D0138458

Microcontroller Engineering with MSP432

Fundamentals and Applications

Microcontroller Engineering with MSP432

Fundamentals and Applications

Ying Bai

CRC Press is an imprint of the
Taylor & Francis Group, an **informa** business

CRC Press
Taylor & Francis Group
6000 Broken Sound Parkway NW, Suite 300
Boca Raton, FL 33487-2742

Printed on acid-free paper
Version Date: 20160916

International Standard Book Number-13: 978-1-4987-7298-3 (Hardback)

Library of Congress Cataloging-in-Publication Data

Names: Bai, Ying, 1956- author.
Title: Microcontroller engineering with MSP432 : fundamentals and applications / author, Ying Bai.
Description: Boca Raton : Taylor & Francis, CRC Press, 2017. | Includes bibliographical references and index.
Identifiers: LCCN 2016020120 | ISBN 9781498772983 (alk. paper)
Subjects: LCSH: Microcontrollers. | Automatic control. | Texas Instruments MSP430 series microprocessors.
Classification: LCC TJ223.M53 B345 2017 | DDC 006.2/2--dc23
LC record available at https://lccn.loc.gov/2016020120

Visit the Taylor & Francis Web site at
http://www.taylorandfrancis.com

and the CRC Press Web site at
http://www.crcpress.com

Printed and bound in the United States of America by Publishers Graphics, LLC on sustainably sourced paper.

This book is dedicated to my wife, Yan Wang, and my daughter, Susan (Xue) Bai.

Contents

Preface

The ARM® Cortex®-M4 and MSP432™ family MCUs are the most popular and updated microcontrollers widely implemented in the education, industrial, and manufacturing fields in recent years. Owing to their relatively simple structures and powerful functions, MCU systems have been applied in many applications in the real world, including automatic, intelligent, and industrial controls, and academic implementations.

The advantages of using microcontrollers include, but are not limited to, the following:

- The ARM® Cortex®-M4 MCU is a 32-bit microcontroller and can work independently as a single controller to provide real-time and multifunction controls to effectively and easily control the most real objectives in our world.
- The internal bus system used in Cortex®-M4 MCU is 32 bits and is based on the so-called advanced microcontroller bus architecture (AMBA) standard. The AMBA standard provides efficient operations and low power cost on the hardware.
- The main bus interface between the MCU and external components is the advanced high-performance bus (AHB), which provides interfaces for memory and system bus, as well as peripheral devices.
- A nested vectored interrupt controller (NVIC) is used to provide support and management to the interrupt responses and processing to all components in the system.
- The Cortex®-M4 MCU also provides standard and extensive debug features and support to enable users to easily check and trace their program with breakpoints and steps.
- The MSP-EXP432P401R EVB provides fundamental and basic peripherals and interfaces to enable users to conveniently communicate to other parallel or serial peripherals via GPIO ports to perform specific control tasks and functions.
- The EduBASE ARM® Trainer provides most popular I/O devices, such as 4-LED, a 4-bit DIP switch, four 7-segment LEDs, a 4×4 keypad working as an input keyboard, a 16×2 LCD connected to HD44780 to work as an output displaying device, two H-Bridge motor drivers, three analog input sensors, a CAN protocol, and other peripheral interfaces. All of these I/O devices and interfaces provide great flexibility to enable users to design and build advanced and professional control units applied in the real world.
- The integrated development environment Keil® ARM-MDK µVersion®5 provides an integrated development environment to enable users to easily create, compile, build, and run professional application projects to control and coordinate the entire control system to perform desired tasks in a short period of time.

The author has tried to provide a full package to cover all components and materials related to the ARM® Cortex®-M4™ and MSP432 family microcontroller systems, including hardware and software as well as practical application notes with real examples. All example projects in the book have been compiled, built, and tested. To help students master the main techniques and ideas, four appendices are also provided to help students overcome some possible learning curves.

Your questions or comments on this book are welcome.

Ying Bai

Acknowledgments

The first and most special thanks to my wife, Yan Wang; I could not have finished this book without her sincere encouragement and support.

Many thanks should be extended to the editor Nora Konopka who made this book available to the public. Without her deep perspective and hard work, it would be difficult to find this book in the marketplace. Thanks are extended to this book's editorial team. Without their contributions, it would be impossible for this book to be published.

Thanks should also be extended to the following book reviewers for their valued opinions:

- Dr. Jiang (Linda) Xie, Professor, Department of Electrical and Computer Engineering, University of North Carolina at Charlotte
- Dr. Daoxi Xiu, Application Analyst Programmer, North Carolina Administrative Office of the Courts
- Dr. Dali Wang, Associate Professor, Department of Physics and Computer Science, Christopher Newport University

Last, but not least, thanks should be extended to all those who supported and enabled me to complete this book.

Author

Dr. Ying Bai is a professor in the Department of Computer Science and Engineering at Johnson C. Smith University located in Charlotte, North Carolina. His special interests include intelligent controls, soft computing, mixed-language programming, fuzzy logic controls, robotic controls, robots calibrations, and fuzzy multicriteria decision-making.

His industry experience includes positions as software and senior software engineers at companies such as Motorola MMS, Schlumberger ATE Technology, Immix Telecom, and Lam Research.

Dr. Bai has published approximately 50 academic research papers in *IEEE Trans.* journals and international conferences. He has also published 13 books with publishers such as Prentice Hall, CRC Press LLC, Springer, Cambridge University Press, and Wiley IEEE Press in recent years. The Russian translation of his first book entitled *Applications Interface Programming Using Multiple Languages* was published by Prentice Hall in 2005. The Chinese translation of his eighth book entitled *Practical Database Programming with Visual C#.NET* was published by Tsinghua University Press in China at the end of 2011. Most of his books are about interfacing software programming, serial port programming, database programming, and fuzzy logic controls in industrial applications as well as microcontroller programming and applications.

Trademarks and references

Trademarks

- ARM®, Cortex®, Keil®, and µVision® are registered trademarks of ARM Limited (or its subsidiaries) in the EU and/or elsewhere.
- ARM7™, ARM9™, and ULINK™ are trademarks of ARM Limited (or its subsidiaries) in the EU and/or elsewhere. All rights reserved.
- Texas Instruments™ is a trademark of Texas Instruments Incorporated.
- MSP432™ is a trademark and product of Texas Instruments Incorporated.
- EnergyTrace™ is a trademark and product of Texas Instruments Incorporated.
- LaunchPad™ is a trademark and product of Texas Instruments Incorporated.
- Code Composer Studio™ is a trademark and product of Texas Instruments Incorporated.

References

- MSP432P4xx Family Technical Reference Manual.pdf (http://www.ti.com/lit/ug/slau356a/slau356a.pdf)
- MSP432P401R MCU.pdf (http://www.ti.com/lit/ds/symlink/msp432p401r.pdf)
- MSP432 EVB Guide.pdf (http://www.ti.com/lit/ug/slau597a/slau597a.pdf)
- Chapters 3, 4, 11, and 12: *Practical Microcontroller Engineering with ARM Technology*, by Ying Bai, Wiley IEEE Press, December 29, 2015

chapter one

Introduction to microcontrollers and this book

Since the development of very large-scale integrated circuits (VLSI) in recent years, more and more advanced semiconductor devices and equipment are built with very high intensity. Over millions of MOSFETs can be integrated in a very small semiconductor chip to generate multi-function processors or **microprocessors**. Microprocessors can be considered as a VLSI that can be programmed to perform specific functions or tasks.

One of the most popular and important microprocessors is the central processing unit (CPU), which is a center of a computer and used to process and coordinate all operations on a computer. Some other popular microprocessors can be categorized into different groups based on their functions. Some popular microprocessors are:

1. CPU or the core of the microprocessor
2. Parallel 8-bit or 16-bit I/O ports
3. Parallel-to-serial converter (UART)
4. Timer and counter
5. Interrupt control unit and priority interrupt controller
6. Random access memory (RAM) chips
7. Erasable programmable read-only memory (EPROM) chips
8. Electrically erasable programmable read-only memory (EEPROM) chips
9. Flash memory chips

By combining microprocessors with memory units and I/O ports, a **microcontroller** system can be built. A microcontroller is also called a microcomputer. In fact, a microcontroller is made by embedding processors, memory unit, and I/O ports into a single semiconductor chip, and this is the current module of a modern microcontroller unit (MCU) used in all aspects in today's society. One of the latest MCU modules is the MSP432™ family MCU—MSP432P401R.

1.1 Microcontroller configuration and structure

By combining some microprocessors with memory units and I/O ports, a microcontroller can be built. In fact, a microcontroller can be built by combining three basic components with three system buses as shown in Figure 1.1.

Three components are CPU, memory, and I/O ports. These three components are connected with three system buses, address bus (**A.B.**), data bus (**D.B.**), and control bus (**C.B.**), to provide the following functions:

- The CPU works as headquarter for the microcontroller to provide all controls to other components and coordinate them to fulfill the desired tasks assigned to the microcontroller.

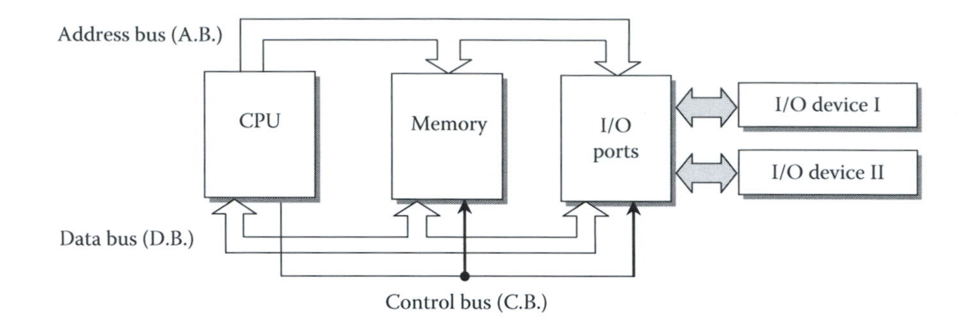

Figure 1.1 Basic structure and configuration of a microcontroller system.

- The memory unit works as a storage unit to store the user's program, including the user's instructions and data. Some system programs and data are also stored in special memory units such as PROM, EPROM, EEPROM, or flash memory.
- The I/O ports work as an interface and provide the communications between the CPU and the peripheral devices.

The communications between these three components are performed via three system buses. The address bus provides a valid address to the memory to enable the CPU to select and pick up the desired instruction or data from the selected memory space. The data bus is used to transfer a valid data item between components. The control bus provides valid operational signals to coordinate the information transfer between components. Some popular control signals are Read/$\overline{\text{Write}}$ ($\mathbf{R}/\overline{\mathbf{W}}$) signal used to read from or write into the memory, Chip Select (**CS**) signal used to decode the address to select the desired microprocessor chip, and Enable signal **E** that is similar to the **CS** signal.

Three components are connected together via three buses in tri-state mode, which means that the connection between any two components is disconnected or high impedance until a valid address is applied and decoded to enable the selected tri-state gates to turn on.

Regularly, a CPU contains three components: (1) a group of registers called register bank, (2) arithmetic and logic unit (ALU), and (3) control signal generator. The registers are used to assist the instruction's decoding and data operations since most operations between the CPU and memory are performed inside registers in the CPU because of the high execution speed of registers. The ALU is used to perform arithmetic and logic operations, and the control unit provides all timing and control signals required to perform all related operations of the CPU.

Generally the memory space is divided into two separate areas, one is the system memory space that is used to store instructions and data related to the system normal operations of the microcontroller, and the other one is the users' memory space that is used to store the users' instructions and data.

The memory spaces also can be divided into the catch and heap areas based on the materials used to build the memory, and the former is made of high-speed static RAM (SRAM) and the latter is made of dynamic RAM (DRAM) with relatively slower accessing speeds. The advantage of using the SRAM is that a higher memory accessing speed can be obtained, but much more MOSFETs are utilized for each SRAM unit and therefore makes the memory structure complicated with higher cost. The advantage of using DRAM is that higher memory densities or integration intensities can be obtained with much simpler

MOSFET structure and lower cost for each DRAM unit, but the working speed is relatively slower because of an additional refresh circuit applied on the DRAM. Because of the cost issue, the size of SRAM or the catch memory is regularly small but the size of the heap or DRAM is huge.

The memory can also be categorized to the RAM, the read only memory (ROM), the EPROM, or EEPROM, and flash memory. Generally, the system instructions and data are stored in the ROM, EPROM, or EEPROM spaces. The users' instructions and data are stored in the RAM space. Based on the functions, the memory can be divided into either volatile memory or non-volatile memory. The RAM belongs to the volatile memory since all information stored in this kind of memory would be gone when the power is off. However, the ROM, EPROM, EEPROM, and flash memory belong to the non-volatile memory since the information stored in this kind of memory would be in there still even if the power is off.

Based on the structure, the RAM can be categorized to SRAM or DRAM.

The I/O ports can be divided into two categories: the parallel and the serial I/O ports. Each I/O port can be mapped to a memory address or each of them can have a special I/O address that is different with a normal memory address. The former is called the I/O memory mapping addressing and the latter is called the direct I/O addressing.

1.2 ARM® Cortex®-M4 microcontroller system

Different embedded systems or MCUs have been developed and built by different vendors in recent years. One of the popular MCUs is the ARM® Cortex®-M MCU family. This kind of MCU provides multi-functions and control abilities, low-power consumptions, high-efficiency signal processing functionality, low-cost, and easy-to-use advantages. The latest product of the ARM® Cortex®-M family is Cortex®-M4 MCU.

The ARM® Cortex-M is a group of 32-bit reduced instruction set computing (**RISC**) ARM® processor cores licensed by ARM® Holdings. The cores are intended for microcontroller use, and consist of the Cortex®-M0, Cortex®-M0+, Cortex®-M1, Cortex®-M3, and Cortex®-M4.

The ARM® Cortex®-M4 processor is the latest embedded processor by ARM® specifically developed to address digital signal control markets that demand an efficient, easy-to-use blend of control and signal processing capabilities. The combination of high-efficiency signal processing functionality with low-power, low-cost, and easy-to-use benefits of the Cortex-M family of processors is designed to satisfy the emerging category of flexible solutions specifically targeting the motor control, automotive, power management, embedded audio, and industrial automation markets.

The ARM® Cortex®-M4 MCU provides the following specific functions

- Although the Cortex-M4 processor is a 32-bit MCU, it can also handle 8-, 16-, and 32-bit data efficiently.
- The Cortex®-M4 MCU itself does not include any memory, but it provides different memory interfaces to the external flash ROMs and SRAMs.
- Due to its 32-bit data length, the maximum searchable memory space is 4 GB.
- In order to effectively manage and access this huge memory space, different regions are created to store system instructions and data, users' instructions, data, and mapped peripheral device registers and related interfaces.
- The internal bus system used in Cortex-M4 MCU is 32-bit and it is based on the so-called advanced microcontroller bus architecture (**AMBA**) standard. The AMBA standard provides efficient operations and low power cost on the hardware.

- The main bus interface between the MCU and external components is the advanced high-performance bus (**AHB**), which provides interfaces for memory and system bus as well as peripheral devices.
- A nested vectored interrupt controller (NVIC) is used to provide all supports and managements to the interrupt responses and processing to all components in the system.
- The Cortex-M4 MCU also provides standard and extensive debug features and supports to enable users to easily check and trace their program with breakpoints and steps.

To assist users to build professional microcontroller application projects, some useful development tools and kits are involved in this book to enable users to develop specific implementations easier and faster.

1.3 MSP432P401R microcontroller and its development tools and kits

In this book, we concentrate on a much updated MCU, MSP432P401R built by Texas Instruments™ with a professional Evaluation Board MSP-EXP432P401R EVB, in which two ARM® Cortex®-M4-based MCUs, TM4C129 and MSP432P401R, are utilized. The related development tools and kits for this book can be categorized into two parts, the hardware part and the software part.

The hardware part includes

- Texas Instruments™ MSP432P401R LaunchPad™ MSP-EXP432P401R EVB
- EduBASE ARM® Trainer (contains most popular peripherals and interfaces)
- Other related peripherals, such as DC motors, LCD, 7-segment LEDs, keypad, photo and temperature sensors, and D/A converters

The software part includes

- Integrated development environment **Keil® ARM®-MDK µVersion®5.15** (IDE).
- MSPWare software driver package **MSPWare_2_21_00_39**.

Appendices A~D in this book provide detailed information and directions for downloading and installing these software tools in your host computers.

1.4 Outstanding features about this book

1. Both ARM® assembly and C codes are provided in the book to assist users to develop professional projects with any language easily and faster.
2. More than 70 real example projects are provided in this book with detailed and line-by-line explanations and illustrations to enable users to understand and learn the programming skills easily and faster. These example projects covered most popular peripherals, such as flash memory and EEPROM, ADC, 4 × 4 keypad, 7-segment LEDs, LCD, DAC, I2C, UART, PWM, Timer_A modules, Timer32 modules, Watchdog timer, analog comparator, FPU, and MPU, in different chapters.

3. Both the direct register access (DRA) model and the software driver (SD) model programming techniques are introduced and discussed with a set of complete real example projects to cover all peripherals in the book.
4. A complete set of homework, including the true/false, multi-choice questions, comprehensive questions, and lab projects, is attached after each chapter. This enables students to understand what they learned better by doing something themselves.
5. A complete set of answers to all homework is provided for the instructors.
6. A complete set of MS PowerPoint teaching slides are provided for the instructors to make teaching this book easy and convenient.
7. Appendices A~D provide a complete set of instructions and directions to enable users to download and install development tools and kits easily and faster.
8. Good textbook for college students, good reference book for programmers, software engineers, and academic researchers.

1.5 Who this book is for

This book is designed for college students and software programmers who want to develop practical and commercial control programming with MSP432P401R MCU and related development tools. Fundamental knowledge and understanding on C language programming is assumed.

1.6 What this book covers

This book comprises 12 chapters with an easy study way to enable students to learn the ARM® Cortex®-M4 microcontroller and MSP432P401R MCU technology and interface implementations easily. Each chapter contains homework and exercises as well as lab projects to enable students to perform necessary exercises to improve their learning and understanding for the related materials and technologies.

Chapter 1: Provides an overview and introduction about the microcontrollers and a global review for the book with highlights on outstanding features and organization of the book.

Chapter 2: Provides detailed discussion and analysis about the MSP432P401R MCU hardware, which includes the architecture of the Cortex®-M4 Core and processor, memory (flash memory, SRAM, and EEPROM), GPIO parallel and serial ports, NVIC, system timer SysTick, system control block (SCB), clock system (CS), ADC_14, eUSCI_A and eUSCI_B modules, port mapping controller (PMC), floating point unit (FPU), and memory protection unit (MPU).

Chapter 3: Provides detailed discussion and analysis about the development tools and kits for the Cortex-M4 MCU. These tools and kits are discussed separately based on the hardware and software sections. The MSP-EXP432P401R EVB and EduBASE ARM® Trainer are discussed as the hardware kits and tools. The Keil® ARM®-MDK µVersion®5 that works as an IDE and the MSPWare software driver package `MSPWare_2_21_00_39` that works as a software suite and driver library are introduced as software tools and kits. The XDS110-ET debugger that works as a system debugger for the MSP-EXP432P401R EVB is also discussed in this chapter.

Chapter 4: Provides detailed and complete discussion about the ARM® Cortex®-M4 microcontroller software and instruction set. These discussions include the ARM®

Cortex®-M4 software development structure, a complete Cortex®-M4 assembly instruction set, the Keil® CMSIS Core-specific intrinsic functions, inline assembler, C programming procedure for the MSP432P401R MCU, and two programming models applied by the MSPWare Peripheral Driver Library. The detailed procedure of building and developing an example MSP432P401R MCU project with C codes is discussed step by step at the end of this chapter.

Chapter 5: Provides detailed discussions about the ARM® Cortex®-M4 and MSP432P401R MCU interrupts and exceptions. These discussions cover the interrupt and exception sources, interrupt handlers, and interrupt and exception vector tables. Most popular control registers involved in the NVIC are introduced in details with related examples. The NVIC macros and NVIC API functions supported by the CMSIS Core software package are also introduced in this chapter. The GPIO interrupts handled by API functions provided by the MSPWare Peripheral Driver Library are discussed at the end of this chapter.

Chapter 6: Provides detailed discussion about ARM® Cortex®-M4 memory system. Specially, the memory system used in the MSP432P401R MCU is extensionally discussed in detail. These discussions include the memory architecture, the entire memory map with accurate addresses for each component, SRAM, flash memory, and internal ROM. Most popular control registers applied on these memory models are introduced and discussed in detail with related example projects. The API functions used for flash memory are also discussed in this chapter. Some special memory implementation techniques, such as bit-band alias and flash memory programming, are introduced with actual example projects.

Chapter 7: Provides detailed discussions about the MSP432P401R MCU parallel I/O ports programming. The major parallel peripherals discussed in this chapter include the on-board keypads, analog to digital converter ADC_14, and analog comparators. All peripherals in the MSP432P401R MCU system, including the parallel and serial, are interfaced to the processor via a group of general-purpose input output (GPIO) ports. All GPIO ports and control registers related to these parallel peripherals are introduced with some real example projects in details. Those peripheral-related API functions provided by the MSPWare Peripheral Driver Library are also discussed in this chapter with actual example projects.

Chapter 8: Provides detailed discussions about the MSP432P401R MCU serial I/O ports programming. The major serial peripherals discussed in this chapter include the serial peripheral interface (SPI), inter-integrated circuit (I2C) interface, and universal asynchronous receivers/transmitters (UARTs). Several real example projects related to these serial peripherals include: on-board LCD interface project, on-board 7-segment LED project, digital to analog converter project, I2C interfacing project, and UART project. The related API functions provided by the MSPWare Peripheral Driver Library are also discussed with some real example projects in this chapter.

Chapter 9: Provides detailed discussions about the MSP432P401R Timer_A modules and their applications. All Timer_A modules-related control registers are introduced and discussed in detail with several real example projects. Different implementations of using Timer_A modules are analyzed and discussed with some real example projects. These implementations include the free-running timer, input edge-count mode, input edge-time mode, and PWM mode. The related API functions supporting those modes and provided by the MSPWare Peripheral Driver Library are also introduced with example projects.

Chapter 10: Provides detailed discussions about the MSP432P401R Timer32 and Watchdog WDT_A modules. Quite a few actual example projects are covered by these discussions and introductions. The related API functions for these peripherals are also introduced in this chapter.

Chapter 11: Provides detailed discussions about the ARM® Cortex®-M4 FPU. First, the single- and double-precision floating point numbers and protocols are introduced. Then the FTP architecture applied in the Cortex-M4 MCU is discussed in detail. All control registers used for the FPU are discussed and illustrated in detail. The FPU-related API functions provided by the MSPWare Peripheral Driver Library are also discussed. Finally, a real example project is provided to illustrate how to use the FPU to perform some sophisticated floating point data operations.

Chapter 12: Provides detailed discussion about the ARM® Cortex®-M4 MPU. An overview about the MPU is provided first in this chapter. All control registers used for the MPU are discussed and illustrated in detail. The MPU-related API functions provided by the MSPWare Peripheral Driver Library are also discussed. Finally, a real example project is provided to illustrate how to use the MPU to perform desired protection functions for the selected memory regions.

Appendix A: Provides instructions about downloading and installing the Keil® MDK-ARM® 5.15 IDE and MSP432 DFP.

Appendix B: Provides instructions about downloading and installing the MSPWare Software Driver Package **`MSPWare_2_21_00_39`**.

Appendix C: Provides the hardware connection configuration between the host computer and the MSP432P401R MCU-based EVB–MSP-EXP432P401R.

Appendix D: Provides a set of CMSIS core-specific intrinsic functions.

1.7 How this book is organized and how to use this book

Chapters 2~10 provide fundamental and professional introductions and discussions about the most popular MSP432P401R MCU applications with the most widely used peripherals, such as flash memory, ADC_14, DAC, ACMP, PWM, UART, I2C, SPI, LCD, 7-segment LEDs, Timer_A, Timer32, and Watchdog WDT_A modules. Two optional components, FPU and MPU, are discussed in Chapters 11 and 12.

Based on the organization of this book we described above, this book can be used as two categories such as Level I and Level II, which is shown in Figure 1.2.

For undergraduate or graduate college students or beginning software programmers, it is highly recommended to learn and understand the contents of Chapters 2~10 since those are fundamental knowledge and techniques used in the MSP432P401R microcontroller programming and implementations (Level I). For the materials in Chapters 4, 6, 11, and 12, they are only related to the ARM® assembly language and memory operations with additional and optional peripherals, FPU and MPU, used in the ARM® Cortex®-M4 microcontroller programming. They are optional to instructors and depend on the time and schedule (Level II).

1.8 How to use the source code and sample projects

All projects in the book can be divided into two parts: the class projects and the lab projects. All source codes for those projects are available in the book. However, all class projects are available to both instructors and students, but the lab projects are only available to instructors since students need to build these lab projects themselves. All source codes for these projects have been debugged and tested, and they are ready to be executed in any MSP432P401R EVB.

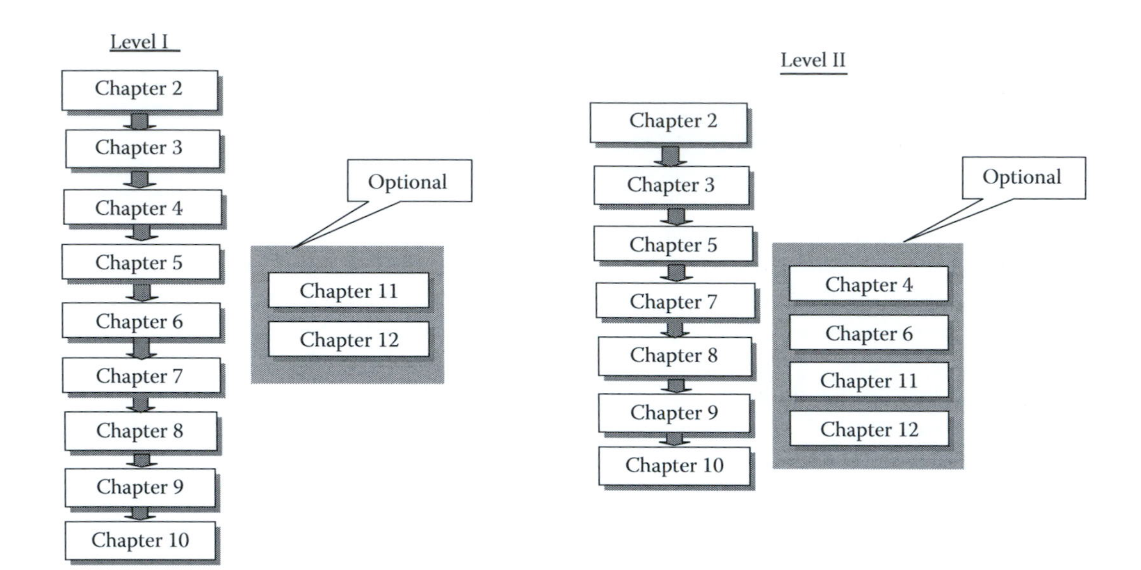

Figure 1.2 Two study levels in the book.

All class projects are categorized into the associated chapters that are located at the folder **MSP432 Class Projects** that is located at the site https://www.crcpress.com/Microcontroller-Engineering-with-MSP432-Fundamentals-and-Applications/Bai/p/book/9781498772983. You need to use either the book **ISBN**, the book **title**, or the **author** name to access and download these projects into your computer and run each project as you like. To successfully run those projects on your computer, the following conditions must be met:

1. The Keil™ MDK-ARM® 5.15 and above IDE must be installed in your computer.
2. The MSPWare Software Package should be installed in your computer, and this package must be installed if you want to use any API function provided by MSPWare Peripheral Driver Library.
3. The MSP-EXP432P401R EVB and EduBASE ARM® Trainer must be installed and connected to your host computer.

Refer to Appendices A~C to complete steps 1~3.

All book-related teaching and learning materials, including the class projects, lab projects, appendices, faculty teaching slides, and homework solutions, can be found in the associated folders located at the CRC Book Support site, as shown in Figure 1.3.

These materials are categorized and stored at different folders in two different sites based on the teaching purpose (for instructors) and learning purpose (for students). For instructors:

1. **MSP432 Class Projects** folder: Contains all class projects for different chapters.
2. **MSP432 Lab Projects** folder: Contains all lab projects included in the homework sections in different chapters. Students need to follow the directions provided in each homework lab section to build and develop these lab projects themselves.

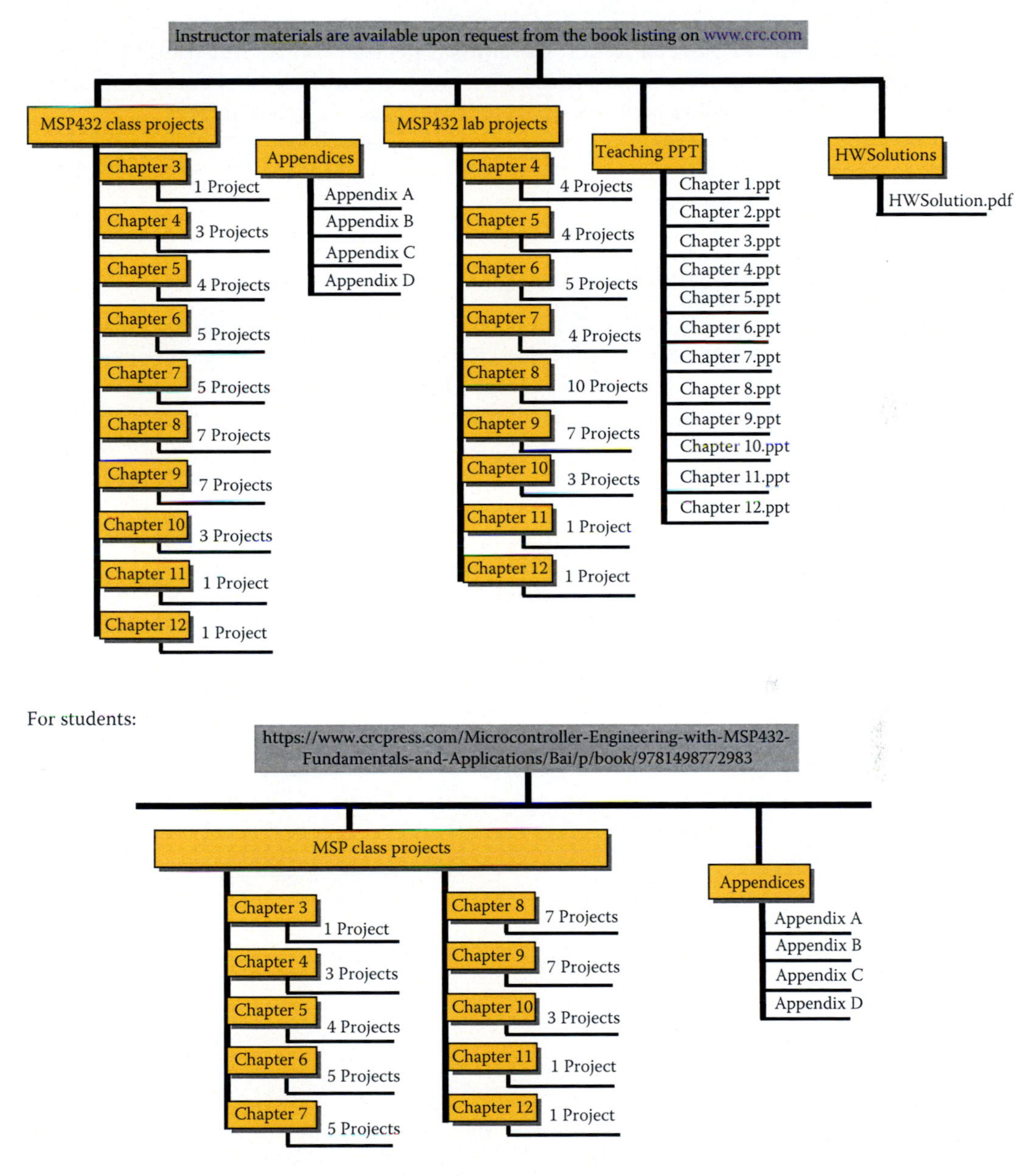

Figure 1.3 Book-related materials on Website.

3. **`Appendices`** folder: Contains all Appendices (**`Appendices A~D`**) that provide useful references and practical instructions to download and install the MSP432P401R MCU-related development tools and kits.
4. **`TeachingPPT`** folder: Contains all MS-PPT teaching slides for each chapter.
5. **`HWSolutions`** folder: Contains a set of complete solutions for the homework developed and used in the book. The solutions are categorized and stored at the different chapter subfolders based on the book chapter sequence.

For students:

1. **MSP432 Class Projects** folder: Contains all class projects in different chapters. Students can download and run these class projects in their host computers after a suitable environment has been set up (refer to conditions listed above).
2. **Appendices** folder: Contains all appendices (**Appendices A~D**) that provide useful references and practical instructions to download and install the MSP432P401R MCU-related development tools and kits.

1.9 Instructors and customers support

The teaching materials for all chapters have been extracted and represented by a sequence of Microsoft PowerPoint files, each file for one chapter. The interested instructors can find those teaching materials from the folder **TeachingPPT** that is located at the site www.crc.com and those instructor materials are available upon request from the book listing on www.crc.com.

A set of complete homework solutions is also available upon request from the book listing on www.crc.com.

E-mail support is available to readers of this book. When you send e-mail to us, please send all questions to the e-mail address: ybai@jcsu.edu.

Detailed structure and distribution of all book-related materials in the CRC site, including the teaching materials for instructors and learning materials for students, are shown in Figure 1.3.

chapter two

MSP432 microcontroller architectures

The main topics to be discussed in this chapter include the architectures and organizations of most popular embedded systems, including the most updated microcontroller ARM® Cortex®-M4, Texas Instruments MSP432P401R MCU, MSP432P401R LaunchPad™ MSP-EXP432P401R evaluation board, and EduBASE ARM® Trainer. All of these components will be used in this book to make our project development process easier and simpler.

2.1 Overview

A so-called embedded system generally comprises a group of programmable devices with some memory and a set of peripheral I/O ports. In fact, an embedded system can be considered an integrated system by embedding some CPUs with some memory subsystems, I/O ports, and maybe several peripheral devices together into a single semiconductor chip to get an intelligent control unit, called an MCU. Therefore, a typical embedded system can be thought of as an MCU. An illustration function block diagram for a general MCU is shown in Figure 2.1.

Most popular important components involved in this MCU include:

- ARM® CPU or processor
- Memory (SRAM and flash memory)
- Parallel input and output (PIO) ports
- Some internal devices (timer/counters) or peripheral devices (analog-to-digital converter [ADC] and UART)

Different embedded systems or MCUs have been developed and built by different vendors in recent years. One of the most popular MCUs is the MSP432™ MCU family. The core of this MCU is one of the latest ARM® Cortex®-M processors or Cortex®-M4 microprocessor.

This chapter is organized in the following sections:

- The architecture of the ARM® Cortex®-M4 processor is discussed first. This includes the architecture of the ARM® Cortex®-M4 Core Processor (CPU).
- The architecture of the MSP432P401R MCU is introduced in the following section. This discussion includes:
  - MSP432P401R microcontroller on-chip memory map
  - The MSP432P401R MCU system peripherals and system controls
  - MSP432P401R microcontroller GPIO modules
- Introduction to EduBASE ARM® Trainer is given at the end of this chapter.

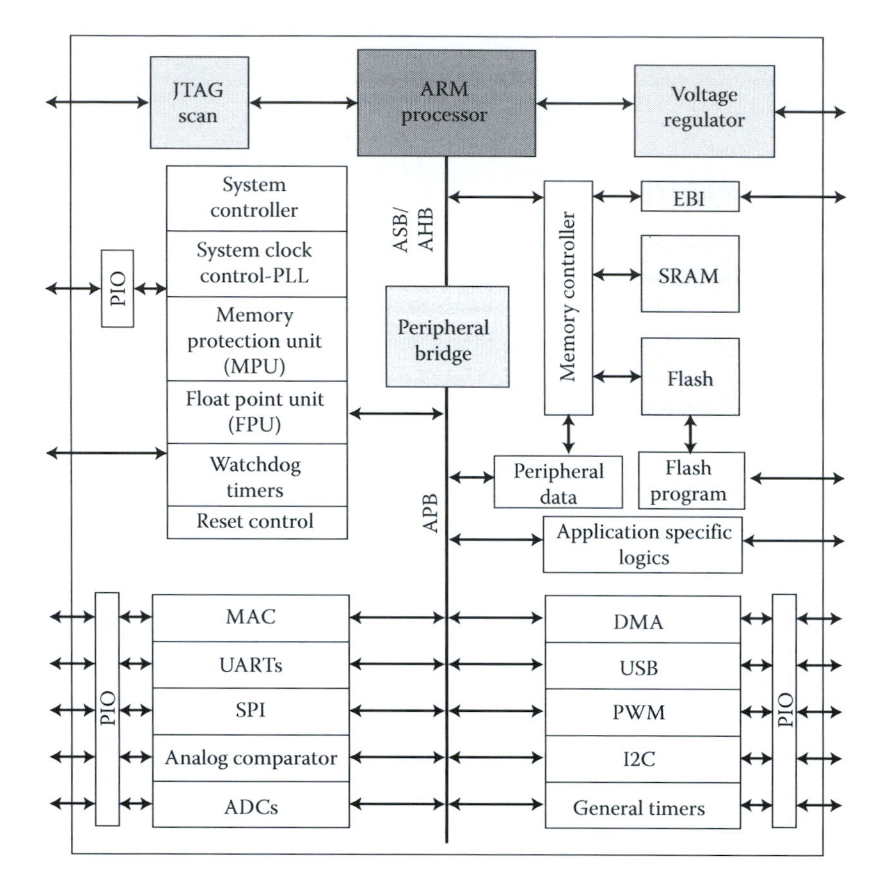

Figure 2.1 Block diagram for a general MCU.

Because the ARM® Cortex®-M4 is the latest MCU with a lot of brand new and advanced techniques involved, therefore, its architecture definitely becomes very complicated if one wants to get a very detailed picture about this product. Fortunately, as a software developer or software engineer, you do not need to understand each piece of the architecture in detail; what you need to learn is the basic functions of the most popular or required components, their block diagrams and interfaces in this MCU.

2.2 *Introduction to ARM® Cortex®-M4 MCU*

The ARM® Cortex®-M is a group of 32-bit RISC ARM® processor cores licensed by ARM® Holdings. The cores are intended for microcontroller use, and consist of the Cortex®-M0, Cortex®-M0+, Cortex®-M1, Cortex®-M3, and Cortex®-M4. Table 2.1 lists a development history for ARM® MCUs family.

It can be seen from Table 2.1 that the ARM® Cortex®-M3 and Cortex®-M4 MCUs belong to the ARMv7 family, and they are 32-bit microcontrollers. However, the early ARM® Cortex®-M MCUs, such as Cortex®-M0~Cortex®-M1, belong to the ARMv6-M family.

The ARM® Cortex®-M4 processor is the latest embedded processor by ARM® specifically developed to address digital signal control markets that demand an efficient,

Table 2.1 Development history of the ARM® MCUs family

Architecture	Bit width	Cores designed by ARM holdings	Cores designed by third parties	Cortex profile
ARMv1	32/26	ARM1		
ARMv2	32/26	ARM2, ARM3	Amber, STORM Open Soft Core	
ARMv3	32	ARM6, ARM7		
ARMv4	32	ARM8	StrongARM, FA526	
ARMv4T	32	ARM7TDMI, ARM9TDMI		
ARMv5	32	ARM7EJ, ARM9E, ARM10E	XScale, FA626TE, Feroceon, PJ1/Mohawk	
ARMv6	32	ARM11		
ARMv6-M	32	ARM Cortex-M0, ARM Cortex-M0+, ARM Cortex-M1		Microcontroller
ARMv7-M	32	ARM Cortex-M3		Microcontroller
ARMv7E-M	32	ARM Cortex-M4		Microcontroller
ARMv7-R	32	ARM Cortex-R4, ARM Cortex-R5, ARM Cortex-R7		Real time
ARMv7-A	32	ARM Cortex-A5, ARM Cortex-A7, ARM Cortex-A8, ARM Cortex-A9, ARM Cortex-A12, ARM Cortex-A15, ARM Cortex-A17	Krait, Scorpion, PJ4/Sheeva, Apple A6/A6X	Application
ARMv8-A	64/32	ARM Cortex-A53, ARM Cortex-A57	X-Gene, Denver, Apple A7 (Cyclone), K12	Application
ARMv8-R	32	No announcements yet		Real time

easy-to-use blend of control and signal processing capabilities. The combination of high-efficiency signal processing functionality with the low-level power, low cost, and ease-to-use benefits of the Cortex®-M family of processors is designed to satisfy the emerging category of flexible solutions specifically targeting the motor control, automotive control, power management, embedded audio, and industrial automation markets.

The ARM® Cortex®-M4 MCU provides the following specific functions:

- The Cortex®-M4 Processor is a 32-bit MCU, but it can also handle 8-bit, 16-bit, and 32-bit data efficiently.
- The Cortex®-M4 MCU itself does not include any memory, but it provides different memory interfaces to the external flash memory, internal ROM, and SRAMs.
- Because of its 32-bit data length, the maximum searchable memory space is 4 GB.
- In order to effectively manage and access this huge memory space, different regions are created to store system instructions and data, user's instructions, data, and mapped peripheral device registers and related interfaces.

- The internal bus system used in Cortex®-M4 MCU is 32 bit, and it is based on the AMBA standard. The AMBA standard provides efficient operations and low power cost on the hardware.
- The main bus interface between the MCU and external components is the advanced AHB, which provides interfaces for memory and system bus, as well as peripheral devices.
- An NVIC is used to provide all supports and managements to the interrupt responses and processing to all components in the system.
- The Cortex®-M4 MCU also provides standard and extensive debug features and supports to enable users to easily check and trace their program with breakpoints and steps.

Overall, the Cortex®-M4 MCU processor incorporates:

- A processor core or CPU
- An NVIC closely integrated with the processor core to achieve low-latency interrupt processing
- Multiple high-performance bus interfaces, including code interface and SRAM and peripheral interface
- A system timer unit SysTick
- A low-cost debug solution with the optional ability, such as debug access port (DAP) and data watchpoint
- An optional MPU
- An optional FPU
- Embedded Trace Macrocell (ETM) interface
- Instrumentation Trace Macrocell (ITM) interface
- The debug and Serial Wire Viewer (SWV) interface

A structure block diagram for ARM® Cortex®-M4 MCU is shown in Figure 2.2.

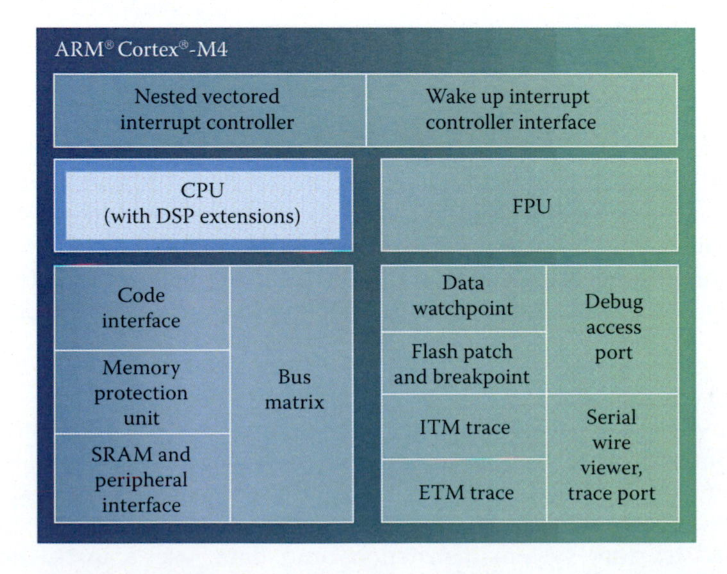

Figure 2.2 Structure block diagram for ARM® Cortex®-M4 MCU. (From ARM Limited. Copyright© ARM® Limited. With permission.)

2.2.1 Architecture of ARM® Cortex®-M4 MCU

Generally, an ARM® Cortex®-M4 MCU is related to the following four architectures:

- ARM® architecture
- Memory architecture
- NVIC architecture
- Debug architecture

In this section, we only concentrate on the ARM MCU architecture since some of them will be discussed in Chapters 5 and 6 later.

2.2.1.1 ARM® MCU architecture block diagram

Let's first introduce and discuss the ARM® architecture. Figure 2.3 shows a functional block diagram for the ARM® Cortex®-M4 MCU.

Cortex®-M4 Core or CPU: This is a central control unit for the entire Cortex®-M4 MCU. All instructions are fetched, decoded, and executed inside this CPU. This CPU or microprocessor consists of three key components: register bank, internal data path, and control unit. This is the core for the normal operation of the entire MCU.

Floating Point Unit: One of the important differences between the Cortex®-M4 MCU and Cortex®-M3 MCU is that an optional FPU is added into the Cortex®-M4 Core to enhance the floating-point data operations. The Cortex®-M4 FPU implements ARMv7E-M architecture with FPv4-SP extensions. It provides floating-point computation functionality that is compliant with the *ANSI/IEEE Std 754-2008, IEEE Standard for Binary Floating-Point Arithmetic.*

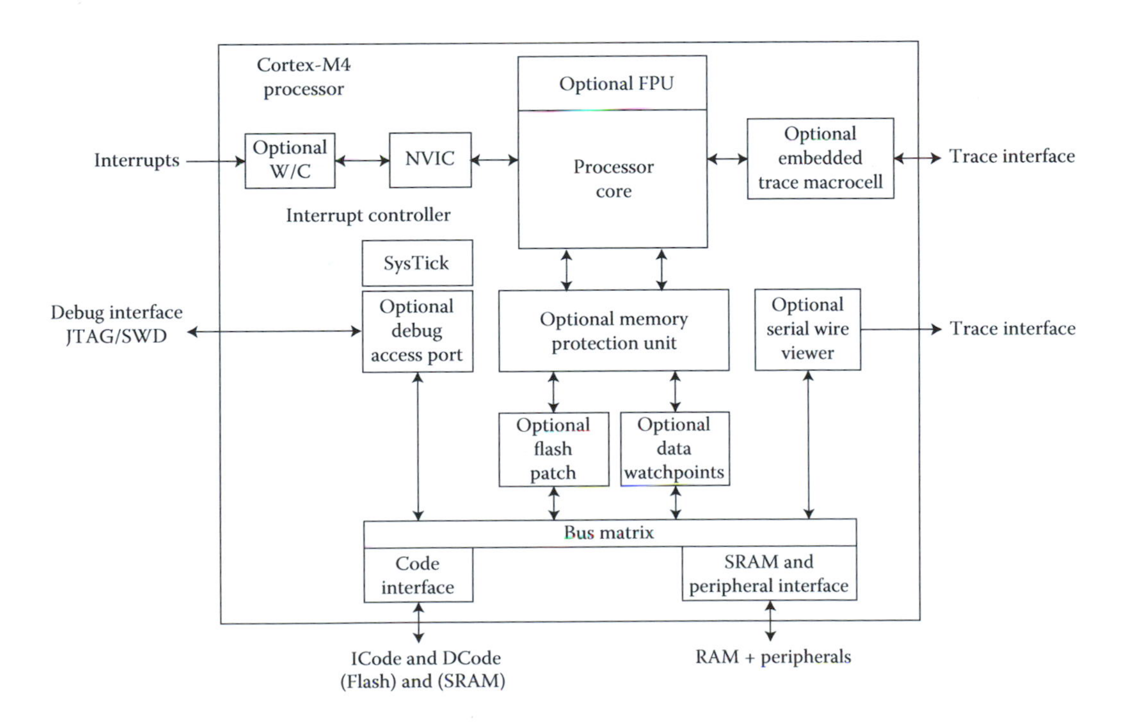

Figure 2.3 Functional block diagram for the ARM® Cortex®-M4 MCU.

System Timer system tick (SysTick): The ARM® Cortex®-M4 MCU provides an integrated system timer, SysTick, which includes a 24-bit counter with clear-on-writing, decrementing, and reloading-on-zero control mechanism. The main purpose of using this timer is to provide a periodic interrupt to ensure that the operating system (OS) kernel can operate regularly.

If you do not need to use any embedded OS in your applications, the SysTick timer can be used to work as a simple timer peripheral for periodic interrupt generator, delay generator, or timing measurement device. The reason for installing this timer inside the MCU is to make software portable, and any program built with Cortex®-M Processor can run any OS written for Cortex®-M4 MCU.

Nested Vectored Interrupt Controller: The NVIC is closely integrated with the processor core to achieve all levels of interrupt processing. These special features include:

- Monitor and preprocess any exception or interrupt occurring during the normal running of the processor. All exceptions and interrupts are categorized into different emergency levels based on their priority levels, from 1 to 15 for exceptions, and from 16 to 240 for interrupts.
- Identify the exception or interrupt source if an exception/interrupt occurred, and direct the main control to the entry point (address) of the related exception/interrupt (interrupt service routine [ISR]) to process the exception/interrupt requests.
- Dynamically manage all exceptions and interrupts based on their priority levels.
- Automatically store the processor states on interrupt entry, and restore it on interrupt exit, with no instruction overhead.
- Optional *Wake-up Interrupt Controller* (WIC) supports the ultra-low-power sleep mode.

Bus Matrix and Bus Interfaces: The Bus Matrix and Bus Interfaces provide:

- Three AHB-Lite interfaces: ICode interface to flash memory, DCode interface to SRAM and peripheral interfaces, as well as system bus interfaces, including the internal control bus and debug components
- Private Peripheral Bus (PPB) interface based on Advanced Peripheral Bus (APB) interface
- Bit-band support that includes atomic bit-band writing and reading operations
- Memory access alignment and write buffer for buffering of write data
- Exclusive access transfers for multiprocessor systems

Memory Protection Unit: This is an optional component for MPU used for memory protection purpose and it includes:

- Eight memory regions
- Subregion disable (SRD), enabling efficient use of memory regions
- The ability to enable a background region that implements the default memory map attributes

System Control Block: This block is located in the system control space (SCS) in the system memory map and it is integrated with the NVIC unit together to provide the following functions:

- Monitor and control the processor configurations, such as low power modes
- Provide fault detection information via fault status register

- Relocate the vector table in the memory map by adjusting the content of the Vector Table Offset Register (VTOR)

Debug Access Port (DAP): The DAP is an optional unit and it provides the following functions:

- Debug access to all memory and registers in the system, including access to memory-mapped devices, access to internal core registers when the core is halted, and access to debug control registers even while **SYSRESETn** is asserted
- Serial Wire Debug Port (SW-DP)/Serial Wire JTAG Debug Port (SWJ-DP) debug access
- Optional Flash Patch and Breakpoint (FPB) unit for implementing breakpoints and code patches
- Optional Data Watchpoint and Trace (DWT) unit for implementing watchpoints, data tracing, and system profiling
- Optional ITM for support of printf() style debugging
- Optional Trace Port Interface Unit (TPIU) for bridging to a Trace Port Analyzer (TPA), including single wire output (SWO) mode
- Optional ETM for instruction trace

2.2.1.2 Architecture of the ARM® Cortex®-M4 core (CPU)

Figure 2.4 shows an architecture block diagram for the ARM® Cortex®-M4 Core or CPU.

The ARM® Cortex®-M4 Core is a 32-bit RISC CPU and it provides the following components:

- Twenty-one 32-bit registers
- 32-bit internal data path
- 32-bit bus interface

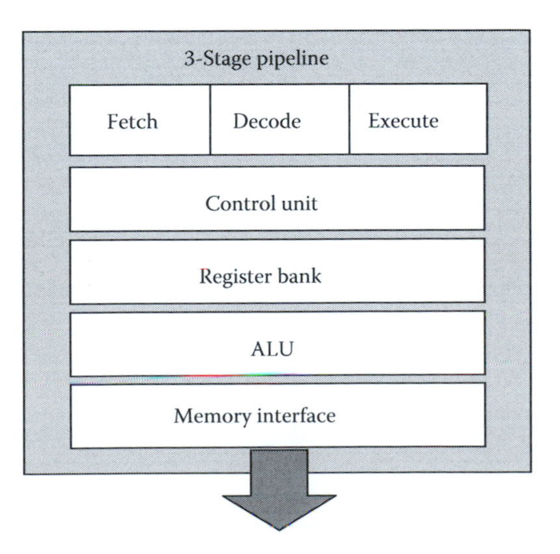

Figure 2.4 Architecture of the ARM® Cortex®-M4 CPU.

Similar to most other microcontrollers' core or CPU, a Cortex®-M4 Core comprises three important components:

- Register bank
- Control unit
- Internal data path (IDP) or ALU

The ARM® Cortex®-M4 Core provides a group of registers and all registers are 32-bit long. These registers can be combined together to form a register group or a register bank.

For each instruction, the Cortex®-M4 Core utilized a three-stage pipeline operation, which is a popular instruction operational style. This operation allows the CPU to get, decode, and execute multiple instructions simultaneously.

The data processing mode in the Cortex®-M4 used a so-called load-store architecture, which means, in order to process the data, the following three steps must be performed:

1. Load data from the memory and write them into registers in the register bank
2. Process data inside the core
3. Write the processed result back to the memory

Let's have a closer look at these registers in the register bank first.

2.2.1.2.1 Register bank in the Cortex®-M4 core Totally, there are twenty-one 32-bit registers located inside the Cortex®-M4 core or microprocessor. These registers can be divided into two groups:

1. Sixteen registers located in the register bank
2. Five special registers located outside of the register bank

A structure block diagram of these registers is shown in Figure 2.5. First, let's take a look at those 16 registers in the register bank.

Inside the register bank, 13 registers, **R0~R12**, belong to general-purpose registers and the other 3 registers, **R13~R15,** are special registers with different specific functions. As we mentioned, all of these registers are 32 bit.

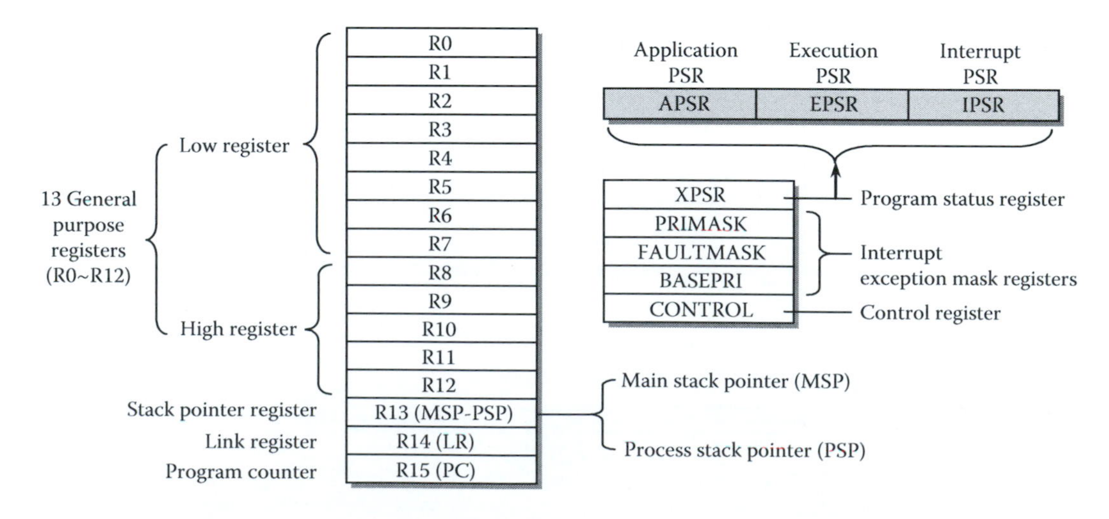

Figure 2.5 Structure block diagram of 21 registers in the Cortex®-M4 Core.

All 13 general-purpose registers can be used to store instructions, data, and addresses. These registers can also be further divided into two subgroups. The first eight registers, **`R0~R7`**, can be categorized to the low register group, and the **`R8~R12`** can be grouped to a high register set. Most 16-bit data operations should be performed in the low register group, and most 32-bit data operations should be processed in the high register group.

The other three registers in the register bank belong to special registers with specific functions. The register **`R13`** is a **`Stack Pointer Register`** (**`SPR`**) used to store the current stack address. In Cortex®-M4 Core, there are two kinds of stack pointers: the **`main stack pointer`** (**`MSP`**) and the **`process stack pointer`** (**`PSP`**). The MSP is used for the system program working in the **`handler mode`** and the PSP is used for the user's program working in the **`thread mode`**. Only one stack pointer is active at a time. The default stack pointer is the MSP after the system is reset.

But this stack pointer can be selected by programming of the **`Control`** register to be discussed later.

The register **`R14`** is a **`link register`** (**`LR`**) and this register provides some linking functions to set up a connection between the main program and the calling functions or subroutines. When a function or subroutine is called, the returning address should be entered into the R14. After the function or subroutine is done, the content of the Link Register R14, which is the returning address to the main program, is fed into the program counter (PC) to enable the processor to continue the work from the address stored in the PC.

A similar situation occurred to the interrupts or exceptions. If some exception or interrupt happened, the returning address to the main program (or the next instruction's address in the main program) should be stored into the link register before the control can be transferred to the **`ISR`**. As the ISR is done, the control can be returned to the main program by popping up the returning address to the PC.

The Register **`R15`** is the **`PC`**. This register keeps the sequence of the program running by automatically updating its content to point to the next instruction's address in the memory. Since the address line is 32-bit wide, each address needs 4-byte space. Therefore, the interval between the neighboring and the adjacent addresses is always four (bytes). In other words, the increment of an address is four. This makes the least significant bit (LSB) of an address always 0.

During the programming process, one can use R0~R12 (or r0~r12) to access general-purpose registers in the register bank. For three special registers used in the register bank, one can use different names to access them, such as R13, r13, SP or sp, for stack pointer register, R14, r14, LR, or lr for the link register, and R15, r15, PC, or pc for the PC.

2.2.1.2.2 Special registers in the Cortex®-M4 core In addition to those registers in the register bank, the Cortex®-M4 Core also includes five special registers located outside the register bank, as shown in Figure 2.5. The purposes of these registers are used to monitor the running status of the CPU, system working states, and interrupt/exception masking. These special registers have the following properties:

- They are not memory mapped, which means that you cannot access them by using any memory-mapped addresses, instead, you must use special register access instructions, such as MSR or MRS.
- They are mainly used for the low-level language programming, such as Assembly, and not for the high-level programming such as C. However, you can access these special registers by using some intrinsic C functions provided by Cortex Microcontroller Software Interface Standard (CMSIS)-Core.

(a)

Bits	31 30 29 28 27	26:25	24	23:20	19:16	15:10	9	8 7 6 5 4 3 2 1 0
APSR	N Z C V Q				GE*	Reserved		
IPSR	Reserved							Exception number
EPSR	Reserved	ICI/IT	T	Reserved		ICI/IT	Reserved	

(b)

Bits	31 30 29 28 27	26:25	24	23:20	19:16	15:10	9	8 7 6 5 4 3 2 1 0
PSR	N Z C V Q	ICI/IT	T		GE*	ICI/IT		Exception number

Figure 2.6 Structure and bit functions in special registers. (a) Tree individual register–APSR, IPSR and EPSR. (b) The combined register PSR.

Now, let's discuss these special registers one by one.

Program Status Register: This register can be divided into three different status registers to show the running status of different units:

- Application Program Status Register (APSR)
- Execution Program Status Register (EPSR)
- Interrupt Program Status Register (IPSR)

These three registers can be accessed individually, such as APSR, EPSR, and IPSR, or in combination as one combined register PSR in your program. Different status is presented or reflected by using different bits on these registers. The meaning and purpose of each bit in these registers are shown in Figure 2.6 and Table 2.2.

Figure 2.6a shows the bit functions on three different status registers, APSR, IPSR, and EPSR, respectively. Figure 2.6b shows the bit functions on a combined status register PSR. The following important points must be kept in mind when we try to use these special registers:

- The running status, **N**, **Z**, **V**, and **C** bits provide information about the execution result of the previous DataPath or ALU operations. The associated bit should be set if a running result is matched. For example, bit Z is set to 1 if the execution result is zero.
- The **Q** bit is used to indicate whether a sticky saturation occurred. It is set by the SSAT and USAT instructions.

Table 2.2 Bit functions in the program status register (PSR)

Bit	Function
N	Negative flag
Z	Zero flag
C	Carry flag
V	Overflow flag
Q	Sticky saturation flag
GE[19:16]	Greater than or equal flag for each byte lane
ICI/IT	Interrupt-continuable instruction (ICI) bits/IF-THEN (IT) instruction status bit for conditional exception
T	Thumb state (always 1)
Exception Number	Indicates which exception occurred and under processed by CPU

- The GE bits are used to indicate whether a greater than or equal result happened when compared to two operands. These bits do not work for Cortex®-M3 processor.
- The EPSR cannot be accessed by user's software codes directly using MRS or MSR instructions.
- The IPSR is a read-only register and can be read with the combined PSR.

Next, let's take care of the Interrupt Exception Mask Registers.

Interrupt Exception Mask Registers: Three special registers, Primary Mask (**PRIMASK**) register, Fault Mask (**FAULTMASK**) register, and Base Priority (**BASEPRI**) register, are mainly used for interrupt or exception masking purposes.

In Cortex®-M4 system, the interrupts and exceptions have the following properties:

- All interrupts and exceptions are categorized into two major groups: Maskable or Non-Maskable, which means those that are maskable interrupts/exceptions can be masked or disabled by the CPU, but those that are nonmaskable interrupts/exceptions cannot be masked or disabled by the CPU.
- Both maskable and nonmaskable interrupts/exceptions are further divided into the different priority levels based on their importance or emergency levels. The smaller number on the priority level, the higher priority the interrupt/exception has.
- Generally, a single bit in a mask register is used to mask (disable) or unmask (enable) certain interrupt/exceptions to occurr. A 1 in that bit is to mask (disable) the selected interrupt/exception, and a 0 is to unmask (enable) the associated interrupt/exception.

The **PRIMASK** register is a Primary Interrupt Mask register, and it uses only one bit (bit 0) to mask (disable) all maskable interrupts/exceptions when this bit is set to 1, and unmask (enable) all maskable interrupts/exceptions when this bit is reset to 0.

Similarly, the **FAULTMASK** register uses its bit 0 to enable or disable all maskable interrupts/exceptions. However, one significant difference between the **PRIMASK** and the **FAULTMASK** registers is that the latter can be used to block the HardFault exception that belongs to the nonmaskable exceptions. By using this register, one can block any further fault by inhibiting the triggering of further faults during fault processing. Another difference is that the **FAULTMASK** register can be reset to 0 automatically as the exception returns.

The **BASEPRI** register provides more flexible interrupt masking strategies. Unlike **PRIMASK** register, the **BASEPRI** register can perform masking or unmasking functions based on the priority levels of related interrupts/exceptions. In this way, if a higher-level interrupt/exception is being executed or handled, any other lower-level interrupt/exception cannot be responded until the current interrupt/exception has been processed.

Unlike the **PRIMASK** and the **FAULTMASK** registers, the **BASEPRI** register uses more than one bit to handle different level interrupts and exceptions. The number of bits is determined by the total number of priority levels defined in a microcontroller system. For instance, in most Cortex®-M4 system, either 8 or 16 priority levels are adopted, and therefore this makes the **BASEPRI** register use either 3 bits or 4 bits to handle those 8 or 16 priority level interrupts.

After resetting, the **BASEPRI** register is always reset to 0, and this disables the operation of the **BASEPRI** register. A nonzero number, which is equivalent to certain priority level, in the **BASEPRI** register can be used to enable all interrupts/exceptions that have higher priority levels, and block or disable all other interrupts/exceptions that have the same or lower priority levels compared with the current interrupt priority level.

	Bits	31~3	2	1	0
Cortex-M4	CONTROL			SPSEL	TMPL
Cortex-M4 with FPU	CONTROL		FPCA	SPSEL	TMPL

Figure 2.7 Bit functions and structures of the CONTROL register.

Regularly one should use assembly language codes to access these interrupt mask registers. However, the CMSIS-Core also provides intrinsic functions to access these registers with the C codes. More detailed discussion about these accessing can be found in Section 4.5.2 in Chapter 4.

CONTROL Register: The CONTROL Register provides the following controllabilities:

- Select the stack pointer to use either main stack pointer or process stack pointer.
- Determine the access level in thread mode to use either privileged or unprivileged level.
- Indicate whether the current executed codes use the FPU or not if the Cortex®-M4 contained an FPU.

Figure 2.7 shows the bit functions and structures for Cortex®-M4 MCU with and without FPU. The function of each bit (bits 0~2) on this CONTROL register is:

- **`TMPL`** (Bit-0): Thread Mode Privilege Level. This bit is used to define the privileged level in the thread mode. Under thread mode, it is privileged level when this bit is 0 (default). It is unprivileged level when this bit is 1. The processor is always in privileged level when it works in the handler mode.
- **`SPSEL`** (Bit-1): Select the Stack Pointer. Under the thread mode, it uses main stack pointer (MSP) when this bit is 0 (default). Otherwise, it uses process stack pointer (PSP) when this bit is 1. When working in the handler mode, this bit is always 0. This means that the handler mode always uses MSP.
- **`FPCA`** (Bit-2): Floating-Point Context Active bit. This bit is only available in the Cortex®-M4 with an FPU involved. The exception handler uses this bit to determine whether registers in the FPU need to be saved when an exception occurred. When this bit is 0 (default), it indicates that no FPU has been used and therefore there is no need to save any register in FPU. However, if this bit is set to 1, an FPU has been used and related registers in the FPU need to be saved. The FPCA bit is automatically set when a floating-point instruction is executed. This bit can be cleared by hardware on exception entry.

Bits 3~31 in this register are reserved for future usage. More detailed discussion about the CONTROL register can be found in Sections 4.5.2 and 4.5.3 in Chapter 4.

2.2.1.3 Architecture of the floating-point registers

The Cortex®-M4 MCU provides an optional FPU. Additional registers are needed to support floating data operations if this FPU is used. These registers include **`Floating-Point Data Processing Register`** (**FPDPR**) and **`Floating-Point Status and Control Register`** (**FPSCR**).

The FPDPR comprises 32 single-precision registers, **`S0~S31`**, or 16 double-precision registers, **`D0~D15`**, respectively. Each of the 32-bit single-precision registers, S0~S31, can be

Floating point unit (FPU)		
S1	S0	D0
S3	S2	D1
S5	S4	D2
S7	S6	D3
S9	S8	D4
S11	S10	D5
S13	S12	D6
S15	S14	D7
S17	S16	D8
S19	S18	D9
S21	S20	D10
S23	S22	D11
S25	S24	D12
S27	S26	D13
S29	S28	D14
S31	S30	D15
FPSCR		

Figure 2.8 Configuration of the floating-point registers.

accessed using floating-point instructions. These registers can also be accessed as a pair or double-precision registers D0~D15 (64-bit). The configuration of these registers is shown in Figure 2.8.

One point to be noted is that the FPU in the Cortex®-M4 does not support double-precision floating-point calculations, but you can still use floating-point instructions to transfer double-precision data.

All floating-point data calculations are under the control of the **FPSCR**. This register provides the following control functions:

- Define the floating-point operation behaviors.
- Provide status information about the floating-point operation results.

Bits functions on the FPSCR are shown in Figure 2.9.

The functions of bits **N**, **Z**, **C**, and **V** are identical with those in the **PSR**. The function of each other bit in the **FPSCR** is (bits 5~6, 8~21 and 27 are reserved)

- **AHP** (Bit-26): The value on this bit defines the alternative half-precision format for the floating-point operations. A 0, which is the default value on this bit, is used to define the IEEE half-precision format. A 1 is to define an alternative half-precision format.

Bits	31 30 29 28	27	26	25	24	23:22	21:8	7	6:5	4	3	2	1	0
FPSCR	N Z C V		AHP	DN	FZ	RMode		IDC		IXC	UFC	OFC	DZC	IOC

Figure 2.9 Bit function and structure on FPSCR.

- **DN** (Bit-25): The value on this bit is used to define the default not a number (NaN) mode. A 0 means that the NaN operands propagate through to the output of a floating-point operation, and this is the default value. A 1 indicates that any operation including one or more NaNs returns the default NaN.
- **FZ** (Bit-24): The value on this bit indicates whether the flush-to-zero model is enabled or disabled. A value of 0, which is the default value, on this bit means that the FZ model is disabled, otherwise if this bit value is 1, it means that the FZ model is enabled.
- **RMode** (Bits 23 and 22): These two bits are used to set up the specified rounding mode that is used by all floating point operational instructions. The values of these bits are:
  - 00—round to nearest (RN) mode (default)
  - 01—round to plus infinity (RP) mode
  - 10—round to minus infinity (RM) mode
  - 11—round to zero (RZ) mode
- **IDC** (Bit-7): This bit is used to monitor whether a floating-point exception has occurred. A 1 indicated that a floating-point exception has happened, and the result is not within the normalized value range. A 0 means that no floating-point exception occurred. This bit can be cleared by writing 0 to it.
- **IXC** (Bit-4): This bit is used to detect whether an inexact cumulative exception occurred. A 1 in this bit indicated that a floating exception has occurred, otherwise a 0 means that no floating-point exception occurred. This bit can be cleared by writing 0 to it.
- **UFC** (Bit-3): This bit is the underflow cumulative exception status bit. A 1 in this bit indicated that an underflow cumulative exception has occurred. Otherwise if this bit is 0, it means that no underflow cumulative exception has occurred. This bit can be cleared by writing 0 to it.
- **OFC** (Bit-2): This bit is the overflow cumulative exception status bit. A 1 in this bit indicated that an overflow cumulative exception has occurred. Otherwise if this bit is 0, it means that no overflow cumulative exception has occurred. This bit can be cleared by writing 0 to it.
- **DZC** (Bit-1): This bit is the divided by zero cumulative exception status bit. A 1 in this bit indicated that a divided by zero cumulative exception has occurred. Otherwise if this bit is 0, it means that no divided by zero cumulative exception has occurred. This bit can be cleared by writing 0 to it.
- **IOC** (Bit-0): This bit is the invalid operation cumulative exception status bit. A 1 in this bit indicated that an invalid operation cumulative exception has occurred. Otherwise if this bit is 0, it means that no invalid operation cumulative exception has occurred. This bit can be cleared by writing 0 to it.

More discussions about the memory architecture in the MSP432P401R MCU system will be given in Chapter 6.

The introduction and discussion about the NVIC features will be provided in Chapter 5.

2.3 *Introduction to MSP-432™ MCU family member—MSP432P401R*

Texas Instrument's MSP-432™ family microcontrollers provide developers a high-performance ARM® Cortex®-M-based architecture with a broad set of integration capabilities and a strong ecosystem of software and development tools. Targeting performance and flexibility, the MSP432P401R MCU architecture offers a 48 MHz Cortex®-M4F with FPU and MPU, a variety of integrated on-chip memory and multiple programmable GPIO.

MSP-432™ family devices offer consumers compelling cost-effective solutions by integrating application-specific peripherals and providing a comprehensive library of software tools, which minimize board costs and design-cycle time. Offering quicker time-to-market and cost savings, the MSP-432™ family microcontrollers are the leading choice in high-performance and low-power 32-bit applications.

This section contains an overview and details about one of the MSP-432™ family microcontrollers, MSP432P401R.

2.3.1 *MSP432P401R microcontroller overview*

The MSP432P401R MCU is a low-power and high-performance embedded controller with multiple functions and advanced features. Unlike ARM® Cortex®-M4, this MCU contains quite of few components, such as on-chip memory and some on-chip peripherals as well as various peripheral device interfaces, and integrates them into this chip. The main components embedded in this MCU include (Figure 2.10):

- A 32-bit ARM® Cortex®-M4F Processor Core with a FPU and an MPU. The CPU speed is 48 MHz with 1.196 DMIPS/MHz performance rate.
- On-chip memory devices include:
  - 256-KB single-cycle flash main memory (2–128 KB banks)
  - 16-KB flash information memory (2–8 KB banks)
  - 64-KB SRAM (including 8-KB backup memory)
  - 32-KB ROM loaded with MSPWare driver libraries
- One system timer (SysTick)
- Eleven GPIO ports (ports 1~10 and Port J)
- One high-speed 14-bit ADC up to 1 MSPS (mega-samples per second)
- Two independent integrated analog comparators (IACs)
- Four 16-bit general-purpose timers (GPTMs) (Timer_A) and pulse width modulation (PWM) modules
- Two 32-bit GPTM modules (Timer32)
- One Watchdog timer module (WDT_A)
- Four Enhanced Universal Serial Communication Interface A (eUSCI_A) modules for UARTs or SPI. Four Enhanced Universal Serial Communication Interface B (eUSCI_B) modules for I2C or SPI
- One eight-channel direct memory access (DMA) and one real-time clock (RTC)
- One internal voltage reference and one internal temperature sensor
- Intelligent ultra-low-power design power consumption as low as 25 nA (LPM4.5 mode)

A functional block diagram of the MSP432P401R MCU is shown in Figure 2.11.

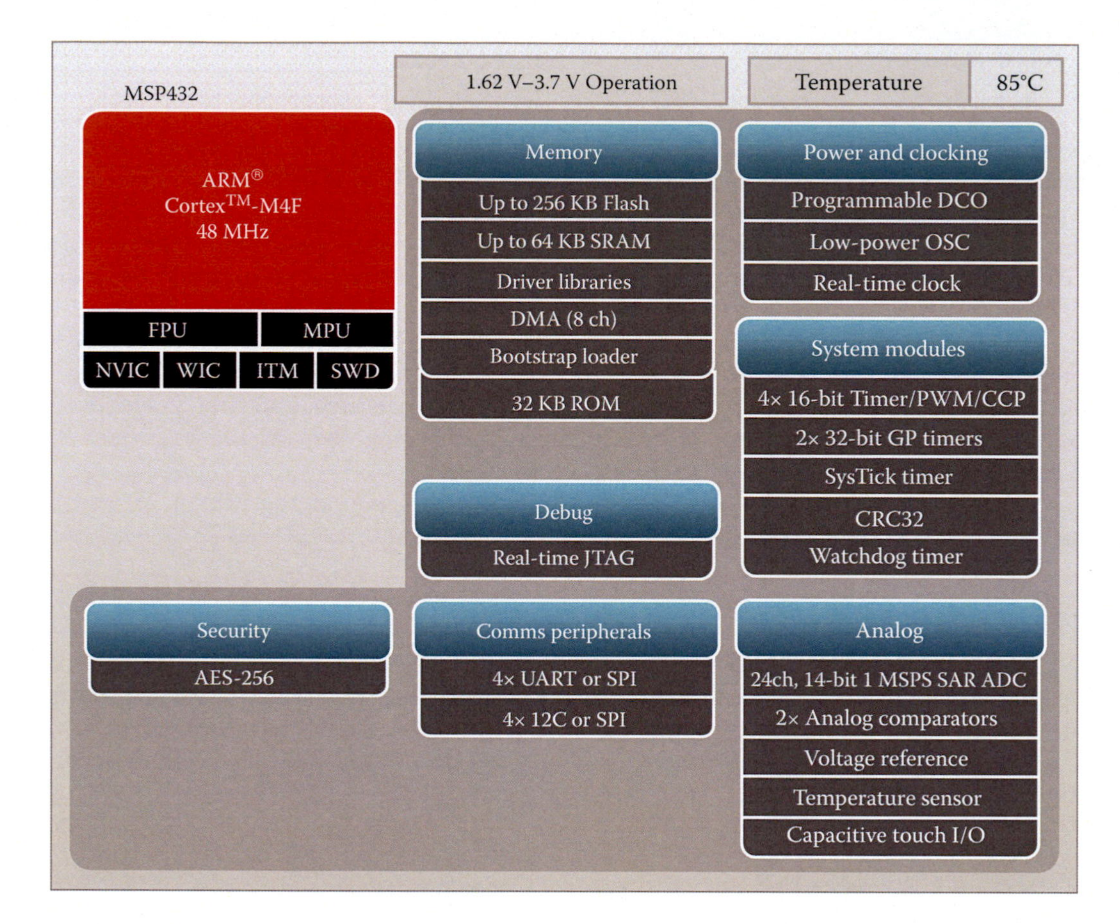

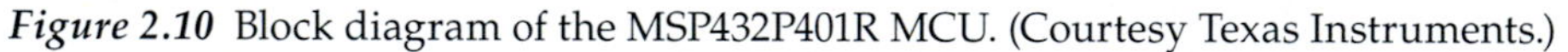
Figure 2.10 Block diagram of the MSP432P401R MCU. (Courtesy Texas Instruments.)

2.3.2 *MSP432P401R microcontroller on-chip memory map*

The MSP432P401R MCU contains the following memory devices on this chip:

- 256 KB single-cycle flash main memory
- 16 KB flash information memory
- 64 KB SRAM (Including 8 KB backup memory)
- 32 KB ROM loaded with MSPWare driver libraries

The MSP432P401R MCU supports a 4-GB address space that is divided into eight 512 MB zones. A global memory map for the MSP432P401R MCU is shown in Figure 2.12.

All on-chip memory devices are controlled by the related control registers, such as flash control registers, ROM control registers, and SRAM control registers. These registers are located at the associated memory spaces.

The **`256 KB flash memory`** is used to store the user's program codes and exception/interrupt vector tables. The exception and interrupt vector tables are located at the lower memory space starting from `0x0000.0000`. To perform any programming for this flash memory, the MSP432™ family devices provide a user-friendly interface with related registers. All reading, erasing, or programming operations are handled via related

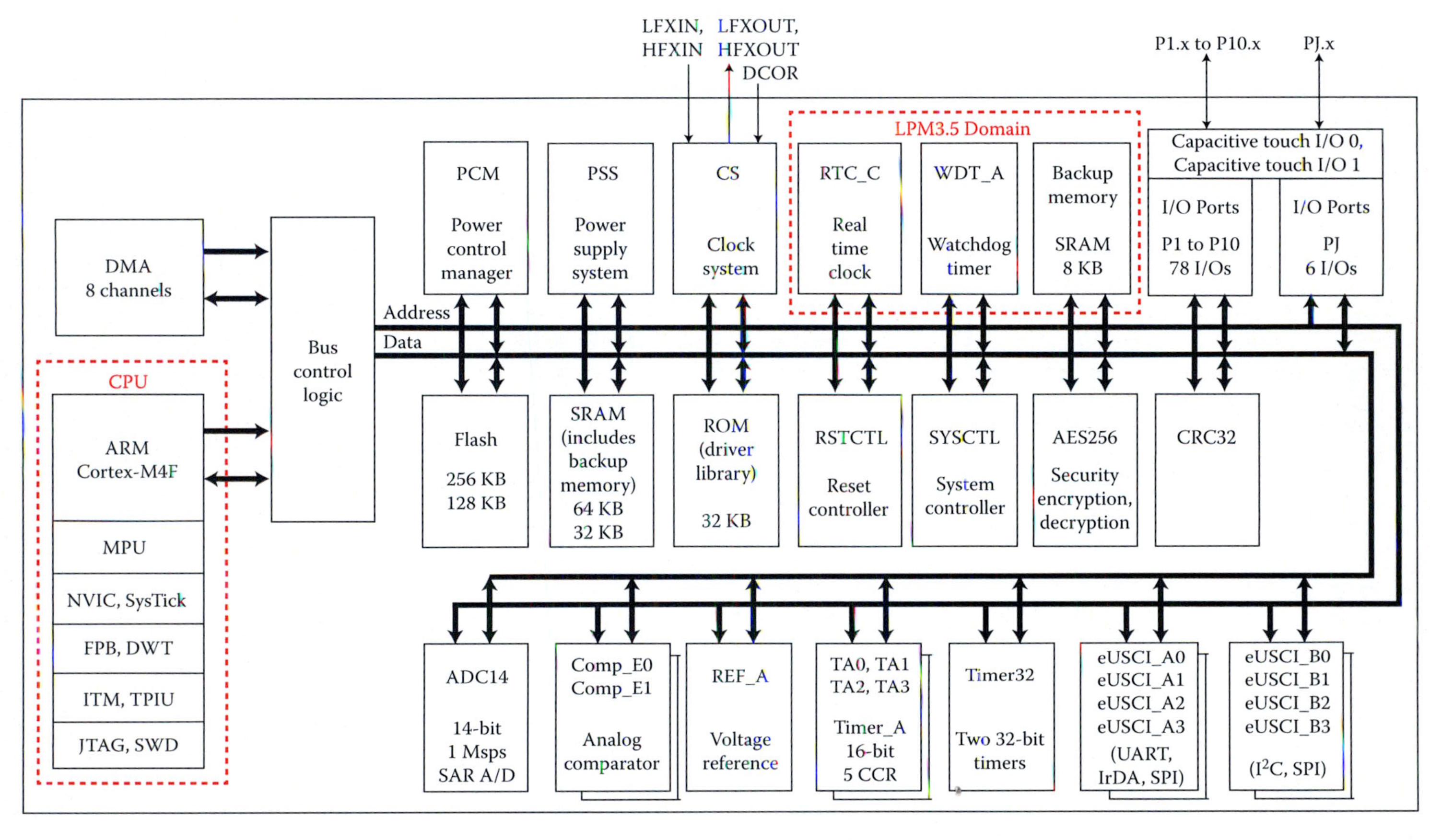

Figure 2.11 Functional block diagram of the MSP432P401R MCU. (Courtesy Texas Instruments.)

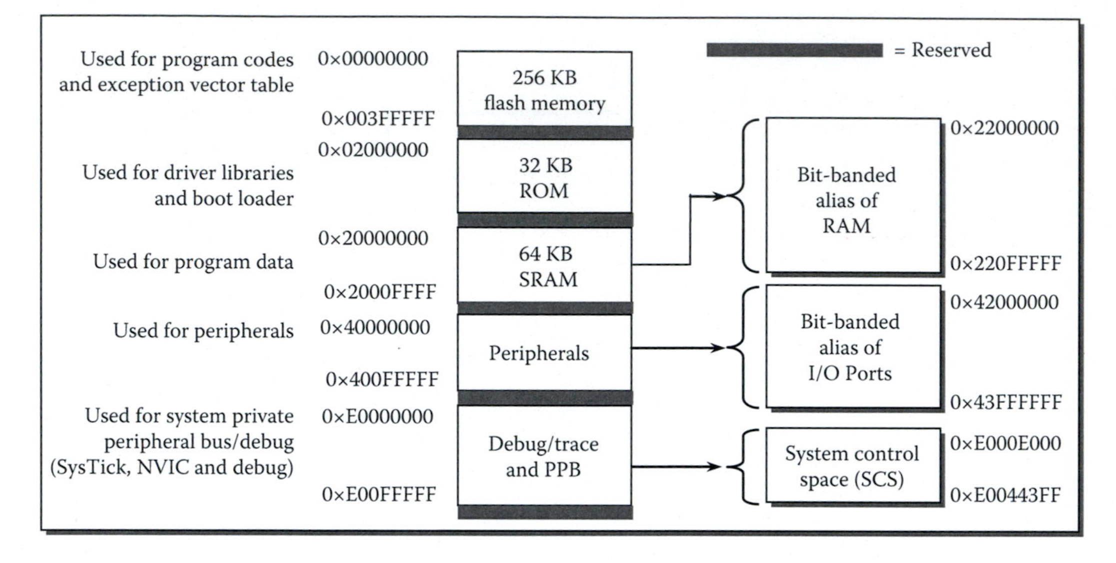

Figure 2.12 Memory map for the MSP432P401R MCU.

operational modes and registers: **flash memory read mode, flash memory program mode**, and **flash memory erase mode**.

The flash memory consists of two independent equal-sized memory banks, each containing the following regions: (1) 256-KB main memory region (two 128 KB banks), this is the primary code memory and is for code/data for the user application; (2) 16-KB information memory region (two 8 KB banks), it is for TI or customer code/data. Some of the information memory sectors could be used by TI and others will be available for users.

The MSP432P401R MCU supports 32 KB of ROM memory. The lower 1 KB of the ROM is reserved for TI internal purposes and accesses to this space will return an error response. The rest of the ROM is used for driver libraries.

The **Internal ROM** can be preprogrammed with the following software and programs:

- MSPW are driver libraries, including the peripheral devices library, USB library and graphical library
- MSP432 MCU bootstrap loader (BSL)
- Advanced encryption standard (AES) cryptography tables
- Cyclic redundancy check (CRC) error-detection functionality

The MSP432 MCU BSL is used to download code to the flash memory of a device without the use of a debug interface. When the core is reset, the user has the opportunity to direct the core to execute the ROM BSL or the application in flash memory.

AES is ideal for applications that can use prearranged keys, such as setup during manufacturing or configuration.

The CRC technique can be used to validate correct receipt of messages (nothing lost or modified in transit), to validate the data after decompression, to validate that flash memory contents have not been changed, and for other cases where the data needs to be validated.

The 1 MB **peripheral** region from **0x4000.0000** to **0x400F.FFFF** is dedicated to the system and application control peripherals of the device. On the MSP432P401x devices, a total of 128 KB of this region is dedicated for peripherals, while the rest is marked as reserved. The main on-chip peripheral devices include:

- Four 16-bit GPTMs
- Two 32-bit timers
- One Watchdog timer
- One ADC
- Two analog comparators
- Flash controller
- One real-time controller (RTC)

The main peripheral interfaces used for the external I/O devices include:

- Four eUSCI_A modules for UART or SPI
- Four eUSCI_B modules for I2C or SPI
- Eleven 8-bit GPIO ports modules (P1~P10, and PJ)
- PMAPC
- Capacitive Touch I/O 0 interface
- Capacitive Touch I/O 1 interface

A global description about the memory mapping regions for the MSP432P401R MCU is shown in Table 2.3.

To get more detailed memory map information and distributions for all the devices and components used in the MSP432P401R MCU, refer to Chapter 6.

Basically, all peripherals in the MSP432P401R MCU system can be divided into the following groups:

1. System peripherals
2. On-chip peripherals
3. Interfaces to external parallel and serial peripherals

For the interfaces to external serial peripherals, they can be further categorized into another two subgroups; synchronous and asynchronous communication mode.

2.3.3 System peripherals

The **`system peripherals`** are related controls to the system peripherals, and the most popular system peripherals involved in this MCU are

- SysTick
- NVIC
- FPU
- MPU

2.3.3.1 On-chip peripherals

The **`on-chip peripherals`** are controls or components integrated on the MSP432P401R chip, which include:

- One 14-bit ADC14
- Four 16-bit and two 32-bit GPTMs

Table 2.3 Memory map for the MSP432P401R MCU

Start address	End address	Descriptions
		256 KB on-chip flash memory
0x0000.0000	0x0003.FFFF	Flash main memory (exceptions/interrupt vector table, system control and user codes)
0x0004.0000	0x001F.FFFF	Reserved
0x0020.0000	0x0020.3FFF	Flash information memory region (flash boot-override mailbox and device descriptor (TLV))
0x0020.4000	0x003F.FFFF	Reserved
		32 KB ROM
0x0200.0000	0x0200.03FF	Reserved for TI for future usage (1 KB)
0x0200.0400	0x020F.FFFF	MSP432 driver libraries, boot loader, AES & CRC
		64 KB SRAM
0x2000.0000	0x200F.FFFF	User data
0x2200.0000	0x23FF.FFFF	Bit-band alias of bit-banded on-chip SRAM starting at 0x2000.0000
		Peripherals
0x4000.0000	0x4000.03FF	Timer_A0
0x4000_0400	0x4000_07FF	Timer_A1
0x4000.0800	0x4000.0BFF	Timer_A2
0x4000.0C00	0x4000.0FFF	Timer_A3
0x4000.1000	0x4000.13FF	eUSCI_A0
0x4000.1400	0x4000.17FF	eUSCI_A1
0x4000.1800	0x4000.1BFF	eUSCI_A2
0x4000.1C00	0x4000.1FFF	eUSCI_A3
0x4000.2000	0x4000.23FF	eUSCI_B0
0x4000.2400	0x4000.27FF	eUSCI_B1
0x4000.2800	0x4000.2BFF	eUSCI_B2
0x4000.2C00	0x4000.2FFF	eUSCI_B3
0x4000.3000	0x4000.33FF	REF_A
0x4000.3400	0x4000.3BFF	COMP_E0 & COMP_E1
0x4000.3C00	0x4000.3FFF	AES256
0x4000.4000	0x4000.43FF	CRC32
0x4000.4400	0x4000.47FF	RTC_C
0x4000.4800	0x4000.4BFF	WDT_A
0x4000.4C00	0x4000.4FFF	Port module
0x4000.5000	0x4000.53FF	Port mapping controller
0x4000.5400	0x4000.57FF	Capacitive touch I/O 0
0x4000.5800	0x4000.5BFF	Capacitive touch I/O 1
0x4000.C000	0x4000.CFFF	Timer32
0x4000.E000	0x4000.FFFF	DMA
0x4001.0000	0x4001.03FF	PCM

(*Continued*)

Table 2.3 (Continued) Memory map for the MSP432P401R MCU

Start address	End address	Descriptions
0x4001.0400	0x4001.07FF	CS
0x4001.0800	0x4001.0FFF	PSS
0x4001.1000	0x4001.17FF	Flash controller
0x4001.2000	0x4001.23FF	ADC14
0x4200.0000	0x43FF.FFFF	Bit-banded alias of 0x4000.0000 through 0x400F.FFFF
Debug trace and private peripheral bus		
0xE000.0000	0xE000.2FFF	ITM, DWT, and FPB
0xE000.E000	0xE000.EFFF	Cortex-M4F system peripherals (SysTick, NVIC, MPU, FPU, and SCB)
0xE004.0000	0xE004.0FFF	Trace port interface unit (TPIU)
0xE004.2000	0xE004.23FF	Reset controller
0xE004.3000	0xE004.43FF	System controller
0xE00F.F000	0xE00F.FFFF	External PPB—ROM table

- Two analog comparators COMP_E0 and COMP_E1
- One shared voltage reference REF_A
- One temperature sensor
- One Watchdog timer
- One DMA controller

2.3.3.2 *Interfaces to external parallel and serial peripherals*

The MSP432P401R MCU provides a set of GPIO port modules. These modules contain eleven GPIO ports (P1~P10 and PJ), each GPIO port can be programmable or configurable to provide multiple or alternate functions to enable the port to handle different tasks.

Nine ports, P1~P9, are 8-bit ports containing eight I/O lines; however, some ports may contain less, such as P10 and PJ, which contain six I/O pins. Totally, 10 ports, P1~P10, provide 78 I/O lines and Port J provides 6 I/O lines. Each port can work as either digital input or output port, analog input port, either parallel or serial port. Furthermore, each I/O pin (line) is individually configurable for input or output direction, and each can be individually read or written. Each I/O line is individually configurable for pull-up or pull-down resistors. Also, each bit or pin on some GPIO ports (P1~P6) can be configured as an interrupt source to create an associate interrupt request to the Cortex®-M4 Core. These interrupt requests can be configured to be triggered by either a rising or falling edge for P1 through P6. Wake-up capability from low-power modes, such as LPM3, LPM4, LPM3.5, and LPM4.5 modes can be programmed over ports P1 through P6.

The GPIO ports are ranged from P1 through P10, and PJ. Also, each two pair or neighboring ports, such as P1 and P2, P3, and P4, until P9 and P10, can be combined together to form a 16-bit port named PA, PB, until PE.

One point to be noted is that although PJ is considered as a general I/O port, in most devices this port will be utilized by JTAG debugger. Therefore, it is recommended to leave this port for the debugging purpose.

Moreover, the MSP432P401R MCU provides a PMAPC to enable some GPIO ports can be reconfigured to perform multiple digital functions dynamically. This PMAPC allows reconfigurable mapping of digital functions over some ports, such as P2, P3, and P7.

A set of port mapping (**PM** or **PMAP**) registers can be used to support the mapping functions performed by the PMAPC.

On the MSP432P401R MCU system, all peripheral devices, including on-chip and external peripherals, are interfaced with the MCU via these GPIO ports, P1~P10, controlled by either the general digital controller (GDC) or the PMAPC.

Because the GPIO plays such a vital role in the interfaces to all peripheral devices, now let's take a closer look at these ports and related interfaces.

2.3.4 MSP432P401R microcontroller GPIO modules

As we mentioned, the GPIO modules comprise 11 physical GPIO ports, each port for Port 1~Port 10 is corresponding to an individual 8-bit or 6-bit I/O port. Most peripherals are interfaced to the MSP432P401R MCU via these GPIO ports. With the help of the GDC or PMAPC, each GPIO port can be configured to perform multiple functions or reconfigured to perform a special function.

To make it convenient for users to perform these interfaces between those GPIO pins and related peripherals, all of these GPIO ports and related pins are connected to four 10-pin jumper connectors J1~J4 on the MSP-EXP432P401R evaluation board.

Figure 2.13 shows a functional block diagram to illustrate these GPIO pins with multiple programmable functions. When using this pin payout, the following points are to be noted:

- In Figure 2.13a, J1 and J3 on the right-hand side are the jumper connectors on the MSP-EXP432P401R EVB, and the most left side are possible connected peripherals.
- In Figure 2.13b, J2 and J4 on the left-hand side are the jumper connectors on the MSP-EXP432P401R EVB, and the most right side are possible connected peripherals.
- The multiple signal boxes with different colors between the EVB and possible peripherals are multiple or alternate functions provided by the same GPIO pins.
- Each GPIO port and pin is defined as **Port_number.Pin_number**. For example, the GPIO Port 1 with pin 0 is defined as P1.0, GPIO Port 4 with pin 3 as P4.3, and so on.
- The pin number on each jumper connector J1~J4 starts at 1 from the top until 10 on the bottom.
- One example is pin **P3.2**. This pin can work as a normal digital I/O pin, or it can be configured to work as an UART RX input (**UCA2RXD**) pin, or an UCA2 Slave Output Master Input (**UCA2SOMI**) pin, by the PMAPC.
- The I/O pins with an exclamation symbol (**!**) indicated that those pins have interrupt capable ability, and an interrupt can be applied on those pins.

Before we can continue to discuss the GPIO programming process, let's first take a look at the system clock since this is an important component in the MCU system.

2.3.4.1 MSP432P401R GPIO ports architecture

Figure 2.14 shows a functional block diagram of the GPIO module used in the MSP432P401R MCU. It can be found that the GPIO module is controlled by two-layer controllers: (1) the regular general digital I/O control and (2) the PM control. All these controls are performed by using two sets of registers.

First, let's take care of the regular general digital I/O control.

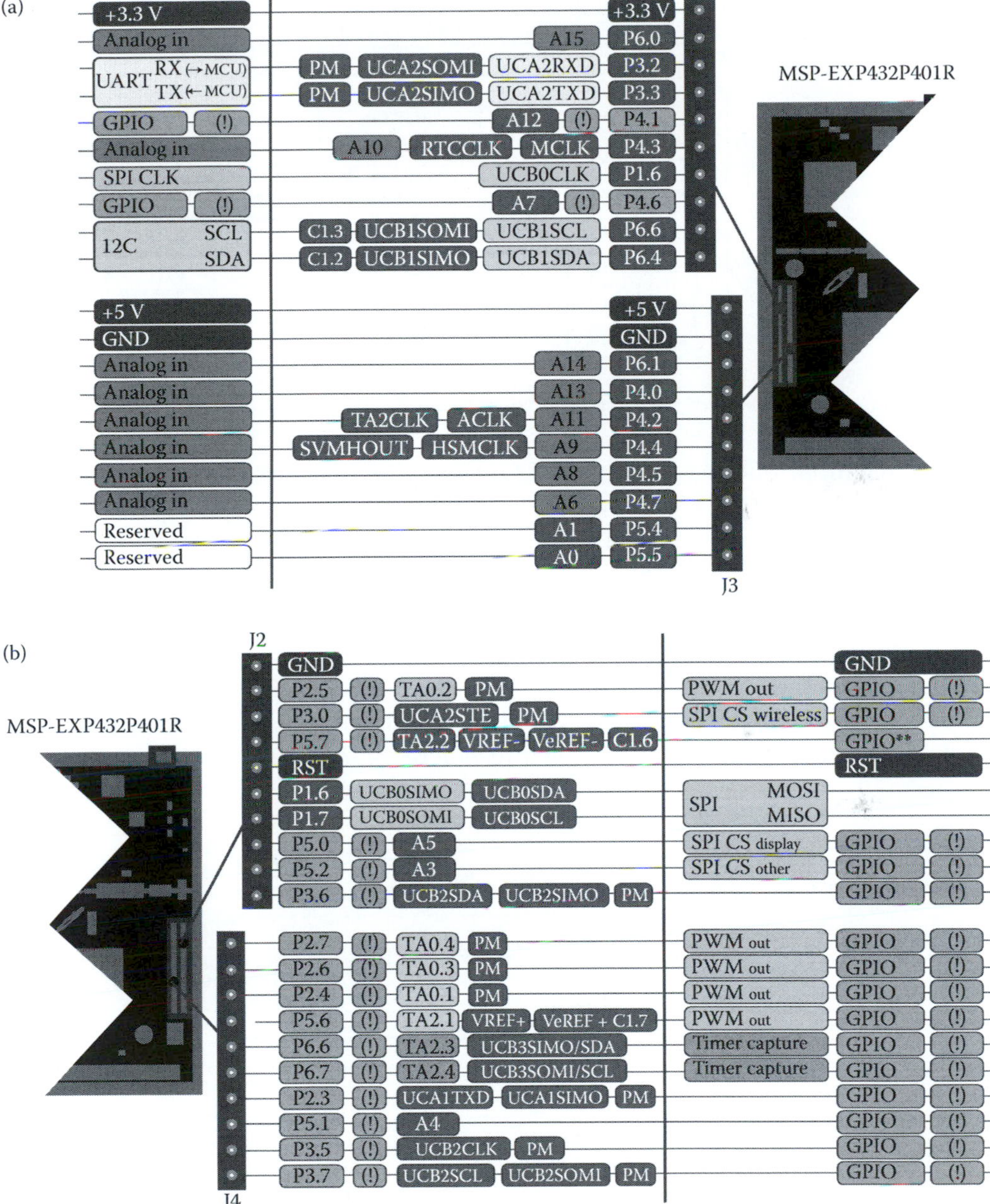

Figure 2.13 20-pin standard layout for MSP-EXP432P401R EVB. (Courtesy of Texas Instruments.)

2.3.4.2 General digital I/O function control

For each GPIO port and GPIO pin, a set of registers are used to configure each of them. The **x** used in the figure indicates the port number, which is arranged from 1 to 10. This means that each port has an individual set of similar registers used to control its function.

For example, **P1IN**, **P1OUT**, **P1DIR**, **P1REN**, and **P1DS**, are used to control input/output functions for GPIO Port 1. The registers **P1SEL0**, **P1SEL1**, and **P1SELC** are used to control

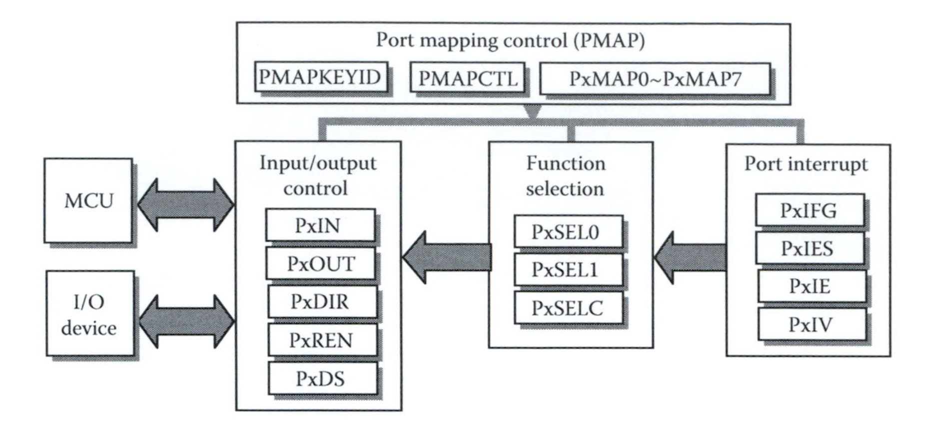

Figure 2.14 Functional block diagram of GPIO ports.

the additional input/output functions for GPIO Port 1. Similarly, the **P1IFG**, **P1IES**, **P1IE**, and **P1IV** registers are used to control the interrupt functions for GPIO Port 1.

To better understand these control registers and their functions, first let's take a look at the bit configuration for each GPIO port as shown in Figure 2.15.

Each GPIO port and pin (on those selected ports and pins) is controlled by a set of related registers shown in Figure 2.15 and each of them is expressed as **PyFunction.x**, where **y** indicates the port number and **x** means the pin number, and **Function** shows the desired operation.

For example, **P1REN.0** means the pull-up or pull-down resistor enable register for the GPIO Port 1 on pin 0. **P2DIR.3** means the direction register for GPIO Port 2 on pin 3.

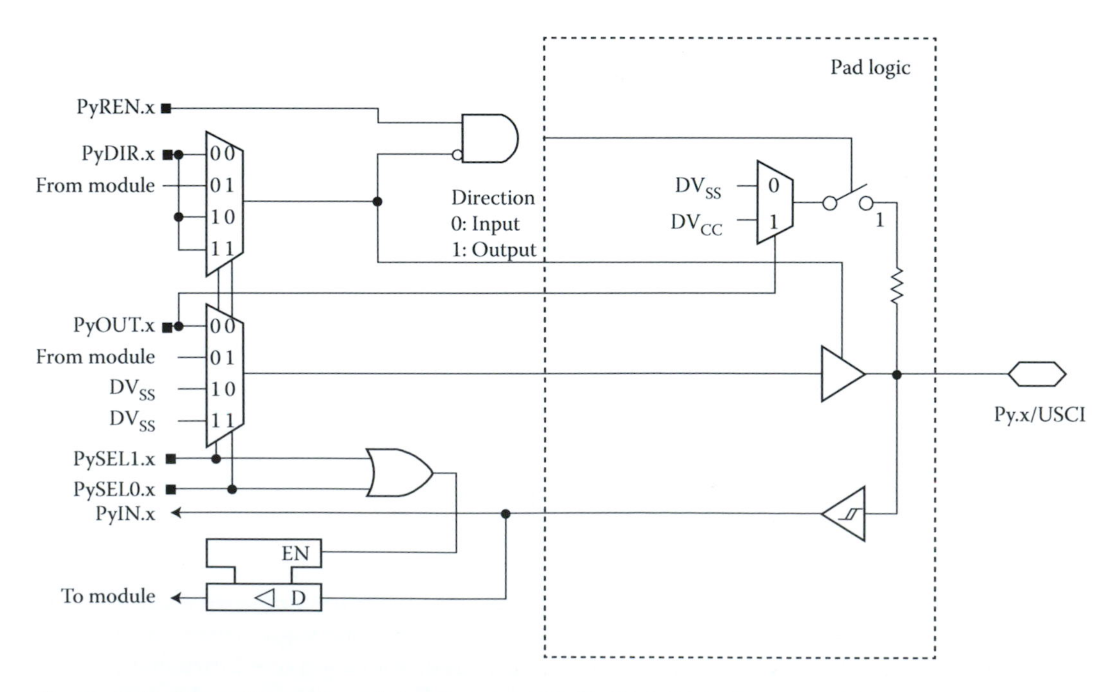

Figure 2.15 Bit configuration of GPIO ports (P1, P3, P2.0~P2.3, P9.4~P9.7, P10.0~P10.3). (Courtesy Texas Instruments.)

There are ten 8-bit I/O ports implemented in the MSP432P401R MCU system with the following functions:

- All individual I/O bits are independently programmable
- Any combination of input, output, and interrupt conditions is possible
- Any pin can be programmable in pull-up or pull-down mode
- Edge-selectable interrupt capability is available on ports P1 through P6
- Wake-up capability from LPM3, LPM4, LPM3.5, and LPM4.5 modes over ports P1 through P6
- Read/write access to port-control registers is supported by all instructions
- Ports can be accessed byte-wise or in pairs (16-bit widths)
- Capacitive touch functionality is supported on all pins of ports P1 through P10 and PJ
- Glitch filtering capability on selected digital I/Os

An import issue in using these registers is that there are two different techniques, port control and Port Mapping, used by the GPIO ports.

The port control used by the port function selection registers, **PxSEL0** and **PxSEL1**, can be used to select one of four different alternate functions;

1. GPIO function
2. The primary module function
3. The secondary module function
4. The tertiary (third) module function

The port control executed by using the **PxSEL0** and **PxSEL1** registers is available to all GPIO ports, P1~P10 even PJ. All functions that can be selected are fixed with definite order and determined by the MCU used in the project.

The Port Mapping is a different technique used to enable users to reconfigure some ports functions dynamically (during the project running), such as Ports 2, 3, and 7. Unlike the port control technique controlled by the **PxSEL0** and **PxSEL1** registers, these mappings are flexible and determined by the user in real time.

Now, let's have a closer look at each of these registers and its function based on Figure 2.15. All of these registers are 8-bit registers with each bit being mapped to each pin on the selected GPIO port.

2.3.4.2.1 Port input registers (PxIN) Each bit in each **PxIN** register stores the value of the input signal at the corresponding I/O pin when the pin is configured as I/O function. These registers are read only. A value of 0 on a bit means that the input bit is low, and a 1 indicates that the bit input is high. The port number **x** = 1~10 and J.

2.3.4.2.2 Port output registers (PxOUT) Each bit in each **PxOUT** register is the value to be output to the corresponding I/O pin when the pin is configured as I/O function. A value of 0 on a bit means that the output bit is low, and a 1 indicates that the bit output is high. The port number **x** = 1~10 and J.

If a pin is configured as an input I/O function (Port Direction Register **PxDIR** = 0), the pin can be configured as a pull-up or pull-down pin with a pull-up or pull-down resistor.

To enable the pull-up or pull-down resistor, the corresponding bit in the **PxOUT** register can be configured to select a pull-up or a pull-down resistor:

- **Bit** = 0: Pin is pulled down
- **Bit** = 1: Pin is pulled up

2.3.4.2.3 Port direction registers (PxDIR) Each bit on each **PxDIR** register selects the transfer direction of the corresponding I/O pin when it is configured as I/O function. In most of the cases, **PxDIR** register also controls the direction of the I/O when it is configured as peripheral functions. **PxDIR** bits for I/O pins that are selected for peripheral functions must be set as required by the peripheral functions. For certain secondary functions, such as **eUSCI**, the I/O direction is controlled by the secondary function itself and not by the **PxDIR** register.

A value of **0** on a bit indicates that the bit works as an **Input** pin, and a **1** means that the bit is an **Output** pin.

2.3.4.2.4 Port pull-up/pull-down resistor enable registers (PxREN) Each bit in each **PxREN** register enables or disables the pull-up or pull-down resistor of the corresponding I/O pin. The corresponding bit in the **PxOUT** register selects if the pin contains a pull-up or pull-down resistor.

- **Bit** = 0: Pull-up or pull-down resistor disabled
- **Bit** = 1: Pull-up or pull-down resistor enabled

This register must be used with the **PxDIR** and **PxOUT** registers together to determine the input bit's pull-up or pull-down status. Table 2.4 lists these combinations.

Refer to Figure 2.15, if **PxSEL0** and **PxSEL1** = 00, and **PxDIR:PxREN:PxOUT** = 010, **PxDIR** = 0 will be transferred to the inverting input of the **AND** gate via first channel (00) on the MUX. This 0 will be inverted to 1 after the inverter and **ANDed** with the **PxREN** = 1 to produce logic 1 to close the switch on the pad logic part. This closed switch will connect the resistor to the ground Vss if **PxOUT** = 0 to make this input pin as a pull-down pin. However, if the **PxOUT** = 1, this closed switch will connect the resistor to the power supply (Vcc) to make the input pin as a pull-up pin. The input signal is transferred to the **PxIN** pin.

2.3.4.2.5 Port function select registers (PxSEL0, PxSEL1) As we mentioned, the port control executed by using the **PxSEL0** and **PxSEL1** registers can select one of four module functions for each I/O pin. Table 2.5 shows all of these four possible functions based on **PxSEL0** and **PxSEL1** registers' value combinations.

The pin configurations shown in Figure 2.15 are only valid for selected ports and pins, such as **P1**, **P3**, **P2.0~P2.3**, **P9.4~P9.7**, and **P10.0~P10.3**. For the rest of ports and pins, refer to Sections 7.3 and 7.4 in Chapter 7.

Table 2.4 Combinations of PxDIR, PxOUT, and PxREN registers

PxDIR	PxREN	PxOUT	I/O description
0	0	x	Input
0	1	0	Input with pull-down resistor
0	1	1	Input with pull-up resistor
1	x	x	Output

Table 2.5 I/O function selection based on PxSEL0 and PxSEL1 registers

PxSEL1	PxSEL0	I/O description
0	0	General purpose I/O function selected
0	1	Primary module function selected
1	0	Secondary module function selected
1	1	Tertiary (third) module function selected

Refer to Figure 2.15, these two bits, **PxSEL1** and **PxSEL0**, control two MUXes to select one of four possible functions. Table 2.6 shows some example module function selections for **P1.0**~**P1.7**. For more function selections, refer to Section 7.3 in Chapter 7.

In Figure 2.15, for the selection **PxSEL1:PxSEL1** = 01, it indicates **from module**. This means that different ports have different secondary module functions.

Setting the **PxSEL1** or **PxSEL0** bits to a module function does not automatically set the pin direction. Other peripheral module functions may require the **PxDIR** bits to be configured according to the direction needed for the module function.

When a port pin is selected as an input to peripheral modules, the input signal to those peripheral modules is a latched representation of the signal at the MCU pin. While **PxSEL1** and **PxSEL0** are other than 00, the internal input signal follows the signal at the pin for all connected modules. However, if **PxSEL1** and **PxSEL0** = 00, the input to the peripherals maintains the value of the input signal at the MCU pin before the **PxSEL1** and **PxSEL0** bits are reset.

Because the **PxSEL1** and **PxSEL0** bits do not reside in contiguous addresses, changing both bits at the same time is not possible. For example, an application might need to change **P1.0** from GPIO to the tertiary module function residing on **P1.0**. Initially, **P1SEL1** = 00 h and **P1SEL0** = 00 h. To change the function, it would be necessary to write both **P1SEL1** = 01 h and **P1SEL0** = 01 h. This is not possible without first passing through an intermediate configuration, and this configuration may not be desirable from an application standpoint. The complement register **PxSELC** can be used to handle such situations.

The **PxSELC** register always reads 0. Each set bit of the **PxSELC** register complements the corresponding respective bit of the **PxSEL1** and **PxSEL0** registers. In the example, with **P1SEL1** = 00 h and **P1SEL0** = 00 h initially, writing **P1SELC** = 01 h causes **P1SEL1** = 01 h and **P1SEL0** = 01 h to be written simultaneously.

Table 2.6 Alternate function selections based on PxSEL1 and PxSEL0

Pin	PxSEL1:PxSEL0			
	00	**01**	**10**	**11**
P1.0	GPIO Port	UCA0STE	DVss	DVss
P1.1	GPIO Port	UCA0CLK	DVss	DVss
P1.2	GPIO Port	UCA0RXD/UCA0SOMI	DVss	DVss
P1.3	GPIO Port	UCA0TXD/UCA0SIMO	DVss	DVss
P1.4	GPIO Port	UCB0STE	DVss	DVss
P1.5	GPIO Port	UCB0CLK	DVss	DVss
P1.6	GPIO Port	UCB0SIMO/UCB0SDA	DVss	DVss
P1.7	GPIO Port	UCB0SOMI/UCB0SCL	DVss	DVss

A point to be noted is that when any **PxSEL** bit is set, the corresponding pin interrupt function is disabled. Therefore, signals on these pins do not generate interrupts, regardless of the state of the corresponding **PxIE** bit.

2.3.4.2.6 Port output drive strength selection registers (PxDS) There are two output types of I/O available. One is with the regular drive strength and the other is with the high drive strength. Most of the I/Os have regular drive strength while some selected I/Os have high drive strength.

PxDS register is used to select the drive strength of the high drive strength I/Os.

- **Bit** = 0: High drive strength I/Os are configured for regular drive strength
- **Bit** = 1: High drive strength I/Os are configured for high drive strength

The **PxDS** register does not have any effect on the I/O with only regular drive strength.

2.3.4.2.7 Port interrupt flag registers (PxIFG) This is an 8-bit register and each bit on each register has an interrupt flag used to indicate that an interrupt has occurred.

Each **PxIFG** bit is the interrupt flag for its corresponding I/O pin, and the flag is set when the selected input signal edge occurs at the pin. All **PxIFG** interrupt flags request an interrupt when their corresponding **PxIE** bit is set. Software can also set each **PxIFG** flag, providing a way to generate a software interrupt. If a bit is set to 1, this means that an interrupt for that pin has occurred and is pending to be responded. Otherwise if bit is 0, it means that no interrupt has occurred.

Only edges transitions, not static levels, cause interrupts.

All interrupt flags for a particular port are prioritized, with **PxIFG.0** being the highest, and combined to source a single interrupt vector. The highest priority-enabled interrupt generates a number in the **PxIV** register. This number can be evaluated or added to the PC to automatically enter the appropriate software routine. Disabled interrupts do not affect the **PxIV** value. The **PxIV** registers are half-word (16 bit) access only.

Any access (read or write) to the **PxIV** register automatically resets the highest pending interrupt flag set in the **PxIFG** register. If another interrupt flag is set, another interrupt is immediately generated after servicing the initial interrupt.

For example, assume that **P1IFG.0** has the highest priority, and both **P1IFG.0** and **P1IFG.2** flags are set in the **P1IFG** register. When the interrupt service routine accesses the **P1IV** register, **P1IFG.0** is reset automatically. After the completion of **P1IFG.0** interrupt service routine, the **P1IFG.2** generates another interrupt.

2.3.4.2.8 Port interrupt edge select registers (PxIES) Each **PxIES** bit selects the interrupt edge, either a rising edge or a falling edge, for the corresponding I/O pin.

- **Bit** = 0: Respective **PxIFG** flag is set on a low-to-high transition (rising edge)
- **Bit** = 1: Respective **PxIFG** flag is set on a high-to-low transition (falling edge)

A point to be noted is that writing to **PxIES** for each corresponding I/O may result in setting the corresponding interrupt flags.

2.3.4.2.9 Port interrupt enable registers (PxIE) Each **PxIE** bit enables the associated **PxIFG** interrupt flag bit.

Table 2.7 PxIV register and related interrupt vector values

Bits	Bit field	PxIV vector values	Interrupt source	Interrupt flag	Priority
15:5	Reserved	0000H			
4:0	PxIV	00H	No interrupt pending		
		02H	Port x.0	PxIFG.0	Highest
		04H	Port x.1	PxIFG.1	
		06H	Port x.2	PxIFG.2	
		08H	Port x.3	PxIFG.3	
		0AH	Port x.4	PxIFG.4	
		0CH	Port x.5	PxIFG.5	
		0EH	Port x.6	PxIFG.6	
		10H	Port x.7	PxIFG.7	Lowest

- **Bit** = 0: The interrupt is disabled
- **Bit** = 1: The interrupt is enabled

2.3.4.2.10 Port interrupt vector registers (PxIV) This is a 16-bit register used to reserve the interrupt vector values for eight interrupt sources via 8 bit or 8 pin on this register. Only the lowest 5 bits (bits 4~0) are used to indicate the interrupt vector values for each source. Table 2.7 shows all possible vector values for eight interrupt sources.

When the **PxIV** bit field = 02H, it means that the interrupt is generated by **Port x.0** pin, and this interrupt has the highest interrupt priority. Similarly, if **PxIV** field = 10H, it means that the interrupt is caused by **Port x.7** pin, and this interrupt has the lowest priority level.

The **PxIV** is the interrupt vector and this value will be combined with some hardware setup code together to be loaded into the PC to direct the program to enter the ISR to execute the task in the interrupt handler.

For example, the following piece of code can be used to configure Port 1 pin 0 (**P1.0**) to work as a general I/O pin:

```
P1DIR  |= BIT0;   // BIT0 = 0x1 → P1.0 as an output pin
P1SEL1 = 0x0;     // P1SEL1:P1SEL0 = 00 → General Digital I/O pin
P1SEL0 = 0x0;     // these two instructions are not necessary since by
                     default they are 0
```

The following piece of codes can be used to configure Port 4 pin 5 (**P4.5**) as an input pin with pull-up resistor:

```
P4DIR = ~BIT5;    // BIT5 = 0x20 = 00100000B. ~BIT5 = invert(BIT5) = 11011111
                     → BIT5 = 0 (input)
P4REN  |= BIT5;   // P4REN = BIT5 → P4REN5 = 1 (connect the resistor on pin5 -
                     see Table 2.4)
P4OUT = BIT5;     // P4OUT = BIT5 → P4OUT5 = 1 (connect the power Vcc - see
                     Table 2.4)
```

In software development process, all GPIO registers can be accessed by using the related macros, such as **P1DIR** → Port 1 Direction Register, **P4REN** → Port 4 Resistor Enable

Register, and so on. By default, after a system reset, all **PxSEL1** and **PxSEL0** are reset to 00, which indicates that all ports are general digital I/O ports.

The following piece of code can be used to configure Port 1 pin 5 (**P1.5**) as a secondary module function pin (**UCB0CLK**):

```
P1SEL1 = 0x0;    // make P1SEL1.5 = 0 (this code line is unnecessary since by
                    default it is 0)
P1SEL0 = 0x20;   // make P1SEL0.5 = 1 → make P1SEL1.5:P1SEL0.5 = 01 (see Table 2.6)
```

Next, let's take care of the PM control.

2.3.4.3 PM control

The **PMAPC** provides a flexible way to enable users to reconfigure some GPIO ports and pins to map them to some special function pins to perform desired functions via those pins.

The PMAPC provides the following features:

- All configurations are protected by write access key.
- A default mapping is provided for each port and pin (this default mapping is device dependent).
- Mapping can be reconfigured during runtime.
- Each output signal can be mapped to several output pins.

All of these mapping operations are performed by using a set of port mapping registers. The mapping operational principle and sequence are

- To enable access to any of the PMAPC registers, the correct key must be written into the PM Key ID Register (**PMAPKEYID**). This is a 16-bit register providing 16-bit key value. The **PMAPKEYID** register always reads **0x96A5**. Writing the key **0x2D52** to this register grants write access to all PMAPC registers. Read access is always possible since this is a read-write register.
- If an invalid key is written while write access is granted, any further write accesses are prevented. It is recommended that the application completes mapping configuration by writing an invalid key, such as 0, into the **PMAPKEYID** register.
- Each port pin, **Px.y**, provides a mapping register, **PxMAPy**. Setting this register to a certain value maps a module's input and output signals to the respective port pin **Px.y**. The port pin itself is switched from a GPIO to the selected peripheral/secondary function by setting the corresponding **PxSEL.y** bit to 1. If the input or the output function of the module is used, it is typically defined by setting the **PxDIR.y** bit.
- There are also peripherals (for example, the eUSCI module) that control the direction or even other functions of the pin (for example, open drain), and these options are documented in the mapping table.
- By default, the PMAPC allows only one configuration after a hard reset, and all mapping registers remain locked. To check the lock status for mapping registers, the **PMAPLOCKED** bit in the PM Control Register (**PMAPCTL**) can be used and this is a read-only bit. A hard reset is required to disable the permanent lock again.
- If it is necessary to reconfigure the mapping during runtime, the **PMAPRECFG** bit in the **PMAPCTL** register must be set during the first write access timeslot. If **PMAPRECFG** is cleared during later configuration sessions, no more configuration sessions are possible.

Figure 2.16 shows the bit configuration and function of the **PMAPCTL** register.

Port mapping control register (PMAPCTL)

15 ... 2	1	0
Reserved Bits 15~2.	PMAPRECFG	PMAPLOCKED

Bit#	Bit-field	Type	Reset value	Bit function
15:0	Reserved	RO	000	Read as 0 always.
1	PMAPRECFG	RW	0	Port mapping reconfiguration control bit. 0: Configuration only once. 1: Allow reconfiguration.
0	PMAPLOCKED	RO	1	Port mapping lock bit (read only). 0: Access to mapping registers is allowed. 1: Access to mapping registers is locked.

Figure 2.16 Bit configurations for PMAPCTL register.

For all mapping registers **PxMAPy**, they are ranged from P1MAP0~P1MAP7, P2MAP0~P2MAP7, until P7MAP0~P7MAP7. All of these mapping registers are 8-bit wide and can be accessed with byte writing or reading. These registers can also be accessed with half-word (16-bit) writing and reading by combining the neighboring registers, such as P1MAP0 and P1MAP1 as P1MAP01, P1MAP2, and P1MAP3 as P1MAP23, and so on.

An example for some PM values stored in the **PxMAPy** registers and mapping functions is shown in Table 2.8.

All mapped port pins provide the function **PM_ANALOG** (**0xFF**). Setting the PM register **PxMAPy** to **PM_ANALOG** disables the output driver and the input Schmitt-trigger to prevent parasitic cross currents when applying analog signals.

Table 2.8 Examples for port mapping values and functions

Value	PxMAPy value	Input pin function	Output pin function
0	PM_NONE	None	DVss
1	PM_UCA0CLK	eUSCI_A0 clock input/output (direction controlled by eUSCI)	
2	PM_UCA0RXD	eUSCI_A0 UART RXD (direction controlled by eUSCI—input)	
	PM_UCA0SOMI	eUSCI_A0 SPI slave out master in (direction controlled by eUSCI)	
3	PM_UCA0TXD	eUSCI_A0 UART TXD (direction controlled by eUSCI—output)	
	PM_UCA0SIMO	eUSCI_A0 SPI slave in master out (direction controlled by eUSCI)	
4	PM_UCB0CLK	eUSCI_B0 clock input/output (direction controlled by eUSCI)	
5	PM_UCB0SDA	eUSCI_B0 I2C data (open drain and direction controlled by eUSCI)	
	PM_UCB0SIMO	eUSCI_B0 SPI slave in master out (direction controlled by eUSCI)	
6	PM_UCB0SCL	eUSCI_B0 I2C clock (open drain and direction controlled by eUSCI)	
	PM_UCB0SOMI	eUSCI_B0 SPI slave out master in (direction controlled by eUSCI)	
7	PM_UCA1STE	eUSCI_A1 SPI slave transmit enable (direction controlled by eUSCI)	
8	PM_UCA1CLK	eUSCI_A1 clock input/output (direction controlled by eUSCI)	
9	PM_UCA1RXD	eUSCI_A1 UART RXD (direction controlled by eUSCI—input)	
	PM_UCA1SOMI	eUSCI_A1 SPI slave out master in (direction controlled by eUSCI)	
10	PM_UCA1TXD	eUSCI_A1 UART TXD (direction controlled by eUSCI—Output)	
	PM_UCA1SIMO	eUSCI_A1 SPI slave in master out (direction controlled by eUSCI)	
31(0xFF)	PM_ANALOG	Disables the output driver as well as the input Schmitt trigger to prevent parasitic cross currents when applying analog signals	

The following piece of code can be used to map Port 2 pin 0 (**P2.0**) as a Timer 0 CCR0 output pin to generate a sequence of square waveform outputs.

```
PMAPKEYID = PMAP_KEYID_VAL;   // Enable Write-access to modify port mapping
                                 registers
PMAPCTL = PMAPRECFG;          // Allow reconfiguration during runtime
P2MAP0 = PM_TA0CCR0A;         // map P2.0 as Timer 0 CCR0 pin by assigning
                                 19 to P2MAP0 register
PMAPKEYID = 0;                // Disable Write-Access to modify port
                                 mapping registers
```

Some macros are used on this piece of code, such as **PMAP_KEYID_VAL** (**0x2D52**), **PMAPRECFG** (**0x02**), and **PM_TA0CCR0A** (**19**), to simplify the coding process.

2.3.4.4 *Comparison between the digital I/O function control and PM control*

It can be found that there is no significant difference between the general digital function control and the PM control on alternate function selections. The only difference between them is

- In general digital function control, for the selected pin, the alternate function selections are performed by setting up the **PxSEL1** and **PxSEL0** registers. These **PxSELn** registers are located inside the digital function input/output module, and they perform these alternate function selections by driving the PMAPC (exactly the MUX). The alternate functions are predefined and fixed.
- In PM control, for the selected pin, the alternate functions are selected by setting up the mapping values in the **PxMAPy** register. The PMAPC (or MUX) is located outside of the digital function control module, exactly between the digital function control module and the alternate function modules. The PMAPC is exactly controlled by the mapping values in the **PxMAPy** register combined with **PxSEL1** and **PxSEL0** registers.
- For all mapping registers **PxMAPy**, from P1MAP0~P7MAP7, only **P2MAPy**, **P3MAPy**, and **P7MAPy** (**y** = 0~7) registers, have default mapping functions and have been mapped to those default alternate functions. Table 2.9 shows those default mapping functions for Port 2, P2. For default mappings of P3 and P7, refer to Chapter 7.

Now, let's take care of the GPIO ports initialization and configuration processes.

Table 2.9 Default mapping values and functions for P2

Pin	PxMAPy value	Input pin function	Output pin function
P2.0	PM_UCA1STE	eUSCI_A1 SPI slave transmit enable (direction controlled by eUSCI)	DVss
P2.1	PM_UCA1CLK	eUSCI_A1 clock input/output (direction controlled by eUSCI)	
P2.2	PM_UCA1RXD	eUSCI_A1 UART RXD (direction controlled by eUSCI—input)	
	PM_UCA1SOMI	eUSCI_A1 SPI slave out master in (direction controlled by eUSCI)	
P2.3	PM_UCA1TXD	eUSCI_A1 UART TXD (direction controlled by eUSCI—output)	
	PM_UCA1SIMO	eUSCI_A1 SPI slave in master out (direction controlled by eUSCI)	
P2.4	PM_TA0.1	TA0 CCR1 capture input CCI1A	TA0 CCR1 compare output Out1
P2.5	PM_TA0.2	TA0 CCR2 capture input CCI2A	TA0 CCR2 compare output Out2
P2.6	PM_TA0.3	TA0 CCR3 capture input CCI3A	TA0 CCR3 compare output Out3
P2.7	PM_TA0.4	TA0 CCR4 capture input CCI4A	TA0 CCR4 compare output Out4

2.3.4.5 Initialization and configuration of GPIO ports

After a system reset, all GPIO port pins are configured as digital inputs with their module functions disabled. To prevent floating inputs, all port pins, including unused ones, should be configured according to the application needs as early as possible during the initialization procedure.

2.3.4.5.1 Configuration of unused GPIO port pins

To prevent a floating input and to reduce power consumption, unused GPIO pins should be configured as general digital I/O function, output direction, and left unconnected on the EVB board. The value of the **PxOUT** bit does not matter, because the pin is unconnected. Alternatively, the integrated pull-up or pull-down resistor can be enabled by setting the **PxREN** bit of the unused pin to prevent a floating input.

A point to be noted is that the application should make sure that port PJ is configured properly to prevent a floating input. Some pins of port PJ are shared with the JTAG TDI and TDO functions, and get initialized to the JTAG functionality on reset. Other pins of Port J are initialized to high impedance inputs by default.

2.3.4.5.2 Configuration of GPIO pins for ultra-low-power modes operation

When the MSP432P401R MCU enters the LPM3, LPM4, LPM3.5, or LPM4.5 low-power operational modes, the state of the GPIO pins gets locked and stored by the MCU through the low-power modes. Upon exit from the low-power modes, this state remains locked until explicitly unlocked by the application.

If the low-power mode is LPM3.5 or LPM4.5, the configuration registers of the digital I/Os get reset; however, the locked state of I/Os ensures that the reset values do not impact the I/O operation. In this case, it is the responsibility of the application to reinitialize the configuration registers appropriately before releasing the lock condition of I/Os.

Before entering LPM3, LPM4, LPM3.5, or LPM4.5 modes, the following operations are required for the I/O devices:

1. Set all GPIO pins to GPIO (**PxSEL0** = 00 h and **PxSEL1** = 00 h) and configure as needed. Each I/O can be set to input high impedance, input with pull-down, input with pull-up, output high, or output low. It is critical that no inputs are left floating in the application; otherwise, excess current may be drawn in the low-power mode.
2. Configuring the I/O in this manner ensures that each pin is in a safe condition prior to entering the low-power mode.
3. Optionally, configure input interrupt pins for wake up from low-power modes. To wake the device from low-power modes, a GPIO port must contain an input port with interrupt and wake-up capability. Not all inputs with interrupt capability offer wake up from low-power modes.
4. To wake up the device, a port pin must be configured properly prior to entering the low-power modes. Each port should be configured as general-purpose input. Pull-downs or pull-ups can be applied if required. Setting the **PxIES** bit of the corresponding register determines the edge transition that wakes up the device. Also, the **PxIE** for the port must be enabled.

A point to be noted is that it is impossible to wake up from a port interrupt if its respective port interrupt flag is already asserted. It is recommended that the flag must be cleared prior to entering the LPM3, LPM4, LPM3.5, or LPM4.5 modes. Any pending flags in this case could then be serviced prior to the low-power mode entry.

Table 2.10 Some popular GPIO Port 1 registers in AHB bus aperture

GPIO register	Base address	Offset	Full address	SW symbolic definition
Port 1 Input Register	0x4000.4C00	0x000	0x4000.4C00	P1IN
Port 1 Output Register	0x4000.4C00	0x002	0x4000.4C02	P1OUT
Port 1 Direction Register	0x4000.4C00	0x004	0x4000.4C04	P1DIR
Port 1 Resistor Enable Register	0x4000.4C00	0x006	0x4000.4C06	P1REN
Port 1 Drive Strength Register	0x4000.4C00	0x008	0x4000.4C08	P1DS
Port 1 Select 0 Register	0x4000.4C00	0x00A	0x4000.4C0A	P1SEL0
Port 1 Select 1 Register	0x4000.4C00	0x00C	0x4000.4C0C	P1SEL1
Port 1 Interrupt Vector Register	0x4000.4C00	0x00E	0x4000.4C0E	P1IV
Port 1 Complement Select Register	0x4000.4C00	0x016	0x4000.4C16	P1SELC
Port 1 Interrupt Edge Select Register	0x4000.4C00	0x018	0x4000.4C18	P1IES
Port 1 Interrupt Enable Register	0x4000.4C00	0x01A	0x4000.4C1A	P1IE
Port 1 Interrupt Flag Register	0x4000.4C00	0x01C	0x4000.4C1C	P1IFG

As mentioned above, during LPM3, LPM4, LPM3.5, or LPM4.5 modes, the I/O pin states are held and locked based on the settings prior to the low-power entry. In the case of LPM3.5 or LPM4.5 modes, only the pin conditions are retained. All other port configuration register settings such as **PxDIR**, **PxREN**, **PxOUT**, **PxIES**, and **PxIE** are lost.

Upon exit from LPM3.5 or LPM4.5 modes, all peripheral registers are set to their default conditions but the I/O pins remain locked while the **LOCKLPM5** bit in the **PCM** is set. Keeping the I/O pins locked ensures that all pin conditions remain stable when returning to the active mode, regardless of the default I/O register settings.

When back in active mode, the I/O configuration and I/O interrupt configuration such as **PxDIR**, **PxREN**, **PxOUT**, and **PxIES** should be restored to the values prior to entering LPM3.5 or LPM4.5. The **LOCKLPM5** bit can then be cleared, which releases the I/O pin conditions and I/O interrupt configuration. Any changes to the port configuration registers while **LOCKLPM5** is set have no effect on the I/O pins.

After enabling the I/O interrupts by configuring **PxIE** and port interrupt enable configuration at NVIC, the I/O interrupt that caused the wake up can be serviced as indicated by the **PxIFG** flags. These flags can be used directly, or the corresponding **PxIV** register may be used. Note that the **PxIFG** flag cannot be cleared until the **LOCKLPM5** bit has been cleared.

A point to be noted is that it is possible that multiple events occurred on various ports. In these cases, multiple **PxIFG** flags are set, and it cannot be determined which port caused the I/O wake up. Table 2.10 shows most popular GPIO Port 1-related control registers, including the register names, base addresses, offset addresses, full addresses, and software macros, used in the real programming codes.

2.3.5 MSP432P401R microcontroller system controls

The MSP432P401R MCU provides a system control unit used to configure and manage the overall operation of the devices and provides necessary control information about the MCU. These system controls and configurations include the MCU system reset control, power control, operating modes control, clock control, and user configuration settings control.

The system control comprises a group of control registers, and most of these registers are read-only registers. Their mapping addresses are located in the internal ROM space in the on-chip memory. To access these registers in the internal ROM space, a set of special application programming interface (API) functions (located in the MSPW are peripheral drive library and BSL) are defined and used in the MSP432P401R MCU system.

The main system control functions included in the MSP432P401R MCU are:

- MCU system reset control
- Power supply system (PSS)
- Power control manager (PCM)
- CS control
- System controller (SYSCTL)
- System initialization and configuration

2.3.5.1 MCU system reset control

The MSP432P401R MCU provides four reset sources and most of them belong to internal reset operations. These reset sources include (1: highest reset priority, 4: lowest reset priority):

1. The power-on reset (POR)
2. The reboot reset (RR)
3. The hard reset (HR)
4. The soft reset (SR)

Except the power-on and reboot resets, all other resets belong to error-triggered resets because of something wrong in the system, including the microcontroller core or other devices.

If a reset occurred, including all of these four types of resets, the reset source can be identified and recorded in a set of reset control registers (**RSTCTL**). Fifteen **RSTCTL** registers are used in the MSP432P10R MCU system to:

- Generate a reset (POR, RR, HR, or SR)
- Clear a reset (POR, RR, HR, or SR)
- Record the reset sources
- Identify the reset sources

Table 2.11 shows one of **RSTCTL** registers, **RSTCTL_HARDRESET_STAT**, which shows all hard reset status as well as the relationship between each bit and each related hard reset source. A value of 1 in a certain bit indicates that the corresponding reset has occurred.

The **RSTCTL_HARDRESET_STAT** is a 32-bit register, but only the lower 16 bits (bits 15:0) are used to indicate the hard reset sources.

After a reset occurred, the user can direct the microcontroller core to execute the BSL stored in the internal ROM space or the user's application in the flash memory.

Almost all resets will trigger the microcontroller core to perform similar jobs in the following sequence:

1. Waiting for the reset to be exited or released.
2. Load PC and SP registers with the initial programer count and initial stack pointer, and begin to execute the instruction starting from the current PC's content.

Table 2.11 RSTCTL_HARDRESET_STAT register description

Bit number	Reset source number	Hard interrupt source
31:16	Reserved	Reserved—Always read 0
15	SRC15	PCM
14	SRC14	CS
13	SRC13	Reserved
12	SRC12	Reserved
11	SRC11	Reserved
10	SRC10	Reserved
9	SRC9	Reserved
8	SRC8	Reserved
7	SRC7	Reserved
6	SRC6	Reserved
5	SRC5	Reserved
4	SRC4	Reserved
3	SRC3	Flash controller
2	SRC2	WDT_A password violation
1	SRC1	WDT_A time-out
0	SRC0	SYSRESETREQ (System reset output of Cortex-M4)

2.3.5.1.1 Power-on reset The POR initiates a complete initialization of the application settings and MCU configuration information. This class of reset may be initiated either by the PSS, the PCM, the RSTn pin, the CS upon digitally controlled oscillator (DCO) external resistor short circuit fault, or the device emulation logic (through the debugger). From an application perspective, all sources of POR return the device to the same state of initialization.

After the system is powered on, a system POR is generated. An internal POR circuit will monitor the power supply voltage (V_{DD}) and generate a reset signal to all of the internal logic including JTAG when the power supply ramp reaches a threshold value (V_{VDD_POK}). The POK means power-OK (POK). The internal POR is only active on the initial power up of the microcontroller or when the microcontroller wakes from low-power mode.

The internal POR monitor circuit is used to keep the analog circuitry in reset until the voltage supply for analog circuits (V_{DDA}) has reached the correct range for the analog circuitry to begin operating. The POK monitor is used to keep the digital circuitry in reset until the V_{DDA} power supply is at an acceptable operational level.

A POR may be required by the MCU in any of the following cases:

- A true power-on or power-off condition (application or removal of power to the device)
- A **`Voltage Exception`** condition that is generated by the **`PSS`**. This condition can be caused either by the supervision logic for the V_{CC} or core domain voltage
- Exit from LPM3.5 or LPM4.5 modes of operation (initiated by the **`PCM`**)
- A user-driven full chip reset. This reset can be initiated either through the **`RSTn`** pin or through the debugger or through the SYSCTL

Table 2.12 Reboot control register (**SYS_REBOOT_CTL**)

Bit	Bit field	Type	Bit function
31:16	Reserved	RO	Reserved and read as 0
15:8	WKEY	W	Key to enable writing to bit 0 WKEY = **0x69** → Enable bit 0 to be written WKEY = **0xxx** → Disable bit 0 to be written
7:1	Reserved	RO	Reserved and read as 0
0	REBOOT	W	**1:** Generate a reboot reset. **0:** No reboot reset

From the user application perspective, all of the above sources result in the same reset state, and are hence classified under one reset category, which is termed as POR.

2.3.5.1.2 Reboot reset The Reboot Reset is identical to the POR, and allows the application to emulate a POR class reset without needing to power cycle the MCU or activating the **RSTn** pin. It can also be initiated through the debugger, and hence does not affect the debug connection to the device. However, a POR will result in a debug disconnection.

The Reboot sources can also be programmed to generate reboot reset by setting the **REBOOT** bit (bit 0) in the Reboot Control Register, **SYS_REBOOT_CTL**. Table 2.12 shows the bit functions for this register. Only when **WKEY** = **0x69**, the **REBOOT** bit can be accessed and modified.

Although **SYS_REBOOT_CTL** is a 32-bit register, only two bit fields, bit-15:8 and bit-0, are used for generating a reboot reset source in this register. Bits 31:16 and 7:1 are reserved.

2.3.5.1.3 Hard Reset The hard reset resets all modules that are set up or modified by the application. This includes all peripherals as well as the nondebug logic of the Cortex-M4. The MSP432P401R MCU supports up to 16 sources of hard reset. Table 2.11 lists the reset source allocation. The reset controller registers can be used to identify the possible source of reset in the device.

From an application perspective, a hard reset performs the following jobs:

- Resets the processor and all application configured peripherals in the system, which includes the bus system, thereby aborting any pending bus transactions
- Returns control to the user code
- Debugger connection to the MCU is maintained
- It does NOT reboot the device
- On-chip SRAM values are retained

A point to be noted is that the hard reset class sets status flag registers that report the exact source of the hard reset. The application can use these registers to select the necessary course of action. See Table 2.11 to get more details about these hard reset sources.

2.3.5.1.4 Soft reset A soft reset is initiated under user application control and is a deterministic event. This class resets only the execution-related components of the system. All other application-related configuration is maintained, thereby retaining the application's view of the device. Peripherals that are configured by the application continue their operation through a soft reset.

Table 2.13 RSTCTL_SOFTRESET_STAT register description

Bit number	Reset source number	Hard interrupt source
31:16	Reserved	Reserved—Always read 0
15	SRC15	Reserved
14	SRC14	Reserved
13	SRC13	Reserved
12	SRC12	Reserved
11	SRC11	Reserved
10	SRC10	Reserved
9	SRC9	Reserved
8	SRC8	Reserved
7	SRC7	Reserved
6	SRC6	Reserved
5	SRC5	Reserved
4	SRC4	Reserved
3	SRC3	Reserved
2	SRC2	WDT_A password violation
1	SRC1	WDT_A time-out
0	SRC0	CPU LOCKUP condition (LOCKUP output of Cortex-M4)

The MSP432P401R MCU supports up to 16 sources of soft reset. Table 2.13 lists the reset source allocation. One of reset controller registers, **`RSTCTL_SOFTRESET_STAT`**, can be used to identify the possible source of soft reset in the design. A value of 1 in a certain bit indicates that the corresponding reset has occurred. This is a 32-bit read-only register but only the lower 16 bits (bits 15:0) are used to record and indicate the soft reset sources, and bits 31:16 are reserved.

The **`WDT_A`**-generated resets can be mapped either as a hard reset or a soft reset.

From an application perspective, a soft reset has the following implications:

- Resets the following execution-related components in the system:
  - **`SYSRESETn`** of the Cortex®-M4. All bus transactions in the M4 (except the debug PPB space) are aborted
  - WDT module
- All system-level bus transactions are maintained
- All peripheral configurations are maintained
- Returns control to the user code
- Debugger connection to the device is maintained
- It does not reboot the device
- On-chip SRAM values are retained

2.3.5.2 *Power supply system (PSS)*

The PSS manages all functions related to the power supply and its supervision for the MCU. Its primary functions are to generate a supply voltage for the core logic and to provide mechanisms for the supervision of the voltage applied to the device (V_{CC}) and the regulated voltage (V_{CORE}).

The PSS uses an integrated voltage regulator to produce a secondary core voltage (V_{CORE}) from the primary voltage that is applied to the device (V_{CC}). In general, V_{CORE} supplies the CPU, memory, and the digital modules, while V_{CC} supplies I/Os and analog modules. On certain devices, there will be one or more separate AV_{CC} pin(s) to supply the analog modules. However, it is assumed that **DV_{CC}** and **AV_{CC}** are shorted on the board or generated from the same source, and no level shifting or isolation is done between these two supplies.

The V_{CORE} output is maintained using a dedicated voltage reference. V_{CORE} voltage level is programmable to allow power savings if the maximum device speed is not required. The core voltage regulators are covered in detail in the PCM section.

The PSS module provides a means for V_{CC} and V_{CORE} to be supervised. The supervisors detect when a voltage falls under a specific threshold and trigger a reset event. The input or primary side of the regulator is referred to as the high side. The output or secondary side is referred as the low side. V_{CC} is supervised and monitored by the high-side supervisor/monitor (**SVSMH**). V_{CORE} is supervised by the low-side supervisor (**SVSL**). The thresholds enforced by these modules are derived from the same voltage reference used by the regulator to generate V_{CORE}.

The PSS controls all the power supply-related functionality of the device. It consists of the following components:

- Supply supervisor and monitor for high side (**SVSMH**)
- Supply supervisor for low side (**SVSL**)
- Core voltage regulator (**CVR**)
- VCC detect (**VCCDET**)

Figure 2.17 shows a functional block diagram of the PSS.

All PSS operations are under the control of a set of PSS-related registers, which includes:

1. PSS Key Register (PSSKEY)
2. PSS Control Register (PSSCTL0)
3. PSS Interrupt Enable Register (PSSIE)
4. PSS Interrupt Flag Register (PSSIFG)
5. PSS Clear Interrupt Flag Register (PSSCLRIFG)

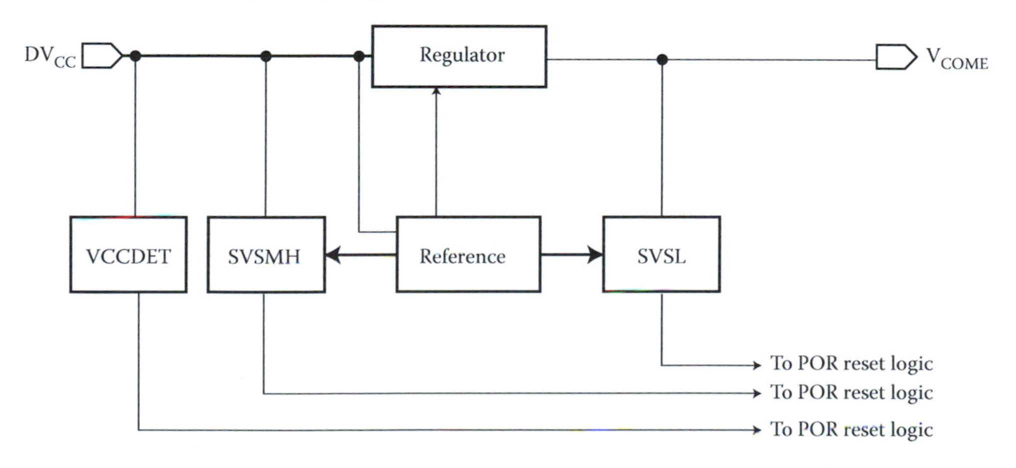

Figure 2.17 Functional block diagram of the PSS. (Courtesy Texas Instruments.)

The **PSSKEY** is an important control register since it controls the accessibility for all other PSS registers. This is a 32-bit register but only lower 16 bits, bits 15:0, are used to store a PSSKEY to lock or unlock all other PSS registers.

To access any other PSS register, a value of **0x695A** must be written into this **PSSKEY** field in the PSSKEY register to unlock all other registers. When reading, this register always returns **0xA596**.

2.3.5.3 Power control manager (PCM)

The MSP432P401R MCU supports several power modes that allow for the optimization of power for a given application. The power modes can be changed dynamically to cover many different power profile requirements across many applications. The **PCM** is responsible for managing power requests from different areas of the system and processing the requests in a controlled manner.

It uses all the information from the system that may affect the power requirements and adjusts the power as required, if possible. The **CS** and the **PSS** settings are the two primary elements that control the power settings of the MCU and hence the power consumption of the microcontroller.

There are several factors determining the power mode setting of the device. This includes existing conditions from the CS settings and the PSS settings. It is possible that a power mode request cannot be safely entered based on the existing conditions in the system. The PCM is an automated subsystem that adjusts the power based on the direct power request settings or indirectly based on the other requests in the system. The PCM is the main interface between the PSS and CS modules.

The PCM works based on different system events. The most common events are:

- PCM Control 0 Register (**PCMCTL0**). This register can be modified directly by the application execution to request that a particular power mode be entered.
- Interrupt and wake-up events. Interrupts from low-power modes cause operation to automatically return to an active mode.
- Reset events. Reset events cause the power mode to be set back to its default setting.
- Debug events. Power mode settings are adapted to support debug hardware requirements.

All of these events, regardless of their source, cause a power change request to the PCM. The power change request is used as an indication that a new power mode setting may be required. The PCM processes all the requests and makes the necessary changes to the system as required. It may not be possible to fulfill all the requests, and the PCM may deny some requests.

DV_{CC} can be powered from a wide input voltage range, but the core logic of the MCU must be kept at a voltage lower than what this range allows. For this reason, a linear regulator or alternately referred to as **LDO** has been integrated into the PSS. The regulator derives the necessary core voltage (V_{CORE}) from DV_{CC}.

Some devices offer an inductor-based DC–DC regulator in addition to an LDO to regulate the core voltage. The DC–DC provides increased efficiency and therefore decreased current consumption from DV_{CC}; however, it requires an external pin and an inductor, as well as a larger capacitor on the V_{CORE} pin. The larger capacitor results in longer wake-up times when the core voltage is changed and at startup.

The device always starts up with the **LDO** and then the user can switch to the DC–DC if so desired by choosing a power mode that enables the DC–DC.

2.3.5.3.1 Power modes Each MCU supports several power modes. Active modes (**AM**) are any power mode under which the CPU execution is possible. Low-power mode 0 (**LPM0**), low-power mode 3 (**LPM3**), low-power mode 4 (**LPM4**), low-power mode 3.5 (**LPM3.5**), and low-power mode 4.5 (**LPM4.5**) do not allow for CPU execution. Each power mode is described in the following sections. Direct transitions between some of the power modes are not possible.

2.3.5.3.1.1 Active modes (AM) All active modes support code execution from flash memory or SRAM. Not all active mode transitions are supported. Active modes are referred as run modes in ARM terminology.

Active modes can be any power mode in which CPU execution is possible. There are six active modes available. Active modes can be logically grouped by the core voltage levels that are supported. There are two core voltage-level settings: core voltage level 0 and core voltage level 1. Three active modes are associated with each core voltage level. The various active modes allow for optimal power and performance across a broad range of application requirements. The core voltage can be supplied either by a low dropout (**LDO**) regulator or a DC–DC regulator.

AM_LDO_VCORE0 is based on the core voltage level 0 with LDO operation.
AM_LDO_VCORE1 is based on the core voltage level 1 with LDO operation.

To use the DC–DC regulator, the application must always transition first to either **AM_LDO_VCORE0** or **AM_LDO_VCORE1**. In addition, the DC–DC regulator requires external components and configuration for proper operation. Because the DC–DC regulator is a switching regulator, it takes time to settle and achieve a stable voltage. During this settling time, the LDO automatically remains enabled. When the DC–DC regulator has settled, the LDO is automatically disabled to reduce power.

2.3.5.3.1.2 Low-power mode 0 (LPM0) LPM0 modes are referred to as sleep modes in ARM terminology.

During LPM0, the processor execution is halted. Halting the processor reduces dynamic power due to reduced switching activities caused by the execution of the processor. In general, there is only one LPM0 setting, and LPM0 can be entered from all active modes. Therefore, LPM0 effectively supports six different modes of operation, corresponding to each active mode:

1. **LPM0_LDO_VCORE0**
2. **LPM0_DCDC_VCORE0**
3. **LPM0_LF_VCORE0**
4. **LPM0_LDO_VCORE1**
5. **LPM0_DCDC_VCORE1**
6. **LPM0_LF_VCORE1**

The maximum frequency requirement in LPM0 is identical to that of the active mode at the time of LPM0 entry.

For example, if LPM0 is entered from **AM_LDO_VCORE0**, the maximum frequency requirement of **AM_LDO_VCORE0** would also apply during **LPM0_LDO_VCORE0**. LPM0 is useful to save power when processor execution is not required, yet very fast wake-up time is necessary. LPM0 exit always takes the device back to the original active mode at the time of LPM0 entry.

2.3.5.3.1.3 Low-power mode 3 (LPM3) and low-power mode 4 (LPM4) LPM3 and LPM4 are referred to as deep sleep modes in ARM terminology. The processor execution is halted during LPM3 and LPM4 operational period.

LPM3 mode restricts maximum frequency of device operation to 32.768 kHz. Only RTC and WDT modules are functional out of the low-frequency clock sources (LFXT, REFO, and VLO) while in LPM3. All other peripherals are disabled and the high-frequency clock sources are turned off in LPM3.

LPM4 mode starts to work when the device is programmed for LPM3 with RTC and WDT modules disabled, and all clock sources are turned off. In LPM4 mode, no peripheral function is available. All SRAM banks that are enabled for data retention during LPM3 and LPM4, and all peripheral registers will retain the data through LPM3 and LPM4 modes.

The device I/O pin states are also latched and retained in LPM3 and LPM4. LPM3 and LPM4 modes are useful for relatively infrequent processor activity followed by long periods of low-frequency activity, better known as low duty cycle applications. The wake-up time from LPM3 and LPM4 is longer than that of LPM0, but the average power consumption is significantly lower. DC–DC regulator operation is not supported in LPM3 and LPM4 modes. LPM3 and LPM4 exit always takes the device back to the original active mode at the time of LPM3 or LPM4 entry.

2.3.5.3.1.4 Low-power mode 3.5 (LPM3.5) and low-power mode 4.5 (LPM4.5) LPM3.5 and LPM4.5 modes are referred to as stop or shutdown modes in ARM terminology. Both modes provide the lowest power consumption possible but at reduced functionality.

In LPM3.5 mode, all peripherals are disabled and powered down except for the RTC and WDT, which can be optionally enabled by the application and clocked out of low-frequency clock sources (LFXT, REFO, and VLO). Furthermore, the MCU is brought down to the core voltage level 0. Wake-up is possible through any of the wake-up events. LPM3.5 mode does not retain any peripheral register data. However, Bank-0 of SRAM is retained for the application's use as a backup memory, and also device I/O pin states are latched and retained.

In LPM4.5 mode, all peripherals and clock sources are powered down and the internal voltage regulator is switched off. Wake up is possible through any of the wake-up events. LPM4.5 mode does not retain any SRAM or peripheral register data but device I/O pin states are latched and retained. Any essential data must be stored to flash prior to entering LPM4.5 mode.

LPM3.5 and LPM4.5 modes are useful for complete power down or simple timekeeping over long periods of time with infrequent wake-up activity.

2.3.5.3.2 Power mode transitions Several power mode transitions are possible under application control, allowing for optimal power performance trade-offs for a wide variety of usage profiles. The relationship between different power modes and high-level representation of different power mode transitions is shown in Figure 2.18.

After power up, upon hard reset or any other higher-class reset, the MCU enters active mode (AM). From there, it is possible to transition into different low-power modes, such as LPM0, LPM3, LPM4, LPM3.5, and LPM4.5 through application programming. Upon defined wake-up events, the MCU returns to the active mode from which the specific

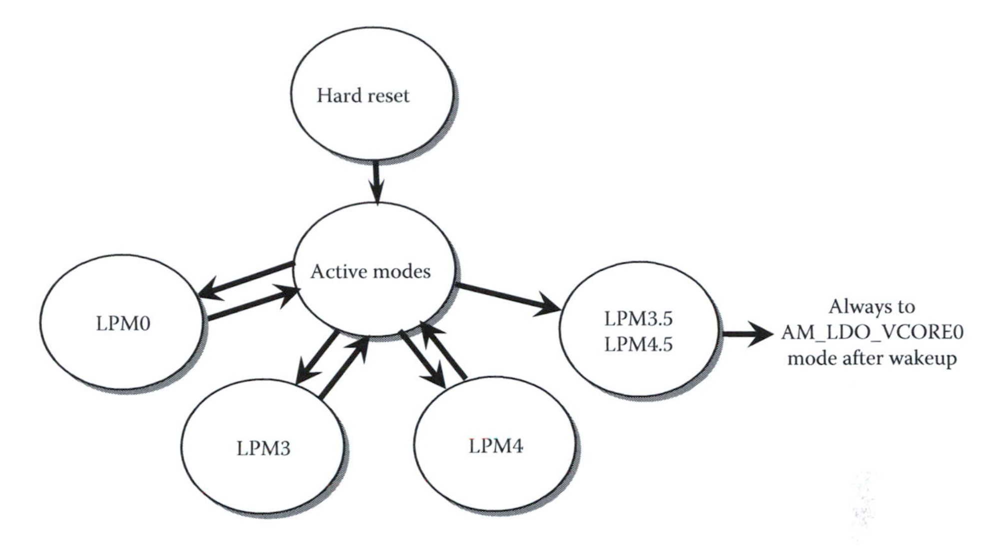

Figure 2.18 High-level power mode transitions.

low-power mode was entered. Only for LPM3.5 and LPM4.5 modes, the MCU always enters **AM_LDO_VCORE0** mode after wake up, as shown in Figure 2.18.

2.3.5.3.3 PCM registers All PCM operations are under the control of a set of PCM-related registers, which includes:

1. PCM Control 0 Register (**PCMCTL0**)
2. PCM Control 1 Register (**PCMCTL1**)
3. PCM Interrupt Enable Register (**PCMIE**)
4. PCM Interrupt Flag Register (**PCMIFG**)
5. PCM Clear Interrupt Flag Register (**PCMCLRIFG**)

All of these registers are read/write registers, except the **PCMIFG** that is a read-only register.

The **PCMCTL0** register is used to set up, control, and process the different power modes. The **PCMCTL1** register is used to monitor the working status of all possible power modes that have been requested and responded, such as whether the power mode is working in transaction or not, or if a power mode is locked or not.

To access these registers, both **PCMCTL0** and **PCMCTL1** registers must be written an appropriate value (**0x695Axxxx**) to the **PCMKEY** field in both registers to unlock all registers. Both are 32-bit registers but the higher 16 bits, bits 31:16, are used to store a **PCMKEY** to lock or unlock all PCM registers. When reading these two registers, a value of **0xA596** is always returned.

The other three registers, **PCMIE**, **PCMIFG,** and **PCMCLRIFG**, are used for the PCM interrupt enabling and processing functions.

2.3.5.4 System clock control

The clock is a timing base and it provides an operational timing standard or criterion to enable computers to perform their jobs step-by-step based on each clock cycle. Similarly, in order to enable microcontrollers to execute their instructions in a defined sequence, a clock source is definitely needed. Without a clock source, no microcontroller or computer can run its instructions properly.

In the MSP432P401R MCU, seven different clock sources are provided:

1. Low-frequency oscillator (**LFXT**): Supports 32.768 KHz low-frequency source.
2. High-frequency oscillator (**HFXT**): Supports high-frequency crystal up to 48 MHz.
3. Very-low-power, low-frequency oscillator (**VLO**): An internal very low-frequency source generating a 10 KHz frequency.
4. Low-frequency reference oscillator (**REFO**): An alternate internal low-power low-frequency oscillator generating selectable 32.768-kHz or 128-kHz typical frequency.
5. Module oscillator (**MODOSC**): An internal low-power oscillator with 25-MHz typical frequency. It is typically used to supply a **Clock On Request** to peripheral modules.
6. Internal **DCO**: The **DCO** is a power-efficient tunable internal oscillator that generates up to 48 MHz. It also supports a high-precision mode when using an external precision resistor.
7. System oscillator (**SYSOSC**): The **SYSOSC** is a lower-frequency version of the **MODOSC** and is calibrated to a frequency of 5 MHz. It drives the ADC sampling clock in the 200-ksps conversion mode. In addition, it is also used for timing of various system-level control and management operations.

By combining these clock sources, the following five system clock signals can be generated from the clock module:

1. **ACLK**: Auxiliary Clock: ACLK is software selectable by individual peripheral modules. ACLK is restricted to maximum frequency of operation of 128 kHz.
2. **MCLK**: Master Clock: MCLK is software selectable as LFXTCLK, VLOCLK, REFOCLK, DCOCLK, MODCLK, or HFXTCLK. MCLK is used by the CPU and peripheral module interfaces, as well as used directly by some peripheral modules.
3. **HSMCLK**: Subsystem Master Clock: HSMCLK is software selectable by individual peripheral modules.
4. **SMCLK**: Low-speed Subsystem Master Clock: SMCLK uses the HSMCLK clock resource selection for its clock resource.
5. **BCLK**: Low-speed Backup Domain Clock: BCLK is software selectable as LFXTCLK and REFOCLK. This is primarily used in the backup domain (domain consisting of a set of peripherals which are available during the low-power modes of the device). BCLK is restricted to a maximum frequency of 32 kHz.

Figure 2.19 shows the functional block diagram of the CS used in the MSP432P401R MCU system. From this block diagram, the following features are provided:

- Two clock sources, **LFXT** and **HFXT**, need the external crystal oscillators and capacitors. However, since the **DCO** can work in either internal or external mode, thus an external resistor (**DCOR**) is needed if it works in the external mode. Therefore, totally three clock sources may need external components to build their clock source function.
- All other four clock sources, including **VLO**, **REFO**, **MODOSC**, and **SYSOSC**, belong to internal clock sources and they do not need any external components to build their clock functions.

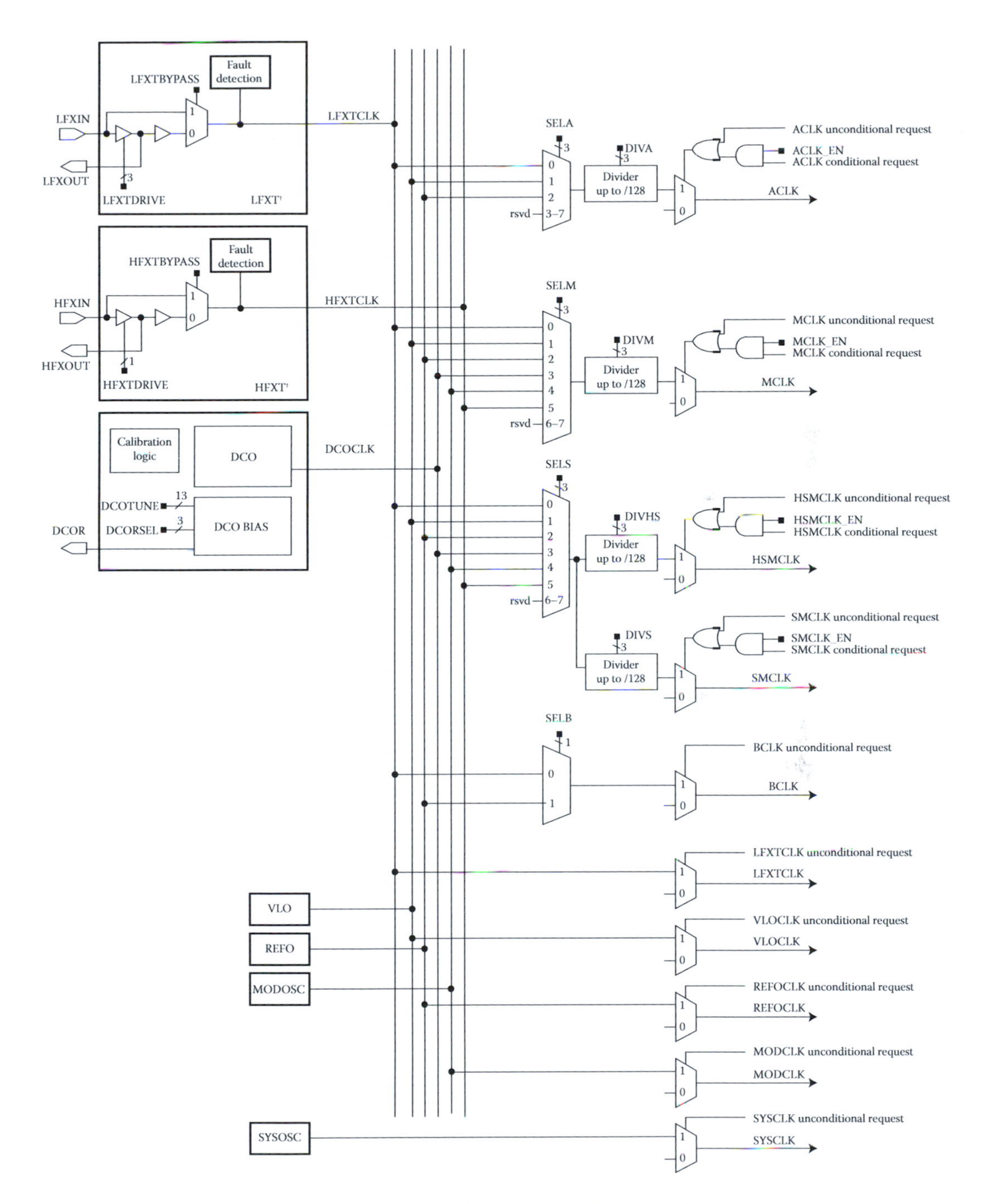

Figure 2.19 Functional block diagram of the clock system. (Courtesy Texas Instruments.)

- For both **LFXT** and **HFXT**, a **BYPASS** function is provided (**LFXTBYPASS** and **HFTXBYPASS**). This function enables both clock sources to bypass those external crystals with an external square input as the clock crystal signals. For example, in the bypass mode, the **LFXIN** and **HFXIN** can accept an external square-wave clock input signal, and **LFXOUT** and **HFXOUT** are configured as a **GPIO**.

- Except the **SYSOSC**, all other clock sources can be controlled and selected by related MUX via corresponding **SELx** signals. These signals are controlled by setting related bits on the corresponding CS registers.
- Also, except the **SYSOSC**, all other clock sources can be divided by related frequency dividers that are controlled by the corresponding **DIVx** signals to get different clocks with lower frequencies. These signals are also controlled by setting related bits on the corresponding CS registers.
- After the clock frequency divider, all clock sources including the **SYSOSC**, another selection MUX is provided and this is used to enable peripherals that want to use a clock source to support their I/O operations to submit a clock-using-request to the CS module. If the following two conditions are met the request clock source is enabled and passed to the selected peripheral device: (1) the selected peripheral is enabled and (2) a clock-using-request is submitted from the enabled peripheral.

Most of these conditions are controlled and performed by using a logic OR gate, and these requests belong to conditional requests. For any conditional requests coming from any devices, it can be disabled or masked by the CS module. However, for all clock sources, including the **VLO**, **REFO**, **MODOSC**, and **SYSOSC**, there are also some unconditional requests available to peripherals. These unconditional requests cannot be disabled or masked, and must be responded by the CS module, if they are coming.

2.3.5.4.1 System clock control registers The clock source selection and control are determined by related bits on the corresponding CS registers. Table 2.14 shows all CS registers. Among them only some registers are important to define the configurations and operations of the clock sources. These registers include:

- Key Register (CSKEY)
- Control 0 Register (CSCTL0)
- Control 1 Register (CSCTL1)
- Control 2 Register (CSCTL2)

Table 2.14 CS control registers

Register name	Macro	Type	Access	Reset value
Key Register	CSKEY	RW	Word	0000A596
Control 0 Register	CSCTL0	RW	Word	00010000
Control 1 Register	CSCTL1	RW	Word	00000033
Control 2 Register	CSCTL2	RW	Word	00070007
Control 3 Register	CSCTL3	RW	Word	000000BB
Clock Enable Register	CSCLKEN	RW	Word	0000000F
Status Register	CSSTAT	RO	Word	00000003
Interrupt Enable Register	CSIE	RW	Word	00000000
Interrupt Flag Register	CSIFG	RO	Word	00000001
Clear Interrupt Flag Register	CSCLRIFG	WO	Word	00000000
Set Interrupt Flag Register	CSSETIFG	WO	Word	00000000
DCO External Resistor (Calibration Resistor)	CSDCOERCAL	RW	Word	02000000

- Clock Enable Register (CSCLKEN)
- Clock Status Register (CSSTAT)

We will have a detailed discussion about these registers in this section. For more detailed descriptions of all rest registers, refer to MSP432 datasheet.

2.3.5.4.1.1 Clock key register (CSKEY) This is a 32-bit CS access control register to control whether all CS registers can be accessed. The upper 16 bits are reserved and only the lower 16 bits, bits 15:0 (**CSKEY**), are used to hold the key value to lock or unlock all CS registers.

To unlock all CS registers, write the password **0x695A** to the lower 16 bits, or bit field **CSKEY**, in this register. To lock all CS registers, write any other value to **CSKEY** field. This register always returns **0xA596** when it is read.

2.3.5.4.1.2 Clock control 0 register (CSCTL0) The CS Control 0 Register (**CSCTL0**) is mainly used to configure and control the DCO clock source, including the enable or disable DCO mode, select internal or external DCO resistor, define the DCO operational frequency and frequency ranges.

This is a 32-bit register but some bits are reserved for future usage. Table 2.15 shows the bit configuration and function of this register.

Two fields are important to most applications: **DCOEN** and **DCORSEL**. The former is used to enable the DCO clock source and the latter is used to select the operational frequency and frequency range of the clock source. After a system reset, the default clock source selected by this register is **DCOCLK** with internal resistor mode and a frequency of 3 MHz (from 2 to 4 MHz).

2.3.5.4.1.3 Clock control 1 register (CSCTL1) The CS Control 1 Register (**CSCTL1**) is mainly used to select a clock source with a frequency divider to get the desired clock source with desired frequency. This is a 32-bit register with some bits reserved. Table 2.16 shows the bit configuration and function of this register.

After a system reset or hard reset, the default clock source selected by this register is:

- For BCLK source, the LFXTCLK is selected (bit-12: SELB).
- For ACLK source, the LFXTCLK is selected (bits-10:8: SELA). Frequency divider is 1 (bits 26:24 DIVA).
- For SMCLK and HSMCLK sources, the DCOCLK is selected (bits-6:4: SELS). Frequency divider is 1 (bits 30:28 DIVS), and (bits 22:20 DIVHS).
- For MCLK source, the DCOCLK is selected (bits-2:0: SELM). Frequency divider is 1 (bits 18:16 DIVM).

2.3.5.4.1.4 Clock control 2 register (CSCTL2) The CS Control 2 Register (**CSCTL2**) is mainly used to configure the **LFXTCLK** and **HFXTCLK** clock sources, including the clock oscillators enabling, frequency selections and oscillator bypass functions. This is a 32-bit register with some bits reserved. Table 2.17 shows the bit configuration and function of this register.

After a system reset, the default clock sources selected by this register are: both LFXT and HFXT use the external crystals, and both LFXTCLK and HFXTCLK modules are enabled.

Table 2.15 CS clock control 0 register (CSCTL0)

Bit	Field	Type	Reset	Function
31:25	Reserved	RO	00	Reserved and always reads as 0
24	DIS_DCO_ DELAY_CNT	RW	0	Enable or disable the DCO settling counter value **0:** Settling delay counter is enabled. The application can change to any frequency desired as long as it is supported by the power mode **1:** Settling delay counter is disabled. The application should ensure that the frequency changes requested are in step size of 5 MHz
23	DCOEN	RW	0	Enables DCO oscillator regardless if used as a clock resource **0:** DCO is on if it is used as a source for MCLK, HSMCLK, or SMCLK and clock is requested, otherwise it is disabled. **1:** DCO is on
22	DCORES	RW	0	Enable the DCO external resistor mode **0:** Internal resistor mode; **1:** External resistor mode
21:19	Reserved	RO	0	Reserved and always reads as 0
18:16	DCORSEL	RW	1	DCO frequency range select **0x0:** Nominal DCO frequency (MHz): 1.5; range: 1–2 MHz **0x1:** Nominal DCO frequency (MHz): 3; range: 2–4 MHz **0x2:** Nominal DCO frequency (MHz): 6; range: 4–8 MHz **0x3:** Nominal DCO frequency (MHz): 12; range: 8–16 MHz **0x4:** Nominal DCO frequency (MHz): 24; range: 16–32 MHz **0x5:** Nominal DCO frequency (MHz): 48; range: 32–64 MHz **0x6** to **0x7**: Nominal DCO frequency (MHz): reserved, defaults to 1.5 when selected; nominal range (MHz): reserved, defaults to 1 to 2 when selected
15:13	Reserved	RO	0	Reserved and always reads as 0
12:0	DCOTUNE	RW	0	DCO frequency tuning select. Two's complement representation Value represents an offset from the calibrated center frequency for the range selected by the **DCORSEL** bits

2.3.5.4.1.5 Clock enable register (CSCLKEN) The CS Enable Register (**CSCLKEN**) is mainly used to configure and set up possible clock sources, including turn on selected clock oscillators and enable-related conditional requests coming from various device and peripheral modules.

This is a 32-bit register with some bits reserved. Table 2.18 shows the bit configuration and function of this register.

After a system reset, the default clock setups by this register are: most internal oscillators are turned on and all system clock conditional requests are enabled.

2.3.5.4.1.6 Clock status register (CSSTAT) This is a 32-bit read-only register used to monitor and indicate the current running status of all clock oscillators and clock modules.

Table 2.16 CS clock control 1 register (CSCTL1)

Bit	Field	Type	Reset	Function
31	Reserved	RO	0	Reserved and always reads as 0
30:28	DIVS	RW	0	SMCLK source divider **0x0:** f(SMCLK)/1; **0x1:** f(SMCLK)/2; **0x2:** f(SMCLK)/4 **0x3:** f(SMCLK)/8; **0x4:** f(SMCLK)/16; **0x5:** f(SMCLK)/32 **0x6:** f(SMCLK)/64; **0x7:** f(SMCLK)/128
27	Reserved	RO	0	Reserved and always reads as 0
26:24	DIVA	RW	0	ACLK source divider **0x0:** f(ACLK)/1; **0x1:** f(ACLK)/2; **0x2:** f(ACLK)/4 **0x3:** f(ACLK)/8; **0x4:** f(ACLK)/16; **0x5:** f(ACLK)/32 **0x6:** f(ACLK)/64; **0x7:** f(ACLK)/128
23	Reserved	RO	0	Reserved and always reads as 0
22:20	DIVHS	RW	0	HSMCLK source divider **0x0:** f(HSMCLK)/1; **0x1:** f(HSMCLK)/2; **0x2:** f(HSMCLK)/4; **0x3:** f(HSMCLK)/8; **0x4:** f(HSMCLK)/16; **0x5:** f(HSMCLK)/32; **0x6:** f(HSMCLK)/64; **0x7:** f(HSMCLK)/128
19	Reserved	RO	0	Reserved and always reads as 0
18:16	DIVM	RW	0	MCLK source divider **0x0:** f(MCLK)/1; **0x1:** f(MCLK)/2; **0x2:** f(MCLK)/4 **0x3:** f(MCLK)/8; **0x4:** f(MCLK)/16; **0x5:** f(MCLK)/32 **0x6:** f(MCLK)/64; **0x7:** f(MCLK)/128
15:13	Reserved	RO	0	Reserved and always reads as 0
12	SELB	RW	0	Select the BCLK source. **0x0:** LFXTCLK; **0x1:** REFOCLK
11	Reserved	RO	0	Reserved and always reads as 0
10:8	SELA	RW	0	Selects the ACLK source **0x0:** LFXTCLK when LFXT available, otherwise REFOCLK **0x1:** VLOCLK; **0x2:** REFOCLK **0x3** - **0x7:** Reserved for future use. Defaults to REFOCLK. Not recommended for use to ensure future compatibilities
7	Reserved	RO	0	Reserved and always reads as 0
6:4	SELS	RW	3	Select the SMCLK and HSMCLK source **0x0:** LFXTCLK when LFXT available, otherwise REFOCLK **0x1:** VLOCLK; **0x2:** REFOCLK; **0x3:** DCOCLK; **0x4:** MODOSC **0x5:** HFXTCLK when HFXT available, otherwise DCOCLK **0x6:** Reserved for future use. Defaults to DCOCLK **0x7:** Reserved for future use. Defaults to DCOCLK
3	Reserved	RO	0	Reserved and always reads as 0
2:0	SELM	RW	3	Select the MCLK source **0x0:** LFXTCLK when LFXT available, otherwise REFOCLK **0x1:** VLOCLK; **0x2:** REFOCLK; **0x3:** DCOCLK; **0x4:** MODOSC **0x5:** HFXTCLK when HFXT available, otherwise DCOCLK **0x7:** Reserved for future use. Defaults to DCOCLK

Table 2.17 CS clock control 2 register (CSCTL2)

Bit	Field	Type	Reset	Function
31:26	Reserved	RO	00	Reserved and always reads as 0
25	HFXTBYPASS	RW	0	HFXT bypass select **0x0:** HFXT is driven by external crystal **0x1:** HFXT is driven by external square wave
24	HFXT_EN	RW	0	Turn on the HFXT oscillator regardless if used as a clock resource. **0x0:** HFXT is on if it is used as a source for MCLK, HSMCLK, or SMCLK, and is selected via the port selection and not in bypass mode **0x1:** HFXT is on if HFXT is selected via the port selection and HFXT is not in bypass mode
23	Reserved	RO	0	Reserved and always reads as 0
22:20	HFXTFREQ	RW	0	HFXT frequency selection These bits must be set to the appropriate value based on the frequency of the crystal connected. These bits are don't care in the HFXT bypass mode **0x0:** 1 MHz~4 MHz; **0x1:** 4 MHz~8 MHz; **0x2:** 8 MHz~16 MHz; **0x3:** 16 MHz~24 MHz; **0x4:** 24 MHz~32 MHz; **0x5:** 32 MHz~40 MHz **0x6:** 40 MHz~48 MHz; **0x7:** Reserved for future use
19	Reserved	RO	0	Reserved and always reads as 0
18:17	Reserved	RW	3	Reserved
16	HFXTDRIVE	RW	1	HFXT oscillator drive selection Reset value is 1 when HFXT available, and 0 when HFXT not available. This bit is a don't care in the HFXT bypass mode **0x0:** To be used for HFXTFREQ setting 000b **0x1:** To be used for HFXTFREQ settings 001b to 110b
15:10	Reserved	RO	0	Reserved and always reads as 0
9	LFXTBYPASS	RW	0	LFXT bypass select **0x0:** LFXT is driven by external crystal **0x1:** LFXT is driven by external square wave
8	LFXT_EN	RW	0	Turn on the LFXT oscillator regardless if used as a clock resource **0x0:** LFXT is on if it is used as a source for ACLK, MCLK, HSMCLK, or SMCLK and is selected via the port selection and not in bypass mode **0x1:** LFXT is on if LFXT is selected via the port selection and LFXT is not in bypass mode
7:3	Reserved	RW	0	Reserved and must be written as 0
2:0	LFXTDRIVE	RW	7	The LFXT oscillator current can be adjusted to its drive needs **0x0:** Lowest current consumption **0x1:** Increased drive strength LFXT oscillator **0x2:** Increased drive strength LFXT oscillator **0x3:** Increased drive strength LFXT oscillator **0x4:** Increased drive strength LFXT oscillator **0x5:** Increased drive strength LFXT oscillator **0x6:** Increased drive strength LFXT oscillator **0x7:** Maximum drive strength LFXT oscillator Reset value is **7** when LFXT available, and **0** when LFXT not available

Table 2.18 CS clock enable register (CSCLKEN)

Bit	Field	Type	Reset	Function
31:16	Reserved	RO	00	Reserved and always reads as 0
15	REFOFSEL	RW	0	Select REFO nominal frequency. **0:** 32.768 kHz; **1:** 128 kHz
14:11	Reserved	RO	0	Reserved and always reads as 0
10	MODOSC_ EN	RW	0	Turn on the MODOSC oscillator regardless if used as a clock resource **0:** MODOSC is on only if it is used as a source for ACLK, MCLK, HSMCLK, or SMCLK; **1:** MODOSC is on
9	REFO_EN	RW	0	Turn on the REFO oscillator regardless if used as a clock resource **0:** REFO is on only if it is used as a source for ACLK, MCLK, HSMCLK, or SMCLK; **1:** REFO is on
8	VLO_EN	RW	0	Turn on the VLO oscillator regardless if used as a clock resource **0:** VLO is on only if it is used as a source for ACLK, MCLK, HSMCLK, or SMCLK; **1:** VLO is on
7:4	Reserved	RO	0	Reserved and always reads as 0
3	SMCLK_EN	RW	1	SMCLK system clock conditional request enable **0:** SMCLK disabled regardless of conditional clock requests **1:** SMCLK enabled based on any conditional clock requests
2	HSMCLK_EN	RW	1	HSMCLK system clock conditional request enable **0:** HSMCLK disabled regardless of conditional clock requests **1:** HSMCLK enabled based on any conditional clock requests
1	MCLK_EN	RW	1	MCLK system clock conditional request enable **0:** MCLK disabled regardless of conditional clock requests **1:** MCLK enabled based on any conditional clock requests
0	ACLK_EN	RW	1	ACLK system clock conditional request enable **0:** ACLK disabled regardless of conditional clock requests **1:** ACLK enabled based on any conditional clock requests

One of main purposes to use this register is to verify whether the clock is stable after a clock-source-changing is performed by checking the related **XCLK_READY** bit. Because a certain period of time is needed for changing clock source or clock frequency, this verification is necessary.

Another purpose of using this register is to confirm whether the desired clock source is active by checking the related **XCLK_ON** bits on this register. After a system reset or power on, all clock sources need some time to make them active. A value of 1 on a bit indicates that the corresponding clock source is stable and active, and ready to be used.

Table 2.19 shows the bit configuration and function for this register.

After a system reset or hardware reset operations, all clock modules are disabled or inactive except the **DCO** clock module. Therefore, the DCO clock source is the only active clock module and available for all peripherals and other on-chip devices.

Refer to Table 2.18, after a system or hardware reset operations, the following clock sources conditional requests are enabled: **MCLK**, **SMCLK**, **HSMCLK**, and **ACLK**.

Refer to Table 2.15, the **DCORSEL** bit-field in the **CSCTL0** register are set to 1 after a system reset. This means that a 3 MHz DCO clock is provided and enabled (Table 2.18)

Table 2.19 CS clock status register (CSSTAT)

Bit	Field	Type	Reset	Function
31:29	Reserved	RO	00	Reserved and always reads as 0
28	BCLK_READY	RO	0	BCLK Ready status (if it is stable after a clock changing) **0:** Not ready; **1:** Ready
27	SMCLK_READY	RO	0	SMCLK Ready status (if it is stable after a clock changing) **0:** Not ready; **1:** Ready
26	HSMCLK_ READY	RO	0	HSMCLK Ready status (if it is stable after a clock changing) **0:** Not ready; **1:** Ready
25	MCLK_READY	RO	0	MCLK Ready status (if it is stable after a clock changing) **0:** Not ready; **1:** Ready
24	ACLK_READY	RO	0	ACLK Ready status (if it is stable after a clock changing) **0:** Not ready; **1:** Ready
23	REFOCLK_ON	RO	0	REFOCLK system clock status **0:** Inactive; **1:** Active
22	LFXTCLK_ON	RO	0	LFXTCLK system clock status **0:** Inactive; **1:** Active
21	VLOCLK_ON	RO	0	VLOCLK system clock status **0:** Inactive; **1:** Active
20	MODCLK_ON	RO	0	MODCLK system clock status **0:** Inactive; **1:** Active
19	SMCLK_ON	RO	0	SMCLK system clock status **0:** Inactive; **1:** Active
18	HSMCLK_ON	RO	0	HSMCLK system clock status **0:** Inactive; **1:** Active
17	MCLK_ON	RO	0	MCLK system clock status **0:** Inactive; **1:** Active
16	ACLK_ON	RO	0	ACLK system clock status **0:** Inactive; **1:** Active
15:8	Reserved	RO	0	Reserved and always reads as 0
7	REFO_ON	RO	0	REFO status **0:** Inactive; **1:** Active
6	LFXT_ON	RO	0	LFXT status. Only available on devices with LFXT **0:** Inactive; **1:** Active
5	VLO_ON	RO	0	VLO status **0:** Inactive; **1:** Active
4	MODOSC_ON	RO	0	MODOSC status **0:** Inactive; **1:** Active
3	Reserved	RO	0	Reserved and always reads as 0
2	HFXT_ON	RO	0	HFXT status. Only available on devices with HFXT **0:** Inactive; **1:** Active
1	DCOBIAS_ON	RO	1	DCO bias status. **0:** Inactive; **1:** Active
0	DCO_ON	RO	1	DCO status. **0:** Inactive; **1:** Active

for all clock-using-request for **MCLK**, **SMCLK**, **HSMCLK**, or **ACLK** for any peripheral device. If any peripheral is enabled, the request-enabled clock source will be applied to that peripheral.

2.3.5.4.2 System clock operations After a system reset, the MCU enters Active Mode 0 (**AM0_LDO**). In **AM0_LDO**, the CS module default configuration is

- For devices that have **LFXT** available:
  - LFXT crystal operation is selected as the clock resource for **LFXTCLK**
  - LFXTCLK is selected for **ACLK** (**SELAx** = 0) and **ACLK** is undivided (**DIVAx** = 0)
  - LFXTCLK is selected for **BCLK** (**SELB** = 0)
  - LFXT remains disabled. The crystal pins (**LFXIN** and **LFXOUT**) are shared with GPIOs. LFXIN and LFXOUT pins are set to GPIOs, and LFXT remains disabled until the I/O ports are configured for LFXT operation. To enable LFXT, the **PSEL** bits associated with the crystal pins must be set.
- For devices that have HFXT available:
  - HFXIN and HFXOUT pins are set to GPIOs, and HFXT is disabled.
- DCOCLK is selected for MCLK, HSMCLK, and SMCLK (**SELMx** = **SELSx** = 3), and each system clock is undivided (**DIVMx** = **DIVSx** = **DIVHSx** = 0), see Table 2.16 for CSCTL1.

2.3.5.4.3 Module clock request system In the MSP432P401R MCU system, for all device modules, including the peripheral modules, they typically use a so-called **Clock On Request** method to obtain their clock sources from the CS module regardless of the current power mode of operation. In other words, all modules need to send a clock-using-request to the CS module to get an appropriate clock source to drive their modules.

As we discussed in Section 2.3.5.4, these clock-using-requests can be divided into two groups: conditional and unconditional. For any of the conditional requests coming from any devices, it can be disabled or masked by the CS module. However, for all internal clock sources, including the **VLO**, **REFO**, **MODOSC**, and **SYSOSC**, there are also some unconditional requests available to peripherals. These unconditional requests cannot be disabled or masked, and must be responded by the CS module if they are coming.

Figure 2.20 shows an example of using conditional and unconditional requests for the ACLK clock source.

As we mentioned, if any device module needs to use any clock source, it needs to send a clock-using-request to the CS module. For conditional requests, two conditions must be met:

1. The selected peripheral is enabled.
2. A clock-using-request is submitted from that enabled peripheral.

The first condition is equivalent to **ACLK selected & module_n enabled** in Figure 2.20, and the second condition is mapped to **ACLK_REQEN_module_n**.

It can be found from Figure 2.20 that if both conditions are true, and if the clock source ACLK is present (**ACLK** is 1) and enabled (**ACLK_EN** = 1), the selected clock source **ACLK_module_n** will be enabled and transferred to the requested peripheral module.

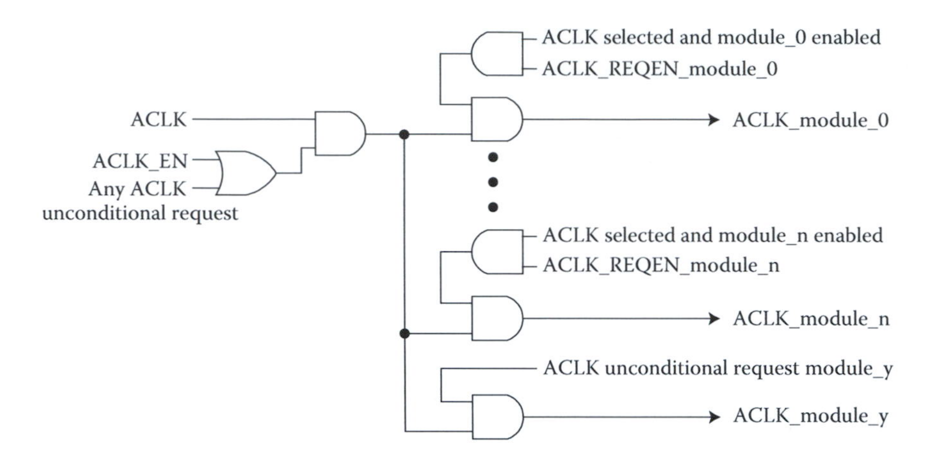

Figure 2.20 Conditional and unconditional requests for ACLK. (Courtesy Texas Instruments.)

For any unconditional request, those two conditions are not needed. The selected clock source **ACLK_module_y** will be transferred to the requested peripheral if this unconditional request is received and the ACLK module is present (**ACLK** = 1) and enabled (**ACLK_EN** = 1), which is shown on the bottom of Figure 2.20.

This example circuit works for all other clock sources, including **MCLK**, **SMCLK**, **HSMCLK**, **BCLK**, and all other internal oscillators. This means that each other clock source can be selected with a similar conditional and unconditional request circuit, as shown in Figure 2.20.

In general, the global clock enable bits (for example, **ACLK_EN**) can be used to enable or disable all conditional system clock requests globally. **By default these bits are set**. Clearing these bits disables the respective system clock even if conditional requests for it are active. If an unconditional request is active, the respective system clock remains active. These bits are useful to disable clocks globally prior to entering a particular power mode. Note that the respective system clock remains disabled when transitioning back to active mode, and the application must reenable it if that system clock is desired.

The system clocks are each distributed to the peripheral modules as individual clock lines as shown in Figure 2.20. This reduces dynamic power to modules that do not require a particular system clock. Only peripheral clocks that request the respective system clock receive it.

A point to be noted is that special care should be taken when enabling or disabling the clock requests to individual modules or globally. Enabling a clock request on an active module immediately causes the requested clock to be presented to the peripheral module. Disabling a clock request on an active module immediately causes any active clock request to be terminated. In both cases, this may cause the status of a module to be affected.

2.3.5.4.4 System clock initialization and configuration The system clock can be any clock sources as we discussed in previous sections. After a system reset, all of these clock sources are enabled (refer to Table 2.18 for **CSCLKEN** register).

By default, the **DCOR** pin works as a GPIO and the DCO mode is internal resistor mode with a default frequency of 3 MHz, which provides clock source for **MCLK**, **HSMCLK**, and

Table 2.20 MSP432P401R MCU default clock sources

Clock	Default clock source	Default clock frequency	Function
MCLK	DCO	3 MHz	Master clock drives CPU and peripherals
HSMCLK	DCO	3 MHz	Subsystem master clock drives peripherals
SMCLK	DCO	3 MHz	Low-speed subsystem master clock drives peripherals
ACLK	LFXT (or REFO if no crystal present)	32.768 KHz	Auxiliary clock drives peripherals
BCLK	LFXT (or REFO if no crystal present)	32.768 KHz	Low-speed backup domain clock drives LPM peripherals

SMCLK clock signals. All default clock sources are shown in Table 2.20. However, only the DCO clock source is enabled and active after a system reset (refer to Table 2.19).

If some applications need higher clock frequency, refer to Section 2.3.5.4.1.2 to change the DCO frequency and check the related bit on the **CSSTAT** register to confirm this changing.

If a peripheral or a device needs to use any of these clocks to support its operation, refer to Section 2.3.5.4 to send a clock-using-request to the desired clock source. Also, the peripheral should be enabled before sending out this request.

2.3.5.5 System controller (SYSCTL)

The SYSCTL is a set of various miscellaneous features of the device, including flash bank and SRAM bank configuration, RSTn/NMI function selection, and peripheral halt control.

In addition, the SYSCTL enables device security features like JTAG and SWD lock and IP protection, which can be used to protect unauthorized accesses either to the entire device memory map or to certain selected regions of the flash.

The SYSCTL on the MSP432P401R MCU provides the following functions:

- Device memory (flash and SRAM) configuration and status
- NMI sources configuration and status
- Watchdog configuration to generate hard reset or soft reset
- Clock run or stop configuration to various modules in debug mode
- Over-ride controls for resets for device debug
- Device security configuration
- Device configuration and peripherals calibration information through device descriptors

Table 2.21 lists related registers that are part of **SYSCTL** and their control functions.

2.4 Introduction to MSP432P401R LaunchPad™ MSP-EXP432P401R evaluation board

The MSP-EXP432P401R LaunchPad™ is an easy-to-use evaluation module (EVM) for the MSP432P401R microcontroller. It contains an ARM® 32-bit Cortex®-M4F MCU, including on-board emulation for programming, debugging, and energy measurements.

Table 2.21 SYSCTL control registers

Address	Register	Macro	Function
0xE0043000	Reboot Control Register	SYS_REBOOT_CTL	Reboot causes the device to reinitialize itself and causes boot-code to execute again
0xE0043004	NMI Control & Status Register	SYS_NMI_CTLSTAT	NMI Control and Status Register Following NMI sources are available on MSP432P401R: • RSTn/NMI device pin in NMI configuration • Clock system (CS) sources • Power supply system (PSS) sources • Power control manager (PCM) sources
0xE0043008	Watchdog Reset Control Register	SYS_WDTRESET_CTL	Configure Watchdog timer module reset sources to hard reset or soft reset based on WDT events: 1. WDT time out 2. WDT violation
0xE004300C	Peripheral Halt Control Register	SYS_PERIHALT_CTL	Enable clock to run or stop on different peripherals while the device is under the control of debugger
0xE0043010	SRAM Size Register	SYS_SRAM_SIZE	Determine the actual SRAM size applied on this special MSP432P401R MCU
0xE0043014	SRAM Bank Enable Register	SYS_SRAM_BANKEN	Select to enable or disable desired SRAM banks to reduce the power consumptions
0xE0043018	SRAM Bank Retention Control Register	SYS_SRAM_BANKRET	Optimize the leakage power consumption of the SRAM in LPM3 and LPM4 modes. In order to enable this, each SRAM bank can be individually configured for retention
0xE0043020	Flash Size Register	SYS_FLASH_SIZE	Indicate the actual size of flash banks used in this MSP432P401R MCU
0xE0043030	Digital I/O Glitch Filter Control Register	SYS_DIO_GLTFLT_CTL	Enable or disable the glitch filter control on selected I/O pins

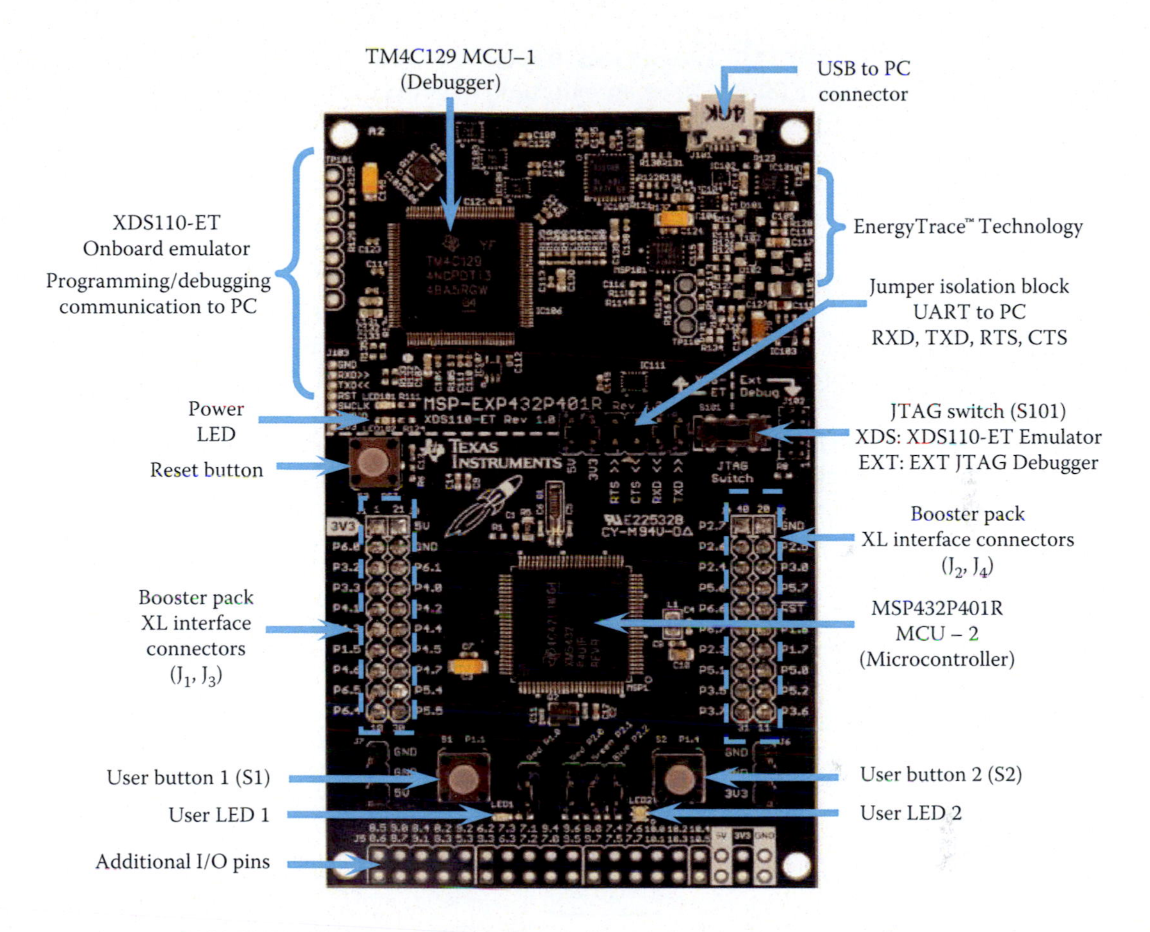

Figure 2.21 MSP-EXP432P401R evaluation board. (Courtesy texas instruments.)

The MSP432P401R MCU supports low-power applications requiring increased CPU speed, memory, analog, and 32-bit performance.

Figure 2.21 shows an illustration photo for the MSP-EXP432P401R evaluation board. The following components and their functions are involved in this board:

- Two microcontrollers, **MCU-1** and **MCU-2**, are included in this EVB. The former MCU **TM4C129** is used as a program loading/debugging controller, and the latter MCU **MSP432P401R** works as a real microcontroller for this board.
- An open-source on-board debugger or emulator, **XDS110-ET emulator**, featuring EnergyTrace+ technology and application UART is installed in the top of this EVB.
- 40-pin LaunchPad Standard (**booster pack XL interface connectors**) that is made of four connectors, **J1~J4**, provides most interface and control signals to peripheral devices.
- One USB connector is provided to support the program development. The **USB to PC connector** (**micro-A/B** cable) is used for users programming download and debugging purpose.
- The **JTAG Switch** (**S101**) selects the **XDS110-ET onboard emulator** or an external JTAG debugger to work as a programming/debugging tool for the MCU. This

switch is a part of **XDS110 Isolation Block**. One can use this switch to select the XDS110-ET emulator (**XDS-ET**) or an external JTAG (**ext debug**) as the program/debug tool for MCU.

- Two user buttons, **S1** and **S2**, are provided to support the users' multiapplication functions. Both user buttons can be used in the preloaded application program to adjust the light spectrum of the LED. The user buttons can also be used for other purposes in the user's custom application.
- Two user LEDs, **LED1** and **LED2**, are provided in the board to enable users to use these LEDs to develop and build different application programs.
- The **reset button** enables users to perform a reset operation for the processor and the entire system. The **power LED** works as an indicator for the MCU working status.
- The **jumper isolation block** allows users to connect or disconnect signals that cross from the XDS110-ET domain into the MSP432P401R target domain. This crossing is shown by the dotted line across the LaunchPad. No other signals cross this domain, so the XDS110-ET can be decoupled from the MSP432P401R target side. This includes XDS110-ET Serial Wire Debug (**SWD**) signals, application UART signals, 3.3 V and 5 V power supplies. A functional block diagram for this block is shown in Figure 2.22.

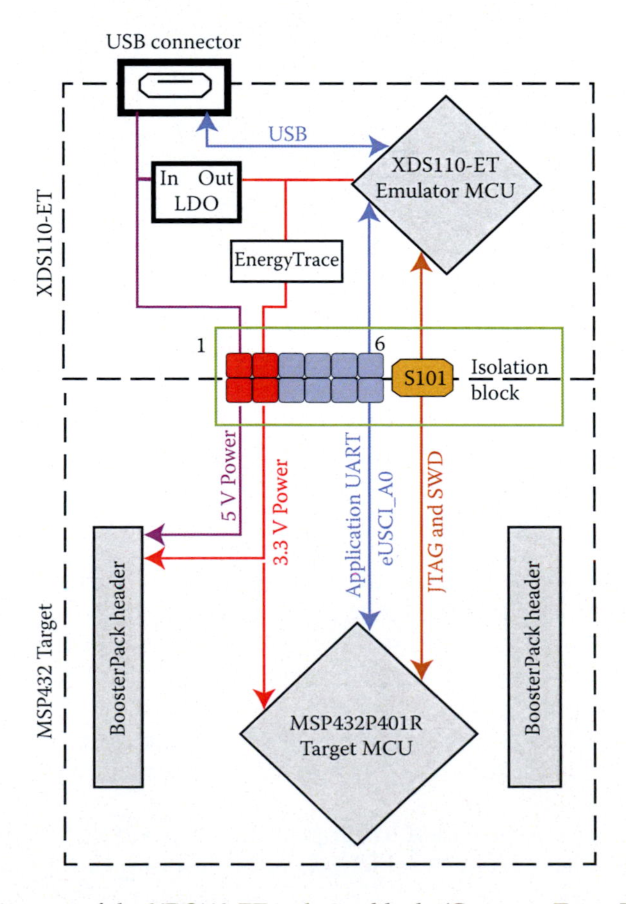

Figure 2.22 Block diagram of the XDS110-ET isolation block. (Courtesy Texas Instruments.)

- **EnergyTrace™ Technology** is an energy-based code analysis tool that measures and displays the application's energy profile and helps to optimize it for ultra-low-power consumption. EnergyTrace™ technology is supported on the LaunchPad™ MSP432P401R device XDS110-ET debugger
- All additional GPIO Pins related to the rest of the internal and external peripherals on the MSP432P401R MCU are connected to another jumper connector **J5** (**additional I/O pins**) that is located at the bottom of this EVB.

This **XDS110-ET Isolation Block** provides flexible control ability for users to enable them to select to either use the target MCU or not on the target board (bottom part) as they like. These control flexible abilities are performed by using a set of jumpers and **JTAG Switch** (**S101**). By using these jumpers to connect the corresponding pins, users will be able to (all jumpers are numbered from the left in Figure 2.22):

- Connect or disconnect 3.3 V power supply to the target board to enable or disable the target MCU MSP432P401R to be driven by this low-level power via the **2nd jumper**.
- Connect or disconnect 5 V power supply to the target board to enable or disable the target MCU MSP432P401R to be driven by this normal-level power via the **1st jumper**.
- Expose the UART interface of the XDS110-ET to enable it to be used for devices other than the onboard MCU, these include:
  - Connect or disconnect **Read-To-Send** (**RTS**—the **3rd jumper**) to the target MCU to enable or disable the MSP432P401R to use this UART **RTS** signal to indicate whether it is ready to receive data from the host PC.
  - Connect or disconnect **Clear-To-Send** (**CTS**—the **4th jumper**) to the target MCU to enable or disable the host PC to use this UART **CTS** signal to indicate whether it is ready to receive data from the MSP432P401R MCU.
  - Connect or disconnect **Receive Data** (**RXD**—the **5th jumper**) to the target MCU to enable or disable the MSP432P401R to use this UART **RXD** signal to indicate whether it is ready to receive data from the host PC.
  - Connect or disconnect **Transmit Data** (**TXD**—the **6th jumper**) to the target MCU to enable or disable the MSP432P401R to use this UART **TXD** signal to indicate whether it is ready to transmit data to the host PC.

By using these jumpers and **JTAG Switch** (**S101**), the top XDS110-ET emulator can be easily disconnected with the lower part, the target MCU (MSP432P401R), to enable users to use any desired external or other debuggers and devices.

By default, all jumpers are connected except **RTS** and **CTS** pins. This means that the target MCU (MSP432P401R) uses both 3.3 V and 5 V power supplies with UART **RXD** and **TXD** signals to communicate with the host PC to enable users to build and download their program via these signal lines.

Figure 2.23 shows the functional block diagram of this MSP-EXP432P401R EVB.

By default, some GPIO pins have been connected to some hardware components on this EVB when shipped, such as user switches **S1** and **S2**, user LEDs, **LED1** and **LED2**. Let's have a closer look at these default connections.

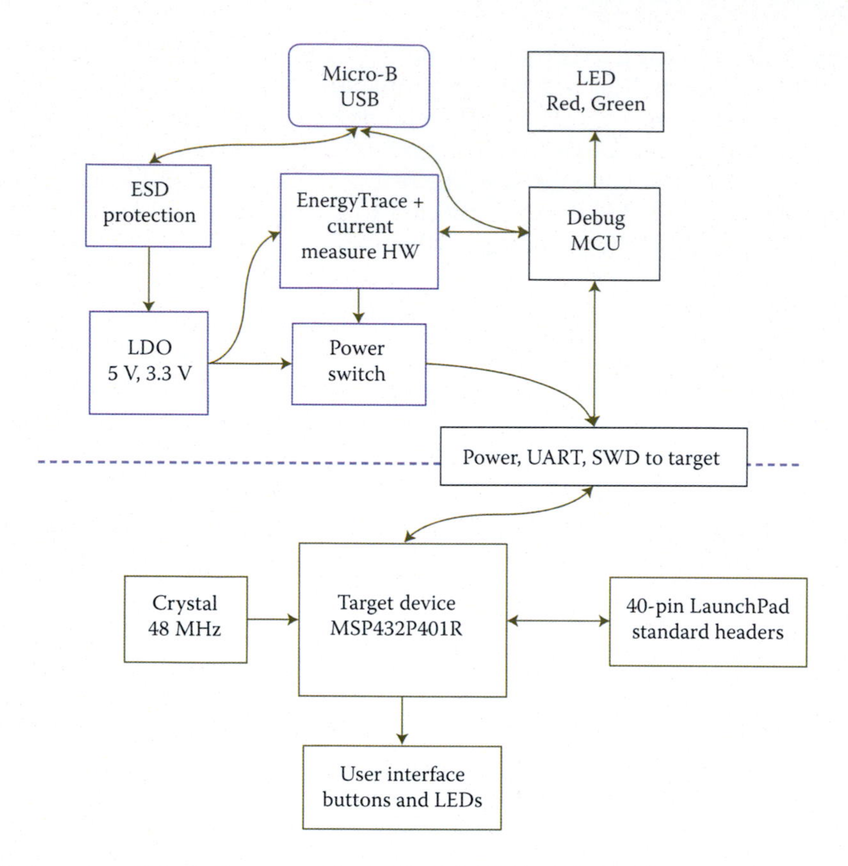

Figure 2.23 Functional block diagram of the MSP-EXP432P401R EVB. (Courtesy Texas Instruments.)

2.4.1 Onboard hardware configurations

For two user LEDs, **LED1** is a single-color (red color) LED, but the second LED, **LED2**, is a three-color LED that can provide red, blue, and green colors when driven by different GPIO pins. These LEDs are connected to different GPIO pins on this EVB. Figure 2.24 shows the hardware connections for these LEDs.

Two user switches, **S1** and **S2**, are also connected to two GPIO pins, respectively. These connections are also shown in Figure 2.24.

JP8~JP11 are four jumpers to enable users to disconnect these peripheral devices if they do not like to use them. Apply a HIGH on GPIO pins **P1.0** or **P2.0~P2.2** to turn related LEDs on. Press any switch button **S1** or **S2** to generate a LOW to GPIO pins **P1.1** or **P1.4**.

2.4.2 GPIO pins configurations on booster pack interface connectors (J1~J4)

Tables 2.22 through 2.25 show the distributions and functions for four jumper connectors, J1~J4 on this EVB.

It can be found from these tables that most GPIO pins provide multiple or alternate functions, and some have interrupt enable abilities to generate some interrupt when an interrupt condition occurs. By using either **PxSEL1/PxSEL0** registers or PMAPC, one can select one of the multiple functions for the related GPIO pin.

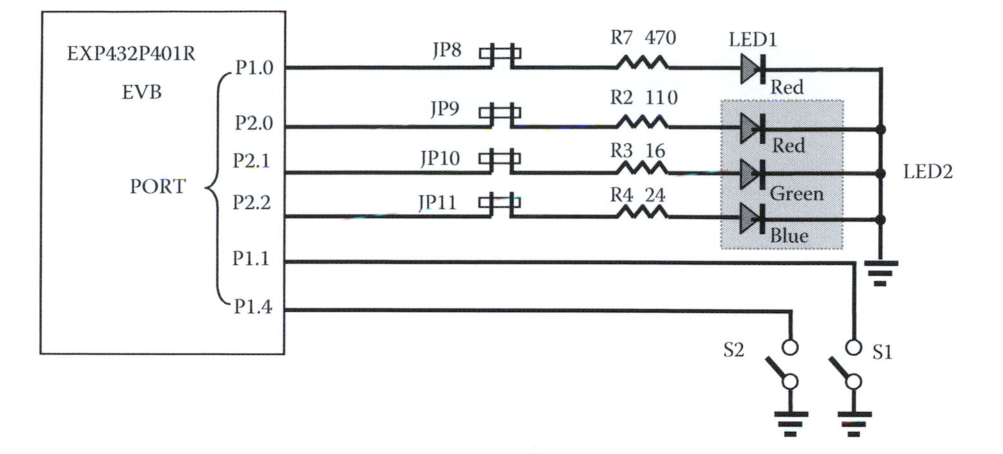

Figure 2.24 Onboard hardware configurations.

Table 2.22 Pin distributions and functions in J1 connector

J1 pin	Digital function	Analog function	Alternate functions	Port mapping	Interrupt enable
1	3.3V	—	—	—	—
2	P6.0	AIN15	—	—	—
3	P3.2	—	UCA2RXD/UCA2SOMI	PM	—
4	P3.3	—	UCA2TXD/UCA2SIMO	PM	—
5	P4.1	AIN12	—	—	Yes
6	P4.3	AIN10	MCLK/RTCCLK	—	—
7	P1.5	—	UCB0CLK	—	—
8	P4.6	AIN7	—	—	Yes
9	P6.5	—	UCB1SCL/UCB1SOMI	PM	—
10	P6.4	—	UCB1SDA/UCB1SIMO	PM	—

Table 2.23 Pin distributions and functions in J2 connector

J2 pin	Digital function	Analog function	Alternate functions	Port mapping	Interrupt enable
1	GND	—	—	—	—
2	P2.5	—	TA0.2	PM	Yes
3	P3.0	—	UCA2STE	PM	Yes
4	P5.7	—	TA2.2/VREF-/VeREF-/C1.6	—	Yes
5	RST	—	—	—	—
6	P1.6	—	UCB0SIMO/UCB0SDA	—	—
7	P1.7	—	UCB0SOMI/UCB0SCL	—	—
8	P5.0	AIN5	—	—	Yes
9	P5.2	AIN3	—	—	Yes
10	P3.6	—	UCB2SDA/UCB2SIMO	PM	—

Table 2.24 Pin distributions and functions in J3 connector

J3 pin	Digital function	Analog function	Alternate functions	Port mapping	Interrupt enable
1	5V	—	—	—	—
2	GND	—	—	—	—
3	P6.1	AIN14	—	—	—
4	P4.0	AIN13	—	—	—
5	P4.2	AIN11	MCLK/TA2CLK	—	—
6	P4.4	AIN9	SVMHOUT/HSMCLK	—	—
7	P4.5	AIN8	—	—	—
8	P4.7	AIN6	—	—	—
9	P5.4	AIN1	—	—	—
10	P5.5	AIN0	—	—	—

Table 2.25 Pin distributions and functions in J4 connector

J4 pin	Digital function	Analog function	Alternate functions	Port mapping	Interrupt enable
1	P2.7	—	TA0.4	PM	Yes
2	P2.6	—	TA0.3	PM	Yes
3	P2.4	—	TA0.1	PM	Yes
4	P5.6	—	TA2.1/VREF + /VeREF + /C1.6	—	Yes
5	P6.6	—	TA2.3/UCB3SIMO/SDA	—	Yes
6	P6.7	—	TA2.4/UCB3SOMI/SCL	—	Yes
7	P2.3	—	UCA1TXD/UCA1SIMO	PM	Yes
8	P5.1	AIN4	—	—	Yes
9	P3.5	—	UCB2CLK	PM	Yes
10	P3.7	—	UCB2SCL/UCB2SOMI	PM	Yes

2.5 *Introduction to EduBASE ARM® trainer*

The EduBASE ARM® Trainer is designed for MSP432P401R LaunchPad™ microcontroller evaluation board MSP-EXP432P401R. Multiple peripheral devices and components are provided by this trainer to enable users to build multiple different applications.

Figure 2.25 shows a photo of this trainer.

The main peripherals and components by this board include:

- 16 × 2 LCD display module with LED backlight
- 4-digit, 7-segment display module for learning multiplexing technique
- 4 × 4 keypad
- Four data LEDs
- A 4-position DIP switch
- Four pushbutton switches

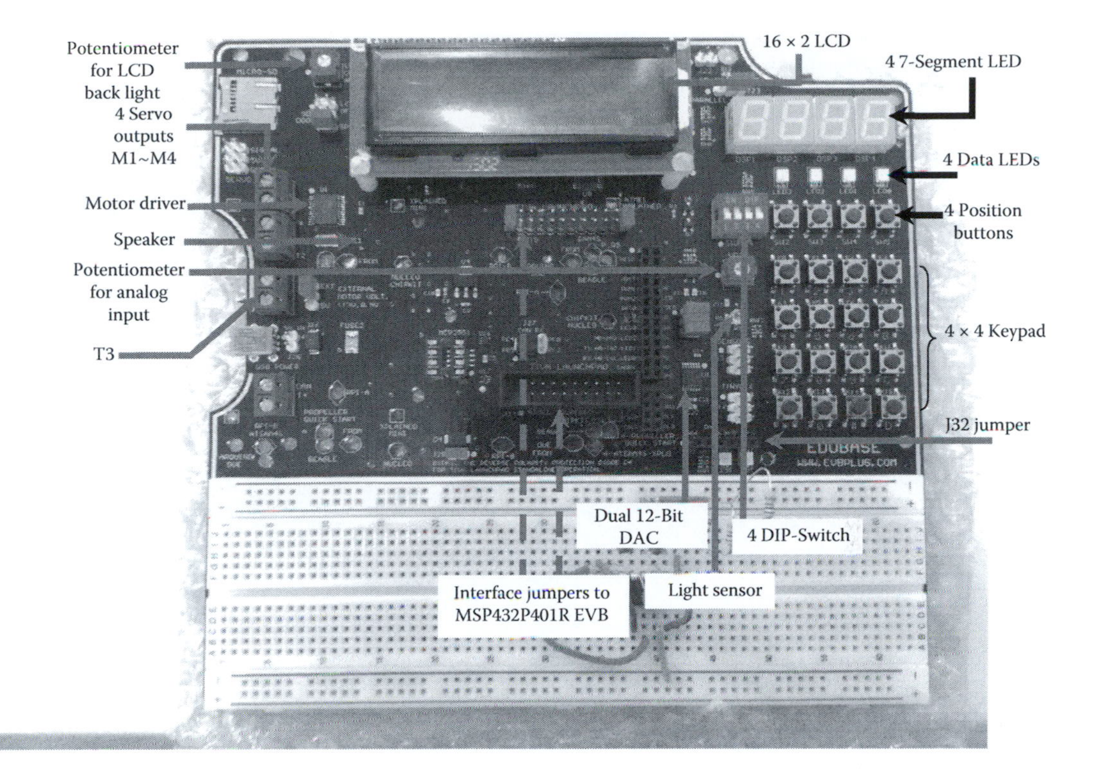

Figure 2.25 EduBASE ARM® trainer.

- Speaker
- Light sensor for home automation applications
- Potentiometer for analog inputs
- X-Y-Z accelerometer module interface header
- Three analog sensor inputs
- Four servo or relay outputs
- SPI-based dual 12-bit DAC for generating analog waveforms
- I²C-based RTC with a capacitor backup
- High efficiency dual H-Bridge for controlling two DC motors or one stepper motor

An interface is provided to allow the MSP-EXP432P401R EVB to be inserted into this trainer. Two dashed lines in Figure 2.25 show these two connectors. A complete connection between the MSP-EXP432P401R EVB and the EduBASE ARM® Trainer is shown in Figure C.1 in Appendix C.

Figure 2.26 shows a complete development system including the EduBASE ARM Trainer with the MSP-EXP432P401R EVB inserted.

2.6 *Summary*

The main topics discussed in this chapter include the architectures and organizations of the most popular embedded systems, including the most updated microcontroller ARM® Cortex®-M4, MSP432P401R MCU, TI™ for MSP432P401R LaunchPad™ MSP-EXP432P401R

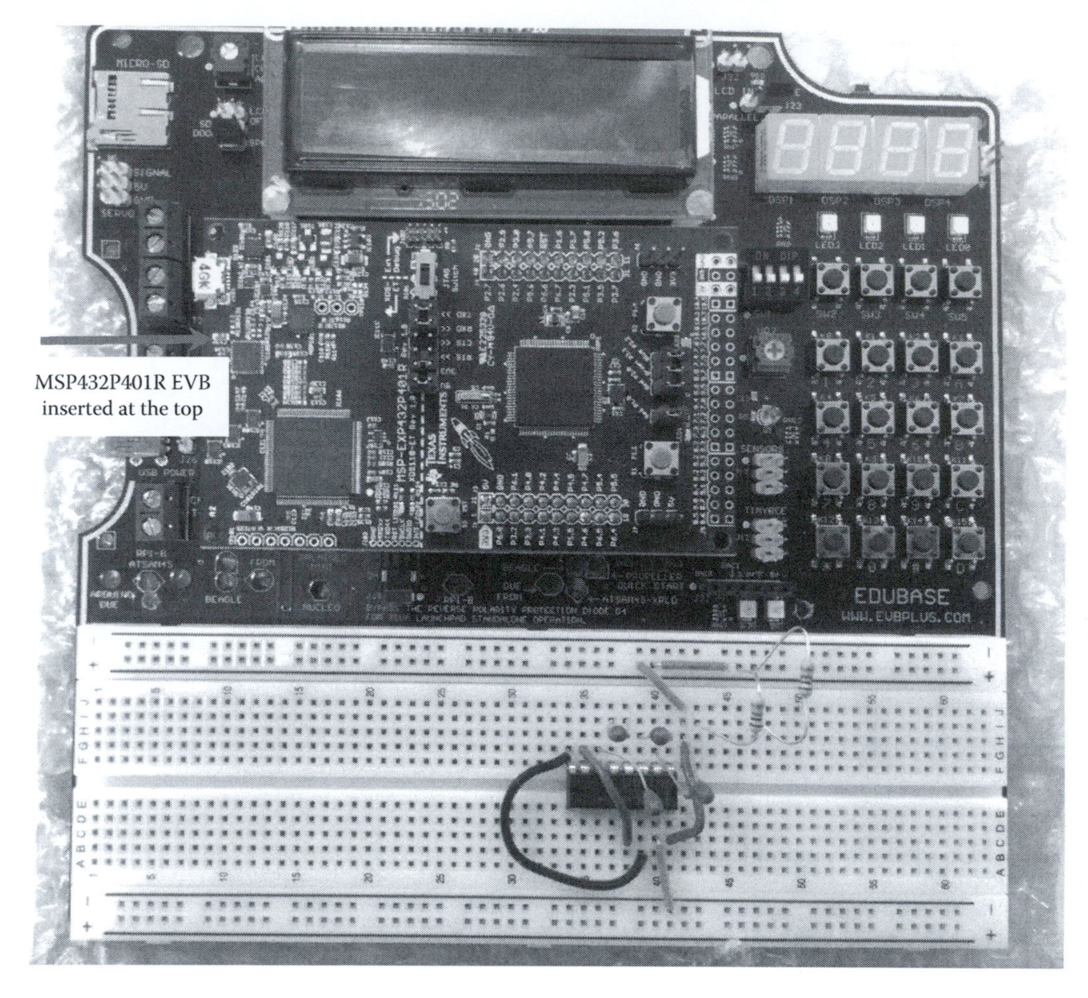

Figure 2.26 EduBASE ARM® trainer with MSP-EXP432P401R EVB.

evaluation board, and EduBASE ARM® Trainer. All of these components will be used in this book to make our project development process easier and simpler.

Starting with an overview of the organizations and architectures of the most popular embedded systems and microcontrollers, one of the most popular and powerful microcontrollers, ARM® Cortex®-M4, is introduced and discussed in detail with the following components:

- An ARM® Cortex®-M4 Processor Core or CPU
- An NVIC closely integrated with the processor core to achieve low-latency interrupt processing
- Multiple high-performance bus interfaces, including code interface and SRAM and peripheral interface
- A system timer unit SysTick
- A low-cost debug solution with the optional ability, such as DAP and data watchpoint
- An optional MPU
- An optional FPU

Then each main component's architectures are discussed in the following sequence:

- ARM® Cortex®-M4 CPU or processor
- FPU
- External Memory Map
- MPU
- NVIC

The TI™ LaunchPad™ for MSP432P4x microcontroller, MSP432P401R, is then discussed in details, which includes:

- On-chip memory map
- GPIO ports architectures
- System controls

Then, the MSP-EXP432P401R evaluation board is discussed with detailed introductions to all components and their functions. The EduBASE ARM® Trainer is also introduced in the last part of this chapter to enable users to have a global picture about the entire development system used in this book.

HOMEWORK

I. True/False Selections

____1. The ARM® Cortex®-M4 processor contains CPU, on-chip memory, and I/O Ports.

____2. The only difference between the ARM® Cortex®-M3 and Cortex®-M4 is that the latter has an optional FPU.

____3. The NVIC is integrated inside the ARM® Cortex®-M4 MCU chip.

____4. Although the Cortex®-M4 processor is a 32-bit MCU, it can also handle 8-bit, 16-bit, and 32-bit data efficiently.

____5. In Cortex®-M4 system, only one peripheral bus system is used and it is called the advanced peripheral bus (APB).

____6. In Cortex®-M4 CPU, there are a total of 21 registers in the register bank. Registers 0~15 are general-purpose registers.

____7. The default clock source of the MSP432P401R MCU is DCO after a system reset.

____8. The MSP432P401R MCU contains 10 GPIO ports, and each port has a set of related control registers.

____9. In the port mapping control, each mapping register **`PxMAPy`** is used to select a specific mapping function for each GPIO pin when it is used.

____10. The MSP432P401R MCU provides seven clock sources and most of them belong to internal clock sources.

II. Multiple Choice

1. All embedded systems or microcontrollers contain ________________ the components.
   a. CPU
   b. Memory device
   c. I/O ports
   d. All of them
2. The ARM® Cortex®-M4 MCU contains ____________.
   a. No memory
   b. 32 KB RAM

c. 25 6KB Flash ROM
d. 2 KB EEPROM

3. One bus interface between the MSP432P401R MCU and external components is ________________________________.
   a. Advanced peripheral bus (APB)
   b. Advanced high-performance bus (AHB)
   c. Advanced microcontroller bus (AMB)
   d. General peripheral bus (GPB)
4. The MSP432P401R MCU memory map system includes ____________________.
   a. 256 KB flash memory
   b. 64 KB SRAM
   c. 32 KB internal ROM
   d. All of them
5. The Internal ROM includes _________________________.
   a. Peripheral device driver library
   b. Graphical library
   c. Bootstrap loader
   d. All of them
6. Any of the following operational modes can be switched between active mode (**AM**) and low-power mode (**LPM**) except _____________________.
   a. LPM0
   b. LPM3
   c. LPM4
   d. LPM3.5 and LPM4.5
7. One prerequisite job of using any system clock register is to ___________________________.
   a. Initialize related clock control register
   b. Enable all related clock control registers
   c. Unlock the Clock Key Register (CSKEY)
   d. Unlock the Clock Control 0 Register (CSCTL0)
8. All GPIO ports are 8-bit ports except _______________, which are 6-bit ports.
   a. Port 1 and Port J
   b. Port 2 and Port 10
   c. Port 9 and Port 10
   d. Port 10 and Port J
9. Most GPIO pins in the MSP432P401R MCU can be accessed to perform multiple functions by using `PxSEL1/PxSEL0` or port mapping, the former is _________________________ and the latter is _________________________.
   a. Direct control, indirect control
   b. Analog control, digital control
   c. Inside the digital function module, outside the digital function module
   d. Inside the peripheral module, outside the peripheral module
10. To save power, the MSP432P401R MCU can run in ____________________ mode(s), it is (they are) _________________________________.
    a. 1, Low-Power Mode
    b. 2, LPM0 and LPM3
    c. 3, LPM0, LPM3 and LPM4
    d. 5, LPM0, LPM3, LPM4, LPM3.5 or LPM4.5

III. Exercises

1. Provide a brief description about basic components used in an embedded system or a microcontroller system.
2. Provide a brief description about basic components used in an ARM® Cortex®-M4 MCU.
3. Explain the functions of 13 general-purpose registers and 3 special registers in the register bank inside an ARM® Cortex®-M4 CPU.
4. Provide a brief description about operational modes available for the MSP432P401R MCU system and their transition relationships.
5. Provide a brief description about the Hard Reset in the MSP432P401R MCU system.

chapter three

MSP432 microcontroller development kits

This chapter provides general information on software development tools and platforms used for the MSP432P401R MCU and MSP-EXP432P401R evaluation board (EVB).

3.1 Overview

Texas Instruments™ MSP432™ Series MCUs offer the industry's most popular ARM® Cortex®-M4 core with scalable memory and package options, unparalleled connectivity peripherals, and advanced analog integration. From Ethernet connectivity to basic UARTs, the MSP432™ Series MCUs offer a variety of solutions for automatic controls, displays, low-power, industrial automation, and much more.

One of the most popular MSP432™ Series microcontroller EVBs is the MSP-EXP432P401R LaunchPad™ evaluation platform from Texas Instruments. These low-cost kits provide developers with everything they need to start designing new applications. The award-winning MSP432P401R Series LaunchPad™ is an ideal introduction to the world of ARM® Cortex®-M4 microcontrollers.

Many commercial development tools, including the IDE, debug adapters, compilers, loaders, and runners, are available for the ARM® Cortex®-M4 microcontrollers. Generally, there are different layers of software used for each different microcontroller system. Figure 3.1 shows an example configuration of using MSP432P401R LaunchPad™ EVB platform.

Basically, the whole development kit can be divided into two layers: (1) the Keil® MDK-ARM® Suite that provides a graphical user interface (GUI) with all general required development tools and (2) the MSP432P401R LaunchPad™ software package that provides specified software and libraries for MSP432P401R EVB.

It can be found from Figure 3.1 that two dash-lines are pointed to the debug adaptor. This means that both Keil® MDK and MSPWare provide a related device driver for the debug adaptor that is connected between the software tools and the EVB. The ARM® developed a vendor-independent hardware abstraction layer for the Cortex-M Processor series called CMSIS. The DAP is a device driver for the debug adaptor involved in the CMSIS, and together it is called CMSIS-DAP. This debug adaptor is used to perform some necessary communications between the software development tools and the microcontroller hardware, in this case, MSP432P401R EVB.

These communications include the debugging user's programs and downloading the compiled programs to the EVB, either to RAM or flash memory.

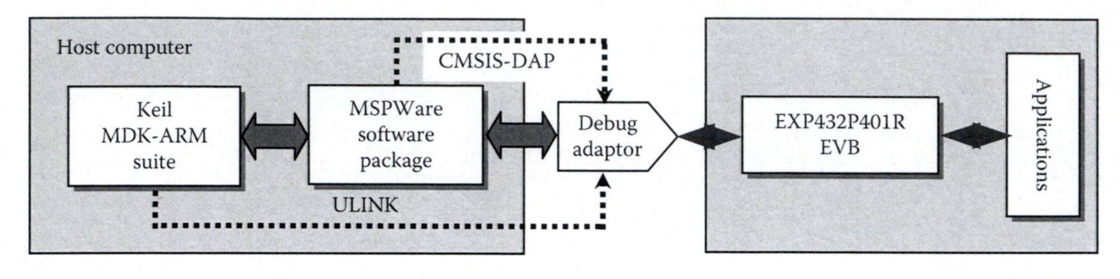

Figure 3.1 Configuration of MSP432P401R-based EVB.

3.2 Entire MSP432P401R-based development system

As we mentioned, we will use an MSP432P401R-based evaluation trainer, EduBASE ARM® Trainer, which comprises MSP432P401R EVB and some other useful peripheral devices, to get a more powerful EVB or Trainer. The hardware setup and connection for the whole evaluation system is shown in Figure 3.2.

It can be found from Figure 3.2 that the entire MSP432P401R-based development system comprises two pieces of important hardware components: the host computer (PC) that works as a control and development unit and the MSP-EXP432P401R EVB with the EduBASE ARM® Trainer EVB. The components and functions of these two pieces of hardware are

- The host computer works as an interface to enable users to create, assemble, debug, and test the user's program in the EduBASE ARM® Trainer using the Keil® IDE and MSPWare software package installed in the host computer. All these functions are performed by accessing various libraries and tools provided by the MSPWare firmware and Keil® IDE installed in the host computer.
- The EduBASE ARM® Trainer and MSP432P401R EVB provide all hardware and software interfacing abilities to facilitate the above operations performed in the host computer.

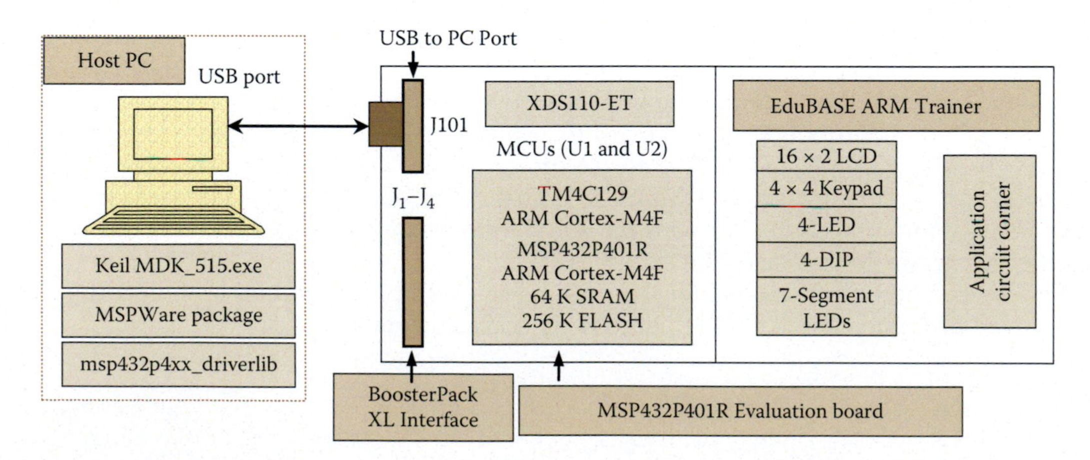

Figure 3.2 Setup and connection of entire MSP432P401R-based development system.

The commands and data communications between the host computer and the MSP-EXP432P401R EVB with EduBASE ARM® Trainer are made through the USB ports, USB port in the host computer and the Debug USB port in the MSP-EXP432P401R EVB, via a USB-to-PC cable.

For the MSP432P401R microcontroller EVB, more than 10 development platforms are available and they are provided by different vendors. However, the following platforms and tools are relatively popular:

- Keil MDK-ARM Microcontroller Development Kit (MDK) IDE
- Texas Instruments' Code Composer Studio™ (CCS) IDE
- IAR Embedded Workbench for ARM®
- Mentor Graphics Sourcery CodeBench
- GNU Compiler Collection (GCC)

Among those tools and platforms, one of the popular choices is the Keil® MDK for ARM®, or MDK-ARM®. This MDK contains all required components and tools to develop application programs for ARM®-related microcontrollers.

The MDK-ARM is a complete software development environment for Cortex™-M, Cortex-R4, ARM7™, and ARM9™ processor-based devices. MDK-ARM® is specifically designed for microcontroller applications; it is easy to learn and use, and powerful enough for the most demanding embedded applications.

MDK-ARM® is available in four editions: MDK-Lite, MDK-Cortex-M, MDK-Standard, and MDK-Professional. All editions provide a complete C/C++ development environment and MDK-Professional includes extensive middleware libraries. Since we are using ARM® Cortex®-M4 MCU, we will concentrate our discussion on MDK-Cortex-M development system.

As we discussed in Chapter 2, the hardware configuration of the entire MSP432P401R-based development system comprises the following components:

- ARM® Cortex®-M4F MCU
- MSP432P401R MCU
- MSP-EXP432P401R EVB
- EduBASE ARM® Trainer
- Host computer

This configuration is presented in Figure 3.3.

Based on the discussion above, the entire development system for ARM® microcontroller comprises the following components or tools:

1. Development kits or suites
2. Debug adaptor and drivers
3. Specified MCU-related software package
4. Program examples
5. Development or EVBs

Since the development EVB has been discussed in the last chapter, now let's discuss the top four components in the following section one by one. However, before we can continue our discussions for these components, we first need to download and install them in the host computer.

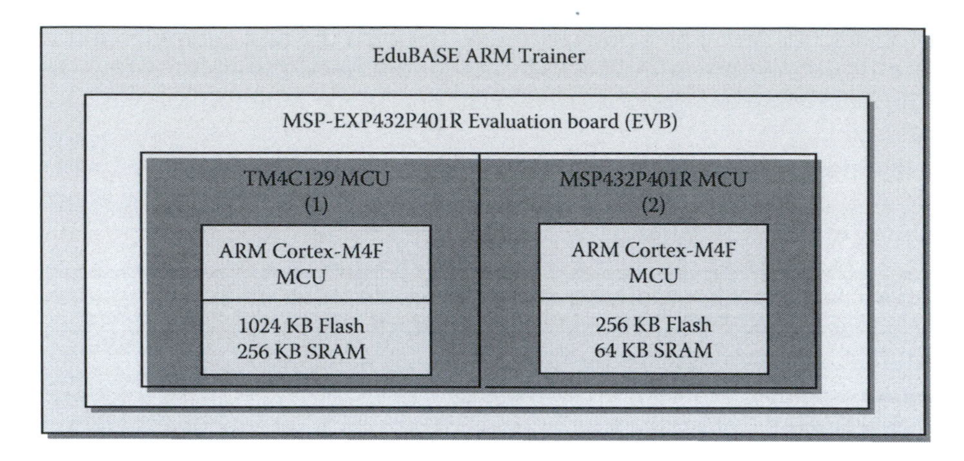

Figure 3.3 Configuration of the EduBASE ARM trainer hardware.

3.3 *Download and install development suite and specified firmware*

Two components, development kits and specified firmware or software package, are the key components and can be downloaded and installed separately. The program examples are MCU-related and they can be installed with the specified firmware together.

Refer to **`Appendix A`** to download and install Keil® MDK-ARM® 5.15. This installation process not only installs the MDK Core but also installs some MSP432P401R-specific interface Software Packs. An icon of installed Keil® MDK, **`Keil µVersion5`**, will be added onto your desktop when the installation is complete. The default installation location of this development suite on your host PC is *C:/Keil_v5*.

Refer to **`Appendix B`** to download and install MSP432™ family-specified firmware package *MSPWare_2_21_00_39*. The default installation location of this firmware in your host computer is *C:/ti/msp*.

Refer to **`Appendix C`** to see the hardware configuration between the MSP432P401R EVB MSP-EXP432P401R and the host computer, and set up the connection between them with the USB-to-PC cable provided by MSP432P401R EVB.

Now, let's discuss these components in the following section one by one. Since the MDK Core contains the debugger, we will discuss these two components together.

3.4 *Introduction to the IDE: Keil® MDK µVersion5*

The Keil® MDK is the most comprehensive software development environment for ARM®-based microcontrollers. MDK Version 5 is now split into the MDK Core and Software Packs which makes new device support and middleware updates independent from the toolchain.

The entire Keil® MDK development system can be divided into the following key components:

- The MDK Core
  - µVersion® IDE with source editor and GUI
  - Pack installer

Figure 3.4 Components included in the MDK core. (Reproduced with permission from ARM® Limited. Copyright© ARM limited.)

    - ARM® C/C++ compiler
    - µVersion® Debugger with trace function
- Software Packs
    - Device drivers for SPI, USB, and Ethernet
    - The CMSIS support, including the CMSIS-CORE, CMSIS-DSP, and CMSIS-RTOS (real-time operating system)
    - MDK middleware support
    - Example programs

MDK Core: The MDK Core contains all development tools, including µVersion IDE, compiler, and debugger. By using the MDK Core, you can create, build, and debug an embedded application for Cortex-M processor-based microcontroller devices. The new Pack Installer adds and updates Software Packs for devices, CMSIS, and middleware. The purpose of the new added Pack Installer is to manage Software Packs that can be added any time to the MDK Core. This makes new device support and middleware updates independent from the toolchain. Software Packs that add support for a complete microcontroller family are called Device Family Packs.

An illustration block diagram of the MDK Core and its components is shown in Figure 3.4.

Software Packs: Software Packs contain device support, CMSIS libraries, middleware, board support, code templates, and example projects. Among all components included in the Software Packs, two components, CMSIS and middleware, need to be explained in more detail.

The CMSIS provides a ground-up software framework for embedded applications that run on Cortex-M-based microcontrollers. The CMSIS enables consistent and simple software interfaces to the processor and the peripherals, simplifying software reuse, reducing the learning curve for microcontroller developers.

The CMSIS application software components include:

- **CMSIS-CORE**: Defines the API for the Cortex-M processor core and peripherals and includes a consistent system startup code. The software components ***CMSIS:CORE*** and ***Device:Startup*** are all you need to create and run applications on the native processor that uses exceptions, interrupts, and device peripherals.
- **CMSIS-RTOS**: Provides standard RTOS and therefore enables software templates, middleware, libraries, and other components that can work across supported RTOS systems.
- **CMSIS-DSP**: Is a library collection for digital signal processing (DSP) with over 60 functions for various data types: fix-point and single- (32-bit) or double-precision floating point (64-bit).

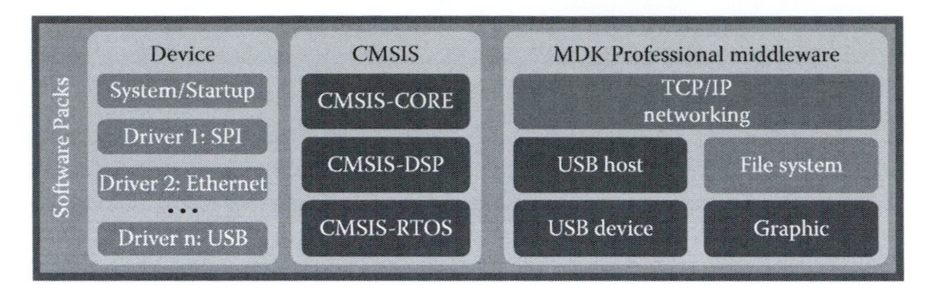

Figure 3.5 Components in the software packs. (Reproduced with permission from ARM® Limited. Copyright© ARM Limited.)

The MDK-Professional Middleware offers a wide range of communication peripherals to meet many embedded design requirements, and it is essential to make efficient use of these complex on-chip peripherals. The MDK-Professional Middleware provides a Software Pack that includes royalty-free middleware with components for TCP/IP networking, USB host, and USB device communication, file system for data storage, and a GUI.

A complete block diagram including all Software Packs components is shown in Figure 3.5.

As the MDK is a powerful development suite with a great amount of components, we will divide our discussions for these components in different chapters. In this chapter, we will concentrate our discussions on the MDK Core, especially on the MDK-Cortex-M family. Two components, µVersion IDE and debugger, are main topics to be discussed in this chapter.

First, let's have a closer look at the Keil® MDK-ARM® works for the MSP432P401R LaunchPad™ EVB.

3.4.1 Keil® MDK-ARM® for the MDK-Cortex-M family

Similar to Keil® MDK-ARM®, the MDK-Cortex-M family contains the following components:

1. **µVision5 IDE**: Provides a GUI with all general required development tools, such as debugger and simulation environment.
2. **ARM compilation tools**: These tools include C/C++ compiler, ARM® assembler, linker, and other utilities.
3. **Debugger**: Provides debug functions for ARM® microcontroller programs.
4. **Simulator**: Provides simulation environment to enable users to build and run program without any real hardware.
5. **Keil RTX RTOS kernel**: Provides a real OS kernel.
6. **TCP/IP networking suite**: Offers multiple protocols and various applications.
7. **USB Device and USB host stacks** are provided with standard driver classes.
8. **ULINKpro** enables on-the-fly analysis of running applications and records every executed Cortex-M instruction.
9. **Complete code coverage** information about your program's execution.
10. Execution Profiler and Performance Analyzer enable program optimization.
11. CMSIS compliant.
12. **Reference start-up codes** for about 1000 microcontrollers.

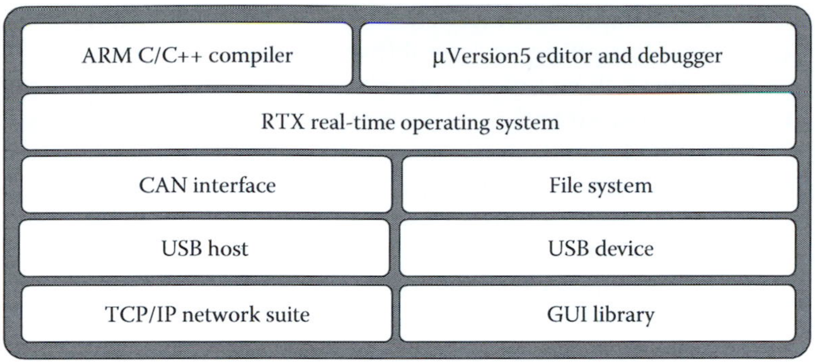

ARM C/C++ compiler | µVersion5 editor and debugger
RTX real-time operating system
CAN interface | File system
USB host | USB device
TCP/IP network suite | GUI library

Figure 3.6 Complete structure of the MDK-Cortex-M development system.

13. Flash programming algorithms.
14. Program examples.

A complete configuration of the MDK-ARM® for Cortex®-M family is shown in Figure 3.6.

To better understand the program development process with Keil® MDK, first let's have a detailed discussion about the general development flow of a user project by using the MDK-Cortex®-M development system.

3.4.2 General development flow with MDK-ARM®

Figure 3.7 shows a general development process of a user project in the Keil® MDK.

Generally, a user project in Keil® MDK can be developed in the following steps:

- A new project is created using the Keil® MDK.
- The user's source files with source codes, either C or ARM assembly codes, are added into the project.
- Depending on your source codes, the armcc (ARM® compiler) or the armasm (ARM® assembler) is called and executed to translate the user's source codes to the object

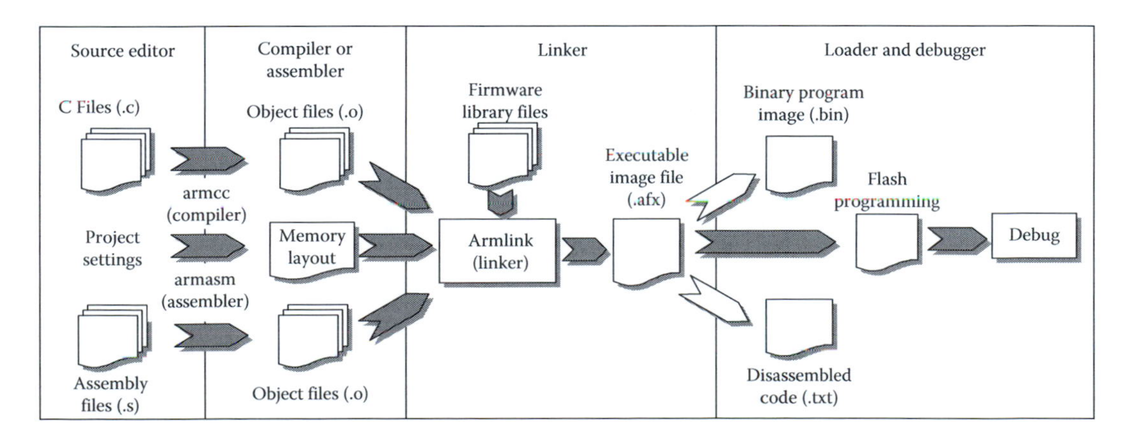

Figure 3.7 Development structure of a user project with MDK.

codes and stored in the host computer. The start-up codes with Project Settings will be involved in this compiling or assembling process.
- The object code files will be linked with all other system library files or MCU-related library files and converted to the executable files, or image file in the ARM® terminology, and downloaded into flash memory or RAM in the EVB.
- Finally, the executable codes can be sent to the debugger to perform the debugging or executing operations. In fact, the compiler, linker, and loader are integrated into one unit, the Builder in the ARM® μVersion® IDE.
- Alternatively, the executable file can also be converted to the binary file or text file for the users' reference.

To successfully build a basic user project with Keil® MDK, one needs to use the following two key components:

1. Keil® MDK Core
2. CMSIS-Core

The MDK Core provides all development tools, including μVersion IDE, compiler, and debugger. The CMSIS-Core defines the API for the Cortex-M processor core and peripherals and includes a consistent system startup code. The software components ***CMSIS:CORE*** and ***Device:Startup*** are all you need to create, build, and run applications on the native processor that uses exceptions, interrupts, and device peripherals in the MDK IDE environment.

All user source codes can be written in C; however, the startup codes that are provided by MCU vendors and generally included in the Keil® MDK installation process are ARM® assembly codes. The users also need to use some library and driver files provided by the MCU vendor, in our case, the firmware MSPWare package provided by Texas Instruments™ to create some system files. This situation is shown in Figure 3.8.

As shown in Figure 3.8, to build and develop an ARM® application project, a native Cortex-M Core with CMSIS-Core should be used. In fact, one CMSIS-Core component ***CMSIS:CORE***, combined with the software component ***Device:Startup*** and MSPWare, can be used to build a successful project. These components provide the following central files:

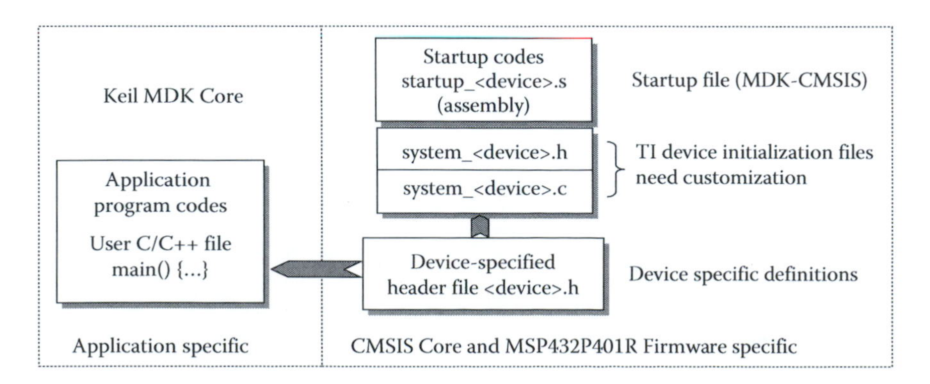

Figure 3.8 Program development with MDK Core and CMSIS-Core.

- The **startup _ <device>.s** file with reset handler and exception vectors
- The **system _ <device>.c** configuration file for basic device setup (clock and memory bus)
- The **system _ <device>.h** includes file for user code accessing to the microcontroller device

The device-specified **<device>.h** header file is included in C source files and defines:

- Peripheral access with standardized register layout
- Access to interrupts and exceptions and the NVIC
- Intrinsic functions to generate special instructions, for example, to activate sleep mode
- Systick Timer (SYSTICK) functions to configure and start a periodic timer interrupt
- Debug access for printf-style I/O and ITM communication via on-chip CoreSight

One point to be noted is that in an actual application file, the *<device>* is the name of the microcontroller device used in the real user project. For example, in our case, this device name should be *<msp432>*. Also, not all these four files can be found at the development stage, and some files, **system _ <device>.h**, and device-specified **<device>.h** cannot be found until the program has been built successfully.

Now that we have some basic understanding about the Keil® MDK-Cortex-M development kit, let's get our feet a little wet with more details to familiarize us with this kit.

3.4.3 Functions of the Keil® MDK-ARM® µVersion®5 GUI

In this section, we will have a detailed discussion about this MDK-ARM® µVision®5 GUI.

The **µVision5.15** (for convenience, we use µVision5 below) is an updated window-based software development platform that combines a robust and modern editor with a project manager and makes facility tool. It integrates all the tools needed to develop embedded applications, including a C/C++ compiler, macro assembler, linker/locator, and a HEX file generator. The µVision5 helps expedite the development process of embedded applications by providing the following:

- Full-featured source code editor
- Device Database® for configuring the development tool
- Project Manager for creating and maintaining your projects
- Integrated Make Utility functionality for assembling, compiling, and linking your embedded applications
- Dialogs for all development environment settings
- True integrated source- and assembler-level debugger with high-speed CPU and peripheral simulator
- Advanced GDI interface for software debugging on target hardware and for connecting to a Keil® ULINK™ debug adapter
- Flash programming utility for downloading the application program into flash memory
- Links to manuals, online help, device datasheets, and user guides

The µ**Vision5 IDE** and **Debugger** are the central part of the Keil® development toolchain and have numerous features that help the programer to develop embedded applications quickly and successfully. The Keil® tools are easy to use, and are guaranteed to help you achieve your design goals in a timely manner.

The µVision5 offers a **`Build Mode`** for creating applications and a **`Debug Mode`** for debugging applications. Applications can be debugged with the integrated µVision5 **`Simulator`** or directly on hardware, for example, with adapters of the Keil® ULINK™ USB-JTAG family. Developers can also use other AGDI adapters or external third-party tools for analyzing applications.

The µVision5 GUI provides menus for selecting commands and toolbars with command buttons. The status bar, at the bottom of the window, displays information and messages about the current µVision5 command. Windows can be relocated and even docked to another physical screen. The window layout is saved for each project automatically and restored the next time the project is used. You can restore the default layout using the menu **`Window|Reset View to Defaults`**.

Before opening the Keil® MDK-ARM® µVersion 5.15 Suite, make sure that the following two important components have been setup with your host computer:

- The MSP432P401R EVB has been connected to your host PC with the USB-to-PC cable. Refer to **`Appendix C`** to complete this hardware setup if you have not.
- All development tools have been downloaded and installed as we did in Section 3.3.

Now open the MDK µVersion5 IDE by double clicking on the icon **`Keil µVision5`** from the desktop. The opened IDE is shown in Figure 3.9. Some important tools in the toolbar have been highlighted. We will discuss all the important menu items one by one in the next section.

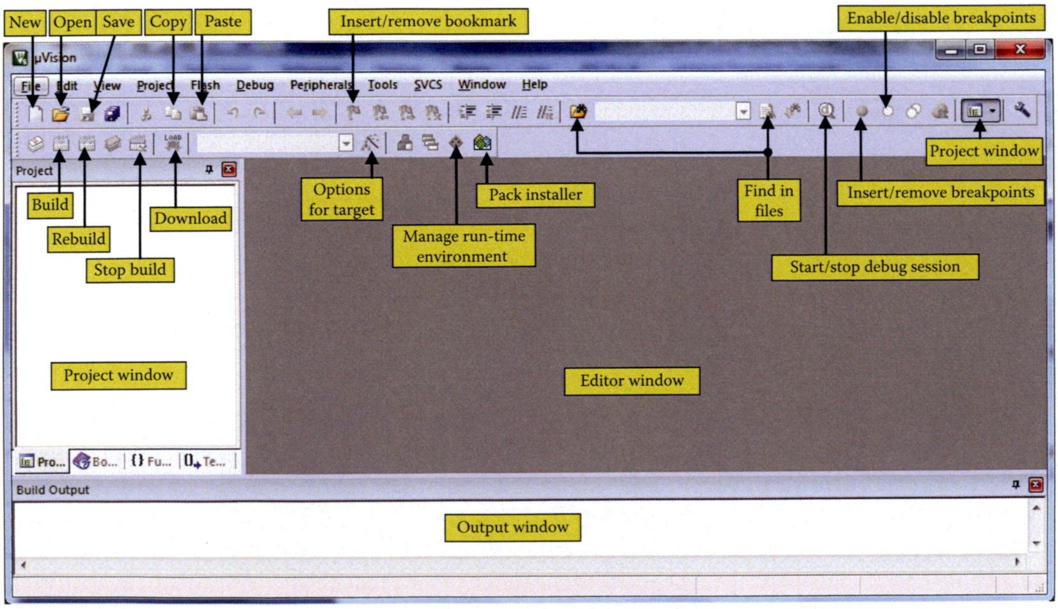

Figure 3.9 Opened MDK-ARM IDE. (Reproduced with permission from ARM® Limited. Copyright© ARM Limited.)

3.4.3.1 File menu

Under the **`File`** menu, there are 12 menu items or submenus, but only two items, **`Device Database`** and **`License Management`**, are new and important to us.

`File`:

- **`Device Database`**
- **`License Management`**

The **`Device Database`** lists all available devices offered by different vendors and provides download access to the related Software Packs. You can also confirm or check the installed device by using this menu item. For example, in our case, we are using an MSP432P401R MCU. To check this device, click on **`Device Database`** item from the **`File`** menu. The Device Database wizard is displayed, as shown in Figure 3.10.

Scroll down the **`Database`** list to find all devices made by Texas Instruments. Expand the **`Texas Instruments`** folder, and find the subfolder **`MSP432 Family`**. Expand this subfolder and **`MSP432P`** folder, then you can find our target device, **`MSP432P401R`**. Click on this device and you can find all pieces of related information about this MCU, as shown in Figure 3.11.

The **`License Management`** is used to manage the license version for the MDK you are using. The following license types are available in the current MDK µVersaion5:

- **`Single-User License`** (Node-Locked) grants the right to use the product by one developer on two computers at the same time.

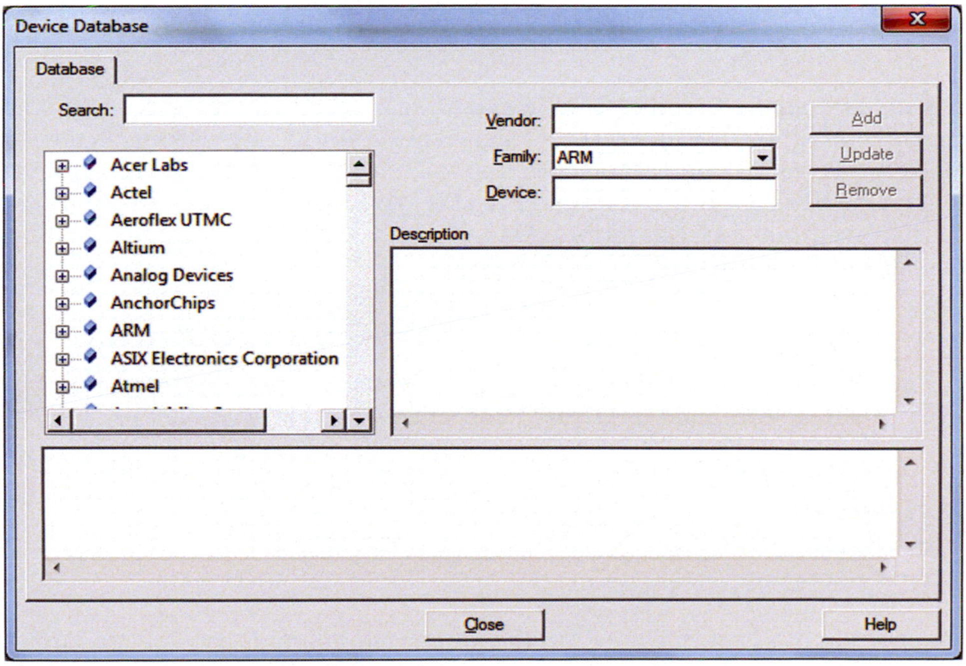

Figure 3.10 Opened device database wizard. (Reproduced with permission from ARM® Limited. Copyright© ARM Limited.)

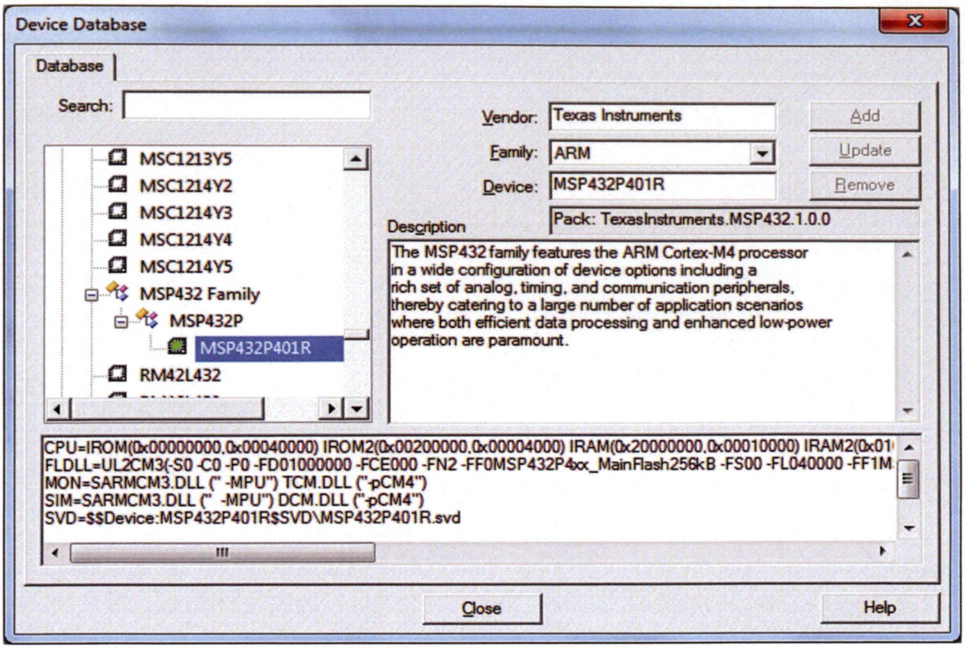

Figure 3.11 Detailed information for the device MSP432P401R. (Reproduced with permission from ARM® Limited. Copyright© ARM Limited.)

- **`Floating-User License`** or **`FlexLM License`** grants the right to use the product on several computers by a number of developers at the same time.

If you select the **`License Management`** item from the file menu, the License Management wizard will be displayed, as shown in Figure 3.12. By clicking on the Single-User License tab, you can find the detailed information for the single user. However, when clicking on the **`Floating License`** and **`FlexLM License`** tabs, the similar information will be displayed, but some controls would be disabled. This means that those controls are not available to the single user.

3.4.3.2 Edit and view menus

There are 18 menu items or submenus under the **`Edit`** menu, but only three items, **`Outlining`**, **`Advanced`**, and **`Configuration`**, are new and important to us.

`Edit`:

- **`Outlining`**
- **`Advanced`**
- **`Configuration`**

The **`Outlining`** is an MDK-ARM Plug-in for Eclipse and it lists the structural elements of a C/C++ file that is currently open in the editor. Developers can sort the list, set filters, and group the elements for viewing. With this menu, you can show or hide all outlining for your source codes. Also you can expand and collapse all definitions, current block, or current procedure by selecting the related item in this menu.

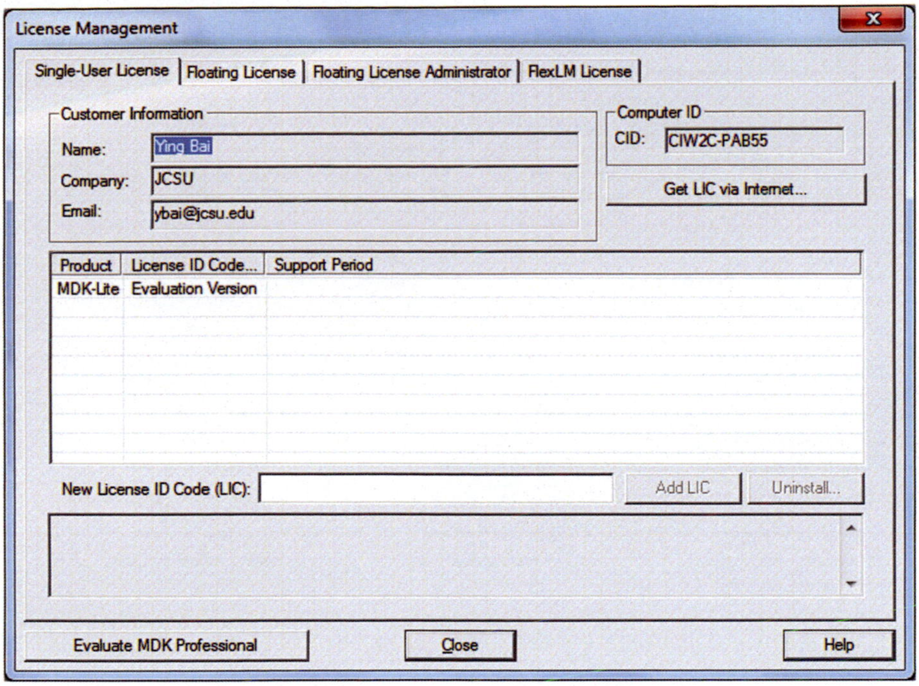

Figure 3.12 License management wizard. (Reproduced with permission from ARM® Limited. Copyright© ARM Limited.)

The **Advanced** is used to assist the editing and formatting of the source codes. These assistants include converting the codes to upper or lower cases, or to comment, increasing or decreasing the line indentation for the selected code line, cutting or deleting the selected lines. When this item is selected, it opens a submenu with extended editor features. The commands are also accessible through the context menu.

The **Configuration** is another editor-assistant and it is used to setup and configure the general settings for the MDK Editor. An example of a Configuration wizard is shown in Figure 3.13.

These configuration tools include:

- Encoding mode
- Color and fonts for the specified editor
- User keywords
- Shortcut keys
- Text completion
- Other settings

The user can define and select desired format and style from this menu for the MDK Editor to get specific codes editing and displaying format.

The **View** menu provides 10 items to enable users to select the desired windows and tools on the toolbar. These windows and tools include:

- **Status Bar**
- **Toolbars**

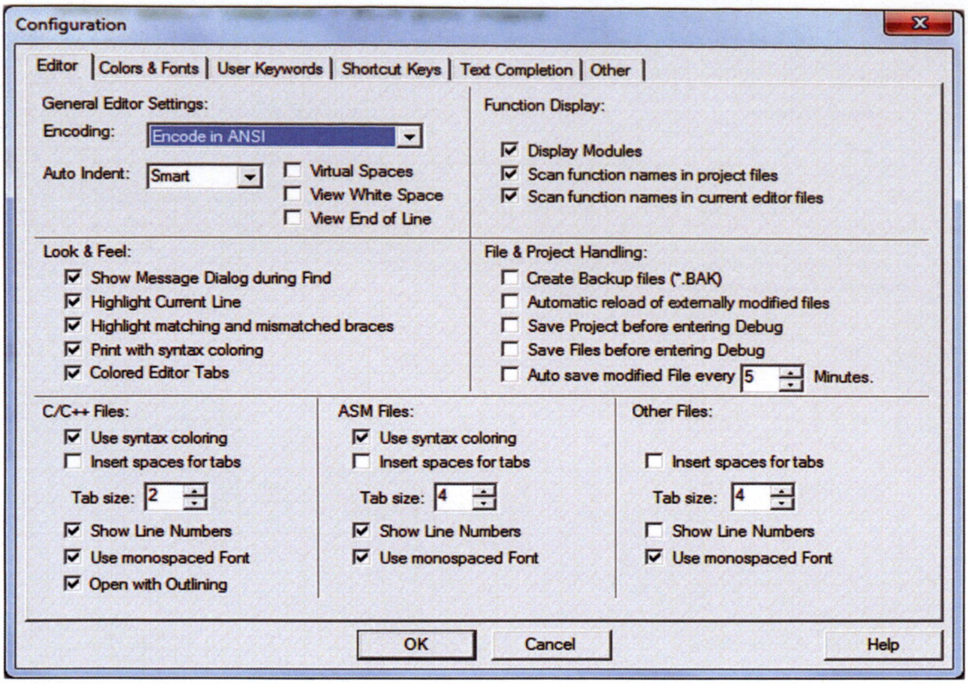

Figure 3.13 Example of the configuration submenu. (Reproduced with permission from ARM® Limited. Copyright© ARM Limited.)

- **`Project Window`**
- **`Books Window`**
- **`Functions Window`**
- **`Templates Window`**
- **`Source Browser Window`**
- **`Build Output Window`**
- **`Error List Window`**
- **`Find In File Window`**

There are two submenu items under the toolbars, **`File Toolbar`** and **`Build Toolbar`**. The former provides all tools used to manage user's files development, and the latter provides all support tools to build the user's projects. You can open or close any of these tools or windows by clicking the selected item from this menu.

3.4.3.3 Project menu

The **`Project`** is a very important menu in the MDK-ARM® μVersion5 IDE. It includes commands to create, open, save, and close project files, **`Export`** the project to a previous version, μVision4, **`Manage`** project components, or set **`Options`** for the target, group, and file, or **`Build`** the project. Multiple projects can be managed through the menu **`Project|Manage|Multi-Project Workspace`**. In total, 16 commands are included in this **`Project`** menu:

`Project`:

- **`New μVersion Project`**
- **`New Multi-Project Workspace`**

- **`Open Project`**
- **`Save Project in µVersion4 Format`**
- **`Close Project`**
- **`Export`**
- **`Manage`**
- **`Select Device for Target Project`**
- **`Remove Items`**
- **`Options for Target`**
- **`Clean Target`**
- **`Build Target`**
- **`Rebuild All Target Files`**
- **`Batch Build`**
- **`Translate`**
- **`Stop Build`**

Let's discuss these menu items one by one with a real new user project **`MyProject`**.

The **`New µVersion Project`** enables users to create a new project in the following sequence:

1. Open the Windows Explorer to create a new folder named **`MyProject`** under the **`C:\MSP432 Class Projects\Chapter 3`** folder.
2. Open the Keil® ARM®-MDK µVersion5 IDE and click on the **`New µVersion Project`** item. On the opened **`Create New Project`** wizard, browse to our new folder **`MyProject`** created above and enter the project name, **`MyProject`** (Figure 3.14). The file extension *.**`uvproj`** is for MDK µVersion4, or *.**`uvprojx`** is for µVersion5. Click on the **`Save`** button to save this new project. It is a good practice to use a separate folder for each project.

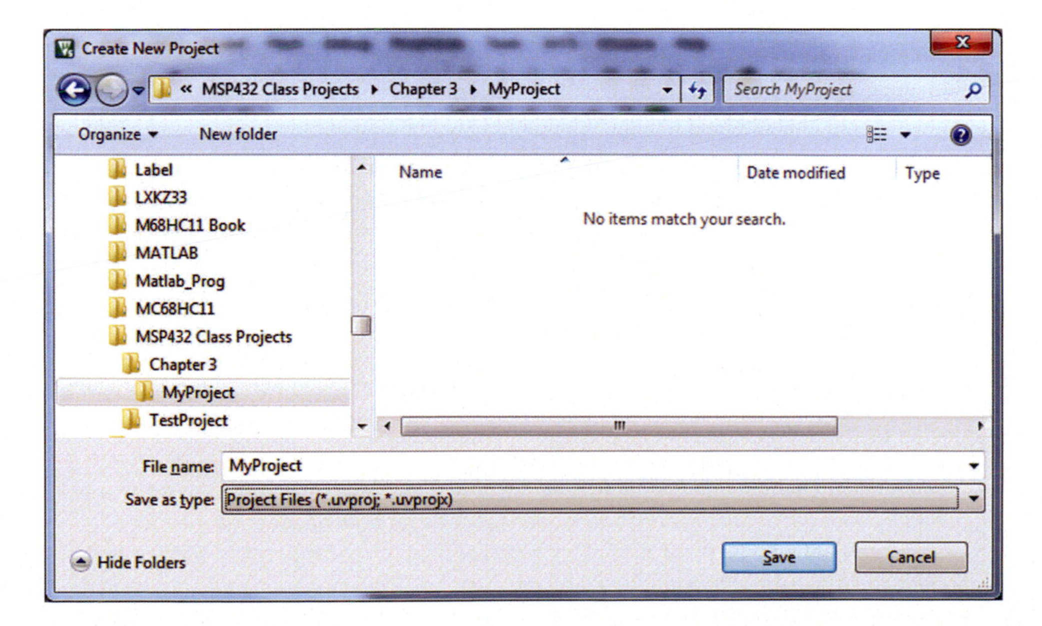

Figure 3.14 Save the new project.

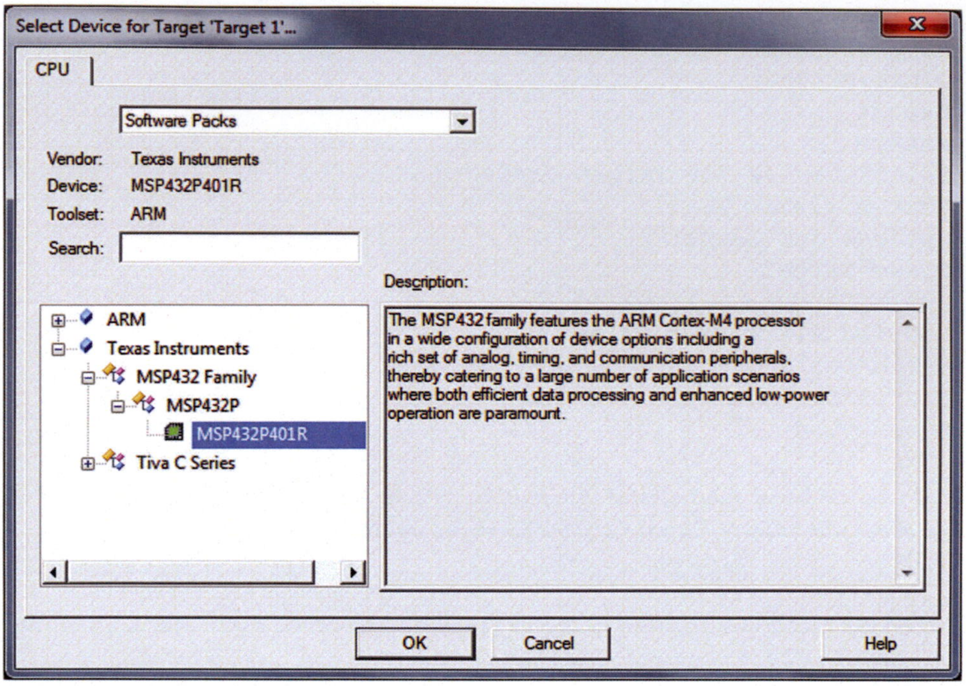

Figure 3.15 Select device for target wizard. (Reproduced with permission from ARM® Limited. Copyright© ARM Limited.)

3. The **Select Device for Target "Target1"** wizard is displayed as shown in Figure 3.15. Expand folders, **Texas Instruments->MSP432 Family->MSP432P**, select our device, **MSP432P401R**. This selection defines essential tool settings such as compiler controls, memory layout for the linker, and the flash programming algorithms. Click on the **OK** button to continue.
4. Immediately the **Manage Run-Time Environment** (**RTE**) wizard is shown, as shown in Figure 3.16, after a new project is created and saved. The **Manage RTE** wizard allows you to manage software components of this new project by adding, deleting, disabling, or updating during the software development process at any time.
5. MDK µVersion5 offers software components for creating applications with a framework called **Manage RTE**. Software components are delivered in Software Packs that get installed independently from the MDK Core. Third-party Software Packs can also be installed to add other middleware libraries.

   When a new project is created, the RTE window opens automatically after you have selected a device. You can open this wizard by going to the menu **Project|Manage|Run-Time Environment**. The **Manage RTE** wizard provides following functions:

   a. Lists prebuilt software components that are installed and available for the selected MCU. Software components can exist in different variants and versions.
   b. Manages software components of a project. Only configurable files are copied to the project folder. Header files, source code, or libraries that need no modification are included directly from the folder structure of the Software Pack. This simplifies the maintenance of different component versions or variants.

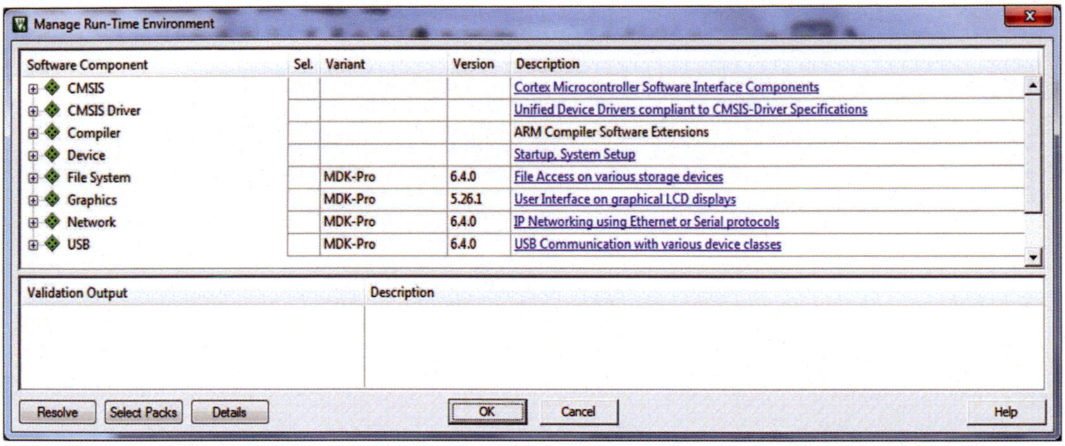

Figure 3.16 Manage run-time environment wizard. (Reproduced with permission from ARM® Limited. Copyright© ARM Limited.)

c. Handles software component versions and variants in a project. Various project targets can use different microcontrollers and/or different versions/variants of a software component. The RTE manager replaces the relevant files of the selected software components automatically.
d. Identifies conflicts between software components. For example, it is not possible to select multiple Liquid Crystal Display (LCD) interfaces for the graphic component.
e. Identifies other required software components. For example, the RTOS kernel or a driver for a device peripheral. The button **`Resolve`** selects other components in case of unambiguous requirements.
f. Provides access to the documentation of a software component.

Now, let's expand the following nodes since we need to use some components and need to setup some of them in this **`Manage RTE`** wizard:

i. **`CMSIS`**
ii. **`Device`**

For a basic user application, we only need the MDK Core and CMSIS-Core. Select and setup the following components by checking the related checkbox in the **`Sel`** column:

i. **`CMSIS:CORE`**
ii. **`Device:Startup`**

Your finished setup and selection for this **`Manage RTE`** wizard should match the one that is shown in Figure 3.17.

One point to be noted is that different colors may be displayed for the different selected components when you do this setup and selection. Also the detailed information will be displayed in the **`Validation Output`** window in the bottom of this wizard to indicate some missed or required components you need to get to complete these setups and selections.

i. **`Green color`**: The software component has been resolved or a software component allowing multiple instances has been resolved. Nothing has been displayed in the validation output window.
ii. **`Yellow color`**: This software component is unresolved. Other components are required for correct operation and are listed in the validation output window.

Manage Run-Time Environment

Software Component	Sel.	Variant	Version	Description
CMSIS				Cortex Microcontroller Software Interface Components
CORE	☑		4.1.0	CMSIS-CORE for Cortex-M, SC000, and SC300
DSP	☐		1.4.5	CMSIS-DSP Library for Cortex-M, SC000, and SC300
RTOS (API)			1.0	CMSIS-RTOS API for Cortex-M, SC000, and SC300
CMSIS Driver				Unified Device Drivers compliant to CMSIS-Driver Specifications
Compiler				ARM Compiler Software Extensions
Device				Startup, System Setup
Startup	☑		1.0.0	System Startup for MSP432 Family
File System		MDK-Pro	6.4.0	File Access on various storage devices
Graphics		MDK-Pro	5.26.1	User Interface on graphical LCD displays
Network		MDK-Pro	6.4.0	IP Networking using Ethernet or Serial protocols
USB		MDK-Pro	6.4.0	USB Communication with various device classes

Validation Output	Description

Resolve | Select Packs | Details | OK | Cancel | Help

Figure 3.17 Selected components in the manage run-time environment wizard. (Reproduced with permission from ARM® Limited. Copyright© ARM Limited.)

iii. **`Red color`**: The software component conflicts with other components or is not installed on the computer. Detailed information is listed in the validation output window.

Click on the **`OK`** button to close this wizard when the setup and selection process is done.

In case you encountered either yellow or red colors, you may need to use the **`Pack Installer`** to install those missed components.

6. Now, our new project **`MyProject`** has been created and necessary environments have been setup using the Manage RTE as we did above. The MDK μVersion5 GUI looks like the one shown in Figure 3.18.

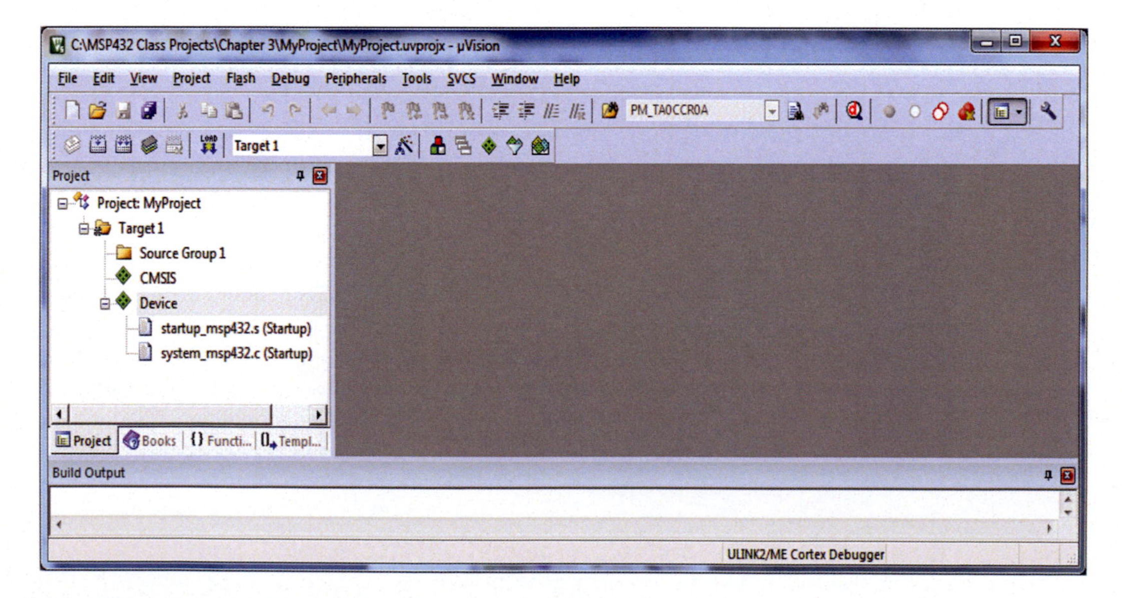

Figure 3.18 New project wizard. (Reproduced with permission from ARM® Limited. Copyright© ARM Limited.)

As we discussed in Section 3.4.2 (Figure 3.8), to make a user project in the MDK environment, the following four files are needed:

a. Startup code file, **startup_msp432.s** (assembly file provided by the MDK)
b. System source file, **system_msp432.c** (C file provided by Texas Instruments)
c. System header file, **system_msp432.h** (header file provided by Texas Instruments)
d. Device-specified definition file, **msp432p401r.h** (header file provided by the Texas Instruments for MSP432P401R MCU)

However, right now we can only find the first two code files, **startup_msp432.s** and **system_msp432.c**, from the **Project** pane in the left in Figure 3.18. Do not worry about this now and you will see all of these four files later as we finish the building of our project.

7. In MDK µVersion5, the projects are organized in a special format, called **Target** → **Source Group** → **Source Files**. All related projects can be collected together to form a group and put into one special group. The default group is **Source Group 1**. You can add additional source files into a source group by double clicking on the selected source group and then using the file browser to add the source file. Different source groups can also be collected together to form a Target, the default target is **Target 1**. You can also add any additional source groups into the Target by right clicking on **Target 1**, and select **Add Group**.
8. Refer to Figure 3.18, you can find that the default group, **Source Group 1**, has been added into the **Target 1** under the **Project** pane in the left. To add our project source file, **MyProject.c**, into this default source group, perform the following operations:
   a. Right click on the **Source Group 1** and select **Add New Item to Group 'Source Group 1'** to open the Add New Item to Group wizard, as shown in Figure 3.19.

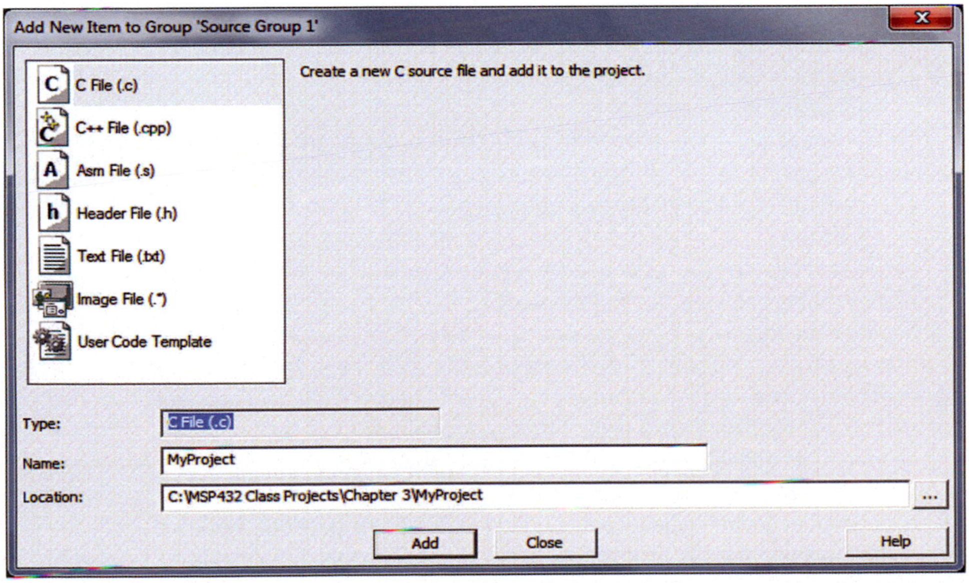

Figure 3.19 Add new item to group wizard. (Reproduced with permission from ARM® Limited. Copyright© ARM Limited.)

```
1  //*****************************************************************
2  // MyProject.c - Main Application File for MyProject
3  //*****************************************************************
4  #include "msp.h"
5  int main(void)                              // Default clock: DCO (3 MHz)
6  {
7    volatile uint32_t i;
8    WDTCTL = WDTPW | WDTHOLD;                 // Stop watchdog timer
9    // The following code toggles P1.0 port
10   P1DIR |= BIT0;                            // Configure P1.0 as output
11   while(1)
12   {
13     P1OUT ^= BIT0;                          // Toggle P1.0
14     for(i = 0; i < 10000; i++);             // Time delay
15   }
16 }
```

Figure 3.20 Source codes in the MyProject.c file.

b. Select **C File (.c)** as the template and enter **MyProject** into the **Name** box. Click on the **Add** button to add this source file into our new project.
c. Add the codes shown in Figure 3.20 into this source file.
d. Go to **File|Save** menu item to save this source file.
e. The function of this piece of code is to setup and enable the GPIO P1.0 pin that is connected to the red-color **LED1** in the EXP432P401R EVB (Figure 2.24 in Chapter 2) to turn on and off (toggle) this LED. The detailed functions are
   i. The code in line 8 is used to disable the Watchdog timer to avoid unnecessary Watchdog timer time-out interrupt errors.
   ii. In line 10, the GPIO pin **P1.0** direction register **P1DIR** is assigned to 1 (**BIT0** = **0x1**) to make pin **P1.0** an output pin (refer to Section 2.3.4.2.3 in Chapter 2).
   iii. A **while()** loop is used to make this project run in an infinitive loop.
   iv. In line 13, the P1.0 pin is toggled by using the **XOR** operator (∧) to make its output register to output an alternate HIGH and LOW sequence to flash the LED1.
   v. The code in line 14 is used to delay a period of time for the program.

Your finished source file **MyProject.c** is shown in Figure 3.21.

9. Now, let's build our project by going to **Project|Build target** menu item. The building process begins and the detailed building steps are shown in the **Build Output** window in the bottom, as shown in Figure 3.22. You can find that our project is built successfully with **0** errors and **0** warnings.
10. Now if you expand our project source file, **MyProject.c**, and the system-specified source file, **system _ msp432.c**, from the **Project** pane in the left, you can find all four code files as we mentioned in Section 3.4.2 (Figure 3.8), which are:
    a. Startup code file, **startup _ msp432.s** (assembly file provided by the MDK)
    b. System source file, **system _ msp432.c** (C file provided by Texas Instruments)
    c. System header file, **msp432p401r.h** (header file provided by Texas Instruments), which is located under System source file, **system _ msp432.c**, and our project file.

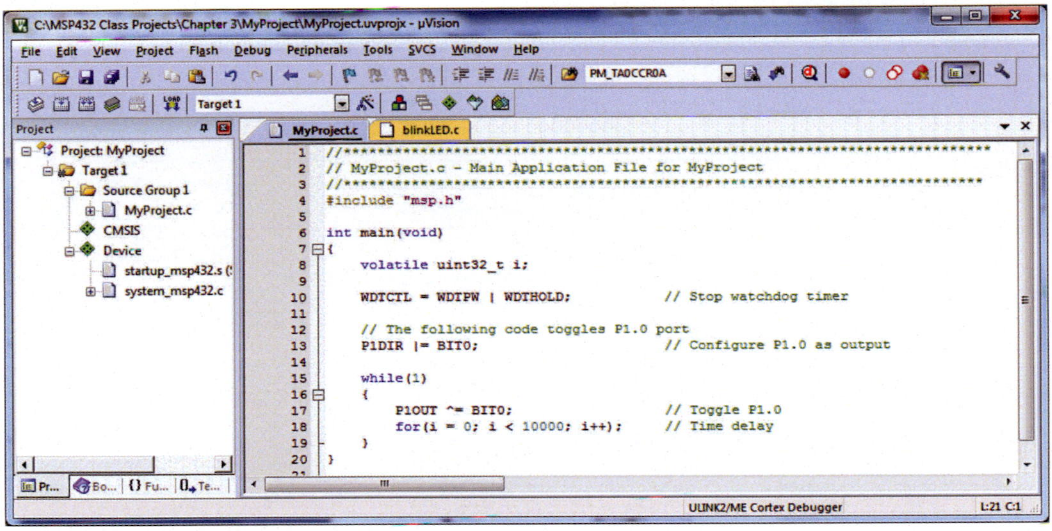

```
//*****************************************************************************
// MyProject.c - Main Application File for MyProject
//*****************************************************************************
#include "msp.h"

int main(void)
{
    volatile uint32_t i;

    WDTCTL = WDTPW | WDTHOLD;            // Stop watchdog timer

    // The following code toggles P1.0 port
    P1DIR |= BIT0;                       // Configure P1.0 as output

    while(1)
    {
        P1OUT ^= BIT0;                   // Toggle P1.0
        for(i = 0; i < 10000; i++);      // Time delay
    }
}
```

Figure 3.21 Finished codes for the source file MyProject.c. (Reproduced with permission from ARM® Limited. Copyright© ARM Limited.)

```
Build Output
*** Using Compiler 'V5.05 update 2 (build 169)', folder: 'C:\Keil_v5\ARM\ARMCC\Bin'
Build target 'Target 1'
".\Objects\MyProject.axf" - 0 Error(s), 0 Warning(s).
Build Time Elapsed:  00:00:00
```

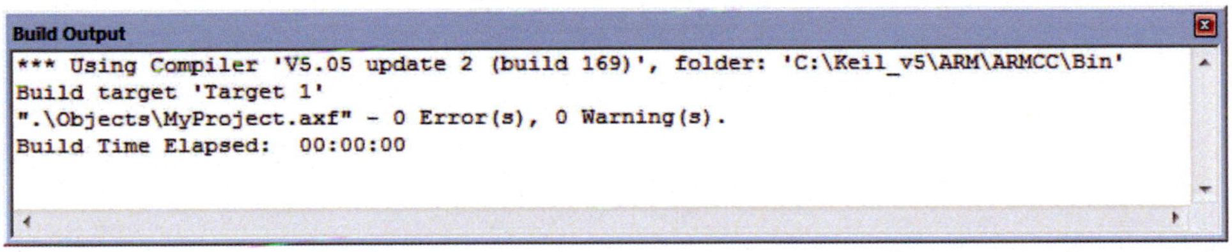

Figure 3.22 Building result of the project MyProject. (Reproduced with permission from ARM® Limited. Copyright© ARM Limited.)

d. Device-specified definition file, **msp.h** (header file provided by the Texas Instruments for MSP432 Family MSP432P401R), which is also located under our project source file, **MyProject.c**

This means that some system code files will not be available until the project has been built successfully.

11. Now, we need to download our project into the flash memory in the MSP432 Family LaunchPad EVB, MSP-EXP432P401R. This download process needs a debug adaptor and related driver. Before we can do this, we need to confirm that the debugger provided by the MDK is compatible with MSP432P401R MCU system. As we discussed in Section 3.1, a debugger **CMSIS-DAP** in MDK IDE is compatible with XDS110-ET debugger and can be used in the MSP432P401R MCU system.
12. Perform the following operations to finish this debugger selection:
    a. Go to **Project|Options for Target "Target 1"** item to open this wizard.
    b. Click on the **Debug** tab to open its settings.
    c. On the upper right part of this wizard, click on the dropdown arrow for the **Use** combo box and select **CMSIS-DAP Debugger** from the list, as shown in Figure 3.23.

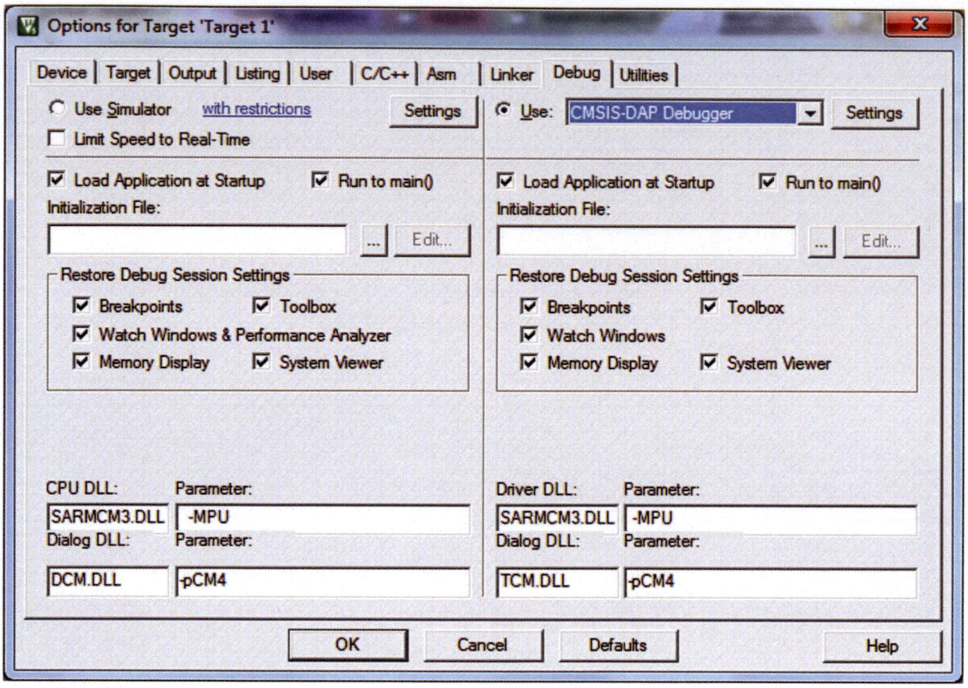

Figure 3.23 Select the CMSIS-DAP debugger. (Reproduced with permission from ARM® Limited. Copyright© ARM Limited.)

d. Now, click on the **Settings** button on the right of this selected debugger to open the **Cortex-M Target Driver Setup** wizard. Make sure to check the following two checkboxes (Figure 3.24):
   i. Verify code download
   ii. Download to flash

   Then click on the **Flash Download** tab on this wizard to open the download function page. Check the **Erase Full Chip** radio button, as shown in Figure 3.25. Click **OK** buttons to close these wizards.

13. Now select the **Download** item from the **Flash** menu, or click the **Download** button (icon). The process takes a few seconds. A sequence of erase, program, and verify operations will be shown at the bottom of the IDE window as the device is programmed. When it is finished, the **Build** window will show that the device was erased, programmed, and **Verified OK**, as shown in Figure 3.26. Our project application, **MyProject.axf**, which is an image file, is now programmed and downloaded into the flash memory on the EVB.
14. We can now start to debug our project by going to **Debug|Start/Stop Debug Session** menu item. Click on the **OK** button for the popup memory limitation message for the evaluation version to begin this debug process. Your finished debug process for our project is shown in Figure 3.27.
15. To run our project, go to **Debug|Run** menu item to run it. The user LED, **LED1**, in the MSP432P401R EVB will be flashed periodically. Click on the **Stop** item from the **Debug** menu to stop our project. Click **Debug|Start/Stop Debug Session** menu to stop debug.

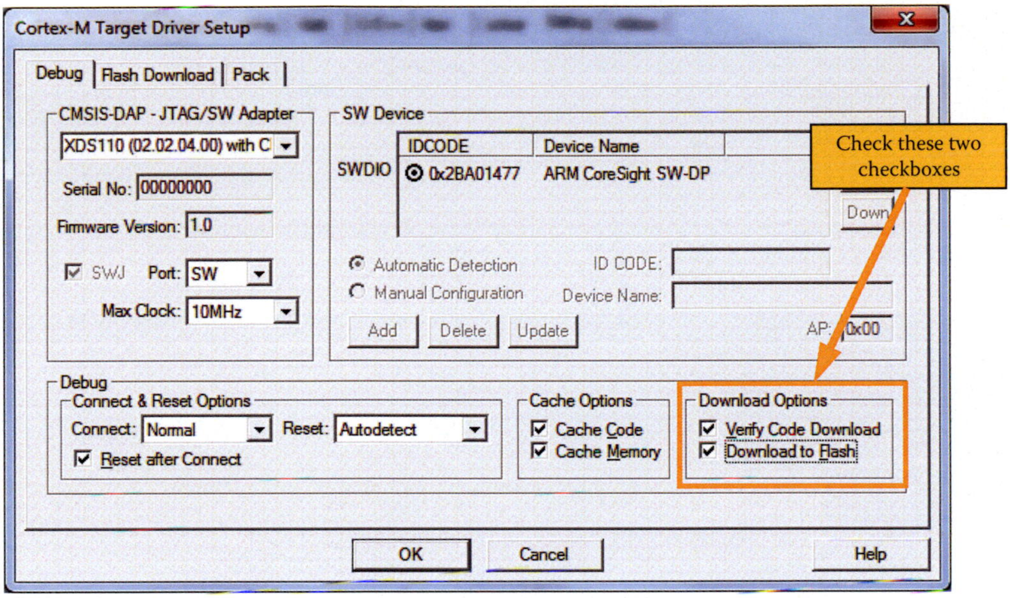

Figure 3.24 Check verify code download and download to flash checkboxes. (Reproduced with permission from ARM® Limited. Copyright© ARM Limited.)

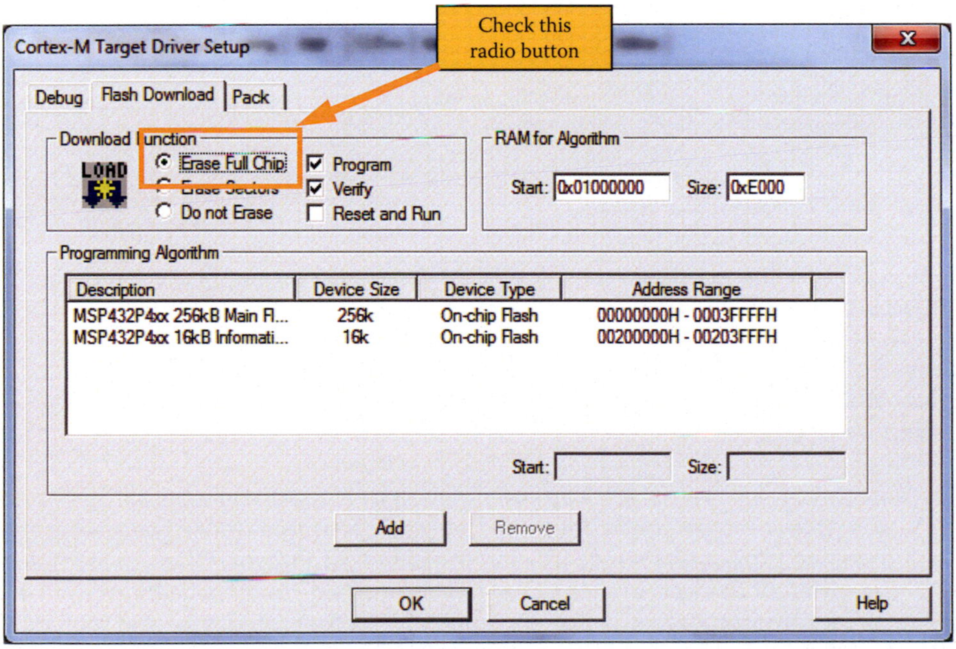

Figure 3.25 Check the erase full chip radio button. (Reproduced with permission from ARM® Limited. Copyright© ARM Limited.)

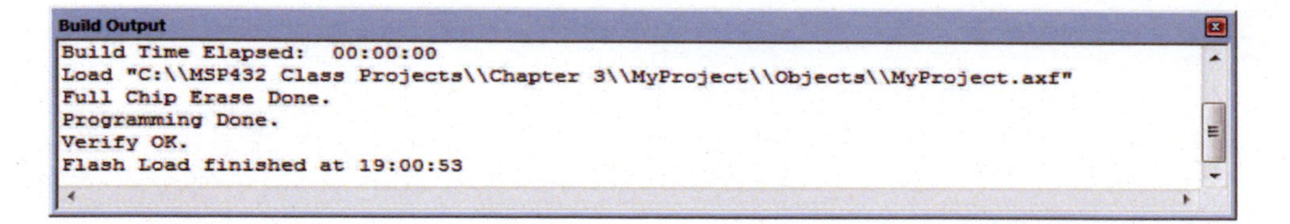

Figure 3.26 Download process for our project. (Reproduced with permission from ARM® Limited. Copyright© ARM Limited.)

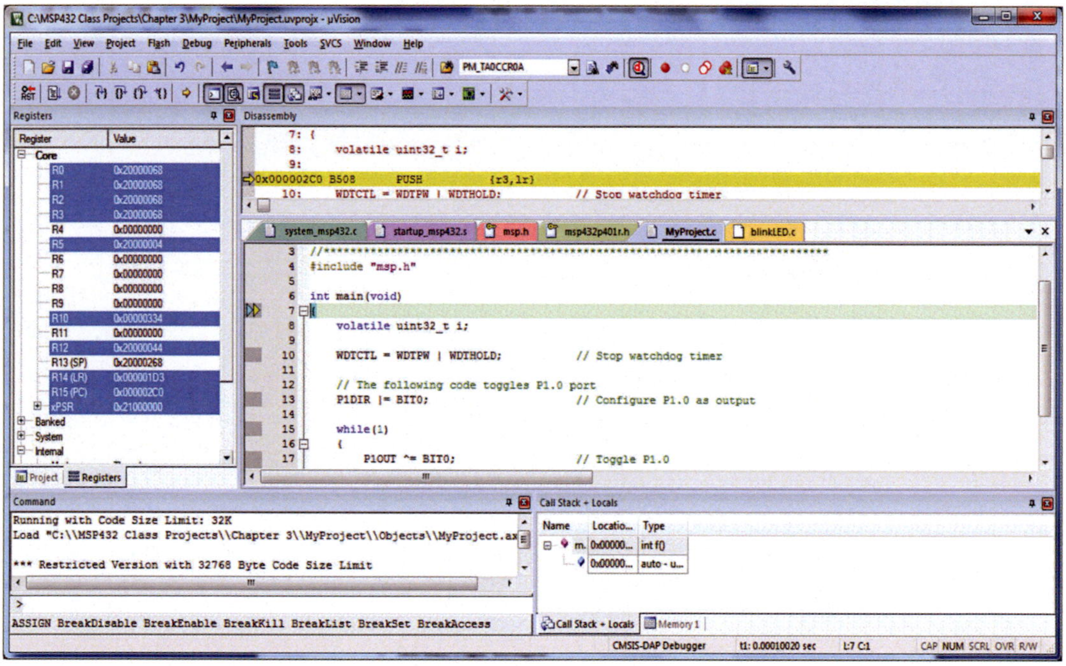

Figure 3.27 Debug process for our project. (Reproduced with permission from ARM® Limited. Copyright© ARM Limited.)

Now, let's continue to discuss the following menu items under the **`Project`** menu, the **`New Multi-Project Workspace`** and **`Open Project`**.

`Project`
- **`New Multi-Project Workspace`**
- **`Open Project`**

The **`New Multi-Project Workspace`** menu item is used to create multiple projects in the µVision project environment. Multiple Projects is a simple-to-use feature for managing more than one project in a single µVision project environment. Often, system designs are targeting different devices. In such cases, it is comfortable to manage the system design using one project environment. Create a µVision5 project for each device and include them into a multiple project.

Multiple project files have the extension **`.UVMPW`**. You can use **`Project|Open Project`** menu item to open a single project or a multiple projects file.

The rest menu items under the Project menu are:

Project:
- **Save Project in μVersion4 Format**
- **Close Project**
- **Export**
- **Manage**
- **Select Device for Target Project**
- **Options for Target Project**
- **Clean Targets**
- **Build Target**
- **Rebuild All Target Files**
- **Batch Build**
- **Translate**
- **Stop Build**

Save Project in μVersion4 Format is used to save the project created in μVersion5 or later to the early μVersion format. The condition is: when no RTE is used, you can save projects in μVision4 format (extension ***.uvproj**). Software components must be removed using **Project|Manage|Run-Time Environment** before you do this saving operation.

Close Project is to close the current project.

Export is used to export the active project, or the current multiproject, to the μVision3 format. Options specific to later μVision versions, such as μVersion5, are not converted. The original project file will still exist untouched.

Under the **Manage** menu item, seven submenu items exist:

Manage
- **Project Items**
- **Multi-Project Workspace**
- **RTE**
- **Select Software Packs**
- **Reload Software Packs**
- **Pack Installer**
- **Migrate to Version 5 Format**

The **Project Items** configures targets, groups, and files; sets file extensions and tool paths; selects development tools; and configures books and manuals. Figure 3.28 shows an example of using this item to manage target, group, and files in our sample project, **MyProject**. You can add additional files from any location in your host computer by using the **Add Files** button.

The **Multi-Project Workspace** is used to add, delete, or rearrange the μVision project files in a MultiProject file.

The **RTE** wizard allows you to manage the software components of a project. Software components can be added, deleted, disabled, or updated during the software development process at any time. When a new project is started, the RTE window opens automatically after you have selected a device. The menu **Project|Manage|Run-Time Environment** opens the wizard.

We have provided a very detailed discussion about the Manage RTE wizard when we created our sample project **MyProject** using the **New μVersion Project** menu item,

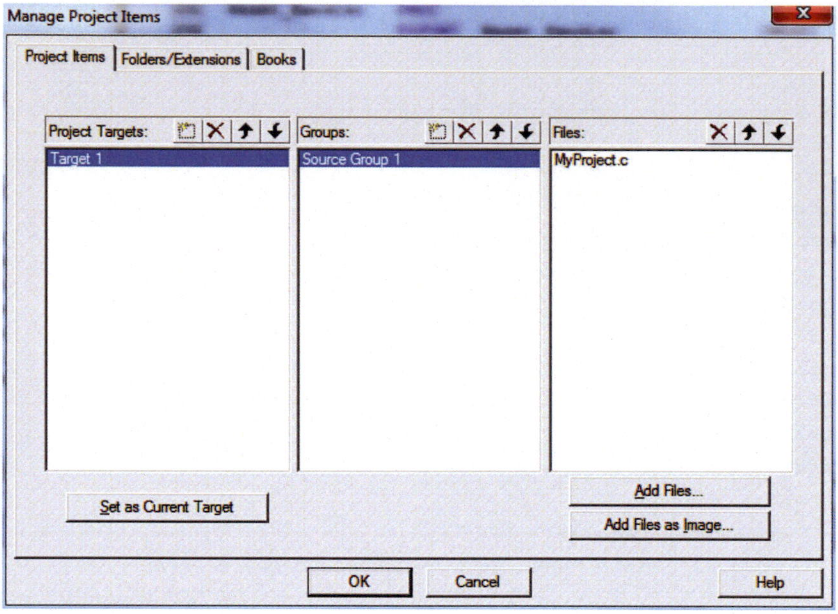

Figure 3.28 Components, environment, books wizard. (Reproduced with permission from ARM® Limited. Copyright© ARM Limited.)

refer to that part for this tool. For your convenience, an example of using this tool to manage our sample project is redisplayed in Figure 3.29.

The **`Select Software Packs`** and **`Reload Software Packs`** menu items can be used to select and refresh Software Pack information in the Pack Installer window.

The **`Pack Installer`** is a utility program that allows users to install, update, and remove Software Packs. This wizard can be launched from within µVision or standalone, outside of µVision. The Pack Installer window offers the following functionality:

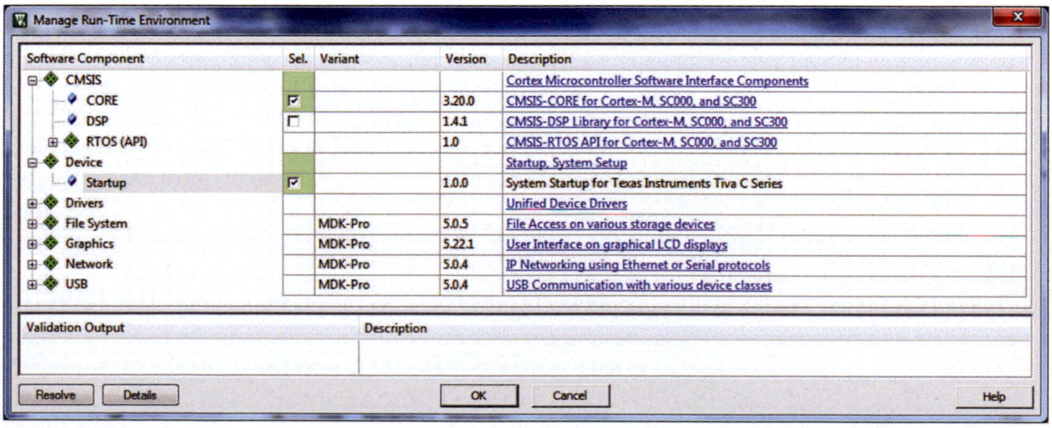

Figure 3.29 Manage run-time environment wizard for our sample project. (Reproduced with permission from ARM® Limited. Copyright© ARM Limited.)

- Installs, updates, or removes Software Packs, and thus software components
- Lists installed Software Packs and checks for updates on the Internet. A brief release history might be displayed before updating a Software Pack
- Lists example projects available from installed Software Packs
- Offers filters to narrow the list of Software Packs or example projects
- Displays the progress of the executed function in the status bar at the bottom of the window

The **`Migrate to Version 5 Format`** menu item can be used to convert early μVersion project to the updated μVersion format.

The **`Select Device for Target Project`** allows users to select the target device (MCU) for the current project. Depending on the EVB and MCU used in your project, you can find and select desired MCU/CPU for your project via this tool. We have provided a very detailed discussion about this wizard when we created our sample project **`MyProject`** using **`New μVersion Project`** menu item (Figure 3.15), refer to that part for this tool and its applications.

The **`Options for Target Project`** is a very important tool to support the project development process with multiple functions. We will pay more attention to this tool and provide more detailed discussions about it.

This tool contains 10 functions or options, and each function is presented with a tab, as shown in Figure 3.30.

- The **`Device`** option allows users to select the desired device (MCU) for the current project. This is similar to the function of the **`Select Device for Target Project`** wizard. When a device is selected, all related settings, including the complier,

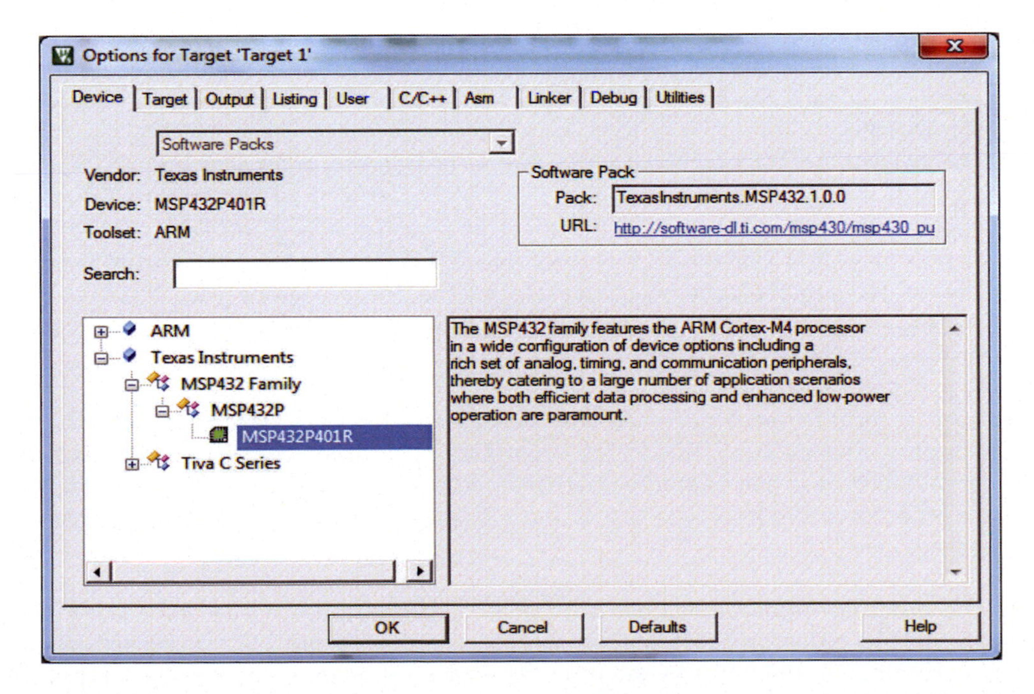

Figure 3.30 Functions provided by the options for target project wizard. (Reproduced with permission from ARM® Limited. Copyright© ARM Limited.)

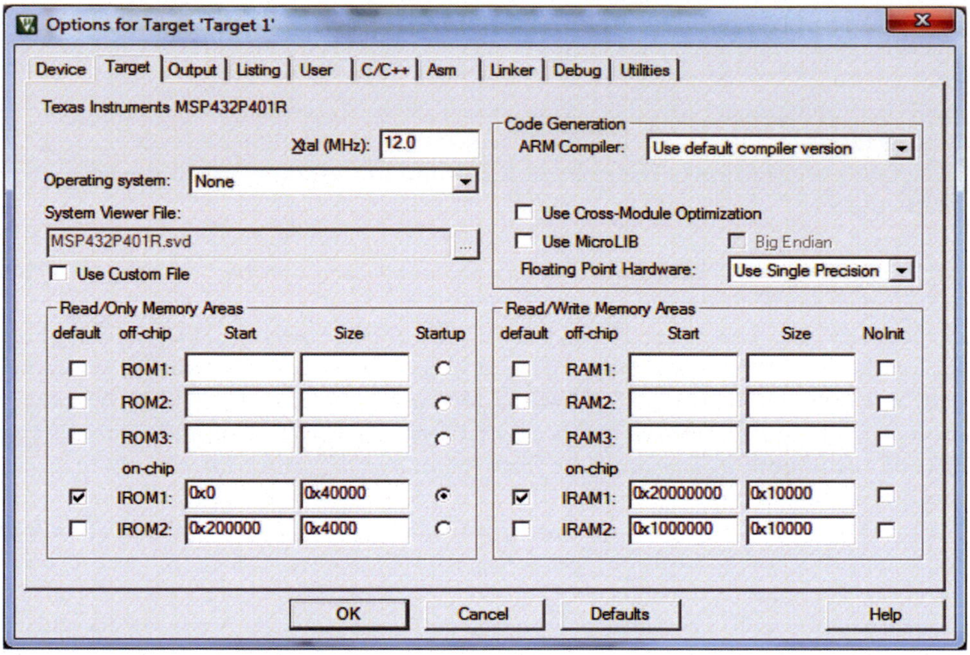

Figure 3.31 Target option for the sample project MyProject. (Reproduced with permission from ARM® Limited. Copyright© ARM Limited.)

memory map, and flash algorithms, are configured for the device. If the MCU that you are using is not listed in the list, you can still select Cortex-M4 under the ARM® section and manually set the configuration options. An example of **`Device`** option selection wizard for our sample project **`MyProject`** we built above is shown in Figure 3.30.

- The **`Target`** option enables users to define the memory map of the device, such as the address ranges of the ROM and the RAM used in the EVB, options to use the FPU on the Cortex-M4 MCU if installed on your device, option to utilize the RTX Kernel, and a RTOS that comes with CMSIS-RTOS. The memory map setup is generally created automatically when you selected the desired device (MCU). An example of **`Target`** option selection wizard for our sample project **`MyProject`** we built above is shown in Figure 3.31.

It can be found from Figure 3.31 that no operating system (RTOS) is used for this target project. The memory ROM and RAM are configured from addresses **`0x0`** and **`0x20000000`** with a size of **`0x40000`** (256 KB) and **`0x10000`** (64 KB), respectively. The **`Use Single-Precision`** item is selected as the FPU in the **`Floating-Point Hardware`** combo box.

- The **`Output`** option allows users to select a different location to save the project output file, either an executable image file or a library file. Object and listing files are created in subfolders of the project folder by default. However, the output can be redirected to other folders. Thus, each project target can have its own output folder. To do this, you can use the **`Select Folder for Objects`** wizard to set

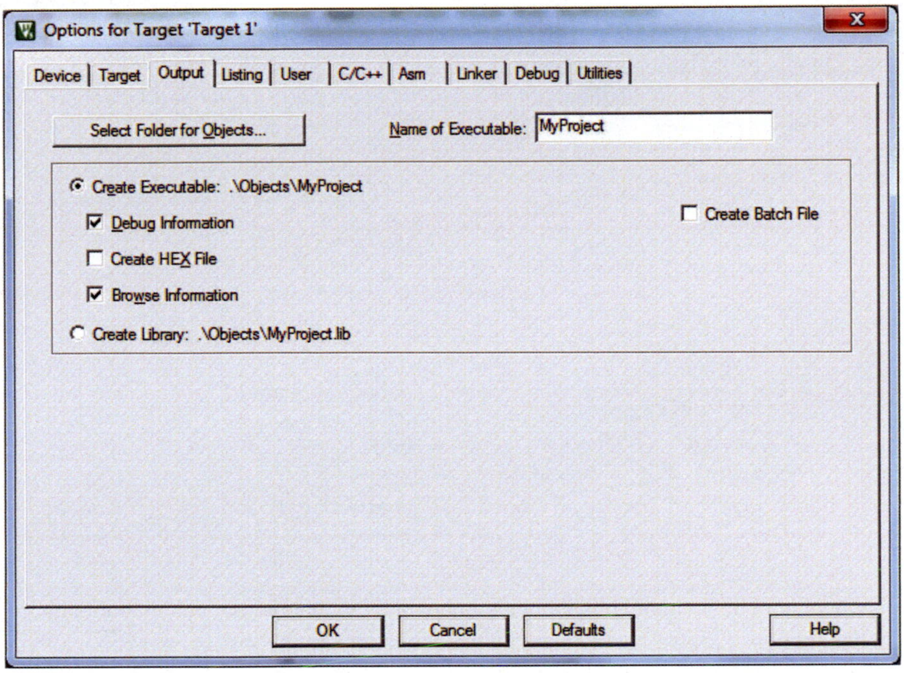

Figure 3.32 Output option for the sample project MyProject. (Reproduced with permission from ARM® Limited. Copyright© ARM Limited.)

the output directory to a folder you created in your desired folder. Furthermore, object and listing files from previous build processes can be preserved. An example of using the **`Output`** option wizard for our sample project **`MyProject`** is shown in Figure 3.32.

- The **`Listing`** option, similar to the **`Output`** option, also allows users to select a different folder to store the output listing files by using the **`Select Folder for Listing`** wizard. This option also allows users to enable or disable assembly listing files. The C compiler listing file is turned off by default. You can turn this option on during the debug process to monitor the generating sequence of the assembly instructions. An example of using the **`Listing`** option wizard for our sample project **`MyProject`** is shown in Figure 3.33.
- The **`User`** option enables users to use some external tools during the process of an application. This option allows users to execute external programs in three ways:
  - Before compiling a C or C++ file (checking for MISRA compliance)
  - Before building an application (invoking a data management utility)
  - After building an application (invoking data converters or debuggers)

    You need to enter the desired external tools or commands into one of the three boxes, which are related to one of the three ways shown above.
  - The **`C/C++`** option allows users to define optimization options, including path, and misc controls for the C/C++ compiler. By default, some include file directories that are automatically included in the project compiler string list. You can check the **`No Auto Includes`** checkbox if you do not want to include these files to be included automatically.

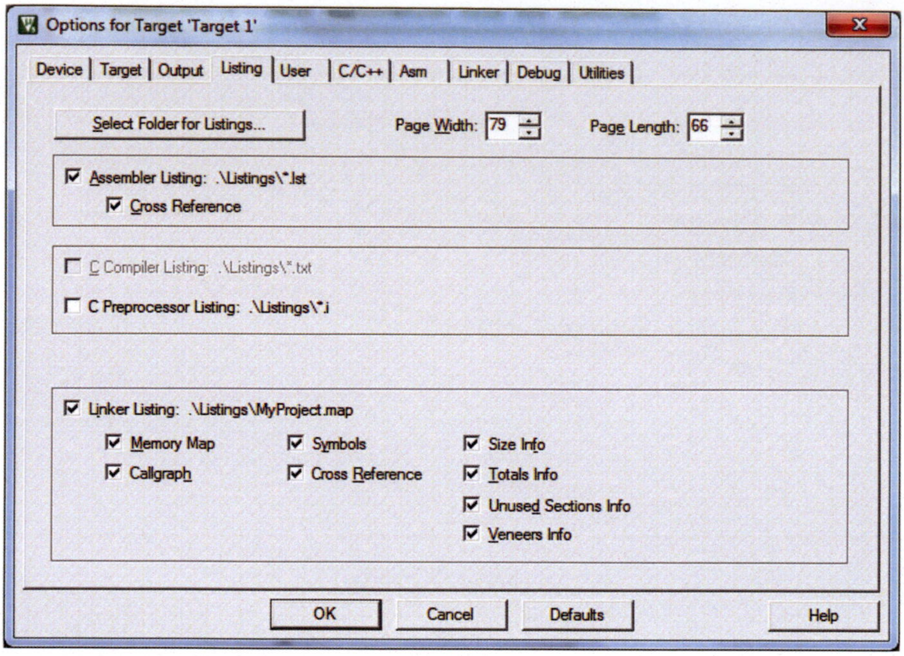

Figure 3.33 Listing option for the sample project MyProject. (Reproduced with permission from ARM® Limited. Copyright© ARM Limited.)

- The **Asm** option, or assembler option, enables users to define preprocessing directives, include paths, and additional assembler command switches.
- The **Linker** option allows users to use the memory layout from the Target wizard. When enabled, µVision creates a linker scatter file from the memory information supplied in the Target wizard. When disabled, the X/O Base, R/O Base, R/W Base, and the scatter file can be set manually. The scatter file defines the memory layout and allows assigning modules to specific memory areas.
- The **Debug** option defines options that apply when a debugging session is started. The screen is split into options for the simulator and for the target driver. This means that the debug option provides two selections for the user, either using an instruction simulator or using an actual hardware device with a debug adaptor. This option also provides the following functions:
  - The type of debug adaptor
  - The type of debug driver
  - Breakpoint setup
  - Watch windows and performance analyzer
  - Memory display
  - System Viewer and toolbox

An example of using the **Debug** option wizard for our sample project **MyProject** is shown in Figure 3.34. One point to be noted is that the debug driver selection tool, which is located at the **Use** part at the upper right corner on this wizard, is very important. Since we are using the **CMSIS-DAP** as our debugger, we therefore need to select this driver from the **Use** list. Otherwise we may encounter some link and download error if the default debugger driver is selected and used.

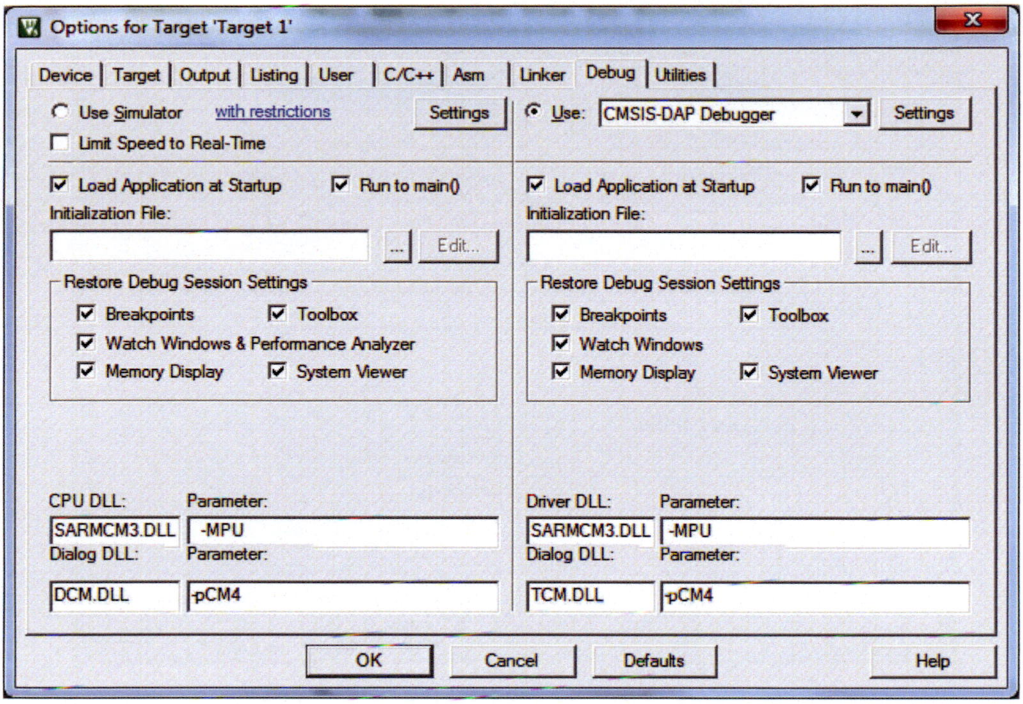

Figure 3.34 Debug option for the sample project MyProject. (Reproduced with permission from ARM® Limited. Copyright© ARM Limited.)

- The **Utilities** option enables users to select the target debug adaptor for flash programming. µVision5 supports several flash programming utilities, for example:
  - Adapters of the Keil® ULINK USB-JTAG family that also offers debugging and tracing capabilities
  - Third-party adapters that can be selected from the configuration dialog
  - External, command-line-driven utilities that are provided by chip vendors

    Using this **Utilities** wizard, applications can be downloaded to flash memory:
  - Manually—through the menu **Flash|Download**
  - Automatically—by enabling the checkbox **Utilities|Update Target before Debugging**

    An example of using the **Utilities** option wizard for our sample project **MyProject** is shown in Figure 3.35. The following utilities options have been selected for our project:
    - Use target driver for flash programming
    - Use debug driver
    - Update target before debugging

  As the **Update Target before Debugging** checkbox is checked, our program will be automatically downloaded into the flash memory as the debug process starts.

  If you click on the **Settings** button, more details about this debug configuration are displayed, as shown in Figure 3.36. The debug adaptor is **XDS110 with CMSIS**, which is displayed in the **CMSIS-DAP-JTAG/SW Adapter** group combo box on the top. The debug port and max clock are also included in this group. The

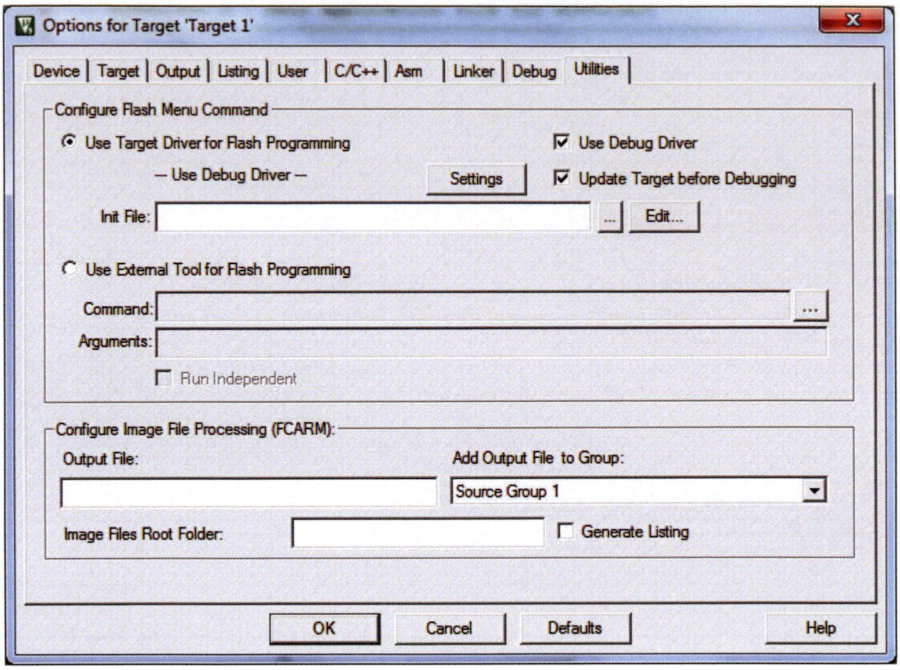

Figure 3.35 Utilities option for the sample project MyProject. (Reproduced with permission from ARM® Limited. Copyright© ARM Limited.)

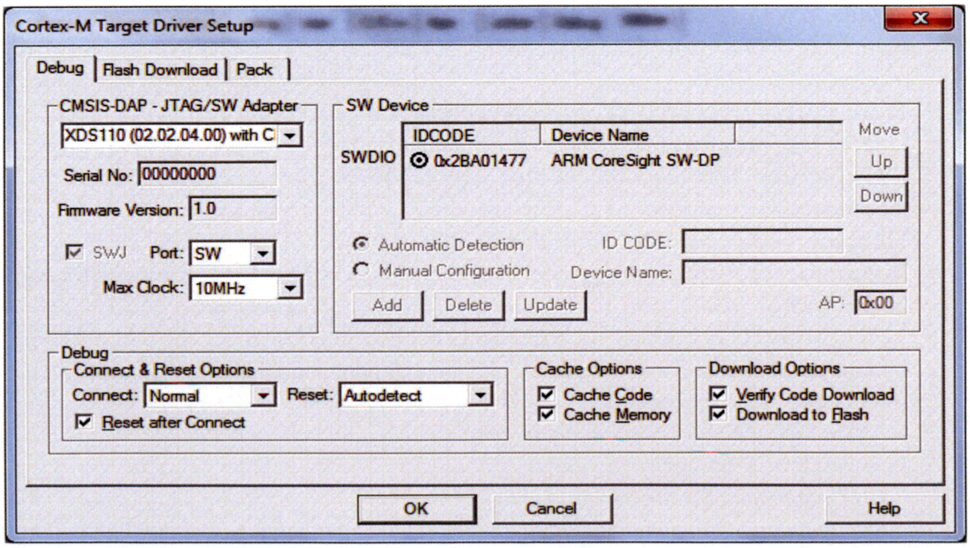

Figure 3.36 Opened settings wizard. (Reproduced with permission from ARM® Limited. Copyright© ARM Limited.)

Debug Connect and Reset information are shown in the **Debug** group box in the bottom.

At this point, we complete the introduction to all tabs under the **Options for Target Project** menu item in the **Project** menu. The rest of menu items under the **Project** are relatively simple. Let's take a quick look at these items.

Project:

- **Clean Targets**
- **Build Target**
- **Rebuild All Target Files**
- **Batch Build**
- **Translate**
- **Stop Build**

The **Clean Targets** is used to delete the intermediate files of the project target. These intermediate files include all source files with the extension .obj, .o, ._ii, ._ia, ._i, .map, and .list.

The **Build Target** is used to build the project, which includes the compile and link processes.

The **Rebuild All Target Files** is similar to Build Target. It will retranslate all source files and build the application.

The **Batch Build** provides a convenient way to enable users to build multiple project targets in one working step.

The **Translate** is used to translate all active files in the project.

The **Stop Build** is to stop the building process.

3.4.3.4 Flash menu

This menu is used to download the user's executable program into the flash memory. Three submenu items under this menu are:

Flash

- **Download**
- **Erase**
- **Configure Flash Tools**

The **Download** and **Erase** items are simple, and the functions of these menus just download the user's executable program to the flash memory in the EVB, or delete it from the flash memory in the EVB. The function of the **Configure Flash Tools** is exactly identical with the **Utilities** option under the **Options for Target Project** in the **Project** menu, which we have discussed above.

Some useful functions can be executed after these flash tools are configured:

- Enable **Use Debug Driver** when using the debugger adapter as a debug and flash programming unit
- Enable **Update Target before Debugging** to download the application to flash whenever a new debugging session is started
- Enter a predownload script into the field **Init File** to specify commands which prepare the device for flash programming. For example, to configure the bus

- Enable **Use External Tool for Flash Programming** to enter options for third-party command-line-based utilities that are not in the list of target drivers. You can use Key Sequences
- Enter options to **Configure Image File Processing (FCARM)** by invoking FCARM during the build process. This converts image files into C source code. Refer to Using FCARM with µVision

3.4.3.5 Debug menu

This menu is used to perform debugging and running functions for the user program. Most submenu items are used to debug the program, which include **Step**, **Step Over**, **Step Out**, **Insert/Remove Breakpoint**. We will concentrate on some new and important commands for this menu; especially, we will provide our discussions for the following items:

Debug

- **Start/Stop Debug Session**
- **Reset CPU**
- **Run**
- **Stop**
- **OS Support**
- **Execution Profiling**
- **Memory Map**
- **Inline Assembly**
- **Function Editor**

The **Start/Stop Debug Session** command enables users to start or stop the debug process.

The **Reset CPU** command sets the CPU to the RESET state.

The **Run** item starts to run the user's program until it hits a breakpoint.

The **Stop** command is used to stop the running of the user's program.

The **OS Support** command is used to access kernel-aware debug information. Define an RTOS in the field **Options for Target Project|Target|Operating System** to activate the menu items. Two submenu items under this command are:

- System and Thread Viewer: Opens the dialog **RTX Tasks and System**
- Event Viewer: Opens the dialog **Event Viewer**

The first command is used to monitor all system and thread-related tasks running with the user program, and the second is to watch all related events.

The menu **Execution Profiling** enables the **Execution Profiler** to record and display execution statistics for each instruction in the user program. Users can view the instruction time and calls in the **Disassembly Window** and in the **Editor**. The values are cumulative numbers. The **Execution Profiler** records timing and execution statistics about instructions for the complete program code. To view the values in the **Editor** or **Disassembly Window**, you need to use **Show Time** or **Show Calls** from this menu.

The **Memory Map** provides some functions to enable users to

- Display the currently mapped memory ranges
- Remove the selected mapped range from the list

- Read or write memory ranges
- Identify the specified memory range as **von Neumann** memory. When specified, μVision5 overlaps the external data memory (**XDATA**) range and code memory (0xFFxxxx). Write access to external data memory also changes code memory.

These operations can be performed via a memory map configuration dialog.

The **Inline Assembly** command can be used to modify instructions while debugging to allow correcting the code or making temporary changes. The **Function Editor** is used to create, modify, and compile debug functions using the built-in debug function editor.

3.4.3.6 Peripherals menu

The **Peripherals** menu includes dialogs to view and change on-chip peripheral settings. The content of this menu is tailored to show specific peripherals of the CPU selected for the application. This menu is active only in **Debug Mode**.

Two submenu items under this menu are

- System Viewer
- Core Peripherals

The **System Viewer** can be used to open the related monitor window for all peripheral devices connected to the MCU EVB. The real-time running status of selected peripheral devices can be traced or inspected in these monitor windows. You can select any desired device by checking on it from the System Viewer submenu. An example of the System Viewer, **TIMER _ A0**, is shown in Figure 3.37. Table 3.1 shows some typical dialogs provided by this menu.

Under the **Core Peripherals** menu, there are five submenu items. Each of them is used to open a related window to display the running status of the selected device. You can also set or change the related configuration for the selected device.

- **NVIC**
- **System Control and Configurations**
- **SysTick Timer**
- **Fault Reports**
- **MPU**

One point to be noted is that all of these five items are available only when the program has been debugged. An example of using the **Core Peripherals** for the NVIC is shown in Figure 3.38.

3.4.3.7 Tools menu

The **Tools** menu allows you to configure and run Gimpel PC-Lint and other custom programs. Six submenu items under this menu are

Tools
- **Set-up PC-Lint**
- **Lint**
- **Lint All C-Source Files**

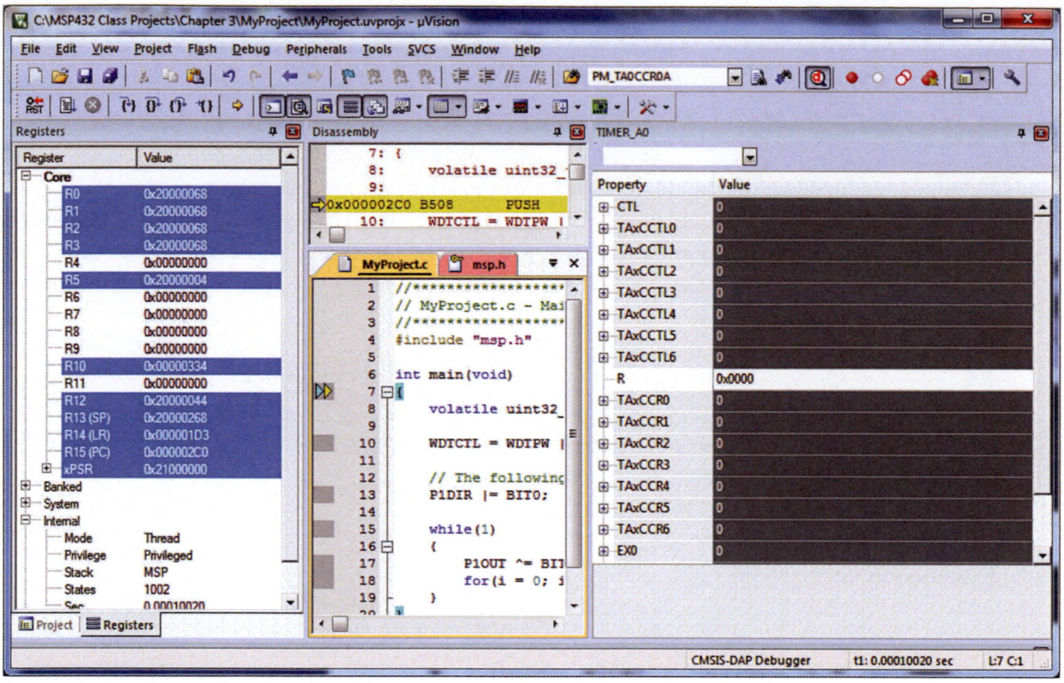

Figure 3.37 Example of using system viewer for TIMER_A0 device. (Reproduced with permission from ARM® Limited. Copyright© ARM Limited.)

- **`Configure Merge Tool`**
- **`Customize Tools Menu`**
- **`Call Notepad`**

The **`Set-up PC-Lint`** is used to configure PC-Lint from Gimpel Software.

The **`Lint`** is to run PC-Lint on the current editor file. PC-Lint from Gimpel Software checks the syntax and semantics of C codes, and reports possible bugs, inconsistencies, and locates unclear, erroneous, or invalid C codes. PC-Lint could reduce debugging efforts considerably.

Table 3.1 Typical system viewers

Menu items	Descriptions
WDT_A	Watchdog timer A
REF_A	Reference voltage A controller
EUSCI	Extended universal SCI controller
COMP	Analog comparator controller
TIMER	Timers/counters
ADC14	Analog-to-digital converter
RTC_C	Real-time control controller
PMAP	Port mapping controller
DIO	Digital input output controller
SYSCTL	System control registers

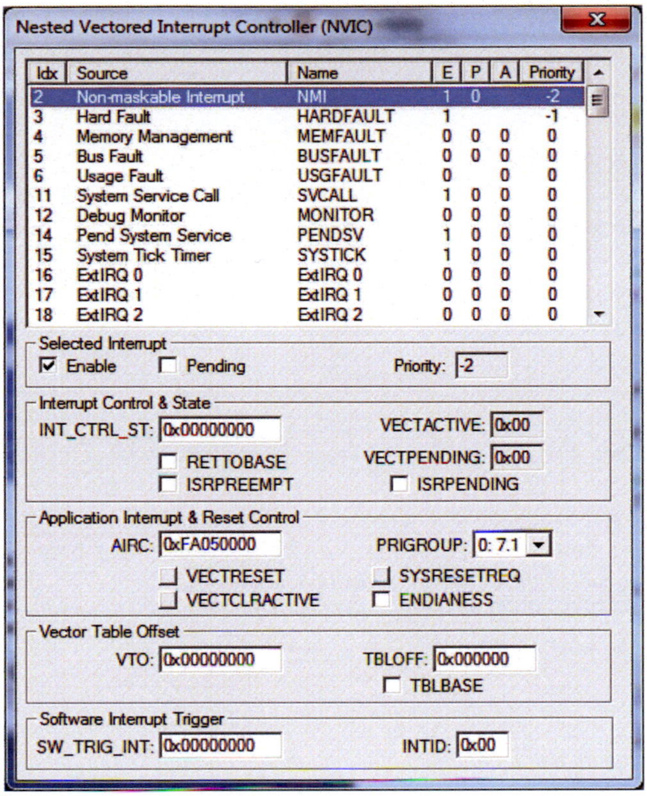

Figure 3.38 Nested vectored interrupt controller configuration dialog. (Reproduced with permission from ARM® Limited. Copyright© ARM Limited.)

Configuration templates are provided for each toolset and various device types, including PC-LINT configuration file for ARM7™, ARM9™, and Cortex®-Mx devices (**..\ARM\BIN\CO-RV.LNT**). The files have the extension ***.LNT**. It is strongly recommended to use the template configuration files as they contain options required by various compilers.

The **Lint All C-Source Files** item can be used to run PC-Lint on all C source files in your project.

The **Configure Merge Tool** item sets up a merge tool (e.g., WinMerge) and gives you the opportunity to merge application-specific configuration settings from a previous file version into the current file version.

The **Customize Tools Menu** is used to add user programs to the **Tools** menu. External programs can be integrated into and run from the Tools menu. To do this, you need to add menu items for running external programs through the **Tools|Customize Tools Menu**.

The **Call Notepad** item enables users to call a user program written in Notepad file from the Keil MDK environment.

3.4.3.8 Software version control system (SVCS) menu

The **SVCS** menu is used to configure the **SVCS**. In fact, the SVCS menu allows you to configure and add commands for an SVCS. Commands that steer the SVCS can be added to and executed from the menu. The command output is shown in the µVision5 window build output after the command finishes executing.

Template files are provided for: Intersolv PVCS, Microsoft SourceSafe, MKS Source Integrity, and Rational ClearCase. The files have the extension **`*.SVCS`** and are located in the folder **`C:\Keil _ v5\UV4`**. Adapt the templates to your needs.

3.4.3.9 Window menu

The **`Window`** menu includes commands to control text editor files and windows format. Three submenu items are:

`Window`
- **`Reset View to Defaults`**
- **`Split`**
- **`Close All`**

The **`Reset View to Defaults`** is to reset the window layout to the μVision5 default look and feel.

The **`Split`** is used to divide the active editor file into two horizontal or vertical panes.

The **`Close All`** item can be used to close all opened editor files. A dialog box is displayed for files that have been changed but not saved yet.

In the bottom of this Window menu, all user projects that had been opened before are listed in the order they are opened. This provides a good record for all projects you opened and built.

3.4.3.10 Help menu

The **`Help`** menu includes commands to start the online help system, to list information about on-chip peripherals, to access the knowledge base, to contact the technical support team, to check for product updates, and to display product version information.

Six submenu items listed under this menu are:

`Help`
- **`μVision Help`**
- **`Open Books Window`**
- **`Simulated Peripherals for Specified MCU`**
- **`Contact Support`**
- **`Check for Update`**
- **`About μVision`**

The functions for all of these menu items are simple and easy to be understood, except the **`Simulated Peripherals for Specified MCU`**, which provides information about the simulated peripherals of the selected device.

The μVision5 Debugger can simulate the behavior of the target application. This enables developers to test applications prior to having the hardware.

The logic behavior of communication peripherals is reflected through virtual registers that can be listed with the command **`DIR VTREG`**. Thus, debug functions that simulate complex peripherals can be written easily. The μVision5 Simulator mimes the timing and logical behavior of serial communication protocols such as UART, I^2C, SPI, and CAN, but does not simulate the I/O port toggling of the physical communication pins on the I/O port.

To perform the simulation task using the simulator, you need to select the **`Use Simulator`** checkbox in the **`Debug`** tab under the **`Options for Target Project`** submenu in the **Project** menu. This simulation function is also dependent on whether the vendor provides the Simulator.

At this point, we finished introductions to all menu items for MDK μVersion5. Now, let's have a summarization about the topics we discussed in those sections.

3.5 Embedded software development procedure

Generally, to successfully develop and build an embedded software project, one needs the following components:

1. Development kits or suites
2. Debug adaptor and debug drivers
3. Development boards or EVBs
4. Specified software device drivers

Specially, to successfully develop and build a user project with MDK μVersion5 IDE, the following operational sequence is needed:

1. Create a new user project using the **New μVersion Project** menu item.
2. Select the target device (MCU) installed in your EVB using the **Select Device for Target "Target1"** wizard
3. Configure and setup the required software components you will use in your project using the **Manage RTE** wizard such as **CMSIS:CORE** and **Device:Startup**. The **MDK:CORE** is automatically setup and involved in your project.
4. Add the user's source files, either assembly or C codes files, into the project using the **Add New Item to Group 'Source Group 1'** wizard. If you prefer to use only C codes, you may also need to add some header files.
5. Build your project using the Project|Build target or rebuild all target files menu item.
6. Download your executable or image file to the flash memory in your EVB using the **Download** item from the **Flash** menu. If the checkbox **Update Target before Debugging** under the **Utilities** menu has been checked, your image file will be automatically downloaded into the flash memory when the debug process starts.
7. Debug your project by going to **Debug|Start/Stop Debug Session** menu item.
8. Run your project using the **Debug|Run** menu item.
9. Stop your project using **Debug|Stop** or click the **Start/Stop Debug Session** menu item again.
10. Erase your project from the flash memory using the **Erase** item from the **Flash** menu.

Figure 3.39 shows an illustration block diagram for this project development, building, debugging, and running process.

Next, let's have a closer look at the debug process in Cortex-M microcontrollers, including the debug adaptor, debug driver, and debug process.

3.6 ARM®-MDK μVision5 debugger and debug process

We have discussed and implemented quite a few debug functions during which we developed and built our sample project **MyProject** in the last section; however, due to the importance of this component, we will introduce this component in more detail in this part.

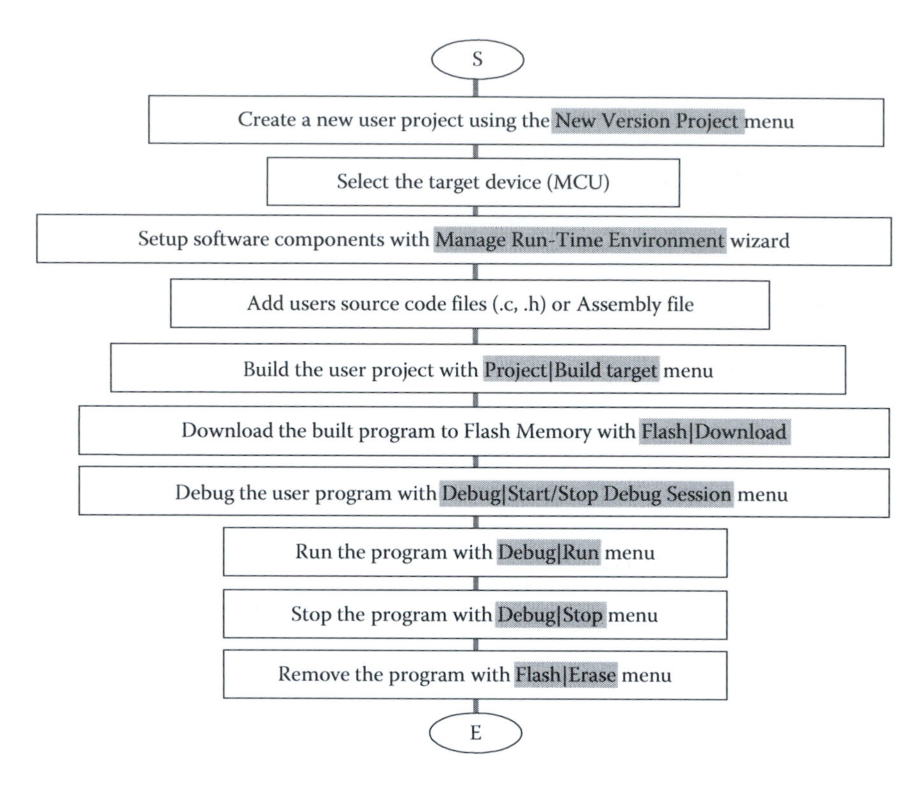

Figure 3.39 Project development steps.

The basic functions of a debugger are to find, locate, identify, and trace any possible bugs or errors that existed in your project. Different strategies and technologies are implemented in different debuggers.

The ARM® CoreSight™ technology integrated into the ARM® Cortex®-M processor-based devices provides powerful debug and trace capabilities. It enables run-control to start and stop programs, breakpoints, memory access, and flash programming. Features such as PC sampling, data trace, exceptions including interrupts, and instrumentation trace are available in most devices. Devices integrate instruction trace using ETM, Embedded Trace Buffer (ETB), or Micro Trace Buffer (MTB) to enable analysis of the program execution.

The processor implementation determines the debug configuration, including whether debug is implemented. If the processor does not implement debug, no ROM table is present and the halt, breakpoint, and watchpoint functionality is not present.

Basic debug functionality includes processor halt, single-step, processor core register access, vector catch, unlimited software breakpoints, and full system memory access.

The debug option might include:

- A breakpoint unit supporting two literal comparators and six instruction comparators, or only two instruction comparators
- A watchpoint unit supporting one or four watchpoints

For processors that implement debug, ARM® recommends that a debugger identifies and connects to the debug components using the CoreSight debug infrastructure.

3.6.1 ARM® μVision5 debug architecture

The μVision5 Debugger from Keil® supports simulation using only your PC or laptop, and debugging using your target system and a debugger interface. The μVision5 includes traditional features such as simple and complex breakpoints, watch windows, and execution control as well as sophisticated features such as trace capture, execution profiler, code coverage, and logic analyzer.

Most popular debug features involved in the μVision5 Debugger include:

- Run-control of the processor allowing you to start and stop programs
- Single-step one source or assembler line
- Set breakpoints while the processor is running
- Read/write memory contents and peripheral registers on-the-fly
- Program internal and external flash memory

The MDK contains the μVision5 Debugger that can be connected to various debug/trace adapters, and allows you to program the flash memory. It supports traditional features such as simple and complex breakpoints, watch windows, and execution control. Using trace, additional features such as event/exception viewers, logic analyzer, execution profiler, and code coverage are supported.

The ARM® debugger comprises two pieces of elements, the hardware—debug adapter and the software—debug adapter driver. These two pieces of elements make up a complete debug system. Different interface protocols are developed to meet the needs of different debug processes. Most popular debug interfaces include JTAG, SWD, and SWV.

Let's first have a clear picture about these terminologies used in the debug adaptor interfaces.

- **`JTAG`** (Join Test Action Group)
  JTAG is the industry-standard interface used to download and debug programs on a target processor, as well as many other functions. It offers a convenient and easy way to connect to devices and is available on all ARM® processor-based devices. The JTAG interface can be used with Cortex-M devices to access the CoreSight debug capabilities.
- **`SWD`**
  The SWD mode is an alternative to the standard JTAG interface. SWD uses two pins to provide the same debug functionality as JTAG with no performance penalty, and introduces data trace capabilities with the SWV. The SWD interface pins can be overlayed with the JTAG signals, allowing the standard target connectors to be used. These pins include:
  - TCLK—SWCLK (Serial Wire Clock)
  - TMS—SWDIO (SWD data I/O)
  - TDO—SWO (Serial Wire Output) (output pin for SWV)

  JTAG and SWD modes are fully supported by ULINK2, ULINK-ME, and ULINK*Pro*.
- **`SWV`**
  Cortex®-M3 and Cortex-M4-based devices are able to provide high-speed data trace information in a number of ways depending on the type of information or analysis you require. The SWV provides real-time data trace information from various

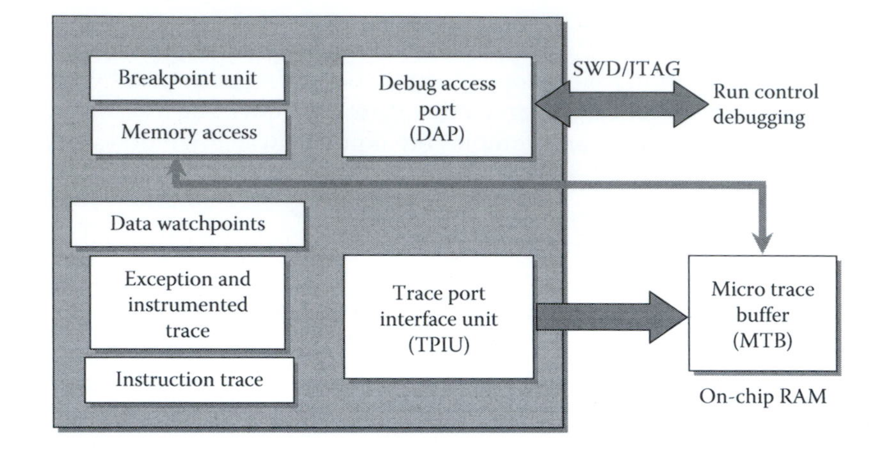

Figure 3.40 Illustration block diagram of the μVision5 Debugger.

sources within the Cortex®-M3/M4 device. This is the output via the single SWO pin while your system processor continues running at full speed. Information is available from the ITM and DWT units, providing:

- PC sampling
- Event counters that show CPU cycle statistics
- Exception and interrupt execution with timing statistics
- Trace data—data reads and writes used for timing analysis
- ITM trace information used for simple *printf*-style debugging
- SWV Data trace is available via the SWO pin in two output formats:
  - UART style (1 Mb/s)—supported by ULINK2 and ULINK-ME
  - Manchester Encoded (100 Mb/s)—supported by ULINK*Pro*

Figure 3.40 shows an illustration block diagram of the μVision5 Debugger used for ARM® Cortex-M3/M4 microcontrollers.

All debug components, including the Breakpoint unit, memory access, and data watchpoints, are embedded into this debugger. Two trace components, exception and instrumented trace, and instruction trace, are also included in this debugger. Two different interfaces, DAP and TPIU, are used to interface to the run-time debugging control via a JTAG or SWD debug adaptor interfaces and the MTB. The former is to perform run-time debug processes and the latter is used to perform trace operations.

3.6.2 ARM® debug adaptor and debug adaptor driver

The **`Debug Adapter`** is a piece of hardware or a hardware interface connected from your host computer to your microcontroller EVB. The debug adapter needs to work with a **`Debug Adapter Driver`**, a piece of software installed in your host computer, to perform all debug functions, including downloading your program to the flash in the EVB, debugging any possible bugs in your program, and tracing the performances of related components in your system.

The μVision5 Debugger can be configured as a **`Simulator`** or as a **`Target Debugger`**. Go to the **`Debug`** tab of the **`Options for Target project`** dialog to switch between these two debug modes and to configure each mode.

The **Simulator** is a software-only product that simulates most features of a microcontroller without the need for target hardware. By using the simulator, you can test and debug your embedded application before any target hardware or EVB is available. μVision also simulates a wide variety of peripherals, including the serial port, external I/O, timers, and interrupts. Peripheral simulation capabilities vary depending on the device you have selected.

The **Target Debugger** is a hybrid product that combines μVision5 with a hardware debugger interfacing to your target system. The following debug devices are supported:

- **JTAG/OCDS Adapters** that connect to on-chip debugging systems such as the ARM® Embedded ICE
- **Target Monitors** that are integrated with user hardware and that are available on many EVBs
- **Emulators** that connect to the MCU pins of the target hardware
- **In-System Debuggers** that are part of the user application program and provide basic test functions

Third-party tool developers may use the Keil® Advanced GDI to interface μVision5 to their own hardware debuggers.

No matter whether you choose to debug with the simulator or with a target debugger, the μVision5 IDE implements a single user interface that is easy to learn and master.

The Keil® ULINK™ Debug Adapters family provides several types of debug adapters, they are

- J-LINK/J-TRACE Cortex
- ULINKpro Cortex
- CMSIS-DAP
- UNLink
- ST-Link

All of these adapters provide support to ARM7®, ARM9®, and Cortex®-M family microcontroller devices, to perform the following tasks:

- Download programs to your target hardware
- Examine memory and registers
- Single-step through programs and insert multiple breakpoints
- Run programs in real-time
- Program flash memory
- Connect using JTAG or Serial Wire modes
- On-the-fly debug of ARM® Cortex®-M-based devices
- Examine trace information from ARM® Cortex®-M3 and Cortex®-M4 devices

The following debug adaptor interfaces are popular and compatible with the μVision5 Debugger:

- The ULINK2 and ULINK-ME Debug adapters interface to JTAG/SWD debug connectors and support trace with the SWO. The ULINK*pro* Debug/Trace adapter also interfaces to ETM trace connectors and uses streaming trace technology to capture the complete instruction trace for code coverage and execution profiling.

- The CMSIS-DAP-based USB JTAG/SWD debug interfaces are typically part of an EVB or starter kit and offer integrated debug features. In addition, several proprietary interfaces that offer a similar technology are supported.
- The MDK supports third-party debug solutions such as Segger J-Link or J-Trace. Some starter kit boards provide the J-Link Lite technology as an on-board solution.

The ULINK2 may be also used for:

- On-chip debugging (using on-chip JTAG, SWD, or SWV)
- Flash memory programming (using user-configurable flash programming algorithms)

Using the ULINK2 adapter together with the Keil® µVision5 IDE/Debugger, you can easily create, download, and test embedded applications on target hardware.

3.6.3 MSP432™ Family Launchpad™ debug adaptor and debug adaptor driver

Since we use the MSP432 Family LaunchPad EVB MSP-EXP432P401R with this book, therefore we need to emphasize the specified debugger used in this system.

The MSP432™ Family LaunchPad™ EVB comes with an on-board XDS110-ET emulator or debugger. This emulator mainly comprises a TM4C129 MCU, and can be considered as a debug controller with a debug adapter. Therefore, we do not need to use any other debug adapter since it has been built on this EVB.

As you know, there are two pieces of MCUs, TM4C129 and MSP432P401R, on the MSP432™ Family LaunchPad™ EVB MSP-EXP432P401R. The one MCU located at the top of this EVB works as a debug controller/adapter. This debugger, combined with other related debug elements in this EVB, is called an emulator.

The XDS110-ET allows for the programming and debug of the MSP432P401R using any of the supported tool chains. Note that the XDS110-ET supports only JTAG and SWD debugging. An external debug interface can be connected for SWV and SWO (trace).

The XDS110-ET also supports the EnergyTrace+ technology to scan and trace the power consumptions on the MCU and most peripherals.

To access the on-board XDS110-ET emulator in the MSP432P401R EVB, a debug adapter driver is needed. A good choice is the CMSIS-DAP Driver. Depending on the different OS the user used, different debug drivers are provided.

A point to be noted is that the XDS110 firmware is still being developed at the time of this writing, and the XDS110 has not yet been optimized. This results in **SLOWER** debug and flash programming speeds. But the debug operations work correctly with slow processing time. For full debugging speed, until the XDS110 is optimized, use any Keil ULINK or a J-Link adapter.

Some other debug adapters are also available to be used to access the XDS110-ET debugger, such as Keil ULINK2, ULINK-ME, and Keil ULINK*pro*. An illustration block diagram using the Keil® ULINK2 and XDS110-ET debug adapters and drivers for the MSP-EXP432P401R EVBs are shown in Figure 3.41.

Next, let's discuss the detailed debug process.

3.6.4 ARM® µVersion5 debug process

Based on the discussions above, we can now start to do some debug jobs for our project. We will use our sample project **MyProject** built in the last section as an example to illustrate this debug process.

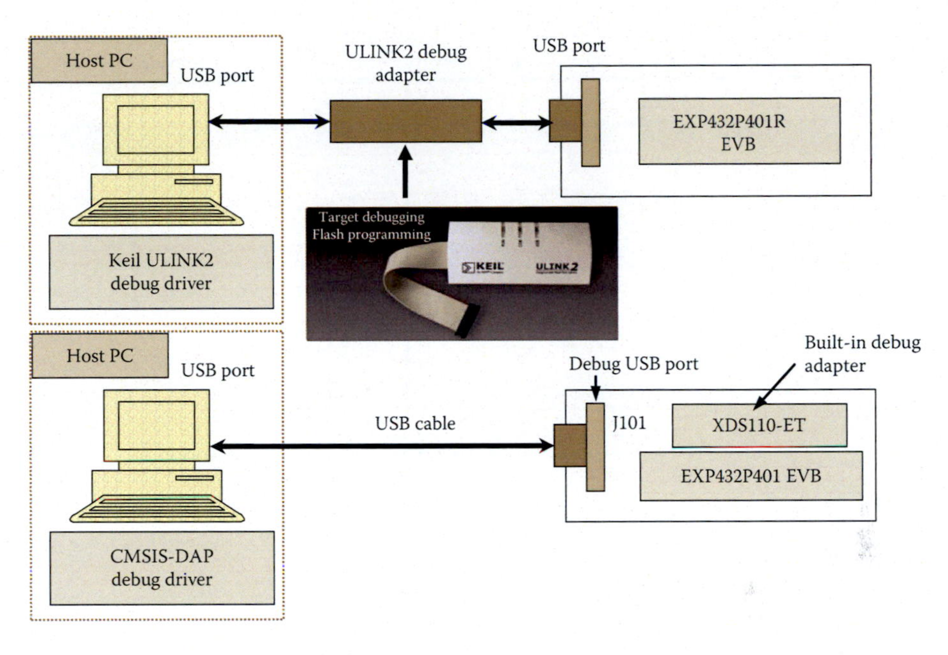

Figure 3.41 Debug system connections for ULINK2 and XDS110-ET.

Go to the **Debug** menu and click on the **Start/Stop Debug Session** item to start the debug process for this sample project. The debug result is shown in Figure 3.42.

From this debug window, you can perform the following debug functions:

- Examine and modify memory via Memory Window
- Program variables and processor registers via Register Window
- Set breakpoints via **Debug|Insert/Remove Breakpoint** menu item
- Single-step through a program using **Debug|Step** menu item
- Perform other typical debugging activities

Since we did not setup any break point and watchpoint in this project, therefore, the debug automatically stopped at the entry point of the main program, **main()** function body, as a blue and a yellow arrow pointed to in the source editor.

The contents for all registers inside the ARM® Cortex®-M4 MCU are displayed in the **Register Window** in the left. In the **Command Window**, the limitation of the used memory size for this EVB, which is 32 KB (32768 Bytes), is displayed. The current used memory size (824 Bytes) for this sample project is also shown in this window.

The **Call Stack** window shows the function nesting and variables of the current program location. The addresses for all modules and names of variables used in this project are shown in the **Symbols Window**. Go to the **Debug|Start/Stop Debug Session** menu to stop this debug process.

As we did not use any debug function, some debug-related windows and functions, such as Memory Window and Peripheral Window, did not show up in our debug process.

3.7 MSP432™ family software suite

Texas Instruments™ MSPWare software for MSP432™ family is an extensive suite of software tools designed to simplify and speed development of MSP432™ family-based MCU

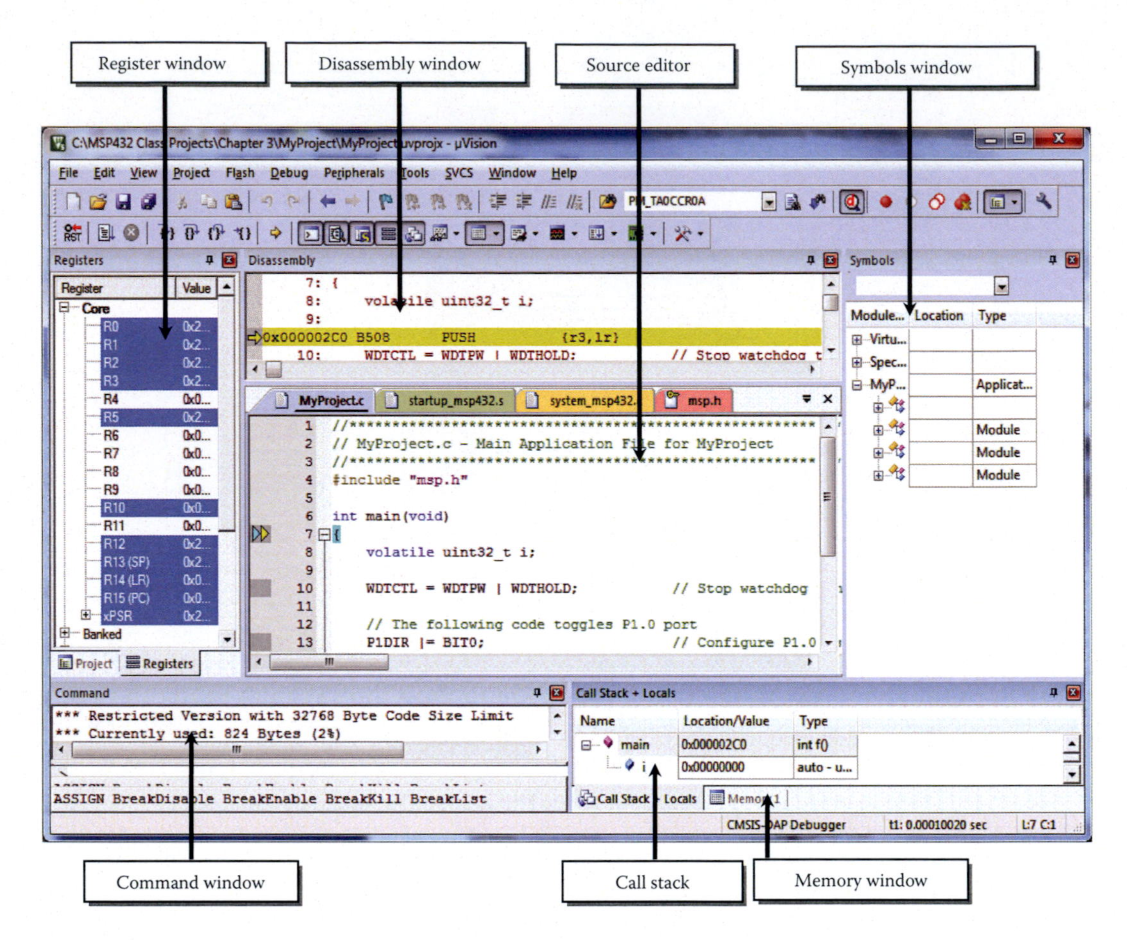

Figure 3.42 Debug example for our sample project MyProject. (Reproduced with permission from ARM® Limited. Copyright© ARM Limited.)

applications. All MSPWare for MSP432™ family software has a free license, and allows royalty-free use so that users can create and build full-function, easy-to-maintain code. MSPWare for MSP432™ family software is written entirely in C to make development and deployment efficient and easy.

The complete MSPWare for MSP432™ Series software suite includes:

- Royalty-free libraries for most popular peripherals
- Powerful interfaces provided by set of API functions
- Kit-and peripheral-specific code examples for MSP432P401R MCU
- Release notes and related documentation
- Written entirely in C
- Everything you need to use your MSP432™ family kits or boards

The MSPWare for MSP432™ Series software supports different development tools and IDEs, the following development tools and IDEs are involved:

- Texas Instruments Code Composer Studio 6.1 (XDS100v3)
- IAR Embedded Workbench for ARM 7.30 (SEGGER J-LINK)

- GNU C Compiler 4.8 (gcc) (SEGGER J-LINK)
- Keil Embedded Development Tools for ARM 5.13 (KEIL U-LINK Pro)

The XDS110-ET provides a built-in debugger to work with these IDEs via the JTAG/SWD interface protocol.

3.7.1 MSPWare for MSP432™ series software package

Texas Instruments™ MSPWare for MSP432™ Series software provides supports to most popular MCUs, such as ARM® Cortex®-M4 core, with scalable memory and package options, unparalleled connectivity peripherals, and advanced analog integration.

The MSPWare for MSP432™ Series suite contains and integrates all user-required source code functions and object libraries, which include:

- Peripheral driver library (**`driverlib`**)
- Graphic library (**`grlib`**)
- USB library (**`usblib430`**)
- Utilities
- Boot loader and in-system programming support
- Example codes
- Third-party examples

Figure 3.43 shows a block diagram for these source code functions and libraries included in the MSPWare for MSP432™ Series suite. MSPWare for MSP432™ libraries offer users the flexibility of working with sample applications or the freedom to create their own projects.

Different source code functions and libraries are provided by this suite, which include:

- The **`Peripheral Driver Library`** offers an extensive set of API functions for controlling the peripherals found on various MSP432™ devices.
- The **`Graphic Library`** includes a set of graphics primitives and a widget set for creating graphical user interfaces on MSPWare for MSP432™ Series-based microcontroller boards that have a graphical display.
- The **`USB Library`** provides an MSPWare for MSP432™ Series royalty-free USB stack to enable efficient USB host, device, and on-the-go operations.

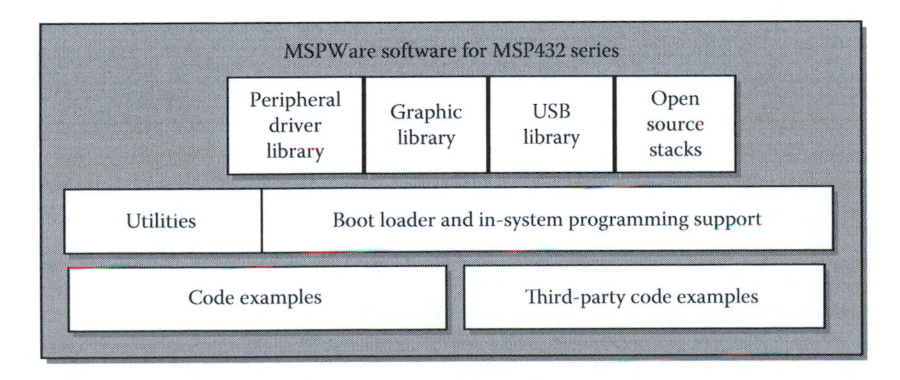

Figure 3.43 Tools included in the MSPWare for MSP432 Series suite.

- The **Utilities** provide all required developing tools and user-friendly functions to make the user's program development easier and simpler.
- The **Boot Loader & In-System Programming** support users to build the startup codes, install them at the beginning of the flash memory in the EVB, and run them when the user program starts.
- The **Code Examples** offer some useful coding guides to help users to start and speed up their coding developments.
- The **Third-Party Code Examples** provide some codes developed by different venders.

Among those code functions and object libraries, some of them are very important and useful in the user's program development. We need to highlight those functions and libraries.

3.7.1.1 Peripheral driver library (driverlib)

Basically, this library provides a collection of source code (**.c**) files and related header (**.h**) files, and those files should be installed on or integrated with the related development IDE to facilitate the user's program development.

All of these source and header files should be located at the related folders when the MSPWare for MSP432™ Series software is installed in your host computer. These folders are

- **C:/ti/msp/MSPWare _ <version>/inc**: Contains all hardware- or device-related specified header files. These files include:
  - Peripheral-specific definitions
  - Required type definitions
  - Macros
- **C:/ti/msp/MSPWare _ <version>/driverlib/driverlib/MSP432P4xx/keil**: Contains all project library files and compiler output directory, which include:
  - C source and header files peripheral-specific functionality
  - Compiler-specific project file for building the driver library
  - Compiler-specific output directories and files for the used compiler

A point to be noted is that the MSPWare for MSP432™ Series peripheral driver library is also preprogrammed into the internal ROM space on all MSP432 Series MCUs.

A set of high-level API interfaces is provided by this library to enable users to access and select all related peripheral devices during the program building process. This library is compatible with most popular IDEs such as CCS, ARM®/Keil® MDK, IAR, and GNU.

3.7.1.2 Boot loader

The Texas Instruments™ MSPWare boot loader is a small piece of code that can be programmed at the beginning of flash to act as an application loader as well as an update mechanism for applications running on an MSP432 ARM® Cortex®-M4-based microcontroller. The boot loader can be built to use the UART0, SPI0, I2C0, Ethernet, or USB ports to update the code on the microcontroller. The boot loader is customizable via source code modifications, or simply deciding at compile time which routines to include. Since full source code is provided, the boot loader can be completely customized.

Three update protocols are utilized. On UART0, SPI0, and I2C0, a custom protocol is used to communicate with the download utility to transfer the firmware image and program it into flash. When using Ethernet or USB Device Firmware Upgrade (DFU), however,

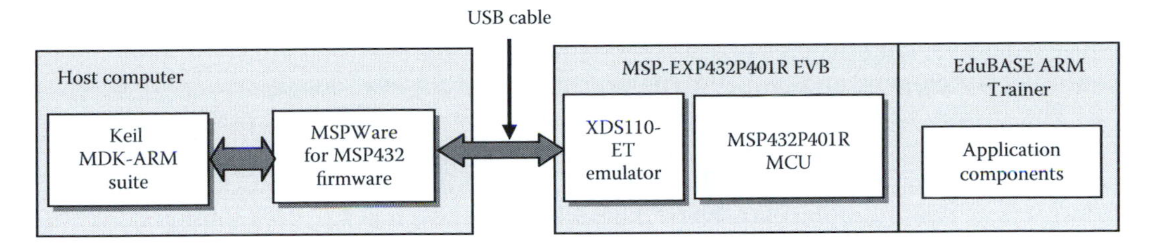

Figure 3.44 Complete development system for our project.

different protocols are employed. On Ethernet, the standard bootstrap protocol (BOOTP) is used, and for USB DFU, updates are performed via the standard DFU class.

When configured to use UART0, Ethernet, or USB, the LM Flash Programer GUI can be used to download an application via the boot loader.

Some other functions provided by the boot loader are

- Download codes to the flash memory for firmware updates
- Interface options include USB, UART, I2C, and SPI

After the MSPWare for MSP432 Series software package is installed in your host computer, all boot loader-related codes, including the source code files, header files, and assembly code files, are located at: **`C:/ti/msp/MSPWare_<version>/boot_loader`**.

3.7.1.3 Code examples

A complete set of program examples is provided by this MSPWare software package and it includes the code examples for the following microcontrollers:

- EZ430
- MSP430
- MSP432

All of these examples are categorized into different groups based on either the EVBs (**`boards`**) or MCUs (**`devices`**) they are built and implemented on.

For our target MCU used in this book, related example codes can be found at the folders:

- **`C:/ti/msp/MSPWare_<version>/examples/boards/MSP432P401R`**
- **`C:/ti/msp/MSPWare_<version>/examples/devices/MSP432P4xx`**

An illustration block diagram that contains complete components, including hardware and software, for our project development system is shown in Figure 3.44.

Next, let's discuss one of the most important properties applied on the MSP432™ Series LaunchPad™ system, MSP432™ Series CMSIS support.

3.8 MSP432™ series CMSIS support

ARM® Cortex® CMSIS, a standardized hardware abstraction layer for the Cortex-M processor series, provides supports for most MCUs built by different venders, including the TI MSP432™ Series devices. The CMSIS enables consistent and simple software interfaces to the processor core and simple basic MCU peripherals for silicon vendors and middleware providers, simplifying software reuse, reducing the learning curve for new microcontroller developers, and reducing the time to market for new devices.

The CMSIS DSP library includes source code and example applications, and saves time by including common DSP algorithms such as complex arithmetic, vector operations, and filters and control functions. The ARM® Cortex®-M4 core uses the DSP SIMD instruction set and floating-point hardware that enhances the Tiva™ C Series microcontroller algorithm capabilities for digital signal control applications.

A standardized software interface allows developers to make the switch from a competitive MCU to TI's MSP432™ Series microcontrollers and more easily to transfer the existing software to any MSP4xx Series microcontroller.

Many microcontroller-based applications can benefit from the use of an efficient DSP library. To that end, ARM® has developed a set of functions called the CMSIS DSP library that is compatible with all Cortex®-M3 and M4 processors and that is specifically designed to use ARM® assembly instructions to quickly and easily handle various complex DSP functions. Currently, ARM® supplies example projects for use in their Keil® µVision IDE that are meant to show how to build their CMSIS DSP libraries and run them on an M3 or M4.

The basic idea of using this MSP432™ Series support to CMSIS is to develop a set of interface functions to access the CMSIS DSP library files developed by the ARM® in the Keil® ARM-MDK environment from the MSP432™ Series EVBs.

A complete CMSIS DSP library is under development by TI and some application examples can be found in the CCS IDE for MSP432 Series EVBs.

3.9 Summary

The main topics in this chapter are about the MSP432P401R microcontroller device development environments and kits.

To successfully develop and build your embedded application software for specified MCU, you need the following kits or tools:

- MDKs
- Application-specified software development package
- Software examples

Since we used the Keil® ARM® MDKs and MSP432 Series EVB in this book, we concentrated our discussions on these two components.

The contents of this chapter include:

1. An overview and introduction about the MDK and related kits
2. The related development suite and specified firmware
3. An introduction to the IDE—Keil® MDK µVersion5
4. Embedded software development procedure
5. The ARM-MDK µVision5 Debugger and debug process
6. The MSP432™ Series software suite MSPWare
7. The MSP432™ Series libraries and other support
8. Program examples

In Section 3.4, detailed introductions and discussions about the Keil® ARM-MDK µVersion5 IDE are given since this kit is a key to build our user projects. These discussions include:

- A step-by-step introduction to each menu and menu items in the Keil® ARM-MDK µVersion5 IDE
- A real sample project **`MyProject`** is used to facilitate the introductions to those menus

In Section 3.7, a detailed discussion about the MSP432™ Series software development package and related firmware is provided with the EXP432P401R EVB used in this book.

HOMEWORK

I. True/False Selections

_____1. To successfully develop a microcontroller project, one needs the general-purpose development kits and the specified EVB-related development software package.

_____2. The built-in XDS110-ET Emulator in the MSP-EXP432P401R EVB is not compatible with the JTAG interface protocol.

_____3. The Keil MDK µVersion5 kit includes an IDE and a debug driver.

_____4. The debugger is a piece of software used to perform the debug functions only.

_____5. One needs to use both debug adapter and debug driver to perform debug-related functions.

_____6. In MSP432 Series LaunchPad EVB, the debug adapter is a built-in unit on the EVB.

_____7. Starting Keil MDK µVersion5, the kit is split into MDK Core and Software Packs. The MDK Core includes a µVersion IDE with editor, a C/C++ compiler and a debugger.

_____8. One needs to use MDK Core, CMSIS-Core, and Device Startup to successfully build a basic ARM-MDK project.

_____9. JTAG is the industry-standard interface used to download and debug programs on a target processor, as well as many other functions.

____10. The MSP432 Series software package MSPWare only includes two libraries, Peripheral Driver Library and USB Library.

II. Multiple Choices

1. The MSP432 Series EXP432P401R EVB comprises _______________ MCU(s).
   a. 1
   b. 2
   c. 3
   d. 4
2. To build a basic microcontroller project, one needs _________________________.
   a. Development kits
   b. Debug adaptor and debugger
   c. Specified MCU-related firmware
   d. Development or EVB
   e. All of above
3. The Keil MDK Core contains ____________________________________.
   a. µVersion IDE with source editor and GUI
   b. ARM® C/C++ compiler and Pack Installer
   c. µVersion Debugger
   d. All of them
4. Two important components in the MDK Software Packs are __________________.
   a. Device supports and middleware
   b. CMSIS libraries and middleware

   c. Board support and code templates
   d. CMSIS-Core and device supports
5. The CMSIS application software components include ________________.
   a. CMSIS:CORE, CMSIS:RTOS
   b. CMSIS:CORE, CMSIS:DSP
   c. Device:Startup, CMSIS:CORE
   d. CMSIS:CORE, CMSIS:RTOS, CMSIS:DSP
6. The µVersion Debugger in the Keil MDK Core can ________________.
   a. Perform the debugging functions for the user's program
   b. Download the user's image files into the flash memory in the EVB
   c. Both a and b
   d. None of them
7. The µVision5 offers a ________________ for creating applications and a ________________ for debugging applications.
   a. Debug mode and Running mode
   b. Debug mode and Download mode
   c. Build mode and Debug mode
   d. Build mode and Running mode
8. When the **`Load Application at Startup`** checkbox in the **`Debug`** tab under the **`Options for Target Project`** menu is selected, your image file will be automatically download to the flash memory as the ________________.
   a. Flash|Download menu item is selected
   b. Debug process begins
   c. Debug process is done
   d. Debug|Run menu item is selected
9. The MSP432 Series Software Package MSPWare includes ________________.
   a. Code functions and object libraries
   b. Utilities
   c. Boot loader and in-system programming
   d. All of them
10. To speed up the executing speed of downloading and debugging the users' program on the EXP432P401R EVB, one can use ________________.
   a. XDS110-ET Emulator
   b. CMSIS-DAP debug adapter protocol (debug driver)
   c. Keil ULINK or a J-Link adapter
   d. Any of them

III. Exercises

1. Provide a brief description about components to be used to develop a basic microcontroller project.
2. Provide a brief description about components used to build a microcontroller project with Keil® MDK µVersion IDE and running in the MSP432 Series EVB MSP-EXP432P401R.
3. Explain the debug components used in Keil® µVersion5 Debugger and EXP432P401R built-in XDS110-ET emulator in the EXP432P401R EVB.
4. Explain the relationship between the general-purpose development kit (Keil MDK) and specified MSPWare for MSP432™ Series LaunchPad™ software package.
5. Explain the development steps used to build a basic microcontroller project with Keil® MDK µVersion5 IDE.

chapter four

ARM® microcontroller software and instruction set

This chapter provides general information on the ARM® Cortex®-M4 microcontroller software and instruction set. The discussion is divided into two major parts: first the ARM® Cortex®-M4 Assembly instruction set is introduced, and then the related intrinsic C functions are discussed with necessary peripheral device driver libraries provided by the vendor.

4.1 Overview

To make any microcontroller or computer work, some directions should be provided to enable the microcontroller to follow those directions to perform certain tasks as the user desired. In fact, those directions can be translated to a sequence of instructions that can be integrated into a group of instructions called a program. Furthermore, in order to enable a microcontroller to understand the user's directions, a programming language is necessary to perform a valid communication tool between the users and the microcontroller. Like human beings, different languages have been used for their communications. In the computer world, the computing languages can be categorized into two levels:

- *High-Level Language*: This kind of language is very similar to the English language, and it is easy to be understood by human beings. However, it cannot be understood by the microcontrollers or computers at all. Some popular high-level languages are C/C++, Visual C++, Visual Basic.NET, Visual C#, and Java. The high-level language is computer or machine independent, which means that this kind of language can be understood by any computer with any OS. A translator or compiler is needed to convert the high-level language instructions into the low-level instructions to enable microcontrollers to understand and execute them.
- *Low-Level Language*: This kind of language comprises binary code or machine code sequence, like `01101110`. The low-level language is a computer- or machine-dependent language, which means that different microcontrollers have their own different languages and cannot be recognized by other microcontrollers or computers.

To enable human beings to use certain languages to communicate for each other, only a language is not enough. A word set written by that language is needed to allow people to use those words to represent their meaning to communicate for each other. Similarly, in order to use certain high-level programming languages to develop the user's application program, an instruction set written by that language is also needed. Unlike other 8-bit or 16-bit microcontrollers, the ARM® defines its instruction set as architecture; exactly it is called **`instruction set architecture`** (**`ISA`**).

In the early days, most ARM® processors could only handle 32-bit instructions. However, as the development of new mobile and microcontroller technologies has taken place, more and more devices only need 8-bit or 16-bit instruction sets. The advantage of using short bit size is that both the size and capacity of the applications in most modern devices become small and the running speed is also faster compared with those 32-bit instruction sets. Therefore, a contradiction arises between the instruction lengths.

To solve this problem, ARM® developed two different instruction set architectures, (1) a traditional 32-bit instruction set that is called ARM® instruction set and (2) a 16-bit instruction set that is called Thumb instruction set.

However, these two instruction sets are not compatible at all. To make them compatible, ARM® developed Thumb-2 technology in 2003. This technology enables a mixture of 16-bit and 32-bit instructions to be executed within one operating state. All the ARM® Cortex®-M processors are based on Thumb-2 technology.

The advantage of using this Thumb-2 instruction set is that during the operation, the ARM® processor can switch between the ARM® state (ARM® instruction set) and the Thumb state (Thumb-2 instruction set) under the software control. Therefore, some codes are compiled with ARM® instruction set for better performance, and some other codes are compiled as Thumb instructions to make the program size smaller.

Both ARM® Cortex®-M3 and Cortex®-M4 processors are based on Thumb-2 technology, however, they can only support Thumb instruction set.

All ARM® Cortex®-M4 instructions are developed by using the ARM® Cortex®-M4 assembly language, which is a low-level and processor-dependent language. There are large numbers of assembly instructions used by the ARM® Cortex®-M4 processors, but you do not need to learn all of them in detail because a C compiler is available to enable users to develop their codes in the C language and translate them to the assembly and machine codes. It is easy to develop a user application with the C codes if you have some knowledge of the C language.

4.2 Introduction to ARM® Cortex®-M4 software development structure

We have provided some discussions about the software development process under Keil® ARM® MDK environment in Chapter 3. Since we will use the Keil® ARM® MDK as our application development IDE, we just copy and redisplay Figure 3.7 in this chapter to illustrate the software development process under this environment.

Figure 4.1 shows a general software development process for ARM® Cortex®-M4 processor.

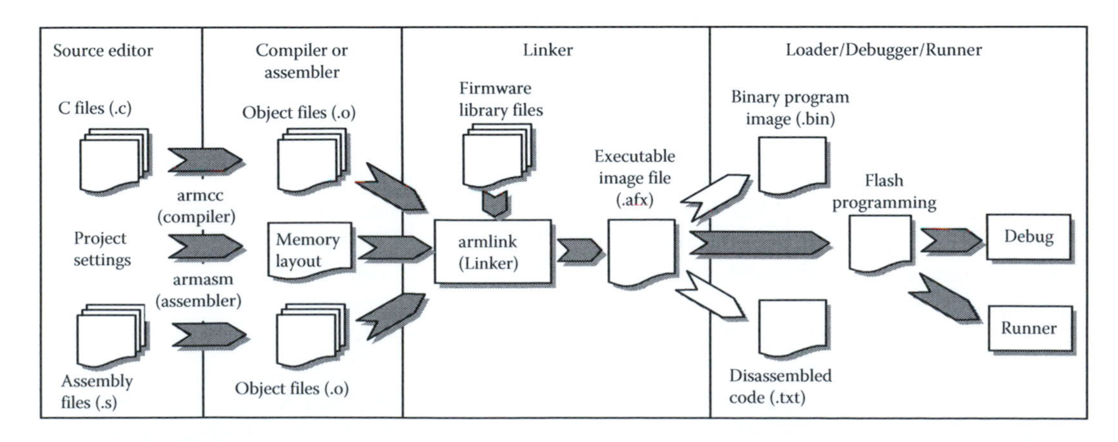

Figure 4.1 ARM Cortex-M4 software development procedure.

Generally, a user ARM® Cortex®-M4 application program can be developed and built with the following five steps:

1. **`Source Editor`**: Create user's source code files, either C codes (**`.c`**) or assembly codes (**`.s`**).
2. **`Compiler or Assembler`**: Translate the user's codes to the machine object codes (**`.o`**).
3. **`Linker`**: Connect the system library files with the user's object files to create the final executable file (image files - **`.afx`**).
4. **`Loader and Debugger`**: Load or download the user's executable or image files to the flash ROM in the development board and perform debugging functions.
5. **`Runner`**: Run the user's program in the flash memory.

The Keil® ARM® MDK µVersion5 provides all of these five components. The compiler and assembler are **`armcc`** and **`armasm`**. The linker is **`armlink`**. The user's executable or image file (**`.afx`**) can be downloaded into the 256-KB flash memory space. Then two options can be selected by the users. A debugging function is available in the MDK µVersion5 IDE, which can be used to debug and test the user's program. The users can also run the program in the flash memory space.

As we mentioned, over hundreds of assembly instructions are provided by ARM® Cortex®-M4 processor. However, you do not need to know them in detail because all of these instructions can be replaced by the related C codes. You can easily develop your program using the C language and convert it to the object file with the help of the ARM® Compiler **`armcc`**.

In the following sections, we will provide a brief introduction and illustration about the ARM® Cortex®-M4 instruction set. The main purpose of introducing these assembly instructions is to enable users to have a basic idea about them and use some of them during the debugging process if it is necessary.

4.3 *ARM® Cortex®-M4 assembly instruction set*

There are more than 200 instructions in the ARM® Cortex®-M4 instruction set excluding the FPU-related instructions. All of these instructions can be categorized into 14 groups based on the functions of those instructions. These groups include:

1. Data moving instructions
2. Arithmetic instructions
3. Logic instructions
4. Shift and rotate instructions
5. Data conversion instructions
6. Bit field processing instructions
7. Compare and test instructions
8. Program flow control instructions
9. Saturation instructions
10. Exception-related instructions
11. Sleep mode instructions
12. Memory barrier instructions
13. Miscellaneous instructions
14. Unsupported instructions

There are more than 60 FPU-related instructions if the FPU is available and enabled in the Cortex-M4 microcontroller system. All these instructions are Cortex-M4 assembly instructions with certain format or syntax. To enable the ARM® Cortex®-M4 processor to execute these instructions, an assembler is used to convert the assembly codes into the binary codes. In order for the assembler to do this conversion job correctly, all of these assembly instructions must follow a rule or syntax to enable the assembler to recognize them. So, first let's take a look at the format or syntax of these assembly instructions.

4.3.1 ARM® Cortex®-M4 assembly language syntax

As we mentioned in the last section, the user's source codes developed by assembly language must follow certain rules and meet some requirements to enable the assembler to recognize and understand those source codes, and furthermore, to enable the assembler to correctly convert them to the target or object codes.

Generally, each user's instruction in the user's source code file comprises four fields:

1. **`Label field`**
2. **`Operation field`**
3. **`Operands field`**
4. **`Comment field`**

Each field can be considered as a column, and each column is separated with the **`Tab`** key in the user's source code file, as shown in Figure 4.2.

Let's have a closer look at these fields by discussing them one by one.

- **`Label field`**: A label is an identifier to mark a line in the user's program. It is optional but when used it can provide a symbolic memory reference, such as a branch instruction's address or a symbol for a constant.

  A label must be located at the first column and started in the first character position in the instruction line. The length of each label is limited to 15 characters, which include upper- or lower-case letters (a~z), digits 0~9, the period, dollar sign, and underscore. The first character must be alphabetic, a period or an underscore. A label may end with a colon. Refer to Figure 4.2 for an example of using the label **`START`**.
- **`Operation field`**: This filed is also called **`mnemonic field`**. Two types of instructions may appear in this field, the opcode and the pseudo instruction, which we will discuss in the next section. The opcode is the mnemonic form of a Cortex-M4 instruction and it tells the microprocessor what kind of operation should be performed, such as **`MOV`**, **`STR`**, and so on. The pseudo instruction tells the assembler what

Label field	Operation field	Operands field	Comment field
NVIC_IRQ_SETEN	EQU	0×E000E100	*; define NVIC SETEN*
NVIC_IRQ_ENABLE	EQU	0×1	*; define the NVIC Enable*
START:	MOVS	R0, #0×56	; set R0 = 0×56
	MOVS	R1, #NVIC_IRQ_ENABLE	; set R1 = 0×1
	STR	R1, [R0]	; set 0×1 => 0×56

Figure 4.2 Syntax of the ARM Cortex-M4 assembly instructions.

to do. An example of using this pseudo instruction is **EQU**. Refer to Figure 4.2 for an example of using the operation **EQU**.

- **Operands field**: The operands field contains the data or an address for its corresponding instruction to be operated or performed. Both opcodes and pseudo codes can have operands. The number of operands depends on the type of instructions. Some instructions do not necessarily need any operand, such as **NOP**, **SEV**, and **WFI** instructions. However, some other instructions may need one, two, or three operands, such as **SADD8 R0, R1, and R5**.

  The operands must follow the opcode.
- **Comment field**: This field enables users to place some comments for each instruction to illustrate the function or purpose of the related instruction line. All comments must start with the semicolon (;) to tell the assembler that this is a comment, and the assembler will not convert any comment to the object codes. Refer to Figure 4.2 for an example of using the comments.

Since the operands field may contain multiple operands and, the following rules are useful to define the related operand:

- For data processing instructions written for the ARM® assembler, the first operand is the destination of the operation.
- For a memory read instruction, excluding multiple load instructions, the first operand is the destination register that data is to be loaded into.
- For a memory write instruction, excluding multiple store instructions, the first operand is the source register that holds the data to be written into the memory.
- For a multiple load instruction, the register list which is the third operand is the destination operand that the data will be loaded into.
- For a multiple store instruction, the register list which is the third operand is the source operand that the stored data will be written into the memory.

When using ARM® Cortex®-M4 assembly instructions to build the user's applications, the following points must be paid attention to:

1. For each instruction, the label and comment fields are optional. However, the operation and operands fields are necessary parts.
2. An instruction operand can be an ARM® register, a constant, or another instruction-specific parameter. When there is a destination register in the instruction, it is usually specified before the operands. Operands in some instructions are flexible and they can either be a register or a constant.
3. Bit-0 of any address written to the PC with a BX, BLX, LDM, LDR, or POP instruction must be 1 for the correct execution, because this bit indicates the required instruction set, and the Cortex-M4 processor only supports thumb instructions.

In order to enable the assembler to successfully convert the user's source codes into the object codes, the user's program must provide a full description about the user's source codes to enable the assembler to recognize and understand each instruction in the user's program, and furthermore to transfer it to the target or object codes.

To meet this requirement, we need to introduce the pseudo assembly instructions provided by the ARM® Cortex®-M4 assembler. The so-called pseudo assembly instructions mean that these codes are not real Cortex-M4 program assembly instructions. Instead, they are a set of assembler-helping codes to assist the assembler to assign and define memory spaces for the user's variables, constants, and program. These pseudo assembly instructions will not be converted to the object codes when the assembler is executed.

4.3.2 ARM® Cortex®-M4 pseudo instructions

Generally, the function of the pseudo assembly instructions is to help and assist the ARM® Cortex®-M4 assembler to convert and build the user's object code file. These assistants include the final real flash memory addresses in which the user's target program, data, constants, and labels should be located and stored, the memory spaces reserved and the ending point for the user's object code program. Table 4.1 shows a list of the most popular pseudo assembly instructions used by ARM® Cortex®-M4 assembler working for Keil® MDK-ARM®. Different IDE vendors may provide different assemblers with different pseudo instruction sets.

In ARM® Cortex®-M4 system, all data are categorized into the following data types:

- A 64-bit data item is called a double word and its length is 8 bytes.
- A 32-bit data item is called a word and its length is 4 bytes.

Table 4.1 Popular pseudo instructions used in Cortex-M4 MCU

Pseudo instruction	Function	Example
DCB n	Reserve 1-byte space (8 bits) for data	DCB 1, DCB 0x25
DCW n	Reserve half-word space (16 bits) for data	DCW 2, DCW 0x1234
DCD n	Reserve 1 word space (32 bits) for data	DCD 0x12345678
DCQ n	Reserve 2 words space (64 bits) for data	DCQ 0x0123456789ABCDEF
DCFS n	Define a single-precision floating-point data	DCFS 1E3
DCFD n	Define a single-precision floating-point data	DCFD 3.141593
DCB s	Reserve a space for a string	DCB "Hello World\n" 0
DCI n	Reserve a space for an instruction	DCI 0xBE00
ALIGN n	Align the current location to a specified boundary by padding with 0 or NOP instructions. The following number **n** is the number of bytes to be aligned	ALIGN 4: align the next data or instruction to 4 bytes or a word boundary
AREA section_name	Ask assembler to assemble a new code or data section	
CODE16	Specify assembly codes as Thumb codes (16 bits)	
EXPORT symbol	Declare a symbol that can be used by the linker to resolve symbol references in separate object or library files	
FILL num_of_bytes	Reserve a block of memory and fill it with the specific value	
IMPORT symbol	Declare a symbol reference in separate object or library files that is to be resolved by linker	
LTORG	Ask the assembler to assemble the current literal pool immediately	
THUMB	Specify assembly codes as Thumb codes in UAL format	

- A 16-bit data item is called a half-word and its length is 2 bytes.
- An 8-bit data item is called a byte and its length is 1 byte.

One point to be noted for these pseudo instructions is the different roles played by THUMB and CODE16. Both instructions are used to ask the assembler to treat the following codes as Thumb (16 bits) codes. However, the THUMB indicates to the assembler that the following codes should be Thumb codes in a **`unified assembly language`** (**`UAL`**) format, but the CODE16 asks the assembler to treat the following codes as Thumb codes with a **`pre-UAL`** format. The difference between the UAL and pre-UAL format is:

- The pre-UAL is an early Thumb syntax and only suitable to the Thumb instruction set.
- The UAL is a new Thumb-2 syntax, and it is more powerful in cross-platform or cross-architecture applications.

The pre-UAL syntax is still accepted by most IEDs, including the Keil® MDK-ARM®. But it is highly recommended to use the UAL syntax for new projects.

Pseudo instructions are very useful to help users to build and develop their application programs by providing more controls and managements to the memory space, variables, and constants used in the program.

Before we can start our discussion about the detailed instruction set, first let's have a clear picture about the addressing modes used in the Cortex-M4 assembler system.

4.3.3 ARM® Cortex®-M4 addressing modes

Generally, there are seven addressing modes used in the ARM® Cortex®-M4 assembly language system, and they are:

1. Immediate offset addressing mode
2. Register offset addressing mode
3. PC-relative addressing mode
4. Load and store multiple registers addressing mode
5. PUSH and POP register addressing mode
6. Load and store register exclusive addressing mode
7. Inherent addressing mode

Let's take a closer look at each of these addressing modes one by one.

4.3.3.1 Immediate offset addressing mode

To access a memory space to load or store a data item, a valid target memory address is needed to do this loading and storing operation. In Cortex-M4 system, this target address is obtained by adding an offset to a base address that is stored in a register. The offset value can be positive or negative, and the register used to store a base address can be any general-purpose register R0~R12. Depending on the different ways to calculate the target memory address, four different immediate offset addressing modes are available:

1. Regular immediate offset addressing mode
2. Preindexed immediate offset addressing mode
3. Postindexed immediate offset addressing mode
4. Regular immediate offset addressing mode with unprivileged access

4.3.3.1.1 Regular immediate offset addressing mode The syntax of this addressing mode is:

LDR{type} Rd, [Rn, {#Offset}]; for example **LDRB R0, [R2, #0x5];**
STR{type} Rt, [Rn, {#Offset}]; for example **STRSB R0, [R2, #0x10];**

Where the **{type}** is one of the following:

- **B**: Unsigned byte, zero extends to 32 bits on loads
- **SB**: Signed byte, sign extends to 32 bits (LDR only)
- **H**: Unsigned half-word, zero extends to 32 bits on loads
- **SH**: Signed half-word, sign extends to 32 bits (LDR only)
- -: Omit, for word

The **Rd** is the destination register for the load data from the memory into the Register (LDR), but the **Rt** is the source register for the store data from the register into the memory (STR). The **Rn** is the register that contains a base memory address. The offset must be prefixed with a **#** sign to indicate that it is an immediate number or offset.

The square bracket [] covering the Rn and an offset indicates that the combination of the content of the Rn and the offset is a valid memory address. In general, the square bracket sign [] means the content of the operand inside the [], not the operand itself. For instance, if the register R2 stores a hexadecimal number 0x01234567, the [R2] means the content of the R2, which is equivalent to a memory address 0x1234567, not R2 itself.

The running result of executing the **LDRB R0, [R2, #0x5]**, is to load an unsigned byte located at a memory address that is equal to the sum of the content of the R2 and 0x5, 0x1234567 + 0x5 = 0x123456C, into the register R0. In this mode, the offset is an immediate number and the content of the register R2 is kept unchanged after the execution of this instruction.

The range of the offset depends on the operation mode used in the instruction. For the regular immediate offset mode, the offset is ranged –255~4095.

4.3.3.1.2 Pre-indexed immediate offset addressing mode Similarly to the immediate offset addressing mode we discussed above, the operation procedure for this mode is:

1. The target memory address is first calculated by summing the content of the register Rn and the offset (pre-indexed).
2. Then, the data item stored in that target address will be loaded into the destination register Rt (for LDR instructions). For the STR operations, the data item stored in the source register Rt will be written into the target memory address.

The only difference between this mode and the regular immediate offset mode is that the content of the register Rn will be modified after running this instruction. For example, after the instruction **LDRB R0, [R2, #0x5]** is executed, the content of the register R2 will be changed to 0x123456C. However, the content of this register will be kept unchanged after running the regular immediate offset mode instructions.

4.3.3.1.3 Post-indexed immediate offset addressing mode The only difference between this mode and the previous modes is that the target memory address is directly obtained from the register Rn without using the offset value. Then, either the loading or storing instructions use this address as the target memory address to access the memory to perform either loading or storing operations. After the instruction is executed, a combination of the content of the register Rn and the offset is performed and this sum is sent back to the register Rn. This means that the content of the register Rn is modified or changed *after* (*post*) this mode's instruction is executed. The difference between this mode and the pre-indexed immediate offset mode is that the target address is first calculated based on the Rn and the offset before accessing the memory in the preindexed mode, however, this mode directly uses the content of the register Rn as the target memory address to access the memory without using the offset value.

For example, to run **`LDRB R0, [R2, #0x5]`** instruction in this mode, the target memory address is the content of the register R2, which is 0x1234567. The data item located at this address will be loaded into the register R0. Then, the content of the register R2 is added with the offset 0x5 to get the target address 0x123456C and this target address is sent to the R2. After running this instruction, the content of the register R2 is changed to 0x123456C, not 0x1234567.

4.3.3.1.4 Regular immediate offset addressing mode with unprivileged access The syntax of this addressing mode is:

```
LDR{type}T Rd, [Rn, {#Offset}]; for example LDRBT  R0, [R2, #0x5];
STR{type}T Rt, [Rn, {#Offset}]; for example STRSBT R0, [R2, #0x10];
```

where the **`{type}`** is one of the following:

- **`B`**: Unsigned byte, zero extends to 32 bits on loads
- **`SB`**: Signed byte, sign extends to 32 bits (LDR only)
- **`H`**: Unsigned half-word, zero extends to 32 bits on loads
- **`SH`**: Signed half-word, sign extends to 32 bits (LDR only)

These load and store instructions perform the same function as the regular immediate offset addressing mode. The difference is that these instructions have only unprivileged access even when used in the privileged software. When used in the unprivileged software, these instructions perform exactly the same function as regular memory access instructions with immediate offset.

Table 4.2 shows the ranges of the offset values used for the different operation modes. It can be found that the regular immediate offset addressing mode has a larger offset range.

4.3.3.2 *Register offset addressing mode*

The operation of this addressing mode is similar to that of the immediate offset addressing mode. The only difference between this mode and the immediate offset mode is that this mode uses a register to replace the immediate offset. In other words, the content of one register works as an offset for the target memory address.

Table 4.2 Offset range for different addressing modes

Instruction type	Immediate offset	Pre-indexed	Post-indexed
Word, half-word, signed half-word, byte, signed byte	–255~4095	–255~255	–255~255
Double words	4 × (–1020~1020)	4 × (–1020~1020)	4 × (–1020~1020)

The syntax of this kind of instruction is:

```
LDR{type}   Rd, [Rn, Rm, { LSL #n }]; for example LDRB   R0, [R2, R5];
STR{type}   Rt, [Rn, Rm, { LSL #n }]; for example STRSB  R0, [R2, R5, LSL #2];
```

where the `{type}` is one of the following:

- `B`: Unsigned byte, zero extends to 32 bits on loads
- `SB`: Signed byte, sign extends to 32 bits (LDR only)
- `H`: Unsigned half-word, zero extends to 32 bits on loads
- `SH`: Signed half-word, sign extends to 32 bits (LDR only)
- `-`: Omit, for word

Three registers Rt, Rd, Rn, and Rm are used in this mode and a logic shift left (LSL) instruction with the number of shift bits is an optional operation. The function of this mode is as follows:

For loading data from the memory and putting it into the destination register, the Rd is the destination register. The register Rn stores a base address and the register Rm stores an offset. Before performing the sum operation to combine the Rn (base address) and the Rm (offset) to get the target memory address, an **`LSL #n`** instruction could be used to shift Rm left up to 3 bits, which is equivalent to multiple Rm by 8. The range of the shifting bit **`n`** is 0~3. Since the LSL #n shift instruction is optional, it is covered by a brace sign {}.

The target memory address can be obtained by using:

$$\text{Target Memory Address} = \frac{\text{The content of Rn}[\mathbf{Rn}] + \text{The content of Rm}[\mathbf{Rm}] \times \mathbf{2}^{n}, n = 0 \sim 3.}{(\mathbf{Base\ address}) \quad + \quad (\mathbf{Offset}) \quad \times \mathbf{2}^{\mathbf{Shift\ bits}}}$$

Then, the instruction goes to the target memory address to pick the data item stored in that address and load it into the register Rt.

For storing data from the register to the memory, calculate the target memory address as above. Then, write the data item stored in the source register Rt into the target memory address.

For instance, the instruction **`LDRSB R0, [R5, R1, LSL #1];`** is to read a byte value from a target address that is equal to sum of R5 and two times R1, sign extend it to a word value and put it into the register R0. If the R5 contains a hexadecimal value: 0x12345678 and the R1 contains another hexadecimal value: 0x00000002, then the target address is 0x12345678 + 2 × 0x00000002 = 0x1234567C. The data byte stored in the address 0x1234567C is loaded and sign extended to 32 bits and put into the register R0.

Another example is the instruction **`STR R0, [R5, R1];`** this instruction is to store the content of R0 into an address that is equal to the sum of R5 and R1. Still using the contents of R5 and R1 in the above example, the target address is 0x12345678 + 0x00000002 = 0x123 4567A. The content of the source register R0 will be stored into the target memory address 0x1234567A.

One condition of using this mode is that the registers Rn and Rm must not be PC or SP.

4.3.3.3 PC-relative addressing mode

This mode is similar with the immediate offset addressing mode and the only difference for this mode is that the current PC value is used to replace the content of the base address register Rn. In other words, the current PC value works a base address for the target memory address. The final target memory address is a sum of the PC value and an offset. The offset can be considered as a relative index to the current PC value.

The syntax of this kind of instruction is:

```
LDR{type}  Rd, [PC, #Offset ];        for example  LDRB R0, [PC, #0x30];
LDRD{type} Rd, Rd2, [PC, #Offset ];   for example  LDRD R0, R1, [PC, #0x200];
```

where the `{type}` is one of the following:

- `B`: Unsigned byte, zero extends to 32 bits on loads
- `SB`: Signed byte, sign extends to 32 bits
- `H`: Unsigned half-word, zero extends to 32 bits on loads
- `SH`: Signed half-word, sign extends to 32 bits
- `-`: Omit, for word

The LDR instruction loads the destination register Rd with a data item whose address is a sum of the current PC value and the offset. The loaded byte and half-word will be extended to 32 bits.

The LDRD instruction loads the destination registers Rd and Rd2 with double words whose address is a sum of the current PC value and the offset. The lower word is loaded into the Rd and the higher word is loaded into the Rd2.

The value to be loaded can be a byte, half-word, or word. The loaded bytes and half-words can either be signed or unsigned. But both the loaded byte and half-word must be extended to 32 bits.

Three points must be noted when using this addressing mode:

1. The current PC value exactly should be equal to the current PC value plus 4.
2. The offset must be within a limited range relative to the current PC value. An error may be generated by the assembler if the offset is beyond this limitation. Table 4.3 shows some offset limitations used for the different instructions.
3. The bit-1 or bit[1] of the target address must be cleared to 0 to make it word aligned.

The reason why the current PC value is PC + 4 is because the so-called current PC value exactly should be the address of the next instruction, not the current instruction. This is due to the calculation process used by the assembler. When the assembler calculates the target address, the PC value used by the assembler is the address of the next instruction (PC + 4), not the address of the current instruction (PC) since this instruction takes 4-byte space itself. To correctly calculate the relative interval or offset between the current instruction's address and the target address, the starting address should not be the current

Table 4.3 PC relative offset range for different instruction types

Instruction type	Offset range
Word, half-word, signed half-word, byte, signed byte	–4095~4095
Double words	–1020~1020

instruction's address since it already takes 4-byte space. Therefore, the correct starting address should be the next instruction's address, which is PC + 4, since each instruction takes 32-bit or 4-byte space.

The reason why the bit-1 or bit[1] in the result target address must be 0 is because each instruction takes 4-byte space in ARM® Cortex®-M4 system. To align each 8-bit or 16-bit instruction to a word (32 bits) instruction, the bit-1 should be 0 since the last two bits (bit-1 and bit-0) on each instruction should be a sequence of 00 → 01 → 10 → 11. The next instruction should repeat this sequence to get another 4-byte space, 00 → 01 → 10 → 11. In this way, the starting address of each instruction on the last two bits, bit1 and bit 0, is always 00. Therefore, the bit-1 should be always 0 to make sure that each instruction is a word (32 bits).

Some restrictions may be applied when using these instructions:

- Rd2 must be neither SP nor PC
- Rd must be different from Rd2
- Rd can be SP or PC only for word loads

4.3.3.4 Load and store multiple registers addressing mode

The LDM and STM instructions enable users to load and store multiple data items that are contiguous in the memory space. These instructions only support 32-bit data.

The syntax of these kinds of instructions is:

```
LDM{mode}  Rn{!}, { reglist };    for example   LDM R0!, {R1, R2, R3};
STM{mode}  Rn{!}, { reglist };    for example   STMDB R1!, {R3-R6, R11, R12};
```

where the **{mode}** is one of the following:

- **IA**: **I**ncrement the target address **A**fter the load or store operation
- **DB**: **D**ecrement the target address **B**efore the load or store operation

The **Rn** is the address register used to store a valid memory address. The instructions can use this address to load from or store into any data items starting from that address.

The exclamation sign **!** is an optional writeback suffix. If it is present, the target address that is loaded from or stored to is written back into the address register Rn.

The **reglist** is a list of multiple registers to be loaded or stored, and it should be separated with a comma if more register is used or multiple ranges are used. The reglist can contain a register range, such as R0~R3.

The **LDM** instructions load the registers in reglist with word values from memory addresses stored in the Rn. The **STM** instructions store the word values in the registers in the reglist to memory addresses stored in the Rn.

For the **LDM**, **LDMIA**, **STM**, and **STMIA** instructions, the memory addresses to be accessed are at 4-byte intervals ranging from **Rn** to **Rn** + **4** × (**n−1**), where **n** is the order number of registers in **reglist**. The accesses happen in an order of increasing register numbers, with the lowest numbered register using the lowest memory address and the highest number register using the highest memory address. If the exclamation sign ! is present, the final address **Rn** + **4** × **n** is written back to **Rn**.

For example, the instruction **LDMIA R0!, {R1, R2, R3}**; is to load three-word data starting from the memory address indicated in the register R0 and put them into three registers R1, R2, and R3. Three registers, R1, R2 and R3 in the **reglist** are numbered as

1, 2, and 3. If the address in the R0 is 0x12345678, the following operational sequence will be executed to fulfill this instruction:

- For the first register located in the reglist R1, which is numbered **`1`**. The first word located in the starting address **`R0`** + **`4`** × **`(n-1)`** = 0x12345678 + 4 × (**1** – 1) = 0 x12345678 is loaded into the register R1. This word takes 4 bytes with a range of 0x12345678~0x1234567B.
- For the second register located in the reglist R2, which is numbered **`2`**. The second word located in the starting address **`R0`** + **`4`** × **`(n-1)`** = 0x12345678 + 4 × (**2**–1) = 0x123 45678 + 0x4 = 0x1234567C is loaded into the register R2. This word takes 4 bytes with a range of 0x1234567C~0x1234567F.
- For the third register located in the reglist R3, which is numbered **`3`**. The third word located in the starting address **`R0`** + **`4`** × **`(n-1)`** = 0x12345678 + 4 × (**3**–1) = 0x123456 78 + 0x8 = 0x12345680 is loaded into the register R3. This word takes 4 bytes with a range of 0x12345680~0x12345683.
- Finally, the final address **`R0`** + **`4`** × **`n`** = 0x12345678 + 4 × **3** = 0x12345678 + 12D = 0x123 45678 + 0x0C = 0x12345684 is written back into the R0 since an exclamation sign ! is present following the register R0 in this instruction.

Figure 4.3 shows the operation sequence for this instruction.

For the **`LDMDB`** and **`STMDB`** instructions, the memory addresses to be accessed are at 4-byte intervals ranging from **`Rn`** to **`Rn - 4*n`**, where **`n`** is the order number of registers in **`reglist`**. The accesses happen in an order of decreasing register numbers, with the lowest numbered register using the highest memory address and the highest number register using the lowest memory address. If the exclamation sign ! is present, the value **`Rn - 4`** × **`n`** is written back to **`Rn`**.

Another example is the instruction **`STMDB R1!, {R3-R6, R11, R12}`**. This instruction is to store six registers' contents, including R3, R4, R5, R6, R11, and R12, (six words) into the memory address starting at R1 – 4 × n. The order numbers for these six registers are: R3 → **`1`**, R4 → **`2`**, R5 → **`3`**, R6 → **`4`**, R11 → **`5`**, R12 → **`6`**. Assume that the R1 contains a hexadecimal value 0x00008000, the following sequence will be executed to fulfill this instruction:

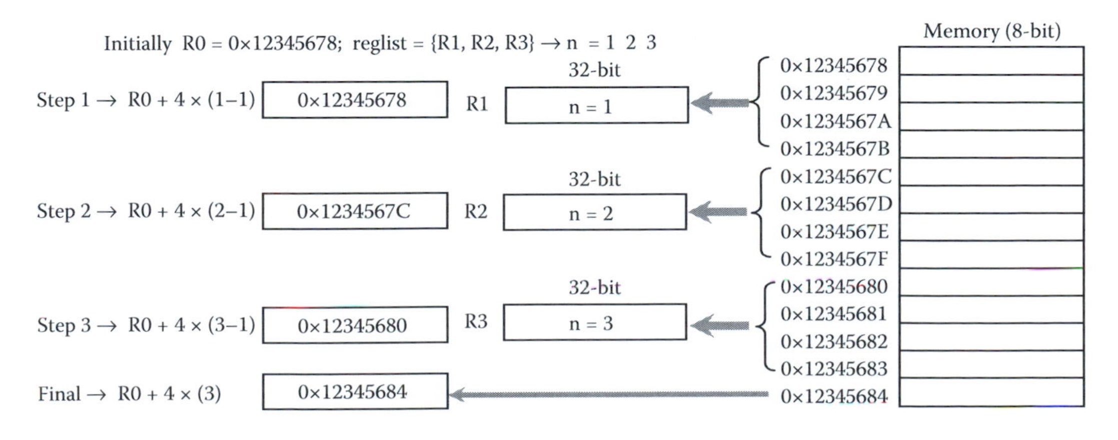

Figure 4.3 Operation sequence of the instruction LDMIA R0!, {R1, R2, R3}.

- For the first register R3 whose order number is **1**, its content is stored into the memory address starting at R1 – 4 × n = 0x8000 – 4 × **1** = 0x7FFC. This word takes 4 bytes with a range of 0x7FFC~0x8000. Since the mode is DB, therefore a decrement is performed first before the data storing.
- For the second register R4 whose order number is **2**, its content is stored into the memory address starting at R1 – 4 × n = 0x8000 – 4 × **2** = 0x8000 – 0x8 = 0x7FF8. This word takes 4 bytes with a range of 0x7FF8~0x7FFB.
- For the third register R5 whose order number is **3**, its content is stored into the memory address starting at R1 – 4 × n = 0x8000 – 4 × **3** = 0x8000 – 12D = 0x8000 – 0x0C = 0x7FF4. This word takes 4 bytes with a range of 0x7FF4~0x7FF7.
- For the fourth register R6 whose order number is **4**, its content is stored into the memory address starting at R1 – 4 × n = 0x8000 – 4 × **4** = 0x8000 – 0x10 = 0x7FF0. This word takes 4 bytes with a range of 0x7FF0~0x7FF3.
- For the fifth register R11 whose order number is **5**, its content is stored into the memory address starting at R1 – 4 × n = 0x8000 – 4 × **5** = 0x8000 – 0x14 = 0x7FEC. This word takes 4 bytes with a range of 0x7FEC~0x7FEF.
- For the sixth register R12 whose order number is **6**, its content is stored into the memory address starting at R1 – 4 × n = 0x8000 – 4 × **6** = 0x8000 – 24D = 0x8000 – 0x18 = 0x7FE8. This word takes 4 bytes with a range of 0x7FE8~0x7FEB.
- The final address **R1–4** × **n** = 0x8000 – 4 × 6 = 0x8000 – 24D = 0x8000 – 0x18 = 0x7FE8 is written back to the R1 since a ! sign is present after the register R1.

Figure 4.4 shows the operation sequence for this instruction.

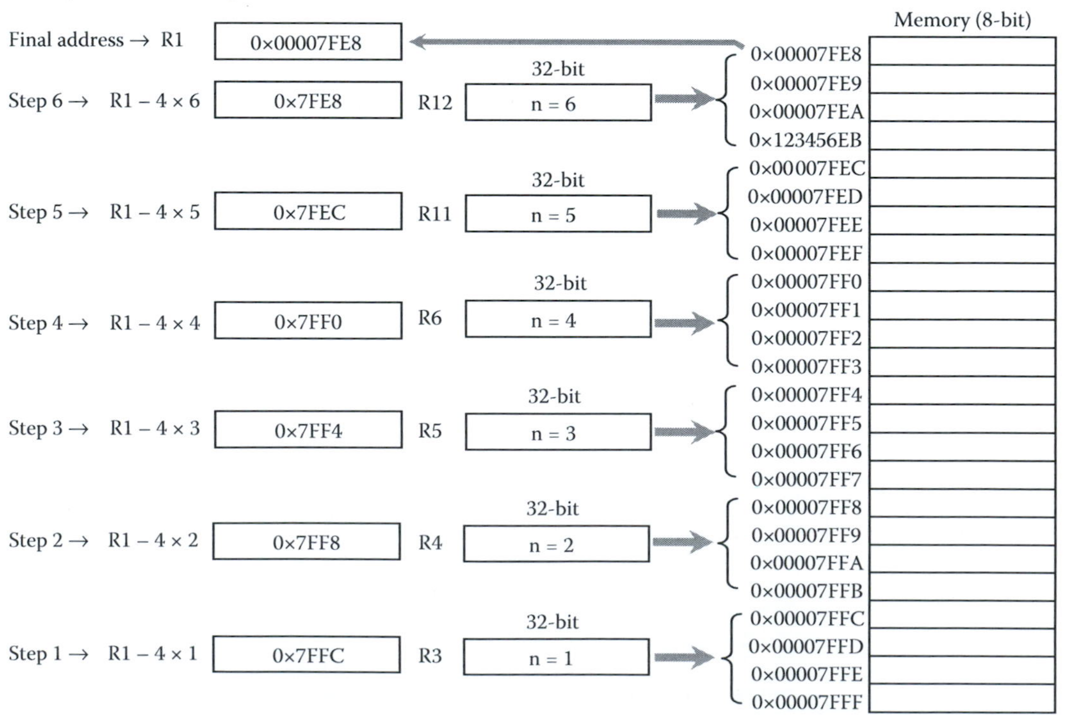

Figure 4.4 Running sequence of the instruction STMDB R1!, {R3-R6, R11, R12}.

Some restrictions are applied when using these instructions:

- The register Rn cannot be PC
- The reglist cannot contain SP
- In any STM instruction, the reglist cannot contain PC
- In any LDM instruction, the reglist cannot contain PC if it contains LR
- The reglist must not contain Rn if you specify the writeback suffix when PC is in the reglist in an LDM instruction
    - Bit[0] of the value loaded to the PC must be 1 for correct execution, and a branch occurs to this half-word-aligned address
    - If the instruction is conditional, it must be the last instruction in the IT block

Next, let's take care of the PUSH and POP register addressing mode.

4.3.3.5 PUSH and POP register addressing mode

The **PUSH** instruction stores the register's content into the memory, and the **POP** instruction performs an opposite operation; it is to load the data into the registers from the memory space.

The function of the PUSH instruction is similar to that of the **STMDB** instruction discussed in the last section, and the function of the POP instruction is similar to that of the **LDM** (or **LDMIA**) instruction discussed in the last section. The syntaxes of these instructions are:

```
PUSH { reglist };   for example,   PUSH {R0, R3-R5, R9};
POP { reglist };    for example,   POP {R2, R3};
```

The definition for the reglist is same as those used for the LDM and STM instructions. Multiple registers can be used in the reglist with each register being separated with comma, and the register ranges can also be used.

Some restrictions are applied when using these instructions, they are:

- The reglist cannot contain **SP**.
- For the PUSH instruction, the reglist cannot contain **PC**.
- For the POP instruction, the reglist cannot contain **PC** if it contains **LR**.

When performing the PUSH operation, the SP should be first decremented by 4 to reserve 4-byte space for the pushed-in data. For the POP operations, the SP is incremented by 4 after the data are pop up.

4.3.3.6 Load and store register exclusive addressing mode

Both the load register exclusive mode and the store register exclusive mode are a special group of memory accessing instructions. Normally, they are used for a system or environment that contains multiple sources to access the memory devices and those sources can share a common memory space. In that case, a control strategy is used to select and enable only one source to access the memory device at a time. Different control strategies have been developed to do this kind of selection, such as semaphores or mutual exclusive (Mutex).

In load register exclusive and store register exclusive modes, a destination register Rd is used to perform this kind of control and selection.

The syntaxes of these instructions are:

```
LDREX  Rt, [Rn, #Offset];       for example: LDREX R0, [R1, #0x12];
STREX  Rd, Rt, [Rn, #Offset];   for example: SREX R0, R1, [R2, #0x0B];
LDREXB Rt, [Rn];                for example: LDREXB R0, [R1];
STREXB Rd, Rt, [Rn];            for example: STREXB R0, R5, [R1];
LDREXH Rt, [Rn];                for example: LDREXH R0, [R1];
STREXH Rd, Rt, [Rn];            for example: STREXH R0, R1, [R2];
```

where

- **Rd**: The destination register for returned status
- **Rt**: The register to load or store
- **Rn**: The register stored the memory address to be accessed to load or store
- **Offset**: Is an optional value to be applied with the address in Rn to get the target address.

The LDREX, LDREXB, and LDREXH can be used to load a word, byte, and half-word, respectively, from a memory address and put them into the register **Rt**.

The STREX, STREXB, and STREXH instructions are used to store a word, byte, and half-word, respectively, from the register **Rt** to a memory address.

The address used in any store-exclusive instruction must be the same as the address in the most recently executed load-exclusive instruction. The value stored by the store-exclusive instruction must also have the same data size as the value loaded by the preceding load-exclusive instruction. This means that both LDRXX and STRXX instructions must be used in pair with the LDRXX instruction being first and the STRXX instruction being second.

The destination register **Rd** is used to store the current memory accessing status. When it is reset to 0 by a STRXX instruction, it indicated that no other source is accessing the memory device between the current load-exclusive and store-exclusive instructions.

Some restrictions are applied when using these instructions:

- Do not use **PC** for any register
- Do not use **SP** for Rd and Rt
- For the STREX instruction, Rd should be different from both Rt and Rn
- The value of the offset must be in the range 4 × (0–1020)

Finally, let's take a look at the inherent addressing mode.

4.3.3.7 Inherent addressing mode

The so-called inherent addressing mode means that both operands are registers and all operational data are located inside registers without needing to access the memory space.

Most arithmetic and logic as well as shift and rotate instructions use this addressing mode, such as the instructions ADD Rd, Rn, Rm; SBC Rd, Rn, Rm; AND Rd, Rn; ASR Rd, Rn. Some other instructions using this mode include:

```
Compare Instruction: CMP.
Compare Negative Instruction: CMN.
Signed and Unsigned Extension Instructions: SXTB, SXTH, UXTB.
Reversing Data Instructions: REV, REVSH.
Test Instruction: TST.
```

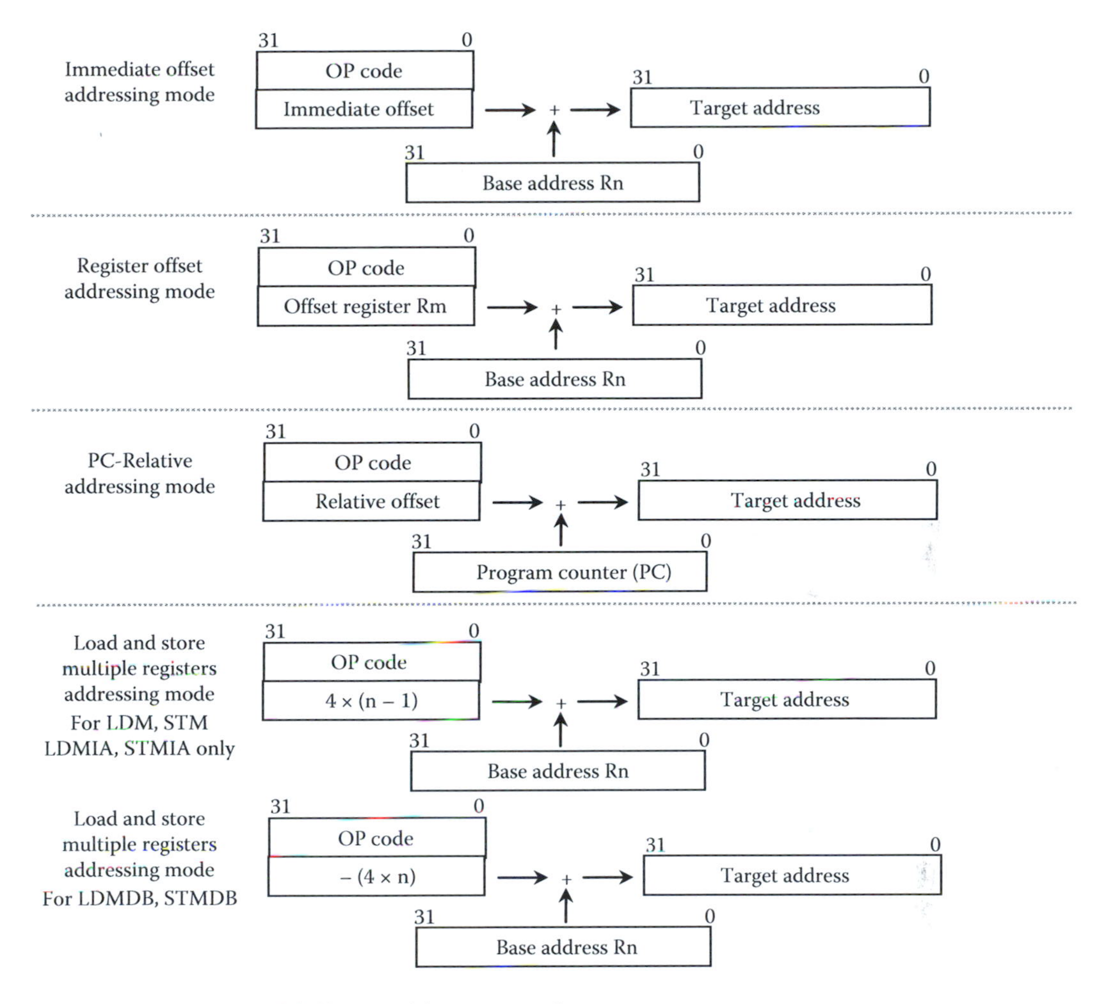

Figure 4.5 Summary of different addressing modes.

Now, let's give a quick summary about these addressing modes.

4.3.3.8 Addressing mode summary

Figure 4.5 provides a graphic summary presentation of most memory accessing addressing modes we discussed. The PUSH, POP, and load and store register exclusive addressing modes are similar to STMDB, LDM, and immediate offset addressing modes. The inherent addressing mode is very easy since this mode only uses registers without memory accessing.

4.3.4 ARM® Cortex®-M4 instruction set categories

As we mentioned, the ARM® Cortex®-M4 assembly language instructions can be divided into 14 groups based on their functions. We will provide a quick review for all of these instructions in the following sections one by one.

4.3.4.1 Data moving instructions

The data moving operation is one of the most popular operations used in any microcontroller programming. In Cortex-M4 instruction set, the data moving instructions include:

1. Move data from one register to another register
2. Move data between a general-purpose register and a special register
3. Move an immediate data into a register
4. Move data from registers to memory space
5. Move data from memory space to registers

The syntax for the first three (1~3) data moving instructions are:

- **MOV Rd, Operand2**
- **MOVS Rd, Operand2**
- **MOVT Rd, Operand2**
- **MOV Rd, #imm16**
- **MOVW Rd, #imm16**
- **MRS Rd, Rs**
- **MSR Rs, Rt**
- **MVN Rd, Operand2**
- **MVNS Rd, Operand2**

Basically, these instructions have the syntax of **MOV Destination, Source**. The function of these instructions is to move a data value from the source to the destination register. The second operand, **Operand2**, is a flexible operand and it can be an immediate number or a source register with a shifting operation. The **#imm16** is a half-word or a 16-bit data. The registers **Rd** and **Rt** represent the destination and the source registers. The **Rs** means a special register.

The suffix **S** and **W** are used to indicate that these instructions will affect the flag bits after it is executed. The suffix **T** indicates that a 16-bit data item is sent to the upper 16 bits of the destination register. Table 4.4 shows these data moving instructions and their functions.

The syntaxes for the last two (4~5) data moving instructions are those syntaxes of the top six addressing modes instructions we discussed in the previous sections. Table 4.5 shows these data moving instructions and their functions.

Figure 4.6 shows an example of using these data moving instructions. The exclamation sign ! means that the execution of the related instruction affects the flag bits, NZVC. If all flag bits are 0, this means that the execution of the related instruction has no effect on any flag.

Table 4.4 Data moving instructions with no memory access

Instruction	Function	Flags
MOV Rd, Rt	Rd ← Rt	—
MOVS Rd, Rt	Rd ← Rt & Update APSR flag	N, Z, C
MOV Rd, #Offset	Rd ← #Offset	—
MOVS Rd, #Offset	Rd ← #Offset & Update APSR flag	N, Z, C
MOVW Rd, #Imm16	Rd ← #Imm16 & Update APSR flag	N, Z, C
MOVT Rd, #Imm16	Upper 16-bit of Rd ← #Imm16	—
MRS Rd, Rs	Rd ← A special register Rs	—
MSR Rs, Rt	A special register Rs ← Rt	N, Z, C
MVN Rd, Rt	Rd ← Negative value of Rt	N, Z, C

Table 4.5 Data moving instructions with memory access

Instruction	Function	Flags
LDR Rd, [Rn, #Offset]	Rd ← [Rn + #Offset]	—
LDRB Rd, [Rn, #Offset]	Rd ← Extended to 32-bit of [Rn + #Offset]	—
LDRH Rd, [Rn, #Offset]	Rd ← Extended to 32-bit of [Rn + #Offset]	—
LDREX Rd, [Rn, #Offset]	Rd ← [Rn + #Offset]	—
STR Rt, [Rn, #Offset]	[Rn + #Offset] ← Rt	—
STRD Rt, Rt2, [Rn, #Offset]	[Rn + #Offset] ← Rt, [Rn + #Offset + 4] ← Rt2	—
LDM Rn! {Reglist}	Reglist ← [Rn + 4 × (n-1)]	—
LDMDB Rn! {Reglist}	Reglist ← [Rn - 4 × n]	—
STM Rn! {Reglist}	[Rn + 4 × (n-1)] ← Reglist	—
STMDB Rn! {Reglist}	[Rn - 4 × n] ← Reglist	—

```
1                       MovingDataExample.s
2   ; DEMO PROGRAM TO USE DATA MOVING INSTRUCTIONS
3
4   ————————————— No Memory Access Modes —————————————
5   START:      MOV     R3, R0                  ;R3 ← R0                 NZVC = 000-
6               MOVS    R4, R0                  ;R4 ← R0 & update APSR   NZVC = !!!!-
7               MOV     R0, #0x25               ;R0 ← #0x25              NZVC = 000-
8               MRS     R6, PRIMASK             ;R6 ← PRIMASK            NZVC = !!!!-
9
10  ————————— Immediate Offset Addressing Mode —————————
11  PLOOP:      LDR     R5, [R0, #0x20]         ;R5 ← [R0 + #0x20]       NZVC = 000-
12              STR     R0, [R5, #0x0C]         ;[R5 + #0x0C] ← R0       NZVC = 000-
13
14  ————————— Register Offset Addressing Mode —————————
15  SLOOP:      LDR     R3, [R0, R2]            ; R3 ← [R0 + R2]         NZVC = 000-
16              STR     R5, [R0, R7]            ;[R0 + R7] ← R5          NZVC = 000-
17              LDREX   R0, [R2, #0x10]         ;R0 ← [R2 + #0x10]       NZVC = 000-
18
19  ——— Load & Store Multiple Register Addressing Mode ———
20  LOOP:       LDM     R1!, {R3-R5}            ; R3,R4,R5 ← [R1+4x(n-1)] NZVC = 000-
21              STM     R0, {R1, R6}            ; [R0+4x(n-1)] ← R1 & R6 NZVC = 000-
22              STMDB   R0!, {R5, R8}           ;[R0-4xn] ← R5 & R8      NZVC = 000-
23              LDRH    R1, {R2, R5, LSL #2}    ;R1 ← [R2 + R5 × 4]      NZVC = 0000-
24              STR     R0, {R1, R2, LSL #1}    ;R0 ← [R1 + R2 × 2]      NZVC = 0000-
25              LDMDB   R7, {R1-R2, R5}         ;(R1, R2, R5) ← [R7 – 4 × n] NZVC = 0000-
```

Figure 4.6 Some example codes for the data moving instructions.

4.3.4.2 Arithmetic instructions

The Cortex-M4 processor provides many different arithmetic operation instructions, including the addition, subtraction, multiplication, and division. The addition instruction group contains addition with carry bit, and the subtraction instruction group includes subtraction with borrow bit.

The ARM® Cortex®-M4 arithmetic instructions include the following popular operations:

- Addition (ADD, ADC)
- Subtraction (SUB, SBC)

- Reverse Subtraction (RSB)
- Multiplication (MUL)
- Division (SDIV, UDIV)
- Multiplication with accumulation (MLA)
- Multiplication with subtraction (MLS)

The syntaxes and functions of the most popular arithmetic instructions are shown in Table 4.6.

The following points should be noted when using Table 4.6:

- The **S** is an optional suffix and if the **S** is specified, the condition codes on the related flag bits are updated when this instruction is done.
- The **Rd** is an optional destination register and if this register is omitted, the **Rn** will work as a destination register. The {} means that the variable inside is an optional one.
- The **Rm** is any other register that can be used to store some operating data.
- The **Imm12** is a 12-bit offset constant with a range of 0~4095.
- The **Operand2** is a flexible operand and it can be a register or an offset.
- The running results of most of these arithmetic instructions affect the condition codes in the flag register, such as negative, zero, overflow, and carry (NZVC) bits.

There are some other addition and subtraction instructions that are available in the Cortex-M4 assembly instruction set, which include signed addition, unsigned addition,

Table 4.6 Syntaxes for most popular arithmetic instructions

Instruction	Function	Flags
ADD{S} {Rd}, Rn, Operand2;	Rd ← Rn + Operand2 (if Rd omits, Rn ← Rn + Operand2)	N, Z, V, C
ADD{S} {Rd}, Rn, Imm12;	Rd ← Rn + Imm12 (if Rd omits, Rn ← Rn + Imm12)	N, Z, V, C
ADC{S} {Rd}, Rn, Operand2;	Rd ← Rn + Operand2 + Carry (if Rd omits, result → Rn)	N, Z, V, C
SUB{S} {Rd}, Rn, Operand2;	Rd ← Rn - Operand2 (if Rd omits, Rn ← Rn - Operand2)	N, Z, V, C
SUB{S} {Rd}, Rn, Imm12;	Rd ← Rn - Imm12 (if Rd omits, Rn ← Rn - Imm12)	N, Z, V, C
SBC{S} {Rd}, Rn, Operand2;	Rd ← Rn - Operand2 - Carry (if Rd omits, Rn ← Rn - Operand2 – Carry)	N, Z, V, C
RSB{S} {Rd}, Rn, Operand2;	Rd ← Operand2 – Rn (if Rd omits, Rn ← Operand2 – Rn)	N, Z, V, C
RSB{S} {Rd}, Rn, Rm;	Rd ← Rm – Rn (if Rd omits, Rn ← Rm – Rn)	N, Z, V, C
MUL{S} {Rd}, Rn, Rm;	Rd ← Rn × Rm (if Rd omits, Rn ← Rn × Rm)	N, Z, V, C
MLA Rd, Rn, Rm, Ra;	Rd ← Rn × Rm + Ra	—
MLS Rd, Rn, Rm, Ra;	Rd ← Ra - Rn × Rm	—
SDIV {Rd}, Rn, Rm;	Rd ← Rn/Rm (if Rd omits, Rn ← Rn/Rm) Signed Division	—
UDIV {Rd}, Rn, Rm;	Rd ← Rn/Rm (if Rd omits, Rn ← Rn/Rm) Unsigned Division	—

Table 4.7 Some other additional arithmetic instructions

Instruction	Function	Flags
SADD8 {Rd}, Rn, Rm;	Rd ← Rn + Rm; Perform four 8-bit signed additions	N, Z, V, C
SADD16 {Rd}, Rn, Rm;	Rd ← Rn + Rm; Perform two 16-bit signed additions	N, Z, V, C
UADD8 {Rd}, Rn, Rm;	Rd ← Rn + Rm; Perform four 8-bit unsigned additions	N, Z, V, C
UADD16 {Rd}, Rn, Rm;	Rd ← Rn + Rm; Perform two 16-bit unsigned additions	N, Z, V, C
SSUB8 {Rd}, Rn, Rm;	Rd ← Rn - Rm; Perform four 8-bit signed integer subtractions	N, Z, V, C
SSUB16 {Rd}, Rn, Rm;	Rd ← Rn - Rm; Perform two 16-bit signed integer subtractions	N, Z, V, C
USUB8 {Rd}, Rn, Rm;	Rd ← Rn - Rm; Perform four 8-bit unsigned subtractions	N, Z, V, C
USUB16 {Rd}, Rn, Rm;	Rd ← Rn + Rm; Perform two 16-bit unsigned subtractions	N, Z, V, C

8-bit or 16-bit additions, signed subtraction, unsigned subtraction, 8-bit and 16-bit subtractions. Table 4.7 shows these additional arithmetic instructions.

Figure 4.7 shows an example of using some popular arithmetic instructions to perform addition, subtraction, multiplication, and division operations.

All data items used for this example program are first defined and related memory spaces are reserved. Then, all data items are assigned to the different registers based on their assigned addresses. Four different arithmetic operations are executed using related arithmetic instructions, such as addition, subtraction with carry, multiplication, and division. The running results are assigned to the related assigned memory space.

```
1                        ARITHMETIC.s
2   ; DEMO PROGRAM TO USE ARITHMETIC INSTRUCTIONS TO PERFORM SOME OPERATIONS
3   ---------------   Define and Reserve Memory Spaces for Source Data   ---------------
4   DATA1       DCD   1             ;Reserve 1-word memory space for data source 1
5   DATA2       DCD   1             ;Reserve 1-word memory space for data source 2
6   SUM         DCD   1             ;Reserve 1-word memory space for sum
7   MINUS       DCD   1             ;Reserve 1-word memory space for minus
8   PRODUCT     DCQ   1             ;Reserve 2-word memory spaces for product
9   QUOTIENT    DCD   1             ;Reserve 1-word memory spaces for quotient
10
11  ---------------   Perform Data Assignment Operations   ---------------
12              LDR   R0, = DATA1    ;R0 ← [DATA1], Set R0 = Address of DATA1
13              LDR   R1, = DATA2    ;R1 ← [DATA2], Set R1 = Address of DATA2
14              LDR   R2, = SUM      ;R2 ← [SUM], Set R2 = Address of SUM
15              LDR   R3, = MINUS    ;R3 ← [MINUS], Set R3 = Address of MINUS
16              LDR   R4, = PRODUCT  ;R4 ← [PRODUCT], Set R4 = Address of PRODUCT
17              LDR   R5, = QUOTIENT ;R5 ← [QUOTIENT], Set R5 = Address of QUOTIENT
18  ---------------   Perform Data Arithmetic Operations   ---------------
19
20              ADD   R0, R0, R1     ;R0 ← DATA1 + DATA2, Get the SUM
21              STR   R0, [R2, #0]   ;SUM ← R0, Save the SUM
22              SBC   R0, R0, R1     ;R0 ← (DATA1 + DATA2) – DATA2, Get the MINUS
23              STR   R0, [R3, #0]   ;MINUS ← R0, Save the MINUS
24              MUL   R0, R0, R1     ;R0 ← DATA1 × DATA2, Get the PRODUCT
25              STR   R0, [R4, #0]   ; PRODUCT ← R0, Save the PRODUCT
26              UDIV  R0, R0, R1     ;R0 ← (DATA1 × DATA2) / DATA2, Get the QUOTIENT
27              STR   R0, [R5, #0]   ;QUOTIENT ← R0, Save the QUOTIENT
```

Figure 4.7 Example of using some instructions to do arithmetic operations.

4.3.4.3 Logic instructions

The ARM® Cortex®-M4 logic instructions include the following popular operations:

- AND (bitwise AND operations)
- ORR (bitwise OR operations)
- ORN (bitwise OR NOT operations)
- EOR (bitwise Exclusive OR operations)
- BIC (logic AND NOT or bit clear operations)

The syntaxes and functions of the most popular logic instructions are shown in Table 4.8. The following points should be noted when using Table 4.8:

- The **S** is an optional suffix and if the **S** is specified, the condition codes on the related flag bits are updated when this instruction is done.
- The **Rd** is an optional destination register and if this register is omitted, the **Rn** will work as a destination register. The {} means that the variable inside is an optional one.
- The **Rm** is any other register that can be used to store some operating data.
- The **Immed** is a 32-bit immediate constant.
- The **Operand2** is a flexible operand and it can be a register or an offset.
- The running results of most of these arithmetic instructions affect the condition codes in the flag register, such as negative, zero, and carry (NZC) bits.
- The **AND** instruction is to perform a bitwise (bit-by-bit) AND operation between two registers or between one register and an Operand2 or an immediate constant.
- The **ORR** instruction is to perform a bitwise (bit-by-bit) OR operation between two registers or between one register and an Operand2 or an immediate constant.
- The **ORN** instruction is to perform a bitwise (bit-by-bit) OR operation in the register **Rn** with the complementary of the corresponding bits in the value of Operand2 or the

Table 4.8 Syntaxes for most popular logic instructions

Instruction	Function	Flags
AND{S} {Rd}, Rn, Operand2;	Rd ← Rn & Operand2 (if Rd omits, Rn ← Rn & Operand2)	N, Z, C
AND{S} {Rd}, Rn, #Immed;	Rd ← Rn & Immed (if Rd omits, Rn ← Rn & Immed)	N, Z, C
ORR{S} {Rd}, Rn, Operand2;	Rd ← Rn	Operand2 (if Rd omits, Rn ← Rn | Operand2)	N, Z, C
ORR{S} {Rd}, Rn, #Immed;	Rd ← Rn	Immed (if Rd omits, Rn ← Rn | Immed)	N, Z, C
ORN{S} {Rd}, Rn, Operand2;	Rd ← Rn	~ Operand2 (if Rd omits, Rn ← Rn | ~ Operand2)	N, Z, C
ORN{S} {Rd}, Rn, #Immed;	Rd ← Rn	~ Immed (if Rd omits, Rn ← Rn | ~ Immed)	N, Z, C
EOR{S} {Rd}, Rn, Operand2;	Rd ← Rn ⊕ Operand2 (if Rd omits, Rn ← Rn ⊕ Operand2)	N, Z, C
EOR{S} {Rd}, Rn, #Immed;	Rd ← Rn ⊕ Immed (if Rd omits, Rn ← Rn ⊕ Immed)	N, Z, C
BIC{S} {Rd}, Rn, #Immed;	Rd ← Rn & (~Immed), (if Rd omits, Rn ← Rn & (~Immed))	N, Z, C
BIC{S} {Rd}, Rn, Rm;	Rd ← Rn & (~Rm), (if Rd omits, Rn ← Rn & (~Rm))	N, Z, C
BIC{S} {Rd}, Rn, Operand2;	Rd ← Rn & (~ Operand2), (if Rd omits, Rn ← Rn & (~ Operand2))	N, Z, C

complementary of an immediate constant. The sign ~ means a complementary or an inverse of the bits in a variable. For example, if Rn = 0x1110111000110010, the complementary or inverse of the Rn is: 0x0001000111001101. Each bit's value gets its inverse value, from 1 to 0, and from 0 to 1.

- The **EOR** instruction is to perform a bitwise (bit-by-bit) exclusive OR (⊕) between the registers or between one register and an Operand2 or an immediate constant.
- The **BIC** instruction is to perform a bitwise (bit-by-bit) AND operation on the bits in the **Rn** with the complements of the corresponding bits in the value of Operand2 or a register. For example, if Rn = 0x1110111000110010 and an instruction **BIC Rn, Rn** will make Rn = 0x00000000, since the complements of Rn or~Rn = 0x0001000111001101, which is just inverse of the original value in the Rn. **0 AND 1** must be 0 and this is equivalent to clear Rn bit-by-bit.

Figure 4.8 shows an example of using some popular logic instructions to perform AND, ORR, EOR, ORN, and BIC operations.

Next, let's take care of the shift and rotation instructions.

4.3.4.4 *Shift and rotate instructions*

The Cortex-M4 processor provides five different shift and rotation operation instructions, including the arithmetic shift right (ASR), LSL, logic shift right (LSR), rotate right (ROR), and rotate right with extended (RRX).

```
1                      LOGIC.s
2   ; DEMO PROGRAM TO USE LOGIC INSTRUCTIONS TO PERFORM SOME OPERATIONS
3   ----------- Define and Reserve Memory Spaces for Source Data -----------
4   DATA1      DCD   1               ;Reserve 1-word memory space for data source 1
5   DATA2      DCD   1               ;Reserve 1-word memory space for data source 2
6   ARESULT    DCD   1               ;Reserve 1-word memory space for AND Result
7   ORESULT    DCD   1               ;Reserve 1-word memory space for ORR Result
8   NRESULT    DCD   1               ;Reserve 1-word memory space for ORN Result
9   ERESULT    DCD   1               ;Reserve 1-word memory spaces for EOR Result
10  BRESULT    DCD   1               ;Reserve 1-word memory spaces for BIC Result
11
12  ----------- Perform Data Assignment Operations -----------
13             LDR   R0, = DATA1     ;R0 ← [DATA1], Set R0 = Address of DATA1
14             LDR   R1, = DATA2     ;R1 ← [DATA2], Set R1 = Address of DATA2
15             LDR   R2, = ARESULT   ;R2 ← [ARESULT], Set R2 = Address of ARESULT
16             LDR   R3, = ORESULT   ;R3 ← [ORESULT], Set R3 = Address of ORESULT
17             LDR   R4, = NRESULT   ;R4 ← [NRESULT], Set R4 = Address of NRESULT
18             LDR   R5, = ERESULT   ;R5 ← [ERESULT], Set R5 = Address of ERESULT
19             LDR   R6, = BRESULT   ;R5 ← [BRESULT], Set R5 = Address of BRESULT
20  ----------- Perform Data Logic Operations -----------
21
22             AND   R0, R0, R1      ;R0 ← DATA1 & DATA2, Get the AND Result
23             STR   R0, [R2, #0]    ;ARESULT ← R0, Save the AND Result
24             ORR   R0, R0, R1      ;R0 ← (DATA1 & DATA2) | DATA2, Get the ORR Result
25             STR   R0, [R3, #0]    ;ORESULT ← R0, Save the ORR Result
26             ORN   R0, R0, R1      ;R0 ← ((DATA1 & DATA2) | DATA2) | (~ DATA2), Get the ORN
27             STR   R0, [R4, #0]    ;NRESULT ← R0, Save the ORN Result
28             EOR   R0, R0, R1      ;R0 ← (R0 ⊕ R1), Get the EOR Result
29             STR   R0, [R5, #0]    ;ERESULT ← R0, Save the EOR Result
30             BIC   R0, R0, R1      ;R0 ← R0 & ( ~ R1), Get the BIC Result
31             STR   R0, [R6, #0]    ;BRESULT ← R0, Save the BIC Result
```

Figure 4.8 Example of using some popular logic instructions.

The ARM® Cortex®-M4 instructions contain the following five popular operations:

- ASR
- LSL
- LSR
- ROR
- RRX

The syntaxes and functions of these shift and rotate instructions are shown in Table 4.9. The following points should be noted when using Table 4.9:

- The `S` is an optional suffix and if the `S` is specified, the condition codes on the related flag bits are updated when this instruction is done.
- The `Rd` is a destination register used to store the shifting result.
- The `Rn` is any other register that contains the value to be shifted.
- The `Rs` is a register holding the shift length to be applied to the value in the register Rn. Only the least significant byte is used and its value can be in the range 0 to 255.
- The `n` is a constant representing the shift length, and its range depends on the instruction.
- The running results of most of these arithmetic instructions affect the condition codes in the flag register, such as NZC bits.
- In all of these instructions, the shifting result is written into the destination register Rd, but the content of the register Rn remains unchanged.
- To use a 16-bit version of these instructions, all registers used must be low registers (R0~R7). The RRX instruction is not available in a 16-bit version.
- For arithmetic shift and rotate instructions, only shift right instructions are available. The reason for that is: (1) the LSL instruction is equivalent to arithmetic shift left (ASL) instruction and both of them perform the same function and (2) the ROR instruction can work as a rotate left (ROL) instruction, since an ROR is equivalent to rotate to left as this rotate is a cycle or a closed loop operation.

Table 4.9 Syntaxes for shift and rotate instructions

Instruction	Function	Flags
ASR{S} Rd, Rn, Rs;	Rd ← Rn ≫ Rs; Rn is shifted right by number of bits stored in Rs	N, Z, C
ASR{S} Rd, Rn, #n;	Rd ← Rn ≫ n; Rn is shifted right by n bits. The range of n is: 1 ~ 32	N, Z, C
ASR{S} Rd, Rn;	Rd ← Rn ≫ Rn; Rn is shifted right by number of bits stored in Rn	N, Z, C
LSL{S} Rd, Rn, Rs;	Rd ← Rn ≪ Rs; Rn is shifted left by number of bits stored in Rs	N, Z, C
LSL{S} Rd, Rn, #n;	Rd ← Rn ≪ n; Rn is shifted left by n bits. The range of n is: 0 ~ 31	N, Z, C
LSL{S} Rd, Rn;	Rd ← Rn ≪ Rn; Rn is shifted left by number of bits stored in Rn	N, Z, C
LSR{S} Rd, Rn, Rs;	Rd ← Rn ≫ Rs; Rn is shifted right by number of bits stored in Rs	N, Z, C
LSR{S} Rd, Rn, #n;	Rd ← Rn ≫ n; Rn is shifted right by n bits. The range of n is: 1 ~ 32	N, Z, C
LSR{S} Rd, Rn;	Rd ← Rn ≫ Rn; Rn is shifted right by number of bits stored in Rn	N, Z, C
ROR{S} Rd, Rn, Rs;	Rd ← Rn rotate right by Rs	N, Z, C
ROR{S} Rd, Rn, #n;	Rd ← Rn rotate right by n bits. The range of n is: 1 ~ 31	N, Z, C
ROR{S} Rd, Rn;	Rd ← Rn rotate right by Rn	N, Z, C
RRX{S} Rd, Rn;	{C, Rd} = {Rn, C}; Rn is shifted right with Carry bit by 1 bit	N, Z, C

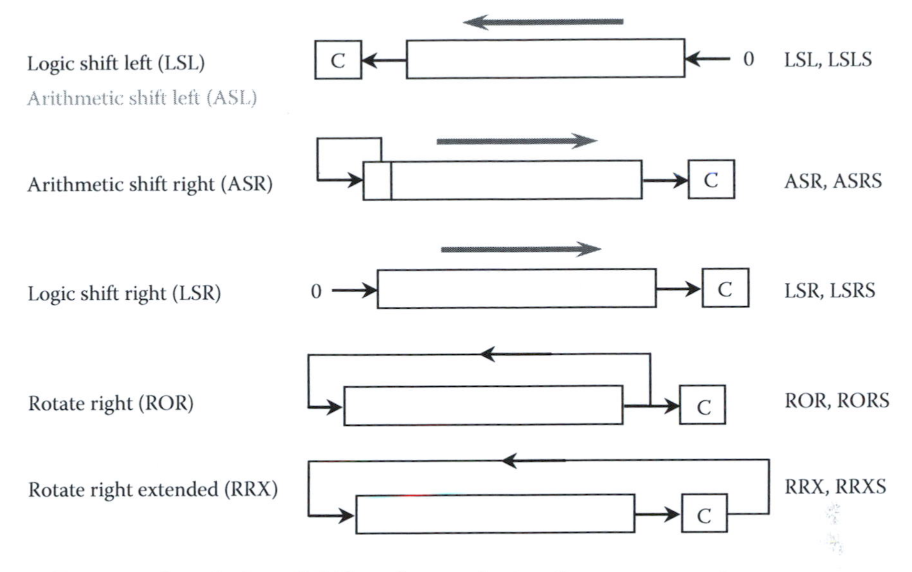

Figure 4.9 Function description of shift and rotate instructions.

Figure 4.9 shows a function description of using these shift and rotate instructions to perform related operations. Figure 4.10 shows an example of using the shift and rotate instructions to perform related operations.

4.3.4.5 Data conversion instructions

The Cortex-M4 processor provides some data conversion and reversion instructions to help users to convert and reverse some data items during the program development to facilitate the coding process. Generally, these instructions enable users to convert 8-bit and 16-bit data items to 32-bit data in either signed or unsigned format. Some instructions provide prerotate function to enable the data items to be rotated before the data conversion. For the data reversion instructions, these instructions allow users to get inverse versions of bytes, half-words or words.

```
1                        SHIFT_ROTATE.s
2   ; DEMO PROGRAM TO USE SHIFT & ROTATE INSTRUCTIONS TO PERFORM SOME OPERATIONS
3
4   ----------          Perform Data Assignment Operations          ----------
5
6                   MOVS    R0, #0x8        ;R0 ← 0x8, Set R0 initial value = 0x8
7                   MOVS    R1, #0x2        ;R1 ← 0x2, Set R1 = number of shift bits
8   ----------          Perform Data Shift & Rotate Operations      ----------
9                   ASRS    R2, R0, R1      ;R2 ← R0 >> 2 bits (R2 = 0x2 now)
10                  ASR     R0, R1          ;R0 ← R0 >> 2 bits (R0 = 0x2 now)
11                  LSL     R3, R0, #0x3    ;R3 ← R0 << 3 bits (R3 = 0x10 now)
12                  LSLS    R4, R1, R1      ;R4 ← R1 << 2 bits (R4 = 0x8 now)
13                  LSRS    R5, R0, R1      ;R5 ← R0 >> 2 bits (R5 = 0x0 now)
14                  ROR     R0, R1, R1      ;R0 ← R1 rotate right by 2-bit (R0 = 0x10000000 now)
15                  RORS    R0, R0, R1      ;R0 ← R0 rotate right by 2-bit (R0 = 0x20000000 now)
16                  RRX     R0, R1          ;R0 ← R1 is rotated right with C by 2-bit (R0 = 0x08000000)
```

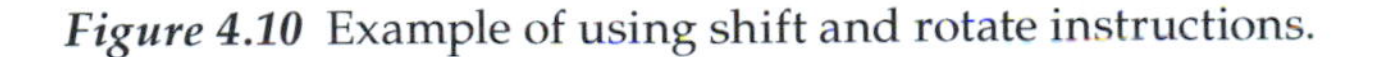

Figure 4.10 Example of using shift and rotate instructions.

The ARM® Cortex®-M4 instructions contain the following popular conversion and reversion operations:

- Signed extended byte to word (SXTB)
- Signed extended half-word to word (SXTH)
- Unsigned extended byte to word (UXTB)
- Unsigned extended half-word to word (UXTH)
- Reverse bytes in word (REV)
- Reverse bytes in each half-word (REV16)
- Reverse bytes in lower half-word and sign extend the result (REVSH)
- Reverse bit order in a 32-bit word (RBIT)

The syntaxes and functions of these shift and rotate instructions are shown in Table 4.10.

The following points should be noted when using Table 4.10:

- The **Rd** is an optional destination register. If it is omitted, the **Rn** replaces the Rd and works as the destination register.
- The **ROR #n** is an optional operation to perform a ROR #n function before the Rn can be converted. The available value for the rotation number **n** is: 8, 16, and 24, which means that you can rotate Rn right with 8, 16, or 24 bits before the conversion.
- The **SXTB** instruction extracts the lowest 8-bit value of Rn, Rn [7:0], and sign extends to a 32-bit value. The so-called sign-extended means that when sign extends an 8-bit data to a 32-bit data, copy the bit value on the MSB of the 8-bit data, bit[7] on Rn, and paste this

Table 4.10 Syntaxes for data conversion and reversion instructions

Instruction	Function	Flags
SXTB {Rd}, Rn, OR #n};	Rd ← Signed_Extend (Rn[7:0]) to 32-bit. ROR #n for Rn. Rd = Rn if Rd missed	N, Z, C
SXTH {Rd}, Rn, {ROR #n};	Rd ← Signed_Extend (Rn[15:0]) to 32-bit. ROR #n for Rn. Rd = Rn if Rd missed	N, Z, C
SXTB16 {Rd}, Rn, {ROR #n};	Rd[15:0] ← Signed_Extend (Rn[7:0]); ROR #n for Rn. Rd = Rn if Rd omits Rd[31:16] ← Signed_Extend (Rn[23:16])	N, Z, C
UXTB {Rd}, Rn, {ROR #n};	Rd ← Unsigned_Extend (Rn[7:0]) to 32-bit. ROR #n for Rn. Rd = Rn if Rd omits	N, Z, C
UXTH {Rd}, Rn, {ROR #n};	Rd ← Unsigned_Extend (Rn[15:0]) to 32-bit. ROR #n for Rn. Rd = Rn if Rd omits	N, Z, C
UXTB16 {Rd}, Rn, {ROR #n};	Rd[15:0] ← Unsigned_Extend (Rn[7:0]); ROR #n for Rn. Rd = Rn if Rd omits Rd[31:16] ← Unsigned_Extend (Rn[23:16])	N, Z, C
REV Rd, Rn;	Rd ← REV (Rn); Reverse byte order in word	N, Z, C
REV16 Rd, Rn;	Rd ← REV16 (Rn); Reverse byte order in each half-word independently	N, Z, C
REVSH Rd, Rn;	Rd ← REVSH (Rn); Reverse byte order in lower half-word and sign extend result	N, Z, C
RBIT Rd, Rn;	Rd ← RBIT (Rn); Reverse the bit order in a 32-bit word in Rn	N, Z, C

value to the upper 24 bits. For example, if Rn = 0x23456789, after the execution of the instruction **SXTB Rd, Rn;** the Rd = 0xFFFFFF89, since the MSB of the Rn, bit[7] is 1.

- The **SXTH** instruction extracts the lower 16-bit value of Rn, Rn [15:0], and sign extends to a 32-bit value. The so-called sign-extended means that when extends a 16-bit data to a 32-bit data, copy the bit value on the MSB of the 16-bit data, bit[15] on Rn, and paste this value to the upper 16 bits. Still using the above example in SXTB, after the execution of the instruction **SXTH Rd, Rn;** the Rd = 0x00006789 since the MSB of the 16-bit data on Rn, bit[15] is 0 (6 → 0110).
- The **UXTB** instruction extracts the lowest 8-bit value of Rn, Rn [7:0], and zero extends to a 32-bit value. The so-called zero-extended means that when zero extends an 8-bit data to a 32-bit data, put 24 zeros to the upper 24 bits. Still using the above example in SXTB, after the execution of the instruction **UXTB Rd, Rn;** the Rd = 0x00000089, since 24 zeros are put in the upper 24 bits in the Rd even the MSB of the lowest 8-bit on the Rn is 1.
- The **UXTH** instruction extracts the lower 16-bit value of Rn, Rn [15:0], and zero extends to a 32-bit value. The so-called zero-extended means that when zero extends a 16-bit data to a 32-bit data, put 16 zeros to the upper 16 bits. If still using the above example in SXTB, where Rn = 0x23456789. After executing: **UXTH Rd, Rn;** the Rd = 0x00006789.
- The **SXTB16** instruction extracts bits[7:0] on Rn and sign extends to 16 bits (Rd[15:0]), and extracts bits [23:16] on Rn and sign extends to 16 bits (Rd[31:16]).
- The **UXTB16** instruction extracts bits[7:0] on Rn and zero extends to 16 bits (Rd[15:0]), and extracts bits [23:16] and zero extends to 16 bits (Rd[31:16]).
- The **REV** instruction reverses the byte order in a word stored in the register Rn. For example, if Rn = 0x11223344, which is a word. After the execution **REV Rd, Rn;** the Rd = 0x44332211. Only the order of each byte, 11, 22, 33, 44, is reversed to: 44, 33, 22,11.
- The **REV16** instruction only reverses the byte order in each half-word stored in the register Rn independently. For example, if Rn = 0x12345678, after executing the instruction: **REV16 Rd, Rn;** the Rd = 0x34127856. It only reverses 2 bytes in each half-word.
- The **REVSH** instruction reverses byte order in the lower half-word, and sign extends to 32 bits. For example, if Rn = 0x23456789, after the instruction: **REVSH Rd, Rn;** is executed, the Rd = 0xFFFF8967. First, the lower half-word in Rn, 6789, is reversed in the byte order to: 8967. Then, it is sign extended to 32-bit word by putting 16 "1" into the upper 16 bits since the MSB value of the lower half-word in Rn, bit[15], is 1 (8 → 0x1000).
- The **RBIT** instruction reverses the bit order in a 32-bit word. For example, if Rn = 0x00000001. After the execution of the instruction: **RBIT Rd, Rn;** the Rd = 0x80000000.

Figure 4.11 shows a function description of using these conversion instructions to perform related operations.

Figure 4.12 shows a function description about these reverse instructions. Figure 4.13 shows an example of using the shift and rotate instructions to perform related operations.

4.3.4.6 Bit field processing instructions

The Cortex-M4 processor provides five different bit field operation instructions. These instructions are used to help users to manipulate and control a single bit or a range of bits in a register. The bit field operation is a brand new technology in the Cortex-M4 microcontroller.

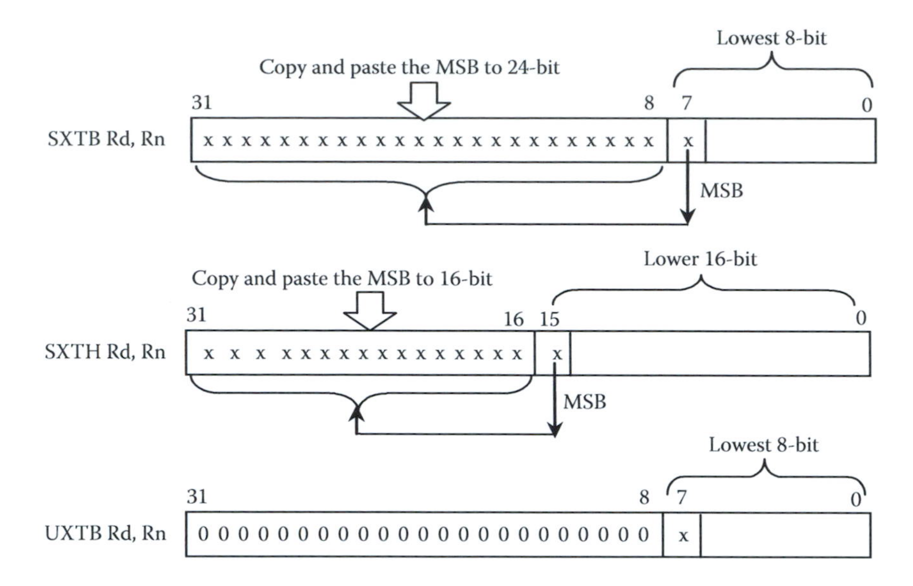

Figure 4.11 Function description about some conversion instructions.

The ARM® Cortex®-M4 processor contains the following five popular bit field instructions:

- Bit field clear (BFC)
- Bit field insert (BFI)
- Count leading zero (CLZ)
- Signed bit field extract (SBFX)
- Unsigned bit field extract (UBFX)

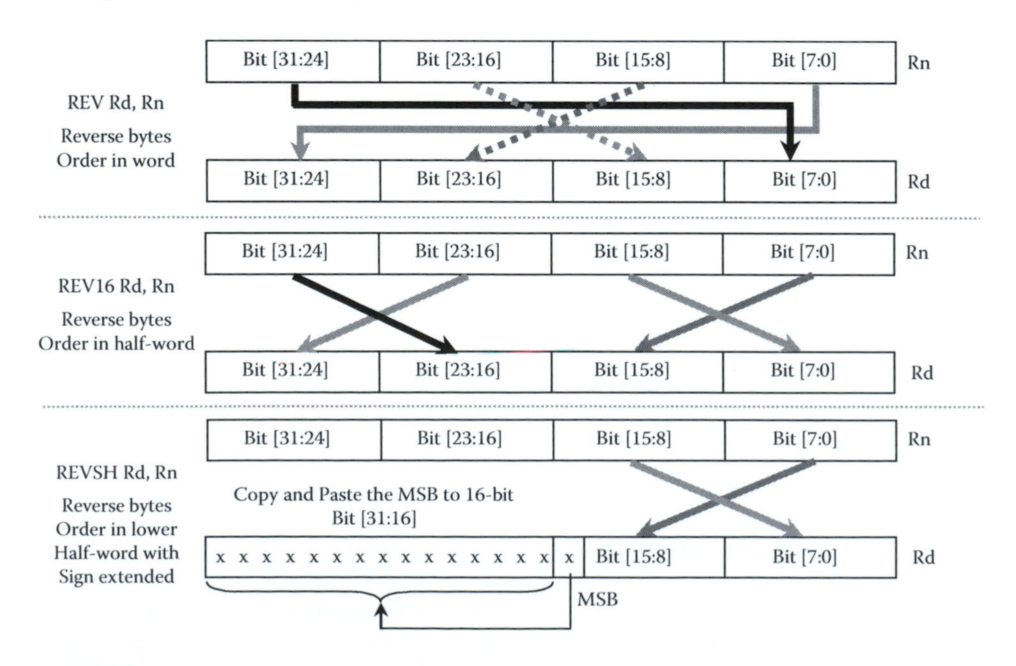

Figure 4.12 Function description about the reverse instructions.

```
1                        CONVERT_REVERSE.s
2  ; DEMO PROGRAM TO USE CONVERT & REVERSE INSTRUCTIONS TO PERFORM SOME OPERATIONS
3
4  ———————   Perform Data Assignment Operations   ———————
5
6          MOVS    R0, #0x12345678      ;R0 ← 0x12345678, Set R0 initial value = 0x12345678
7          MOVS    R1, #0x11223344      ;R1 ← 0x11223344, Set R1 initial value = 0x11223344
8  ———————   Perform Data Conversion & Reverse Operations   ———————
9          SXTB    R2, R0          ;R2 ← Sign extends R0[7:0] to 32-bit
10         SXTH    R3, R1          ;R3 ← Sign extends R1[15:0] to 32-bit
11         UXTB    R4, R0          ;R4 ← Unsign extends R0[7:0] to 32-bit
12         UXTH    R4, R1, ROR #2  ;Rotate R1 right 2 bits, R4 ← Unsign extends R1[15:0] to 32-bit
13         REV     R5, R0          ;R5 ← Reverse bytes order in Word for R0
14         REV16   R0, R1          ;R0 ← Reverse bytes order in the lower half-word in R1
15         REVSH   R1, R0          ;R1 ← Reverse bytes order in lower halfword with sign extension
16         RBIT    R0, R1          ;R0 ← Reverse bit order in Word for R1
```

Figure 4.13 Example of using some reversion instructions.

The syntaxes and functions of these shift and rotate instructions are shown in Table 4.11.

The following points should be noted when using Table 4.11:

- The **Rd** is the destination register.
- The **Rn** is the operand register.
- The **1stb** is the first bit or starting bit position of the LSB of the bit field. For example, if a **1stb** is **5** and a **width** is **3**, which means that the starting position in the Rd should be bit [5] and the ending bit should be bit [7] (1stb + width – 1 = 5 + 3 – 1 = 7). All bits in this field (bit [5]~bit [7]) belong to the selected or target bit field. The **1stb** must be in the range 0 to 31, and the **width** must be in the range 1~32.
- The **BFC** instruction is used to clear a bit field in the register Rd. The range of selected bit field starts from the **1stb** bit and ends at **1stb** + **width** – **1** bit. Refer to Figure 4.14 to get more details about this instruction.

Table 4.11 Syntaxes for bit field processing instructions

Instruction	Function	Flags
BFC Rd, #1stb, #width;	Clears **width** bits in Rd, starting at the low bit position **lstb** (1st bit) to the (*1stb* + *width* -1) bit. Other bits in Rd are unchanged	—
BFI Rd, Rn, #1stb, #width;	Replaces **width** bits in Rd starting at the low bit position **lstb** (1st bit), with width bits from Rn starting at bit[0]. Other bits in Rd keep unchanged	—
CLZ Rd, Rn;	Count the number of leading zeros in Rn and return result to Rd. If all bits in Rn are 0, it returns 32. If bit [31] in Rn is 1, it returns 0.	—
SBFX Rd, Rn; #1stb, #width;	Rd ← Extract a bit field from Rn and sign extends it to 32 bits	—
UBFX Rd, Rn; #1stb, #width;	Rd ← Extract a bit field from Rn and zero extends it to 32 bits	—

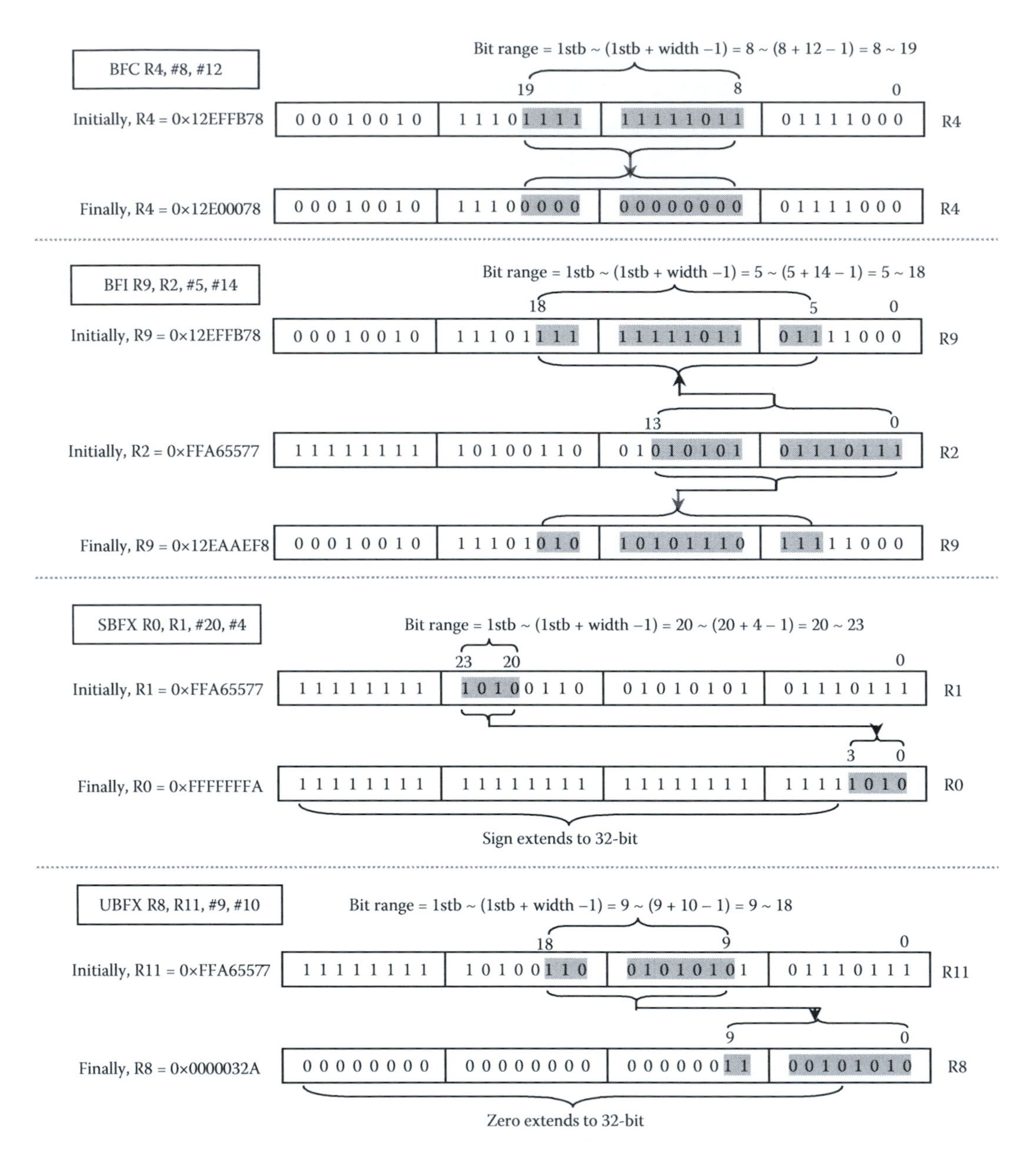

Figure 4.14 Illustrations for 4-bit field instructions.

- The **BFI** instruction is used to replace a bit field in the Rd with the content of a bit field in the Rn starting at bit [0]. The range of the bit field is from **1stb** to **1stb + width − 1**.
- The **CLZ** instruction is used to count the leading zeros before the first bit whose value is 1 in the operand register Rn and return the result to the destination register Rd.
- The **SBFX** instruction is used to extract a bit field from the register Rn, sign extends it to 32 bits, and writes the result to the destination register Rd.

- The **UBFX** instruction is to extract a bit field from the register Rn, zero extends it to 32 bits, and writes the result to the destination register Rd.

For example, the following instructions perform the related clear and insert functions:

1. **BFC R4, #8, #12;** Clear bit-8 to bit-19 (12 bits) of R4 to 0.
2. **BFI R9, R2, #5, #14;** Replace bit-5 to bit-18 (14 bits) of R9 with bit-0 to bit-13 from R2.
3. **SBFX R0, R1, #20, #4;** Extract bit-20 to bit-23 (4 bits) from R1 and sign extend to 32 bits and then write the result to R0.
4. **UBFX R8, R11, #9, #10;** Extract bit-9 to bit-18 (10 bits) from R11 and zero extend to 32 bits and then write the result to R8.

It can be found from above examples that the range of the bit field is always starting from bit **lstb** and ending at bit **lstb + width – 1**. The reason to minus 1 is because the LSB starts at 0, not 1. Figure 4.14 shows the execution process for these four instructions.

4.3.4.7 Compare and test instructions

The Cortex-M4 processor provides some compare and bit testing instructions. These instructions are used to help users to make comparisons between registers, and between register and immediate data. The comparison result is used to update the condition codes in the flag bits, but the result is not reserved. The bit testing instructions are used to perform bit-by-bit checking.

The ARM® Cortex®-M4 processor contains the following two compare instructions and two bit testing instructions:

- Compare (CMP)
- Compare Negative (CMN)
- Bitwise and Test (TST)
- Bitwise XOR Test (TEQ)

The syntaxes and functions of these compare and bit testing instructions are shown in Table 4.12. The following points should be noted when using Table 4.12:

- The **Rn** is the operand register.
- The **Operand2** is a flexible operand and it can be a register or an immediate constant.
- The **CMP** instruction subtracts the value of Operand2 from the value in the Rn. This is the same as a **SUBS** instruction, except that the result is discarded.
- The **CMN** instruction compares the negative number with the Rn by adding the value of Operand2 to the value in the Rn. This is the same as an ADDS instruction, except that the result is discarded. If this addition result is negative, the flag bit N is set to 1.
- The **TST** instruction performs a bitwise AND operation on the value of Rn and the value of Operand2. This is the same as the ANDS instruction, except that it discards the result.
- The **TEQ** instruction performs a bitwise exclusive OR operation on the value in the Rn and the value of the Operand2. This is the same as the EORS instruction, except that it discards the result. This instruction can be used to test if two values are equal without affecting the V or C flags.

Table 4.12 Syntaxes for compare and bit testing instructions

Instruction	Function	Flags
CMP Rn, Operand2;	Compare the value in the Rn with Operand2. Then update the condition flags on the result, but do not write the result to the Rn	N, Z, C, V
CMN Rn, Operand2;	Compare the value in the Rn with Operand2 by adding the value of Operand2 to the value in Rn. The result updates the flag bits but does not write to the Rn	N, Z, C, V
TST Rn, Operand2;	Perform a bitwise AND operation on the value in Rn and the value of the Operand2. The result do not write to Rn.	N, Z, C, V
TEQ Rn, Operand2;	Perform a bitwise exclusive OR operation on the value in Rn and the value of the Operand2. The result does not write to Rn	N, Z, C, V

The following example instructions perform related compare and bit test functions:

1. CMP R0, R1;
2. CMN R2, #1;
3. TST R0, 0x101;
4. TEQ R1, 0x80;

The first instruction **`CMP R0, R1`** is to compare the content of R0 with the content of R1. The related flag bit will be set if R0 > R1 (N = 0), and if R0 < R1 (N = 1) and if R0 = R1 (Z = 1).

The second instruction **`CMN R2, #1`** is to compare if the R2 equal to –1. In fact, the so-called compare with negative is to subtract the NOT value of the Operand2 from the Rn. Here, it is to subtract ~1 from the R2, or R2 – (NOT 1) = R2 – (–1) = R2 + 1.

The third instruction **`TST R0, 0x101`** is to test whether bit [0] and bit [8] in the R0 is zeros since an AND operation is performed between these two bits in the R0 and the immediate constant 0x101 = 000100000001B. If the running result of this instruction is 0, which means that both bit[0] and bit[8] in the R0 is 0. The flag bit **`Z`** is set to 1. But the content of the R0 is kept unchanged.

The fourth instruction **`TEQ R1, 0x80`** is to test whether the content of the R1 is equal to 0x80. The flag bits Z and N will be affected based on the running result. Still the content of the R1 is kept unchanged.

4.3.4.8 Program flow control instructions

The Cortex-M4 processor provides 10 program branch and control instructions. These instructions are used to control the program flows and executions with certain conditions or with no condition. So, these instructions provide intelligent control ability for the microcontrollers. These instructions can be divided into five groups based on their functions:

1. Unconditional branches
2. Conditional branches
3. Compare and branches
4. Table branches (TBB, TBH)
5. Conditional executions (If-Then or IT)

The unconditional branch instructions include:

- Branch to label (B <label>)
- Branch with link (BL <label>)
- Branch indirect (BX Rn)
- Branch indirect with link (BLX Rn)
- Branch wider (B.W <label>)

The conditional branch instructions are the same as the unconditional branch instructions, and the only difference is that a condition <cond> is attached with each related instruction for the conditional branch instructions.

In fact, all B, BL, BLX, BX, and B.W can be either unconditional or conditional branch instructions depending on whether the optional condition <cond> item is attached with each branch instruction.

The compare and branch instructions include:

- Compare and branch if zero (CBZ)
- Compare and branch if nonzero (CBNZ)

The table branch instructions include:

- Table branch byte (TBB)
- Table branch half-word (TBH)

The conditional execution has only one instruction, If-Then or IT block instruction. By using the IT block instruction, up to four subsequent instructions can be conditionally executed based on the condition provided by the IT instruction and the APSR value.

The syntaxes and functions of these branch instructions are shown in Table 4.13. The following points should be noted when using Table 4.13:

- The **Label** is a PC-relative address and it can be expressed as an address label. Different branch instructions have different ranges for this address label. Refer to Table 4.13 to get more details about the range for each branch instruction.
- The **{cond}** is one of the 14 possible condition suffixes shown in Table 4.14. All branch instructions use these suffixes to compare with the current condition values in the APSR, exactly the values on four flag bits: N, Z, C, and V. If the comparison result is true, the branch occurs. Otherwise if the comparison result is false, the branch is not occurred and the program continues running the next instruction.
- For branch instructions, the **Rn** is a register that contains the target address to be branched to. For CBZ and CBNZ instructions, the **Rn** is an operand register holding an operand. For TBB and TBH instructions, the **Rn** contains the address of the table and the **Rm** contains an index into the table.
- Both **BL** and **BLX** instructions perform a branch and a link function, which means that before branching the program to an address label (directly) or an address stored in a register Rn (indirect), write the address of the next instruction to the **LR**, or register R14, to reserve the returning point for this branch operation. These two instructions are mostly used for calling functions or subroutines since they execute the branch and save the return address (next instruction's address) to the LR at the same time, so the processor can branch back to the original program after the function or subroutine call is finished.

Table 4.13 Syntaxes for branch instructions

Instruction	Function	Flags
B Label;	Branch program to Label unconditionally. Branch range: – 16~16 MB	—
B {cond}, Label	Branch program to Label if the {cond} is True. Branch range: – 1~16 MB (inside IT block); –1 MB~1 MB (outside IT block)	—
B.W {cond}, Label	Branch program widely to Label if the {cond} is True. The W indicates that a 32-bit version of branch instruction is used for wider range	—
BL {cond} Label;	Branch and Link program to Label either conditionally (with {cond} item) or unconditionally (without {cond} item). Branch range: – 16~16 MB	—
BX {cond} Rn;	Branch (indirect) program to the address indicated in the Rn if the {cond} is True. This instruction also creates a UsageFault exception if bit[0] of the Rn is 0	—
BLX {cond} Rn;	Branch (indirect) and Link program to the address indicated in the Rn if the {cond} is True. This instruction also creates a UsageFault exception if bit[0] of the Rn is 0	—
CBZ Rn, Label	Compare the content of the Rn with 0, branch the program to Label if the Rn is 0. Rn must be lower registers (R0~R7)	—
CBNZ Rn, Label	Compare the content of the Rn with 0, branch the program to Label if the Rn is not 0. Rn must be lower registers (R0~R7)	—
TBB [Rn, Rm];	Branch program to [Rn + Rm]. The Rn contains the address of the table and the Rm contains an index into the table. The offset in the Rm is a single byte	—
TBH [Rn, Rm, LSL #n]	Branch program to [Rn + Rm]. The Rn contains the address of the table and the Rm contains an index into the table. The offset in the Rm is a half-word	—
IT <X> <Y> <Z> cond	IT block instruction allows up to four subsequent instructions to be executed or not based on the condition (cond) provided by the IT instruction	—

- Both **`BX`** and **`BLX`** are indirect branch instructions, which mean that the program is branched to an address that is stored in the register Rn, not a direct address label.
- For the **`IT`** instruction, the <**X**><**Y**><**Z**> indicates the number of possible executable subsequent instructions included in this IT block. Each of <X>, <Y>, and <Z> can either be **`T`** (true) or **`E`** (else) for the condition (**`cond`**). The number of T or E appeared in an IT instruction block determined the number of subsequent instructions involved in this IT block. The <X> indicates the execution condition for the 2nd instruction in the IT block, and the <Y> indicates the execution condition for the 3rd instruction in the IT block, and the <Z> indicates the execution condition for the 4th instruction in the IT block.
- The **`IT`** block instruction can have up to four subsequent instructions, even a branch instruction can be involved into an IT block. However, the **`B {cond}, Label`** is the only conditional instruction that can be either inside or outside an IT block. All other branch instructions must be conditional inside an IT block, and must be unconditional outside the IT block.

Table 4.14 Condition suffixes for conditional branch (execution) instructions

Suffix	Branch condition	Flags (APSR)
EQ	Equal	Z = 1 if equal
NE	Not equal	Z = 0 if not equal
CS/HS	Carry set/unsigned higher or same	C = 1 if unsigned higher or same
CC/LO	Carry clear/unsigned lower	C = 0 if unsigned lower
MI	Minus/negative	N = 1 if minus
PL	Plus/positive or zero	N = 0 if positive or zero
VS	Overflow	V = 1 if overflow
VC	No overflow	V = 0 if no overflow
HI	Unsigned higher	C = 1 & Z = 0
LS	Unsigned lower or same	C = 0 & Z = 1
GE	Signed greater than or equal	N = 0 & V = 0
LT	Signed less than	N = 1 & V = 0
GT	Signed greater than	Z = 0, N = V
LE	Signed less than or equal	Z = 1, N ! = V

Figure 4.15 shows some branch and compare branch instruction examples.

The TBB instruction is used to branch the program to a location inside a branch table and all entries in that table are arranged as a byte array. The maximum offset from the base address is $2 \times 2^8 = 512$B, which is stored in the Rm. The TBH instruction is used to branch program to a location inside the branch table and all entries in that table are arranged as a half-word (16-bit) array. The maximum offset from the base address is $2 \times 2^{16} = 128$ KB, which is stored in the Rm.

The base table address stored in the Rn could be the current PC value, in other words, the Rn could be PC, or an address stored in some other register. Since all instructions in the Cortex-M4 must be aligned to a word, the current PC value should be PC + 4 if the PC value works as a base address of the branch table.

```
1                     BRANCH.s
2   ; DEMO PROGRAM TO USE BRANCH & COMPARE BRANCH INSTRUCTIONS TO PERFORM SOME OPERATIONS
3
4   ---------- Perform Branch & Compare and Branch Operations ----------
5
6         B      LOOP          ;Branch program to a label address LOOP
7         BLE    DONE          ;Branch & Link program to DONE if the result is Signed Less Than or Equal.
                               ;The next instruction's address is written into the register Link Register (LR).
8         B.W    TARGET        ;Branch program to TARGET within 16MB range.
9         BEQ    EXIT          ;Branch program to EXIT if the last operation result is equal.
10        BL     FUNC          ;Branch with Link (Call) to a function FUNC, the return address (next
11                             ;instruction's address) is written into the Link Register (LR).
12        BX     LR            ;Return from the function call.
13        BXNE   R0            ;Branch program to an address stored in the R0 if the last operation result is
14                             ;not equal.
15        BLX    R0            ;Branch with Link program to an address stored in the R0.
16        CBZ    R2, TARGET    ;Compare R2 with 0. If R2 = 0, branch program to TARGET.
17        CBNZ   R0, EXIT      ;Compare R0 with 0, If R0 ≠ 0, branch program to EXIT.
```

Figure 4.15 Some coding examples of using branch and compare branch instructions.

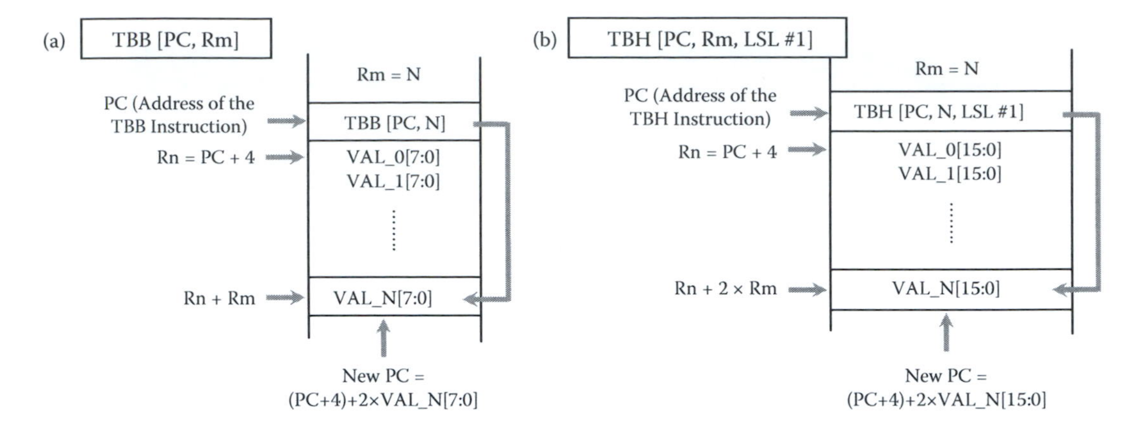

Figure 4.16 Illustrations for two table branch instructions.

Figure 4.16 provides a functional illustration for execution of two table branch instructions, **TBB [Rn, Rm];** and **TBH [Rn, Rm, LSL #1]**. These two examples suppose that the table base address is the current PC value (Rn = PC) and the branch table starts immediately after these instructions.

For the first instruction (Figure 4.16a), the Rn = PC and the Rm = N, where N is an offset or an index. The offset value is located in the memory unit [Rn + Rm], or VAL_N[7:0]. Since the TBB branch offset is twice the unsigned value of the byte returned from the table (align to word), the target table address is: PC + 4 + 2 × VAL_N[7:0]. The program is branched to that location.

Similarly, the second instruction TBH performs a similar table branch operation with half-word data arrangement (Figure 4.16b). A similar result can be obtained.

Figure 4.17 shows an example of using an IT instruction block to control two subsequent instructions to be executed conditionally.

4.3.4.9 Saturation instructions

As we experienced in most analog amplifier circuits, sometimes the gain of an amplifier is so large, such as an ideal operational amplifier (OP), to make the output of the amplifier either too small in negative or too large in positive. This is called saturations. This means

```
1                          IF_THEN _BLOCK.s
2  ; DEMO PROGRAM TO USE IF-THEN INSTRUCTIONS TO PERFORM SOME OPERATIONS
3
4  ————————————   Perform IT Block Instruction Operations   ————————————
5
6          CMP     R0, #0        ;Compare R0 with 0.
7          ITTE    EQ            ;ITTE means: The 1st T means that if cond is True (Equal to 0), the 1st
8                                ;instruction executes. The 2nd T means that if cond is True (Equal to 0),
9                                ;the 2nd instruction executes. The 3rd E means that if cond is Else (Not
10                               ;Equal to 0), the 3rd instruction executes.
11         MOVEQ   R3, #5        ;The 1st instruction is executed (R3 ← 0x5) if R0 = 0.
12         MOVEQ   R5, #8        ;The 2nd instruction is executed (R5 ← 0x8) if R0 = 0.
13         MOVNE   R5, #2        ;The 3rd instruction is executed (R5 ← 0x2) if R0 ≠ 0.
```

Figure 4.17 Example of using IF-THEN branch instructions.

Table 4.15 Syntaxes for saturation instructions

Instruction	Function	Flags
SSAT Rd, #n, Rn, {shift #s};	Rd ← Apply the specified shift, then saturates data to the signed range $-2^{n-1} \le x \le 2^{n-1} - 1$.	—
USAT Rd, #n, Rn, {shift #s};	Rd ← Apply the specified shift, then saturates data to the unsigned range $0 \le x \le 2^{n}-1$	—

that the positive peak value of the output is beyond the positive voltage V^+ and the negative peak value of the output is below the negative voltage V^-.

Similarly, in Cortex-M4 and other microcontrollers, the situations have also existed and happened in some situations. They are similar to overflows for arithmetic operations. To reduce any possible saturation, Cortex-M4 provides some instructions, such as SSAT to handle the signed saturations, USAT to handle unsigned saturations, QADD to handle saturating addition, QSUB to handle saturating subtraction, and so on. In order to save space, we only introduce the SSAT and USAT instructions since they are popular in the signal processing.

The syntaxes and functions of these saturation instructions are shown in Table 4.15. The following points should be noted when using Table 4.15:

- The **Rd** is the destination register
- The **n** specifies the bit position to saturate to:
  - For SSAT, n = 1~32
  - For USAT, n = 0~31
- The **Rn** is the register containing the value to be saturated
- The **shift #s** is an optional shift applied to **Rn** before saturating. It must be one of the following:
  - **ASR #s** (where *s* is in the range 1~31)
  - **LSL #s** (where *s* is in the range 0~31)

Generally, the SSAT and USAT perform the following operations to reduce any possible saturation to be occurred:

SSAT (for signed n-bit saturation):

1. If the value to be saturated is less than -2^{n-1}, the result returned is -2^{n-1}.
2. If the value to be saturated is greater than $2^{n-1} - 1$, the result returned is $2^{n-1} - 1$.
3. Otherwise, the result returned is the same as the value to be saturated.

USAT (for unsigned n-bit saturation):

1. If the value to be saturated is less than 0, the result returned is 0.
2. If the value to be saturated is greater than 2^{n-1}, the result returned is 2^{n-1}.
3. Otherwise, the result returned is the same as the value to be saturated.

If the returned result is different from the original value to be saturated, it is called saturation. If saturation occurs, the **Q** flag bit is set to 1 in the APSR by the instruction. Otherwise, it leaves the Q flag unchanged. To clear the Q flag to 0, one must use the **MSR** instruction. To access the Q flag bit, you need to use the **MRS** instruction to read the state of the Q flag.

Two examples of using SSAT and USAT instructions to process the saturations are:

```
SSAT R1, #16, R7, LSL #2    ;Logical shift left value in R7 by 2, then saturate it as a
                            ;signed 16-bit value and write it back to R1.
USATNE R0, #8, R5           ;Conditionally saturate value in R5 as an unsigned 8 bit
                            ;value and write it to R0.
```

When processing saturation with SSAT and USAT, each bit on the original register should be checked and adjusted if any of them is beyond the required ranges.

4.3.4.10 Exception-related instructions

The Cortex-M4 processor provides quite a few of instructions related to system exceptions, such as change processor state (CPS), send event (SEV), supervisor call (SVC), wait for event (WFE), and wait for interrupt (WFI). Among them, two instructions, CPS and SVC, are more important and popular in most implementations. We will introduce these instructions with more details in this section.

The **CPS** instruction is used to change the processor state, exactly to change the state or value of some interrupt masking registers, such as **PRIMASK** and **FAULTMASK**. These registers can also be accessed by using the MSR and MRS instructions.

The syntaxes and functions of the CPS instruction are shown in Table 4.16. The following points should be noted when using Table 4.16:

- The CPS can only be used for privileged software, and it has no effect if used in unprivileged software.
- The CPS cannot be used in any conditional block, including the IT block.

The SVC instruction is used to create an SVC exception. Mostly, this instruction is used for an RTOS or embedded OS environment. Specially, when an application is running in the unprivileged level, it can send a request to ask some services that are running in the privileged state from the OS.

The SVC also enables applications to access various system services without knowing the actual program memory address of the service. The only information an application needs to know is the SVC service number, the input parameters, and the returned results.

The syntax of the SVC instruction is simple, it looks like:

```
SVC #immed;
```

Table 4.16 Syntaxes and functions for CPS instruction

Instruction	Function	Flags
CPSIE I;	Clear the PRIMASK register to enable interrupts (same as _enable_irq(); function)	—
CPSID I;	Set the PRIMASK register to disable interrupts (same as _disable_irq(); function). The NMI and HardFault are not affected	—
CPSIE F;	Clear the FAULTMASK register to enable fault interrupt (same as _enable_fault_irq();)	—
CPSID F;	Set the FAULTMASK register to disable fault interrupt (same as _disable_fault_irq(); function). The NMI is not affected	—

The **immed** is an 8-bit-value parameter to be evaluated to an integer with the range 0~255. This parameter is regularly ignored by the processor. If required, it can be retrieved by the exception handler to determine what service is being requested.

No flag bit would be affected by running this instruction.

An example of using this SVC instruction is:

```
SVC 0x32;    SVC handler can extract the immediate value by locating it
             via the stacked PC.
```

Next, let's take a look at the sleep mode instructions.

4.3.4.11 Sleep mode instructions

The ARM® Cortex®-M4 processor provides two major sleep mode instructions to allow the processor to enter the sleep mode: **wait for event** (**WFE**) and **wait for interrupt** (**WFI**). The syntaxes for these two instructions are very simple as shown below:

```
WFE;   Wait for Events
WFI;   Wait for Interrupts
```

Depending on the value in the single-bit event register, the processor can enter the sleep mode or not when a WFE instruction is executed:

- If the single-bit event register is reset to 0, the processor enters the sleep mode and will be woken up by the next event
- If the single-bit event register is set to 1, the processor will not enter the sleep mode and continue executing the next instruction.

Unlike the WFE instruction, the WFI instruction causes the processor to enter the sleep mode immediately. The processor will be woken up by the next interrupt or reset.

The CMSIS-Compliant device driver provides similar C functions for these two instructions: **_WFE();** and **_WFI();** By using these functions, users can perform the same functionalities in C codes as the assembly codes did.

4.3.4.12 Memory barrier instructions

The ARM® architecture provides some parallel data processing abilities, which means that the memory accessing can occur at the same time as other nonmemory-accessing instructions are executed. In order to coordinate these parallel data processes and protect memory transfer, three memory barrier instructions are provided in the Cortex-M4 system:

- Data memory barrier (DMB)
- Data synchronization barrier (DSB)
- Instruction synchronization barrier (ISB)

The function of the DMB is to ensure that all memory accesses that appear before the DMB instruction are completed before any new memory access that appears after the DMB instruction. The DMB instructions do not affect the ordering or execution of instructions that do not access memory.

The function of the DSB is to ensure that all memory accesses are completed before the next instruction is executed. In other words, all instructions that come after DSB (in program order) do not execute until the DSB instruction completes.

Table 4.17 Syntaxes and functions for miscellaneous instruction

Instruction	Function	Flags
BKPT #immed;	Set a breakpoint for a program to enable program to enter debug mode. The immed is an expression evaluating to an integer in the range 0–255. Generally, the immed indicates the address line in which the instruction is stopped and debugged	—
NOP;	No operation and just do nothing. The time delay caused by the NOP instruction is not guaranteed and the processor may remove it from the pipeline before it can be executed	—
SEV;	Send an event to all processors, and set the single-bit event register to 1	—

The function of the ISB is to ensure that all previous instructions are completed before executing any new instruction. In other words, all instructions following the ISB instruction are fetched from the cache or memory, but cannot be executed until the ISB is completed.

The CMSIS-Compliant device driver provides similar C functions for these three instructions:

1. `void_DMB(void);`
2. `void_DSB(void);`
3. `void_ISB(void);`

By using these functions, users can perform the same functionalities in C codes as the assembly codes did.

4.3.4.13 Miscellaneous instructions

In addition to the instruction set we discussed above, the Cortex-M4 processor also provides some other useful instructions, such as **`Breakpoint`** (**`BKPT`**), **`No Operation`** (**`NOP`**), and **`Send Event`** (**`SEV`**). These instructions cannot be clearly categorized into certain related instruction groups. Therefore, they can be considered as miscellaneous instructions.

The syntaxes for these miscellaneous instructions are shown in Table 4.17.

The CMSIS-Compliant device driver provides similar C functions for these three instructions:

1. `_BKPT(immed);`
2. `_NOP();`
3. `_SEV();`

By using these functions, users can perform the same functionalities in C codes as the assembly codes did.

4.4 ARM® Cortex®-M4 software development procedures

In Chapter 3, we provided detailed introduction about the software development tools and kits for the ARM® Cortex®-M4 MCU. The Keil® ARM®-MDK provides a full set of development kits and tools, including the source code editor, assembler, compiler, linker, loader, debugger, and runner, to enable users to design and build their applications easily and quickly in an IDE with various GUIs.

To successfully develop the user applications, it is not enough by using only those IDE and GUIs. Some peripheral device drivers included in certain related driver libraries are needed to enable users to combine their application object codes with those drivers located in the peripheral libraries provided by the vendors to build a complete executable file. When we use the MSP432™ Series LaunchPad™ Evaluation Board, MSP-EXP432P401R, these libraries are located in the internal ROM space in the on-chip memory map discussed in Section 2.3.2 in Chapter 2. These libraries and system files include:

- MSPWare driver libraries, including the peripheral devices library, USB library and graphical library
- BSL
- AES cryptography tables
- Cyclic Redundancy Check (CRC) error-detection functionality

In fact, each library is a collection of interfacing functions written by C, and different libraries provide different control and interfacing functions. One of the most popular libraries is the driver library, which provides all control and interfacing functions for peripheral devices in the MSP-EXP432P401R EVB.

As we discussed in Section 3.5 in Chapter 3, to successfully design and build user application programs, the following software and tools are needed:

1. **`The Keil ARM-MDK Core`** (including µVersion5 IDE, compiler, and debugger)
2. **`CMSIS:CORE`** (including the API for the Cortex-M processor core and peripherals, and a consistent system startup code—Device:Startup)
3. MSPWare for MSP432 Series Software, which includes:
   a. MSPW are driver libraries, including the peripheral devices library, USB library, and graphical library
   b. BSL
   c. AES cryptography tables
   d. CRC error-detection functionality

For all of these software and tools, refer to Section 3.3 in Chapter 3 to download and install these development tools and software.

To successfully develop and build user application programs, the following hardware and components are needed, since we are using MSP432™ Series LaunchPad™ MSP-EXP432P401R EVB and EduBASE Trainer:

1. A MSP432™ Series LaunchPad™ Evaluation Board—MSP-EXP432P401R
2. A USB with Micro-A/-B Cable
3. An EduBASE ARM Trainer
4. A host computer

Important points to be noted for the MSP432™ Series software package MSPWare are

1. MSP432™ Series Software Package MSPWare contains all driver libraries and will be downloaded and installed by the users to their host computer.
2. However, the same libraries are also installed in an internal ROM space in the on-chip memory in the MSP432P401R MCU.

The second library is located at the internal ROM space in the on-chip memory unit in the MSP432P401R MCU, which is included in the MSP-EXP432P401R Evaluation Board.

All interfacing functions in this library can also be used or called by the users as read-only functions from the user's application program. A prefix **ROM_** must be added before each function to distinguish them from those same functions included in the libraries installed in the users' host computer.

In order to make our application development process easy and simple, we will use C language to replace assembly codes to build our application program in this book. This will significantly reduce the learning curves for assembly language and greatly simplify our project building and developing process.

4.5 Using C language to develop MSP432P401R microcontroller applications

Today, most software program developers prefer to use high-level language, such as C or C++, not low-level language, such as ARM® Assembly, to build and develop their applications. The reason for that is obvious since the high-level language is easy to learn and understand, and the most important point is that the programmers are not required to have detailed knowledge about the hardware and components they are using in their development system. Of course, it would be much better if the developers have some solid and deep understanding and knowledge about the hardware components they used, and in this way they can develop and build more professional and highly efficient applications to be implemented in most applications.

Possible barriers or problems when using high-level language, such as C or C++, to build and develop users' applications are

1. How to use C or C++ codes to directly access hardware components, such as general and special registers in the register bank in the Cortex-M4 Core?
2. How to use high-level language to access all special registers?
3. How to use C/C++ codes to replace and simulate the assembly instructions to perform the same functions?

The answers to these questions are

1. Use the intrinsic functions provided by C/C++ compiler related to the vendor-dependent IDE, such as Keil® ARM® C/C++ Compiler-specific intrinsic functions.
2. Use special software packages, such as Keil® CMSIS Core-specific intrinsic functions.
3. Use inline assembler, such as embedded assembler in Keil® ARM® toolchains, to insert the required assembly instructions in the C/C++ codes.
4. Use compiler-specific features such as keywords or idiom recognitions.

For most applications, the top two answers are most widely adopted: The Keil® ARM® C/C++ Compiler-specific intrinsic functions and the Keil® CMSIS Core intrinsic functions.

Before we can go deep on these two sources, first let's have a clear picture about the standard data-type definitions about data items used in these two source systems.

4.5.1 Standard data types used in intrinsic functions

Different standard data types are defined for the data used for those intrinsic functions. Table 4.18 shows an example of some standard data-type definitions for intrinsic functions

Table 4.18 Standard data types used in Keil® ARM® C/C++ compiler

Data type	Description
int8_t	Signed char (8 bits)
int16_t	Signed short integer (16 bits)
int32_t	Signed integer (32 bits)
uint8_t	Unsigned char (8 bits)
uint16_t	Unsigned short integer (16 bits)
uint32_t	Unsigned integer (32 bits)
INT8_MIN	–128
INT8_MAX	127

Table 4.19 Standard data types used in Keil® CMSIS core

Data type	Description
int8_t	Signed 8-bit integer
int16_t	Signed 16-bit integer
int32_t	Signed 32-bit integer
uint8_t	Unsigned 8-bit integer
uint16_t	Unsigned 16-bit integer
uint32_t	Unsigned 32-bit integer

provided by the Keil® ARM® C/C++ Compiler (**stdint.h**), and Table 4.19 shows another example of standard data-type definitions for intrinsic functions provided by the CMSIS Core (**stdint.h**).

It can be found from these two tables that there are tiny differences between these two kinds of standard data-type definitions. However, in some applications, there is no difference between a signed char and signed integer.

Like object-oriented programming (OOP) style, in order to organize data types better and make them to an integrated format or a collection format, some data structures are used in most intrinsic functions. A data structure, or **struct**, is exactly a data collection with different data items and related data types integrated into this collection to make data well organized.

For example, to use a union type **APSR_TYPE** to access the **APSR**, a data structure shown in Figure 4.18 is used.

To use this **APSR_TYPE** structure to access each item defined in this structure, one needs to create a variable **apsr** based on this structure, such as:

```
struct  APSR_TYPE  apsr;
apsr.Q = 0;
apsr.Z = 0;
apsr.C = 1;
```

Another example is the **CONTROL_TYPE** data structure shown in Figure 4.19.

```
typedef union
{
    struct
    {
            uint32_t  _reserved0:26;          /*bit: 0 ~ 26 Reserved */
            uint32_t  Q:1;                    /* bit: 27 Saturation condition flag */
            uint32_t  V:1;                    /* bit: 28 Overflow condition code flag */
            uint32_t  C:1;                    /* bit: 29 Carry condition code flag */
            uint32_t  Z:1;                    /* bit: 30 Zero condition code flag */
            uint32_t  N:1;                    /* bit: 31 Negative condition code flag */
    } b;                                      /*Structure used for bit access. */
     uint32_t w;                              /* Type used for word access. */
} APSR_TYPE;
```

Figure 4.18 APSR_TYPE data structure.

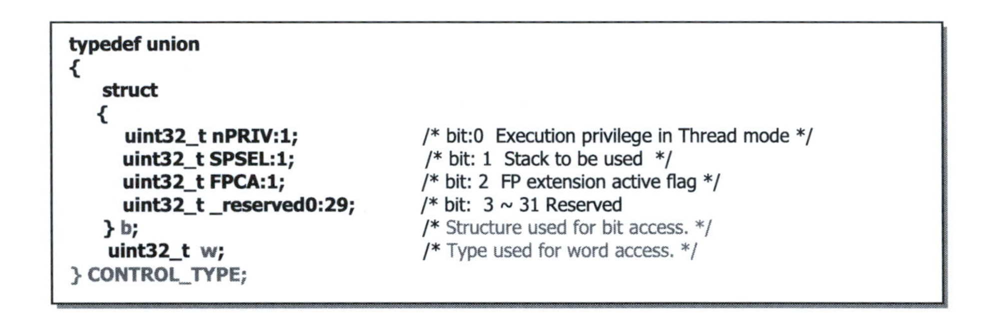

```
typedef union
{
    struct
    {
        uint32_t nPRIV:1;              /* bit:0  Execution privilege in Thread mode */
        uint32_t SPSEL:1;              /* bit: 1  Stack to be used  */
        uint32_t FPCA:1;               /* bit: 2  FP extension active flag */
        uint32_t _reserved0:29;        /* bit:  3 ~ 31 Reserved
    } b;                               /* Structure used for bit access. */
     uint32_t  w;                      /* Type used for word access. */
} CONTROL_TYPE;
```

Figure 4.19 CONTROL_TYPE data structure.

4.5.2 *CMSIS core-specific intrinsic functions*

As we discussed in Chapter 3, the CMSIS-Core provides a basic run-time system for the Cortex-M4 device and enables the user to access the Cortex-M4 processor and the related device peripherals. The system includes:

1. **`Hardware Abstraction Layer`** (**`HAL`**) for Cortex-M4 processor registers with standardized definitions for the SysTick, NVIC, SCB registers, MPU registers, FPU registers, and processor access functions.
2. **`System Exception Names`** to interface to system exceptions without having compatibility issues.
3. **`Methods to Organize Header Files`**. These methods make it easy to learn new Cortex-M4 microcontroller products and improve software portability. This includes naming conventions for device-specific interrupts.
4. **`Methods for System Initialization`** to be used by each MCU vendor. For example, the standardized **`SystemInit()`** function is essential for configuring the CS of the device.
5. **`Intrinsic Functions`** used to generate CPU and memory access instructions that are not supported by some standard C functions. These intrinsic functions also contain a group of Core peripheral access functions.
6. A **`variable`** is used to determine the system clock frequency which simplifies the setup the SysTick timer.

All of these features are included in related intrinsic functions and these intrinsic functions and core peripheral access functions are defined in the following C/C++ header files:

- **`core_cm4.h`**—CMSIS Cortex-M4 Core peripheral access layer header file
- **`core_cmInstr.h`**—CMSIS Cortex-M Core instruction access header file
- **`core_cmFunc.h`**—CMSIS Cortex-M Core function access header file
- **`core_cmSimd.h`**—CMSIS Cortex-M single instruction multidata (SIMD) header file.

Table 4.20 shows these header files and related intrinsic functions involved in these header files. These intrinsic functions can be further divided into the following eight groups based on their functions:

1. **`Peripheral access`**: Define all I/O naming conventions, requirements, and optional features to access peripherals. These definitions include the I/O data type, peripheral device name, constants, or arguments.
2. **`System and clock configuration`**: Two intrinsic functions and one variable are included in this group. SystemInit(), SystemCoreClockUpdate(), and SystemCoreClock. Two functions are used to initialize a microcontroller system and update a device-specific system clock. The **`SystemCoreClock`** is a global variable used to store the system clock frequency.
3. **`Interrupts and exceptions`**: Define the conventions and features of all interrupts and exceptions used in the MCU system. These include the interrupt numbers, priority levels, and IRQ types. These intrinsic functions allow users to access related interrupt registers to enable or disable any interrupt, set or clear pending for selected interrupt source, and even perform a system reset function via NVIC mechanism.
4. **`Core register access`**: Provides all intrinsic functions to simulate related assembly instructions to access special registers used in the Cortex-M4 processor. Most

Table 4.20 Header files and related intrinsic functions

Header file	Description
core_cm4.h	This is the HAL definition file for Cortex-M4 processor registers with standardized definitions for the SysTick, NVIC, SCB registers, MPU registers, FPU registers, and processor access functions
core_cmInstr.h	This is the definition file for all intrinsic functions used to simulate most ARM assembly instructions, including data moving, memory access, arithmetic, logic, shift and rotate, exception related and miscellaneous instructions
core_cmFunc.h	This is the definition file for all intrinsic functions used to access all special registers in the Cortex-M4 Core, such as CONTROL, APSR, IPSR, PSP, MSP, PRIMASK, FAULTMASK. These intrinsic functions allow users to get and set related registers, enable and disable interrupts
core_cm4_simd.h	This is the definition file for intrinsic functions used to simulate most ARM assembly arithmetic instructions, including SADD8, SADD16, SSUB8, USUB8, SSAX, and SEL. These intrinsic functions allow users to perform similar arithmetic operations as those assembly instructions did

popular registers include CONTROL, PRIMASK, APSR, IPSR, PSP, MSP, BASEPRI, and FAULTMASK. Four functions can be used to enable or disable general IRQs and FAULT_IRQs interrupts.

5. **`Intrinsic functions for CPU instructions`**: These intrinsic functions generate specific Cortex-M4 instructions that cannot be directly accessed by using the ARM® C/C++ Compiler. The related functions can be used to perform the same function as the assembly instructions did, such as NOP, WFI, WFE, SEV, BKPT, ISB, DSB, REV, and CLZ.
6. **`Intrinsic functions for SIMD instructions`**: This is the definition file for intrinsic functions used to simulate most ARM® assembly arithmetic instructions, including SADD8, SADD16, SSUB8, USUB8, SSAX, and SEL. These intrinsic functions allow users to perform similar arithmetic operations as those assembly instructions did.
7. **`SysTick timer`**: One intrinsic function—**`SysTick_Config(uint32_t tick)`** is defined in this group and it is used to initialize and start the SysTick timer.
8. **`Debug access`**: Three intrinsic functions and one external variable are defined in this group to enable users to enter the debug mode to perform debugging function.

Table 4.21 shows some popular intrinsic functions included in the CMSIS Core. Refer to Appendix D to get a complete list of CMSIS Core-specific intrinsic functions.

Table 4.21 Some popular intrinsic functions provided by CMSIS core

Instructions	CMSIS core intrinsic function	Functions
NOP	void __NOP (void);	No operation
SEV	void __SEV(void);	Send event
WFI	void __WFI(void);	Wait for interrupt (enter sleep mode)
WFE	void __WFE(void);	Wait for event
BKPT	void __BKPT(uint8_t value);	Set a software breakpoint
LDREXB	uint8_t __LDREXB (volatile uint8_t addr*);	Exclusive load byte
LDREX	uint32_t __LDREXW (volatile uint32_t addr*);	Exclusive load word
STREXB	uint32_t __STREXB (uint8_t value, volatile uint8_t addr*);	Exclusive store byte
CLZ	uint8_t __CLZ (unsigned int val);	Count leading zeros
RBIT	uint32_t __RBIT (uint32_t val);	Reverse bits order in word
ROR	unit32_t __ROR (uint32_t value, uint32_t shift);	Rotate shift right by n bits
SADD8	uint32_t __SADD8 (uint32_t val1, uint32_t val2);	Perform four 8-bit signed addition
SSUB8	uint32_t __SSUB8 (uint32_t val1, uint32_t val2);	Perform four 8-bit signed subtraction
MRS	uint32_t __get_CONTROL (void);	Read the CONTROL register
MSR	uint32_t __set_CONTROL (uint32_t control);	Set the CONTROL register
MRS	uint32_t __get_APSR (void);	Read the APSR register
MSR	uint32_t __set_PRIMASK (uint32_t priMask);	Set the PRIMASK register
CPSIE I	void __enable_irq (void);	Globally enable the IRQ interrupts
CPSID I	void __disable_irq (void);	Globally disable the IRQ interrupts

4.5.3 Keil® ARM® compiler-specific intrinsic functions

The intrinsic functions provided by the Keil® C/C++ Compiler are very similar to those defined in the CMSIS Core discussed above. Although the functions are very similar, some parameters and arguments may be differently defined in two systems. These intrinsic functions are included with the compiler itself, and therefore they are platform or vendor dependent. This means that different microcontrollers developed and built by the various vendors may have different compilers with different intrinsic functions. In other words, these intrinsic functions cannot be portable across platforms or tools.

Table 4.22 shows some example intrinsic functions provided by the Keil® C/C++ Compiler (**c55x.h** and **dspfns.h**).

Compare Table 4.22 with the CMSIS Core intrinsic functions shown in Table 4.21, it can be found that there are some small differences between the intrinsic functions provided by the CMSIS Core and the Keil® C/C++ Compiler, especially for the data types and arguments used for these two systems. In some microcontroller development kits or toolchains, the CMSIS Core intrinsic functions are fully supported by the compilers. However, in Keil® ARM-MDK μVersion5, these intrinsic functions are different and cannot be supported for each other.

The Keil® ARM®-MDK C/C++ Compiler, **armcc**, also provides a method called **named register variables** to enable users to access special registers in the Cortex-M4 processor. To use this method to access any special register, you first need to declare the special register as a variable.

The named register variables are declared by combining the **register** keyword with the **__asm** keyword. The **__asm** keyword takes one parameter, a character string, which is the name of the special register. The syntax is:

```
register int regname _asm ("regname");
```

where the **regname** is the name of the special register to be accessed in the Cortex-M4 processor.

For example, to declare the general register R0 as a named register variable, use:

```
register int R0 _asm ("r0");
```

A typical use of named register variables is to access bits in the **APSR**.

Recall that we introduced a data structure, **struct APSR_TYPE**, in Section 4.5.1 and used that structure as the data type for the APSR register. Figure 4.20 shows an example of using that structure and the named register variables to set the saturation flag Q in the APSR.

Table 4.23 lists most often used special registers in the Cortex-M4 processor and their related reference names. You can use these names to generate associate named register variables to access each of these registers.

4.5.4 Inline assembler

Another way to access the Cortex-M4 processor registers from the C/C++ codes is to use the inline assembler method. This means that you can insert some assembly instructions in your C/C++ programs to execute them when your program runs. The Keil® ARM® C/C++ Compiler supports the inline assembler feature with Thumb-2 technology.

Table 4.22 Some example intrinsic functions provided by Keil® C/C++ compiler

Instructions	C/C++ compiler intrinsic function	Functions
NOP	void __nop(void);	No operation
SEV	void __sev(void);	Send event
WFI	void __wfi(void);	Wait for interrupt (enter sleep mode)
WFE	void __wfe(void);	Wait for event
BKPT	void __breakpoint(int value);	Set a software breakpoint
LDREXB	unsigned int __ldrex (volatile void *ptr);	Exclusive load byte
LDREXH	unsigned int __ldrex (volatile void *ptr);	Exclusive load half-word (16 bits)
LDREX	unsigned int __ldrex (volatile void *ptr);	Exclusive load word
STREXB	int __strex (unsigned int value, volatile void *ptr);	Exclusive store byte
STREXH	int __strex (unsigned int value, volatile void *ptr);	Exclusive store half-word (16 bits)
STREX	int __strex (unsigned int value, volatile void *ptr);	Exclusive store word
CLZ	unsigned char __clz (unsigned int val);	Count leading zeros
RBIT	unsigned int __rbit (unsigned int val);	Reverse bits order in word
ROR	unsigned int __ror (unsigned int val, unsigned int shift);	Rotate shift right by n bits
SADD16	_ARM_INTRINSIC int16_t _sadd(int16_t src1, int16_t src2);	Perform four 16-bit signed addition
SSUB8	_ARM_INTRINSIC int16_t _ssub(int16_t src1, int16_t src2);	Perform four 16-bit signed subtraction
ADD	__ARM_INTRINSIC int16_t add(int16_t x, int16_t y);	Add two signed 16-bit data
SUB	__ARM_INTRINSIC int16_t sub(int16_t x, int16_t y);	Subtract two signed 16-bit data
ADC	__ARM_INTRINSIC int32_t L_add_c(int32_t x, int32_t y);	Add two 32-bit signed data with carry
SBC	__ARM_INTRINSIC int32_t L_sub_c(int32_t x, int32_t y);	Subtract two 32-bit signed data with carry
SDIV	__ARM_INTRINSIC int16_t div_s(int16_t x, int16_t y);	Divide two 16-bit signed data
MUL	__ARM_INTRINSIC int32_t L_mult(int16_t x, int16_t y);	Multiply two 16-bit data to get 32-bit result
ASR (LSL)	__ARM_INTRINSIC int32_t L_shr(int32_t x, int16_t shift);	Shift 32-bit data to right (to left if shift < 0)
CLREX	void __clrex(void);	Clear exclusive
CPSID I	void __disable_irq(void);	Disable IRQ interrupts & fault handlers
CPSIE I	void __enable_irq(void);	Enable IRQ interrupts & fault handlers
CPSID F	void __disable_fiq(void);	Disable fault interrupts
CPSIE F	void __enable_fiq(void);	Enable fault interrupts
USAT	int __usat(unsigned int val, unsigned int sat);	Unsigned saturation
SSAT	int __ssat(int val, unsigned int sal);	Signed saturation

```
typedef union
{
    struct
   {
         uint32_t  _reserved0:27;          /*bit: 0..26 Reserved */
         uint32_t  Q:1;                    /* bit: 27 Saturation condition flag */
         uint32_t  V:1;                    /* bit: 28 Overflow condition code flag */
         uint32_t  C:1;                    /* bit: 29 Carry condition code flag */
         uint32_t  Z:1;                    /* bit: 30 Zero condition code flag */
         uint32_t  N:1;                    /* bit: 31 Negative condition code flag */
   } b;                                    /*structure used for bit access. */
    uint32_t w;                            /* type used for word access. */
} APSR_TYPE;
register APSR_TYPE apsr _asm("apsr");      /* declare apsr as a named register variable */
void set_Q(void)                           /* the set_Q() method */
{
    apsr.b.Q = 1;                          /* access bit Q to set it to 1   */
}
```

Figure 4.20 Example of using named register variables to access a bit in the APSR.

Table 4.23 Most often used special registers in the Cortex-M4 processor

Register	Named register string name in __asm()
APSR	"apsr"
BASEPRI	"basepri"
BASEPRI_MAX	"basepri_max"
CONTROL	"control"
EAPSR (EPSR + APSR)	"eapsr"
EPSR	"epsr"
FAULTMASK	"faultmask"
IAPSR (IPSR + APSR)	"iapsr"
IEPSR (IPSR + EPSR)	"iepsr"
IPSR	"ipsr"
MSP	"msp"
PRIMASK	"primask"
PSP	"psp"
PSR	"psr"
R0 ~ R12	"r0" ~ "r12"
R13	"r13" or "sp"
R14	"r14" or "lr"
R15	"r15" or "pc"
XPSR	"xpsr"

However, some instructions are not supported and cannot be executed in the inline assembler way:

- The instructions **TBB**, **TBH**, **CBZ**, and **CBNZ**
- The instruction **SETEND**

```
/* call inline assembler to perform addition with s1 and s2, returned result is in res */
int ADD32(int s1, int s2)              /* C code function header */
{
   int res;                            /* C code returned variable */
   __asm                               /* call inline assembler */
  {
     ADD res, s1, s2                   /* { assembly instruction } */
  }
   return res;                         /* C code */
}

/* call inline assembler to enable the CPU to enter the sleep mode */
void CPUSleep(void)                    /* C code function header */
{
   __asm (" WFI\n");                   /* call inline assembler */
   return;                             /* C code */
}
```

Figure 4.21 Some examples of using inline assembler from C codes.

Although for these limitations, the inline assembler is still a powerful tool to facilitate programmers to build their mix-code applications easily and quickly. The inline assembler provides great flexibility for users to access the low-level components from their high-level code program. In addition to the inline assembler, the Keil® ARM® C/C++ Compiler also supports a feature called embedded assembler and that feature enables user to create some assembly functions in a C/C++ program.

Figure 4.21 shows some examples of using the inline assembler to call assembly instructions to perform special functions from the C code program.

Similar to an inline assembler, the opposite way works. This means that you can call a C function from assembly codes. You can use **IMPORT** instruction to select and import your C function into the assembly codes and use **BL** to call this C function.

Figure 4.22 shows an example of using these instructions to call a C function from assembly codes to perform some addition function, and return result to the assembly codes.

```
/* call C function from assembly codes to perform addition with s1 and s2, returned result is in res */
/* definition of the C function ADD_C
int ADD_C(int s1, int s2)              /* C code function header */
{
   int res;                            /* C code returned variable */
   res = s1 + s2;                      /* perform addition in C code */
   return res;                         /* C code */
}
; call C function ADD_C to perform addition and get result from R0 = res
MOVS    R0, #0x3                       ; assembly codes, R0 = 0x3
MOVS    R1, #0x2                       ; assembly codes, R1 = 0x2
IMPORT  ADD_C                          ; import C function ADD_C()
BL      ADD_C                          ; call C function ADD_C to do the addition
```

Figure 4.22 Example of calling a C function from the assembly codes.

4.5.5 Naming convention and definition in C programming developments

Now, we have a clear picture about the ARM® Cortex®-M4 Microcontroller assembly instruction set and related intrinsic functions provided by the Keil® ARM® C/C++ Compiler and CMSIS Core. It looks like we are ready to use C codes to develop and build our applications to access Cortex-M4 MCU and peripherals to fulfill our desired tasks. However, before we can do that, we need to understand some special features about C program development procedure, as well as special definitions used by the C codes for the Cortex-M4 processor and actual registers, and peripherals.

In C programming, some definitions are necessary and they are closely related to the peripherals or I/O devices used in the related MCU. In Section 2.3.4 in Chapter 2, we have discussed most GPIO ports used in this MCU and their addresses.

In order to use those ports, you need to access related registers located at those ports. To facilitate users to use these peripherals to develop their software in C codes, the following rules have been built to help developers to build their program easily and quickly:

1. Each GPIO port contains a group of related registers with unique addresses. To access each register in C code, a symbolic definition is applied to each register based on its address. Refer to Section 2.3.4 to get more details about these GPIO ports and related addresses. Most often used registers for each GPIO port are: Port Input Register (**PxIN**), Port Output Register (**PxOUT**), Port Direction Register (**PxDIR**), Port Function Select Registers (**PxSEL0**, **PxSEL1**), and Port Interrupt Enable Registers (**PxIE**). In your C program, you need to use these symbolic definitions as the related registers' names to access them.
2. To use these registers, you should first declare them by using the **#define** statement. In fact, you need to define each register in your C codes based on its address. Since the address in C code can be mapped a pointer, you can define each register by using a pointer pointed to this register. One easy way to do this definition is to use **#define** statement to define each register one by one. A better way is to define a structure for a port that contains all registers. An example of this definition for Port 1 Input register is:

```
#define  P1IN   (* ((volatile uint8_t *) 0x40004C00))
```

   It looks like that this is a double pointer definition. The (**volatile uint8_t ***) indicates that this is a pointer or address with the **volatile uint8_t** data type, and the keyword **volatile** means that the data written into this address can be modified, which is opposite to the keyword **const**. The whole address is defined as **(*((volatile uint8_t *) 0x40004C00))**, and the first * points to the starting bit on this address (totally 32 bits on this register). Only the lowest 8 bits in the Port 1 Input register is used even if it has 32 bits. The first * is pointed to the entire 8 bits of P1IN register, exactly pointing to the starting bit. This register can be considered as a 2D array, such as **P1IN** = **{{bit31}, {bit30}**, ... **{bit1}, {bit0}};** since each single bit (pin) of this port may be accessed during the programming stage.
3. Similarly, you need to use the **#define** statement to define and declare all port registers, interrupt numbers, and bit fields used for GPIO-related components. Two examples of these kind of definitions are:

```
#define   BIT0        0x0001
#define   INT_PORT1   51
```

The first code line is to define specific bit (bit-0) on the P1IN register. The second line is to define the interrupt number for the Port 1.

In fact, you do not need to do these definitions yourself in your program since all of these definitions have been made by the vender in the system header file **msp432p401r.h**. Later on during your program development process, you can directly use these symbolic definitions to access all registers on related ports. You can use these registers to do inputs, outputs, and other functions by using logic AND or logic OR operations.

4.5.6 MSPWare peripheral driver library

The MSP432™ Series LaunchPad™ software package provides a set of complete software development tools and libraries to support the users to build their applications.

4.5.6.1 Programming models

In MSP432™ Series software driver library MSPWare, two programming models are provided to support for the user's program development: the **DRA** model and the **SD** model. Each model can be used independently or combined, based on the users' needs for their applications or the programming environment desired by the developer.

The advantage of using the DRA model is that the target program can be developed in smaller size with higher efficiency. But the developers need to get very detailed knowledge about the processor and peripheral hardware architectures and interfacing configurations as well as detailed interfacing parameters. By using the SD model, the users do not need to know too much about the peripheral hardware architecture and interfacing configurations, and can easily build their interfacing functions with the help of the driver library to control and interface to peripherals to get the desired objectives.

In the SD model, a related application program interface (API) is provided by the peripheral driver library, and the API enables users to access and control the peripherals via a set of interfacing functions involved in that API. Because these drivers provide complete control of the peripherals in their normal mode of operation, it is possible to write an entire application without direct access to the hardware. This method provides for rapid development of the application without requiring detailed knowledge of how to program the peripherals. Because we have provided very detailed introductions and discussions about the Cortex-M4 MCU and related architecture as well as assembly instruction set, therefore we can handle either DRA or SD programming model with our applications without any problem.

In the following sections and chapters, we will use a combining model to build and develop our application projects. In other words, we will combine the DRA model and SD model together to build our applications. In part of our programs, we can use the DRA model to directly access the hardware, including the MCU-related registers, memory, and peripherals. In some other parts, we can call the API functions provided by the driver library to indirectly access and control the hardware and peripherals to realize our objectives.

4.5.6.2 DRA model

In the DRA model, the users can access and control peripherals by writing values directly into the peripheral's registers. The users need to use a set of register macros or register symbolic definitions provided by the peripheral driver library to simplify their coding process. These register symbolic definitions are stored in MCU-specific header files contained in the **`inc`** directory. The name of the header file matches the MCU number. For example, the header file for the MSP432P401R MCU is in the **`inc/msp432p401r.h`** directory under the **`C:\ti\msp\MSPWare_2_21_00_39\driverlib`** in your host computer. By including this header file that matches the MCU used, all related macros or symbolic definitions are available for accessing all registers on that MCU, as well as all bit fields within those registers.

Before we can continue our discussion about how to use the DRA model to build our sample project, first let's have a closer look at the hardware architecture used for this example.

4.5.6.2.1 Hardware architecture of the example project We try to use an example project named **`DRA LED Project`** to illustrate how to use DRA model to build a simple project. The MSP-EXP432P401R Evaluation Board is used as the hardware base to support our project development. In fact, we will use a three-color LED, **`LED2`**, which is connected to three pins on the Port 2, exactly pins **`P2.0`**, **`P2.1`**, and **`P2.2`**, in that EVB to test our project. The hardware configuration for this connection is shown in Figure 2.24 in Chapter 2. To make it convenient, we duplicate it and show here as Figure 4.23.

As shown in Figure 4.23, the pins **`P1.1`** and **`P1.4`** on **`PORT1`** are connected to two user switches or user buttons, **`S1`** and **`S2`**, respectively. The rest three pins, **`P2.0`**, **`P2.1`**, and **`P2.2`** on Port2, are connected to a three-color LED, in which three LEDs are integrated into one three-color LED in this evaluation board and can be individually driven by three pins in the Port 2.

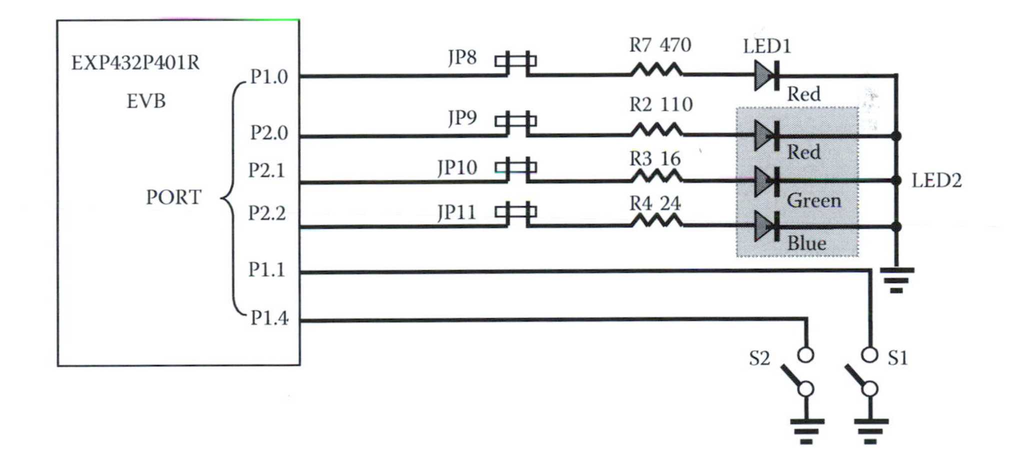

Figure 4.23 GPIO PORT 2 configuration in EXP432P401R EVB.

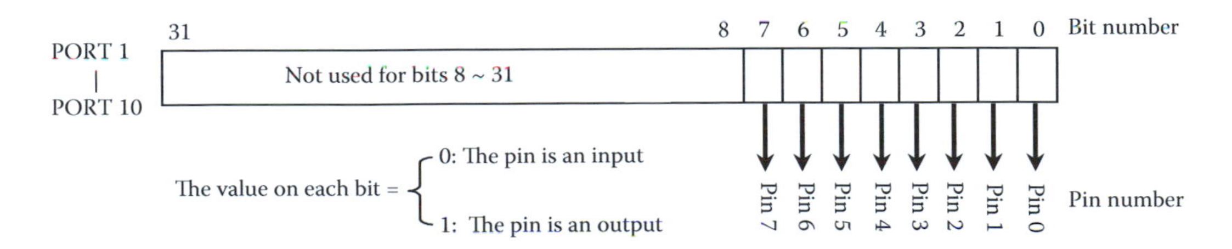

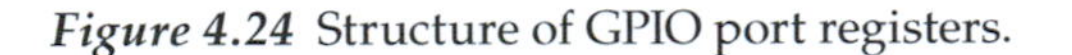

Figure 4.24 Structure of GPIO port registers.

In this example, we use three pins, **P2.0~P2.2**, to drive three-color LED to periodically turn it on and off to make a three-color flashing action.

As we discussed in Section 2.3.4.5 in Chapter 2, to enable a GPIO port to work properly, you need firstly to initialize and configure the GPIO port in the following sequence:

1. Configure all GPIO Port 2 pins to GPIO pins (**P2SEL0** = 00 h and **P2SEL1** = 00 h). In fact, this step can be ignored since by default, both **P2SEL1** and **P2SEL0** are 00
2. Setup the direction for three pins on the Port 2 by programming the **P2DIR** register
3. Turn on and off three-color LED periodically by assigning appropriate values to GPIO Port 2 three pins via Port 2 Output register (**P2OUT**)
4. Make this turn on and off operations as an infinitive loop.

4.5.6.2.2 Structure and bit function of GPIO registers In order to effectively using symbolic definitions to define and use those registers to successfully initialize and control the GPIO ports, let's have a closer look at the structure and bit function for these registers used by the Port 2.

These registers include:

- The GPIO PORT2 Direction Register (**P2DIR**)
- The GPIO PORT2 Output Register (**P2OUT**)

A point to be noted is that in EXP432P401R EVB, the Watchdog timer keeps its working after a system reset to monitor and detect any possible error caused by the microcontroller and other devices. To avoid unnecessary interrupts produced by this Watchdog timer, we need to stop its operations before we can build and run our sample project.

4.5.6.2.3 GPIO ports and port control registers As we mentioned in the last section, in this sample project we need to use the following registers to perform this three-color LED flashing operation:

- The GPIO PORT2 Direction Register (**P2DIR**)
- The GPIO PORT2 Output Register (**P2OUT**)

Although the registers used in each GPIO Port are 32 bits, only the lowest 8 bits is used, in which each bit is corresponding to each pin, as shown in Figure 4.24.

When working in the AHB aperture, the memory mapping addresses for these registers are:

- The GPIO PORT2 Direction Register (**P2DIR**): **0x40004C05**
- The GPIO PORT2 Output Register (**P2OUT**): **0x40004C03**

The functions of these registers are: the output register is used to output data to the related pins; the direction register is used to define the transferring direction, either input or output, for the data to be transferred in the **IN** or **OUT** Register.

When using these registers, the following points should be noted:

- The output register is used to store an 8-bit output data. Although this is 32-bit register, only the lowest 8 bits is used with one bit matching to one pin as shown in Figure 4.24.

- Each bit (pin) in this register can be operated separately, in other words, each bit can be read out or written in, and can be setup as an input or output bit (pin) separately by configuring the GPIO direction Register.
- The GPIO direction register is used to define the data transferring direction, either input or output, for each bit in the **IN** or **OUT** Register. A 0 in a bit in this register indicates that the corresponding pin in the **IN** or **OUT** register is an input pin, and a 1 in a bit in this register means that the corresponding pin in the **IN** or **OUT** Register is an output pin.

Now that we have a clear picture about these registers, let's start to define these registers in our software to use them to run our sample project.

4.5.6.2.4 Symbolic definitions and macros The **#define** statements used by the DRA model follow a naming convention that makes it easier to know how to use a particular symbolic definition.

The GPIO modules have many registers that do not have bit field definitions. For those registers, the register bits represent the individual GPIO pins. Therefore, the bit 0 in these registers corresponds to the **Px.0** pin on the port, where **x** represents a GPIO module number, such as **1** for **PORT1** and **2** for **PORT2**, and bit 1 corresponds to the **Px.1** pin, and so on.

A complete register symbolic definitions or macros for the MSP432P401R MCU can be found in the related system header file provided by the MSPWare for MSP432™ Series software package. In most applications, it should be located in the file **inc/msp432p401r.h** that is under the folder **C:\ti\msp\MSPWare_2_21_00_39\driverlib** in your host computer.

Table 4.24 shows related Port 2 registers, including their names, base address, offset, and software symbolic definitions or macros, to be used in this sample project.

4.5.6.2.5 Programming operations for symbolic definitions To use those symbolic definitions to access each register and perform some data assignment operations in our program, we need to introduce some popular programming operations for the symbolic definitions in the C codes. Most popular operations include:

- Set or reset a bit to 1 or 0: Using direct assignments to assign a **1** (HIGH) or a **0** (LOW) to a specified bit or pin. For example, using the following assignments to set and reset bit 3 (pin 3 – **P2.3**) in the Port 2 Direction Register (**P2DIR**):

```
P2DIR = 0x08;      Set P2.3 to 1 to make this pin output.
P2DIR = 0x0;       Reset P2.3 to 0 to make this pin input.
```

- Set a bit to **1** (HIGH): Using a logic **OR** operation to set a bit to **1** (HIGH). For example, using the following logic OR operation, one can set bit 5 in the Port 2 Output register:

```
P2OUT  |= BIT5;   → P2OUT = P2OUT | 0x20;
```

Table 4.24 GPIO port 2 related registers in AHB bus aperture

GPIO register	Base address	Offset	Full address	SW symbolic definition
GPIO Port 2 Input Register	0x4000.4C00	0x01	0x4000.4C01	P2IN
GPIO Port 2 Output Register	0x4000.4C00	0x03	0x4000.4C03	P2OUT
GPIO Port 2 Direction Register	0x4000.4C00	0x05	0x4000.4C05	P2DIR

No matter what is the original value on bit 5 (either 0 or 1) in Port 2, the bit 5 would be definitely set to 1 after this OR operation.

- Reset a bit to **0** (LOW): Using logic **AND Complement** operation to reset a bit to 0. For example, using the following logic AND Complement operation, one can reset bit 5 in the Port 2 Output register to 0:

```
P2OUT &= ~BIT5;    → P2OUT = P2OUT &~0x20
```

No matter what is the original value on bit 5 (either 0 or 1) in Port 2, the bit 5 would be definitely reset to 0 after this AND complement operation.

- Toggle a bit value from 0 to 1 or from 1 to 0: Using logic Exclusive OR (**EOR**) operation to toggle a bit value. The following EOR operation can toggle bit 5 on Port 2 Output register:

```
P2OUT ^ = BIT5;      → P2OUT =~ 0x20;
```

- Get a bit value by shifting operation: By using shift left or right operations, one can get desired value for a data bit. For example, the following shift operations enable users to get desired bit result:

```
#define  P2.0 (1 << 0)   → P2.0 = 0x00000001 (1 in bit 0 is shifted left by 0 bit)
#define  P2.1 (1 << 1)   → P2.1 = 0x00000002 (1 in bit 0 is shifted left by 1 bit)
#define  P2.2 (1 << 2)   → P2.2 = 0x00000004 (1 in bit 0 is shifted left by 2 bit)
#define  P2.3 (1 << 3)   → P2.3 = 0x00000008 (1 in bit 0 is shifted left by 3 bit)
```

The instruction **return (P2OUT >> 4);** is equivalent to shift right all 32-bit data in Port 2 output register by 4 bits to move **P2.7~P2.4** to bits **P2.3~P2.0**.

Now, we have everything done and we are ready to build our sample project.

4.5.6.2.6 Develop a sample project using the DRA model In this sample project, we use three pins, **P2.0~P2.2**, to drive a three-color LED to periodically turn it on and off to make a three-color flashing action (Figure 4.23).

Since this project is very simple, we only use a header file and an application file with the names of "**DRALED.h**" and "**DRALED.c.**"

The general steps to develop a microcontroller application with Keil® ARM®-MDK include:

1. Create a new µVersion5 project with selecting of the MCU and MDK Core/CMSIS Core
2. Create the user's header files and add them into the project
3. Create the user's C source files and add them into the project
4. Include the system header files and other header files into the project by either adding them into the project or using the **Include Paths** function in the ARM®-MDK µVersion5
5. Include the external functions, variables and other C source files into the project by placing them under the current project folder
6. Link the library (static or dynamic library) provided by the vendor with the project by adding it as a project file into the project

7. Compile and link project to create the image or executable file
8. Download the image file into the flash memory by using **`Flash|Download`** menu item
9. Run the project by using the **`Debug|Run`** menu item.

Now, let's follow these nine steps to build our first sample project **`DRA LED Project`**.

4.5.6.2.6.1 Create a new μVersion5 project—DRA LED project Perform the following operations to create this new project **`DRA LED Project`**:

1. Open the Windows Explorer to create a new folder named **`DRA LED Project`** under the **`C:\MSP432 Class Projects\Chapter 4`** folder
2. Open the Keil® ARM®-MDK μVersion5 and go to **`Project|New μVersion Project`** menu item to create a new μVersion Project. On the opened wizard, browse to our new folder **`DRA LED Project`** that is created in step 1 above. Enter **`DRA LED Project`** into the **File name** box and click on the **Save** button to create this project
3. On the next wizard, you need to select the device (MCU) for this project. Expand three icons, **`Texas Instruments`**, **`MSP432 Family`** and **`MSP432P`**, and select the target device **`MSP432P401R`** from the list by clicking on it. Click on the **OK** to close this wizard
4. Next, the Software Components wizard is opened, and you need to setup the software development environment for your project with this wizard. Expand two icons, **`CMSIS`** and **`Device`**, and check the **`CORE`** and **`Startup`** checkboxes in the **Sel.** column, and click on the **OK** button since we need these two components to build our project.

4.5.6.2.6.2 Create the user's header file Perform the following operations to create this new header file **`DRALED.h`**:

1. In the **Project** pane, expand the **`Target`** folder and right click on the **`Source Group 1`** folder and select the **`Add New Item to Group "Source Group 1."`**
2. Select the **`Header File (.h)`** and enter **`DRALED`** into the **Name:** box, and click on the **Add** button to add this file into the project
3. Enter the codes shown in Figure 4.25 into this header file, and click on the **`File|Save`** menu item to save this file.

```
1  //****************************************************************************
2  // DRALED.h - Header file for the project - DRA LED Project
3  //****************************************************************************
4  #include <stdint.h>
5  #include <stdbool.h>
6  #include "msp.h"
7
8  void delayTime(int time);
```

Figure 4.25 Header file DRALED.h.

The codes for this header file are straightforward and easy to be understood. Three system header files are declared first in lines 4~6, and a user-defined function **delayTime()** is declared here to perform some time delay function for this project.

4.5.6.2.6.3 Create the user's source file Perform the following operations to create this new source file **DRALED.c**:

1. In the **Project** pane, expand the **Target** folder and right click on the **Source Group 1** folder and select the **Add New Item to Group "Source Group 1."**
2. Select the **C File (.c)** and enter **DRALED** into the **Name:** box, and click on the **Add** button to add this source file into the project
3. Enter the codes shown in Figure 4.26 into this source file, and click on the **File | Save** menu item to save this file.

Let's have a closer look at this piece of codes to see how it works.

- The user-defined header file **DRALED.h** is included first at line 4 on this file since we need to use related macros and functions defined in that header file.
- The function body for the user-defined function **delayTime()** is presented following the user header file in lines 5~8. A simple **for()** loop is used to delay a period of time for the project to make the three-color LED to stop flashing periodically.

```
1  //****************************************************************************
2  // DRALED.c - Main application file for the project - DRA LED Project
3  //****************************************************************************
4  #include "DRALED.h"
5  void delayTime(int time)
6  {
7     int count;
8     for(count = 0; count < time; count++);    // time delay
9  }
10 int main(void)
11 {
12    uint32_t num = 100000;
13    WDTCTL = WDTPW | WDTHOLD;                  // stop watchdog timer
14    P2DIR |= BIT0 | BIT1 | BIT2;               // set P2.0 ~ P2.2 as output pins
15    while(1)
16    {
17    P2OUT ^= BIT0;                             // toggle P2.0
18    delayTime(num);                            // time delay
19    P2OUT ^= BIT1;                             // toggle P2.1
20    delayTime(num);                            // time delay
21    P2OUT ^= BIT2;                             // toggle P2.2
22    delayTime(num);                            // time delay
23    }
24 }
```

Figure 4.26 Application file for the sample project DRA LED Project.

- The main program starts at line 9. First, a local **`uint32_t`** variable **`num`** is declared and assigned with a value of **`100000`** in line 11, and this variable will work as a terminate value for the loop counter for the **`delayTime()`** function later.
- In line 12, the Watchdog timer is terminated to avoid any unnecessary interrupts.
- Then the Direction Register on the GPIO Port 2, **`P2DIR`**, is initialized and configured to enable **`P2.0~P2.2`** pins to work as output pins in line 13. Three macros, **`BIT0`**, **`BIT1`**, and **`BIT2`**, which are defined in the system header file **`msp432p401r.h`**, are **`ORed`** together and assigned to the **`P2DIR`** register to complete this output configuration. A point to be noted is that the system header file **`msp432p401r.h`** is covered and included in another system header file **`msp.h`** that has been included in our user-defined header file. Thus, you do not need to clearly include it in this project as long as you include the **`msp.h`** header file.
- Starting line 14, an infinitive **`while()`** loop is used to repeatedly toggle the three-color LED via three pins, **`P2.0~P2.2`**. After each toggling, a time delay is performed by calling the user-defined function **`delayTime()`** to make the flashing evenly and periodically. The codes between lines 16 and 21 perform these operations.

In fact, the **`num`** value can be used to determine the length or the period of the time delay performed by the **`delayTime()`** function.

4.5.6.2.6.4 Include the other header files into the project Because we did not use any other header files in this sample project, we need to do nothing in this step. However, in some other complicated projects, you must do this step otherwise your project cannot work properly.

4.5.6.2.6.5 Include the external functions and other C source files into the project Because we did not use any other C source file and external function in this sample project, we need to do nothing in this step. However, in some other complicated projects, you must do this step otherwise your project cannot work properly.

4.5.6.2.6.6 Link the static or dynamic library with the project In this sample project, we do not use any static or dynamic link library since we used DRA model for this project. Therefore, we need to do nothing with this step. However, in some other projects where either a static or dynamic link library, such as MSPWare peripheral driver library, is used, you must do this step. Otherwise you cannot run your project at all.

4.5.6.2.6.7 Compile and link project to create the image or executable file Go to **`Project|Build Target`** menu item to compile and link this sample project with system files to make our project image or executable file. A message displaying **`0 Error(s)`** with **`0 Warning(s)`** should be displayed in the **`Build Output`** window at the bottom if nothing is wrong with this sample project.

4.5.6.2.6.8 Download the image file into the flash memory and run the project Before we can continue to perform downloading our project file into the flash memory, we must make sure that a correct debugger has been selected for this project. Refer to Section 3.4.3.3 in Chapter 3 to get more details about this debugger selection process. To make it simple, just check this from the **`Debug`** tab under the **`Project|Options for Target "Target 1"`** menu item, and make sure that the debugger used is **`CMSIS-DAP Debugger`** from the **`Use`** box. The **`CMSIS-DAP`** is a compatible debugger with XDS110-ET emulator in the MSP-EXP432P401R EVB.

Now, using the **Flash|Download** menu item, you can download the project image file into the flash memory. You can also run this program by going to **Debug|Start/Stop Debug Session** and **Debug|Run** menu item. As the program running, you can find that the three-color LED will be flashing periodically with red, green, and blue colors.

Next, let's discuss how to use the SD model to call the driver library functions to perform this LED flashing function.

4.5.6.3 Peripheral driver library and API functions

The driver library is installed in the internal ROM space in the on-chip memory, and it is also included in the MSPWare for MSP432™ Series software package and can be downloaded and installed in the user's host computer.

This library contains about 26 groups API functions for related peripherals. The most popular groups included in this library are

1. GPIO ports
2. Analog comparators
3. ADC
4. Advanced encryption standard (AES256)
5. Flash memory
6. FPU
7. MPU
8. Interrupt controller (NVIC)
9. SPI
10. I2C
11. Port map control (PMAP)
12. System control
13. SysTick
14. Timer
15. UART
16. Real-time clock (RTC_C)
17. Watchdog timer

We will discuss most of these peripherals in the following chapters. However, in this section, we want to use the GPIO library API functions as an example to illustrate how to use these library functions to access the GPIO ports to develop our next sample project.

As we discussed in Section 2.3.3.2 in Chapter 2, the MSP432P401R MCU provides a set of GPIO modules. These modules contain 11 GPIO ports, Ports 1~10 and Port J. Each module is related to an individual GPIO port, and each GPIO port is programmable or configurable to provide multiple functions to enable the port to handle different tasks. Not all 78 pins on Ports 1~10 are configurable or programmable since some pins are not available. Depending on the peripherals being used, each GPIO pin provides the following capabilities:

- Can be configured as a digital I/O pin with input or output function. On reset, they default to be an input.
- Can be configured as an analog input pin to receive an analog signal as input.
- In the input mode, each pin can generate interrupts on rising or falling edges from P1 through P6. Wake-up capability from low-power modes, such as LPM3, LPM4, LPM3.5, and LPM4.5 modes can be programmed over ports P1 through P6.

- In the output mode, each pin can be configured to regular or high drive strength. On reset, they default to regular drive strength.
- Provide input pull-up or pull-down resistors selection function.
- Can be configured to be a GPIO or a mapped peripheral function pin. On reset, they default to GPIO pins.
- Most pins on most ports have multiple peripheral functions, in other words, when it is configured for special mapped peripheral function, the pin will do special functions.

First, let's have a closer look at some important API functions for GPIO ports.

4.5.6.3.1 GPIO API functions Most of the GPIO API functions in this library can operate on more than one pin at a time. The argument parameter **selectedPins** in these functions is used to specify the pins that are affected; only the pins corresponding to the related bits in this parameter that are set are affected. There is a one-to-one relationship between each bit and the corresponding pin on each port, where bit 0 is pin 0, bit 1 is pin 1, and so on. For example, if **selectedPins** is **0x09** (**0x0000**1001B), then only pins 0 and 3 are affected by the function.

The most useful API functions for our following sample projects are:

- **GPIO_setAsInputPin()**
- **GPIO_setAsOutputPin()**
- **GPIO_setOutputHighOnPin()**
- **GPIO_setOutputLowOnPin()**
- **GPIO_toggleOutputOnPin()**

The **GPIO_setAsInputPin()** and **GPIO_setAsOutputPin()** functions configure the related pins as either input or output pins. The **GPIO_setOutputHighOnPin()** and **GPIO_setOutputLowOnPin()** functions output HIGH or LOW on the selected pins. The **GPIO_toggleOutputOnPin()** function perform the toggle operations for the selected pins.

There are about 20 API functions provided by the GPIO module and these functions are used to control all GPIO-related peripherals in MSP432P401R MCU system. We will select the above three functions as an example to illustrate how to use them to access GPIO Port 2 to control the three-color LED to perform a flashing operation.

4.5.6.3.1.1 The GPIO_setAsInputPin() and GPIO_setAsOutputPin() functions In order to perform a reading or writing operation from/to the selected pins on a GPIO port, the selected pins must be configured as input or output pins by using the **GPIO_setAsInputPin()** and **GPIO_setAsOutputPin()** functions. The protocols of these two functions are similar as:

```
GPIO_setAsInputPin(uint_fast8_t selectedPort, uint_fast16_t selectedPins)
GPIO_setAsOutputPin(uint_fast8_t selectedPort, uint_fast16_t
  selectedPins)
```

Both functions have two arguments, **selectedPort** and **selectedPins**. The first is used to select the desired GPIO port and the second is used to define the desired pins.

Table 4.25 Macros used for GPIO port and pins in the gpio.h header file

GPIO port	Macro	GPIO pin	Macro
GPIO Port 1	GPIO_PORT_P1	0	GPIO_PIN0
GPIO Port 2	GPIO_PORT_P2	1	GPIO_PIN1
GPIO Port 3	GPIO_PORT_P3	2	GPIO_PIN2
GPIO Port 4	GPIO_PORT_P4	3	GPIO_PIN3
GPIO Port 5	GPIO_PORT_P5	4	GPIO_PIN4
GPIO Port 6	GPIO_PORT_P6	5	GPIO_PIN5
GPIO Port 7	GPIO_PORT_P7	6	GPIO_PIN6
GPIO Port 8	GPIO_PORT_P8	7	GPIO_PIN7
GPIO Port 9	GPIO_PORT_P9	8	GPIO_PIN8
GPIO Port 10	GPIO_PORT_P10	9	GPIO_PIN9
GPIO Port 11	GPIO_PORT_P11	10	GPIO_PIN10
GPIO Port J	GPIO_PORT_PJ	11–15	GPIO_PIN11 ~ GPIO_PIN15

When using these two functions in your program, one can directly use the number of ports or pins for these arguments. However, one can also use macros defined in the system header file **gpio.h** to work as port and pin numbers. Table 4.25 shows these macros used for ports and pins.

A point to be noted is that the pin numbers shown in Table 4.25 are more than 7. In fact, these pin numbers are defined from 0 through 15, or from **GPIO_PIN0~GPIO_PIN15**, in the **gpio.h** header file. As we mentioned in Section 2.3.3.2 in Chapter 2, each two pair or neighboring ports, such as P1 and P2, P3 and P4, until P9 and P10, can be combined together to form a 16-bit port named PA, PB, until PE. These 16 pin macros are defined for those 16-bit ports, not for only single 8-bit port. One can also use macros **PIN_ALL8** or **PIN_ALL16** defined in the **gpio.h** header file to access all 8 bits or all 16 bits on any 8-bit or 16-bit registers.

To read or write multiple pins by using these functions, an **OR** operator can be used to combine all related bits or pins together.

4.5.6.3.1.2 GPIO_setOutputHighOnPin() and GPIO_setOutputLowOnPin() functions These two functions are used to output either HIGH or LOW on the selected GPIO ports and pins. Both functions have two identical arguments as those defined in **GPIO_setAsInputPin()** and **GPIO_setAsOutputPin()** functions.

One can use the same macros for these ports and pins defined in Table 4.25.

4.5.6.3.1.3 GPIO_toggleOutputOnPin() function This function allows users to toggle the output for the selected GPIO port and pins. The protocol of this function is:

```
void GPIO_toggleOutputOnPin(uint_fast8_t selectedPort, uint_fast16_t
  selectedPins);
```

Two arguments are used in this function and the meanings for these arguments are

1. The first argument **selectedPort** is the number of the GPIO port.
2. The second argument **selectedPins** is the number of the pins to be affected.

One can use the same macros for these ports and pins defined in Table 4.25.

For example, the following two functions perform the same function: To toggle outputs of pins 0 and 3 on the GPIO Port 2:

```
GPIO_toggleOutputOnPin(2, GPIO_PIN0|GPIO_PIN3);
GPIO_toggleOutputOnPin(GPIO_PORT_P2, GPIO_PIN0|GPIO_PIN3);
```

4.5.6.3.1.4 GPIO_setAsPeripheralModuleFunctionInputPin() function This function can be used to configure the GPIO pins to perform alternate peripheral module functions, either primary, secondary, or ternary module function modes, in the input direction for the selected pin. The protocol of this function is:

```
void GPIO_setAsPeripheralModuleFunctionInputPin(uint_fast8_t
    selectedPort, uint_fast16_t selectedPins, uint_fast8_t mode);
```

For the first two arguments, **selectedPort** and **selectedPins**, the same macros shown in Table 4.25 can be used for this function. The third argument is the **mode** that the pin should be configured to perform an alternate function, and it can be one of the following macros:

- **GPIO_PRIMARY_MODULE_FUNCTION**
- **GPIO_SECONDARY_MODULE_FUNCTION**
- **GPIO_TERTIARY_MODULE_FUNCTION**

4.5.6.3.1.5 GPIO_setAsPeripheralModuleFunctionOutputPin() function This function can be used to configure the GPIO pins to perform alternate peripheral module functions, either primary, secondary, or ternary module function modes, in the output direction for the selected pin. The protocol of this function is:

```
void GPIO_setAsPeripheralModuleFunctionOutputPin(uint_fast8_t
    selectedPort, uint_fast16_t selectedPins, uint_fast8_t mode);
```

For the first two arguments, **selectedPort** and **selectedPins**, the same macros shown in Table 4.25 can be used for this function. The third argument is the **mode** that the pin should be configured to perform an alternate function and it can be one of the following macros:

- **GPIO_PRIMARY_MODULE_FUNCTION**
- **GPIO_SECONDARY_MODULE_FUNCTION**
- **GPIO_TERTIARY_MODULE_FUNCTION**

These above two GPIO API functions are very useful and very important to configure the GPIO pins to work as alternate functions.

More other GPIO library API functions will be discussed in the following related chapters when more GPIO-related peripherals are introduced.

Now, let's use the GPIO Port 2 as our target GPIO port to illustrate how to call these related GPIO API functions to initialize, configure, access and control this port to perform a three-color LED flashing function on the MSP-ESP432P401R Evaluation Board.

4.5.6.3.2 Develop a sample project using the SD model Refer to Figure 4.23 to get details about the hardware configuration of the GPIO Port 2 in the MSP-ESP432P401R Evaluation Board.

In this sample project, **SD LED Project**, we still use **P2.0~P2.2** pins which are connected to a three-color LED as an output pin to drive those LEDs to on and off to make a flashing action.

Since this project is very simple, so we only use a header file and an application file with the names of **SDLED.h** and **SDLED.c**. The MSPWare peripheral driver library is used in the sample project to allow us to use those GPIO-related API functions to access GPIO ports and pins.

Perform the following steps to build our sample project **SD LED Project**.

4.5.6.3.2.1 Create a new µVersion5 project - SD LED project Perform the following operations to create this new project **SD LED Project**:

1. Open the Windows Explorer to create a new folder named **SD LED Project** under the **C:\MSP432 Class Projects\Chapter 4** folder.
2. Open the Keil® ARM®-MDK µVersion5 and go to **Project|New µVersion Project** menu item to create a new µVersion Project. On the opened wizard, browse to our new folder **SD LED Project** that is created in step 1 above. Enter **SD LED Project** into the **File name** box and click on the **Save** button to create this project.
3. On the next wizard, you need to select the device (MCU) for this project. Expand three icons, **Texas Instruments**, **MSP432 Family** and **MSP432P**, and select the target device **MSP432P401R** from the list by clicking on it. Click on the **OK** to close this wizard.
4. Next, the software components wizard is opened, and you need to setup the software development environment for your project with this wizard. Expand two icons, **CMSIS** and **Device**, and check the **CORE** and **Startup** checkboxes in the **Sel.** column, and click on the **OK** button since we need these two components to build our project.

4.5.6.3.2.2 Create the user's header file Perform the following operations to create this new header file **SDLED.h**:

1. In the **Project** pane, expand the **Target** folder and right click on the **Source Group 1** folder and select the **Add New Item to Group "Source Group 1."**
2. Select the **Header File (.h)** and enter **SDLED** into the **Name:** box, and click on the **Add** button to add this file into the project.
3. Enter the codes shown in Figure 4.27 into this header file, and click on the **File|Save** menu item to save this file. The function for each code line is:
   a. The code lines 4~8 are used to include five system header files to be used in this project. The <**stdint.h**> and <**stdbool.h**> are standard C type data definition files, and these two header files defined most standard integer data types and Boolean data types used in C/C++ programs. Two files are located at the default folder of installing the Keil® ARM®-MDK µVersion5 IDE, **C:\Keil_v5\ARM\ARMCC\include**. The other three header files, <**msp.h**>, <**gpio.h**>, and <**wdt_a.h**>, are TI system header files used to define the GPIO and Watchdog API

```
1  //*******************************************************************
2  // SDLED.h - Header File for the project - SD LED Project
3  //*******************************************************************
4  #include <stdint.h>
5  #include <stdbool.h>
6  #include <msp.h>
7  #include <MSP432P4xx/gpio.h>
8  #include <MSP432P4xx/wdt_a.h>
9  void delayTime(int time);
```

Figure 4.27 Header file SDLED.h.

functions used in this project. The last two header files are located at the folder: **C:\ti\msp\MSPWare_2_21_00_39\driverlib\driverlib\MSP432P4xx**. We will explain the included path for these header files later.

b. The code line 9 is used to declare a user-defined time delay function.

Now, let's take care of the C source file **SDLED.c**.

4.5.6.3.2.3 Create the user's source file Perform the following operations to create this new source file **SDLED.c**:

1. In the **Project** pane, expand the **Target** folder and right click on the **Source Group 1** folder and select the **Add New Item to Group "Source Group 1."**
2. Select the **C File (.c)** and enter **SDLED** into the **Name:** box, and click on the **Add** button to add this source file into the project.
3. Enter the codes shown in Figure 4.28 into this source file, and click on the **File|Save** menu item to save this file.

Let's have a closer look at this piece of codes to see how it works.

- In coding line 4, our user-defined header file **SDLED.h** is included since it contains all system header files and user-defined time delay function to be used in this project
- The coding lines between 5 and 9 are used to define the body for the user-defined time delay function.
- The program entry point starts from the **main()** function in line 10.
- In line 12, a local variable **num** is declared since we need to use it as the time delay length value when we call the **delayTime()** function later.
- The API function **WDT_A_holdTimer()** is called in line 13 to terminate the function of the Watchdog timer to avoid undesired time interrupts.
- The GPIO API function **GPIO_setAsOutputPin()** is called in line 15 to configure **P2.0~P2.2** pins to work as output pins. Three pins are **ORed** together via their macros to enable them to work as output pins.
- An infinitive **while()** loop is used in line 16 to enable the codes between lines 18 and 23 to be executed repeatedly.
- The GPIO API function **GPIO_toggleOutputOnPin()** is called and executed in line 18 to toggle the output on **P2.0** pin to flash the red-color LED. Then, a time delay is performed by calling the **delayTime()** function in line 19.

```
1  //*****************************************************************************
2  //SDLED.c - Main application file for the project - SD LED Project
3  //*****************************************************************************
4  #include "SDLED.h"
5  void delayTime(int time)
6  {
7    int count;
8    for(count = 0; count < time; count++);      // time delay
9  }
10 int main(void)
11 {
12   uint32_t  num = 100000;
13   WDT_A_holdTimer();                           // stop watchdog timer
14   // set P2.0 ~ P2.2 as output pins
15   GPIO_setAsOutputPin(GPIO_PORT_P2, GPIO_PIN0|GPIO_PIN1|GPIO_PIN2);
16   while(1)
17   {
18     GPIO_toggleOutputOnPin(GPIO_PORT_P2, GPIO_PIN0);         // toggle P2.0
19     delayTime(num);                                          // time delay
20     GPIO_toggleOutputOnPin(GPIO_PORT_P2, GPIO_PIN1);         // toggle P2.1
21     delayTime(num);                                          // time delay
22     GPIO_toggleOutputOnPin(GPIO_PORT_P2, GPIO_PIN2);         // toggle P2.2
23     delayTime(num);                                          // time delay
24   }
25 }
```

Figure 4.28 C source codes for the application file SDLED.c.

- Similarly, the GPIO API function **GPIO_toggleOutputOnPin()** is called and executed in line 20 to toggle the output on **P2.1** pin to flash the green-color LED. Then, a time delay is performed by calling the **delayTime()** function in line 21.
- Finally, the GPIO API function **GPIO_toggleOutputOnPin()** is called and executed in line 22 to toggle the output on **P2.2** pin to flash the blue-color LED. Then, a time delay is performed by calling the **delayTime()** function in line 23.

4.5.6.3.2.4 Include the system header files path and other header files We used two system header files in our application C source file, <**MSP432P4xx/gpio.h**> and <**MSP432P4xx/wdt_a.h**>. There are two ways to include or add them into our project.

- Use the **Include Paths** function provided by the Keil® ARM®-MDK C/C++ Compiler to include a path for those header files to enable the compiler to know where to find these header file when the project is compiled
- Add all of these header files into the project.

The advantage of using the first method is that it is very easy for the users to include these header files with one path. However, the shortcoming is that the project developed in

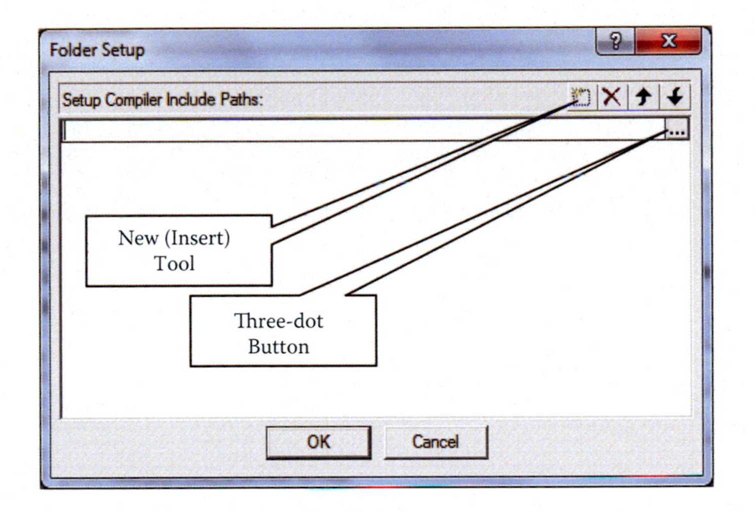

Figure 4.29 Finished folder setup wizard.

this way cannot be portable since all header files may not be in the same paths in another computer. In other words, those header files are not included in your project and may be in some other location in your computer.

The advantage of using the second way is that these header files are included in your project and the project can be portable. The bad thing is that you need to spend your time to add those header files one by one into your project.

To save time, in this project we used the first way. Perform the following operations to include these system header files into the project:

1. When the project is open, go to **`Project|Options for Target "Target 1"`** menu item to open the options wizard
2. Then click on the **`C/C++`** tab and go to the **`Include Paths`** box.

Click on the three-dot-button on the right of this box to open the **Folder Setup** wizard, as shown in Figure 4.29. Click on the **`New`** (**`Insert`**) tool to open a new textbox. Click on the three-dot-button on the right of this textbox to browse to the location where these header files are located, which is **`C:\ti\msp\MSPWare_2_21_00_39\driverlib\driverlib\MSP432P4xx`**. Here, two options come: you can include the whole path as listed above, or you can include a part of this whole path. In our case, we include a part of this entire path in here (**`C:\ti\msp\MSPWare_2_21_00_39\driverlib`**) and leave another part involved in our header file declaration (**`MSP432P4xx\gpio.h`**). Then click on the **OK** button to select this path. Your finished include path wizard should look like one that is shown in Figure 4.30. Click on the **OK** button to close this wizard.

Your finished **`Options`** wizard is shown in Figure 4.30. Click the **OK** to close this wizard.

4.5.6.3.2.5 Include the external functions and other C source files Because we did not use any other C source file and external function in this sample project, we need to do nothing in this step. However, in some other complicated projects, you must do this step otherwise your project cannot work properly.

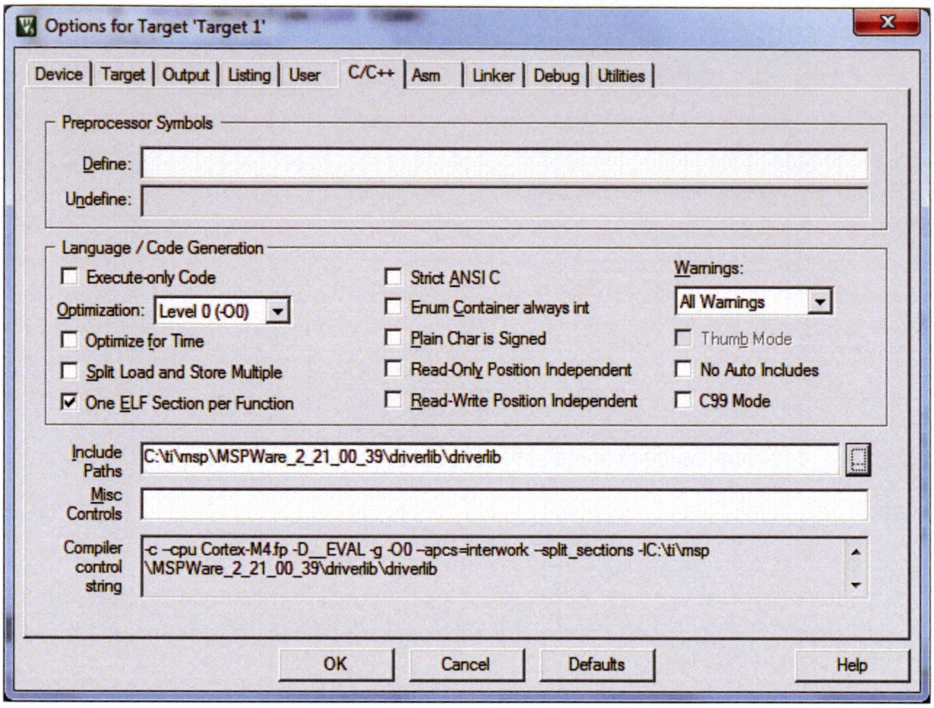

Figure 4.30 Finished options wizard.

4.5.6.3.2.6 Link the static or dynamic library with the project Generally, there are two kinds of libraries you can link to the user's project: the static library (**`.lib`**) and dynamic link library (**`.dll`**).

The dynamic link library means that you will not add this library into your project until your project runs, which means that the library is dynamically added or linked with your project when the project is running. In this way, compiler will not consider that this library is a part of your project, and will not link it until your project runs. Therefore, the size of your target file is much smaller since it does not contain the library but only links the library when it runs.

The static library comprises a collection of files with related header files. Depending on the type of the files included in a static library, a static library can be further divided into two categories: (1) the library comprises a collection of C code files with related header files, and (2) the library is made up of a binary file that is the compiling result of those C code files with the related header files. The difference between these two static libraries is that the first one is only a group of C files with related header files, but the second is the compiling result of the first one and it is a binary file.

To link the first kind of static library, you need to add all of those C files and header files into your project. This method is a little time consuming since you need time to do those file additions. To link the second kind of static library, you only need to add a single binary library file with related header files. We will use the second way since it is simple and easy.

The compiler considered that this library is with your project and compiles this library with your project together to get the target file. Because the library is added into your project before your project runs, it is called a static link or a static library.

In this sample project, we need to use the MSPWare peripheral driver library, which is a static binary library, to access GPIO ports. Therefore, we need to link that library with our project. Perform the following operations to link this library with our project:

- In the **Project** pane, expand the **Target** folder and right click on the **Source Group 1** folder and select the **Add Existing Files to Group "Source Group 1."**
- Browse to find the library, the MSPWare peripheral driver library, which is located at: **C:\ti\msp\MSPWare_2_21_00_39\driverlib\driverlib\MSP432P4xx\keil** in your host computer and the library is named **msp432p4xx_driverlib.lib**.
- Select this library by clicking on it and click on the **Add** button to add it into our project.

4.5.6.3.2.7 Compile and link project to create the image or executable file Before you can build the project, make sure that the debugger you are using is **CMSIS-DAP** and the debugger settings are correct. You can do these checks by:

- Going to **Project|Options for Target "Target 1"** menu item to open the options wizard
- On the opened options wizard, click on the **Debug** tab
- Making sure that the debugger shown in the **Use:** box is **CMSIS-DAP Debugger**. Otherwise you can click on the drop-down arrow to select this debugger from the list
- Making sure that the debugger settings are correct by clicking on the **Settings** button located at the right of the debugger box. Then click on the **Debug** tab to check both checkboxes: **Verify Code Download** and **Download to Flash** inside the **Download Options** group box. Then click on the **Flash Download** tab to check **Erase Full Chip** radio button inside the **Download Function** group box.

Now, go to **Project|Build Target** menu item to compile and link this sample project with system files to make the project image or executable file. A message displaying **0 Error(s)** with **0 Warning(s)** should be displayed in the **Build Output** window at the bottom if nothing wrong with this sample project.

4.5.6.3.2.8 Download the image file into the flash memory and run the project By using **Flash|Download** menu item, you can download the project image file into the flash memory. Now you can begin to debug this project by going to **Debug|Start/Stop Debug Session**. Click on the **OK** button for the memory limitation for the evaluation version message, and run this program by going to **Debug|Run** menu item. As the program running, you can find that the three-color LED will be flashing forever. Go to **Debug|Stop** menu to stop the project running.

4.5.6.4 Comparison between two programming models

Comparing for these sample projects with two programming models, we can get the following conclusions:

1. The sample project developed with the DRA model has small size and less program coding lines. But good and solid knowledge and understanding about the microcontroller hardware is a necessary requirement to build the project in this model.

2. The sample project developed with the SD model has relatively large size and more program coding lines. More system header files and a static link library are needed to build this kind of project. But the developers do not need to have deep and detailed understanding about the microcontroller hardware to build this kind of project.

An interesting question is, how to develop a project by combining these two programming models together? The following section is an answer to this question.

4.5.6.5 Combined programming model example

In this section, we try to use the Wayne EduBASE Trainer and a combined programming model to build a project with the following functions:

- Use GPIO Port 4 bit 4 (**P4.4**) as an input pin to receive the state of a position switch **SW2** installed on the EduBASE Trainer since the pins **P4.4**, **P4.2**, **P4.0**, and **P6.1** in **J3** connector on the EXP432P401R EVB are connected to **SW2~SW5** position switches in the EduBASE Trainer. Figure 4.31 shows this configuration of this Trainer.
- Use GPIO Port 2 bit 2 (**P2.2**) as an output pin to turn on or off the blue-color LED installed in the EXP432P401R EVB (refer to Figure 2.24 in Chapter 2).
- When the position switch **SW2** is pressed, the blue-color LED is ON.
- When the position switch **SW2** is released, the blue-color LED is OFF.

The interfacing jack **J3** in the EXP432P401R EVB is connected to the interfacing jack J14 in the EduBASE Trainer. A total of 4 bits or pins, **P4.4**, **P4.2**, **P4.0**, and **P6.1**, are connected to four position switches, **SW5~SW2**, in the EduBASE Trainer, respectively. These four pins are also connected to a DIP switch **SW1** and four rows, **ROW0~ROW3**, in a 4 × 4 keypad on the EduBASE Trainer.

A logic LOW is applied to a pin if no associate position switch is pressed since a 4.7 K resistor that is connected to the ground is connected to these pins.

A logic HIGH would be appeared on a pin if the related position switch is pressed since a voltage divider applied on the resistors R3 and RN10 generates a HIGH ($V_{pin} = 4.7/(4.7 + 1) \times 5\ V = 4.1\ V$). One point to be noted is that all switches on the DIP switch **SW1** should be in the OFF position when using these position switches **SW2~SW5** to interface to any pin to do the related program controls.

Four pins on GPIO ports, **P3.2**, **P3.3**, **P2.5**, and **P2.4**, in the MSP-EXP432P401R EVB are connected to 4 LEDs (**PB0~PB3**) in the EduBASE Trainer via pins 3 and 4 in J1, pin 2 in J2, and pin3 in J4 in the EXP432P401R EVB. Also, the **P5.6** is connected to a speaker via pin 4 in J4.

Based on this hardware configuration, let's build our combining model project **CP LED Project**.

4.5.6.5.1 Create the header file Create a new folder **CP LED Project** under the folder **C:\MSP432 Class Projects\Chapter 4** in the Windows Explorer. Then create a new μVersion5 project **CP LED Project** in the Keil® ARM®-MDK μVersion5 IDE and add it into the folder **CP LED Project**.

Create a new header file named **CPLED.h** and add it into the project **CP LED Project**. Enter the codes shown in Figure 4.32 into this header file.

Two system header files, <**stdint.h**> and <**stdbool.h**>, are included since these files contain definitions for all integer data types and Boolean data types in ARM® C/C++ compiler system.

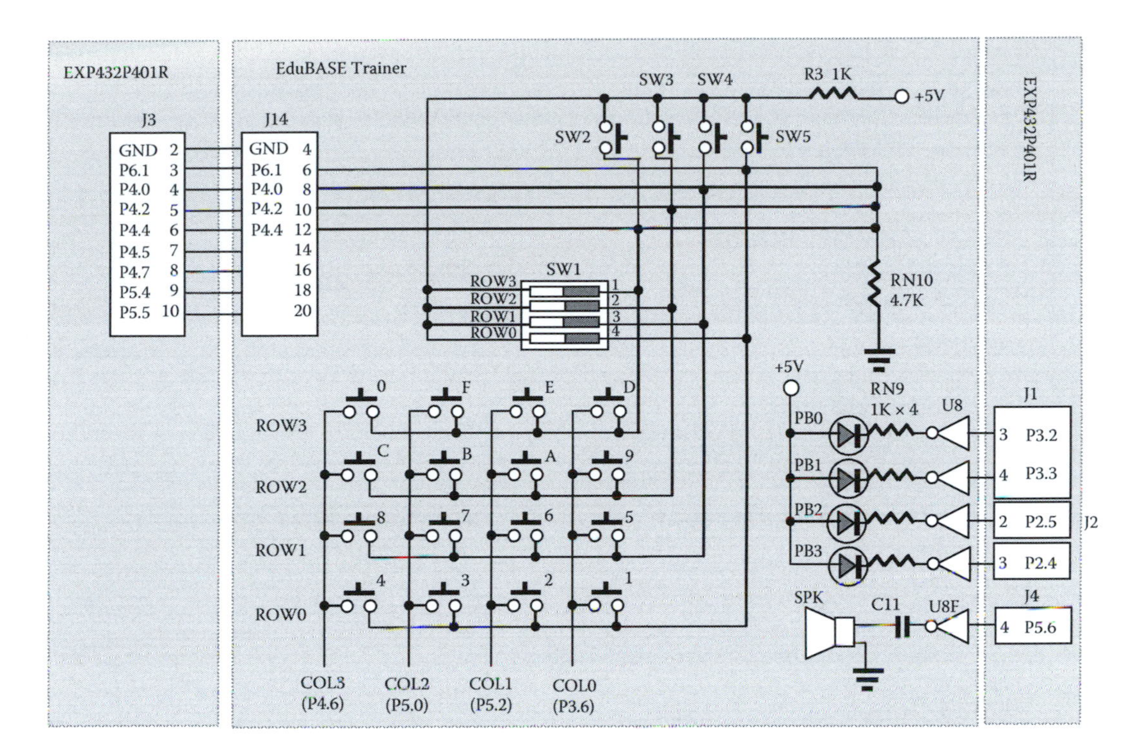

Figure 4.31 Some hardware configurations for EduBASE Trainer.

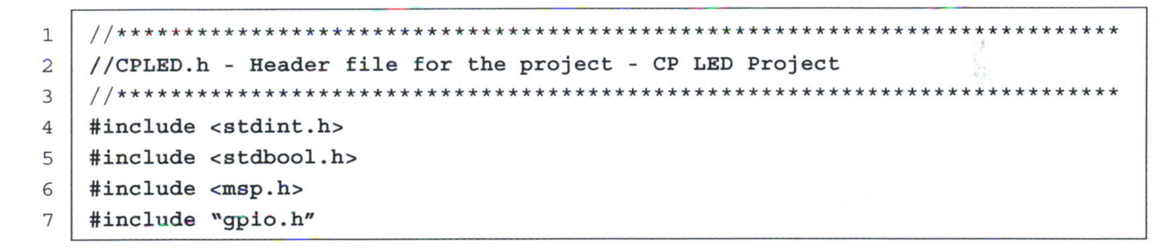

```
1  //****************************************************************************
2  //CPLED.h - Header file for the project - CP LED Project
3  //****************************************************************************
4  #include <stdint.h>
5  #include <stdbool.h>
6  #include <msp.h>
7  #include "gpio.h"
```

Figure 4.32 Header file CPLED.h.

The TI system header file <**msp.h**> contains all definitions and macros for control registers used for peripherals. The header file "**gpio.h**" includes definitions for all GPIO API functions defined in the MSPWare driver library. Since we want to add all system files, including system header files, into this project to make it portable, here we use **gpio.h** as a user-defined header file and cover it with double quotation marks.

4.5.6.5.2 Create the C source file Now, create a new C code file **CPLED.c** and add it into the project. Enter the codes shown in Figure 4.33 into this source file.

Let's have a closer look at this piece of code to see how it works.

- The code in line 4 is used to include the user-defined header file in which all system header files are included.
- The program starts at **main()** entry point at line 5.

```
1  //****************************************************************************
2  //CPLED.c - Main application file for the project - CP LED Project
3  //****************************************************************************
4  #include "CPLED.h"
5  int main(void)
6  {
7    WDTCTL = WDTPW | WDTHOLD;              // stop watchdog timer
8    P4DIR &= ~BIT4;                        // set P4.4 as input pin
9    P2DIR |= BIT2;                         // set P2.2 as output pin
10   while(1)
11   {
12     if (P4IN & BIT4)                     // if SW2 pressed, blue LED on
13       GPIO_setOutputHighOnPin(2, 4);     // P2OUT = BIT2;  BIT2 = 0x4
14     else
15       GPIO_setOutputLowOnPin(2, 4);      // P2OUT &= ~BIT2; BIT2 = 0x4
16   }
17 }
```

Figure 4.33 C source codes for the project CPModel.

- The Watchdog timer is terminated in line 7 to avoid unnecessary timer interrupts.
- In line 8, the GPIO pin **P4.4** is configured as an input pin by using an inverse of **BIT4**, **~BIT4** (BIT4 = **00010000** and~BIT4 = **11101111**), with an **AND** of **P4DIR** register to make sure that BIT4 on **P4DIR** is 0 (input).
- The **P2.2** pin is configured as an output pin in line 9.
- An infinitive **while()** loop is used starting at line 10 to make the following codes to be executed infinitely.
- To check whether the button switch **SW2** on the EduBASE Trainer is pressed or not, the **P4IN** register is **ANDed** with **BIT4** to check if this bit has an input HIGH, which means that the **SW2** is pressed in line 12.
- If the returned value is **True**, the GPIO API function **GPIO_setOutputHighOnPin()** is executed to turn the blue-color LED on in line 13.
- Otherwise the **SW2** is not pressed, the API function **GPIO_setOutputLowOnPin()** is called to turn the blue-color LED off in line 15.
- Both API functions can be replaced by two DRA model instructions, **P2OUT** = **BIT2** and **P2OUT &** = **~BIT2**, to turn the blue-color LED on and off, respectively.

Go to **File | Save** menu item to save this source file and add it into our project.

4.5.6.5.3 Include system header files and add static library into the project Instead of including the system header file path, in this project we want to add all system header files into our project to make it portable. The system header files we need to use include:

- **msp.h**
- **gpio.h**

Since the header file **msp.h** can be automatically added by the MDK-ARM compiler (see Section 3.4.3.3 in Chapter 3) during the compiling time, therefore, we only need to take care of the system header file **gpio.h**.

This header file is located at the folder **C:\ti\msp\MSPWare_2_21_00_39\driverlib\driverlib\ MSP432P4xx**. Perform the following operations to add this header file into our project:

1. Open the Windows Explorer and browse to our project folder, which is **C:\MSP432 Class Projects\Chapter 4\CP LED Project**
2. Go to the folder **C:\ti\msp\MSPWare_2_21_00_39\driverlib\driverlib\ MSP432P4xx** to find, copy, and paste **gpio.h** header file into our project folder listed in step 1.

In this way, we consider the **gpio.h** as a user-defined header file stored in our current project folder, therefore, we have to use double quotation marks " " to cover it in our user-defined header file **CPLED.h** shown in Figure 4.32. It would not work if you try to add this header file directly into the project folder via the Keil® ARM®-MDK µVersion5 IDE.

To add the MSPWare driver library into our project, perform the following operations:

- In the **Project** pane, expand the **Target** folder and right click on the **Source Group 1** folder and select the **Add Existing Files to Group "Source Group 1"**
- Browse to find the library, **msp432p4xx_driverlib.lib**, which is located at the folder: **C:\ti\msp\MSPWare_2_21_00_39\driverlib\driverlib\MSP432P4xx\keil** in your host computer
- Select this library by clicking on it and click on the **Add** button to add it into our project.

Now, we are almost ready to compile, build, and run our project. Before you can do that, make sure that the debugger you are using is the **CMSIS-DAP**. You can do this checking by:

- Going to **Project|Options for Target "Target 1"** menu item to open the options wizard.
- On the opened options wizard, click on the **Debug** tab.
- Make sure that the debugger shown in the **Use:** box is **CMSIS-DAP Debugger**. Otherwise you can click on the drop-down arrow to select this debugger from the list. Also, make sure that the **Settings** for the debugger and flash functions are correct. Refer to Section 4.5.6.3.2.7 for correct settings.

Now, we are ready to compile, load, and run our project.

4.5.6.5.4 Compile and link project to create the image or executable file Go to **Project|Build Target** menu item to compile and link this sample project with system files to make our project image or executable file. A message displaying **0 Error(s)** with **0 Warning(s)** should be displayed in the **Build Output** window at the bottom if nothing is wrong with this sample project.

By using **Flash|Download** menu item, you can download the project image file into the flash memory. Now you can begin to debug this project by going to **Debug|Start/Stop Debug Session**. Click on the **OK** button for the memory limitation for the evaluation version message, and run this program by going to **Debug|Run** menu item. As the program is running, you can find that the blue-color LED will be on as soon as you pressed the **SW2** position switch in the EduBASE Trainer. Otherwise, no LED can be on at all.

Go to **Debug|Stop** menu item to stop the project from running.

At this point, we have finished introductions and discussions about the software, instruction set, and C code programming for this MSP-EXP432P401R Evaluation Board with the EduBASE Trainer.

One more step you can do is: can you use some position switches in the EduBASE Trainer to control the flashing of some LEDs in the EduBASE Trainer, not in the EXP432P401R EVB? This may be considered as one of your homework or labs in this chapter.

4.6 Summary

This chapter is mainly about the introduction and discussion about the software development and application of using MSP432 Series LaunchPad™ ARM® Microcontroller and evaluation board EXP432P401R. The EduBASE Trainer is also a useful tool and development emulator to be used in these software implementations.

Following the overview section, the development architecture and procedure of using the ARM® Cortex®-M4 MCU to build a microcontroller system is introduced. This chapter is basically divided into the following three parts:

1. Introductions and discussions about the ARM® Cortex®-M4 Assembly instruction set, including 14 groups of instructions that are categorized based on their functions and implementations. The instruction set includes:
   a. ARM® Cortex®-M4 assembly instruction structure and syntax
   b. ARM® Cortex®-M4 pseudo assembly instructions
   c. ARM® Cortex®-M4 complete assembly instruction set
   d. ARM® Cortex®-M4 assembly instruction addressing modes
   e. ARM® Cortex®-M4 unsupported assembly instructions
2. Discussions and introductions to ARM® Cortex®-M4 software development procedure, including the Keil® ARM®-MDK µVersion5 IDE, CMSIS Core, device drivers, and MSPWare Software Package and MSP-EXP432P401R Evaluation Board. The MSPWare driver library, including the peripheral driver library, graphical library, USB library, and boot loader, which are provided by MSPWare software package, is also introduced in this part. An introduction to intrinsic instructions is included in this part, which contains:
   a. The standard data types used in intrinsic instructions program
   b. The CMSIS Core-specific intrinsic functions
   c. The Keil® ARM® Compiler-specific intrinsic functions
   d. Inline assembler
3. Using the C codes to build and develop ARM® Cortex®-M4 MCU application programs. This part includes:
   a. An introduction to C programming development guidelines and procedures, especially for the name definitions and conversions in C programming.
   b. Introduction to MSPWare driver library, mainly about the peripheral driver library and GPIO API functions.

c. Introduction to two popular application programming models used in the MSPW are driver library, DRA model, and SD model.
d. Introduction to use the MSPWare driver library to build a sample project DRA LED Project, in which only the DRA method is used.
e. Introduction to use the MSPPDL to build a sample project SD LED Project, in which the static library is used.
f. Introduction to use a combining model method to build a sample project CP LED Project, in which both the MSPWare driver library API functions and the DRA method are used.
g. Introduction to the EduBASE Trainer, including the interface and connection between the Jacks, J1~J4, in the MSP-EXP432P401R EVB and some peripheral devices provided in the EduBASE Trainer.

When finished with this chapter, users can develop and build most simple ARM® Cortex®-M4 MCU application projects by using any models or methods discussed in this chapter.

HOMEWORK

I. True/False Selections

_____1. The *low-level language* codes comprise programming codes similar to English, and are easy to understand by humans.

_____2. The Thumb-2 technology enables a mixture of 16-bit and 32-bit instructions to be executed within one operating state.

_____3. The different addressing modes used in ARM® Cortex®-M4 instructions are to access the memory using the different methods to calculate the target memory address.

_____4. The instruction SXTB extracts the lowest 8-bit value of Rn, Rn [7:0], and sign extends to a 32-bit value.

_____5. The CMP instruction performs a subtraction between two operands, and modifies the contents of two operands when the instruction is done.

_____6. The Keil® ARM-MDK Core includes a μVersion5 IDE, a compiler, and a debugger.

_____7. To access some ARM® Cortex®-M4 CPU registers, one can use intrinsic functions provided by CMSIS Core and Keil® ARM® C/C++ Compiler.

_____8. To develop a standard C application project, one needs to create exactly two files, the header file and application file.

_____9. Two programming models are used in the MSPWare driver library, the DRA model and sequence development (SD) model.

____10. Although all GPIO port registers are 32-bit registers, only the lowest 8 bits in those registers is used.

II. Multiple Choices

1. The following instruction(s), ________________, is (are) ARM® Cortex®-M4 assembly instructions.
   a. ui32int uiLoop
   b. a = b + c
   c. void __WFI(void)
   d. WFI

2. The ARM® Cortex®-M4 assembly instructions contain _______________.
   a. Label field
   b. Operation field and operands field
   c. Operation field, operands field, and comment field
   d. Both a and c
3. The pseudo instruction **DCW 2** is to _______________.
   a. Reserve 2-byte space for a data item in the program
   b. Reserve 2 half-word space for a data item in the program
   c. Reserve 4-byte space for a data item in the program
   d. Reserve 2-word space for a data item in the program
4. The **LDRB R0, [R2, #0x5]** is to load an unsigned byte located at ___________ into the register R0 if the R2 contains 0x40004000.
   a. 0x40000005
   b. 0x40004000
   c. 0x40004005
   d. 0x40000000
5. The instruction **BFC R4, #8, #12;** is to _______________.
   a. Set bits 8–12 in the register R4 to 1
   b. Reset bits 8–12 in the register R4 to 0
   c. Set bits 8–19 in the register R4 to 1
   d. Reset bits 8–19 in the register R4 to 0
6. The MSPWare driver libraries include _______________.
   a. Peripheral driver library
   b. Graphical library
   c. USB library
   d. All of above
7. Two sets of popular intrinsic functions are provided by _______________.
   a. Keil® ARM C/C++ Compiler and CMSIS Core
   b. MSPWare driver library and Keil® ARM-MDK µVersion5 IDE
   c. Keil® ARM® C/C++ Compiler and MSPWare driver library
   d. MSPWare driver library and Keil® ARM® C/C++ Compiler
8. To build a C code project, regularly one needs to develop _______________ file(s).
   a. The implementation and application
   b. The header and application
   c. The header and implementation
   d. The application
9. To use macros or symbolic definitions to define the GPIO Port 2 Output Register, one can use _______________ macro.
   a. **GPIO_PORT2_OUT**
   b. **GPIO_PORT2_OUT_REG**
   c. **P2OUT**
   d. **PORT2_OUT**
10. To use **P1DIR** register to set **P1.0** and **P1.2** pins as output pins, one needs to set the **P1DIR** register with a hexadecimal data value ___________.
    a. 0x3
    b. 0x4
    c. 0x5
    d. 0x6

III. Exercises
1. Provide a brief description about the ARM® Cortex®-M4 assembly instruction syntax
2. Provide a brief description about seven addressing modes used in the ARM® Cortex®-M4 assembly instructions.
3. Explain the difference between the data conversion instructions SXTB and UXTB.
4. Provide a description about the software and tools to be used to build an ARM® Cortex®-M4 MCU application project with MSP432 Series LaunchPad EXP432P401R EVB.
5. Provide a brief description about two programming models used in MSPWare peripheral driver library.

IV. Practical Laboratory

LABORATORY 4 ARM AND MSP432 SOFTWARE AND C PROGRAMMING

4.0 GOALS

This laboratory exercise allows students to practice ARM® Cortex®-M4 instructions and C programming through developing four labs.

1. Program **`Lab4_1(InLineASM)`** lets you practice several ARM® Cortex®-M4 assembly instructions such as **LDR**, **STR**, and **MOV** with inline assembler.
2. Program **`Lab4_2(DIPLED)`** enables you to build a C code program to access and control a three-color LED with three DIP switches in the EduBASE Trainer.
3. Program **`Lab4_3(LEDSPK)`** allows you to control a single LED with one position switch combined with a sound output driven by the speaker in the EduBASE Trainer.
4. Program **`Lab4_4(LED4Buttons)`** provides you a chance to control four LEDs with four position switches in the EduBASE Trainer.

After completion of these programs, you should understand some popular ARM® Cortex®-M4 assembly instructions and the architecture of the inline assembler, as well as architectures and configurations of some popular peripherals installed in the EduBASE Trainer. You should be able to code some interfacing and controlling programs to access those peripherals to perform some closed-loop control functions.

4.1 LAB4_1

4.1.1 Goal

The **`Lab4_1`** project builds a C code program with inline assembler to use some popular ARM® Cortex®-M4 assembly instructions to perform a simulated PUSH and POP operation (Keil® C/C++ compiler does not allow to directly use PUSH and POP instructions in the inline assembler). The tasks of the program are as follows:

- Push 10 data 0, 1, 2,…, 9 into the stack in the order from 0x0 to 0x9.
- Pop these data back to another memory space in the same order.

4.1.2 Data assignment

The 10 data items should be first written into a memory space **`0x20002000`** and read back from another memory space **`0x20003000`**. This memory space change is made by the PUSH and POP instructions. These two memory spaces can be defined by using the **#define** macro:

- **#define WADDRESS 0x20002000**
- **#define RADDRESS 0x20003000**

4.1.3 Development of the source code

Follow the steps below to develop this C source code. Only a C code file is used in this project since it is a simple application without needing any header file.

1. Refer to Section 4.5.7.2.6.1 to create a new folder **Lab4_1(InLineASM)** under the folder **C:\MSP432 Lab Projects\Chapter 4** in the Windows Explorer.
2. Open the Keil® ARM-MDK µVersion5 and go to **Project|New µVersion Project** menu item to create a new µVersion Project. On the opened wizard, browse to our new created folder **Lab4_1(InLineASM)** in step 1 above. Enter **Lab4_1** into the **File name** box and click on the **Save** button to create this project in the folder **Lab4_1(InLineASM)**.
3. On the next wizard, you need to select the device (MCU) for this project. Expand three icons, **Texas Instruments**, **MSP432 Family,** and **MSP432P**, and select the target device **MSP432P401R** from the list by clicking on it. Click on the **OK** to close this wizard.
4. Next the software components wizard is opened, and you need to setup the software development environment for your project with this wizard. Expand two icons, **CMSIS** and **Device**, and check the **CORE** and **Startup** checkboxes in the **Sel.** column, and click on the **OK** button since we need these two components to build our project.
5. In the **Project** pane, expand the **Target** folder and right click on the **Source Group 1** folder and select the **Add New Item to Group "Source Group 1."**
6. Select the **C File (.c)** and enter **Lab4_1** into the **Name:** box, and click on the **Add** button to add this source file into the project.
7. On the top of this C source file, include two system header files, <**stdint.h**> and <**stdbool.h**>
8. Using the **#define** macro to define two memory addresses, **WADDRESS 0x20002000**, and **RADDRESS 0x20003000**.
9. Since we need to use inline assembler to include some ARM® Cortex®-M4 assembly instructions to perform PUSH and POP operations, we need to create two functions to cover these inline assemblers since all inline assemblers must be included inside functions in the Keil® ARM-MDK C/C++ Compiler.
10. Create the first C function named **pushdata()** with one argument **int uidata** and returning a **void** value.
11. Inside the function **pushdata()**, create three local integer variables, **R0**, **R2**, and **SP**, which are three pseudo registers representing three real corresponding registers in the ARM® CPU. You have to use those pseudo registers, not real registers, to perform any data operation and the compiler can perform data transferring between those pseudo and real registers later.
12. Put an inline assembler sign, **__asm**, to tell the compiler that the following codes are assembly codes, not C codes. The point to be noted is that the underscore before the **asm** keyword is a double underscore, not a single one.
13. Put a starting brace under the __asm inline assembler sign to start this assembler block.
14. Use the **MOV** instruction to move the **WADDRESS** into the register **R0** (**#WADDRESS**).

15. Use the **MOV** instruction to move the **R0** into the **SP** to setup the SP to point to the written address **WADDRESS**.
16. Use the **MOV** instruction to move the argument **#uidata** to the register **R2**.
17. Use the **STR** instruction to store the data stored in the **R2** into the **SP** with an immediate offset **#uidata** multiplied by **4**. The reason to multiply by 4 is to align each data to a word (4 bytes). If you do not do this alignment, each data item would be a byte to be stored in each memory address.
18. Put an ending brace to finish this inline assembler block.
19. Put another ending brace to finish the first function.
20. Create the second C function named **popdata()** with one argument **int uidata** and returning a **void** value.
21. Inside the function **popdata()**, create four local integer variables, **R0**, **R1**, **R3**, and **SP**.
22. Put an inline assembler sign, **__asm**, to tell the compiler that the following codes are assembly codes, not C codes.
23. Put a starting brace under the __asm inline assembler sign to start this assembler block.
24. Use the **MOV** instruction to move the **WADDRESS** into the register **R0** (**#WADDRESS**).
25. Use the **MOV** instruction to move the **RADDRESS** into the register **R1** (**#RADDRESS**).
26. Use the **MOV** instruction to move the **R0** into the **SP** to setup the SP to point to the written address **WADDRESS**.
27. Use the **LDR** instruction to load data from the **SP** with an immediate offset **#uidata** multiplied by **4** into the register **R3**.
28. Use the **STR** instruction to store the data in the **R3** into the **R1** with an immediate offset **#uidata** multiplied by **4**.
29. Put an ending brace to finish this inline assembler block. Put another ending brace to finish the second function.
30. Put **int main(void)** to start our main program and put a starting brace under this main function to begin our main program.
31. Declare a **uint32_t** local variable **uiData**.
32. Use a **for** loop with **uiData** as the loop counter, starting at **0** and ending at less than **10**. Put a starting brace under this for loop to start this loop.
33. Call **pushdata()** function to perform the data pushing operation with the argument **uiData**.
34. Put an ending brace to finish this for loop block.
35. Use another for loop with the same argument **uiData** to call **popdata()** function to perform data popping operation.
36. Put another ending brace to finish the main program.

4.1.4 Demonstrate your program by testing the running result

Now, you can compile your program by going to **Project|Build target** menu item. However, before you can download your program into the flash memory in the EVB, make sure that the debugger you are using is the **CMSIS-DAP**. You can do this checking by:

- Going to **Project|Options for Target "Target 1"** menu item to open the options wizard.
- On the opened options wizard, click on the **Debug** tab
- Making sure that the debugger shown in the **Use:** box is **CMSIS-DAP Debugger**. Also, make sure that all **Settings** for the debugger and flash function are correct.

Perform the following operations to run your program and check the running results:

- Go to **Flash|Download** menu item to download your program into the flash ROM.
- Go to **Debug|Start/Stop Debug Session** to begin debugging your program. Click on the **OK** button on the 32-KB memory size limitation message box to continue.
- Then go to **Debug|Run** menu item to run your program.
- To check your running result, do the following checks:
  - Click on the **Memory 1** tab on the Memory 1 window located at the lower-right corner of your screen.
  - Enter the address **0x20002000** into the **Address:** box and press the Enter key.
  - You can find that all 10 data items, from 0 to 9, have been stored in this piece of memory space (**00 00 00 00 01 00 00 00 02 00 00 00 03 00 00 00 04 00 00 00**...).
  - Enter another address **0x20003000** into the **Address:** box and press the Enter key again, you would find the same result in that piece of memory space.
  - Go to Debug|Start/Stop Debug Session to stop your program.

Based on these results, try to answer the following questions:

- Can you tell me whether these results are correct? If they are right, why are these data items arranged in this way?
- Can you modify your codes to remove all multiplied by 4 for each immediate offset? What is the result of your program after this modification?
- What did you learn from this project?

4.2 LAB4_2

4.2.1 Goal

This project is to build a C code program to access and control a three-color LED with three DIP switches in the EduBASE Trainer. Refer to Figure 4.31 to get more details about the hardware configuration for some popular peripherals installed in the EduBASE Trainer. In this lab, we try to use DIP switches 1–3 in the EduBASE Trainer to control a three-color LED, LED2 on the EXP432P401R EVB, with a DRA programming model.

4.2.2 Data assignment

It can be found from Figure 4.31 that the DIP (**SW1**) switches 1~3 are connected to the **P4.4**, **P4.2**, and **P4.0** pins and GPIO pins **P2.0~P2.2** are connected to a three-color LED in the MSP-EXP432P401R EVB. To enable and set the appropriate pins to turn on and off three-color LED, the following data and registers should be used:

- **P4DIR** register should be reset to 0 for bits 4, 2, and 0 to enable **P4.4**, **P4.2**, and **P4.0** to work as input pins to receive the DIP switches pressing status. If any DIP switch is off, the corresponding pin receives a LOW input. Otherwise, a HIGH is received by the corresponding pin if the related DIP switch is on.
- **P2DIR** register should be configured to enable **P2.0~P2.2** as output pins to control the three-color LED to turn it on and off based on the DIP switch status.

4.2.3 Development of the project

Follow the steps below to develop this project. Only a C source file is used in this project since this project is simple. Create the project with the following steps:

1. Create a new folder **`Lab4_2(DIPLED)`** under the folder **`C:\MSP432 Lab Projects\Chapter 4`** in the Windows Explorer.
2. Open the Keil® ARM-MDK µVersion5 and go to **`Project|New µVersion Project`** menu item to create a new µVersion Project. On the opened wizard, browse to our new created folder **`Lab4_2(DIPLED)`** that is created in step 1 above. Enter **`Lab4_2`** into the **File name** box and click on the **Save** button to create this project in the folder **`Lab4_2(DIPLED)`**.
3. On the next wizard, you need to select the device (MCU) for this project. Expand three icons, **`Texas Instruments`**, **`MSP432 Family`**, and **`MSP432P`**, and select the target device **`MSP432P401R`** from the list by clicking on it. Click on the **OK** to close this wizard.
4. Next the Software Components wizard is opened, and you need to setup the software development environment for your project with this wizard. Expand two icons, **`CMSIS`** and **`Device`**, and check the **`CORE`** and **`Startup`** checkboxes in the **Sel.** column, and click on the **OK** button since we need these two components to build our project.

4.2.4 Development of the C source file

1. In the **Project** pane, expand the **`Target`** folder and right click on the **`Source Group 1`** folder and select the **`Add New Item to Group "Source Group 1."`**
2. Select the **`C File (.c)`** and enter **`Lab4_2`** into the **Name:** box, and click on the **Add** button to add this source file into the project.
3. Include the following system header files into this source file first:
   a. **`<stdint.h>`**
   b. **`<stdbool.h>`**
   c. **`<msp.h>`**
4. Place the **`int main(void)`** to start our main program and put a starting brace under this main function to begin our main program.
5. Write a code line to stop the Watchdog timer (refer to Figure 4.26).
6. Configure **`P4.4`**, **`P4.2`**, and **`P4.0`** as input pins.
7. Configure **`P2.0~P2.2`** as output pins.
8. Use an infinitive **`while()`** loop to perform checking DIP switches and turning three-color LED on and off repeatedly.
9. Inside the **`while()`** loop, use (**`P4IN & BIT4`**) to check whether DIP switch 1 is on via **`P4.4`** pin. When DIP switch 1 is on, a HIGH is received on **`P4.4`** pin.
10. If the DIP switch 1 is on, send a 1 to the output register of the GPIO Port 2 to make **`P2.0`** pin HIGH to turn on the red-color LED.
11. Inside the **`while()`** loop, use (**`P4IN & BIT2`**) to check whether DIP switch 2 is on via **`P4.2`** pin. When DIP switch 2 is on, a HIGH is received on **`P4.2`** pin.
12. If the DIP switch 2 is on, send a 1 to the output register of the GPIO Port 2 to make **`P2.1`** pin HIGH to turn on the green-color LED.
13. Inside the **`while()`** loop, use (**`P4IN & BIT0`**) to check whether DIP switch 3 is on via **`P4.0`** pin. When DIP switch 3 is on, a HIGH is received on **`P4.0`** pin.

14. If the DIP switch 3 is on, send a 1 to the output register of the GPIO Port 2 to make **`P2.2`** pin HIGH to turn on the blue-color LED.
15. Some necessary ending braces should be added to finish the **`while()`** loop and the main program.

4.2.5 Setup of the include path and linking of the static library

Since you do not need to use any other system header files and any library in this project, therefore you need to do nothing for this step.

Now you can compile your program by going to **`Project`**|**`Build target`** menu item. However, before you can download your program into the flash memory in the EVB, make sure that the debugger you are using is the **`CMSIS-DAP`**. You can do this checking by:

- Going to **`Project`**|**`Options for Target "Target 1"`** menu item to open the options wizard
- On the opened options wizard, click on the **`Debug`** tab.
- Make sure that the debugger shown in the **`Use:`** box is **`CMSIS-DAP Debugger`**. Also make sure that all **`Settings`** for the debugger and flash function are correct.

4.2.6 Demonstrate your program

Perform the following operations to run your program and check the running results:

- Go to **`Flash`**|**`Download`** menu item to download your program into the flash memory
- Go to **`Debug`**|**`Start/Stop Debug Session`** to begin debugging your program. Click on the **OK** button on the 32KB memory size limitation message box to continue.
- Then go to **`Debug`**|**`Run`** menu item to run your program.

As the project runs, the three-color LED will be on with different colors, such as red, green, and blue, when you slide the DIP (**`SW1`**) switches 1~3 to the ON position. You can also press three button switches, **`SW2`**, **`SW3`**, and **`SW4`** on the EduBASE Trainer to get the same running results (Why?).

Based on these results, try to answer the following questions:

- Can you modify your codes to control any other four LEDs, such as PB0~PB3, which are installed on the EduBASE Trainer by checking the four button switches, such as **`SW2~SW5`** in the EduBASE Trainer?
- What will happen if you slide one DIP switch to the ON position and then slide another DIP switch to the ON position without recovering the first DIP switch back to its OFF location?
- What did you learn from this project?

4.3 LAB4_3

4.3.1 Goal

This project is to build a C code program to access and control a single LED with one position or a DIP switch combined with a sound output driven by the speaker in the EduBASE Trainer. Refer to Figure 4.31 to get more details about the hardware configuration for some popular peripherals installed in the EduBASE Trainer. In this lab, we try to use the position switch **`SW2`** or switch-1 in the DIP-SW1 switch to control the LED PB0 and send the speaker in the EduBASE Trainer with a 400 Hz frequency with a combined programming model.

4.3.2 Data assignment

It can be found from Figure 4.31 that the DIP switch 1 or button switch **SW2** is connected to **P4.4** pin and the LED PB0 is connected to the **P3.2** pin of the GPIO ports in the EXP432P401R EVB. Also, the **P5.6** pin is connected to a speaker via pin 4 in the J4 in the EXP432P401R EVB. To enable and set the clock for these three GPIO ports and pins, the following data and registers should be used:

- Pin 4 on the GPIO Port 4 (**P4.4**) should be configured as an input pin since it is connected to the button switch **SW2** on the EduBASE Trainer.
- Pin 2 on the GPIO Port 3 (**P3.2**) and pin 6 on the GPIO Port 5 (**P5.6**) are set to output pins since they are connected to the LED PB0 and a speaker in the EduBASE Trainer.

4.3.3 Development of the project

Use the steps given below to develop this project. Only a C source file is used in this project since this project is simple. Create the project with the following steps:

1. Create a new folder **Lab4_3(LEDSPK)** under the folder **C:\MSP432 Lab Projects\Chapter 4** in the Windows Explorer.
2. Open the Keil® ARM-MDK µVersion5 and go to **Project|New µVersion Project** menu item to create a new µVersion Project. On the opened wizard, browse to our new created folder **Lab4_3(LEDSPK)** in step 1 above. Enter **Lab4_3** into the **File name** box and click on the **Save** button to save this project in the folder **Lab4_3(LEDSPK)** created in step 1.
3. On the next wizard, you need to select the device (MCU) for this project. Expand three icons, **Texas Instruments**, **MSP432 Family**, and **MSP432P**, and select the target device **MSP432P401R** from the list by clicking on it. Click on the **OK** to close this wizard.
4. Next, the Software Components wizard is opened, and you need to setup the software development environment for your project with this wizard. Expand two icons, **CMSIS** and **Device**, and check the **CORE** and **Startup** checkboxes in the **Sel.** column, and click on the **OK** button since we need these two components to build our project.

4.3.4 Development of the C source file

1. In the **Project** pane, expand the **Target** folder and right click on the **Source Group 1** folder and select the **Add New Item to Group "Source Group 1."**
2. Select the **C File (.c)** and enter **Lab4_3** into the **Name:** box, and click on the **Add** button to add this source file into the project.
3. Include the following header files into this source file first:
   a. **<stdint.h>**
   b. **<stdbool.h>**
   c. **<msp.h>**
   d. **<MSP432Pxx/gpio.h>**
4. Declare the user-defined time delay function **delayTime()**. Refer to Section 4.5.7.2.6.3 to get detailed codes for this function body.

5. Place the **int main(void)** to start our main program and put a starting brace under this main function to begin our main program.
6. Write a code line to stop the Watchdog timer (refer to Figure 4.26).
7. Configure **P4.4** as an input pin.
8. Configure **P3.2** as an output pin.
9. Configure **P5.6** as an output pin.
10. Use an infinitive **while()** loop to perform checking **SW2** switch and (1) turning the single-color LED **PB0** on and off, and (2) sending 400 Hz audio signal to the speaker repeatedly.
11. Inside the **while()** loop, use (**P4IN & BIT4**) to check whether the **SW2** is on via **P4.4** pin. When it is on, a HIGH is received on **P4.4** pin.
12. If the **SW2** is on, send a 1 to the output register of the GPIO Port 3 to make **P3.2** pin HIGH to turn on the red-color LED PB0.
13. Also, use **GPIO_setOutputHighOnPin()** to send a HIGH to the GPIO Port 5 to make **P5.6** pin output a HIGH.
14. Call user-defined time delay function **delayTime()** to delay program by 20000 clocks.
15. Use **GPIO_setOutputLowOnPin()** to send a Low to the GPIO Port 5 to make **P5.6** pin output a Low.
16. Call user-defined time delay function **delayTime()** to delay program by 20000 clocks.
17. If the **SW2** is off, turn LED **PB0** off by sending a 0 to the pin **P3.2**.

4.3.5 Setup of the include path and linking of the static library

Now, you need to include the system header files by adding the include path and link the static MSPWare peripheral driver library by adding that library into the project.

Refer to Section 4.5.7.3.2.4 to add this include path into the project. Perform the following operations to link this library with our project:

- In the **Project** pane, expand the **Target** folder and right click on the **Source Group 1** folder and select the **Add Existing Files to Group "Source Group 1."**
- Browse to find the library, **msp432p4xx_driverlib.lib**, which is located at the folder: **C:\ti\msp\MSPWare_2_21_00_39\driverlib\driverlib\MSP432P4xx\keil** in your host computer.
- Select this library by clicking on it and click on the **Add** button to add it into our project.

Now, you can compile your program by going to **Project|Build target** menu item. However, before you can download your program into the flash ROM in the EVB, make sure that the debugger you are using is the **CMSIS-DAP**. You can do this checking by:

- Going to **Project|Options for Target "Target 1"** menu item to open the options wizard.
- On the opened Options wizard, click on the **Debug** tab.
- Making sure that the debugger shown in the **Use:** box is **CMSIS-DAP Debugger**. Also, make sure that all **Settings** for the debugger and flash function are correct.

4.3.6 Demonstrate your program

Perform the following operations to run your program and check the running results:

- Go to **Flash|Download** menu item to download your program into the flash memory.
- Go to **Debug|Start/Stop Debug Session** to begin debugging your program. Click on the **OK** button on the 32KB memory size limitation message box to continue.
- Then go to **Debug|Run** menu item to run your program.

As the project runs, the LED **PB0** will be ON when you press the **SW2** position switch. Also, a 400 Hz signal is sent to the speaker during the **SW2** switch is pressed or DIP-SW1 is kept in the ON position. The LED **PB0** will be off when you release the **SW2** button, and no signal is sent to the speaker.

Based on these results, try to answer the following questions:

- Can you modify your codes to control any other LED, such as **PB1~PB3**, by checking the other SW switches, such as SW3~SW5 in the EduBASE Trainer?
- Can you modify your codes to send the speaker with other frequency signal?
- What did you learn from this project?

4.4 LAB4_4

4.4.1 Goal

This project is to build a C code program to access and control all four LEDs **PB0~PB3** with all position buttons **SW2~SW5** combined with different sound outputs driven by the speaker in the EduBASE Trainer. Refer to Figure 4.31 to get more details about the hardware configuration for some popular peripherals installed in the EduBASE Trainer. In this lab, we try to build this project with the DRA programming model.

4.4.2 Data assignment

It can be found from Figure 4.31 that four position buttons, **SW2~SW5**, are connected to four pins in GPIO ports, **P4.4**, **P4.2**, **P4.0**, and **P6.1**. Four LEDs, **PB0~PB3**, are connected to four pins in GPIO ports, **P3.2**, **P3.3**, **P2.5**, and **P2.4** in the J1 and J2 at the MSP-EXP432P401R EVB. Also, the **P5.6** pin is connected to a speaker via pin 4 in the J4 in the EXP432P401R EVB. To enable and set related pins on GPIO ports, the following data and registers should be used:

- GPIO pins **P4.4**, **P4.2**, **P4.0**, and **P6.1** should be configured as input pins to receive and check four position buttons **SW2~SW5** status.
- GPIO pins **P3.2**, **P3.3**, **P2.5**, and **P2.4** are configured as output pins to control four LEDs, **PB0~PB3**, work status.
- GPIO pin **P5.6** should be configured as output pin to output 400 Hz audio signal to the speaker installed in the EduBASE Trainer.

The project is designed to perform the following functions:

1. The main clock source is DOC with 3 MHz clock frequency provided after a system reset. Therefore, the clock cycle for this 3 MHz clock is 0.33 μs.

2. When different position buttons is pressed, the related LED will be ON and a related sound with certain frequency is sent to the speaker in the EduBASE Trainer. The relationship between each position button and related sound is
   a. If **SW2** is pressed: A 400 Hz signal is sent to the speaker.
   b. If **SW3** is pressed: A 600 Hz signal is sent to the speaker.
   c. If **SW4** is pressed: An 800 Hz signal is sent to the speaker.
   d. If **SW5** is pressed: A 1000 Hz signal is sent to the speaker.
   e. For all other situations, a 200 Hz signal is sent to the speaker.

The sound signal is sent out with a square waveform, half-cycle is HIGH and another half-cycle is LOW. Therefore, we need to send a HIGH signal with half-cycle time, and send a LOW with another half-cycle to the speaker. The half-cycle time of the sound signal for each frequency can be calculated in this way:

1. Determine the whole cycle time by dividing the desired frequency by 1.
2. Divide the half-cycle time by the system clock cycle time to get the delay time count.
3. Use this delay time count to call a related subroutine to send that signal with that frequency.

For example, for a 400 Hz signal, the calculation steps are

1. The whole cycle time for 400 Hz is: 1/400 = 0.0025 s = 2.5 ms = 2500 μs.
2. The half-cycle time is 1250 μs, which is divided by the system clock cycle time 0.33 μs, the result is 3788.
3. Use this 3788 to call the related subroutine to send this signal to the speaker to get a 400 Hz frequency sound signal.

Table 4.26 shows this relationship between each frequency signal and related half-cycle time, and delay time counts used in this project.

4.4.3 Development of the project and the header file

Using the steps given below to develop this project. Both the header and C source files are used in this project since this project is a little more complicated. Create the project and develop the header file with the following steps:

1. Create a new folder **Lab4_4(LED4Buttons)** under the folder **C:\MSP432 Lab Projects\Chapter 4** in the Windows Explorer.

Table 4.26 Half-cycle delay time for different frequency signals

Desired frequency	Half cycle time (μs)	Delay time count
400 Hz	1250	3788
600 Hz	833	2500
800 Hz	625	1875
1000 Hz	500	1500
1200 Hz	417	1250
1400 Hz	357	1071
200 Hz	2500	7500

2. Open the Keil® ARM-MDK µVersion5 and go to **Project|New µVersion Project** menu item to create a new µVersion Project. On the opened wizard, browse to our new created folder **Lab4_4(LED4Buttons)** in step 1. Enter **Lab4_4** into the **File name** box and click on the **Save** button to create and save this project in the folder **Lab4_4(LED4Buttons)**.
3. On the next wizard, you need to select the device (MCU) for this project. Expand three icons, **Texas Instruments**, **MSP432 Family**, and **MSP432P**, and select the target device **MSP432P401R** from the list by clicking on it. Click on the **OK** to close this wizard.
4. Next, the software components wizard is opened, and you need to setup the software development environment for your project with this wizard. Expand two icons, **CMSIS** and **Device**, and check the **CORE** and **Startup** checkboxes in the **Sel.** column, and click on the **OK** button since we need these two components to build our project.
5. In the **Project** pane, expand the **Target** folder and right click on the **Source Group 1** folder and select the **Add New Item to Group "Source Group 1."**
6. Select the **Header File (.h)** and enter **Lab4_4** into the **Name:** box, and click on the **Add** button to add this header file into the project.
7. Enter the following codes into this header file:

```
#include <stdint.h>
#include <stdbool.h>
#include <msp.h>
void Delay(int time);
void SetSound(int period);
```

The top three are system header files to be used in this project. The following two user-defined functions are used to delay a period for the program and send various audio signals to the speaker when the different position buttons are pressed.

4.4.4 Development of the C source file

1. In the **Project** pane, expand the **Target** folder and right click on the **Source Group 1** folder and select the **Add New Item to Group "Source Group 1."**
2. Select the **C File (.c)** and enter **Lab4_4** into the **Name:** box, and click on the **Add** button to add this source file into the project.
3. Include the following header file into this source file first:

```
"Lab4_4.h"
```

4. Define the first user function **Delay()** with an integer variable **time** as the argument for this function. This function returns a **void**. The function body includes a local integer variable **count** that works as a loop counter and a **for()** loop. The lower bound of the for( ) loop counter is **0**, and the upper bound is the **time**.
5. Define the second user function **SetSound()** with an integer variable **period** as the argument for this function. This function returns a **void**. The function body includes:
   a. Assign logic HIGH to the GPIO pin **P5.6** to send half-cycle HIGH to the speaker.
   b. Call the **Delay()** function we defined in step 4 to delay a **period** of time to keep **P5.6** as HIGH for half-cycle.

   c. Assign logic LOW to the GPIO pin **P5.6** to send a LOW to the speaker.
   d. Call the **Delay()** function we defined in step 4 to delay a **period** of time to keep **P5.6** as LOW for half-cycle.
6. Place the **int main(void)** to start our main program and put a starting brace under this main function to begin our main program.
7. Write a code line to terminate the Watchdog timer.
8. Set GPIO pins **P4.4**, **P4.2,** and **P4.0** as input pins.
9. Configure GPIO pin **P6.1** as an input pin.
10. Set GPIO pins **P3.2** and **P3.3** as output pins.
11. Set GPIO pins **P2.4** and **P2.5** as output pins.
12. Set GPIO pin **P5.6** as an output pin.
13. Use an infinitive **while()** loop to repeatedly turn ON- and OFF-related LED and send various audio signals when related position button **SW** is pressed.
14. Use **(P4IN & BIT4)** to check if the position button **SW2** is pressed.
15. If it is, turn on the **PB0** LED and send a 400 Hz audio signal to the speaker via **P5.6** pin by calling the user-defined function **SetSound()**.
16. Use **(P4IN & BIT2)** to check if the position button **SW3** is pressed.
17. If it is, turn on the **PB1** LED and send a 600 Hz audio signal to the speaker via **P5.6** pin by calling the user-defined function **SetSound()**.
18. Use **(P4IN & BIT0)** to check if the position button **SW4** is pressed.
19. If it is, turn on the **PB2** LED and send a 800 Hz audio signal to the speaker via **P5.6** pin by calling the user-defined function **SetSound()**.
20. Use **(P6IN & BIT1)** to check if the position button **SW5** is pressed.
21. If it is, turn on the **PB3** LED and send a 1000 Hz audio signal to the speaker via **P5.6** pin by calling the user-defined function **SetSound()**.

4.4.5 Setup of the include path and linking of the static library

Since we did not use any system header files and static library in this project, we do not need to include any header file and link any library.

Now, you can compile your program by going to **Project|Build target** menu item. However, before you can download your program into the flash memory in the EVB, make sure that the debugger you are using is the **CMSIS-DAP**. You can do this checking by:

- Going to **Project|Options for Target "Target 1"** menu item to open the options wizard.
- On the opened options wizard, click on the **Debug** tab.
- Making sure that the debugger shown in the **Use:** box is **CMSIS-DAP Debugger**. Also, make sure that all **Settings** for the debugger and flash function are correct.

4.4.6 Demonstrate your program

Perform the following operations to run your program and check the running results:

- Go to **Flash|Download** menu item to download your program into the flash ROM.
- Go to **Debug|Start/Stop Debug Session** to begin debugging your program. Click on the **OK** button on the 32-KB memory size limitation message box to continue.
- Then go to **Debug|Run** menu item to run your program.

As the project runs, the corresponding LEDs will be ON when you press a position button **SW**. Also, different sound signal with different frequency is sent to the speaker during the position button **SW** is pressed.

Based on these results, try to answer the following questions:

- Can you modify your codes to change the control order for each **SW** button and related LED, such as **SW2** to **PB3** LED, **SW3** to **PB2** LED, **SW4** to **PB1** LED, and **SW5** to **PB0** LED in the EduBASE Trainer?
- Can you modify your codes to inverse the frequency range (from high to low) to be sent to the speaker as the related position button (**SW2~SW5**) is pressed?
- What did you learn from this project?

chapter five

MSP432 microcontroller interrupts and exceptions

This chapter provides general information about exceptions and interrupts occurred and handled in the MSP432 and ARM®Cortex®-M4 microcontrollers. The discussion is divided into several parts, which include: the exception and interrupt sources, the priority levels of different exceptions and interrupts, the interrupt vector table used to direct the processor to response and handle the accepted exceptions and interrupts, and the ISR used to process the exception and interrupt requests. Two programming models in MSPWare Peripheral Driver Library (MSPPDL) are covered in these discussions. Both CMSIS-Core and MSPPDL API functions used to access NVIC registers to process exceptions and interrupts are also introduced.

5.1 Overview

Generally, a user's application program comprises a sequence of assembly or high-level instructions stored in the user memory space in most microcontroller systems. The CPU in a microcontroller sequentially executes the user's program by fetching them one by one from the memory to the instructor register (IR) in the CPU, and decoding each of the instructions in the CPU and running them in that sequence. However, in many applications, it is necessary to stop the execution of the current instructions and execute sets of instructions in response to requests from various events or peripheral devices during the user's program running. These requests, called exceptions or interrupts, are often asynchronous to the execution of the user's program or called the main program. Exceptions and interrupts provide a way to temporarily suspend the user's program execution so the CPU can be free to service these exceptions or interrupts requests. After an exception/interrupt has been serviced, the main program resumes as if there had been no interruption. Generally, an exception/interrupt processing system contains three components:

- **`Exception/Interrupt Request`** coming from exceptions or interrupts sources
- **`Exception/Interrupt Monitor and Masking`** installed in the CPU to monitor, enable, or disable any exception or interrupt request. This includes both global and local masking systems
- **`Exception/Interrupt Response`** allowing the CPU to response and handle accepted exception or interrupt request

The exception/interrupt source works as an exception/interrupt request and it is provided by an internal or an external device that needs the interrupting service with an interrupt request sending to the CPU. This exception/interrupt request is sent to the CPU with an interrupt flag signal and waiting for the CPU's response. When the CPU finishes the current instruction, exactly at the end of the third clock cycle of the execution of the current instruction, it checks whether there is any exception or interrupt request by

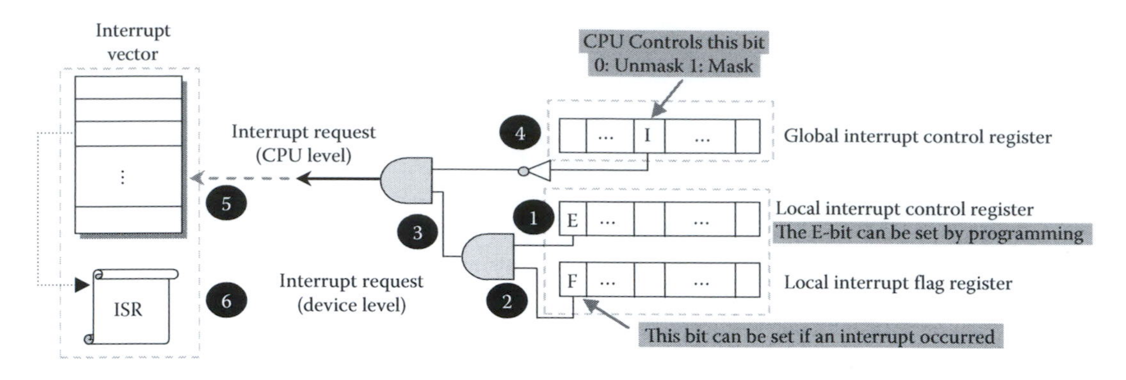

Figure 5.1 Illustration for exception/interrupt processing.

inspecting all interrupt flags. If an interrupt occurs and the interrupt flag has been set with no other higher priority interrupt occurring, the CPU will transfer the control to the ISR to execute the service subroutine to process the request sent by the associated exception or interrupt source.

To successfully identify, accept, and respond to an exception or interrupt request sent by either an internal or an external device, the following working units are necessary:

- A *global exception/interrupt control register*, which is under the control of the CPU, is needed. By resetting or setting certain bits in this global interrupt control register, the CPU can unmask (enable) or mask (disable) all maskable interrupt requests. In ARM® Cortex®-M4 system, the primary interrupt mask (**PRIMASK**) register is equivalent to this register.
- A *local exception/interrupt control register* is related to some associated events or devices. Certain bits on this register can be accessed and controlled by the programer by setting (enabling) or resetting (disabling) the interrupt created by the associated device by programming. In ARM® Cortex®-M4 system, the **BASEPRI** register is equivalent to this register and provides this priority control function.
- An *exception or interrupt vector table* or a collection of the entry addresses of the ISR is needed. This vector table is device dependent or interrupt source dependent, which means that each different exception or interrupt source has a different unique vector. In the ARM® Cortex®-M4 system, a vector table is provided to collect and store entry addresses for all exceptions and interrupts to occur.

Figure 5.1 provides an illustration for exceptions and interrupts processing structure in most microcontroller systems.

5.2 Exceptions and interrupts in the ARM® Cortex®-M4 MCU system

As we discussed in Section 2.2.1.1 in Chapter 2, all exceptions and interrupts are controlled and managed by a NVIC in the ARM® Cortex®-M4 system. The NVIC is a control unit that is embedded in the Cortex-M4 MCU and is used to handle and preprocess all exceptions and interrupts, including maskable and unmaskable interrupts, occurred during the normal running of application codes in the ARM® Cortex®-M4 system.

Like any other interrupt processing unit (IPU), the NVIC processes any exception or interrupt in the following sequence:

1. An exception or interrupt is first created by an exception/interrupt source and an interrupt service request is sent to the Cortex-M4 CPU.
2. Based on the mask register's content (**`PRIMASK`**) and the interrupt priority level (**`BASEPRI`**), the CPU will determine whether to respond or process the interrupt request.
3. If the interrupt request is accepted, the associated hardware will provide interrupt-related information, such as the interrupt source number and related ISR entry point, in a vector table.
4. Before the control can be transferred to the ISR, all related registers, including R0~R3, R12, LR, PSR, and PC, are pushed into the stack to reserve their contents. During this protection process, all other interrupts or exceptions are masked or disabled to avoid any data to be lost.
5. Then the control will be directed to the entry point (entry address of the ISR) stored in the vector table to run the ISR to perform the required interrupt service. During this process, all other interrupts and exceptions are unmasked or enabled to allow higher level priority interrupts or exceptions to be requested and responded.
6. After the ISR is done and before the control can be transferred back to the main program, (a) all other interrupts or exceptions are masked or disabled to avoid any data to be lost, (b) all related registers protected in step 4, including the PC, will be recovered by popping them back to the related registers.
7. Then the control can be directed to the main program to continue executing the normal application codes based on the old PC content. At this time, all interrupts and exceptions are unmasked or enabled to allow any interrupt or exception to be requested and responded.

In fact, there is no significant difference between an exception and an interrupt in the ARM® Cortex®-M4 system. Regularly, an exception is generated by the ARM® processor or the CPU and it has a higher priority level compared with other interrupts. An exception is often generated by an event, such as the system fault event and other system mis-operation event. An interrupt is generally created by an internal or an external peripheral with lower priority level. Figure 5.2 shows an illustration for all exceptions and interrupts used in the ARM® Cortex®-M4 microcontroller system.

It can be found from Figure 5.2 that all exceptions, except the SysTick Time exception and nonmaskable interrupt (NMI), are related to the ARM® processor and generated by the related event in this CPU. All Interrupt Requests (IRQs), including one NMI that can be considered as an exception, are generated either by internal peripherals located in the ARM® Cortex®-M4 EVB or external peripherals via some I/O ports.

Table 5.1 lists the most popular used exceptions and interrupts, including the exception types, numbers, and priority levels. Three exceptions, Reset, NMI, and Hard Fault, have fixed priority levels and cannot be modified by programming.

5.2.1 Exception and interrupt types

It can be found from Table 5.1 that the NIVC in the Cortex-M4 system can handle up to 240 exceptions and interrupt inputs. However, in most real applications, they do not have so many interrupt or exception sources available.

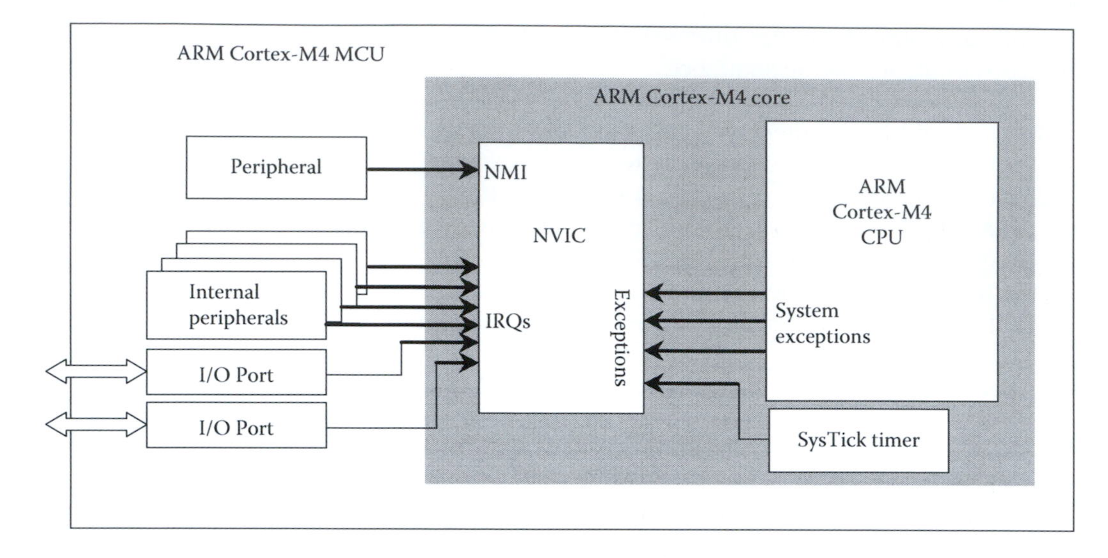

Figure 5.2 Illustration for all exceptions and interrupts in the ARM® NVIC.

There are 15 system exceptions in the Cortex-M4 system, and their exception numbers range from 1 to 15. The exceptions with number greater than 15 and above (until up to 240) are considered as interrupts. The smaller the number, the higher the priority level of the exception or interrupt, and this is true for both exceptions and interrupts.

The top three exceptions, **`Reset`**, **`NMI`**, and **`Hard Fault`**, have fixed priority levels with minus numbers to indicate that these exceptions cannot be masked or disabled by using **`PRIMASK`** register and they have the highest priority level and must be handled first if they are coming. Starting from exception four, the priority levels for all the following exceptions and interrupts can be programmed to the different priority levels based on those exceptions and peripherals applied in your applications. Also, these exceptions and interrupts can be masked or disabled by using **`PRIMASK`** register in your application programs.

To respond to an exception or an interrupt, a related exception handler (EXH) or a related interrupt service subroutine (ISR) is developed and they comprise a piece of program code to process the exception or interrupt.

5.2.2 *Exceptions and interrupts management*

All exceptions and interrupts in the Cortex-M4 MCU are handled by the NVIC. In order to manage and control all of those exceptions and interrupts, the NVIC provides a group of programmable registers. Most of these registers are located inside the NVIC and the system control block (SCB). The SCB can be considered as an assistant to the NVIC to help it to control and manage exceptions and interrupts in the Cortex-M4 interrupt system. The special registers inside the ARM® Cortex®-M4 processor, such as **`PRIMASK`**, **`FAULTMASK`**, and **`BASEPRI`**, also provide mask and unmask, priority level selections for most exceptions and interrupts to support their processing.

Besides NVIC and SCB, the Keil® CMSIS-Core also provides a set of registers, definitions, and functions to support the handling and processing of exceptions and interrupts.

Table 5.1 Exception and interrupt types in the Cortex-M4 system

Exception number	Exception type	Priority level	Vector address	Description
0	—	—	0x0000.0000	Initial SP value
1	Reset	−3 (Highest)	0x0000.0004	Reset
2	NMI	−2	0x0000.0008	Nonmaskable interrupt
3	Hard Fault	−1	0x0000.000C	Hardware-related fault
4	Memory Manage Fault	Programmable	0x0000.0010	Memory Management Fault. MPU violations or program address faults
5	BusFault	Programmable	0x0000.0014	Bus error
6	Usage fault	Programmable	0x0000.0018	Program error
7–10	Reserved	N/A	—	—
11	SVC	Programmable	0x0000.002C	SuperVisor call
12	Debug Monitor	Programmable	0x0000.0030	Debug-related exceptions, such as brealpoints
13	Reserved	N/A	—	—
14	PendSV	Programmable	0x0000.0038	Pendable service call
15	SYSTICK	Programmable	0x0000.003C	System tick timer
16	Interrupt 0	Programmable	0x0000.0040	These interrupts can be generated by on-chip internal peripherals or external peripherals
17	Interrupt 1	Programmable	0x0000.0044	
18	Interrupt 3	Programmable	0x0000.0048	
...	...	...	...	
240	Interrupt 224	Programmable	0x0000.0384	

The CMSIS-Core defines a set of different priority levels called enumeration values for exceptions and interrupts with a set of macros or symbolic definitions for related handlers. Table 5.2 lists the most popular macros for these exceptions and interrupts. Table 5.3 lists the most popular CMSIS-Core functions used to access related special registers to mask, unmask, and set priority levels for selected exceptions and interrupts.

It can be found from Table 5.2 that both exceptions and interrupts are all considered as IRQn and each IRQ has a different priority level. The **`Enumeration (IRQn)`** and **`EXH`** columns defined macros for these IRQn names and handler names, and the users can use these macros to define related exceptions and call-related handlers in their program.

The exception numbers and EXHs for all interrupts are device dependent, and are defined in related device-specified header files in a **`typedef`** structure called **`IRQn`**, which are provided by the different microcontroller vendors.

Table 5.3 shows some exceptions and interrupts accessing functions defined in the CMSIS-Core. The users can use and call these functions to access related registers to configure exception and interrupt in their program. Of course, the users can also directly access those registers in NVIC and SCB in their program to configure related interrupts.

Table 5.2 CMSIS-core exception definitions in the Cortex-M4 system

Exception number	Exception type	CMSIS-core enumeration (IRQn)	Enumeration value	Exception handler
1	Reset	—	—	Reset_Handler
2	NMI	NonMaskableInt_IRQn	−14	NMI_Handler
3	Hard fault	HardFault_IRQ	−13	HardFault_Handler
4	Memory manage fault	MemoryManagement_ IRQn	−12	MemManage_ Handler
5	BusFault	BusFault_IRQn	−11	BusFault_Handler
6	Usage fault	UsageFault_IRQn	−10	UsageFault_ Handler
11	SVC	SVCall_IRQn	−5	SVC_Handler
12	Debug monitor	DebugMonitor_IRQn	−4	DebugMon_ Handler
14	PendSV	PendSV_IRQn	−2	PendSV_Handler
15	SYSTICK	SysTick_IRQn	−1	SysTick_Handler
16	Interrupt 0	(device-specified)	0	(device-specified)
17	Interrupt 1–239	(device-specified)	1–239	(device-specified)

Table 5.3 CMSIS-core functions for interrupt controls

CMSIS-core function	Description
`void NVIC_EnableIRQ(IRQn_Type IRQn)`	Enable an external interrupt
`void NVIC_DisableIRQ(IRQn_Type IRQn)`	Disable an external interrupt
`void NVIC_SetPriority(IRQn_Type IRQn, uint32_t priority)`	Set the priority for an interrupt
`void __enable_irq(void)`	Clear PRIMASK to enable all interrupts
`void __disable_irq(void)`	Set PRIMASK to disable all interrupts
`void NVIC_SetPriorityGrouping(uint32_t PriorityGroup)`	Set priority grouping structure

However, a limitation of using this method is the portable issue, which means that if the porting codes are defined based on different processors, the users' codes may not be compatible with another Cortex-M processor.

For the MSP432P401R MCU, most definitions for related exceptions and interrupts are defined in the system header files, such as **interrupt.h** and **msp432p401r.h**.

The MSPPDL also provides a group of interrupt controller API functions to facilitate users to directly access related NVIC registers to configure and control selected exceptions

Table 5.4 NVIC API functions in the MSPWare peripheral driver library

API function	Description
`void Interrupt_enableInterrupt` (uint32_t interruptNumber)	Enable an interrupt
`void Interrupt_disableInterrupt` (uint32_t interruptNumber)	Disable an interrupt
`bool Interrupt_isEnabled` (uint32_t interruptNumber)	Check if an interrupt has been enabled. Returning a nonzero indicates that the interrupt is enabled
`bool Interrupt_enableMaster` (void)	Enable processor to receive any interrupt. Returning a True means that all interrupts are enabled
`bool Interrupt_disableMaster` (void)	Prevent processor from receiving any interrupt. Returning a True means all interrupts are disabled
`void Interrupt_unpendInterrupt` (uint32_t interruptNumber)	Clear a specified pending interrupt
`void Interrupt_pendInterrupt` (uint32_t interruptNumber)	Set a specified interrupt to be pending status
`void Interrupt_setVectorTableAddress` (uint32_t addr)	Sets the address of the vector table
`uint32_t Interrupt_getVectorTableAddress` (void)	Returns the address of the interrupt vector table
`void Interrupt_enableSleepOnIsrExit` (void)	Enables the processor to sleep when exiting an ISR
`uint8_t Interrupt_getPriority` (uint32_t interruptNumber)	Get the priority level for a specified interrupt
`void Interrupt_setPriority` (uint32_t interruptNumber, uint8_t priority)	Set the priority level for a specified interrupt
`uint32_t Interrupt_getPriorityGrouping` (void)	Get the priority grouping configuration
`void Interrupt_setPriorityGrouping (uint32_t bits)`	Set the priority grouping configuration
`uint8_t Interrupt_getPriorityMask` (void)	Get the current priority masking bits
`void Interrupt_setPriorityMask` (uint8_t priorityMask)	Set the current priority masking level
`void Interrupt_registerInterrupt` (uint32_t interruptNumber, void(*intHandler)(void))	Register an interrupt handler
`void Interrupt_unregisterInterrupt` (uint32_t interruptNumber)	Unregister a specified interrupt

and interrupts. Table 5.4 shows all of these NVIC API functions. When using these NVIC API functions, the following points should be noted:

- If two interrupts with the same priority are asserted at the same time, the one with the lower interrupt number is processed first. The NVIC keeps track of the nesting of interrupt handlers, allowing the processor to return from interrupt context only once all nested and pending interrupts have been handled.

- ISR can be configured in two ways; statically at compile time or dynamically at run time. Static configuration of an ISR is accomplished by editing the interrupt vector table in the application's startup code. When statically configured, the interrupts must be explicitly enabled via the API function **`Interrupt_enableInterrupt()`** before the processor can respond to the interrupt. Statically configuring, the interrupt vector table provides the fastest interrupt response time.
- Alternatively, interrupts can be configured at run-time using another NVIC API function **`Interrupt_registerInterrupt()`**. When using this function, the interrupt must also be enabled as before. Run-time configuration of interrupts adds a small latency to the interrupt response time because the stacking operation (a write to SRAM) and the interrupt handler table fetch (a read from SRAM) must be performed sequentially.
- For the API function **`Interrupt_setVectorTableAddress()`**, this function is designed for advanced users who might want to switch between multiple instances of vector tables, such as between flash memory and SRAM. The argument **`addr`** is the new address of the vector table.

For most general applications and implementations, the users can use the NVIC API functions provided by the MSPPDL to handle any exception and interrupt implemented in their applications. The users can also select to use the CMSIS-Core accessing functions shown in Table 5.3 to access related registers in the NVIC to process those exceptions and interrupts. The point to be noted is that you must select the CMSIS-Core tool when you create your new µVersion5 project with the Keil® ARM®-MDK IDE if you want to use the CMSIS-Core access functions to handle the exceptions and interrupts in applications.

As we discussed in Section 2.3.2 in Chapter 2, the NVIC and **`SCB`** are located in the **`SCS`** at the on-chip memory map with a memory range of **`0xE000E000~0xE000EFFF`**.

5.2.3 *Exception and interrupt processing*

After a system reset, all maskable exceptions and interrupts are disabled with a priority level of zero. In order to use any exception or interrupt, you need to:

1. Set up the priority level for the required exception or interrupt
2. Enable the exception or interrupt generation mechanism in the processor or the peripheral
3. Enable the interrupt in the related register in the NVIC

When these preparation jobs have been done, the processor will run the user's program in a normal sequence and wait for an exception or interrupt coming. The ARM® Cortex®-M4 NVIC provides the following registers to support to access, configure, and respond to all different exceptions and interrupts:

- Interrupt Priority Level Register (0xE000E400~0xE000E4EF)
- Interrupt Set Enable Register (ISER)(0xE000E100~0xE000E11C)
- Interrupt Clear Enable Register (ICER) (0xE000E180~0xE000E19C)
- Interrupt Set Pending Register (0xE000E200~ 0xE000E21C)
- Interrupt Clear Pending Register (0xE000E280~ 0xE000E29C)
- Interrupt Active Status Register (0xE000E300~ 0xE000E31C)

Table 5.5 Most popular NVIC registers used for interrupt controls

NVIC register	CMSIS-core macros	CMSIS-core function	Function
Interrupt Set Enable Registers	NVIC→ISER[0] to NVIC→ISER[7]	void NVIC_EnableIRQ(IRQn)	1→enable
Interrupt Clear Enable Registers	NVIC→ICER[0] to NVIC→ICER[7]	void NVIC_DisableIRQ(IRQn)	1→clear enable
Interrupt Set Pending Registers	NVIC→ISPR[0] to NVIC→ISPR[7]	void NVIC_SetPendingIRQ(IRQn)	1→set pending
Interrupt Clear Pending Registers	NVIC→ICPR[0] to NVIC→ICPR[7]	void NVIC_ ClearPendingIRQ(IRQn)	1→clear pending
Interrupt Active Bits Registers	NVIC→IABR[0] to NVIC→IABR[7]	uint32_t NVIC_GetActive(IRQn)	Get active bits
Interrupt Priority Registers	NVIC→IP[0] to NVIC→IP[239]	void NVIC_SetPriority(IRQn, Pri) uint32_t NVIC_GetPriority(IRQn)	Priority level (8-bit wide) for each INT
Software Trigger Interrupt Register	NVIC→STIR	Write an interrupt number to set its pending status	Write an interrupt number to set its pending status

- Interrupt Controller Type Register (0xE000E004)
- Software Trigger Interrupt Register (0xE000EF00)

Table 5.5 shows the most popular NVIC registers, CMSIS-Core macros, and functions used for exceptions and interrupts in the ARM® Cortex®-M4 system.

Each register takes 32-bit or 4 bytes. For example, the ISER **NVIC→ISER[0]** is a 32-bit register and it takes an address range of **0xE000E100~0xE000E103**, and another ISER **NVIC→ISER[1]** is also a 32-bit register and it takes an address range of **0xE000E104~0xE000E107**. Since the register **NVIC→ISER[0]** is a 32-bit register, it can enable 32 interrupts, each bit is for 1 interrupt source. This means that the bit-0 is for Interrupt 0, bit-1 is for Interrupt 1, and so on. Therefore, this register can be used to enable the first 32 interrupts, from Interrupt 0 through to Interrupt 31. Similarly, the **NVIC→ISER[1]** can be used to enable Interrupt 32~Interrupt 63. For example, if you want to enable the Interrupt 12 using this register, just set bit-12 in the **NVIC→ISER[0]** to 1. You can write the code likes: **NVIC→ISER[0] = 0x00001000** or **NVIC→ISER[0] = 1 << 12**.

5.2.3.1 Exception and interrupt inputs and pending status

When any exception or interrupt occurs, the processor will respond to the exception or interrupt source in one of the following possible ways:

- The interrupt can be masked (disabled) or unmasked (enabled)
- The interrupt can be pending or not pending
- The interrupt can be in an active or inactive state

All of these functions can be achieved by using several NVIC programmable registers, such as Interrupt Set (Clear) Enable Register, Interrupt Set (Clear) Pending Register, and read-only Interrupt Active Status Register listed in Table 5.5.

An exception or an interrupt request can be accepted by the processor if:

- The processor is in the normal running status (not reset or halt state)
- The exception or interrupt is enabled or is in the pending status (exclude Reset, NMI, and Hard Fault)
- The exception or interrupt has higher priority level compared with the current running program or handler

After an exception/interrupt request has been accepted, the NVIC will help the CPU to identify the exact exception/interrupt source and determine the entry address of the related ISR to further process this exception/interrupt. In the early ARM® Cortex®-M MCUs, these jobs were handled by using software, which was low in efficiency and slow in processing speed. Starting from Cortex-M3, these jobs can be handled by hardware with the help of the NVIC. The entry address of the related ISR for the identified exception/interrupt can be easily and quickly located from a so-called interrupt vector table, in which all exception and interrupt sources are located according to their priority levels in order.

5.2.3.2 Exception and interrupt vector table

The vector table is exactly a word collection of all exception and interrupt sources used in the ARM® Cortex®-M4 system. This collection is distributed in a block of memory with a table format and it is in the order of the priority levels of all exception and interrupt sources. Each source in this collection, which takes 4-byte or one word, represents the starting address of an EXH or an ISR for related exception or interrupt. Since the address bus used in the Cortex-M4 is 32-bit wide and the width of each byte in the memory is 8-bit, so 4-byte is needed to store one starting address for the handler or ISR.

This vector table can be located at any area of the system memory. However, the default location for this table is the bottom of the memory. The exact location of the vector table can be determined by a programmable register in the NVIC, the **VTOR**. The content of this register is 0 when a reset operation is performed. Therefore, the vector table is located at address 0x00000000 after a system reset.

The starting address of each EXH or interrupt ISR can also be called a vector since it provides not only a direction to the related handler or ISR but also a detailed entry address of the handler or ISR. As we mentioned, each vector takes 4-byte space in the memory. To access the desired vector, the selected exception number or vector number in the vector table must be multiplied by 4, and then go to that resulted address to pick up the desired vector for the selected exception/interrupt. For example, the system reset is considered as an exception and its exception number is 1 (the highest priority level). To get its vector or the starting address of its ISR, the target address for this vector in the vector table is: 1×4 = 0x00000004. Inside this address following with 4 continuous bytes, or from 0x00000004~0x00000007, a 32-bit starting address of the ISR related to the reset is stored.

The LSB of each vector indicates if the exception is to be processed in the Thumb state. Because all Cortex-M processors support only Thumb instructions, the LSB of all exception and interrupt vectors should be 1. Table 5.6 shows a typical vector table with related starting addresses for most popular EXHs and interrupt ISRs.

Since the actual vector for each EXH or interrupt ISR can be modified either statically or dynamically by programming via **VTOR**, therefore the real value for each vector cannot be indicated at this moment. For detailed default vector table and interrupt related vectors, refer to **startup _ msp432.s** file and this file contains the definitions for all exceptions and interrupts used in the MSP432P401R microcontroller system. This file should be

Table 5.6 Vector table used in the Cortex-M4 microcontroller

Exception number	Memory address	Related vector	LSB
......	......	......	
19	0x0000004C	Interrupt 3 vector	[1]
18	0x00000048	Interrupt 2 vector	[1]
17	0x00000044	Interrupt 1 vector	[1]
16	0x00000040	Interrupt 0 vector	[1]
15	0x0000003C	SysTick vector	[1]
14	0x00000038	PendSV vector	[1]
13	0x00000034	Reserved	
12	0x00000030	Debug monitor vector	[1]
11	0x0000002C	SVC Vector	[1]
10	0x00000028	Reserved	
9	0x00000024	Reserved	
8	0x00000020	Reserved	
7	0x0000001C	Reserved	
6	0x00000018	Usage fault vector	[1]
5	0x00000014	BusFault vector	[1]
4	0x00000010	MemManage vector	[1]
3	0x0000000C	HardFault vector	[1]
2	0x00000008	NMI vector	[1]
1	0x00000004	Reset vector	[1]
0	0x00000000	MSP initial value	

loaded into your project when you selected the **`Device|Startup`** tool as you created your new project. The LSB for each vector must be 1 to indicate that this is compatible for the Thumb instructions.

5.2.3.3 Definitions of the priority levels

All exceptions and interrupts in the Cortex-M4 system have certain priority levels, either maskable or unmaskable sources. Most maskable interrupts have programmable priority levels, but all NMI have fixed priority levels. When an exception or interrupt occurs, the NVIC performs a comparison between the priority level of current exception or interrupt and the priority level of the new coming exception/interrupt. The current running task will be suspended and the control will be transferred to the service routine of the new coming exception/interrupt if the priority level of the new coming exception/interrupt is higher.

In the ARM® Cortex®-M4 system, the interrupt priority levels are controlled by the Interrupt Priority Registers (IPRs), as shown in Table 5.5. Each priority register can use 3 bits, 4 bits, or 8 bits to cover all priority levels used in the priority control system. Totally, 8 priority levels can be used if 3 bits are used in this register, and 16 priority levels can be obtained if 4 bits are used in this register. Devices in the MSP432 family support up to 64 interrupt sources and 8 priority levels, which means that 3 bits are used in the priority register in the MSP432P401R MCU.

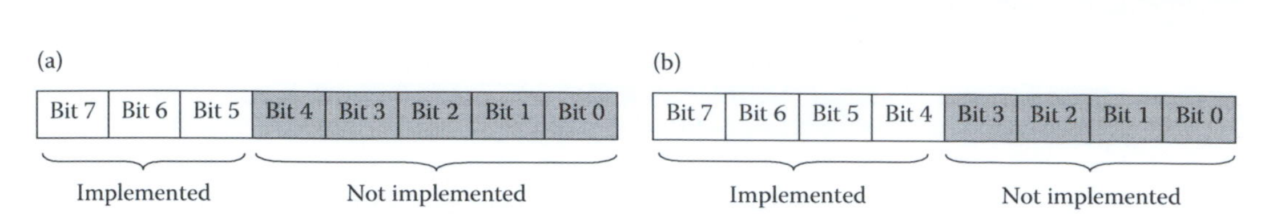

Figure 5.3 Priority levels used in the priority register.

Figure 5.3 shows an illustration of using 3 bits and 4 bits in the priority level configuration register. Totally, 8 and 16 priority levels are generated and available to the NVIC unit.

In Figure 5.3a, three bits (bits 7~5) are used in the priority register to provide eight priority levels. Since bits 4~0 are not used, they are always read out as 0. Therefore, the eight priority levels are: 0x00, 0x20, 0x40, 0x60, 0x80, 0xA0, 0xC0, and 0xE0.

Similarly, four bits are used in the priority register in Figure 5.3b. Therefore, 16 priority levels are provided by this kind of bits setting. The reason for using only the MSB, not LSB, in the priority register is that a 4-bit priority configuration can run on devices with 3-bit priority configuration register.

In some more complex applications, the **`group priority`** and **`sub-priority`** configuration may be used to extend the functions of the priority levels. For example, for an 8-bit priority configuration register, the upper 4-bit is defined as the group priority and the lower 4-bit is considered as the subpriority. Of course, it is unnecessary to divide the group and subpriority in this even way, any combinations, such as 7-bit for group and 1-bit for subpriority for an 8-bit priority register configuration, and 2-bit for group and 1-bit for subpriority for a 3-bit priority configuration, are acceptable and work.

Table 5.7 shows an example of bit fields and their values for the different combinations of group priority and subpriority. In MSP432P401R MCU system, only the last four groups, from group priority 4 to group priority 7, are available since only three bits (bits 7~5) are used for the priority levels definitions in this system.

The reason for using the group priority level is to determine whether an interrupt that has the same priority level as the one that is currently being processed by the processor can be accepted. If this happened, the processor will further check the subpriority levels for both interrupts to determine whether the new coming interrupt should be accepted or not. Generally, the subpriority that has higher level (lower value) should be accepted and processed. Tables 5.3 and 5.4 provide some popular functions, either CMSIS-Core supported functions or the MSPPDL supported API functions, to facilitate to set up and access these group priority level configurations.

Table 5.7 Definitions of group priority and subpriority fields

Priority group	Group priority field	Subpriority field
0 (default)	Bits 7~1	Bit 0
1	Bits 7~2	Bits 1~0
2	Bits 7~3	Bits 2~0
3	Bits 7~4	Bits 3~0
4	Bits 7~5	Bits 4~0
5	Bits 7~6	Bits 5~0
6	Bit 7	Bits 6~0
7	None	Bits 7~0

Refer to Table 5.1 to get a detailed description about the priority levels for most exceptions and interrupts used in the ARM® Cortex®-M4 microcontroller system. Next, let's concentrate on the exceptions and interrupts used in the MSP432P401R MCU system.

5.3 Exceptions and interrupts in the MSP432P401R microcontroller system

In addition to 15 exceptions, the Cortex-M4 processor on MSP432P401R MCUs implements a NVIC with 64 external interrupt lines and 8 levels of priority. From an application perspective, the interrupt sources at the MCU level are divided into two classes, the NMI and the User Interrupts. Internally, the CPU exception model handles the various exceptions (internal and external events, including CPU instruction, memory, and BusFault conditions) in a fixed and configurable order of priority.

The NMI input of the NVIC has the following possible interrupt sources:

- External NMI pin (if configured in NMI mode)-**PIN_FLG|PIN_SRC**
- Oscillator fault condition-**CS_FLG|CS_SRC**
- PSS generated interrupts-**PSS_FLG|PSS_SRC**
- PCM generated interrupts-**PCM_FLG|PCM_SRC**

The source that finally feeds the NMI of the NVIC is configured through the NMI Control and Status Register (**SYS_NMI_CTLSTAT**). The bit field and bit function for this register are shown in Table 5.8.

Table 5.8 Bit field and function of the NMI control and status register

Bit	Field	Type	Reset	Function
31:20	Reserved	RO	00	Reserved and reads as 0
19	PIN_FLG	RW	0	0: The RSTn/NMI pin was not the source of NMI 1: The RSTn/NMI pin was the source of NMI
18	PCM_FLG	R	0	0: The PCM interrupt was not the source of NMI 1: The PCM interrupt was the source of NMI
17	PSS_FLG	R	0	0: The PSS interrupt was not the source of NMI 1: The PSS interrupt was the source of NMI
16	CS_FLG	R	0	0: The CS interrupt was not the source of NMI 1: The CS interrupt was the source of NMI
15:4	Reserved	RO	00	Reserved and reads as 0
3	PIN_SRC	RW	0	0: Configure RSTn/NMI pin as a source of POR Class Reset 1: Configure RSTn/NMI pin as a source of NMI Note: Setting this bit to 1 prevents the RSTn pin from being used as a reset. An NMI is triggered by the pin only if a negative edge is detected
2	PCM_SRC	RW	1	0: Disable the PCM interrupt as a source of NMI 1: Enable the PCM interrupt as a source of NMI
1	PSS_SRC	RW	1	0: Disable the PSS interrupt as a source of NMI 1: Enable the PSS interrupt as a source of NMI
0	CS_SRC	RW	1	0: Disable the CS interrupt as a source of NMI 1: Enable the CS interrupt as a source of NMI

It can be found from Table 5.8 that the top four bits (bits 19~16) are NMI interrupt source status bits used to monitor any of the four NMI interrupt status. A value of 1 on the related bit indicates that a corresponding NMI interrupt is triggered by a NMI source (_**FLG**). The lower four bits (bits 3~0) are used to control (enable or disable) four NMI interrupt sources (_**SRC**). By default, the **RSTn/NMI** pin has been set as a source of Power-On-Reset (POR) interrupt source, which means that as soon as the power is on, a reset is executed and a POR interrupt is generated and sent to this pin. Also the other three NMI interrupts, PCM, PSS, and CS, are enabled by default.

MSP432P401R MCU supports 15 exceptions and up to 64 interrupt sources. Figures 5.4 and 5.5 show the vector table for these exceptions and interrupt sources, as well as the exception and interrupt processing handlers.

Generally, users can directly use these handler macros in their applications. However, alternatively users can also modify these handlers' names to match their applications. The key issue is that you need to make sure that both handlers' names are identical in both vector tables shown in Figures 5.4, 5.5, and your applications.

As shown in Figures 5.4 and 5.5, all exceptions and interrupts are handled by different handlers or ISR based on the exception and interrupt sources. This can be summarized as follows:

1. All maskable Interrupt Requests (IRQs) are handled by the related ISRs.
2. All faults, including the Hard Fault, Memory Management Fault, Usage Fault, and BusFault are handled by the fault handlers.
3. All other exceptions, including NMI, PendSV, SVCall, SysTick, and the fault exceptions, belong to system exceptions and are handled by system handlers.

```
__Vectors   DCD   __initial_sp          ;Top of Stack
            DCD   Reset_Handler         ;Reset Handler
            DCD   NMI_Handler           ;NMI Handler
            DCD   HardFault_Handler     ;Hard Fault Handler
            DCD   MemManage_Handler     ;MPU Fault Handler
            DCD   BusFault_Handler      ;Bus Fault Handler
            DCD   UsageFault_Handler    ;Usage Fault Handler
            DCD   0                     ;Reserved
            DCD   0                     ;Reserved
            DCD   0                     ;Reserved
            DCD   0                     ;Reserved
            DCD   SVC_Handler           ;SVCall Handler
            DCD   DebugMon_Handler      ;Debug Monitor Handler
            DCD   0                     ;Reserved
            DCD   PendSV_Handler        ;PendSV Handler
            DCD   SysTick_Handler       ;SysTick Handler
```

Figure 5.4 Exceptions and their handlers in MSP432P401R MCU system.

	Handler_Name	Interrupt_Number
DCD	PSS_IRQHandler	;0: PSS Interrupt
DCD	CS_IRQHandler	;1: CS Interrupt
DCD	PCM_IRQHandler	;2: PCM Interrupt
DCD	WDT_A_IRQHandler	;3: WDT_A Interrupt
DCD	FPU_IRQHandler	;4: FPU Interrupt
DCD	FLCTL_IRQHandler	;5: FLCTL Interrupt
DCD	COMP_E0_IRQHandler	;6: COMP_E0 Interrupt
DCD	COMP_E1_IRQHandler	;7: COMP_E1 Interrupt
DCD	TA0_0_IRQHandler	;8: TA0_0 Interrupt
DCD	TA0_N_IRQHandler	;9: TA0_N Interrupt
DCD	TA1_0_IRQHandler	;10: TA1_0 Interrupt
DCD	TA1_N_IRQHandler	;11: TA1_N Interrupt
DCD	TA2_0_IRQHandler	;12: TA2_0 Interrupt
DCD	TA2_N_IRQHandler	;13: TA2_N Interrupt
DCD	TA3_0_IRQHandler	;14: TA3_0 Interrupt
DCD	TA3_N_IRQHandler	;15: TA3_N Interrupt
DCD	EUSCIA0_IRQHandler	;16: EUSCIA0 Interrupt
DCD	EUSCIA1_IRQHandler	;17: EUSCIA1 Interrupt
DCD	EUSCIA2_IRQHandler	;18: EUSCIA2 Interrupt
DCD	EUSCIA3_IRQHandler	;19: EUSCIA3 Interrupt
DCD	EUSCIB0_IRQHandler	;20: EUSCIB0 Interrupt
DCD	EUSCIB1_IRQHandler	;21: EUSCIB1 Interrupt
DCD	EUSCIB2_IRQHandler	;22: EUSCIB2 Interrupt
DCD	EUSCIB3_IRQHandler	;23: EUSCIB3 Interrupt
DCD	ADC14_IRQHandler	;24: ADC14 Interrupt
DCD	T32_INT1_IRQHandler	;25: T32_INT1 Interrupt
DCD	T32_INT2_IRQHandler	;26: T32_INT2 Interrupt
DCD	T32_INTC_IRQHandler	;27: T32_INTC Interrupt
DCD	AES256_IRQHandler	;28: AES256 Interrupt
DCD	RTC_C_IRQHandler	;29: RTC_C Interrupt
DCD	DMA_ERR_IRQHandler	;30: DMA_ERR Interrupt
DCD	DMA_INT3_IRQHandler	;31: DMA_INT3 Interrupt
DCD	DMA_INT2_IRQHandler	;32: DMA_INT2 Interrupt
DCD	DMA_INT1_IRQHandler	;33: DMA_INT1 Interrupt
DCD	DMA_INT0_IRQHandler	;34: DMA_INT0 Interrupt
DCD	PORT1_IRQHandler	;35: PORT1 Interrupt
DCD	PORT2_IRQHandler	;36: PORT2 Interrupt
DCD	PORT3_IRQHandler	;37: PORT3 Interrupt
DCD	PORT4_IRQHandler	;38: PORT4 Interrupt
DCD	PORT5_IRQHandler	;39: PORT5 Interrupt
DCD	PORT6_IRQHandler	;40: PORT6 Interrupt
DCD	0	;41: Reserved
DCD	…	;…: Reserved
DCD	0	;64: Reserved

Figure 5.5 Interrupts and their handlers in MSP432P401R MCU system.

Generally, both the second and the third handlers are developed and controlled by the microcontroller vendors. Therefore in this section, we only pay our attention to the first one.

The interrupts are widely applied in all events and peripherals in the MSP432P401R MCU system, but one of the most popular peripherals is the GPIO. In this section, we try to introduce the interrupts used by the GPIO ports to illustrate how to use GPIO-related interrupt mechanism to handle different interrupts in this system since all other peripherals in this MCU system used the similar interrupt handling configurations and procedures. In this way, we do not need to discuss the interrupt configurations and procedures for all other peripherals one by one.

So we will concentrate on the interrupts used in the GPIO ports at this part.

In MSP432P401R MCU system, the GPIO-related interrupts are controlled by three layers or three levels with different control components:

1. Local interrupt configurations and controls for each GPIO pin (four interrupt control registers)
2. Local interrupt configurations and controls for each GPIO port (NVIC interrupt control registers)
3. Global interrupt configurations and controls for all peripherals by the processor (PRIMASK and BASEPRI registers)

These interrupt controls and configurations are performed by different registers that belong to the various components. Let's discuss these one by one in details in the following sections.

5.3.1 *Local interrupts configurations and controls for GPIO pins*

As we discussed in Section 2.3.4.2 in Chapter 2, four interrupt control and status registers are used to process interrupts occurred in the GPIO ports and pins. These registers include:

- **`GPIO Interrupt Enable Register`** (**`PxIE`**): Allows selected pins to cause interrupts
- **`GPIO Interrupt Edge Select Register`** (**`PxIES`**): Determines the interrupt signal edges (rising or falling) to trigger the interrupt mechanism
- **`GPIO Interrupt Vector Register`** (**`PxIV`**): Stores a vector identifier to the GPIO pin that generates an interrupt. This vector identifier is combined with the port interrupt vector in the vector table together to form the target vector (entry of the ISR)
- **`GPIO Interrupt Flag Register`** (**`PxIFG`**): Indicates the raw interrupt status for a pin

All of these registers are 32-bit, but only the lowest 8-bit is used and each bit is corresponding to each pin in the selected GPIO port, bit 0 is for pin 0, bit 1 is for pin 1, and so on. Table 5.9 shows the bit values and their functions for these registers. Only the lowest 5 bits are used for the **`PxIV`** registers.

In Sections 2.3.4.2.7 through 2.3.4.2.10 in Chapter 2, we have provided detailed discussions about these registers. Refer to those sections to get more details about these registers.

Table 5.9 Bit values and functions for GPIO interrupt *control registers*

GPIO register	Each bit value (lowest 8-bit) and each pin function
`PxIE`	`0:` Interrupt for the selected pin is disabled `1:` Interrupt for the selected pin is enabled
`PxIES`	`0:` Interrupt is triggered by a low-to-high pulse (Rising edge) `1:` Interrupt is triggered by a high-to-low pulse (Falling edge)
`PxIV`	`0x02~0x10:` Store a vector identifier to the GPIO pin that generates an interrupt
`PxIFG`	Store an interrupt flag to indicate the GPIO pin whose interrupt is pending

When using these registers to configure and process interrupts generated by any GPIO pin, the following key points should be kept in mind:

- In MSP432P401R MCU system, only GPIO ports 1~6 are interrupt enabled, which means that only P1~P6 have the interrupt mechanism to provide interrupt function. All other GPIO ports, including P7~P11 and PJ, have no interrupt function.
- To enable a GPIO pin to provide interrupt function, the corresponding pin must be enabled by setting the related bit in the **`PxIE`** register.
- If multiple GPIO pins in a single GPIO port are set to enable interrupts, the **`PxIFG.0`** pin has the highest priority level and the **`PxIFG.7`** pin has the lowest priority level. The issue is that all of these interrupts will be handled by the **`PORTx_IRQHandler()`** as shown in Figure 5.5 if these interrupts occur. A good way to process these interrupts is to identify all possible interrupt sources one by one first by checking the related **`PxIV`** register, and then response to each identified interrupt by executing the related interrupt handler.
- Any access (read or write) of the **`PxIV`** register automatically resets the highest pending interrupt flag. If another interrupt flag is set, another interrupt is immediately generated after servicing the initial interrupt.
- When using the interrupt flag register **`PxIFG`**, special attention should be paid. The point is that any writing to **`PxOUT`**, **`PxDIR`**, or **`PxREN`** register can result in setting the corresponding **`PxIFG`** flags. Therefore, a good way to avoid this kind of possible error is: All GPIO port initializations and configurations should be completed before setting the **`PxIE`** register for the GPIO pins to enable their interrupts. In other words, to enable GPIO pins, interrupt function is the last step in any application.
- When using **`PxIES`** register to define the interrupt trigger edges, special attention should be paid to the content on the **`PxIN`** register. As shown in Table 5.10, when setting a rising edge on **`PxIES`** register with the **`PxIN`** = 0, the **`PxIFG`** may be set. The same thing happens when setting a falling edge on **`PxIES`** register with the **`PxIN`** = 1. To avoid this kind of error, a good way is to set a falling edge in the **`PxIES`** register with **`PxIN`** = 0, and set a rising edge in the **`PxIES`** with the **`PxIN`** = 1. Another way is to clear **`PxIFG`** register after all configurations are done and before the pin is enabled by setting the **`PxIE`** register.

5.3.2 Local interrupts configurations and controls for GPIO ports

In addition to configure and initialize each GPIO pins, one also needs to configure the priority levels for the selected GPIO ports. In addition to using the **`PxIFG`** registers, these jobs

Table 5.10 Relationship between writing PxIES register and the resulting PxIFG register

PxIES	PxIN	PxIFG
0→1 (Rising Edge)	0	May be set
0→1 (Rising Edge)	1	Unchanged
1→0 (Falling Edge)	0	Unchanged
1→0 (Falling Edge)	1	May be set

can also be handled by using some NVIC interrupt control registers. In fact, two registers are necessary to be configured and initialized to make GPIO ports to correctly generate interrupt requests to be responded by the processor.

In Section 5.2.3, we have provided a brief review about the most popular NVIC interrupt control registers. Among them, two register groups are very useful and they are:

- Interrupt Priority Level Registers (**0xE000E400~0xE000E4EF**)
- ISERs (**0xE000E100~0xE000E11C**)

These two register groups are used to set the priority levels for all peripherals and enable selected peripherals.

5.3.2.1 NVIC interrupt priority level registers

The NVIC Interrupt Control Unit provides **35** Interrupt Priority Level Registers as a group used to configure all peripheral, including both internal and external peripherals, interrupt priority levels. These registers are arranged from **NVIC_IPR0~NVIC_IPR34**, and each register can be considered as a group **IPRn** (**n** = 0~34 is the group number) and is a 32-bit register taking 4 bytes memory space starting from **0xE000E400** to **0xE000E4EF**.

Each of these 32-bit priority register group can be ordered from 0 to 34 and it is divided into four segments with each segment having 8 bits. Each 8-bit segment only used the upper 3-bits as the priority level bits for each different peripheral. Therefore, $35 \times 4 = 140$ different priority levels can be defined and used for 140 peripherals. Figure 5.6 shows an example of this kind of priority level register, group 0 or **NVIC_IPR0**.

It can be found from Figure 5.6 that **IPR0** is a group 0 priority register with four segments:

1. Segment 1—Bits 7~5: **IPR0_0** interrupts priority level control (INTA)
2. Segment 2—Bits 15~13: **IPR0_1** interrupts priority level control (INTB)
3. Segment 3—Bits 23~21: **IPR0_2** interrupts priority level control (INTC)
4. Segment 4—Bits 31~29: **IPR0_3** interrupts priority level control (INTD)

Table 5.11 shows the relationship between each interrupt, its handler, and its priority bits.

	31	30	29	28	27	26	25	24	23	22	21	20	19	18	17	16
		INTD				Reserved				INTC				Reserved		
Type	RW	RW	RW	RO	RO	RO	RO	RO	RW	RW	RW	RO	RO	RO	RO	RO
reset	0	0	0	0	0	0	0	0	0	0	0	0	0	0	0	0

	15	14	13	12	11	10	9	8	7	6	5	4	3	2	1	0
		INTB				Reserved				INTA				Reserved		
Type	RW	RW	RW	RO	RO	RO	RO	RO	RW	RW	RW	RO	RO	RO	RO	RO
reset	0	0	0	0	0	0	0	0	0	0	0	0	0	0	0	0

Figure 5.6 Example of the NVIC priority level register IPR0.

Table 5.11 Relationship between vectors and NVIC definitions

Vector address	Exception number	IRQ number	ISR name in startup_TM4C123.s	NVIC macros for priority register	Priority bits
0x00000038	14	−2	PendSV_Handler	NVIC_SYS_PRI3_R	23–21
0x0000003C	15	−1	SysTick_Handler	NVIC_SYS_PRI3_R	31–29
0x00000040	16	0	PSS_IRQHandler	NVIC_IPR0	7–5
0x00000044	17	1	CS_IRQHandler	NVIC_IPR0	15–13
0x00000048	18	2	PCM_IRQHandler	NVIC_IPR0	23–21
0x0000004C	19	3	WDT_A_IRQHandler	NVIC_IPR0	31–29
0x00000050	20	4	FPU_IRQHandler	NVIC_IPR1	7–5
0x00000054	21	5	FLCTL_IRQHandler	NVIC_IPR1	15–13
0x00000058	22	6	COMP_E0_IRQHandler	NVIC_IPR1	23–21
0x0000005C	23	7	COMP_E1_IRQHandler	NVIC_IPR1	31–29
0x00000060	24	8	TA0_0_IRQHandler	NVIC_IPR2	7–5
0x00000064	25	9	TA0_N_IRQHandler	NVIC_IPR2	15–13
0x00000068	26	10	TA1_0_IRQHandler	NVIC_IPR2	23–21
0x0000006C	27	11	TA1_N_IRQHandler	NVIC_IPR2	31–29
0x00000070	28	12	TA2_0_IRQHandler	NVIC_IPR3	7–5
0x00000074	29	13	TA2_N_IRQHandler	NVIC_IPR3	15–13
0x00000078	30	14	TA3_0_IRQHandler	NVIC_IPR3	23–21
0x0000007C	31	15	TA3_N_IRQHandler	NVIC_IPR3	31–29
0x00000080	32	16	EUSCIA0_IRQHandler	NVIC_IPR4	7–5
0x00000084	33	17	EUSCIA1_IRQHandler	NVIC_IPR4	15–13
0x00000088	34	18	EUSCIA2_IRQHandler	NVIC_IPR4	23–21
0x0000008C	35	19	EUSCIA3_IRQHandler	NVIC_IPR4	31–29
0x00000090	36	20	EUSCIB0_IRQHandler	NVIC_IPR5	7–5
0x00000094	37	21	EUSCIB1_IRQHandler	NVIC_IPR5	15–13
0x00000098	38	22	EUSCIB2_IRQHandler	NVIC_IPR5	23–21
0x0000009C	39	23	EUSCIB3_IRQHandler	NVIC_IPR5	31–29
0x000000A0	40	24	ADC14_IRQHandler	NVIC_IPR6	7–5
0x000000A4	41	25	T32_INT1_IRQHandler	NVIC_IPR6	15–13
0x000000A8	42	26	T32_INT2_IRQHandler	NVIC_IPR6	23–21
0x000000AC	43	27	T32_INTC_IRQHandler	NVIC_IPR6	31–29
0x000000B0	44	28	AES256_IRQHandler	NVIC_IPR7	7–5
0x000000B4	45	29	RTC_C_IRQHandler	NVIC_IPR7	15–13
0x000000B8	46	30	DMA_ERR_IRQHandler	NVIC_IPR7	23–21
0x000000BC	47	31	DMA_INT3_IRQHandler	NVIC_IPR7	31–29
0x000000C0	48	32	DMA_INT2_IRQHandler	NVIC_IPR8	7–5
0x000000C4	49	33	DMA_INT1_IRQHandler	NVIC_IPR8	15–13
0x000000C8	50	34	DMA_INT0_IRQHandler	NVIC_IPR8	23–21
0x000000CC	51	35	PORT1_IRQHandler	NVIC_IPR8	31–29
0x000000D0	52	36	PORT2_IRQHandler	NVIC_IPR9	7–5

(Continued)

Table 5.11 (Continued) Relationship between vectors and NVIC definitions

Vector address	Exception number	IRQ number	ISR name in startup_TM4C123.s	NVIC macros for priority register	Priority bits
0x000000D4	53	37	PORT3_IRQHandler	NVIC_IPR9	15–13
0x000000D8	54	38	PORT4_IRQHandler	NVIC_IPR9	23–21
0x000000DC	55	39	PORT5_IRQHandler	NVIC_IPR9	31–29
0x000000E0	56	40	PORT6_IRQHandler	NVIC_IPR10	7–5
0x000000E4	57	41	Reserved	NVIC_IPR10	15–13
0x000000E8	58	42	Reserved	NVIC_IPR10	23–21
...	...	...	...	...	...
0x00000140	80	64	Reserved	NVIC_IPR11	31–29

When using Table 5.11 to build the user's program to access NVIC registers directly to configure and set up related exceptions and interrupts, the following points should be noted:

1. The third column in Table 5.11 lists the interrupt numbers or IRQ numbers for all interrupts. Note that the interrupts that have IRQ numbers 0~3 are controlled by the same priority level register (group **0**) **NVIC_IPR0**, and this means that each group **n** register **NVIC_IPRn** can control up to four peripherals (priority levels) with the interrupt numbers from 4n to (4**n** + 3). This relationship is shown in Table 5.12. The related priority bits filed for each **IPRn** are shown in column 6 on Table 5.11. These IRQ numbers are closely related to the associated peripherals to be enabled by using two NVIC Interrupt Enable Registers, **NVIC_ISER0**~**NVIC_ISER1**. These registers are similar to those ISERs **NVIC→ISER[]** defined in the CMSIS-Core macros in Table 5.5. Table 5.14 shows the relationship between each bit on these two registers and each related peripheral to be enabled.
2. The fifth column lists the macro or symbolic definition for each NVIC priority register for the related interrupt that is associated with each interrupt handler in column 4. Each NVIC priority register (**NVIC_IPRn**) is a 32-bit register. Table 5.13 shows the most popular priority registers used in the MSP432P401R system. For example, the priority register **NVIC_IPR0** is corresponding to the peripherals that have interrupt number from 4 × **0**~[4 × **0** + 3] = 0~3. It can be found from Table 5.11 that four peripherals have these interrupt numbers. Therefore, the priority levels of these four peripherals should be configured by using the same priority level register, **NVIC_IPR0**, which is: bits 31~29 define the priority levels for the Watchdog timer A, and bits 23~21 define the priority levels for the PCM, and so on. Do not be confused about the address shown in the first column in Table 5.11, which is only for the handler's vector address, not for the priority register's address.

Table 5.12 Bit filed of priority levels and related interrupt priority group

PRIn register bit field	Interrupt source	Priority register macros
`Bits 31:29`	Interrupt[IRQ] = Interrupt[4n + 3]	NVIC_IPRn NVIC→IP[4**n**]~NVIC→IP[4**n** + 3]
`Bits 23:21`	Interrupt[IRQ] = Interrupt[4n + 2]	
`Bits 15:13`	Interrupt[IRQ] = Interrupt[4n + 1]	
`Bits 7:5`	Interrupt[IRQ] = Interrupt[4n]	

Table 5.13 Most popular priority registers used in the MSP432P401R NVIC

Priority register	Priority bits				Address
	31–29	23–21	15–13	7–5	
NVIC_IPR0	WDT_A_IRQ	PCM_IRQ	CS_IRQ	PSS_IRQ	0xE000E400
NVIC_IPR1	COMP_E1_IRQ	COMP_E0_IRQ	FLCTL_IRQ	FPU_IRQ	0xE000E404
NVIC_IPR2	TA1_N_IRQ	TA1_0_IRQ	TA0_N_IRQ	TA0_0_IRQ	0xE000E408
NVIC_IPR3	TA3_N_IRQ	TA3_0_IRQ	TA2_N_IRQ	TA2_0_IRQ	0xE000E40C
NVIC_IPR4	EUSCIA3_IRQ	EUSCIA2_IRQ	EUSCIA1_IRQ	EUSCIA0_IRQ	0xE000E410
NVIC_IPR5	EUSCIB3_IRQ	EUSCIB2_IRQ	EUSCIB1_IRQ	EUSCIB0_IRQ	0xE000E414
NVIC_IPR6	T32_INTC_IRQ	T32_INT2_IRQ	T32_INT1_IRQ	ADC14_IRQ	0xE000E418
NVIC_IPR7	DMA_INT3_IRQ	DMA_ERR_IRQ	RTC_C_IRQ	AES256_IRQ	0xE000E41C
NVIC_IPR8	PORT1_IRQ	DMA_INT0_IRQ	DMA_INT1_IRQ	DMA_INT2_IRQ	0xE000E420
NVIC_IPR9	PORT5_IRQ	PORT4_IRQ	PORT3_IRQ	PORT2_IRQ	0xE000E424
NVIC_IPR10	Reserved	Reserved	Reserved	PORT6_IRQ	0xE000E428
NVIC_IPR11	Reserved	Reserved	Reserved	Reserved	0xE000E42C
NVIC_SYS_PRI3	SysTick	PendSV	—	Debug	0xE000ED20

3. The first column lists the vector locations for related interrupts in the vector table. Each vector is exactly a 32-bit entry address of the related interrupt handler and takes 4 bytes in the on-chip memory space.
4. Corresponding to each vector location in column 1, the related interrupt handler's name defined in the **`Startup_msp432.s`** file is shown in column 4. The users are highly recommended to use these handler's names defined in this startup file in their program without any modifications. Of course, you can change these handlers' names (statically or dynamically) to meet the needs of your special applications. However, the name you changed in this startup file must be identical with the name you used in your program.

Table 5.12 shows the relationship for each group number (**`n`**) on priority register, the related priority bit field and IRQ number. The macro **`NVIC_IPRn`** is defined by the MSPWare software system, but the **`NVIC→IP[]`** is defined by the CMSIS-Core system. Table 5.13 shows some most popular NVIC Priority Level Registers and bit fields used for each different peripheral in the MSP432P401R MCU system.

In fact, Tables 5.11, 5.12, and 5.13 can be used together to get a clear picture about the peripheral devices, their interrupt numbers, their interrupt vectors, their interrupt handlers, and their related priority registers and bit fields.

For example, the **`EUSCIA0_IRQHandler`** that is corresponding to the Extended Universal SCI A0 peripheral in Table 5.11, and the IRQ number for this peripheral is 16. From Table 5.12, it can be found that in order to make the equation Interrupt [4**n**] = 16 hold, the group number of the priority register, **n** = **4**. Therefore, the corresponded Priority Level Register should be NVIC_IPR**4** and it can configure priority levels for 4**n**~4**n** + 3 peripherals whose IRQ numbers are ranged 4 × 4~4 × 4 + 3 = 16~19. It can be found from Table 5.11 that the peripherals whose IRQ numbers are 16~19 are: **`EUSCIA0_IRQHandler`**, **`EUSCIA1_IRQHandler`**, **`EUSCIA2_IRQHandler`**, and **`EUSCIA3_IRQHandler`**, respectively. Take a look at Table 5.13, the Priority Level Register **NVIC_IPR4** that is located at row 5 can exactly handle the priority levels for these four peripherals in four segments, bits 7~5, bits 15~13, bits 23~21, and bits 31~29.

Since only upper 3-bit is used for each segment, the available priority levels for group 0 priority register **NVIC_IPR0** are:

- 0x00, 0x20, 0x40, 0x60, 0x80, 0xA0, 0xC0 and 0xE0 for segment 1
- 0x0000, 0x2000, 0x4000, 0x6000, 0x8000, 0xA000, 0xC000 and 0xE000 for segment 2
- 0x000000, 0x200000, 0x400000, 0x600000, 0x800000, 0xA00000, 0xC00000 and 0xE00000 for segment 3
- 0x00000000, 0x20000000, 0x40000000, 0x60000000, 0x80000000, 0xA0000000, 0xC0000000 and 0xE0000000 for segment 4

During the programming process, users can directly use various macros defined for all Priority Level Registers, **NVIC_IPR0~NVIC_IPR15**, in their program to access them to perform priority level configurations for selected peripherals.

5.3.2.2 NVIC Interrupt set enable registers (ISERs)

All NVIC registers can be fully accessed from privileged mode, but interrupts can be pended while they are in the unprivileged mode by enabling the **Configuration and Control (CFGCTRL)** register. Any other unprivileged mode access may cause a BusFault.

In MSP432P401R MCU system, the NVIC provides two ISERs, **ISER0~ISER1**. The symbolic definitions or macros for these registers are: **NVIC_ISER0~NVIC_ISER1**.

All these two set enable registers are 32-bit registers and all these registers use the full length, 32-bit, to enable related peripherals. These registers, **ISER0~ISER1**, use single bit, or bit by bit, to enable related 64 peripheral with the following functions:

- **NVIC_ISER0**: Provides 32 bit-by-bit enable control ability to 32 peripherals whose IRQ numbers are 0~31 (see Table 5.11 for IRQ numbers). This means that bit-0 controls the peripheral whose IRQ number is 0, bit-1 controls the peripheral whose IRQ number is 1, and bit-31 controls the peripheral whose IRQ number is 31.
- **NVIC_ISER1**: Provides 32 bit-by-bit enable control ability to 32 peripherals whose IRQ numbers are 32~63.

These registers are only used to set enables to related peripherals by writing 1 to the related bits. Writing zeros to any bits has no effects. To disable any peripherals, one needs to use the corresponding bits in the NVIC Clear Enable Registers, **NVIC_ICER0** through **NVIC_ICER1**.

Each bit on the set enable register is to enable one peripheral, and the bit number is equal to the IRQ number of the peripheral to be enabled. For example, to enable the DMA Error Interrupt whose IRQ number is 30 (Table 5.11), the bit-30 on the **NVIC _ ISER0** should be set to 1. To enable DMA INT1 Interrupt whose IRQ number is 33, the bit-1 on the **NVIC _ ISER1** should be set to 1. Since each **NVIC _ ISERn** register can only handle 32 peripherals (0~31), if the IRQ number is greater than 31, the target bit number should be calculated as

Target bit number = IRQ number − 32 × (n − 1). **n = 1** if IRQ ≤ 31, **n = 2** if (64 > IRQ > 31).

Table 5.14 Relationship between each bit on interrupt enable register and related peripheral

Enable register	32 Enable bits 0	1	2	3	4	5	6–29	30	31	Address
NVIC_ ISER0	PSS	CS	PCM	WDT_A	FPU	FLCTL	...	DMA_ ERR	DMA_ INT3	0xE000E100
NVIC_ ISER1	DMA_ INT2	DMA_ INT1	DMA_ INT0	PORT1	PORT2	PORT3	...	Reserved	Reserved	0xE000E104

Table 5.14 lists the most popularly used peripherals and their bit numbers in the NVIC set enable registers. Each bit is associated with a peripheral and a setting to a bit enables the selected peripheral. Each 32-bit set enable register can be used to enable 32 peripherals.

During the programming process, users can directly use these macros defined for all ISERs, **NVIC_ISER0~NVIC_ISER1**, in their program to access them to perform the enable configurations for selected peripherals.

Now, let's take care of the last interrupt configuration process.

5.3.3 Global interrupts configurations and controls

To enable the ARM® Cortex®-M4 processor to globally control all interrupts for all peripherals used in the system, two special registers, **PRIMASK** and **BASEPRI**, are provided inside the Cortex-M4 CPU to support this control function. We have discussed these two registers in detail in Section 2.2.1.2.2 in Chapter 2.

The **PRIMASK** register is a Primary Interrupt Mask register and it uses one bit (bit 0) to mask or disable all maskable interrupts/exceptions in the Cortex-M4 MCU system. When this bit is set to 1, it masks or disables all maskable interrupts. When this bit is reset to 0, it will unmask or enable all maskable interrupts used in the system. This function provides the global interrupt control ability for the Cortex-M4 processor.

The **BASEPRI** register provides more flexible interrupt masking strategies and it can perform masking or unmasking functions based on the priority levels of related interrupts/exceptions. In this way, if a higher level interrupt/exception is being executed or handled, any other lower level interrupt/exception will not be responded until the current interrupt/exception has been processed. As we mentioned, although this register is a 32-bit register, only three bits (bits 7~5) are used to configure eight different priority levels for peripherals.

The **BASEPRI** can be considered as a global priority control register, and its contents on bits 7~5 should be compared with the contents of all local priority registers, such as NVIC_IPRn, to determine whether the peripherals whose priority levels are configured by the NVIC_IPRn can be accepted or not by the processor. If the NVIC_IPRn includes higher priority (smaller number) than that of the BASEPRI, the interrupt requested by the peripheral can be accepted. Otherwise the interrupt may be pended until the current task is done.

As we discussed in Chapter 2, these two special registers can only be accessed by using ARM® Assembly instructions **MSR** or **MRS**. Fortunately, some intrinsic functions are provided by the ARM® MDK C/C++ Compiler and the CMSIS-Core to help users to avoid using the assembly instructions to fulfill these control abilities. Three options can be adopted by the users when they build their applications to access these special registers:

1. Use inline assembler
2. Use intrinsic function **__enable_irq()** provided by the ARM® MDK C/C++ Compiler
3. Use intrinsic function **__enable_irq()** provided by CMSIS-Core

Although it looks like that both intrinsic functions provided by two different sources are same, they are different in definitions and in codes, but perform the same function. Refer to Sections 4.5.2 and 4.5.3 in Chapter 4 to get more details for these intrinsic functions.

Now we have completed the interrupts initialization and configuration process. However, before we can accept and respond to any interrupt, we need to know how to direct the accepted interrupt to the correct interrupt handler or ISR to process that interrupt. The vector table is used for this purpose. Let's have a clear picture about the vector table and vectors located in that table.

5.3.4 GPIO interrupt handling and processing procedure

When all related interrupt control registers for a GPIO port and pins are initialized and configured, the system is ready to accept and respond to any interrupt. Figure 5.7 shows a block diagram for responding to a rising-edge-triggered interrupt occurred on **P6.4** pin.

- First, two GPIO port 6 interrupt control registers, **P6IE** and **P6IES** should be initialized and configured to setup pin **P6.4** to generate an appropriate interrupt request when a rising-edge-triggered interrupt is applied on **P6.4** pin.
- Then the NVIC priority register **NVIC_IPR10** and the NVIC ISER **NVIC_ISER1** should be configured to set up the priority level and set to enable the **P6.4** pin. Refer to Tables 5.11 through 5.14, bits 7~5 in the **NVIC_IPR10** should be configured to set up a priority level 3 to port 6 and bit-8 on the **NVIC_ISER1** should be set to enable this port (**NVIC_IPR10 = 0x00000060, NVIC_ISER1 = 0x08**).
- Two CPU-controlled special registers, **PRIMASK** and **BASEPRI**, should be programmed to globally enable all interrupts and set up the appropriate priority level

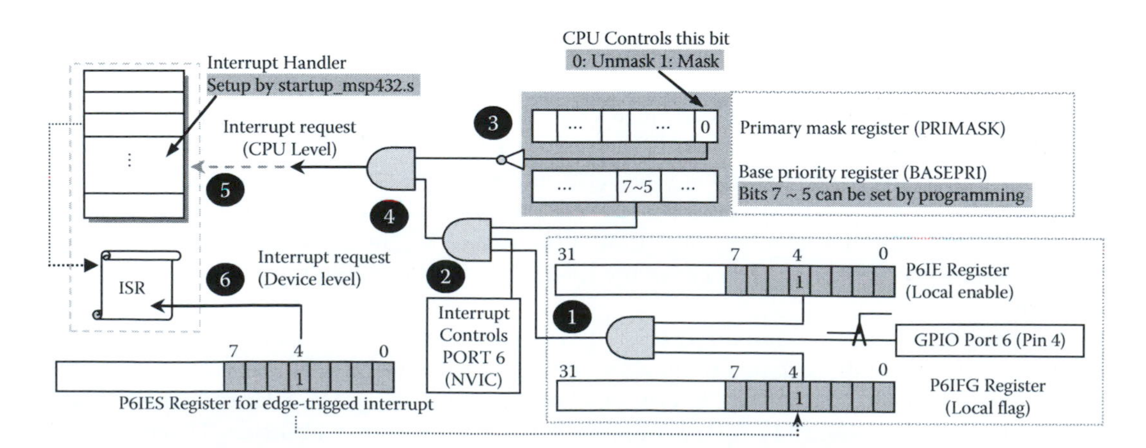

Figure 5.7 Interrupt handling and processing procedure.

to enable CPU to accept and response to any interrupts whose priority level is higher than that setting in the BASEPRI. Generally, the BASEPRI is reset to 0 after a system reset to enable the processor to accept any level's interrupt.

- When **P6.4** pin is triggered by a rising edge pulse, an interrupt request is sent to the NVIC controller and directed to the ISR entry address via the vector (**PORT6_IRQHandler**) in the vector table. The users can build their codes to process this interrupt request inside this handler in their program.
- Inside the interrupt handler (**PORT6_IRQHandler**), the user needs to reset the processed interrupt by reading the **P6IV** register to clear that interrupt flag set by **P6IFG** in pin **P6.4**. Refer to Section 2.3.4.2.7 in Chapter 2 to get more details about this interrupt clear.

Now that we have enough knowledge about the interrupts in MSP432P401R MCU system, let's build some projects to test the interrupt functions.

5.4 Developing GPIO port interrupt projects to handle GPIO interrupts

Generally, there are four ways to build GPIO-related interrupt programs to configure and response all GPIO-related interrupts:

1. Use DRA model to configure GPIO port and pins, as we discussed in Sections 5.3.1 and 5.3.2, to accept and handle related interrupt occurred on the desired pins with the selected GPIO port.
2. Use CMSIS-Core macros defined for NVIC interrupt control registers, as we discussed in Section 5.2.3, to configure and handle interrupts occurred on the desired pins with the selected GPIO ports.
3. Use MSPPDL API functions, as we discussed in Section 5.2.2, to configure and handle interrupts occurred on desired pins on the selected GPIO ports.
4. Use CMSIS-Core functions, as we discussed in Section 5.2.2, to configure and handle interrupts occurred on the desired pins with the selected GPIO ports.

Basically, the first two ways belong to the DRA programming model, and the third and fourth ways belong to the SD programming model. We will introduce and discuss all of these four ways to build our interrupt applications in this chapter. The point to be noted is that the first method works only for GPIO port interrupts, but the other three methods work for all other peripheral interrupts. We may combine the first two methods together as the DRA model to handle the GPIO-related interrupts in this chapter.

5.4.1 Using DRA programming model to handle GPIO interrupts

Similar to all other interrupt and exception processing procedure, to accept and handle an interrupt in the MSP432P401R MCU system using the DRA model, the following operations should be performed:

1. Set up and configure the interrupt for GPIO pins using GPIO interrupt control registers
2. Configure the priority level and enable the GPIO port using NVIC-related registers

3. Set up the global interrupt control using two special registers in the processor
4. Connect the interrupt handler with the users' project
5. Respond and process to the interrupt as it occurred, and reset and clear the active flag bit set in the **PxIFG** register

In this section, we will use a sample project **DRAInt** to illustrate how to build a GPIO-related interrupt project to access, response, and process an interrupt triggered by a GPIO pin.

The hardware we will use is the EduBASE ARM® Trainer and some configurations for this trainer were shown in Figure 4.31 in Chapter 4. For your convenience, we redrew this configuration in this section as shown in Figure 5.8.

In this project, we try to use **P4.4** pin that is connected to a position switch **SW2** in the EduBASE Trainer as a rising-edge-triggered interrupt as this switch is pressed. It can be found from Figure 5.8 that the **P4.4** pin gets a LOW level as the **SW2** is released and a HIGH level if the **SW2** is pressed. As this interrupt occurs, an ISR is executed to toggle the LED **PB0** installed in the EduBASE Trainer via **P3.2** pin in the GPIO port 3.

5.4.1.1 *Create a new project DRAInt and the header file*

Open the Windows Explorer to create a new folder **DRAInt** under the folder **C:\MSP432 Class Projects\Chapter 5**. Then refer to Sections 4.5.6.2.6.1 and 4.5.6.2.6.2 in Chapter 4 to create a new µVersion5 project **DRAInt** and add a new header file **DRAInt.h** into this project. Enter the codes shown in Figure 5.9 into this header file.

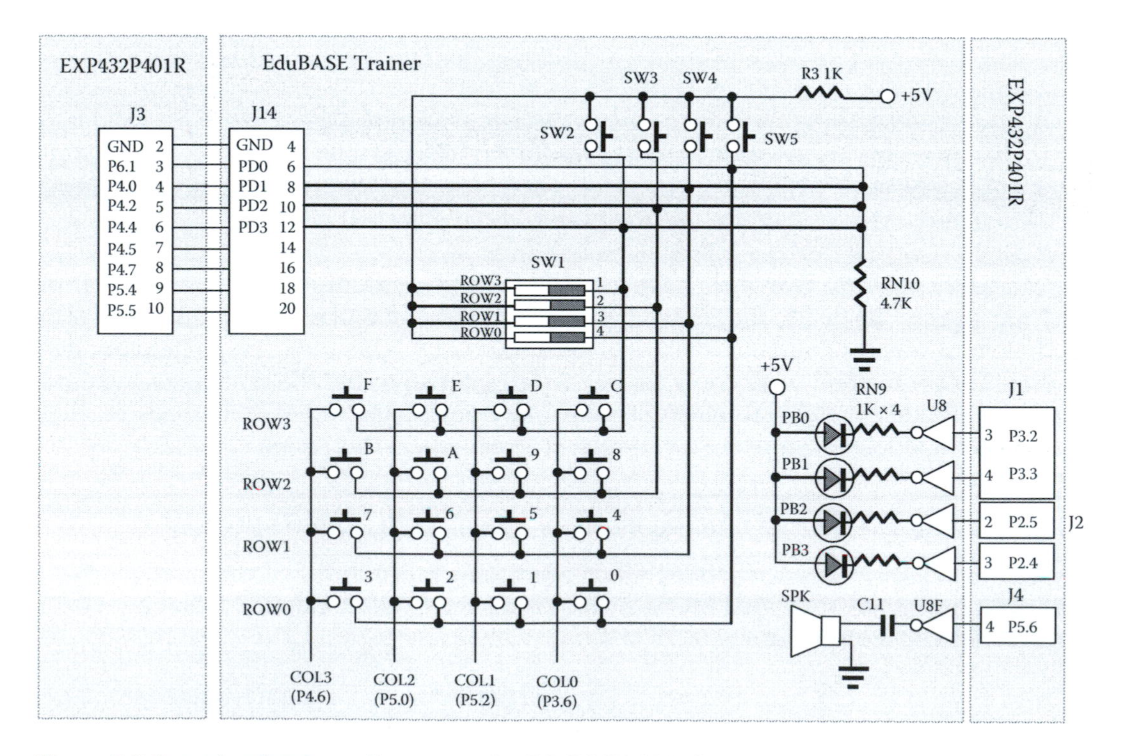

Figure 5.8 Some hardware configurations for EduBASE ARM® Trainer.

```
1  //*****************************************************
2  // DRAInt.h - Header file for the project - DRAInt
3  //*****************************************************
4  #include <stdint.h>
5  #include <stdbool.h>
6  #include <msp.h>
7  uint16_t ret = 0;
8  void Enable_IRQ(void);
```

Figure 5.9 Project header file DRAInt.h.

Three system header files, <**stdint.h**>, <**stdbool.h**>, and <**msp.h**> are declared first to enable the compiler to know most integer, Boolean data types, and related macros used in this project. Then a global variable **ret** is declared and this variable works as a holder to receive and keep the returned value from reading the **P4IV** register after an interrupt coming from **P4.4** pin is handled.

5.4.1.2 Create a new C code file DRAInt and add it into the project

Refer to Section 4.5.6.2.6.3 in Chapter 4 to create a new C File **DRAInt.c** and add it into the project **DRAInt**. Enter the codes shown in Figure 5.10 into this file.

Let's have a closer look at this piece of code to see how it works.

- In coding line 4, the user-defined header files, **DRAInt.h**, are included into this file. This file contains all macros or symbolic definitions for GPIO-related registers and NVIC-related registers to handle the interrupt request.
- A user-defined C function, **Enable_IRQ()**, is defined in lines 5 and 9. An inline assembler is used to insert some ARM® Cortex®-M4 assembly instructions to configure the PRIMASK and the BASEPRI registers to globally enable interrupts and reset the global priority register BASEPRI to 0 to allow any interrupt to be accepted by the processor. The **MSR** instruction is used to configure BASEPRI register and the **CPSIE** is to configure the PRIMASK register to unmask all interrupts. In fact, the BASEPRI is reset to 0 after a system reset operation.
- Inside the **main()** program, the Watchdog timer is first disabled to avoid any unnecessary interrupt caused by the Watchdog timer in line 12.
- In line 13, the GPIO pin **P4.4** is configured as an input pin since it is connected to the position button **SW2**. When **SW2** is pressed, a rising edge interrupt is generated.
- The GPIO pin **P3.2** is set as an output pin in line 14 since it is connected to the **PB0** LED in the EduBASE Trainer.
- In lines 15 through to 17, three GPIO interrupt registers, **P4IES**, **P4IFG**, and **P4IE**, are configured to initialize the pin **P4.4** as a rising-edge-triggered interrupt pin. One point to be noted is that the order of these instructions is very important. The **P4IFG** must be cleared after the **P4IES** has been configured to make sure that all previous sets on **P4IFG** register caused by setting the **P4IES** can be removed. The final action is to enable the interrupt function on the pin **P4.4**, and this step must be executed in the last step to avoid any possible wrong interrupt to be triggered.
- In lines 18 through 20, both NVIC registers, NVIC priority register (**NVIC_IPR9**), and NVIC set enable register (**NVIC_ISER1**), are used to set the priority level and enable the interrupt function on the GPIO pin **P4.4**. Since the IRQ number of the GPIO port 4 is **38** (refer to Table 5.11), a level 3 (**0x3** = **011B**) priority level is selected

```
 1  //****************************************************************************
 2  // DRAInt.c - Main application file for project - DRAInt
 3  //****************************************************************************
 4  #include "DRAInt.h"
 5  void Enable_IRQ(void)
 6  {
 7    int R2;
 8    __asm { MOV  R2, 0x0;  MSR  BASEPRI, R2;  CPSIE  I }
 9  }
10  int main(void)
11  {
12    WDTCTL = WDTPW | WDTHOLD;   // stop watchdog timer
13    P4DIR &= ~BIT4;             // set P4.4 as input pin to SW2
14    P3DIR |= BIT2;              // set P3.2 as output pin to PB0 LED
15    P4IES &= ~BIT4;             // reset P4IES.4 to 0 to set a rising-edge-trigger
16    P4IFG = 0;                  // clear all flags to avoid false interrupts
17    P4IE |= BIT4;               // enable P4.4 to receive interrupt
18    NVIC_IPR9 = 0x00600000;     // set P4 priority level as 3 (bits 23-21)
19    NVIC_ISER1 = 0x40;          // set bit 38 (bit 6) on ISER1 to enable P4 interrupt
20    Enable_IRQ();               // globally enable the interrupt
21    while(1) {}                 // wait for interrupt happen...
22  }
23  void PORT4_IRQHandler(void)
24  {
25    ret = P4IV;                 // clear P4IFG register
26    P3OUT ^= BIT2;              // toggle PB0 LED on & off
27  }
```

Figure 5.10 The codes for the C file DRAInt.c.

for this port. Because only three bits, bits 23~21, are used in the **`NVIC_IPR9`** register (refer to Table 5.13), therefore the priority bits value should be **`0000.0000.0110.0000.0000.0000.0000.0000B`** = **`0x00600000`**. All other bits are not used for this priority setting and therefore can be considered as 0. If you use the CMSIS-Core macros defined for these NVIC registers in your codes (see Table 5.5), you can use the code **`NVIC→IP[9]`** = **`0x60`** or **`NVIC→IP[9]`** = **`6`** ≪ **`20`** to replace that lone code line above. The reason for that is because in CMSIS-Core definitions, you can access each 8-bit segment on **`NVIC_IPR0`** register individually without considering other 24 bits. Bits 23~21 belong to the 3rd segment and it can be considered as an individual 8-bit segment without any relationship with other 24 bits in this register. The expression **`6≪20`** means that a number 6 (**`0x6`** = **`0110B`**) is shifted left by 20 bits to make it become 0110B and locating it at bits 23~21. Similar idea is used for the **`NVIC_ISER0`** register to enable pin **`P4.4`**.

- In line 20, a user-defined C-function **`Enable_IRQ()`** is called to use an inline assembler to execute some Cortex-M4 assembly instructions to globally enable all interrupts and set priority level to 0 to enable all interrupts to be accepted. An option is that you can call an intrinsic function, **`__enable_irq()`**, which is defined by either Keil® ARM® MDK C/C++ Compiler or CMSIS-Core, to replace this user-defined function to perform this global interrupt enable function.
- An infinitive **`while()`** loop is executed to wait for an interrupt to occur on pin **`P4.4`** in line 21.

- The code lines 23~27 contain the P4.4 interrupt handler and the interrupt processing codes. The name of this handler, **PORT4_IRQHandler**, is defined in the system startup file, **startup_msp432.s**, which has been added into this project. When using this handler, you must make sure that the name of this handler used in your program must be identical with that defined in the startup file. Otherwise you may encounter some compiling errors if you used different name for this handler. It is highly recommended to use the default handler name defined in this startup file. Refer to Table 5.11 to get the handler names you want to use in your program. The processing codes for this interrupt is simple, just toggle the **PB0** LED in the EduBASE Trainer via pin **P3.2** in the GPIO port 3.

Now we are ready to compile, link, download, and run this project to test the interrupt function via pin **P4.4** in the GPIO port 4. However, before we can do these jobs, make sure that the environment used by our project meets the requirements of the MSP432P401R MCU.

5.4.1.3 Set up environment to compile and run the project

To set up the correct environment for this project, you need to:

1. Open the **Debug** tab in the **Project|Options for Target "Target 1"** menu to make sure that the debugger you are using is **CMSIS-DAP Debugger** in the **Use** box. Also make sure that all settings related to the debugger and the flash functions are correct.

Now it is time to compile, download, and run the project. During the project runs, press the position switch **SW2** in the EduBASE Trainer, and a rising-edge-triggered interrupt occurred in the pin **P4.4**, and it is detected by the interrupt hardware and NVIC. This interrupt is then directed to the related handler, **PORT4_IRQHandler**. Inside the handler, the **PB0** LED in the EduBASE Trainer is toggled. Each time you press the **SW2** switch, the LED is on and off periodically.

5.4.2 Using CMSIS-core macros for NVIC registers to handle GPIO interrupts

In this section, we will use the CMSIS-Core macros defined for the NVIC interrupt control registers to set up and configure some GPIO ports and pins to perform interrupt

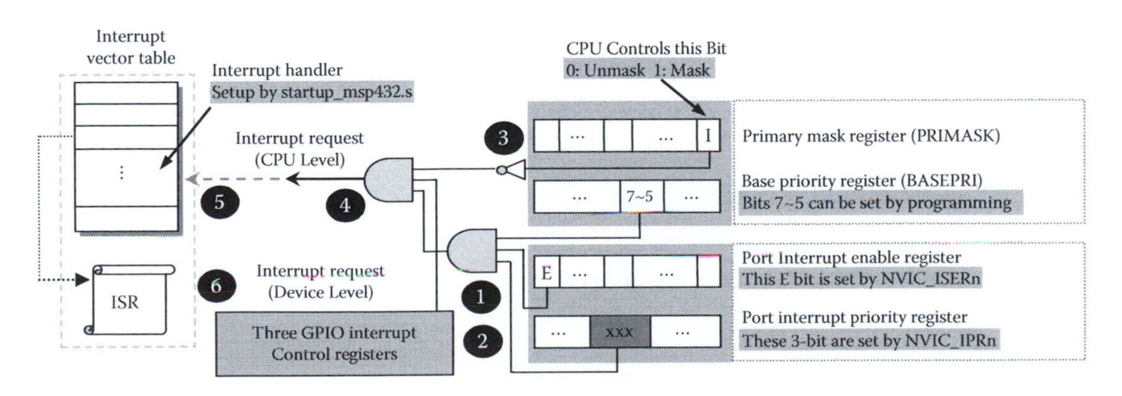

Figure 5.11 Block diagram using NVIC macros to handle GPIO interrupts.

generations, priority level detections, interrupt enables, and interrupt processing process.

We have provided a brief discussion and introduction about the CMSIS-Core macros defined for NVIC interrupt control registers in Section 5.2.3. Refer to Table 5.5 to get more details about these macro definitions.

Figure 5.11 shows a block diagram of using the CMSIS-Core macros for NVIC interrupt control registers to handle GPIO interrupts. Two control registers, **NVIC_IPRn** and **NVIC_ISERn**, are used to control those related GPIO interrupt registers to configure the priority level for the interrupt and enable the selected GPIO port to generate an interrupt request to the processor.

All of these NVIC registers and related macros are defined in the **core_cm4.h** file that is provided by the CMSIS-Core package. This header file provides all structure definitions for the NVIC registers.

5.4.2.1 Popular data structures defined in the MSP432P401R header file

In order to facilitate and help users to build and develop their application programs, the MSPWare software package provides a complete set of definitions for the control registers and structures in the header file **msp432p401r.h**. Specially, this file defined a set of special structures for all peripheral registers and IRQ numbers to make the users' program neat and simple.

These structures are defined based on all peripherals, both internal and external peripherals, NVIC registers and system control registers, used in the MSP432P401R MCU system with one-to-one relationship. This means that each peripheral has a corresponding structure and it is defined in the following ways:

- First the structure type for each peripheral is defined with the peripheral name and all registers and constants used by that peripheral
- Then the peripheral is defined as a pointer of the structure type of the peripheral

Some example structure types for the peripheral **WatchDog A** and **DIO** are shown in Figures 5.12 and 5.13.

After the structure type for the peripheral is defined, each peripheral can be further defined as a pointer of the defined structure type of the peripheral. Figure 5.14 shows an

```
typedef struct {
  uint8_t  RESERVED0[12];
  union {                          /* WDTCTL Register */
    __IO uint16_t r;
   struct {                        /* WDTCTL Bits */
    __IO uint16_t  bIS      :    3;/* Watchdog timer interval select */
    __O  uint16_t  bCNTCL   :    1;/* Watchdog timer counter clear */
    __IO uint16_t  bTMSEL   :    1;/* Watchdog timer mode select */
    __IO uint16_t  bSSEL    :    2;/* Watchdog timer clock source select */
    __IO uint16_t  bHOLD    :    1;/* Watchdog timer hold */
    __IO uint16_t  bPW      :    8;/* Watchdog timer password */
   } b;
  } rCTL;
} WDT_A_Type;
```

Figure 5.12 Structure type example—WDT_A_Type.

```
typedef struct {
 union {                          /* PAIN Register */
   __I uint16_t r;
   struct {                       /* PAIN Bits */
    __Iuint16_t bP1IN     :  8;  /* Port 1 Input */
    __Iuint16_t bP2IN     :  8;  /* Port 2 Input */
   } b;
 } rPAIN;
 union {                          /* PAOUT Register */
   __IO uint16_t r;`
   struct {                       /* PAOUT Bits */
    __IO uint16_t bP1OUT :  8;  /* Port 1 Output */
    __IO uint16_t bP2OUT :  8;  /* Port 2 Output */
   } b;
 } rPAOUT;
 union {                          /* PADIR Register */
   __IO uint16_t r;
   struct {                       /* PADIR Bits */
    __IO uint16_t bP1DIR :  8;  /* Port 1 Direction */
    __IO uint16_t bP2DIR :  8;  /* Port 2 Direction */
   } b;
 } rPADIR;
 union {                          /* PAREN Register */
   __IO uint16_t r;
   struct {                       /* PAREN Bits */
    __IO uint16_t bP1REN  :  8;  /* Port 1 Resistor Enable */
    __IO uint16_t bP2REN  :  8;  /* Port 2 Resistor Enable */
    } b;
   } rPAREN;
   ......
 } DIO_Type;
```

Figure 5.13 A structure type example – DIO_Type.

example of using the peripheral DIO structure **DIO_Type** to define all GPIO ports **DIO** as a pointer of the **DIO_Type** with the base address of each port. Using the **WDT_A_Type** structure type to define the Watchdog A as a pointer of that type is also shown here.

The meaning of each definition, such as **#define DIO ((DIO_Type *) DIO_BASE)**, is to define the peripheral DIO as a pointer of the **DIO_Type** structure with the starting address as **DIO_BASE**. In this way, the peripheral macro DIO now is a structure (not a constant) and it contains all registers involved in this peripheral. To access each register in this peripheral, a pointer format→should be used. The starting address of this structure is **DIO_BASE** (**0x40004C00**), which is also defined in this header file. Similarly, all peripherals' registers can be accessed and initialized in this way by using this kind of pointer definition.

```
#define WDT_A      ((WDT_A_Type *) WDT_A_BASE)
#define DIO        ((DIO_Type *) DIO_BASE)
#define TIMER_A0   ((TIMER_A0_Type *) TIMER_A0_BASE)
#define TIMER_A1   ((TIMER_A1_Type *) TIMER_A1_BASE)
#define TIMER_A2   ((TIMER_A2_Type *) TIMER_A2_BASE)
#define TIMER_A3   ((TIMER_A3_Type *) TIMER_A3_BASE)
......
```

Figure 5.14 Some examples of using structure type to define the peripheral.

One issue of using the peripheral structure type to define the peripheral as a pointer of that type is that the users' program codes may be longer, and therefore make them a little difficult to be coded and understood. For example, by using this definition, to access each register inside a peripheral, such as GPIO ports 1 and 3, the pointer format can be used and the following code line is a valid one:

- DIO->rPAOUT.b.bP1OUT = 0x1;
- DIO->rPADIR.b.bP2DIR = 0x2;
- DIO->rPCIE.b.bP5IE = 0x3;

Since a **union** and a **struct** are combined together to be used in these structure definitions, it makes them difficult to be accessed. Also these registers are grouped as 16-bit, not 8-bit, registers, such as **P1OUT** and **P2OUT** are grouped together as **PAOUT**, **P3IES**, and **P4IES** are grouped together to form **PBIES**, and so on. To access each desired 8-bit register, one needs to first access the union, and then the struct, and finally the GPIO pin. For example, to access **P1OUT** register, you need to:

1. First access the union **rPAOUT** (**P1OUT** is in **PAOUT** group)
2. Then you can access the struct **b**
3. Finally you can access the **bP1OUT** register and related pins

Table 5.15 shows a comparison between the register accessing with normal peripheral register macros and peripheral macros defined with the pointer of the peripheral structure type. It can be found that the pointer of structure type definition has longer code line and is not easy to remember and program.

In MSP432P401R MCU system, all peripherals have been defined in the normal and this structure type pointer format. Table 5.16 shows most popular peripheral structure type definitions. All of these structure definitions can be found from the system header file **msp432p401r.h** provided by the MSPWare software package.

5.4.2.2 IRQ numbers defined in the MSPWare system header file

Besides all peripherals and related registers, the system header file **msp432p401r.h** also contains the macro definitions for all exceptions and interrupts, especially for the IRQ number definitions for all exceptions and interrupts used in the MSP432P401R MCU system.

Figure 5.15 shows an example of structure definitions (**IRQn_Type**) for all exceptions involved in this header file and used in the MSP432P401R MCU system.

Table 5.15 Comparison between normal and structure *definitions for register accessing*

Normal definition accessing	Structure type definition accessing
P1OUT = 0x01	DIO→rPAOUT.b.bP1OUT = 0x1
P1DIR = 0x01	DIO→rPADIR.b.bP1DIR = 0x1
P1IFG = 0x01	DIO→rPAIFG.b.bP1IFG = 0x1
P1IES = 0x01	DIO→rPAIES.b.bP1IES = 0x1
P3IE = 0x01	DIO→rPBIE.b.bP3IE = 0x1
P3SEL0 = 0x01	DIO→rPBSEL0.b.bP3SEL0 = 0x1
P3SEL1 = 0x01	DIO→rPBSEL1.b.bP3SEL1 = 0x1

Table 5.16 Popular peripherals structure type definitions

Peripheral	Structure type definition
Watch Dog A	WDT_A_Type
GPIO PORT 1~GPIO PORT 11	DIO_TYPE
GPIO PORT J	DIO_Type
SYSCTL	SYSCTL_Type
TIMER A0~TIMER A3	TIMER_A0_Type~TIMER_A3_Type
RTC_C	RTC_C_Type
REF_A	REF_A_Type
ADC14	ADC14_Type
AES256	AES256_Type
CAPTIO0	CAPTIO0_Type
CAPTIO1	CAPTIO1_Type
COMP_E0	COMP_E0_Type
COMP_E1	COMP_E1_Type
CS Registers	CS_Type
DMA	DMA_Type
EUSCI A0~EUSCI A3	EUSCI_A0_Type~EUSCI_A3_Type
EUSCI B0~EUSCI B3	EUSCI_B0_Type~EUSCI_B3_Type
PCM	PCM_Type
PMAP	PMAP_Type
PSS	PSS_Type
TIMER32	TIMER32_Type

Figure 5.16 shows an example of structure definitions (**IRQn_Type**) for all interrupts involved in this header file and used in the MSP432P401R MCU system.

In this header file, the exception numbers are also defined with direct mapping method, such as:

```
#define INT_PSS  16  // PSS Exception Number = 16
```

```
typedef enum IRQn {
/* ------------------- Cortex-M4 Processor Exceptions Numbers ------------------- */
 Reset_IRQn              = -15,   /*!<  1 Reset Vector, invoked on Power up and warm reset  */
 NonMaskableInt_IRQn     = -14,   /*!<  2 Non maskable Interrupt, cannot be stopped      */
 HardFault_IRQn          = -13,   /*!<  3 Hard Fault, all classes of Fault     */
 MemoryManagement_IRQn   = -12,   /*!<  4 Memory Management, MPU mismatch   */
 BusFault_IRQn           = -11,   /*!<  5 Bus Fault, Pre-Fetch-, Memory Access Fault */
 UsageFault_IRQn         = -10,   /*!<  6 Usage Fault, i.e. Undef Instruction, Illegal State */
 SVCall_IRQn             = -5,    /*!< 11 System Service Call via SVC instruction */
 DebugMonitor_IRQn       = -4,    /*!< 12 Debug Monitor   */
 PendSV_IRQn             = -2,    /*!< 14 Pendable request for system service   */
 SysTick_IRQn            = -1,    /*!< 15 System Tick Timer */
 ......                  ......       ......
} IRQn_Type;
```

Figure 5.15 Structure definitions for exceptions.

```
typedef enum {
/* ---------------- MSP432P401R Specific Interrupt Numbers --------------- */
 PSS_IRQn           =  0,     /* 16 PSS Interrupt */
 CS_IRQn            =  1,     /* 17 CS Interrupt */
 PCM_IRQn           =  2,     /* 18 PCM Interrupt */
 WDT_A_IRQn         =  3,     /* 19 WDT_A Interrupt */
 FPU_IRQn           =  4,     /* 20 FPU Interrupt */
 FLCTL_IRQn         =  5,     /* 21 FLCTL Interrupt */
 COMP_E0_IRQn       =  6,     /* 22 COMP_E0 Interrupt */
 COMP_E1_IRQn       =  7,     /* 23 COMP_E1 Interrupt */
 TA0_0_IRQn         =  8,     /* 24 TA0_0 Interrupt */
 TA0_N_IRQn         =  9,     /* 25 TA0_N Interrupt */
 TA1_0_IRQn         = 10,     /* 26 TA1_0 Interrupt */
 TA1_N_IRQn         = 11,     /* 27 TA1_N Interrupt */
 TA2_0_IRQn         = 12,     /* 28 TA2_0 Interrupt */
 TA2_N_IRQn         = 13,     /* 29 TA2_N Interrupt */
 TA3_0_IRQn         = 14,     /* 30 TA3_0 Interrupt */
 TA3_N_IRQn         = 15,     /* 31 TA3_N Interrupt */
 EUSCIA0_IRQn       = 16,     /* 32 EUSCIA0 Interrupt */
 EUSCIA1_IRQn       = 17,     /* 33 EUSCIA1 Interrupt */
 EUSCIA2_IRQn       = 18,     /* 34 EUSCIA2 Interrupt */
 EUSCIA3_IRQn       = 19,     /* 35 EUSCIA3 Interrupt */
 EUSCIB0_IRQn       = 20,     /* 36 EUSCIB0 Interrupt */
 EUSCIB1_IRQn       = 21,     /* 37 EUSCIB1 Interrupt */
 EUSCIB2_IRQn       = 22,     /* 38 EUSCIB2 Interrupt */
 EUSCIB3_IRQn       = 23,     /* 39 EUSCIB3 Interrupt */
 ADC14_IRQn         = 24,     /* 40 ADC14 Interrupt */
 T32_INT1_IRQn      = 25,     /* 41 T32_INT1 Interrupt */
 T32_INT2_IRQn      = 26,     /* 42 T32_INT2 Interrupt */
 T32_INTC_IRQn      = 27,     /* 43 T32_INTC Interrupt */
 AES256_IRQn        = 28,     /* 44 AES256 Interrupt */
 RTC_C_IRQn         = 29,     /* 45 RTC_C Interrupt */
 DMA_ERR_IRQn       = 30,     /* 46 DMA_ERR Interrupt */
 DMA_INT3_IRQn      = 31,     /* 47 DMA_INT3 Interrupt */
 DMA_INT2_IRQn      = 32,     /* 48 DMA_INT2 Interrupt */
 DMA_INT1_IRQn      = 33,     /* 49 DMA_INT1 Interrupt */
 DMA_INT0_IRQn      = 34,     /* 50 DMA_INT0 Interrupt */
 PORT1_IRQn         = 35,     /* 51 PORT1 Interrupt */
 PORT2_IRQn         = 36,     /* 52 PORT2 Interrupt */
 PORT3_IRQn         = 37,     /* 53 PORT3 Interrupt */
 PORT4_IRQn         = 38,     /* 54 PORT4 Interrupt */
 PORT5_IRQn         = 39,     /* 55 PORT5 Interrupt */
 PORT6_IRQn         = 40      /* 56 PORT6 Interrupt */
}IRQn_Type;
```

Figure 5.16 Structure definitions for interrupts.

One point to be noted is that the IRQ numbers are different from exception numbers. Most of the time we use the IRQ numbers, not the exception numbers, in our program. The exception numbers just combine both exception and IRQ numbers together starting from 0.

Before we can build our sample interrupt project using the CMSIS-Core macros for NVIC interrupt registers to handle GPIO-related interrupts, let's have a closer look at these macros.

5.4.2.3 NVIC macros defined in the MSPWare system header files

In addition to data structures and IRQ numbers, the MSP432P401R MCU system also defined a set of macros used for the NVIC interrupt control register accessing and configuration. These macros are defined based on the actual addresses of each NVIC register and the users can directly access these registers and assign values to them.

Figure 5.17 shows some popular NVIC registers defined in the header file **msp432p401r.h**.

For example, to enable the **CSS** and **DMA_ERR** interrupts (refer to Table 5.14), the **NVIC_ISER0** should be initialized with the following code line (bits 30 and 1 are set to 1):

NVIC_ISER0 = **0100.0000.0000.0000.0000.0000.0000.0010B** = **0x40000002**

To set up the priority for the GPIO port 1 to level 3, the **NVIC_IPR8** register should be coded (refer to Table 5.13) since the IRQ number of the GPIO port 1 is 35 (refer to Table 5.11):

NVIC_IPR8 = **0110.0000.0000.0000.0000.0000.0000.0000B** = **0x60000000**

The macro **HWREG32()** is used to convert a named constant to a double pointer address, which is defined in this header file as:

```
#define HWREG32(x) (*((volatile uint32_t *)(x)))
```

```
//*****************************************************************************
// NVIC Registers (NVIC)
//*****************************************************************************
#define NVIC_ISER0 (HWREG32(0xE000E100)) /* IRQ 0 to 31 Set Enable Register */
#define NVIC_ISER1 (HWREG32(0xE000E104)) /* IRQ 32 to 63 Set Enable Register */
#define NVIC_ICER0 (HWREG32(0xE000E180)) /* IRQ 0 to 31 Clear Enable Register */
#define NVIC_ICER1 (HWREG32(0xE000E184)) /* IRQ 32 to 63 Clear Enable Register */
#define NVIC_ISPR0 (HWREG32(0xE000E200)) /* IRQ 0 to 31 Set Pending Register */
#define NVIC_ISPR1 (HWREG32(0xE000E204)) /* IRQ 32 to 63 Set Pending Register */
#define NVIC_ICPR0 (HWREG32(0xE000E280)) /* IRQ 0 to 31 Clear Pending Register */
#define NVIC_ICPR1 (HWREG32(0xE000E284)) /* IRQ 32 to 63 Clear Pending Register */
#define NVIC_IABR0 (HWREG32(0xE000E300)) /* IRQ 0 to 31 Active Bit Register */
#define NVIC_IABR1 (HWREG32(0xE000E304)) /* IRQ 32 to 63 Active Bit Register */
#define NVIC_IPR0  (HWREG32(0xE000E400)) /* IRQ 0 to 3 Priority Register */
#define NVIC_IPR1  (HWREG32(0xE000E404)) /* IRQ 4 to 7 Priority Register */
#define NVIC_IPR2  (HWREG32(0xE000E408)) /* IRQ 8 to 11 Priority Register */
#define NVIC_IPR3  (HWREG32(0xE000E40C)) /* IRQ 12 to 15 Priority Register */
#define NVIC_IPR4  (HWREG32(0xE000E410)) /* IRQ 16 to 19 Priority Register */
#define NVIC_IPR5  (HWREG32(0xE000E414)) /* IRQ 20 to 23 Priority Register */
#define NVIC_IPR6  (HWREG32(0xE000E418)) /* IRQ 24 to 27 Priority Register */
#define NVIC_IPR7  (HWREG32(0xE000E41C)) /* IRQ 28 to 31 Priority Register */
#define NVIC_IPR8  (HWREG32(0xE000E420)) /* IRQ 32 to 35 Priority Register */
#define NVIC_IPR9  (HWREG32(0xE000E424)) /* IRQ 36 to 39 Priority Register */
#define NVIC_IPR10 (HWREG32(0xE000E428)) /* IRQ 40 to 43 Priority Register */
#define NVIC_IPR11 (HWREG32(0xE000E42C)) /* IRQ 44 to 47 Priority Register */
#define NVIC_IPR12 (HWREG32(0xE000E430)) /* IRQ 48 to 51 Priority Register */
#define NVIC_IPR13 (HWREG32(0xE000E434)) /* IRQ 52 to 55 Priority Register */
#define NVIC_IPR14 (HWREG32(0xE000E438)) /* IRQ 56 to 59 Priority Register */
#define NVIC_IPR15 (HWREG32(0xE000E43C)) /* IRQ 60 to 63 Priority Register */
#define NVIC_STIR  (HWREG32(0xE000EF00)) /* Software Trigger Interrupt Register */
```

Figure 5.17 NVIC register macros defined by MSP432P401R MCU system.

For example, the `#define NVIC_IPR8 (HWREG32(0xE000E420))` is exactly equal to:

```
#define NVIC_IPR8 (*((volatile uint32_t *) 0xE000E420))
```

5.4.2.4 NVIC structure defined in the CMSIS-core header file

As we discussed in Section 5.2.3, to help users to develop the interrupt application programs, the CMSIS-Core provides a set of structure type definitions for the system control and NVIC interrupt control registers as well as some NVIC interfacing functions. These definitions and functions include:

- Structure type definitions for system special control and status registers, including ASPR, ISPR, CONTROL, SysTick, and SCB
- Structure type definition for the NVIC registers
- NVIC interface functions such as **NVIC_EnableIRQ(), NVIC_DisableIRQ(), NVIC_GetPriority(), and NVIC_SetPriority()**

Exactly these structure type macros and NVIC functions are defined in a CMSIS-Core system header file **core_cm4.h**. Figure 5.18 shows an example of the structure type definition **NVIC_Type** and the pointer structure definition for NVIC.

It can be found from Figure 5.18 that the **NVIC_Type** is first defined as a structure and this structure contains all registers used in the NVIC, including the **ISER[]**, **IPR[]**, and **ICER[]** registers. Then the **NVIC** variable is defined as a pointer of the **NVIC_Type** structure with its base address **0xE000E100UL** (UL means Unsigned Long). If you need to access any NVIC register by using this pointer structure definition in your program, you must use the pointer operator (→). For example, you want to configure the NVIC ISER **ISER[0]** to enable an interrupt source in your program, then you need to use **NVIC→ISER[0]** to do this job. With a similar way, you can access any other register in this NVIC structure.

```
typedef struct
{
  __IO uint32_t  ISER[8];       /*!< Offset: 0x000 (R/W)  Interrupt Set Enable Register     */
     uint32_t RESERVED0[24];
  __IO uint32_t  ICER[8];       /*!< Offset: 0x080 (R/W)  Interrupt Clear Enable Register   */
     uint32_t RSERVED1[24];
  __IO uint32_t  ISPR[8];       /*!< Offset: 0x100 (R/W)  Interrupt Set Pending Register    */
     uint32_t RESERVED2[24];
  __IO uint32_t  ICPR[8];       /*!< Offset: 0x180 (R/W)  Interrupt Clear Pending Register  */
     uint32_t RESERVED3[24];
  __IO uint32_t  IABR[8];       /*!< Offset: 0x200 (R/W)  Interrupt Active bit Register     */
     uint32_t RESERVED4[56];
  __IO uint32_t  IP[240];       /*!< Offset: 0x300 (R/W)  Interrupt Priority Register       */
     uint32_t RESERVED5[644];
  __O  uint32_t  STIR;          /*!< Offset: 0xE00 ( /W)  Software Trigger Interrupt Register  */

} NVIC_Type;

#define  SCS_BASE     (0xE000E000UL)              /* System Control Space Base Address */
#define  NVIC_BASE    (SCS_BASE +  0x0100UL)      /* NVIC Base Address  */
#define  NVIC    ((NVIC_Type  *)  NVIC_BASE )     /* NVIC configuration struct */
```

Figure 5.18 Structure type definition for the NVIC.

Recall that in Section 5.2.3 we discussed the CMSIS-Core macros for NVIC registers and CMSIS-Core functions to interface these NVIC registers. The point to be noted is that each ISER, from **ISER[0]** to **ISER[7]**, is a 32-bit register and can be used to enable 32 interrupts bit by bit. For example, the **ISER[0]** is a 32-bit register and it can enable 32 interrupt sources with one bit for one interrupt source. Each **ISER[]** can be used to enable different 32 interrupt sources. Table 5.17 shows the range of interrupts that can be enabled by the different **ISER[]** registers. The bit order is equivalent to the IRQ number of the interrupt. Bit-0 is for interrupt whose IRQ number is 0, bit-1 is for interrupt whose IRQ number is 1, and so on.

To enable one interrupt, the corresponding bit in the **ISER[]** should be set to 1. If **ISER[0]** = **0x00000001**, it enabled the interrupt whose IRQ number is 0, and **ISER[0]** = **0x00000008**, it enabled the interrupt whose IRQ number is 3, and so on. The **ISER[0]** is equivalent to **NVIC_ISER0** as shown in Table 5.14. The **NVIC_ISERn** and **NVIC_IPRn** macros are defined by the MSP432P401R MCU system. Refer to Table 5.14 to get more details for this register.

However, the **IPR[]**, which can be considered as an 8-bit register, is used to set up the priority level for one interrupt source. In fact, only three bits on each segment are used to set up the priority level for one interrupt source. The **IP[0]**, which is equivalent to 7~5 bits in the register **NVIC_IPR0** as shown in Table 5.13, is an 8-bit register and it is used to set up priority level for the interrupt whose IRQ number is 0. Figure 5.19 shows a mapping

Table 5.17 NVIC registers used for enable and priority controls

CMSIS-core macros	Enabled interrupt source	MSP432P401R macros
NVIC→ISER[0]	Interrupt sources 0~31	NVIC_ISER0
NVIC→ISER[1]	Interrupt sources 32~63	NVIC_ISER1
NVIC→ISER[2]	Interrupt sources 64~95	
NVIC→ISER[3]	Interrupt sources 96~127	
CMSIS-core macros	**Set interrupt priority level**	**MSP432P401R macros**
NVIC→IP[0]~NVIC→IP[3]	Interrupt sources with IRQ0~IRQ3	NVIC_IPR0
NVIC→IP[4]~NVIC→IP[7]	Interrupt sources with IRQ4~IRQ7	NVIC_IPR1
NVIC→IP[8]~NVIC→IP[11]	Interrupt sources with IRQ8~IRQ11	NVIC_IPR2
NVIC→IP[12]~NVIC→IP[15]	Interrupt sources with IRQ12~IRQ15	NVIC_IPR3
NVIC→IP[16]~NVIC→IP[19]	Interrupt sources with IRQ16~IRQ19	NVIC_IPR4
NVIC→IP[20]~NVIC→IP[23]	Interrupt sources with IRQ20~IRQ23	NVIC_IPR5
NVIC→IP[24]~NVIC→IP[27]	Interrupt sources with IRQ24~IRQ27	NVIC_IPR6
NVIC→IP[28]~NVIC→IP[31]	Interrupt sources with IRQ28~IRQ31	NVIC_IPR7
NVIC→IP[32]~NVIC→IP[35]	Interrupt sources with IRQ32~IRQ35	NVIC_IPR8
NVIC→IP[36]~NVIC→IP[39]	Interrupt sources with IRQ36~IRQ39	NVIC_IPR9
NVIC→IP[40]~NVIC→IP[43]	Interrupt sources with IRQ40~IRQ43	NVIC_IPR10
NVIC→IP[44]~NVIC→IP[47]	Interrupt sources with IRQ44~IRQ47	NVIC_IPR11
NVIC→IP[48]~NVIC→IP[51]	Interrupt sources with IRQ48~IRQ51	NVIC_IPR12
NVIC→IP[52]~NVIC→IP[55]	Interrupt sources with IRQ52~IRQ55	NVIC_IPR13
NVIC→IP[56]~NVIC→IP[59]	Interrupt sources with IRQ56~IRQ59	NVIC_IPR14
NVIC→IP[60]~NVIC→IP[63]	Interrupt sources with IRQ60~IRQ63	NVIC_IPR15

	31~29		23~21		15~13		7~5	
NVIC_IPR0	SegD		SegC		SegB		SegA	
	IP[3]		IP[2]		IP[1]		IP[0]	

Figure 5.19 Mapping relationship between NVIC_IPR0 and IP[].

relationship between each **`IP[]`** register and each **`NVIC_IPR0`** register. In this mapping way, the **`NVIC_IPR1`** includes **`IP[4]~IP[7]`**, **`NVIC_IPR2`** contains **`IP[8]~IP[11]`**, and **`NVIC_IPR3`** includes **`IP[12]~IP[15]`**, as shown in Table 5.17.

A point to be noted from Table 5.17 is that in MSP432P401R MCU system, only 64 interrupts sources exist, therefore only two 32-bit ISERs, **`NVIC_ISER0`** and **`NVIC_ISER1`**, are available.

5.4.2.5 Building sample project to use CMSIS-core macros for NVIC to handle interrupts

In this section, we want to use a sample project **`NVICInt`** to use CMSIS-Core macros for NVIC interrupt registers to handle a falling edge detected interrupt at pin **`P4.4`** at the GPIO port 4 since this pin is connected to a position switch **`SW2`** in the EduBASE Trainer. Refer to Figure 5.8 to get details about the hardware configuration for this connection. As the **`SW2`** switch is released, a falling edge interrupt occurs at **`P4.4`**, and four LEDs, **`PB0~PB3`**, installed in the EduBASE Trainer will be toggled in this interrupt handler.

Perform three steps to establish and configure the interrupt on pin **`P4.4`** at GPIO port 4:

1. Use three GPIO interrupt control registers to initialize and configure pin **`P4.4`** at the GPIO port 4 to detect a falling edge interrupt request signal.
2. Use two NVIC interrupt control registers to configure priority for the GPIO port 4 and enable this port.
3. Use an intrinsic function **`__enable_irq()`** to globally enable all interrupts.

Table 5.18 shows a set of configuration interrupt parameters used for the pin **`P4.4`** at the GPIO port 4 to enable this pin to perform the interrupt functions listed in step 1.

Table 5.18 Interrupt setup parameters for pin P4.4 on GPIO port 4

		GPIO PORT4 interrupt control registers							
Register	**Pin function**	**7**	**6**	**5**	**4**	**3**	**2**	**1**	**0**
P4OUT	0: Output LOW 1: Output HIGH	x	x	x	x	x	x	x	x
P4DIR	0: Input 1: Output	x	x	x	0	x	x	x	x
P4IES	0: Rising-Edge Trigger 1: Falling-Edge Trigger	x	x	x	1	x	x	x	x
P4IE	0: Disable Interrupt 1: Enable Interrupt	x	x	x	1	x	x	x	x
P4IFG	0: No Interrupt Occurred 1: Interrupt Occurred	x	x	x	0	x	x	x	x

Table 5.19 NVIC macros for GPIO port 4

NVIC register	CMSIS-core macros	Function
Interrupt Set Enable Registers	NVIC→ISER[1] = 0x00000040	Write 1 to enable the interrupt
Interrupt Priority Registers	NVIC→IP[38] = 0x60	Priority level 3 (8-bit wide) for the P4.4

Table 5.19 lists NVIC macro configuration parameters for the GPIO port 4 to set up the priority level and enable this port to perform the interrupt functions shown in step 2 above.

Now, we are ready to develop our sample interrupt project **NVICInt**.

5.4.2.6 Create a new project NVICInt and add the C code file

Open the Windows Explorer to create a new folder **NVICInt** under the folder **C:\MSP432 Class Projects\Chapter 5**. Then refer to Sections 4.5.6.2.6.1 and 4.5.6.2.6.3 in Chapter 4 to create a new µVersion5 project **NVICInt** and add a new C source file **NVICInt.c** into this project.

```
1  //*****************************************************************************
2  // NVICInt.c - Main application file for project - NVICInt
3  //*****************************************************************************
4  #include <stdint.h>
5  #include <stdbool.h>
6  #include <msp.h>
7  uint16_t ret;
8  int main(void)
9  {
10   WDTCTL = WDTPW | WDTHOLD;  // stop watchdog timer
11   P4DIR &= ~BIT4;            // set P4.4 as input pin to SW2
12   P3DIR |= BIT2 | BIT3;      // set P3.2 & P3.3 as output pins to PB0-PB1 LEDs
13   P2DIR |= BIT4 | BIT5;      // set P2.4 & P2.5 as output pins to PB2-PB3 LEDs
14   P4IES |= BIT4;             // set P4IES.4 to 1 to set a falling-edge-trigger
15   P4IFG = 0;                 // clear all flags to avoid false interrupts
16   P4IE |= BIT4;              // enable P4.4 to receive interrupt
17   NVIC->IP[38] = 3 << 5;     // set P4 priority level as 3 (bits 23-21)
18   NVIC->ISER[1] = 0x40;      // set bit 38 (bit 6) on ISER[1] to enable P4
19   __enable_irq();            // globally enable the interrupt
20   while(1) {}                // wait for interrupt happen...
21 }
22 void PORT4_IRQHandler(void)
23 {
24   ret = P4IV;                // clear P4IFG register
25   P3OUT ^= (BIT2 | BIT3);    // toggle PB0-PB1 LEDs on & off
26   P2OUT ^= (BIT4 | BIT5);    // toggle PB2-PB3 LEDs on & off
27 }
```

Figure 5.20 The C code file for the NVICInt project.

Table 5.20 Most popular NVIC API functions in the MSPWare peripheral driver library

API function	Description
void Interrupt_enableInterrupt (uint32_t interruptNumber)	Enable an interrupt
bool Interrupt_isEnabled (uint32_t interruptNumber)	Check if an interrupt has been enabled. Returning a non-zero indicates that the interrupt is enabled
bool Interrupt_enableMaster (void)	Enable processor to receive any interrupt. Returning a True means that all interrupts are enabled
void Interrupt_setPriority (uint32_t interruptNumber, uint8_t priority)	Set the priority level for a specified interrupt
void Interrupt_registerInterrupt (uint32_t interruptNumber, void(*intHandler)(void))	Register an interrupt handler

Enter the codes shown in Figure 5.20 into this source file. Let's have a closer look at this file to see how it works.

- Three system header files, **<stdint.h>**, **<stdbool.h>**, and **<msp.h>**, are declared first to enable the compiler to know the most integer and Boolean data types as well as all interrupt control registers and their macros used in this project. All GPIO ports and all CMSIS-Core macros of the NVIC interrupt control registers are defined in the third header file and we need to use them to initialize and configure GPIO ports 4, 3, and 2.
- A global variable **ret** is defined in line 7 and this variable works as a data holder to receive the returned value from reading the **P4IV** register inside the interrupt handler in the program. Another purpose of reading this **P4IV** register is to clear the highest interrupt flag in the **P4IFG** register if the current interrupt has been responded and processed.
- The **main()** program starts at line 8.
- The Watchdog timer is terminated in line 10 to avoid unnecessary interrupts caused by the Watchdog timer.
- The codes in lines 11 and 13 are used to configure GPIO ports 4, 3, and 2 and related pins to make pin **P4.4** as input pin, pins **P3.3** and **P3.2**, as well pins **P2.4** and **P2.5**, to work as output pins.
- In coding lines 14 through 16, the GPIO pin **P4.4** is configured to receive a falling-edge-triggered interrupt when the **SW2** is released. Before this pin can be interrupt-enabled, all flags in the **P4IFG** register must be cleared to remove all previous interrupt flags. Finally, this pin is interrupt-enabled by setting the **P4IE** register in line 16.
- The NVIC IPR **NVIC->IP[38]** is used to set up the priority level for the pin **P4.4**. Refer to Tables 5.11 and 5.17 to get more details about using the this **IP[]** register to set up the priority level for the interrupt source. Here we use **3 ≪ 5** to replace **0x60** since after 3 (**011B**) is shifted left with 5 bits, it becomes **01100000B** and it is equivalent to **0x60**.
- In line 18, the NVIC set enable register is used to set the enable bit (bit-38) for the GPIO port 4, which is equivalent to bit-6 (38–32) in the **ISER[1]** register. The CMSIS-Core macro, **NVIC→ISER[1]**, is used to set bit-6 to do this enabling job.

- An CMSIS-Core intrinsic function **__enable_irq()** is called in line 19 to perform the global interrupt enable job.
- Then an infinitive **while()** loop is executed to wait for the **P4.4** interrupt to occur.
- The code lines between 22 and 27 are the codes for the GPIO port 4 interrupt handler. First, the **P4IV** register is read to clear the falling-edge-triggered interrupt on **P4.4** pin, and then four LEDs on the EduBASE Trainer are toggled with two instructions in lines 25 and 26. The purpose of using this toggle action is to turn on four LEDs when the interrupt is accepted and handled in the odd time, and turn off four LEDs when the interrupt is accepted and handled in the even time.

Now you can compile, download, and run the program to test the interrupt function as the project runs. Make sure that the debugger in the **Debug** tab under the **Project|Options for Target "Target 1"** is **CMSIS-DAP Debugger** before you compile your project. Also you need to set up the correct settings for the debugger and the flash function. Refer to Section 3.4.3.3 in Chapter 3 to get more details about these settings.

As the project runs, press the position switch **SW2** firmly and then release it. All four LEDs on the EduBASE Trainer should be ON, and they should be OFF when you press and release that position switch again the second time.

5.4.3 *Using MSPPDL API functions to handle GPIO interrupts*

In the MSP432P401R MCU system, the MSPPDL provides two layers of API functions to handle all peripheral interrupts. One is the peripheral-specified interrupt API functions, and another is the NVIC-related API functions. Similarly, for the GPIO system, it also provides two layers of Interrupt Control API functions: the NVIC Interrupt Control API functions (Section 5.2.2) and the GPIO Interrupt Control API functions, to support users to access both NVIC control and GPIO port registers to perform configuration, setup, and control functions to all interrupts used in the GPIO ports system.

The MSPPDL provides all API functions for most popular and often used peripherals, including the internal and external peripherals, as we discussed in Section 4.5.6.3 in Chapter 4. In this section, we try to use the GPIO interrupt mechanism as an example to illustrate how to call these API functions to perform and handle GPIO-related interrupts.

We provided detailed discussions for some popular GPIO API functions in Section 4.5.6.3.1 in Chapter 4. In the following sections, we will concentrate on the NVIC and GPIO interrupt-related API functions in detail since we need to use both layers of API functions to develop a sample project to illustrate how to use these two layers of API functions to access related registers to initialize and configure them to perform our desired interrupt control functions.

Since both NVIC and GPIO interrupt-related API functions will be used in our project, we need to discuss both of them. First, let's take a look at the NVIC API functions.

5.4.3.1 *NVIC API functions defined in the MSPPDL*

As we mentioned in Section 5.2.2, the MSPPDL provides a group of interrupt controller API functions to facilitate users to directly access related NVIC registers to configure and control selected exceptions and interrupts. For the MSP432P401R MCU, most definitions for related exceptions and interrupts are defined in the system header files, such as **interrupt.h**, and **msp432p401r.h**.

Table 5.4 in Section 5.2.2 shows a complete set of NVIC API functions provided by the Peripheral Driver Library and used in the MSP432P401R MCU system. Table 5.20 shows the most popular NVIC API functions to be used in general interrupt control functions.

The NVIC is closely coupled with the Cortex-M4 microprocessor. When the processor accepts and responds to an interrupt, the NVIC provides the address of the ISR to handle the interrupt directly to the processor. This action eliminates the need for a global interrupt handler that queries the interrupt controller to determine the source of the interrupt and branch to the appropriate handler, reducing interrupt response time.

The interrupt handler or the entry address of the ISR can be configured or modified in two ways, either statically or dynamically. The API function **Interrupt_enableInterrupt()** must be used to enable the processor to know the existence of the interrupt handler when an interrupt handler is statically configured. Similarly, the API function **Interrupt_registerInterrupt()** must be applied to enable the processor to respond to the interrupt when the interrupt is dynamically configured.

The so-called static configuration is to modify the handler's name before the project can be compiled. This modification includes the editing or changing of the interrupt handler's name in both the system startup file, such as **startup_msp432.s**, and the user's program, and both names must be identical.

The so-called dynamic configuration is to change the handler's name after the user's project runs. The API function **Interrupt_registerInterrupt()** is used to enable this change to be acknowledged by the processor, and furthermore it can be accepted and responded to.

Each interrupt can be locally enabled by using the **Interrupt_enableInterrupt()** API function. However, to allow the processor to know and respond to each interrupt, two special registers, **PRIMASK** and **BASEPRI**, must be globally configured by using another API function, **Interrupt_enableMaster()**, to enable the CPU to respond to each interrupt.

Let's have a closer look at these NVIC API functions.

- The enable state of each interrupt source can be checked by using the API function **Interrupt_isEnabled()**. A nonzero value is returned if the tested interrupt source is enabled.
- Each interrupt source can be set up with certain priority level with **Interrupt_setPriority()** API function. In MSP432Px MCU system, only the upper 3-bit for each 8-bit segment is used to configure the priority levels for all interrupts. The smaller the priority number, the higher the priority level is. Priority 0 is the highest and priority 7 is the lowest level.
- Each interrupt source and handler can be dynamically configured via the API function **Interrupt_registerInterrupt()**. This means that an interrupt handler can be modified when the project runs. The function of the **Interrupt_registerInterrupt()** is to move the vector table from the flash memory to the SRAM space to enable this modification to occur. A point to be noted is that you need to make sure that the vector table is moved and located at the beginning of the SRAM space when linking your project with other system libraries and files.

5.4.3.2 GPIO interrupt-related API functions in the MSPPDL

In this section, we will introduce and discuss some GPIO interrupt-related API functions. These functions are very important and useful for processing of the GPIO-related interrupts. Because these API functions are specially used for the GPIO interrupts, they are therefore not included in those GPIO API functions we discussed in Section 4.5.6.3.1 in Chapter 4.

Table 5.21 lists the most popular and often used GPIO interrupt-related API functions. For all of these functions, the meanings of used arguments are

- **`uint_fast8_t selectedPort:`** Indicates the GPIO port to be accessed. Generally this is the macro address of the port and can be expressed by using the macros **`GPIO_PORT_P1`** (for GPIO port 1), or **`GPIO_PORT_P2`** (GPIO port 2) in the program.

Table 5.21 Most popular GPIO interrupt related API functions

GPIO interrupt-related API function	Description
void **GPIO_enableInterrupt**(uint_fast8_t selectedPort, uint_fast16_t selectedPins)	Enable an interrupt. **selectedPort** is the port to be enabled, and **selectedPins** is the pins to be enabled. **selectedPort**: GPIO_PORT_P1, GPIO_PORT_P2, GPIO_PORT_PA. **selectedPins:** GPIO_PIN_0, ... GPIO_PIN_15
void **GPIO_disableInterrupt**(uint_fast8_t selectedPort, uint_fast16_t selectedPins)	Disable an interrupt. **selectedPort** is the port to be disabled, and **selectedPins** is the pins to be disabled (same as above)
void **GPIO_clearInterruptFlag**(uint_fast8_t selectedPort, uint_fast16_t selectedPins)	Clear an interrupt flag. **selectedPort** is the port to be cleared, and **selectedPins** is the pins to be cleared. **selectedPort**: GPIO_PORT_P1, GPIO_PORT_P2, GPIO_PORT_PA. **selectedPins:** GPIO_PIN_0, ... GPIO_PIN_15
void **GPIO_registerInterrupt**(uint_fast8_t selectedPort, void(*)(void) intHandler)	Register an interrupt with its handler. **selectedPort** is the port to be registered and **intHandler** is the name of the interrupt handler. Specific GPIO interrupts must be enabled via GPIO Enable API func **GPIO_enableInterrupt()**. It is the interrupt handler's duty to clear the interrupt source via **GPIO_clearInterruptFlag()**
uint_fast16_t **GPIO_getInterruptStatus**(uint_fast8_t selectedPort, uint_fast16_t selectedPins)	Get the interrupt status for selected port and pins. **selectedPort**: GPIO_PORT_P1, GPIO_PORT_P2, GPIO_PORT_PA. **selectedPins:** GPIO_PIN_0, ... GPIO_PIN_15
void **GPIO_interruptEdgeSelect**(uint_fast8_t selectedPort, uint_fast16_t selectedPins, uint_fast8_t edgeSelect)	Select on what edge the port interrupt flag should be set for a transition. **selectedPort**: GPIO_PORT_P1,... GPIO_PORT_P11, GPIO_PORT_PJ. **selectedPins:** GPIO_PIN_0, ... GPIO_PIN_15. **edgeSelect:** GPIO_HIGH_TO_LOW_TRANSITION, GPIO_LOW_TO_HIGH_TRANSITION
void **GPIO_unregisterInterrupt**(uint_fast8_t selectedPort)	Unregister the interrupt handler for the port. **selectedPort**: GPIO_PORT_P1,... GPIO_PORT_P11, GPIO_PORT_PJ.

- **`uint_fast16_t selectedPins:`** Indicate the pin number(s) to be accessed. For multiple pins, the OR operator can be used to make them work. Generally this or these pins can be expressed by using the macro **`GPIO_PIN_0`**, **`GPIO_PIN_1`** or **`GPIO_PIN_15`**.
- **`uint_fast8_t edgeSelect:`** Indicates the edge, either rising or falling edge, on which the port interrupt flag should be set. Two macros can be used, **`GPIO_LOW_TO_HIGH_TRANSITION`** and **`GPIO_HIGH_TO_LOW_TRANSITION`**.
- **`void(*)(void) intHandler:`** This is a pointer that points to the entry address of the interrupt handler. The actual handler name should be used to replace the nominal name **`intHandler`** in the real application program.

Some points should be noted when using these GPIO interrupt-related API functions to handle GPIO-related interrupts in your program:

1. The **`GPIO _ registerInterrupt()`** function is a very important function and must be used in your program to enable the processor to know the interrupt and its handler. By default, the processor should know the interrupt and its handler if you used the default vector located in the vector table in the startup file. However, you must explicitly indicate this by using this function in your program when you use the API functions provided by the MSPPDL. Otherwise your program may not work properly.
2. Two types of GPIO interrupt-related API functions, Interrupt Master Enable and Interrupt Priority Setup, are not provided by these GPIO interrupt-related API functions. Therefore, you must use another layer of API functions provided by the NVIC API functions, **`Interrupt_enableMaster()`** and **`Interrupt_setPriority()`**, to perform these configuration jobs.
3. Some GPIO-related API interrupt functions can only access limited GPIO ports. Only GPIO ports 1 and 2, or GPIO port A can be accessed by the following functions:
   a. **`GPIO_enableInterrupt()`**
   b. **`GPIO_disableInterrupt()`**
   c. **`GPIO_clearInterruptFlag()`**
   d. **`GPIO_getInterruptStatus()`**

In that case, one can use either the DRA model or the NVIC API functions to access other GPIO ports to perform related interrupt functions.

After finishing these API functions introduction and discussion, now we are ready to develop and build our sample project to use the API functions to handle some GPIO-related interrupts.

5.4.3.3 Building sample project to use API functions to handle interrupts

In this section, we still use a sample project to illustrate how to use the API functions provided by the MSPPDL to handle a GPIO interrupt.

We want to develop a sample project **`SDInt`** to use GPIO and NVIC API functions provided by the MSPPDL to handle a falling edge detected interrupt at pin **`P1.1`** at the GPIO port 1 since this pin is connected to a position switch **`S1`** in the EXP432P401R EVB.

Refer to Figure 5.8 to get details about the hardware configuration for this connection. As the **`S1`** switch is pressed as the project is running, a falling edge interrupt occurs at **`P1.1`**, and four LEDs (**`PB0~PB3`**) installed in the EduBASE Trainer and a three-color LED installed in the MSP432™ LaunchPad™ MSP-EXP432P401R EVB (see Figure 4.23) will be toggled in this interrupt handler.

The coding for this project is divided into 10 parts:

1. Using system control API function to terminate Watchdog timer
2. Using GPIO API functions to configure and initialize GPIO Port 1 and pin **P1.1** to work as an input pin since it is connected to position switch **S1** in the EXP432P401R EVB
3. Using GPIO API functions to configure and initialize GPIO Port 2 and pins **P2.0~P2.2** to work as output pins since they are connected to a three-color LED on EXP432P401R EVB (see Figure 4.23)
4. Using GPIO API functions to configure and initialize GPIO Ports 3 and 2, exactly pins **P3.2** and **P3.3**, **P2.4** and **P2.5**, to work as a output pins since they are connected to four LEDs, **PB0~PB3** in the EduBASE Trainer
5. Using GPIO Interrupt-related API functions to initialize pin **P1.1** as a falling-edge-detected interrupt source
6. Using GPIO Interrupt-related API function **GPIO_registerInterrupt()** to register the interrupt handler to enable the processor to know this handler
7. Using GPIO Interrupt-related API function to enable the pin **P1.1** interrupt source
8. Using the NVIC API functions to set priority level and enable **P1.1** pin to be ready to receive a falling-edge-triggered interrupt when the **S1** is pressed and released
9. Using NVIC API functions to globally enable all interrupts in the system
10. Developing the codes for the interrupt handler to response to pin **P1.1** interrupt to toggle 4 LEDs in the Trainer and a three-color LED in the EXP432P401R EVB. The **P1.1** pin interrupt flag should also be cleared at the beginning of this interrupt handler

Now, let's start our sample project developing process.

5.4.3.4 *Create a new project SDInt and add the C code file*

Open the Windows Explorer to create a new folder **SDInt** under the folder **C:\MSP432 Class Projects\Chapter 5**. Then refer to Sections 4.5.6.2.6.1 and 4.5.6.2.6.3 in Chapter 4 to create a new µVersion5 project **SDInt** and add a new C source file **SDInt.c** into this project. Enter the codes shown in Figure 5.21 into this source file.

Let's have a closer look at this file to see how it works.

- Three system header files, <**stdint.h**>, <**stdbool.h**>, and <**msp.h**>, are declared first to enable the compiler to know most integer and Boolean data types as well as all interrupt control registers and their macros used in this project. All GPIO ports and NVIC-related interrupt control registers are defined in the third header file and we need to use them to initialize and configure GPIO Ports 1, 2, and 3.
- The code lines between 7 and 9 are used to include another three important system header files. The <**gpio.h**> defined all GPIO-related API functions and macros, including the GPIO interrupt APIs. The <**interrupt.h**> contains all NVIC-related API functions and macros. The <**wdt_a.h**> is used to cover the Watchdog-related API functions. The prefix **MSP432P4xx** is the super folder and it contains these header files. Later on, we will set up the **Include Path** to point to this super folder to allow the compiler to locate them correctly.

- In line 10, the interrupt handler that responds and processes the interrupt received by pin **P1.1** is declared. You can use any name for this handler, but this name must be identical in all locations when it is used in your program, and it also must be declared first.

The following codes are straightforward and easy to understand based on the comments. Some coding lines need special attentions and they are

- Three interrupt enable functions are used in this program, **GPIO_enableInterrupt()**, **Interrupt_enableInterrupt()**, and **Interrupt_enableMaster()**. The first one is used to locally enable the GPIO pin **P1.1** and it belongs to the local interrupt control. The second is the NVIC level interrupt control or globally interrupt control for all peripherals, and the third is the CPU level interrupt control or globally interrupt control for the processor.
- The API function **GPIO_registerInterrupt()** must be applied before any interrupt enable functions can be executed. Also the name of the interrupt handler, **GPIO_P1_Handler** used in this function, must be identical with the name when it is declared at the beginning. In this way, it allows the compiler to know where to find this interrupt handler as the project runs. This step is important and your program may not work if you missed it.
- Inside the interrupt handler **GPIO_P1_Handler**, four LEDs, **PB0~PB3**, and the three-color LED are periodically toggled as the **S1** button is pressed.

Now we can compile, link, download, and run this project to test the interrupt function. However, before we can do that, we need to configure the environments to enable our project to be compiled and linked correctly. Three jobs are related to these configurations: (1) set up the include path for all system header files used in the project, (2) add the MSPPDL into our project to enable us to use all related API functions, and (3) select the correct debugger to download and run the project.

5.4.3.5 Configure environments and run the project

Perform the following operations to include all system header files into our project:

- When the project is open, go to **Project|Options for Target "Target 1"** menu item to open the Options wizard
- Then click on the **C/C++** tab and go to the **Include Paths** box
- Click on the three-dot button on the right of this box to open the **Folder Setup** wizard
- Click on the **New** (**Insert**) tool to open a new textbox. Click on the three-dot button on the right to browse to the location where these header files are located, which is **C:\ti\msp\MSPWare_2_21_00_39\driverlib\driverlib**. Then click on the **OK** button to select this location
- Click on the **OK** button to close this wizard. Refer to Section 4.5.6.3.2.4 in Chapter 4 to get more details for this operation

Perform the following operations to add the MSPPDL into the project:

- In the **Project** pane, expand the **Target** folder and right click on the **Source Group 1** folder and select the **Add Existing Files to Group "Source Group 1"**.

```
   //*****************************************************************************
1
2  // SDInt.c - Main application file for project - SDInt
3  //*****************************************************************************
4  #include <stdint.h>
5  #include <stdbool.h>
6  #include <msp.h>
7  #include <MSP432P4xx\gpio.h>
8  #include <MSP432P4xx\interrupt.h>
9  #include <MSP432P4xx\wdt_a.h>

10 void GPIO_P1_Handler(void);            // interrupt handler

11 int main(void)
12 {
13   WDT_A_holdTimer();                   // stop watchdog timer

14   // set P1.1 as input pin with pull-up resistor
15   GPIO_setAsInputPinWithPullUpResistor(GPIO_PORT_P1, GPIO_PIN1);
16   // set P2.0 ~ P2.2 as output pins
17   GPIO_setAsOutputPin(GPIO_PORT_P2, GPIO_PIN0 | GPIO_PIN1|GPIO_PIN2);
18   // set P3.2 ~ P3.3 as output pins
19   GPIO_setAsOutputPin(GPIO_PORT_P3, GPIO_PIN2 | GPIO_PIN3);
20   // set P2.4 ~ P2.5 as output pins
21   GPIO_setAsOutputPin(GPIO_PORT_P2, GPIO_PIN4 | GPIO_PIN5);
22   // select a falling-edge-triggered interrupt for P1.1
23   GPIO_interruptEdgeSelect(GPIO_PORT_P1, GPIO_PIN1, GPIO_HIGH_TO_LOW_
         TRANSITION);
24   // register interrupt handler for P1.1
25   GPIO_registerInterrupt(GPIO_PORT_P1, GPIO_P1_Handler);
26   // locally enable GPIO P1.1 interrupt
27   GPIO_enableInterrupt(GPIO_PORT_P1, GPIO_PIN1);
28   // set priority level for P1.1 interrupt source
29      Interrupt_setPriority(INT_PORT1, 3);
30      // globally enable P1.1 interrupt source
31      Interrupt_enableInterrupt(INT_PORT1);
32      // globally enable interrupt for the processor
33      Interrupt_enableMaster();
34      // reset pins P2.0 ~ 2.3 to turn off 3-color LED
35      GPIO_setOutputLowOnPin(GPIO_PORT_P2, GPIO_PIN0 | GPIO_PIN1 | GPIO_PIN2);

36      while(1) {}                          // wait for P1.1 interrupt coming...
37 }

38 void GPIO_P1_Handler(void)
39 {
40   GPIO_clearInterruptFlag(GPIO_PORT_P1, GPIO_PIN1);  // clear P1.1 interrupt flag
41   GPIO_toggleOutputOnPin(GPIO_PORT_P2,
42                         GPIO_PIN0 | GPIO_PIN1 | GPIO_PIN2 | GPIO_PIN4 | GPIO_PIN5);
43   GPIO_toggleOutputOnPin(GPIO_PORT_P3, GPIO_PIN2 | GPIO_PIN3);
44 }
```

Figure 5.21 Detailed codes for the SDInt.c file.

- Browse to find the library, the MSPPDL, which is located at: **C:\ti\msp\MSPWare_2_21_00_39\driverlib\driverlib\MSP432P4xx\keil** in your host computer and the library is named **msp432p4xx_driverlib.lib**.
- Select this library by clicking on it and click on the **Add** button to add it into our project. Refer to Section 4.5.6.3.2.6 in Chapter 4 to get more details for this operation.

Perform the following operations to select the correct debugger for this project:

- Go to **Project|Options for Target "Target 1"** menu item to open the Options wizard.
- On the opened Options wizard, click on the **Debug** tab.
- Make sure that the debugger shown in the **Use** box is **CMSIS-DAP Debugger**. Otherwise you can click on the dropdown arrow to select this debugger from the list.
- Make sure that the debugger settings are correct by clicking on the **Settings** button located at the right of the debugger box. Then click on the **Debug** tab to check both checkboxes: **Verify Code Download** and **Download to Flash** inside the **Download Options** group box. Then click on the **Flash Download** tab to check **Erase Full Chip** radio button inside the **Download Function** group box.

Now you can compile, download, and run the project to test the interrupt function.

As the project runs, press the position switch **S1**, and you can find that all LEDs in the EduBASE Trainer and one three-color LED in the MSP432P401R EVB are on. But they are all off when pressing **S1** the second time. By pressing this switch repeatedly, you can find that our project responded to each interrupt with no problem.

Next let's take care of the interrupt response with the CMSIS-Core functions.

5.4.4 Using CMSIS-core access functions to handle GPIO interrupts

Like MSPPDL, the CMSIS-Core also provides a set of interrupt API functions to map to the related NVIC registers to perform interrupt configurations and controls for peripherals used in the MSP432P401R MCU system. In the last section, we introduced the interrupt API functions provided by the MSPPDL. In this section, we will discuss another set of interrupt API functions provided by the CMSIS-Core.

We have provided a detailed discussion about the CMSIS-Core functions used to access NVIC interrupt control registers to perform interrupt configurations and controls for peripherals in Section 5.2.2. Table 5.22 lists the most popular and often used CMSIS-Core functions used for interrupt controls and configurations in MSP432P401R MCU system.

It can be found from Table 5.22 that only a limited number of functions are provided by the CMSIS-Core, and some other useful functions, such as setting up the priority for a specified pin, clearing an interrupt for a pin and configuring the interrupt type for a pin, are not provided by this Core package. Therefore, we cannot use these pure CMSIS-Core functions to build our sample interrupt control project; instead, we need to use a combined set of functions, which include both CMSIS-Core functions for NVIC registers and GPIO macros defined for GPIO registers, to develop our sample project to illustrate how to use the CMSIS-Core functions defined for NVIC registers to process and handle a GPIO interrupt.

In order to use these CMSIS-Core functions for NVIC registers and GPIO structures to access and configure related NVIC and GPIO registers, we first need to have a clear picture about header files we should use in this project.

- To use CMSIS-Core functions for NVIC registers, we need to use the header file, **`core_cm4.h`**, since this header file contains all definitions for NVIC interrupt API functions listed in Table 5.22. This header file is provided by the CMSIS-Core software package.
- To use system control and GPIO registers with their structure definitions, we need to use the header file, **`msp.h`**, since this header file provides all structure definitions for GPIO ports (**`DIO_Type`**). This header file is provided by the MSPWare software package.

Now, let's build a sample project to use the CMSIS-Core supported functions and GPIO-related registers to perform interrupt processing.

5.4.4.1 Building a sample project to use CMSIS-core functions to handle interrupts

In this section, we will use a sample project to illustrate how to use the CMSIS-Core NVIC functions provided by the CMSIS-Core software package to handle a GPIO interrupt.

We want to develop a sample project **`CMSISInt`** to use GPIO and NVIC API functions to handle a rising edge detected interrupt at pin **`P6.1`** at the GPIO Port 6 since this pin is connected to a position switch **`SW5`** in the EduBASE Trainer.

Refer to Figure 5.8 to get details about the hardware configuration for this connection. As the **`SW5`** switch is pressed as the project is running, a rising edge interrupt is occurred at **`P6.1`**, and four LEDs (**`PB0~PB3`**) installed in the EduBASE Trainer and a blue-color LED installed in the MSP432 LaunchPad™ MSP432P401R EVB (see Figure 4.23) will be toggled in this interrupt handler.

Table 5.22 Most popular CMSIS-core functions used for interrupt controls

NVIC register	CMSIS-core function	Description
Interrupt Set Enable Registers	void **`NVIC_EnableIRQ`**(IRQn_Type IRQn)	Enable an external interrupt
Interrupt Clear Enable Registers	void **`NVIC_DisableIRQ`**(IRQn_Type IRQn)	Disable an external interrupt
Interrupt Set Pending Registers	void **`NVIC_SetPendingIRQ`**(IRQn_Type IRQn)	Set the pending status for an interrupt
Interrupt Clear Pending Registers	void **`NVIC_ClearPendingIRQ`**(IRQn_Type IRQn)	Clear pending status for an interrupt
Interrupt Active Bits Registers	uint32_t **`NVIC_GetActive`**(IRQn_Type IRQn)	Get the active status for an interrupt
Interrupt Set Priority Register	void **`NVIC_SetPriority`**(IRQn_Type IRQn, uint32_t Priority)	Set the priority for an interrupt
Interrupt Get Priority Register	uint32_t **`NVIC_GetPriority`**(IRQn_Type IRQn)	Get the priority for an interrupt
Global enable an interrupt	void **`__enable_irq`**(void)	Clear PRIMASK to enable all interrupts
Global disable an interrupt	void **`__disable_irq`**(void)	Set PRIMASK to disable all interrupts

The coding for this project is divided into seven steps:

1. Using Watchdog structure pointer to hold the Watchdog timer.
2. Using GPIO structure pointer to configure and initialize GPIO Ports 2 and 3, exactly **P3.2**, **P3.3**, **P2.4**, and **P2.5**, to work as output pins since they are connected to 4 LEDs **PB0**~**PB3** in the ARM® EduBASE Trainer.
3. Using GPIO structure pointer to configure and initialize GPIO Port 2, exactly pin **P2.2**, to work as an output pin since it is connected to a blue-color LED on the EXP432P401R EVB (see Figure 4.23).
4. Using GPIO structure pointer to configure and initialize GPIO Ports 6, exactly pin **P6.1**, to work as an input pin since it is connected to a position switch **SW5** in the Trainer.
5. Using GPIO structure pointer to initialize pin **P6.1** as a rising-edge-detected interrupt source.
6. Using CMSIS-Core functions for NVIC to set up the priority level 3 for GPIO Port 6 and enable its interrupt source.
7. Developing the codes for the interrupt handler to response to the **P6.1** interrupt to toggle 4 LEDs in the Trainer and a blue-color LED in the EXP432P401R EVB. The pin **P6.1** interrupt flag should also be cleared at the beginning of this interrupt handler.

Now, let's start our sample project developing process.

5.4.4.2 Create a new project CMSISInt and add the C code file

Open the Windows Explorer to create a new folder **CMSISInt** under the folder **C:\MSP432 Class Projects\Chapter 5**. Then refer to Sections 4.5.6.2.6.1 and 4.5.6.2.6.3 in Chapter 4 to create a new µVersion5 project **CMSISInt** and add a new C source file **CMSISInt.c** into this project. Enter the codes shown in Figure 5.22 into this source file.

Let's have a closer look at this file to see how it works.

- Three system header files, **<stdint.h>**, **<stdbool.h>**, and **<msp.h>**, are declared first to enable the compiler to know most integer and Boolean data types as well as all interrupt control registers and their macros used in this project. All GPIO ports-related registers and CMSIS-related interrupt control functions are defined in the third header file and we need to use them to initialize and configure GPIO Ports 2, 3, and 6.
- A global variable **ret** is defined in line 7 and this variable works as a data holder to receive the returned value from reading the **P6IV** register inside the interrupt handler in the program. Another purpose of reading this **P6IV** register is to clear the highest interrupt flag in the **P6IFG** register if the current interrupt has been responded to and processed.
- The Watchdog register is accessed by writing the **WDTPW|WDTHOLD** value to hold its function. To access the Watchdog timer register, a password (**PW**) must be written into the **WDTPW** register to enable any writing to be effective. Here, the Watchdog structure pointer **WDT_A** is used with the→operator to access this 16-bit register **r** to write down this hold value in line 10.
- The code lines between 11 and 13 are used to configure **P6.1** as an input pin, **P3.2**, **P3.3**, **P2.4**, and **P2.5** as output pins. Here, similarly the GPIO structure pointer **DIO** is used to access each related register.

```
1  //*****************************************************************************************************
2  // CMSISInt.c - Main application file for project - CMSISInt
3  //*****************************************************************************************************
4  #include <stdint.h>
5  #include <stdbool.h>
6  #include <msp.h>
7  uint16_t ret = 0;
8  int main(void)
9  {
10   WDT_A->rCTL.r = WDTPW | WDTHOLD;       // hold WDT (WDTHOLD = 0x0080)
11   DIO->rPCDIR.b.bP6DIR &= ~BIT1;         // set P6.1 as input pin
12   DIO->rPBDIR.b.bP3DIR |= BIT2|BIT3;     // set P3.2 & 3.3 output pins
13   DIO->rPADIR.b.bP2DIR |= BIT2|BIT4|BIT5; // set P2.2, P2.4 & 2.5 output pins
14   DIO->rPCIES.b.bP6IES &= ~BIT1;         // set P6.1 a rising-edge-trigger interrupt
15   DIO->rPCIFG.b.bP6IFG = 0;              // reset all P6 IFGs to 0
16   DIO->rPCIE.b.bP6IE |= BIT1;            // enable P6.1 interrupt
17   NVIC_SetPriority(PORT6_IRQn, 0x3);     // set P6 priority level as 3 (bits 7-5)
18   NVIC_EnableIRQ(PORT6_IRQn);            // set bit 40 (bit 8) on ISER1 to enable P6
19   __enable_irq();                        // globally enable interrupts for CPU
20   while(1) {}                            // wait for interrupt happen...
21 }
22 void PORT6_IRQHandler(void)
23 {
24   ret = DIO->rP6IV.b.bP6IV;              // clear P4IFG register
25   DIO->rPBOUT.r ^= (BIT2|BIT3);          // toggle PB0-PB1 (P3.2 & P3.3) LEDs on & off
26   DIO->rPAOUT.r ^= (BITA|BITC|BITD);     // toggle P2.2 (blue) , PB2-PB3 (P2.4 & P2.5) LEDs on & off
27 }
```

Figure 5.22 Detailed codes for the CMSISInt.c file.

- In line 14, the GPIO pin **P6.1** is configured as a rising-edge-detective interrupt pin. Any previous interrupt flag is cleared by resetting the **P6IFG** register in line 15. Finally, the pin **P6.1** is locally enabled by setting bit 1 in the GPIO Port 6 in line 16. Since the GPIO port structure pointer is used, therefore a pointer operator→is applied to all related bits in the GPIO-related registers.
- A CMSIS-Core function **NVIC_SetPriority()** is executed in line 17 to set up the priority level for the Port 6 to level 3. As you know, the priority levels in the MSP432P401R system are ranged at 0~7 for each port. The passed argument of this function is the IRQn_Type, **PORT6_IRQn**.
- In line 18, another CMSIS-Core function **NVIC_EnableIRQ()** is called to enable a rising-edge-detective interrupt for the Port 6.
- The processor interrupt is globally enabled in line 19 by calling an intrinsic function **__enable_irq()**.
- An infinitive **while()** loop is used in line 20 to wait for any interrupt to occur.
- The codes for the interrupt handler are shown between lines 22 and 27.
- First, the **P6IV** register is read to clear the current interrupt flag to enable any further interrupt to be occurred in line 24.

- The code lines between 25 and 26 are used to perform toggle operations for all LEDs installed in both MSP432P401R EVB and EduBASE Trainer. Since the GPIO port structure pointer **DIO** is used, therefore a pointer operator→is applied to all related bits in the GPIO-related registers.

A point to be noted in the last coding line (line 26) inside the interrupt handler is: when using GPIO structure pointer to access each register, only register pair, such as **PA** (P1 and P2), **PB** (P3 and P4), until **PE** (P9 and P10), can be accessed. You cannot access any single register directly with this pointer structure. For example, if you want to access GPIO **P2**, you need to first access the **PA** that is a 16-bit register with **P2** as higher 8-bit and **P1** as lower 8-bit. In order to access **P2.2**, **P2.4**, and **P2.5** pins, you must access bits 10, 12, and 13, respectively (not bits 2, 4, and 5 since now **P2** is higher 8-bit for 16-bit register **PA**). These bits (10, 12, and 13) are defined as **BITA**, **BITC**, and **BITD** in the <**msp.h**> header file.

Another point to be noted is that there are two ways to use this structure pointer to access each 16-bit or 8-bit GPIO register or just a single pin in that register; one is to use **DIO→rPAOUT.r** or **WDT_A→rCTL.r** to access entire related 16-bit or 8-bit GPIO register, another way is to use **DIO→rPAOUT.b.bP1OUT** to access only a single pin on the related register. Refer to Figure 5.13, in the **DIO_Type** structure, the **rPAOUT** is defined as a *union* with a 16-bit register **r**, but **b** is defined as a bit *struct* with **bP1OUT** and **bP2OUT** and it can be used to access each single bit.

Now we can compile, link, download, and run this project to test the interrupt function. However, before we can do that, we need to configure the environments to enable our project to be compiled and linked correctly. Only one job is related to these configurations; select the correct debugger to download and run the project.

5.4.4.3 *Configure environments and run the project*

Perform the following operation to select the correct debugger for this project:

1. Go to **Project|Options for Target "Target 1"** menu item to open the Options wizard.
2. On the opened Options wizard, click on the **Debug** tab.
3. Make sure that the debugger shown in the **Use:** box is **CMSIS-DAP Debugger**. Also make sure that all the settings for the debugger and the flash function are correctly configured.

Now you can compile, download, and run the project to test the interrupt function.

As the project runs, press the position switch **SW5**, and you can find that all four LEDs in the EduBASE Trainer and one blue-color LED in the MSP432P401R EVB are on. But they are all off when pressing **SW5** the second time. Again and again pressing this switch, you can find that our project responded to each interrupt with no problem.

5.5 *Comparison among four interrupt programming methods*

By developing and building our sample interrupt projects with four different methods as we did in the last sections, we can compare these methods to get some conclusions as listed below:

1. Comparably speaking, the DRA methods look better in the coding process and interrupt configuration as well as the handling process. Relatively, the project **DRAInt**

developed by using the GPIO interrupt-related registers method is a little better than the project **NVICInt** built by using the CMSIS-Core functions for NVIC method since the DRA model is used in the first method.
2. Another two projects, **SDInt** and **CMSISInt**, are developed by using the SD method, and have longer codes with more coding process. Especially, for the project **SDInt** that is built by using the API functions provided by MSPPDL, it involves a lot of function calls, and makes the project codes much longer compared with the coding process in the other three methods.
3. Summarily, the advantage of using the DRA method to develop interrupt projects is that the project coding process is easier with less code lines. However, the developers must have good understanding and knowledge about the details of the Cortex-M4 MCU and peripheral devices as well as all related registers and the NVIC interrupt mechanism.
4. The advantage of using the SD method to build interrupt projects is that most API functions in the MSPPDL and CMSIS-Core package provide a good path and easy going way to facilitate the coding and development process for the users, and the users do not need solid understanding about the microcontroller hardware and interrupt processing mechanism to build a professional interrupt application.

Developers can use any method to build their interrupt-related project and this depends on the requirements of real applications and developing environments.

5.6 Summary

The chapter mainly concentrates on exceptions and interrupts that belong to MSP432P401R MCU system. All exceptions and interrupts related to the ARM® Cortex®-M4 processor are introduced, but only interrupts are discussed in detail in this chapter since most exceptions belong to system faults or errors, and they should be processed and handled by the system itself.

First, an overview about the general interrupts is given with a function block diagram of the most popular interrupt processing procedure. Then the exceptions and interrupts involved in the ARM® Cortex®-M4 MCU system are introduced. These discussions include the following basic interrupt elements:

- Interrupt sources
- Interrupt masks
- Interrupt priorities
- Interrupt vector table
- Interrupt handler or ISR

Then the fundamental interrupt processing procedure for ARM® Cortex®-M4 processor is provided, which includes:

- Interrupt request is established
- Interrupt is accepted or pended
- Interrupt is responded
- Interrupt is processed

Popular interrupt processing API functions provided by two software packages, MSPWare software package and CMSIS-Core software package, are introduced and discussed in detail. These discussions include the API function protocols, syntax, operational principle, and possible running results. The main interrupt control unit, NVIC, is also introduced in detail.

Following these basic introductions and discussions, all exceptions and interrupts related to the MSP432P401R MCU system are provided. The peripheral GPIO is used as an example to illustrate how to generate, respond, and process interrupts coming from any GPIO port.

Two major interrupt processing modes and functions provided by MSPPDL and CMSIS-Core package are discussed and analyzed with some examples. Both MSPWare library and CMSIS-Core package provide a set of API functions to map NVIC-related registers to facilitate the interrupt configuration and processing.

Four types of interrupt processing methods are introduced with four actual example projects, and these methods include:

1. Use DRA model to configure GPIO port and pins to accept and handle related interrupts that occurred on the desired pins with the selected GPIO port.
2. Use CMSIS-Core macros defined for NVIC interrupt control registers to configure and handle interrupts that occurred on the desired pins with the selected GPIO ports.
3. Use MSPPDL API functions to configure and handle interrupts that occurred on desired pins on the selected GPIO ports.
4. Use CMSIS-Core functions to configure and handle interrupts that occurred on the desired pins with the selected GPIO ports.

The detailed discussions, line-by-line explanations, and illustrations for each project are given to help readers to understand the purpose and function of each coding line. When finishing this chapter, the readers should be able to develop and build general interrupt projects to handle most GPIO-related interrupts with different methods.

HOMEWORK

I. True/False Selections

____1. An exception is often generated by an event, such as the system fault event and other system mis-operation event. An interrupt is generally created by an internal or an external peripheral with lower priority level.

____2. The Reset belongs to an exception and has the highest priority level in the ARM® Cortex®-M4 MCU system.

____3. In the Cortex-M4 system, there are about 240 interrupts that can be handled by this MCU system. Each interrupt has an interrupt number ranging from 0 to 240.

____4. The top three exceptions, **`Reset`**, **`NMI`**, and **`Hard Fault`**, have fixed priority levels with minus numbers to indicate that these exceptions cannot be masked or disabled by using **`PRIMASK`** register and they have the highest priority levels.

____5. Two software packages, the MSPWare software package and CMSIS-Core package, provide different interrupt processing API functions to access, configure, and handle all interrupts via NVIC.

____6. The CMSIS-Core macro **NVIC→ISER[1]** is an 8-bit register and it can be used to enable 8 interrupts, from Interrupt 0 to Interrupt 7.

____7. The starting address of each interrupt ISR can be called a vector since it provides not only a direction to the related handler but also an entry address of the ISR.

____8. In MSP432P401R MCU system, an 8-bit segment is used to define the priority levels for each related interrupt. Therefore, totally 256 priority levels can be used by each interrupt source.

____9. Any GPIO Interrupt Flag Register (**PxIFG**) is used to determine the triggering mode of an interrupt source, either a rising-edge or a falling-edge.

____10. The **PRIMASK** and **BASEPRI** registers are used to globally enable or disable all interrupts in the MSP432P401R MCU system.

II. Multiple Choices

1. Each vector in the vector table takes ________ bytes to make a starting address of the ISR.
   a. 1
   b. 2
   c. 3
   d. 4
2. Most GPIO interrupt-related registers are ____ bits, but only _________ are used.
   a. 32, 16
   b. 32, the highest 8-bit
   c. 32, the lowest 8-bit
   d. 8, 3
3. Each NVIC priority register **NVIC_IPRn** is a 32-bit register and it can configure _______ priority levels for _______ related interrupt sources.
   a. 1, 1
   b. 2, 2
   c. 3, 3
   d. 4, 4
4. In order to configure an interrupt coming from the GPIO Port 2 as a priority level of 5, which of the following codes is correct?
   a. **NVIC_IPR9 = 0x0000000A**
   b. **NVIC_IPR9 = 0x000000A0**
   c. **NVIC_IPR9 = 0x00000A00**
   d. **NVIC_IPR9 = 0x0000A000**
5. To use the **NVIC IPRn** register to configure the priority level for an interrupt coming from the GPIO Port 3, which **NVIC_IPRn** register with which 3 bits should be used?
   a. NVIC_IPR6, 23~21
   b. NVIC_IPR7, 31~29
   c. NVIC_IPR8, 7~5
   d. NVIC_IPR9, 15~13
6. To use NVIC ISER _______________ to enable a GPIO Port 5 interrupt, which bit in this register should be set to 1?
   a. **NVIC_ISER1**, Bit-5
   b. **NVIC_ISER1,** Bit-6
   c. **NVIC_ISER1,** Bit-7
   d. **NVIC_ISER1,** Bit-1

7. To configure the priority level for an interrupt coming from the **ADC14** whose IRQ number is 24, which NVIC priority register group (n) should be used with which 3 bits?
   a. **NVIC_IPR9, 7~5**
   b. **NVIC_IPR6, 7~5**
   c. **NVIC_IPR5, 7~5**
   d. **NVIC_IPR4, 7~5**
8. The difference between functions, **GPIO_enableInterrupt()** and **Interrupt_enableInterrupt()**, is ________________.
   a. The first function is used to globally enable all interrupts, but the second function is to locally enable a GPIO peripheral interrupt.
   b. The first function is to globally enable all interrupts for the processor, but the second function is to enable a specific GPIO interrupt.
   c. The first function is to locally enable a GPIO peripheral interrupt, but the second function is to globally enable all interrupts.
   d. The first function is to locally enable a GPIO peripheral interrupt, but the second function is to globally enable all interrupts for the processor.
9. The intrinsic function **__enable_irq()** is defined by the ________ and it can be used to ________________.
   a. CMSIS-Core software package, enable all interrupts
   b. MSPWare software package, enable all interrupts
   c. MSPWare and CMSIS-Core software packages, enable all interrupts
   d. ARM® MDK μVersion5, enable all interrupts
10. To dynamically register an interrupt handler, one can use ________ API function.
    a. `NVIC_IntRegister()`
    b. `GPIO_registerInterrupt()`
    c. `Interrupt_registerInterrupt()`
    d. Both b and c

III. Exercises

1. Provide a brief description about three components used in the general exceptions and interrupts processing system.
2. Provide a brief description about interrupts processing configuration and procedure in the ARM® Cortex®-M4 MCU system.
3. Explain three layers used in the configurations and responses for the GPIO interrupts in the MSP432P401R MCU system.
4. Provide a description about four methods used to configure and response GPIO-related interrupts in the MSP432P401R MCU system.
5. Provide a brief comparison between two models, **DRA** and **SD**, to illustrate the advantages and disadvantages of these two models in building interrupt related projects.

IV. Practical Laboratory

LABORATORY 5 ARM® Cortex®-M4 and MSP432 exceptions and interrupts

5.0 GOALS

These laboratory exercises allow students to learn and practice ARM® Cortex®-M4 and MSP432 interrupts handling and processing for GPIO-related interrupts by developing four labs.

1. Program **Lab5_1(EdgeINT)** lets you build an interrupt processing project to handle an edge-triggered interrupt on GPIO **P4.0** pin by using the DRA programming model.
2. Program **Lab5 _ 2(NVICINT)** enables students to build a falling-edge-triggered interrupt on GPIO **P4.2** pin by using the CMSIS-Core macros defined for NVIC interrupt control registers to access and control a single LED **PB2** with a DIP switch in the EduBASE Trainer. A 400 Hz signal is also sent to the speaker to make a sound for this interrupt.
3. Program **Lab5 _ 3(2INTPRI)** allows students to use both **P6.1** and **P1.1** pins to handle two edge-triggered interrupts to control two different sets of LEDs, **PB0** and **PB1**, **PB2** and **PB3**, in the EduBASE Trainer. This lab enables students to learn and practice how to set up priority levels for two interrupts coming from two different ports based on their real priority levels and how to control and response these two interrupts asynchronously and synchronously. Any programming model can be used in this project.
4. Program **Lab5_4(INTGroup)** provides you another way to use both **P6.1** and **P1.1** pins to handle two edge-triggered interrupts to control two different sets of LEDs, **PB0** and **PB1**, **PB2** and **PB3**, in the EduBASE Trainer. In this project, students need to use the NVIC Set Priority Grouping and NVIC Encoder Priority functions to configure two interrupts coming from GPIO Ports 6 and 1 to replace the CMSIS **NVIC→IP[]** macros to set up the priority levels for these two interrupts.

After completion of these programs, you should understand the most popular interrupts applied in the GPIO ports, and these interrupt architectures and configurations as well as interrupt handling methods. You should be able to code some interrupt controlling programs to access the NVIC registers to control and response related interrupt requests sent by the peripherals to perform desired interrupt control functions.

5.1 LAB5_1

5.1.1 Goal

In this project, students need to use the **P4.0** pin that is connected to a position switch **SW4** in the EduBASE Trainer to handle a rising-edge-triggered interrupt as this switch is pressed. It can be found from Figure 5.8 that pin **P4.0** gets a LOW level as the **SW4** is released and a HIGH level if the **SW4** is pressed. As this interrupt is occurred, an ISR is executed to toggle a three-color LED in the EXP432P401R EVB via **P2.2~P2.0** pins in the GPIO Port 2.

The DRA model is required to use to build this project.

5.1.2 Data assignment and hardware configuration

It can be found from Figure 5.8 that the position switch **SW4** is connected to the **P4.0** pin and the three-color LED is connected to **P2.2~P2.0** pins of the GPIO ports in the EXP432P401R EVB (refer to Figure 4.23). To enable and set these two GPIO ports, Ports 4 and 2, the following data and registers should be used:

- Bit 0 in the Port 4 direction register should be reset to 0 to enable **P4.0** pin to work as an input pin
- Bits 2~0 in the Port 2 direction register should be set to 1 to enable **P2.2~P2.0** pins to work as output pins

5.1.3 Development of the source code

Refer to Figure 5.8 and follow the steps below to develop this C source code. Only a C code file is used in this project since it is a simple application without needing any header file.

1. Create a new folder `Lab5_1(EdgeINT)` under the folder `C:\MSP432 Lab Projects\Chapter 5` in the Windows Explorer.
2. Create a new µVersion5 project named `Lab5_1` and save this project to the folder `Lab5_1(EdgeINT)` that is created in step 1 above.
3. On the next wizard, you need to select the device (MCU) for this project. Expand three icons, `Texas Instruments`, `MSP432 Family`, and `MSP432P`, and select the target device `MSP432P401R` from the list by clicking on it. Click on the `OK` to close this wizard.
4. Next the Software Components wizard is opened, and you need to set up the software development environment for your project with this wizard. Expand two icons, `CMSIS` and `Device`, and check the `CORE` and `Startup` checkboxes in the `Sel.` column, and click on the `OK` button since we need these two components to build our project.
5. In the `Project` pane, expand the `Target` folder and right click on the `Source Group 1` folder and select the `Add New Item to Group "Source Group 1".`
6. Select the `C File (.c)` and enter `Lab5_1` into the `Name:` box, and click on the `Add` button to add this source file into the project.
7. On the top of this C source file, you need first to include three system header files, `<stdint.h>`, `<stdbool.h>`, and `<msp.h>` since we need to use them in this project.
8. Declare a global `uint16_t` variable `ret` as a data holder to receive the returned value from reading the `P4IV` register inside the interrupt handler later.
9. Start the main program with the code `int main(void)`.
10. Hold the Watchdog timer by using the Watchdog register `WDTCTL`.
11. Set up three pins, `P2.2`~`P2.0` on the GPIO Port 2, to work as output pins using `P2DIR` register.
12. Reset all pins on P2 to turn off the three-color LED using `P2OUT` register.
13. Configure GPIO pin `P4.0` as an input pin using `P4DIR` register.
14. Configure GPIO pin `P4.0` to receive a rising-edge-triggered interrupt by using the `P4IES` register.
15. Clear all previous interrupt flags for the GPIO Port 4 by resetting the `P4IFG` register.
16. Enable pin `P4.0` interrupt function using the `P4IE` register.
17. Use NVIC priority register `NVIC_IPR9` to set up the priority level for the GPIO Port 4 as level 3.
18. Use NVIC set enable register `NVIC_ISER1` to enable the GPIO Port 4 to be triggered by an interrupt source. The IRQ number for GPIO Port 4 is `38` (see Table 5.11) and it should be located at bit-6 in the `NVIC_ISER1` register and this bit should be set to 1.
19. Use an intrinsic function `__enable_irq();` to globally enable all interrupts.
20. Use an infinitive `while()` loop to wait for any interrupt to occur.
21. Now you need to develop the codes for the interrupt handler, `PORT4_IRQHandler()`. Using the code `void PORT4_IRQHandler(void)` to start this handler.
22. Inside this handler, you first need to clear the interrupt flag set in the `P4IFG` register by reading the `P4IV` register, assigning the returned value to the global variable `ret`, and then you need to toggle the three-color LED via `P2.2`~`P2.0` pins with `P2OUT` register.

5.1.4 Demonstrate your program by testing the running result

Now, you can compile your program by going to **Project|Build target** menu item. However, before you can download your program into the flash memory in the EVB, make sure that the debugger you are using is the **CMSIS-DAP Debugger**. You can check this by:

- Going to **Project|Options for Target "Target 1"** menu item to open the Options wizard
- On the opened Options wizard, click on the **Debug** tab
- Making sure that the debugger shown in the **Use:** box is **CMSIS-DAP Debugger**. You also need to set up the correct settings for the debugger and the flash function by clicking on the **Settings** button located at the right of the selected debugger

Perform the following operations to run your program and check the running results:

- Go to **Flash|Download** menu item to download your program into the flash memory.
- Go to **Debug|Start/Stop Debug Session** to begin debugging your program. Click on the **OK** button on the 32 KB memory size limitation message box to continue.
- Then go to **Debug|Run** menu item to run your program.
- As the project runs, you can press the **SW4** position switch and the three-color LED on the EXP432P401R EVB will be on, and it will be off as you press the **SW4** switch again.
- Go to **Debug | Start/Stop Debug Session** to stop your program.

Based on these results, try to answer the following questions:

- Can you tell me what color is displayed on this three-color LED? Why?
- Why there are no codes used to initialize and set up the system clock for this project? What is the default system clock?
- When you press the **SW4** button in some time, the three-color LED may be kept in the ON or OFF status (not toggling), why does this happened?
- What did you learn from this project?

5.2 LAB5_2

5.2.1 Goal

This project enables students to build a falling-edge-triggered interrupt on GPIO **P4.2** pin by using the CMSIS-Core macros defined for NVIC interrupt control registers to access and control a single LED **PB2** with a DIP switch 2 in the EduBASE Trainer. A 1000 Hz signal is also sent to the speaker to make a sound for this interrupt.

5.2.2 Data assignment and hardware configuration

It can be found from Figure 5.8 that the DIP switch 2 is connected to the **P4.2** pin, the LED **PB2** is connected to the **P2.5** pin, and the speaker is connected to the **P5.6** pin of the GPIO ports in the EXP432P401R EVB. To enable and set these three GPIO ports, the following data and registers should be used:

- Bit-2 in the GPIO Port 4 direction register should be reset to 0 to enable pin **P4.2** to work as an input pin

- Bit-5 in the GPIO Port 2 direction register should be set to 1 to enable **P2.5** pin to work as an output pin
- Bit-6 in the GPIO Port 5 direction register should be set to 1 to enable **P5.6** pin to work as an output pin to output a 1000 Hz audio signal (refer to **Lab4_4**) to the speaker installed in the EduBASE ARM Trainer

5.2.3 Development of the project

Follow the steps below to develop this project. Only a C code source file is used in this project since this project is simple. Create the project and develop the C source file with the following steps:

1. Create a new folder **Lab5_2(NVICINT)** under the folder **C:\MSP432 Lab Projects\Chapter 5** in the Windows Explorer.
2. Open the Keil® ARM®-MDK µVersion5, create a new project named **Lab5_2** and save this project into the folder **Lab5_2(NVICINT)** created in step 1.
3. On the next wizard, you need to select the device (MCU) for this project. Expand three icons, **Texas Instruments**, **MSP432 Family**, and **MSP432P**, and select the target device **MSP432P401R** from the list by clicking on it. Click on the **OK** to close this wizard.
4. Next the Software Components wizard is opened, and you need to set up the software development environment for your project with this wizard. Expand two icons, **CMSIS** and **Device**, and check the **CORE** and **Startup** checkboxes in the **Sel.** column, and click on the **OK** button since we need these two components to build our project.

5.2.4 Development of the C source file

1. In the **Project** pane, expand the **Target** folder and right click on the **Source Group 1** folder and select the **Add New Item to Group "Source Group 1."**
2. Select the **C File (.c)** and enter **Lab5_2** into the **Name:** box, and click on the **Add** button to add this source file into the project.
3. Include the following header files into this source file first:
   a. **<stdint.h>**
   b. **<stdbool.h>**
   c. **<msp.h>**
4. Declare a global **uint16_t** variable **ret** as a data holder to receive the returned value from reading the **P4IV** register inside the interrupt handler later.
5. Create two user-defined functions, **void Delay(int time)** and **void SetSound(int period)**, since we need to use these functions to generate a 400 Hz signal.
6. Inside the function **Delay()**, declare an integer local variable **i** that works as a loop count, and use **for()** loop to delay a period of time. The starting value for the loop count is **0**, and the ending value is the input argument **time**.
7. Inside the function **SetSound()**, set **P5.6** to 1 by using **P5OUT** register to generate the first half-cycle of the 400 Hz signal. Then call the function **Delay()** to delay a **period**. Reset the **P5.6** to 0 by using **P5OUT** register to generate the second half-cycle of the 400 Hz signal. Then call the function **Delay()** to delay another **period**.
8. Place the **int main(void)** to start the main program and put a starting brace under this main function to begin the main program.
9. Hold the Watchdog timer by using the Watchdog register **WDTCTL**.

10. Configure **P4.2** as an input pin using the **P4DIR** register.
11. Set **P2.5** as an output pin using the **P2DIR** register.
12. Configure **P5.6** as an output pin using the **P5DIR** register.
13. Clear **P2.5** and **P5.6** pins using the **P2OUT** and **P5OUT** registers to turn off both LED **PB2** and the speaker.
14. Configure GPIO pin **P4.2** to receive a falling-edge-triggered interrupt by using the **P4IES** register.
15. Clear all previous interrupt flags for the GPIO Port 4 by resetting the **P4IFG** register.
16. Enable pin **P4.2** interrupt function using the **P4IE** register.
17. Use the NVIC structure pointer (**NVIC→IP[38]**) to set up the priority level for the Port 4 as priority level 3. Refer to Tables 5.5, 5.11, and 5.13 to get more details about this setting.
18. Use the NVIC structure pointer (**NVIC→ISER[1]**) to enable the Port 4 interrupt. Refer to Tables 5.5, 5.11, and 5.13 to get more details about this setting.
19. Use an intrinsic function **__enable_irq()** to globally enable all interrupts.
20. Use an infinitive **while(1)** loop to wait for any possible interrupt to occur.
21. Generate the interrupt handler **void PORT4_IRQHandler(void)** to handle interrupts coming from **P4.2** pin.
22. Inside the handler, you first need to clear the interrupt flag set in the **P4IFG** register by reading the **P4IV** register and assigning the returned value to the global variable **ret**.
23. Then you can toggle the LED **PB2** by using the **P2OUT** register.
24. Finally, you need to call the user-defined function **SetSound()** with an argument to send out a 1000 Hz signal to the speaker.

5.2.5 Set up the environment and check the debugger

Now you need to perform the following operations to make sure that the debugger you are using is the **CMSIS-DAP Debugger**. You can check this by:

- Going to **Project|Options for Target "Target 1"** menu item to open the Options wizard.
- On the opened Options wizard, click on the **Debug** tab.
- Making sure that the debugger shown in the **Use:** box is **CMSIS-DAP Debugger**. You also need to set up the correct settings for the debugger and the flash function by clicking on the **Settings** button located at the right of the selected debugger.

5.2.6 Demonstrate your program

Perform the following operations to run your program and check the running results:

- Go to **Flash|Download** menu item to download your program into the flash memory.
- Go to **Debug|Start/Stop Debug Session** to begin debugging your program. Click on the **OK** button on the 32 KB memory size limitation message box to continue.
- Then go to **Debug|Run** menu item to run your program.

As the project runs, the LED **PB2** will be ON when you slide the DIP-SW2 to the ON position (or press the position switch **SW3** since both are connected together), and a 1000 Hz signal is sent to the speaker. The LED **PB2** is OFF when you slide the DIP-SW2 back to the OFF position (or release the position button **SW3**).

Based on these results, try to answer the following questions:

- Can you modify your codes to make any other LED, such as **PB1~PB3**, ON or OFF controlled by different interrupts coming from any other pins on the Port 4, which are connected to the other DIP switches, such as DIP-SW2~DIP-SW4 in the EduBASE Trainer?
- What did you learn from this project?

5.3 LAB5_3

5.3.1 Goal

This project allows students to use both **P6.1** and **P1.1** pins to handle two edge-triggered interrupts to control two different sets of LEDs, **PB0** and **PB1**, **PB2** and **PB3**, in the EduBASE Trainer. This lab enables students to learn and practice how to set up priority levels for two interrupts coming from two different ports based on their real priority levels and how to control and respond to these two interrupts asynchronously and synchronously. Any programming model can be used in this project.

5.3.2 Data assignment and hardware configuration

In this project, two interrupts, one comes from **P6.1** and the other one comes from the **P1.1**, are developed with two interrupt handlers, **PORT6_IRQHandler()** and **PORT1_IRQHandler()**. The **P6.1** has a higher priority level (**3**) and the **P1.1** has a lower priority level (**5**). When the **P6.1** interrupt occurred, two LEDs **PB0** and **PB1** are turned ON and a 1000 Hz signal is sent to the speaker. As the **P1.1** interrupt occurs, two LEDs **PB2** and **PB3** are turned ON and an 800 Hz signal is sent to the speaker.

It can be found from Figure 5.8 that the **P6.1** pin is connected to the position switch **SW5** in the EduBASE Trainer and the **P1.1** pin is connected to the user switch **S1** in the EXP432P401R EVB. The LEDs **PB0~PB3** are connected to the **P3.2**, **P3.3**, **P2.5**, and **P2.4** pins on the GPIO Ports 3 and 2 in the EXP432P401R EVB. The **P5.6** pin is connected to the speaker controller in the EduBASE Trainer. To enable and set these five GPIO ports (Ports 1, 2, 3, 5, and 6), the following data and registers should be used:

- Bit-1 in the GPIO Port 6 direction register and bit-1 in the GPIO Port 1 direction register should be reset to 0 to make **P6.1** and **P1.1** pins as input pins.
- Bits 2 and 3 in the GPIO Port 3 direction register should be set to 1 to enable **P3.2** and **P3.3** to work as output pins.
- Bits 4 and 5 in the GPIO Port 2 direction register should be set to 1 to enable **P2.4** and **P2.5** to work as output pins.
- Bit-6 in the GPIO Port 5 direction register should be set to 1 to enable **P5.6** to work as an output pin.

5.3.3 Development of the project

Use the below steps to develop this project. Only the C source file is used in this project since this project is not complicated. Create a new project with the following steps:

1. Create a new folder **Lab5_3(2INTPRI)** under the folder **C:\MSP432 Lab Projects\Chapter 5** in the Windows Explorer.

2. Open the Keil® ARM®-MDK µVersion5, create a new project **Lab5_3** and save this project in the folder **Lab5_3(2INTPRI)** created in step 1.
3. On the next wizard, you need to select the device (MCU) for this project. Expand three icons, **Texas Instruments**, **MSP432 Family**, and **MSP432P**, and select the target device **MSP432P401R** from the list by clicking on it. Click on the **OK** to close this wizard.
4. Next the Software Components wizard is opened, and you need to set up the software development environment for your project with this wizard. Expand two icons, **CMSIS** and **Device**, and check the **CORE** and **Startup** checkboxes in the **Sel.** column, and click on the **OK** button since we need these two components to build our project.

5.3.4 Development of the C source file

1. In the **Project** pane, expand the **Target** folder and right click on the **Source Group 1** folder and select the **Add New Item to Group "Source Group 1".**
2. Select the **C File (.c)** and enter **Lab5_3** into the **Name:** box, and click on the **Add** button to add this source file into the project.
3. Include the following header files into this source file first:
   a. **#include <stdint.h>**
   b. **#include <stdbool.h>**
   c. **#include <msp.h>**
4. Declare a global **uint16_t** variable **ret** as a data holder to receive the returned values from reading the **P1IV** and **P6IV** registers inside the interrupt handlers later.
5. In this project we need to use three user-defined functions, **Ports_Init()**, **Delay()**, and **SetSound()** to initialize the GPIO Ports, 1, 2, 3, 5, and 6, delay a period of time and send two sound signals with two different frequencies, 1000 and 800 Hz, to the speaker controller in the EduBASE these three functions as:
   void Ports_Init(void);
   void Delay(int time);
   void SetSound(int period);
6. Place the **int main(void)** to start our main program and put a starting brace under this main function to begin our main program.
7. Hold the Watchdog timer by using the Watchdog register **WDTCTL**.
8. Call the **Ports_Init()** function to initialize and configure GPIO Ports 1, 2, 3, 5, and 6.
9. Use the NVIC structure pointer to set up the interrupt priority level of Port 6 as 3, and enable this port based on its IRQ number with the following codes:

   a. **NVIC-> IP[40] = 0x06;** // set Port 6 interrupt priority to 3

   b. **NVIC-> ISER[1] |= 0x00000100;** // enable IRQ40

   Refer to Tables 5.5, 5.11 through 5.13 to get more details about these bits values.
10. Use the NVIC structure pointer to set up the interrupt priority level of Port 1 as 5, and enable this port based on its IRQ number with the following codes:

    a. **NVIC-> IP[35] = 0xA0;** // set Port 1 interrupt priority to 5

    b. **NVIC-> ISER[1] |= 0x00000008;** // enable IRQ35

Refer to Tables 5.5, 5.11 through 5.13 to get more details about these bit values.

11. Call the intrinsic function **__enable_irq()** to globally enable all interrupts.
12. Use an infinitive **while()** loop to wait for any interrupt to be occurred.
13. Now, let us develop the codes for our first user-defined function **Ports_Init()**. The purpose of this function is to:
    a. Configure pins **P6.1** and **P1.1** as input pins by using **P6DIR** and **P1DIR** registers.
    b. Configure pins **P3.2** and **P3.3** on GPIO Port 3 as output pins using **P3DIR** register.
    c. Configure pins **P2.4** and **P2.5** on GPIO Port 2 as output pins using **P2DIR** register.
    d. Configure one pin, **P5.6**, on GPIO Port 5 as an output pin using **P5DIR** register and output **0** to this pin using **P5OUT** register.
    e. Reset all bits on GPIO Ports 2 and 3 to turn off all connected LEDs using **P2OUT** and **P3OUT** registers.
    f. Configure pin **P1.1** to work as a pull-up mode to enable the **S1** to be connected to a pull-up resister to work properly to provide a falling-edge input signal (refer to Section 2.3.4.2.4 in Chapter 2 to get more details about this pull-up setting).
    g. Configure pins **P6.1** and **P1.1** as falling-edge-triggered interrupt pins using **P6IES** and **P1IES** registers.
    h. Reset all previous flags set in both **P6IFG** and **P1IFG** registers by resetting these registers one by one.
    i. Enable interrupts on pins **P6.1** and **P1.1** using **P6IE** and **P1IE** registers.

    The point to be noted is step (f): when using the **P1.1** pin that is connected to a user switch **S1** in the EXP432P401R EVB as an input pin to receive any interrupt, the pin **P1.1** by default is in the open-collector-mode, which means that pin **P1.1** is in the tri-state-mode. This pin must be configured by software to work either in the pull-up or the pull-down connection mode to enable it to receive a valid edge signal from this pin.
14. The codes for the second user-defined function **Delay()**, refer to Section 5.2.4 in **Lab5_2**.
15. The codes for the third user-defined function **SetSound()**, refer to Section 5.2.4 in **Lab5_2**.
16. The codes for the GPIO Port 6 handler are:

```
void PORT6_IRQHandler(void)
{
  ret = P6IV;             // clear interrupt flag set in P6IFG register
  P3OUT ^= (BIT3|BIT2);   // toggle LED PB0 & PB1 in Trainer
  SetSound(1500);         // send 1000 Hz signal to speaker
}
```

17. The codes for the GPIO Port 1 handler are:

```
void PORT1_IRQHandler(void)
{
  ret = P1IV;             // clear interrupt flag set in P1IFG register
  P2OUT ^= (BIT4|BIT5);   // turn LED PB2 & PB3 in Trainer
  SetSound(1875);         // send 800 Hz signal to speaker
}
```

5.3.5 Set up the environment and select the debugger

Now, you can compile your program by going to **Project|Build target** menu item. However, before you can download your program into the flash memory in the EVB, you need to perform the following operations to make sure that the debugger you are using is the **CMSIS-DAP Debugger**. You can check this by:

- Going to **Project|Options for Target "Target 1"** menu item to open the Options wizard.
- On the opened Options wizard, click on the **Debug** tab.
- Making sure that the debugger shown in the **Use:** box is **CMSIS-DAP Debugger**. You also need to set up the correct settings for the debugger and the flash function by clicking on the **Settings** button located at the right of the selected debugger.

5.3.6 Demonstrate your program

Perform the following operations to run your program and check the running results:

- Go to **Flash|Download** menu item to download your program into the flash ROM.
- Go to **Debug|Start/Stop Debug Session** to begin debugging your program. Click on the **OK** button on the 32 KB memory size limitation message box to continue.
- Then go to **Debug|Run** menu item to run your program.

As the project runs, the LEDs **PB0** and **PB1** will be ON and a 1000 Hz signal is sent to the speaker when you press and hold down the position switch **SW5** in the EduBASE Trainer. But both LEDs will be OFF when you press the **SW5** the second time. This means that a falling-edge-triggered interrupt occurs when the **SW5** is pressed. The LEDs **PB2** and **PB3** will be ON and an 800 Hz signal is sent to the speaker when you press and hold down the user switch **S1** in the EXP432P401R EVB. This means that a falling-edge-triggered interrupt occurs when the **S1** is pressed. Based on these results, try to answer the following questions:

- Now try to press the press button **SW5** first, and then press the switch button **S1**, what happened? Why?
- Try to press the switch **S1** first, then press the press button **SW5**, what happened? Why?
- Now modify your codes in two interrupt handlers as below:

```
void PORT6_IRQHandler(void)
{
  P3OUT ^= (BIT3|BIT2);   // toggle LED PB0 & PB1 in Trainer
  SetSound(1500);         // send 1000 Hz signal to speaker
  Delay(800000);
  ret = P6IV;             // clear interrupt flag set in P6IFG register
}
void PORT1_IRQHandler(void)
{
 P2OUT ^= (BIT4|BIT5);    // turn LED PB2 & PB3 in Trainer
 SetSound(1875);          // send 800 Hz signal to speaker
 Delay(800000);
 ret = P1IV;              // clear interrupt flag set in P1IFG register
}
```

Now compile and run your project. Try to press **SW5** first, and then press **S1**. What happened? It looks like the interrupt response for **S1** is delayed? Why?

- What did you learn from this project?

5.4 LAB5_4

5.4.1 Goal

This project is similar to **Lab5_3** but it provides students with another way to use both **P6.1** and **P1.1** pins to handle two edge-triggered interrupts to control two different sets of LEDs, **PB0** and **PB1**, **PB2** and **PB2**, in the EduBASE Trainer. In this project, students need to use the NVIC Set Priority Grouping and NVIC Encoder Priority functions to configure two interrupts coming from GPIO Ports 6 and 1 to replace the CMSIS **NVIC→IP[]** macros to set up the priority levels for these two interrupts.

5.4.2 Data assignment and hardware configuration

Refer to project **Lab5_3** above to get details about the hardware configurations for this project.

In this project, students need to use two NVIC API functions to configure and set up the priority levels for GPIO pins **P6.1** and **P1.1**. These functions are

1. **void NVIC_SetPriorityGrouping(uint32 PriorityGroup)**. This function can be used to split the three-priority-bit used in the MSP432P401R MCU system into two regions; the group priority and the subpriority. Only four possible group levels (**PriorityGroup**) are available in the MSP432P401R MCU system (refer to Table 5.7), which are:
   a. Group **4**: Bits 7~5 work as group bits, no bit (0 bit) works as the subpriority bit
   b. Group **5**: Bits 7~6 work as group bits, bit 5 (1 bit) works as the subpriority bit
   c. Group **6**: Bit 7 works as group bit, bits 6~5 (2 bits) work as the subpriority bits
   d. Group **7**: No bit works as group bit, bits 7~5 (3 bits) work as the subpriority bits
   In this project, students are recommended to use group **5** to make 2-bit, bits 7~6, as group bits (4 group options), and 1-bit, bit 5 as subpriority bit (1 subpriority option).
2. **uint32_t NVIC_EncodePriority(uint32_t PriorityGroup, uint32_t PreemptPriority, uint32_t SubPriority)**. This function is used to generate encoded priority levels for GroupPriority and SubPriority under the same group. The smaller the number, the higher the priority level. In this project, students are recommended to use **2** as the group priority level for **P6.1** and **3** as the group priority level for **P1.1**, but both have the same subpriority level of **1**. Then students can change these priority levels to make both pins have the same group priority level but different subpriority levels.

5.4.3 Development of the project

Create a new folder **Lab5_4(INTGroup)** under the folder **C:\MSP432 Lab Projects\Chapter 5** with the Windows Explorer.

Refer to Sections 5.3.3 and 5.3.4 in **Lab5_3** to create a new µVersion5 project **Lab5_4** and add it into the folder **Lab5_4(INTGroup)** created above. Create a new C code file **Lab5_4.c** and add it into this project. Copy all codes in the file **Lab5_3.c** and paste them into the **Lab5_4.c** file. Perform the following modifications to this file:

1. Before the code lines to set priority and enable **P6.1**, using the **NVIC_SetPriorityGrouping()** function to set the priority group as **5** for both **P6.1** and **P1.1** interrupt sources
2. Replace the code line **NVIC→IP[40] = 0x06;** with **NVIC_SetPriority(PORT6_IRQn, NVIC_EncodePriority(5, 2, 1));**
3. Replace the code line **NVIC→IP[35] = 0xA0;** with **NVIC_SetPriority(PORT1_IRQn, NVIC_EncodePriority(5, 3, 1));**
4. Try to understand the meanings of these two new code lines and their functions, especially for three arguments in two functions

5.4.4 Set up environment and check the debugger

Now you can compile your program by going to **Project|Build target** menu item. However, before you can download your program into the flash memory in the EVB, you need to perform the following operations to make sure that the debugger you are using is the **CMSIS-DAP Debugger**. You can check this by:

- Going to **Project|Options for Target "Target 1"** menu item to open the Options wizard
- On the opened Options wizard, click on the **Debug** tab
- Making sure that the debugger shown in the **Use:** box is **CMSIS-DAP Debugger**. You also need to set up the correct settings for the debugger and the flash function by clicking on the **Settings** button located at the right of the selected debugger

5.4.5 Demonstrate your program

Perform the following operations to run your program and check the running results:

- Go to **Flash|Download** menu item to download your program into the flash memory
- Go to **Debug|Start/Stop Debug Session** to begin debugging your program. Click on the **OK** button on the 32 KB memory size limitation message box to continue
- Then go to **Debug|Run** menu item to run your program

As the project runs, what will happen if you press the user switch button **S1** in the EXP432P401R EVB? What will occur if you press and hold down the **S1** and press the position switch **SW5** in the EduBASE Trainer? What will happen if you change the order of pressing these two switch buttons?

Based on these results, try to answer the following questions:

- If you change the second argument's value in two **NVIC_EncodePriority()** functions above to make both equal, what will happen if you press two switch buttons?
- What did you learn from this project?

chapter six

ARM® and MSP432™ MCU memory system

This chapter provides general information about the ARM® Cortex®-M4 and MSP432P4 microcontroller memory system. The discussion is mainly concentrated on the memory system used in an MSP432P401R MCU system. This discussion includes the system memory map specially designed for that MCU, connections between the processor and memory, and the connection between the memory and peripherals, memory architecture and requirements, bit-band principle and operations, memory access attributes, memory endianness, memory access behaviors, and memory programming applications in the MSP432P401R MCU system.

6.1 Overview

The ARM® Cortex®-M4 processor provides an internal bus matrix to enable the processor to access external memory system via two interfaces: the code interface and SRAM/peripheral interface. The address buses used in both inside and outside of the Cortex®-M4 processor are 32 bits to make the maximum searchable space up to 4 GB. The bus interface between the processor and external memory is the AHB, which provides interfaces and connections to various 32/16/8-bit memory devices.

In order to access and control the 4-GB memory space effectively and easily, the entire 4-GB memory space in the Cortex®-M4 system is divided into the different regions for various predefined memory and peripheral devices uses. With the help of the multiple bus interfaces provided by the Cortex®-M4, the Cortex®-M4 processor can access different memory regions, such as from the CODE region stored program codes to DATA region in the SRAM or peripheral regions, simultaneously or at the same time. The following buses can be used to access memory or peripheral devices in parallel:

- ICode Bus: fetch opcode from the flash ROM
- DCode Bus: read constant data from flash ROM
- System Bus: read/write data from SRAM or I/O, fetch opcode from SRAM
- PPB: read/write data from internal peripheral devices (NVIC)
- AHB: read/write data from high-speed I/O and parallel ports

In summary, the ARM® Cortex®-M4 memory system provides the following specifications and advantages:

- The Cortex®-M4 processors can work with either little-endian or big-endian memory systems. Generally, the Cortex®-M4 is designed with just one endian configuration.
- The ARM® Cortex®-M4 provides a bit-band feature to enable read/write access to individual bits in one 1 MB SRAM region (from **0x22000000** to **0x220FFFFF**), and one 256 MB I/O port region (from **0x42000000** to **0x43FFFFFF**) in the memory devices used in the Texas Instruments™ MSP432 Series LaunchPad™ MCU—MSP432P401R. To

use this bit-band feature, two parameters are needed: the target memory address and the target bit number to be accessed. An example of using this bit-band feature will be discussed in the next section.
- In the ARM® Cortex®-M4 MCU, an optional unit, memory protection unit (MPU) is provided to enable users to access different memory regions with certain permissions. The MPU is a programmable unit that defines access permissions for different regions. The MPU supports eight programmable regions.
- In Cortex®-M4 memory systems, the unaligned transfer operations are supported to perform unaligned data transfers.
- To assist to access different memory regions, memory attributes and access permissions are provided to facilitate these accessing.
- The bus interfaces on the ARM® Cortex®-M4 MCU are generic bus interfaces, which mean that these kinds of interfaces enable the processors to connect and interface to different types and sizes of memory with various memory controllers. Generally, two types of memory: flash memory for program codes and SRAM for program data, are widely adopted in most Cortex®-M microcontroller systems.

In Sections 2.2 and 2.3.2 in Chapter 2, we have provided brief discussions about the memory system used for ARM® Cortex®-M4 and memory map system used for an MSP432P401R MCU. In this chapter, we will focus our discussions on the memory system used in the MSP432P401R MCU system. First, let's take a closer look at the memory architecture in MSP432P401R MCU.

6.2 *Memory architecture in the MSP432P401R MCU system*

Figure 6.1 shows an architecture block diagram for MSP432P401R MCU memory system.

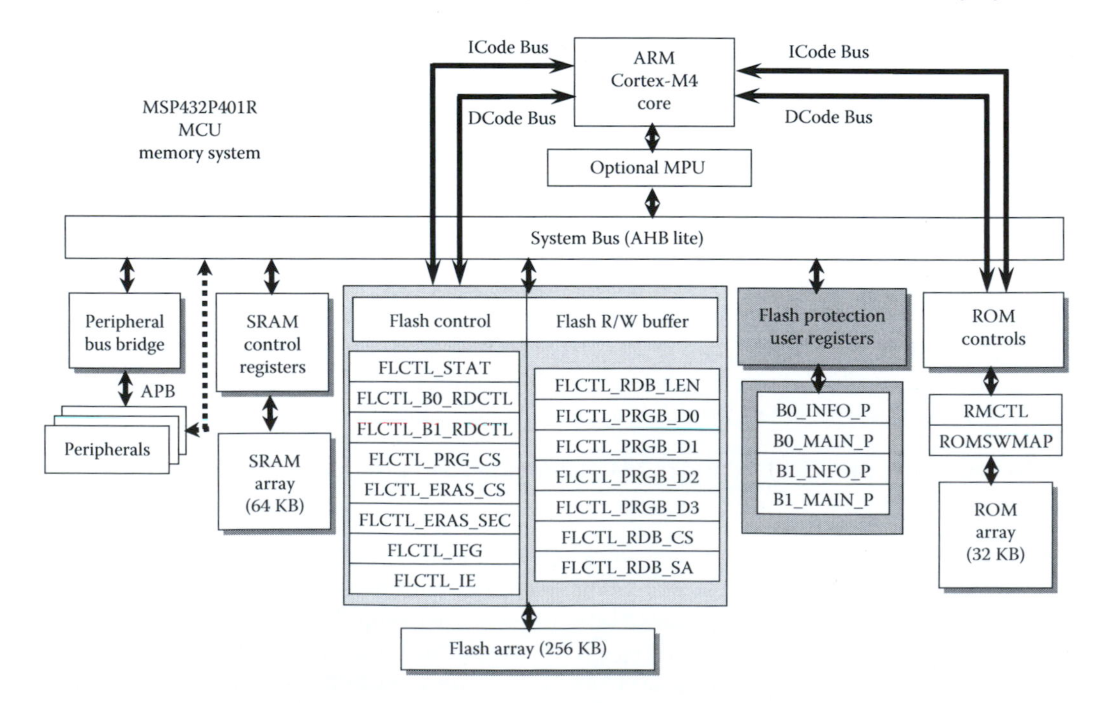

Figure 6.1 Memory systems in the MSP432P401R MCU system.

It can be found from Figure 6.1 that the following components or devices exist in the MSP432P401R MCU memory system

- SRAM (64 KB)
- Flash memory (256 KB)
- Internal ROM (32 KB)
- Flash protection unit
- Optional MPU

Most components or devices inside the memory system are connected to the ARM® Cortex®-M4 processor via system bus, such as SRAM, peripheral bus bridge, flash control, and ROM control. However, the internal ROM and flash memory spaces are connected to the Cortex®-M4 processor via two bus interfaces, ICode and DCode. In simple designs, these two buses can be combined together by using a bus multiplexer to get a simple bus system.

The advantages of using ICode and DCode bus interfaces are:

- The Cortex®-M4 processor can perform data access and fetch the next instruction at the same time, just like a pipeline operation.
- During the interrupt processing period, the processor can access the stack space to reserve or recover all used registers and read the vector table simultaneously.
- Different microcontroller vendors can use these two interfaces to speed up the accessing to the internal ROM space to execute some built-in library functions, such as MSPPDL, boot loader program, graphical library, and USB Library.

The general connection between the Cortex®-M4 processor and peripherals is by AHB protocol directly by the processor using the different macros based on memory mappings in the MSP432P401R MCU system.

In this chapter, we will concentrate on three major components in the MSP432P401R memory system, which are:

- SRAM
- Internal ROM
- Flash memory

The optional MPU will be discussed in Chapter 12. Now, let's take a look at these components or devices one by one.

6.3 Static random-access memory (SRAM)

The **64-KB SRAM** is used to store users' program data and provides system stack and user stack spaces. To reduce the time consuming on read-modify-write (RMW) operations, the ARM® Cortex®-M4 provides bit-banding technology in the processor. With a bit-band-enabled processor, certain regions in the memory map (SRAM and peripheral space) can use address aliases to access individual bits in a single and atomic operation. This 64-KB SRAM space can be extended up to 512 MB.

In order to understand the actual functions and accessing abilities of the SRAM, first let's have a clear picture about the SRAM memory map used in the MSP432P401R MCU system. In Section 2.3.2 in Chapter 2, we have provided a memory map used for this MCU. But we need to discuss this issue in more detail in the following sections.

6.3.1 SRAM memory map in the MSP432P401R MCU system

A more detailed memory map specially designed for 64-KB SRAM is shown in Figure 6.2.

It can be found from Figure 6.2 that a 1-MB memory region from **0x0100.0000** to **0x010F.FFFF** is embedded and defined as an SRAM region in the 256-KB flash memory space. This region is also aliased in the SRAM zone of the device, thereby allowing efficient access to the SRAM, both for instruction fetches as well as data reads.

The 64-KB SRAM space is arranged from **0x2000.0000** to **0x2000.FFFF** with a bit-band alias space arranged from **0x2200.0000** to **0x220F.FFFF**. The rest space is reserved for the future usage.

As we discussed in Section 2.3.2 in Chapter 2, the 64-KB SRAM memory space can be equally divided into eight 8-KB memory spaces or called eight banks arranged from Bank0 to Bank7.

Each bank is controlled by the related control register. A memory map used by the MSP432P401R MCU system is also shown in that section. To get more details about this SRAM device, let's start from its detailed memory map.

Each bank can be individually enabled or disabled, and can also be accessed individually to perform read and write operations. The application can choose to optimize the power consumption of each bank. Banks that are powered down remain powered down in both active as well as low-power modes of operation, thereby limiting any unnecessary inrush current when the device transitions between active and retention-based low-power modes. The application can also choose to disable one (or more) banks for a certain stage in the processing and re-enable it for another stage.

Whenever a particular bank is disabled, reads to its address space return 0, and writes are discarded. To prevent **holes** in the memory map, if a particular bank is enabled; all the lower banks are forced to be enabled as well. This ensures a contiguous memory map through the set of enabled banks instead of a possible disabled bank appearing between enabled banks.

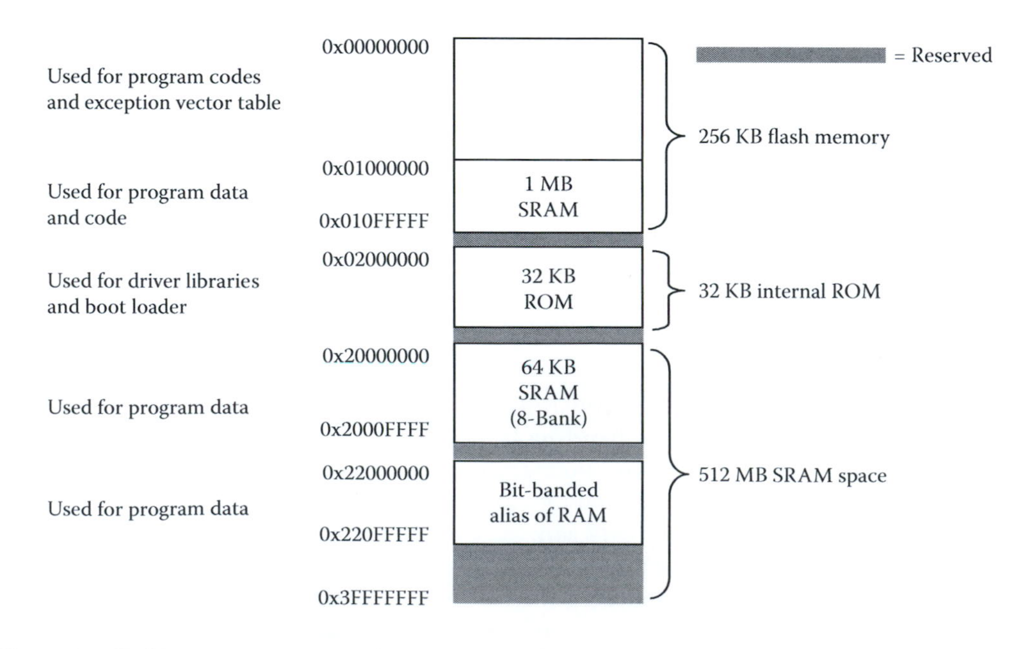

Figure 6.2 SRAM memory map in the MSP432P401R MCU system.

One point to be noted is that Bank0 is always enabled and cannot be powered down by default. Also, when any SRAM bank is enabled or disabled, accesses to the SRAM are temporarily stalled to prevent spurious reads. This is handled transparently and does not require any code intervention.

The application can also choose to optimize the leakage power consumption of the SRAM in LPM3 and LPM4 operation modes. In order to do this, each SRAM bank can be individually configured for retention. Banks that are enabled for retention retain their data through the LPM3 and LPM4 modes. The application can also choose to retain a subset of the enabled banks.

For example, the application may need 32 KB of SRAM for its processing needs, which is four banks are kept enabled. However, for all these four banks, only one bank may contain critical data that must be retained in LPM3 or LPM4 modes while the rest are powered off completely to minimize power consumption.

As we mentioned, Bank0 of SRAM is always retained and cannot be powered down at any operation mode. Therefore, it also operates up as a possible backup memory in the LPM3, LPM4, and LPM3.5 modes of operation.

All of these enable and retain abilities and functions are performed by using related SRAM control and status registers. In order to understand these control functions, let's have a closer look at these registers.

6.3.2 *SRAM control and status registers*

Three system control registers, SRAM Size (**SYS_SRAM_SIZE**) Register, SRAM Bank Enable (**SYS_SRAM_BANKEN**) Register, and SRAM Bank Retention Control (**SYS_SRAM_BANKRET**) Register, are used to control and configure the SRAM operations. All of these three registers belong to the System Control (**SYSCTL**) Module registers.

6.3.2.1 *SRAM size (SYS_SRAM_SIZE) register*

This is a 32-bit read-only register used to indicate the memory size (in bytes) of the SRAM available in the MSP432P401R MCU system. The memory mapping address for this register is **0xE004.3010**, which is located at the **SYSCTL** space range.

6.3.2.2 *SRAM bank enable (SYS_SRAM_BANKEN) register*

This is a 32-bit register but only bit-16 and the lower 8 bits are used to indicate which bank of the SRAM is powered up (enabled) and available for the application. The rest bits are reserved for the future usage.

The application can choose to enable or disable SRAM banks on the fly. When the SRAM banks are being powered up or down, accesses to the SRAM space is temporarily stalled during the powered up or down period until the SRAM banks are finished and ready. Accesses to the rest of the memory map remain unaffected.

Figure 6.3 shows the bits configurations and values of this register.

SRAM bank enable register (SYS_SRAM_BANKEN)

31–17	16	15–8	7	6	5	4	3	2	1	0
Reserved	SRAM_RDY	Reserved	BNK7_EN	BNK6_EN	BNK5_EN	BNK4_EN	BNK3_EN	BNK2_EN	BNK1_EN	BNK0_EN

Figure 6.3 SRAM bank enable register.

Table 6.1 shows the bit field and bit function for this **SYS_SRAM_BANKEN** register. From this table, the following points can be noted:

1. All eight banks (Bank7~Bank0) are enabled and powered up after a system reset.
2. All lower eight bits, including **BNK7_EN~BNK1_EN**, are read-write (RW) accessible except the bit-0, **BNK0_EN**, which is a read-only bit and cannot be modified by any writing. This means that the Bank0 is a default available memory region and it always powered up and enabled to the device, and it cannot be powered down and disabled at any time.

The SRAM Bank Enable Register controls which banks of the SRAM are enabled for read/write accesses. There is one bit for each available bank in the lower eight bits on this register. Banks that are not enabled are powered down to minimize the power consumption. Each bit in this register corresponds to one bank of the SRAM. Banks may only be enabled in a contiguous form, which means that if a bank is enabled, all banks below this bank are all enabled and powered up. For example:

- For a device that has eight banks in the system, values of **00111111** and **00000111** on this **SYS_SRAM_BANKEN** register are acceptable.
- But values such as **00010111** or **10110111** are not valid, and the resultant bank configuration will be set to **00011111** or **11111111**.
- For a 4-bank SRAM, the only allowed values are **0001**, **0011**, **0111**, and **1111**.

6.3.2.3 SRAM bank retention control register (SYS_SRAM_BANKRET)

Similar to SRAM Bank Enable Register, this is also a 32-bit register but only bit-16 and lower eight bits are used to indicate which bank of the SRAM is retained when the MCU entered the LPM3 or LPM4 modes. Any bank that is not enabled for retention will be completely powered down in these modes and will lose its data. The rest bits are reserved for the future usage.

The bit-16 in this register is still a **SRAM_RDY** bit, which is a read-only bit. If this bit is 0, which means that the SRAM banks are being set up for retention. Entry into LPM3 or LPM4 modes should not be attempted until this bit is set to 1. The lower eight bits on this register (bits 7~0), **BNK7_RET~BNK0_RET**, with each bit being corresponding to a bank, Bank7~Bank0, indicates whether the related memory bank is retained when entering the LPM3 or LPM4 mode.

Table 6.2 shows the bit field and bit function for this **SYS_SRAM_BANKRET** register. From this table, the following points can be noted:

1. All seven banks (Bank7~Bank1) are not retained for LPM3 and LPM4 modes after a system reset. But only Bank0 is retained for LPM3 and LPM4 modes by default after a system reset operation.
2. All lower eight bits, including **BNK7_RET~BNK1_RET**, are RW accessible except the bit-0, **BNK0_RET**, which is a read-only bit and cannot be modified by any writing. This means that the Bank0 is a default available memory region that can be automatically retained in LPM3, LPM4, and LPM3.5 operation modes.
3. Write to lower seven bits (**BNK7_RET~BNK1_RET**) are allowed ONLY when the **SRAM_RDY** bit of this register is set to 1. If the **SRAM_RDY** bit is 0, writes to these bits will be ignored.

Table 6.1 Bit value and its function for SYS_SRAM_BANKEN register

Bit	Field	Type	Reset	Function
31:17	**Reserved**	RO	00	Reserved and returns 0 as read
16	**SRAM_RDY**	RO	0	SRAM is ready to be accessed and used: **0:** SRAM is not ready to be accessed; **1:** SRAM is ready to be accessed. During the powered up or down period, this bit is reset to 0 to disable any access to the selected banks. This bit will be set to 1 when the power-up or down is done
15:8	**Reserved**	RO	00	Reserved and returns 0 as read
7	**BNK7_EN**	RW	1	Bank 7 of SRAM enable or disable control: **0:** Bank 7 of SRAM is disabled; **1:** Bank 7 of SRAM is enabled When set to 1, bank enable bits for all banks below this bank are set to 1 as well
6	**BNK6_EN**	RW	1	Bank 6 of SRAM enable or disable control: **0:** Bank 6 of SRAM is disabled; **1:** Bank 6 of SRAM is enabled When set to 1, bank enable bits for all banks below this bank are set to 1 as well
5	**BNK5_EN**	RW	1	Bank 5 of SRAM enable or disable control: **0:** Bank 5 of SRAM is disabled; **1:** Bank 5 of SRAM is enabled When set to 1, bank enable bits for all banks below this bank are set to 1 as well
4	**BNK4_EN**	RW	1	Bank 4 of SRAM enable or disable control: **0:** Bank 4 of SRAM is disabled; **1:** Bank 4 of SRAM is enabled When set to 1, bank enable bits for all banks below this bank are set to 1 as well
3	**BNK3_EN**	RW	1	Bank 3 of SRAM enable or disable control: **0:** Bank 3 of SRAM is disabled; **1:** Bank 3 of SRAM is enabled When set to 1, bank enable bits for all banks below this bank are set to 1 as well
2	**BNK2_EN**	RW	1	Bank 2 of SRAM enable or disable control: **0:** Bank 2 of SRAM is disabled; **1:** Bank 2 of SRAM is enabled When set to 1, bank enable bits for all banks below this bank are set to 1 as well.
1	**BNK1_EN**	RW	1	Bank 1 of SRAM enable or disable control: **0:** Bank 1 of SRAM is disabled; **1:** Bank 1 of SRAM is enabled When set to 1, bank enable bits for all banks below this bank are set to 1 as well
0	**BNK0_EN**	R	1	Bank 0 of SRAM enable control: **1:** Bank 0 of SRAM is enabled

Table 6.2 Bit value and its function for SYS_SRAM_BANKRET register

Bit	Field	Type	Reset	Function
31:17	**Reserved**	RO	00	Reserved and returns 0 as read
16	**SRAM_RDY**	RO	0	SRAM is ready to be accessed and used: **0:** SRAM is not ready to be accessed; **1:** SRAM is ready to be accessed During the retention setup period, this bit is reset to 0 to disable any entering to the LPM3 or LPM4 mode. This bit will be set to 1 when the retention set up is done
15:8	**Reserved**	RO	00	Reserved and returns 0 as read
7	**BNK7_RET**	RW	0	Bank 7 of SRAM retention control: **0:** Bank7 of the SRAM is not retained in LPM3 or LPM4 mode **1:** Bank7 of the SRAM is retained in LPM3 or LPM4 mode
6	**BNK6_RET**	RW	0	Bank 6 of SRAM retention control: **0:** Bank6 of the SRAM is not retained in LPM3 or LPM4 mode **1:** Bank6 of the SRAM is retained in LPM3 or LPM4 mode
5	**BNK5_RET**	RW	0	Bank 5 of SRAM retention control: **0:** Bank5 of the SRAM is not retained in LPM3 or LPM4 mode **1:** Bank5 of the SRAM is retained in LPM3 or LPM4 mode
4	**BNK4_RET**	RW	0	Bank 4 of SRAM retention control: **0:** Bank4 of the SRAM is not retained in LPM3 or LPM4 mode **1:** Bank4 of the SRAM is retained in LPM3 or LPM4 mode
3	**BNK3_RET**	RW	0	Bank 3 of SRAM retention control: **0:** Bank3 of the SRAM is not retained in LPM3 or LPM4 mode **1:** Bank3 of the SRAM is retained in LPM3 or LPM4 mode
2	**BNK2_RET**	RW	0	Bank 2 of SRAM retention control: **0:** Bank2 of the SRAM is not retained in LPM3 or LPM4 mode **1:** Bank2 of the SRAM is retained in LPM3 or LPM4 mode
1	**BNK1_RET**	RW	0	Bank 1 of SRAM retention control: **0:** Bank1 of the SRAM is not retained in LPM3 or LPM4 mode **1:** Bank1 of the SRAM is retained in LPM3 or LPM4 mode
0	**BNK0_RET**	R	1	Bank 0 of SRAM retention control: Bank0 is always retained in LPM3, LPM4, and LPM3.5 modes of operation

6.3.3 Bit-band operations

It can be found from Figure 6.2 that both the 64-KB SRAM and the peripheral memory spaces in the memory map can be mapped to two bit-band alias regions, bit-banded alias of SRAM and bit-banded alias of I/O ports, respectively.

The first 1-MB memory space for SRAM, **0x2000.0000~0x200F.FFFF**, and the first 1-MB space for the peripheral, **0x400.0000~0x400F.FFFF**, in the memory map are called **bit-band regions** and they can be accessed in two different ways. This means that this first 1 MB can be accessed as normal SRAM or peripheral memory mapping spaces, but they can also be accessed via different memory regions called the bit-band alias regions.

A bit-band region maps each word in a **bit-band alias** region to a single bit or the least significant bit (LSB) in the **bit-band region**. Since each word contains 32 bits, therefore the 1-MB bit-band region can be mapped to 32-MB bit-band alias regions. The memory map has two 32 MB alias regions that map to two 1-MB bit-band regions:

- Accesses to the 32-MB SRAM alias region map to the 1-MB SRAM bit-band region
- Accesses to the 32-MB peripheral alias region map to the 1-MB peripheral bit-band region.

The mapping relationships between the bit-band regions and the bit-band alias regions for the SRAM and the peripheral are shown in Table 6.3.

Within the 1-MB bit-band region, each bit is mapped to a 32-bit word in the bit-band alias region; exactly each bit in the bit-band region can be mapped to the LSB of a 32-bit word in the bit-band alias region. In this way, each word in the bit-band region can be mapped to an LSB of 32 words in the bit-band alias region. When using the bit-band alias regions, the following points should be noted:

- When the bit-band alias address is accessed, this address is remapped into a bit-band address. For reading operations, the word is read and the desired bit location is shifted to the LSB of the returned data. For writing operations, the written bit data are shifted to the required bit location, and a RMW operation is executed.

Table 6.3 Bit-banding regions in SRAM and peripheral

Address range	Memory region	Instruction and data accesses
0x20000000~0x200FFFFF	SRAM bit-band region (1 MB)	This SRAM memory space can be accessed as normal SRAM, but this region can also be accessed via bit-band alias region
0x22000000~0x23FFFFFF	SRAM bit-band alias region (32 MB)	Data accesses to this region are remapped to the SRAM bit-band region. A write operation is performed as read-modify-write. Instruction accesses are not remapped
0x40000000~0x400FFFFF	Peripheral bit-band region (1 MB)	This peripheral memory range can be accessed as normal peripheral space, but this region can also be accessed via bit-band alias region
0x42000000~0x43FFFFFF	Peripheral bit-band alias region (32 MB)	Data accesses to this region are remapped to the peripheral bit-band region. A write operation is performed as read-modify-write. Instruction accesses are not permitted

- For the Cortex®-M4 processor, there is no special instruction to access the bit-band alias addresses. However, when data accesses to these regions that are defined as the bit-band alias address ranges, the addresses can be automatically converted to the bit-band regions.
- The SRAM starts at the base address 0x20000000, equivalently the bit-banded SRAM can be mapped to start at the base address **0x2200000**.
- The peripheral space starts at the base address 0x40000000, equivalently the bit-banded peripheral space is mapped to start at the base address **0x42000000**.

Let's take a closer look at the mapping relationship between the bit-band region and bit-band alias region for SRAM and the peripheral memory space.

6.3.3.1 *Mapping relationship between the bit-band region and bit-band alias region*

The relationship between the bit-band region address and the bit-band alias address can be described by the following equation:

```
Bit-band alias = bit-band-base + (byte_offset × 32) + (bit_number × 4)
```

This means that the bit-band alias address can be calculated by using this equation. Where:

- The **bit-band base** is the starting address of the bit-band alias regions. For the SRAM, it is **0x22000000**, and for the peripheral, it is **0x42000000**.
- The **byte_offset** is the number of the bytes in the bit-band region that contains the targeted bit.
- The **bit_number** is the bit position, 0~7, of the targeted bit.

For example, bit-7 (bit_number) at the address 0x22002000 (bit-band-base) is:

Bit-band alias = 0x22002000 + (0x2000 × 32) + (7 × 4) = 0x2204201C

Based on this equation, some remapping bit-band addresses in the SRAM alias region are shown in Table 6.4. The remapping bit-band addresses in the peripheral alias region are shown in Table 6.5.

It can be found from these tables that both the bit-band region of the SRAM and the bit-band region of the peripheral memory space can be accessed via bit-band aliased addresses, which means that as long as an access to the bit-band alias address is created, it can be automatically converted or remapped to the associated bit-band address in either SRAM or the peripheral memory space. In fact, both bit-band region and the bit-band alias region is physically the same memory region but they can be accessed in two different ways.

6.3.3.2 *Advantages of using the bit-band operations*

The advantages of using the bit-band alias to access the memory space include:

- Allows a single load/store operation to access to a single data bit to simplify the data writing operations.
- Allows a single load/store operation to access to a single data bit to simplify the data reading operations.

Table 6.4 Remapping of bit-band addresses in the SRAM alias region

Bit-band range	Bit-band alias region
0x20000000 bit[0]	0x22000000 bit[0]
0x20000000 bit[1]	0x22000004 bit[0]
0x20000000 bit[2]	0x22000008 bit[0]
0x20000000 bit[3]	0x2200000C bit[0]
0x20000000 bit[4]	0x22000010 bit[0]
0x20000000 bit[5]	0x22000014 bit[0]
0x20000000 bit[6]	0x22000018 bit[0]
0x20000000 bit[7]	0x2200001C bit[0]
......	......
0x20000000 bit[31]	0x2200007C bit[0]
0x20000004 bit[0]	0x22000080 bit[0]
0x20000004 bit[1]	0x22000084 bit[0]
......	......
0x20000004 bit[31]	0x220000FC bit[0]
......	......
0x200FFFFC bit[31]	0x23FFFFFC bit[0]

Table 6.5 Remapping of bit-band addresses in the peripheral alias region

Bit-band range	Bit-band alias region
0x40000000 bit[0]	0x42000000 bit[0]
0x40000000 bit[1]	0x42000004 bit[0]
0x40000000 bit[2]	0x42000008 bit[0]
0x40000000 bit[3]	0x4200000C bit[0]
0x40000000 bit[4]	0x42000010 bit[0]
0x40000000 bit[5]	0x42000014 bit[0]
0x40000000 bit[6]	0x42000018 bit[0]
0x40000000 bit[7]	0x4200001C bit[0]
......	......
0x40000000 bit[31]	0x4200007C bit[0]
0x40000004 bit[0]	0x42000080 bit[0]
0x40000004 bit[1]	0x42000084 bit[0]
......	......
0x40000004 bit[31]	0x420000FC bit[0]
......	......
0x400FFFFC bit[31]	0x43FFFFFC bit[0]

```
        Without Bit-Band ——— Write Operation ——— With Bit-Band
LDR     R1, = 0x20000004; Set address          LDR   R1, =0x22000080;  Set address
LDR     R0, [R1];         Read                 MOV   R0, #1;           Set data
ORR.W   R0, #0x1;         Modify bit value     STR   R0, [R1];         Write data
STR     R0, [R1];         Write back data
        Without Bit-Band ——— Read Operation ——— With Bit-Band
LDR     R1, = 0x20000004; Set address          LDR   R1, = 0x22000080; Set address
LDR     R0, [R1];         Read data            LDR   R0, R1;           Read data
UBFX.W  R0, R0, #0, #1;   Extract bit[0]
```

Figure 6.4 Bit data writing and reading with and without bit band.

Figure 6.4 shows an example of performing data bit writing and reading operations with and without bit-band method.

It can be found from Figure 6.4 that in total four assembly instructions are used to perform a data bit writing operations without the bit-band support, but only three instructions are used to do the same job with the bit-band alias method. Similar situation occurred to the reading operations.

Regularly, to modify a data bit value in the ARM® Cortex®-M4 system, a three-step operation, RMW, should be performed. This means that at least three instructions are needed to complete this modification. However, with the bit-band region support, this modification only needs two steps, modify and write (without read), which means that only two instructions are needed to complete this modification.

To get a clearer picture about the bit-band technology, we try to use the following example to illustrate the mapping relationships between the bit-band region and the bit-band alias region with SRAM. Similar idea can be obtained for the peripheral memory regions.

6.3.3.3 *Illustration example of using bit-band alias addresses*

Figure 6.5 shows an example to illustrate how to map 32-bit-band alias word to related 1-MB bit-band region for SRAM.

Based on the mapping equation shown in Section 6.3.3.1, the following bit-band alias words, exactly the LSB or bit[0] of the following bit-band alias words, can be mapped to related bits on the bit-band region:

1. The bit[0] of the alias word at **0x22000000** maps to bit[0] of the bit-band byte at **0x20000000** → **0x22000000** = **0x22000000 + (0*32) + (0 *4)**. In this mapping, the bit-band base is 0x22000000, the byte offset is 0 and the bit number is 0.
2. The bit[0] of the alias word at **0x2200001C** maps to bit[7] of the bit-band byte at **0x20000000** → **0x2200001C** = **0x22000000+ (0*32) + (7*4)**. In this mapping, the bit-band base is 0x22000000, the byte offset is 0 and the bit number is 7.
3. The bit[0] of the alias word at **0x23FFFFE0** maps to bit[0] of the bit-band byte at **0x200FFFFF** → **0x23FFFFE0** = **0x22000000 + (0xFFFFF*32) + (0*4)**. In this mapping, the bit-band base is 0x22000000, the byte offset is 0xFFFFF and the bit number is 0.
4. The bit[0] of the alias word at **0x23FFFFFC** maps to bit[7] of the bit-band byte at **0x200FFFFF** → **0x23FFFFFC** = **0x22000000 +(0xFFFFF*32) + (7*4)**. In this mapping, the bit-band base is 0x22000000, the byte offset is 0xFFFFF and the bit number is 7.

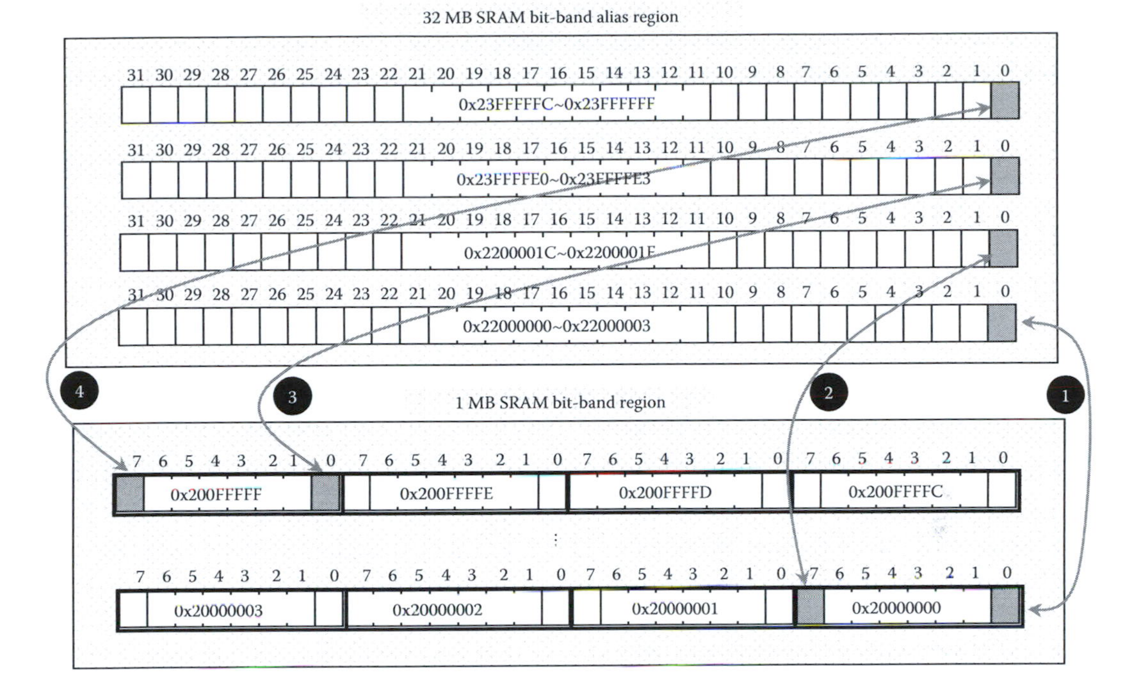

Figure 6.5 Bit-band mapping for SRAM.

Some points to be noted when doing this mapping calculation are:

- When performing the **bit _ number** × **4** calculation, the product should be converted to the hexadecimal number before adding it to the sum. For example, the product of 7 × 4 = 28 should be converted to **0x1C** before it can be added into the sum to get the alias address.
- When performing the **byte_offset** × **32** operation, first convert the **byte_offset** from the hexadecimal number to the binary number, and then attach five 0s at the end of this binary number, which is equivalent to multiply by 32. Finally, convert that binary number back to the hexadecimal number and add it into the sum to get the alias address.
- Each bit[0] on the related bit-band alias word has 4 bytes or 32 bits. Therefore, the calculated address for the alias word is the address of the first byte of that word. For example, in the first mapping in Figure 6.5, the bit[0] of the alias word **0x22000000** maps to bit[0] of the bit-band byte at **0x20000000**. This means that the **0x22000000** is the address of the first byte that is included in the word who takes 4 bytes with addresses ranged from **0x22000000~0x22000003**.

Now we can determine some bit-band alias addresses easily with the equation:

```
Bit-band alias = bit-band-base + (byte_offset × 32) + (bit_number × 4)
```

For example, determine the bit-band alias addresses for the following bits at the bit-band regions: (1) bit[2] of the bit-band region at 0x20000FFD, (2) bit[5] of the bit-band region at 0x20002222, and (3) bit[3] of the bit-band region at 0x20007890.

1. Bit-band alias = 0x22000000 + (0xFFD × 32) + (2 × 4) = 0x2201FFA8.
2. Bit-band alias = 0x22000000 + (0x2222 × 32) + (5 × 4) = 0x22044454.
3. Bit-band alias = 0x22000000 + (0x7890 × 32) + (3 × 4) = 0x220F120C.

Based on these results, the bit[0] of the alias word at 0x2201FFA8 maps to bit[2] of the bit-band region at 0x20000FFD, the bit[0] of the alias word at 0x22044454 maps to bit[5] of the bit-band region at 0x20002222, and the bit[0] of the alias word at 0x220F120C maps to bit[3] of the bit-band region at 0x20007890.

6.3.3.4 Bit-band operations for different data sizes

The bit-band operations are not only limited to map each bit in the bit-band region to a word in the bit-band alias region, it is also available to map bits in the bit-band region to bytes and half-words in the bit-band alias region. For example, when a byte or a half-word access instruction, such as **LDRB/STRB** or **LDRH/STRH**, is used to access a bit-band alias region, the accesses generated to the bit-band region will be in byte size or in half-word size.

Refer to bit mapping example shown in Figure 6.5. For the first mapping, in which the bit[0] of the alias word at **0x22000000** maps to bit[0] of the bit-band byte at **0x20000000**. In fact, this can be considered as the bit[0] of the alias *byte* at **0x22000000** maps to bit[0] of the bit-band region at **0x20000000**. You can also consider that the bit[0] of the alias *half-word* at **0x22000000~0x22000001** maps to bit[0] of the bit-band region at **0x20000000**.

6.3.3.5 Bit-band operations built in C programs

Since the bit-band operations belong to special features used in the assembly instructions in the Cortex®-M4 MCU system, therefore this feature cannot be recognized by any high-level language, such as C/C++. The C/C++ compiler does not know that the same memory space can be accessed by using two different mapping regions, the bit-band region and bit-band alias region. Also, the compiler cannot convert from a bit-band region to the related bit-band alias region. In order to use the bit-band feature in C/C++, the easiest way is to declare the memory addresses related to the bit-band region and bit-band alias region separately.

For example, you can declare the SRAM bit-band region and bit-band alias region as two separate base addresses as shown in Figure 6.6.

The first declaration is to define a normal SRAM memory address starting at **0x20000000**. The data type is a volatile 32-bit pointer. The **volatile** property is used to inform the compiler that the data value in this memory address can be changed and each

```
Define SRAM Bit-Band Address
#define MEM_REG0       *((volatile unsigned long *) (0x20000000)   // bit-band region
#define MEM_REG0_BIT0  *((volatile unsigned long *) (0x22000000)   // bit-band alias region
#define MEM_REG0_BIT1  *((volatile unsigned long *) (0x22000004)   // bit-band alias region
#define MEM_REG0_BIT2  *((volatile unsigned long *) (0x22000008)   // bit-band alias region
......
MEM_REG0 = 0x18;             // access the memory register by normal address
MEM_REG0 = MEM_REG0|0x2;     // set bit without using bit-band feature
MEM_REG0_BIT2 = 0x1;         // set bit using the bit-band feature via bit-band alias address
MEM_REG0_BIT1 = 0x1;         // set bit using the bit-band feature via bit-band alias address
```

Figure 6.6 Declare bit-band and bit-band alias regions in C.

```
Define SRAM Bit-Band Conversion Equation
#define  BIT_BAND (addr, bitnum) ((0x22000000 + ((addr & 0xFFFFF) << 5) + (bitnum << 2))
Convert the Address to a Pointer
#define  MEM_ADDR (addr)  *((volatile unsigned long *) (addr))
Define SRAM Bit-Band Address
#define  MEM_REG0         *((volatile unsigned long *) (0x20000000)       // bit-band region
Data Assignments without and wit Bit-Band Operations
MEM_ADDR (MEM_REG0) = 0x18;                          // access the memory register by normal address
MEM_ADDR (MEM_REG0) = MEM_ADDR (MEM_REG0) | 0x2;     // set bit without using bit-band feature
MEM_ADDR (BIT_BAND(MEM_REG0, 1)) = 0x1;              // set bit using the bit-band feature via bit-band
                                                     // alias address
```

Figure 6.7 C macros used to convert bit-band alias and bit-band operations.

time a variable is accessed, the memory location is accessed instead of a copy of the data inside the processor.

The second through the fourth declarations define bit-band alias addresses. The first two assignments are for normal memory operations, but the last two assignments used the bit-band alias addresses to set the bits values.

You can also develop some C macros to define useful bit-band-related operations and addresses, and this makes the bit-band operations easier in the C programs. For example, the macros defined in Figure 6.7 are used to convert the bit-band address and the bit number into the bit-band alias address, which is the converting equation shown in Section 6.3.3.1, and access the memory bit-band region to set data bit value.

In Figure 6.7, the macro used to define the SRAM bit-band conversion equation is to convert the bit-band region with the bit number to the associated bit-band alias address. The **`bit-band-base`** address for SRAM is **`0x22000000`**. The **`(addr & 0xFFFFF)`** is to **`AND`** the input bit-band address with **`0xFFFFF`** to get the **`byte_offset`** (SRAM bit-band region is: **`0x20000000~0x200FFFFF`**). Then the **`byte_offset`** is shifted left by 5 bits (**`<< 5`**), which is equivalent to multiply the **`byte_offset`** by 32. The (**`bitnum << 2`**) is to shift bitnum left by 2 bits, which is equivalent to multiply the **`bitnum`** by 4. Finally, all of these items are added together to get the bit-band alias address.

The last assignment **`MEM_ADDR(BIT_BAND(MEM_REG0, 1)) = 0x1`**; is to set the LSB of the bit-band alias word at 0x22000004 to 1, which is equivalent to set the bit[1] in the bit-band region at 0x20000000 to 1. The **`BIT_BAND(MEM_REG0, 1)`** is used to get the bit-band alias address 0x22000004. The **`MEM_ADDR(addr)`** is used to define an address pointer.

6.4 Internal ROM

As we discussed in Section 2.3.2 in Chapter 2, the internal ROM is a memory device in MSP432P401R MCU, and this device contains the following software and programs:

- MSPWare driver libraries, including the peripheral devices library, USB library, and graphical library
- MSPWare BSL
- AES cryptography tables
- Cyclic Redundancy Check (CRC) error-detection functionality

The MSPWare BSL is used to download code to the flash memory of a device without the use of a debug interface. When the core is reset, the user has the opportunity to direct the core to execute the ROM BSL or the application in flash memory by using any GPIO serial interface, such as UART, I2C, or SPI, as configured in the BSL Peripheral Interface Instance (**BSL_PER_IF_SEL**) register. More detailed information about the boot loader can be found in Section 3.7.1.2 in Chapter 3. For detailed information about the **BSL_PER_IF_SEL** register, refer to Section 6.4.1 in this chapter.

AES is ideal for applications that can use prearranged keys, such as setup during manufacturing or configuration.

The CRC technique can be used to validate correct receipt of messages (nothing lost or modified in transit), to validate data after decompression, to validate that flash memory contents have not been changed, and for other cases where the data need to be validated.

It can be found from Table 2.3 in Chapter 2 that the internal ROM memory map starts at **0x0200.0000** and ends at **0x020F.FFFF**. The 1-KB memory space between **0x0200.0000** and **0x200.03FF** is reserved for future usage.

6.4.1 Bootstrap loader (BSL)

The Texas Instruments™ boot loader is a small piece of code that can be programmed at the beginning of flash to act as an application loader as well as an update mechanism for applications running on an MSP432 ARM® Cortex®-M4-based microcontroller. The boot loader can be built to use the UART, SPI, or I2C ports to update the code on the microcontroller. The boot loader is customizable via source code modifications, or simply deciding at compile time which routines to include. Since full source code is provided, the boot loader can be completely customized.

Three update protocols are utilized. On UART, SPI, and I2C, a custom protocol is used to communicate with the download utility to transfer the firmware image and program it into flash.

When configured to use UART0, Ethernet, or USB, the LM Flash Programmer GUI can be used to download an application via the boot loader.

Some other functions provided by the boot loader are:

- Download codes to the flash memory for firmware updates
- Interface options include UART, I2C, and SPI

The content of the **BSL_PER_IF_SEL** register helps the BSL to determine and estimate the properties of used interface as below:

1. **Interface multiplexing status**: If the corresponding interface is directly available on device pins or if it is multiplexed with GPIO.
2. **Interface module**: Module used to implement the interface, such as eUSCIA or eUSCIB.
3. Instance of the module used by the BSL.

The bit map and bit function of the **BSL_PER_IF_SEL** register are shown in Table 6.6.

In MSP432™ MCU system, the BSL can automatically select the interface used to communicate with the MCU. The specific instance of the peripheral interfaces to be used depends on the selected MCU and can be found in the device-specific data sheet. This

Table 6.6 Bit value and its function for BSL_PER_IF_SEL register

Bit	Field	Type	Function
31:24	**Reserved**	RO	Reserved for future usage
23	**I2C_MUX**	RO	I2C interface mux status: **1:** I2C is muxed with GPIO; **0:** I2C pins are dedicated on device pin out
22	**Reserved**	RO	Reserved for future usage
21:20	**I2C_MOD**	RO	Module used to implement I2C: **0:** eUSCIB; **1–3:** Reserved for future IPs implementing I2C
19:16	**I2C_INST**	RO	Instance number of the IP specifically used to implement I2C: **0x0:** Use instance 0 of the IP used for implementing I2C (for example, eUSCIB0 when I2C_MOD = 00) **0x1:** Use instance 1 of the IP used for implementing I2C (for example, eUSCIB1 when I2C_MOD = 00) .... **0xF:** Use instance 15 of the IP used for implementing I2C (for example, eUSCIB15, when I2C_MOD = 00)
15	**SPI_MUX**	RO	SPI interface mux status: **0:** SPI pins are dedicated on device pin out; **1:** SPI is muxed with GPIO
14	**Reserved**	RO	Reserved for future usage
13:12	**SPI_MOD**	RO	Module used to implement SPI: **0:** eUSCIA; **1:** eUSCIB; **2–3:** Reserved for future IPs implementing SPI
11:8	**SPI_INST**	RO	Instance number of the IP specifically used to implement SPI: **0x0:** Use instance 0 of the IP used for implementing SPI (for example, eUSCIB0 when SPI_MOD = 01) **0x1:** Use instance 1 of the IP used for implementing SPI (for example, eUSCIB1 when SPI_MOD = 01) .... **0xF:** Use instance 15 of the IP used for implementing SPI (for example, eUSCIB15, when SPI_MOD = 01)
7	**UART_MUX**	RO	UART interface mux status: **0:** UART pins are dedicated on device pin out; **1:** UART is muxed with GPIO
6	**Reserved**	RO	Reserved for future usage
5:4	**UART_MOD**	RO	Module used to implement UART: **0:** eUSCIA; **1–3:** Reserved for future IPs implementing UART
3:0	**UART_INST**	RO	Instance number of the IP specifically used to implement UART: **0x0:** Use instance 0 of the IP used for implementing UART (for example, eUSCIA0, when UART_MOD = 00) **0x1:** Use instance 1 of the IP used for implementing UART (for example, eUSCIA1, when UART_MOD = 00) .... **0xF:** Use instance 15 of the IP used for implementing UART (for example, eUSCIA15, when UART_MOD = 00)

information is also part of the device descriptor table (TLV), which is used by the BSL to select the correct instance of the interface.

To use the BSL, a user-selectable BSL entry sequence must be applied. An added sequence of commands initiates the desired function. A boot-loading session can be exit by continuing operation at a defined user program address or by the reset condition.

If the device is secured by disabling JTAG, it is still possible to use the BSL. Access to the MSP432™ memory through the BSL is protected against misuse by a user-defined password.

To invoke the BSL, the BSL entry sequence must be applied to dedicated pins. After that, the BSL header character, followed by the data frame of a specific command, initiates the desired function.

Prior to calling the BSL, the BSL needs the GPIO configuration details, such as the module type, module instance, and mux status of the interface being used. These configuration details can be obtained from the **BSL_PER_IF_SEL** register shown in Table 6.6. To get actual interfacing operational parameters for each interface peripheral to be used, the related port configuration registers, such as **BSL Port Configuration for UART** (**BSL_PORTCNF_UART**), **BSL Port Configuration for SPI** (**BSL_PORTCNF_SPI**), and **BSL Port Configuration for I2C** (**BSL_PORTCNF_I2C**), should be used to get these detailed implementation parameters.

After initialization, the BSL uses RAM addresses between **0x2000:0000** and **0x2000:07FF** for data buffer and local variables. When invoking the BSL from a main application, the contents of RAM may be lost.

The MSP432™ BSL can be invoked by any of the three following approaches. Any one of the conditions is a sufficient condition for BSL entry.

1. The BSL is called by the bootcode when the application memory is erased. Bootcode reads out addresses **0x0** and **0x4** from the application flash memory and compares it with **0xFFFF.FFFF** to determine whether or not the application memory is erased.
2. The BSL is called by application software from calling the API table located at address **0x0020:2000**. The table contains the address of the function that starts the BSL execution.
3. The BSL is called by the device bootcode by applying a hardware entry sequence involved in the boot-override command.

For more detailed information about these configuration registers and BSL invocation process, refer to MSP432P401R BSL User Guide.

After the MSPWare Series software package is installed in your host computer, all boot loader-related codes, including the source code files, header files, and assembly code files, are located at: **C:/ti/TivaWare_C_Series-<version>/boot_loader**.

6.4.2 MSPWare peripheral driver library

The MSPWare Series suite contains and integrates all user-required source-code functions and object libraries, which include:

- The **Peripheral Driver Library** offers an extensive set of API functions for controlling the peripherals found on various MSP432 devices.

- The **graphic library** includes a set of graphics primitives and a widget set for creating GIUs on MSPWare Series-based microcontroller boards that have a graphical display.
- The **USB Library** provides an MSPWare royalty-free USB stack to enable efficient USB host, device, and on-the-go operations.
- The **utilities** provide all required developing tools and user-friendly functions to make the user program development easier and simpler.
- The **boot loader and in-system programming** support users to build the startup codes, install them at the beginning of the flash memory in the EXP432P401R EVB, and run them when the user program starts.
- The **open source stacks** offers different stacks for most popular host and peripheral devices.
- The **code examples** offer some useful coding guides to help users to start and speed up their coding developments.
- The **third-party code examples** provide some codes developed by different venders.

In this section, we only pay our attention to the Peripheral Driver Library.

Basically, the Peripheral Driver Library provides a set of high-level API interfaces to enable users to access and interface all related peripheral devices to build their applications. This set of API interfaces is a collection of source code (**.c**) files and related header (**.h**) files used to support users to develop and build their application projects by calling those API interface functions. This library is compatible with most popular IDEs), such as CCS, ARM®/Keil® MDK, IAR, and GNU.

This library should be installed on or integrated with the related development IDE to facilitate the user's program development. All source and header files in the library are located at the related folders when the MSPWare Series software is installed in your host computer. These folders are:

- **C:/ti/msp/MSPWare_<version>/driverlib/inc**: Contains all hardware or device-related specified header files. These files include:
  - Peripheral specific definitions
  - Required type definitions
  - Macros
- **C:/ti/msp/MSPWare_<version>/driverlib/driverlib/MSP432P4xx/keil**: Contains all project library files (object codes) and compiler output directory, which include:
  - C source and header files peripheral specific functionality
  - Compiler-specific project file for building the driver library "libraries"
  - Compiler-specific output directories and files for the used compiler. It should be: **C:/ti/msp/MSPWare_<version>/driverlib/driverlib/MSP432P4xx/keil** for the ARM-MDK μVersion5.

A point to be noted is that this MSPWare Series Peripheral Driver Library is also installed in the internal ROM space on all MSP432 Series MCUs. Two header files, **rom.h** and **rom_map.h**, provide assistance to the users to enable them to access and use these API functions in the library. These two header files are located at the folder: **C:/ti/msp/MSPWare_<version>/driverlib/driverlib/MSP432P4xx** in your host computer.

Several tables at the beginning of the internal ROM point to the entry points for the API functions that are provided in the ROM. Accessing the API functions through these tables provides scalability; while the API locations may be changed in future versions of the ROM, but the API tables will not. The tables are split into two levels; the main table contains one pointer per peripheral which points to a secondary table that contains one pointer per API function that is associated with that peripheral. This means that one peripheral can have multiple API functions. The main table is located at **0x0100.0010**, right after the Cortex®-M4F vector table in the ROM.

To access or use these API functions, a prefix **ROM_** must be added before each function to distinguish them from those same functions included in the library that is installed in the users' host computer. For example, to access or use the API function, **GPIO_setAsOutputPin()**, you must use: **ROM_GPIO_setAsOutputPin()**, to distinguish the same function in the Peripheral Driver Library installed in your host computer.

Sometimes for some reasons, bugs may exist in this ROM library. If those bugs are fixed and some API functions in the library are corrected, these corrected API functions can only be reloaded into the static library stored in the users' host computers, not in the internal ROM space, because of the timing issue. In order to solve this problem, the users can use a prefix **MAP_** before any API function to call it. When this happened, the MCU can automatically determine from which library to find that function, either the ROM or the host computer where the library is stored. For example, to access or use the API function, **GPIO_setAsOutputPin()**, you may use: **MAP_GPIO_setAsOutputPin()** to enable the MCU to automatically select the correct function from any library.

6.5 Flash memory

The **256 KB flash ROM** memory is used to store the user's program codes, exceptions and interrupts vector tables. The exception and interrupt vector tables are located at the lower memory space starting from **0x0000.0000**. To perform any programming for this flash memory, the MSP432 Series devices provide a user-friendly interface with a group of control and status registers.

As we discussed in Section 2.3.2 in Chapter 2, the flash memory consists of two independent equal-sized memory banks, each containing the following regions:

- 256 KB main memory region (two 128 KB banks): this is the primary code memory and is for code/data used by the user application.
- 16 KB information memory region (two 8 KB banks): it is for TI or customer code/data. Some of the sectors could be used by TI and others will be available for users.

Figure 6.8 shows the detailed memory map for the flash memory used in the MSP432P401R MCU system. The 256 KB main flash memory space can be divided into two independent 128 KB banks, Bank0 and Bank1. These two equally independent banks allow simultaneous read and execute from one bank while the other bank is undergoing program and erase operation. The flash main memory can also be divided into 64 sectors of 4 KB each, with a minimum erase granularity of 4 KB (1 sector).

The flash information memory region is 16 KB. The flash information memory consists of four sectors of 4 KB each, with a minimum erase granularity of 4 KB (one sector). The information memory can be viewed as two independent banks of 8 KB each, which allows read or execute from one bank while the other bank is undergoing a program or erase operation.

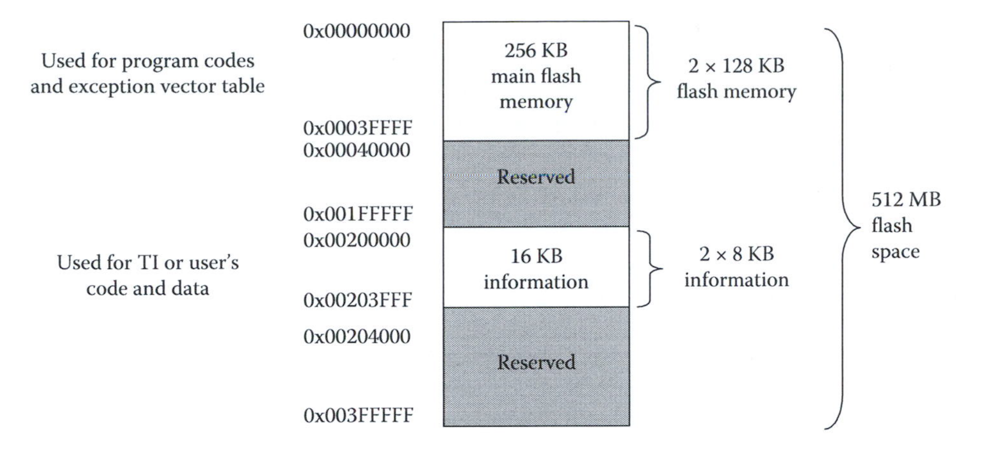

Figure 6.8 Flash memory map in the MSP432P401R MCU system.

The flash information memory region that contains the device descriptor (TLV) is factory configured for protection against write or erase operations. The Bank1 in both sectors is an 8 KB space used to work as a part of BSL operational space.

Generally, the flash memory comprises the following operational units:

1. General flash controls
2. Flash burst read and program (write) controls
3. Flash protection unit
4. 256 KB flash memory array

The 256 KB flash memory array is controlled and managed by three units listed in 1~3.

The flash control unit is used to control the flash memory programming process, including the reading, writing (programming), and erasing operations. When performing the writing or erasing operations, the following points need to be noted:

- Only an erasing operation can change bits values from 0 to 1.
- A writing operation can only change bits from 1 to 0. If the write attempts to change a 0 to a 1, the write fails and no bits are changed.
- A flash operation can be started before entering the sleep or deep-sleep mode (by using instruction **WFI**). It can also be completed while in sleep or deep-sleep period.

During a flash memory operation (writing, burst erasing, or mass erasing), any access to the flash memory is inhibited. As a result, instruction and literal fetches are held off until the flash memory operation is done. If an instruction execution is required during a flash memory operation, that instruction must be placed in the SRAM space and executed from there while the flash operation is in progress.

The basic operations of a flash memory include the reading, writing (programming), and erasing, which includes the burst reading, burst writing, and burst erasing.

In addition to the basic operations, the flash memory also provides multiple read and program modes of operation that the application can deploy. Up to 128 bits (memory word width) can be programmed (set from 1 to 0) in a single program operation. Although the CPU data buses are 32-bit wide, the flash can buffer 128-bit write data (four words) before initiating flash programming, thereby making it more seamless and power efficient for software to program large blocks of data at one time. Also, the flash memory supports a burst read and write mode that takes less time when compared to reading or programming words individually.

Both flash main and information memory regions offer write/erase protection control at a sector granularity to enable software to optimize operations such as mass erase while protecting certain regions of the flash. In low-power modes of operation, the flash memory is disabled and put in a power-down state to minimize the leakage.

6.5.1 Basic operational components of the flash memory

The basic operational components for a flash memory include:

- Flash controller clocking
- Flash controller address mapping
- Flash controller access privileges

First, let's take care of the flash controller clocking.

6.5.1.1 Flash controller clocking

The flash controller can be controlled by two kinds of clocks; the clock used to drive the bus interface, and clock used to drive the flash operations or flash timing clock.

The bus interface clock is used for all software interactions with the flash memory, which happen over the memory or configuration bus interface. The bus interface clock has a varying frequency and is controlled as per application needs but typically running at the same frequency as the CPU. However, the memory and configuration bus interfaces may end up working at frequencies that are integral multiples of each other. This is typical of multilayer bus systems where the memory bus interface is treated as a high bandwidth interface and therefore runs off a higher frequency (typically the CPU frequency). The configuration bus interface is typically a low bandwidth interface and may run off a slower or divided clock.

The flash timing clock is used for all flash operations that require time-controlled sequences. It is typically a request-based clock derived from a CS resource that powers up on request (refer to Section 2.3.5.4.3 in Chapter 2), and provides a predetermined frequency (within accuracy constraints). This allows the flash controller to use the timing clock for controlling the following flash internal operations:

- Flash power up/down sequencing
- Flash program/erase operations
- Switching between different modes of operation
- Any other operation that needs controlled timing (TBD further)

In the MSP432P401R MCU system, the frequency of the flash timing clock will typically be 5 MHz, and sourced from the **SYSOSC** internal oscillator. Refer to Section 2.3.5.4 in Chapter 2 to get details about this kind of oscillator source.

6.5.1.2 Flash controller address mapping

As we mentioned, the 256 KB main flash memory space can be divided into two independent 128 KB banks, Bank0 and Bank1. These two equally independent banks allow simultaneous read and execute from one bank while the other bank is undergoing program and

Table 6.7 Flash 256-KB main memory banks

Banks	Address range	Content	Write/erase protected?
Bank 0–32 Sectors	0x0000_0000–0x0001_FFFF	User Codes–128 KB	NO
Bank 1–32 Sectors	0x0002_0000–0x0003_FFFF	User Codes–128 KB	NO

erase operation. The flash main memory can also be divided into 64 sectors of 4 KB each, with a minimum erase granularity of 4 KB (1 sector). Table 6.7 shows these two banks and their addresses ranges.

The flash information memory region is 16 KB. Flash information memory consists of four sectors of 4 KB each, with a minimum erase granularity of 4 KB (one sector). The information memory can be viewed as two independent banks of 8 KB each, which allows read or execute from one bank while the other bank is undergoing a program or erase operation. Table 6.8 lists these two 8 KB flash information memory banks and their addresses ranges as well the contents.

During the flash memory normal operations, which include the flash memory reading, programming, and erasing, all of these operations are performed based on banks. This means that each bank has its own configuration and operation registers, and each operation is performed inside each bank. For the burst reading, burst programming, and burst erasing, either banks or sectors can be applied to perform these operations.

For each flash memory operation, an appropriate memory address range should be determined to decide which memory bank should be accessed or used to perform the desired operation.

6.5.1.3 Flash controller access privileges

The flash memory space on the MSP432P401R MCU system can be accessed by three components; by the CPU, the DMA, or via the debugger (JTAG/SW)

- CPU (instruction and data buses)
  - The CPU can issue instruction fetches to the entire flash memory region.
  - The CPU can issue data reads/writes to the entire flash memory region (unless the access is blocked by the device security architecture).
- DMA
  - The DMA has conditional access read permissions to the flash memory. If the device is nonsecure, or JTAG and SWD lock-based security is active, the DMA will have full permissions to the entire flash memory space. If the IP protection is active on the device, the DMA will be allowed reads/writes *only* to Bank1. In this case, DMA accesses to Bank0 will return an error response.

Table 6.8 Flash information memory banks

Bank and sector	Address range	Content	Write/erase protected?
Bank 0, Sector 0	0x0020_0000–0x0020_0FFF	Flash Boot-override Mailbox–4 KB	NO
Bank 0, Sector 1	0x0020_1000–0x0020_1FFF	Device Descriptor (TLV)–4 KB	YES
Bank 1, Sector 0	0x0020_2000–0x0020_2FFF	TI BSL–4 KB	NO
Bank 1, Sector 1	0x0020_3000–0x0020_3FFF	TI BSL–4 KB	NO

    - DMA writes to the flash memory will be processed as per the program mode of the flash.
- Debugger
    - A debugger can initiate accesses to the flash memory. If the device is nonsecure, all debugger accesses are permitted. However, if the device is enabled for any form of code security, debugger accesses to the flash memory will be denied by the device security architecture.

Next, let's take a closer look at the various operational modes of the flash memory.

6.5.2 Operational modes of the flash memory

The basic operations of a flash memory include the reading, writing (programming), and erasing, which includes the burst reading, burst writing, and mass erasing.

6.5.2.1 Flash memory read mode and features

In the MSP432P401R MCU system, the following read modes are available for the flash memory:

- Normal read mode
- Marginal read mode (Margin 0 or Margin 1 read mode)
- Program verify read mode
- Erase verify read mode
- Leakage verify read mode

The normal read mode is used to read any data or instruction from the valid flash memory addresses. The so-called marginal read mode is used to check and make sure that programmed bits or cells (0) in the flash memory are really programmed zeros, and erased cells or bits (1) in the flash memory are really ones.

As we know, the flash memory comprises arrays of MOSFETs that have floating gates. Before a flash memory cell (a single MOSFET with a floating gate) can be programmed, it must be first erased to 1 by charging more electrons in the floating gate for that MOSFET. Similarly, during the programming process, each flash cell must be programmed or written from 1 to 0 by removing those electrons from that floating gate. If the erasing process is not perfect, the erased 1 would be not sufficient and less electrons are charged to some floating gates. An error may be occurred or those erased cells (1) may be read as 0. Similarly, if the programming process is not as good as expected, some programmed cells (0) may have more electrons on some floating gates to make those flash cells to be read as 1.

To overcome these insufficient charging or discharging operations because of un-perfect erasing or programming processes, the marginal read mode is used to recharge or release those insufficient floating gates to make sure that all flash memory cells are correctly erased or programmed.

The program verify read mode is a read mode which helps check if the bit(s) has/have been programmed with sufficient electrons margin. Similarly, the erase-verify read mode is a read mode which helps check if the bit(s) has/have been erased with sufficient electrons margin. The leakage-verify read mode is a read mode which helps check if a bit is at/near depletion.

A point to be noted is that each bank has its own individual read mode settings, and all read accesses, including instruction fetches, to a bank are carried out as per its configured read mode.

Another point to be noted is that due to mode transitions, there will be a time delay between configuration of the new mode setting for a bank and the time it is actually ready to process reads in that mode. In addition, the flash controller may stall the transition of a bank into a particular mode because of a simultaneous operation on the other bank. After changing the mode setting, it is the application's responsibility to ensure that the read mode status for the bank reflects the new mode as well. This will ensure that all subsequent reads happen in the targeted mode, else a few of the reads to the bank may continue to be serviced in the old mode.

All reading operations are controlled by related flash read mode control registers. These registers not only control the data flowing direction, but also the timing relationship for each reading. These registers are configurable in terms of the number of memory bus cycles they take to service any read command. This allows a decoupling of the CPU's execution frequency from the maximum read frequency supported by the flash memory. If the bus clock speed is higher than the timing clock frequency of the flash, the access will be stalled for the configured number of wait states, allowing data from the flash to be accessed reliably. The **FLCTL_BANK0_RDCTL** and **FLCTL_BANK1_RDCTL** registers can be used to control the number of wait states that are inserted by the flash controller for each flash access.

Figure 6.9 shows the bit map for the **FLCTL_BANK0_RDCTL** register. The **FLCTL_BANK1_RDCTL** register has the identical bit map and structure as those of the **FLCTL_BANK0_RDCTL** register.

Table 6.9 shows the bit value and function for the **FLCTL_BANK0_RDCTL** register. The memory mapping address for this register is **0x40011010**. The **FLCTL_BANK1_RDCTL** register has the same bit value and function, but its memory mapping address is **0x40011014**.

When we configure and define the flash memory operation modes using these read control registers, the following points must be noted:

- Bits **19:16** on these registers are used to reflect the current reading mode for the selected flash bank. Therefore, these bits are read-only bits without writing ability. After a system reset, the default read mode is normal read mode (0). Also, these bits will be forced to 0 (normal read mode) when the device is in low-frequency active and low-frequency LPM0 modes of operation.
- Bits **3:0** on these registers are used to control or set the read mode for the selected flash bank. Therefore, these bits are write/read bits. Similarly, the default read mode is the normal read mode after a system reset. Besides, these bits have the following special control functions:
  - These bits will be forced to 0 (normal read mode) when the device is in low-frequency active and low-frequency LPM0 modes of operation.
  - These bits will be forced to 0 when the device is in the 2 T mode of operation.
  - These bits are writable *ONLY* when the burst status bit field (**17:16**) on the flash read burst/compare control and status register (**FLCTL_RDBRST_CTLSTAT**) shows the **Idle** state. In all other cases, the bits will remain locked so as to not disrupt an operation that is in progress.

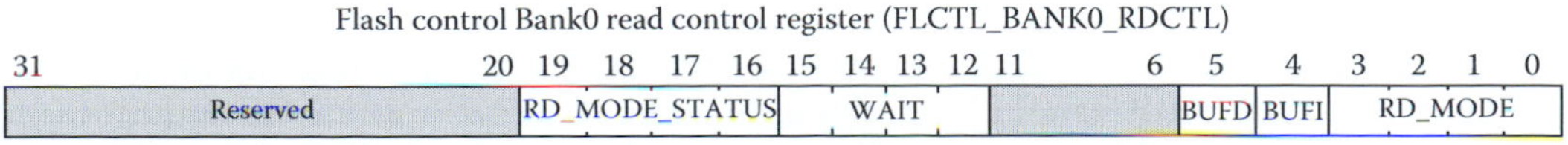

Flash control Bank0 read control register (FLCTL_BANK0_RDCTL)

31–20	19–16	15–12	11–6	5	4	3–0
Reserved	RD_MODE_STATUS	WAIT		BUFD	BUFI	RD_MODE

Figure 6.9 Bit map and structure of the FLCTL_BANK0_RDCTL register.

Table 6.9 Bit value and bit function for FLCTL_BANK0_RDCTL register

Bit	Field	Type	Reset	Function
31:20	**Reserved**	RO	00	Reserved and returns 0 as read
19:16	**RD_MODE_ STATUS**	RO	0	Reflects the current read mode of the bank: **0000** = Normal read mode; **0001** = Read Margin 0 **0010** = Read Margin 1; **0011** = Program Verify **0100** = Erase Verify; **0101** = Leakage Verify **1001** = Read Margin 0B; **1010** = Read Margin 1B All others = Reserved, bank is set to Normal read mode
15:12	**WAIT**	RW	3	Defines the number of wait states required for a read to the bank: **0000** = 0 wait states; **0001** = 1 wait states **0010** = 2 wait states; **0011** = 3 wait states **0100** = 4 wait states; **0101** = 5 wait states **0110** = 6 wait states; **0111** = 7 wait states **1000** = 8 wait states; **1001** = 9 wait states **1010** = 10 wait states; **1011** = 11 wait states **1100** = 12 wait states; **1101** = 13 wait states **1110** = 14 wait states; **1111** = 15 wait states
11:8	**Reserved**	RW	00	Reserved
7:6	**Reserved**	R	NA	Reserved and returns 0 as read
5	**BUFD**	RW	0	Enables read buffering feature for data reads to this bank
4	**BUFI**	RW	0	Enables read buffering feature for instruction fetches to this bank
3:0	**RD_MODE**	RW	00	Flash read mode control setting for bank: **0000** = Normal read mode; **0001** = Read Margin 0 **0010** = Read Margin 1; **0011** = Program Verify **0100** = Erase Verify; **0101** = Leakage Verify **1001** = Read Margin 0B; **1010** = Read Margin 1B All others = Reserved, bank is set to Normal read mode

- Bit field (**15:12**) is used to coordinate the asynchronous read/write operations between the MCU and the flash memory by adding the desired time delay or the number of bus cycles (waiting states) for each read mode. The default waiting state is three bus cycles after a system reset.
- Bits **5** (buffering read data) and **4** (buffering read instructions) are used to enable or disable the read buffering features for data and instructions. The default settings for these features are disabled (0).

The **FLCTL_BANK0_RDCTL** register is used to control the first 128 KB flash main memory, but the **FLCTL_BANK1_RDCTL** register is used for the second 128 KB flash main memory.

The following piece of codes illustrate how to use the normal read mode to read a group of data (in words) starting from the flash memory address **srcAddr** and save into the flash destination address, **destAddr** in Bank0. The length of data to be read out is **dataLength**.

```
FLCTL_BANK0_RDCTL = 0x00003000;       // normal read mode with 3 wait states
for (n = 0; n < dataLength; n++)      // repeat reading for dataLength times
{
 HWREG32(destAddr) =                  // read and save a data word
  HWREG32(srcAddr);
 srcAddr += 4;                        // update the start address
 destAddr += 4;                       // update the destination address
}
```

6.5.2.1.1 *Flash memory read buffering feature* In order to provide optimal power consumption and performance across predominantly contiguous memory accesses, the flash controller offers a read buffering feature. If the read buffering feature is enabled, the flash memory is always read in entire 128-bit chunks, even though the read access may only be 8, 16, or 32-bit wide. The 128-bit data and its associated address is buffered by the flash controller, so subsequent accesses (expected to be contiguous in nature) within the same 128-bit address boundary are serviced by the buffer. Using this scheme, the flash accesses will need wait states only when the 128-bit boundary is crossed, while read accesses within the buffer's range will be serviced without any bus stalls.

If the read buffering feature is disabled, accesses to the flash will bypass the buffer and the data read from the flash will be limited to the width of the access (8, 16, or 32 bits). Each bank will have independent settings for the read buffer. In addition, within each bank, the application will have the flexibility of enabling read buffering either for instruction fetches only, or data fetches only, or both.

A point to be noted is that the read buffers will always be bypassed (even if enabled) if the flash is currently performing any type of erase or program operation. This is to ensure data coherence through operations that may change the contents of the flash.

To perform reading 128-bit data with the read buffering feature via the flash memory Bank1, the **FLCTL_BANK1_RDCTL** register should be configured as:

```
FLCTL_BANK1_RDCTL = 0x20;             // normal read mode with read
                                         buffering feature
for (n = 0; n < dataLength; n++)      // repeat reading for dataLength times
{
HWREG32(destAddr)= HWREG32(srcAddr); // read and save a 128-bit data
}
```

6.5.2.1.2 *Flash memory burst read-compare feature* In addition to read buffering feature, the flash controller supports a burst read/compare feature. When working with this feature, it permits a fast read/compare operation on a contiguous section of the flash memory. Implementing a burst read permits the flash controller to optimize the time taken for the operation by comparing all 128 bits at a time to avoid the additional setup/active/hold parameters for each read iteration operation.

The control for the read burst operation is implemented using the following registers:

- Burst Read/Compare Control/Status Register (**FLCTL_RDBRST_CTLSTAT**).
  - Set up the memory region type:
    - Information memory
    - Engineering memory
    - Main memory
  - Enable or disable generation of an interrupt on completion

    - Report status of read operation
    - Enable or disable stop of burst operation on first failure. If disabled, burst runs through its entire length
    - Trigger off the read operation through a start bit
- Burst Read/Compare Start Address Register (**FLCTL_RDBRST_STARTADDR**)
    - Contain the start address for the read operation. This is the direct byte offset address of the memory type (information memory/engineering memory/main memory) chosen and *must* be a 128-bit boundary
- Burst Read/Compare Length Register (**FLCTL_RDBRST_LEN**)
    - Contain the number of 128-bit flash IP addresses to be compared
- Data compare value—Can be either of the following:
    - `0000_0000_0000_0000_0000_0000_0000_0000`
    - **FFFF_FFFF_FFFF_FFFF_FFFF_FFFF_FFFF_FFFF**
- Burst Read/Compare Fail Address Register (**FLCTL_RDBRST_FAILADDR**).
    - Store the first flash address where compare failure occurs. This is a byte address, but will reflect the 128-bit boundary where the last fail occurred
- Burst Read/Compare Fail Count Register (**FLCTL_RDBRST_FAILCNT**)
    - Store the number of failures encountered from start through stop

When using the burst read-compare feature, the following points must be noted:

1. The burst read/compare feature is useful only when there is a predetermined (fixed) value stored in each 128-bit location of the flash. In other words, this is typically useful for erase verify and leakage verify operations on large sections of the memory.
2. If the burst/compare is configured to continue on fails, this register will reflect the address of the last failed comparison.
3. Certain burst read operations such as program verify will be split into multiple burst operations if the address crosses a sector boundary. This will be handled transparently by the flash controller, but the overall time taken for the read/compare operations will involve additional setup/hold times due to the increase in number of entry/exit sequences into/from the read mode.
4. If the burst starts at, or enters (during its operation) a reserved location in the flash memory space, the burst will terminate immediately with an error condition.

Now, let's have a closer look at these control and state registers.

6.5.2.1.2.1 Burst read-compare control-status register (FLCTL_RDBRST_CTLSTAT) This is a 32-bit register but only 12 bits are used to set up the control functions for the burst read and compare operation and check the running status of this function. Figure 6.10 shows the bit fields for this register and Table 6.10 lists the bit values and functions for this register.

Flash control burst read-compare control-status register (FLCTL_RDBRST_CTLSTAT)

31	23		19	18	17 16		6	5	4	3	2 1	0
Reserved	CLR_STAT		ADDR_ERR	CMP_ERR	BRST_STAT		TEST_EN		DATA_CMP	STOP_FAIL	MEM_TYPE	START

Figure 6.10 Bit fields for the FLCTL_RDBRST_CTLSTAT register.

It can be found from Table 6.10 that the bit fields on this **FLCTL_RDBRST_CTLSTAT** register have the following functions:

1. Bits 19~16 are read-only bits and used to display the running status of these data and instruction burst/compare operations. Bits 19 and 18 are used to monitor if any possible address error or comparison error occurred. A **0** on these bits indicates that no error is encountered. A value of **1** on any of these bits shows that an error has occurred.

Table 6.10 Bit value and bit function for FLCTL_RDBRST_CTLSTAT register

Bit	Field	Type	Reset	Function
31:24	**Reserved**	RO	NA	Reserved and returns 0 as read
23	**CLR_STAT**	W	NA	Write **1** to clear status bits 19–16 of this register. Write **0** has no effect
22:20	**Reserved**	RO	NA	Reserved and returns 0 as read
19	**ADDR_ERR**	RO	0	Memory Address Access Error: **0:** No address error occurred **1:** Burst/Compare Operation was terminated due to access to reserved memory
18	**CMP_ERR**	RO	0	Burst/Compare Has Data Compare Error: **0:** No data compare error occurred **1:** Burst/Compare Operation encountered at least one data comparison error
17:16	**BRST_STAT**	RO	0	Status of Burst/Compare operation: **00** = Idle **01** = Burst/Compare START bit written, but operation pending **10** = Burst/Compare in progress **11** = Burst complete (status of completed burst remains in this state unless explicitly cleared by SW)
15:7	**Reserved**	RO	NA	Reserved and returns 0 as read
6	**TEST_EN**	RW	0	**1:** Enable comparison against test data compare registers
5	**Reserved**	RW	0	Reserved
4	**DATA_CMP**	RW	0	Data pattern used for comparison against memory read data: **0** = 0000_0000_0000_0000_0000_0000_0000_0000 **1** = FFFF_FFFF_FFFF_FFFF_FFFF_FFFF_FFFF_FFFF
3	**STOP_FAIL**	RW	0	**0:** Continue the Burst/Compare operation even encounter a mismatch error **1:** Burst/Compare operation terminates on the first compare error
2:1	**MEM_TYPE**	RW	00	**00** = Main Memory; **01** = Information memory; **10** = Reserved; **11** = Engineering
0	**START**	W	0	**1:** Trigger start of Burst/Compare operation. **0:** No effect

2. Bits 17~16 are used to check the current running status of the burst/compare operations. A value of **0x01** on these bits indicates that the burst/compare operation is going on, but a value of **0x11** shows that the operation is done.
3. By setting bit-23, **CLR_STAT**, all status values on bits 19~16 can be cleared to 0.
4. Bit-6 and bits 4~0 are used to control the burst/compare normal operations. These bits values defined the data pattern to be compared, whether the operation should stop if encountered any error, memory type to be compared and trigger the starting of this operation.

Next, let's take care of the FLCTL_RDBRST_STARTADDR register.

6.5.2.1.2.2 Burst read-compare start address register (FLCTL_RDBRST_STARTADDR) This is a 32-bit register but only the lower 21 bits (bits 20~0 = bit field **START_ADDRESS**) are used to set up an initial or starting address for the burst compare operations. Totally, a 1 MB memory space (2^{20} = 1024 KB) can be used to access 64 KB 128-bit (64 × 1024 × 16 bytes) memory space to perform the burst/compare operations (128 bit = 16 bytes). When using this register to store or set up an initial memory address for the burst/compare operations, the following points must be noted:

1. The offset is **0x0** starting from the type of memory region selected.
2. Bits 3~0 on this register are always 0 to get a 128-bit boundary. Since each 128-bit can be mapped to 16 bytes in address, in other words, each 128-bit data can be stored in 16 continuous addresses, such as from **0x0000** to **0x000F**. The next 128-bit data are stored starting from **0x0010** to **0x001F**, and the next are from **0x0020** to **0x002F**, and so on, until **0xFFFF**. Since each data is 128-bit long and its starting address must be **0x0000**, **0x0010**, **0x0020**, **0x0030**, and continue in this way until **0xFFF0**. Therefore, the last four bits, from bits 3~0, must always be 0.
3. If the amount of memory available is less than 2 MB, the upper bits of the **START_ADDRESS** will behave as reserved.
4. This bit field is writable *ONLY* when burst status (**17:16**) on the **FLCTL_RDBRST_CTLSTAT** register shows the **Idle** (**00**) state. In all other cases, the bits will remain locked so as to not disrupt an operation that is in progress.

6.5.2.1.2.3 Burst read-compare length register (FLCTL_RDBRST_LEN) Similarly, to **FLCTL_RDBRST_STARTADDR**, this is also a 32-bit register but only the lower 21 bits (bits 20~0 = bit field **BURST_LENGTH**) are used to set up the length of the burst operations (in 16 bytes). The maximum length can be set up is 64 KB.

When using this register to set up the length of the burst operations, two points must be noted:

1. If amount of memory available is less than 2 MB, the upper bits of the **BURST_LENGTH** will behave as reserved.
2. This bit field is writable *ONLY* when burst status (**17:16**) on the **FLCTL_RDBRST_CTLSTAT** register shows the **Idle** (**00**) state. In all other cases, the bits will remain locked so as to not disrupt an operation that is in progress.

6.5.2.1.2.4 Burst read-compare fail address register (FLCTL_RDBRST_FAILADDR) This is also a 32-bit register but only the lower 21 bits (bits 20~0 = bit field **FAIL_ADDR**) are used to set up the address of the last failed comparison operation.

This register can be used together with the **FLCTL_RDBRST_CTLSTAT** register to track and identify the running status and result of the burst/compare operations. When bit 3 (**STOP_FAIL**) on the **FLCTL_RDBRST_CTLSTAT** register is set to 1, the burst/compare operation will be stopped immediately if any error is encountered. The memory address of that failed compare can be found from the **FLCTL_RDBRST_FAILADDR** register. However, if bit-3 (**STOP_FAIL**) on the **FLCTL_RDBRST_CTLSTAT** register is 0, the burst compare operations will be continued even when some comparisons have failed. When the burst compare operations are done, the address of the last failed compare can be found from the **FLCTL_RDBRST_FAILADDR** register.

When using this register, the following points must be noted:

1. Application may choose to reset this register to 0 before starting any new burst compare operation.
2. If the amount of memory available is less than 2 MB, the upper bits of the **FAIL_ADDR** will behave as reserved.
3. This bit field is writable *ONLY* when burst status (17:16) of the **FLCTL_RDBRST_CTLSTAT** register shows the **Idle** (**00**) state. In all other cases, the bits will remain locked so as to not disrupt an operation that is in progress.

6.5.2.1.2.5 Burst read-compare fail count register (FLCTL_RDBRST_FAILCNT) This is a 32-bit register but only the lower 17 bits (bits 16~0 = bit field **FAIL_COUNT**) are used to record the number of failures encountered in the burst operation.

When using this register to record the number of failed burst operations, the following points must be noted:

1. Application may need to reset this register to 0 before starting any new burst compare operation. Else, it will increment from the current value each time a new burst is started.
2. The **FAIL_COUNT** may be as high as 128 K for a 2 MB flash memory size. In case the size of memory on device is less than 2 MB, upper bits will behave as reserved.
3. This bit field is writable *ONLY* when burst status (**17:16**) of the **FLCTL_RDBRST_CTLSTAT** register shows the **Idle** (**00**) state. In all other cases, the bits will remain locked so as to not disrupt an operation that is in progress.

The following two examples show how the read burst/compare feature can be deployed.

A. *Erase Verify for Bank0 in information memory sector:*
   - Read mode: erase verify
   - Memory type: information memory
   - Burst/compare start address: start address of sector boundary (4 KB boundary) from the offset 0 of the information memory space
   - Burst/compare data_**x** (**x** = **0** through **3**): **0xFFFF_FFFF**
   - Burst/compare length: **256** (total of 4 KB at 128 bits = 16 bytes per read).

B. *Erase verify after mass erase of Bank1 in 256 KB main memory:*
   - Read mode: erase verify
   - Memory type: main flash
   - Burst/compare start address: **0x20000**
   - Burst/compare data_**x** (**x** = **0** through **3**): **0xFFFF_FFFF**
   - Burst/Compare Length: **8192** (total of 128 KB at 128 bits = 16 bytes per read)

A piece of codes shown in Figure 6.11 can be used for example B.

Now, let's take care of the flash memory programming operations.

```
1   uint32_t ret;
2   ret = FLCTL_RDBRST_CTLSTAT;          // get the current content of FLCTL_RDBRST_CTLSTAT
                                         register
3   while(!(ret && 0xFFFCFFFF)) {}       // bits 17:16 are zero (idle)? If not, wait here......
4   FLCTL_BANK1_RDCTL |= 0x4;            // erase verify operation with 3 wait states for Bank1
5   FLCTL_RDBRST_CTLSTAT = 0x00800010;   // bit-23 = 1 to clear 19:16 bits, bit-4 = 1 → data:
                                         0xFF....FF
6   ret = FLCTL_RDBRST_CTLSTAT;          // get the current content of FLCTL_RDBRST_CTLSTAT
                                         register
7   while(!(ret && 0xFFFCFFFF)) {}       // bits 17:16 are zero (idle)? If not, wait here......
8   FLCTL_RDBRST_STARTADDR = 0x20000;    // assign starting address to FLCTL_RDBRST_STARTADDR
9   FLCTL_RDBRST_LEN = 8192;             // set burst/compare length
10  FLCTL_RDBRST_CTLSTAT |= 0x1;         // start the burst/compare operation
```

Figure 6.11 Piece of code used for example B.

6.5.2.2 Flash memory program (write) mode

As we discussed in Section 6.5, a flash programming or writing operation can only change bits from 1 to 0. In the MSP432P401R MCU system, all programming bits can be written into the flash memory, either the main or the information banks, as written into a SRAM space with quite a few of CPU cycles. The flash controller stores these writes without stalling the CPU, and transparently handles the programming of the flash memory.

In addition, the flash memory architecture supports programming of bits from a single bit up to 128 bits in one program operation. To provide further flexibility, the application can choose to enable single 128-bit programming in one of two flavors: immediate writes or full-word (128 bits) writes. The so-called immediate writes is to program the current bits required by the application as soon as a write command is issued no matter whether the required bits are equal to or less than 128 bits. However, the full-word 128-bit writes cannot start any programming operation until a whole 128-bit data is ready.

Let's take a closer look at these two writing modes.

6.5.2.2.1 Immediate write mode In this write mode, the flash controller initiates a program operation immediately upon receiving a write command on the memory bus. This can be executed in the following way:

- When a write command is issued, the flash controller can perform writing operation to immediately write current data into the flash, or the controller can compose the current data to a 128-bit word by using the following way:
  - If the data is 32-bit wide, the 32 bits are placed in the appropriate section of the 128-bit word, and the rest of the bits in the composed word are set to 1.
  - If the data is 16-bit wide, the 16 bits are placed in the appropriate section of the 128-bit word, and the rest of the bits in the composed word are set to 1.
  - If the data is 8-bit wide, the 8 bits are placed in the appropriate section of the 128-bit word, and the rest of the bits in the composed word are set to 1.
- The composed word is now used as the target data input for the flash program operation (at the targeted address). All bits set to 1 are masked for programming to 0.

6.5.2.2.2 Full-word write mode In order to optimize write operations and the power consumption during flash program processes, the flash controller can be configured to buffer multiple writes from the CPU and initiate the program operation only after a complete

128-bit word has been composed. This is extremely useful in cases where the application is working with larger word sizes and prefers to initiate writes only when it has at least 16 bytes (128 bits) of data ready for programming.

The following operational sequence explains how this mode is applied:

- The data **must** be written in an incremental address fashion, starting with a 128-bit aligned LSB (bits 3~0 must be 0)
    - Can be written as 4 × 32-bit data, starting from the least significant 32-bit word.
    - Can be written as 8 × 16-bit data, starting from the least significant 16-bit word.
    - Can be written as 16 × 8-bit data, starting from the least significant byte.
    - Can be written as a combination of the above, but must start with the LSB getting loaded with the first write and end with the most significant byte being loaded with the last write.
- Once the above is completed, the flash controller now has a complete 128-bit write word that is used for the program operation.

Depending on the application selection, it can write a single byte to the LSB to start the word composition, and a single byte to the MSB of the 128-bit address to end the word composition. This will also result in the flash controller composing a full 128-bit word in which only the least and most significant bytes have actual data, but the intermediate data bits are all filled with 1 s to mask them for programming.

In addition to these two programming modes, the MSP432P401R MCU system also provides an auto verify feature to automatically perform a program verify and compare function to check to make sure that the programming process is correct and all data bits have been sufficiently programmed into the desired flash memory space.

6.5.2.2.3 Auto verify feature The following functions can be performed by executing this auto verify feature:

1. Compose the 128-bit word with either in immediate write mode, full-word write mode, or burst write mode.
2. Read the 128-bit data from the target address in program verify mode.
3. Compare the 128-bit read data (from the flash) with the 128-bit composed write data using the following rules:
    a. If a bit in the composed write data is 1, it is masked from comparison, irrespective of the value of the corresponding bit in the read data from the flash.
    b. If a bit in the composed write data is 0 *and* the corresponding bit in the read data is also 0, it is considered to be already programmed.
        i. Detection of this condition causes the controller to abort the write/program operation with the appropriate interrupt flag set.
        ii. The application can choose to correct the data to be written to the flash, or decide to clear the preprogram auto-verify option and proceed.
        iii. If the preprogram auto-verify option is disabled, or it is enabled and the compare passes, the controller now proceeds to the next step.
4. A program operation is initiated to the target address with the write data value.
5. After the program operation, the address is now read again in program-verify mode.
6. This is compared with the original composed write data, with bits that are written as "1" being masked from comparison.
    a. Detection of a mismatch causes the controller to abort the write/program operation with the appropriate interrupt flag set.

Flash control program control-status register (FLCTL_PRG_CTLSTAT)

31–19	18	17–16	15–4	3	2	1	0
Reserved	BNK_ACT	STATUS		VER_PST	VER_PRE	MODE	ENABLE

Figure 6.12 Bit fields for the FLCTL_PRG_CTLSTAT register.

The flash programming operations, either the immediate mode or the full-word mode, are under control of the Flash Program Control and Status Register (**FLCTL_PRG_CTLSTAT**). Figure 6.12 shows the bit map and field for this register. Table 6.11 lists the bit values and functions for this register.

It can be found from Figure 6.12 and Table 6.11 that this **FLCTL_PRG_CTLSTAT** register has 32 bits but only seven bits, bits 18~16 and bits 3~0, are used to set up and monitor the flash memory programming operations. All other bits are reserved for the future usage.

The higher 3 bits, bit-18 (**BNK_ACT**) and bits 17~16 (**STATUS**), are used to show the current flash programming banks (Bank0 or Bank1) and the running status of the single word

Table 6.11 Bit value and bit function for FLCTL_PRG_CTLSTAT register

Bit	Field	Type	Reset	Function
31:19	**Reserved**	RO	NA	Reserved and returns 0 as read
18	**BNK_ACT**	RO	0	Indicate which bank is currently undergoing a program operation (valid only if the **STATUS** shown in bits 17–16 are NOT **idle**) **0:** Word in **Bank0** being programmed; **1:** Word in **Bank1** being programmed
17:16	**STATUS**	RO	0	Indicate the status of program operations in the Flash memory: **00:** Idle (no program operation currently active) **01:** Single word program operation triggered, but pending **10:** Single word program in progress **11:** Reserved (Idle)
15:4	**Reserved**	RO	NA	Reserved and returns 0 as read
3	**VER_PST**	RW	1	Control automatic postprogram-verify operations: **0:** No postprogram verification **1:** Postverify feature automatically invoked for each write operation (any mode)
2	**VER_PRE**	RW	1	Control automatic preprogram verify operations: **0:** No preprogram verification **1:** Preverify feature automatically invoked for each write operation (any mode)
1	**MODE**	RW	0	Control write mode selected by application: **0:** Write Immediate Mode **1:** Write Full-Word Mode
0	**ENABLE**	RW	0	Master control for all word program operations: **0:** Disable Word program operation; **1:** Enable Word program operation

program (in pending or in progress). By default, the Bank0 is selected as the programming space after a system reset.

The lower 4 bits are used to control and set up the operational write mode and auto-verify functions. Bits 3 (**VER_PST**) and 2 (**VER_PRE**) are used to control whether a post or preprogram verification function is selected for any mode of writing operation. Both verification functions are selected by default after a system reset. Bit 1 (**MODE**) is used to select and set up the write mode, either immediate mode or full-word mode. By default, the immediate mode is selected after a system reset. Finally, bit 0 (**ENABLE**) can be used to initiate or start a write operation.

When using this register to control flash memory programming operations, the following points must be noted:

1. All four control bits (bits 3~0), including the **VER_PST**, **VER_PRE**, **MODE**, and **ENABLE**, can be modified and configured *ONLY* when the flash memory running status indicated on bit-field **STATUS** (bits 17:16) is Idle (**00**). In all other cases, these bits will remain locked so as to not disrupt an operation that is in progress.
2. When the programming mode is full-word mode (bit-1 = **1**), the application must ensure that the writes in this mode follow the LSB/MSB loading order within the 128-bit address boundary. The flash controller needs to collate data over multiple writes to compose the full 128-bit word before initiating this program mode operation.
3. The bit-0 (**ENABLE** bit) will be forced to 0 when the device is in the low-frequency active and low-frequency LPM0 modes of operation.

Figure 6.13 shows a piece of code used to perform an immediate write mode for programming sixteen 32-bit words and write them from the starting address **src** to the destination address **dest** in the flash memory. In this piece of code, both **src** and **dest** are addresses.

Before performing any programming operation, the **STATUS** bit field in the **FLCTL_PRG_CTLSTAT** register must be checked to make sure that it is in the **Idle** status in lines

```
1   uint32_t n, ret;
2   uint32_t dataLength = 16;                 // write 16 32-bit words from src to dest in
                                              flash memory
3   ret = FLCTL_PRG_CTLSTAT;                  // get the current content of FLCTL_PRG_CTLSTAT
4   register while(!(ret & 0xFFFCFFFF)) {}    // bits 17:16 are zero (idle)? If not, wait here……
5   FLCTL_PRG_CTLSTAT = 0x0D;                 // Immediate mode with auto-verify (enable the
                                              write)
6   FLCTL_PRGBRST_CTLSTAT &= ~0x000000C0;     // clear burst program auto verify functions
7   for (n = 0; n < dataLength; n++)          // repeat writing for dataLength data items
8   {
9     FLCTL_BANK0_MAIN_WEPROT = 0;            // clear all write-erase protection functions
10    FLCTL_CLRIFG = 0x0000020E;              // Clear IFG errors (PRG Done |PreV | PostV |PRG
                                              Error)
11    HWREG32(dest) = HWREG32(src);           // Write a 32-bit word from src to dest in flash
                                              memory
12    while(FLCTL_IFG & 0x08){}               // Wait for the PRG to be done……
13    //check all possible errors. If no error,
14    src += 4;                               // update the start address
15    dest += 4;                              // update the destination address
16  }
```

Figure 6.13 Piece of code used to perform immediate write mode.

3 and 4. If it is in the idle status, program the **FLCTL_PRG_CTLSTAT** register to set up the programming mode (immediate mode with default post and preverifications in this example) and enable the program operation (**0x0D** = **00001101**—see Table 6.11) in line 5.

In order to perform any program operation, the auto-verify functions for both post and preprogramming in the flash burst program register should also be cleared to disable their functions. This is done in line 6.

Then a write loop is performed to continuously write sixteen 32-bit words in the immediate mode. Before each write loop, the flash Bank0 main memory Write-Erase (**WE**) Protection register must be cleared to disable all **WE** protection functions (see Section 6.5.2.4). Also, the **FLCTL_CLRIFG** register is used to clear any previous error and program-complete flag by writing 1 to the associate bits in line 10. Then, a write instruction is executed to write a 32-bit word from the source address **src** to the destination address **dest** in the flash memory in line 11. Both addresses should be defined by the users before this piece of codes can be executed.

After each word is written into the destination address, the **FLCTL_IFG** register is used to check whether this write operation is complete in line 12. Also, some error checking should be performed after each write (codes are not shown in here). If no error has occurred, the source and the destination addresses are updated to point to the next address.

A piece of modified code shown in Figure 6.14 can be used to perform eight full-words write mode operation to continuously program eight 128-bit (32 × 32-bit) words into the flash memory. The full mode writing will not happen until a complete 128-bit data are composed by the flash controller, which means that only after four 32-bit data have been composed to 128-bit data item, the write operation starts its operation to program 128-bit word data into the flash memory.

As shown in Figure 6.14, eight full words (thirty-two 32-bit) data are declared in line 2 by defining the **dataLength** variable. Then, a **while()** loop is executed to repeatedly

```
1   uint32_t n = 0, ret;
2   uint32_t dataLength = 32;                  // write 32 32-bit words from src to dest in
                                                  flash memory
3   ret = FLCTL_PRG_CTLSTAT;                   // get the current content of FLCTL_PRG_CTLSTAT
                                                  register
4   while(!(ret &                              // bits 17:16 are zero (idle)? If not, wait
    0xFFFCFFFF)) {}                               here......
5   FLCTL_PRG_CTLSTAT |= 0x0F;                 // Full Word mode with auto-verify (enable the
                                                  write)
6   FLCTL_PRGBRST_CTLSTAT &= ~0x000000C0;      // clear burst program auto verify functions
7   while (n < dataLength)                     // repeat reading for dataLength times
8   {
9     FLCTL_BANK0_MAIN_WEPROT = 0;             // clear all write/erase protection functions
10    FLCTL_CLRIFG = 0x0000020E;               // Clear IFG errors (PRG Done |PreV | PostV |PRG
                                                  Error)
11    HWREG32(dest) = HWREG32(src);            // Write a 32-bit word from src to dest in flash
                                                  memory
12    while(FLCTL_IFG & 0x08){}                // Wait for the PRG to be done....
13    if (FLCTL_IFG & 0x00000206)              // check all possible errors (PreV|PostV|PRG
                                                  Error)
14      return false;                          // if any error occurred, return false
15    dest += 4;                               // update the destination address
16    n++;
17  }
```

Figure 6.14 Piece of code used to perform full-word write mode.

write eight full words from the source (**src**) to the destination (**dest**) address in the main flash memory Bank0 in line 7. A point to be noted is that this full-word write or program would not occur until a 128-bit or four 32-bit data stored in the **src** is complete in line 11. You can test this piece of code for the full-word write mode by reducing the **dataLength** to 3. In that case, no data can be written into the flash.

6.5.2.2.4 Burst program feature A more advantaged program feature provided by the MCP432P401R MCU system is the Burst program. The so-called Burst program feature adds further enhancement over the full-word write mode of operation by permitting multiple (up to 4) 128-bit data words to be written in one single burst writing command. This feature is very valuable in the case of applications where a large number of bytes, or a block of data, is required to be programmed to contiguous addresses in the flash in one quick operation. Implementing a burst writing operation provides a lower overall write latency since the set up and hold times related to flash programming operation are not repeated for each iteration.

The Burst program feature is enabled through the following components:

- **Data Input Buffer**: This is a 4-chunk 128-bit wide buffer that can be pre-loaded with data either by the application code or by the DMA. The buffer will actually be implemented as sixteen 32-bit registers to facilitate direct writes from the CPU/DMA
- **Start Address Register**: Contain and indicate the start address of the program burst into the flash memory array. This address must be a 128-bit boundary (bits 3~0 must be 0)
- **Burst Program Length**: Contain and indicate the number of 128-bit words to be written in sequence. The maximum length is 4, which means that up to four words can be written in succession.
- **Auto Verify**: As with single word programming, the application can choose to enable auto program verification before and/or after the program operation.

In this mode of operation, if all the writes are within the same sector, the set up and hold delays of the program operation are incurred just once through the burst. If the write happens to cross a sector boundary, the operation will be broken into two bursts.

A point to be noted is that the Burst program will be controlled by the **START** bit (bit-0) in the Flash Program Burst Control and Status Register (**FLCTL_PRGBRST_CTLSTAT**). If the Burst program feature is used, it is the responsibility of the application to ensure that the buffer registers are loaded with appropriate data to be programmed and this loading must be completed before the burst operation is started. Bits that are not to be programmed should be set to 1 to mask them from the program pulse.

The following control and status registers are used in the burst write operation:

- Flash Program Burst Control and Status Register (**FLCTL_PRGBRST_CTLSTAT**)
- Flash Program Burst Start Address Register (**FLCTL_PRGBRST_STARTADDR**)
- Flash Program Burst Data Registers (**FLCTL_PRGBRST_DATA0_0~FLCTL_PRGBRST_DATA3_3**) sixteen 32-bit registers used to store or buffer four 128-bit data to be programmed.

Now, let's have a closer look at these burst writing control and status registers.

Flash program burst control-status register (FLCTL_PRGBRST_CTLSTAT)

31	23	22	21	20	19	18 17 16		7	6	5 4 3	2 1	0
Reserved	CLR_STAT		ADDR_ERR	PST_ERR	PRE_ERR	BURST_STATUS		AUTO_PST	AUTO_PRE	LEN	TYPE	START

Figure 6.15 Bit fields for the FLCTL_PRGBRST_CTLSTAT register.

6.5.2.2.4.1 Flash program burst control and status register (FLCTL_PRGBRST_CTLSTAT) This is a 32-bit register but only 15 bits are used to configure the burst programming control and monitor the burst operational status. Figure 6.15 shows the bit map and field for this register. Table 6.12 lists the bit values and functions for this register.

It can be found from Table 6.12 that this register includes both control and status functions for the burst program operations. The higher bits, bits 21~16, are used to monitor and check the running status of the burst program operations, but the lower 8 bits, bits 7~0, are used to control and configure the burst program operations.

When using this register to configure and set up the burst program operations, the following points must be noted:

1. Write **1** to the bit field **CLR_STAT** (bit-23) will clear the status bits 21:16 *ONLY* when the bit field **BURST_STATUS** (bits 18:16) shows the completion state (**111**). In all other cases, write **1** to this bit will have no effect. This is to allow deterministic behavior.
2. All control and configure bits (8 lower bits), bits 7~0, cannot be written or modified until the bit field **BURST_STATUS** (bits 18:16) shows the **Idle** state (**000**). In all other cases, the bits will remain locked so as to not disrupt an operation that is in progress.
3. Writes to the **START** bit will be ignored if the device is in low-frequency active and low-frequency LPM0 modes of operation.

Next, let's take a look at the Flash Program Burst Start Address Register.

6.5.2.2.4.2 Flash program burst start address register (FLCTL_PRGBRST_STARTADDR) This is a 32-bit register but only the lower 22 bits (bits 21~0 = bit field **START_ADDRESS**) are used to set up an initial or starting address for the burst program operations. Totally, a 2 MB memory space (2^{21} = 2048 KB) can be used to access 128 KB 128 bits (128 × 1024 × 16 bytes) memory space to perform the burst program operations (128 bits = 16 bytes). When using this register to store or set up an initial memory address for the burst program operations, the following points must be noted:

1. The offset is **0x0** starting from the type of memory region selected.
2. Bits 3~0 on this register are always 0 to get a 128-bit boundary.
3. This bit field is writable *ONLY* when the bit field **BURST_STATUS** (**18:16**) on the register **FLCTL_PRGBRST_CTLSTAT** shows the **Idle** (**000**) state. In all other cases, the bits will remain locked so as to not disrupt an operation that is in progress.

Bits 31~22 on this register are reserved for future usage.

6.5.2.2.4.3 Flash program burst data registers Four chunks 128-bit wide buffer are used to preload the data to be written into the target flash memory. This preload operation with data can either be performed by the application code or by the DMA. The buffer will actually be implemented as sixteen 32-bit registers to facilitate direct writes from the CPU or DMA.

Table 6.12 Bit value and bit function for FLCTL_PRGBRST_CTLSTAT register

Bit	Field	Type	Reset	Function
31:24	**Reserved**	RO	NA	Reserved and returns 0 as read
23	**CLR_STAT**	W	NA	Write **1** to clear status bits 21–16 of this register. Write **0** has no effect
22	**Reserved**	RO	NA	Reserved and returns 0 as read
21	**ADDR_ERR**	RO	0	Memory Address Access Error: **0:** No address error occurred **1:** Burst operation was terminated due to access to reserved memory
20	**PST_ERR**	RO	0	Postprogram auto-verify error: **0:** No postprogram auto-verify errors occurred **1:** Burst operation encountered postprogram auto-verify errors
19	**PRE_ERR**	RO	0	Preprogram auto-verify error: **0:** No preprogram auto-verify errors occurred **1:** Burst operation encountered preprogram auto-verify errors
18:16	**BURST_STATUS**	RO	0	Burst operation status: **000** = Idle (Burst not active) **001** = Burst program started but pending **010** = Burst active, with 1st 128-bit word being written into Flash **011** = Burst active, with 2nd 128-bit word being written into Flash **100** = Burst active, with 3rd 128-bit word being written into Flash **101** = Burst active, with 4th 128-bit word being written into Flash **110** = Reserved (Idle) **111** = Burst Complete (status of completed burst remains in this state unless explicitly cleared by SW)
15:8	**Reserved**	RO	NA	Reserved and returns 0 as read
7	**AUTO_PST**	RW	1	Control the Auto-verify operation after the Burst Program: **0:** No program-verify operations enabled **1:** Cause an automatic Burst Program Verify after the Burst Program Operation
6	**AUTO_PRE**	RW	1	Control the Auto-Verify operation before the Burst Program: **0:** No program-verify operations enabled **1:** Cause an automatic Burst Program Verify before the Burst Program Operation

(*Continued*)

Table 6.12 (Continued) Bit value and bit function for FLCTL_PRGBRST_CTLSTAT register

Bit	Field	Type	Reset	Function
5:3	LEN	RW	00	Length of burst (in 128 bits): **000:** No burst operation **001:** 1* 128 bits burst write starting from address in the STARTADDR Register **010:** 2*128 bits burst write starting at address in the STARTADDR Register **011:** 3*128 bits burst write starting at address in the STARTADDR Register **100:** 4*128 bits burst write starting at address in the STARTADDR Register **101~111:** Reserved. No burst operation
2:1	TYPE	RW	00	Type of memory that burst program is carried out on: **00:** Main memory **01:** Information memory **10:** Reserved **11:** Engineering memory
0	START	W	NA	**1:** Trigger start of Burst Program operation. **0:** Disable Burst Program

Each 128-bit chunk buffer comprises four 32-bit burst data registers arranged from **Program Burst Datan Register0** to **Program Burst Datan Register3** (**n** = 0~3). In fact, the **n** is the chunk number arranged from 0 to 3.

The actual names for these 16 registers are

- FLCTL_PRGBRST_DATA0_0~FLCTL_PRGBRST_DATA0_3 (first chunk)
- FLCTL_PRGBRST_DATA1_0~FLCTL_PRGBRST_DATA1_3 (second chunk)
- FLCTL_PRGBRST_DATA2_0~FLCTL_PRGBRST_DATA2_3 (third chunk)
- FLCTL_PRGBRST_DATA3_0~FLCTL_PRGBRST_DATA3_3 (fourth chunk)

Each register **FLCTL_PRGBRST_DATAn_x** can be used to preload a 32-bit data to be programmed into the flash memory with the following bit order:

```
Bits (32*(x+1)-1) to 32 * x; where x=0, 1, 2, 3
```

So, for each chunk **n**, four 32-bit registers are arranged **FLCTL_PRGBRST_DATAn_x** (**x** = **0**~**3**).

For example, the first 32-bit register in chunk **0**, **FLCTL_PRGBRST_DATA0_0** (**x** = **0**), holds the lowest 32-bit data for a 128-bit data with the data bits between 31 and 0. The second 32-bit register **FLCTL_PRGBRST_DATA0_1** (**x** = **1**) holds the less-lower 32-bit data between bits 63 and 32. Finally, the fourth 32-bit register **FLCTL_PRGBRST_DATA0_3** (**x** = **3**) holds the highest data word with the data bits between 127 and 96.

One point to be noted when using these registers to preload 128-bit data is that *ONLY* when the bit field **BURST_STATUS** (bits **18:16**) on the **FLCTL_PRGBRST_CTLSTAT** register shows the **Idle** state (**000**), a writing operation is valid. In all other cases, these data bits will remain locked so as to not disrupt an operation that is in progress.

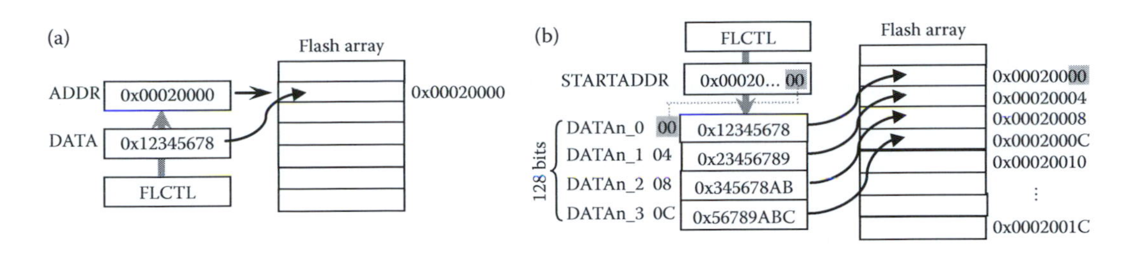

Figure 6.16 Write a 32-bit word and a burst program data into flash memory.

Figure 6.16 shows a functional block diagram to compare writing a single word and performing a burst program data operation.

When performing a burst program operation, the first 128-bit data are preloaded into four 32-bit registers arranged from **DATAn_0** to **DATAn_3**. Then, the content of these burst data registers are written to the address stored in the Burst Start Address Register when a burst program command is executed. The address range for this 128-bit data is from **0x00020000** to **0x0002000C**. Then, the next 128-bit data will be written to the next memory block starting from the address **0x00020010**. The address difference between each 128-bit data is a 128-bit boundary, or the bits 3~0 are 0 on the **FLCTL_PRGBRST_STARTADDR** register.

A piece of example codes shown in Figure 6.17 can be used to perform a burst program operation to write four 128-bit data into the main flash memory space at Bank0 with starting address of **0x2000**.

A user-defined function **InitFBPData()** is used in this piece of codes to help to fill sixteen 32-bit registers **FLCTL_PRGBRST_DATAn_0~FLCTL_PRGBRST_DATAn_3** (**n** = 0~3) with a 32-bit initial data word **0x0**. Each of the following fifteen 32-bit data is incremented by **4 x uiLoop** number.

Let's have a closer look at this piece of code to see how it works.

- First, the base address of all the 16 program burst data registers, **0x40011060**, is defined as a constant in line 3, and this constant will be used with the macro **HWREG32()** together to convert it to an address in the **InitFBPData()** function to fill all sixteen 32-bit registers.
- Then, the starting address for this burst program operation, **0x2000** that is a 128-bit boundary address, is assigned to the **FLCTL_PRGBRST_STARTADDR** register in line 4.
- The user-defined function **InitFBPData()** is called in line 5 to fill sixteen 32-bit data registers **FLCTL_PRGBRST_DATAn_0~FLCTL_PRGBRST_DATAn_3** (**n** = 0~3).
- In line 6, the WE Protection register of the main flash memory Bank0 is cleared to remove protections for all 32 sectors in that bank to enable the program operations.
- All error flags in the Flash Interrupt Flag Register (**FLCTL_IFG**) are cleared in line 7 to make it ready for this burst programming operation.
- Also, the auto-verify bits, including the pre and postprogram verify in the flash program control-status register **FLCTL_PRG_CTLSTAT**, are cleared in line 8 since we do not need these verifications on this burst program operation.
- The **FLCTL_PRGBRST_CTLSTAT** register is configured to perform four 128-bit burst program operation with auto pre and postprogram verifications in line 9. Also, this operation is enabled in this configuration.
- In line 10, the content of the **FLCTL_IFG** register is read back to check if this burst program is complete.

```
1   uint32_t ret;
2   void InitFBPData(void);                      // declare user defined function InitFBPData()
3   #define FL_PRGBRST_D_0 0x40011060            // define the base addr for the DATA0 register

4   FLCTL_PRGBRST_STARTADDR = 0x2000;            // setup the starting address as 0x2000.
5   InitFBPData();                               // call the user defined function InitFBPData() to
                                                    fill 16 32-bit
                                                 // registers to get 4 128-bit data to be
                                                    programmed to flash.
6   FLCTL_BANK0_MAIN_WEPROT = 0;                 // clear all write/erase protection functions
7   FLCTL_CLRIFG = 0x0000021E;                   // clear any IFG errors (PRGB Done |PreV | PostV
                                                    |PRG Error)
8   FLCTL_PRG_CTLSTAT &= ~0x000000C0;            // clear program auto verify functions
9   FLCTL_PRGBRST_CTLSTAT = 0xE1;                // enable burst program with 4 128-bit data
                                                    writing.
10  while(FLCTL_IFG & 0x10){}                    // Wait for the PRGB to be done......
11  if (FLCTL_IFG & 0x00000206)                  // check all possible errors (PreV|PostV|PRG
                                                    Error)
12    return false;
13  ret = FLCTL_PRGBRST_CTLSTAT;                 // Burst program is done. Get its running status
14  if (ret & 0x00380000)                        // check any error (ADDR_ERR, PST_ERR & PRE_ERR)
                                                    occurred?
15    return false;
16  FLCTL_PRGBRST_CTLSTAT |= 0x00800000;         // write 1 to CLR_STAT bit to clear 21-16 bits
                                                    status.
17  ......

18  void InitFBPData(void)
19  {
20    uint32_t uiLoop, wData = 0x0;

21    for(uiLoop = 0; uiLoop < 16; uiLoop++)
22    {
23      HWREG32(FL_PRGBRST_D_0 + 4 * uiLoop) = wData;
24      wData += 0x11111111;
25    }
26  }
```

Figure 6.17 Burst program to write four 128-bit data into flash Bank0 space.

- If the burst program operation is done. Next, we need to check if any error is encountered during that operation in line 11. If any error has occurred, a **false** is returned to the main program in line 12 to indicate this error.
- The codes in lines 13~15 are used to check if any local error has occurred for that burst program operation by checking the **FLCTL_PRGBRST_CTLSTAT** register. These include the **ADDR_ERR**, **PRE_ERR**, and **PST_ERR**. A false is returned to the main program if any error is occurred.
- In line 16, a 1 is written to the **CLR_STAT** bit in the **FLCTL_PRGBRST_CTLSTAT** register to clear 21–16 bits status.
- The codes between lines 18~26 are the body of the user-defined function **InitFBPData()**. The only point to be noted is the code line 23, where the base address constant of the burst program data registers **FL_PRGBRST_D_0**, is converted to the starting address with the system macro **HWREG32()**. All of the following 15 data registers are created by using this base address by adding 4 bytes offset and assigned with the related loop data values.

6.5.2.3 Flash memory erase mode

As we mentioned in Section 6.5, the erasing of bits in the flash memory involves setting the targeted bits to the value 1. Erase operations are permitted in a minimum granularity of the sector size (4 KB). Therefore, there are 32 sectors (Sector0~Sector31) in each main flash memory bank (32×4 KB = 128 KB) and 2 sectors (Sector0~Sector1) in each information flash memory bank (2×4 KB = 8 KB).

There are two erase modes that exists in the flash memory: **`Sector Erase`** and **`Mass Erase`**.

- **`Sector Erase Mode`**: In the sector erase mode, the flash controller can be configured to erase a targeted sector. This sector can be in either bank, and in any of the three memory regions. This kind of erase operation is controlled by a group of flash erase control registers. Once the erase is started, its status can be polled to check its completion, and the controller can also be configured to generate an interrupt on its completion.
- **`Mass Erase Mode`**: In the mass erase mode, the flash controller is set up to erase the entire flash memory. Mass erase is simultaneously applied to both banks. As with sector erase, its status can be polled to check its completion, and the controller can also be configured to generate an interrupt on its completion.

Two registers are used to control, configure, and monitor erase operations, either sector or mass, for any flash memory banks. These two registers are:

- Flash Erase Control and Status Register (**`FLCTL_ERASE_CTLSTAT`**)
- Flash Erase Sector Address Register (**`FLCTL_ERASE_SECTADDR`**)

The first register is used to control and monitor both erase operations, and the second register is mainly used to define the starting address for the sectors to be erased.

6.5.2.3.1 Flash erase control and status register (FLCTL_ERASE_CTLSTAT) This is a 32-bit register but only 8 bits are used to configure the erase operation and monitor the erase operational status. Figure 6.18 shows the bit map and field for this register. Table 6.13 lists the bit field values and functions for this register.

It can be found from Figure 6.18 that the higher 3 bits, bits 18~16, are running status bits used to monitor and indicate any possible running errors encountered during the erase operations. The lower 4 bits, bits 3~0, are used to control the erase operations, such as setting the erase memory type (the main, the information, or the engineering flash memory), the erase mode (sector or mass) and the enable as well as disable the erase operations.

It also can be found from Table 6.13 that after a system reset, the default erase setup is:

- The main flash memory is selected as the target memory to be erased in the **`TYPE`** field.
- The sector erase operation mode is selected as the erase mode in the **`MODE`** field.
- The erase operation is disabled by clearing the **`START`** bit.

Flash erase control-status register (FLCTL_ERASE_CTLSTAT)

31–20	19	18	17–16	15–4	3–2	1	0
Reserved	CLR_STAT	ADDR_ERR	STATUS	Reserved	TYPE	MODE	START

Figure 6.18 Bit fields for the FLCTL_ERASE_CTLSTAT register.

Table 6.13 Bit value and bit function for FLCTL_ERASE_CTLSTAT register

Bit	Field	Type	Reset	Function
31:20	**Reserved**	RO	NA	Reserved and returns 0 as read
19	**CLR_STAT**	W	NA	Running status of the erase operation: **0:** No effect; **1:** Clear status bits **18–16** on this register
18	**ADDR_ ERR**	RO	0	Erase Address Error: **0:** No address error occurred; **1:** Erase Operation was terminated due to attempted erase of reserved memory address
17:16	**STATUS**	RO	00	Indicate the status of erase operations in the flash memory: **00:** Idle (no erase operation currently active) **01:** Erase operation triggered to START but pending **10:** Erase operation in progress **11:** Erase operation completed (status of completed erase remains in this state unless explicitly cleared by SW)
15:4	**Reserved**	RO	NA	Reserved and returns 0 as read
3:2	**TYPE**	RW	0	Type of memory that erase operation is carried out on (either mode): **00:** Main memory **01:** Information memory **10:** Reserved **11:** Engineering memory
1	**MODE**	RW	0	Control Erase Mode Selected by Application: **0:** Sector Erase (controlled by **FLTCTL_ERASE_SECTA-DDR**) **1:** Mass Erase (includes all main and information memory sectors that do not have corresponding WE bits set)
0	**START**	W	0	Master control for erase operations: **0:** Disable erase operation; **1:** Enable erase operation

When using this register to configure the erase operation and monitor the running status of the erase operation, the following points must be noted:

1. Writing a 1 to bit-19, the **`CLR_STAT`** bit, will clear the running status bits `18:16` *ONLY* when the erase status shown in bit field **`STATUS`** (`17:16`) is complete (`11`). In all other cases, this writing-1 operation will have no effect.
2. To configure or modify any control bits (bits 3~0), one needs to write some values (either 0 or 1) to these bits. However, these bit fields are writable *ONLY* when status bits (`17:16`) show the `Idle` state (`00`). In all other cases, the bits will remain locked so as to not disrupt an operation that is in progress.
3. When performing sector erase operation by resetting the **`MODE`** bit to 0, the application must ensure that the writes in this mode follow the LSB-MSB loading order within the 128-bit address boundary in the **`FLCTL_ERASE_SECTADDR`** register.
4. Write to the **`START`** bit will be ignored if the device is in low-frequency active and low-frequency LPM0 modes of operation.

6.5.2.3.2 Flash erase sector address register (FLCTL_ERASE_SECTADDR) This is a 32-bit register but only the lower 22 bits (bits 21~0 = bit field **SECT_ADDRESS**) are used to set up an initial or starting address for the sector erase operations. Totally, a 2 MB memory space (2^{21} = 2048 KB) can be accessed to perform the sector erase operations. When using this register to store or set up an initial memory address for the sector erase operations, the following points must be noted:

1. The offset is **0x0** starting from the type of memory region selected.
2. If memory type is set to information/main memory, bits 11~0 are always 0 (forced sector boundary of 4 KB). If memory type is set to engineering memory, bits 10~0 are always 0 (forced sector boundary of 2 KB).
3. This bit field is writable *ONLY* when the **STATUS** bits (**17:16**) on the **FLCTL_ERASE_CTLSTAT** register shows the **Idle** (**00**) state. In all other cases, the bits will remain locked so as to not disrupt an operation that is in progress.

Although the flash controller may perform either sector or mass erase modes, the erase operation may not have an effect on any sector that has its **Write-Erase (WE)** protection enabled. This will be explained in more detail in the following section.

6.5.2.4 Flash memory Write-Erase protection

In the MSP432P401R MCU system, the flash memory controller provides a Write-Erase (WE) Protection function for each sector at both main and information flash memory arrays. These protection functions are performed by setting related WE registers, with 1 bit dedicated for one flash sector. If a bit is set to 1, the corresponding sector is protected and the entire sector can be considered as a read-only sector without any writing and erasing functions. Using these bits, the application can protect sectors from inadvertent program/erase operations. In addition, these bits can also be used to optimize erase timings as follows:

- Consider a case where the application wishes to erase 128 KB of a 256 KB main memory
- This can be carried out as 32 sector erase operations

Four 32-bit control registers are used to perform these **WE** protection functions:

1. Main Memory Bank0 WE Protection Register (**FLCTL_BANK0_MAIN_WEPROT**)
2. Main Memory Bank1 WE Protection Register (**FLCTL_BANK1_MAIN_WEPROT**)
3. Information Memory Bank0 WE Protection Register (**FLCTL_BANK0_INFO_WEPROT**)
4. Information Memory Bank1 WE Protection Register (**FLCTL_BANK1_INFO_WEPROT**)

Since both main flash memory arrays, Bank0 and Bank1, have 128-KB capacity with 32 sectors, therefore, both registers, **FLCTL_BANK0_MAIN_WEPROT** and **FLCTL_BANK1_MAIN_WEPROT**, are 32-bit registers with each bit providing the **WE** protection for one corresponding sector. But for both information flash memory array, each of them only has 8-KB capacity with 2 sectors, therefore, both registers, **FLCTL_BANK0_INFO_WEPROT** and **FLCTL_BANK1_INFO_WEPROT**, only use the lowest two bits, bits 1 and 0, to provide **WE** protection function even when both registers are 32 bits.

A point to be noted is that after a system reset, all sectors on all flash memory spaces, both main and information, are protected by default. Therefore, before a write or an erase

```
1  uint32_t sector, secAddr;
2  secAddr = 0x1000;                          // sector starting address
3  for(sector = 1; sector < 5; sector++)
4  {
5    FLCTL_CLRIFG = 0x00000226;               // clear IFG errors (ERASE Done |PreV | PostV
                                              |PRG Error)
6    FLCTL_BANK0_MAIN_WEPROT = 0;             // clear all write/erase protection functions
7    FLCTL_ERASE_SECTADDR = secAddr * sector; // set the starting address to be erased for
                                              4KB as interval.
8    FLCTL_ERASE_CTLSTAT |= 0x01;             // start the sector erase operation.
9    while(FLCTL_IFG & 0x20){}                // Wait for ERASE to be done......
10   if (FLCTL_ERASE_CTLSTAT & 0x00040000)    // sector erase is complete. Now check the
                                              error on bit-18.
11     return false;                          // ADDR_ERR is set to 1, an address error
                                              occurred. Return.
12 }
```

Figure 6.19 Erase four sectors in the flash Bank0 space.

operation can be performed, the applications have the responsibility to reset related bits on related registers to remove those **WE** protections from those sectors in which the write or erase operations are to be performed. Otherwise the applications may encounter some problems and the desired operations cannot be executed at all.

A piece of example code shown in Figure 6.19 can be used to perform a sector erase operation to erase four sectors in the main flash memory space at Bank0 with starting address of **0x1000**.

Let's have a closer look at this piece of code to see how it works.

- Two local variables, **sector** and **secAddr**, are declared at line 1 and the starting address of the flash memory (**0x1000**) to be erased is assigned to the **secAddr** variable in line 2.
- A **for** loop is used to repeat executing erase operations for four sectors starting at line 3.
- Any possible errors appeared in the **FLCTL_IFG** register are cleared in line 5 to make it ready for this erase operation.
- The **WE** protection bits for all sectors in the main flash memory Bank0 are reset in line 6 to remove those protections to enable the erase operations to be executed.
- In line 7, the starting sector address **0x1000** stored in the variable **secAddr** that is multiplied by the loop counter **sector** is assigned to the **FLCTL_ERASE_SECTADDR** register.
- The bit-0 (**START** bit) in the **FLCTL_ERASE_CTLSTAT** register is set to 1 in line 8 to start this sector erase operation.
- Then a conditional **while()** loop is applied in line 9 to monitor whether the first sector erase operation is complete by checking bit-5 (**ERASE**) in the **FLCTL_IFG** register. This **while()** loop will continue this checking until the bit-5 is **1**, which means that the first sector erase operation is complete.
- After the first sector erase is done, next we need to check if any error related to the sector address has occurred in line 10.
- If an address error is encountered, a **false** is returned to the main program in line 11.

This for loop will be executed four times to erase four sectors starting from **0x1000** and ending at **0x4FFF** in the first bank (Bank0) in the main flash memory space. Each loop erases 1 sector (4 KB) with address interval as **0x1000** (4 KB). The first sector is addressed between **0x1000** and **0x1FFF**, the second sector is between **0x2000** and **0x2FFF**, and so on.

6.5.3 *Flash memory interrupt handling*

In the MSP432P401R MCU system, the flash controller can be configured to trigger and generate interrupts based on some flash operations and events. These interrupts can be generated and indicated by related flags set in the **FLCTL_IFG** when those interrupt sources are enabled by setting the Flash Interrupt Enable Register (**FLCTL_IE**). After the triggered interrupts are accepted and processed, the applications have the responsibility to clear and reset those handled interrupts via the Flash Clear Interrupt Flag Register (**FLCTL_CLRIFG**). The following operations and events can trigger an interrupt:

- Word composition error in the full-word write mode
- Pre-program auto-verify operation error
- Post-program auto-verify mismatch error
- Completion of burst program mode operation
- Completion of erase operation
- Completion of burst read-compare operation (can also stop due to compare mismatch)
- A match on the benchmark counter

Among these interrupts, no priority level has existed and these interrupts can be generated at the same time if multiple operations are complete and multiple events have occurred.

Four control registers are used to configure, control, and monitor these flash operations-related interrupts:

1. Flash Interrupt Enable Register (**FLCTL_IE**)
2. Flash Interrupt Flag Register (**FLCTL_IFG**)
3. Flash Clear Interrupt Flag Register (**FLCTL_CLRIFG**)
4. Flash Set Interrupt Flag Register (**FLCTL_SETIFG**)

All of these registers are 32 bits, but only the lower 8 bits, bits 9:8 and 5:0, are used to set up and monitor flash-related interrupts shown above. Since the similarity between these registers, we only show one register, **FLCTL_IE**, to illustrate the bit fields for these registers. Figure 6.20 shows the bit fields for this register. If any bit on this register is set to 1, it enables the controller to generate an interrupt based on the corresponding flash memory operation or event shown in the **FLCTL_IFG**. Whenever a related flash operation is enabled for interrupt and complete or an event occurs, the related bits on the **FLCTL_IFG** register is set to 1.

All of these registers have the same bit fields. Table 6.14 shows the bit functions for the **FLCTL_IFG** register.

When a flash-operation-related interrupt is enabled and generated, it should be handled by a flash interrupt handler, **FLCTL_IRQHandler()**, which is a default flash interrupt handler in the MSP432P401R MCU system. The applications have the responsibility to build and code this handler to respond to and process the accepted interrupt.

After any interrupt that flagged in the **FLCTL_IFG** register has been accepted and processed by the processor, the applications need to use the **FLCTL_CLRIFG** to clear or reset

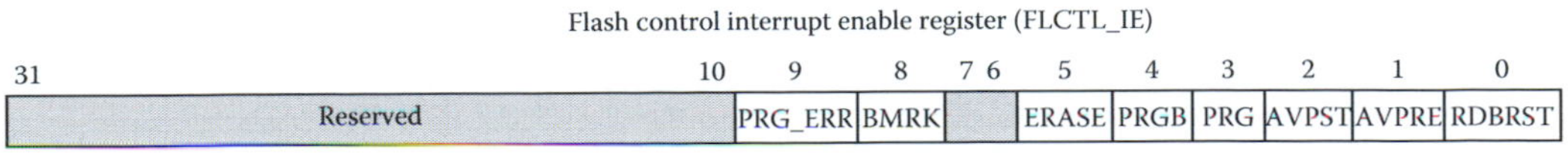

Flash control interrupt enable register (FLCTL_IE)

31 … 10	9	8	7 6	5	4	3	2	1	0
Reserved	PRG_ERR	BMRK		ERASE	PRGB	PRG	AVPST	AVPRE	RDBRST

Figure 6.20 Bit fields for the FLCTL_IE register.

Table 6.14 Bit value and bit function for FLCTL_IFG register

Bit	Field	Type	Reset	Function
31:10	**Reserved**	RO	NA	Reserved and returns 0 as read
9	**PRG_ERR**	RO	0	Word Composition Error for Full-Word Program: **0:** No error. **1:** A word composition error in full-word write mode (data loss due to writes crossing over to a new 128-bit boundary before full word has been composed)
8	**BMRK**	RO	0	Benchmark Compare Match: **0:** No benchmark match; **1:** A Benchmark Compare match occurred
7:6	**Reserved**	RO	00	Reserved
5	**ERASE**	RO	0	Erase Operation Status: **0:** Erase is working; **1:** Erase operation is complete
4	**PRGB**	RO	0	Burst Program Operation Status: **0:** Burst Program is working **1:** Burst Program operation is complete
3	**PRG**	RO	0	Word Program Operation Status: **0:** Word program is working **1:** Word program operation is complete
2	**AVPST**	RO	0	Postprogram verify operation: **0:** Postprogram verify operation is OK **1:** Postprogram verify operation has failed comparison
1	**AVPRE**	RO	0	Preprogram verify operation: **0:** Preprogram verify operation is OK **1:** Preprogram verify operation has failed comparison
0	**RDBRST**	RO	0	Read burst/compare operation: **0:** Read burst/compare operation is working **1:** Read burst/compare operation is complete

those flags in the **FLCTL_IFG** register inside the **FLCTL_IRQHandler()**. The bit fields and functions of the **FLCTL_CLRIFG** are identical with those in the **FLCLT_IFG**. Setting a bit to 1 in the **FLCTL_CLRIFG** register is equivalent to clearing the corresponding flag set in the **FLCTL_IFG** register.

The purpose of the **FLCTL_SETIFG** is to set up some desired flags in the **FLCTL_IFG** to simulate generating some flash-related interrupts.

A point to be noted is that before enable any flash-operation-related interrupt, the register **FLCTL_CLRIFG** should be used to clear any possible previous interrupt flags set in the **FLCTL_IFG** to avoid any possible mis-interrupt operation.

6.5.4 *Low-frequency active and low-frequency LPM0 modes support*

The MSP432P401R MCU devices support low-frequency active and low-frequency LPM0 modes. In these modes, the MCU runs in a very low frequency, low leakage mode, with the maximum bus clock frequency restricted to 128 KHz. When operating in this mode, the flash controller has the following functionality:

- Read will be carried out in normal read mode only. The read mode setting for both banks will be automatically set to normal read when the flash controller detects the low-frequency active or low-frequency LPM0 modes (**AM_LF_VCOREx** or **LPM0_LF_VCOREx** modes).
- Read burst and compare operation will not be enabled or permitted.
- Any form of program or erase operation will not be enabled or permitted.

A point to be noted when working in the low-frequency active and low-frequency LPM0 mode is that it is the application's responsibility to ensure that only read operations can be performed to the flash when the device is in low-frequency active or low-frequency LPM0 modes. All other operations will be ignored or discarded without generating an exception. If the flash is enabled for full-word write mode, and an entry to low-frequency active or low-frequency LPM0 mode is initiated, any partially composed data in the 128-bit write word will be discarded without generating any exception.

Another point to be noted is the working status of the flash controller when a system reset is issued, including a SR, a HR, or a POR action.

For the SR, it will have no impact on the flash controller functionality. However, for a hardware reset, it will have the following impact on flash controller functionality:

- All current or outstanding read operations will be terminated.
- Any program or erase operation that is currently active will complete as normal. This ensures that bits are not left in a marginal state due to HR conditions.
- Any outstanding program or erase operation will be terminated.
- All application settings in the flash control registers will be reset.

A POR will have the following impact on flash controller functionality:

- All current or outstanding read operations will be terminated.
- All current or outstanding program or erase operations will be terminated.
- All application settings in the flash control registers will be reset.

6.5.5 Flash controller benchmarking features

The flash controller also offers two counters for application benchmarking purposes. This feature is extremely useful when monitoring the number of flash accesses is a key issue, primarily because the flash access power is the main contributor to the overall active power of the device. These two 32-bit counters can be described as below:

- Instruction Fetch Benchmark Counter (**FLCTL_BMRK_IFETCH**)
  - Readable/writable by software
  - Increments on each instruction fetch to the flash
- Data Fetch Benchmark Counter (**FLCTL_BMRK_DREAD**)
  - Readable/writable by software
  - Increments on each data fetch to the flash

In addition, the flash controller also implements a compare-based interrupt generation capability with a Flash Benchmark Count Compare Register (**FLCTL_BMRK_CMP**). The

Flash benchmark control and status register (FLCTL_BMRK_CTLSTAT)

31–4	3	2	1	0
Reserved	CMP_SEL	CMP_EN	D_BMRK	I_BMRK

Figure 6.21 Bit fields for the FLCTL_BMRK_CTLSTAT register.

interrupt generation logic can be configured to monitor either of the benchmark counters and fire an event when the counter in consideration reaches a particular value. This event may trigger a related **BMRK** interrupt by setting a flag in the **FLCTL_IFG** register.

Four registers are used to configure, control, and monitor these benchmark operations for instructions and data:

1. Flash Benchmark Control and Status Register (**FLCTL_BMRK_CTLSTAT**)
2. Flash Benchmark Instruction Fetch Count Register (**FLCTL_BMRK_IFETCH**)
3. Flash Benchmark Data Read Count Register (**FLCTL_BMRK_DREAD**)
4. Flash Benchmark Count Compare Register (**FLCTL_BMRK_CMP**)

Figure 6.21 shows the bit fields for the **FLCTL_BMRK_CTLSTAT** register. Table 6.15 shows the bit functions for this register.

It can be found from Table 6.15 that the lower 4 bits on this register control the benchmark operations via either the instruction or the data benchmark register. Bit-3 (**CMP_SEL**) is to select to compare either the Instruction Benchmark Register (0) or the Data Benchmark Register (1) with a threshold value set in the **FLCTL_BMRK_CMP**.

Bit-2 (**CMP_EN**) is used to enable (1) or disable (0) the comparison of the Instruction or Data Benchmark registers against the threshold value set in the **FLCTL_BMRK_CMP**

Table 6.15 Bit value and bit function for FLCTL_BMRK_CTLSTAT register

Bit	Field	Type	Reset	Function
31:4	**Reserved**	RO	NA	Reserved and returns 0 as read
3	**CMP_SEL**	RW	0	Select which benchmark register should be compared against the threshold: **0:** Compare the Instruction Benchmark Register against the threshold value **1:** Compare the Data Benchmark Register against the threshold value
2	**CMP_EN**	RW	0	Benchmark Compare Enable: **0:** No benchmark compare; **1:** Enable comparison of the Instruction or Data Benchmark Registers against the threshold value
1	**D_BMRK**	RW	0	Data Benchmark Counter Increment: **0:** No increment **1:** Increment Data Benchmark count register on each data read access to the Flash
0	**I_BMRK**	RW	0	Instruction Benchmark Counter Increment: **0:** No increment **1:** Increment Instruction Benchmark count register on each instruction fetch to the Flash

register. Bit-1 (**D_BMRK**) and bit-0 (**I_BMRK**) are used to enable (**1**) or disable (**0**) to increment the Data Benchmark Count register or the Instruction Benchmark Count register when a data reading or an instruction fetch operation is occurred.

Both **FLCTL_BMRK_IFETCH** and **FLCTL_BMRK_DREAD** are 32-bit registers and used to record the number of instruction fetches to the flash and the number of data read operations to the flash. Each of these registers would be increment by 1 if an instruction is fetched to or a data item is read out from the flash memory.

The **FLCTL_BMRK_CMP** register is also a 32-bit register and it is used to store the threshold value that is compared against either the **IFETCH** or **DREAD** Benchmark Counters.

6.5.6 Recommended settings and flow for program and erase operations

When performing flash program or erase operations, some operational procedure and sequence should be followed to enable users to build and develop reliable applications successfully. These include the correct register settings and software flow as shown below:

Flash Program Operations (immediate full-word program or Burst program):

1. Any program operation must enable **PRE-VERIFY** and **POST-VERIFY** options to ensure a reliable programming of the flash bit-cell.
   a. Writing **0** to a bit which is already **0** will lead to over-programming of the bit-cell and will have impact on the flash memory reliability. Having **PRE-VERIFY** enabled prevents this occur.
   b. Application may choose to disable **PRE-VERIFY** option if it is programming a location which is already erased.
2. **POST-VERIFY** is required for any program operation and if **POST-VERIFY** fails, the application should do the following for the failed data:
   a. Read the word in program-verify mode and compare with the expected data to find out which bits failed the postverify.
   b. Initiate a program to that address with only the failed bits set as **0**.
3. For immediate and full-word program operations, the Burst program auto-verify functions must be disabled by clearing **AUTO_PST** and **AUTO_PRE** bits in the **FLCTL_PRGBRST_CTLSTAT** register. This disable action can avoid the disturbance of the burst auto-verify functions to the normal program operations.
4. For Burst program operations, the program auto-verify functions must be disabled by clearing **VER_PST** and **VER_PRE** bits in the **FLCTL_PRG_CTLSTAT** register. This disable action can avoid the disturbance of the normal program auto-verify functions to the burst program operations.

Flash Erase Operations (sector and mass erase):

1. An erase operation must be followed by a read of the erased locations to confirm that erase was successful.
2. Verification of erase operation should be done in software and no hardware support is available for automatic verification (unlike program operation).
3. The read for verification of erase operation must be done by setting the bank in erase verify read mode.
4. An efficient software implementation of erase verification should use the burst read feature of the FLCTL and compare the data against an all **1** pattern.
5. If burst read gives an error, then reinitiate erase function until the verification pass.

A point to be noted is that for either write or erase operations, the related Write-Erase **`Protection Register`** must be cleared to enable both operations to be performed. Now, let's build some projects to perform desired operations to the flash memory to test our study.

6.6 Memory system programming methods

As we discussed in Section 6.2 in this chapter, the MSP432P401R MCU memory system contains the following memory components or devices:

- SRAM (64 KB)
- Flash memory (256 KB)
- Internal ROM
- Flash protection unit
- Optional MPU

Because the internal ROM is used by the vendor to store necessary peripheral driver library and BSL, we do not need to touch that component. The optional MPU will be discussed in Chapter 12. The SRAM bit-band operation programming has been discussed in Sections 6.3.3.2 through 6.3.3.5. Therefore, in this section we will focus our discussions on the flash memory operations, which include:

1. Flash memory reading operation
2. Flash memory burst reading-comparing operation
3. Flash memory writing (immediate word and full-word programming)
4. Flash memory burst programming operation
5. Flash memory erasing operation (sector and mass erasing)

There are two ways to program the flash memory with C programming language. These two popular ways are

- Programming flash memory with DRA model
- Programming flash memory with SD model provided by MSPWare Peripheral Driver Library

In the previous sections, we have provided detailed discussions about all registers used for flash memory programming in the memory architecture and memory map parts. These provided good bases for programming the flash memory with the DRA method. In the following sections, we will first concentrate on the flash memory programming operations, including the flash reading, programming, and erasing with the DRA model. Then, we will move our attention to the MSPPDL since it provides all API functions to enable us to program flash memory with the SD method.

6.6.1 Flash memory projects with DRA model

In this section, we concentrate on the flash memory projects developed by using the DRA model, including the flash memory reading, programming, and erasing operations.

6.6.1.1 Flash memory reading projects with DRA model

In Section 6.5.2.1, we have discussed in detail about the flash memory read mode, which includes the normal read mode and burst read-compare mode. In this section, we will provide an example project to illustrate how to perform a normal flash read operation to read 32 words from a piece of flash memory space. For the burst read-compare mode, we leave this kind of project in the flash erase mode to enable users to perform a burst read-compare operation to confirm the correctness of the sector erase operations.

As we discussed in Section 6.5.2.1, the flash memory reading operations are controlled by two registers; the **FLCTL_BANK0_RDCTL** and **FLCTL_BANK1_RDCTL**, where the former is used to control the first 128-KB flash main memory, but the **FLCTL_BANK1_RDCTL** register is used for the second 128-KB flash main memory.

In this example project, we try to use the **FLCTL_BANK0_RDCTL** register to read 32 words (each word is a 32-bit data item) from the main flash memory starting at address **0**. Refer to Figure 6.9 and Table 6.9 to get more details about this register, include the bit field and functions.

Perform the following operations to create and build a new Keil µVersion5 project **DRAFlashREWords** in the location **C:\MSP432 Class Projects\Chapter 6** at your computer:

- Open the Windows Explorer and create a new folder **DRAFlashREWords** under the folder **C:\MSP432 Class Projects\Chapter 6** at your computer
- Open the Keil µVersion5 MDK IDE and create a new µVersion5 project **DRAFlashREWords** and save this project to the folder **DRAFlashREWords** created above
- Create a new C File **DRAFlashREWords.c** and add it into our project **DRAFlashREWords**
- Add the code shown in Figure 6.22 into this C File.

Now, let's have a closer look at this piece of code to see how it works.

- Three system header files are declared first at code lines 4~6 since we need to use some system macros defined for the flash control registers in this project.
- Some local variables and array are declared inside the main program body in lines 9 and 10. The array **destAddr[]** is a 32-word array used to store the read back data words from the flash memory. The variable **n** works as a loop counter for the for loop,

```
1  //****************************************************************************
2  // DRAFlashREWords.c - Main application file for project DRAFlashREWords
3  //****************************************************************************
4  #include <stdint.h>
5  #include <stdbool.h>
6  #include <msp.h>
7  int main(void)
8  {
9    uint32_t destAddr[32];
10   uint32_t n, dataLength = 32, srcAddr = 0x0;
11   FLCTL_BANK0_RDCTL = 0x00003000;          // normal read mode with 3 wait states
12   for (n = 0; n < dataLength; n++)         // repeat reading for dataLength times
13   {
14     destAddr[n] = HWREG32(srcAddr);        // read and save a data word
15     srcAddr += 4;                          // update the start address
16   }
17   while(1);
18 }
```

Figure 6.22 Detailed code for the project DRAFlashREWords.

and the **dataLength** defined the length of this flash reading. The **srcAddr** is the starting address of the flash main memory from where the 32-word data can be read.
- In line 11, the Flash Control Bank0 Read Control Register, **FLCTL_BANK0_RDCTL**, is configured to perform a normal read operation from the main flash memory (the first 128 KB) with three wait states.
- Then, a **for** loop is used to repeatedly read 32 words starting from the flash memory address **0** in lines 12 through 16.
- In line 14, the starting address **0** is converted to an address pointer by using the system macro **HWREG32()**. The content of that address is read out and assigned to the data array **destAddr[]**. The variable **n** works as the loop counter to scan all 32 words until the upper bound **dataLength** is touched.
- After each reading and assigning, the starting address is updated by incrementing 4 to point to the next address in the Bank0 flash memory in line 15.
- An infinitive **while(1)** loop in line 17 is used to temporarily halt the program to enable users to check this reading result.

Before building and running this project, make sure:

1. The debugger used in this project is **CMSIS-DAP** (**Project**|**Options for Target "Target 1"**|**Debug** tab). All debugger-related settings are correct.

We need to set up a breakpoint at the **while()** loop line to enable us to check the contents of the Bank0 flash memory space to confirm our reading results. Go to **Debug**|**Insert/Remove Breakpoint** menu item to set up this breakpoint.

Now you can build, download, and run the project, and the project will be halted at the **while()** loop line. Let's check the running results of this project by performing the following operations:

1. Open the flash Bank0 main memory by clicking the **Memory 1** tab located at the lower right corner, and enter **0** into the **Address:** box. Press the Enter key from your keyboard. The contents of the Bank0 main memory are displayed, as shown in Figure 6.23.
2. Now, click on the **Call Stack + Locals** tab that is on the left of the **Memory 1** tab to open that window. Then, expand the data array **destAddr** by clicking on the left plus icon. All 32 words read from the flash Bank0 main memory space are shown in this window, as shown in Figure 6.24.

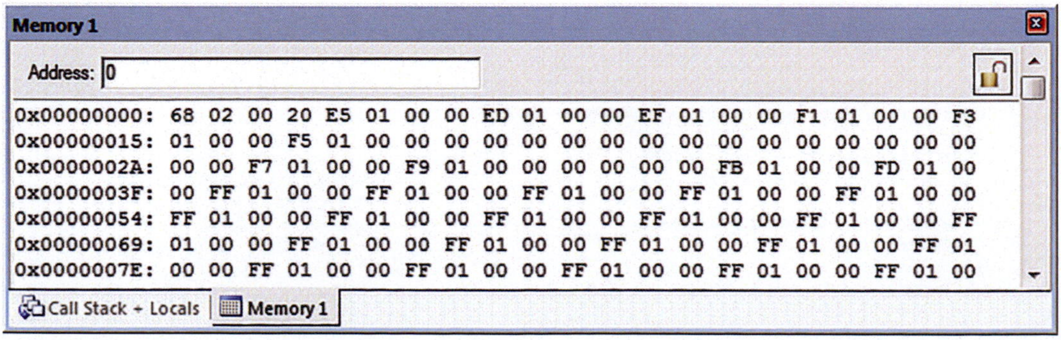

```
Memory 1
Address: 0
0x00000000: 68 02 00 20 E5 01 00 00 ED 01 00 00 EF 01 00 00 F1 01 00 00 F3
0x00000015: 01 00 00 F5 01 00 00 00 00 00 00 00 00 00 00 00 00 00 00 00 00
0x0000002A: 00 00 F7 01 00 00 F9 01 00 00 00 00 00 00 FB 01 00 00 FD 01 00
0x0000003F: 00 FF 01 00 00 FF 01 00 00 FF 01 00 00 FF 01 00 00 FF 01 00 00
0x00000054: FF 01 00 00 FF 01 00 00 FF 01 00 00 FF 01 00 00 FF 01 00 00 FF
0x00000069: 01 00 00 FF 01 00 00 FF 01 00 00 FF 01 00 00 FF 01 00 00 FF 01
0x0000007E: 00 00 FF 01 00 00 FF 01 00 00 FF 01 00 00 FF 01 00 00 FF 01 00
Call Stack + Locals | Memory 1
```

Figure 6.23 Contents on the flash Bank0 main memory.

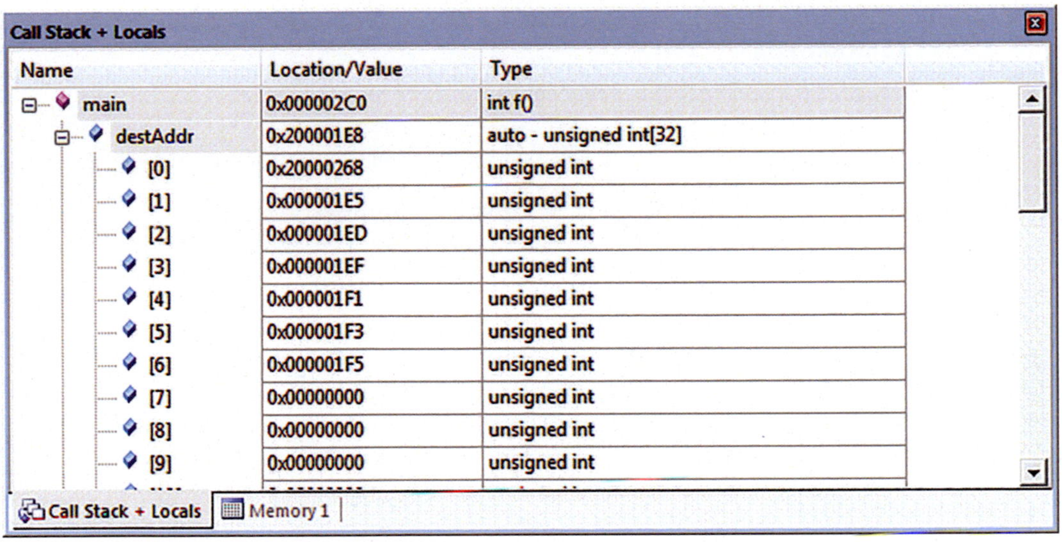

Call Stack + Locals

Name	Location/Value	Type
main	0x000002C0	int f()
destAddr	0x200001E8	auto - unsigned int[32]
[0]	0x20000268	unsigned int
[1]	0x000001E5	unsigned int
[2]	0x000001ED	unsigned int
[3]	0x000001EF	unsigned int
[4]	0x000001F1	unsigned int
[5]	0x000001F3	unsigned int
[6]	0x000001F5	unsigned int
[7]	0x00000000	unsigned int
[8]	0x00000000	unsigned int
[9]	0x00000000	unsigned int

Call Stack + Locals | Memory 1

Figure 6.24 Read results stored in the data array destAddr[].

3. Compare these 32 words in this array with those contents shown in Figure 6.23, one can find that they are identical, right? One issue is the order of these words arranged in this data array and stored in the flash memory, it is different! The **MSB** and the **LSB** in this data array and in the memory are just opposite. In the memory, the first word is `68 02 00 20` starting from address `0`. But in the data array, the first word is `20 00 02 68`. Can you see the difference? Why?
4. Click on the **Debug**|**Start/Stop Debug Session** menu item to stop the running of the project.

6.6.1.2 *Flash memory immediate writing projects with DRA model*

In this section, we will build a flash memory write (program) project to perform writing immediate words function. We have provided detailed discussions about this kind of operational mode in Section 6.5.2.2. Two operational modes, the immediate word and full-word programming, are provided by the MSP432P401R MCU system. One control register, Flash Program Control and Status Register (**FLCTL_PRG_CTLSTAT**), is used to configure and control both word programming operations. Refer to Figure 6.12 and Table 6.11 to get more details about this register used for this mode.

We try to program three 32-bit words into the flash main memory Bank0 space with the immediate mode since this mode provides the flexibility to write any length of words into the flash memory with immediate response.

Perform the following operations to create and build a new Keil µVersion5 project **DRAFlashWEWords** in the location **C:\MSP432 Class Projects\Chapter 6** at your computer:

- Open the Windows Explorer and create a new folder **DRAFlashWEWords** under the folder **C:\MSP432 Class Projects\Chapter 6** at your computer
- Open the Keil µVersion5 MDK IDE and create a new µVersion5 project **DRAFlashWEWords** and save this project to the folder **DRAFlashWEWords** created above

- Create a new C File **DRAFlashWEWords.c** and add it into our project **DRAFlashWEWords**
- Add the code shown in Figure 6.25 into this C File.

Now, let's have a closer look at this piece of codes to see how it works.

- Three system header files are declared first at code lines 4~6 since we need to use some system macros defined for the flash control registers in this project.

```
1  //*****************************************************************************************
2  // DRAFlashWEWords.c - Main application file for project DRAFlashWEWords
3  //*****************************************************************************************
4  #include <stdint.h>
5  #include <stdbool.h>
6  #include <msp.h>
7  uint32_t srcData[4];
8  void InitFWData(void);
9  int main(void)
10 {
11  uint32_t n, ret, dest = 0x1000;
12  uint32_t dataLength = 3;              // write 3 32-bit words to dest in flash memory
13  InitFWData();                         // call InitFWData() to fill 4 32-bit data
                                          into srcData[]
14  ret = FLCTL_PRG_CTLSTAT;              // get the current content of FLCTL_PRG_
                                          CTLSTAT register
15  while(!(ret & 0xFFFCFFFF)) {}         // bits 17:16 are zero (idle)? If not, wait
                                          here......
16  FLCTL_PRG_CTLSTAT = 0x0D;             // Immediate mode with auto-verify (enable the
                                          write)
17  FLCTL_PRGBRST_CTLSTAT &= ~0x000000C0; // clear burst program auto verify functions
18  for (n = 0; n < dataLength; n++)      // repeat writing for dataLength data
19  {
20    FLCTL_BANK0_MAIN_WEPROT = 0;        // clear all write/erase protection functions
21    FLCTL_CLRIFG = 0x0000020E;          // Clear IFG errors (PRG Done |PreV | PostV
                                          |PRG Error)
22    HWREG32(dest) = srcData[n];         // Write a 32-bit word from src to dest in
                                          flash memory
23    while(FLCTL_IFG & 0x8){}            // Wait for the PRG to be done....
24    if (FLCTL_IFG & 0x00000206)         // check all possible errors (PreV|PostV|PRG
                                          Error)
25      return false;                     // if any error occurred, return false
26    dest += 4;                          // update the destination address
27  }
28  while(1);                             // used for checking the writing result......
29 }
30 void InitFWData(void)
31 {
32  uint32_t uiLoop, wData = 0x0;
33  for(uiLoop = 0; uiLoop < 4; uiLoop++)
34  {
35    srcData[uiLoop] = wData;
36    wData += 0x11111111;
37  }
38 }
```

Figure 6.25 Detailed code for the project DRAFlashWEWords.

- A global array **srcData[]** and a global function **InitFWData()** are declared in lines 7 and 8. The data array **srcData[]** is used to store three 32-bit data to be written into the flash memory starting at **0x1000** in the main flash Bank0 space. The user-defined function **InitFWData()** is used to fill out three 32-bit data into the data array **srcData[]**.
- Some local variables **n**, **ret**, **dest**, and **dataLength**, are declared inside the main program body in lines 11 and 12. The variable **n** will work as a loop counter for the for loop, the **ret** works as a reading holder to keep the running status of this write operation, and the **dataLength** defined the number of words to be written into the flash. The **dest** is the starting address of the flash main memory from where three 32-bit words can be written into the flash.
- In line 13, the user-defined function **InitFWData()** is called to fill out three 32-bit data in the **scrData[]** array.
- The codes in lines 14 and 15 are used to read bits 17:16 in the Flash Program Control and Status Register (**FLCTL_PRG_CTLSTAT**) to check whether the Flash Controller is in the idle status. If both bits are 0, which means that the flash controller is in the idle status and the next instruction can be executed. Otherwise a **while()** loop is used to continue checking and wait until both bits are 0 in line 15.
- If the flash controller is in the idle status, the **FLCTL_PRG_CTLSTAT** register is configured to start an immediate word programming operation in line 16 with the post and preverify function to be activated by assigning **0x0D** (1101)—refer to Table 6.11.
- In line 17, the Burst program auto-verify bits (**AUTO_PST** and **AUTO_PRE**) in the **FLCTL_PRGBRST_CTLSTAT** register are cleared to avoid any possible disturbance for those bits' functions (refer to Table 6.12).
- The codes in lines 18 through 27 are used to continuously write three 32-bit words into the flash memory starting at **0x1000**.
- Before any WE operation can be performed, the related flash **WE** Protection Register must be cleared to remove those protections since all 32 bits on those protection registers are set to 1 (protected) after a system reset. Since we try to write data into the Bank0 in the main flash memory, the related **WE** protection register for Bank0, **FLCTL_BANK0_MAIN_WEPROT**, is cleared in line 20.
- In order to check this writing result, we need first to clear all previous setup bits on the **FLCTL_IFG** in line 21 by setting the related bits on the **FLCTL_CLRIFG** register. Bit-3 (**PRG**) on the **FLCTL_IFG** register will be set to 1 when a program operation is complete. Also, bits 9 (**PRG_ERR**), 2 (**AVPST**), and 1 (**AVPRE**) on the **FLCTL_IFG** register would be set to 1 if any error related to programming operation (**PRG_ERR** = 1), or any error related to either post auto-verify (**AVPST** = 1) or pre auto-verify (**AVPRE** = 1) have occurred (refer to Table 6.14).
- In line 22, the source data stored in the **srcData[]** array is written into the flash memory starting at the address of **dest** (**0x1000**). The address constant **dest** is converted to the pointer by using the system macro **HWREG32()** to facilitate this data writing.
- Then a conditional **while()** loop is used in line 23 to check whether this data has been written into the flash by inspecting bit-3 in the **FLCTL_IFG** register. If not, this **while()** loop will wait there until this programming is complete.
- If this writing is complete, we need to check if any error has been encountered by inspecting bits 9 (**PRG_ERR**), 2 (**AVPST**), and 1 (**AVPRE**) on the **FLCTL_IFG** register in line 24.
- If any error has occurred, a false is returned to the main program in line 25.

- Otherwise the address **dest** is updated by adding it by 4 to point to the next address in the flash memory to make it ready for the next writing in line 26.
- In line 28, an infinitive **while()** loop is used to temporarily halt the running of the program to enable users to check these writing results via reading the contents of the related addresses in the flash memory.
- The codes in lines 30 through 38 are the function body of the user-defined function **InitFWData()** and these codes are used to fill out four 32-bit words into the **srcData[]** array. These four data are valued from **0x00000000**, **0x11111111**, **0x22222222** and **0x33333333**.

Now, perform the following operations to build this project:

1. Go to **Project|Build target** menu item to build the project to get the project map file, **DRAFlashWEWords.map**. The reason we need this map file is to determine the starting address for writing data in the flash memory block. Because in order to run our project, our project image file will also be downloaded into this flash memory. Therefore, we must make sure that the flash memory block we try to write is not the same block in which our project image file will be downloaded.
2. Open the Windows Explorer and browse to our project folder, which is **C:/MSP432 Class Projects/Chapter 6/DRAFlashWEWords/Listings**, and open the **DRAFlashWEWords.map** file that is under this folder.
3. Scroll down this map file until you find the memory map for our image file, which is indicated by: **Memory Map of the image** line, as shown in Figure 6.26.
4. It can be found from this map file that the download starting address of our project image file is **0x000001e5**, the loading and the running regions of our project image file start from base address **0x0** with the sizes of **0x038C** and **0x0388**. Therefore, it should not have any conflict if we select the **0x1000** as the starting address of the flash memory block to perform the writing operations since it is greater than **0x38C** that is the ending address of our stored project image file in the flash memory.

Now, we can download and run the project. However, before downloading and running this project, we must make sure:

1. The debugger used in this project is **CMSIS-DAP (Project|Options for Target "Target 1"|Debug** tab).
2. All debugger-related settings are correct.

We need to set up a breakpoint at the **while()** loop line (28) to enable us to check the contents of the Bank0 flash memory space to confirm our writing results. Go to **Debug|Insert/Remove Breakpoint** menu item to set up this breakpoint.

```
==============================================================================
Memory Map of the image

  Image Entry point: 0x000001e5
  Load Region LR_1 (Base: 0x00000000, Size: 0x0000038C, Max: 0xffffffff, ABSOLUTE)
    Execution Region ER_RO (Base: 0x00000000, Size: 0x00000388, Max: 0xffffffff, ABSOLUTE)
    ......
```

Figure 6.26 Opened DRAFlashWEWords.map file.

```
Memory 1
Address: 1000
0x00000FE2: FF FF FF FF FF FF FF FF FF FF FF FF FF FF FF FF FF FF FF FF FF
0x00000FF7: FF FF FF FF FF FF FF FF FF 00 00 00 00 11 11 11 11 22 22 22 22
0x0000100C: FF FF FF FF FF FF FF FF FF FF FF FF FF FF FF FF FF FF FF FF FF
0x00001021: FF FF FF FF FF FF FF FF FF FF FF FF FF FF FF FF FF FF FF FF FF
0x00001036: FF FF FF FF FF FF FF FF FF FF FF FF FF FF FF FF FF FF FF FF FF
0x0000104B: FF FF FF FF FF FF FF FF FF FF FF FF FF FF FF FF FF FF FF FF FF
0x00001060: FF FF FF FF FF FF FF FF FF FF FF FF FF FF FF FF FF FF FF FF FF
Call Stack + Locals | Memory 1
```

Figure 6.27 Contents on the flash Bank0 main memory.

Now you can build, download, and run the project, and the project will be run and halted at the **`while()`** loop line (28). Let's check the running results by performing the following operations:

1. Open the flash Bank0 main memory by clicking the **`Memory 1`** tab located at the lower right corner, and enter **`1000`** into the **`Address:`** box. Press the Enter key from your keyboard. The contents of the flash Bank0 main memory are displayed starting from some address, as shown in Figure 6.27.
2. Now scroll down the flash memory addresses until you find the address that is closed to **`0x1000`**, such as **`0x00000FF7`**. Then count to the real address **`0x1000`**, and you can find that three 32-bit words, **`0x00000000`**, **`0x11111111`**, and **`0x22222222`**, have been written into related addresses, as shown in Figure 6.27.
3. Click on the **`Debug|Start/Stop Debug Session`** menu item to stop the running of the project.

Our flash programming project is successful.

6.6.1.3 Flash memory burst programming projects with DRA model

In this section, we will build a flash memory write (program) project to perform burst programming function. We have provided detailed discussions about this kind of operational feature in Section 6.5.2.2.4. Two major control registers, Flash Program Burst Control and Status Register (**`FLCTL_PRGBRST_CTLSTAT`**) and Flash Program Burst Start Address Register (**`FLCTL_PRGBRST_STARTADDR`**), are used to configure and control this burst word programming operations. Refer to Figure 6.15 and Table 6.12 to get more details about these registers used for this mode.

In addition to those two control and address registers, 16 Flash Program Burst Data Input Registers (**`FLCTL_PRGBRST_DATAn_0~FLCTL_PRGBRST_DATAn_3`**, **`n`** = 0~3) are used to store sixteen 32-bit data or four 128-bit words to make it ready for this burst program operation.

In this project, we try to program sixteen 32-bit data into the flash main memory Bank0 space with the burst program method since this mode provides the ability to enable us to write a huge block of words (4 × 128 bits) into the flash memory with one operation.

Perform the following operations to create and build a new Keil µVersion5 project **DRAFlashBurstPrgm** in the location **C:\MSP432 Class Projects\Chapter 6** at your computer:

- Open the Windows Explorer and create a new folder **DRAFlashBurstPrgm** under the folder **C:\MSP432 Class Projects\Chapter 6** at your computer.
- Open the Keil µVersion5 MDK IDE and create a new µVersion5 project **DRAFlashBurstPrgm** and save this project to the folder **DRAFlashBurstPrgm** created above.
- Create a new C File **DRAFlashBurstPrgm.c** and add it into our project **DRAFlashBurstPrgm**.
- Add the code shown in Figure 6.28 into this C File.

```
1  //*****************************************************************************************
2  // DRAFlashBurstPrgm.c - Main application file for project DRAFlashBurstPrgm
3  //*****************************************************************************************
4  #include <stdint.h>
5  #include <stdbool.h>
6  #include <msp.h>
7  #define FL_PRGBRST_DATA0_0 0x40011060
8  void InitFBPData(void);
9  int main(void)
10 {
11   uint32_t ret;
12   FLCTL_PRGBRST_STARTADDR = 0x2000;      // setup the starting address as 0x2000
13   InitFBPData();                         // call InitFBPData() to fill 16 32-bit reg to
                                             get 4 128-bit data
14   FLCTL_BANK0_MAIN_WEPROT = 0;           // clear all write/erase protection functions
15   FLCTL_CLRIFG = 0x00000216;             // clear any IFG errors (PRGB Done |PreV | PostV
                                             |PRG Error)
16   FLCTL_PRG_CTLSTAT &= ~0x000000C0;      // clear program auto verify functions
17   FLCTL_PRGBRST_CTLSTAT = 0xE1;          // enable burst program with 4 128-bit data writing
18   while(FLCTL_IFG & 0x10){}              // Wait for the PRGB to be done......
19   if (FLCTL_IFG & 0x00000206)            // check all possible errors (PreV|PostV|PRG Error)
20     return false;                        // return false if any error occurred
21   ret = FLCTL_PRGBRST_CTLSTAT;           // Burst program is done. Get its running status
22   if (ret & 0x00380000)                  // check any error (ADDR_ERR, PST_ERR & PRE_ERR)
                                             occurred?
23     return false;                        // return false if any error occurred
24   FLCTL_PRGBRST_CTLSTAT |= 0x00800000;   // write 1 to CLR_STAT bit to clear 21-16 bits
                                             status
25     while(1);
26 }
27 void InitFBPData(void)
28 {
29   uint32_t uiLoop, wData = 0x0;
30   for(uiLoop = 0; uiLoop < 16; uiLoop++)
31   {
32       HWREG32(FL_PRGBRST_DATA0_0 + 4 * uiLoop) = wData;
33       wData += 0x11111111;
34   }
35 }
```

Figure 6.28 Detailed code for the project DRAFlashBurstPrgm.

Now, let's have a closer look at this piece of code to see how it works.

- Three system header files are declared first at code lines 4~6 since we need to use some system macros defined for the flash control registers in this project.
- A global macro **FL_PRGBRST_DATA0_0** and a global function **InitFBPData()** are declared in lines 7 and 8. The macro **FL_PRGBRST_DATA0_0** is defined as a constant starting address **0x40011060**, which is a nonpointer address of the first Flash Program Burst Data Input Register **FLCTL_PRGBRST_DATA0_0**. The reason we defined this nonpointer address is to make us easier to get and access all following 15 burst data registers by adding a 4-byte offset based on this starting nonpointer address. Then, we can use the system macro **HWREG32()** to convert these addresses to the pointers to assign sixteen 32-bit data to them. The user-defined function **InitFBPData()** is used to fill out sixteen 32-bit data to all 16 Flash Program Burst Data Input Registers.
- A local variable **ret** is declared inside the main program body in line 11. The variable **ret** works as a reading holder to keep the running status of this burst program operation.
- In line 12, the address **0x2000** in the flash memory is assigned to the Flash Program Burst Start Address Register (**FLCTL_PRGBRST_STARTADDR**) to enable it to work as the starting address from where the burst program will be performed.
- In line 13, the user-defined function **InitFBPData()** is called to fill out all 16 Flash Program Burst Data Input Registers with sixteen 32-bit data.
- Before any WE operation can be performed, the related Flash **WE** Protection Register must be cleared to remove those protections, since all 32 bits on those protection registers are set to 1 (protected) after a system reset. Since we try to write data into the Bank0 in the main flash memory, the related **WE** protection register for Bank0, **FLCTL_BANK0_MAIN_WEPROT**, must be cleared in line 14.
- In order to check this burst programming result, we need first to clear all previous setup bits on the **FLCTL_IFG** in line 15 by setting the related bits on the **FLCTL_CLRIFG** register. Bit-4 (**PRGB**) on the **FLCTL_IFG** register will be set to 1 when a burst program operation is complete. Also, bits 9 (**PRG_ERR**), 2 (**AVPST**), and 1 (**AVPRE**) on the **FLCTL_IFG** register would be set to 1 if any error related to programming operation (**PRG_ERR** = 1), or any error related to either post auto-verify (**AVPST** = 1) or pre auto-verify (**AVPRE** = 1) have occurred (refer to Table 6.14). Therefore, the command data **0x00000216** is assigned to the **FLCTL_CLRIFG** register to clear all previous possible errors.
- In line 16, the program auto-verify bits (**VER_PST** and **VER_PRE**) in the **FLCTL_PRG_CTLSTAT** register are cleared to avoid any possible disturbance for those bits' functions (refer to Table 6.11).
- The flash burst program operation is started at line 17 by assigning **0xE1** to the Flash Program Burst Control and Status Register (**FLCTL_PRGBRST_CTLSTAT**). The command data **0xE1** indicates that this burst program uses both auto-verify functions (**AUTO_PST** and **AUTO_PRE**) with programming function of four 128-bit data words and starts this burst program operation immediately.
- Then a conditional **while()** loop is used in line 18 to check whether this burst program has been completed by inspecting bit-4 (**PRGB**) in the **FLCTL_IFG** register. If this bit is 0, which means that this burst program has not been done, this **while()** loop will wait there until this burst programming is complete.
- If this burst writing is complete, we need to check if any error has been encountered by inspecting bits 9 (**PRG_ERR**), 2 (**AVPST**), and 1 (**AVPRE**) on the **FLCTL_IFG** register in line 19.

- If any error has occurred, a false is returned to the main program in line 20.
- The codes in lines 21 and 22 are used to check whether any other errors related to this burst programming operation have occurred. These errors include the address error **ADDR_ERR** (going to reserved memory space), postverify error **PST_ERR** and preverify error **PRE_ERR** in the **FLCTL_PRGBRST_CTLSTAT** register. The command code **0x00380000** covers these error bits.
- A **false** is returned if any of these errors is encountered in line 23.
- In line 24, a **1** is written to bit-23 (**CLR_STAT**) in the **FLCTL_PRGBRST_CTLSTAT** register to clear all six status bits (bits 21~16).
- An infinitive **while()** loop is used in line 25 to temporarily halt the running of the program to enable users to check these burst programming results via reading the contents of the related addresses in the flash memory.
- The codes in lines 27 through 35 are the function body of the user-defined function **InitFBPData()** and these codes are used to fill out all 16 Flash Program Burst Data Input Registers with sixteen 32-bit data. These 16 data are valued from **0x00000000**, **0x11111111**, **0x22222222** until **0xFFFFFFFF**. The address range for these data is **0x2000~0x203F**, totally 64-Byte space.

A point to be noted is line 32, where the system macro **HWREG32()** is used to convert a nonpointer address to a pointer address. The advantage of using this kind of nonpointer address is that each following sequential program burst data register's address can be obtained by adding a 32-bit address offset (4 × uiLoop) with the base address of the Flash Program Burst Data0 Input Register (**FL_PRGBRST_DATA0_0**). The coding job is greatly reduced by using this coding strategy.

Now, we can download and run the project. However, before downloading and running this project, we must make sure:

1. The debugger used in this project is **CMSIS-DAP (Project|Options for Target "Target 1"|Debug** tab).
2. All debugger-related settings are correct.

We need to set up a breakpoint at the **while()** loop line (25) to enable us to check our burst writing results. Go to **Debug|Insert/Remove Breakpoint** menu item to set up this breakpoint.

Now you can build, download, and run the project, and the project will be run and halted at the **while()** loop line (25). Let's check the running results by performing the following operations:

1. Open the flash Bank0 main memory by clicking the **Memory 1** tab located at the lower right corner, and enter **2000** into the **Address:** box. Press the Enter key from your keyboard. The contents of the flash Bank0 main memory are displayed starting from some address.
2. Now, scroll down the flash memory addresses until you find the address that is close to **0x2000**, such as **0x00001FF6**. Then, count to the real address **0x2000**, and you can find that sixteen 32-bit words, **0x00000000**, **0x11111111**, **0x22222222**...... until **0xFFFFFFFF** have been written into related addresses from **0x2000~0x203F**, as shown in Figure 6.29.
3. Click on the **Debug|Start/Stop Debug Session** menu item to stop the running of the project.

Our flash burst programming project is successful.

```
Memory 1
Address: 2000
0x00001FCA: FF FF FF FF FF FF FF FF FF FF FF FF FF FF FF FF FF FF FF FF FF FF
0x00001FE0: FF FF FF FF FF FF FF FF FF FF FF FF FF FF FF FF FF FF FF FF FF FF
0x00001FF6: FF FF FF FF FF FF FF FF FF FF 00 00 00 00 11 11 11 11 22 22 22 22
0x0000200C: 33 33 33 33 44 44 44 44 55 55 55 55 66 66 66 66 77 77 77 77 88 88
0x00002022: 88 88 99 99 99 99 AA AA AA AA BB BB BB BB CC CC CC CC DD DD DD DD
0x00002038: EE EE EE EE FF FF FF FF FF FF FF FF FF FF FF FF FF FF FF FF FF FF
0x0000204E: FF FF FF FF FF FF FF FF FF FF FF FF FF FF FF FF FF FF FF FF FF FF
0x00002064: FF FF FF FF FF FF FF FF FF FF FF FF FF FF FF FF FF FF FF FF FF FF
Call Stack + Locals | Memory 1
```

Figure 6.29 Contents on the flash Bank0 main memory.

6.6.1.4 Flash memory erasing projects with DRA model

In this section, we will build a flash memory sector erase project to perform four sectors erasing operations for the flash main memory space (Bank0 has 32 sectors = 128 KB).

We have provided detailed discussions about erasing operational mode in Section 6.5.2.3. Two erasing modes, the sector and mass erase, are provided by the MSP432P401R MCU system. Two registers, Flash Erase Control and Status Register (**FLCTL_ERASE_CTLSTAT**) and Flash Erase Sector Address Register (**FLCTL_ERASE_SECTADDR**), are used to configure and control both erasing operations. Refer to Figure 6.18 and Table 6.13 to get more details about these registers used for this mode.

We try to erase four sectors (16 KB with each sector = 4 KB) in the flash main memory Bank0 with the sector erase mode starting from **0x1000** and ending at **0x4FFF**.

Perform the following operations to create and build a new Keil µVersion5 project **DRAFlashErase** in the location **C:\MSP432 Class Projects\Chapter 6** at your computer:

- Open the Windows Explorer and create a new folder **DRAFlashErase** under the folder **C:\MSP432 Class Projects\Chapter 6** at your computer.
- Open the Keil µVersion5 MDK IDE and create a new µVersion5 project **DRAFlashErase** and save this project to the folder **DRAFlashErase** created above.
- Create a new C File **DRAFlashErase.c** and add it into our project **DRAFlashErase**.
- Add the code shown in Figure 6.30 into this C File.

Now, let's have a closer look at this piece of code to see how it works.

- Three system header files are declared first at code lines 4~6 since we need to use some system macros defined for the flash control registers in this project.
- Some local variables, **sector** and **secAddr**, are declared inside the main program body in line 9. The variable **sector** will work as a loop counter for the **for** loop to scan all four sectors starting from address **0x1000**, and the **secAddr** works as the starting address holder to keep the starting address (**0x1000**) to perform this erase operation.
- A **for** loop is used in line 10 to repeatedly perform erase operation for each sector, from sector 1 to sector 4.
- In order to check this sector erasing result, we need first to clear all previous setup bits on the **FLCTL_IFG** in line 12 by setting the related bits on the **FLCTL_CLRIFG**

```
1   //****************************************************************************
2   // DRAFlashErase.c - Main application file for project DRAFlashErase
3   //****************************************************************************
4   #include <stdint.h>
5   #include <stdbool.h>
6   #include <msp.h>

7   int main(void)
8   {
9    uint32_t sector, secAddr = 0x1000;

10   for(sector = 1; sector < 5; sector++)
11   {
12     FLCTL_CLRIFG = 0x00000226;                // Clear IFG errors (ERASE Done |PreV |
                                                  PostV |PRG Error)
13     FLCTL_BANK0_MAIN_WEPROT = 0;              // clear all write/erase protection
                                                  functions
14     FLCTL_ERASE_SECTADDR = secAddr * sector;  // set the starting address to be erased
                                                  for 4KB as interval.
15     FLCTL_ERASE_CTLSTAT |= 0x01;              // start the sector erase operation
16     while(FLCTL_IFG & 0x20){}                 // Wait for ERASE to be done......
17     if (FLCTL_ERASE_CTLSTAT & 0x00040000)     // sector erase is complete. Now check the
                                                  error on bit-18
18      return false;                            // ADDR_ERR is 1, an address error
                                                  occurred. Return false
19   }
20   FLCTL_ERASE_CTLSTAT =0x00080000 ;           // set CLR_STAT bit to clear 18-16 status
                                                  & disable erase

21   FLCTL_CLRIFG = 0x1;                         // clear RDBRST bit (bit-0) in IFG
22   FLCTL_RDBRST_FAILCNT = 0x0;                 // clear burst read/compare fail counter
                                                  to 0
23   FLCTL_RDBRST_STARTADDR = secAddr;           // start burst read/compare function to
                                                  verify erase
24   FLCTL_RDBRST_LEN = 0x4000;                  // set the compare length
25   FLCTL_RDBRST_CTLSTAT = 0x19;                // set flash - pattern: FFF...F & start
                                                  burst read/compare
26   while(FLCTL_IFG & 0x1){}                    // wait for burst read/compare to be
                                                  done......
27   if (FLCTL_RDBRST_CTLSTAT & 0x000C0000)      // check ADDR_ERR or CMP_ERR bits
28     return false;
29   FLCTL_RDBRST_CTLSTAT |= 0x00800000;         // set CLR_STAT bit to clear status
30   if (FLCTL_RDBRST_FAILCNT != 0)              // return false if any read/compare failed
31     return false;

32   while(1);
33  }
```

Figure 6.30 Detailed code for the project DRAFlashErase.

register. Bit-5 (**ERASE**) on the **FLCTL_IFG** register will be set to 1 when an erase operation is complete. Also, bits 9 (**PRG_ERR**), 2 (**AVPST**), and 1 (**AVPRE**) on the **FLCTL_IFG** register would be set to 1 if any error related to erasing operation (**PRG_ERR** = 1), or any error related to either post auto-verify (**AVPST** = 1) or pre auto-verify (**AVPRE** = 1) have occurred (refer to Table 6.14). Therefore, the command data **0x00000226** is assigned to the **FLCTL_CLRIFG** register to clear all previous possible errors and status.

- Before any WE operation can be performed, the related Flash **WE** Protection Register must be cleared to remove those protections, since all 32 bits on those protection registers are set to 1 (protected) after a system reset. Since we try to erase data from the Bank0 in the main flash memory, the related **WE** protection register for Bank0, **FLCTL_BANK0_MAIN_WEPROT**, must be cleared in line 13.

- In line 14, the starting address for each sector from which all sector data will be erased is assigned to the **FLCTL_ERASE_SECTADDR** register. Because the capacity of each sector is 4 KB, the starting address for each sector can be obtained by multiplying the initial starting address **0x1000** stored in the variable **secAddr** with the loop counter **sector**.
- Then the erase operation is started by assigning **0x01** to the **FLCTL_ERASE_CTLSTAT** register in line 15.
- A conditional **while()** loop is used to check whether this 4-sector erasing operation is complete by inspecting the **FLCTL_IFG** register in line 16, exactly the bit-5 (**ERASE**) in this register.
- If this erase operation is done, we need further to check if any address-related error has occurred for this erasing in line 17. This checking can be performed by inspecting bit-18 (**ADDR_ERR**) in the **FLCTL_ERASE_CTLSTAT** register. If this bit is set to 1, this means that the erasing operation is terminated due to attempted erase of reserved memory address. A **false** is returned to the main program if this error occurred in line 18.
- The codes between lines 21 and 31 are used to confirm this sector-erase operation by performing a burst read and compare operation. A detailed discussion about this operation mode has been given in Section 6.5.2.1.2. Refer to that part to get more details about this mode.
- In line 21, the bit-0 (**RDBRST**) in the **FLCTL_IFG** register is cleared by setting that bit via the **FLCTL_CLRIFG** register since this bit will be set to 1 when a burst read-compare operation is complete.
- Then, the Flash Read Burst-Compare Fail Count Register (**FLCTL_RDBRST_FAILCNT**) is cleared in line 22 to make sure that no previous burst read-compare error has existed. This step is important since we need to check this register to make sure that no error has been encountered after this burst read-compare operation. This register is used to count how many errors have occurred for this operation mode. A **0** in this register means that no error has occurred.
- In line 23, the starting address **0x1000** is assigned to the Flash Read Burst-Compare Start Address Register (**FLCTL_RDBRST_STARTADDR**) to make it ready for this operation mode.
- Similarly, the erasing length **0x5000** (bytes) is assigned to the Flash Read Burst/Compare Length Register (**FLCTL_RDBRST_LEN**) in line 24 as the amount of bytes to be erased. A point is that the erase starting address is **0x1000**, and the ending address should be **0x4FFF** since from **0x1000** to **0x4FFF**, it contains four sectors = 4 × 4 KB = 16 KB. However, the length for the four sectors is **0x4000** because **0x4000** = 16 KB.
- In line 25, the Flash Read Burst-Compare Control and Status Register (**FLCTL_RDBRST_CTLSTAT**) is configured by setting **0x19** to start this burst read-compare operation with the following properties:
  - Data compare pattern = **0xFFFFFFFF** (bit **DATA_CMP** = 1) since after an erase operation, all bits must be changed from 0 to 1
  - This burst read-compare operation will stop for the first mismatch (bit **STOP_FAIL** = 1), since we want this burst read-compare to stop if any error has occurred
  - Start this burst read-compare operation immediately (bit **START** = 1).
- Another conditional **while()** loop is used to check whether this burst read-compare operation is complete by inspecting the **FLCTL_IFG** register in line 26, exactly the bit-0 (**RDBRST**) in this register.

- If this burst read-compare operation is done, we need to further check if any address and compare-related errors have occurred for this operation in line 27. This checking can be performed by inspecting bit-19 (**ADDR_ERR**) and bit-18 (**CMP_ERR**) in the **FLCTL_RDBRST_CTLSTAT** register. If any of these bits is set to 1, which means that the burst read-compare operation is terminated due to either attempted erase of reserved memory address or a compare error. A **false** is returned to the main program if any error is occurred in line 28.
- In line 29, a **1** is written to bit-23 (**CLR_STAT**) in the **FLCTL_RDBRST_CTLSTAT** register to clear all four status bits (bits 19~16).
- Then, we need to check if any error or fail operation has been encountered for this burst read-compare operation by inspecting the content of the Flash Read Burst-Compare Fail Count Register (**FLCTL_RDBRST_FAILCNT**) in line 30. If this register contains a nonzero value, which means that at least one error has occurred, a **false** is returned to the main program to indicate this situation in line 31.
- An infinitive **while()** loop is used in line 32 to temporarily halt the running of the program to enable users to check these erase and burst read-compare results via reading the contents of the related addresses in the flash memory.

Before downloading and running this project, we must make sure:

1. The debugger used in this project is **CMSIS-DAP (Project|Options for Target "Target 1"|Debug** tab).
2. All debugger-related settings are correct.

We need to set up a breakpoint at the **while()** loop line (32) to enable us to check our running results. Go to **Debug | Insert/Remove Breakpoint** menu item to set up this breakpoint.

Now build, download, and run the project. The project will be run and halted at the **while()** loop line (32). You can check the running results by opening the contents of the flash memory addressed in the range of **0x1000~0x4FFF**. All bytes in this addresses range should be **0xFF**.

So far, all of these flash memory example projects are built with the DRA model. However, the MSPWare driver library also provides a set of API functions to assist users to build similar projects with the SD model. Now, let's take a look at these API functions.

6.6.2 *Use API functions in the MSPWare library to perform flash memory programming*

To facilitate developers to build and develop flash memory projects to read, program, and erase flash memory blocks, the MSPWare library provides a set of API functions. These APIs provide 25 functions for dealing with the on-chip flash. Functions are provided to read, program and erase the flash, configure the flash protection, and handle the flash interrupts.

The flash memory banks can be erased based on sector space interval or mass erase for entire flash memory space. When erasing, all sectors or entire flash memory space are set or changed from all 0s to all 1s. Each bank can be protected or marked as read only or execute only. The read-only banks cannot be erased or programmed, protecting the contents of those banks from being modified. Execute-only banks cannot be erased or programmed, and can only be read by the processor.

The flash memory banks can be programmed on immediate or full-word basis. Programming process can only make a bit from 1 to 0 (not from 0 to 1). With a flash burst program function, up to four 128-bit words can be programmed or written in a flash memory block at one operation.

The timing for the flash is automatically handled by the flash memory controller.

The flash controller can generate an interrupt when an invalid access is attempted, such as reading from execute-only flash. The flash memory controller can also generate an interrupt when a read, erase, or programming operation has completed.

All of these flash-related API functions are located in the MSPWare Peripheral Driver Library and contained at the file **flash.c** that is located at: **C:\ti\msp\MSPWare_2_21_1_00_39\driverlib\driverlib\MSP432P4xx**, with the header file **flash.h** containing the API declarations for those functions.

Most popular flash memory API functions can be divided into the following four groups:

1. Flash reading-related API functions
2. Flash programming-related API functions
3. Flash erasing-related API functions
4. Flash program-erase protection-related functions
5. Flash interrupt processing-related API functions

We will discuss these API functions group by group in the following sections.

6.6.2.1 Flash reading-related API functions

Seven API functions are involved in this flash reading group:

1. FlashCtl_setWaitState()
2. FlashCtl_getWaitState()
3. FlashCtl_setReadMode()
4. FlashCtl_getReadMode()
5. FlashCtl_enableReadBuffering()
6. FlashCtl_disableReadBuffering()
7. FlashCtl_verifyMemory()

The operational sequence of performing a read operation, either normal or buffering read, from a flash bank, should be:

- The appropriate wait states must be set using the **FlashCtl_setWaitState()** function before any reading operation can be performed. The current states can be retrieved by using the API function **FlashCtl_getWaitState()**.
- Then, the **FlashCtl_setReadMode()** function can be executed to configure the desired reading mode for the desired reading operation. The current read mode can be obtained by calling the **FlashCtl_getReadMode()** function.
- The API function **FlashCtl_enableReadBuffering()** should be called if a buffer reading operation is executed.
- The function **FlashCtl_verifyMemory()** can be used to perform a burst read-compare operation to verify the correctness of any erasing operation.

Table 6.16 lists these flash memory API functions.

Table 6.16 Flash reading-related API functions

API function	Parameter	Description
void **FlashCtl_setWait State(**uint32_t bank, uint32_t waitState**)**	**bank** is the flash bank to set wait state for **waitState** is the number of wait states to set	Valid values for the **bank** are: **FLASH_BANK0; FLASH_BANK1.** Valid values for **waitState** are: **0x0000~0x1111.**
uint32_t **FlashCtl_getWait State(**uint32_t bank**)**	**bank** is the flash bank to get wait state for	Return the set number of flash wait states for the given flash bank. Valid values for the **bank** are: **FLASH_BANK0; FLASH_BANK1**
bool **FlashCtl_setRead Mode(**uint32_t bank, uint32_t readMode**)**	**bank** is the flash bank to set read mode for **readMode** is the mode to be set for the flash bank Return a **True** means that the set mode is successful	Set the read mode to be used by flash read operations. Valid values for the **bank** are: **FLASH_BANK0; FLASH_BANK1.** Valid **readMode** values are: **FLASH_NORMAL_READ_MODE, FLASH_MARGIN0_READ_MODE, FLASH_MARGIN1_READ_MODE, FLASH_PROGRAM_VERIFY_ READ_MODE, FLASH_ERASE_VERIFY_READ_ MODE, FLASH_LEAKAGE_VERIFY_ READ_MODE, FLASH_MARGIN0B_READ_ MODE, FLASH_MARGIN1B_READ_MODE**
uint32_t **FlashCtl_getRead Mode(**uint32_t flashBank**)**	**flashbank** is the flash bank to get read mode from	Return the read mode to set. Valid returned mode values are identical with those used in **setMode()** above
void **FlashCtl_enableRead Buffering(**uint_fast8_t bank, uint_fast8_t method**)**	**bank** is the memory bank to enable read buffering **method** is the access type to enable read buffering	Enable read buffering on accesses to a specified bank of flash memory. Valid values for the **bank** are: **FLASH_BANK0; FLASH_BANK1.** Valid value for the **method** are: **FLASH_DATA_READ, FLASH_ INSTRUCTION_FETCH.**
void **FlashCtl_disableRead Buffering(**uint_fast8_t bank, uint_fast8_t method**)**	**bank** is the memory bank to disable read buffering **method** is the access type to disable read buffering	Disable read buffering on accesses to a specified bank of flash memory. Valid values for **bank** and **method** are identical with those used in the **enableReadBuffering()**
bool **FlashCtl_verify Memory(**void * verifyAddr, uint32_t length, uint_fast8_t pattern**)**	**verifyAddr** is the start address where verification will begin **length** is in bytes to verify based on the pattern **pattern** is the format which verification will check versus	Verify a given segment of memory based on either a high (1) or low (0) state. Valid values for pattern are: **FLASH_0_PATTERN, FLASH_1_ PATTERN**

When using these API functions to access and perform various flash operations, especially for the reading operations, the **`FlashCtl_setWaitState()`** should be called first to set up the wait states for the reading operations. The default wait state is 3 and it would be used as the wait state if this function is not executed before any reading operation is performed.

6.6.2.2 Flash programming-related API functions

Four API functions are involved in this flash programming group:

1. **`FlashCtl_enableWordProgramming()`**
2. **`FlashCtl_disableWordProgramming()`**
3. **`FlashCtl_ isWordProgrammingEnabled()`**
4. **`FlashCtl_programMemory()`**

Table 6.17 lists these flash API functions.

The operational sequence of performing a programming operation (only immediate word) to a flash bank should be:

- The API function **`FlashCtl_enableWordProgramming()`** should be called first to enable the program operation and set up the appropriate writing mode, either the immediate or the full-word mode, to the program operations.
- The API function **`FlashCtl_isWordProgrammingEnabled()`** is used to check and confirm the flash controller current working status to make sure that the desired flash memory bank has been enabled and ready to be programmed.
- The API function **`FlashCtl_programMemory()`** is used to perform immediate programming or writing a group of data into the flash memory banks. A point to be

Table 6.17 Flash programming-related API functions

API function	Parameter	Description
void **FlashCtl_ enableWordProgramming(**uint32_t mode**)**	**mode** is the programming mode to be performed	Enable word programming of the flash memory and set the mode of behavior when the flash write occurs. Valid values for the mode are: **FLASH_IMMEDIATE_WRITE_MODE** **FLASH_COLLATED_WRITE_MODE**
void **FlashCtl_ disableWord Programming(**void**)**	**void**	Disables word programming of flash memory
uint32_t **FlashCtl_ isWordProgrammingEnabled(**void**)**	**void** Returning a **0** means that the Word programming is disabled	Return if word programming mode is enabled. Valid returning values are: **FLASH_IMMEDIATE_WRITE_MODE** **FLASH_COLLATED_WRITE_MODE**
bool **FlashCtl_ programMemory (**void *src, void *dest, uint32_t length**)**	**src** is a pointer to the data source to program into flash **dest** is a pointer to the destination in flash to program **length** is the bytes to program	Program a portion of flash memory with the provided data. Return a **true** means the program is successful. Otherwise if a **false** is returned, the program is failed

noted when using this function to program data into the flash memory is that this function body has already included to call the following four other API functions:

- **FlashCtl_enableWordProgramming**() with the **FLASH_IMMEDIATE_WRITE_MODE** as the selected program mode.
- **FlashCtl_setProgramVerification()** with **FLASH_REGPRE|FLASH_REGPOST** as the arguments to set auto pre and postverify functions.
- **FlashCtl_clearProgramVerification()** with **FLASH_BURSTPOST|FLASH_BURSTPRE** as arguments to clear burst auto pre and postverify functions.
- **FlashCtl_clearInterruptFlag()** with **FLASH_PROGRAM_ERROR|FLASH_POSTVERIFY_FAILED|FLASH_PREVERIFY_FAILED|FLASH_WRDPRGM_COMPLETE** as arguments to clear previous interrupt flags set by program error, auto pre and postverify errors and word program complete actions.

Therefore, these four functions should not be used again when calling this function to perform programming a group of data to the flash memory.

6.6.2.3 Flash erasing-related API functions

Two API functions are involved in this flash erasing group:

1. **FlashCtl_eraseSector()**
2. **FlashCtl_performMassErase()**

The first function is used to sectors erasing and the second one is for entire flash memory space erasing operation. A point to be noted is that for the sector erase operation, the starting address must be a 128-bit boundary address (bits 3~0 must be 0).

Table 6.18 lists these flash memory erasing API functions.

6.6.2.4 Flash program-erase protection-related functions

Three API functions are involved in this flash program-erase protection group:

1. **FlashCtl_protectSector()**
2. **FlashCtl_unprotectSector()**
3. **FlashCtl_isSectorProtected()**

These functions are used to set or enable, reset or disable WE protection functions for the flash memory banks in the MSP432P401R MCU system.

Table 6.18 Flash erasing-related API functions

API function	Parameter	Description
bool **FlashCtl_eraseSector** (uint32_t addr)	**addr** is the starting address from which the sector erasing is performed	Erase a sector of MAIN or INFO flash memory. Note that the minimum allowed size that can be erased is a flash sector (4 KB on the MSP432 family). If an address is provided to this function which is not on a 4-KB boundary, the entire sector will still be erased A **true** is returned if this erasing is successful
bool **FlashCtl_performMassErase**(void)	**void**	Perform a mass erase on all unprotected flash sectors Protected sectors are ignored A **true** is returned if the erase is successful

A point to be noted is that before any flash program or erase operation can be performed, the **`FlashCtl_unprotectSector()`** API function should be executed to remove any protection for the desired flash banks or sectors from which the programming or erasing operations to be performed. Otherwise the desired program or erase operations cannot be executed as you desired because of the protected functions applied on those flash banks and sectors.

Table 6.19 lists these flash memory protection-related API functions.

Table 6.19 Flash program-erase protection-related API functions

API function	Parameter	Description
bool **FlashCtl_protectSector(**uint_fast8_t memSpace, uint32_t secMask**)**	**memSpace** is the value of the memory bank to enable program protection. Must be one of the following values: **FLASH_MAIN_MEMORY_SPACE_BANK0, FLASH_MAIN_MEMORY_SPACE_BANK1, FLASH_INFO_MEMORY_SPACE_BANK0, FLASH_INFO_MEMORY_SPACE_BANK1** **secMask** is a bit mask of the sectors to enable program protection. Must be a bitfield of the following values: **FLASH_SECTOR0,.... FLASH_SECTOR31**	Enable program protection on the given sector mask. This setting can be applied on a sector-wise bases on a given memory space (INFO or MAIN) A **true** is returned if the sectors protection is enabled. Otherwise a **false** is returned If multiple sectors are to be protected, an **OR** can be used to combine multiple **FLASH_SECTORn** together
bool **FlashCtl_unprotectSector(**uint_fast8_t memSpace, uint32_t secMask**)**	**memSpace** and **secMask** are identical with those used in the function **FlashCtl_protectSector().**	Disable program protection on the given sector mask. This setting can be applied on a sector-wise bases on a given memory space (INFO or MAIN) A **true** is returned if the sectors protection is disabled. Otherwise a **false** is returned If multiple sectors are to be unprotected, an **OR** can be used to combine multiple **FLASH_SECTORn** together
bool **FlashCtl_isSectorProtected(**uint_fast8_t memSpace, uint32_t secMask**)**	**memSpace** and **secMask** are identical with those used in the function **FlashCtl_protectSector()**.	Return the sector protection for given sector mask and memory space A **true** is returned if the sectors protection is enabled. Otherwise a **false** is returned

6.6.2.5 Flash interrupt processing-related API functions

Seven API functions are involved in this flash interrupt processing group:

1. **FlashCtl_registerInterrupt()**
2. **FlashCtl_unregisterInterrupt()**
3. **FlashCtl_enableInterrupt()**
4. **FlashCtl_disableInterrupt()**
5. **FlashCtl_clearInterruptFlag()**
6. **FlashCtl_getInterruptStatus()**
7. **FlashCtl_getEnabledInterruptStatus()**

Table 6.20 lists these flash memory interrupt processing-related API functions.

When using these functions to handle flash memory interrupt-related operations, the following points should be noted:

1. The flash memory interrupt handler should be registered first with the API function **FlashCtl_registerInterrupt()**. The handler can be built by the user in any name and declared in the project file.
2. The flash memory interrupt sources must be then enabled by using the API function **FlashCtl_enableInterrupt()** to allow the processor to unmask that interrupt source and to respond it.
3. All previous interrupt flags set in the **FLCTL_IFG** should be cleared by using the API function **FlashCtl_clearInterruptFlag()** to make the program ready for the new coming interrupts.
4. An infinite **while()** loop may be used in the program to wait for the interrupt to be occurred and processed by the interrupt handler.
5. It is the users' responsibility to build and code the interrupt handler to respond to and process the interrupt if it is occurred.
6. After an interrupt has been processed, the related interrupt flag set in the **FLCTL_IFG** register must be cleared by the users' program inside the interrupt handler.

A point to be noted is that when using the function **FlashCtl_registerInterrupt()** to register an interrupt handler to process any flash interrupt, this function has already included another two global interrupt functions, **Interrupt_registerInterrupt()** and **Interrupt_enableInterrupt()**. These two global interrupt register and interrupt enable functions are used to register and enable global interrupt registers to the NVIC to make the processor ready to respond to those interrupts. Therefore, the users do not need to call them again in their applications.

We have provided a detailed discussion about the API functions used in the MSPPDL, now let's build some example projects to illustrate how to use these functions to perform related flash memory operations.

6.6.3 Use API functions to build flash memory example projects with SD model

The so-called SD Model is to call API functions built in the MSPWare Peripheral Driver Library to perform related flash memory operations.

Table 6.20 Flash memory interrupt processing-related API functions

API function	Parameter	Description
void **FlashCtl_registerInterrupt(**void(*)(void) intHandler**)**	**intHandler** is a pointer to the function to be called when the flash operation interrupt occurs	Register an interrupt handler for flash memory operation system interrupt
void **FlashCtl_unregisterInterrupt(**void**)**	**void**	Unregisters the interrupt handler for the flash system
void **FlashCtl_enableInterrupt (**uint32_t flags**)**	**flags** is a bit mask of the interrupt sources to be enabled. Must be a logical OR of: **FLASH_PROGRAM_ERROR, FLASH_BENCHMARK_INT, FLASH_ERASE_COMPLETE, FLASH_BRSTPRGM_COMPLETE, FLASH_WRDPRGM_COMPLETE, FLASH_POSTVERIFY_FAILED, FLASH_PREVERIFY_FAILED, FLASH_BRSTRDCMP_COMPLETE**	Enable individual flash control interrupt sources
void **FlashCtl_disableInterrupt (**uint32_t flags**)**	**flags** is identical with those used in **enableInterrupt()**	Disable individual flash system interrupt sources
void **FlashCtl_clearInterruptFlag (**uint32_t flags**)**	**flags** is identical with those used in **enableInterrupt()**	Clear flash system interrupt sources. The processed interrupt must be cleared by calling this function inside the interrupt handler
uint32_t **FlashCtl_getInterruptStatus (**void**)**	**void**	Get the current interrupt status. The returned value can be any one that is identical with the **flags** used in **enableInterrupt()**
uint32_t **FlashCtl_getEnabledInterruptStatus(**void**)**	**void**	Get the current interrupt status masked with the enabled interrupts. This function is useful to call in ISRs to get a list of pending interrupts that are actually enabled and could have caused the ISR The returned value is same as those in the last function

Some important points must be kept in mind when using this model to build projects:

1. The API function-related header files, such as **`flash.h`**, must be included in the project to enable the compiler to locate these functions when it compiles the project.
2. The MSPWare Peripheral Driver Library file, **`driverlib.lib`**, must be added into the project to enable the linker to find those API function bodies when it links the project.

Now, let's build an example SD Model project **SDFlashWEWords** to perform flash memory write-word operation by using the related API functions in the MSPWare Driver Library. In this project, we try to use the immediate mode to write four 32-bit words located at the starting address **0x0** into the flash memory Bank1, exactly into the Sector0 space, at starting address **0x20000**.

Perform the following operations to create and build a new Keil μVersion5 project **SDFlashWEWords** in the location **C:\MSP432 Class Projects\Chapter 6** at your computer:

- Open the Windows Explorer and create a new folder **SDFlashWEWords** under the folder **C:\MSP432 Class Projects\Chapter 6** at your computer.
- Open the Keil μVersion5 MDK IDE and create a new μVersion5 project **SDFlashWEWords** and save this project to the folder **SDFlashWEWords** created above.
- Create a new C File **SDFlashWEWords.c** and add it into our project **SDFlashWEWords**.
- Add the codes shown in Figure 6.31 into this C File.

Now, let's have a closer look at this piece of code to see how it works.

- Five system header files are declared first at code lines 4~8 since we need to use some system macros defined for the flash control registers and some API functions defined in the related library in this project.

```
1   //*****************************************************************************
2   // SDFlashWEWords.c - Main application file for project SDFlashWEWords
3   //*****************************************************************************
4   #include <stdint.h>
5   #include <stdbool.h>
6   #include <msp.h>
7   #include <MSP432P4xx\flash.h>
8   #include <MSP432P4xx\wdt_a.h>
9   int main(void)
10  {
11    bool ret;
12    uint32_t src = 0x0, dest = 0x20000, dataLength = 16;// write 4 32-bit words to dest
13    void *srcAddr, *destAddr;
14    srcAddr = (void*)src;
15    destAddr = (void*)dest;
16    WDT_A_holdTimer();                  // stop WDT
17    P2DIR |= BIT0|BIT2;                 // configure P2.0 & P2.2 as output pins
18    P2OUT &= ~(BIT0|BIT2);              // disable P2.0 & P2.2 output
19    ret = FlashCtl_unprotectSector(FLASH_MAIN_MEMORY_SPACE_BANK1, FLASH_SECTOR0);
20    if (!ret)
21      return false;
22    ret = FlashCtl_programMemory(srcAddr, destAddr, dataLength);
23    if (!ret)
24      P2OUT |= BIT0;                    // turn on red LED P2.0 to indicate this error
25    else
26      P2OUT |= BIT2;                    // turn on blue LED P2.2 to indicate success
27    while(1);
28  }
```

Figure 6.31 Detailed code for the project SDFlashWEWords.

- Some local variables are declared inside the main program body in lines 11~13. The variable **`ret`** is a bool variable and it works as a status holder to receive and hold the running status of API functions. The **`src`** and **`dest`** are the source and destination addresses (**`0x0`** and **`0x20000`**) for the data to be read from and written into the flash memory. The **`dataLength`** is the number of bytes of data to be written into the flash sector0 area (four 32-bit words = 16 bytes).
- The code in line 13 is very important since the API function **`FlashCtl_programMemory()`** can only accept the void pointers as the source and the destination addresses, thus we need to create these two void pointers here.
- In lines 14 and 15, two addresses, **`src`** and **`dest`**, are converted to two void pointer variables, **`srcAddr`** and **`destAddr`**, respectively. These conversions are critical otherwise the API function **`FlashCtl_programMemory()`** cannot work without these conversions.
- A Watchdog timer-related API function **`WDT_A_holdTimer()`** is executed in line 16 to hold the Watchdog timer temporarily to avoid any unnecessary disturbance from it. The protocol of this function is defined in the header file **`wdt_a.h`**.
- The code line in 17 is used to configure GPIO Port 2, especially for **`P2.0`** and **`P2.2`** pins, to be output pins. Also, these two pins are disabled in line 18. We need to use these two pins to control two LEDs on the EXP-432P401R EVB to test our programming result later.
- As we discussed in Section 6.6.2.2, before any data can be programmed or written into the flash memory, the related flash memory banks and sectors must be unprotected first to remove those protections. For this purpose, the API function **`FlashCtl_unprotectSector()`** is called in line 19 to remove the protections for Bank1 and Sector0.
- The codes in lines 20 and 21 are used to check the status holder **`ret`** to make sure that the unprotect function is executed successfully.
- In line 22, the API function **`FlashCtl_programMemory()`** is called with the starting and destination addresses as well as the data length as input arguments to perform this four 32-bit word programming operation. Since this function will call another four API functions, including **`FlashCtl_enableWordProgramming()`**,**`FlashCtl_setProgramVerification()`**, **`FlashCtl_clearProgramVerification()`** and **`FlashCtl_clearInterruptFlag()`** (refer to Section 6.6.2.2), therefore we do not need to call these functions again. By the way, these functions are necessary to be executed before this program operation can be performed.
- The codes in lines 23~26 are used to check the execution of this program operation. The red-color LED would be turned on if an error is returned by executing this operation. Otherwise the blue-color LED would be on if that program operation is successful.
- The infinitive **`while(1)`** loop in line 27 is used to temporarily hold the program to enable users to check the programming results by inspecting the contents on the destination address in the flash memory.

Now we can compile, link, download, and run this project to test the programming function. However, before we can do that, we need to configure the environments to enable our project to be compiled and linked correctly. Three jobs are related to these configurations: (1) set up the include path for all system header files used in the project, (2) add the MSPPDL file into our project to enable us to use all related API functions, and (3) select the correct debugger to download and run the project.

6.6.3.1 Configuring the environments and run the project

Perform the following operations to include all system header files into our project:

- Go to **Project|Options for Target "Target 1"** menu item to open the options wizard.
- Then click on the **C/C++** tab and go to the **Include Paths** box.
- Click on the three-dot-button on the right of this box to open the **Folder Setup** wizard.
- Click on the **New** (**Insert**) tool to open a new textbox. Click on the three-dot-button on the right to browse to the location where these header files are located, which is **C:\ti\msp\MSPWare_2_21_00_39\driverlib\driverlib**. Then click on the **OK** button to select this location.
- Click on the **OK** button to close this wizard.

Perform the following operations to add the MSPPDL file **driverlib.lib** into the project:

- In the **Project** pane, expand the **Target** folder and right click on the **Source Group 1** folder and select the **Add Existing Files to Group "Source Group 1."**
- Browse to find the library, the MSPPDL, which is located at: **C:\ti\msp\MSPWare_2_21_00_39\driverlib\driverlib\MSP432P4xx\keil** in your host computer and the library is named **msp432p4xx_driverlib.lib**.
- Select this library by clicking on it and click on the **Add** button to add it into our project.

Perform the following operations to select the correct debugger for this project:

- Go to **Project|Options for Target "Target 1"** menu item to open the options wizard
- On the opened options wizard, click on the **Debug** tab.
- Make sure that the debugger shown in the **Use:** box is **CMSIS-DAP Debugger**. Otherwise you can click on the drop-down arrow to select this debugger from the list.
- Make sure that the debugger settings are correct by clicking on the **Settings** button located at the right of the debugger box. Then, click on the **Debug** tab to check both checkboxes: **Verify Code Download** and **Download to Flash** inside the **Download Options** group box. Then, click on the **Flash Download** tab to check **Erase Full Chip** radio button inside the **Download Function** group box.

We need to set up a breakpoint at the **while()** loop line (27) to enable us to check our running results. Go to **Debug|Insert/Remove Breakpoint** menu item to set up this breakpoint.

Now build, download, and run the project. The project will be run and halted at the **while()** loop line (27). You can check the running results by opening the contents of the related flash memory addresses. First, let's check the contents of the source data address located at **0x0000**, from which four words or 16 bytes are read and then written into the flash memory Bank1–Sector0 starting at **0x20000**.

Click on the **Memory 1** tab to open the flash memory and type **0** into the **Address:** box to check the contents of the source data, as shown in Figure 6.32.

```
Memory 1
Address: 0
0x00000000: 68 02 00 20 E5 01 00 00 ED 01 00 00 EF 01 00 00 F1 01 00 00 F3
0x00000015: 01 00 00 F5 01 00 00 00 00 00 00 00 00 00 00 00 00 00 00 00 00
0x0000002A: 00 00 F7 01 00 00 F9 01 00 00 00 00 00 00 FB 01 00 00 FD 01 00
0x0000003F: 00 FF 01 00 00 FF 01 00 00 FF 01 00 00 FF 01 00 00 FF 01 00 00
0x00000054: FF 01 00 00 FF 01 00 00 FF 01 00 00 FF 01 00 00 FF 01 00 00 FF
Call Stack + Locals | Memory 1
```

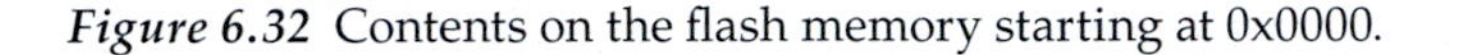

Figure 6.32 Contents on the flash memory starting at 0x0000.

```
Memory 1
Address: 99999
0x0001FFE9: FF FF FF FF FF FF FF FF FF FF FF FF FF FF FF FF FF FF
0x0001FFFB: FF FF FF FF FF 68 02 00 20 E5 01 00 00 ED 01 00 00 EF
0x0002000D: 01 00 00 FF FF FF FF FF FF FF FF FF FF FF FF FF FF FF
0x0002001F: FF FF FF FF FF FF FF FF FF FF FF FF FF FF FF FF FF FF
0x00020031: FF FF FF FF FF FF FF FF FF FF FF FF FF FF FF FF FF FF
0x00020043: FF FF FF FF FF FF FF FF FF FF FF FF FF FF FF FF FF FF
Call Stack + Locals | Memory 1
```

Figure 6.33 Contents on the flash memory starting at 0x20000.

Try to remember the first 16 bytes data in the flash memory shown in Figure 6.32. Then type **99999** into the **Address:** box and scroll down until the address is close to **0x20000**. Now you can check and compare the four 32-bit words or 16 bytes data starting at **0x20000** with those shown in Figure 6.33. They are identical, right? What conclusion you can get from this comparison?

Our SD Model project is successful.

We will leave some other flash operation-related projects developed by using the SD Model as lab projects for readers in the home work section that is located at the end of this chapter. Those projects involve flash interrupt processes functions and handlers.

Next, let's take a look at the entire memory map and some other memory properties.

6.7 Complete memory map in the MSP432P401R MCU system

We have provided detailed discussions about the memory map used in the MSP432P401R MCU system in Sections 2.3.2 in Chapter 2. Therefore, in this section, we just highlight some important sections in this map and devices mapped to this system.

Figure 6.34 shows a system memory map for the MSP432P401R MCU system.

It can be found from this memory map that the default flash memory capacity is 16 MB with an address range of **0x0000.0000~0x00FF.FFFF**. Currently, only a 256 KB flash memory space (**0x0000.0000~0x0003.FFFF**) is used for user's program codes and

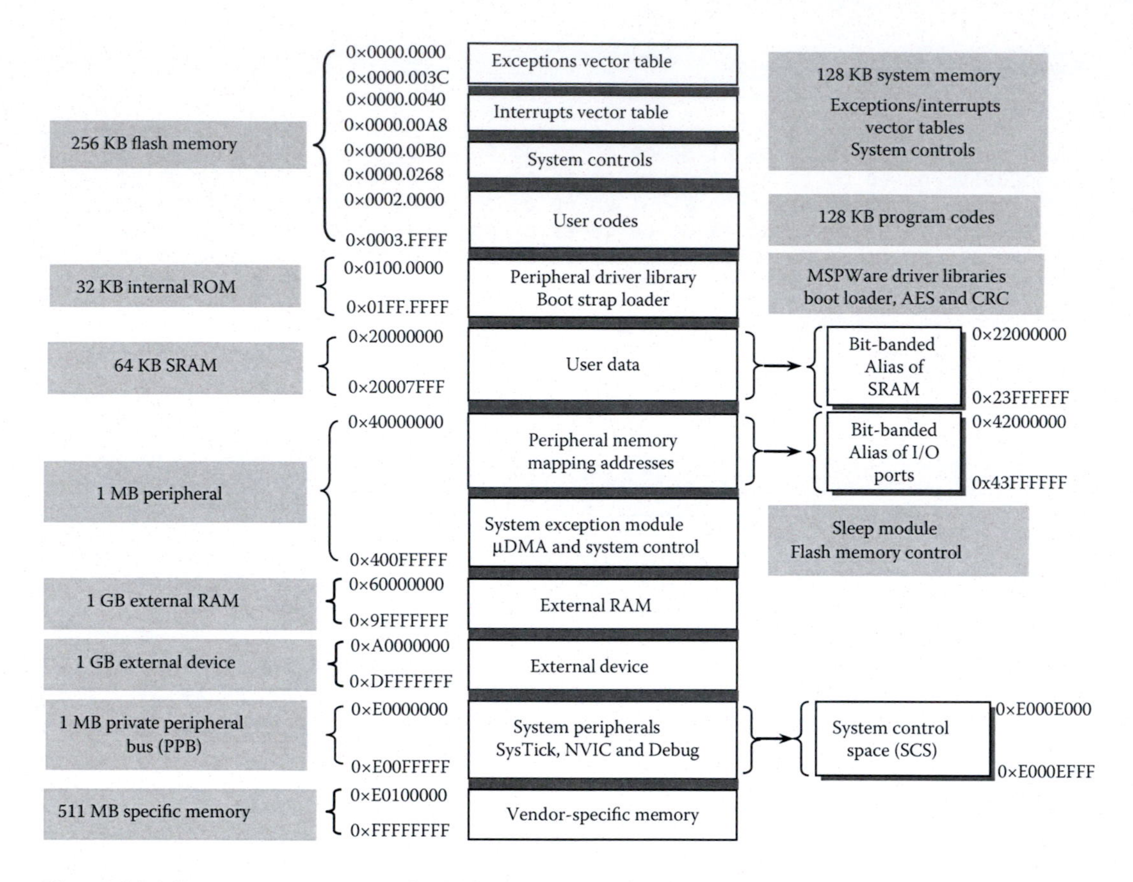

Figure 6.34 System memory map for MSP432P401R MCU.

exception/interrupt vector tables. Similarly, the default SRAM memory capacity is 512 KB, but currently only a 64 KB space is available to the users in this map. Additional flash memory and SRAM can be added if more memory spaces are needed for special applications. The memory space used for peripherals has the similar situations.

Regularly users should use flash memory to store their program codes and use SRAM to save their program data since the Cortex®-M4 processor uses the different system buses to fetch instructions and data synchronously. In this way, the application programs can be executed fast and effectively.

The 32 KB internal ROM is used to store MSPWare Peripheral Driver Library and BSL, AES, and CRC. The microcontroller can use this BSL as an initial program loader to load and run the user's program if the flash memory is empty.

Both 64 KB SRAM and 1 MB Peripheral memory spaces include bit-band regions. Bit banding provides atomic operations to bit data. A bit-band region maps each word in a **`bit-band alias`** region to a single bit in the **`bit-band region`**. The bit-band regions occupy the lowest 1 MB of the SRAM and peripheral memory regions. More detailed discussions about the bit-band operations can be found in Section 6.3.3.

Most system control-related registers are located at the memory space ranged from **`0x400F.9000`** to **`0xE00F.FFFF`** in this memory map. The last 511 MB space is reserved for the vendor-specific memory usage.

Table 6.21 provides a more detailed memory map for this MSP432P401R MCU system.

Table 6.21 Detailed memory map for MSP432P401R MCU

Start address	End address	Descriptions
		256 KB On-chip flash memory
0x0000.0000	0x0003.FFFF	Flash main memory (exceptions/interrupt vector table, system control and user codes)
0x0004.0000	0x001F.FFFF	Reserved
0x0020.0000	0x0020.3FFF	Flash information memory region (flash boot-override mailbox and device descriptor (TLV))
0x0020.4000	0x003F.FFFF	Reserved
		32 KB ROM
0x0200.0000	0x0200.03FF	Reserved for TI for future usage (1 KB)
0x0200.0400	0x020F.FFFF	MSP432 Driver Libraries, Boot Loader, AES & CRC
		64 KB SRAM
0x2000.0000	0x200F.FFFF	User data
0x2200.0000	0x23FF.FFFF	Bit-band alias of bit-banded on-chip SRAM starting at 0x2000.0000
		Peripherals
0x4000.0000	0x4000.03FF	Timer_A0
0x4000_0400	0x4000_07FF	Timer_A1
0x4000.0800	0x4000.0BFF	Timer_A2
0x4000.0C00	0x4000.0FFF	Timer_A3
0x4000.1000	0x4000.13FF	eUSCI_A0
0x4000.1400	0x4000.17FF	eUSCI_A1
0x4000.1800	0x4000.1BFF	eUSCI_A2
0x4000.1C00	0x4000.1FFF	eUSCI_A3
0x4000.2000	0x4000.23FF	eUSCI_B0
0x4000.2400	0x4000.27FF	eUSCI_B1
0x4000.2800	0x4000.2BFF	eUSCI_B2
0x4000.2C00	0x4000.2FFF	eUSCI_B3
0x4000.3000	0x4000.33FF	REF_A
0x4000.3400	0x4000.3BFF	COMP_E0 & COMP_E1
0x4000.3C00	0x4000.3FFF	AES256
0x4000.4000	0x4000.43FF	CRC32
0x4000.4400	0x4000.47FF	RTC_C
0x4000.4800	0x4000.4BFF	WDT_A
0x4000.4C00	0x4000.4FFF	Port module
0x4000.5000	0x4000.53FF	Port mapping controller

(*Continued*)

Table 6.21 (Continued) Detailed memory map for MSP432P401R MCU

Start address	End Address	Descriptions
0x4000.5400	0x4000.57FF	Capacitive touch I/O 0
0x4000.5800	0x4000.5BFF	Capacitive touch I/O 1
0x4000.C000	0x4000.CFFF	Timer32
0x4000.E000	0x4000.FFFF	DMA
0x4001.0000	0x4001.03FF	PCM
0x4001.0400	0x4001.07FF	CS
0x4001.0800	0x4001.0FFF	PSS
0x4001.1000	0x4001.17FF	Flash controller
0x4001.2000	0x4001.23FF	ADC14
0x4200.0000	0x43FF.FFFF	Bit-banded alias of 0x4000.0000 through 0x400F.FFFF
Debug-trace and private peripheral bus		
0xE000.0000	0xE000.2FFF	ITM, DWT, and FPB
0xE000.E000	0xE000.EFFF	Cortex-M4F system peripherals (SysTick, NVIC, MPU, FPU, and SCB)
0xE004.0000	0xE004.0FFF	Trace port interface unit (TPIU)
0xE004.2000	0xE004.23FF	Reset controller
0xE004.3000	0xE004.43FF	System controller
0xE00F.F000	0xE00F.FFFF	External PPB—ROM Table

6.8 Memory requirements and memory properties

The memory map and the programming of the MSP432P401R MPU split the memory map into different regions, as shown in Figure 6.34 and Table 6.21. Each region has a defined memory type, and some regions have additional memory attributes. The memory type and attributes determine the behavior of accesses to the region.

In order to correctly and effectively access the desired memory region to perform the selected operations, we need to have a clear and a full picture about the memory requirements and some important memory properties.

There are so many different memory properties applied on memory systems in most popular microcontroller systems. However, in this section we only limit our discussions on some most important properties, such as memory endianness, memory access attributes, and behavior of memory accesses. For some memory properties listed below, we will not cover them in this part since they are not very important from the point of view of most applications:

- Memory data alignment and unaligned data access
- Memory exclusive accesses
- Memory barriers
- Memory access permissions

First, let's take care of the memory requirements.

6.8.1 Memory requirements

In Section 6.2, we have provided a detailed discussion about the memory architecture used in the MSP432P401R MCU system. From Figure 6.1, it can be found that different types of memory can be connected to the processor and peripherals via different buses. Although the bus size is 32 bits in the Cortex®-M4 MCU, the memory can also be connected with other width in the bus, such as 8 bits, 16 bits, 64 bits, and 128 bits, and so on, if related conversion hardware exists.

In the MSP432P401R MCU system, the used memory are categorized as SRAM, flash memory and internal ROM, as shown in Figure 6.1. However, there is no real limitation on what kind of memory can be used in this system. For instance, the SRAM can be replaced by DRAM, SDRAM (synchronous DRAM) and DDR SDRAM (double data rate SDRAM), and so on. Also, the program codes could be in the flash memory, RAM, EEPROM, and even ROM.

The memory size used in a microcontroller system is also flexible. Some low-cost Cortex®-M MCUs have only 8 KB of flash memory and 4-KB on-chip SRAM. In the MSP432P401R MCU system, the default memory capacity for the SRAM is 512 MB, but only 64 KB SRAM is used for this system. Similarly, the default size for the flash memory is also 512 MB, however, only 256 KB has been used by the MSP432P401R MCU system.

The only requirement for the data memory, such as SRAM, is that the memory must be byte addressable and the memory interface needs to support byte, half-word, and word conversions.

For the memory spaces used for external RAM and external device, as shown in Figure 6.34, the user needs to refer to the specifications of the memory chips and peripherals connected to those spaces.

The most popular memory types include:

- **`Normal`**—The processor can change the order of transactions to improve the efficiency, or perform speculative reads.
- **`Device`**—The processor keeps transaction order relative to other transactions to the device or strongly ordered memory.
- **`Strongly Ordered`**—The processor preserves transaction order relative to all other transactions.

The different ordering requirements for the device and strongly ordered memory mean that the memory system can buffer a write to device memory, but must not buffer a write to strongly ordered memory.

The additional memory attribute is the **`Execute Never`** (**`XN`**). This means that the processor prevents instruction accesses. Any attempt to fetch an instruction from an XN region causes a memory management fault exception.

Table 6.22 shows the different requirements for the different memory regions used in the MSP432P401R MCU memory system.

The Code, SRAM, and external RAM regions can hold user's programs and data. These regions can be accessed by instructions. However, it is recommended that programs always use the code region. This is because the processor has separate buses that enable instruction fetches and data accesses to be occurred simultaneously.

6.8.2 Memory access attributes

In the last section we discussed the memory requirements, and different memory regions can be accessed with various methods. Besides these requirements, the memory map also

Table 6.22 Requirements to use different memory regions

Address range	Memory region	Memory type	XN	Description
0x00000000~ 0x1FFFFFFF	Code	Normal	—	Executable region for program code Can also put data here
0x20000000~ 0x3FFFFFFF	SRAM	Normal	—	Executable region for data. Can also put code here This region includes bit band and bit-band alias areas
0x40000000~ 0x5FFFFFFF	Peripheral	Device	XN	This region includes peripheral registers mapping & bit band and bit-band alias areas
0x60000000~ 0xDFFFFFFF	Reserved	Normal	—	Reserved in MSP432P401R
0xE0000000~ 0xE00FFFFF	Private Peripheral Bus	Strongly ordered	XN	This region includes the NVIC, System timer, and system control block
0xE0100000~ 0xFFFFFFFF	Reserved	—	—	This region is reserved in MSP432P401R

defines the memory attributes of the access. The most popular memory attributes used in the Cortex®-M4 processor system are:

- **`Bufferable`**—The instructions or data written to the memory can be carried out by a writing buffer, and the processor can continue to execute the next instruction.
- **`Cacheable`**—The data read out from the memory can be copied to a memory cache and it can be used when next time it is accessed. In this way, the data processing can be speeded up to improve the execution efficiency of instructions.
- **`Executable`**—The processor can fetch and execute instructions from this memory region.
- **`Sharable`**—Data in this memory space can be shared by multiple bus systems. The memory system needs to ensure coherency of data among different bus systems in the sharable memory space.

The processor bus interfaces send out the memory access attributes information to the memory controller for each instruction and data processing. If the MPU is present and the MPU region configurations are reprogrammed, the default memory attributes can be overridden.

Only the executable and bufferable attributes affect the operations of the most applications in Cortex®-M4 microcontroller systems. The cacheable and sharable attributes are generally used by a cache controller, as shown in Table 6.23.

However, the sharable attribute is often needed and used in systems where multiple cores or processors and multiple cache units work with cache coherency control. The cache controller needs to make sure that the value is coherent with other cache units when a sharable data is used.

Regularly, there is not any cache memory or cache controller available for Cortex®-M4 microcontroller. However, a cache unit can be added into the microcontroller to improve the memory accessing performances and behaviors.

Table 6.23 Memory attributes related to different memory types

Bufferable	Cacheable	Memory type
0	0	Strongly ordered. The processor must wait until the current instruction is complete before the next instruction can be executed
1	0	Device. The processor can use the write buffer to execute the current instruction while continuing to the next instructions unless the next instruction is also needs a memory access
0	1	Normal memory access with write through (WT) cache
1	1	Normal memory access with write back (WB) cache

6.8.3 Memory endianness

The so-called memory endianness is to define how to store or arrange data bytes into words in the memory system and registers. As you know, each data word comprises four data bytes. In Cortex®-M4 microcontroller system, there are two memory endian formats; the little endian and the big endian.

6.8.3.1 Little-endian format

In general, the Cortex®-M4 processor views memory as a linear collection of bytes numbered in ascending order starting from zero. For example, bytes 0~3 hold the first stored word, and bytes 4~7 hold the second stored word. In other words, the lowest byte should be located on the top of the memory space and the highest byte should be in the bottom of the memory space, just as shown in Figure 6.35a.

However, when each word is loaded from the memory and stored into a register, the lowest byte should be loaded to the lowest byte in a register, and the higher byte should be loaded into the higher byte in the register, as shown in Figure 6.35b.

It can be found from Figure 6.35a that the lowest byte (**byte0**) is located at the top or lowest memory address (**A + 0**), the low byte (**byte1**) is in the low address (**A + 1**), the high byte (**byte2**) is in the high address (**A + 2**) and the highest byte (**byte3**) is located at the bottom memory address (**A + 3**). These four bytes make up a single 32-bit word and stored in the memory in this way, which is called memory little-endian format.

When loading this word to a register, each byte is loaded into the corresponding byte location shown in Figure 6.35b.

In summary, in the little endian format, the processor stores the least significant byte (LSB) of a word at the lowest-numbered byte, and the most significant byte (MSB) at the highest-numbered byte.

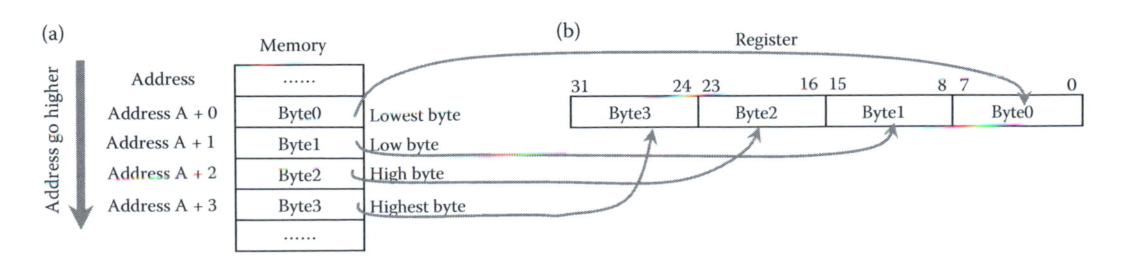

Figure 6.35 Memory little-endian format.

For most microcontroller systems, including the Cortex®-M4, the little-endian format is used.

6.8.3.2 Big-endian format

Optionally, the data can be stored in the memory in the big-endian format, which means that the bytes 4~7 hold the first stored word, and bytes 0~3 hold the second stored word. In other words, the highest byte is located on the top of the memory space (lowest address) and the lowest byte is in the bottom of the memory space (highest address), just as shown in Figure 6.36a.

However, when each word is loaded from the memory and stored into a register, the highest byte should be loaded to the highest byte in a register, and the lower byte should be loaded into the lower byte in the register, as shown in Figure 6.36b.

It can be found from Figure 6.36a that the highest byte (**byte3**) is located at the top or lowest memory address (**A + 0**), the high byte (**byte2**) is in the high address (**A + 1**), the low byte (**byte1**) is in the high address (**A + 2**), and the lowest byte (**byte0**) is located at the highest memory address (**A + 3**). These four bytes make up a single 32-bit word and stored in the memory in this way, which is called memory big-endian format.

When loading this word to a register, each byte is loaded into the corresponding byte location shown in Figure 6.36b.

In summary, in the big-endian format, the processor stores the MSB of a word at the lowest-numbered address, and the Least Significant Byte (LSB) at the highest-numbered address.

6.9 Summary

In this chapter, we discussed and analyzed all kinds of memory devices used in the MSP432P401R MCU system. The memory devices we discussed include:

- SRAM (64 KB)
- Internal ROM (32 KB)
- Flash memory (256 KB)

Starting with an introduction to the memory architecture used in the MSP432P401R MCU system, each memory device is discussed and analyzed in details with all related control registers used by those devices.

The SRAM is discussed first starting from Section 6.3. The SRAM memory map used in the MSP432P401R MCU system is discussed in Section 6.3.1. All SRAM-related control and status registers are introduced in Section 6.3.2. The bit-band operations implemented in the SRAM are introduced in Section 6.3.3.

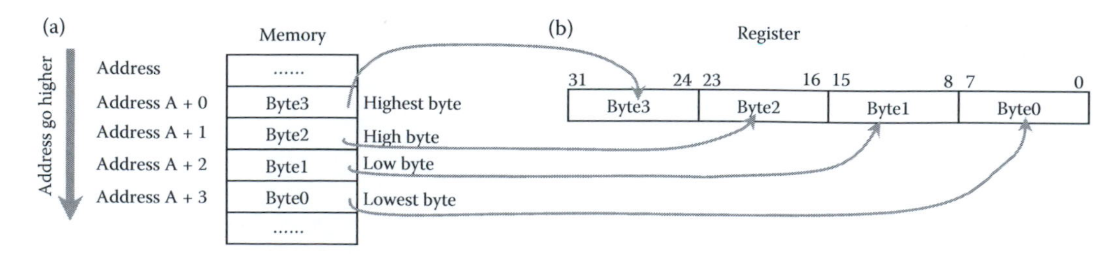

Figure 6.36 Memory big-endian format.

Starting from Section 6.4, a detailed discussion about the 32-KB internal ROM is given with the related materials stored in this region, including the MSPPDL and BSL.

The flash memory is introduced in Section 6.5. The details about the flash memory controller, including the flash controller operational clocking and addressing mode, all different flash memory operation modes, and related control and status registers, are introduced in the following sections. Various flash memory programming methods and projects are introduced in Section 6.6. These programming projects include:

- Flash reading project
- Flash programming project
- Flash burst programming project
- Flash erasing project
- Flash burst reading and comparison project

Using API functions provided by the MSPWare peripheral driver library to access and perform various flash memory operations are also discussed in Section 6.6.

A complete and detailed system memory map is then provided in Section 6.7 to give users a clear and global picture about the address ranges and locations for all memory components used in this MCU system.

The memory requirements and memory properties are introduced in Section 6.8, which include:

- Memory requirements
- Memory access attributes
- Memory little-endian and big-endian format

Five example projects related to most flash memory operations are included and discussed in detail in this chapter with line-by-line illustrations.

HOMEWORK

I. True/False Selections

____1. In the MSP432P401R MCU system, memory system works with little-endian format.

____2. The ARM® Cortex®-M4 provides a bit-band feature to enable read/write access to individual bits in one 1-MB SRAM region.

____3. The internal ROM and flash memory spaces are connected to the Cortex®-M4 processor via two bus interfaces, ICode and DCode.

____4. In the MSP432P401R MCU system, the memory system comprises 32-KB SRAM, 256-KB flash memory, and an internal 32-KB ROM device.

____5. The flash main memory is organized as 2 banks of 128-KB blocks that can be individually erased for each 4-KB sector, and both erasing and writing operations can change bits' values from 0 to 1.

____6. For an immediate data word writing operation, the flash memory address can be any address as long as it is located in the flash memory map range.

____7. Before any program or erase operation can be performed for the flash memory, the target flash memory space must be unprotected via the related **WEPROT** registers.

____8. The full-word 128-bit writes cannot start any programming operation until a whole 128-bit data is ready.

_____9. After a system reset, all flash memory spaces are unprotected, which means that all related **`WEPROT`** registers are reset to 0.

_____10. A bit-band region maps each word in a **`bit-band alias`** region to a single bit or the least significant bit (LSB) in the **`bit-band region`**. Since each word contains 32 bits, therefore the 1-MB bit-band region can be mapped to 32-MB bit-band alias regions.

II. Multiple Choices

1. The following buses can be used to access memory or peripheral devices in parallel ____________________________.
   a. ICode Bus, DCode Bus, System Bus
   b. DCode Bus, System Bus, AHB
   c. ICode Bus, DCode Bus, System Bus, AHB
   d. None of them
2. Which of the following units is an optional device in the Cortex®-M4 processor ______________?
   a. NVIC
   b. SysTick
   c. FPU
   d. MPU
3. Which of the following two memory devices can be accessed by using the ICode Bus and DCode Bus besides the system bus in the MSP432P401R MCU system ____________________________?
   a. Peripherals and SRAM
   b. Flash memory and internal ROM
   c. SRAM and flash memory
   d. Flash memory and EEPROM
4. The general connection between the Cortex®-M4 processor and peripherals is by ____________ protocol. However, the peripherals can also be accessed via the ____________ protocol directly by the processor.
   a. PPB, APB
   b. System Bus, APB
   c. APB, AHB
   d. Bus Bridge, AHB
5. When an erase operation is complete, the ______ bit in the ______ register is set.
   a. 3, FLCTL_CLRIFG
   b. 4, FLCTL_SETIFG
   c. 5, FLCTL_IFG
   d. 6, FLCTL_IFG
6. It is the application's responsibility to __________________ after a flash-related interrupt has been handled or processed.
   a. Enable that interrupt
   b. Register that interrupt
   c. Set that interrupt
   d. Clear that interrupt
7. When performing a burst program operation, the ____________________ functions should be cleared.
   a. Burst read compare
   b. Program auto pre and postverify

   c. Burst program auto pre and postverify
   d. Erase auto pre and postverify
8. Based on the bit-band equation in Section 6.3.3.1, the **0x20000000** bit[31] in the bit-band region can be mapped to ______________ in the bit-band alias region.
   a. **0x2000007C** bit[3]
   b. **0x2000007C** bit[2]
   c. **0x2000007C** bit[1]
   d. **0x2000007C** bit[0]
9. To clear a flash memory word program completion interrupt, one needs to set the ________________________ bit in the ________________________ register.
   a. 3, FLCTL_IFG
   b. 4, FLCTL_IFG
   c. 3, FLCTL_CLRIFG
   d. 4, FLCTL_CLRIFG
10. To erase the third sector, sector2, in the first bank of the flash main memory, the starting address in the **FLCTL_ERASE_SECTADDR** register should be ________________________.
   a. **0x0000**
   b. **0x1000**
   c. **0x2000**
   d. **0x3000**

III. Exercises
1. Provide a brief description about the memory architecture implemented in the MSP432P401R MCU system.
2. Provide a brief description about the flash memory used in the MSP432P401R MCU system.
3. Explain the operational process of the flash memory in the MSP432P401R MCU system.
4. Provide a description about the basic operations of the flash memory implemented in the MSP432P401R MCU system.
5. Provide a brief explanation about the flash memory interrupt process, including how to generate a flash memory interrupt and how to enable, handle and clear that interrupt.

IV. Practical Laboratory

Laboratory 6 MSP432P401R MCU memory system and device

6.0 GOALS

This laboratory exercise allows students to learn and practice ARM® Cortex®-M4 and MSP432P401R MCU memory system programming by developing five labs.

1. Program **Lab6_1**(**DRAFlashWEFullWords**) lets you build a flash memory full-word program project to use DRA model to write eight 128-bit (thirty-two 32-bit) words to the flash main memory Bank0 space.
2. Program **Lab6_2**(**DRAFlashIntBurstPrgm**) enables students to build a flash burst program project with the interrupt mechanism to write four 128-bit data into the flash main memory Bank0 space.

3. Program **Lab6_3(DRAFlashIntErase)** enables students to use DRA model to build a flash memory sector erase project with interrupt mechanism to erase 3 sectors in the flash main memory space.
4. Program **Lab6_4(SDFlashIntErase)** allows students to familiar and utilize the SD Model to build a flash memory erase project with interrupt function to erase 4 sectors from the flash main memory space.
5. Program **Lab6_5(SRAMBitBand)** allows students to familiar with the bit-band technology by building a bit-band related project to perform some bit-band operations.

After completion of these programs, students should be able to understand the fundamental architecture and operational procedure for most popular memory devices, including the flash memory and SRAM, installed in the MSP432P401R MCU system. Students should be able to code some basic programs to access the desired memory devices to perform popular operations, such as erasing, writing, and reading, on flash memory.

6.1 LAB6_1

6.1.1 Goal

In this project, students need to use DRA model to build a flash memory full-word program project to write eight 128-bit (thirty-two 32-bit) words to the flash main memory Bank0 space.

The Flash Program Control and Status Register (**FLCTL_PRG_CTLSTAT**) will be used in this lab project to facilitate the development of this program.

6.1.2 Data assignment and hardware configuration

No hardware configuration is needed in this project.

To use the DRA model to access related flash control register to perform full-word program operation, some related system header file, such as **msp.h**, should be included in the project.

6.1.3 Development of the project

Use the steps given below to develop this project. Only the C source file is used in this project since this project is not complicated. Create a new project with the following steps:

1. Create a new folder **Lab6_1(DRAFlashWEFullWords)** under the folder **C:\MSP432 Lab Projects\Chapter 6** in the Windows Explorer.
2. Create a new µVersion5 project named **DRAFlashWEFullWords** and save this project to the folder **Lab6_1(DRAFlashWEFullWords)** that is created in step 1 above.
3. On the next wizard, you need to select the device (MCU) for this project. Expand three icons, **Texas Instruments**, **MSP432 Family** and **MSP432P**, and select the target device **MSP432P401R** from the list by clicking on it. Click on the **OK** to close this wizard.
4. Next the software components wizard is opened, and you need to set up the software development environment for your project with this wizard. Expand two icons, **CMSIS** and **Device**, and check the **CORE** and **Startup** checkboxes in the **Sel.** column, and click on the **OK** button since we need these two components to build our project.

6.1.4 Development of the C source file

1. In the **Project** pane, expand the **Target** folder and right click on the **Source Group 1** folder and select the **Add New Item to Group "Source Group 1."**

2. Select the **C File (.c)** and enter **DRAFlashWEFullWords** into the **Name:** box, and click on the **Add** button to add this source file into the project.
3. On the top of this C source file, you need first to include three system header files, **<stdint.h>**, **<stdbool.h>**, and **<msp.h>** since we need to use them in this project.
4. Declare one global data array **uint32_t srcData[32]** and one global user-defined function **void InitFWData(void)**.
5. Start the main program with the code **int main(void)**.
6. Create the following local variables:

```
uint32_t n, ret, dest = 0x1000, dataLength = 32;
```

   where variable **n** works as the loop counter, **ret** as the value-holder, **dest** as the starting address from which eight 128-bit data will be written into the flash and **dataLength** defined the length of the data to be written in the flash Bank0 space.
7. The user-defined function **InitFWData()** should be called now to fill the data array **srcData[]** with thirty-two 32-bit predefined data.
8. Check the working status of the register **FLCTL_PRG_CTLSTAT** by reading its value and assigned it to the variable **ret**.
9. Use this read value stored in the **ret** with a **while()** loop to check whether bits 17:16 in the **FLCTL_PRG_CTLSTAT** register are 00 (the flash controller is idle). If not, wait here until both bits are 00.
10. After the **FLCTL_PRG_CTLSTAT** register is idle, configure it to perform:
    a. Auto-postprogram-verify
    b. Auto-preprogram-verify
    c. Full-word program mode
    d. Enable the full-word program to start immediately
11. Clear the burst program auto-verify functions by assigning an appropriate value to the **FLCTL_PRGBRST_CTLSTAT** register.
12. Return **1** if the returned **ret** value is nonzero. This means that the erase operation has something wrong.
13. A **while()** or a **for()** loop can be used to repeatedly write thirty-two 32-bit data into the flash main memory starting at **dest = 0x1000**. If a **while()** loop is used, the variable **n** can work as the loop condition (**n < dataLength**) to complete this loop writing.
14. Inside this **while()** loop, first the **WE** protecting function for all Bank0 in the flash main memory space should be disabled by assigning an appropriate value to the **FLCTL_BANK0_MAIN_WEPROT** register.
15. Then, all previous interrupt flags set in the **FLCTL_IFG** register should be cleared by assigning an appropriate value to the **FLCTL_CLRIFG** register.
16. Next, write the current 32-bit data stored in the data array **srcData[n]** to the flash main memory starting at the address **dest**. To perform this writing, the system macro **HWREG32(dest)** should be used to convert this constant address **dest** to an address pointer. Then the data in the data array **scrData[n]** can be assigned to the macro **HWREG32(dest)**.
17. Finally, another **while()** loop can be used to monitor and wait for this full-word writing to be complete. One can use the **FLCTL_IFG & 0x8** as the condition for this **while()** loop to wait for this writing to be done.
18. Still inside the **while()** loop, if a full-word program is done, we need to check all possible errors related to this writing operation. An **if()** selection structure can be used to check these errors by the condition **FLCTL_IFG & 0x00000206**.

19. If this condition is true, which means that some errors have occurred, a **false** is returned to the main program to indicate this situation. Otherwise, no error has occurred, then the target address **dest** is updated by adding 4 to it and the loop counter **n** is also updated by adding 1 to it for the next loop. In this way, the coding for the **while()** loop is complete.
20. Put another infinitive **while(1)** loop here and it can be used to check the full-word writing operation results.
21. The **main()** program is done at this point. Next, let's see the codes for the user-defined function **InitFWData()**.
22. Put the function header **void InitFWData(void)** to start this function.
23. Declare two local variables, **uiLoop** and **wData = 0x0**, both are **uint32_t** type.
24. Use a **for()** loop with the **uiLoop** as the loop counter. The loop counter starts from 0 but less than 32.
25. Inside the **for()** loop, create 32 data and fill them into the data array **srcData[]** by using the instruction likes **srcData[uiLoop] = wData;** Then update the **wData** by 1.

6.1.5 Set up the environment to build and run the project

To build and run the project, one needs to perform the following operations to set up the environments:

- Set up a breakpoint at the infinitive **while(1)** loop line (20). We need this breakpoint to temporarily stop the program to check our programming results.
- Select the **CMSIS-DAP Debugger** in the **Debug** tab under the **Project|Options for Target "Target 1"** menu item. Also make sure that the correct debugger setting and flash download setting are selected by checking related checkboxes and radio button (refer to Section 3.4.3.3 with Figures 3.24 and 3.25 in Chapter 3 to get more details about these settings).
- Now let's build the project by going to **Project|Build target** menu item.
- Then go to **Flash|Download** menu item to download the image file of this project into the flash memory in the MSP-EXP432P401R Evaluation Board.
- Now go to **Debug|Start/Stop Debug Session** to ready to run the project.
- Go to **Debug|Run** menu item to run the project.

6.1.6 Demonstrate your program by checking the running result

After the project runs, it will be temporarily stopped at the infinitive **while(1)** loop line. Now open the **Memory 1** window by clicking on it at the bottom of this screen. Type **3000** into the Address box, press the Enter key from your keyboard and scroll down to the address **0x1000**. Then you can find that our thirty-two 32-bit words, starting from **0x00000000** to **0x0000001F**, have been written into the address starting at **0x00001000** and ending at **0x0000107F** in the flash main memory space. But the order for these bytes is opposite to our normal word bytes.

Go to **Debug|Start/Stop Debug Session** to stop your program.

Based on these results, try to answer the following questions:

- By comparing the original 32 words created in the data array **srcData[32]** in our main program with the resulting 32 words written into the flash memory, what you can say? Why is the order of each byte opposite?
- In the flash main memory space, why have all other bytes have written as **FF**?
- What did you learn from this project?

6.2 LAB6_2

6.2.1 Goal

This project enables students to use the DRA model to build a flash burst program project with the interrupt mechanism to write four 128-bit data into the flash main memory Bank0 space.

6.2.2 Data assignment and hardware configuration

No hardware configurations are needed for this project.

In this project, we need to use the red and the blue-color LEDs installed on the MSP-EXP432P401R EVB as indicators to monitor and display the running error and result for this burst programming operation. Therefore, we need to correctly configure and enable these LEDs to work for these purposes. Also, the Watchdog timer should be temporarily stopped to avoid any unnecessary disturbance to our project.

Since the interrupt mechanism is used in this burst programming project, we also need to configure the related registers and develop the codes for the flash interrupt handler.

6.2.3 Development of the project

Follow the steps below to create and develop this project. Only a C code source file is used in this project since this project is simple. Create the project with the following steps:

1. Create a new folder **Lab6_2(DRAFlashIntBurstPrgm)** under the folder **C:\MSP432 Lab Projects\Chapter 6** in the Windows Explorer.
2. Open the Keil® ARM-MDK µVersion5, create a new project **DRAFlashIntBurstPrgm** and save this project into the folder **Lab6_2(DRAFlashIntBurstPrgm)** created in step 1.
3. On the next wizard, you need to select the device (MCU) for this project. Expand three icons, **Texas Instruments**, **MSP432 Family** and **MSP432P**, and select the target device **MSP432P401R** from the list by clicking on it. Click on the **OK** to close this wizard.
4. Next the software components wizard is opened, and you need to set up the software development environment for your project with this wizard. Expand two icons, **CMSIS** and **Device**, and check the **CORE** and **Startup** checkboxes in the **Sel.** column, and click on the **OK** button since we need these two components to build our project.

6.2.4 Development of the C source file

1. In the **Project** pane, expand the **Target** folder and right click on the **Source Group 1** folder and select the **Add New Item to Group "Source Group 1."**
2. Select the **C File (.c)** and enter **DRAFlashIntBurstPrgm** into the **Name:** box, and click on the **Add** button to add this source file into the project.
3. Include the following three system header files into this source file:
   a. **#include <stdint.h>**
   b. **#include <stdbool.h>**
   c. **#include <msp.h>**
4. First, define the base constant address for the flash burst program data registers using **#define FL_PRGBRST_DATA0_0 0x40011060**. There are 16 flash burst program registers arranged from **FLCTL_PRGBRST_DATA0_0** to **FLCTL_PRGBRST_DATA3_3**. Since we need to fill all of these 16 data registers with our sixteen 32-bit

target data to be programmed into the flash later, thus we need first to define the base address for these registers.

5. Next, we also need to define the user-defined function **InitFBPData()** since we need to use this function to fill sixteen 32-bit data into 16 flash burst program data registers. This function has a **void** input argument and returns the **void** data type.
6. Inside the **main()** program, first use the instruction **WDTCTL = WDTPW|WDTHOLD;** to temporarily stop the Watchdog timer's function to avoid any unnecessary disturbance.
7. Since we need to use the red- and the blue-color LEDs, thus we need to configure **P2.0** and **P2.2** as output pins since both pins are connected to these two LEDs.
8. Then you may disable pins **P2.0** and **P2.2** to block any default outputs.
9. Now assign the **FLCTL_PRGBRST_STARTADDR** register with appropriate value to set up the starting address as **0x2000** to that register.
10. Call the user-defined function **InitFBPData()** to fill sixteen 32-bit flash burst program data registers with 16 predefined 32-bit data.
11. Clear all WE protection functions for flash Bank0 by assigning an appropriate value to the **FLCTL_BANK0_MAIN_WEPROT** register.
12. Clear related error or status flags set on the **FLCTL _ IFG** register by assigning an appropriate value to it, these flags include **PRGB Done**, **PreV**, **PostV** and **PRG Error**. An **OR** operator (|) can be used to combined all of these flags together to reset them.
13. Clear both program auto-verify functions, **VER_POST** and **VER_PRE**, by assigning an appropriate value to the **FLCTL_PRG_CTLSTAT** register.
14. Enable burst program related interrupts by assigning an appropriate value to the FLCTL_IE register. These interrupts include **PRGB Done**, **PreV**, **PostV** and **PRG Error**.
15. Enable the burst program with four 128-bit data writing by assigning an appropriate value to the **FLCTL_PRGBRST_CTLSTAT** register.
16. Set flash interrupt request level as 3 (**011**) by assigning an appropriate value to the IPR **NVIC_IPR1**.
17. Enable the flash interrupt sources by assigning an appropriate value to the interrupt enable register **NVIC_ISER0** (the flash interrupt number is 5—refer to Table 5.11).
18. Globally, enable the flash Interrupt Request (IRQ) by calling the CMSIS intrinsic function **__enable_irq()**.
19. Use an infinitive **while(1)** loop to wait for any flash interrupt to be occurred.

The following codes are for the user-defined function **void InitFBPData(void)**:

20. Inside the function, first declare two **uint32_t** variables, **uiLoop** and **wData = 0x0**. The first variable is used as the loop counter and the second works as a value holder to hold the data for 16 flash burst program data registers.
21. Then a **for()** loop is used to repeatedly assign each data item to the 16 flash burst program data registers. The variable **uiLoop** works as the loop counter starting with 0 and ending with 16.
22. Inside the **for()** loop, each of 16 flash burst program data registers is filled with a data item stored in the variable **wData** by using the system macro as:

```
HWREG32(FL_PRGBRST_DATA0_0 + 4 * uiLoop) = wData;
```

23. Then each data item in **wData** is incremented by **0x11111111**. Thus, the total sixteen 32-bit data are arranged from **0x00000000~0xFFFFFFFF**.

The following codes are for the flash interrupt handler—**void FLCTL_IRQHandler (void)**:

24. Inside the interrupt handler, first we need to check whether any interrupt is caused by an error during the flash programming process by using an **if()** selection structure. The condition for this **if()** structure is the **AND** of the **FLCTL_IFG** register and an appropriate value (**FLCTL_IFG & Value**). The possible errors include the auto-post and auto-preverify (**PostV** and **PreV**) errors and program error (**PRG ERR**).
25. If this interrupt is triggered by any of these three errors, the related error flag set in the **FLCTL_IFG** register is cleared by assigning an appropriate value to the **FLCTL_CLRIFG** register. Also, the red-color LED is turned on to indicate this situation by using the instruction **P2OUT = BIT0;**
26. If this interrupt is not triggered by any error, next we need to check if this interrupt is caused by the completion of a flash burst program by using another **if()** selection structure. The condition used for this **if()** structure is the **AND** of the **FLCTL_IFG** register with an appropriate value.
27. If this interrupt is triggered by the completion of a flash burst program operation, the flag set for this interrupt on the **FLCTL_IFG** should be cleared by assigning an appropriate value to the **FLCTL_CLRIFG** register. Also, the blue-color LED should be turned on by using the instruction **P2OUT = BIT2;**
28. Finally, a 1 is written to the **CLR_STAT** bit in the **FLCTL_PRGBRST_CTLSTAT** register to clear 21–16 bits status on this register.

6.2.5 Set up the environment to build and run the project

To build and run the project, one needs to perform the following operations to set up the environments:

- Select the **CMSIS-DAP Debugger** in the **Debug** tab under the **Project|Options for Target "Target 1"** menu item. Also, make sure that correct debugger setting and flash download setting are selected by checking related checkboxes and radio button (refer to Section 3.4.3.3 with Figures 3.24 and 3.25 in Chapter 3 to get more detail about these settings).
- Now, let's build the project by going to **Project|Build target** menu item.
- Then go to **Flash|Download** menu item to download the image file of this project into the flash memory in the MSP-EXP432P401R Evaluation Board.
- Now go to **Debug|Start/Stop Debug Session** to ready to run the project.
- Go to **Debug|Run** menu item to run the project.

6.2.6 Demonstrate your program by checking the running result

After the project running, the blue-color LED will be on to indicate that an interrupt has been triggered by the completion of the flash burst program and sixteen 32-bit data have been written into the flash main memory space starting at address **0x2000**.

Now go to **Debug** menu item and click the **Stop** item to halt the project.

Next, we need to check the burst programming results by opening the flash memory. Click on the **Memory 1** tab and enter **8000** into the **Address** box, and then scroll down to find the address **0x00002000**. You can find that all sixteen 32-bit data, from **0x00000000** to **0xFFFFFFFF**, have been successfully programmed into the flash main memory with an address range of **0x000020000~0x0000203F**, totally 64 bytes or 4 × 128 bits.

Go to **Debug|Start/Stop Debug Session** menu item to stop the project.
Based on these results, try to answer the following questions:

- What is the meaning of the **HWREG32(FL_PRGBRST_DATA0_0+4*uiLoop)=wData;** instruction in the user-defined function **InitFBPData()**? Why each **uiLoop** must be multiplied by 4?
- What is the advantage of using this system macro **HWREG32()** to assign each data to each flash burst program data register?
- What did you learn from this project?

6.3 LAB6_3

6.3.1 Goal

This project allows students to use DRA model to build a flash memory sector erase project with interrupt mechanism to erase 3 sectors in the flash main memory space.

6.3.2 Data assignment and hardware configuration

No hardware configuration and data assignment are needed for this project.

6.3.3 Development of the project

Follow the steps given below to create and develop this project. Only a C code source file is used in this project since this project is simple. Create the project with the following steps:

1. Create a new folder **Lab6_3(DRAFlashIntErase)** under the folder **C:\MSP432 Lab Projects\ Chapter 6** in the Windows Explorer.
2. Open the Keil® ARM-MDK µVersion5, create a new project **DRAFlashIntErase** and save this project into the folder **Lab6_3(DRAFlashIntErase)** created in step 1.
3. On the next wizard, you need to select the device (MCU) for this project. Expand three icons, **Texas Instruments**, **MSP432 Family** and **MSP432P**, and select the target device **MSP432P401R** from the list by clicking on it. Click on the **OK** to close this wizard.
4. Next, the software components wizard is opened, and you need to set up the software development environment for your project with this wizard. Expand two icons, **CMSIS** and **Device**, and check the **CORE** and **Startup** checkboxes in the **Sel.** column, and click on the **OK** button since we need these two components to build our project.

6.3.4 Development of the C source file

1. In the **Project** pane, expand the **Target** folder and right click on the **Source Group 1** folder and select the **Add New Item to Group "Source Group 1."**
2. Select the **C File (.c)** and enter **DRAFlashIntErase** into the **Name:** box, and click on the **Add** button to add this source file into the project.
3. Include the following header files into this source file first:
   a. **#include <stdint.h>**
   b. **#include <stdbool.h>**
   c. **#include <msp.h>**
4. Place the **int main(void)** to start our main program.
5. Declare one unsigned 32-bit integer local variable, **secAddr = 0x1000**. This variable stored the starting address from which three sectors in the flash main memory will

be erased. In fact, this address **0x1000** is the starting address of the first sector (sector1) in the flash main memory since we want to erase three sectors, Sector1~Sector3, in this project.

6. Stop the Watchdog timer by using the instruction **WDTCTL = WDTPW|WDTHOLD;**
7. Since we need to use red-color and green-color LEDs in this project to indicate the running status of this project, configure GPIO pins **P2.0** and **P2.1** as output pins (**P2DIR|=BIT0|BIT1;**)
8. Disable the outputs of the GPIO pins **P2.0** and **P2.1** (**P2OUT &=~(BIT0|BIT1);**)
9. Clear previous error flags (**ERASE Done |PreV | PostV |PRG Error**) in the **FLCTL_IFG** register by assigning an appropriate value to the **FLCTL_CLRIFG** register.
10. Clear WE protections for Sector1~Sector3 in the flash main memory by assigning an appropriate value to the **FLCTL_BANK0_MAIN_WEPROT** register.
11. Start a mass erase operation by assigning an appropriate value to the register **FLCTL_ERASE_CTLSTAT**.
12. Use a conditional **while()** loop to wait for the mass erase operation to be completed. The condition is an **AND** of the **FLCTL_IFG** with a value (**FLCTL_IFG & Value**). If this condition is **false**, which means that the erase operation has not been done, this **while()** loop will keep its waiting until this condition becomes to true (the erase is done).
13. If the erase operation is done, next we need to check if any error is involved in this erase. Use an **if()** structure to check all possible errors, including the PreV, PostV, and PRG error. The condition for this **if()** structure should be an **AND** of the **FLCTL_IFG** with a value that contains all possible errors (**FLCTL_IFG & Value**).
14. If any error has occurred, the related error flag set in the **FLCTL_IFG** register must be cleared by assigning an appropriate value to the **FLCTLCLR_IFG** register. Also, the red-color LED is turned on to indicate this error by using the instruction **P2OUT = BIT0;**
15. A 1 is written to the **CLR_STAT** bit in the **FLCTL_ERASE_CTLSTAT** register to clear 18–16 bits status in that register.

Next, we need to use the flash burst read-compare function to read, compare, and verify our erase operation performed above. The so-called read-compare function is exactly to read the contents of all 3 sectors and compare them with **0xFF** for each byte since after each erase, each bit is changed from 0 to 1. Therefore, all bytes in those selected three sectors should be **0xFF** after a successful erase operation.

16. First, clear the **RDBRST** bit in **FLCTL_IFG** register by assigning an appropriate value to the **FLCTLCLR_IFG** register to make it ready for this read-compare operation.
17. Then, clear the Burst Read-Compare Fail Counter Register (**FLCTL_RDBRST_FAILCNT**) to 0.
18. Assign the starting erase address stored in the **secAddr** variable to the Flash Read Burst-Compare Start Address Register (**FLCTL_RDBRST_STARTADDR**).
19. Assign the compare length **0x3000** to the Flash Read Burst-Compare Length Register (**FLCTL_RDBRST_LEN**) since we need to check three erased sectors, Sector1~Sector3, with an address range of **0x1000~0x3FFF**.
20. Set the main flash compare pattern as **0xFFF...FFFFF** and start this burst read-compare operation by assigning an appropriate value to the **FLCTL_RDBRST_CTLSTAT** register. Since after all bytes are erased in three sectors, the contents of all bytes should be **0xFF**.

21. Enable the burst read-compare and error-related interrupts by assigning an appropriate value to the **FLCTL_IE** register.
22. Set the flash IRQ level as 3 by using the instruction **NVIC_IPR1 = 0x00006000;**
23. Enable the flash IRQ by using the instruction **NVIC_ISER0=0x20;**
24. Globally enable all interrupts using CMSIS intrinsic function **__enable_irq();**
25. Use an infinitive **while()** loop to wait for any interrupt to be occurred.

The following codes are for the flash interrupt handler:

26. First, we need to check all possible errors (**PreV|PostV|PRG Error**) to identify whether this interrupt is triggered by an error by using an **if()** structure. The condition for this **if()** structure is an **AND** of the **FLCTL_IFG** with a **Value** (**FLCTL_IFG & Value**) related to those errors.
27. If this interrupt is caused by any error, the related error flag set in the **FLCTL_IFG** register must be cleared by assigning an appropriate value to the **FLCTL_CLRIFG** register.
28. The red-color LED is turned on by using the instruction **P2OUT = BIT0;** to indicate this error.
29. If this interrupt is not triggered by an error, we need furthermore to check whether this interrupt is caused by a completion of the burst read-compare operation. Similarly, an **if()** structure is used with an **AND** of the **FLCTL_IFG** register with a **Value** as the condition.
30. If this interrupt is triggered by a completion of the burst read-compare operation, the following five operations should be performed:
    a. Clear the flag set by this completion of the burst read-compare operation by assigning an appropriate value to the **FLCTL_CLRIFG** register.
    b. Check if an **ADDR_ERR** or a **CMP_ERR** have occurred by **AND**ing an appropriate value with the **FLCTL_RDBRST_CTLSTAT** register (**FLCTL_RDBRST_CTLSTAT & Value**) via an **if()** structure. If any error has occurred, turn on the red-color LED to indicate this.
    c. Use another **if()** structure to check if the Burst Read-Compare Fail Counter Register (**FLCTL_RDBRST_FAILCNT**) is equal to 0. If not, at least one compare error has been met, the red-color LED is turned on to indicate this error.
    d. Otherwise no error occurred, the green-color LED is turned on (**P2OUT = BIT1;**) to indicate that our burst read-compare operation is successful, and therefore our erase operation is successful, too.
    e. Set the **CLR_STAT** bit in **FLCTL_RDBRST_CTLSTAT** to clear all status in that register.

6.3.5 Set up the environment to build and run the project

Refer to Lab6_2 (**6.2.5**) to complete this environment setup and run the project.

6.3.6 Demonstrate your program

After the project running, the green-color LED will be on to indicate that an interrupt has been triggered by the completion of the flash burst read-compare operation and three sectors in the flash main memory have been successfully erased starting at the address **0x1000**.

Now go to **Debug** menu item and click the **Stop** item to halt the project.

Next, we need to check the erasing results by opening the flash memory. Click on the **Memory 1** tab and enter **4000** into the **Address** box, and then scroll down to find the address **0x00001000**. You can find that all bytes in three sectors addressed from **0x00001000** to **0x00003FFF**, have been successfully erased from 0s to 1s the flash main memory.

Go to **Debug|Start/Stop Debug Session** menu item to stop the project.

Based on these results, try to answer the following questions:

- Did you pay attention to our codes used to erase the sectors? Why did we not directly erase those three sectors, instead we used a mass erase operation?
- How do we configure the **FLCTL_BANK0_MAIN_WEPROT** register to only remove the WE protection functions for Sector1~Sector3 in the flash main memory?
- What did you learn from this project?

6.4 LAB6_4

6.4.1 Goal

In this project, students need to use API functions provided by the MSPPDL to build a flash erase project with interrupt function to erase four sectors from the flash main memory space by using the SD programming model.

These API functions are located at the MSPPDL **driverlib.lib** file. Refer to Table 6.18 in Section 6.6.2.3 and Table 6.19 in Section 6.6.2.5 to get more details about these functions.

6.4.2 Data assignment and hardware configuration

No hardware configuration is needed in this project.

To use these API functions, one needs to include the related system header files, **flash.h**, which contains all definitions for macros and functions used in those API functions.

6.4.3 Development of the project

Follow steps below to create and develop this project. Only a C code source file is used in this project since this project is simple. Create the project with the following steps:

1. Create a new folder **Lab6_4(SDFlashIntErase)** under the folder **C:\MSP432 Lab Projects\ Chapter 6** in the Windows Explorer.
2. Open the Keil® ARM-MDK µVersion5, create a new project **SDFlashIntErase** and save this project into the folder **Lab6_4(SDFlashIntErase)** created in step 1.
3. On the next wizard, you need to select the device (MCU) for this project. Expand three icons, **Texas Instruments**, **MSP432 Family** and **MSP432P**, and select the target device **MSP432P401R** from the list by clicking on it. Click on the **OK** to close this wizard.
4. Next the software components wizard is opened, and you need to set up the software development environment for your project with this wizard. Expand two icons, **CMSIS** and **Device**, and check the **CORE** and **Startup** checkboxes in the **Sel.** column, and click on the **OK** button since we need these two components to build our project.

6.4.4 Development of the C source file

1. In the **Project** pane, expand the **Target** folder and right click on the **Source Group 1** folder and select the **Add New Item to Group "Source Group 1."**

2. Select the **C File (.c)** and enter **SDFlashIntErase** into the **Name:** box, and click on the **Add** button to add this source file into the project.
3. Include the following header files into this source file first:
   a. **#include <stdint.h>**
   b. **#include <stdbool.h>**
   c. **#include <msp.h>**
   d. **#include <MSP432P4xx\flash.h>**
4. Declare one global **bool** variable **waitErase** and the flash interrupt handler **void Flash_IRQHandler(void);** The **waitErase** will be used as a condition for a **while()** loop to wait for the erase-complete interrupt to occur.
5. Place the **int main(void)** to start our main program.
6. Declare one **bool** variable **ret** and one unsigned 32-bit integer variable, **dest = 0x1000**. The variable **ret** works as a testing value holder and the variable **dest** stored the starting address from which four sectors in the flash main memory (Sector1~Sector4) will be erased. In fact, this address **0x1000** is the starting address of the first sector (Sector1) in the flash main memory since we want to erase four continuous sectors, Sector1~Sector4, in this project.
7. Stop the Watchdog timer by using the instruction **WDTCTL=WDTPW|WDTHOLD;**
8. Since we need to use red-color, green-color, and blue-color LEDs in this project to indicate the running status of this erasing operation, configure GPIO pins **P2.0~P2.2** as output pins (**P2DIR|=BIT0|BIT1|BIT2;**).
9. Disable the outputs of the GPIO pins **P2.0~P2.2** (**P2OUT &=~(BIT0|BIT1|BIT2);**).
10. Remove any WE protection for four sectors in the flash main memory space by calling the API function **FlashCtl_unprotectSector()**. Two arguments are involved in this function, the **FLASH_MAIN_MEMORY_SPACE_BANK0** and the OR of four sectors, **FLASH_SECTOR1~FLASH_SECTOR4**. The returning value of calling this function is assigned to the variable **ret**.
11. If this function calling is successful, a **true** is returned and assigned to the **ret**. Otherwise, a **false** is returned if this function calling is failed. Use an **if()** structure to check this **ret** and returns a false to the main program if this function is failed.
12. Clear previous error flags (**ERASE Done** |**PreV** | **PostV** |**PRG Error**) in the **FLCTL_IFG** register by calling the API function **FlashCtl_clearInterrupt-Flag()**. Only one argument is involved to this function, and one can use the **OR** operator to or all of these four flags together to form a single argument.
13. Register the flash memory interrupt handler, **Flash_IRQHandler**, by calling the API function **FlashCtl_registerInterrupt()**. The argument of this function is a pointer to the flash interrupt handler, which is **Flash_IRQHandler**.
14. Enable the flash interrupt sources by calling the API function **FlashCtl_enableIn-terrupt()**. This function contains only one argument, and one can use the **OR** operator or all four flags shown in step 12 together to make them as one single argument.
15. Call the API function **FlashCtl_eraseSector(dest)** to start this four sectors erase operation.
16. Use a conditional **while()** loop to wait for this erase operation to be done. The condition is the global bool variable **waitErase**. Recall that in step 4 when we declared this variable, we did not assign any value to it. By default, a **false** should be assigned to this bool variable by the system if it is declared but without assigning any value to it.
17. Inside this **while()** loop, the API function **FlashCtl_verifyMemory((void*) dest, 0x4000, FLASH_1_PATTERN);** will be called to perform a burst

read-compare operation to verify the correctness of our erase operation if it is done (when the **waitErase** becomes to true). The returning value of calling this function is a bool value and it is assigned to the variable **ret**.

18. Use an **if()** structure to check this **ret** value and the red-color LED is turned on (**P2OUT|=BIT0;**) if it is false, which means that the burst read-compare verify function is failed.
19. Otherwise the green-color LED is turned on (**P2OUT=BIT1;**) if a true is returned to indicate that no error has occurred for this burst read-compare verification and our erase operation is successful.

The following codes are for the flash interrupt handler **void Flash_IRQHandler (void)**:

20. Inside this handler, first create a **uint32_t** local variable **intStatus**.
21. Get the current interrupt sources by calling the API function **FlashCtl_getInterruptStatus()**, and the returned interrupt sources are assigned to the local variable **intStatus**.
22. Use an **if()** structure to check and test this **intStatus** to see whether any error has occurred during the erase operation. These errors include **PreV | PostV |PRG ERR**. One can use an **OR** operator or all of these three errors together to compare it with the **intStatus**.
23. If any or all of these errors have occurred, clear them by calling the API function **FlashCtl_clearInterruptFlag()**. One can use an OR operator or all of these three errors together to form a single argument and put it in that clear function above. One can use three system macros to present these errors. Then, turn on the red-color LED to indicate this error or these errors.
24. If no error has been encountered, use another **if()** structure to check and test the **intStatus** to see whether this interrupt is triggered by the completion of this erase operation (**FLASH_ERASE_COMPLETE**).
25. If it is, clear this interrupt flag by calling the API function **FlashCtl_clearInterruptFlag()** with the system macro **FLASH_ERASE_COMPLETE** as the argument for this function. Then turn on the blue-color LED, and set the global bool variable **waitErase** to **true** to stop the **while(waitErase)** loop in the main program and allow the burst read-compare operation to be executed inside that loop.

6.4.5 Set up the environment to build and run the project

To build and run the project, one needs to perform the following operations to set up the environments:

- Set up the path for the selected system header files in the **Include Paths** box in the **C/C++** tab under the **Project|Options for Target "Target 1"** menu item. Go to that box, and browse to the folder **C:\ti\msp\MSPWare_2_21_00_39\driverlib\driverlib**. Then click on the **OK** button to complete this setting.
- Add the MSPPDL **msp432p4xx_driverlib.lib** into the project. Go to the **Project** pane and right click on the **Source Group 1** item, and select the **Add Existing Files to Group "Source Group 1"** menu item. Browse to the folder where the library file is located, **C:\ti\msp\MSPWare_2_21_00_39\driverlib\driverlib\MSP432P4xx\keil**. Then select that library file and click on the **Add** button. Click on the **Close** button to finish this step.

- Select the **CMSIS-DAP Debugger** in the **Debug** tab under the **Project|Options for Target "Target 1"** menu item and set up the correct settings for the debugger and the flash download function.
- Now, let's build the project by going to **Project|Build target** menu item.
- Then go to **Flash|Download** menu item to download the image file of this project into the flash memory in the MSP-EXP432P401R Evaluation Board.

6.4.6 Demonstrate your program by checking the running result

- Now go to **Debug|Start/Stop Debug Session** to ready to run the project.
- Go to **Debug|Run** menu item to run the project. The blue-color LED is on, which means that the erase operation is complete.
- Now go to **Debug|Stop** menu to halt the project.
- Go to **Debug|Start/Stop Debug Session** menu to stop the project. Now the green-color LED is turned on, which means that the burst read-compare verify operation is completed successfully.

Based on these results, try to answer the following questions:

- By comparing the previous projects built with DRA model and this project built with SD Model, what are the advantages and disadvantages for both models?
- How do you confirm that this erase operation is successful by checking bytes in the target flash memory?
- What did you learn from this project?

6.5 LAB6_5

6.5.1 Goal

This project allows students to become familiar with the bit-band technology by building a bit-band-related project to perform some bit-band operations. This project enables students to write 32 bit-band values into 32 words in the bit-band alias range.

6.5.2 Data assignment and hardware configuration

No hardware configuration and data assignment are needed for this project.

6.5.3 Development of the project

Use the steps below to develop this project. Only the C source file is used in this project since this project is not complicated. Create a new project with the following steps:

1. Create a new folder **Lab6_5(SRAMBitBand)** under the folder **C:\MSP432 Lab Projects\ Chapter 6** in the Windows Explorer.
2. Open the Keil ARM-MDK µVersion5, create a new project named **SRAMBitBand** and save this project in the folder **Lab6_5(SRAMBitBand)** created in step 1.
3. On the next wizard, you need to select the device (MCU) for this project. Expand three icons, **Texas Instruments**, **MSP432 Family** and **MSP432P**, and select the target device **MSP432P401R** from the list by clicking on it. Click on the **OK** to close this wizard.
4. Next the software components wizard is opened, and you need to set up the software development environment for your project with this wizard. Expand two icons,

CMSIS and **Device**, and check the **CORE** and **Startup** checkboxes in the **Sel.** column, and click on the **OK** button, since we need these two components to build our project.

6.5.4 Development of the C source file

1. In the **Project** pane, expand the **Target** folder and right click on the **Source Group 1** folder and select the **Add New Item to Group "Source Group 1"**
2. Select the **C File (.c)** and enter **SRAMBitBand** into the **Name:** box, and click on the **Add** button to add this source file into the project
3. Include the following header files into this source file first:
   a. **#include <stdint.h>**
   b. **#include <stdbool.h>**
   c. **#include <msp.h>**
4. In this project, we need to use one user-defined macro:

```
#define MEM_REG0    0x20000000
```

   This macro defined the starting address of the bit-band region.
5. In this project we need to use one user-defined functions, **Mem_Addr()** to calculate the bit-band alias address. The declaration for this function is:

```
uint32_t Mem_Addr(uint32_t addr, uint32_t bitnum);
```

6. Place the **int main(void)** to start our main program.
7. Declare three unsigned 32-bit integer local variables, **bit_alias**, **uiValue** and **uiAddr**. The first variable is used to accept and hold the returned calculated bit-alias address by calling the function **Mem_Addr()**. The second one works as a loop number and the third variable **uiAddr** is the starting address of the bit-band region, which is **0x20000000**. Assign this address to the variable **uiAddr**.
8. Declare two unsigned 32-bit integer data array, **prData[32]** and **prAddr[32]**. The first data array is used to hold and store 32 bit-band values into the bit-band alias range, and the second data array is to hold and store 32 bit-band alias addresses.
9. Use a **for()** loop to repeatedly call the function **Mem_Addr()** to calculate 32 bit-band alias addresses, store those addresses into the data array **prAddr[]**. Assign bit-band value **0x1** to each address, and also store each value into the data array **prData[]**.
10. Use a **while(1)** loop to keep the project runs forever.
11. In the function **Mem_Addr()**, first create a local 32-bit unsigned integer variable **bit_band**.
12. Then, put the equation: **bit_band = 0x22000000 + ((addr & 0xFFFFF) << 5) + (bitnum << 2)**; under that local variable.
13. Return the local variable **bit_band**.

6.5.5 Set up the environment to build and run the project

- You need to set up a breakpoint at the code line that includes a **while(1)** loop. Now go to **Project|Build target** menu item to build the project.
- Select the **CMSIS-DAP Debugger** in the **Debug** tab under the **Project|Options for Target "Target 1"** menu item and set up the correct settings for the debugger and the flash download function.

6.5.6 Demonstrate your program

Perform the following operations to run your program and check the running results:

- Go to **Flash|Download** menu item to download your program into the flash memory.
- Go to **Debug|Start/Stop Debug Session** to begin debugging your program. Click on the **OK** button on the 32-KB memory size limitation message box to continue.
- Then go to **Debug|Run** menu item to run your program.

As the project runs, it will stop at the breakpoint, the **while(1)** loop. Now open the **Call Stack + Locals** window, expand the **prData[]** array. You can find that all 32 bit-band alias range have been set up by 1, exactly in the LSB of each address. Expand the **prAddr[]** array, you can find that the calculated bit-band alias addresses have been stored in this array.

Based on these results, try to answer the following questions:

- Can you find a way to change the starting address (**0x20000000**) on the bit-band region in this source file to allow this project to calculate any other desired bit-band alias address? If so, how?
- What did you learn from this project?

chapter seven

MSP432™ parallel I/O ports programming

This chapter provides general information about MSP432P401R microcontroller general-purpose I/O (GPIO) port programming. The discussion is mainly concentrated on the GPIO ports-related peripherals used in the MSP432P401R MCU system. This discussion includes the GPIO ports and peripherals specially designed for the MSP432P401R MCU, GPIO port architecture and configurations, general and special or PM control functions for different GPIO ports and pins, and GPIO ports and peripheral programming applications in the MSP432P401R MCU system.

7.1 Overview

As we discussed in Section 2.3.4 in Chapter 2, in the MSP432P401R MCU system, the GPIO module provides interfaces for multiple peripherals or I/O devices. Generally, the GPIO modules comprise 11 physical GPIO ports, Port 1~Port 10 and Port J, and each port is made of an individual 8-bit or 6-bit I/O port. With the help of the GDC or PMAPC techniques, each GPIO port can be configured to perform multiple functions or reconfigured to perform a special function. In addition, all GPIO ports provide various interfaces to access most system or on-chip peripherals as well as external peripherals to perform input/output functions. These system and on-chip peripherals can be categorized into the following groups:

1. System peripherals (peripherals 6~9 belong to ARM® Cortex®-M4 processor)
   a. One System Timer (**SysTick**)
   b. One Watchdog timer (**WDT_A**)
   c. Four 16-bit General Purpose Timers (**TA0~TA3**) and PWM modules (**PWM**)
   d. Two 32-bit General Purpose Timers (GPTMs)
   e. A CRC32 Module (**CRC32**)
   f. Instrumentation Trace Macrocell (**ITM**)
   g. Data Watchpoint and Trace (**DWT**)
   h. Flash Patch and Breakpoint (**FPB**)
   i. Trace Port Interface Unit (**TPIU**)
2. On-chip peripherals
   a. One 14-bit ADC (**ADC14**)
   b. Two analog comparators (**COMP_E0** and **COMP_E1**)
   c. One voltage reference (**REF_A**)
   d. One temperature sensor
   e. One capacitive touch IO module
   f. One eight-channel **DMA** and one real-time clock (**RTC**)
3. Interfaces to external peripherals
   a. Four enhanced universal serial communications interface A (eUSCI_A) modules for 4 × UART or SPI
   b. Four enhanced universal serial communication interface B (eUSCI_B) Modules for 4 × I2C or SPI

As we discussed in Section 2.3.4.1 in Chapter 2, each GPIO port can be mapped to an I/O block, and each block contains eight bits or eight pins. Each bit or each pin can be configured as either input or output pin. Also with the help of the general digital I/O control and the PM control techniques, each pin can be configured to perform multiple or different functions, which are called alternative functions.

The advantage of using the Port Mapping (PM) in the GPIO module is that multiple functions can be configured and fulfilled dynamically by using a limited number of GPIO ports and pins. The shortcoming is that this will make the GPIO port configurations and programming more complicated and more difficult.

Most of the peripherals, including system, on-chip, and interfaces to the external peripherals listed above, are connected and controlled to the GPIO ports with related pins. We will divide these peripherals into different groups and discuss them one by one in the following different chapters. In this chapter, we will concentrate on the following peripherals since these peripherals belong to the parallel I/O peripherals:

- On-board keypads interface programming project
- ADC programming project
- Analog comparator programming project

In Chapter 8, we will discuss SPI, UART, and I²C since these devices belong to series I/O peripherals. The following projects will be built in that chapter since these peripherals are connected to the SPI:

- On-board 7 segment LED interface programming project
- On-board LCD interface programming project
- Digital-to-analog converter (DAC) programming project

The timer in the MSP432P401R MCU system is a big topic and we will discuss this peripheral in Chapters 9 and 10. The discussion includes six real timers and one Watchdog timer.

7.2 GPIO module architecture and GPIO port configuration

We have provided detailed discussions and introductions about the GPIO architecture and related registers in Section 2.3.4 in Chapter 2. In this part, we will discuss more about the GPIO ports and pin configurations used in the MSP432P401R MCU system.

Figure 7.1 shows a functional block diagram for GPIO ports 1~10 and all related pins used in the MSP432P401R MCU system. One point to be noted is that although PJ is considered as a general I/O port, in most devices this port will be utilized by JTAG Debugger. Therefore, it is recommended to leave this port for the debugging purpose.

It can be found from Figure 7.1 that each pin in each GPIO port can provide multiple or alternative functions. The actual function for each pin is determined by two control techniques:

1. The port function select registers (**`PxSEL0`**, **`PxSEL1`**) in the general digital I/O control
2. The Port Mapping Controller (PMC)

In Sections 2.3.4.2 and 2.3.4.3 in Chapter 2, we have provided detailed discussions about these two control techniques. There is no significant difference between these two methods and one can use any one technique to define and map each pin on the selected

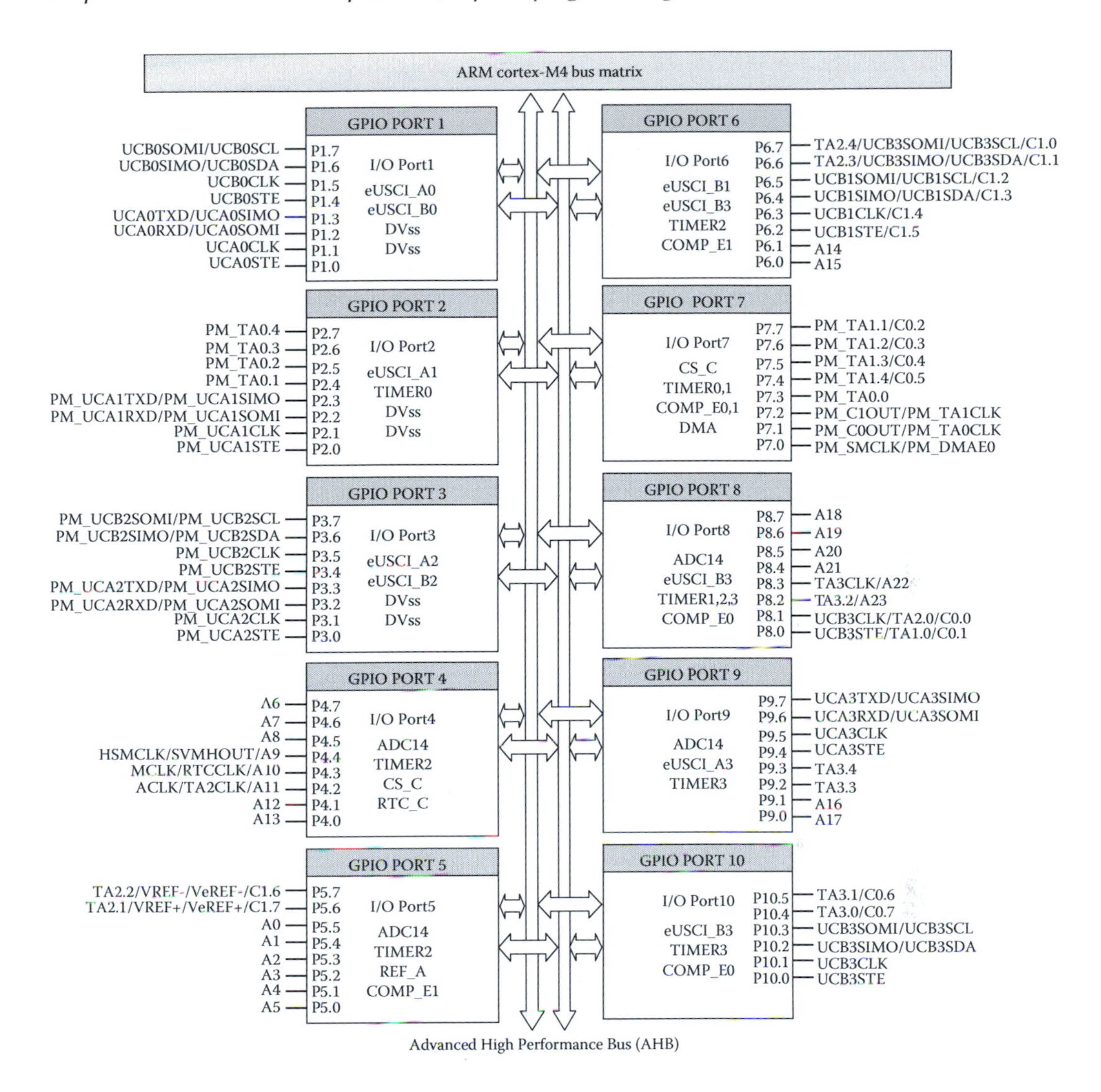

Figure 7.1 GPIO port and pin configuration.

GPIO port to perform a specific function. Refer to Section 2.3.4.4 in Chapter 2 to get the actual difference between these two methods.

It can also be found from Figure 7.1 that in the MSP432P401R MCU system, the PM technique can only be applied on 24 pins on 3 GPIO ports, `P2`, `P3`, and `P7`. This can be confirmed by checking all input/output signals connected to 24 pins on those 3 ports, in which all signals are preceded by `PM_` prefix to indicate that these signals are multiple function signals and must be activated by using the PMC via these pins to make them effect.

Based on these two control techniques, in the following sections, we will divide our discussions into two major parts:

1. The GPIO ports controlled by using the general digital I/O control
2. The GPIO ports controlled by using the PMC

7.3 GPIO ports controlled by using the general digital I/O control

The GPIO ports belonging to this group include GPIO Ports **P1, P4~P6, P8~P10**. All GPIO pins on these ports can be configured to perform either general digital functions or multifunction by setting two port function select registers **PxSEL0** and **PxSEL1**, respectively, via the software before the program runs.

A basic configuration for each GPIO pin on these ports was shown in Figure 2.15 in Chapter 2. For the convenience, we redisplay it in this chapter as Figure 7.2.

Two control bits, **PySEL1.x** and **PySEL0.x**, can be combined together to select one input from multi-input applied on a 4-to-1 MUX as shown in Figure 7.2. In fact, for these GPIO ports working in this mode, only two possible functions can be selected and performed:

- Default general digital I/O function when **PySEL1.x:PySEL0.x = 00.**
- Specified module function coming from the predefined module when **PySEL1.x: PySEL0.x = 01, 10 or 11.**

When these two control bits are other values, such as **10** or **11**, the **DVss** signal is selected.

For all available module functions, refer to related tables. Table 7.1 lists all available alternate functions for GPIO Port 1. Tables 7.2 through 7.4 list related alternate functions for P4~P6, and Tables 7.5 through 7.7 show all available alternate functions for P8~P10.

As different module functions may have different configurations for the related GPIO pins, we only discuss the most popular configurations with selected GPIO pins. For more details about these configurations, refer to related manufacturing documents.

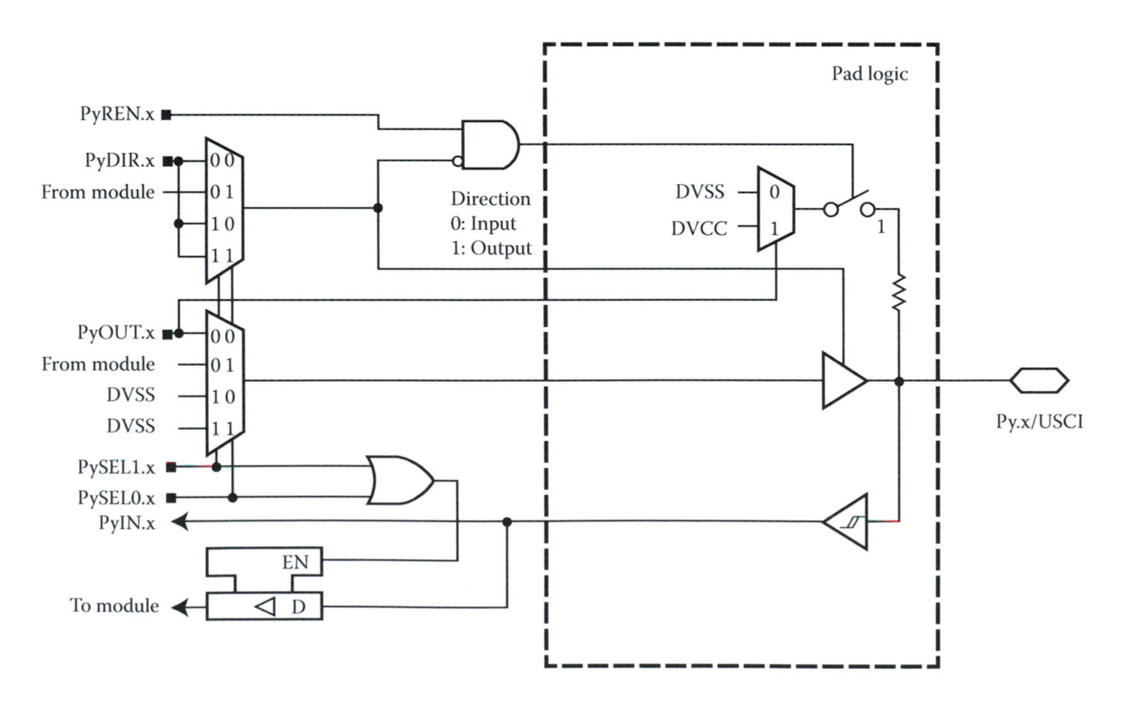

Figure 7.2 Bit configuration of GPIO ports (P1, P3, P2.0~P2.3, P9.4~P9.7, P10.0~P10.3). (Courtesy Texas Instruments.)

Table 7.1 GPIO port 1 (P1.0~P1.7) pin functions

Pin	Function	Control bits or signals		
		P1DIR.x	**P1SEL1.x**	**P1SEL0.x**
P1.0	Digital I/O	0:I, 1:O	0	0
	UCA0STE	X	0	1
	DVSS	1	1	0
	DVSS	1	1	1
P1.1	Digital I/O	0:I, 1:O	0	0
	UCA0CLK	X	0	1
	DVSS	1	1	0
	DVSS	1	1	1
P1.2	Digital I/O	0:I, 1:O	0	0
	UCA0RXD/UCA0SOMI	X	0	1
	DVSS	1	1	0
	DVSS	1	1	1
P1.3	Digital I/O	0:I, 1:O	0	0
	UCA0TXD/UCA0SIMO	X	0	1
	DVSS	1	1	0
	DVSS	1	1	1
P1.4	Digital I/O	0:I, 1:O	0	0
	UCB0STE	X	0	1
	DVSS	1	1	0
	DVSS	1	1	1
P1.5	Digital I/O	0:I, 1:O	0	0
	UCB0CLK	X	0	1
	DVSS	1	1	0
	DVSS	1	1	1
P1.6	Digital I/O	0:I, 1:O	0	0
	UCB0SIMO/UCB0SDA	X	0	1
	DVSS	1	1	0
	DVSS	1	1	1
P1.7	Digital I/O	0:I, 1:O	0	0
	UCB0SOMI/UCB0SCL	X	0	1
	DVSS	1	1	0
	DVSS	1	1	1

The following important points can be found from Table 7.1:

- Each GPIO pin, **P1.0**~**P1.7**, can perform at least two functions: the normal digital I/O function and an alternate function, such as **UCA0STE**, **UCA0CLK**, **UCA0RXD**, or **UCA0SOMI**. These alternate functions are selected and controlled by setting two registers, **P1SEL1.x** and **P1SEL0.x**, as 01, respectively.
- The bit value in the **P1DIR.x** Register determines the data transfer directions for the normal digital I/O functions. When **P1DIR.x** = 0, the corresponding pin works as an input pin. If **P1DIR.x** = 1, the related pin works as an output pin.

Table 7.2 GPIO port 4 (P4.0~P4.7) pin functions

Pin	Function	Control bits or signals		
		P4DIR.x	P4SEL1.x	P4SEL0.x
P4.0	Digital I/O	0:I, 1:O	0	0
	DVSS	1	0	1
	DVSS	1	1	0
	A13	X	1	1
P4.1	Digital I/O	0:I, 1:O	0	0
	DVSS	1	0	1
	DVSS	1	1	0
	A12	X	1	1
P4.2	Digital I/O	0:I, 1:O	0	0
	N/A	0	0	1
	ACLK	1		
	TA2CLK	0	1	0
	DVSS	1		
	A11	X	1	1
P4.3	Digital I/O	0:I, 1:O	0	0
	N/A	0	0	1
	MCLK	1		
	N/A	0	1	0
	RTCCLK	1		
	A10	X	1	1
P4.4	Digital I/O	0:I, 1:O	0	0
	N/A	0	0	1
	HSMCLK	1		
	N/A	0	1	0
	SVMHOUT	1		
	A9	X	1	1
P4.5	Digital I/O	0:I, 1:O	0	0
	DVSS	1	0	1
	DVSS	1	1	0
	A8	X	1	1
P4.6	Digital I/O	0:I, 1:O	0	0
	DVSS	1	0	1
	DVSS	1	1	0
	A7	X	1	1
P4.7	Digital I/O	0:I, 1:O	0	0
	DVSS	1	0	1
	DVSS	1	1	0
	A6	X	1	1

Table 7.3 GPIO port 5 (P5.0~P5.7) pin functions

Pin	Function	Control bits or signals		
		P5DIR.x	P5SEL1.x	P5SEL0.x
P5.0	Digital I/O	0:I, 1:O	0	0
	DVSS	1	0	1
	DVSS	1	1	0
	A5	X	1	1
P5.1	Digital I/O	0:I, 1:O	0	0
	DVSS	1	0	1
	DVSS	1	1	0
	A4	X	1	1
P5.2	Digital I/O	0:I, 1:O	0	0
	DVSS	1	0	1
	DVSS	1	1	0
	A3	X	1	1
P5.3	Digital I/O	0:I, 1:O	0	0
	DVSS	1	0	1
	DVSS	1	1	0
	A2	X	1	1
P5.4	Digital I/O	0:I, 1:O	0	0
	DVSS	1	0	1
	DVSS	1	1	0
	A1	X	1	1
P5.5	Digital I/O	0:I, 1:O	0	0
	DVSS	1	0	1
	DVSS	1	1	0
	A0	X	1	1
P5.6	Digital I/O	0:I, 1:O	0	0
	TA2.CCI1A	0	0	1
	TA2.1	1		
	N/A	0	1	0
	DVSS	1		
	VREF+, VeREF+, C1.7	X	1	1
P5.7	Digital I/O	0:I, 1:O	0	0
	TA2.CCI2A	0	0	1
	TA2.2	1		
	N/A	0	1	0
	DVSS	1		
	VREF−, VeREF−, C1.6	X	1	1

Table 7.4 GPIO port 6 (P6.0~P6.7) pin functions

Pin	Function	Control bits or signals		
		P6DIR.x	P6SEL1.x	P6SEL0.x
P6.0	Digital I/O	0:I, 1:O	0	0
	DVSS	1	0	1
	DVSS	1	1	0
	A15	X	1	1
P6.1	Digital I/O	0:I, 1:O	0	0
	DVSS	1	0	1
	DVSS	1	1	0
	A1 4	X	1	1
P6.2	Digital I/O	0:I, 1:O	0	0
	USB1STE	X	0	1
	N/A	0	1	0
	DVSS	1		
	C1.5	X	1	1
P6.3	Digital I/O	0:I, 1:O	0	0
	UCB1CLK	X	0	1
	N/A	0	1	0
	DVSS	1		
	C1.4	X	1	1
P6.4	Digital I/O	0:I, 1:O	0	0
	UCB1SIMO/UCB1SDA	X	0	1
	N/A	0	1	0
	DVSS	1		
	C1.3	X	1	1
P6.5	Digital I/O	0:I, 1:O	0	0
	UCB1SOMI/UCB1SCL	X	0	1
	N/A	0	1	0
	DVSS	1		
	C1.2	X	1	1
P6.6	Digital I/O	0:I, 1:O	0	0
	TA2.CCI3A	0	0	1
	TA2.3	1		
	UCB3SIMO/UCB3SDA	X	1	0
	C1.1	X	1	1
P6.7	Digital I/O	0:I, 1:O	0	0
	TA2.CCI4A	0	0	1
	TA2.4	1		
	UCB3SOMI/UCB3SCL	X	1	0
	C1.0	X	1	1

Table 7.5 GPIO port 8 (P8.0~P8.7) pin functions

Pin	Function	Control bits or signals		
		P8DIR.x	P8SEL1.x	P8SEL0.x
P8.0	Digital I/O	0:I, 1:O	0	0
	UCB3STE	X	0	1
	TA1.CCI0A	0	1	0
	TA1.0	1		
	C0.1	X	1	1
P8.1	Digital I/O	0:I, 1:O	0	0
	UCB3CLK	X	0	1
	TA2.CCI0A	0	1	0
	TA2.0	1		
	C0.0	X	1	1
P8.2	Digital I/O	0:I, 1:O	0	0
	TA3.CCI2A	0	0	1
	TA3.2	1		
	N/A	0	1	0
	DVSS	1		
	A23	X	1	1
P8.3	Digital I/O	0:I, 1:O	0	0
	TA3CLK	0	0	1
	DVSS	1		
	N/A	0	1	0
	DVSS	1		
	A24	X	1	1
P8.4	Digital I/O	0:I, 1:O	0	0
	N/A	0	0	1
	DVSS	1		
	N/A	0	1	0
	DVSS	1		
	A21	X	1	1
P8.5	Digital I/O	0:I, 1:O	0	0
	N/A	0	0	1
	DVSS	1		
	N/A	0	1	0
	DVSS	1		
	A20	X	1	1
P8.6	Digital I/O	0:I, 1:O	0	0
	N/A	0	0	1
	DVSS	1		
	N/A	0	1	0
	DVSS	1		
	A19	X	1	1
P8.7	Digital I/O	0:I, 1:O	0	0
	N/A	0	0	1
	DVSS	1		
	N/A	0	1	0
	DVSS	1		
	A18	X	1	1

However, when the desired pin has been selected to work as an alternate function pin (**P1SEL1.x:P1SEL0.x** = 01), the data transfer directions are determined by the selected alternate functions, and have nothing to do with the **P1DIR.x** Register (this is shown by a **X** mark in Table 7.1).

- When **P1SEL1.x:P1SEL0.x** registers equal to other values, such as 10 and 11, if **P1DIR.x** is set to 1, the **DVSS** signal is transferred as an output via that pin. Otherwise if **P1DIR.x** = 0, no function/signal is available to that pin (N/A).

Table 7.2 shows the alternate functions for the GPIO Port 4, **P4.0~P4.7**.

It can be found from Table 7.2 that pins **P4.2~P4.4** can perform more alternate functions, and therefore make the situation more complicated.

When **P4SEL1.x:P4SEL0.x** = 01, depending on the value set in the **P4DIR.x** Register, pins **P4.2~P4.4** can perform more alternate functions. Similarly, when **P4SEL1.x:P4SEL0.x** = 10, pins **P4.2~P4.4** can perform some other alternate functions based on the different values set in the **P4DIR.x** Register.

When **P4SEL1.x:P4SEL0.x** = 11, no matter what value is set in the **P4DIR.x** Register, the selected pins work as analog input pins and this setup works for all GPIO Port 4 pins, **P4.0~P4.7**.

Table 7.3 lists the alternate functions for the GPIO Port 5, **P5.0~P5.7**.

It can be found from Table 7.3 that pins **P5.0~P5.5** can perform two functions, either normal digital I/O or analog inputs, depending on the combination values set in the **P5SEL1.x:P5SEL0.x** registers (00 or 11). But pins **P5.6** and **P5.7** can perform more alternate functions, and therefore make the situation more complicated. When the values set in the **P5SEL1.x:P5SEL0.x** registers are equal to 01 or 10, depending on the **P5DIR.x** Register, pins **P5.6** and **P5.7** can perform totally four different functions.

Table 7.4 shows the alternate functions for the GPIO Port 6, **P6.0~P6.7**.

It can be found from Table 7.4 that pins **P6.0~P6.5** can perform two functions, either normal digital I/O or analog inputs, depends on the combination values set in the **P6SEL1.x:P6SEL0.x** registers (00 or 11). But pins **P6.6** and **P6.7** can perform more alternate functions, and therefore make the situation more complicated. When the values set in the **P6SEL1.x:P6SEL0.x** registers are equal to 01 or 10, depends on the **P6DIR.x** Register, pins **P6.6** and **P6.7** can perform totally four different functions.

The signals **C1.0~C1.7** indicate the Analog Comparator_E1 inputs 0~7. The **TA2.0~TA2.4** signals indicate five timer capture inputs for Timer A: TA2 CCR0 capture-CCI0A~TA2 CCR4 capture-CCI4A inputs.

Table 7.5 shows the alternate functions for the GPIO Port 8, **P8.0~P8.7**.

It can be found from Table 7.5 that GPIO pins **P8.0~P8.3** can perform multiple functions, but pins **P8.4~P8.7** can only perform two functions, digital I/O or analog inputs (**A18~A21**).

The signals **C0.0~C0.7** indicate the Analog Comparator_E0 inputs 0~7. The signal **TAx.CCIyA** indicates the **yth** capture input for Timer A channel **x**.

Table 7.6 lists the alternate functions for the GPIO Port 9, **P9.0~P9.7**.

It can be found from Table 7.6 that GPIO pins **P9.0~P9.1**, **P9.4~P9.7** can only perform two functions, but pins **P9.2** and **P9.3** can perform multiple functions.

Table 7.7 lists the alternate functions for the GPIO Port 10, **P10.0~P10.5**.

It can be found from Table 7.7 that GPIO pins **P10.0~P10.3** can only perform two functions, but pins **P10.4** and **P10.5** can perform multiple functions.

The signals **C0.6~C0.7** indicate the Analog Comparator_E0 inputs 6~7. The signal **TAx.CCIyA** indicates the **yth** capture input for timer A channel **x**.

Table 7.6 GPIO port 9 (P9.0~P9.7) pin functions

Pin	Function	Control bits or signals		
		P9DIR.x	P9SEL1.x	P9SEL0.x
P9.0	Digital I/O	0:I, 1:O	0	0
	DVSS	1	0	1
	DVSS	1	1	0
	A17	X	1	1
P9.1	Digital I/O	0:I, 1:O	0	0
	DVSS	1	0	1
	DVSS	1	1	0
	A16	X	1	1
P9.2	Digital I/O	0:I, 1:O	0	0
	TA3.CCI3A	0	0	1
	TA3.3	1		
	N/A	0	1	0
	DVSS	1		
	N/A	0	1	1
	DVSS	1		
P9.3	Digital I/O	0:I, 1:O	0	0
	TA3.CCI4A	0	0	1
	TA3.4	1		
	N/A	0	1	0
	DVSS	1		
	N/A	0	1	1
	DVSS	1		
P9.4	Digital I/O	0:I, 1:O	0	0
	UCA3STE	X	0	1
	N/A	0	1	0
	DVSS	1		
	N/A	0	1	1
	DVSS	1		
P9.5	Digital I/O	0:I, 1:O	0	0
	UCA3CLK	X	0	1
	N/A	0	1	0
	DVSS	1		
	N/A	0	1	1
	DVSS	1		
P9.6	Digital I/O	0:I, 1:O	0	0
	UCA3RXD/ UCA3SOMI	X	0	1
	N/A	0	1	0
	DVSS	1		
	N/A	0	1	1
	DVSS	1		

(*Continued*)

Table 7.6 (Continued) GPIO port 9 (P9.0~P9.7) pin functions

Pin	Function	Control bits or signals		
		P9DIR.x	**P9SEL1.x**	**P9SEL0.x**
P9.7	Digital I/O	0:I, 1:O	0	0
	UCA3TXD/ UCA3SIMO	X	0	1
	N/A	0	1	0
	DVSS	1		
	N/A	0	1	1
	DVSS	1		

From these tables, all related alternate functions that can be performed by GPIO Ports, P1, P4~P6, P8~P10, which are determined by the values set in the **PxSEL1.y:PxSEL0.y** registers, are clearly shown in the above table. We need to use these tables to get detailed GPIO pin function mapping relationships to help us to build GPIO example projects in the following sections.

Next, let's take care of the rest GPIO ports that are controlled by the PMC to perform alternate functions.

7.4 GPIO ports controlled by using the PMC

Three GPIO ports are involved in this group: GPIO Ports P2, P3, and P7.

We have provided detailed discussions about the PM control in Section 2.3.4.3 in Chapter 2. Regularly, a PMC provides the following features:

- All configurations are protected by write access key
- A default mapping is provided for each port and pin (this default mapping is device-dependent)
- Mapping can be reconfigured during runtime
- Each output signal can be mapped to several output pins

All of these mapping operations are performed by using a set of PM registers. The mapping operational principle and sequence are:

1. To enable access to any of the PMC registers, the correct key must be written into the PM Key ID Register (**PMAPKEYID**). This is a 16-bit register providing 16-bit key value. The **PMAPKEYID** Register always reads **0x96A5**. Writing the key **0x2D52** to this register grants write access to all PMC registers. Read access is always possible since this is a read–write register.
2. If an invalid key is written while write access is granted, any further write accesses are prevented. It is recommended that the application completes mapping configuration by writing an invalid key, such as 0, into the **PMAPKEYID** Register.
3. Each port pin, **Px.y**, provides a mapping register, **PxMAPy**. Setting this register to a certain value maps a module's input and output signals to the respective port pin Px.y.
4. By default, the PMC allows only one configuration after a hard reset (HR), and all mapping registers remain locked.

Table 7.7 GPIO port 10 (P10.0~P10.5) pin functions

Pin	Function	Control bits or signals		
		P10DIR.x	**P10SEL1.x**	**P10SEL0.x**
P10.0	Digital I/O	0:I, 1:O	0	0
	UCB3STE	X	0	1
	N/A	0	1	0
	DVSS	1		
	N/A	0	1	1
	DVSS	1		
P10.1	Digital I/O	0:I, 1:O	0	0
	UCB3CLK	X	0	1
	N/A	0	1	0
	DVSS	1		
	N/A	0	1	1
	DVSS	1		
P10.2	Digital I/O	0:I, 1:O	0	0
	UCB3SIMO/UCB3SDA	X	0	1
	N/A	0	1	0
	DVSS	1		
	N/A	0	1	1
	DVSS	1		
P10.3	Digital I/O	0:I, 1:O	0	0
	UCB3SOMI/UCB3SCL	X	0	1
	N/A	0	1	0
	DVSS	1		
	N/A	0	1	1
	DVSS	1		
P10.4	Digital I/O	0:I, 1:O	0	0
	TA3.CCI0A	0	0	1
	TA3.0	1		
	N/A	0	1	0
	DVSS	1		
	C0.7	X	1	1
P10.5	Digital I/O	0:I, 1:O	0	0
	TA3.CCI1A	0	0	1
	TA3.1	1		
	N/A	0	1	0
	DVSS	1		
	C0.6	X	1	1

5. If it is necessary to reconfigure the mapping during runtime, the **PMAPRECFG** bit in the **PMAPCTL** Register must be set during the first write access timeslot. If **PMAPRECFG** is cleared during later configuration sessions, no more configuration sessions are possible.

All possible PM values, **PxMAPy**, are shown in Table 7.8. In the following sections, we need to show the PM values for GPIO Ports P2, P3, and P7.

7.4.1 PM control for GPIO P2

Table 7.8 shows the default PM values and functions for the GPIO P2.

When using these PM values to configure the GPIO P2, P3, and P7 to perform the desired PM functions, the following points must be kept in mind:

- Each pin has a unique **PxMAPy** value, which is the default mapping value. For example, the default **P2MAPy** values shown in Table 7.8 for P2.
- These PM values must be combined with the values set in the **PxSEL1.y** and **PxSEL0.y** registers together to map the selected pin to perform mapped functions.

Table 7.9 shows the pin functions for GPIO **P2** by using the combinations of the **P2MAPy** values and values set in the **P2SEL1.y** and **P2SEL0.y** registers.

7.4.2 PM control for GPIO P3

Table 7.10 shows the default PM values and functions for the GPIO P3.

Table 7.11 shows the pin functions for GPIO **P3** by using the combinations of the **P3MAPx** values and values set in the **P3SEL1.y** and **P3SEL0.y** registers.

It can be found from Table 7.11 that each default PM value plays its role only when the values set in the **P3SEL1.x** and **P3SEL0.x** registers are equal to **01**. For all other values, there is no effect for the selected pin even if the default PM value is applied. The default values shown in the sixth column, **P3MAPx** in Table 7.11, are exactly the **P3MAPy** values shown in Table 7.10.

Table 7.8 Default mapping values and functions for GPIO P2

<table>
<tr><th>Pin</th><th>P2MAPy value</th><th>Input pin function</th><th>Output pin function</th></tr>
<tr><td>P2.0</td><td>PM_UCA1STE</td><td colspan="2">eUSCI_A1 SPI slave transmit enable (direction controlled by eUSCI)</td></tr>
<tr><td>P2.1</td><td>PM_UCA1CLK</td><td colspan="2">eUSCI_A1 clock input/output (direction controlled by eUSCI)</td></tr>
<tr><td>P2.2</td><td>PM_UCA1RXD</td><td colspan="2">eUSCI_A1 UART RXD (direction controlled by eUSCI—input)</td></tr>
<tr><td></td><td>PM_UCA1SOMI</td><td colspan="2">eUSCI_A1 SPI slave out master in (direction controlled by eUSCI)</td></tr>
<tr><td>P2.3</td><td>PM_UCA1TXD</td><td colspan="2">eUSCI_A1 UART TXD (direction controlled by eUSCI—output)</td></tr>
<tr><td></td><td>PM_UCA1SIMO</td><td colspan="2">eUSCI_A1 SPI slave in master out (direction controlled by eUSCI)</td></tr>
<tr><td>P2.4</td><td>PM_TA0.1</td><td>TA0 CCR1 capture input CCI1A</td><td>TA0 CCR1 compare output Out1</td></tr>
<tr><td>P2.5</td><td>PM_TA0.2</td><td>TA0 CCR2 capture input CCI2A</td><td>TA0 CCR2 compare output Out2</td></tr>
<tr><td>P2.6</td><td>PM_TA0.3</td><td>TA0 CCR3 capture input CCI3A</td><td>TA0 CCR3 compare output Out3</td></tr>
<tr><td>P2.7</td><td>PM_TA0.4</td><td>TA0 CCR4 capture input CCI4A</td><td>TA0 CCR4 compare output Out4</td></tr>
</table>

Table 7.9 GPIO port 2 (P2.0~P2.7) pin mapping functions

Pin	Function	Control bits or signals			
		P2DIR.x	P2SEL1.x	P2SEL0.x	P2MAPx
P2.0	Digital I/O	0:I, 1:O	0	0	X
	UCA1STE	X	0	1	PM_UCA1STE
	N/A	0	1	0	X
	DVSS	1			
	N/A	0	1	1	X
	DVSS	1			
P2.1	Digital I/O	0:I, 1:O	0	0	X
	UCA1CLK	X	0	1	PM_UCA1CLK
	N/A	0	1	0	X
	DVSS	1			
	N/A	0	1	1	X
	DVSS	1			
P2.2	Digital I/O	0:I, 1:O	0	0	X
	UCA1RXD/UCA1SOMI	X	0	1	Default value
	N/A	0	1	0	X
	DVSS	1			
	N/A	0	1	1	X
	DVSS	1			
P2.3	Digital I/O	0:I, 1:O	0	0	X
	UCA1TXD/UCA1SIMO	X	0	1	Default value
	N/A	0	1	0	X
	DVSS	1			
	N/A	0	1	1	X
	DVSS	1			
P2.4	Digital I/O	0:I, 1:O	0	0	X
	TA0.CCI1A	0	0	1	Default value (Table 7.8)
	TA0.1	1			
	N/A	0	1	0	X
	DVSS	1			
	N/A	0	1	1	X
	DVSS	1			
P2.5	Digital I/O	0:I, 1:O	0	0	X
	TA0.CCI2A	0	0	1	Default value (Table 7.8)
	TA0.2	1			
	N/A	0	1	0	X
	DVSS	1			
	N/A	0	1	1	X
	DVSS	1			
P2.6	Digital I/O	0:I, 1:O	0	0	X
	TA0.CCI3A	0	0	1	Default value (Table 7.8)
	TA0.3	1			
	N/A	0	1	0	X
	DVSS	1			
	N/A	0	1	1	X
	DVSS	1			

(*Continued*)

Table 7.9 (Continued) GPIO port 2 (P2.0~P2.7) pin mapping functions

Pin	Function	Control bits or signals			
		P2DIR.x	P2SEL1.x	P2SEL0.x	P2MAPx
P2.7	Digital I/O	0:I, 1:O	0	0	X
	TA0.CCI4A	0	0	1	Default value (Table 7.8)
	TA0.4	1			
	N/A	0	1	0	X
	DVSS	1			
	N/A	0	1	1	X
	DVSS	1			

7.4.3 PM control for GPIO P7

Table 7.12 shows the default PM values and functions for the GPIO P7.

Table 7.13 shows the pin functions for GPIO **P7** by using the combinations of the **P7MAPx** values and values set in the **P7SEL1.y** and **P7SEL0.y** registers.

It can be found from Table 7.13 that each default PM value plays its role only when the values set in the **P7SEL1.x** and **P7SEL0.x** registers equal to **01**. For all other values, there is no effect for the selected pin even the default PM value is applied. The default values

Table 7.10 Default mapping values and functions for GPIO P3

Pin	P3MAPy value	Input pin function	Output pin function
P3.0	PM_UCA2STE	eUSCI_A2 SPI slave transmit enable (direction controlled by eUSCI)	
P3.1	PM_UCA2CLK	eUSCI_A2 clock input/output (direction controlled by eUSCI)	
P3.2	PM_UCA2RXD	eUSCI_A2 UART RXD (direction controlled by eUSCI—input)	
	PM_UCA2SOMI	eUSCI_A2 SPI slave out master in (direction controlled by eUSCI)	
P3.3	PM_UCA2TXD	eUSCI_A2 UART TXD (direction controlled by eUSCI—output)	
	PM_UCA2SIMO	eUSCI_A2 SPI slave in master out (direction controlled by eUSCI)	
P3.4	PM_UCB2STE	eUSCI_B2 SPI slave transmit enable (direction controlled by eUSCI)	
P3.5	PM_UCB2CLK	eUSCI_B2 clock input/output (direction controlled by eUSCI)	
P3.6	PM_UCB2SIMO	eUSCI_B2 SPI slave in master out (direction controlled by eUSCI)	
	PM_UCB2SDA	eUSCI_B2 I2C data (open drain and direction controlled by eUSCI)	
P3.7	PM_UCB2SOMI	eUSCI_B2 SPI slave out master in (direction controlled by eUSCI)	
	PM_UCB2SCL	eUSCI_B2 I2C clock (open drain and direction controlled by eUSCI)	

Table 7.11 GPIO port 3 (P3.0~P3.7) pin mapping functions

Pin	Function	Control bits or signals			
		P3DIR.x	P3SEL1.x	P3SEL0.x	P3MAPx
P3.0	Digital I/O	0:I, 1:O	0	0	X
	UCA2STE	X	0	1	PM_UCA2STE
	N/A	0	1	0	X
	DVSS	1			
	N/A	0	1	1	X
	DVSS	1			
P3.1	Digital I/O	0:I, 1:O	0	0	X
	UCA2CLK	X	0	1	PM_UCA2CLK
	N/A	0	1	0	X
	DVSS	1			
	N/A	0	1	1	X
	DVSS	1			
P3.2	Digital I/O	0:I, 1:O	0	0	X
	UCA2RXD/UCA2SOMI	X	0	1	Default value
	N/A	0	1	0	X
	DVSS	1			
	N/A	0	1	1	X
	DVSS	1			
P3.3	Digital I/O	0:I, 1:O	0	0	X
	UCA2TXD/UCA2SIMO	X	0	1	Default value
	N/A	0	1	0	X
	DVSS	1			
	N/A	0	1	1	X
	DVSS	1			
P3.4	Digital I/O	0:I, 1:O	0	0	X
	UCB2STE	X	0	1	Default value
	N/A	0	1	0	X
	DVSS	1			
	N/A	0	1	1	X
	DVSS	1			
P3.5	Digital I/O	0:I, 1:O	0	0	X
	UCB2CLK	X	0	1	Default value
	N/A	0	1	0	X
	DVSS	1			
	N/A	0	1	1	X
	DVSS	1			
P3.6	Digital I/O	0:I, 1:O	0	0	X
	UCB2SIMO/UCB2SDA	X	0	1	Default value
	N/A	0	1	0	X
	DVSS	1			
	N/A	0	1	1	X
	DVSS	1			

(*Continued*)

Table 7.11 (Continued) GPIO port 3 (P3.0~P3.7) pin mapping functions

Pin	Function	Control bits or signals			
		P3DIR.x	P3SEL1.x	P3SEL0.x	P3MAPx
P3.7	Digital I/O	0:I, 1:O	0	0	X
	UCB2SOMI/UCB2SCL	X	0	1	Default value
	N/A	0	1	0	X
	DVSS	1			
	N/A	0	1	1	X
	DVSS	1			

shown in the sixth column, **P7MAPx** in Table 7.13, are exactly the **P7MAPy** values shown in Table 7.12.

A detailed discussion about the GPIO port initialization and configuration has been given in Section 2.3.4.5 in Chapter 2. All GPIO port control registers are also introduced in Section 2.3.4.2 in Chapter 2. Refer to those sections to get more details about these registers. Next, we need to use these registers to interface some GPIO ports to perform a few sophisticated projects to illustrate how to use GPIO modules to perform general digital and alternate functions.

7.5 On-board keypads interface programming project

In the EduBASE ARM® Trainer, there is a 4 × 4 keypad installed and connected to four GPIO ports, P3, P4, P5, and P6. As shown in Figure 7.3, four rows (rows 0~3) are connected to four lower bits, P6.1, P4.0, P4.2, and P4.4 on GPIO P4 and P6. Four columns (columns 3~0) are connected to four bits, P4.6, P5.0, P5.2, and P3.6 on the GPIO P3, P4, and P5, respectively.

Table 7.12 Default mapping values and functions for GPIO P7

Pin	P7MAPy value	Input pin function	Output pin function
P7.0	PM_SMCLK/ PM_DMAE0	DMAE0 input	SMCLK
P7.1	PM_C0OUT/ PM_TA0CLK	Timer_A0 external clock input	Comparator-E0 output
P7.2	PM_C1OUT/ PM_TA1CLK	Timer_A1 external clock input	Comparator-E1 output
P7.3	PM_TA0.0	TA0 CCR0 capture input CCI0A	TA0 CCR0 compare output Out0
P7.4	PM_TA1.4	TA1 CCR4 capture input CCI4A	TA1 CCR4 compare output Out4
P7.5	PM_TA1.3	TA1 CCR3 capture input CCI3A	TA1 CCR3 compare output Out3
P7.6	PM_TA1.2	TA1 CCR2 capture input CCI2A	TA1 CCR2 compare output Out2
P7.7	PM_TA1.1	TA1 CCR1 capture input CCI1A	TA1 CCR1 compare output Out1

Table 7.13 GPIO port 7 (P7.0~P7.7) pin mapping functions

Pin	Function	Control bits or signals			
		P7DIR.x	P7SEL1.x	P7SEL0.x	P7MAPx
P7.0	Digital I/O	0:I, 1:O	0	0	X
	DMAE0	0	0	1	Default value (Table 7.12)
	SMCLK	1			
	N/A	0	1	0	X
	DVSS	1			
	N/A	0	1	1	X
	DVSS	1			
P7.1	Digital I/O	0:I, 1:O	0	0	X
	TA0CLK	0	0	1	Default value (Table 7.12)
	C0OUT	1			
	N/A	0	1	0	X
	DVSS	1			
	N/A	0	1	1	X
	DVSS	1			
P7.2	Digital I/O	0:I, 1:O	0	0	X
	TA1CLK	0	0	1	Default value (Table 7.12)
	C1OUT	1			
	N/A	0	1	0	X
	DVSS	1			
	N/A	0	1	1	X
	DVSS	1			
P7.3	Digital I/O	0:I, 1:O	0	0	X
	TA0.CCI0A	0	0	1	Default value (Table 7.12)
	TA0.0	1			
	N/A	0	1	0	X
	DVSS	1			
	N/A	0	1	1	X
	DVSS	1			
P7.4	Digital I/O	0:I, 1:O	0	0	X
	TA1.CCI4A	0	0	1	Default value (Table 7.12)
	TA1.4	1			
	N/A	0	1	0	X
	DVSS	1			
	C0.5	X	1	1	X
P7.5	Digital I/O	0:I, 1:O	0	0	X
	TA1.CCI3A	0	0	1	Default value (Table 7.12)
	TA1.3	1			
	N/A	0	1	0	X
	DVSS	1			
	C0.4	X	1	1	X

(*Continued*)

Table 7.13 (Continued) GPIO port 7 (P7.0~P7.7) pin mapping functions

Pin	Function	Control bits or signals			
		P7DIR.x	P7SEL1.x	P7SEL0.x	P7MAPx
P7.6	Digital I/O	0:I, 1:O	0	0	X
	TA1.CCI2A	0	0	1	Default value (Table 7.12)
	TA1.2	1			
	N/A	0	1	0	X
	DVSS	1			
	C0.3	X	1	1	X
P7.7	Digital I/O	0:I, 1:O	0	0	X
	TA1.CCI1A	0	0	1	Default value (Table 7.12)
	TA1.1	1			
	N/A	0	1	0	X
	DVSS	1			
	C0.2	X	1	1	X

Since we do not need to use any special or alternate function for these GPIO ports, therefore we do not need to take care of configuring the **PxSEL1.y** and **PxSEL0.y** registers for this project, since by default all of these bits are reset to 0 to enable all GPIO ports to work as general digital I/O mode.

It can be found from this keypad hardware configuration that both four pins on these GPIO ports are pulled down by pull down resistors, **RN10** and **RN13**. Therefore, all of these pins are driven as LOW if no logic signal is applied to these pins.

Generally, any keypad interface can be developed by either hardware or software. The former has higher running speed with high cost, and the latter has lower running speed with low cost. In this part, we will use software to build a keypad interfacing program to scan and identify the pressed key and display it in four LEDs, **PB3~PB0**, which are installed in the EduBASE ARM® Trainer, via the GPIO pins, P3.2, P3.3, P2.5, and P2.4 (refer to Figure 4.31 in Chapter 4).

To check and identify a pressed key, two methods can be applied on the software:

1. Using the poll-up method to repeatedly scan and check all keypad key arrays
2. Using the interrupt method to check the pressed key

The second method is better and more efficient as compared with the first method. However, in this section we will discuss the first method since some GPIO pins have no interrupt generation functions in the MSP-EXP432P401R EVB.

7.5.1 Keypad interfacing programming structure

From the hardware configuration of this keypad device shown in Figure 7.3, it can be seen that four GPIO pins, P3.6, P5.2, P5.0, and P4.6, are connected to four keypad columns, columns 0~3. These four columns are generally in the LOW status since they are pulled down by four resistors **RN13**.

The so-called poll-up method is to scan all four columns by outputting HIGH to each row via each GPIO pin, P6.1, P4.0, P4.2, and P4.4. If any key on any column is pressed, all four columns are still in the LOW status. These can be checked by getting

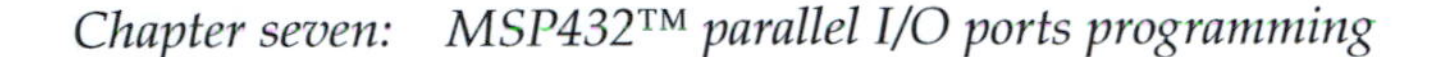

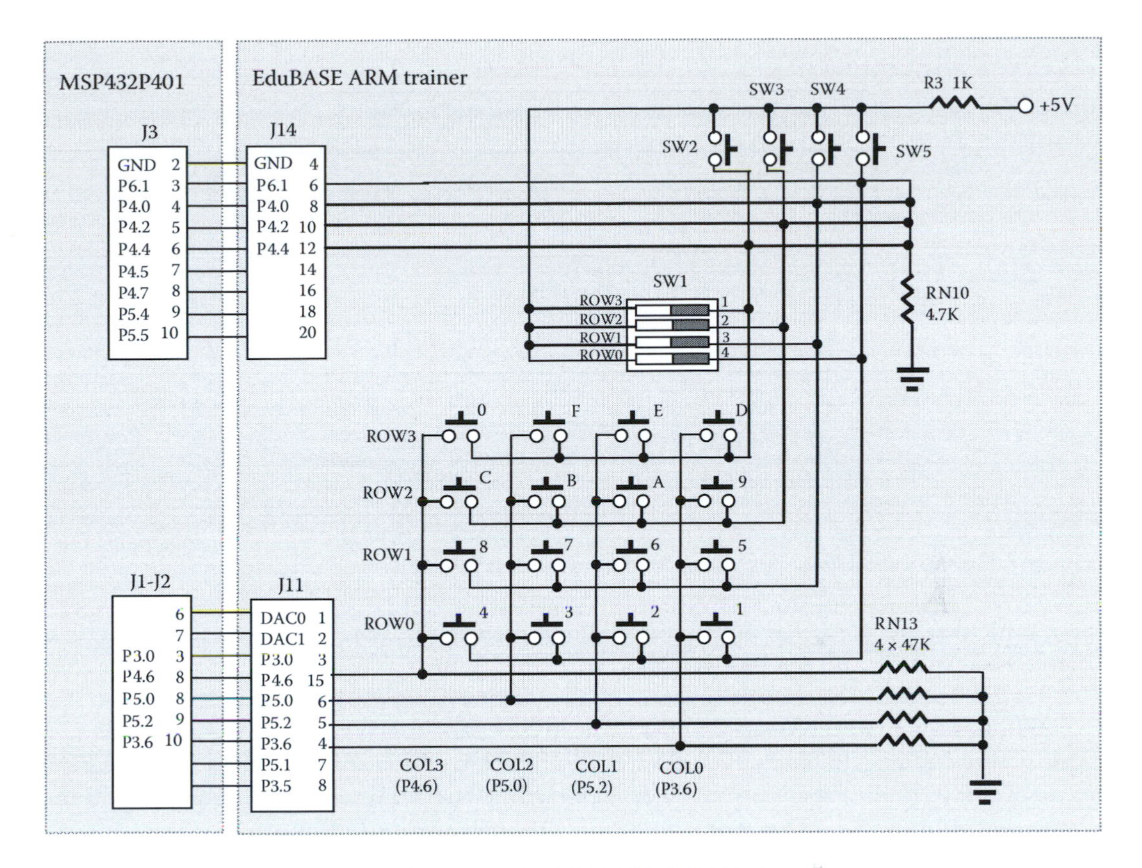

Figure 7.3 Keypad hardware configuration for EduBASE ARM Trainer.

them back via four pins, P3.6, P5.2, P5.0, and P4.6. If a key is pressed, this makes the corresponding column to HIGH. By checking the row number and column number, we can identify which key has been pressed.

In this way, we repeatedly send HIGH to each row one by one, and read back all columns via P3.6, P5.2, P5.0, and P4.6. Then we can check if any column is HIGH to identify the pressed key.

This method is easy but it is time consuming with low efficiency because most of the time the program is performing scanning and reading back functions without any chance to handle other possible jobs.

The functional block diagram for this program is shown in Figure 7.4.

Three user-defined functions are involved in this project, the **`InitKeypad()`**, **`ReadKeypad()`**, and **`GetKey()`**, to perform the initialization process and reading key process, respectively.

In the main program, the **`InitKeypad()`** function is called first to initialize and configure GPIO pins, P6.1, P4.0, P4.2, and P4.4, to make them as output pins and GPIO pins, P3.6, P5.2, P5.0, and P4.6, as input pins. Then a **`while(1)`** loop is used to repeatedly scan the keypad to try to read a key if any of them is pressed. This scanning and reading process is performed by calling the function **`ReadKey()`** in two times. The first-time reading result is saved in the **`KeyCode`** variable, and the second-time reading result is compared with the first-time result to make sure that the same key is pressed. This two-time reading process is necessary since the keypad pressing may cause some hardware vibrations and

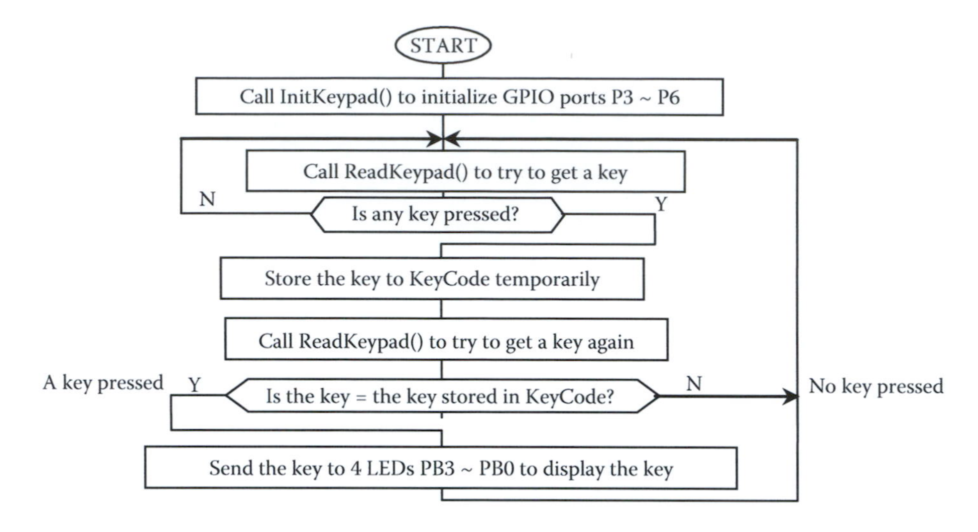

Figure 7.4 Functional block diagram of keypad scanning process.

bouncing to create some wrong key-press signal. In order to avoid these possible mistakes, two-time scanning and reading is a good solution.

Figure 7.5 shows the bouncing actions caused by pressing a key from the keypad.

If both time reading results are the same, this means that a key is pressed and the key code is sent to the GPIO pins, P3.2, P3.3, P2.5, and P2.4, which are connected to four LEDs, to display it (refer to Figure 4.31 in Chapter 4). Otherwise no key has been pressed and the program continues to perform loop scanning and reading for the future keys.

Another issue is the keypad connection and key number displayed in the EduBASE ARM® Trainer. The current keypad with the key numbers is only a simulation to a standard telephone keypad and numbers, as shown in Figure 7.6a. The actual connection for these keys and key numbers is shown in Figure 7.6b. Our keypad project is developed

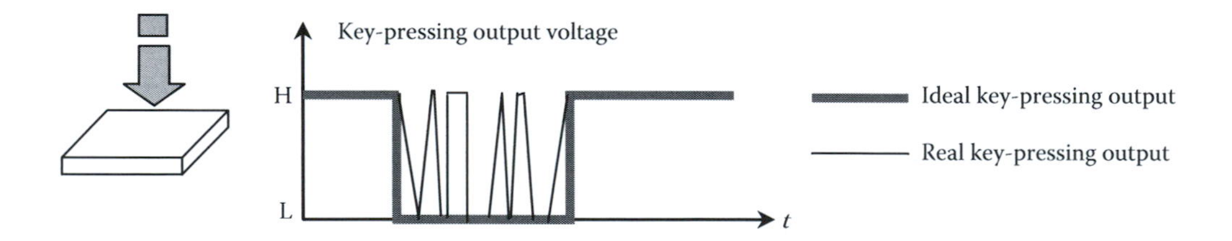

Figure 7.5 Bouncing actions caused by pressing a key from the keypad.

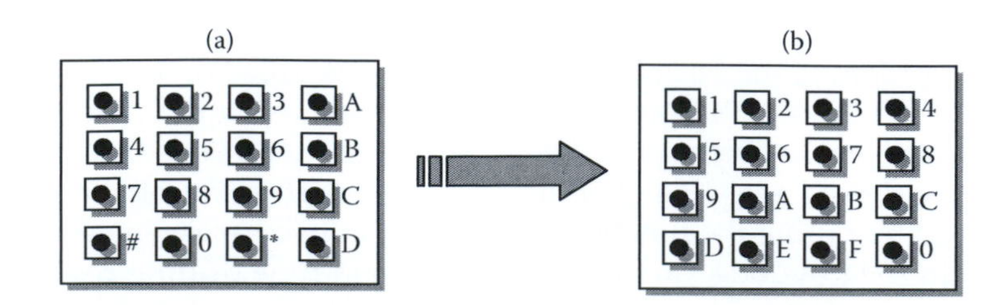

Figure 7.6 Two configurations of the keys and key numbers in the keypad.

based on the actual keypad and key numbers shown in Figure 7.6b. Therefore when we test our project with this keypad later, we should use the correct keypad with their key numbers shown in Figure 7.6b.

7.5.2 Create the keypad interfacing programming project

Now, let's create our keypad project **DRAKeyPad**. This project is built based on the **DRA** method with the polling strategy.

Perform the following operations to create a new project **DRAKeyPad**:

1. Open the Windows Explorer to create a new folder **DRAKeyPad** under the **C:\MSP432 Class Projects\Chapter 7** folder.
2. Open the Keil® ARM®-MDK µVersion5 and go to **Project|New µVersion Project** menu item to create a new µVersion Project. On the opened wizard, browse to our new folder **DRAKeyPad** that is created in step 1 above. Enter **DRAKeyPad** into the **File name** box and click on the **Save** button to create this project.
3. On the next wizard, you need to select the device (MCU) for this project. Expand three icons, **Texas Instruments**, **MSP432 Family**, and **MSP432P**, and select the target device **MSP432P401R** from the list by clicking on it. Click on the **OK** to close this wizard.
4. Next the Software Components wizard is opened, and you need to set up the software development environment for your project with this wizard. Expand two icons, **CMSIS** and **Device**, and check the **CORE** and **Startup** checkboxes in the **Sel.** column, and click on the **OK** button since we need these two components to build our project.

Since this project is relatively simple, therefore only a C source file is good enough.

7.5.2.1 Create the C source file DRAKeyPad.c

Perform the following operations to create a new C source file **DRAKeyPad.c**:

1. In the **Project** pane, expand the **Target** folder and right click on the **Source Group 1** folder and select the **Add New Item to Group "Source Group 1"**
2. Select the **C File (.c)** and enter **DRAKeyPad** into the **Name:** box, and click on the **Add** button to add this source file into the project
3. Enter the first part codes shown in Figure 7.7 into this source file

Let's have a closer look at the first part source codes to see how they work.

- Three system header files, **stdint.h**, **stdbool.h**, and **msp.h**, are included first at lines 4~6 in this source file since we need to use them in this project. In fact, we need to use most system macros defined in the header file **msp.h**.
- In lines 7~9, three user-defined functions, **InitKeypad()**, **ReadKeypad()**, and **GetKey()**, are declared since we need to use them to perform initialization and read-key processes later.
- Two local char variables, **keyCode** and **keyNum**, are created at lines 12. The first variable is used to accept and hold the returned key code by calling the **ReadKeypad()** function. The second variable works as a temporary key code holder to save the key code obtained from the first-reading process.

```
1  //*****************************************************************************************
2  // DRAKeyPad.c - Main application file for project DRAKeyPad (First Part Codes)
3  //*****************************************************************************************
4  #include <stdint.h>
5  #include <stdbool.h>
6  #include <msp.h>

7  void InitKeypad(void);
8  char ReadKeypad(void);
9  char GetKey(char row);

10 int main(void)
11 {
12  char keyCode, keyNum;

13  InitKeypad();                        // initialize GPIO pins
14  while(1)
15  {
16   keyCode = ReadKeypad();             // try to read a key from keypad
17   if (keyCode != 0)                   // if a key is pressed...
18   {
19    keyNum = keyCode;                  // reserve the key code
20    keyCode = ReadKeypad();            // try to get the key again
21    if (keyCode == keyNum)             // is the same key?
22    {
23     P3DIR = BIT2|BIT3;                // set P3.2 & P3.3 as output pins (LEDs - PB0 & PB1)
24     P2DIR = BIT5|BIT4;                // set P2.5 & P2.4 as output pins (LEDs - PB2 & PB3)
25     P3OUT = (keyNum & 0x0F) << 2;     // display keycode on LEDs PB1-PB0
26     P2OUT = 0;                        // clean P2 output pins to 0
27     P2OUT |= (keyCode & BIT2) << 3;   // display bit2 on keyCode in PB2 (P2.5 pin)
28     P2OUT |= (keyCode & BIT3) << 1;   // display bit3 on keyCode in PB3 (P2.4 pin)
29    }
30   }
31  }
32 }

33 void InitKeypad(void)
34 {
35  WDTCTL = WDTPW | WDTHOLD;            // stop WDT
36  P6DIR = BIT1;                        // configure P6.1 as output pin
37  P4DIR = BIT0|BIT2|BIT4;              // set P4.0, P4.2 & P4.4 as output pins
38  P3DIR &= ~BIT6;                      // set P3.6 as input pin
39  P5DIR &= ~(BIT0|BIT2);               // set P5.0 & P5.2 as input pins
40  P4DIR &= ~BIT6;                      // set P4.6 as input pin
41  P3REN = BIT6;
42  P3OUT &= ~BIT6;                      // pull-down resistor input on P3.6
43  P5REN = BIT0|BIT2;
44  P5OUT &= ~(BIT0|BIT2);               // pull-down resistor input on P5.0 & P5.2
45  P4REN = BIT6;
46  P4OUT &= ~BIT6;                      // pull-down resistor input on P4.6
47 }
```

Figure 7.7 First part codes for the source file DRAKeyPad.c.

- In line 13, the **InitKeypad()** function is called to initialize and configure the GPIO Ports P3, P4, P5, and P6 to make P6.1, P4.0, P4.2, and P4.4 as output pins and P3.6, P5.2, P5.0, and P4.6 as input pins.
- An infinitive **while()** loop is then executed to repeatedly scan and read the keypad to try to get the pressed key and display the key code on the four LEDs in line 14.
- In line 16, the **ReadKeypad()** function is called to try to get the pressed key, and the pressed and returned key code is assigned to the variable **keyCode**.
- If the **keyCode** is not a 0, which means that a key has been pressed, its value is then stored into the variable **keyNum** in line 19. This value will be compared with the next reading key code to confirm whether the same key is pressed later.
- In line 20, the **ReadKeypad()** function is called again to try to get the second pressed key.
- Now, compare the second reading key code with the first one. If both are same, which means that a key has been really pressed and it is not a keypad bouncing or a vibration. Then the key code is assigned to four LEDs to display it starting at line 21.
- A point to be noted is that four LEDs, **PB3~PB0**, which are installed in the EduBASE ARM Trainer, are not connected to four GPIO pins in a sequence format, and this makes the displaying the key code a little trouble. Refer to Figure 4.31 in Chapter 4, it can be found: P3.2 → PB0; P3.3 → PB1; P2.5 → PB2 and P2.4 → PB3. These connections need us to do some modifications on key code to display it on these four LEDs.
- In lines 23~24, these four pins are defined as output pins by assigning the logic HIGH to the related DIR registers.
- Since pins P3.2 (**PB0**) and P3.3 (**PB1**) are in normal connection format, we can just shift left 2-bit for the lowest two digits, bits 1 and 0, on the key code to align them to fit P3.3 and P3.2 pins to display them on these two LEDs in line 25.
- In line 26, the GPIO Port P2 is first reset to 0 to make it ready to display key code.
- Since pins P2.5 (**PB2**) and P2.4 (**PB3**) are not in normal connection format, and they are just inversion in the displaying-order. Therefore, we need to exchange these two bits, bits 2 and 3, on the key code to make them align to P2.5 and P2.4, respectively. In line 27, we pick up bit 2 on the key code and shift it left by three bits to align it to fit P2.5 (**PB2**).
- In line 28, we pick up bit 3 on the key code and shift it left by 1-bit to align it to fit P2.4 (**PB3**).
- The **InitKeypad()** function starts at line 33.
- In line 35, the Watchdog timer is temporarily terminated to avoid its disturbance.
- The codes in lines 36~37 are used to configure GPIO pins, P6.1, P4.0, P4.2, and P4.4 as output pins.
- The GPIO pins, P3.6, P5.2, P5.0, and P4.6, are configured as input pins in lines 38~40.
- The codes in lines 41~46 are used to configure input pins P3.6, P5.2, P5.0, and P4.6 to be connected with related pull down resistors.

Now, enter the second part codes shown in Figure 7.8 into this source file. Let's have a closer look at the second part source codes to see how they work.

- The **ReadKeypad()** function starts at line 51.
- First configure GPIO port P4 to output 0 to make rows 1 (**P4.0**), 2 (**P4.2**), and 3 (**P4.4**) to LOW in line 53.
- Then output a HIGH to GPIO pin P6.1 to make row0 as HIGH in line 54.

```
48 //*****************************************************************************
49 // DRAKeyPad.c - Main application file for project DRAKeyPad (Second Part Codes)
50 //*****************************************************************************
51 char ReadKeypad(void)
52 {
53  P4OUT = 0;                              // output Low to all P4 pins
54  P6OUT = BIT1;                           // output High to P6.1 (row0)
55  if (GetKey(0)) {return GetKey(0);}      // get the key on row0
56  P6OUT &= ~BIT1;                         // output LOW to P6.1 (row0)
57  P4OUT = BIT0;                           // output High to P4.0 (row1)
58  if (GetKey(1)) {return GetKey(1);}      // get the key on row1
59  P4OUT &= ~BIT0;                         // output LOW to P4.0 (row1)
60  P4OUT = BIT2;                           // output High to P4.2 (row2)
61  if (GetKey(2)) {return GetKey(2);}      // get the key on row2
62  P4OUT &= ~BIT2;                         // output LOW to P4.2 (row2)
63  P4OUT = BIT4;                           // output High to P4.4 (row3)
64  if (GetKey(3)) {return GetKey(3);}      // get the key on row3
65  P4OUT &= ~BIT4;                         // output LOW to P4.4 (row3)
66  return 0;
67 }

68 char GetKey(char row)
69 {
70  if (P3IN & BIT6) return (row * 4 + 1);  // key in column 0 (P3.6)
71  if (P5IN & BIT2) return (row * 4 + 2);  // key in column 1 (P5.2)
72  if (P5IN & BIT0) return (row * 4 + 3);  // key in column 2 (P5.0)
73  if (P4IN & BIT6) return (row * 4 + 4);  // key in column 3 (P4.6)

74  return 0;
75 }
```

Figure 7.8 Second part codes for the source file DRAKeyPad.c.

- In line 55, the user-defined function **GetKey(0)** is executed to try to find a pressed key that is connected to row0. The argument of this function is exactly the row number. If this function returned a nonzero value, which means that a key connected to row0 has been pressed, and this key value is returned to the **main()** function.
- The GPIO pin P6.1 is then deselected by assigning a 0 to it in line 56.
- The GPIO pin P4.0 is selected by assigning it to HIGH to make row1 as HIGH in line 57.
- In line 58, the user-defined function **GetKey(1)** is executed to try to find a pressed key that is connected to row1. If this function returned a nonzero value, which means that a key connected to row1 has been pressed, and this key value is returned to the **main()** function.
- The GPIO pin P4.0 is then deselected by assigning a 0 to it in line 59.
- The GPIO pin P4.2 is selected by assigning it to HIGH to make row2 as HIGH in line 60.
- In line 61, the user-defined function **GetKey(2)** is executed to try to find a pressed key that is connected to row2. If this function returned a nonzero value, which means that a key connected to row2 has been pressed, and this key value is returned to the **main()** function.
- The GPIO pin P4.2 is then deselected by assigning a 0 to it in line 62.

- The GPIO pin P4.4 is selected by assigning it to HIGH to make row3 as HIGH in line 63.
- In line 64, the user-defined function **GetKey(3)** is executed to try to find a pressed key that is connected to row3. If this function returned a nonzero value, which means that a key connected to row3 has been pressed, and this key value is returned to the **main()** function.
- The GPIO pin P4.4 is then deselected by assigning a 0 to it in line 65.
- Otherwise, a zero is returned to the **main()** function to indicate that no key has been pressed in line 66.
- The **GetKey()** function starts at line 68. The input argument of this function is the row number that has been selected and set to HIGH.
    - The codes in lines 70~73 are used to check four input pins, P3.6, P5.0, P5.2, and P4.6, in a sequence to see whether any of these columns is set to HIGH. If this happened, it means that a key has been pressed, and the pressed key value is calculated using the equation: **row** × **4** + **1**, and the key value is returned to the **ReadKeypad()** function.
- If no key is pressed, a zero is returned in line 74.

Now, let's build and run the project to test the function of this project. Before we can do this, we first need to set up the environment for this project to enable the compiler and debugger to locate our software sources.

7.5.3 Set up the environment to build and run the project

The only environment setup is to select the correct debugger for this project, which is **CMSIS-DAP Debugger**. Go to **Project**|**Options for Target "Target 1"** menu item to finish this setup. Also, make sure that the Debugger settings and the Flash Download setting are correct.

Now, build the project and download the image file of this project to the flash memory. Run the project by going to **Debug**|**Start/Stop Debug Session**, and then **Debug**|**Run** menu item.

During the project runs, press any key on the keypad and you can find that the corresponding LED with the related key number should be ON. Press the 0 key to turn off all IEDs.

Go to the **Debug**|**Stop** and **Debug**|**Start/Stop Debug Session** to stop the project. Our keypad project is very successful.

7.6 ADC programming project

There is one ADC module in the MSP432P401R MCU system, **ADC14**. The ADC14 module supports fast 14-bit analog-to-digital conversions. The module implements a 14-bit SAR core, sample select control, and enables up to 24 independent external analog inputs and two internal analog inputs to be converted simultaneously. The conversion-and-control buffer allows up to 32 independent ADC samples to be converted and stored without any CPU intervention.

In the EduBASE ARM® Trainer, the first channel (**AIN0**) in the module ADC0 is connected to a potentiometer **VR2** and the second channel (**AIN1**) is connected to a photo sensor **Q1**.

In order to have a clear picture about the ADC modules in the MSP432P401R MCU system, we have provided a detailed discussion about this module.

7.6.1 ADC module in the MSP432P401R MCU system

The MSP432P401R microcontroller contains a 14-bit ADC module. This ADC module provides the following features:

- 24 analog input channels (including the internal and external analog inputs)
- 14-bit precision ADC
- Single-ended and differential-input configurations
- On-chip internal temperature sensor
- Maximum sample rate of one million samples/second (**MSPS**)
- Sample-and-hold with programmable sampling periods controlled by software or timers
- Conversion initiation by software or timers
- Software-selectable on-chip reference voltage generation (1.2 V, 1.45 V, or 2.5 V) with option to make externally available
- Software-selectable internal or external reference
- Selectable conversion clock source
- Four conversion modes: single-channel, repeat-single-channel, sequence (auto-scan), and repeat-sequence (repeated auto-scan)
- Interrupt vector register for fast decoding of 38 ADC interrupts

This ADC14 module mainly comprises the following eight key components to coordinate the ADC to perform each conversion:

1. External and internal analog input selection
2. Internal and external reference voltage selection
3. ADC clock selection and clock predivision
4. ADC triggering source and sample-hold mode selection
5. ADC Conversion mode selection
6. ADC Conversion result memory-register selection
7. ADC Power Management
8. ADC Conversion interrupt-source selection and processing

Now, let's have a closer look at the ADC14 module architecture and organization.

7.6.2 ADC module architecture and functional block diagram

A simplified ADC module architecture and functional block diagram is shown in Figure 7.9.

Let's discuss these components one by one based on this architecture in the following sections.

7.6.2.1 External and internal analog input selection

In the MSP432P401R MCU system, there are 24 analog input channels available in total: 22 external analog inputs and 2 internal analog inputs. Two internal analog inputs include the Battery Monitor coming from analog input channel 23 and Internal Temperature Sensor coming from analog input channel 22.

An internal MUX is used and controlled by the **ADC14INCHx** bit field in the **ADC14 Conversion Memory Control x Register—ADC14MCTLx** (**x** = 0~31) to select an

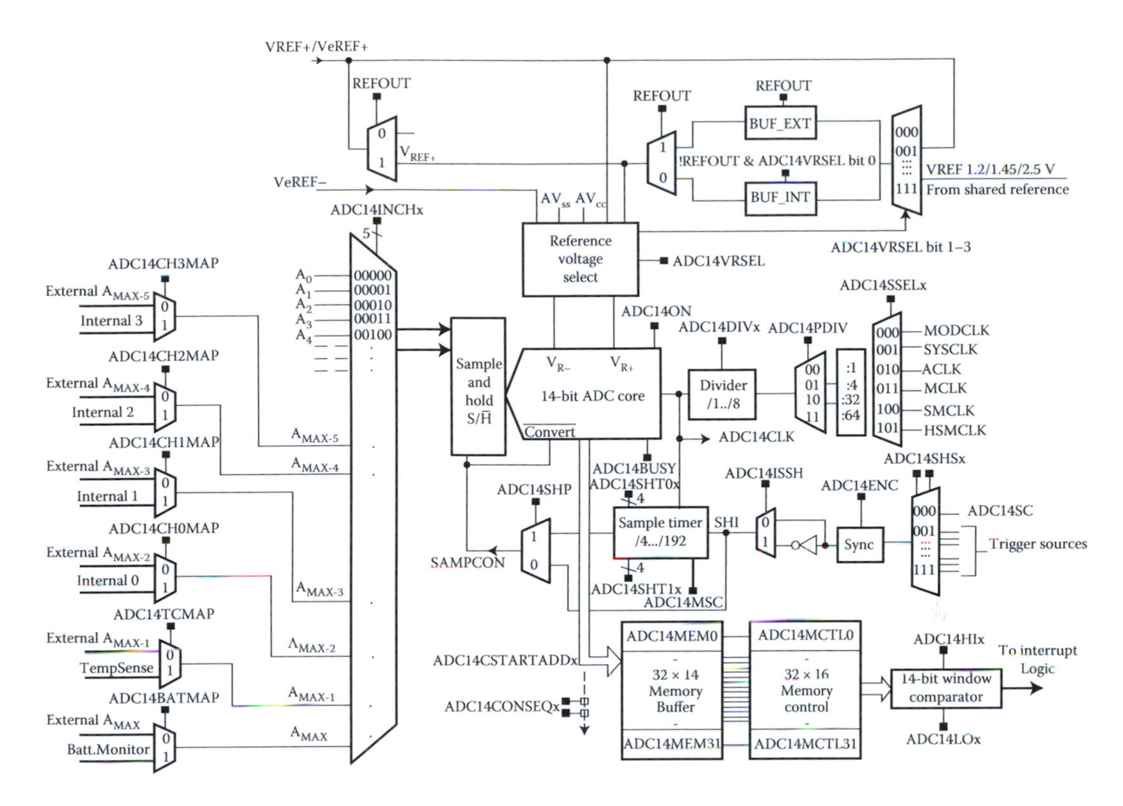

Figure 7.9 Simplified ADC module functional block diagram. (Courtesy Texas Instruments.)

external analog source as the input to the ADC. Totally, there are 32 control registers (**ADC14MCTL0~ADC14MCTL31**) used and each of them is used to select 1 analog input via an input channel. However, in the MSP432P401R MCU system, only 22 control registers (**ADC14MCTL0~ADC14MCTL21**) are utilized to select 22 external analog sources as inputs.

The first internal analog input source, **ADC14BATMAP**, is exactly to monitor the half of AVCC (1/2 AVCC) power supply to make sure that the MCU is working under a normal power supply. The second internal analog input source, **ADC14TCMAP**, is coming from an internal temperature sensor integrated with the ADC14 module. These two internal analog inputs can be selected by two bits (bits 22 and 23), **ADC14BATMAP** and **ADC14TCMAP**, in the ADC14 Control 1 Register (**ADC14CTL1**).

All 22 external analog inputs are defined as macros, **ADC14INCH_0~ADC14INCH_21**. All of these macros can be found from the system header file **msp.h**.

7.6.2.2 Internal and external reference voltage selection

As shown in Figure 7.9, various different reference voltage sources can be selected as the analog-to-digital conversion voltage references. Basically, the internal **VREF+** and **VREF−** pair as well as external **VeREF+** and **VeREF−** pair can work as references for this ADC14 module. These reference selections can be controlled by the **ADC14VRSEL** bit field (bits 11~8) in the **ADC14MCTLx** (**x** = 0~31) registers for certain selected conversions.

It can also be found from Figure 7.9 that there are two buffers, **BUF_INT** and **BUF_EXT**, available for the selected reference. When the **REFOUT** bit (bit-1) in the **REF Control**

Register 0 (**REFCTL0**) is 0, the selected reference voltage is buffered by the internal buffer (**BUF_INT**). However, if the **REFOUT** = 1, the selected reference is buffered by the external buffer (**BUF_EXT**).

In addition to internal and external references, one of three other reference voltages, 1.2, 1.45, or 2.5 V, which comes from the output of the REF_A module, can also be selected as a reference for the ADC module. When the **ADC14VRSEL** bit-file = 1111, one of three voltage references, 1.2, 1.45, or 2.5 V, can be selected as the ADC reference voltage. Which voltage can be selected is determined by the output of the REF_A module, which is based on the **REFVSEL** bit field (bits 5~4) value in the **REFCTL0** Register as shown below:

- **REFVSEL** = 00: 1.2 V (when **REFON** bit in **REFCTL0** is set to 1)
- **REFVSEL** = 01: 1.45 V (when **REFON** bit in **REFCTL0** is set to 1)
- **REFVSEL** = 11: 2.5 V (when **REFON** bit in **REFCTL0** is set to 1)

7.6.2.3 ADC clock selection and clock predivision

It can be found from Figure 7.9 that six possible clock sources can be selected to work as the conversion clock ADC14CLK. In the pulse sampling mode, this selected clock can work as both the conversion clock and can generate the sampling period.

The ADC14 source clock can be selected using the **ADC14SSELx** (**x** = 0~31) bit in the ADC14 Control 0 Register (**ADC14CTL0**). The input clock can be predivided by 1, 4, 32, or 64 by using the **ADC14PDIV** bits (bits 31~30) and then subsequently divided by 1~8 using the **ADC14DIVx** bits (bits 24~22) in the **ADC14CTL0** Register.

Possible ADC14CLK sources are MODCLK, SYSCLK, ACLK, MCLK, SMCLK, and HSMCLK.

7.6.2.4 ADC triggering source and sample-hold mode selection

An analog-to-digital conversion can be initiated with a rising edge of the sample input signal SHI. The source for SHI is selected with the **ADC14SHSx** bits (bits 29~27) in the **ADC14CTL0** Register. Table 7.14 shows all possible values in the **ADC14SHSx** bit field and related trigger sources. This means that the ADC14 module can be either triggered by the software or by one of timer comparison outputs.

The ADC14 module requires 9, 11, 14, and 16 ADC14CLK cycles for 8-bit, 10-bit, 12-bit, and 14-bit resolution modes respectively. The actual resolution is controlled by the bit field

Table 7.14 ADC14 trigger signals

ADC14SHSx (bits 29~27)		Trigger sources
Decimal	Binary	
0	000	Software (ADC14SC)
1	001	TA0_C1
2	010	TA0_C2
3	011	TA1_C1
4	100	TA1_C2
5	101	TA2_C1
6	110	TA2_C2
7	111	TA3_C1

ADC14RES (bits 5~4) in the **ADC14CTL1** Register. The polarity of the SHI signal source can be inverted by setting the **ADC14ISSH** bit (bit 25) in the **ADC14CTL0** Register. The Sample Control (**SAMPCON**) signal controls the sample period and start of conversion.

Basically, the ADC14 module performs two steps to complete its one analog-to-digital conversion:

1. The sampling-and-hold mode with the Sample and Hold module (SHM) for one analog input. During this sampling period, the input analog signal is connected to the sampling capacitor of the ADC module and charges the sampling capacitor.
2. Then the input analog signal is disconnected from the sampling capacitor of the ADC module to hold the input sample value on that capacitor to enable the ADC14 to perform its conversion.

These two steps are controlled by the SAMPCON signal shown in Figure 7.9. When the SAMPCON is high, sampling starts and executes. When the sample process is done, a high-to-low SAMPCON transition enables the hold process to begin to start the analog-to-digital conversion. The longer the sampling period, the more accurate ADC result can be obtained. However, the lower conversion frequency would be achieved.

In the MSP432P401R MCU system, two bit fields, **ADC14SHT1x** and **ADC14SHT0x**, in the **ADC14CTL0** Register are used to configure the sampling period. The **ADC14SHT1x** bit field (bits 15~12) is used to configure the sampling length for the registers **ADC14MEM8~ADC14MEM23**, but the **ADC14SHT0x** bit field (bits 11~8) is to configure the sampling period for the registers **ADC14MEM0~ADC14MEM7** as well as **ADC14MEM24~ADC14MEM31**.

Table 7.15 lists all possible sampling lengths and related values.

In the MSP432P401R MCU system, two different sample-and-hold modes are defined by control bit **ADC14SHP** (bit 26) in the **ADC14CTL0** Register:

- Extended Sample Mode
- Pulse Mode

7.6.2.4.1 Extended sample mode The extended sample mode is selected when **ADC14SHP** = 0. In this mode, the ADC trigger signal is coming from the analog input. The input analog signal controls the SAMPCON signal and defines the sampling period t_{sample}.

As shown in Figure 7.10, when the input analog signal works as the SHI, a low-to-high or a rising edge is generated as the SHI is greater than a threshold value. As

Table 7.15 ADC14 sampling period selection

ADC14SHT1x/ADC14SHT0x		Sampling length (ADC14CLK periods)
Decimal	Binary	
0	0000	4
1	0001	8
2	0010	16
3	0011	32
4	0100	64
5	0101	96
6	0110	128
7	0111	192

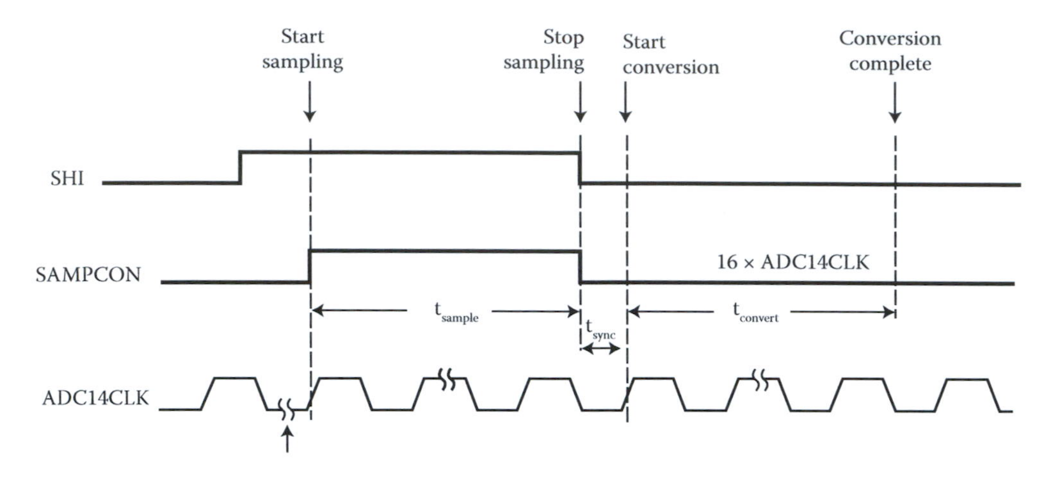

Figure 7.10 Extended sample mode for 14-bit conversion. (Courtesy Texas Instruments.)

the next ADC14CLK comes, the SAMPCON is triggered to high to begin the sampling process with the sampling length t_{sample}. When the sampling period is complete, a high-to-low transition on SAMPCON stops the sampling period and starts the hold and conversion process. For 14-bit resolution, after 16 ADC14CLK periods ($t_{convert}$), the conversion is complete and the result is stored to the related output register **ADC14MEMx**.

If an ADC internal buffer is used, the user should assert the sample trigger, wait for the **ADC14RDYIFG** flag to be set (indicating the ADC14 local buffered reference is settled) and then keep the sample trigger asserted for the desired sample period before deasserting.

One point to be noted is that the maximum sampling time must not exceed 420 μs.

7.6.2.4.2 Pulse mode The pulse sample mode is selected when **ADC14SHP** = 1. The ADC sample trigger signal is coming from one of the timer's comparator outputs. The SHI signal is used to trigger the sampling timer. The **ADC14SHT0x** and **ADC14SHT1x** bits in the **ADC14CTL0** Register control the interval of the sampling timer that defines the SAMPCON sample period t_{sample}. The sampling timer keeps SAMPCON high while waiting for reference and internal buffer to settle if the internal reference is used, synchronization with ADC14CLK and for the programmed interval t_{sample}.

In this mode, the sampling length is determined by the timer. During the timer performs the counting process, the SAMPCON signal should be kept in High to enable the sampling process to be executed. Once the counting process is done and the time in the timer is up, the output of the comparator of the timer triggers the SAMPCON to become Low to stop the sampling period, and begin the hold-conversion process, as shown in Figure 7.11.

7.6.2.5 ADC conversion mode selection

In the MSP432P401R MCU system, the ADC14 module provides four different conversion modes and these modes can be selected by the values in the bit field **ADC14CONSEQx** (bits 18~17) in the **ADC14CTL0** Register.

Table 7.16 shows these conversion modes and associated values defined in the bit field **ADC14CONSEQx** in the **ADC14CTL0** Register.

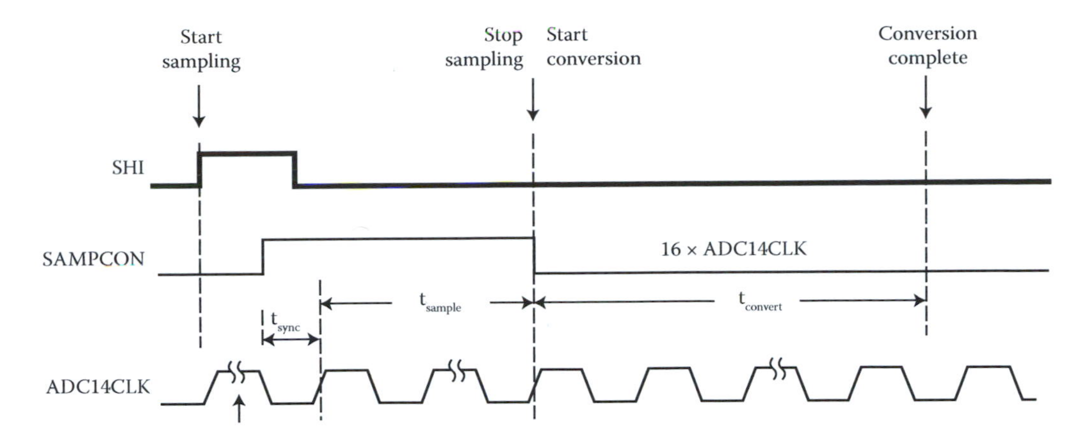

Figure 7.11 Pulse mode for 14-bit resolution conversion. (Courtesy Texas Instruments.)

Table 7.16 ADC14 conversion mode selection

ADC14CONSEQx		ADC14 conversion mode
Decimal	**Binary**	
0	**00**	Single-channel single-conversion
1	**01**	Sequence-of-channels (auto-scan)
2	**10**	Repeat-single-channel
3	**11**	Repeat-sequence-of-channels (repeated auto-scan)

The ***Single-Channel Single-Conversion Mode*** means that a single channel is converted by one time, and then the ADC stops its conversion. The result is stored in the related **ADC14MEMx** Register defined by the **ADC14CSTARTADDx** bit field (bits 20~16) in the **ADC14CTL1** Register.

An analog-to-digital conversion can be triggered by software when the **ADC14SC** bit (bit-0) in the **ADC14CTL0** Register is set to 1. When **ADC14SC** bit triggers a conversion, successive conversions can be triggered by the **ADC14SC** bit. When any other trigger source is used, the **ADC14ENC** bit must be toggled between each conversion, and the **ADC14ENC** low pulse duration must be at least three **ADC14CLK** cycles.

The ***Sequence-Of-Channels or Auto-Scan Mode*** means that a sequence of channels is converted by one time, and then the conversion stops. The sequence of channels can be selected by setting the **ADC14EOS** (ADC14 End of Sequence) bit in the ADC14 Conversion Memory Control **x** Registers (**ADC14MCTL0~ADC14MCTL31**). The ADC results are written to the conversion memories starting with the **ADC14MEMx** registers defined by the **ADC14CSTARTADDx** bit field. The sequence automatically stops after the measurement of the channel with a set **ADC14EOS** bit.

For example, if we want to convert a sequence of channels, from channel 0 to channel 7, all **ADC14EOS** bits in **ADC14MCTL0~ADC14MCTL6** should be reset to 0, but the **ADC14EOS** bit in the **ADC14MCTL7** must be set to 1 to indicate that channel 7 is the last channel for this sequence.

Similar to single-channel single-conversion, when using the **ADC14SC** bit to start a sequence, additional sequences can also be started by the **ADC14SC** bit.

The ***Repeat-Single-Channel Mode*** indicates that a single channel should be converted repeatedly. This kind of conversion works as a loop until the **ADC14ENC** bit (bit 1) in the **ADC14CTL0** Register is cleared to 0 to disable the ADC module. The ADC results are written to the **ADC14MEMx** Register defined by the **ADC14CSTARTADDx** bit field. When each single channel conversion is complete, the related **ADC14IFGx** bit in the **ADC14IFG0** Register should be set to indicate this completion. It is necessary to read the result after the completed conversion because only one **ADC14MEMx** memory is used and is overwritten by the next conversion.

The ***Repeat-Sequence-Of-Channels Mode*** means that a sequence of channels is converted repeatedly. Similar to repeat-single-channel, this mode enables the ADC module to repeatedly convert a sequence of channels until the **ADC14ENC** bit (bit 1) in the **ADC14CTL0** Register is cleared to 0 to disable the ADC module. The ADC results are written to the conversion memories starting with the **ADC14MEMx** registers defined by the **ADC14CSTARTADDx** bit field. The sequence ends after the measurement of the channel with a set **ADC14EOS** bit and the next trigger signal restarts the sequence.

Stopping ADC14 conversion depends on the conversion mode, however the following operational sequence is recommended:

- Reset the **ADC14ENC** bit in single-channel single-conversion mode to stop a conversion immediately. The results may be unreliable. For reliable results, poll the ADC busy bit, **ADC14BUSY** (bit 16) in the **ADC14CTL0** Register until it is reset before clearing the bit **ADC14ENC**.
- Reset the **ADC14ENC** bit during repeat-single-channel operation to stop the converter at the end of the current conversion.
- Reset the **ADC14ENC** bit during a sequence or repeat-sequence mode to stop the converter at the end of the current conversion.
- To stop conversion immediately in any mode, reset the bit **ADC14CONSEQx** to 0 and reset the **ADC14ENC** bit. In this case, however the conversion results may unreliable.

A point to be noted is that if no **ADC14EOS** bit has been set and a sequence mode is selected, resetting the **ADC14ENC** bit does not stop the sequence. To stop the sequence, first select a single-channel mode and then reset **ADC14ENC**.

To configure the ADC to perform successive conversions automatically and as quickly as possible, a multiple sample and convert function is available. When the **ADC14MSC** bit (bit 7) is set to 1, and the bit **ADC14CONSEQx** > 0, and a sample timer is used, the first rising edge of the SHI signal triggers the first conversion. Successive conversions are triggered automatically as soon as the prior conversion is completed. Additional rising edges on SHI are ignored until the sequence is completed in the single-sequence mode, or until the **ADC14ENC** bit is toggled in repeat-single-channel or repeated-sequence modes. The function of the **ADC14ENC** bit is unchanged when using the **ADC14MSC** bit.

7.6.2.6 ADC conversion result memory-register selection

Totally, there are 32 **ADC14MEMx** conversion memory registers available to store conversion results. Each **ADC14MEMx** Register is configured with an associated **ADC14MCTLx** control register. The **ADC14INCHx** and **ADC14DIF** bits in the **ADC14MCTLx** Register select the input channels. The **ADC14EOS** bit defines the end of sequence when a sequential conversion mode is used. A sequence rolls over from **ADC14MEM31** to **ADC14MEM0** when the **ADC14EOS** bit in the register **ADC14MCTL31** is not set.

The **ADC14CSTARTADDx** bit field defined the first **ADC14MCTLx** used for any conversion. If the conversion mode is single-channel or repeat-single-channel, the **ADC14CSTARTADDx** points to the single **ADC14MCTLx** Register to be used. If the conversion mode is either sequence-of-channels or repeat-sequence-of-channels, the **ADC14CSTARTADDx** bit field points to the first **ADC14MCTLx** Register to be used in a sequence. A pointer, not visible to software, is incremented automatically to the next **ADC14MCTLx** Register in a sequence when each conversion completes. The sequence continues until an **ADC14EOS** bit in **ADC14MCTLx** Register is set to 1; this is the last control byte processed.

When a conversion result is written to a selected **ADC14MEMx** Register, the corresponding flag bit **ADC14IFGx** in the **ADC14IFGR0** Register is set.

7.6.2.7 ADC power management

The ADC14 module supports two power modes selected through **ADC14PWRMD** bit field (bits 1~0) in **ADC14CTL1** Register. When **ADC14PWRMD** bits are reset to 00, the Regular Power Mode is selected but when the **ADC14PWRMD** bits are set to 10, the Low Power Mode is selected. The ADC14 supports 8-bit, 10-bit, 12-bit, and 14-bit resolution settings selected through **ADC14RES** bit field (bits 5~4) in **ADC14CTL1** Register.

The Regular Power Mode (when the **ADC14PWRMD** = 00) supports sampling rates up to 1 MSPS and can be used with any of the resolutions settings. The Low Power Mode (**ADC14PWRMD** = 10) is a power saving mode recommended for 12-bit, 10-bit, or 8-bit resolution settings with sampling rates not exceeding 200 KSPS.

The ADC14 module is designed for low-power applications. When the ADC14 module is not actively converting, the core is automatically disabled and automatically re-enabled when it is needed. The **MODOSC** or **SYSOSC** are also automatically enabled to provide MODCLK or SYSCLK to ADC14 when needed and disabled when not needed for ADC14.

7.6.2.8 ADC conversion interrupt-source selection and processing

Theoretically, there are 38 interrupt sources in the ADC14 module, which include 32 analog conversions completion and six internal interrupt sources.

The 32 analog conversions completion interrupt sources include:

- **ADC14IFG0~ADC14IFG31** bits in the **ADC14IFGR0** Register: All of the related **ADC14IFGx** bits are set when their corresponding **ADC14MEMx** memory registers are loaded with the conversion results. An ADC interrupt request is generated if the corresponding **ADC14IEx** bit in the **ADC14IER0** Register is set.

  The six internal interrupt sources contain:
- **ADC14OV** (ADC14MEMx Register Overflow Interrupt): The **ADC14OV** interrupt occurs when a conversion result is written to the related **ADC14MEMx** Register before its previous conversion result was read. The **ADC14OVIFG** bit (bit 4) in the **ADC14IFGR1** Register is set to 1 if this overflow happened and an **ADC14OV** interrupt is generated if the bit **ADC14OVIE** (bit 4) in the **ADC14IER1** Register is set to 1 to enable this interrupt.
- **ADC14TOV** (ADC14 Conversion Time Overflow Interrupt): The **ADC14TOV** condition is generated when another sample-and-conversion is requested before the current conversion is completed. The **ADC14TOVIFG** bit (bit 5) in the **ADC14IFGR1** Register is set to 1 if this condition happened and an **ADC14TOV** interrupt is generated if the bit **ADC14TOVIE** (bit 5) in the **ADC14IER1** Register is set to 1 to enable this interrupt. The DMA is triggered after the conversion in single-channel conversion mode or after the completion of a sequence of channel conversions in sequence of channels conversion mode.

- **ADC14RDY** (ADC14 Local Buffered Reference Ready Interrupt): The **ADC14RDY** condition is created when the local buffer reference is ready to be used. The **ADC14RDYIFG** bit (bit 6) in the **ADC14IFGR1** Register is set if the local buffered reference is ready. An ADC14RDY interrupt can be generated if the **ADC14RDYIE** bit (bit 6) in the **ADC14IER1** Register is set to 1 to enable this interrupt. It can be used during extended sample mode instead of adding the maximum ADC14 local buffered reference settle time to the sample signal time.
- **ADC14HI** (ADC14MEMx Register's content is greater than the upper limit of the window comparator interrupt): The **ADC14HI** event is generated if the conversion result stored in an ADC14MEMx is greater than the upper limit set by the window comparator. The **ADC14HIIFG** bit (bit 3) in the **ADC14IFGR1** Register is set to 1 if this event happened and an **ADC14HI** interrupt is generated if the bit **ADC14HIIE** (bit 3) in the **ADC14IER1** Register is set to 1 to enable this interrupt.
- **ADC14LO** (ADC14MEMx Register's content is smaller than the lower limit of the window comparator interrupt): The **ADC14LO** event is generated if the conversion result stored in an ADC14MEMx is smaller than the lower limit set by the window comparator. The **ADC14LOIFG** bit (bit 2) in the **ADC14IFGR1** Register is set to 1 if this event happened and an **ADC14LO** interrupt is generated if the bit **ADC14LOIE** (bit 2) in the **ADC14IER1** Register is set to 1 to enable this interrupt.
- **ADC14IN** (ADC14MEMx Register's content is greater than the lower limit and smaller than the upper limit of the window comparator interrupt): The **ADC14IN** event is generated if the conversion result stored in an ADC14MEMx is greater than the lower limit and smaller than the upper limit set by the window comparator. The **ADC14INIFG** bit (bit 1) in the **ADC14IFGR1** Register is set to 1 if this event happened and an **ADC14IN** interrupt is generated if the bit **ADC14INIE** (bit 1) in the **ADC14IER1** Register is set to 1 to enable this interrupt.

Generally, an ADC interrupt can be enabled, generated, and handled in the following sequence:

1. Each desired ADC interrupt source should be first enabled by setting the corresponding bit in the **ADC14IER0** (for 32 analog inputs) or **ADC14IER1** (for 6 internal events) Register.
2. If an ADC conversion is complete and that conversion has the highest priority, the conversion result is written into the corresponding **ADC14MEMx** Register and the related **ADC14IFGx** bit in the **ADC14IFGR0** Register is set to 1. Then a related ADC14 conversion complete interrupt is generated and sent to the NVIC to be further processed. If an ADC internal event occurs, the related bit in the **ADC14IFGR1** Register is set and a corresponding interrupt is generated and sent to the NVIC, too.
3. This highest-priority enabled ADC14 interrupt then generates a number in the **ADC14IV** (ADC14 Interrupt Vector) Register. This number can be evaluated or loaded into the PC to automatically enter the appropriate ISR to handle and process this interrupt. Disabled ADC14 interrupts do not affect the **ADC14IV** value.
4. As soon as entering the ISR, the responded interrupt should be cleared immediately to enable other higher priority interrupts to be generated and handled. A reading of the **ADC14IV** Register automatically resets the highest-pending Interrupt condition and flag except the **ADC14IFGx** flags. The **ADC14IFGx** bits are reset automatically by accessing their associated **ADC14MEMx** Register or may be reset with software.

Writing to the **ADC14IV** Register clears all pending interrupt conditions and flags. An optional way to clear an interrupt flag is to set the corresponding bit on the **ADC14CLRIFGR0** Register (for 32 analog input sources) or set the related bit on the **ADC14CLRIFGR1** Register (for 6 internal ADC events).

A point to be noted is: If another interrupt is pending after the servicing of an interrupt, another interrupt is generated. For example, if both the **ADC14OV** and the **ADC14IFG3** interrupts are pending when the ISR accesses the **ADC14IV** Register, the **ADC14OV** Interrupt condition is reset automatically. After the **ADC14OV** interrupt service is completed, the **ADC14IFG3** generates another interrupt.

7.6.3 ADC14 core and conversion result format

The ADC14 core is used to convert an analog input to a 14-bit digital value. The core uses two programmable-selectable voltage references (VR+ and VR–) to define the upper and lower limits of the conversion. The digital output (N_{ADC}) is full scale (**0x3FFFh**) when the input signal is equal to or higher than VR+, and is 0 when the input signal is equal to or is lower than VR–. The input channel and the reference voltage levels (VR+ and VR–) are defined in the conversion-control memory.

For the single-ended mode conversion, the conversion resolution is ($2^{14} = 16{,}384$)

$$1\text{LSB} = \frac{V_{R+} - V_{R-}}{16{,}384}$$

The conversion input and output equation for a single-ended conversion is

$$\text{N}_{\text{ADC}} = 2^{14} \times \frac{V_{IN+} - V_{R-}}{V_{R+} - V_{R-}} = 16{,}384 \frac{V_{IN+} - V_{R-}}{V_{R+} - V_{R-}}$$

For the differential mode conversion, the conversion resolution is

$$1\text{LSB} = \frac{V_{R+} - V_{R-}}{8192}$$

The conversion input and output equation for a differential mode conversion is

$$\text{N}_{\text{ADC}} = 2^{13} \times \frac{V_{IN+} - V_{IN-}}{V_{R+} - V_{R-}} + 2^{13} = 8192 \times \frac{V_{IN+} - V_{IN-}}{V_{R+} - V_{R-}} + 8192$$

The ADC14 core can be configured by two control registers, ADC14 Control Register 0 (**ADC14CTL0**) and ADC14 Control Register 1 (**ADC14CTL1**). The core is reset when the **ADC14ON** bit (bit 4) in the **ADC14CTL0** Register is 0. When **ADC14ON** = 1, reset is removed and the core is ready to power up when a valid conversion is triggered.

The ADC14 can be turned off by resetting the **ADC14ON** bit to 0 when not in use to save the power. If during a conversion the **ADC14ON** bit is cleared to 0, the conversion is abruptly stopped and everything is powered down. With few exceptions, all ADC14 control bits or bit field can only be modified when **ADC14ENC** = 0.

Table 7.17 ADC14 conversion result formats

Analog input voltage range	ADC14DIF	ADC14DF	ADC14RES	Ideal conversion results (ADC14DIF = 1)	ADC14MEMx read value
VIN+–VR– (–VREF~+VREF)	0	0	00	0~255	0000h~00FFh
	0	0	01	0~1023	0000h~03FFh
	0	0	10	0~4095	0000h~0FFFh
	0	0	11	0~16386	0000h~3FFFh
	0	1	00	–128~127	8000h~7F00h
	0	1	01	–512~511	8000h~7FC0h
	0	1	10	–2048~2047	8000h~7FF0h
	0	1	11	–8192~8191	8000h~7FFCh
VIN+–VIN– (–VREF~+VREF)	1	0	00	–128~127 [0~255]	0000h~00FFh
	1	0	01	–512~511 [0~1023]	0000h~03FFh
	1	0	10	–2048~2047 [0~4095]	0000h~0FFFh
	1	0	11	–8192~8191 [0~16383]	0000h~3FFFh
	1	1	00	–128~127	8000h~7F00h
	1	1	01	–512~511	8000h~7FC0h
	1	1	10	–2048~2047	8000h~7FF0h
	1	1	11	–8192~8191	8000h~7FFCh

The **ADC14ENC** bit must be set to 1 before any conversion can take place.

The conversion results are always stored in binary unsigned format. For differential inputs, the result has an offset of 8192 added to it to make the number positive. The data format bit (bit 3), ADC14 Differential Format (**ADC14DF**) in the ADC14CTL1 Register allows the user to read the conversion results as binary unsigned or signed binary in the 2's compliment format.

The ADC14 Differential Mode (**ADC14DIF**) bit (bit 13) in the **ADC14MCTLx** Register defined the conversion mode for the selected channel. When this bit is set to 1, a differential mode conversion is selected for the input channel. Otherwise a 0 means that this conversion is a single-ended mode. Table 7.17 shows the relationships among these control bits and converting result formats.

7.6.4 ADC14 registers

The ADC14 module is mainly controlled by the following registers:

1. ADC14 Control Register 0 (**ADC14CTL0**)
2. ADC14 Control Register 1 (**ADC14CTL1**)
3. ADC14 Conversion Memory Control Registers (**ADC14MCTL0**~**ADC14MCTL31**)
4. ADC14 Conversion Memory Registers (**ADC14MEM0**~**ADC14MEM31**)
5. ADC14 Interrupt Enable Registers (**ADC14IER0**~**ADC14IER1**)
6. ADC14 Interrupt Flag Registers (**ADC14IFGR0**~**ADC14IFGR1**)
7. ADC14 Clear Interrupt Flag Registers (**ADC14CLRIFGR0**~**ADC14CLRIFGR1**)
8. ADC14 Interrupt Vector Register (**ADC14IV**)

31	30	29	28	27	26	25	24
ADC14PDIV		ADC14SHSx			ADC14SHP	ADC14ISSH	ADC14DIVx
23	**22**	**21**	**20**	**19**	**18**	**17**	**16**
ADC14DIVx		ADC14SSELx			ADC14CONSEQx		ADC14BUSY
15	**14**	**13**	**12**	**11**	**10**	**9**	**8**
ADC14SHT1x				ADC14SHT0x			
7	**6**	**5**	**4**	**3**	**2**	**1**	**0**
ADC14MSC	Reserved		ADC14ON	Reserved		ADC14ENC	ADC14SC

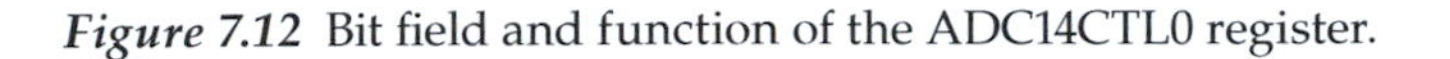

Figure 7.12 Bit field and function of the ADC14CTL0 register.

7.6.4.1 ADC14 control register 0 (ADC14CTL0)

This register provides the major controls for the ADC module, such as the sample-and-hold mode selection, input channel selection, ADC14CLK selection, conversion mode selection, sample-and-hold time selection, ADC enable, and start controls.

This is a 32-bit register and Figure 7.12 shows the bits configuration for this register. Table 7.18 shows the bit field and bit function for this register.

A point to be noted is that no modification or change for any bit or bit field on this register is allowed unless the **ADC14ENC** bit is reset to 0. When this bit is 0, the ADC14 module is disabled and all initializations and configurations can be performed under this condition. As soon as the **ADC14ENC** = 1, the ADC14 module is enabled and no modification can be made.

Another point to be noted is that the **ADC14BUSY** bit (bit 16) in this register can be used as a polled-up signal to detect whether a conversion is complete.

7.6.4.2 ADC14 control register 1 (ADC14CTL1)

This is a 32-bit register that provides additional control and configure functions to the ADC14 module. These control signals include six internal analog input channels selection, start conversion address, ADC conversion resolutions, conversion mode, and ADC power mode. Figure 7.13 shows the bit configurations of this register. Table 7.19 lists the bits or bit field values and functions.

7.6.4.3 ADC14 conversion memory control registers (ADC14MCTL0~ADC14MCTL31)

These 32 conversion memory control registers are used to configure and set up the ADC operational modes and analog input channels as well as the sample sequence. Each of these registers is used to configure the corresponding **ADC14MEMx** (**x** = 0~31) register. Equivalently, each register is used to set up and configure one related analog channel to enable the selected analog input to be converted and stored in the corresponding **ADC14MEMx** Register correctly.

Figure 7.14 shows the bit configurations of this register. Table 7.20 lists the bits or bit field values and functions.

7.6.4.4 ADC14 conversion memory registers (ADC14MEM0~ADC14MEM31)

These are 32-bit registers but only the lower 16-bit are used to store the ADC conversion results. Based on the conversion mode, the conversion results are arranged in these registers in the following different ways:

Table 7.18 Bit value and function for ADC14CTL0 register

Bit	Field	Type	Reset	Function
31:30	**ADC14PDIV**	RW	00	ADC14 Predivider. This bit predivides the selected ADC14 clock source **00** = Predivide by 1; **01** = Predivide by 4; **10** = Predivide by 32; **11** = Predivide by 64
29:27	**ADC14SHSx**	RW	000	ADC14 Sample-And-Hold Source Select **000** = ADC14SC bit (Software) **001** = TA0_C1; **010** = TA0_C2; **011** = TA1_C1; **100** = TA1_C2 **101** = TA2_C1; **110** = TA2_C2; **111** = TA3_C1
26	**ADC14SHP**	RW	0	ADC14 Sample-And-Hold Pulse-Mode Select **0** = SAMPCON signal is sourced from the sample-input signal **1** = SAMPCON signal is sourced from the sampling timer
25	**ADC14ISSH**	RW	0	ADC14 invert signal sample-and-hold **0** = The sample-input signal is not inverted **1** = The sample-input signal is inverted
24:22	**ADC14DIVx**	RW	000	ADC14 clock divider **000** = ÷1; **001** = ÷2; **010** = ÷3; **011** = ÷4; **100** = ÷5; **101** = ÷6 **110** = ÷7; **111** = ÷8
21:19	**ADC14SSELx**	RW	000	ADC14 clock source select **000** = MODCLK; **001** = SYSCLK; **010** = ACLK; **011** = MCLK **100** = SMCLK; **101** = HSMCLK; **110** = Reserved; **111** = Reserved
18:17	**ADC14CON-SEQx**	RW	00	ADC14 conversion sequence mode select **00** = Single-channel, single-conversion **01** = Sequence-of-channels **10** = Repeat-single-channel **11** = Repeat-sequence-of-channels
16	**ADC14BUSY**	RW	0	ADC14 busy. This bit indicates an active sample or conversion operation **0** = ADC has no operation **1** = A sequence, sample, or conversion is active
15:12	**ADC14SHT1x**	RW	0000	ADC14 sample-and-hold time. Define the number of ADC14CLK cycles in the sampling period for registers ADC14MEM8 to ADC14MEM23 **0000** = 4; **0001** = 8; **0010** = 16; **0011** = 32; **0100** = 64 **0101** = 96; **0110** = 128; **0111** = 192; **1000** to **1111** = Reserved
11:8	**ADC14SHT0x**	RW	0000	ADC14 sample-and-hold time. Define the number of ADC14CLK cycles in the sampling period for registers ADC14MEM0 to ADC14MEM7 and ADC14MEM24 to ADC14MEM31 **0000** = 4; **0001** = 8; **0010** = 16; **0011** = 32; **0100** = 64 **0101** = 96; **0110** = 128; **0111** = 192; **1000** to **1111** = Reserved

(*Continued*)

Table 7.18 (Continued) Bit value and function for ADC14CTL0 register

Bit	Field	Type	Reset	Function
7	**ADC14MSC**	RW	0	Multiple sample-conversion. Valid only for sequence or repeated modes **0** = The sampling timer requires a rising edge of the SHI signal to trigger each sample and convert **1** = The first rising edge of the SHI signal triggers the sampling timer, but further sample-and-conversions are performed automatically as soon as the prior conversion is completed
6:5	**Reserved**	RO	00	Reserved. Always reads as 0
4	**ADC14ON**	RW	0	ADC14 ON **0** = ADC14 off. **1** = ADC14 on. ADC core is ready to power up when a valid conversion is triggered
3:2	**Reserved**	RO	00	Reserved. Always reads as 0
1	**ADC14ENC**	RW	0	ADC14 enable conversion **0** = ADC14 disabled; **1** = ADC14 enabled
0	**ADC14SC**	RW	0	ADC14 start conversion. Software-controlled sample-and-conversion start ADC14SC and ADC14ENC may be set together with one instruction ADC14SC is reset automatically **0** = No sample-and-conversion-start. **1** = Start sample-and-conversion

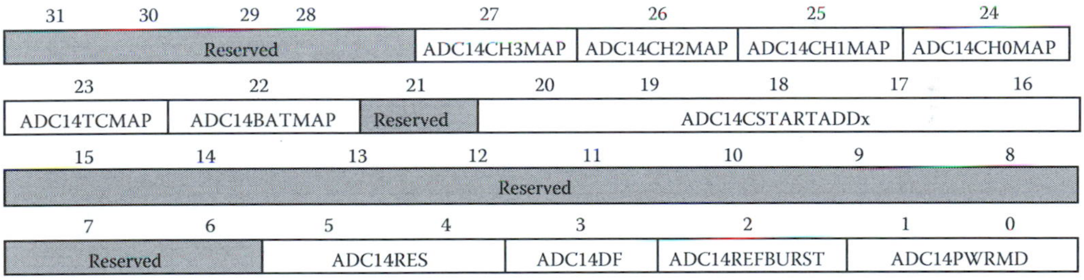

31–28	27	26	25	24
Reserved	ADC14CH3MAP	ADC14CH2MAP	ADC14CH1MAP	ADC14CH0MAP

23	22	21	20–16
ADC14TCMAP	ADC14BATMAP	Reserved	ADC14CSTARTADDx

15–8
Reserved

7–6	5–4	3	2	1–0
Reserved	ADC14RES	ADC14DF	ADC14REFBURST	ADC14PWRMD

Figure 7.13 Bit field and function of the ADC14CTL1 Register.

- If **ADC14DF** = 0, the 14-bit conversion results are right aligned with unsigned binary:
  - In 14-bit mode, bit 13 is the MSB and bits 15~14 are 0
  - In 12-bit mode, bits 15~12 are 0
  - In 10-bit mode, bits 15~10 are 0
  - In 8-bit mode, bits 15~8 are 0
- If **ADC14DF** = 1, the 14-bit conversion results are left aligned with 2's-complement format:
  - In 14-bit mode, bit 15 is the MSB and bits 1~0 are 0
  - In 12-bit mode, bits 3~0 are 0
  - In 10-bit mode, bits 5~0 are 0
  - In 8-bit mode, bits 7~0 are 0

Table 7.19 Bit value and function for ADC14CTL1 register

Bit	Field	Type	Reset	Function
31:28	**Reserved**	RO	00	Reserved. Always reads as 0
27	**ADC14CH3MAP**	RW	0	Control internal channel 3 selection to ADC input channel MAX-5 **0** = ADC input channel internal 3 is not selected **1** = ADC input channel internal 3 is selected for ADC input channel MAX-5
26	**ADC14CH2MAP**	RW	0	Control internal channel 2 selection to ADC input channel MAX-4 **0** = ADC input channel internal 2 is not selected **1** = ADC input channel internal 2 is selected for ADC input channel MAX-4
25	**ADC14CH1MAP**	RW	0	Control internal channel 1 selection to ADC input channel MAX-3 **0** = ADC input channel internal 1 is not selected **1** = ADC input channel internal 1 is selected for ADC input channel MAX-3
24	**ADC14CH0MAP**	RW	0	Control internal channel 0 selection to ADC input channel MAX-2 **0** = ADC input channel internal 0 is not selected. **1** = ADC input channel internal 0 is selected for ADC input channel MAX-2
23	**ADC14TCMAP**	RW	0	Control temperature sensor ADC input channel selection **0** = ADC internal temperature sensor channel is not selected for ADC **1** = ADC internal temperature sensor channel is selected for ADC input channel MAX-1
22	**ADC14BATMAP**	RW	0	Control 1/2 AVCC ADC input channel selection **0** = ADC internal 1/2 AVCC channel is not selected for ADC **1** = ADC internal 1/2 AVCC channel is selected for ADC input channel MAX
21	**Reserved**	RO	0	Reserved. Always reads as 0
20:16	**ADC14C-STARTADDx**	RW	00000	ADC14 conversion start address. select which ADC14 conversion memory register is used for a single conversion or for the first conversion in a sequence. The value of **ADC14C-STARTADDx** is 0 to 1F, corresponding to ADC14MEM0 to ADC14MEM31
15:6	**Reserved**	RO	0	Reserved. Always reads as 0

(*Continued*)

Table 7.19 (Continued) Bit value and function for ADC14CTL1 register

Bit	Field	Type	Reset	Function
5:4	**ADC14RES**	RW	3	ADC14 Resolution **00** = 8 bits; **01** = 10 bits; 10 = 12 bits; **11** = 14 bits
3	**ADC14DF**	RW	0	ADC14 data read-back format **0** = Binary unsigned; **1** = Signed binary
2	**ADC14REFBURST**	RW	0	ADC reference buffer burst **0** = ADC reference buffer on continuously **1** = ADC reference buffer on only during sample-and-conversion
1:0	**ADC14PWRMD**	RW	0	ADC power modes **00** = Regular power mode for use with any resolution setting **10** = Low-power mode for 12-bit, 10-bit, and 8-bit resolution settings **01** = Reserved; 11 = Reserved

31	30	29	28	27	26	25	24
Reserved							
23	**22**	**21**	**20**	**19**	**18**	**17**	**16**
Reserved							
15	**14**	**13**	**12**	**11**	**10**	**9**	**8**
ADC14WINCTH	ADC14WINC	ADC14DIF	Reserved	ADC14VRSEL			
7	**6**	**5**	**4**	**3**	**2**	**1**	**0**
ADC14EOS	Reserved		ADC14INCHx				

Figure 7.14 Bit field and function of the ADC14MCTLx register.

If the user writes to the conversion memory registers, the results are corrupted.

The data is stored in the right-justified format and is converted to the left-justified 2's-complement format during read back. Reading this register clears the corresponding bit in the **ADC14IFGR0** Register.

A point to be noted is that the **ADC14MEMx** Register read by debugger does not clear the corresponding interrupt flag in **ADC14IFGR0** Register.

7.6.4.5 ADC14 interrupt enable registers (ADC14IER0~ADC14IER1)

Two registers, **ADC14IER0** and **ADC14IER1**, are used to enable the ADC-related interrupts for all external and internal analog input conversions.

Both are 32-bit register, but the **ADC14IER0** used its all 32 bits (**ADC14IE31~ADC14IE0**), each bit is for each related external analog conversion channel, to enable an interrupt to be generated if the corresponding **ADC14IFGx** bit in the **ADC14IFGR0** Register is set to 1. This means that an external analog input has been converted to the related digital value. If an **ADC14IEx** bit is set to 1, which means that the related external analog input in channel **x** is enabled for its interrupt when it is converted to a digital value.

Table 7.20 Bit value and function for ADC14MCTL0~ADC14MCTL31 registers

Bit	Field	Type	Reset	Function
31:16	**Reserved**	RO	00	Reserved. Always reads as 0
15	**ADC-14WINCTH**	RW	0	Window comparator threshold register selection **0** = Use window comparator thresholds 0, ADC14LO0 and ADC14HI0 **1** = Use window comparator thresholds 1, ADC14LO1 and ADC14HI1
14	**ADC14WINC**	RW	0	Comparator window enable **0** = Comparator window disabled; 1 = Comparator window enabled
13	**ADC14DIF**	RW	0	Differential mode **0** = Single-ended mode enabled; 1 = Differential mode enabled
12	**Reserved**	RO	0	Reserved. Always reads as 0
11:8	**ADC14VRSEL**	RW	0	Select combinations of V(R+) and V(R−) sources as well as the buffer selection and buffer on or off **0000** = V(R+) = AVCC, V(R−) = AVSS **0001** = V(R+) = VREF buffered, V(R−) = AVSS **0010** to **1101** = Reserved **1110** = V(R+) = VeREF+, V(R−) = VeREF− **1111** = V(R+) = VeREF+ buffered, V(R−) = VeREF− It is recommended to connect VeREF− to on-board ground when VeREF− is selected for V(R−)
7	**ADC14EOS**	RW	0	End of sequence. Indicates the last conversion in a sequence **0** = Not end of sequence **1** = End of sequence (last sample)
6:5	**Reserved**	RO	0	Reserved. Always reads as 0
4:0	**ADC14INCHx**	RW	0	Input channel select. If even channels are set as differential then odd channel configuration is ignored **00000** = If ADC14DIF = 0: A0; If ADC14DIF = 1: Ain+ = A0, Ain− = A1 **00001** = If ADC14DIF = 0: A1; If ADC14DIF = 1: Ain+ = A0, Ain− = A1 **00010** = If ADC14DIF = 0: A2; If ADC14DIF = 1: Ain+ = A2, Ain− = A3 **00011** = If ADC14DIF = 0: A3; If ADC14DIF = 1: Ain+ = A2, Ain− = A3 **00100** = If ADC14DIF = 0: A4; If ADC14DIF = 1: Ain+ = A4, Ain− = A5 .... **11111** = If ADC14DIF = 0: A31; If ADC14DIF = 1: Ain+ = A30, Ain− = A31

Table 7.21 Bit value and function for ADC14IER1 register

Bit	Field	Type	Reset	Function
31:7	**Reserved**	RO	00	Reserved. Always reads as 0
6	**ADC14RDYIE**	RW	0	ADC14 local buffered reference ready interrupt enable 0 = Interrupt disabled; 1 = Interrupt enabled
5	**ADC14TOVIE**	RW	0	ADC14 conversion-time-overflow interrupt enable 0 = Interrupt disabled; 1 = Interrupt enabled
4	**ADC14OVIE**	RW	0	ADC14MEMx overflow interrupt enable 0 = Interrupt disabled; 1 = Interrupt enabled
3	**ADC14HIIE**	RW	0	Interrupt enable for the exceeding the upper limit of the window comparator for ADC14MEMx result register 0 = Interrupt disabled; 1 = Interrupt enabled
2	**ADC14LOIE**	RW	0	Interrupt enable for the falling short of the lower limit of the window comparator for the ADC14MEMx result register 0 = Interrupt disabled; 1 = Interrupt enabled
1	**ADC14INIE**	RW	0	Interrupt enable for the ADC14MEMx result register being greater than the ADC14LO threshold and below the ADC14HI threshold 0 = Interrupt disabled; 1 = Interrupt enabled
0	**Reserved**	RO	0	Reserved. Always reads as 0

The **ADC14IER1** only used its lower six bits, bits 6~1, to enable six conversion-related events to generate an interrupt if a condition occurs. Table 7.21 shows all of these six bit values and related functions. If one of these six bits is set to 1, which means that the selected interrupt is enabled.

7.6.4.6 ADC14 interrupt flag registers (ADC14IFGR0~ADC14IFGR1)

These two 32-bit registers are used to monitor 32 external analog input conversion results and 6 conversion-related events to provide a flag when the monitored external analog input conversion result is written into the related **ADC14MEMx** Register or an internal event occurred.

The **ADC14IFGR0** Register uses its 32 bits (**ADC14IFG31~ADC14IFG0**) to monitor all 32 external analog input conversions, each bit for each related input channel. The related **ADC14IFGx** (**x** = 0~31) bit in this register is set to 1 when the corresponding **ADC14MEMx** Register is loaded with a conversion result. This bit is reset to 0 when the **ADC14MEMx** Register is read, or when the **ADC14IV** Register is read, or when the corresponding bit in the **ADC14CLRIFGR0** Register is set to 1.

Unlike the **ADC14IFGR0** Register, the **ADC14IFGR1** Register only used its lower six bits to monitor six conversion-related events to be occurred. If the monitored event occurs, the related bit is set to 1 to indicate that the event has occurred. That bit is reset to 0 by reading the **ADC14IV** Register or when corresponding bit in **ADC14CLRIFGR1** is set to 1.

Table 7.22 shows these bit values and functions.

7.6.4.7 ADC14 clear interrupt flag registers (ADC14CLRIFGR0~ADC14CLRIFGR1)

These two 32-bit registers are used to clear flags set by **ADC14IFGR0** and **ADC14IFGR1** registers when related interrupts or events have been processed.

Table 7.22 Bit value and function for ADC14IFGR1 register

Bit	Field	Type	Reset	Function
31:7	**Reserved**	RO	00	Reserved. Always reads as 0
6	**ADC14RDYIFG**	RO	0	ADC14 local buffered reference ready interrupt flag **0** = No interrupt pending; **1** = Interrupt pending
5	**ADC14TOVIFG**	RO	0	ADC14 conversion time overflow interrupt flag **0** = No interrupt pending; **1** = Interrupt pending
4	**ADC14OVIFG**	RO	0	ADC14MEMx overflow interrupt flag **0** = No interrupt pending; **1** = Interrupt pending
3	**ADC14HIIFG**	RO	0	Interrupt flag for the exceeding the upper limit of the window comparator for ADC14MEMx result register **0** = No interrupt pending; **1** = Interrupt pending
2	**ADC14LOIFG**	RO	0	Interrupt flag for the falling short of the lower limit of the window comparator for the ADC14MEMx result register **0** = No interrupt pending; **1** = Interrupt pending
1	**ADC14INIFG**	RO	0	Interrupt flag for the ADC14MEMx result register being greater than the ADC14LO threshold and below the ADC14HI threshold **0** = No interrupt pending; **1** = Interrupt pending
0	**Reserved**	RO	0	Reserved. Always reads as 0

The **ADC14CLRIFGR0** Register used its all 32 bits, **CLRADC14IFG31~CLRADC14IFG0**, to clear the corresponding flags set by the **ADC14IFGR0** Register. When a bit in this register is set to 1, the related flag is cleared.

The **ADC14CLRIFGR1** Register only uses its lower six bits to clear flags set by the **ADC14IFGR1** Register. When a bit in this register is set to 1, the related flag is cleared. Table 7.23 lists these bit values and functions.

7.6.4.8 ADC14 interrupt vector register (ADC14IV)

This 32-bit register is used to store or hold a valid interrupt vector to direct the program to enter a related ISR to process the pended interrupt if it has been enabled. The

Table 7.23 Bit value and function for ADC14CLRIFGR1 register

Bit	Field	Type	Reset	Function
31:7	**Reserved**	RO	00	Reserved. Always reads as 0
6	**CLRADC14RDYIFG**	W	0	Clear ADC14RDYIFG Flag **0** = No effect; **1** = Clear the pending interrupt flag
5	**CLRADC14TOVIFG**	W	0	Clear ADC14TOVIFG Flag **0** = No effect; **1** = Clear the pending interrupt flag
4	**CLRADC14OVIFG**	W	0	Clear ADC14OVIFG Flag **0** = No effect; **1** = Clear the pending interrupt flag
3	**CLRADC14HIIFG**	W	0	Clear ADC14HIIFG Flag **0** = No effect; **1** = Clear the pending interrupt flag
2	**CLRADC14LOIFG**	W	0	Clear ADC14LOIFG Flag 0 = No effect; 1 = Clear the pending interrupt flag
1	**CLRADC14INIFG**	W	0	Clear ADC14INIFG Flag **0** = No effect; **1** = Clear the pending interrupt flag
0	**Reserved**	RO	0	Reserved. Always reads as 0

highest-priority enabled ADC14 interrupt generates a number in the **ADC14IV** Register. This number can be evaluated or loaded into the PC to automatically enter the appropriate ISR to handle and process this interrupt. Disabled ADC14 interrupts do not affect the **ADC14IV** value.

A reading of the **ADC14IV** Register automatically resets the highest-pending interrupt condition and flag. The flags set in the **ADC14IFGx** bits are reset automatically by reading their associated **ADC14MEMx** registers or by setting the **ADC14CLRIFGR0** Register. Writing to the **ADC14IV** Register also clears all pending interrupt conditions and flags.

7.6.5 Analog input signal channels and GPIO pins

As we discussed in Section 7.2, most peripheral devices in the MSP432P401R MCU system are connected to GPIO ports. This means that all inputs and outputs of peripherals are transferred by using related GPIO pins. This is also true to ADC14 module. The ADC14 module is closely related to GPIO ports and all ADC14 input channels are connected to the related GPIO port and pins.

In MSP432P401RIPZ MCU system, 24 external and 2 internal analog input channels are available in total. The 24 external analog inputs are connected to 24 GPIO pins and 2 internal analog inputs are directly connected to ADC MUX input (Figure 7.9) by setting bits 23 (**ADC14TCMAP**) and 22 (**ADC14BATMAP**) in the **ADC14CTL1** Register.

Refer to Figure 7.1 in Section 7.2, only five GPIO Ports, P4, P5, P6, P8, and P9, are connected to 24 external analog inputs. To make this clear, we collect those analog channels and GPIO pins and show their relationships in Table 7.24.

In MSP-EXP432P401R EVB, only limited or 13 external analog input channels (A3~A15) are provided and connected to the GPIO pins as highlighted in Table 7.24.

Refer to Tables 7.2 through 7.6 in Section 7.3, it can be found that in order to make those GPIO pins shown in Table 7.24 to work as analog input pins, the corresponding bits **PySEL1x** and **PySEL0x** must be set to 11.

7.6.6 ADC14 module initialization

In order for the ADC14 module to be used as an ADC, the appropriate initialization and configuration should be performed to enable it to work properly. The general initialization sequence for the ADC14 module is as follows:

1. Disable the ADC module by clearing the **ADC14ENC** bit to 0 in the **ADC14CTL0** Register
2. Configure the ADC module by setting the following control bits on the **ADC14CTL0** Register:
   a. Configure ADC14 clock predivider by setting the **ADC14PDIV** bit
   b. Configure ADC14 sample-and-hold source by setting the **ADC14SHSx** bit
   c. Configure ADC14 to use extended sample or pulse mode by setting the **ADC14SHP** bit
   d. Configure ADC14CLK divider by setting the **ADC14DIVx** bit
   e. Select the ADC14CLK source by setting the **ADC14SSELx** bit field
   f. Determine the ADC14 conversion mode by setting the **ADC14CONSEQx** bit field
   g. Select the ADC14 sample-and–hold time length by setting the **ADC14SHT1x** and **ADC14SHT0x** bit fields

Table 7.24 ADC input channels and GPIO pins distributions

ADC channel	GPIO pin	Pin type	Pin function
A0	**P5.5**	Input	ADC analog input channel 0 (Reserved for USB)
A1	**P5.4**	Input	ADC analog input channel 1 (Reserved for USB)
A2	**P5.3**	Input	ADC analog input channel 2
A3	**P5.2**	Input	ADC analog input channel 3
A4	**P5.1**	Input	ADC analog input channel 4
A5	**P5.0**	Input	ADC analog input channel 5
A6	**P4.7**	Input	ADC analog input channel 6
A7	**P4.6**	Input	ADC analog input channel 7
A8	**P4.5**	Input	ADC analog input channel 8
A9	**P4.4**	Input	ADC analog input channel 9
A10	**P4.3**	Input	ADC analog input channel 10
A11	**P4.2**	Input	ADC analog input channel 11
A12	**P4.1**	Input	ADC analog input channel 12
A13	**P4.0**	Input	ADC analog input channel 13
A14	**P6.1**	Input	ADC analog input channel 14
A15	**P6.0**	Input	ADC analog input channel 15
A16	**P9.1**	Input	ADC analog input channel 16
A17	**P9.0**	Input	ADC analog input channel 17
A18	**P8.7**	Input	ADC analog input channel 18
A19	**P8.6**	Input	ADC analog input channel 19
A20	**P8.5**	Input	ADC analog input channel 20
A21	**P8.4**	Input	ADC analog input channel 21
A22	**P8.3**	Input	ADC analog input channel 22/TA3CLK
A23	**P8.2**	Input	ADC analog input channel 23/TA3.2

h. Optionally, one can configure the trigger method for the ADC14 multiple sample and conversions by setting the **ADC14MSC** bit if the conversion is a sequence or a repeated sequence mode

3. Power the ADC14 module on by setting the **ADC14ON** bit to 1
4. Enable and trigger the ADC14 module to be ready for the conversion by setting both the **ADC14ENC** and **ADC14SC** bits to 1

Now, the ADC14 module is ready to perform any analog-to-digital conversion.

7.6.7 Build the ADC programming project

Three analog channels are interfaced between the MSP-EXP432P401R EVB and the EduBASE ARM® Trainer, they are **A6**, **A8**, and **A10**. In order to build some implementation applications, first let's have a closer look at these analog interfaces.

7.6.7.1 ADC module in EduBASE ARM® trainer

Functional block diagram for ADC module, GPIO pins, **A6**, **A8**, and **A10** input channels in the EduBASE ARM® Trainer is shown in Figure 7.15.

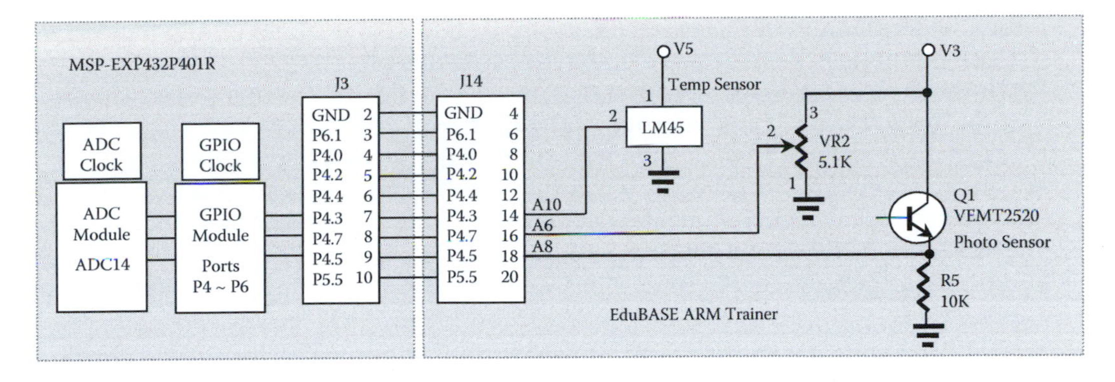

Figure 7.15 Functional block diagram of ADC module in EduBASE trainer.

Three analog input channels, **A6**, **A8**, and **A10**, are connected to three peripherals:

- **A10** is connected to the output of a temperature sensor **LM45** via **P4.3** and this channel collects an analog voltage converted by this temperature sensor.
- **A6** is connected to a potentiometer **VR2** via **P4.7** and this channel collects an analog input voltage from this potentiometer (0~3.3 V).
- **A8** is connected to a photo sensor **VEMT2520** via **P4.5** and this channel collects an analog input signal from the output of this photo transistor.

Three analog channels, **A6**, **A8**, and **A10**, are connected to the EduBASE ARM® Trainer via three GPIO pins, **P4.7**, **P4.5**, and **P4.3**.

In the following we will create an ADC module-related project **DRAADC14** by using the DRA method. This project contains the following features:

- The ADC14 module with sequence sample mode is used for this project to collect three analog signals coming from potentiometer **VR2**, photo sensor **Q1** and temperature sensor **LM45** via three channels, **A6**, **A8**, and **A10**.
- The polling-driven method will be used to monitor and check the **ADC14IFGx** status.
- **A6** collects the output voltage from the potentiometer **VR2**. **A8** gets the analog voltage from the photo sensor **Q1** and **A10** samples the output voltage from the **LM45** temperature sensor.
- The conversion result for channel **A6** is displayed in the three-color LED via GPIO pins **P2.0**~**P2.2** in the MSP-EXP432P401R EVB. The conversion result for **A8** is displayed in four LEDs, **PB0**~**PB3**, in the EduBASE ARM® Trainer, and the conversion result for **A10** is sent to the speaker in the EduBASE ARM® Trainer via GPIO pin **P5.6**.

Now, let's create our ADC project **DRAADC14**.

7.6.7.2 Create the ADC programming project (polling-driven)

Perform the following operations to create a new project **DRAADC14**:

1. Open the Windows Explorer to create a new folder named **DRAADC14** under the folder **C:\MSP432 Class Projects\Chapter 7**.
2. Open the Keil® ARM-MDK µVersion5 and go to **Project**|**New µVersion Project** menu item to create a new µVersion Project. On the opened wizard, browse to our

new folder **DRAADC14** that is created in step 1 above. Enter **DRAADC14** into the **File name** box and click on the **Save** button to create this project.
3. On the next wizard, you need to select the device (MCU) for this project. Expand three icons, **Texas Instruments**, **MSP432 Family**, and **MSP432P**, and select the target device **MSP432P401R** from the list. Click on the **OK** to close this wizard.
4. Next the Software Components wizard is opened, and you need to set up the software development environment for your project with this wizard. Expand two icons, **CMSIS** and **Device**, and check the **CORE** and **Startup** checkboxes in the **Sel.** column, and click on the **OK** button since we need these two components to build our project.

Since this project is relatively simple, therefore only a C source file is good enough.

7.6.7.3 Create the source file DRAADC14.c

Perform the following operations to create a new C source file **DRAADC14.c**:

1. In the **Project** pane, expand the **Target** folder and right click on the **Source Group 1** folder and select the **Add New Item to Group "Source Group 1"**
2. Select the **C File (.c)** and enter **DRAADC14** into the **Name:** box, and click on the **Add** button to add this source file into the project
3. Enter the code shown in Figure 7.16 into this source file

Let's have a closer look at this source file to see how it works.

1. Since we need to use them in this project. In fact, we need to use the **msp.h** system header file to access all the macros defined for GPIO ports and ADC controls to utilize related registers and components in this project.
2. A user-defined function **Delay()** is declared in lines 7~11 and it is used to delay some period of time during the project runs.
3. In line 14, three local variables, **pSensor**, **pMeter**, and **pTemp**, are declared in the beginning of the **main()** program. These variables are used to receive and hold the ADC conversion results for three peripherals, potentiometer, photo sensor, and temperature sensor.
4. In lines 15~18, three GPIO ports and related pins, **P5.6**, **P2.0**~**P2.2**, **P3.2,** and **P3.3**, as well as **P2.4** and **P2.5**, are defined as output pins since these pins are connected to the different LEDs and a speaker, and we need to use them to display the analog conversion results. **P2.0**~**P2.2** are connected to the three-color LED in the MSP-EXP432P401R EVB, and **P3.2**, **P3.3**, **P2.4,** and **P2.5** are connected to four LEDs in the EduBASE Trainer. **P5.6** is connected to the speaker in the EduBASE Trainer (refer to Figure 4.31 in Chapter 4).
5. The Watchdog timer is disabled in line 19.
6. The codes in lines 20~21 are used to configure the GPIO pins **P4.3**, **P4.5,** and **P4.7** as analog input pins by setting the related **P4SEL1x** and **P4SEL0x** registers to 11.
7. In line 22, the ADC14 module is configured by using the following control parameters:
   a. ADC14 module is turned on (**ADC14ON** = 1).
   b. ADC14 multisample and conversion mode is enabled (**ADC14MSC** = 1).
   c. ADC14 sample-and-hold length for **A6** is defined as 96 ADC14CLK period (**ADC14SHT0_96**).
   d. ADC14 sample-and-hold length for **A8** and **A10** are defined as 96 ADC14CLK period (**ADC14SHT1_96**).

```
1   //****************************************************************************
2   // DRAADC14.c - Main application file for project DARADC14
3   //****************************************************************************
4   #include <stdint.h>
5   #include <stdbool.h>
6   #include <msp.h>
7   void Delay(uint32_t time)
8   {
9    uint32_t Loop;
10   for (Loop = 0; Loop < time; Loop++) {}
11  }
12  int main(void)
13  {
14   int pSensor, pMeter, pTemp;
15   P5DIR = BIT6;                    // set P5.6 as output pin
16   P2DIR = BIT2|BIT1|BIT0;          // set P2.0 ~ P2.2 as output pins (3-color LED)
17   P3DIR = BIT2|BIT3;               // set P3.2 & P3.3 as output pins (LEDs PB0 & PB1)
18   P2DIR |= BIT5|BIT4;              // set P2.5 & P2.4 as output pins (LEDs PB2 & PB3)
19   WDTCTL = WDTPW|WDTHOLD;          // stop watchdog timer
20   P4SEL1 |= BIT3|BIT5|BIT7;        // P4 - analog input: P4SEL1x:P4SEL0x = 11
21   P4SEL0 |= BIT3|BIT5| BIT7;       // enable A-D input channel A6, A8 & A10
22   ADC14CTL0 = ADC14ON|ADC14MSC|ADC14SHT0__96|
23               ADC14SHT1__96|ADC14SHP|ADC14CONSEQ_3; // Turn on ADC14
24   ADC14MCTL6 = ADC14INCH_6;                // ref+=AVcc, channel = A6
25   ADC14MCTL8 = ADC14INCH_8;                // ref+=AVcc, channel = A8
26   ADC14MCTL10 = ADC14INCH_10|ADC14EOS;     // ref+=AVcc, channel = A10, end seq.
27   while(1)
28   {
29    ADC14CTL0 |= ADC14ENC | ADC14SC;            // start conversion-software trigger
30    while((ADC14IFGR0 & ADC14IFG10) == 0);     // waiting for three conversions done
31    pMeter = ADC14MEM6;                        // get result from A6 (Potentiometer)
32    pSensor = ADC14MEM8;                       // get result from A8 (Photo Sensor)
33    pTemp = ADC14MEM10;                        // get result from A10 (Temp)
34    P2OUT = pMeter >> 8;                       // send 4 MSB in pMeter to 3-color LED
35    P3OUT = pSensor >> 8;                      // send 4 MSB in pSensor to LEDs PB1 & PB0
36    P5OUT = pTemp << 2;                        // send pTemp * 4 to speaker
37    Delay(50000);
38   }
39  }
```

Figure 7.16 Source code for the DARADC.c.

   e. ADC14 source of the sampling signal is configured as the sampling timer (**ADC14SHP** = 1).
   f. ADC14 conversion sequence mode is selected to enable the ADC to work in the repeat-sequence-of-channels mode (**ADC14CONSEQx** = **11** ⇒ **ADC14CONSEQ_3**).

8. The code in line 24~26, three analog conversion result control memory registers, **ADC14MCTL6**, **ADC14MCTL8**, and **ADC14MCTL10**, are configured to select and convert input analog channels **A6**, **A8**, and **A10** with the default conversion voltage references.
9. In line 27, an infinitive **while()** loop is used to repeatedly collect three analog conversion results and assign them to the related LEDs and the speaker to display the conversion results.
10. Then the ADC14 module is enabled and started with the software trigger in line 29.

11. Another **`while()`** loop is used in line 30 with the **`ADC14IFGR0`** and **`ADC14IFG10`** as the loop condition to detect the **`ADC14IFG10`** conversion complete flag. If the conversion is not complete, this loop will keep its checking and wait for its completion. Since the channel **`A10`** is the last channel to be converted, thus checking the **`ADC14IFG10`** flag is a correct way to determine whether the repeat-sequence-of-channel (**`A6-A8-A10`**) mode is done.
12. If the repeat-sequence-of-channel conversion is complete, three conversion results stored in three **`ADC14MEMx`** registers, **`ADC14MEM6`**, **`ADC14MEM8`**, and **`ADC14MEM10`**, are read and assigned to three local variables, **`pMeter`**, **`pSensor`**, and **`pTemp`** in lines 31~33.
13. Then these conversion results are assigned to related LEDs and speaker to display them in lines 34~36.
14. In line 37, the user-defined function **`Delay()`** is called to delay the program a period of time to enable the displayed results stable.

Now, let's set up the environment to build and run the project to test the ADC functions.

7.6.7.4 Set up the environment to build and run the project

The only environment setup is to select the correct debugger adapter for this project, which is **`CMSIS-DAP Debugger`**. Go to **`Project|Options for Target "Target 1"`** menu item to finish this setup. Also make sure that the Debugger Settings and the Flash Download setting are correct.

Now, build the project by going to **`Project|Build target`** menu item. If everything is fine, go to **`Flash|Download`** item to download the image file of the project into the flash memory. Now, go to **`Debug|Start/Stop Debug Session`** and **`Debug|Run`** item to run the project.

During the project running, you can test it by:

- Tuning the potentiometer **`VR2`** with a small screw driver, you can find that the intensity and color of the three-color LED on the MSP-EXP432P401R EVB are also changed. This is because the analog input to the channel **`A6`** is modified as you tune the potentiometer.
- Either covering the photo sensor by hand or exposing it under some light sources, you can find that two LEDs, **`PB1~PB0`**, on the Trainer will also be changed to ON or OFF.
- Hearing a noise coming from the speaker since the conversion result of the temperature sensor is sent to it. Because the change of the temperature needs a long time to be responded to by the temperature sensor, the noise seems constant.

Now, let's have a quick look at the API functions provided by MSPWare Peripheral Driver Library since we can use some functions to perform ADC-related operations in applications.

7.6.8 ADC module API functions provided in the TivaWare peripheral driver library

The library provides over 30 API functions to support the ADC operations, and these functions can be divided into the following three groups:

1. Configure and set up the ADC14 module
2. Enable and disable the ADC14 module

3. Get the running status and receive the ADC14 conversion results
4. Configure and process ADC14-related interrupts

The first group contains the following nine functions:

- **ADC14_initModule()**
- **ADC14_configureSingleSampleMode()**
- **ADC14_configureMultiSequenceMode()**
- **ADC14_configureConversionMemory()**
- **ADC14_setPowerMode()**
- **ADC14_setResolution()**
- **ADC14_setResultFormat()**
- **ADC14_setSampleHoldTime()**
- **ADC14_setSampleHoldTrigger()**

The second group contains the following seven API functions:

- **ADC14_enableModule()**
- **ADC14_enableSampleTimer()**
- **ADC14_enableConversion()**
- **ADC14_disableModule()**
- **ADC14_disableConversion()**
- **ADC14_disableSampleTimer()**
- **ADC14_toggleConversionTrigger()**

The third group contains the following five API functions:

- **ADC14_isBusy()**
- **ADC14_getResult()**
- **ADC14_getResultArray()**
- **ADC14_getResolution()**
- **ADC14_getMultiSequenceResult()**

The fourth group contains the following five ADC interrupt-related functions:

- **ADC14_registerInterrupt()**
- **ADC14_enableInterrupt()**
- **ADC14_getInterruptStatus()**
- **ADC14_clearInterruptFlag()**
- **ADC14_disableInterrupt()**

These API functions are contained in a C file **adc14.c** with a header file **adc14.h**. Both files are located at the folder **C:\ti\msp\MSPWare_2_21_00_39\driverlib\driverlib\MSP432P4xx** in your host computer.

Let's discuss these API functions one by one based on their groups.

7.6.8.1 Configure and set up the ADC14 module API functions

Table 7.25 shows the API functions used to configure and handle sample sequencers.

The **ADC14_initModule()** is the first API function to be executed to initialize the ADC14 module before it can be operated properly. In most cases, the default setup

Table 7.25 API functions used to configure and set up the ADC14 module

API function	Parameters	Description
bool **ADC14_initModule(**uint32_t clockSource, uint32_t clockPredivider, uint32_t clockDivider, uint32_t internalChannelMask**)**	**clockSourse** is the clock source to use for the ADC **clockPredivider** is the clock predivider **clockDivider** is the clock divider **internalChannelMask** is a mapping to the external pins or internal components	Clock source can be any of the following sources: • **ADC_CLOCKSOURCE_ADCOSC** [DEFAULT] • **ADC_CLOCKSOURCE_SYSOSC** • **ADC_CLOCKSOURCE_ACLK** • **ADC_CLOCKSOURCE_MCLK** • **ADC_CLOCKSOURCE_SMCLK** • **ADC_CLOCKSOURCE_HSMCLK** Predivider can be any of the following one: • **ADC_PREDIVIDER_1** [DEFAULT] • **ADC_PREDIVIDER_4** • **ADC_PREDIVIDER_32** • **ADC_PREDIVIDER_64** Clock divider can be any of the following one: • **ADC_DIVIDER_1** [Default] • **ADC_DIVIDER_2; ADC_DIVIDER_3** • **ADC_DIVIDER_4; ADC_DIVIDER_5** • **ADC_DIVIDER_6; ADC_DIVIDER_7** • **ADC_DIVIDER_8** internalChannelMask can be any following one: • **ADC_MAPINTCH3; ADC_MAPINTCH2** • **ADC_MAPINTCH1; ADC_MAPINTCH0** • **ADC_TEMPSENSEMAP; ADC_BATTMAP** • **ADC_NOROUTE If internalChannelMask** is not desired, pass **ADC_NOROUTE** in lieu of this parameter
bool **ADC14_configureSingleSampleMode(**uint32_t memoryDestination, bool repeatMode)	**memoryDestination** is the memory location to store sample/conversion value **repeatMode** indicates whether to repeat the conversion cycle	**The memoryDestination** can be: **ADC_MEM0** through **ADC_MEM31** The **repeatMode** can be **true** or **false**

(*Continued*)

Table 7.25 (Continued) API functions used to configure and set up the ADC14 module

API function	Parameters	Description
bool **ADC14_configureMultiSequenceMode(**uint32_t memoryStart, uint32_t memoryEnd, bool repeat-Mode**)**	**memoryStart** is the starting location to store the first conversion result **memoryEnd** is the last location to store the last conversion result **repeatMode** indicates whether to repeat the conversion cycle	**memoryStart** can be: **ADC_MEM0** through **ADC_MEM31** **memoryEnd** can be: **ADC_MEM0** through **ADC_MEM31** **repeatMode** can be **true** or **false**
bool **ADC14_configureConversionMemory(**uint32_t memorySelect, uint32_t refSelect, uint32_t chan-nelSelect, bool differntial-Mode**)**	**memorySelect** is the memory to be selected **refSelect** is the voltage reference to be selected **channelSelect** is the analog input channel **differentialMode** is to select differential mode	**memorySelect** can be (OR is used for multimem): **ADC_MEM0** through **ADC_MEM31** **refSelect** can be any of the following: • **ADC_VREFPOS_AVCC_VREF-NEG_VSS** [DEFAULT] • **ADC_VREFPOS_INTBUF_VRE-FNEG_VSS** • **ADC_VREFPOS_EXTPOS_VRE-FNEG_EXTNEG** • **ADC_VREFPOS_EXTBUF_VRE-FNEG_EXTNEG** **channelSelect** can be any of the following one: **ADC_INPUT_A0** through **ADC_INPUT_A31** **differentialMode** can be **true** or **false**
bool **ADC14_setPowerMode(**uint32_t powerMode**)**	**powerMode** is the desired power mode to be set	The possible **powerMode** value is: **ADC_UNRESTRICTED_POWER_MODE** (no restriction) **ADC_LOW_POWER_MODE** (500ksps restriction) **ADC_ULTRA_LOW_POWER_MODE** (200ksps restriction) **ADC_EXTREME_LOW_POWER_MODE** (50ksps restrict)
void **ADC14_setResolution** (uint32_t resolution)	**resolution** is the desired resolution to be set	The possible **resolution** value is: **ADC_8BIT** (10 clock cycle conversion time) **ADC_10BIT** (12 clock cycle conversion time) **ADC_12BIT** (14 clock cycle conversion time) **ADC_14BIT** (16 clock cycle conversion time)[DEFAULT]

(*Continued*)

Table 7.25 (Continued) API functions used to configure and set up the ADC14 module

API function	Parameters	Description
bool **ADC14_setResult-Format(**uint32_t result-Format)	**resultFormat** is the format of the conversion result in ADC14MEMx	The **resultFormat** can be: **ADC_UNSIGNED_BINARY** [DEFAULT] **ADC_SIGNED_BINARY**
bool **ADC14_setSample-HoldTime(**uint32_t firstPulseWidth, uint32_t secondPulseWidth**)**	The **firstPulseWidth** controls ADC memory **ADC_MEMORY_0** through **ADC_MEM-ORY_7** and **ADC_MEM-ORY_24** through **ADC_MEMORY_31** The **secondPulseWidth** controls memory loca-tions **ADC_ MEMORY_8~ADC_ MEMORY_23**	Both the **firstPulseWidth** and **secondPulseWidth** can be any of the following values: • **ADC_PULSE_WIDTH_4** [DEFAULT] • **ADC_PULSE_WIDTH_8** • **ADC_PULSE_WIDTH_16** • **ADC_PULSE_WIDTH_32** • **ADC_PULSE_WIDTH_64** • **ADC_PULSE_WIDTH_96** • **ADC_PULSE_WIDTH_128** • **ADC_PULSE_WIDTH_192**
bool **ADC14_setSample HoldTrigger(**uint32_t source, bool invertSignal**)**	**source** is the source for sampling **invertSignal** is a Boolean variable used to indicate whether the trigger signal is inverted	Possible values for **source** include: **ADC_TRIGGER_ADCSC** [DEFAULT] **ADC_TRIGGER_SOURCE1** **ADC_TRIGGER_SOURCE2** **ADC_TRIGGER_SOURCE3** **ADC_TRIGGER_SOURCE4** **ADC_TRIGGER_SOURCE5** **ADC_TRIGGER_SOURCE6** **ADC_TRIGGER_SOURCE7**

parameters can be adopted for most parameters used in this function. However, for the **internalChannelMask**, if no internal analog component is used and the ADC14 is only used for external analog channel conversions, this parameter should be replaced by **ADC_NOROUTE** to enable the ADC14 module to convert only external analog inputs.

For the API function **ADC14_configureConversionMemory()**, multiple memory registers can be selected and used for sequence or repeat-sequence mode conversions. In that case, the target memory registers, **ADC_MEM0~ADC_MEM31**, can be **ORed** together to enable them to be selected.

The function **ADC14_setSampleHoldTime()** is used to configure the length of the sampling and hold time. Both the **firstPulseWidth** and the **secondPulseWidth** can be defined as any of value of **ADC_PULSE_WIDTH_8~ADC_PULSE_WIDTH_192**.

For the **ADC14_setSampleHoldTrigger()** function, the trigger source **ADC_TRIGGER_SOURCE1~ADC_TRIGGER_SOURCE7**, should be selected based on the MCU model used. In the MSP432P401R MCU system, there are seven internal trigger sources, **TA0_C1**, **TA0_C2**, **TA1_C1**, **TA1_C2**, **TA2_C1**, **TA2_C2**, and **TA3_C1**, are available, and these seven trigger sources can be used to replace the **ADC_TRIGGER_SOURCE1~ADC_TRIGGER_SOURCE7** used in this function. Refer to Section 7.6.2.3 and Table 7.14 to get more details about these trigger sources used in the MSP432P401R MCU system.

Table 7.26 API functions used to enable and disable the ADC14 module

API function	Parameters	Description
void **ADC14_enableModule(**void**)**	None	Enables the ADC block This will enable operation of the ADC block
bool **ADC14_enableSampleTimer(**uint32_t multiSampleConvert**)**	**multiSampleConvert** is used to switch between manual and automatic iteration when using the sample timer	Valid values for **multiSampleConvert** are: **ADC_MANUAL_ITERATION**—The user have to manually set SHI signal (by ADC14_toggleConversionTrigger) at the end of each sample/conversion cycle **ADC_AUTOMATIC_ITERATION**—After one sample/ convert is finished, the ADC module will automatically continue on to the next sample
bool **ADC14_enableConversion(**void**)**	None	Enable conversion of ADC data. Note that this only enables conversion. To trigger the conversion, you have to call the **ADC14_toggleConversionTrigger()** or use the source trigger configured in **ADC14_setSampleHoldTrigger()**
bool **ADC14_disableModule(**void**)**	None	Disables the ADC block This will disable operation of the ADC block
bool **ADC14_disableSampleTimer(**void**)**	None	Disable SAMPCON from being sourced from the sample timer
void **ADC14_disableConversion(**void**)**	None	Halts conversion of the ADC module. Note that the software bit for triggering conversions will also be cleared with this function
bool **ADC14_toggleConversionTrigger(**void**)**	None	Toggle the trigger for conversion of the ADC module by toggling the trigger software bit. Note that this will cause the ADC to start conversion regardless if the software bit was set as the trigger using ADC14_setSampleHoldTrigger

7.6.8.2 Enable and disable the ADC14 module API functions

Table 7.26 shows the API function used to enable and disable the ADC14 module.

The `ADC14_enableConversion()` function is only used to enable the analog-to-digital conversion. To start an ADC operation, one needs to call the `ADC14_toggleConversionTrigger()` function or use the trigger source obtained from calling of the `ADC14_setSampleHoldTrigger()` function to trigger the ADC to begin its conversion job.

Table 7.27 API functions used to get running status and conversion results

API function	Parameters	Description
bool **ADC14_isBusy(**void**)**	None	Return a Boolean value that tells if a conversion/sample is in progress. Returned a **true** means that ADC is working
uint_fast16_t **ADC14_get Result(**uint32_t memory Select**)**	**memorySelect** is the memory location to get the conversion result.	Valid values for **memorySe-lect** are: **ADC_MEM0** through **ADC_ MEM31**
void `ADC14_ getResultArray`(uint32_t memoryStart, uint32_t memoryEnd, uint16_t * res)	**memoryStart** is the starting location to get the conversion result **memoryEnd** is the ending location to get the result **res** is a pointer point to the location to store the result	Valid values for **memoryStart** and **memoryEnd** is: **ADC_MEM0** through **ADC_ MEM31** **res** is the starting address including the conversion result of the last multisequence sample in an array of unsigned 16-bit integers
uint_fast32_t **ADC14_ getResolution(**void**)**	None	Get the resolution of the ADC module. Valid value is: **ADC_8BIT** (10 clock cycle conversion time) **ADC_10BIT** (12 clock cycle conversion time) **ADC_12BIT** (14 clock cycle conversion time) **ADC_14BIT** (16 clock cycle conversion time)
`void` **ADC14_getMulti-Sequ enceResult(**uint16_t*res**)**	**res** is a pointer point to the location to store the result	**res** is the starting address from which the conversion result of the last multisequence sample in an array of unsigned 16-bit integers

7.6.8.3 *Get the running status and receive the ADC14 conversion results API functions*

Five API functions are included in this group. Table 7.27 lists these functions and their parameters. Both API functions, **`ADC14_getResultArray()`** and **`ADC14_getMulti-SequenceResult()`**, contain a pointer variable, **`*res`**, which is a pointer point to the starting address from which the multiconversion results are stored. Before calling these two functions, a blank memory space should be assigned and the starting address of that space should be assigned to **`res`**.

7.6.8.4 *Configure and process the ADC14 interrupts API functions*

Five ADC14 Interrupt-related API functions are included in this group. Table 7.28 shows these functions and their arguments.

Table 7.28 API functions used to configure and handle ADC14 interrupts

API function	Parameters	Description
void **ADC14_registerInterrupt(**void(*)(void) intHandler**)**	**intHandler** is a pointer to the function to be called when the ADC interrupt occurs	Register the handler to be called when an ADC interrupt occurs. This function enables the global interrupt in the interrupt controller Specific ADC14 interrupts must be enabled via the function **ADC14_enableInterrupt()**. It is the interrupt handler's responsibility to clear the interrupt source via the function **ADC14_clearInterruptFlag()**
void `ADC14_enableInterrupt`(uint_fast64_t mask)	`mask` is the bit mask of interrupts to enable	Enable the indicated ADCC interrupt sources. The **ADC_INT0~ADC_INT31** parameters correspond to a completion event of the corresponding memory location **mask** value can be a bitwise OR of the following values: • **ADC_INT0** through **ADC_INT31** • **ADC_IN_INT**—Interrupt enables a conversion in the result register that is either greater than the ADCLO or lower than the ADCHI threshold • **ADC_LO_INT**—Interrupt enable for the falling short of the lower limit interrupt of the window comparator for the result register • **ADC_HI_INT**—Interrupt enable for the exceeding the upper limit of the window comparator for the result register • **ADC_OV_INT**—Interrupt enable for a conversion that is about to save to a memory buffer that has not been read out yet • **ADC_TOV_INT**—Interrupt enable for a conversion that is about to start before the previous conversion has been completed • **ADC_RDY_INT**—Interrupt enable for the local buffered reference ready signal
uint_fast64_t **ADC14_getInterruptStatus(**void**)**	None	Return the status of a the ADC interrupt register Returned value is a bitwise OR of the **mask** value used in the `ADC14_enableInterrupt()`
void **ADC14_clearInterruptFlag(**uint_fast64_t mask**)**	**mask** is the bit mask of interrupts to clear	Clear the indicated ADCC interrupt sources. The `mask` is identical with that used in the **ADC14_enableInterrupt()**
void **ADC14_disableInterrupt(**uint_fast64_t mask**)**	**mask** is the bit mask of interrupts to disable	Disable the indicated ADCC interrupt sources. The `mask` is identical with that used in **ADC14_enableInterrupt()**

When using these functions to configure and handle any ADC14-related interrupt, the following operational sequence must be followed:

- The **ADC14_registerInterrupt()** function must be called or executed first to set up an interrupt handler to handle any ADC14-related interrupt to occur.
- The **ADC14_enableInterrupt()** function must be called or executed second to enable the specific ADC14 interrupt source.
- The **ADC14_clearInterruptFlag()** function must be executed inside the interrupt handler to clear the current interrupt to enable the next one to occur.

The parameter **mask** used in three functions, **ADC14_enableInterrupt()**, **ADC14_disableInterrupt()**, and **ADC14_clearInterruptFlag()**, is identical as shown in Table 7.28.

There are some other API functions available to the ADC modules. Here, we only discussed the most often used functions. Refer to **MSP432P401R Microcontroller Data Sheet** to get more details for those functions.

Next, let's build an example project to use these API functions to access and control the ADC14 module to collect data from channel 1 to test the functions of these API functions.

7.6.8.5 Build an example ADC14 project using API functions

In this project, we use a repeat-single-channel mode to collect the analog input coming from the potentiometer VR2 via **A6** that is connected to the GPIO pin **P4.7**. Since this project is very simple and easy, therefore we only need to create a C file to include all codes. In order to use those API functions, the following operations are needed to set up the environments to build and run this project:

- The MSPPDL file, **msp432p4xx_driverlib.lib**, must be added into the project. This file is located at: **C:\ti\msp\MSPWare_2_21_00_39\driverlib\driverlib\MSP432P4xx\keil** folder in your host computer.
- Some system header files must be included, and the related path to access those header files should also be added into the project by using **C/C++** tab under the **Project|Options for Target "Target 1"** menu item.
- The debug adaptor **CMSIS-DAP Debugger** must be selected by using the **Debug** tab under the **Project|Options for Target "Target 1"** menu item. Also make sure that the Debugger settings and the Flash Download setting are correct.

Now, let's create a new folder **SDADC14** under the **C:\MSP432 Class Projects\Chapter 7** folder.

Then create a new Keil µVersion5 project **SDADC14**, and add a new C file **SDADC14.c** into this project. Enter the code shown in Figure 7.17 into this source file. Let's have a closer look at this piece of code to see how it works.

1. The codes in lines 4~9 are used to include some useful system header files for this project. The lower three are specific header files for GPIO, Watchdog timer, and ADC14.
2. The local variable **cResult** is declared in line 12 and this variable is used to get and hold the converted result from the **ADC14MEM6** Register. The data type of this variable is **uint16_t**, which is a 16-bit unsigned integer variable matched to the requirement of the API function **ADC14_getResult()**.
3. In line 13, the Watchdog timer is temporarily terminated to avoid its disturbance.

```
1  //***************************************************************************
2  // SDADC14.c - Main application file for project SDADC14
3  //***************************************************************************
4  #include <stdint.h>
5  #include <stdbool.h>
6  #include <msp.h>
7  #include <MSP432P4xx/gpio.h>
8  #include <MSP432P4xx/wdt_a.h>
9  #include <MSP432P4xx/adc14.h>

10 int main(void)
11 {
12  uint16_t cResult;

13  WDT_A_holdTimer();                      // stop watchdog timer
14  // set P2.0 ~ P2.2 as output pins
15  GPIO_setAsOutputPin(GPIO_PORT_P2, GPIO_PIN0|GPIO_PIN1|GPIO_PIN2);
16  GPIO_setAsPeripheralModuleFunctionInputPin(GPIO_PORT_P4, GPIO_PIN7,
17                                            GPIO_TERTIARY_MODULE_FUNCTION);

18  ADC14_initModule(ADC_CLOCKSOURCE_MCLK, ADC_PREDIVIDER_1,
19                                            ADC_DIVIDER_1, ADC_NOROUTE);
20  ADC14_configureSingleSampleMode(ADC_MEM6, true);
21  ADC14_configureConversionMemory(ADC_MEM6, ADC_VREFPOS_AVCC_VREFNEG_VSS,
22                                            ADC_INPUT_A6, false);
23  ADC14_setSampleHoldTime(ADC_PULSE_WIDTH_64, ADC_PULSE_WIDTH_64);
24  ADC14_enableSampleTimer(ADC_AUTOMATIC_ITERATION);

25  ADC14_enableModule();
26  ADC14_enableConversion();
27  ADC14_toggleConversionTrigger();

28  while(1)
29  {
30   while(!ADC14_isBusy());
31   cResult = ADC14_getResult(ADC_MEM6);
32   P2OUT = cResult >> 8;
33   ADC14_toggleConversionTrigger();
34  }
35 }
```

Figure 7.17 Code for the source file SDADC.c.

4. The code line 15 is used to configure the GPIO **P2** to enable **P2.0~P2.2** pins to work as output pins since these pins are connected to a three-color LED in the MSP-EXP432P401R EVB, and we need them to display the conversion results later.
5. In line 16, a GPIO-related API function **GPIO_setAsPeripheralModuleFunctionInputPin()** is used to configure the GPIO **P4.7** pin as an analog input pin by setting **P4SEL1.7:P4SEL0.7** to **11** since this pin works as an analog input **A6**.
6. The codes in lines 18~24 are used to initialize and configure the ADC14 module and analog input channel **A6**. The API function **ADC14_initModule()** is called first in line 18 to initialize the ADC14 module with desired ADC14 clock, clock predivider, and clock divider. The last parameter **ADC_NOROUTE** for this function indicates that no internal analog source is used for this conversion.

7. In line 20, the API function **ADC14_configureSingleSampleMode()** is executed with the **ADC14MEM6** as the target conversion memory register for the input channel **A6**. The Boolean value **true** means that this single conversion is a repeat-single-channel mode.
8. The API function **ADC14_configureConversionMemory()** is called in line 21 to set up the conversion memory register as **ADC14MEM6** for the input analog channel **A6**. The ADC references are default reference voltages. The last Boolean value **false** indicates that this conversion is not a differential conversion, instead it is a single-ended conversion.
9. In line 23, the API function **ADC14_setSampleHoldTime()** is used to set up the length of sample-and-hold during the ADC operations. The first parameter controls the registers **ADC14MEM0~ADC14MEM7** and **ADC14MEM24~ADC14MEM31**. Since **A6** belongs to this range, we set the length as 64 ADC14CLK cycles. The second parameter is not used in this project, but it is also set to this value.
10. The API function **ADC14_enableSampleTimer()** is called in line 24 with the parameter **ADC_AUTOMATIC_ITERATION** to enable the sample timer to repeat its trigger itself.
11. The codes in lines 25~27 are used to enable and start trigger the ADC14 module to make it ready for the conversion job. The API function **ADC14_enableConversion()** in line 26 only enables the conversion, but the conversion cannot be started until it is triggered by a trigger source. Generally, two API functions, **ADC14_setSampleHoldTrigger()** and **ADC14_toggleConversionTrigger()**, can handle this trigger job. However, the latter is more powerful to trigger and start an ADC conversion. Therefore, the second function is used in this project in line 27.
12. An infinitive **while()** loop is used starting at line 28 to repeatedly check the conversion results and assign them to the GPIO P2 pins **P2.0~P2.3** to display them in a three-color LED.
13. In line 30, the API function **ADC14_isBusy()** is tested with another conditional **while()** loop to see whether the ADC conversion is complete. A true is returned if the ADC is busy in conversion. This **while()** loop will not be released until a **false** is returned, which means that the ADC is not busy and a conversion has been done.
14. The API function **ADC14_getResult()** is called in line 31 to pick up the conversion result stored in the **ADC14MEM6** Register if a false is returned. Then the result is shift left by eight bits and assigned to the GPIO P2 to display it on the three-color LED in line 32.
15. Finally, the API function **ADC14_toggleConversionTrigger()** is called again to trigger the ADC module to start the next conversion.

After building the project, you can download it into the flash memory. As the project runs, you can tune the potentiometer **VR2** since it is connected to channel 6 (**A6**) via **P4.7**. As you tune the **VR2**, the intensity and color on the three-color LED will be changed and four LEDs on the Trainer will also be changed.

7.7 Analog comparator project

In the MSP432P401R MCU system, two identical analog comparator modules, **COMP_E**, are provided to assist users to perform precision analog signals comparison functions. These two modules are called **COMP_E0** and **COMP_E1**, respectively.

Because of the identity of two modules, we will concentrate on one common module **COMP_E** to discuss this peripheral in this chapter.

The core of the **COMP_E** module is an analog comparator used to compare two analog inputs applied on two input terminals, V_+ and V_-. If the V_+ terminal is more positive than the V_- terminal, the comparator output **COUT** is high. Otherwise, it outputs a low. The comparator can be turned on or off by using the control bit **CEON** (bit 10) in the Comparator Control Register 1 (**CExCTL1**: **x** = 0 or 1). The comparator should be turned off when not in use to reduce current consumption. When the comparator is turned off, **COUT** is always low. The bias current of the comparator is programmable.

The **COMP_E** module supports precision slope analog-to-digital conversions, supply voltage supervision, and monitoring of external analog signals. Features of the **COMP_E** include:

- Inverting and noninverting terminal input multiplexer
- Software-selectable RC filter for the comparator output
- Output provided to Timer_A capture input
- Software control of the port input buffer
- Interrupt capability
- Selectable reference voltage generator, voltage hysteresis generator
- Reference voltage input from shared reference
- Ultra-low power comparator mode
- Interrupt-driven measurement system for low-power operation support

Let's first have a closer look at the architecture and organization of the **COMP_E** module.

7.7.1 Architecture and organization of the COMP_E Module

Figure 7.18 shows the functional block diagram of the **COMP_E** module.

The function of the **COMP_E** module can be divided into the following four sections:

1. The compared analog inputs selection
2. The reference voltages selection
3. The comparator output controls
4. The comparator interrupt controls

All of these selections and controls are performed by using a set of **COMP_E** module registers. In the following sections, we will discuss these components one by one with the related control registers.

All of the **x** used in related registers in the following sections, such as **CExCTL0**, **CExCTL1**, **CExCTL2**, **CExCTL3**, **CExINT**, and **CExIV**, has a value of 0 or 1, which represents the registers used for **COMP_E0** or **COMP_E1** module, respectively.

In MSP432P401R MCU, each **COMP_E** module can compare up to eight analog input signal pairs.

7.7.2 Compared analog inputs selection

The components in this part are used to select two analog inputs from multiple analog input signals, and connect them to two terminals of the **COMP_E** comparator.

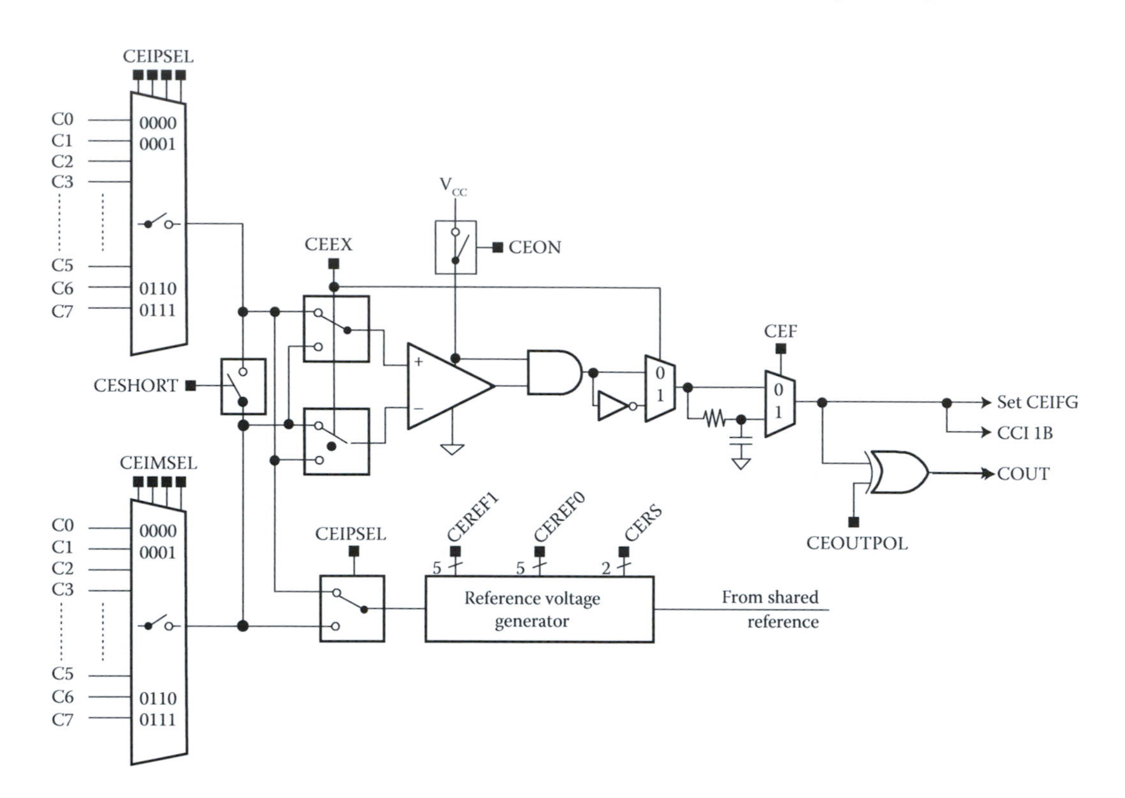

Figure 7.18 Functional block diagram of the COMP_E module. (Courtesy Texas Instruments.)

Two 8-to-1 multiplexers are used to perform this selection function with two control bit field **CEIPSELx** and **CEIMSELx** in the Comparator Control Register 0 (**CExCTL0**). The **CEIPSELx** bit field (bits 3~0) in the **CExCTL0** Register is used to control the upper multiplexer to select one analog input to be connected to the positive terminal (V_+) of the comparator. The **CEIMSELx** bit field (bits 11~8) in the **CExCTL0** Register is used to control the lower multiplexer to select an analog input to be connected to the minus terminal (V_-) of the comparator. In fact, these two bit fields provide the following control functions:

- Route an internal reference voltage to an associated output port pin
- Apply an external current source (for example, resistor) to the V+ or V– terminal of the comparator
- Map both terminals of the internal multiplexer to the outside

Two control bits, **CEIPEN** (bit 7) and **CEIMEN** (bit 15), in the **CExCTL0** Register can be used to control whether enabling the selected analog input from the upper multiplexer to be connected to the V_+ on the comparator, and the selected analog input from the lower multiplexer to be connected to the V_- terminal on the comparator.

The **CESHORT** bit (bit 4) in the **CExCTL1** Register can be used to short-connect two analog inputs, and this can be used to build a simple sample-and-holder for the comparator.

It can be found from Figure 7.18 that The **CEEX** bit (bit 5) in the **CExCTL1** Register can be used to exchange the input signals on the comparator's V_+ and V_- terminals and invert the comparator's output. This control function is performed by switching two

31	30	29	28	27	26	25	24
Reserved							
23	**22**	**21**	**20**	**19**	**18**	**17**	**16**
Reserved							
15	**14**	**13**	**12**	**11**	**10**	**9**	**8**
CEIMEN	Reserved			CEIMSEL			
7	**6**	**5**	**4**	**3**	**2**	**1**	**0**
CEIPEN	Reserved			CEIPSEL			

Figure 7.19 Bit field and function of the CExCTL0 Register.

electronic-switches connected to two input terminals on the comparator. This allows the user to determine or compensate for the comparator input offset voltage.

Figure 7.19 shows the bit configurations of the **CExCTL0** Register. Table 7.29 lists the bit value and function of this register.

7.7.3 Reference voltages selection

A functional block diagram of the reference generator in the **COMP_E** module is shown in Figure 7.20.

It can be found from Figure 7.20 that there are two sources to work as the reference voltages to the comparator; 1) the internal resistor ladder with internal voltage source V_{CC},

Table 7.29 Bit value and function for CExCTL0 register

Bit	Field	Type	Reset	Function
31:16	**Reserved**	RO	00	Reserved. Always reads as 0
15	**CEIMEN**	RW	0	Channel input enable for the V− terminal of the comparator **0:** Disable selected analog input channel to be connected to V− terminal **1:** Enable selected analog input channel to be connected to V− terminal
14:12	**Reserved**	RO	00	Reserved. Always reads as 0
11:8	**CEIMSEL**	RW	0000	Select an analog input to be connected to the minus terminal (V−) of the comparator **0000~0111:** Select one analog input from analog inputs C0~C7
7	**CEIPEN**	RW	0	Channel input enable for the V_+ terminal of the comparator **0:** Disable selected analog input channel to be connected to V_+ terminal **1:** Enable selected analog input channel to be connected to V_+ terminal
6:4	**Reserved**	RO	00	Reserved. Always reads as 0
3:0	**CEIPSEL**	RW	0000	Select an analog input to be connected to the positive terminal (V_+) of the comparator **0000~0111:** Select one analog input from analog inputs C0~C7

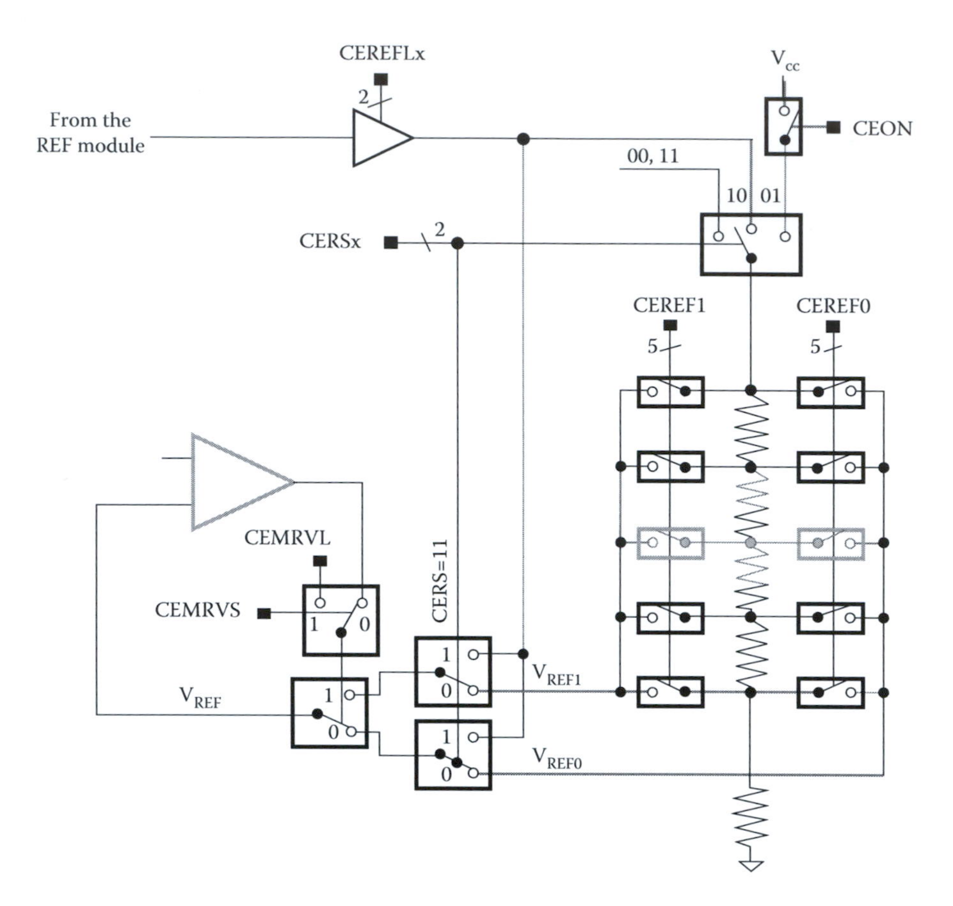

Figure 7.20 Functional block diagram of the reference voltage generator. (Courtesy texas instruments.)

and 2) an external reference coming from the precision **REF_A** module. The following registers control the operation of this reference voltage generator and select the reference sources.

The **CERSx** bit fields (bits 7~6) in the Comparator Control Register 2 (**CExCTL2**) are used to control and determine which reference source should be used and applied to the comparator.

- When **CERSx** = 01, the internal voltage source V_{CC} is used and connected to the internal resistor ladder (the **COMP_E** module must have been enabled by setting **CEON** to 1). The actual reference voltage values are determined by the **CEREF1** and **CEREF0** bit fields in the **CExCTL2** Register.
- When **CERSx** = 10, the precision reference voltage coming from **REF_A** module is applied on the internal resistor ladder. The actual reference voltage values are determined by the **CEREF1** and **CEREF0** bit fields in the **CExCTL2** Register.
- When **CERSx** = 11, the precision reference voltage coming from **REF_A** module works as the reference voltage and is directly applied on the input terminals of the comparator.

The **CEREFLx** bit fields (bits 14~13) in the **CExCTL2** Register select the reference voltage level for the reference coming from the **REF_A** module when it is used as the reference source.

- When **CEREFLx** = 01, the 1.2 V reference voltage is selected
- When **CEREFLx** = 10, the 2.0 V reference voltage is selected
- When **CEREFLx** = 11, the 2.5 V reference voltage is selected

The actual reference voltage values applied on the comparator input terminals are determined by two bit fields, **CEREF1** and **CEREF0**, in the **CExCTL2** Register when the reference source (V_{CC} or REF_A) is determined. Each of these two bit fields control different switches to make them on or off to select different resistor tap or path to get different voltage-divider outputs. Both **CEREF1** and **CEREF0** bit fields are five bits with a value range of 0~31. These two bit fields determine the upper and lower reference limits for a hysteresis reference effect. If a static mode (normal mode) reference is used, these two limits should be identical.

The value in these two bit fields is exactly a ***Numerator*** for the equation:

VBase × (*Numerator*/32) (**VBase** = V_{CC} or REF_A)

This equation is used to calculate the actual reference voltage. The **VBase** can be either the internal V_{CC} or REF_A shared voltage reference. Therefore the value in both **CEREF1** and **CEREF0** can be ranged 0~32. This provides a reference voltage range in **0.03125 × VBase~VBase**.

The **VREF1** is used while **COUT** is 1, and **VREF0** is used while **COUT** is 0. This allows the generation of a hysteresis without using external components. If the hysteresis is not needed for a normal analog voltages comparison, the value in both **CEREF1** and **CEREF0** should be same.

The interrupt flags of the comparator and the comparator output are kept unchanged when the reference voltage coming from the **REF_A** module is settling. If **CEREFLx** is changed from a nonzero value to another nonzero value, the interrupt flags may be triggered and set with some unpredictable behaviors. Therefore it is recommended to set **CEREFLx** = 00 prior to changing the **CEREFLx** settings.

The **CERSEL** bit (bit 5) in the **CExCTL2** Register selects the comparator terminal to which VREF is applied.

- When **CERSEL** = 0 and the **CEEX** = 0, VREF is applied to the V_+ terminal; When **CEEX** = 1, VREF is applied to the V− terminal
- When **CERSEL** = 1 and the **CEEX** = 0, VREF is applied to the V− terminal; When **CEEX** = 1, VREF is applied to the V+ terminal

If external signals are applied to both comparator input terminals, the internal reference generator should be turned off to reduce current consumption.

Figure 7.21 shows the bit configuration for the **CExCTL2** Register, and Table 7.30 shows the bit value and function of this register.

The **CEREFACC** bit (bit 15) in the **CExCTL2** Register is used to determine the reference voltage accuracy if **CEREFL** bit is not 0. When this bit is 0, a static or a normal reference voltage with high accuracy is required and used. When this bit is 1, a dynamic or a low power and low accuracy reference voltage is needed and used. The default value for this bit is static mode.

7.7.4 Comparator output controls

The comparator's output voltage level can be selected between the reference voltages, either **VREF0** or **VREF1**. This is controlled by the bits, **CEMRVS** and **CEMRVL**, in the **CExCTL1** Register.

31	30	29	28	27	26	25	24
Reserved							
23	**22**	**21**	**20**	**19**	**18**	**17**	**16**
Reserved							
15	**14**	**13**	**12**	**11**	**10**	**9**	**8**
CEREFACC	CEREFL		CEREF1				
7	**6**	**5**	**4**	**3**	**2**	**1**	**0**
CERS		CERSEL	CEREF0				

Figure 7.21 Bit field and function of the CExCTL2 register.

Table 7.30 Bit value and function for CExCTL2 register

Bit	Field	Type	Reset	Function
31:16	**Reserved**	RO	00	Reserved. Always reads as 0
15	**CEREFACC**	RW	0	Reference accuracy. A REF_A voltage is requested only if CEREFL > 0 **0:** Static mode (normal mode) **1:** Dynamic mode (low power, low accuracy)
14:13	**CEREFL**	RW	00	REF_A reference voltage level **00:** No reference voltage is requested **01:** 1.2 V is selected from REF_A reference voltage input **10:** 2.0 V is selected from REF_A reference voltage input **11:** 2.5 V is selected from REF_A reference voltage input
12:8	**CEREF1**	RW	00	Reference resistor tap 1. This register defines the tap of the resistor string while CEOUT = 1 and the reference V_{REF1} is used
7:6	**CERS**	RW	00	Define whether the reference voltage is coming from VCC or from the precise shared reference REF_A **00:** No current is drawn by the reference circuitry **01:** VCC applied to the resistor ladder **10:** Shared reference voltage REF_A applied to the resistor ladder **11:** Shared reference voltage supplied to V(CREF). Resistor ladder is off
5	**CERSEL**	RW	0	Select which terminal the VCCREF is applied to **0:** When CEEX = 0, VREF is applied to the V+ terminal; When CEEX = 1, VREF is applied to the V− terminal **1:** When CEEX = 0, VREF is applied to the V− terminal; When CEEX = 1, VREF is applied to the V+ terminal
4:0	**CEREF0**	RW	0	Reference resistor tap 0. This register defines the tap of the resistor string while CEOUT = 0 and the reference V_{REF0} is used

The **CEMRVS** bit in the **CExCTL1** Register defines whether the comparator output selects the **VREF0** or **VREF1** if **CERSx** = 00, 01, or 10. If this bit is 0, the comparator output state selects either **VREF0** or **VREF1**. However, if this is 1, the **CEMRVL** bit in the **CExCTL1** Register will determine the comparator output, either **VREF0** (if **CEMRVL** = 0) or **VREF1** (if **CEMRVL** = 1).

31	30	29	28	27	26	25	24
Reserved							
23	22	21	20	19	18	17	16
Reserved							
15	14	13	12	11	10	9	8
Reserved			CEMRVS	CEMRVL	CEON	CEPWRMD	
7	6	5	4	3	2	1	0
CEFDLY		CEEX	CESHORT	CEIES	CEF	CEOUTPOL	CEOUT

Figure 7.22 Bit field and function of the CExCTL1 Register.

In the MSP432P401R MCU system, an internal low-pass RC filter is available to reduce the output ripple and it is connected to the output of the comparator. When the control bit **CEF** (bit 2) in the **CExCTL1** Register is set, the comparator output is filtered with this on-chip RC filter. The delay of the filter can be adjusted in four different steps with the value set in the bit fields **CEFDLY** (bits 7~6) in the **CExCTL1** Register.

All comparator outputs contain some kinds of oscillating or ripple if the voltage difference across the input terminals is small. Internal and external parasitic effects and cross coupling on and between signal lines, power supply lines, and other parts of the system are the sources for this behavior. The comparator output oscillation reduces the accuracy and resolution of the comparison result. Therefore selecting the output filter is a good way to reduce errors associated with comparator oscillation and ripple.

Figure 7.22 shows the bit configuration for the **CExCTL1** Register, and Table 7.31 shows the bit value and function of this register.

The **CEOUTPOL** bit (bit 1) in this register can be used to set the comparator output **CEOUT** as a normal (noninverted) output or an inverted output.

7.7.5 Comparator interrupt controls

In the MSP432P401R MCU system, two comparator's actions can be used as the interrupt sources:

- Comparator is ready to perform its comparison function
- Comparator completes its comparison with a valid output

One interrupt flag and one interrupt vector is associated with the comparator. The interrupt flag and interrupt vector are included in two registers, the **CExINT** and **CExIV**. Figure 7.23 shows the bit configuration for the **CExINT** Register, and Table 7.32 shows the bit value and function of this register.

The interrupt flag **CEIFG** bit (bit 0) in the **CExINT** Register is set on either the rising or falling edge of the comparator output, selected by the **CEIES** bit in the **CExCTL1** Register. The comparator generates an interrupt when both **CEIFG** and **CEIE** bits are set. The comparator interrupt can be serviced by the CPU when the comparator interrupt is enabled appropriately at the NVIC.

It can be found from Figure 7.23 and Table 7.32 that two sets of events are existed for each comparator module, and these events can be categorized into two pairs:

Table 7.31 Bit value and function for CExCTL1 register

Bit	Field	Type	Reset	Function
31:13	**Reserved**	RO	00	Reserved. Always reads as 0
12	**CEMRVS**	RW	0	Defines whether the comparator output selects either VREF0 or VREF1 if CERS = 00, 01, or 10 **0:** Comparator output state selects either VREF0 or VREF1 **1:** CEMRVL selects either VREF0 or VREF1
11	**CEMRVL**	RW	0	This bit is valid of CEMRVS is set to 1 **0:** VREF0 is selected if CERS = 00, 01, or 10 **1:** VREF1 is selected if CERS = 00, 01, or 10
10	**CEON**	RW	0	Turn the comparator on. When the comparator is turned off, the comparator consumes no power **0:** Off; **1:** On
9:8	**CEPWRMD**	RW	00	Comparator operational power mode **00:** High-speed mode **01:** Normal mode **10:** Ultra-low power mode **11:** Reserved
7:6	**CEFDLY**	RW	00	Filter delay. This delay can be selected in four steps **00:** Typical filter delay of TBD (450) ns **01:** Typical filter delay of TBD (900) ns **10:** Typical filter delay of TBD (1800) ns **11:** Typical filter delay of TBD (3600) ns
5	**CEEX**	RW	0	Exchange the input signals on the comparator's V_+ and V_- terminals and invert the comparator's output **0:** No exchange **1:** Exchange inputs on comparator's V+ and V– terminals, invert the output
4	**CESHORT**	RW	0	Input short. This bit shorts the V+ and V– input terminals **0:** Inputs not shorted; **1:** Inputs shorted
3	**CEIES**	RW	0	Interrupt edge select for CEIIFG and CEIFG **0:** Rising edge for CEIFG, falling edge for CEIIFG **1:** Falling edge for CEIFG, rising edge for CEIIFG
2	**CEF**	RW	0	Output filter. Available if CEPWRMD = 00, 01 **0:** Comparator output is not filtered; **1:** Comparator output is filtered
1	**CEOUTPOL**	RW	0	Output polarity. This bit defines the CEOUT polarity **0:** Noninverted; **1:** Inverted
0	**CEOUT**	RW	0	Output value. This bit reflects the value of the comparator output. Writing this bit has no effect on the comparator output

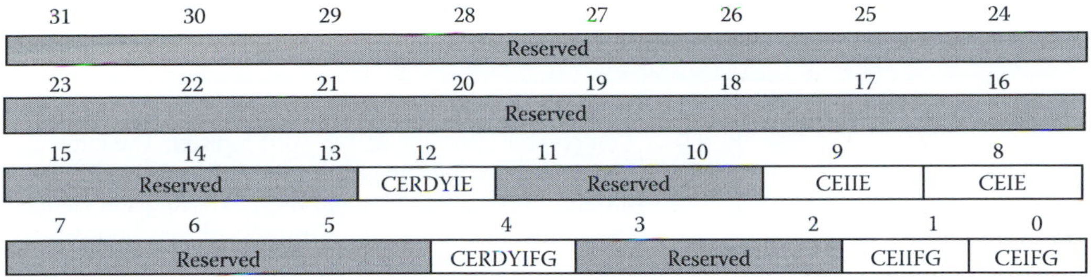

31	30	29	28	27	26	25	24
Reserved							
23	**22**	**21**	**20**	**19**	**18**	**17**	**16**
Reserved							
15	**14**	**13**	**12**	**11**	**10**	**9**	**8**
Reserved			CERDYIE	Reserved		CEIIE	CEIE
7	**6**	**5**	**4**	**3**	**2**	**1**	**0**
Reserved			CERDYIFG	Reserved		CEIIFG	CEIFG

Figure 7.23 Bit field and function of the CExINT register.

Table 7.32 Bit value and function for CExINT register

Bit	Field	Type	Reset	Function
31:13	**Reserved**	RO	00	Reserved. Always reads as 0
12	**CERDYIE**	RW	0	Comparator ready interrupt enable signal **0:** Interrupt is disabled; **1:** Interrupt is enabled
11:10	**Reserved**	RO	0	Reserved. Always reads as 0
9	**CEIIE**	RW	0	Comparator output interrupt enable inverted polarity **0:** Interrupt is disabled; **1:** Interrupt is enabled
8	**CEIE**	RW	0	Comparator output interrupt enable **0:** Interrupt is disabled; **1:** Interrupt is enabled
7:5	**Reserved**	RO	0	Reserved. Always reads as 0
4	**CERDYIFG**	RW	0	Comparator ready interrupt flag. This bit is set if the comparator reference sources are settled and the comparator module is operational. This bit has to be cleared by software **0:** No interrupt is pending; **1:** A ready interrupt is pending
3:2	**Reserved**	RO	0	Reserved. Always reads as 0
1	**CEIIFG**	RW	0	Comparator output inverted interrupt flag. The bit CEIES defines the transition of the output setting this bit **0:** No interrupt pending; **1:** An output inverted interrupt is pending
0	**CEIFG**	RW	0	Comparator output interrupt flag. The bit CEIES defines the transition of the output setting this bit **0:** No interrupt pending; **1:** A comparator output interrupt is pending

- The comparator is ready event. This event can be controlled and reflected by a pair of control bits: the **`CERDYIE`** and **`CERDYIFG`** in the **`CExINT`** Register. If the **`CERDYIE`** is set to 1 to enable this event, the **`CERDYIFG`** is also set to 1 to indicate this event if the comparator is ready to perform its comparison operation.
- The comparator is done event. This event can be controlled and reflected by another pair of control bits: the **`CEIE`** and **`CEIFG`** in the **`CExINT`** Register. If the **`CEIE`** bit is set to 1 to enable this event, the **`CEIFG`** will be set to 1 to indicate the occurrence of this event if the comparator completes its comparison operation. The transition edge of this interrupt is controlled by the **`CEIES`** bit in the **`CExCTL1`** Register.

Table 7.33 Bit value and function for CExIV register

Bit	Field	Type	Reset	Function
31:16	**Reserved**	RO	00	Reserved. Always reads as 0
15:0	**CEIV**	RO	00	Comparator interrupt vector word register. The interrupt vector register reflects only interrupt flags whose interrupt is enabled. Reading the CExIV Register clears the pending interrupt flag with the highest priority **00:** No interrupt pending **02:** CEOUT interrupt; Interrupt Flag: CEIFG; Interrupt Priority: Highest **04:** CEOUT interrupt inverted polarity; Interrupt Flag: CEIIFG **06~08:** Reserved **0A:** Comparator ready interrupt; Interrupt Flag: CERDY-IFG; Interrupt Priority: Lowest

The control bits **CEIIE** and **CEIIFG** in this register just work as the inverting status of the **CEIE** and **CEIFG** bits.

If a comparator related interrupt occurred, a related interrupt vector is sent to the **CExIV** Register. This is a 32-bit register but the lower 16-bit is used to store an interrupt vector.

Table 7.33 lists the bit value and function for the **CExIV** Register.

A point to be noted when using this register is that any reading this register will clears the pending interrupt flag with the highest priority.

7.7.6 Analog comparator interfacing signals and GPIO pins

In the MSP432P401R MCU system, totally there are two groups of 8 analog input signals can be compared with the comparator. These analog input signals are connected to the MSP-EXP432P401R EVB via related GPIO pins. Refer to Figure 7.1, it can be found that only GPIO Ports, **P5~P8** and **P10**, contain related pins to be connected to the corresponding analog inputs. Table 7.34 shows all of these analog inputs and related GPIO pins for **COMP_E0** and **COMP_E1** modules used in the MSP432P401R MCU system.

A point to be noted when using these GPIO pins is: special attention should be paid to two **COMP_E** output pins. Refer to Tables 7.12 and 7.13, GPIO pins **P7.1** and **P7.2** can be mapped to **PM_C0OUT** and **PM_C1OUT** function pins only when the value in the **P7SEL1x**:**P7SEL0x** = 01.

However, in MSP-EXP432P401R EVB, only limited GPIO pins are exposed to the users via four Jumper Connectors, J1~J4. These GPIO pins include:

- **P6.4~P6.7** (C1.3~C1.0)
- **P5.6~P5.7** (C1.7~C1.6)

Therefore, only **COMP_E1** module is available if we want to use GPIO pins provided by J1~J4 connectors to compare external analog signals and connect them via these pins. Of course you can use extra GPIO pins provided by J5 connector in the MSP-EXP432P401R EVB but all of those pins are only prototype pins without any connector available.

Table 7.34 COMP_E External Control Signals and GPIO Pin Distributions

COMP_E Pin	GPIO Pin	Pin Type	Buffer Type	Pin Function
C0.0	**P8.1**	I	Analog	Comparator_E0 analog input 0
C0.1	**P8.0**	I	Analog	Comparator_E0 analog input 1
C0.2	**P7.7**	I	Analog	Comparator_E0 analog input 2
C0.3	**P7.6**	I	Analog	Comparator_E0 analog input 3
C0.4	**P7.5**	I	Analog	Comparator_E0 analog input 4
C0.5	**P7.4**	I	Analog	Comparator_E0 analog input 5
C0.6	**P10.5**	I	Analog	Comparator_E0 analog input 6
C0.7	**P10.4**	I	Analog	Comparator_E0 analog input 7
C1.0	**P6.7**	I	Analog	Comparator_E1 analog input 0
C1.1	**P6.6**	I	Analog	Comparator_E1 analog input 1
C1.2	**P6.5**	I	Analog	Comparator_E1 analog input 2
C1.3	**P6.4**	I	Analog	Comparator_E1 analog input 3
C1.4	**P6.3**	I	Analog	Comparator_E1 analog input 4
C1.5	**P6.2**	I	Analog	Comparator_E1 analog input 5
C1.6	**P5.7**	I	Analog	Comparator_E1 analog input 6
C1.7	**P5.6**	I	Analog	Comparator_E1 analog input 7
PM_C0OUT	**P7.1**	O	Digital	Comparator_E0 output
PM_C1OUT	**P7.2**	O	Digital	Comparator_E1 output

Next, let's have a closer look at the initialization and configuration process for the COMP_E module.

7.7.7 Initialization and configuration process for the analog comparator

Prior to using any COMP_E module to perform analog voltage comparison function, the COMP_E module must be initialized and configured with the following steps:

1. Reset COMP_E **`CExCTL1`** and **`CExINT`** registers for initialization
2. Select the compared analog input signal by configuring the **`CExCTL0`** Register
3. Set the selected analog input to be connected to the V_+ or V_- terminal of the comparator by configuring the **`CExCTL0`** Register
4. Select the reference source (either the V_{CC} or the REF_A shared reference, either using the internal resistor ladder or not) by configuring the **`CExCTL2`** Register
5. Determine whether the selected reference to be connected to the V_+ or V_- terminal of the comparator by configuring the **`CExCTL2`** Register
6. Select the REF_A reference voltage level by configuring the **`CExCTL2`** Register if the REF_A reference is used
7. Set the power mode for the comparator by configuring the **`CExCTL1`** Register
8. Enable comparator related interrupt by configuring the **`CExCTL1`** and **`CExINT`** registers if interrupt mechanism is used
9. Optionally one can set up the output selection (**`VREF1`** or **`VREF0`**), output polarity and using the output filter by configuring the **`CExCTL1`** Register
10. Enable the comparator module to make it ready to perform the comparison by configuring the **`CExCTL1`** Register

Now, let's build a simple project to test the functions of the COMP_E1 module.

7.7.8 Develop an example analog comparator project

Now, let's build a comparator project using the **COMP_E1** module to perform a comparison between an analog input and the internal reference.

In this simple project, we try to use **COMP_E1** module to compare an external voltage coming from the potentiometer **VR2** installed in the EduBASE ARM® Trainer with a programmable internal reference voltage $\mathbf{V_{REF1}} = \mathbf{V_{REF0}} = \mathbf{0.25V_{CC}}$. The hardware configuration for this project is shown in Figure 7.24.

One needs to use some jump wires to perform the following connections to finish this hardware and software configuration:

- Connect the potentiometer output that is **A6** input to the ADC to the **V-** terminal of the **COMP_E1** module as an external analog input to the comparator. To do this connection, just connect the **P4.7** pin on the J3 connector with the **P5.6** pin on the J4 connector in the MSP-EXP432P401R EVB.
- To monitor and check the comparison result **C1OUT**, one needs to build this project to use the interrupt mechanism to monitor and handle a COMP_E1 compare output interrupt to get the output of this comparator. If this C1OUT is set to 1, which means that the Vin+ > Vin–, four LEDs **PB3~PB0** installed in the ARM® Trainer are turned on. Otherwise if this bit is cleared to 0, which means that the Vin+ < Vin–, four LEDs are off.

The only hardware connection has been highlighted in Figure 7.24. All other connections have been completed by the vendor. Your finished connection should match one that is shown in Figure 7.24. Now, let's build the project **DRACOMPE1** to complete this analog comparator function.

Create a new Keil® µVersion®5 project **DRACOMPE1** under the folder **C:\MSP432 Class Projects\Chapter 7**. Then create a new C source file **DRACOMPE1.c** and enter the code shown in Figure 7.25 into this source file.

Let's have a closer look at this source file to see how it works.

1. The codes in lines 4~6 are used to include all system header files used in this project.
2. In code line 9, the Watchdog timer is temporarily stopped to avoid its disturbance to this project.
3. The code lines 10~11 are used to configure GPIO **P2.4** and **P2.5** as output pins since these pins are connected to two LEDs, **PB3** and **PB2**, in the EduBASE ARM Trainer.

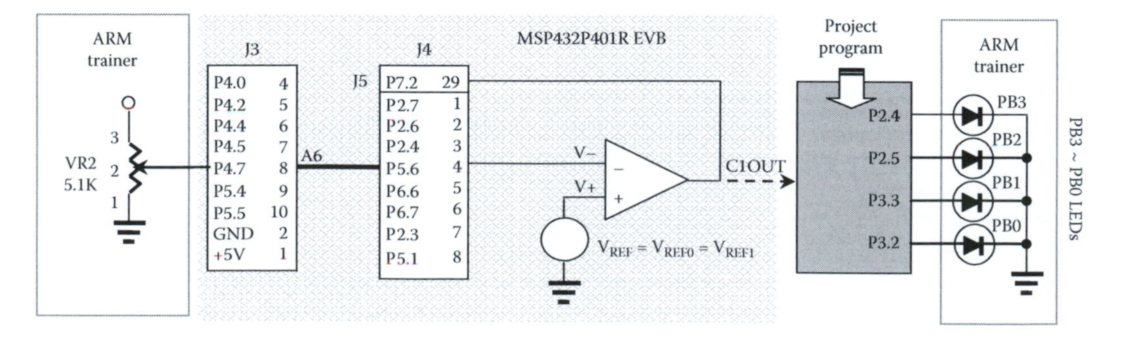

Figure 7.24 Hardware and software configuration for the COMP_E1.

```
1  //*****************************************************************************
2  // DRACOMPE1.c - Main application file for project DRACOMPE1
3  //*****************************************************************************
4  #include <stdint.h>
5  #include <stdbool.h>
6  #include <msp.h>
7  int main(void)
8  {
9   WDTCTL = WDTPW | WDTHOLD;       // stop WDT
10  P2DIR = BIT4|BIT5;              // set P2.4 & P2.5 as output pins (PB3 & PB2)
11  P3DIR = BIT2|BIT3;              // set P3.3 & P3.2 as output pins (PB1 & PB0)
12  P7DIR |= BIT2;                  // C1OUT output direction (p7.2)
13  P7SEL0 |= BIT2;                 // map P7.2 as C1OUT function (P7SEL1 = 0 by default)
14  CE1CTL0 = CEIMEN | CEIMSEL_7;   // enable V- to be connected to input channel 7
15  CE1CTL1 = CEPWRMD_1;            // normal power mode (CEPWRMD_1 = 0x1)
16  CE1CTL2 = 0x747;                // CERS = 01 (Vcc is applied to the resistor ladder)
17                                  // CERSEL = 0: VREF is applied to V+ terminal
18                                  // CEREF1 = CEREF0 = 8: 8/32 = 0.25Vcc = 0.25*3.3V = 0.825V
19  CE1INT = CEIE;                  // enable COMP_E1 module output interrupt
20  CE1CTL1 |= CEON;                // turn on Comparator_E1 module
21  while(1)                        // when A6 (V-) < V+ (0.825V), C1OUT = High
22  {                               // when A6 (V-) > V+ (0.825V), C1OUT = Low
23   if (CE1CTL1 & CEOUT)           // check if C1OUT outputs high
24   {
25    P2OUT = BIT4|BIT5;            // if it is high, turn on LEDs PB3 ~ PB0
26    P3OUT = BIT2|BIT3;
27   }
28   else                           // if C1OUT is low, turn off LEDs PB3 ~ PB0
29   {
30    P2OUT = 0;
31    P3OUT = 0;
32   }
33  }
34 }
```

Figure 7.25 Detailed codes for the project DRACOMPE1.

4. The GPIO pin **P7.2** is configured as an output pin in line 12 since this pin is connected to the **C1OUT** pin (refer to Tables 7.12 and 7.13) when this pin is mapped to an alternate function pin by using the mapping controller.
5. The GPIO pin **P7.2** is configured to work as a comparator output pin **C1OUT** in line 13 by setting **P7SEL1.2**:**P7SEL0.2** = **01** (refer to Tables 7.12 and 7.13). After a system reset, the **P7SEL1.2** is reset to 0, therefore you do not need to code for this register.
6. The code lines 14~16 are used to initialize and configure the **COMP_E1** module to enable it to work as an analog comparator to compare one external analog input coming from **A6** with an internal reference. The external analog input from **A6** is configured to be connected to the V- terminal of the comparator via the comparator input channel **C1.7**, and the internal reference coming from the internal resistor ladder is configured to be connected to the V_+ terminal of the comparator. The detailed coding process for these initializations are listed below in steps 7~9.
7. In line 14, the **CE1CTL0** Register is configured to enable the V- terminal of the comparator to be connected to the analog input via channel **C1.7** (**CEIMEN** = **1** and **CEIMSEL** = **0x7**). The system macro **CEIMSEL_7** defined in the system header file **msp432p401r.h** is equal to **0x7**.
8. The **CE1CTL1** Register is configured in line 15 to enable the **COMP_E1** module to use the normal power mode. The system macro **CEPWRMD_1** is defined as **0x1**.

9. In line 16, the **CE1CTL2** Register is configured to perform the following functions:
   a. Enable the Vcc to be connected to the internal resistor ladder by setting **CERS** (bits 7:6) to **0x01** since we need to use the internal power supply Vcc.
   b. Enable the internal reference VREF to be connected to the V+ terminal of the comparator by resetting the **CERSEL** bit (bit 5) to **0**.
   c. Enable two references, VREF1 and VREF0, to be applied on the comparator to be equal by setting the bit fields **CEREF1** (bits 11:8) and **CEREF0** (bits 4:0) to **8** since our target reference voltage level is $(\mathbf{8}/32) \times \text{Vcc} = 0.25\text{Vcc} = 0.25 \times 3.3\text{V} = 0.825$ V. However, the starting value for both references is 0, not 1, therefore we must subtract 1 from this **8** to get and place **7** into these two bit fields.

   Figure 7.26 shows the detailed bits configurations and values for this initialization on the **CE1CTL2** Register. The initialized value is: **0x0747**.

10. In line 19, the **CE1INT** Register is configured to enable the **COMP_E1** module to have the interrupt ability by setting the **CEIE** bit to 1.
11. The **COMP_E1** module is enabled in line 20 to make it begin to perform the comparison function by setting the **CEON** bit in the **CE1CTL1** Register to 1.
12. Starting line 21, an infinitive **while()** loop is used to repeatedly check the **COMP_E1** module to see whether the comparison function is done and whether the **C1OUT** outputs a High. If it is, four LEDs, **PB3~PB0**, will be turned on. Otherwise they are off.

After setting up the environments, such as checking and making sure that the debugger to be used for this project is **CMSIS-DAP** and the download settings are correct, you can build, download, and run the project.

As the project runs, rotate the potentiometer **VR2** with a small screwdriver, and you can find that four LEDs are on if the analog input voltage is lower than 0.825 V, which can be monitored by using a multimeter. Four LEDs will be off when the input voltage is greater than 0.825 V.

7.7.9 Analog comparator API functions in the MSPPDL

To assist users to build and develop sophisticated analog comparator projects, a set of API functions is provided by the MSPPDL.

A special struct **COMP_E_Config** is designed for the COMP_E module. This struct is:

```
typedef struct _COMP_E_Config
{
 uint_fast16_t positiveTerminalInput;
 uint_fast16_t negativeTerminalInput;
```

15	14	13	12	11	10	9	8
CEREFACC	CEREFL		CEREF1				
0	0	0	0	0	1	1	1

7	6	5	4	3	2	1	0
CERS		CERSEL	CEREF0				
0	1	0	0	0	1	1	1

Figure 7.26 Bits configurations and values for the initialization of CE1CTL2.

```
 uint_fast8_t outputFilterEnableAndDelayLevel;
 uint_fast8_t invertedOutputPolarity;
 uint_fast16_t powerMode;
} COMP_E_Config;
```

This struct is used in some API functions to simplify the configuration process for the COMP_E module programming in applications.

Totally 21 API functions are available to users and these functions are used to:

1. Initialize the COMP_E modules
2. Set up reference voltages
3. Other COMP_E module controls (enable module, set the power mode and output)
4. Manage interrupts for the COMP_E modules

7.7.9.1 API functions used to initialize COMP_E modules

The following API functions are involved in this group:

- **COMP_E_initModule()**
- **COMP_E_shortInputs()**
- **COMP_E_swapIO()**
- **COMP_E_unshortInputs()**

Table 7.35 shows the prototypes and arguments of these functions.

When using the **COMP_E_initModule()** to initialize the COMP_E modules, this function will reset all necessary register bits and set the given options in the registers. To actually use the comparator module, the **COMP_E_enableModule()** function must be explicitly called before use. If a reference voltage is set to a terminal, the function **COMP_E_setReferenceVoltage()** should be called to set the desired reference voltage.

7.7.9.2 API functions used to set up reference voltages

The following API functions are involved in this group:

- **COMP_E_setReferenceVoltage()**
- **COMP_E_setReferenceAccuracy()**

Table 7.36 shows the prototypes and arguments of these functions.

The arguments **supplyVoltageReferenceBase** is used to determine which reference source should be used for the comparator. When the **COMP_E_REFERENCE_AMPLFIER_DISABLED** is selected, it means that no shared REF_A is needed for this reference. Otherwise a voltage coming from the shared REF_A source is used.

The **lowerLimitSupplyVoltageFraction32** and the **upperLimitSupplyVoltageFraction32** are used to provide a number as a **Numerator** that is ranged from 0~32 and worked for an equation **VBase** × (**Numerator/32**) to calculate the lower and upper reference voltages to be applied on two terminals of the comparator. For a normal comparator, these two parameters should be identical. However, if a hysteresis effect is needed, these two parameters should be different.

Table 7.35 API functions used to initialize COMP_E Modules

API Function	Parameters	Description
bool **COMP_E_init-Module(**uint32_t comparator, const **COMP_E_Config** * config**)** Returns a **true:** Initialization is successful Returns a **false:** Initialization is failed	**Comparator** is the instance of the comparator module. Value can be: **COMP_E0_MODULE or COMP_E1_MODULE** **config** is a pointer to the struct **COMP_E_Config** Valid values for both **positiveTerminal-Input** and **Nega-tiveTerminalInput** can be: **COMP_E_INPUT0~COMP_E_INPUT7**, and **COMP_E_VREF**. The default is **COMP_E_INPUT0** Modified bits are **CEIPSEL** and **CEIPEN** of **CECTL0** Register, **CERSEL** of **CECTL2** Register	Initialize the COMP_E module Valid values for **outputFilterEnableAndDelayLevel** can be: **COMP_E_FILTEROUTPUT_OFF** [Default] **COMP_E_FILTEROUTPUT_DLYLVL1** **COMP_E_FILTEROUTPUT_DLYLVL2** **COMP_E_FILTEROUTPUT_DLYLVL3** **COMP_E_FILTEROUTPUT_DLYLVL4** Modified bits are **CEF** & **CEFDLY** of **CECTL1** Register Valid values for the **invertedOutputPolarity** can be: **COMP_E_NORMALOUTPUTPOLARITY**—indi-cates the output should be normal [Default] **COMP_E_INVERTEDOUTPUTPOLARITY**—the output should be inverted. Modified bits are **CEOUTPOL** of **CECTL1** Register Valid values for the **powerMode** can be: **COMP_E_HIGH_SPEED_MODE** [default] **COMP_E_NORMAL_MODE COMP_E_ULTRA_LOW_POWER_MODE**
void **COMP_E_shortInputs(**uint32_t comparator**)**	**Comparator** is the instance of the comparator module. Value can be: **COMP_E0_MODULE or COMP_E1_MOD-ULE**	Short two input pins chosen during initialization Modified bits are **CESHORT** of **CECTL1** Register Return: None
void **COMP_E_swapIO(**uint32_t comparator**)**	**Comparator** is the instance of the comparator module. Value can be: **COMP_E0_MODULE or COMP_E1_MOD-ULE**	Toggle the bit that swaps which terminals the inputs go to, also invert the output of the comparator Return: None
void **COMP_E_unshort Inputs(**uint32_t comparator**)**	**Comparator** is the instance of the comparator module. Value can be: **COMP_E0_MODULE or COMP_E1_MOD-ULE**	Disable the short of the two input pins chosen during initialization Return: None

Table 7.36 API functions used to set up references for COMP_E Modules

API Function	Parameters	Description
void **COMP_E_setReferenceVoltage(** uint32_t comparator, uint_fast16_t supplyVoltageReferenceBase, uint_fast16_t lowerLimitSupplyVoltageFractionOf32, uint_fast16_t upperLimitSupplyVoltageFractionOf32**)**	**Comparator** is the instance of the comparator module. Value can be: **COMP_E0_MODULE** or **COMP_E1_MODULE** Valid values for the argument **supplyVoltageReferenceBase** can be: **COMP_E_REFERENCE_AMPLIFIER_DISABLED** [Default] **COMP_E_VREFBASE1_2V** **COMP_E_VREFBASE2_0V** **COMP_E_VREFBASE2_5 V**	Generate a voltage to serve as a reference to the terminal selected at initialization. The voltage is equal to: **VBase*(Numerator/32)**. If the upper and lower limit voltage numerators are equal, then a static reference is defined, whereas when they are different then a hysteresis effect is generated Valid values for **lowerLimitSupplyVoltageFractionOf32** can be a numerator of the equation to generate the reference voltage for the lower limit reference voltage. Valid values are between **0** and **32** Modified bits are **CEREF0** of **CECTL2** Register Valid values for **upperLimitSupplyVoltageFractionOf32** can be a numerator of the equation to generate the reference voltage for the upper limit reference voltage. Valid values are between **0** and **32** Modified bits are **CEREF1** of **CECTL2** Register
void **COMP_E_setReferenceAccuracy(** uint32_t comparator, uint_fast16_t referenceAccuracy**)**	**Comparator** is the instance of the comparator module. Value can be: **COMP_E0_MODULE** or **COMP_E1_MODULE**	Set the reference accuracy Valid values for the argument **referenceAccuracy** can be: **COMP_E_ACCURACY_STATIC** or **COMP_E_ACCURACY_CLOCKED** Modified bits are **CEREFACC** of **CECTL2** Register

For the API function `setReferenceAccuracy()`, the argument `referenceAccuracy` can be either the `COMP_E_ACCURACY_STATIC` with high accuracy or `COMP_E_ACCURACY_CLOCKED` with lower power mode and lower accuracy.

7.7.9.3 API functions used to control and manage the COMP_E modules

The following API functions are included in this group:

- `COMP_E_enableModule()`
- `COMP_E_disableModule()`
- `COMP_E_setPowerMode()`
- `COMP_E_outputValue()`

Table 7.37 shows the prototypes and arguments of these functions.

A point to be noted is that when the COMP_E module is initialized and configured, the module must be disabled, which is a default situation.

7.7.9.4 API functions used to manage interrupts related to the COMP_E modules

The following API functions are included in this group:

- `COMP_E_registerInterrupt()`
- `COMP_E_enableInterrupt()`

Table 7.37 API functions used to control and manage the COMP_E modules

API function	Parameters	Description
void **COMP_E_enableModule(**uint32_t comparator**)**	**Comparator** is the instance of the comparator module. Value can be: **COMP_E0_MODULE** or **COMP_E1_MODULE**	Set the **CEON** bit to enable the operation of the comparator module
void **COMP_E_disableModule(**uint32_t comparator**)**	**Comparator** is the instance of the comparator module. Value can be: **COMP_E0_MODULE** or **COMP_E1_MODULE**	Clear the **CEON** bit to disable the operation of the comparator module
void **COMP_E_setPowerMode(**uint32_t comparator, uint_fast16_t powerMode**)**	**Comparator** is the instance of the comparator module. Value can be: **COMP_E0_MODULE** or **COMP_E1_MODULE** Valid value for the **powerMode**: **COMP_E_HIGH_SPEED_MODE** **COMP_E_NORMAL_MODE** **COMP_E_ULTRA_LOW_POWER_MODE** Modified bits are **CEPWRMD** of **CECTL1** Register	Set the power mode for the comparator
uint8_t **COMP_E_outputValue(**uint32_t comparator**)**	**Comparator** is the instance of the comparator module. Value can be: **COMP_E0_MODULE** or **COMP_E1_MODULE**	Return the output value of the comparator module Valid returned value is: **COMP_E_HIGH or COMP_E_LOW**

- `COMP_E_getEnabledInterruptStatus()`
- `COMP_E_getInterruptStatus()`
- `COMP_E_setInterruptEdgeDirection()`
- `COMP_E_toggleInterruptEdgeDirection()`
- `COMP_E_clearInterruptFlag()`

Table 7.38 shows the prototypes and arguments of these functions.

When using these API functions to perform the COMP_E module related interrupts, the following points should be kept in mind:

1. Any COMP_E module related interrupt source must be enabled before it can be responded.
2. Any COMP_E module related interrupt handler must be registered first before it can be called when the related interrupt occurred.
3. After an interrupt is responded, it must be cleared by calling the API function `COMP_E_clearInterruptFlag()` inside the interrupt handler to enable the further interrupt to be occurred.

7.7.10 Use API functions to build a comparator project with the SD model

Now we can use the API functions provided by MSPPDL to build an analog comparator project **SDCOMPE1**. This project has the following features:

- The external analog input comes from **C1.7** via the GPIO pin **P5.6** and is connected to the V-terminal of the COMP_E1 module.
- The internal reference comes from the interval Vcc = 3.3 V and the resistor ladder, which is connected to the V+ terminal of the COMP_E1 module.
- Two different references are applied to the terminals of the comparator, **VREF1** and **VREF0**, to obtain a hysteresis effect. The upper limitation of the reference is **VREF1** = $0.75Vcc = 0.75 \times 3.3V = 2.475$ V and the lower limitation of the reference is **VREF0** = $0.25Vcc = 0.25 \times 3.3V = 0.825$ V.
- The comparator out is High when $V- < V+ = 0.25Vcc$ (0.825 V) and Low when $V- > V+ = 0.75Vcc = 2.475$ V. When the comparator outputs High, the **VREF1** (2.475 V) is used as the reference applied in the V+ terminal, but when the comparator outputs Low, the **VREF0** (0.825 V) is used as the reference applied on the V+ terminal.
- Four LEDs, **PB3~PB0** installed in the EduBASE ARM Trainer, will be on when the comparator outputs High, and off when the comparator outputs Low.

Refer to Figure 7.24 to complete the hardware configuration for this project (exactly connect the GPIO pin **P4.7** (**A6**) with GPIO pin **P5.6** (**C1.7**) on J3 and J4 in the MSP-EXP432P401R EVB to enable the analog input coming from the **VR2** potentiometer to be connected to the comparator input module 1 at channel 7 (**C1.7**).

Create a new folder **SDCOMPE1** under the folder **C:\MSP432 Class Projects\Chapter 7** in the Windows Explorer. Then create a new µVersion5 project **SDCOMPE1** and add it into the folder **SDCOMPE1** created above.

Create a new C file **SDCOMPE1.c** and enter the following code shown in Figure 7.27 into this C file. Let's have a closer look at this piece of code to see how it works.

1. The codes in lines 4~9 are system header files to be used in this project.
2. The codes in lines 10~17 are used to set up the comparator struct **COMP_E_Config** to configure the initial parameters. The positive input terminal is connected to the internal VREF voltage, and the negative input terminal is connected to an external analog input coming from the **VR2** potentiometer via comparator module 1 input channel 7. All other parameters are default set up values.
3. The Watchdog timer is temporarily terminated in line 21.
4. In lines 22~25, GPIO pins **P2.4**, **P2.5**, **P3.2**, and **P3.3**, are configured as output pins since these pins are connected to four LEDs, **PB3~PB0** installed in the EduBASE ARM Trainer, and we need to use them to display the output status of the comparator.
5. In lines 26~27, the GPIO pin **P7.2** is configured as a digital output pin to enable it to work as the output pin for the COMP_E1 module.

   The code lines 29~31 are used to configure the GPIO pin **P5.6** that is the analog output of the potentiometer **VR2** as the external analog input to the comparator via COMP_E1 module channel 7 (**C1.7**).
6. The COMP_E1 module is initialized by calling the API function **COMP_E_initModule()** with the **compConfig** as the struct argument in line 33.
7. In line 39, the COMP_E1 module is configured by setting up the reference voltages for both positive and negative terminals of the comparator. The system macro constant **COMP_E_REFERENCE_AMPLIFIER_DISABLED** indicates that the shared reference source coming from the REF_A is disabled and the Vcc with the internal resistor ladder are to be used for both references. The lower and upper numerators used for these references are 8 and 24, respectively, to create related lower (0.825 V) and upper (2.475 V) reference limitations to generate a hysteresis effect.

Table 7.38 API functions used to manage interrupts related to the COMP_E modules

API Function	Parameters	Description
void **COMP_E_registerInterrupt(**uint32_t comparator, void(*)(void) intHandler**)**	**Comparator** is the instance of the comparator module. Value can be: **COMP_E0_MODULE** or **COMP_E1_MODULE** The **intHandler** is a pointer to the function to be called when the comparator interrupt occurs	Register the handler to be called when a comparator interrupt occurs. This function enables the global interrupt in the interrupt controller The specific comparator interrupts must be enabled via **COMP_E_enableInterrupt()**. It is the interrupt handler's responsibility to clear the interrupt source via **COMP_E_clearInterruptFlag()**
void **COMP_E_enableInterrupt(**uint32_t comparator, uint_fast16_t mask**)**	**Comparator** is the instance of the comparator module. Value can be: **COMP_E0_MODULE** or **COMP_E1_MODULE** **mask** is the bit mask of the interrupt sources to be enabled. Mask value is the logical or of any of the following: **COMP_E_OUTPUT_INTERRUPT**—Output interrupt **COMP_E_INVERTED_ POLARITY_INTERRUPT**— Output interrupt inverted polarity **COMP_E_READY_ INTERRUPT**— Ready interrupt	Enable selected comparator interrupt sources Any interrupt source must be enabled first before it can be responded or handled by the related interrupt handler
uint_fast16_t **COMP_E_getEnabledInterruptStatus(**uint32_t comparator**)**	**Comparator** is the instance of the comparator module. Value can be: **COMP_E0_MODULE** or **COMP_E1_MODULE**	This function is useful to call in ISRs to get all pending interrupts that are actually enabled and could have caused the ISR
uint_fast16_t **COMP_E_getInterruptStatus** **(**uint32_t comparator**)**	**Comparator** is the instance of the comparator module. Value can be: **COMP_E0_MODULE** or **COMP_E1_MODULE**	Get the current comparator interrupt status only
void **COMP_E_setInterruptEdgeDirection(**uint32_t comparator, uint_fast8_t edgeDirection**)**	**Comparator** is the instance of the comparator module. Value can be: **COMP_E0_MODULE** or **COMP_E1_MODULE** Valid value of the **edgeDirection:** **COMP_E_FALLINGEDGE** [Default] **COMP_E_RISINGEDGE**	Set the edge direction that would trigger an interrupt

(*Continued*)

Table 7.38 ***(Continued)*** API functions used to manage interrupts related to the COMP_E modules

API Function	Parameters	Description
void **COMP_E_toggle InterruptEdge Direction(**uint32_t comparator**)**	**Comparator** is the instance of the Comparator module. Value can be: **COMP_E0_MODULE** or **COMP_E1_MODULE**	Toggle the edge direction that would trigger an interrupt
void **COMP_E_clear InterruptFlag(**uint32_t comparator, uint_fast16_t mask**)**	**Comparator** is the instance of the comparator module. Value can be: **COMP_E0_ MODULE** or **COMP_E1_MODULE** **mask** is a bit mask of the interrupt sources to be cleared. It can be the logical OR of any of the following: **COMP_E_INTERRUPT_ FLAG**—Output interrupt flag **COMP_E_ INTERRUPT_FLAG_INVERTED_ POLARITY**—Output interrupt flag inverted polarity **COMP_E_ INTERRUPT_FLAG_ READY**—Ready interrupt flag	Clear Comparator interrupt flags

8. The COMP_E1 module is enabled in line 41.
9. Starting at line 42, an infinitive **while()** loop is used to repeatedly check the output status of the comparator. If the output is High, four LEDs, **PB3~PB0**, will be turned on and the reference will be changed to 2.475 V. If the output is Low, four LEDs will be off and the reference will be changed to 0.825 V. For these changed references, the output will be High if V– < V+ (0.825 V) and Low when V– > V+ (2.475 V).

Before you can build and run the project, you need to set up the environment for this project. Follow the steps below to complete this setup:

- Set the header file path by going to **Project|Options for Target "Target 1"** menu item. Then select the **C/C++** tab and go to the **Include Paths** box. Click the three-dot button and select **C:\ti\msp\MSPWare_2_21_00_39\driverlib\driverlib** as the system header file path.
- Select the correct debugger by clicking on the **Debug** tab and select the **CMSIS-DAP Debugger** from the **Use** box. Also make sure that all settings for the debugger and the flash download are correct by clicking on the Settings button.
- Add the MSPPDL into this project by right clicking on the **Source Group 1** folder in the **Project** panel on the left, and select the **Add Existing Files to Group "Source Group 1"** menu item. On the opened box, select the driver library file path **C:\ti\msp\MSPWare_2_21_00_39\driverlib\driverlib\MSP432P4xx\keil** and then select the driver library file **msp432p4xx_driverlib.lib**. Click the **Add** button to add it into the project.

Now you can build and run the project to check its comparison function. As the project runs, use a small screw driver to rotate the **VR2** potentiometer. Four LEDs, **PB3~PB0**,

```
1  //****************************************************************************
2  // SDCOMPE1.c - Main application file for project SDCOMPE1
3  //****************************************************************************
4  #include <stdint.h>
5  #include <stdbool.h>
6  #include <msp.h>
7  #include <MSP432P4xx\comp_e.h>
8  #include <MSP432P4xx\gpio.h>
9  #include <MSP432P4xx\wdt_a.h>
10 COMP_E_Config compConfig =
11 {
12  COMP_E_VREF,                          // positive Input Terminal
13  COMP_E_INPUT7,                        // negative Input Terminal
14  COMP_E_FILTEROUTPUT_OFF,              // filter off
15  COMP_E_NORMALOUTPUTPOLARITY,          // normal Output Polarity
16  COMP_E_NORMAL_MODE                    // normal power mode
17 };
18 int main(void)
19 {
20  uint8_t res;
21  WDT_A_holdTimer();                    // stop watchdog timer
22  // set P2.4 ~ P2.5 as output pins (LEDs PB3 & PB2)
23  GPIO_setAsOutputPin(GPIO_PORT_P2, GPIO_PIN4|GPIO_PIN5);
24  // set P3.2 ~ P3.3 as output pins (LEDs PB1 & PB0)
25  GPIO_setAsOutputPin(GPIO_PORT_P3, GPIO_PIN2|GPIO_PIN3);
26  // set P7.2 to output primary module function, (C1OUT)
27  GPIO_setAsPeripheralModuleFunctionOutputPin(GPIO_PORT_P7, GPIO_PIN2,
28                                              GPIO_PRIMARY_MODULE_FUNCTION);
29  // set P5.6 to be comparator V-in (C1.7)
30  GPIO_setAsPeripheralModuleFunctionInputPin(GPIO_PORT_P5, GPIO_PIN6,
31                                              GPIO_TERTIARY_MODULE_FUNCTION);
32  // initialize COMP_E1 module
33  COMP_E_initModule(COMP_E1_MODULE, &compConfig);
34  /* set the reference voltage that is being supplied to the V+ terminal
35   * comparator instance 1,
36   * reference voltage = internal Vcc & resistor ladder,
37   * lower limit = Vcc*(8/32) = 0.25Vcc, (= 0.825V at Vcc = 3.3V)
38   * upper limit = Vcc*(24/32) = 0.75Vcc (= 2.475V at Vcc = 3.3V) */
39  COMP_E_setReferenceVoltage(COMP_E1_MODULE, COMP_E_REFERENCE_AMPLIFIER_DISABLED, 8, 24);
40  // enable the COMP_E1 module
41  COMP_E_enableModule(COMP_E1_MODULE);
42  while(1)                              // when A6 (V-) < V+ (0.825V), C1OUT = High
43  {                                     // when A6 (V-) > V+ (2.475V), C1OUT = Low
44   res = COMP_E_outputValue(COMP_E1_MODULE);
45   if (res == COMP_E_HIGH)              // check if C1OUT outputs high
46   {
47    // if it is high, turn on LEDs PB3 ~ PB0
48    GPIO_setOutputHighOnPin(GPIO_PORT_P2, GPIO_PIN4|GPIO_PIN5);
49    GPIO_setOutputHighOnPin(GPIO_PORT_P3, GPIO_PIN2|GPIO_PIN3);
50   }
51   else                                 // if C1OUT is low, turn off LEDs PB3 ~ PB0
52   {
53    GPIO_setOutputLowOnPin(GPIO_PORT_P2, GPIO_PIN4|GPIO_PIN5);
54    GPIO_setOutputLowOnPin(GPIO_PORT_P3, GPIO_PIN2|GPIO_PIN3);
55    }
56   }
57 }
```

Figure 7.27 Detailed code for the project SDCOMPE1.

will be on when the analog input voltage is lower than 0.825 V, but will be off when the analog input is greater than 2.475 V. By using a multimeter, you can check and monitor these voltage values.

We leave a similar project as a lab project for readers in the homework section.

7.8 Summary

This chapter is concentrated on the MSP432 parallel I/O interface programming. Three parallel I/O ports programming projects are included in this chapter, they are:

- On-board keypads interface programming project
- ADC programming project
- Analog comparator project

All of these parallel I/O port programming are closely related to the GPIO ports since most GPIO ports provide multiple functions to interface to different peripheral devices in the MSP432P401R MCU system. All peripherals in this system are configured to be connected to the different GPIO pins to interface to the MCU and other control components. All GPIO ports are programmable ports with a set of control registers to be programmed to perform various important and vital interfacing functions.

The on-board keypads interface programming project is directly related to GPIO ports with GPIO control. Therefore, the GPIO module and its multiple function property are first discussed with its architecture. For example the keypad project is built following that discussion.

As for the analog-to-digital conversion project, the related GPIO control pins and registers are introduced and discussed since these registers provide the controllability about the alternate functions for the GPIO ports.

Following the discussions about the GPIO module and related registers, an example project used to control and interface the ADC14 module is provided. All important and related to ADC14 control registers provided by the MSP432P401R MCU system are introduced and discussed in detail. The popular API functions related to the ADC14 module and provided by the MSPWare Peripheral Driver Library are also introduced and discussed in detail.

The third part of this chapter is about the analog comparator modules and controls implemented in the MSP432P401R MCU system. A quick introduction about two analog comparator modules is provided first. Then a detailed introduction about the most control and configuration registers used in the COMP_E modules is provided with a discussion about the architecture of the COMP_E modules. An example project that implemented the COMP module 1 to compare an external analog input with the internal reference is introduced and analyzed.

Finally, some popular API functions related to COMP_E modules provided by the MSPPDL are introduced. An example project that uses those API functions to perform analog comparison function is also discussed in detail.

HOMEWORK

I. True/False Selections

_____1. With the GPIO **`PM Control`** technique, some GPIO pins can be configured to perform multiple functions, which are called alternative functions.

_____2. The port function select registers (**PxSEL0**, **PxSEL1**) can be used to select multiple different functions for some GPIO pins.

_____3. After a system reset, all GPIO pins are configured to be in the output state.

_____4. The difference between using the port function select registers (**PxSEL0**, **PxSEL1**) and **PMC** technique to select alternate functions for some specific GPIO pins is that these two methods are used for different GPIO ports.

_____5. There are two ADC modules in the MSP432P401R MCU system, ADC0 and ADC1.

_____6. The ADC module must be clocked by selecting one of possible six clock sources.

_____7. After the ADC conversion is complete, the conversion results can be obtained from the related **ADC14MEMx** registers.

_____8. The MSP432P401R MCU contains two analog comparator modules, **COMP_E0** and **COMP_E1**. Each module has 16 analog inputs with selectable reference sources.

_____9. Each COMP_E module can select to use internal power supply Vcc with internal resistor ladder or a shared reference coming from REF_A as the reference source.

_____10. Each COMP_E module can directly use the shared reference REF_A as the comparison reference voltages without the internal resistor ladder.

II. Multiple Choices

1. The following initialization steps should be performed before using a GPIO port ____.
   a. Enable and clock the selected GPIO port
   b. Set up the direction for each pin on the GPIO port
   c. Enable GPIO pins as digital I/O or an alternate function pin
   d. All of them
2. Before the ADC14 module can perform normal ADC conversions, ___________.
   a. The module must be clocked by selecting a clock source
   b. The reference sources must be determined
   c. The analog input must be selected
   d. All of them
3. In order to enable a GPIO pin to work as ADC14 input pin, one the following configurations should be performed ______________________________.
   a. Enable the clock to the appropriate GPIO module
   b. The related bit in the **PxDEN** register must be set to enable the digital function
   c. The related bit in the **PxDIR** register must be set to enable it as output pin
   d. The related **PySEL1x** and **PySEL0.x** registers should set to 11
4. The bit field ____________________ in the **ADC14 Conversion Memory Control x Register**: **ADC14MCTLx** (**x** = 0~31) is to select an external analog source as the input to the ADC.
   a. **ADC14SHSx**
   b. **ADC14MSCx**
   c. **ADC14INCHx**
   d. **ADC14DIFx**
5. If we want to perform a three-channel sequence sampling for channels 0~2, the __________ bit in the _________________ register must be set to 1.
   a. **ADC14EOS, ADC14MCTL1**

   b. **ADC14EOS, ADC14MCTL2**
   c. **ADC14EOS, ADC14MCTL3**
   d. **ADC14EOS, ADC14MCTL4**
6. If we want to stop a single-channel single-conversion mode, we can reset the bit ________________ in the ________________ register.
   a. **ADC14ENC, ADC14CTL0**
   b. **ADC14BUSY, ADC14CTL0**
   c. **ADC14SC, ADC14CTL0**
   d. **ADC14SHP, ADC14CTL0**
7. During the ADC14 initialization process, the bit ________ in the ________ register should be reset to 0 until the initialization and configuration process is done.
   a. **ADC14ENC, ADC14CTL0**
   b. **ADC14BUSY, ADC14CTL0**
   c. **ADC14SC, ADC14CTL0**
   d. **ADC14SHP, ADC14CTL0**
8. When the ADC14 operates in a low-power mode, the bits ________________________ in the ________________________ register should be set to 10.
   a. **ADC14BATMAP, ADC14CTL0**
   b. **ADC14BATMAP, ADC14CTL1**
   c. **ADC14PWRMD, ADC14CTL1**
   d. **ADC14PWRMD, ADC14CTL0**
9. If we want to exchange the input signals on the comparator's V_+ and V_- terminals, the bit ____________________ in the ____________________ register should be set to 1.
   a. **CEMRVS, CExCTL1**
   b. **CEMRVL, CExCTL1**
   c. **CEON, CExCTL1**
   d. **CEEX, CExCTL1**
10. To set two different reference voltages on the V+ terminal of the **COMP_E1** module, **VREF1** = 0.25Vcc and **VREF0** = 0.75Vcc, these two bit fields should be set to ______________________.
   a. 12, 20
   b. 8, 22
   c. 8, 24
   d. 10, 24

III. Exercises

1. Provide a brief description about how to configure the **GPIO** pin **P4.7** to work as an analog input pin.
2. Provide a brief description about the operational procedure of the ADC14 module used in the MSP432P401R MCU system.
3. Provide a brief explanation about the procedure of generating and handling an ADC14 interrupt.
4. Provide a brief description about the COMP_E modules in the MSP432P401R MCU system.
5. Provide a brief discussion about the initialization process for the **COMP_E1** module.

IV. Practical Laboratory

LABORATORY 7 MSP432 MCU parallel port programming

7.0 GOALS

This laboratory exercise allows students to learn and practice MSP432 MCU GPIO parallel ports programming by developing four labs.

1. Program **Lab7_1** let students to modify one of class projects, **DARADC14** built in Section 7.6.7, to use the DRA model and ADC14 interrupt facilities to perform a repeat-sequence-of-channel mode conversion for three analog channels **A6**, **A8**, and **A10**, and display conversion results in three peripherals on the EduBASE ARM Trainer.
2. Program **Lab7_2** enables students to modify another class project **SDADC14** to use the SD model and ADC14 interrupt functions to handle a single channel **A6** conversion, and display the conversion result on a three-color LED installed in the MSP-EXP432P401R EVB.
3. Program **Lab7_3** allows students to become familiar with COMP_E1 module to build a **COMPE1** project by modifying a class project **DRACOMPE1** with the DRA method. Two reference voltages VREF1 and VREF0 are used in this project to enable the comparator to have a hysteresis effect to change the output on two references.
4. Program **Lab7_4** enables students to modify one class project **SDCOMPE1** to use the **COMP_E1** module to compare an external analog input coming from the **VR2** potentiometer with a shared reference coming from REF_A.

After completion of these programs, students should be able to understand the basic architecture and operational procedure for most popular GPIO-related peripheral devices, including the ADC14 and COMP modules installed in the MSP432P401R MCU system. They also should be able to code some sophisticated programs to access the desired peripheral devices to perform specific control operations via GPIO parallel ports.

7.1 LAB7_1

7.1.1 Goal

In this project, students need to modify one of the class projects, **DARADC14** built in Section 7.6.7, to use the DRA model and ADC14 interrupt facilities to perform a repeat-sequence-of-channel mode conversion for three analog channels **A6**, **A8**, and **A10**, and display conversion results in three peripherals on the EduBASE ARM Trainer.

7.1.2 Data assignment and hardware configuration

No hardware configuration is needed in this project. Refer to Figure 7.15 to get a detailed hardware interface configuration about ADC14 and connection to the EduBASE Trainer.

7.1.3 Development of the project and modification of the source code

Perform the following operations to create this lab project.

1. Create a new folder **Lab7_1(DARADC14INT)** under the folder **C:\MSP432 Lab Projects\Chapter 7** in the Windows Explorer.
2. Create a new µVersion5 project **DRAADC14INT** and save this project to the folder **Lab7_1(DARADC14INT)** that is created in step 1 above.

3. On the next wizard, you need to select the device (MCU) for this project. Expand three icons, **Texas Instruments**, **MSP432 Family**, and **MSP432P**, and select the target device **MSP432P401R** from the list. Click on the **OK** to close this wizard.
4. Next the Software Components wizard is opened, and you need to set up the software development environment for your project with this wizard. Expand two icons, **CMSIS** and **Device**, and check the **CORE** and **Startup** checkboxes in the **Sel.** column, and click on the **OK** button since we need these two components to build our project.
5. In the **Project** pane, expand the **Target** folder and right click on the **Source Group 1** folder and select the **Add New Item to Group "Source Group 1."**
6. Select the **C File (.c)** and enter **DRAADC14INT** into the **Name:** box, and click on the **Add** button to add this source file into the project.
7. Open the class project **DRAADC14** and its C file **DRAADC14.c**, copy all codes from that file and paste them to this new C file **DRAADC14INT.c**.

Perform the following modifications to this C file to make it to use ADC14 interrupts:

1. Refer to Figure 7.16, add one code line under line 26 to enable the interrupt source, analog input channel **A6** (**ADC14IE10**), by setting the **ADC14IER0** Register.
2. Configure **NVIC_IPR6** Register to set **A6** interrupt source as a priority level of 3.
3. Enable this interrupt source by configuring the **NVIC_ISER0** Register.
4. Call the intrinsic function **__enable_irq()** to globally enable the interrupt mechanism.
5. Delete the original **while(1)** loop and create a new infinitive **while()** loop to repeatedly enable the ADC14 module and start the trigger of the ADC14 module to perform the conversions. Then call the **Delay()** function to delay the program by 50,000.

Declare the ADC14 Interrupt Handler **void ADC14_IRQHandler(void)**. Inside the handler, perform the following code development.

1. Move three local variables, **pMeter**, **pSensor,** and **pTemp** from the **main()** program to this interrupt handler.
2. Use an **if()** selection structure to check whether the ADC14 channel 10 has completed its conversion. This inspection can be performed by checking the **ADC14IFG10** flag bit in the **ADC14IFGR0** Register. An **&** operation may be needed for this checking.
3. If this flag is set, the conversion has been done. Pick up these results by reading three conversion result registers: **ADC14MEM6**, **ADC14MEM8**, and **ADC14MEM10**, and assign these conversion results to three variables: **pMeter**, **pSensor**, and **pTemp**. Keep the original shifting bits for each assignment.

7.1.4 Set Up the environment to build and run the project

To build and run the project, one needs to perform the following operations to set up the environments:

- Select the **CMSIS-DAP Debugger** in the **Debug** tab under the **Project|Options for Target "Target 1"** menu item. Make sure the debugger and flash download settings are correct.
- Build the project by going to **Project|Build target** menu item.

- Then go to **Flash|Download** menu item to download the image file of this project into the flash memory in the MSP-EXP432P401R EVB.
- Now go to **Debug|Start/Stop Debug Session** and **Debug|Run** menu item to run the project.

7.1.5 Demonstrate your program by checking the running result

After the project runs, tune the VR2 potentiometer and cover the photo sensor. One can see that the three-color LED will be flashing with different colors, and the **PB0** and **PB1** LEDs on the Trainer will also be changed. A noise is coming from the speaker.

Go to **Debug|Start/Stop Debug Session** to stop your program.

Based on these results, try to answer the following questions:

- When you tune the VR2 potentiometer, the three-color LED will be flashing and changing its colors. Is this changing continuous in the color order (red-green-blue)? If not, why?
- Why can't you use PB2 and PB3 LEDs?
- What did you learn from this project?

7.2 LAB7_2

7.2.1 Goal

This project enables students to modify another class project **SDADC14** to use the SD model and ADC14 interrupt functions to handle a single channel **A6** conversion, and display the conversion result on a three-color LED installed in the MSP-EXP432P401R EVB.

7.2.2 Data assignment and hardware configuration

No hardware configuration is needed in this project. Refer to Figure 7.15 to get a detailed hardware interface configuration about ADC14 and connection to the EduBASE Trainer.

7.2.3 Development of the project and modification of the source codes

Perform the following operations to create this lab project.

1. Create a new folder **Lab7_2(SDADC14INT)** under the folder **C:\MSP432 Lab Projects\Chapter 7** in the Windows Explorer.
2. Create a new µVersion5 project **SDADC14INT** and save this project to the folder **Lab7_2(SDADC14INT)** that is created in step 1 above.
3. On the next wizard, you need to select the device (MCU) for this project. Expand three icons, **Texas Instruments**, **MSP432 Family** and **MSP432P**, and select the target device **MSP432P401R** from the list. Click on the **OK** to close this wizard.
4. Next the Software Components wizard is opened, and you need to set up the software development environment for your project with this wizard. Expand two icons, **CMSIS** and **Device**, and check the **CORE** and **Startup** checkboxes in the **Sel.** column, and click on the **OK** button since we need these two components to build our project.
5. Add a new C file **SDADC14INT.c** into the project.
6. Open the class project **SDADC14** and its C file **SDADC14.c**, copy all codes from that file and paste them to this new C file **SDADC14INT.c**.

Perform the following modifications to this C file to make it to use ADC14 interrupt API functions to use ADC14 interrupt mechanism:

7.2.4 Modification of the C source file

1. Refer to Figure 7.17, add two API functions, **`ADC14_registerInterrupt()`** and **`ADC14_enableInterrupt()`**, under the coding line 27 to register the interrupt handler **`ADC14_IntHandler()`** and enable the ADC channel 6 (**`A6`**) to work as an interrupt source when its conversion is complete.
2. Remove and delete all code inside the original **`while()`** loop, but keep this empty **`while()`** loop and we need to use this loop to wait for **`ADC_INT6`** interrupt to occur later.

Declare the ADC14 Interrupt Handler **`void ADC14_IntHandler(void)`**. Inside the handler, perform the following code development.

1. Move the local variable **`uint16_t cResult`** from the **`main()`** program and place it in the interrupt handler.
2. Add another local variable **`uint64_t cStatus`** since we need to use it to get and hold the interrupt status later.
3. Call the API function **`ADC14_getInterruptStatus()`** and assign the returned value to the local variable **`cStatus`**.
4. Call the API function **`ADC14_clearInterruptFlag()`** with the **`cStatus`** as the argument to clear the **`ADC_INT6`** interrupt.
5. Use an **`if()`** selection structure to check whether the returned status is **`ADC_INT6`**. An **`&`** operation may be used for this checking for **`ADC_INT6`** and **`cStatus`**.
6. If this interrupt is generated by the ADC_INT6, the API function **`ADC14_getResult()`** should be called with the **`ADC_MEM6`** as the argument to pick up the conversion result. This conversion result should be assigned to the variable **`cResult`**.
7. Then this result is shift right by eight bits and assigned to the GPIO **`P2`** to display it on the three-color LED.
8. Finally, the API function **`ADC14_toggleConversionTrigger()`** is called again to trigger and start the next conversion.

7.2.5 Set up the environment to build and run the project

To build and run the project, one needs to perform the following operations to set up the environments:

- Select the **`CMSIS-DAP Debugger`** in the **`Debug`** tab under the **`Project|Options for Target "Target 1"`** menu item. Make sure the debugger and flash download settings are correct.
- Build the project by going to **`Project|Build target`** menu item.
- Then go to **`Flash|Download`** menu item to download the image file of this project into the flash memory in the MSP-EXP432P401R EVB.
- Now go to **`Debug|Start/Stop Debug Session`** and **`Debug|Run`** menu item to run the project.

7.2.6 Build and demonstrate your program

As the project runs, tune the VR2 potentiometer. One can see that the three-color LED will flash with different colors, and four LEDs, **PB3~PB0** installed in the EduBASE ARM Trainer will also be flashing.

Based on these results, try to answer the following questions:

- Why is the color on the three-color LED changed and why are four LEDs, **PB3~PB0** changed? Is the change continuous from low to high?
- What did you learn from this project?

7.3 LAB7_3

7.3.1 Goal

This project allows students to become familiar with the COMP_E1 module to build a **DRACOMPE1** project by modifying a class project **DRACOMPE1** with the DRA method. Two reference voltages VREF1 and VREF0 are used in this project to enable the comparator to have a hysteresis effect to change the output on two references, VREF1 and VREF0, with different voltage values.

7.3.2 Data assignment and hardware configuration

No hardware configuration is needed in this project. Refer to Figure 7.24 to get a detailed hardware interface configuration about the COMP_E1 module and connection to the EduBASE Trainer. The only difference is that two references are applied in the comparator inputs in this project to make VERF1 ≠ VREF0 to generate a hysteresis effect.

7.3.3 Development of the project

Using the steps below to develop this project, create a new project with the following steps:

1. Create a new folder **Lab7_3(DRACOMPE1)** under the folder **C:\MSP432 Lab Projects\Chapter 7** with the Windows Explorer.
2. Open the Keil® ARM-MDK µVersion5, create a new project named **DRACOMPE1** and save this project in the folder **Lab7_3(DRACOMPE1)** created in step 1.

7.3.4 Modification of the C source file

You need to modify the C source file we built in one of our class projects, **DRACOMPE1**, to make it as the source file for this project.

The only modification is to modify the code in line 16 in Figure 7.25. This line is used to configure the **CE1CTL2** Register to enable the comparator to use two different references, VERF1 and VREF0. The original VERF0 is still $0.25Vcc = 0.25 \times 3.3V = 0.825$ V, which is the lower limitation reference voltage to be applied on the V+ input terminal of the comparator. However, the upper limitation VREF1 should be set as $0.75Vcc = 0.75 \times 3.3V = 2.475$ V.

You need to figure out which value should be assigned to the **CE1CTL2** Register. A hint is that to make VREF1 = 0.75Vcc, a numerator value **24** (**24**/32 = 0.75), exactly a numerator value of **24** − 1 = **23** (**0x17**) should be assigned to the bit field **CEREF1** (bits 12~8) in the **CE1CTL2** Register.

7.3.5 Set up the environment to build and run the project

The only environment to be set is to make sure that the debugger to be used for this project is **CMSIS-DAP**. Also make sure that all settings about the Debugger and the Flash Download are correct. Then you can build the project.

7.3.6 Demonstrate your program

Perform the following operations to run your program and check the running results:

- Go to **Flash|Download** menu item to download your program into the flash memory
- Go to **Debug|Start/Stop Debug Session** to begin debugging your program. Click on the **OK** button on the 32 KB memory size limitation message box to continue
- Then go to **Debug|Run** menu item to run your program

As the project runs, when you rotate the **VR2** potentiometer to change the analog input to the V-terminal of the comparator via **A6**, you can find that when the analog input is lower than 0.825 V, the comparator outputs High and four LEDs are turned on. Then when you inversely rotate the **VR2**, four LEDs will not turn off until the potentiometer output (analog input) is greater than 2.475 V. Therefore, two voltage limitations are set for this project; 0.825 V and 2.475 V. The comparator outputs High when the input analog is less than 0.825 V, but it outputs a Low when the analog input is greater than 2.475 V.

Based on these results, try to answer the following questions:

- Why are the comparator outputs High and Low with two different analog voltage values?
- Can you modify this project to enable the comparator to output High and Low with some other different analog input values?
- What did you learn from this project?

7.4 LAB7_4

7.4.1 Goal

This project allows students to modify one class project **SDCOMPE1** to use the **COMP_E1** module to compare an external analog input coming from the **VR2** potentiometer with a shared reference coming from REF_A. The selected shared reference voltage VREF is 2.5 V and it is applied on the internal resistor ladder. The output of the **COMP_E1** module is High when the input analog that is applied on the V-terminal of the comparator is less than $0.5VREF = 0.5 \times 2.5 = 1.25$ V, but the output of the **COMP_E1** module is Low when the analog input is greater than $0.5VREF = 1.25$ V.

7.4.2 Data Assignment and hardware configuration

No hardware configuration is needed in this project. Refer to Figure 7.24 to get a detailed hardware interface configuration about COMP_E1 module and connection to the EduBASE Trainer. The only difference is that the reference applied on the V+ terminal of the comparator is coming from the shared reference REF_A applied on the internal resistor ladder.

7.4.3 Development of the project

Use the steps below to develop this project. Create a new project with the following steps:

1. Create a new folder **Lab7_4(SDCOMPE1)** under the folder **C:\MSP432 Lab Projects\Chapter 7** with the Windows Explorer.
2. Open the Keil® ARM-MDK µVersion5, create a new project named **SDCOMPE1** and save this project in the folder **Lab7_4(SDCOMPE1)** created in step 1.

7.4.4 Modification of the C source file

You need to modify the C source file we built in one of our class projects, **SDCOMPE1**, to make it the source file for this project.

The only modification is code in line 39 (refer to Figure 7.27), **COMP_E_setReferenceVoltage()**. Modify this line to use a shared reference REF_A (= 2.5 V) and appropriate numerators for both lower limitation and upper limitation of the references, **VERF0** and **VREF1**, and enter both values into this function to enable the COMP_E1 module to output High as the analog input is lower than $0.5 \times 2.5V = 1.25$ V, and output Low when the analog input is greater than $0.5 \times 2.5V = 1.25$ V.

7.4.5 Set up the environment to build and run the project

This setup contains the following three operations:

- Set the header file path by going to **Project|Options for Target "Target 1"** menu item. Then select the **C/C++** tab and go to the **Include Paths** box. Click the three-dot button and select **C:\ti\msp\MSPWare_2_21_00_39\driverlib\driverlib** as the system header file path.
- Select the correct debugger by clicking on the **Debug** tab and select the **CMSIS-DAP Debugger** from the **Use** box. Also make sure that all settings for the Debugger and the Flash Download are correct by clicking on the Settings button.
- Add the MSPPDL into this project by right clicking on the **Source Group 1** folder in the **Project** panel on the left, and select the **Add Existing Files to Group "Source Group 1"** menu item. On the opened box, select the driver library file path **C:\ti\msp\MSPWare_2_21_00_39\driverlib\driverlib\MSP432P4xx\keil** and then select the driver library file **msp432p4xx_driverlib.lib**. Click the **Add** button to add it into the project.

7.4.6 Demonstrate your program

Perform the following operations to run your program and check the running results:

- Go to **Flash|Download** menu item to download your program into the flash memory
- Go to **Debug|Start/Stop Debug Session** to begin debugging your program. Click on the **OK** button on the 32 KB memory size limitation message box to continue
- Then go to **Debug|Run** menu item to run your program

As the project runs, use a small screwdriver to rotate the **VR2** potentiometer. Four LEDs, **PB3~PB0**, will be on when the analog input voltage is lower than 1.25 V, but will be off when the analog input is greater than 1.25 V. By using a multimeter, you can check and monitor these voltage values.

Based on these results, try to answer the following questions:

- Compare **`Lab7_3`** and **`Lab7_4`**, which one is better in coding process and control performance? Why?
- Explain the advantages and disadvantages for both projects.
- What did you learn from this project?

chapter eight

MSP432™ serial I/O ports programming

This chapter provides general information about an MSP432P401R microcontroller GPIO port programming. The discussion is mainly concentrated on the GPIO ports related to serial peripherals used in the MSP432P401R MCU system. This chapter includes the GPIO ports and serial peripherals specially designed for the MSP432P401R MCU, general and special or alternative control functions for different GPIO ports and pins, and GPIO ports and peripheral programming applications in the MSP432P401R MCU system.

8.1 Overview

As we discussed in Section 2.3.4 in Chapter 2, in the MSP432P401R MCU system, the GPIO module provides interfaces for multiple peripherals or I/O devices. Generally, GPIO module provides 11 physical GPIO ports, Port 1~Port 10 and Port J, and each port is made of an individual 8-bit or 6-bit I/O port. With the help of the General Digital Controller (GDC) or Port Mapping Controller (PMAPC) techniques, each GPIO port can be configured to perform multiple functions or reconfigured to perform a special function.

The advantage of using GDC and PMAPC techniques in the GPIO module is that multiple functions can be configured and fulfilled by using a limited number of GPIO ports and pins. The shortcoming is that this will make the GPIO port configurations and programming more complicated and difficult.

Most of the peripherals, including system, on-chip, and interfaces to the external peripherals, are connected and controlled to the GPIO ports with related pins. In this chapter, we will concentrate on the most popular serial peripherals used in the MSP432P401R MCU system:

- Synchronous peripheral interface (SPI)
- Inter-integrated circuit (I2C) interface
- Universal asynchronous receivers/transmitters (UARTs)

Some example projects related to those serial peripherals to be developed include:

- On-board LCD interface programming project
- On-board seven-segment LED interface programming project
- DAC Programming Project
- I2C interfacing programming project
- UART programming project

The first three projects are related to the SPI interfacing projects.

The timer in the MSP432P401R MCU system is a big topic and we will discuss this peripheral in Chapters 9 and 10 with general-purpose timers and Watchdog timers. In Chapter 9, we will concentrate on our discussion on four 16-bit timers with PWM module and related implementations. Two 32-bit timers and Watchdog timers will be discussed in Chapter 10 with related implementation projects.

In Chapters 11 and 12, the FPU and MPU will be discussed in detail with related implementation projects.

8.2 Introduction to SPI mode and SPI operations

Three terminologies are widely applied in most serial communication systems: (1) *serial communication interface* (SCI), (2) *synchronous serial interface* (SSI), and (3) *serial peripheral interface* (SPI). Generally, the SCI belongs to the asynchronous SCI but SSI and SPI belong to the synchronous SCI. However, even though both SSI and SPI are SSIs, there are some differences that exist between them. These differences include:

- The SSI is a differential, simplex, nonmultiplexed interface, and it relies on a time-out to frame the data.
- The SPI is a single-ended, duplex, multiplex interface and it uses a select line to frame the data.

However, SPI peripherals on most microcontrollers can implement SSI with external differential driver ICs and program-controlled timing.

8.2.1 Architecture and organization of the SPI mode

The SPI exactly provides a bus system that is a synchronous SCI specification used four-signal-line for short-distance communication. The SPI devices communicate in full-duplex mode using a master-slave architecture with a single master. The master device originates the frame for reading and writing. Multiple slave devices are supported through selection with individual slave select (**SS**) signal lines.

The full-duplex communication mode uses two data lines (input and output data lines) with one line carrying serial data from peripheral to microcontroller (input) and the other line carrying serial data from microcontroller to peripheral (output).

These four signal lines include:

- **SCLK**: Serial clock (output from master)
- **MOSI**: Master output and slave input (output from master)
- **MISO**: Master input and slave output (output from slave)
- **SS**: Slave select (output from master and the active level is low)

Figure 8.1 shows an example of an SPI device using a single master with a single slave. A frame is exactly a data transmit/receiving protocol or format to enable both master and slave to understand and follow it to perform a synchronous data communication correctly.

Figure 8.1 SPI illustration of using a single master and a slave.

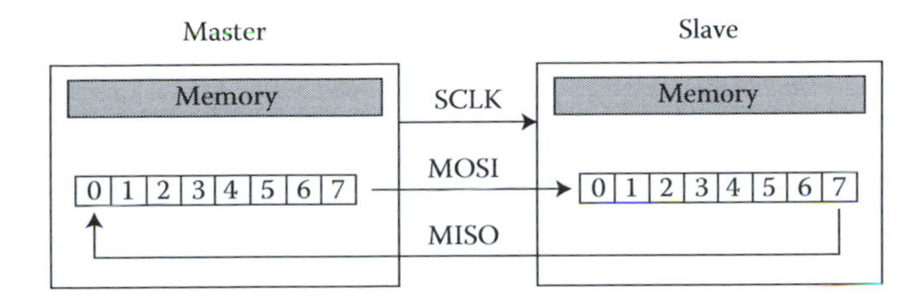

Figure 8.2 SPI data transmitting and receiving operation mode.

A normal operational sequence of an SPI bus is

1. To begin communication, the master configures the clock **SCLK** by using a frequency supported by the slave device.
2. The master then selects the slave device with a logic low on the select line (**SS**).
3. During each following SPI clock cycle, a full-duplex data transmission occurs.
4. The master needs to determine the frame by configuring clock polarity (CKPL) and clock phase (CKPH).
5. The master sends a bit on the **MOSI** line and the slave reads it, while the slave sends a bit on the **MISO** line and the master reads it. This sequence is maintained even when only one-directional data transfer is intended.

As shown in Figure 8.2, the SPI transmissions normally involve two shift registers by using a given word size, such as 8 bits, one in the master and one in the slave side. They are connected in a virtual ring topology. Data are usually shifted out with the most significant bit first, while shifting a new less significant bit into the same register. At the same time, data from the counterpart are shifted into the least significant bit register. After the register bits have been shifted out and in, the master and slave have exchanged register values. If more data need to be exchanged, the shift registers are reloaded and the process repeats. Transmission may continue for any number of clock cycles. When complete, the master stops toggling the clock signal and typically deselects the slave.

8.2.2 CKPL and CKPH (transmit–receive frame)

In the SPI mode, the transmit/receiving frame is determined by the **CKPL** and **CKPH**. Before a normal SPI bus can perform its work, the master must set up the appropriate CKPL and CKPH with respect to the data to select a desired frame.

The **CKPL** determines the active level of the clock **SCLK** and the **CKPH** determines the initial phase of the first data to be transmitted–received. They are defined as below:

- If the **CKPL** = 0, means that the base value or the idle level of the clock **SCLK** is low (0), but the active level of the clock **SCLK** is high (1).
  - For **CKPH** = 0, the data are captured on the rising edge of the clock, but the data are output on a falling edge of each clock.
  - For **CKPH** = 1, the data are captured on the falling edge of each clock, but the data are output on the rising edge of each clock.
- If the **CKPL** = 1, means that the base value or the idle level of the clock **SCLK** is high (1), but the active level of the clock **SCLK** is low (0).

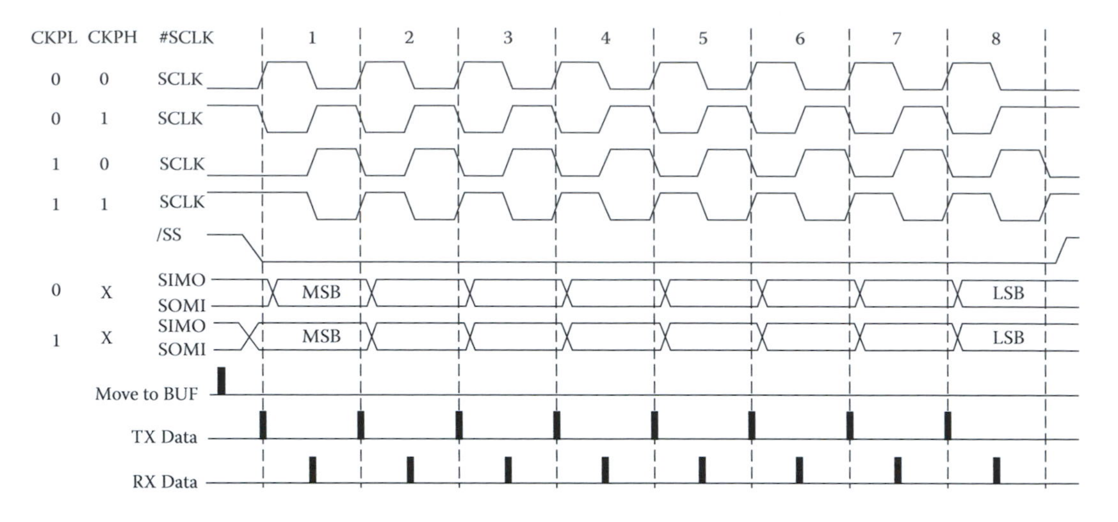

Figure 8.3 Relationship between clock setups and data frames. (Courtesy Texas Instruments.)

- For **CKPH** = 0, the data are captured on the falling edge of each clock, but the data are output on the rising edge of each clock cycle.
- For **CKPH** = 1, the data are captured on the rising edge of each clock, but the data are output on the falling edge of each clock cycle.

An example frame is shown in Figure 8.3 with the selected **CKPL** and **CKPH** values.

When the **CKPL:CKPH** = 10 or 11, the first data to be transmitted–received are delayed by a half cycle, which is equivalent to having a 180° phase difference with the normal data transmission when the **CKPL:CKPH** = 00 or 01.

The first data bit to be transmitted–received can be controlled by configuring a related control registers, either the MSB or the LSB can be transmitted/received first.

8.3 *Enhanced universal serial communication interfaces eUSCI_A and eUSCI_B*

In the MSP432P401R MCU system, all serial or synchronous peripheral interfaces are supported by one hardware module. This hardware module supports two SCIs, called **eUSCI_A** and **eUSCI_B**. Both the **eUSCI_A** and the **eUSCI_B** support multiple serial communication modes, including the SPI, I2C, and UART.

In the MSP432P401R MCU system, each eUSCI interface supports four serial communication modules, **eUSCI_A0~eUSCI_A3** and **eUSCI_B0~eUSCI_B3**. Each module can support one of the three eUSCI modes, SPI, I2C, or UART. In this chapter, we will concentrate on the eUSCI for the SPI mode.

All GPIO ports that are related to **eUSCI_A** and **eUSCI_B** interfaces in the MSP432P401R MCU system are GPIO **P1~P3**, **P6**, and **P8~P10**, as shown in Figure 8.4. It can be found from Figure 8.4 that some GPIO pins are multifunction pins and they can be used as SPI, I2C, or UART function pins.

It can also be found from Figure 8.4 that GPIO ports 2 and 3 belong to PMC Ports and all **eUSCI** interface-related pins are prefixed with the keyword **PM**. All other ports can be

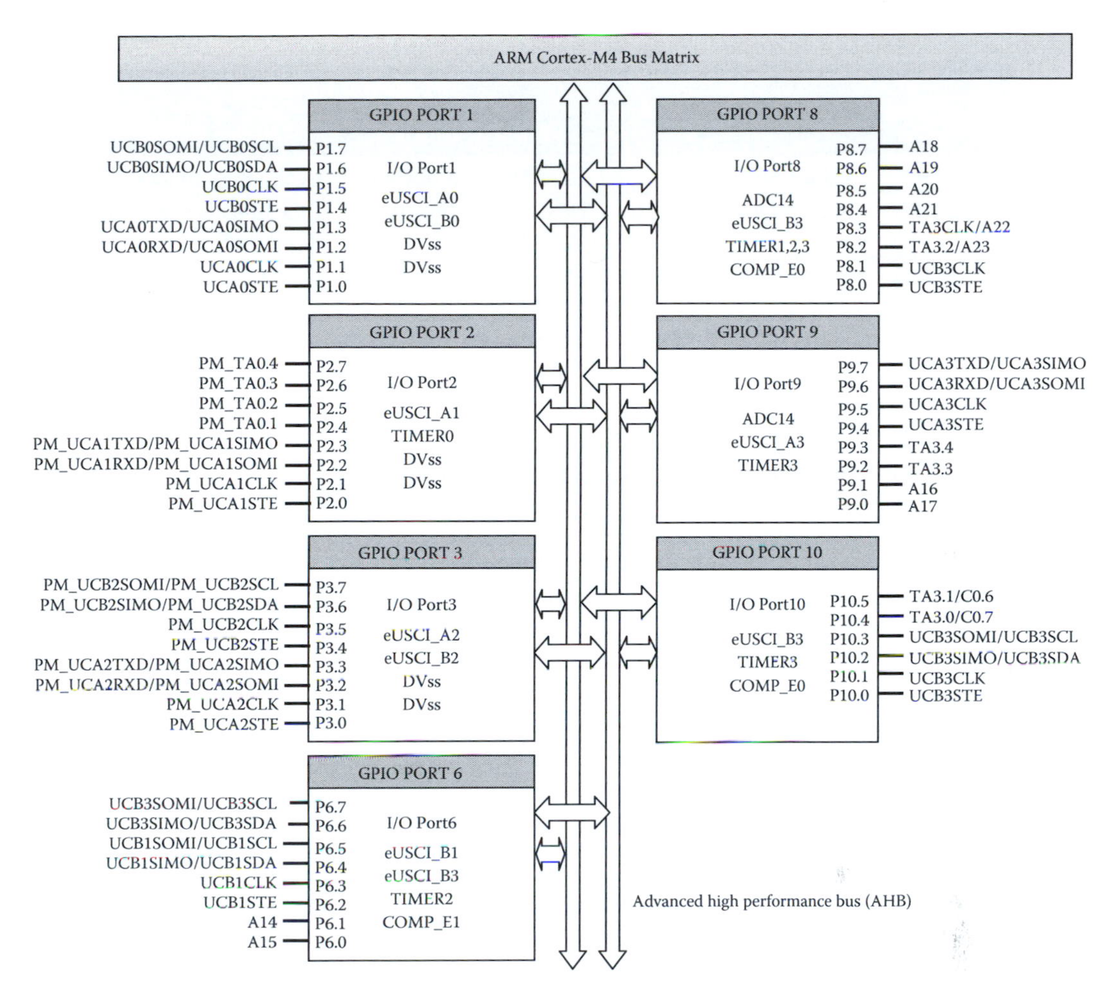

Figure 8.4 GPIO port and pin configuration.

controlled by configuring **PySEL1.x** and **PySEL0.x** registers to set up the related **eUSCI** function pins.

8.4 eUSCI interface used for SPI mode

In the MSP432P401R MCU system, when an eUSCI interface is selected to work as an SPI mode, the eUSCI interface connects the device to an external system via three or four pins: **UCxCLK**, **UCxSOMI**, **UCxSIMO**, and **UCxSTE** (**x** = 0~3). These signals are equivalent to **SCLK**, **SOMI**, **SIMO**, and **SS** signals in the SPI mode. Since both interfaces, **eUSCI_A** and **eUSCI_B**, are identical and can be selected, configured to work as an SPI mode, in the following discussions, we will use the **eUSCI_A** interface as an example to illustrate how to use it to work as an SPI mode.

Table 8.1 lists SPI control signals used for 3-pin or 4-pin mode and their functions.

Unlike traditional SPI bus systems, the eUSCI interface used in the MSP432P401R MCU system allows users to select and control the active level of the slave transmit enable signal **UCxSTE**, which is equivalent to the **SS** signal, by setting the **UCMODEx** bit field in the **eUSCI_Ax Control Word 0** (**UCAxCTLW0**) Register.

Table 8.1 SPI control signals and their functions

SPI signal	SPI signal name	SPI signal function	
		Master mode	Slave mode
UCxCLK	eUSCI SPI Clock	UCxCLK is an output	UCxCLK is an input
UCxSIMO	Master Out & Slave In	UCxSIMO is the data output line	UCxSIMO is the data input line
UCxSOMI	Master In & Slave Out	UCxSOMI is the data input line	UCxSOMI is the data output line
UCxSTE	Slave Transmit Enable	Not used in 3-pin mode Used in 4-pin mode to allow multiple masters on a single bus	

If the SPI mode is selected, it has the following features:

- 7-bit or 8-bit data length
- LSB-first or MSB-first data transmit and receive
- 3-pin and 4-pin SPI operation
- Master or slave modes
- Independent transmit and receive shift (RXS) registers
- Separate transmit and receive buffer registers
- Continuous transmit and receive operation
- Selectable CKPL and phase control
- Programmable clock frequency in master mode
- Independent interrupt capability for receive and transmit

All of these features provide the flexibility in using the eUSCI interface as the SPI mode, and these features are controlled and configured by using a set of SPI-related registers. Let's have a closer look at the architecture and organization of the eUSCI interface used for the SPI mode.

8.4.1 Architecture and functional block diagram of the eUSCI used for SPI mode

Figure 8.5 shows the architecture and functional block diagram for an eUSCI used in the SPI mode. For all eUSCI interfaces, each of them contains an identical and independent set of these registers to control and configure each SPI mode's data transmitting and receiving functions.

Based on Figure 8.5, each SPI mode can be controlled by four sets of registers:

1. General controls and bit clock generation registers
2. SPI data transmit control registers
3. SPI data receiving control registers
4. SPI interrupt-related control registers

All SPI data transmit and receiving processes are under the control of these registers.

8.4.2 General controls and bit clock generation registers

Two registers are used in this section to select the clock source (CKSE) for the SPI mode and generate the bit clock or **SCLK** to assist the data transmit-receive operations.

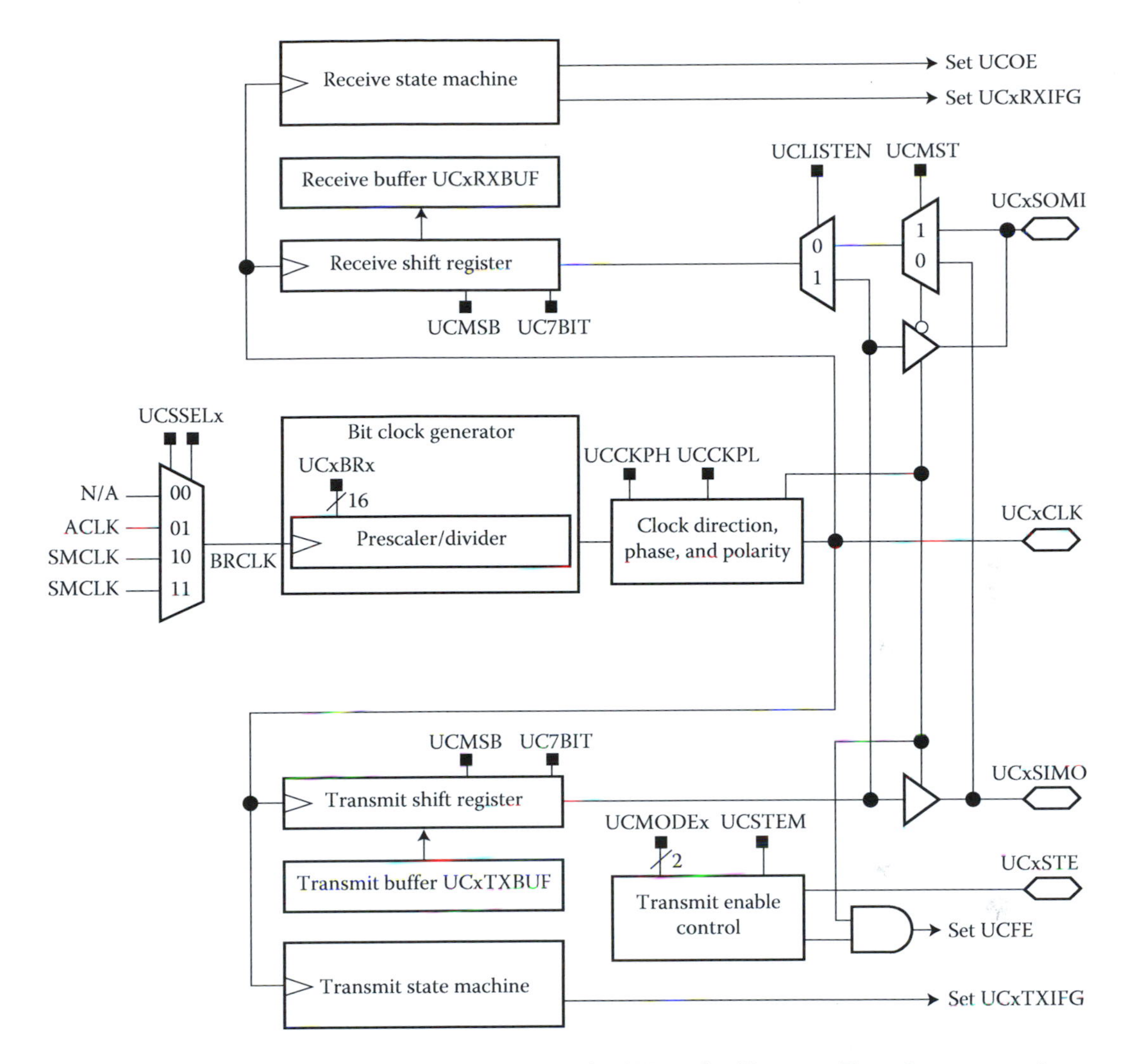

Figure 8.5 Functional block diagram of the eUSCI for SPI mode. (Courtesy Texas Instruments.)

These two registers are

- eUSCI_Ax Control Word 0 Register (**UCAxCTLW0**)
- eUSCI_Ax Bit Rate Control Word Register (**UCAxBRW**)

The **UCAxCTLW0** Register is a 16-bit or half-word register and it is used to provide all controls to the SPI mode, including the CKSE selection, synchronous or asynchronous mode selection, master or slave mode selection, MSB- or LSB-first selection, 7 or 8-bit data length selection, 3-pin or 4-pin mode selection, CKPL and phase selection, and active-level selection for the **UCxSTE** control signal. The bit configuration for the **UCAxCTLW0** Register is shown in Figure 8.6. The bit values of this register are shown in Table 8.2.

By checking Figures 8.5, 8.6, and Table 8.2, it can be found that the related signals in this **UCAxCTLW0** Register control the following eUSCI for SPI components:

15	14	13	12	11	10	9	8
UCCKPH	UCCKPL	UCMSB	UC7BIT	UCMST	UCMODEx		UCSYNC
7	**6**	**5**	**4**	**3**	**2**	**1**	**0**
UCSSELx		RESERVED				UCSTEM	UCSWRST

Figure 8.6 Bit configurations for the UCAxCTLW0 register.

Table 8.2 Bit value and function for UCAxCTLW0 register

Bit	Field	Type	Reset	Function
15	**UCCKPH**	RW	0	USCI clock phase select **0:** Data are changed on the first clock edge and captured on the following edge **1:** Data are captured on the first clock edge and changed on the following edge
14	**UCCKPL**	RW	0	Clock polarity select **0:** The active level is high but the inactive state is low **1:** The active level is low but the inactive state is high
13	**UCMSB**	RW	0	MSB-first select. Controls the direction of receive and transmit shift register **0:** LSB is first. **1:** MSB is first
12	**UC7BIT**	RW	0	Transfer data character length. Selects 7-bit or 8-bit length **0:** 8-bit data length; **1:** 7-bit data length
11	**UCMST**	RW	0	Master mode selection **0:** Slave mode; **1:** Master mode
10:9	**UCMODEx**	RW	0	The eUSCI 3-pin or 4-pin mode selection **00:** 3-pin SPI mode **01:** 4-pin SPI with UCxSTE active high: Slave enabled when UCxSTE = 1 **10:** 4-pin SPI with UCxSTE active low: Slave enabled when UCxSTE = 0 **11:** I2C mode
8	**UCSYNC**	RW	1	Synchronous mode enable **0:** Asynchronous mode is selected; **1:** Synchronous mode is selected
7:6	**UCSSELx**	RW	3	The eUSCI clock source select. These bits select the BRCLK source clock in master mode. UCxCLK is always used in slave mode **00:** Reserved; **01:** ACLK; **10:** SMCLK; **11:** SMCLK
5:2	**Reserved**	RO	0	Reserved
1	**UCSTEM**	RW	0	STE mode select in master mode. This byte is ignored in slave or 3-wire mode **0:** STE pin is used to prevent conflicts with other masters **1:** STE pin is used to generate the enable signal for a 4-wire slave
0	**UCSWRST**	RW	1	Software reset enable **0:** Reset is disabled. The eUSCI is working in normal operation **1:** Reset is enabled. The eUSCI is in reset state (only in this state, initialization and configuration of the eUSCI can be performed)

- The CKSE selection and UCLK generation:
  - **UCSSELx**—select the clock source
  - **UCCKPL** and **UCCKPH**—determine the CKPL and CKPH
- The data transmit and receive controls:
  - **UCMSB** and **UC7BIT**—determine the first data bit (MSB or LSB) and data length for both transmit shift (TXS) and RXS registers
  - **UCMODEx** and **UCSTEM**—control the transmit enable controller to select 3-pin or 4-pin SPI mode and determine the active level for the **UCSTE** signal for 4-pin mode

The **UCAxBRW** register is also a 16-bit register and its 16-bit value **UCBRx** is used to store a prescaler or a divider to divide the selected CKSE **BRCLK** (Figure 8.5) to calculate the final desired SPI clock UCLK. The frequency of the final UCLK is determined by the following equation

$$f_{UCLK} = f_{BitClock} = f_{BRCLK} / \mathbf{UCBRx}$$

This clock rate is exactly a bit transmit–receive rate and regularly it should be smaller or equal to the selected CKSE **BRCLK**.

8.4.3 SPI data transmit control registers

There are four components that are involved in the SPI data transmit part. These four components are controlled by the following related registers:

1.	Transmit state machine	The eUSCI_Ax Status Word (**UCAxSTATW**) register
2.	Transmit buffer	The eUSCI_Ax Transmit Buffer (**UCAxTXBUF**) register
3.	Transmit shift (TSX) register	The **UCMSB** and **UC7BIT** in **UCAxCTLW0** Register
4.	Transmit enable control	The **UCMODEx** and **UCSTEM** in **UCAxCTLW0** Register

The **UCAxTXBUF** register is a 16-bit register, but only the lower 8 bits (bits 7~0) are used to hold the data waiting to be moved into the TXS register and transmitted. Writing to this register clears the **UCTXIFG** bit in the **UCAxIFG** register. The MSB of **UCAxTXBUF** is not used for 7-bit data transfer and is reset to 0.

The **UCAxSTATW** register is also a 16-bit register, but only the lower 8 bits (bits 7~0) are used to monitor and display the working status of the SPI data transmit. Figure 8.7 shows the bit configuration of this register, and Table 8.3 lists the bit value and bit function of this register.

A point to be noted when using the **UCAxSTATW** register to monitor the working status of the SPI, the **UCFE** flag is not used for 3-pin SPI data transmission or any

15	14	13	12	11	10	9	8
RESERVED							
7	**6**	**5**	**4**	**3**	**2**	**1**	**0**
UCLISTEN	UCFE	UCOE	RESERVED				UCBUSY

Figure 8.7 Bit value and configuration for the UCAxSTATW register.

Table 8.3 Bit value and function for UCAxSTATW register

Bit	Field	Type	Reset	Function
15:8	**Reserved**	RO	0	Reserved
7	**UCLISTEN**	RW	0	Listen enable. When this bit is 1, the loopback mode is selected **0:** Disabled **1:** Enabled. The transmitter output is internally fed back to the receiver
6	**UCFE**	RW	0	Framing error flag. This bit indicates a bus conflict in 4-wire master mode. The **UCFE** is not used in 3-wire master or any slave mode **0:** No error; **1:** Bus conflict error occurred
5	**UCOE**	RW	0	Overrun error flag. This bit is set when a character is transferred into **UCxRXBUF** before the previous character was read. UCOE is cleared automatically when **UCxRXBUF** is read, and must not be cleared by software. Otherwise, it does not function correctly **0:** No error; **1:** Overrun error occurred
4:1	**Reserved**	RO	0	Reserved
0	**UCBUSY**	RO	0	The eUSCI is busy. This bit indicate if a transmit or receive is in progress **0:** The eUSCI is inactive **1:** The eUSCI is transmitting or receiving data

slave mode. Also the **UCBUSY** bit is a read-only bit and cannot be written any value to this bit.

8.4.4 SPI data receiving control registers

There are three components involved in the SPI data receiving part. These three components are controlled by the following related registers:

1.	Receive state machine	The **UCAxSTATW** register
2.	Receive buffer	The eUSCI_Ax Receive Buffer (**UCAxRXBUF**) register
3.	RXS register	The **UCMSB** and **UC7BIT** in **UCAxCTLW0** Register

The **UCAxRXBUF** register is a 16-bit register, but only the lower 8 bits (bits 7~0) are used to contain the last received character from the RXS register. Reading the **UCAxRXBUF** register resets the receive error bits in the **UCAxSTATW** register and **UCRXIFG** flag bit in the **UCAxIFG** register. If a 7-bit data mode is used, the **UCAxRXBUF** register is LSB justified and the MSB is always reset to 0.

The **UCAxSTATW** register has been discussed in the last section. This register is used to monitor and display the working status of the SPI data transmit and receive. Two major possible error sources are the framing error (**UCFE**) and overrun error (**UCOE**), as shown in Figure 8.7 and Table 8.3.

A point to be noted when using the bit **UCMST** in the **UCAxCTLW0** Register to control and select the master or the slave mode is that when this bit is 1, the eUSCI_Ax module works as a master mode (master out—MO and master in—MI). As this bit is 0, the eUSCI_Ax module works as a slave mode (slave out—SO and slave in—SI). Refer to Figure 8.5 to get more details.

8.4.5 SPI interrupt-related control registers

Three registers are used to control and process any interrupt related to the SPI operations:

1. The eUSCI_Ax Interrupt Enable (**UCAxIE**) register
2. The eUSCI_Ax Interrupt Flag (**UCAxIFG**) register
3. The eUSCI_Ax Interrupt Vector (**UCAxIV**) register

Both **UCAxIE** and **UCAxIFG** registers are 16-bit registers, but only the lowest two bits, bit-1 and 0, are used to enable and monitor the SPI data transmit and receive interrupts.

In **UCAxIE** register, two lowest bits, **UCTXIE** (bit-1) and **UCRXIE** (bit-0), are used to enable (1) or disable (0) the SPI data transmit and receive interrupts.

In **UCAxIFG** register, two lowest bits, **UCTXIFG** (bit-1) and **UCRXIFG** (bit-0), are used to monitor whether an SPI data transmit interrupt (**UCTXIFG** = 1) or an SPI data receive interrupt (**UCRXIFG** = 1) has occurred. When the **UCAxTXBUF** register is empty, the **UCTXIFG** bit is set to 1 to indicate that an SPI data transmit interrupt is occurred. Similarly, when the **UCAxRXBUF** register has received a complete character, the **UCRXIFG** bit is set to 1 to indicate that an SPI data receive interrupt is occurred.

The **UCAxIV** register is a 16-bit register and it is used to store and hold the current interrupt source vector. Table 8.4 lists bit value and function for this register. Any access, read or write, to the **UCAxIV** register can automatically reset the highest pending interrupt flag. If another interrupt flag is set, another interrupt is immediately generated after servicing the initial interrupt.

It can be found from Table 8.4 that the SPI data receiving has the highest priority interrupt level, but the SPI data transmit buffer empty has a lower priority level.

8.4.5.1 SPI transmit interrupt operation

When the **UCAxTXBUF** register is empty, the **UCTXIFG** interrupt flag in the **UCAxIFG** register is set by the transmitter to indicate that the **UCAxTXBUF** register is ready to accept another character. Also an interrupt request is generated if the **UCTXIE** bit in the **UCAxIE** register is set. The **UCTXIFG** bit is automatically reset to 0 if a character is written to the **UCAxTXBUF** register. The **UCTXIFG** bit can be set after a HR or when the bit **UCSWRST** = 1. The **UCTXIE** bit can be reset after a HR or when **UCSWRST** = 1.

A point to be noted is that the data written to the **UCAxTXBUF** register when **UCTXIFG** = 0 may result in erroneous data transmission. Therefore, the best way to avoid this possible mistake is not to write any data into the **UCAxTXBUF** register until the **UCTXIFG** is set to 1.

Table 8.4 Bit value and function for UCAxIV register

Bit	Field	Type	Reset	Function
15:0	**UCIVx**	RO	0	The eUSCI interrupt vector value **000:** No interrupt pending **002:** Interrupt source: Data received; Interrupt Flag: **UCRXIFG**; Priority: Highest **004:** Interrupt source: Transmit buffer empty; Interrupt Flag: **UCTXIFG**; Priority: Lowest

8.4.5.2 SPI receive interrupt operation

The **UCRXIFG** interrupt flag is set each time as a character is received and loaded into the **UCAxRXBUF** register. An interrupt request is also generated if the **UCRXIE** bit is set. The **UCRXIFG** and the **UCRXIE** bits can be reset by a HR or when the **UCSWRST** = 1. The **UCRXIFG** bit is automatically reset to 0 when the **UCAxRXBUF** register is read.

In summary, as long as the **UCAxTXBUF** register is empty, the **UCTXIFG** flag bit is set to 1. As soon as the **UCAxRXBUF** register received a complete character, the **UCRXIFG** flag bit is also set to 1. No matter whether the **UCTXIE** bit or the **UCRXIE** bit has been set to enable interrupts.

8.4.6 eUSCI_Ax module operational principle

When working at the different modes, master or slave, an eUSCI_Ax module has different operational principle and sequence. The eUSCI module in the SPI mode supports 7-bit and 8-bit character lengths selected by the UC7BIT bit.

In 7-bit data mode, the **UCAxRXBUF** register is LSB justified and the MSB is always reset. The **UCMSB** bit controls the direction of the transfer and selects the LSB- or the MSB-first. The default SPI character transmission is LSB-first. For communication with other SPI interfaces, MSB-first mode may be required. However, to make it convenient, in this chapter we always use the MSB-first format for our discussions.

8.4.6.1 Master mode

Figure 8.8 shows the eUSCI as a master in both 3-pin and 4-pin configurations. The operation sequence as an SPI master is

1. When data are moved to the transmit data buffer **UCxTXBUF** register, the eUSCI initiates a data transfer operation.
2. The data in the **UCxTXBUF** register are then moved to the TXS register when the TXS register is empty, initiating data transfer on the **UCxSIMO** line to start it with either the MSB- or LSB-first, depending on the bit **UCMSB** setting.
3. At the same time, the data on the **UCxSOMI** line are shifted into the RXS register on the opposite clock edge.

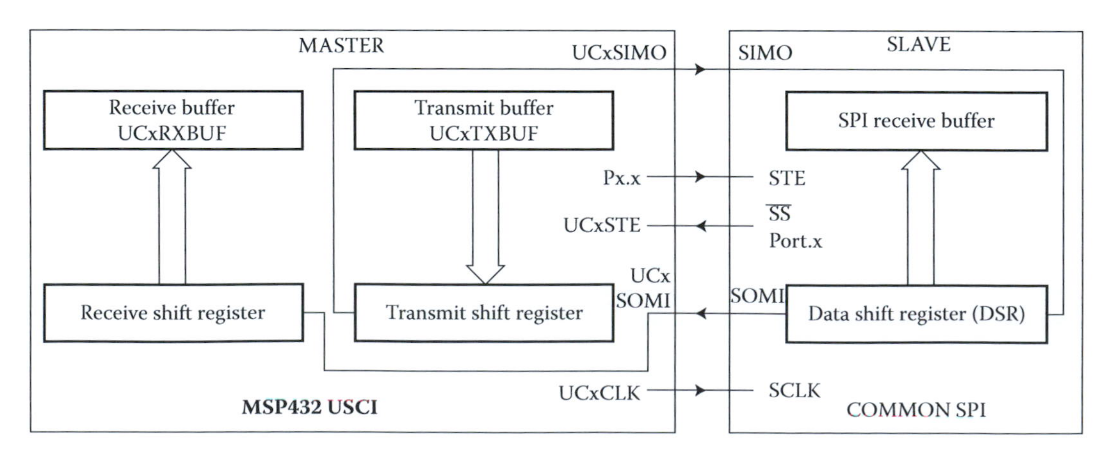

Figure 8.8 Configuration of the eUSCI working as a master mode (UCSTEM = 0). (Courtesy Texas Instruments.)

4. When a complete character is received, the received data are moved from the RXS register to the received data buffer **UCxRXBUF** register and the receive interrupt flag **UCRXIFG** is set, indicating a RX/TX operation is complete.

A set on the transmit interrupt flag, **UCTXIFG**, only indicates that data have been moved from the **UCxTXBUF** to the TXS register and the **UCxTXBUF** register is ready for new data. It does not indicate a RX/TX transition completion. To receive data into the eUSCI in the master mode, data must be written to the **UCxRXBUF** register because both receive and transmit operations operate concurrently.

For a 3-pin (3-wire) SPI transition mode, the fourth line **UCxSTE** is not used. However, for a 4-pin (4-wire) SPI transition mode, two different options to configure the eUSCI as a 4-pin master mode are available, which are

- The fourth pin is used as input to prevent conflicts with other masters (**UCSTEM** = 0).
- The fourth pin is also used as output to generate a slave enable signal (**UCSTEM** = 1).

The bit **UCSTEM** in the **UCAxCTLW0** Register is used to select the corresponding mode.

8.4.6.2 Slave mode

Figure 8.9 shows the eUSCI as a slave mode in both 3-pin and 4-pin configurations. The operation sequence as an SPI slave is

1. The **UCxCLK** is used as the input for the SPI clock and must be supplied by an SPI master. The data transfer rate is determined by this clock and not by the internal bit clock generator in the slave side.
2. The data written to the **UCxTXBUF** register and moved to the TXS register before a start of the **UCxCLK** is transmitted on the **UCxSOMI** line.
3. Then the data on the **UCxSIMO** line are shifted into the RXS register on the opposite edge of the **UCxCLK** and moved to the **UCxRXBUF** register when the set numbers of bits are received.
4. When data are moved from the RXS register to the **UCxRXBUF** register, the **UCRXIFG** interrupt flag is set, indicating that data have been received. The overrun error bit **UCOE** is set when the previously received data are not read from the **UCxRXBUF** register before new data are moved into the **UCxRXBUF** register.

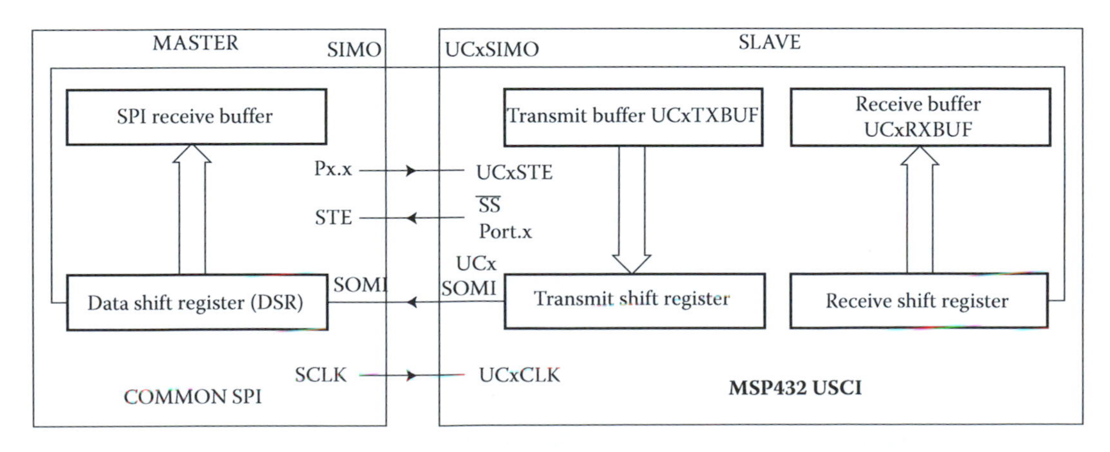

Figure 8.9 Configuration of the eUSCI working as a slave mode. (Courtesy Texas Instruments.)

For a 3-pin (3-wire) slave mode, the fourth pin **UCxSTE** is not used.

However, for a 4-pin slave mode, the **UCxSTE** is a digital input used by the slave to enable the transmission and receiving operations and is driven by the SPI master. When the **UCxSTE** bit is in the slave-active state, the slave operates normally. When the **UCxSTE** is in the slave-inactive state:

- Any receive operation in progress on the **UCxSIMO** line is halted.
- The **UCxSOMI** line is set to the input direction.
- The shift operation is halted until the **UCxSTE** line transitions into the slave transmit active state.

To allow an SPI bus to start a data transmit–receive operation, the SPI mode must be enabled.

8.4.6.3 SPI mode enable

An eUSCI module can be enabled by clearing the **UCSWRST** bit in the **UCAxCTLW0** Register. After it is enabled, it is ready to receive and transmit data. In master mode, the bit clock generator is ready, but is not clocked to produce any clocks. In slave mode, the bit clock generator is disabled and the clock is provided by and comes from the master side.

A transmit or receive operation is indicated by the bit **UCBUSY** = 1.

An Hard Reset or set the **UCSWRST** bit disables the eUSCI immediately and any active transfer is terminated.

Transmit Enable: In master mode, writing to the **UCxTXBUF** register activates the bit clock generator and the data begin to transmit. In slave mode, transmission begins when a master provides a clock, and in 4-pin mode when the **UCxSTE** is in the slave-active state.

Receive Enable: The SPI receives data when a transmission is active. Receive and transmit operations operate concurrently.

8.4.6.4 SPI mode working in the low-power modes

The eUSCI module is not functional when the MCU works in the LPM3, LPM4, or LPMx.5 of operation. However, the application can make use of the **FORCE_LPM_ENTRY** bit in the **PCMCTL1** Register to determine whether the entry to LPM should be aborted if the eUSCI is active, or if the device can continue low-power entry regardless. The latter option is useful if the eUSCI is transmitting–receiving data at a very slow rate and the application will prefer entry to LPM at the expense of a packet of data being lost.

8.4.6.5 Initialization and configuration of the eUSCI module

After a system reset, the **UCSWRST** bit in the **UCAxCTLW0** Register is automatically set to 1 to keep the eUSCI in the reset condition. During this reset period, the **UCSWRST** bit resets most interrupt-related bits and error flag bits, such as the **UCRXIE**, **UCTXIE**, **UCRXIFG**, **UCOE**, and **UCFE** bits, and sets the **UCTXIFG** flag. Clearing the **UCSWRST** bit is to terminate the reset state to release the eUSCI to the normal operation state.

Configuring and reconfiguring the eUSCI module should be performed during the eUSCI is in the reset state, which is when the **UCSWRST** bit is set to avoid unpredictable behavior.

Generally, a recommended eUSCI initialization and configuration process includes:

1. Set the **UCSWRST** bit in the **UCAxCTLW0** Register to allow the eUSCI to be in the reset state.

2. Initialize and configure related eUSCI registers, such as:
   a. The **UCAxCTLW0** Register to define the data operation mode, SCLK CKSE, MSB- or LSB-first, 3-pin or 4-pin mode, data length, CKPL, and phase.
   b. The **UCxBRW** register to define the clock prescaler.
3. Initialize and configure the eUSCI-related GPIO ports and pins.
4. Reset the **UCSWRST** bit in the **UCAxCTLW0** Register to terminate the reset period and enable the eUSCI module to be ready for its normal operation.
5. Enable eUSCI-related interrupts by setting the UCTXIE and UCRXIE bits in the **UCAxIE** register. This step is optional and not necessary if the poll-up the **UCAxIFG** is used.
6. Start the eUSCI data transmit operation by writing new data to the **UCTXBUF** register in the master mode or enable and start the SCLK to begin to receive the data in the slave mode.

8.4.7 eUSCI for SPI mode control signals and GPIO pins in MSP432P401R EVB

As we discussed in the previous sections, most peripheral devices in the MSP432P401R MCU system are related to GPIO Ports. This means that all control signals and I/O of peripherals are transferred by using related GPIO pins. This is also true to eUSCI for SPI mode used in the MSP-EXP432P401R evaluation board. All eUSCI for SPI modes are closely related to GPIO ports and all SPI control signals are connected to related GPIO port and pins, as shown in Table 8.5.

Compare Table 8.5 and Figure 8.4, it can be found that only some selected eUSCI for SPI mode control signals are connected to the J1~J4 jumper connectors in the MSP-EXP432P401R EVB, as shown in Table 8.5. All others are connected to the extended connector J5.

In order to enable the corresponding GPIO pins shown in Table 8.5 to work as SPI module pins to transfer related SPI control signals, the pins must be configured in the following ways:

- For those GPIO ports that are not controlled by the port map controller (**PMAPC**), such as **P1** and **P6**, the related **P1SEL1.x** and **P1SEL0.x** registers must be set up to 01, and **P6SEL1.x** and **P6SEL0.x** registers must be set up to 01 (**x** = 2~5), but **P6SEL1.x**

Table 8.5 eUSCI for SPI mode control signals and GPIO pins distributions

SPI pin	GPIO pin	Pin type	Pin function
UCA2STE	**P3.0**	I/O	SPI Module UCA2 STE Control Signal
UCA2SOMI	**P3.2**	I/O	SPI Module UCA2 SOMI Line
UCA2SIMO	**P3.3**	I/O	SPI Module UCA2 SIMO Line
UCB0CLK	**P1.5**	I/O	SPI Module UCB0 Clock—SCLK
UCB0SIMO	**P1.6**	I/O	SPI Module UCB0 SIMO Line
UCB0SOMI	**P1.7**	I/O	SPI Module UCB0 SOMI Line
UCB1SIMO	**P6.4**	I/O	SPI Module UCB1SIMO Line
UCB1SOMI	**P6.5**	I/O	SPI Module UCB1SOMI Line
UCB2CLK	**P3.5**	I/O	SPI Module UCB2 Clock—SCLK
UCB2SOMI	**P3.7**	I/O	SPI Module UCB2SOMI Line
UCB2SIMO	**P3.6**	I/O	SPI Module UCB2 SIMO Line
UCB3SIMO	**P6.6**	I/O	SPI Module UCB3 SIMO Line
UCB3SOMI	**P6.7**	I/O	SPI Module UCB3 SOMI Line

and **P6SEL0.x** registers must be set to 10 (**x** = 6~7), to enable the selected pins to work as SPI mode pins.

- For those GPIO ports that are controlled by the **PMAPC**, such as **P3**, in addition to setting the **P3SEL1.x** and **P3SEL0.x** registers to 01, the related mapping value must be written into the port mapping register **P3MAPx**.

To configure GPIO **P1** and **P6**, one can directly access related registers to do these settings. But to configure GPIO **P3**, an access key must be used to access all mapping control registers to configure the selected GPIO port. Refer to Sections 7.3 and 7.4 in Chapter 7 to get more details for these settings.

Based on these discussions, now let's build our example project to interface some peripherals via the SPI modules provided by the MSP432P401R MCU system.

8.4.8 Build the on-board LCD interface programming project

In this example project, we use the eUSCI for SPI Module, **eUSCI_B0**, to interface an on-board 16 × 2 LCD device installed in the EduBASE ARM® Trainer to display some interesting letters on it. First let's have a closer look at the hardware configuration and connection for this LCD via **eUSCI_B0**.

8.4.8.1 eUSCI_B0 for SPI mode interface for the LCD in EduBASE ARM® trainer

A high-speed CMOS shift register 74VHCT595 is used as an interface to the LCD from the **eUSCI_B0** in the MSP432P401R MCU system. A functional block diagram of this interface circuit connected to the LCD installed in the EduBASE ARM® Trainer is shown in Figure 8.10.

It can be found from Figure 8.10 that the **eUSCI_B0** module is connected to a HCOMS shift register 74VHCT595 via 3-pin (3-wire) mode, GPIO **P1.6**, **P1.7**, and **P1.5** pins. The GPIO pin **P6.7** just works as a digital function pin to provide a trigger signal. The functions of these pins are:

- **P1.6**: **UCB0SIMO**—masters outputs serial data to the peripheral 74VHCT595
- **P1.5**: **UCB0CLK**—provides SCLK signal to coordinate the serial data transmission
- **P1.7**: **UCB0SOMI**—receives serial data from the peripheral (not used in this example)
- **P6.7**: **CS_LCD**—provides the LCD chip select signal. Here it works as a trigger clock to start a serial-to-parallel data conversion to convert 8-bit serial data stored in the serial shift register to the 8-bit parallel data to be stored in the 8-bit output registers in the 74VHCT595 register

In this hardware configuration, the **eUSCI_B0** module works as a master to use the **UCB0SIMO** pin to output serial data to the LCD via a serial shift register 74VHCT595 and an LCD controller SPLC780. The data size to be transmitted from the **eUSCI_B0** to the LCD is 4 bits, not 8 bits. Therefore, an 8-bit data must be broken into two sections with 4 bits for each section to be transmitted. The higher 4 bit (**DB7~DB4**) are LCD data and the lower 3 bit are LCD control signals, **RS** (register select), **E** (enable), and **BL** (back light enable). The read/write (**R/W̄**) control signal is connected to the ground to make the LCD controller work in the writing mode.

We will discuss these components one by one in the following sections.

8.4.8.2 Serial shift register 74VHCT595

74VHCT595 is an 8-bit serial-in serial-out or serial-in parallel-out register with output latches. In this example, we use it as a serial-in parallel-out register to convert the serial

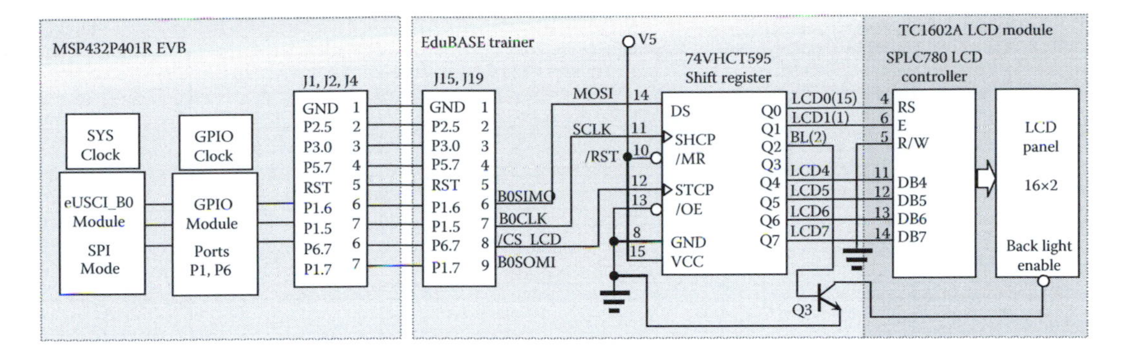

Figure 8.10 Hardware configuration of the LCD interfacing circuit.

input coming from the **eUSCI_B0** module to the parallel output, and the latter is feed into the LCD controller to display the result on the LCD panel.

The operational principle and sequence of this register is:

- Data are shifted into the serial input pin (**DS**) on the positive-going edges of the shift register clock input (**SHCP**).
- The data in each flip-flop register are transferred to the storage register on a positive-going edge of the storage register clock input (**STCP**). If both clocks are connected together, the shift register will always be one clock pulse ahead of the storage register.
- The shift register has a serial input (**DS**) and a serial standard output (**Q7S**) for cascading.
- It is also provided with asynchronous master reset (**MR**-active LOW) for all eight shift flip-flop registers. The storage register has eight parallel three-state bus driver outputs. Data in the storage register appear at the output whenever the output enable (**OE**) input is LOW.

Figure 8.11 shows the functional block diagram of this register.

The **STAGES 1 TO 6** in this figure indicates that this stage comprises six shift flip-flops. All eight parallel outputs are latched and controlled by the **OE** signal.

In our example application, each time after the **eUSCI_B0** module transmits 8 bits data to this shift register via **DS** (**UCB0SIMO-P1.6**) pin and **SHCP** (**UCB0CLK-P1.5**) pin, a positive-going edge (low-to-high) signal should be applied on the **STCP** (**CS_LCD-P6.7**) pin to transfer 8-bit serial data in the shift register to the 8-bit parallel output buffer or latches. The **OE** pin is connected to the ground to enable eight latched outputs to be directly transferred to the output pins (Figure 8.10).

8.4.8.3 LCD module TC1602A and LCD controller SPLC780

The TC1602A LCD module is controlled and driven by an LCD controller SPLC780, which is a dot-matrix LCD controller and driver to control and display alpha-numeric, Japanese kana characters, and symbols in a 16 × 2 LCD device. It can be configured to drive a dot-matrix LCD under the control of a 4- or 8-bit microprocessor. Since all the functions such as display RAM, character generator, and liquid crystal driver, required for driving a dot-matrix LCD are internally provided on one chip, a minimal system can be interfaced with this controller/driver.

In the MSP432P401R MCU system, the SPLC780 is configured to interface with the MCU with 4-bit style; therefore, we will concentrate on this style in our following discussions.

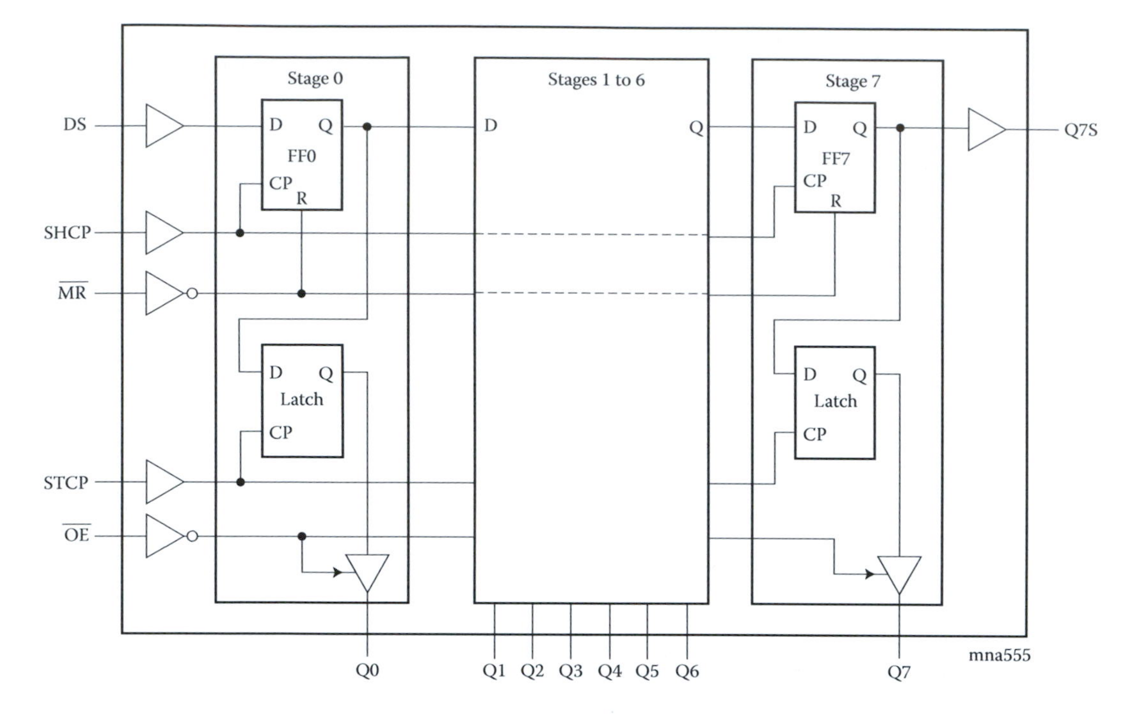

Figure 8.11 Functional block diagram of 74VHCT595 shift register.

An internal structure block diagram of the SPLC780 is shown in Figure 8.12.

The SPLC780 provides a full controllability to the LCD. Some most important and popular components with the associated controlling and interfacing abilities to the LCD are listed below:

- **`Instruction register (IR) and data register (DR)`**
  The SPLC780 has two 8-bit registers, an **`IR`** and a **`DR`**. The IR stores instruction codes, such as display clear and cursor shift, and address information for display data RAM

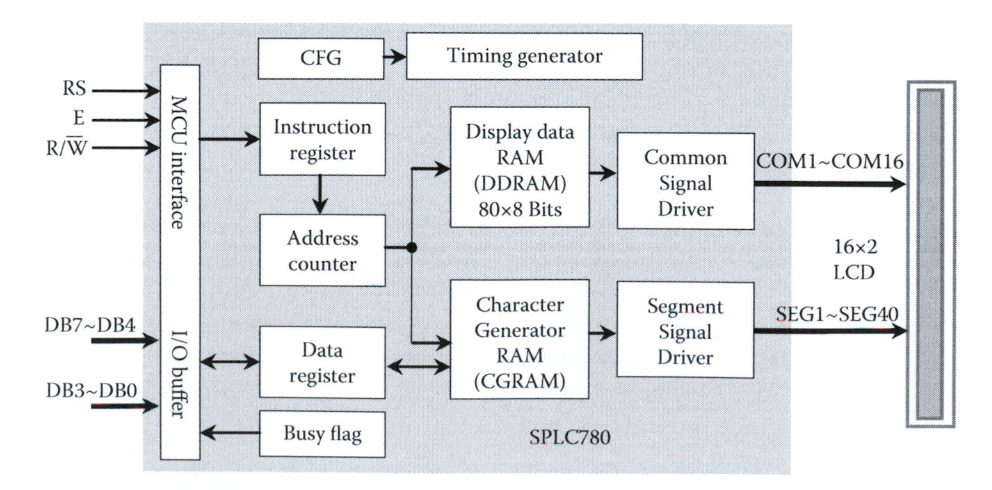

Figure 8.12 Functional block diagram of the LCD controller SPLC780.

(**DDRAM**) and character generator RAM (**CGRAM**). The IR can only be written from the MCU. The DR temporarily stores data to be written into **DDRAM** or **CGRAM** and temporarily stores data to be read from **DDRAM** or **CGRAM**. Data written into the DR from the MCU are automatically written into **DDRAM** or **CGRAM** by an internal operation. The DR is also used for data storage when reading data from **DDRAM** or **CGRAM**. When address information is written into the IR, data are read and then stored into the DR from DDRAM or CGRAM by an internal operation. Data transfer between the MCU is then completed when the MCU reads the DR. After the read, data in **DDRAM** or **CGRAM** at the next address are sent to the DR for the next read from the MCU. By using the register select (**RS**) signal, these two registers can be selected as shown in Table 8.6.

- **Busy flag (BF)**
  When the BF is 1, the SPLC780 is in the internal operation mode, and the next instruction will not be accepted. When RS = 0 and R/$\overline{W}$ = 1 (Table 8.6), the BF is output to DB7. The next instruction must be written after ensuring that the BF is 0.
- **Address counter (AC)**
  The AC assigns addresses to both DDRAM and CGRAM. When an address of an instruction is written into the IR, the address information is sent from the IR to the AC. Selection of either DDRAM or CGRAM is also determined concurrently by the instruction. After writing into (reading from) DDRAM or CGRAM, the AC is automatically incremented by 1 (decremented by 1). The AC contents are then output to DB0 to DB6 when RS = 0 and R/W = 1 (Table 8.6).
- **Display Data RAM (DDRAM)**
  DDRAM stores display data represented in 8-bit character codes. Its extended capacity is 80 × 8 dots or bits or 80 characters. If each character comprises 5 × 8 dots, each line on the LCD can display 80/5 = 16 characters. The area in DDRAM that is not used for display can be used as general data RAM. Figure 8.13 shows the relationships between DDRAM addresses and positions on the 2-line by 16-characters LCD. The DDRAM address (ADD) is set in the AC as hexadecimal. In the MSP432P401R MCU system, a 16 × 2 LCD is used, which means that this LCD can display 2 lines with 16 characters being displayed at each line.
- **Character generator ROM (CGROM)**
  The CGROM generates 5 × 8 dots or 5 × 10 dots character patterns from 8-bit character codes. It can generate 208 5 × 8 dots character patterns and 32 5 × 10 dots character patterns. User-defined character patterns are also available by mask-programmed ROM.
- **Character Generator RAM (CGRAM)**
  In the CGRAM, the user can rewrite character patterns by programming. For 5 × 8 dots, eight character patterns can be written, and for 5 × 10 dots, four character patterns can be written.

Some other components, including the timing generation circuit, LCD-driven circuit, and cursor blink control circuit, are not discussed in this section.

Table 8.6 Register selection

RS	R/W	Function
0	0	Write an instruction to the SPLC780
0	1	Read an instruction from the SPLC780
1	0	Write a data item to the SPLC780
1	1	Read a data item from the SPLC780

	1	2	3	4	5	6	7	8	9	10	11	12	13	14	15	16
Display position	00	01	02	03	04	05	06	07	08	09	0A	0B	0C	0D	0E	0F
DDRAM address	40	41	42	43	44	45	46	47	48	49	4A	4B	4C	4D	4E	4F

Figure 8.13 Relationship between the display positions and DDRAM addresses.

8.4.8.3.1 Interfacing control signals between the MCU and the SPLC780 Based on the discussion and analysis in the last section, we have had a clear and global picture about the SPLC780 LCD controller. In summary, three important signals are widely used in the interfacing process between a MCU and the SPLC780:

- Register selection **RS** signal: When **RS** = 0, it means that currently an instruction is transferred from the MCU to the SPLC780. If **RS** = 1, it means that currently a data item is reading from or writing into the SPLC780.
- **R/$\overline{\text{W}}$** signal: When this signal is 0, it means that a writing operation is performed from the MCU to the SPLC780. If this bit is 1, it means that a reading operation is performed by reading a data item, including the AC from the SPLC780 to the MCU.
- **E** signal: When this signal is 0, it means that the SPLC780 is disabled and no matter what kind of instructions or data are read from or written into the SPLC780, no information can be obtained from the SPLC780 since it is disabled. If this bit is 1, it means that the SPLC780 is enabled and it can accept any control signal with the appropriate responses.

The SPLC780 can send data in either two 4-bit operations or one 8-bit operation, thus allowing interfacing with 4-bit or 8-bit MCUs.

- For 4-bit interface data, only four higher bits (**DB4~DB7**) are used for transfer. Bus lines **DB0~DB3** are disabled. The data transfer between the SPLC780 and the MCU is completed after the 4-bit data have been transferred twice. As for the order of data transfer, the four high-order bits (**DB4~DB7**) are transferred before the four low-order bits (**DB0~DB3**) can be transferred. The BF must be checked after the 4-bit data have been transferred twice. Two more 4-bit operations then transfer the BF and AC data.
- For 8-bit interface data, all eight bus lines (**DB0~DB7**) are used.

Since we are using the MSP432P401R MCU system with the EduBASE ARM® Trainer, in which a 4-bit interfacing style is used, therefore we will concentrate on this 4-bit interfacing technique.

The operational sequence of interfacing between the MCU and the SPLC780 is

1. The LCD must be initialized first by performing a reset operation. Regularly this reset operation can be performed by the SPLC780 itself via an internal reset process when it is powered up. However, if this internal reset operation is not as good as desired, the user must perform an external reset operation by programming the SPLC780 to do that reset operation. This reset process includes the following operations:
   a. Clear the display area in the LCD
   b. Set the display function by setting the related bit with the appropriate values:

i. **DL** = **0**:4-bit interface data, **DL** = **1**:8-bit interface data
ii. **N** = **0**:1-line display, **N** = **1**:2-line display
iii. **F** = **0**:5 × 8 dots character font, **F** = **1**:5 × 10 dots character font

c. Set the display on/off control by setting the related bit with the appropriate values:
i. **D** = **0**: display off, **D** = **1**: display on
ii. **C** = **0**: cursor off, **C** = **1**: cursor on
iii. **B** = **0**: blinking off, **B** = **1**: blinking on

d. Select the entry mode by setting the related bit with the appropriate values:
i. **I/D** = **0**: AC is decremented by 1 after each operation, **I/D** = **1**: AC is incremented by 1 after each operation
ii. **S** = **0**: no shift, **S** = **1**: shift character
iii. **S/C** = **0**: cursor move, **S/C** = **1**: display shift
iv. **R/L** = **0**: shift to the left, **R/L** = **1**: shift to the right

2. After the reset process, perform the instruction writing operation to set the internal DDRAM address from which the data can be written into or read out for SPLC780
3. To display characters in the LCD, perform the data writing operations to send the data to the DDRAM in the SPLC780. After each 8-bit character writing (two 4-bit writings), an instruction read (**IR**) should be performed to check the **BF** before the next data can be sent to the SPLC780

All of these initialization parameters and their values are shown in Table 8.7. A block diagram of this operational sequence is shown in Figure 8.14.

8.4.8.3.2 Control and interface programming for SPLC780 Based on the introduction of the interface between the MCU and the SPLC780 in the last section, now we can begin to build a control and interfacing program to control the TC1602A LCD module to display desired characters in the LCD panel.

All operation steps described in the operational sequence in the last section can be performed by executing a sequence of instructions. For actual reasons, we only concentrate on the 4-bit data transfer since this style is used in our example project.

When the SPLC780 is connected to a MCU by using 4-bit interface, an 8-bit instruction or data must be broken into two parts to be transferred. The higher 4-bit instruction or

Table 8.7 LCD initialization functions

Bit	Value	Functions	Bit	Value	Functions
DL	0	4-bit interface data	**C**	0	Cursor off
	1	8-bit interface data		1	Cursor on
N	0	1-line display	**B**	0	Blinking off
	1	2-line display		1	Blinking on
F	0	5 × 8 dots character font	**I/D**	0	Decrement by 1
	1	5 × 10 dots character font		1	Increment by 1
D	0	Display off	**S**	0	No shift
	1	Display on		1	Shift character
S/C	0	Cursor move	**R/L**	0	Shift to the left
	1	Display shift		1	Shift to the right

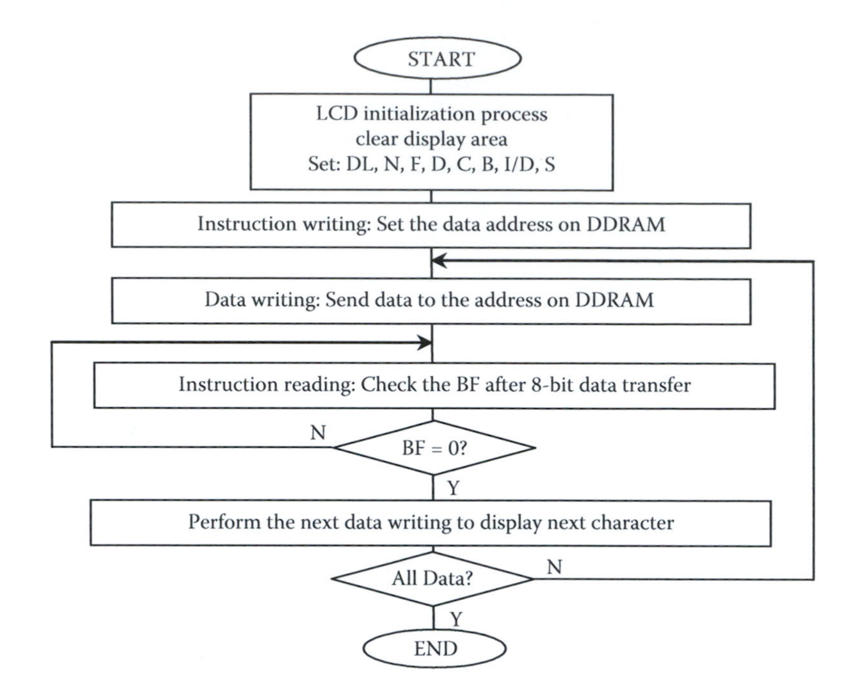

Figure 8.14 Block diagram of the operational sequence of interfacing to SPLC780.

data are transferred first via **DB7~DB4** in the SPLC780. Then the lower 4-bit is transferred via **DB7~DB4** again. The relationship for the lower 4 bit in the 74VHCT595 and SPLC780 is

- **Q7 = DB7**
- **Q6 = DB6**
- **Q5 = DB5**
- **Q4 = DB4**

Tables 8.8 and 8.9 list the general LCD instructions and the actual LCD instructions specially applied in the example project for 4-bit data transfer format implemented in SPLC780 LCD controller (**X** = don't care).

One important point to be noted is the operational steps 3 and 4. After the LCD initialization process, the SPLC780 uses the default data transfer format, 8-bit data format. Therefore, in step 3, even the SPLC780 is configured as 4-bit mode, the SPLC780 may still be considered as 8-bit mode. In order to enable the SPLC780 to know this 4-bit configuration, the SPLC780 needs to be reconfigured by rewriting **DB7~DB4** = 0010 again in step 4. Then the lower 4 bit is written to the SPLC780 by performing the second writing operation. This rewriting is very important and necessary. Otherwise the SPLC780 may perform some abnormal operations later.

Another point to be noted is the **ENABLE** signal **E** on the SPLC780. In the instruction sequence listed above, we did not list the **ENABLE** signal. In fact, this control signal is very important and all operations listed in both tables need this signal to be active (HIGH) when any instruction is written and executed.

8.4.8.3.3 LCD programming instruction structure and sequence As we mentioned in Section 8.4.8.1, in this EduBASE ARM® Trainer hardware configuration, the **eUSCI_B0** module's output is connected to the LCD via a serial shift register 74VHCT595 and an

Table 8.8 Most popular LCD instructions for 4-bit data transfer

	Items	Code						Description
	Instruction	**RS**	**R/$\overline{W}$**	**DB7**	**DB6**	**DB5**	**DB4**	
1	**Clear Display**	**0** **0**	**0** **0**	**0** **0**	**0** **0**	**0** **0**	**0** **1**	Clear display area and set DDRAM address 0 in the address counter (AC)
2	**Return Home**	**0** **0**	**0** **0**	**0** **0**	**0** **0**	**0** **1**	**0** **X**	Return the display to the original position and set DDRAM address to 0
3	**Set Function**	**0**	**0**	**0**	**0**	**1**	**DL**	Set to 4-bit operation. In this case, operation is handled as 8 bits by initialization, and only this instruction is performed with one write
4	**Set Function**	**0** **0**	**0** **0**	**0** **N**	**0** **F**	**1** **X**	**DL** **X**	Set 4-bit operation and select 2-line display and 5 × 8 dot font. 4-bit operation starts from this step and resetting is necessary. (Number of display lines and character fonts cannot be changed after this step)
5	**Display On/Off**	**0** **0**	**0** **0**	**0** **1**	**0** **D**	**0** **C**	**0** **B**	Turns on display (D = 1) and cursor (C = 1). Turn off blinking (B = 0). Entire display is in space mode because of initialization
6	**Set Entry Mode**	**0** **0**	**0** **0**	**0** **0**	**0** **1**	**0** **I/D**	**0** **S**	Sets mode to increment the address by one (I/D = 1) and to shift the cursor to the right at the time of write to the DD/CGRAM. Display is not shifted (S = 0)
7	**Set DDRAM Address**	**0** **0**	**0** **0**	**1** **ADD**	**ADD** **ADD**	**ADD** **ADD**	**ADD** **ADD**	Set DDRAM address (DB7 = 1). ADD is the address bit
8	**Write Data to DDRAM**	**1** **1**	**0** **0**	**8-BIT WRITE-IN DATA** **(ASCII CODE)**				Write data into DDRAM and display them in the LCD
9	**Read Data from DDRAM**	**1** **1**	**1** **1**	**8-BIT READ-OUT DATA** **(ASCII CODE)**				Read data from DDRAM

LCD controller SPLC780. The data size to be transmitted from the **eUSCI_B0** module to the LCD is 4 bits not 8 bits. Therefore, an 8-bit data must be broken into two sections with 4 bits for each section to be transmitted. The higher 4 bits (**DB7~DB4**) are LCD data and the lower 3 bits are LCD control signals, **RS** (register select), **E** (enable), and **BL** (back light enable). The **R/$\overline{W}$** control signal is connected to the ground to make the LCD controller work in the writing mode.

Each **eUSCI_B0** module serial data output is an 8-bit serial data sequence transmitted from the SPI transmit buffer (**UCxTXBUF**) to the **UCB0SIMO** pin via TXS register. Since each 8-bit data must be broken into two 4-bit sections and transmitted by two times to the serial shift register 74VHCT595, each 8-bit data should contain LCD data (upper 4-bit **DB7~DB7**) and LCD commands (lower 3-bit **RS**, **E**, and **BL**). Figure 8.15 shows an 8-bit serial data configuration that is sent to the **UCB0SIMO** pin and transmitted to the **DS** serial input on the serial shift register 74VHCT595.

Table 8.9 Actual LCD instructions for 4-bit data transfer in the example project

	Items	Code						Description
	Instruction	RS	R/$\overline{W}$	DB7	DB6	DB5	DB4	
1	**Clear Display**	**0** **0**	**0** **0**	**0** **0**	**0** **0**	**0** **0**	**0** **1**	Clear display area and set DDRAM address 0 in the address counter (AC)
2	**Return Home**	**0** **0**	**0** **0**	**0** **0**	**0** **0**	**0** **1**	**0** **X**	Return the display to the original position and set DDRAM address to 0
3	**Set Function**	**0**	**0**	**0**	**0**	**1**	**0**	Set to 4-bit operation. In this case, operation is handled as 8 bits by initialization, and only this instruction is performed with one write
4	**Set Function**	**0** **0**	**0** **0**	**0** **1**	**0** **0**	**1** **X**	**0** **X**	Set 4-bit operation and select 2-line display and 5 × 8 dot font. 4-bit operation starts from this step and resetting is necessary. (Number of display lines and character fonts cannot be changed after this step)
5	**Display On/ Off**	**0** **0**	**0** **0**	**0** **1**	**0** **1**	**0** **1**	**0** **1**	Turns on display (D = 1) and cursor (C = 1)Turn on blinking (B = 1). Entire display is in space mode because of initialization
6	**Set Entry Mode**	**0** **0**	**0** **0**	**0** **0**	**0** **1**	**0** **1**	**0** **0**	Sets mode to increment the address by one (I/D = 1) and to shift the cursor to the right at the time of write to the DD/CGRAM. Display is not shifted (S = 0)
7	**Set DDRAM Address**	**0** **0**	**0** **0**	**1** **ADD**	**ADD** **ADD**	**ADD** **ADD**	**ADD** **ADD**	Set DDRAM address (DB7 = 1). ADD is the address bit
8	**Write Data to DDRAM**	**1** **1**	**0** **0**	**8-BIT WRITE-IN DATA** **(ASCII CODE)**				Write data into DDRAM and display them in the LCD
9	**Read Data from DDRAM**	**1** **1**	**1** **1**	**8-BIT READ-OUT DATA** **(ASCII CODE)**				Read data from DDRAM

Refer to Table 8.9. For example, if we want to clear display on LCD panel, we need to first send **DB7~DB4** = **0000** and then **DB7~DB4** = **0001** as LCD data to the LCD controller in two times. In both times, we should also include the LCD command **BL:E:RS** = **110** with those LCD data together to send out. The resulted instruction of this 8-bit code is 00000110 (**0x06**) for the first 8-bit code and 00010110 (**0x16**) for the second 8-bit code.

In our actual project, we can build some subroutines to perform these data-splitting operations, and therefore we can directly send 00000001 (**0x01**) to the subroutine to allow that subroutine to break this 8-bit data into two pieces of 4-bit data and send them out with the corresponding LCD command bits together.

Generally, the LCD programming process includes three sections:

1. LCD initialization (if internal initialization is not as good as desired, the programming initialization built by using the user's program is necessary)

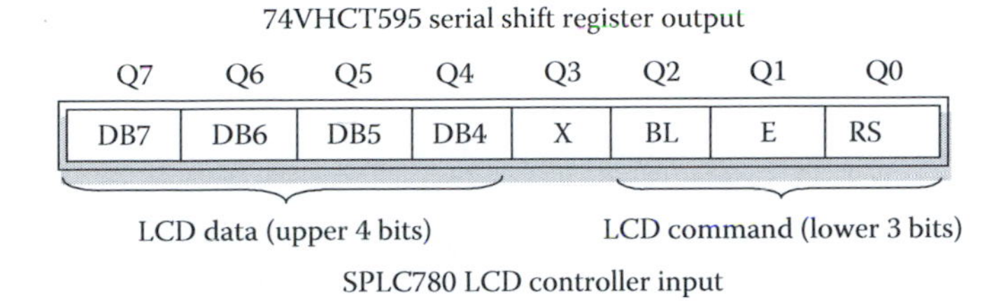

Figure 8.15 8-bit serial data configuration to be transmitted to LCD controller.

2. LCD function setup and configuration
3. LCD displaying data output and display

Now, let's have a closer look at the LCD initialization and function setup instruction structure and sequence. First let's handle the LCD initialization process.

8.4.8.3.3.1 LCD initialization process The operational sequence of the LCD manually initialization process is shown in Figure 8.16.

Since we are using 4-bit data transfer mode to interface to SPLC780 LCD controller via serial shift register 74VHCT595, only 4 data bits **DB7~DB4** are used. However, the data transfer between the 74HCT595 and the SPLC780 is 8-bit mode via **Q7~Q0** (refer to Figure 8.10), which means that besides the 4 data bits transferred in **DB7~DB4**, three control signals, **E**, **RS**, and **BL**, are also transferred via **Q2~Q0**, respectively. Therefore, we divide each instruction into two pieces: the upper 4 bits are instructions to be transferred and the lower 3 bits are **RS**, **E**, and **BL** commands to be transferred, as shown in Figure 8.16.

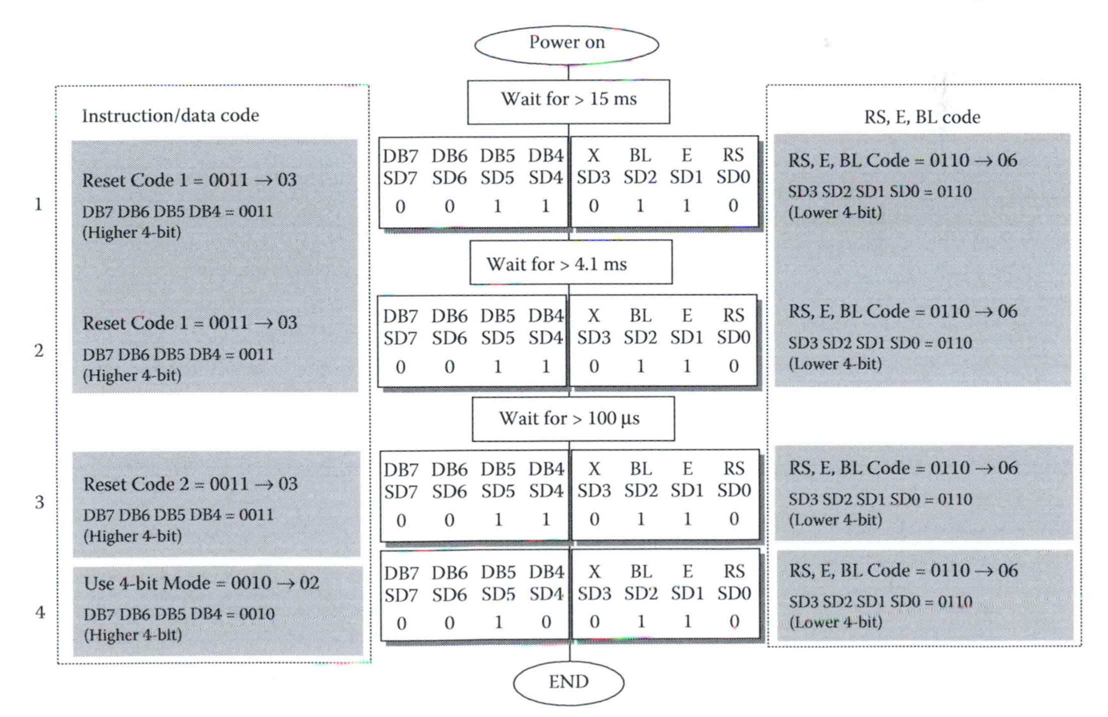

Figure 8.16 Coding process for the LCD initialization operation.

Two reset codes are used for this LCD initialization process. Reset code 1 is sent to the SPLC780 two times (called rewritten for steps **1** and **2** in Figure 8.16). The combined single 8-bit code is **0x36** (**DB7-DB4** = 0011, **BL**:**E**:**RS** = 0110). Some time delays are needed between these two reset codes to be transferred to the SPLC780. The **SD7~SD0** is 8-bit data sent by the **eUSCI_B0** module.

Similarly, if we combine reset code 2 together to form a single reset code, it is **0x36** (step **3** in Figure 8.16). After these reset codes are sent and executed by the SPLC780, the LCD should be initialized properly and can be configured to perform our desired data display process.

Step **4** in Figure 8.16 is to set the data mode for the LCD module. The higher 4 bit is 0010 and the lower 4 bit is 0110. Combining these two pieces of nibbles together, we can get **0x26**.

The reason we combine these two codes together to get a single 8-bit code is that we will use this kind of 8-bit code to do the LCD initialization in our program. With the help of some subroutines, we can decompose these combinations into two separate 4-bit codes and transfer them in 8-bit format with the higher 4-bit as **DB7~DB4** and lower 4-bit as **RS, E,** and **BL** codes.

One point to be noted when transmitting these two-piece codes by using subroutines is that the LCD controller can only accept these instruction/data when it is enabled. Therefore, in order to make the LCD controller enabled to accept these instructions/data, we need to simulate an **E** pulse by using the software codes in our subroutines.

For example, to send the reset code 1 to the SPLC780, one needs to use the code shown in Figure 8.17 to send **0x36** with a simulated **E** pulse.

The codes in lines 6~8 are used to send a combined instruction/command code **0x34** (**DB7~DB4** = 0011, **BL**:**E**:**RS** = 100) to the LCD controller. However, in order to allow this code to be accepted by the SPLC780, the bit **E** must be high. The code in line 7 does this job. The codes in lines 6 and 8 are used to simulate a low before and after the high pulse of **E** in line 7. The function **SPIB_Write()** is used to send serial data to the SPLC780 via 74VHCT595.

8.4.8.3.3.2 LCD function setup and configuration The LCD function setup and configuration process are equivalent to perform operations in steps 1~6 shown in Table 8.9.

The operational sequence of the LCD function setup and configuration process is shown in Figure 8.18. Similar to the initialization, each 8-bit code is broken into two pieces (higher 4-bit and lower 4-bit) and transferred to the SPLC780 in two times with the **RS**, **E**, and **BL** code.

```
1   LCD_write(0x30, 0);                 // call LCD_Write() function...

2   void LCD_write(char data, unsigned char cmd)
3   {
4     data &= 0xF0;                     // clear lower nibble for data
5     cmd &= 0x0F;                      // clear upper nibble for control
6     SPIB_Write (data|cmd|BL);         // RS = 0, R/W = 0, E = 0 ┐
7     SPIB_Write (data|cmd|EN|BL);      // RS = 0, R/W = 0, E = 1 │ A simulated E pulse
8     delay();                          //                        │
9     SPIB_Write (data|BL);             // RS = 0, R/W = 0, E = 0 ┘
10    SPIB_Write (BL);                  // RS = 0, R/W = 0, BL = 1
    }
```

Figure 8.17 Piece of code to create a simulated E pulse.

After initialization

Step	Instruction/data code	DB7 DB6 DB5 DB4 / SD7 SD6 SD5 SD4	X BL E RS / SD3 SD2 SD1 SD0	RS & E code
1	Function set code = 0010 → 02 DB7 DB6 DB5 DB4 = 0010 (High 4-bit)	0 0 1 0	0 1 1 0	RS, E, BL Code = 0110 → 06 SD3 SD2 SD1 SD0 = 0110
	Function set code = 1000 → 08 DB7 DB6 DB5 DB4 = 1000 (High 4-bit) N = 1: 2-Line Display F = 0: 5×8 dots	1 0 0 0 N F - -	0 1 1 0	RS, E, BL Code = 0110 → 06 SD3 SD2 SD1 SD0 = 0110
2	Entry Mode Code = 0000 → 00 DB7 DB6 DB5 DB4 = 0000 (High 4-bit)	0 0 0 0	0 1 1 0	RS, E, BL Code = 0110 → 06 SD3 SD2 SD1 SD0 = 0110
	Entry Mode Code = 0110 → 06 DB7 DB6 DB5 DB4 = 0000 (High 4-bit) I/D = 1: Increment by 1 after moving. S = 0: No shift.	0 1 1 0 I/D S	0 1 1 0	RS, E, BL Code = 0110 → 06 SD3 SD2 SD1 SD0 = 0110
3	Display On/Off = 0000 → 00 DB7 DB6 DB5 DB4 = 0000 (High 4-bit)	0 0 0 0	0 1 1 0	RS, E, BL Code = 0110 → 06 SD3 SD2 SD1 SD0 = 0110
	Display On/Off = 1100 → 0C DB7 DB6 DB5 DB4 = 1100 (High 4-bit) D = 1: Display On; C = 0: Cursor Off B = 0: Blinking Off.	1 1 0 0 D C B	0 1 1 0	RS, E, BL Code = 0110 → 06 SD3 SD2 SD1 SD0 = 0110
4	Clear Display = 0000 → 00 DB7 DB6 DB5 DB4 = 0000 (High 4-bit)	0 0 0 0	0 1 1 0	RS, E, BL Code = 0110 → 06 SD3 SD2 SD1 SD0 = 0110
	Clear Display = 0001 → 01 DB7 DB6 DB5 DB4 = 0001 (High 4-bit)	0 0 0 1	0 1 1 0	RS, E, BL Code = 0110 → 06 SD3 SD2 SD1 SD0 = 0110

END

Figure 8.18 Operational sequence of LCD function set and configuration.

For example, to transfer the function set code **0x28** at step 1 in Figure 8.18, this 8-bit code **0x28** is broken into two pieces, **0x2** and **0x8**, as shown in coding steps 4 and 5 in Figure 8.17.

The higher 4 bit **0x2** is sent first with the **RS**, **E**, and **BL** code **0x7** together to form an 8-bit code **0x27**. Then the lower 4 bit **0x8** is transferred with the **RS**, **E**, and **BL** code **0x7** to form an 8-bit code **0x87** in the second time. All of these breaking and combining jobs as well as a simulated **E** pulse are performed by the software codes in a subroutine similar to one shown in Figure 8.17.

Step 2 in Figure 8.18 is used to set the entry mode code, which is 0006 or **0x06**. This 8-bit code is also broken to two pieces of 4-bit codes: **0000** and **0110** and sent in two times with the **BL**, **E**, and **RS** signal together to the SPLC780 via the 74VHCT595.

Step 3 is to set up the display mode with the code of 00001111 = **0x0F** and step 4 is to clear the display area with the code of 00000001 = **0x01**. These 8-bit codes are also broken into two pieces of 4-bit codes and sent out in two times with the **BL**, **E**, and **RS** signal together.

8.4.8.4 Build the example LCD interfacing project

Now we have enough knowledge and skill to build a practical LCD interfacing project to serially transmit our desired letters or numbers to the LCD to display them on it.

The hardware configuration of this project is shown in Figure 8.10. In the EduBASE ARM® Trainer, it has been configured to use the **eUSCI_B0** module to serially transmit instructions and data to LCD controller SPLC780 via the serial shift register 74VHTC595. The GPIO pins used in this project include:

- **P1.6**: **UCB0SIMO**—masters outputs serial data to the peripheral 74VHCT595
- **P1.5**: **UCB0CLK**—provides SCLK signal to coordinate the serial data transmission
- **P1.7**: **UCB0SOMI**—receives serial data from the peripheral (not used in this example)
- **P6.7**: **CS_LCD**—provides the LCD chip select signal. Here it works as a trigger clock to start a serial-to-parallel data conversion to convert 8-bit serial data stored in the serial shift register to the 8-bit parallel data to be stored in the 8-bit output registers in the 74VHCT595 register

In our example application, each time after the **eUSCI_B0** module transmits 8-bit data to the shift register via **DS** (**UCB0SIMO-P1.6**) pin and clock **SHCP** (**UCB0CLK-P1.5**) pin, a positive-going edge (low-to-high) signal should be applied on the **STCP** (**CS_LCD-P6.7**) pin to transfer 8-bit serial data in the shift register to the 8-bit parallel output buffer or latches. The **OE** pin is connected to the ground to enable eight latched outputs to be directly transferred to the output pins (Figure 8.10).

We try to use the DRA model to build this project. The entire coding process includes the following four sections:

1. Initialize and configure **eUSCI_B0** module-related GPIO ports **P1** and **P6**
2. Initialize and configure **eUSCI_B0** module and related registers (**UCB0CTLW0**, **UCB0BRW**, and **UCB0IE**)
3. Build four subroutines to break, combine, and transmit 8-bit instructions/data to the LCD controller to display desired letters and numbers on the LCD panel
4. Build some time delay functions to delay different periods of time to make the LCD data transmissions stable and reliable

Now, let's first create a new LCD project **DRALCD**.

8.4.8.4.1 Create a DRA LCD project DRALCD Perform the following operations to create a new project **DRALCD**:

1. Open the Windows Explorer window to create a new folder named **DRALCD** under the **C:\MSP432 Class Projects\Chapter 8** folder
2. Open the Keil® ARM MDK µVersion5 and go to **Project|New µVersion Project** menu item to create a new µVersion Project. On the opened wizard, browse to our new folder **DRALCD** that is created in step 1 above. Enter **DRALCD** into the **File name** box and click on the **Save** button to create this project
3. Select the correct processor core **MSP432P401R** from the next wizard and click on the **OK** to close this wizard
4. Next the software components wizard is opened, and you need to set up the software development environment for your project with this wizard. Expand two

icons, **CMSIS** and **Device**, and check the **CORE** and **Startup** checkboxes in the **Sel.** column, and click on the **OK** button since we need these two components to build our project

Since this project is a little complex, therefore both a header file and a C file are needed.

8.4.8.4.2 Create the header file DRALCD.h Create a new header file named **DRALCD.h** and enter the code shown in Figure 8.19 into this header file. Let's have a closer look at this header file to see how it works.

1. Three system header files are included in lines 4~6.
2. The codes in lines 7~9 are used to define three LCD commands as constants. These LCD commands, **RS**, **EN**, and **BL** are defined based on their bit values on the output of the serial shift register 74VHCT595. The **RS** is bit-0 (**Q0**) whose value is 1, the **EN** is bit-1 (**Q1**) whose value is 2, and the **BL** is bit-2 (**Q2**) whose value is 4 (Figure 8.10).
3. The codes in lines 10~16 are used to declare seven user-defined subroutines to perform different functions to be used in the project.

8.4.8.4.3 Create the C source file DRALCD.c Create a new C file named **DRALCD.c** and enter the first part code shown in Figure 8.20 into this C file. Let's have a closer look at this first part source file to see how it works.

1. The user-defined header file **DRALCD.h** is first included in this source file in line 4.
2. Before we can do any writing to the LCD, first we need to reset and initialize the LCD via the LCD controller SPLC780. A user-defined subroutine **LCD_Init()** is used for this purpose in line 8.
3. Then another user-defined subroutine **LCD_Comd(1)** is called to clear the LCD and set the cursor at the home position on the LCD panel in line 9.
4. The codes between lines 10 and 13 are used to write some letters, such as **WELCOME TO JCSU!**, to the LCD panel by calling a user-defined subroutine **LCD_Data()**.

```
1   //*****************************************************************************
2   // DRALCD.h - Header Files for the LCD Project - DRALCD
3   //*****************************************************************************
4   #include <stdint.h>
5   #include <stdbool.h>
6   #include <msp.h>
7   #define RS    1                          //74VHCT595 Q0 bit for RS (Reg Select)
8   #define EN    2                          //74VHCT595 Q1 bit for E (Enable LCD)
9   #define BL    4                          //74VHCT595 Q2 bit for BL (Backlight)
10  void delay_ms(int time);
11  void delay_us(int time);
12  void LCD_cd_Write(char data, unsigned char control);
13  void LCD_Comd(unsigned char cmd);
14  void LCD_Data(char data);
15  void LCD_Init(void);
16  void SPIB_Write(unsigned char data);
```

Figure 8.19 Codes for the header file DRALCD.h.

```
1   //***************************************************************************************
2   // DRALCD.c - Main Application File for LCD Project - The First Part Codes
3   //***************************************************************************************
4   #include "DRALCD.h"
5   int main(void)
6   {
7      WDTCTL = WDTPW | WDTHOLD;      // Stop watchdog timer
8      LCD_Init();                    // initialize LCD controller
9      LCD_Comd(1);                   // clear screen, move cursor to home
10     // Write "WELCOME TO JCSU!" on LCD
11     LCD_Data('W'); LCD_Data('E'); LCD_Data('L'); LCD_Data('C'); LCD_Data('O');
12     LCD_Data('M'); LCD_Data('E'); LCD_Data(' '); LCD_Data('T'); LCD_Data('O');
13     LCD_Data(' '); LCD_Data('J'); LCD_Data('C'); LCD_Data('S'); LCD_Data('U'); ;LCD_Data('!');
14  }
15
16  void LCD_Init(void)               // initialize eUSCI_B0 then initialize LCD controller
17  {
18     P1SEL0 |= BIT5|BIT6|BIT7;      // set SPI 3-pin P1.5, P1.6 & P1.7 as second function
19     P6DIR |= BIT7;                 // set P6.7 as output pin
20     UCB0CTLW0 |= UCSWRST;          // set UCSWRST = 1 to reset eUSCI_B0 module
21     // 3-pin master mode, 8-bit data length, clock polarity high, MSB first
22     UCB0CTLW0 |= UCMST|UCSYNC|UCCKPL|UCMSB;
23     UCB0CTLW0 |= UCSSEL__ACLK;     // clock source = ACLK
24     UCB0BRW = 0x01;                // fBitClock = fBRCLK/(UCBRx+1)= fBRCLK/2
25     UCB0CTLW0 &= ~UCSWRST;         // reset UCSWRST to enable USCIB0 to begin normal work
26     delay_ms(20);                  // LCD controller reset sequence
27     LCD_cd_Write(0x30, 0);         // send reset code 1 two times to SPLC780
28     delay_ms(5);
29     LCD_cd_Write(0x30, 0);
30     delay_ms(1);
31     LCD_cd_Write(0x30, 0);         // send reset code 2 to SPLC780
32     delay_ms(1);
33     LCD_cd_Write(0x20, 0);         // use 4-bit data mode
34     delay_ms(1);
35     LCD_Comd(0x28);                // set 4-bit data, 2-line, 5x7 font
36     LCD_Comd(0x06);                // move cursor right
37     LCD_Comd(0x0C);                // turn on display, cursor off - no blinking
38     LCD_Comd(0x01);                // clear screen, move cursor to home
39  }
40  void SPIB_Write(unsigned char data)
41  {
42     P6OUT = 0;                     // clear STCP (CS_LCD) in 74VHCT595 to Low (P6.7)
43     UCB0TXBUF = data;              // write serial data into 74VHCT595
44     while (UCB0STATW & UCBUSY);    // wait for 74VHCT595 serial data shift done
45     P6OUT |= BIT7;                 // set CS_LCD (STCP) to High to simulate a
                                         positive-going-edge
46  }
47  void LCD_cd_Write(char data, unsigned char control)
48  {
49     data &= 0xF0;                            // clear lower nibble for data
50     control &= 0x0F;                         // clear upper nibble for control
51     SPIB_Write (data | control | BL);        // RS = 0, R/W = 0
52     SPIB_Write (data | control | EN | BL);   // pulse E
53     delay_ms(0);
54     SPIB_Write (data | BL);
55     SPIB_Write (BL);
56  }
```

Figure 8.20 First part source codes for the project DRALCD.

5. The user-defined subroutine **LCD_Init()** starts at line 16.
6. As we discussed in Section 8.4.6.5, to initialize the **eUSCI_B0** module, first we need to configure the related GPIO pins, **P1.6**, **P1.7**, and **P1.5** as **eUSCI_B0** function pins, such as **UCB0SIMO**, **UCB0SOMI**, and **UCB0CLK**. This can be done by setting bits 5~7 in the **P1SEL0** Register in line 18. Since by default, all bits in the **P1SEL1** Register are 0 after a system reset, and this makes **P1SEL1.x** and **P1SEL0.x** equal to **01** (**x** = 5~7) to select **eUSCI** mode.
7. In line 19, the GPIO pin **P6.7** is configured as an output pin since we need it to output a rising-edge signal **STCP** to trigger 74VHCT595 to enable all 8-bit serial data stored in eight serial flip-flops to be converted to 8-bit parallel data and stored in the output latches.
8. The codes between lines 21 and 24 are used to configure the **eUSCI_B0** module to enable it to work as an SPI mode to transmit control commands and data to the SPLC780 LCD controller via the shift register 74VHCT595.
9. In line 20, the **eUSCI_B0** module is first reset by setting the **UCSWRST** bit in the **UCB0CTLW0** Register to enable this module to be configured and initialized.
10. The **eUSCI_B0** module is then configured by setting the related bits in the **UCB0CTLW0** Register with different system macros in line 22:
    a. **UCMST** = 3-pin master mode
    b. **UCSYNC** = synchronous (SPI) mode
    c. **UCCKPL** = the inactive state of the SCLK is high
    d. **UCMSB** = MSB is to be transmitted first
11. In line 23, the bit field **UCSSELx** in the **UCB0CTLW0** Register is configured to select the **ACLK** as the CKSE for the data transitions.
12. The **UCB0BRW** register is configured in line 24 by setting 1 as a prescaler to divide the **ACLK** by 2 to get the bit clock.
13. After the **eUSCI_B0** module is initialized, the **UCSWRST** bit in the **UCB0CTLW0** Register is reset in line 25 to enable the **eUSCI_B0** module to be ready for its normal data transfer.
14. The codes in lines 26~32 are used to reset the LCD controller by sending two reset codes, reset code 1 and reset code 2, respectively. Refer to Section 8.4.8.3.3.1 and Figure 8.16, the LCD reset process needs two reset codes. This reset job is fulfilled by calling a user-defined subroutine **LCD_cd_Write()**. This **cd** means that both LCD commands and data can be processed by using this function. The first argument of this subroutine is an LCD data and the second is an LCD command. These two arguments will be broken and then recombined together to be sent out via that subroutine.
15. In line 33, this subroutine is called again to transfer **0x20** to the LCD controller to configure LCD to use 4-bit data mode. Refer to step 4 in Figure 8.16 in Section 8.4.8.3.3.1 to get a clearer picture for this configuration.
16. The codes in lines 35~38 are used to set function mode, select display mode, set entry mode, and clear the display for the LCD. Refer to Section 8.4.8.3.3.2 and Figure 8.18 to get the meaning for each coding line in this part. A user-defined subroutine **LCD_Comd()** is used to process and transfer these commands to the LCD controller one by one. Each command is first broken into two pieces or two 4-bit commands; each piece is combined with the **RS**, **E**, and **BL** control bits together to form a final 8-bit command to be sent to the LCD controller via the serial shift register 74VHCT595.

17. The codes in lines 40~46 are used to perform writing an 8-bit serial data via **UCB0SIMO** pin to the SPLC780 via 74VHTC595. A user-defined subroutine **SPIB_Write()** is used to fulfill this function.
18. In line 42, a 0 is sent to the GPIO pin **P6.7** to clear **STCP** to make it low via **CS_LCD** signal. Then the transmitted data are sent to the **UCB0SIMO** pin via **UCB0TXBUF** register in line 43.
19. In line 44, a conditional **while()** loop is used to wait for this serial data transmission to be completed. This can be done by monitoring the **UCBUSY** bit (bit-0) in the **UCB0STATW** register (**UCB0STATW & UCBUSY**). When the data transmission is done, this bit is cleared to 0. As soon as this happens, it means that all 8-bit serial data have been shifted into the 8-bit flip-flop registers in the 74VHCT595. Then in line 45, a macro **BIT7** is sent to the GPIO **P6** to make pin **P6.7** to high and to set the **STCP** to high to get a positive-going edge pulse to trigger the transferring of 8-bit data in the serial shift flip-flops into the 8-bit parallel output registers, and furthermore to output to the LCD controller SPLC780 via Q4~Q7 (**DB7~DB4**) and Q0~Q2 (**RS**, **E**, and **BL**).
20. The user-defined subroutine **LCD_cd_Write()** starts at line 47. Two arguments are involved in this subroutine, the first one is the LCD data and the second is the LCD command.
21. The codes in lines 49~50 are used to get the upper 4-bit data and the lower 4-bit control from two arguments by **ANDing 0xF0** and **0x0F**, respectively.
22. The codes in lines 51~55 are used to combine each piece of data with related control commands, such as **RS**, **E**, and **BL**, to form an 8-bit serial data and then send it out via **UCB0SIMO** pin. An **OR** operator is used to combine data and commands together. These codes are also used to create a simulated low-to-high pulse for the **E** control signal to enable the LCD controller SPLC780 to get that serial data (see Figure 8.10).

Now, let's take care of the second part of the source code, which is shown in Figure 8.21.

23. Another user-defined subroutine **LCD_Comd()** starts at line 60. This subroutine is used to break the input command into two pieces of 4-bit command and combine with the related control signal, **RS**, **E**, and **BL** to form an 8-bit serial data to be sent out via the **LCD_cd_Write()** subroutine in two times.
24. In line 64, if the command is <4, which means that if the command is 1 (clear display), 2 (set function), or 3 (return home), delay those commands by 2 ms. Otherwise delay 1 ms for all other commands. Refer to steps 2 and 3 in Table 8.9 to get more details for those LCD commands.
25. The codes in lines 69~74 are used to transfer the LCD data. This transmission is completed by calling a user-defined subroutine **LCD_Data()**. The LCD data to be transferred is first broken into two pieces of 4-bit data, the upper 4 bit and the lower 4 bit. Then each piece of data is combined with the **RS** command to form an 8-bit serial data to be sent via the **LCD_cd_Write()** subroutine. Refer to Section 8.4.8.3.1 to get more details about the **RS** command and its function (**RS** = 1 means that LCD data are transmitted).
26. The codes in lines 75~86 are used to build two time delay subroutines to delay certain ms and μs to make the LCD controller work properly.

Now, let's set up the environment to build and run our project to test the LCD functions.

```
57  //**************************************************************************
58  // DRALCD.c - Main Application File for LCD Project - The Second Part Codes
59  //**************************************************************************
60  void LCD_Comd(unsigned char cmd)
61  {
62    LCD_cd_Write(cmd & 0xF0, 0);      // upper nibble first
63    LCD_cd_Write(cmd << 4, 0);        // then lower nibble
64    if (cmd < 4)
65        delay_ms(2);                  // command 1 and 2 needs up to 1.64ms
66    else
67        delay_ms(1);                  // all others 40 µs
68  }
69  void LCD_Data(char data)
70  {
71    LCD_cd_Write(data & 0xF0, RS);    // upper nibble first
72    LCD_cd_Write(data << 4, RS);      // then lower nibble
73    delay_us(40);
74  }
75  void delay_ms(int time)             // delay n milliseconds (16 MHz CPU clock)
76  {
77    int i, j;
78    for(i = 0 ; i < time; i++)
79        for(j = 0; j < 3180; j++) {}  // do nothing for 1 ms
80  }
81  void delay_us(int time)             // delay n microseconds (16 MHz CPU clock)
82  {
83    int i, j;
84    for(i = 0 ; i < time; i++)
85        for(j = 0; j < 3; j++) {}     // do nothing for 1 µs
86  }
```

Figure 8.21 Second part source code for the project DRALCD.

8.4.8.4.4 Set up the environment to build and run the project Perform the following operations to set up the environment for this project:

- Select the **CMSIS-DAP Debugger** in the **Debug** tab under the **Project|Options for Target "Target 1"** menu item. Make sure that all settings for the debugger and the flash download are correct.

Now build the project by going to **Project|Build target** menu item. If everything is fine, go to **Flash|Download** item to download the image file of the project into the flash memory. Then go to **Debug|Start/Stop Debug Session** and **Debug|Run** item to run the project.

After the project running, you can find that the letter sequence: **WELCOME TO JCSU!** is displayed in the LCD panel. Our project is successful!

Can we display something on the second line in this LCD with the desired starting position? Yes, but you need to refer to step 7 in Table 8.9 and Figure 8.13 to figure out the coding process. Figure 8.22 shows the relationship between the LCD instruction and the DDRAM address.

You can use the subroutine **LCD_Comd(0xC0)** to select the first position on the second line in the LCD. How about other positions? Think about it.

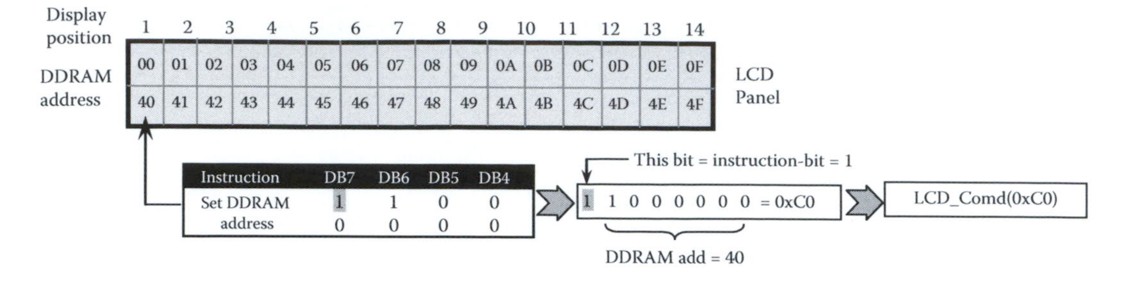

Figure 8.22 Relationship between the LCD instruction and the DDRAM address.

8.4.9 Build on-board seven-segment LED interface programming project

In this section, we illustrate how to use the **eUSCI_B0** module to interface four on-board seven-segment LEDs installed in the EduBASE ARM® Trainer. First let's have a closer look at the hardware configuration and connection for these LEDs via **eUSCI_B0** module.

In order to enable us to develop our programs to control and display our desired data on these four seven-segment LEDs, let's have a clear picture about the hardware configuration for these four seven-segment LEDs and their control functions.

8.4.9.1 Structure of seven-segment LEDs

Figure 8.23a shows a one-digit seven-segment common cathode LED and its structure.

To make any segment, such as **a**, **b**, **c**, and so on to light or ON, just apply logic HIGH to the associated input bit D_i since the cathode terminals of all of these seven-segment LEDs are connected together to the ground. In the EduBASE ARM® Trainer, these 8 input bits (D_0~D_7) are connected to 8 latched output bits (**Q0**~**Q7**) on the 74VHCT595. Figure 8.23b shows the layout of a single bit on seven-segment LED.

8.4.9.2 USCIB0 module interface to the seven-segment LED in the EduBASE ARM® trainer

Similar to LCD interface, four seven-segment LEDs are connected to the **eUSCI_B0** module via two cascaded serial shift registers 74VHCT595. A functional block diagram of this interface circuit connected to the LED installed in the EduBASE ARM® Trainer is shown in Figure 8.24.

It can be found from Figure 8.24 that two serial shift registers 74VHCT595 are cascaded or serial connected together to form two 8-bit parallel outputs to perform the LED digit

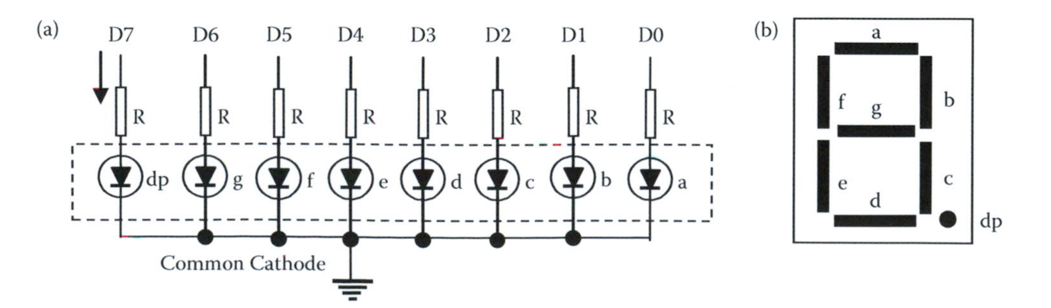

Figure 8.23 Structure of common cathode LED.

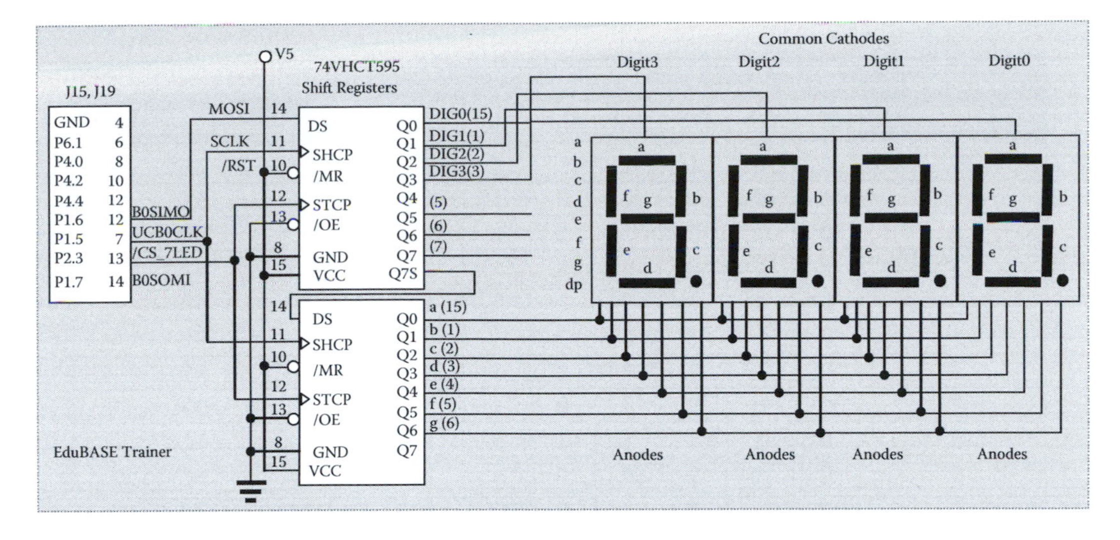

Figure 8.24 Hardware configuration of 74VHCT595 and seven-segment LEDs.

selection and seven-segment selection. The serial input and controls to the 74VHCT595 are still from **eUSCI_B0** module control pins (**P1.6** = **UCB0SIMO**, **P1.5** = **UCB0CLK**). The GPIO pin **P2.3** now works as a general digital function pin and it is used to provide the slave selection and triggering signal to convert eight serial data into eight parallel data in the 74VHCT595 when a positive-going edge signal is provided by this pin.

When working this mode, the first serial 8-bit data to be transferred by **eUSCI_B0** module should be seven-segment control signal used to select the required segments to display desired number on all four LEDs. The second serial 8-bit data (only the lower 4 bits are used) should be used to select the desired LED digit.

In order to display desired numbers or letters on these seven-segment LEDs, a code conversion job is necessary to translate the binary or hexadecimal codes output from the 74VHCT595 to the seven-segment codes to be displayed on those LEDs. Table 8.10 shows the relationship between these code translations. In the real program, a subroutine can be developed and used to perform this coding conversion before the number or letter can be displayed on these LEDs.

In addition to converting the data sent to the 74VHCT595 form binary or hexadecimal codes to seven-segment codes, an LED digit selection code should also be sent to the 74VHCT595 to select the desired LED digit. Table 8.11 shows a relationship between the codes to be sent to the 74VHCT595 and the selected LED digit.

For example, to display a number **0** in the first digit of seven-segment LED, **Digit0**, the following data should be created and sent to the 74VHCT595, respectively:

- A **0x3F** or **00111111B** should be sent to the second 74VHCT595 to make **Q7~Q0** = **00111111** on the second 74VHCT595. This will make segments (anodes) **a**, **b**, **c**, **d**, **e**, and **f** ON, and segments **g** and **dp** OFF.
- A **0xFE** or **11111110B** should be sent to the first 74VHCT595 to make **Q3~Q0** = **1110**. This will make the cathode of the **Digit0** to low to select this digit (the higher 4 bit can be any values). The key is: a low or 0 must be sent to the cathode of the selected LED digit to enable it to be ON to display it.

Now, let's build an example project to access and interface these seven-segment LEDs.

Table 8.10 Relation between the 7-segment data code and normal code

Number/ letter	dp	g	f	e	d	c	b	a	Seven-segment code
	Q7	Q6	Q5	Q4	Q3	Q2	Q1	Q0	
0	0	0	1	1	1	1	1	1	3F
1	0	0	0	0	0	1	1	0	06
2	0	1	0	1	1	0	1	1	5B
3	0	1	0	0	1	1	1	1	4F
4	0	1	1	0	0	1	1	0	66
5	0	1	1	0	1	1	0	1	6D
6	0	1	1	1	1	1	0	1	7D
7	0	0	0	0	0	1	1	1	07
8	0	1	1	1	1	1	1	1	7F
9	0	1	1	0	1	1	1	1	6F
A	0	1	1	1	0	1	1	1	77
b	0	1	1	1	1	1	0	0	7C
C	0	0	1	1	1	0	0	1	39
d	0	1	0	1	1	1	1	0	5E
E	0	1	1	1	1	0	0	1	79
F	0	1	1	1	0	0	0	1	71
H	0	1	1	1	0	1	1	0	76
h	0	1	1	1	0	1	0	0	74
J	0	0	0	1	1	1	1	0	1E
L	0	0	1	1	1	0	0	0	38
S	0	1	1	0	1	1	0	1	6D
P	0	1	1	1	0	0	1	1	73
U	0	0	1	1	1	1	1	0	3E

Table 8.11 Codes sent to the second 74VHCT595 and the selected LED digits

Qn CODE / LED digits	Q7	Q6	Q5	Q4	Q3	Q2	Q1	Q0	HEX CODE
DIGIT0	X	X	X	X	1	1	1	0	$FE
DIGIT1	X	X	X	X	1	1	0	1	$FD
DIGIT2	X	X	X	X	1	0	1	1	$FB
DIGIT3	X	X	X	X	0	1	1	1	$F7

8.4.9.3 Build an example LED interfacing project

Now we have enough knowledge and skill to build a practical LED interfacing project to serially transmit our desired letters or numbers to the LED to display them on it.

The hardware configuration of this project is shown in Figure 8.24. In the EduBASE ARM® Trainer, it has been configured to use the **`eUSCI_B0`** module to serially transmit instructions and data to LEDs via two cascaded serial shift registers 74VHTC595. The GPIO pins used in this project include:

- **`P1.6`**: **`UCB0SIMO`**—transmits instructions/data to the peripheral 74VHCT595
- **`P1.5`**: **`UCB0CLK`**—provides the SCLK to conduct the serial data transmission

- **P1.7**: **UCB0SOMI**—receives serial data from the peripheral (not used in this example)
- **P2.3**: **CS_7LED**—provides the LED chip select signal. Here it works as a trigger clock to start a serial-to-parallel data conversion to convert 8-bit serial data stored in the serial shift register to 8-bit parallel data to be stored in the 8-bit output registers in the 74VHCT595 register

In our example application, each time after **eUSCI_B0** module transmits 8 bits data to the shift register via **DS** (**UCB0SIMO-P1.6**) pin and clock **SHCP** (**UCB0CLK-P1.5**) pin, a positive-going edge (low-to-high) signal should be applied on the **STCP** (**CS_7LED-P2.3**) pin to transfer 8-bit serial data in the shift register into the 8-bit parallel output buffer or latches. The **OE** pin is connected to the ground to enable eight latched outputs to be directly transferred to the output pins (Figure 8.24).

We try to use the DRA model to build this project. The entire coding process includes the following four sections:

1. Initialize and configure **eUSCI_B0** module-related GPIO pins (**P1.6**, **P1.5**) and **P2.3**
2. Initialize and configure **eUSCI_B0** module-related registers (**UCB0CTLW0** and **UCB0BRW**)
3. Build one subroutine to break, combine, and transmit 8-bit instructions/data to the LED to display desired letters and numbers on the selected LEDs
4. Build some time delay functions to delay different periods of time to make the LED data transmissions stable and reliable

Now, let's first create a new LED project **DRALED**.

8.4.9.3.1 Create a DRA LED project DRALED Perform the following operations to create a new project **DRALED**:

1. Open the Windows Explorer window to create a new folder named **DRALED** under the **C:\MSP432 Class Projects\Chapter 8** folder
2. Open the Keil® ARM MDK µVersion5 IDE, create a new µVersion5 project **DRALED** and save it to our new folder **DRALED** that is created in step 1 above
3. On the next wizard, you need to select the correct MCU **MSP432P401R**. Click on the **OK** to close this wizard
4. Next the software components wizard is opened, and you need to set up the software development environment for your project with this wizard. Expand two icons, **CMSIS** and **Device**, and check the **CORE** and **Startup** checkboxes in the **Sel.** column, and click on the **OK** button since we need these two components to build our project

Since this project is very simple, therefore only a C source file is good enough.

8.4.9.3.2 Create the C source file DRALED.c Create a new C file named **DRALED.c** and enter the code shown in Figure 8.25 into this C file. Let's have a closer look at this source file to see how it works.

1. Three system header files, **stdint.h**, **stdbool.h**, and **msp.h**, are included at the top three coding lines, lines 4~6, since all system macros we will use in this project are included in those files.
2. Three user-defined subroutines, **delay()**, **LED_Init()**, and **SPIB_Write()**, are declared at coding lines 7~9.

```
1  //***************************************************************************************
2  // DRALED.c - Main Application File for the LED Class Project - DRALED
3  //***************************************************************************************
4  #include <stdint.h>
5  #include <stdbool.h>
6  #include <msp.h>
7  void delay(int time);
8  void LED_Init(void);
9  void SPIB_Write(unsigned char data);
10 int main(void)
11 {
12    uint32_t dtime = 2;
13    WDTCTL = WDTPW | WDTHOLD;          // Stop watchdog timer
14    LED_Init();                        // initialize UCB0 that connects to the shift registers
15    while(1)
16    {
17     SPIB_Write(0x5B);                 // write num 2 to the seven segments
18     SPIB_Write(0xF7);                 // select digit 3
19     delay(dtime);
20     SPIB_Write(0x3F);                 // write num 0 to the seven segments
21     SPIB_Write(0xFB);                 // select digit 2
22     delay(dtime);
23     SPIB_Write(0x06);                 // write num 1 to the seven segments
24     SPIB_Write(0xFD);                 // select digit 1
25     delay(dtime);
26     SPIB_Write(0x7D);                 // write num 6 to the seven segments
27     SPIB_Write(0xFE);                 // select digit 0
28     delay(dtime);
29    }
30 }
31 void LED_Init(void)                   // configure UCB0 and associated GPIO pins
32 {
33    P1SEL0 |= BIT5|BIT6|BIT7;          // set SPI 3-pin P1.5, P1.6 & P1.7 as second function
34    P2DIR = BIT3;                      // set P2.3 as output pin
35
36    UCB0CTLW0 |= UCSWRST;              // set UCSWRST = 1 to reset eUSCI_B0 module
37    UCB0CTLW0 |= UCMST|UCSYNC|UCCKPL|UCMSB;    // 3-pin master mode, 8-bit data length
38                                               // clock polarity high, MSB first
39    UCB0CTLW0 |= UCSSEL__ACLK;         // clock source = ACLK
40    UCB0BRW = 0x01;                    // fBitClock = fBRCLK/(UCBRx+1)= fBRCLK/2
41    UCB0CTLW0 &= ~UCSWRST;             // reset UCSWRST to enable USCIB0 to begin normal work
42 }
43 void SPIB_Write(unsigned char data)
44 {
45    P2OUT = 0;                         // clear STCP (CS_LED) in 74VHCT595 to Low (P2.3)
46    UCB0TXBUF = data;                  // write serial data into 74VHCT595
47    while (UCB0STATW & UCBUSY);        // wait for 74VHCT595 serial data shift done
48    P2OUT |= BIT3;                     // set CS_LED (STCP) to High to simulate a
                                         positive-going-edge
49 }
50 void delay(int time)                  // delay some time
51 {
52    int i, j;
53    for(i = 0; i < time; i++)
54     for(j = 0; j < 100; j++) {}
55 }
```

Figure 8.25 C source file for the DRALED project.

3. A local variable **dtime** is declared at line 12 and it is used as a time delay constant to delay the program by certain period of time.
4. The Watchdog timer is temporarily terminated in line 13.
5. The **LED_Init()** subroutine is called to initialize and configure the **eUSCI_B0** module-related registers and GPIO pins to enable **P1.6** and **P1.5** pins to work as **UCB0SIMO** and **UCB0CLK** pins in line 14.
6. An infinitive **while()** loop is used to repeatedly send desired numbers to the LEDs starting at line 15.
7. The **SPIB_Write()** subroutine is called to transmit a number **2** (**0x5B**) to the third LED (Digit3) in lines 17 and 18 to display it.
8. In line 19, the user-defined subroutine **delay()** is called to delay the program by **dtime** period.
9. Similarly, the codes in lines 20~27 are used to transmit numbers **0**, **1**, and **6** to the second, first, and zeroth LEDs, respectively, by using the **SPIB_Write()** subroutine.
10. The subroutine **LED_Init()** starts at line 31. The codes for this subroutine are identical with those on the subroutine **LCD_Init()** we discussed in the last project **DRALCD**. The only difference is that the GPIO pin **P2.3** (**CS_7LED**) is used to replace pin **P6.7** used in the last project to provide a triggering signal for the 74VHCT595 to start the serial-to-parallel conversion.
11. The codes in lines 43~49 are built for the subroutine **SPIB_Write()**. In line 45, the **P2.3** is reset to low to select the slave device 74VHCT595 and clear the **STCP** input signal.
12. The serial data are written into the **UCB0TXBUF** register and transmitted to the **UCB0SIMO** pin in line 46.
13. In line 47, a **while()** loop is used to check the **UCBUSY** bit in the **UCB0STATW** register to wait for this data transmission to be done.
14. As soon as that data are transmitted, the GPIO pin **P2.3** outputs a high to provide a positive-going edge signal to convert serial data to the parallel data and output to the LEDs device via 74VHCT595 in line 48.
15. The codes in lines 50~55 are used to build the user-defined subroutine **delay()** to delay certain period of time to make the LED displaying stable.

8.4.9.3.3 Set up the environment to build and run the project Perform the following operations to set up the environment for this project:

- Select the correct debugger **CMSIS-DAP Debugger** from the **Debug** tab under the **Project|Options for Target "Target 1"** menu item
- Make sure that all settings for the debugger and the flash download are correct

Now build the project by going to **Project|Build target** menu item. If everything is fine, go to **Flash|Download** item to download the image file of the project into the flash memory. Then go to **Debug|Start/Stop Debug Session** and **Debug|Run** item to run the project.

After the project running, you can find that four numbers: **2016** are displayed in four LEDs. Our project is successful!

We leave a similar LED project for the readers in the lab exercise part as the homework. The function of that project is to repeatedly display each segment of the LED on each digit.

Next, let's take care of the DAC Programming Project by using the EduBASE ARM Trainer.

8.4.10 Build DAC programming project

In this section, we illustrate how to use the **eUSCI_B0** module to interface to a DAC-MCP4922 installed in the EduBASE ARM® Trainer. First let's have a closer look at the hardware configuration and connection for this DAC via **eUSCI_B0** module.

8.4.10.1 eUSCI_B0 module interface to the MCP4922 in the EduBASE ARM® trainer

Different with the LCD and LED interfaces, the DAC-MCP4922 is directly connected to the **eUSCI_B0** module without using any serial shift register. This is because the DAC-MCP4922 provides a serial data input. A functional block diagram of this interface circuit connected to the DAC-MCP4922 installed in the EduBASE ARM® Trainer is shown in Figure 8.26.

It can be found from Figure 8.26 that two **eUSCI_B0** module control signals, **UCB0SIMO** and **UCB0CLK**, are connected to the DAC-MCP4922 to control the DAC conversions and serial data transmission. Optionally, a serial ADC-TLC548 can be connected to the output of the **DAC0** to receive and check the DAC transferring results. An oscilloscope can also be connected to **DAC0** (**J32-1**) and **DAC1** (**J32-3**) pins to monitor and check the converted analog waveform outputs.

Now, let's have a closer look at these two peripherals, the DAC-MCP4922 and the ADC-TLC548.

8.4.10.2 Operations and programming for MCP4922 DAC

The MCP4922 device is a dual 12-bit buffered voltage output DAC. The devices operate from a single 2.7~5.5 V supply with SPI compatible serial peripheral interface. This DAC is a programmable device and needs to be programmed to perform the desired DAC operations.

A functional block diagram of the MCP4922 DAC is shown in Figure 8.27.

The operational principle and sequence of this DAC are

- The **SDI** is the serial data input pin to accept serial digital data as input.
- The **SCK** terminal is used to accept an external clock and the DAC performs its conversions based on this clock as a timing base.
- The **/CS** is a chip select signal used to start the DAC operations. Both DAC modules use the same CKSE with the same starting signal. The active level is low.

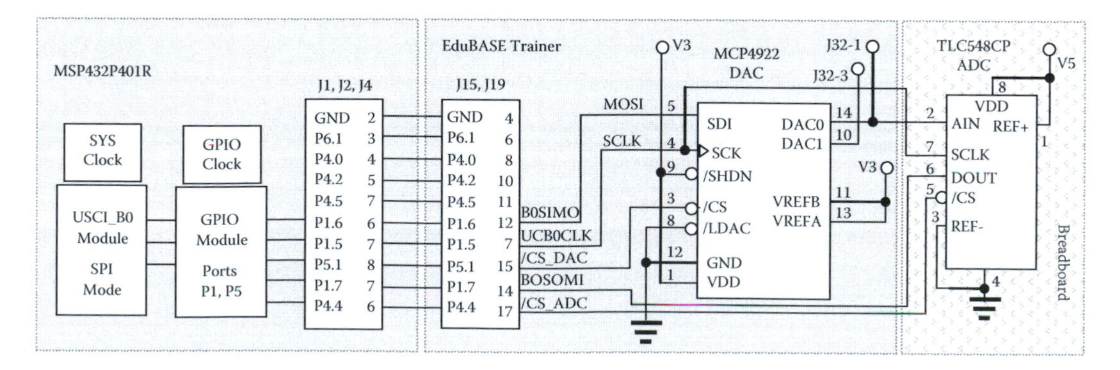

Figure 8.26 Hardware configuration of the DAC-MCP4492 and eUSCI_B0 module.

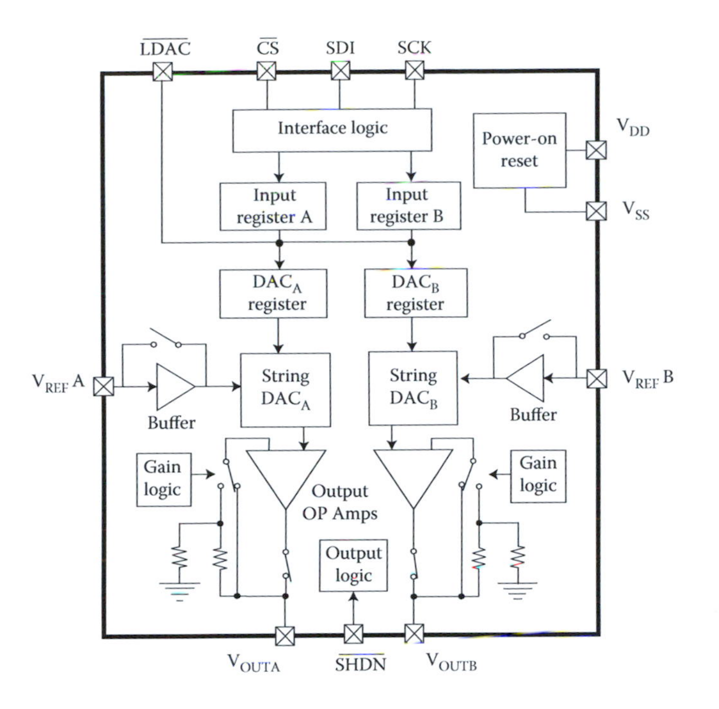

Figure 8.27 Functional block diagram of the DAC-MCP4922.

- The **/LDAC** signal is used to enable both buffered conversion results to be updated in two DAC outputs. The active level of this signal is low.
- The dual DAC can be shut down for their outputs if the **/SHDN** signal is low.
- The dual DAC outputs can be obtained from $\mathbf{V}_{OUTA}$ and $\mathbf{V}_{OUTB}$ pins in serial format. Both outputs can be ranged up to $\mathbf{V}_{REF}$ (× 1 = 2.048 V) or doubled $\mathbf{V}_{REF}$ (× 2 = 4.096 V).

In the EduBASE ARM® Trainer, the **/LDAC** is connected to the ground and this is equivalent to setting the **/LDAC** pin to low. This connection enables both outputs to be updated as a rising edge pulse appeared on the **/CS** signal (Figure 8.26).

The **/SHDN** pin is connected to the power supply **V3**, and this is equivalent to setting this pin to high. This configuration enables both DAC outputs to be available at any time. Both reference pins, $\mathbf{V}_{REFA}$ and $\mathbf{V}_{REFB}$, are tied to the power supply **V3**. This connection makes both DAC use the same reference voltage.

To configure the MCP4922 to perform any DAC operation, the DAC needs first to be programmed. Each programming instruction is 16-bit long and comprises commands part (4 MSB bits) and data part (following 12 bits).

The write command is initiated by driving the **CS** pin to low, followed by clocking the 4-MSB configuration bits and the 12 data bits into the **SDI** pin on the rising edge of **SCK**. The **CS** pin is then raised to high to cause the data to be latched into the selected DAC input registers. The MCP4922 utilizes a double-buffered latch structure to allow both **DACA** and **DACB** outputs to be synchronized with the **LDAC** pin if desired. Upon the **LDAC** pin achieving a low state, the values held in the DAC input registers are transferred into the DAC output registers. The outputs will transition to the value and held in the **DACA** (**DAC0**) or **DACB** (**DAC1**) register.

All writes to the MCP4922 are 16-bit instructions. Any clocks past the 16th clock will be ignored. The most significant 4 bits are configuration bits. The remaining 12 bits are

$\overline{A}$/B	BUF	$\overline{GA}$	$\overline{SHDN}$	D11	D10	D9	D8	D7	D6	D5	D4	D3	D2	D1	D0
bit-15															bit-0

Figure 8.28 Example of the command data structure for the MCP4922 DAC.

data bits. No data can be transferred into the device with **CS** high. Figure 8.28 shows an example of the command data structure to be written into the MCP4922 DAC.

As we mentioned, the 4 MSBs are configuration or command bits. The values for these configuration bits and related functions are shown in Table 8.12.

For example, if we want to configure the MCP4922 to transfer a 12-bit serial data with the following configurations:

- Select the **DACA** (**DAC0**) channel as the converting channel
- Enable the buffered input and output
- Select the output gain as 1
- Do not shut down the DAC channel to enable it active

The configuration bits should be: **0x7xxx** = **0111xxxxxxxxxxxx**. The lower 12 data bits are **x**, which means that they do not matter and the real values depend on the application data.

8.4.10.3 Analog-to-digital converter TLC-548

The TLC-548 is an 8-bit serial-in serial-out ADC with 3-wire serial interface, such as SPI and SSI interfaces. The conversion results of the DAC can be feed into the TLC-548 serial ADC input via pin2 (**AIN**). The **CS_ADC** (**P4.4**) provides a chip select (**/CS**) signal to start the ADC conversion and the ADC conversion process is controlled by the **UCB0CLK** clock signal in a timing sequence. The ADC conversion results can be received by the **UCB0SOMI** pin (**P1.7**) since the ADC output (**DOUT**) is connected to the **UCB0SOMI** pin.

The reference pin (**REF+**) in the TLC-548 is connected to the power supply and **REF-** is connected to the ground. With this configuration, the ADC used the internal reference to allow the output up to be 4.096 V. Figure 8.29 shows the functional block diagram of the ADC-TLC548.

Table 8.12 Bit value and its function for MCP4922 DAC

Bit	Name	Function
15	**$\overline{A}$/B**	DACA or DACB selection bit **0:** Write to DACA (DAC0); **1:** Write to DACB (DAC1)
14	**BUF**	VREF input buffer control bit **0:** Unbuffered **1:** Buffered
13	**$\overline{GA}$**	Output gain selection bit **0:** 2x (VOUT = 2*VREF*D/4096) **1:** 1x (VOUT = VREF*D/4096)
12	**$\overline{SHDN}$**	Output shutdown control bit **0:** Shutdown the selected DAC channel. Analog output is not available at the channel that was shut down. VOUT pin is connected to 500 k **1:** Active mode operation. VOUT is available
11:0	**D11:D0**	DAC input data bits

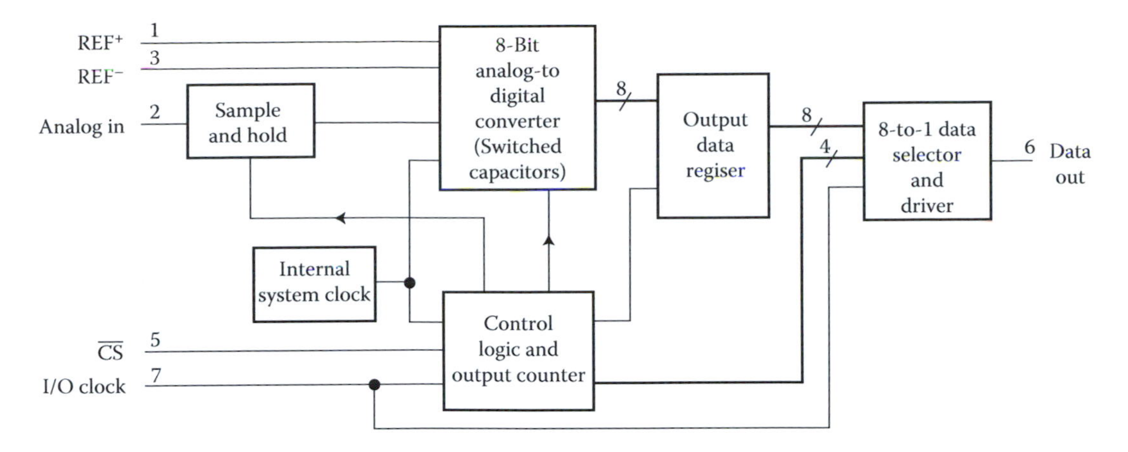

Figure 8.29 Functional block diagram of the ADC-TLC548.

The serial 8-bit output can be obtained from **DOUT** pin (pin 6) on the TLC548, which is controlled by the input clock sequence **SCLK**. The data can be shifted out bit by bit as each falling edge of the **SCLK** is coming.

Now, let's build an example project to access and interface to the DAC-MCP4922.

8.4.10.4 Build the example DAC interfacing project

The hardware configuration of this project is shown in Figure 8.26.

In the EduBASE ARM® Trainer, it has been configured to use the **eUSCI_B0** module to serially transmit digital data to the DAC-MCP4922. You can use an oscilloscope to monitor **DAC0** (**J32-1**) and **DAC1** (**J32-3**) pins on the EduBASE Trainer to check the analog out results. The jumper connector **J32** is located at the lower right corner of the EduBASE ARM Trainer.

Optionally, you can build the ADC part by using the TLC548 with the breadboard provided by the EduBASE ARM® Trainer. Refer to Figure 8.26 to finish this circuit building process. For the power supply **V5** and the ground signals, you can use **5V** and **GND** pins on the **J32** connector in the EduBASE ARM® Trainer, respectively.

The GPIO pins used in this project include:

- **P1.6**: **UCB0SIMO**—transmits digital data to the peripheral DAC-MCP4922
- **P1.5**: **UCB0CLK**—provides the SCLK to conduct the serial data transmission
- **P1.7**: **UCB0SOMI**—receives the serial analog data from the peripheral ADC-TLC548
- **P5.1**: **CS_DAC**—provides the DAC Chip Select signal. Here it works as a trigger signal to start the digital-to-analog conversion to convert 12-bit serial digital data to 12-bit serial analog data, which optionally can be sent to the serial input of the TLC548
- **P4.4**: **CS_ADC**—provides the ADC chip select signal. Here it works as a trigger signal to start the analog-to-digital conversion to convert 12-bit serial analog data back to 8-bit serial digital data, which optionally can be received by the **UCB0SOMI** pin

In our example application, each time after the **eUSCI_B0** module transmits 12 bits digital data to the MCP4922, a low-level signal should be applied on the **/CS_DAC** (**P5.1**) pin to start a DAC operation. The DAC conversion results can be obtained serially from the first conversion channel V_{OUTA} or **DAC0** as each rising edge of the **UCB0CLK** coming (Figure 8.26). While this serial analog output can be optionally sent to the serial input of the ADC-TLC548 to perform an analog-to-digital conversion to convert it back to the digital data. Then the digital data can be received by the **UCB0SOMI** pin to enable us to collect and then confirm the correctness of these transmission and conversions. A point to be noted is that before this ADC can start, a low-level signal must be applied on the **/CS_ADC** (**P4.4**) to enable the TLC548 to start its conversions.

We try to use the DRA model to build this project. The entire coding process includes the following four sections:

1. Initialize and configure **eUSCI_B0** module-related GPIO Ports (**P1**, **P5**, and **P4**)
2. Initialize and configure **eUSCI_B0** module and related registers (**UCB0CTLW0**, **UCB0BRW**, and **UCB0IE**)
3. Build a subroutine to perform data transmission from the **eUSCI_B0** module to the DAC-MCP4922 via **UCB0SIMO** pin
4. Build a subroutine to perform data receiving from the ADC-TLC548 to the **eUSCI_B0** module via **UCB0SOMI** pin (optional)

Now, let's first create a new DAC project **DRADAC**.

8.4.10.4.1 Create a DRA DAC project DRADAC Perform the following operations to create a new project **DRADAC**:

1. Open the Windows Explorer window to create a new folder **DRADAC** under the **C:\MSP432 Class Projects\Chapter 8** folder
2. Open the Keil® ARM MDK µVersion5 to create a new µVersion Project. On the opened wizard, browse to our new folder **DRADAC** that is created in step 1 above. Enter **DRADAC** into the **File name** box and click on the **Save** button to create this project
3. On the next wizard, you need to select the device (MCU) **MSP432P401R**
4. Next, the software components wizard is opened, and you need to set up the software development environment for your project with this wizard. Expand two icons, **CMSIS** and **Device**, and check the **CORE** and **Startup** checkboxes in the **Sel.** column, and click on the **OK** button since we need these two components to build our project

Since this project is a little longer, therefore a header file and a C source file are needed.

8.4.10.4.2 Create the header file DRADAC.h Create a new header file **DRADAC.h** and enter the code shown in Figure 8.30 into this file. Since the codes are easy and straightforward, no explanation is needed.

```
1  //*****************************************************************************
2  // DRADAC.h - Header File for DAC-MCP4922 (eUSCI_B0)
3  //*****************************************************************************
4  #include <stdint.h>
5  #include <stdbool.h>
6  #include <msp.h>
7  #include <math.h>

8  void DAC_Init(void);
9  void DAC_Write(uint16_t value, int channel);
10 void ADC_Read(void);
```

Figure 8.30 Header file for the project DRADAC.

8.4.10.4.3 Create the C source file DRADAC.c Create a new C file **DRADAC.c** and enter the code shown in Figure 8.31 into this C file. Let's have a closer look at this source file to see how it works.

1. The main program starts at line 5.
2. Two variables, the **uint16_t** variable **i_data** and **d_data**, are declared first in the main program in line 7. Both the **i_data** and the **d_data** have a range of 0~4095 and they are used to create an incremented and a decremented sawtooth waveform later.
3. In line 9, the **DAC_Init()** function is called to initialize the DAC-related registers and the **eUSCI_B0** module-related GPIO ports and pins.
4. An infinitive **while()** loop is used to repeatedly call the user-defined functions **DAC_Write()** to create two kinds of sawtooth waveforms in line 10. Optionally, you can read back the analog data coming from the ADC-TLC548.
5. In lines 12 and 13, the **DAC_Write()** functions are called to create an incremented and a decremented sawtooth waveform with the **i_data** as the argument. Each loop the **i_data** is incremented or decremented by 1 and sent to the DAC MCP4922 via **UCB0SIMO** pin to generate two sawtooth signals until the **i_data** gets its upper bound 4095 or lower bound 0. Then the **i_data** is reset to 0 and 4095 to restart its increment and decrement action from there. The second argument is the channel number, 0 means the DAC channel 0 or **DAC0** and 1 means the DAC channel 1 **DAC1**.
6. Optionally, one can get the ADC result by calling a user-defined function **ADC_Read()** from the TLC-548 in line 14.
7. The **DAC_Init()** function starts at line 17. This function is used to initialize and configure the **eUSCI_B0** module-related registers and GPIO pins to enable **P1.6**, **P1.5,** and **P1.7** pins to work as **UCB0SIMO**, **UCB0CLK**, and **UCB0SOMI** pins in line 19.
8. In lines 20 and 21, both **P5.1** and **P4.4** pins are configured as the output pins to trigger the **CS_DAC** and **CS_ADC** pins for both DAC-MCP4922 and ADC TLC-548 later.
9. In lines 22 and 23, the GPIO pins **P5.1** and **P4.4** are set to high to disable **CS_DAC** and **CS_ADC** two trigger signals to deselect DAC-MAP4922 and ADC TLC-548.
10. The code lines between 24 and 29 are used to configure the **eUSCI_B0** module by setting the related bits in the **UCB0CTLW0** Register with different system macros:
    a. **UCMST** = 3-pin master mode
    b. **UCSYNC** = synchronous (SPI) mode
    c. **UCCKPL** = inactive state of the SCLK is high
    d. **UCMSB** = MSB is to be transmitted first
11. In line 27, the bit field **UCSSELx** in the **UCB0CTLW0** Register is configured to select the **ACLK** as the CKSE for the data transitions.
12. The **UCB0BRW** register is configured in line 28 by setting 1 as a prescaler to divide the **ACLK** by 2 to get the bit clock.
13. After the **eUSCI_B0** module is initialized, the **UCSWRST** bit in the **UCB0CTLW0** Register is reset in line 29 to enable the **eUSCI_B0** module to be ready for its normal data transfer.
14. The function **DAC_Write()** starts at line 31.
15. Some local variables, **config**, **cmd_data**, and **data**, are declared in line 33 since we need to use them to reconfigure the higher 8-bit command data and lower 8-bit data. The reason for this is because the **eUSCI_B0** module working in the SPI mode can only transfer 8-bit data, but the MCP4922 DAC module needs to use 16-bit word (4-MSB work as the configuration command and the following 12-bit work as the data). Therefore, we must break this 16-bit command data into two 8-bit format to transfer them in two times.

```
1  //****************************************************************************
2  // DRADAC.c - Main Applicatiopn File for MCP4922 DAC
3  //****************************************************************************
4  #include "DRADAC.h"
5  int main(void)
6  {
7     uint16_t i_data = 0, d_data = 4095;
8     WDTCTL = WDTPW | WDTHOLD;         // stop watchdog timer
9     DAC_Init();                       // initialize eUSCI_B0 module for DAC
10    while(1)
11    {
12       DAC_Write(i_data++, 0);        // create an incremented sawtooth waveform in DAC0
13       DAC_Write(d_data--, 1);        // create a decremented sawtooth waveform in DAC1
14       ADC_Read();                    // receive the ADC result from TLC-548
15    }
16 }
17 void DAC_Init(void)                  // initialize UCB0 that connects to the DAC
18 {
19    P1SEL0 |= BIT5|BIT6|BIT7;         // set SPI 3-pin P1.5, P1.6 & P1.7 as primary
                                           function
20    P5DIR = BIT1;                     // set P5.1 as output pin (CS_DAC)
21    P4DIR = BIT4;                     // set P4.4 as output pin (CS_ADC)
22    P5OUT |= BIT1;                    // set P5.1 to High to deselect DAC
23    P4OUT |= BIT4;                    // set P4.4 to High to deselect ADC
24    UCB0CTLW0 |= UCSWRST;             // set UCSWRST = 1 to reset eUSCI_B0 module
25    UCB0CTLW0 |= UCMST|UCSYNC|UCCKPL|UCMSB;  // 3-pin master mode, 8-bit data length
                                               // clock polarity high, MSB first
26    UCB0CTLW0 |= UCSSEL ACLK;         // clock source = ACLK
27    UCB0BRW = 0x01;                   // fBitClock = fBRCLK/(UCBRx+1)= fBRCLK/2
28    UCB0CTLW0 &= ~UCSWRST;            // reset UCSWRST to enable USCIB0 to begin normal
29                                         work
   }
30
   void DAC_Write(uint16_t value, int channel)    // write 2 8-bit values to DAC through
31                                                   UCB0
   {
32    uint8_t config, cmd_data, data;
33
      if (channel == 0) {config = 0x70;}          // set config (4 MSB) as 7 (0111) for
34                                                   channel 0
      else if (channel == 1) {config = 0xF0;}     // set config (4 MSB) as F (1111) for
35                                                   channel 1
      data = value & 0x00FF;                      // get lower 8-bit data value
36    cmd_data = (value >> 8) & 0x0F;             // get upper 4-bit data
37    cmd_data = cmd_data | config;               // get 4 MSB config & upper 4-bit data
38    P5OUT &= ~BIT1;                             // setup CS_DAC to Low to select DAC
39    UCB0TXBUF = cmd_data;                       // transfer the configure & upper 4-bit
40                                                   data
      while (UCB0STATW & UCBUSY);                 // wait for first 8-bit transmit done
41    UCB0TXBUF = data;                           // transfer the lower 8-bit data
42    while (UCB0STATW & UCBUSY);                 // wait for second 8-bit transmit done
43    P5OUT |= BIT1;                              // set CS_DAC to High to deselect DAC
44 }
45
   void ADC_Read(void)                            // read the ADC result from TLC-548
46 {
47    uint8_t adc;
48    P2DIR = BIT0|BIT1|BIT2;                     // set P2.0~P2.2 as output pins
49
      P4OUT &= ~BIT4;                             // reset P4.4 as Low to CS_ADC to enable
50                                                   ADC TLC-548
      while (!(UCB0IFG & UCRXIFG)){}              // wait for UCB0 data receiving done
51    adc = UCB0RXBUF;                            // get the ADC data
52    P2OUT = adc;                                // display ADC result on 3-color LED P2.0
53                                                   ~ P2.2
      adc = UCB0IV;                               // clear UCRXIFG flag
54 }
55
```

Figure 8.31 Source file for the project DRADAC.

16. In lines 34 and 35, two **if()** selection structures are used to check which channel has been selected by the user to interface to this DAC. If the first channel **DAC0** is selected, the 4-MSB configuration parameters should be **0x70** (see Table 8.12). If the second channel **DAC1** is selected, the 4-MSB configuration parameters should be **0xF0** (Table 8.12).
17. The lower 8-bit data stored in the lower 8-bit on the **value** variable is assigned to the local variable **data** in line 36. A trick is that when 16-bit data are assigned to an 8-bit variable, only the lower 8-bit of that 16-bit data are assigned and the upper 8-bit are ignored. You can directly perform this assignment by using: **data** = **value**; without using that **AND** operation, like we did here. To use this **AND** is to make it safer for this assignment.
18. In line 37, the upper 4-bit data in bits 12~9 in the **value** variable are assigned to the lower 4-bit on the local variable **cmd_data**. This assignment is executed by first shifting the **value** variable right by eight times to move the upper 8-bit to the lower 8-bit and then **ANDing** the shifting result with **0x0F**.
19. The resulted **cmd_data** is then **ORed** with the configuration parameter **config** obtained in step 16 to get the final format: 4 MSB is the **config** and lower 4-bit are upper data bits in line 38.
20. In line 39, the GPIO pin **P5.1** is reset to 0 to set **CS_DAC** to low to select the DAC.
21. Then the **cmd_data** (4-MSB **config** + upper 4-bit data) is assigned to the **UCB0TXBUF** register to be transferred to the MCP4922 DAC via **UCB0SIMO** pin in line 40.
22. A **while()** loop is used in line 41 to wait for the **UCBUSY** bit in the **UCB0STATW** register to be clear, which means that the **eUSCI_B0** module has completed its current data transferring.
23. In line 42, the lower 8-bit data are assigned to the **UCB0TXBUF** register again to make the second transferring to the MCP4922 DAC. Similarly, another **while()** loop is used to wait for this transfer to be completed in line 43.
24. After a complete 16-bit word is transferred to the MCP4922 DAC, the GPIO pin **P5.1** is set to high to set up the **CS_DAC** to high to deselect the DAC.
25. The function **ADC_Read()** starts at line 46. A local 8-bit integer variable **adc** is created in line 48 and this variable is used to hold the returned ADC result later.
26. The GPIO pins **P2.0~P2.2** are configured as the output pins since these pins are connected to a three-color LED and we need to display the ADC result on those LEDs.
27. The GPIO pin **P4.4** is reset to low in line 50 to set up the **CS_ADC** to low to select the ADC TLC-548 chip to begin the analog-to-digital conversion for the analog input coming from the output of the DAC-MCP4922.
28. A **while()** loop is used in line 51 to wait for UCB0 data receiving done.
29. The received ADC result is sent to the local variable **adc** to be reserved in line 52.
30. This result is further sent to the three-color LED via **P2.0~P2.2** pins in line 53.
31. The **UCB0IV** register is read in line 54 to reset the **UCB0RXIFG** flag to make it ready for the next data receiving. This reading result is assigned to the variable **adc**. In fact, this is not necessary since this coding line is only to read the **UCB0IV** and the result is not useful.

Now, let's set up the environment to build and run the project to test the desired functions.

8.4.10.4.4 Set up the environment to build and run the project Perform the following operations to set up the environment for this project:

- Select the correct debugger **CMSIS-DAP Debugger** from the **Debug** tab under the **Project|Options for Target "Target 1"** menu item
- Make sure that all settings for the debugger and the flash download are correct

Now build, download, and run the project. As the project runs, you can use an oscilloscope to monitor **DAC0** and **DAC1** outputs via two pins, **J32-1** and **J32-3** on the **J32** connector, to get two saw-tooth waveforms (0~3.3 V). The period may be different with our calculation values since this program repeatedly generates two waveforms in a **while()** loop and the period may be doubled. Also you may find that the intensity and color on the three-color LED on the MSP EXP-432P401R EVB are periodically changed based on the ADC result if you build the ADC TLC-548 circuit on this project.

8.4.11 eUSCI SPI API functions provided by MSPWare Peripheral Driver Library

The MSPPDL provides three sets of API functions used for the eUSCI modules in the MSP432P401R MCU system. These three sets of functions are specially designed for **eUSCI_Ax**, **eUSCI_Bx**, and general SPI modules. Since the general SPI module covered the first two modules, we will discuss this module in this section.

These SPI API functions can be categorized into the following groups:

1. The initialization and configuration functions:
   a. **SPI_initMaster()**
   b. **SPI_initSlave()**
   c. **SPI_selectFourPinFunctionality()**
   d. **SPI_changeClockPhasePolarity()**
   e. **SPI_changeMasterClock()**
2. The control and status functions:
   a. **SPI_enableModule()**
   b. **SPI_disableModule()**
   c. **SPI_isBusy()**
3. The data processing functions:
   a. **SPI_transmitData()**
   b. **SPI_receiveData()**
4. The interrupt source and processing functions:
   a. **SPI_registerInterrupt()**
   b. **SPI_enableInterrupt()**
   c. **SPI_disableInterrupt()**
   d. **SPI_getInterruptStatus()**
   e. **SPI_clearInterruptFlag()**

In addition to these API functions, the MSPPDL also provides two data structs, **eUSCI_SPI_MasterConfig** and **eUSCI_SPI_SlaveConfig**, to assist the users to build SPI-related projects.

These API functions and structs are contained in a C file **spi.c** with a header file **spi.h**. Both files are located at the folder **C:\ti\msp\MSPWare_2_21_00_39\driverlib\driverlib\MSP432P4xx** in your host computer.

8.4.11.1 eUSCI_SPI_MasterConfig and eUSCI_SPI_SlaveConfig structs

These two structs are used to facilitate the initialization of the master and the slave SPI mode. The struct **eUSCI_SPI_MasterConfig** contains all initialization parameters used to configure a master mode with the following format:

```
typedef struct_eUSCI_SPI_MasterConfig
{
  uint_fast8_t          selectClockSource;
  uint32_t              clockSourceFrequency;
  uint32_t              desiredSpiClock;
  uint_fast16_t         msbFirst;
  uint_fast16_t         clockPhase;
  uint_fast16_t         clockPolarity;
  uint_fast16_t         spiMode;
} eUSCI_SPI_MasterConfig;
```

Similarly, the struct **eUSCI_SPI_SlaveConfig** contains all initialization parameters used to configure a slave mode with the following format:

```
typedef struct_eUSCI_SPI_SlaveConfig
{
  uint_fast16_t         msbFirst;
  uint_fast16_t         clockPhase;
  uint_fast16_t         clockPolarity;
  uint_fast16_t         spiMode;
} eUSCI_SPI_SlaveConfig;
```

8.4.11.2 SPI module initialization and configuration functions

Five API functions are included in this group and Table 8.13 shows these SPI module initialization and configuration functions. The API function **SPI_initMaster()** is a very important and useful function used to set up and configure the entire SPI module. Only after the SPI module is correctly configured by using this function, it can work properly to provide the normal functions. The argument **config** is a pointer of the struct **eUSCI_SPI_MasterConfig**. All parameters included inside this struct can be represented by related macros shown in Table 8.13. These parameters must be configured correctly to make the SPI master initialization successful.

8.4.11.3 SPI module control and status functions

Three functions are included in this group. Table 8.14 shows these SPI module control and status API functions.

After the SPI module is initialized by executing the **SPI_initMaster()** or **SPI_initSlave()** function, the SPI module must be enabled by calling the **SPI_enableModule()** to enable it to begin to perform the data transfer and receiving functions. The **SPI_isBusy()** is exactly to check the **UCBUSY** bit in the control register to see if the SPI is busy in data transferring or in idle status.

8.4.11.4 SPI module data processing functions

Two functions are involved in this group. Table 8.15 shows these API functions.

A point to be noted is that when using the **SPI_receiveData()**, a precheck should be performed first to make sure that a new data item has been placed into the **UCAxRXBUF** register.

Table 8.13 SPI module initialization and configuration functions

API function	Parameter	Description
bool **SPI_initMaster(**uint32_t mInst, const eUSCI_SPI_MasterConfig * config**)**	**mInst** is the instance of the eUSCI A/B module. Valid parameters can be: **EUSCI_A0_MODULE** **EUSCI_A1_MODULE** **EUSCI_A2_MODULE** **EUSCI_A3_MODULE** **EUSCI_B0_MODULE** **EUSCI_B1_MODULE** **EUSCI_B2_MODULE** **EUSCI_B3_MODULE** **config** is a struct and its parameters can be values shown in Description column	**selectClockSource**. Valid values can be: **EUSCI_SPI_CLOCKSOURCE_ACLK** **EUSCI_SPI_CLOCKSOURCE_SMCLK** **clockSourceFrequency**. Any desired value in Hz **desiredSpiClock**. Desired clock rate for SPI communication **msbFirst**. Valid values can be: **EUSCI_SPI_MSB_FIRST** **EUSCI_SPI_LSB_FIRST** [Default] **clockPhase**. Valid values can be: **EUSCI_SPI_PHASE_DATA_CHANGED_ONFIRST_CAPTURED_ON_NEXT** [Default Value] **EUSCI_SPI_PHASE_DATA_CAPTURED_ONFIRST_CHANGED_ON_NEXT** **clockPolarity**. Valid values can be: **EUSCI_SPI_CLOCKPOLARITY_INACTIVITY_HIGH** **EUSCI_SPI_CLOCKPOLARITY_INACTIVITY_LOW** [Default] **spiMode**. Valid values can be: **EUSCI_SPI_3PIN** [Default] **EUSCI_SPI_4PIN_UCxSTE_ACTIVE_HIGH** **EUSCI_SPI_4PIN_UCxSTE_ACTIVE_LOW**
bool **SPI_initSlave(**uint32_t mInst, const eUSCI_SPI_SlaveConfig * config**)**	**mInst** is the instance of the eUSCI A/B module. Valid parameters are identical with those in **SPI_initMaster()**. **config** is a struct and its parameters are identical with those used in the **SPI_initMaster()**	Modified bits are UCMSB, UC7BIT, UCMST, UCCKPL, UCCKPH, UCMODE, UCSWRST bits of **UCAxCTLW0** Reg Returns **true** if successful
void **SPI_selectFourPinFunctionality(**uint32_t mInst, uint_fast8_t 4PinFunction**)**	**mInst** is the instance of the eUSCI A/B module. Valid parameters are identical with those in **SPI_initMaster()** **4PinFunction** is the parameter selecting 4-pin function (see Description)	Valid values for **4PinFunction** are: **EUSCI_SPI_PREVENT_CONFLICTS_WITH_OTHER_MASTERS** **EUSCI_SPI_ENABLE_SIGNAL_FOR_4WIRE_SLAVE** This function should be invoked only in 4-wire mode. Invoking this function has no effect in 3-wire mode

(*Continued*)

Table 8.13 (Continued) SPI module initialization and configuration functions

API function	Parameter	Description
void **SPI_changeClockPhasePolarity(**uint32_t mInst, uint_fast16_t clockPhase, uint_fast16_t clockPolarity**)**	**mInst** is the instance of the eUSCI A/B module. Valid parameters are identical with those in **SPI_initMaster()** **clockPhase** and **clockPolarity** are identical with those used in **config** in **SPI_initMaster()**	Modified bits are **UCSWRST, UCCKPH, UCCKPL, UCSWRST** bits of **UCAxCTLW0** Register **Returns** none
void **SPI_changeMasterClock(**uint32_t mInst, uint32_t clockFrequency, uint32_t spiClock**)**	**mInst** is the instance of the eUSCI A/B module. Valid parameters are identical with those in **SPI_initMaster()**	**clockFrequency** is the frequency of the selected clock source in Hz **spiClock** is the desired clock rate in Hz for SPI communication

8.4.11.5 SPI module interrupt source and processing functions

Five API functions are involved in this group. Table 8.16 shows these SPI module interrupt processing API functions.

Before any interrupt can be handled and responded by the selected SPI module, the API function `SPI_registerInterrupt()` must be executed first to register the related

Table 8.14 SPI module control and status functions

API function	Parameter	Description
void **SPI_enableModule** **(**uint32_t mInst**)**	**mInst** is the instance of the eUSCI A/B module. Valid parameters can be: **EUSCI_A0_MODULE** **EUSCI_A1_MODULE** **EUSCI_A2_MODULE** **EUSCI_A3_MODULE** **EUSCI_B0_MODULE** **EUSCI_B1_MODULE** **EUSCI_B2_MODULE** **EUSCI_B3_MODULE**	Enable operation of the selected SPI block Modified bits are **UCSWRST** bit of **UCBxCTLW0** Register Returns none
void **SPI_disableModule** **(**uint32_t mInst**)**	**mInst** is the instance of the eUSCI A/B module. Valid parameters are identical with those in **SPI_enableModule()**	Disable operation of the selected SPI block Modified bits are **UCSWRST** bit of **UCAxCTLW0** Register Returns none
uint_fast8_t **SPI_isBusy** **(**uint32_t mInst**)**	**mInst** is the instance of the eUSCI A/B module. Valid parameters are identical with those in **SPI_enableModule()**	Return an indication of whether or not the SPI bus is busy. This function checks the status of the bus via **UCBBUSY** bit Returns **EUSCI_SPI_BUSY** if the SPI module transmitting or receiving is busy; otherwise, returns **EUSCI_SPI_NOT_BUSY**

Table 8.15 SPI module data processing functions

API function	Parameter	Description
void **SPI_transmitData (**uint32_t mInst, uint_fast8_t tData**)**	**mInst** is the instance of the eUSCI A/B module. Values are identical with those in **SPI_enableModule()** **tData** is data to be transmitted from the SPI module	Place the supplied data into SPI transmit data register to start transmission. Modified register is **UCAxTXBUF** Returns none
uint8_t **SPI_receiveData(**uint32_t mInst**)**	**mInst** is the instance of the eUSCI A/B module. Values are identical with those in **SPI_enableModule()**	Read a byte of data from the SPI receive data register Return the byte received from by the SPI module, cast as an uint8_t

interrupt handler to response the interrupt if it is occurred. The second argument of this function is a pointer point to the handler subroutine or an entry address of the ISR.

The argument **mask** for **SPI_enableInterrupt()**, **SPI_disableInterrupt()**, **SPI_clearInterruptFlag()**, and **SPI_getInterruptStatus()** API functions is a bit mask for the selected interrupt sources. A corresponding bit value of 1 indicates that the related interrupt source is selected, otherwise if a bit is 0, and the related interrupt source is not included. Some system macros, such as **EUSCI_SPI_RECEIVE_INTERRUPT** and **EUSCI_SPI_TRANSMIT_INTERRUPT**, should be used for those interrupt sources.

8.4.11.6 Build an example project to interface serial peripherals using the SPI module

In this section, we use the SPI-related API functions and some SPI registers to build a mixed interfacing project to access and display desired letters and numbers on the LCD device. We can modify the project **DRALCD** project we built in Section 8.4.8.4 and use the related API functions to replace DRA methods to access and interface LCD controller SPLC780 via 74VHCT595.

8.4.11.6.1 Create an SD model project SDLCD Perform the following operations to create a new project **SDLCD**:

1. Open the Windows Explorer window to create a new folder named **SDLCD** under the **C:\MSP432 Class Projects\Chapter 8** folder
2. Open the Keil® ARM MDK µVersion5 and create a new µVersion Project. On the opened wizard, browse to our new folder **SDLCD** that is created in step 1 above. Enter **SDLCD** into the **File name** box and click on the **Save** button to create this project
3. On the next wizard, you need to select the device (MCU) **MSP432P401R**. Click on the **OK** to close this wizard
4. Next the software components wizard is opened, and you need to set up the software development environment for your project with this wizard. Expand two icons, **CMSIS** and **Device**, and check the **CORE** and **Startup** checkboxes in the **Sel.** column, and click on the **OK** button since we need these two components to build our project

Since this project is a little longer, therefore a header file and a C source file are needed.

Table 8.16 SPI module interrupt source and processing functions

API function	Parameter	Description
void **SPI_registerInterrupt (**uint32_t mInst, void(*)(void) intHandler**)**	**mInst** is the instance of the eUSCI A/B module. Values are identical with those in **SPI_ enable-Module()** **intHandler** is a pointer to the function to be called when the interrupt occurs	Register the handler to be called when an interrupt occurs This function enables the global interrupt in the interrupt controller; specific SPI interrupts must be enabled via **SPI_ enableInterrupt()** It is the interrupt handler's responsibility to clear the interrupt source via **SPI_clear-InterruptFlag()**
void **SPI_enableInterrupt (**uint32_t mInst, uint_fast8_t mask**)**	**mInst** is the instance of the eUSCI A/B module. Values are identical with those in **SPI_ enable-Module()** **mask** is the bit mask of the interrupt sources to be enabled. Valid values are shown in Description column	Enable the indicated SPI interrupt sources Valid values for the **mask** can be logical **OR** of the following: **EUSCI_SPI_RECEIVE_ INTERRUPT** Receive interrupt **EUSCI_SPI_TRANSMIT_ INTERRUPT** transmit interrupt Modified registers are **UCAxIFG** and **UCAxIE** **Returns** none
void **SPI_disableInterrupt (**uint32_t mInst, uint_fast8_t mask**)**	**mInst** is the instance of the eUSCI A/B module. Values are identical with those in **SPI_enable-Module()** **mask** is the bit mask of the interrupt sources to be disabled. Valid values are shown in Description column	Disable individual SPI interrupt sources The **mask** is the logical **OR** of any of the following: **EUSCI_SPI_RECEIVE_ INTERRUPT** receive interrupt **EUSCI_SPI_TRANSMIT_ INTERRUPT** transmit interrupt Modified register is **UCAxIE** **Returns** none
uint_fast8_t **SPI_getInterrupt Status(**uint32_t mInst, uint16_t mask**)**	**mInst** is the instance of the eUSCI A/B module. Values are identical with those in **SPI_enable-Module()** **mask** is the bit mask of the interrupt sources have been enabled. Valid values are shown in Description column	Check the current SPI interrupt pending sources Valid values for the **mask** can be any of the following: **EUSCI_SPI_RECEIVE_ INTERRUPT** receive interrupt **EUSCI_SPI_TRANSMIT_ INTERRUPT** transmit interrupt **Returns**: any value of the following: **EUSCI_SPI_RECEIVE_ INTERRUPT** receive interrupt **EUSCI_SPI_TRANSMIT_ INTERRUPT** transmit interrupt

(*Continued*)

Table 8.16 (Continued) SPI module interrupt source and processing functions

API function	Parameter	Description
void **SPI_clearInterruptFlag (**uint32_t mInst, uint_fast8_t mask**)**	**mInst** is the instance of the eUSCI A/B module. Values are identical with those in **SPI_enable-Module()** **mask** is the bit mask of the interrupt sources occurred. Valid values are shown in the Description column	Clear the selected SPI interrupt status flag The **mask** parameter is the logical **OR** of the following: **EUSCI_SPI_RECEIVE_INTERRUPT** receive interrupt **EUSCI_SPI_TRANSMIT_INTERRUPT** transmit interrupt Modified registers are **UCAxIFG** **Returns** none

8.4.11.6.2 Create the header file SDLCD.h Create a new header file **SDLCD.h** and enter the code shown in Figure 8.32 into this file.

Since the codes are easy and straightforward, no explanation is needed. Five header files are system header files used to define the protocols of related API functions to be used in this project.

8.4.11.6.3 Create the C source file SDLCD.c In this project, we want to display "**WELCOME TO JCSU**" in the first line on the LCD, and "**GOOD JOB**" in the center of the second line on the LCD. Therefore, the function **LCD_Comd(0xC4)** is used to select the position or address of the DDRAM for the first letter in the second line to make the second line's letters located at the center. Refer to Figure 8.22 to get more details about this start position or the address of the DDRAM and this function.

Create a new C file named **SDLCD.c** and enter the code shown in Figure 8.33 into this C file.

```
1  //****************************************************************************
2  // SDLCD.h - Header File for the Main Application File SDLCD.c
3  //****************************************************************************
4  #include <stdint.h>
5  #include <stdbool.h>
6  #include <MSP432P4xx\gpio.h>
7  #include <MSP432P4xx\spi.h>
8  #include <MSP432P4xx\wdt_a.h>
9  #define RS    1                               // Q0 bit for RS (Reg Select)
10 #define EN    2                               // Q1 bit for E (Enable LCD)
11 #define BL    4                               // Q2 bit for BL (Backlight)
12 void delay_ms(int time);
13 void delay_us(int time);
14 void LCD_cd_Write(char data, unsigned char control);
15 void LCD_Comd(unsigned char cmd);
16 void LCD_Data(char data);
17 void LCD_Init(void);
18 void SPIB_Write(unsigned char data);
```

Figure 8.32 Codes for the header file SDLCD.h.

Now, let's have a closer look at this piece of code to see how it works. For those codes that are duplicated from the project **DRALCD.c**, we will skip them and only pay attention to those new added API functions.

1. Two new **char** arrays, **Line1[]** and **Line2[]**, are generated in lines 8 and 9 to hold two strings to be displayed in two lines in the LCD.
2. The **LCD_Init()** function is called in line 11 to initialize and configure the LCD controller to make it ready to interface and control the LCD panel.
3. In line 12, the **LCD_Comd(1)** is executed to clear the LCD panel and move the cursor to the home position on the LCD.
4. The codes in lines 13~15 are used to set up and send desired letters to the LCD.
5. The user-defined subroutine **LCD_Init()** starts at line 17.
6. To initialize the **eUSCI_B0** module, first we need to configure the related GPIO pins, **P1.6**, **P1.7**, and **P1.5** as **eUSCI_B0** function pins, such as **UCB0SIMO**, **UCB0SOMI**, and **UCB0CLK**. This can be done by calling an API function **GPIO_setAsPeripheralModuleFunctionOutputPin()** to set GPIO pins **P1.5~P1.7** as the primary function pins in line 19.
7. In line 21, the GPIO pin **P6.7** is configured as an output pin by executing an API function **GPIO_setAsOutputPin()** since we need it to output a rising-edge signal **STCP** to trigger 74VHCT595 to enable all 8-bit serial data stored in eight serial flip-flops to be converted to 8-bit parallel data and stored in the output latches.
8. The GPIO pin **P6.7** is set to high by calling an API function **GPIO_setOutputHighOnPin()** in line 22 to deselect the STCP signal.
9. The codes between lines 23 and 27 are used to configure the **eUSCI_B0** module to enable it to work as an SPI mode to transmit control commands and data to the SPLC780 LCD controller via the shift register 74VHCT595. Some SPI registers are directly accessed and configured to make this initialization process easy and fast.
10. The **eUSCI_B0** module is first disabled by calling the API function **SPI_disableModule()** in line 23 to enable this module to be configured and initialized.
11. The **eUSCI_B0** module is then configured by setting the related bits in the **UCB0CTLW0** Register with different system macros in line 24:
    a. **UCMST** = 3-pin master mode
    b. **UCSYNC** = synchronous (SPI) mode
    c. **UCCKPL** = inactive state of the SCLK is high
    d. **UCMSB** = MSB is to be transmitted first
12. This configuration is continued in lines 26 and 27.
13. The **eUSCI_B0** module is enabled by calling the API function **SPI_enableModule()** in line 28 to make it ready to perform data transfer operations.
14. The codes between lines 29 and 38 are identical with those codes in the **DRALCD** project.
15. The codes for the subroutine **SPIB_Write()** are totally new. First in line 42, the API function **GPIO_setOutputLowOnPin()** is called to reset the **CS_LCD** (**P6.7**) signal to low to prepare a low-to-high (positive-going edge) signal to the STCP on the serial shift register 74VHCT595.
16. In line 43, the API function **SPI_transmitData()** is used to send data item to the transmit register and then to the **UCB0SIMO** pin.
17. Then a **while()** loop is used to wait for the data sent to the transmit register to be transmitted in line 44. The API function **SPI_isBusy()** works as the loop condition for this **while()** loop and the subroutine will not continue to execute the next instruction until this loop condition becomes **False**, which means that the

```
1  //*******************************************************************************************
2  // SDLCD.c - Main Application File for LCD Project SDLCD
3  //*******************************************************************************************
4  #include "SDLCD.h"
5  int main(void)
6  {
7       uint8_t n;
8       char Line1[] = {'W', 'E', 'L', 'C', 'O', 'M', 'E', ' ', 'T', 'O', ' ', 'J', 'C', 'S', 'U', '!'};
9       char Line2[] = {'G', 'O', 'O', 'D', ' ', 'J', 'O', 'B'};
10      WDT_A_holdTimer();                  // stop the watchdog timer
11      LCD_Init();                         // initialize LCD controller
12      LCD_Comd(1);                        // clear screen, move cursor to home
13      for (n = 0; n < 17; n++) {LCD_Data(Line1[n]);} // write "WELCOME TO JCSU!" on the 1st line.
14      LCD_Comd(0xC4);
15      for (n = 0; n < 8; n++) {LCD_Data(Line2[n]);}   // write "GOOD JOB" in the center of the 2nd line in LCD
16 }
17 void LCD_Init(void)
18 {
19      GPIO_setAsPeripheralModuleFunctionOutputPin(GPIO_PORT_P1, GPIO_PIN5|GPIO_PIN6|GPIO_PIN7,
20      GPIO_PRIMARY_MODULE_FUNCTION);
21      GPIO_setAsOutputPin(GPIO_PORT_P6, GPIO_PIN7);
22      GPIO_setOutputHighOnPin(GPIO_PORT_P6, GPIO_PIN7);      // output a High on pin P6.7 for STCP
23      SPI_disableModule(EUSCI_B0_MODULE);                     // reset the UCB0 module
24      UCB0CTLW0 |= UCMST|UCSYNC|UCCKPL|UCMSB;                 // 3-pin master mode, 8-bit data length
25                                                              // clock polarity high, MSB first
26      UCB0CTLW0 |= UCSSEL__ACLK;                              // clock source = ACLK
27      UCB0BRW = 0x01;                                         // fBitClock = fBRCLK/(UCBRx+1)= fBRCLK/2
28      SPI_enableModule(EUSCI_B0_MODULE);                      // enable UCB0 module
29      delay_ms(20);                                           // LCD controller reset sequence
30      LCD_cd_Write(0x30, 0);                                  // send reset code 1 two times to SPLC780
31      LCD_cd_Write(0x30, 0);
32      LCD_cd_Write(0x30, 0);                                  // send reset code 2 to SPLC780
33      LCD_cd_Write(0x20, 0);                                  // use 4-bit data mode
34      delay_ms(1);
35      LCD_Comd(0x28);                                         // set 4-bit data, 2-line, 5x7 font
36      LCD_Comd(0x06);                                         // move cursor right
37      LCD_Comd(0x0C);                                         // turn on display, cursor off - no blinking
38      LCD_Comd(0x01);                                         // clear screen, move cursor to home
39 }
40      void SPIB_Write(unsigned char data)
41 {
42      GPIO_setOutputLowOnPin(GPIO_PORT_P6, GPIO_PIN7);       // output a Low on pin P6.7 for STCP
43      SPI_transmitData(EUSCI_B0_MODULE, data);                // transmit data into UCB0SIMO pin
44      while(SPI_isBusy(EUSCI_B0_MODULE)) {};                  // wait for transmit done
45      GPIO_setOutputHighOnPin(GPIO_PORT_P6, GPIO_PIN7);      // output a High on P6.7 for STCP
46 }
```

Figure 8.33 Source code for the project SDLCD.

eUSCI_B0 module has completed this data transmission and a **False** is returned from the function **SPI_isBusy()**.

18. In line 45, the function **GPIO_setOutputHighOnPin()** is executed to set the **CS_LCD** (**P6.7**) to high to create a positive-going-edge signal to trigger the 74VHCT595 to start a serial-to-parallel data conversion and output the conversion result to the LCD controller SPLC780.

All codes for other subroutines are identical with those codes in the project **DRALCD**.

8.4.11.6.4 Set up the environment to build and run the project To build and run this project, perform the following operations to set up the environment:

1. Go to **C/C++** tab in the **Project|Options for Target "Target 1"** to set up the **Include Path** as: **C:\ti\msp\MSPWare_2_21_00_39\driverlib\driverlib.**
2. Add the MSPPDL into this project by right clicking on the **Source Group 1** folder in the **Project** panel on the left, and select the **Add Existing Files**

to Group "**Source Group 1**" menu item. On the opened box, select the driver library file path **C:\ti\msp\MSPWare_2_21_00_39\driverlib\driverlib\MSP432P4xx\keil** and then select the driver library file **msp432p4xx_driverlib.lib**. Click the **Add** button to add it into the project.
3. Select the correct debugger by clicking on the **Debug** tab and select the **CMSIS-DAP Debugger** from the **Use** box. Also make sure that all settings for the debugger and the flash download are correct by clicking on the Settings button.

Now you can build and run the project. As the project runs, the letters **WEICOME TO JCSU!** are appeared on the first line, and **GOOD JOB** are appeared in the second line in the LCD.

We leave some projects as homework for the readers, and you can modify projects, such as **DRAADC** and **DRALED**, by using the interrupt mechanism and API functions to rebuild those projects. These projects are included in **Lab8_1** and **Lab8_3**, respectively.

8.5 Inter-Integrated Circuit (I2C) interface

The I2C module is exactly a serial communication bus system to enable I2C devices or peripherals to perform high-speed and bidirectional data transfer via two wire design, a serial data line **SDA** and a SCLK line **SCL**. By using this bus system, the microcontroller and other I2C compatible devices can interface to external I2C devices such as serial memory (RAMs and ROMs), networking devices, LCDs, tone generators, external clock, and so on. The I2C bus may also be used for system testing and diagnostic purposes in product development and manufacturing. The MSP432P401R microcontrollers include and provide the ability to communicate (both transmit and receive) with other I2C devices on the bus.

In the MSP432P401R MCU system, only four (4) **eUSCI_Bn** modules, **eUSCI_B0~eUSCI_B3**, are provided to support the I2C-related data operations. Each **eUSCI_Bn** module can work as a master or slave to perform data transmit or receive operation via two-wire bus.

The I2C master and slave modules provide the ability to communicate to other I2C devices over an I2C bus. The I2C bus is specified to support devices that can both transmit and receive (write and read) data. Also, devices on the I2C bus can be designated as either a master or a slave. The **eUSCI_Bn** modules support both sending and receiving data as either a master or a slave, and also support the simultaneous operation as both a master and a slave. Finally, the **eUSCI_Bn** modules can operate at two speeds: Standard (100 kbps) and Fast (400 kbps).

Both the master and slave I2C modules can generate interrupts. The I2C master module generates interrupts when a transmit or a receive operation is completed or aborted due to an error, and on some devices when a clock low timeout (CLTO) has occurred. The I2C slave module generates interrupts when data have been sent or requested by a master, and on some devices, when a **START** or **STOP** condition is present.

8.5.1 I2C module bus configuration and operational status

Each I2C module comprises both master and slave units and can be identified by a unique address. A master-initiated communication generates the clock signal, **SCL**. For proper operation, the **SDA** pin must be configured as an open-drain signal. Due to the internal circuitry that supports high-speed operation, the SCL pin must not be configured as an open-drain signal. Both **SDA** and **SCL** signals must be connected to a positive power supply voltage using a pull-up resistor. When both wires are in this status, it is called that the bus is idle. A typical I2C bus configuration is shown in Figure 8.34.

As we mentioned, the I2C bus uses only two wires, **SDA** and **SCL**, named **I2CSDA** and **I2CSCL** to perform bidirectional data communications on MSP432P401R microcontrollers. **SDA** is the bidirectional serial data line and **SCL** is the bidirectional SCLK line. The bus is considered idle when both lines are high.

Every transaction on the I2C bus is nine (**9**) bits long, consisting of 8 data bits and a single acknowledge bit. The number of bytes per transfer that defined as the time period between a valid **START** and **STOP** condition, is unrestricted, but each data byte has to be followed by an acknowledge bit, and data must be transferred in MSB-first.

The **START** and **STOP** are two conditions or two states defined in the protocol of the I2C module. A high-to-low transition on the **SDA** line while the **SCL** is high is defined as a **START** condition, and a low-to-high transition on the **SDA** line while **SCL** is high is defined as a **STOP** condition. The bus is considered busy after a **START** condition and free after a **STOP** condition. Figure 8.35 shows an illustration for both **START** and **STOP** conditions.

Regularly, the data on the **SDA** line must be stable during the high period of the **SCL** line, and can only be changed during the low period of the **SCL** line. In this way, the **START** and **STOP** state can be clearly distinguished with the normal data signals on the **SDA** line.

The **STOP** bit determines if the cycle stops at the end of the data cycle or continues on to a repeated **START** condition.

8.5.2 I2C module architecture and functional block diagram

A functional block diagram of each I2C module is shown in Figure 8.36.

It can be found from Figure 8.36 that each **eUSCI_B** module for I2C mode is similar to each **eUSCI_B** module for SPI mode. Each mode can work either an I2C master or an I2C slave core. Each core controls and coordinates the data transmit and receive operations via related registers. All I2C control registers are divided into two groups, the master and the slave group.

Three major components are included in each mode:

1. The clock source selection and bit clock generation part
2. The data transmit part (transmit shift and transmit buffer registers)
3. The data receive part (receive shift and receive buffer registers)

Each part comprises and is controlled by a group of registers.

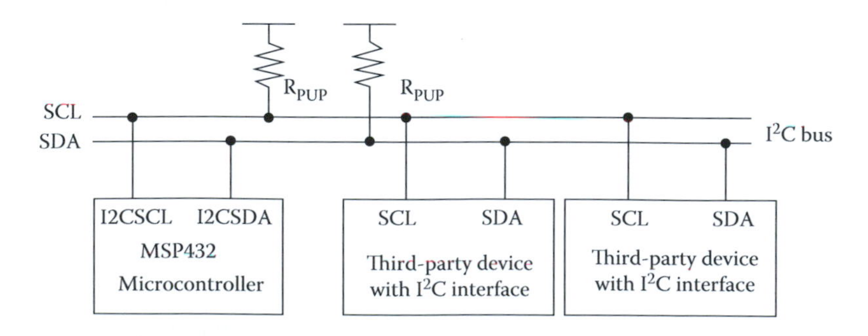

Figure 8.34 I2C bus configuration and status.

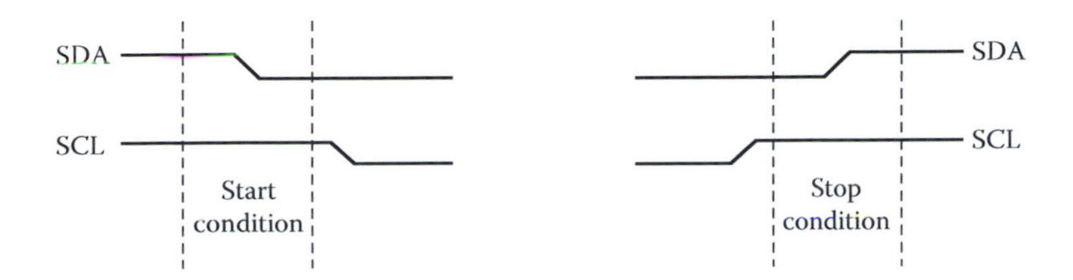

Figure 8.35 Definition of START and STOP conditions.

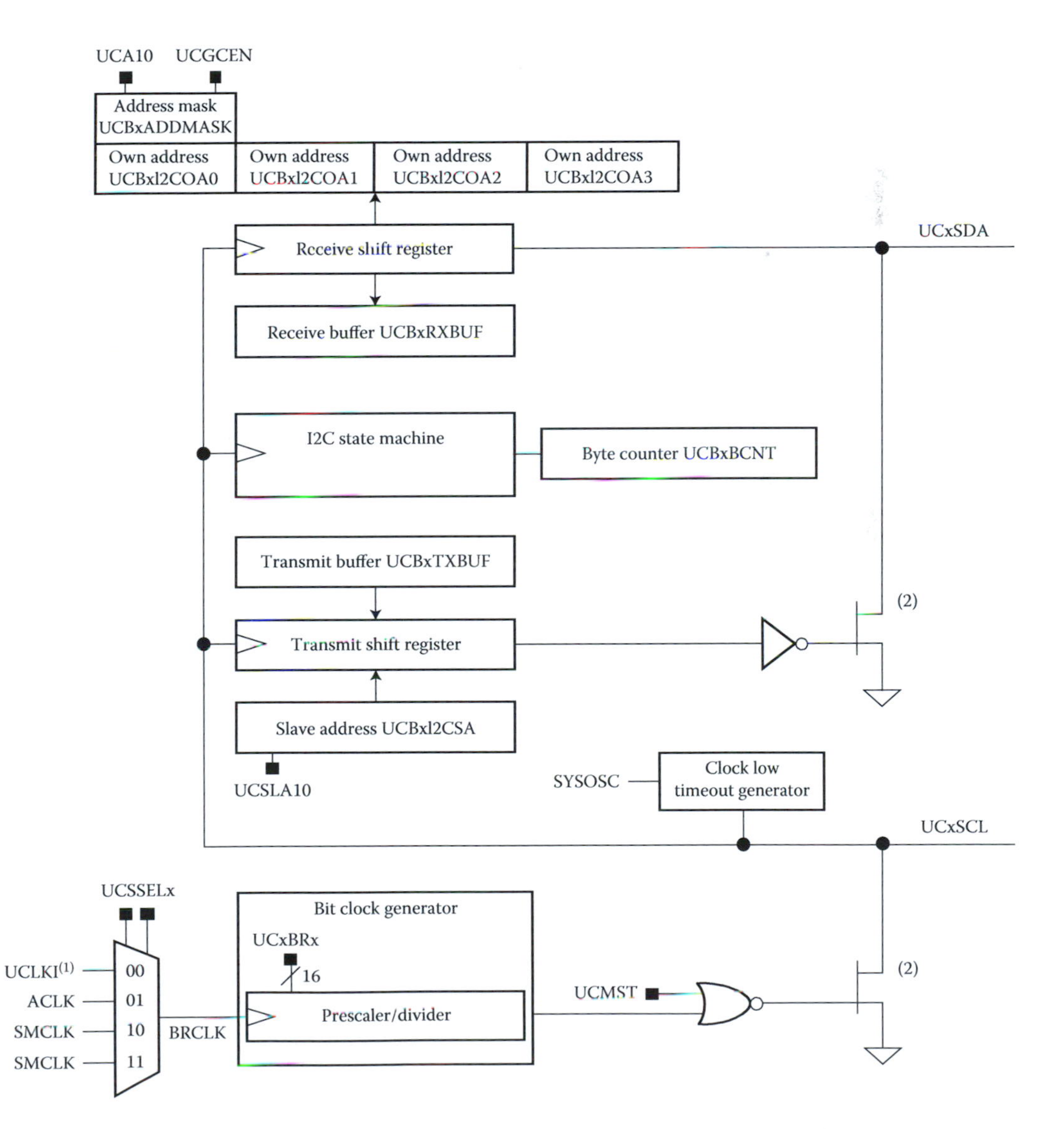

Figure 8.36 Functional block diagram of each eUSCI_B module for I2C mode. (Courtesy Texas Instruments.)

It can also be found from Figure 8.36 that two additional components, byte counter register (**UCBxBCNT**) and CLTO generator (bit field **UCCLTO**) in the **UCBxCTLW1** Register, are added to reflect the current number of bytes to be transmitted or received by I2C mode and monitor the clock working status. A CLTO interrupt can be generated if the clock works in an abnormal status.

On the top of Figure 8.36, four address registers, **UCBxI2COA0**~**UCBxI2COA3**, are used to store four slave addresses. These addresses will be compared with addresses received from the I2C bus to identify the correct I2C slaves if multiple slaves are used. The address mask register **UCBxADDMASK** is used to facilitate this identification.

8.5.3 I2C module data transfer format and frame

In the I2C module, the data transfers follow the format shown in Figure 8.37.

After the **START** condition, a slave address is first transmitted. This address is 7-bit long followed by an **R/S** bit, which is a data direction bit. If the **R/S** bit is 0, the master transmits data to a slave. If it is 1, the master receives data from a slave.

A data transfer is always terminated by a **STOP** condition generated by the master; however, a master can initiate communications with another device on the bus by generating a repeated **START** condition and addressing another slave without first generating a **STOP** condition. Various combinations of receive/transmit formats are then possible within a single transfer.

The first 7 bits of the first byte make up the slave address. The eighth bit determines the direction of the message. The ninth bit is the **ACK** bit that is generated by the master. When the I2C module operates in master receiver mode, the **ACK** bit is normally set causing the I2C bus controller to transmit an acknowledge automatically after each byte. This bit must be cleared when the I2C bus controller requires no further data to be transmitted from the slave transmitter.

Both a slave address and the following data must be transferred starting with the MSB and each piece of address or each data must comprise 9 bits.

All bus transactions have a required acknowledge clock cycle that is generated by the master. During the acknowledge (**ACK**) cycle, the transmitter that can be the master or slave releases the **SDA** line as shown in Figure 8.37. To acknowledge the transaction, the receiver must pull down **SDA** during the acknowledge clock cycle.

8.5.4 I2C module operational sequence

The I2C module can work as either a master or a slave with data transmitting and receiving mode. The master data transmission and receiving operations are controlled by the I2C master group registers, and the slave data transmission and receiving operations are controlled by the I2C slave group registers.

The operational sequences for each different mode are listed below.

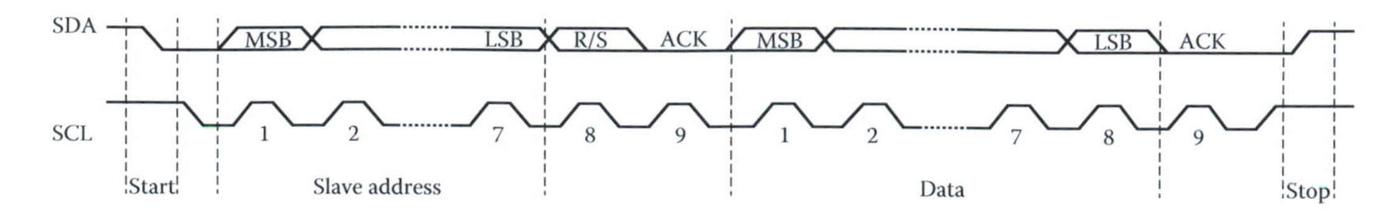

Figure 8.37 I2C data transfer format and frame.

8.5.4.1 I2C module works in the master transmit mode

To configure the **eUSCI_B** module to work as an I2C master mode, perform the following configurations for the **UCBxCTLW0** Register:

1. Select the I2C mode by setting the bit field **UCMODEx** = 11, **UCSYNC** = 1, and **UCMST** = 1.
2. When the master is part of a multimaster system, **UCMM** bit must be set and its own address must be programmed into the **UCBxI2COA0** Register. Support for multiple slave addresses is explained in Section 8.5.5.
3. When **UCA10** = 0, 7-bit addressing is selected. When **UCA10** = 1, 10-bit addressing is selected.
4. The **UCGCEN** bit selects if the **eUSCI_B** module responds to a general call.

When the **eUSCI_B** module works in the master transmit mode, it performs the following operations:

1. After initialization, master transmitter mode is initiated by writing the desired slave address to the **UCBxI2CSA** register.
2. Select the size of the slave address as 7-bit or 10-bit by configuring the **UCSLA10** bit, setting **UCTR** bit for transmitter mode, and setting **UCTXSTT** bit to automatically generate a **START** condition. All of these bits are located in the **UCBxCTLW0** Register.
3. The **eUSCI_B** module waits until the bus is not busy by checking the **UCBBUSY** bit in the **UCBxSTATW** register, then it generates the **START** condition and transmits the slave address.
4. The **UCTXIFG0** bit in the **UCBxIFG** register is set when the **START** condition is generated and the first data to be transmitted are written into **UCBxTXBUF** register. The **UCTXSTT** flag bit in the **UCBxCTLW0** Register is cleared as soon as the complete address is sent.
5. The data written into the **UCBxTXBUF** register are transmitted if arbitration is not lost during transmission of the slave address. The **UCTXIFG0** bit in the **UCBxIFG** register is set again as soon as the data are transferred from the TX buffer into the TX shift register. If there is no data loaded to **UCBxTXBUF** register before the acknowledge cycle, the bus is held during the acknowledge cycle with **SCL** low until data are written into **UCBxTXBUF** register. Data are transmitted or the bus is held, as long as:
   a. No automatic **STOP** is generated
   b. The **UCTXSTP** bit is not set
   c. The **UCTXSTT** bit is not set
6. Setting the **UCTXSTP** bit in the **UCBxCTLW0** Register generates a **STOP** condition after the next acknowledge signal is received from the slave. If the **UCTXSTP** bit is set during the transmission of the slave address or while the eUSCI_B module waits for data to be written into the **UCBxTXBUF** register, a **STOP** condition is generated, even if no data were transmitted to the slave. In this case, the **UCSTPIFG** bit in the **UCBxIFG** register is set to indicate that a **STOP** interrupt is pending.
7. When transmitting a single byte of data, the **UCTXSTP** bit must be set while the byte is being transmitted or any time after transmission begins, without writing new data into the **UCBxTXBUF** register. Otherwise, only the address is transmitted.
8. When the data are transferred from the buffer to the shift register, the **UCTXIFG0** bit is set, indicating data transmission has begun, and the **UCTXSTP** bit may be set for a single byte data transmission. When **UCASTPx** = 10 is set in the **UCBxCTLW1** Register,

a **STOP** condition can be generated automatically after the byte counter value reached the threshold value set in the **UCBxTBCNT** register. Also the **UCBCNTIFG** bit in the **UCBxIFG** register is set when the byte counter reaches this threshold value. In that case, the user does not need to set the **UCTXSTP** bit to generate any **STOP** condition. This is recommended when transmitting only a single byte.

9. Setting the **UCTXSTT** bit generates a repeated **START** condition. In this case, the **UCTR** bit in the **UCBxCTLW0** Register can be set or reset to configure data transmitter or receiver, and a different slave address may be written into the **UCBxI2CSA** register, if desired.
10. If the slave does not acknowledge the transmitted data, the not-acknowledge (**NAK**) interrupt flag bit **UCNACKIFG** in the **UCBxIFG** register is set. The master must react with either a **STOP** condition or a repeated **START** condition. If data were already written into the **UCBxTXBUF** register, it is discarded. If this data should be transmitted after a repeated **START**, it must be written into the **UCBxTXBUF** register again. Any set **UCTXSTT** or **UCTXSTP** bit is also discarded.

Figure 8.38 shows the operational sequence for the I2C master working as a transmitter.

8.5.4.2 I2C module works in the master receive mode

When the **eUSCI_B** module works in the master receive mode, it perform the following operations:

1. After initialization, the master receiver mode is initiated by writing the desired slave address to the **UCBxI2CSA** register.
2. Select the size of the slave address as 7 bit or 10 bit by configuring the **UCSLA10** bit, clearing the **UCTR** bit for the receiver mode, and setting the **UCTXSTT** bit to generate a **START** condition.

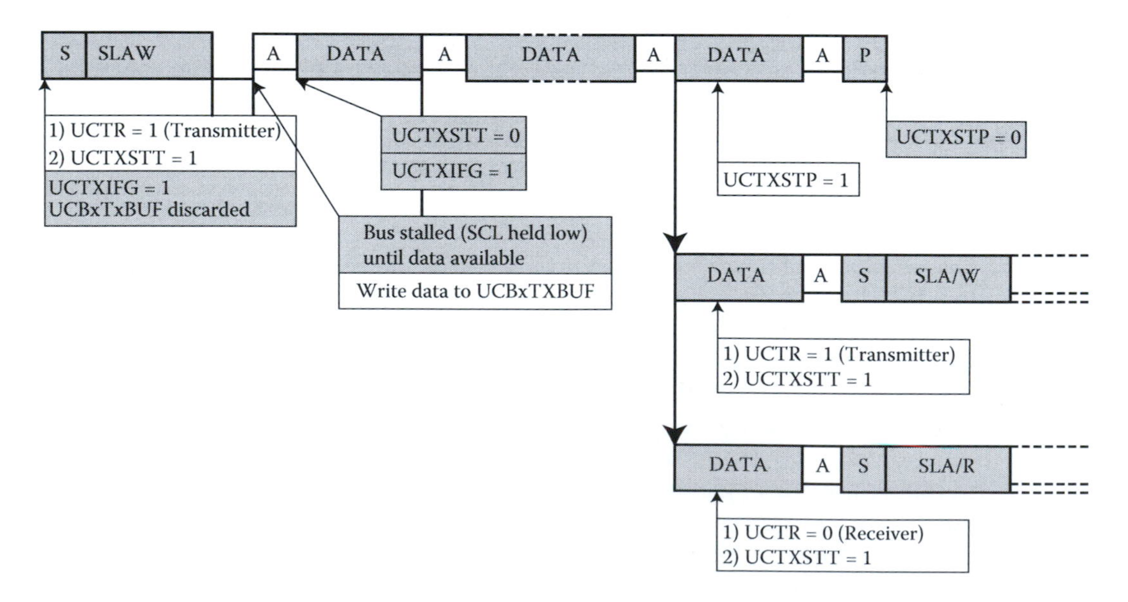

Figure 8.38 Operational sequence of the master working in the transmit mode. (Courtesy Texas Instruments.)

3. The **eUSCI_B** module waits until the bus is not busy by checking the **UCBBUSY** bit in the **UCBxSTATW** register, then it generates the **START** condition and transmits the slave address.
4. The **UCTXSTT** flag is cleared as soon as the complete slave address is sent. After an acknowledge of the address is received from the slave, the first data byte from the slave is received and acknowledged and the **UCRXIFG** flag bit in the **UCBxIFG** register is set. Data are received from the slave, as long as:
    a. No automatic **STOP** is generated
    b. The **UCTXSTP** bit is not set
    c. The **UCTXSTT** bit is not set
5. If a **STOP** condition is generated by the eUSCI_B module, the **UCSTPIFG** bit is set. If the **UCBxRXBUF** register is not read, the master holds the bus during reception of the last data bit until the **UCBxRXBUF** register is read.
6. If the slave does not acknowledge the transmitted address, the **NAK** interrupt flag **UCNACKIFG** in the **UCBxIFG** register is set. The master must react with either a **STOP** condition or a repeated **START** condition.
7. A **STOP** condition is either generated by the automatic **STOP** generation or by setting the **UCTXSTP** bit in the **UCBxCTLW0** Register. The next byte received from the slave is followed by a **NAK** and a **STOP** condition. This **NAK** occurs immediately if the **eUSCI_B** module is currently waiting for the **UCBxRXBUF** register to be read.
8. If a **RESTART** is sent, the **UCTR** bit can be set or reset to configure transmitter or receiver operation, and a different slave address may be written into the **UCBxI2CSA** register if desired.

The complete operational sequence of the master working in the receiving mode is shown in Figure 8.39. For these master transmit and receive modes, the error processing is not covered, and this includes the error identification and error services.

8.5.4.3 I2C module works in the slave transmit and receive modes

To configure the **eUSCI_B** module to work as an I2C slave mode, perform the following configurations for the **UCBxCTLW0** Register:

1. Select the I2C slave mode by setting the bit field **UCMODEx** = 11 and the **UCSYNC** = 1, and clearing the **UCMST** bit to 0.
2. By default, the **eUSCI_B** module must be initially configured in the receiver mode by clearing the **UCTR** bit to receive the I2C address. Then transmit and receive operations are controlled automatically, which are dependent on the R/S bit received together with the slave address.
3. The **eUSCI_B** slave address is configured by programming the **UCBxI2COA0** Register. Support for multiple slave addresses is explained in Section 8.5.5. When the bit **UCA10** = 0, 7-bit addressing is selected. Otherwise, 10-bit addressing is selected when **UCA10** = 1. The **UCGCEN** bit selects if the slave responds to a general call.
4. When a **START** condition is detected on the bus, the **eUSCI_B** module receives the transmitted address and compares it against its own address stored in **UCBxI2COA0**. The **UCSTTIFG** flag is set when address received matches the **eUSCI_B** slave address.

When the **eUSCI_B** module works in the slave transmit mode, it performs the following operations (Figure 8.40):

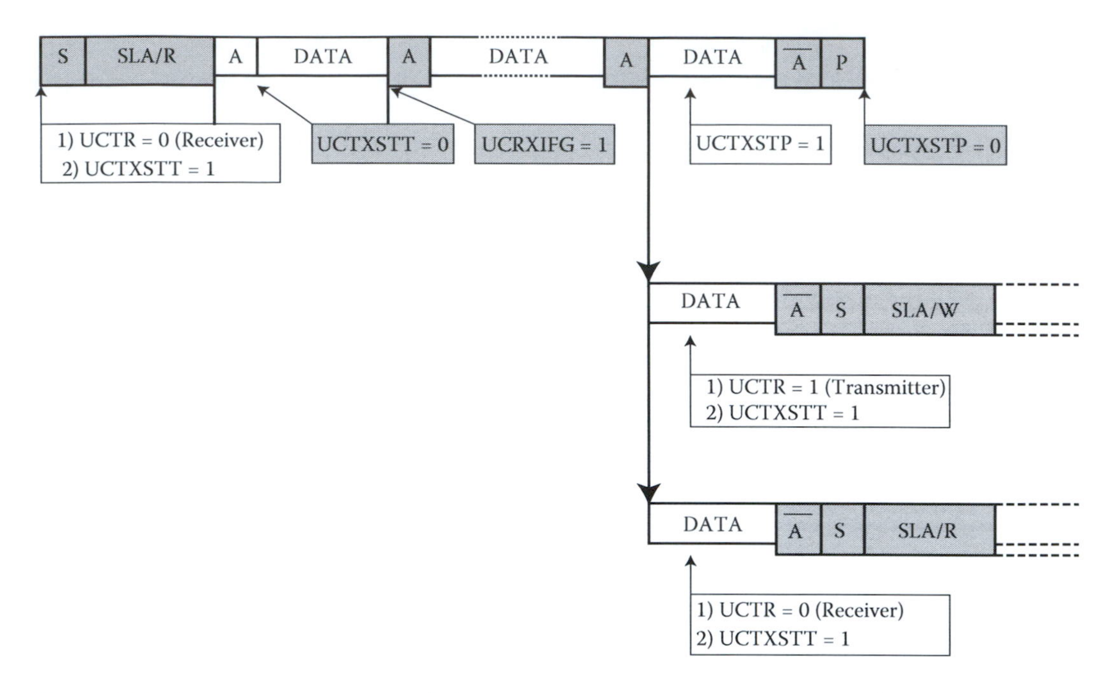

Figure 8.39 Operational sequence of the master working in the receive mode. (Courtesy Texas Instruments.)

1. The slave transmitter mode is started when the slave address transmitted by the master is identical to its own address in the **UCBxI2COA0** Register with a set **R/S** bit.
2. Then the slave transmitter shifts the serial data out on the **SDA** line with the clock pulses that are generated by the master. The slave device does not generate the clock, but it does hold the **SCL** line low when it cannot receive a complete byte to force the transmitter into a wait state. The data transfer continues when the receiver releases the clock **SCL**.
3. If the master requests data from the slave, the **eUSCI_B** module (slave) is automatically configured as a transmitter to set the **UCTR** and the **UCTXIFG0** bits to 1. The **SCL** line is held low until the first data to be sent are written into the transmit buffer **UCBxTXBUF**. Then the address is acknowledged and the data are received.
4. As soon as the data are transferred into the shift register, the **UCTXIFG0** is set again.
5. After the data are acknowledged by the master, the next data byte written into **UCBxTXBUF** is transmitted. However, if the buffer is empty, the bus is stalled during the acknowledge cycle by holding the **SCL** line to low until new data are written into **UCBxTXBUF**.
6. If the master sends a **NAK** followed by a **STOP** condition, the **UCSTPIFG** flag is set. If the **NAK** is followed by a repeated **START** condition, the eUSCI_B I2C state machine returns to its address-reception state.

When the **eUSCI_B** module works in the slave receiver mode, it performs the following operations (Figure 8.41):

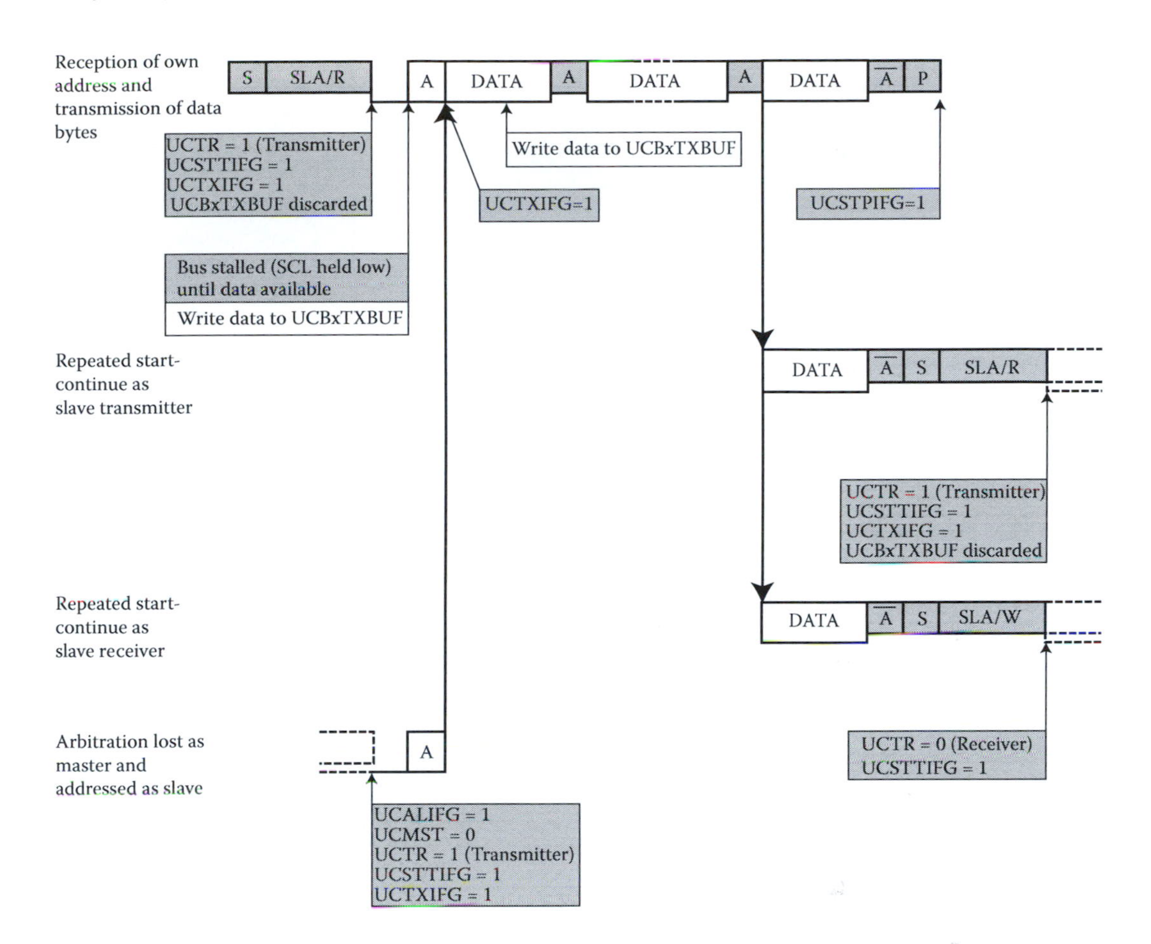

Figure 8.40 Operation sequence of the I2C module working in slave transmit mode. (Courtesy Texas Instruments.)

1. The slave receiver mode is started when the slave address transmitted by the master is identical to its own address in the **UCBxI2COA0** Register and a cleared **R/S** bit is received.
2. Then the serial data bits received on the **SDA** line are shifted into the **UCBxRXBUF** register with the clock pulses generated by the master device. The slave device does not generate the clock, but it can hold the **SCL** line to low if a byte has been received improperly.
3. If the slave receives data from the master, the **eUSCI_B** module is automatically configured as a receiver and the **UCTR** bit is cleared to 0. After the first data byte is received, the receive interrupt flag **UCRXIFG0** is set. The **eUSCI_B** module automatically acknowledges the received data and can receive the next data byte.
4. If the previous data were not read from the receive buffer **UCBxRXBUF** at the end of a reception, the bus is stalled by holding the **SCL** line to low until the **UCBxRXBUF** register is read, then the new data are transferred into the **UCBxRXBUF** register, an acknowledge is sent to the master, and the next data can be received.
5. Setting the **UCTXNACK** bit causes a **NAK** to be transmitted to the master during the next acknowledgment cycle. An **NAK** is sent even if the **UCBxRXBUF** register is not ready to receive the latest data.

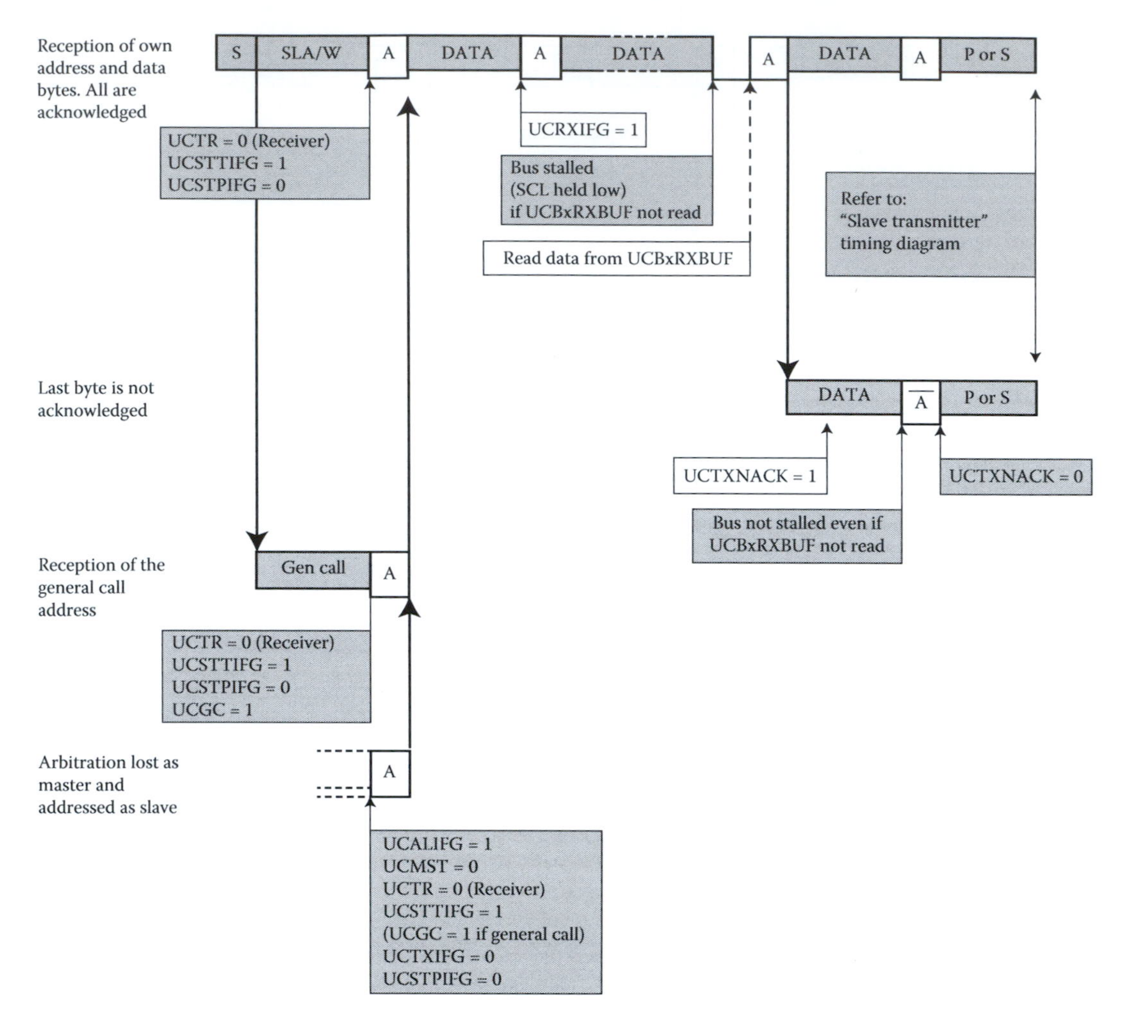

Figure 8.41 Operation sequence of the I2C module working in slave receiver mode. (Courtesy Texas Instruments.)

6. If the **UCTXNACK** bit is set while the **SCL** line is held low, the bus is released with a **NAK** being transmitted immediately, and the **UCBxRXBUF** register is loaded with the last received data. Because the previous data were not read, those data are lost. To avoid loss of data, the **UCBxRXBUF** register must be read before the **UCTXNACK** bit is set.
7. When the master generates a **STOP** condition, the **UCSTPIFG** flag is set.
8. If the master generates a repeated **START** condition, the eUSCI_B I2C state machine returns to its address-reception state.

Now we have a clear picture about the I2C operational sequence. Next let's have a closer look at the I2C major control signals.

8.5.5 I2C module major operational control signals

The following operational control signals are mainly implemented in the I2C module:

8.5.5.1 Acknowledge

All bus transactions have a required acknowledge clock cycle that is generated by the master. During the acknowledge cycle, the transmitter (either master or slave) releases the **SDA** line. To acknowledge the transaction, the receiver must pull down the **SDA** line during the acknowledge clock cycle.

When a slave receiver does not acknowledge the slave address, the **SDA** must be left high by the slave so that the master can generate a **STOP** condition and abort the current transfer. If the master device is acting as a receiver during a transfer, it is responsible for acknowledging each transfer made by the slave. Because the master controls the number of bytes in the transfer, it signals the end of data to the slave transmitter by not generating an **ACK** on the last data byte. The slave transmitter must then release **SDA** to allow the master to generate the **STOP** or a repeated **START** condition.

If the slave is required to provide a manual **ACK** or **NAK**, the **UCBxCTLW0** Register allows the slave to **NAK** for invalid data or command or **ACK** for valid data or command. When this operation is enabled, the MCU slave module I2C clock is pulled low after the last data bit until this register is written with the indicated response.

8.5.5.2 Repeated start

The I2C master module has the capability to execute a repeated **START** (transmit or receive) after an initial transfer has occurred. This means that after data have been transmitted, the master does not generate a **STOP** condition but instead writes another slave address to the **UCBxI2CSA** register and then writes **0x3** to initiate the repeated **START**.

8.5.5.3 Clock low timeout (CLTO)

The I2C slave sometimes can extend the transaction by pulling the clock low periodically to create a slow bit transfer rate. To check and avoid this clock to work in the low status with long time, a CLTO generator is implemented. The CLTO feature is enabled using the **UCCLTO** bits in the **UCBxCTLW1** Register. It is possible to select one of three predefined times for the CLTO. If the clock has been low longer than the time defined with the **UCCLTO** bits and the eUSCI_B was actively receiving or transmitting, the **UCCLTOIFG** is set and an interrupt request is generated if **UCCLTOIE** bit is set as well. The **UCCLTOIFG** flag is set only once, even if the clock is stretched a multiple of the time defined in **UCCLTO** bits.

In the event of an **UCCLTO** condition, application software must choose a way to try to recover it. Most applications may attempt to manually toggle the I2C pins to force the slave to let go of the clock signal (a common solution is to attempt to force a **STOP** on the bus).

8.5.5.4 Multiple slave addresses

The **eUSCI_B** for I2C interface supports multiple addresses capability for the slave. The additional programmable addresses are provided and can be matched if enabled. In this mode, the I2C slave provides an **ACK** on the bus if the address in one the **UCBxI2COA0~UCBxI2COA3** registers is matched to the address in the **UCBxI2CSA** register. The enable for multiple addresses is programmable through the **UCOAEN** bits in the **UCBxI2COA0~UCBxI2COA3** registers.

8.5.5.5 Arbitration

A master may start a transfer only if the bus is idle. However, it is possible for two or more masters to generate a **START** condition within minimum hold time of the **START** condition. In these situations, an arbitration scheme takes place on the **SDA** line while **SCL** is high.

During arbitration, the first of the competing master devices places a 1 (high) on **SDA**, while another master transmits a 0 (low), switches off its data output stage and retires until the bus is idle again.

Arbitration can take place over several bits. Its first stage is a comparison of address bits, and if both masters are trying to address the same device, arbitration continues on to the comparison of data bits.

8.5.5.6 Glitch suppression in multimaster configuration

When a multimaster configuration is used, the **UCGLITx** bits in the **UCBxCTLW1** Register can be configured and set up to enable glitch suppression on the **SCL** and **SDA** lines to assure proper signal values. The length of this glitch filter can be programmed to different filter widths using the **UCGLITx** bits in the **UCBxCTLW1** Register. The glitch suppression value is based on the buffered system clocks. For example, if **UCGLITx** is set to **0x00**, the maximum length 50 ns can be applied. For **UCGLITx** = 11, the minimum length of 6.25 ns can be obtained. Note that all signals will be delayed internally when glitch suppression is applied.

8.5.6 I2C module running speeds (clock rates) and interrupts

As we mentioned, the I2C clock **SCL** is provided by the master on the I2C bus. When the **eUSCI_B** module works in master mode, the **BITCLK** is provided by the **eUSCI_B** bit clock generator and the CKSE is selected with the **UCSSELx** bits in the **UCBxCTLW0** Register. In the slave mode, the bit clock generator is not used and the **UCSSELx** bits are neglected.

The 16-bit value in the **UCBRx** register is the division factor of the **eUSCI_B** CKSE, **BRCLK**. The maximum bit clock used in the single-master mode is **fBRCLK/4**. In the multimaster mode, the maximum bit clock is **fBRCLK/8**. The **BITCLK** frequency is:

```
fBitClock = fBRCLK/UCBRx
```

The minimum high and low periods of the generated **SCL** are:

$t_{LOW,MIN} = t_{HIGH,MIN} =$ (UCBRx/2)/fBRCLK   when UCBRx is even
$t_{LOW,MIN} = t_{HIGH,MIN} =$ ((UCBRx - 1)/2)/fBRCLK   when UCBRx is odd

The **eUSCI_B** module clock source frequency and the prescaler setting in the **UCBRx** register must be chosen to meet the minimum low and high period times of the I2C specification.

During the arbitration procedure the clocks from the different masters must be synchronized. A device that first generates a low period on the **SCL** overrules the other devices, forcing them to start their own low periods. The **SCL** is then held low by the device with the longest low period. The other devices must wait for the **SCL** to be released before starting their high periods. This allows a slow slave to slow down a fast master.

8.5.6.1 I2C module interrupts generation and processing

The I2C module can generate interrupts when the following conditions are observed:

- Master transmit complete
- Master receive complete
- Slave transmit complete
- Slave receive complete
- State change occur

The I2C master and slave modules have separate interrupt signals. While both modules can generate interrupts for multiple conditions, only a single interrupt signal can be sent to the interrupt controller.

Each interrupt flag has its own interrupt enable bit. When an interrupt is enabled by configuring the related bit on the **UCBxIE** register, the interrupt flag bit in the **UCBxIFG** register generates an interrupt request. DMA transfers are controlled by the **UCTXIFGx** and **UCRXIFGx** flags with a DMA controller. It is possible to react on each slave address with an individual DMA channel.

All interrupt flags cannot be cleared automatically, and they need to be cleared together by the user program (e.g., by reading the **UCRXBUF** register to clear the **UCRXIFGx** flag bit). If users want to use an interrupt flag, they need to ensure that the flag has the correct state before the corresponding interrupt is enabled.

8.5.6.1.1 I2C transmit interrupts The **UCTXIFG0** interrupt flag in the **UCBxIFG** register is set whenever the transmitter is able to accept a new byte (the **UCBxTXBUF** register is empty). When operating as a slave with multiple slave addresses, the **UCTXIFGx** flags are set corresponding to which address was received before. If, for example, the slave address specified in register **UCBxI2COA3** did match the address seen on the bus, the **UCTXIFG3** flag bit indicates that the **UCBxTXBUF** register is ready to accept a new byte.

When operating in master mode with automatic **STOP** generation (**UCASTPx** = 10), the **UCTXIFG0** bit is set as many times as defined in **UCBxTBCNT** register.

An interrupt request is generated if the **UCTXIEx** bit in the **UCBxIE** register is set. The **UCTXIFGx** bit can be automatically reset if a write to **UCBxTXBUF** register occurs or if the **UCALIFG** is cleared. The **UCTXIFGx** flag bit can be set when:

- Master mode: the **UCTXSTT** bit is set by the user (transmit START)
- Slave mode: the own address is received (**UCETXINT** = 0) or a **START** is received (**UCETXINT** = 1). The **UCETXINT** means the early TX interrupt flag **UCTXIFG0**

The **UCTXIEx** bit is reset after a HR or when the bit **UCSWRST** = 1.

If the application does not need to use interrupts, the raw interrupt status is always visible in the **UCBxIFG** register.

8.5.6.1.2 I2C receive interrupts The **UCRXIFG0** interrupt flag is set when a character is received and loaded into the **UCBxRXBUF** register. When operating as a slave with multiple slave addresses, the **UCRXIFGx** flag is set corresponding to which slave address was received before.

An interrupt request is generated if the bit **UCRXIEx** in the **UCBxIFG** register is set. The **UCRXIFGx** and the **UCRXIEx** are reset after a HR signal or when the bit **UCSWRST** = 1. The **UCRXIFGx** flag bit is automatically reset when the **UCxRXBUF** register is read.

8.5.6.1.3 I2C state change interrupts If one of the following states or events occurred, a related interrupt can be generated:

1. Arbitration lost: Enabled by the **UCALIE** bit in the **UCBxIE** register and indicated by the **UCALIFG** flag bit in the **UCBxIFG** register
2. NAK: Enabled by the **UCNACKIE** bit in the **UCBxIE** register and indicated by the **UCNACKIFG** flag bit in the **UCBxIFG** register

3. CLTO: Enabled by the **UCCLTOIE** bit in the **UCBxIE** register and indicated by the **UCCLTOIFG** flag bit in the **UCBxIFG** register
4. Byte counter interrupt: Enabled by the **UCBCNTIE** bit in the **UCBxIE** register and indicated by the **UCBCNTIFG** flag bit in the **UCBxIFG** register
5. **START** condition detected: Enabled by the **UCSTTIE** bit in the **UCBxIE** register and indicated by the **UCSTTIFG** flag bit in the **UCBxIFG** register
6. STOP condition detected: Enabled by the **UCSTPIE** bit in the **UCBxIE** register and indicated by the **UCSTPIFG** flag bit in the **UCBxIFG** register
7. Transferring ninth clock cycle of a byte of data: Enabled by the **UCBIT9IE** bit in the **UCBxIE** register and indicated by the **UCBIT9IFG** flag bit in the **UCBxIFG** register

When a related interrupt source is enabled by setting the corresponding bit on the **UCBxIE** register, the selected flag bit on the **UCBxIFG** register is set when that interrupt is occurred.

The **eUSCI_B** interrupt flags are prioritized and combined the source to a single interrupt vector stored in the interrupt vector register **UCBxIV**. This register is used to determine which flag requested an interrupt. The highest priority-enabled interrupt generates a number in the **UCBxIV** register, which can be evaluated or loaded into the PC register to correctly select and enter the appropriate ISR. Disabled interrupts do not affect the **UCBxIV** value.

Reading of the **UCBxIV** register automatically resets the highest pending interrupt flag. If another interrupt flag is set, another interrupt is immediately generated after servicing the initial interrupt. Writing to the **UCBxIV** register clears all pending interrupt conditions and flags.

8.5.7 I2C interface control signals and GPIO I2C control registers

All peripheral devices used in the MSP432P401R MCU system are interfaced to MCU via related GPIO ports and pins, and this is also true to the I2C modules. Table 8.17 lists the external signals of the **eUSCI_B** module for I2C interface and related function.

All of these interfacing signals shown in Table 8.17 are connected to the connectors J1~J4 in the MSP-EXP432P401R EVB except the last two signals on GPIO pins **P10.2** and **P10.3** that are connected to the additional connector J5 on this EVB.

Table 8.17 eUSCI for I2C mode control signals and GPIO pins distributions

I2C pin	GPIO pin	Pin type	Pin function
UCB0SDA	**P1.6**	I/O	I2C Module Signal Line (UCB0 Module)
UCB0SCL	**P1.7**	I/O	I2C Module Clock Line (UCB0 Module)
UCB2SDA (PM)	**P3.6**	I/O	I2C Module Signal Line (UCB2 Module)
UCB2SCL (PM)	**P3.7**	I/O	I2C Module Clock Line (UCB2 Module)
UCB1SDA	**P6.4**	I/O	I2C Module Signal Line (UCB1 Module)
UCB1SCL	**P6.5**	I/O	I2C Module Clock Line (UCB1 Module)
UCB3SDA	**P6.6**	I/O	I2C Module Signal Line (UCB3 Module)
UCB3SCL	**P6.7**	I/O	I2C Module Clock Line (UCB3 Module)
UCB3SDA	**P10.2**	I/O	I2C Module Signal Line (UCB3 Module)
UCB3SCL	**P10.3**	I/O	I2C Module Clock Line (UCB3 Module)

8.5.8 I2C module control registers and their functions

As shown in Figure 8.36, all **eUSCI_B** modules for I2C components and their control functions are globally illustrated by this block diagram. However, in order to get more detail about these components and control signals, we need to discuss them one by one with more detail based on each group of registers and their functions.

These registers can be divided into the following five groups:

1. The eUSCI_B module for i2c mode control registers
2. The eUSCI_B module for I2C mode transmit and receive registers
3. The eUSCI_B module for I2C mode status and error control registers
4. The eUSCI_B module for I2C mode addresses control registers
5. The eUSCI_B module for I2C mode interrupt control registers

8.5.8.1 eUSCI_B module for I2C mode control registers

Three registers are involved in this group:

- The eUSCI_Bx Control Word Register 0 (**UCBxCTLW0**)
- The eUSCI_Bx Control Word Register 1 (**UCBxCTLW1**)
- The eUSCI_Bx Bit Rate Control Word (**UCBxBRW**) register

All of these registers are 16-bit registers with the different control functions.

The **UCBxCTLW0** Register (Figure 8.42) is used to configure and set up all necessary controls for the I2C mode operation, which include:

- Address length selection for the own and slave addresses
- Master or slave mode selection
- Multimaster environment selection
- 3-pin or 4-pin mode selection
- Synchronous or asynchronous selection
- SPI, I2C, or UART mode selection
- Clock source selection
- Transmit or receive mode selection
- Start or stop condition selection
- ACK or NAK transmit selection

The bit configuration for the **UCBxCTLW0** Register is shown in Figure 8.42. The bit values of this register are shown in Table 8.18. One point is that all bits or bit fields that have special pattern background in Figure 8.42 and highlighted in Table 8.18 can only be modified during the eUSCI_B module reset status (**UCSWRST** = 1).

The **UCBxCTLW1** Register is used to set up additional controls to assist the I2C operations.

15	14	13	12	11	10	9	8
UCA10	UCSLA10	UCMM	Reserved	UCMST	UCMODEx		UCSYNC
7	**6**	**5**	**4**	**3**	**2**	**1**	**0**
UCSSELx		UCTXACK	UCTR	UCTXNACK	UCTXSTP	UCTXSTT	UCSWRST

Figure 8.42 Bit configurations for the UCBxCTLW0 Register.

Table 8.18 Bit value and function for UCBxCTLW0 register

Bit	Field	Type	Reset	Function
15	**UCA10**	RW	0	Own addressing mode select **0:** Own address is a 7-bit address. **1:** Own address is a 10-bit address
14	**UCSLA10**	RW	0	Slave addressing mode select **0:** Address slave with 7-bit address. **1:** Address slave with 10-bit address
13	**UCMM**	RW	0	Multimaster environment select **0:** Single master environment. No other master in the system. The address compare unit is disabled. **1:** Multimaster environment
12	**Reserved**	RO	0	Reserved
11	**UCMST**	RW	0	Master mode selection **0:** Slave mode; **1:** Master mode
10:9	**UCMODEx**	RW	0	The eUSCI 3-pin or 4-pin mode selection **00:** 3-pin SPI mode **01:** 4-pin SPI with UCxSTE active high: Slave enabled when UCxSTE = 1 **10:** 4-pin SPI with UCxSTE active low: Slave enabled when UCxSTE = 0 **11:** I2C mode
8	**UCSYNC**	RW	1	Synchronous mode enable **0:** Asynchronous mode is selected; **1:** Synchronous mode is selected
7:6	**UCSSELx**	RW	3	The eUSCI clock source select. These bits select the BRCLK source clock in master mode. **UCLK** is always used in slave mode **00:** UCLK1; **01:** ACLK; **10:** SMCLK; **11:** SMCLK
5	**UCTXACK**	RW	0	Transmit **ACK** condition in slave mode with enabled address mask register. After the **UCSTTIFG** has been set, the user needs to set or reset the **UCTXACK** flag to continue with the I2C protocol. The clock is stretched until the **UCBxCTL1** Register is written. This bit is cleared automatically after the ACK has been send **0:** Do not acknowledge the slave address. **1:** Acknowledge the slave address
4	**UCTR**	RW	0	Transmitter/receiver operation **0:** Receiver. **1:** Transmitter
3	**UCTXNACK**	RW	0	Transmit a **NAK**. **UCTXNACK** is automatically cleared after a NAK is sent. Only for slave receiver mode **0:** Acknowledge normally. **1:** Generate NAK
2	**UCTXSTP**	RW	0	Transmit **STOP** condition in master mode. Ignored in slave mode In master receiver mode, the **STOP** condition is preceded by a **NAK**. **UCTXSTP** is automatically cleared after **STOP** is generated. This bit is a don't care, if automatic **UCASTPx** is different from 01 or 10 **0:** No STOP generated. **1:** Generate STOP

(*Continued*)

Table 8.18 (Continued) Bit value and function for UCBxCTLW0 register

Bit	Field	Type	Reset	Function
1	**UCTXSTT**	RW	0	Transmit **START** condition in master mode. Ignored in slave mode In master receiver mode, a repeated **START** condition is preceded by a **NAK**. **UCTXSTT** is automatically cleared after **START** condition and address information is transmitted. Ignored in slave mode **0:** Do not generate START condition. **1:** Generate START condition
0	**UCSWRST**	RW	1	Software reset enable **0:** Reset is disabled. The eUSCI is working in normal operation **1:** Reset is enabled. The eUSCI is in reset state (only in this state, initialization and configuration of the eUSCI can be performed)

The bit configuration for the **UCBxCTLW1** Register is shown in Figure 8.43. The bit values of this register are shown in Table 8.19. One point is that all bits or bit fields that have special pattern background in Figure 8.43 and highlighted in Table 8.19 can only be modified during the eUSCI_B module reset status (**UCSWRST** = 1 in the **UCBxCTLW0** Register).

The **UCBxBRW** register is a 16-bit register and it is used to define a prescaler as the clock divider to divide the selected CKSE BRCLK to get the desired I2C clock SCL.

8.5.8.2 eUSCI_B module for I2C mode transmit and receive registers

Two registers are included in this group: the eUSCI_Bx Transmit Buffer (**UCBxTXBUF**) register and the eUSCI_Bx Receive Buffer (**UCBxRXBUF**) register.

Both registers are 16-bit registers but only the lower 8-bit are used to store either a transmit or receive data byte. When transmit, the data byte to be transmitted is first sent to the **UCBxTXBUF** register, and then it is moved into the shift register. The **UCTXIFGx** flag bit is set as soon as the **UCBxTXBUF** register is empty.

When receive, the received data byte is first moved into the shift register. Then it is shifted to the **UCRXBUF** register. The **UCRXIFGx** flag bit is set as soon as the **UCRXBUF** register is received and loaded with a data byte.

8.5.8.3 eUSCI_B module for I2C mode status and error control registers

Two registers are involved in this group: the eUSCI_Bx Status Word (**UCBxSTATW**) register and the eUSCI_Bx Byte Counter Threshold (**UCBxTBCNT**) register.

The **UCBxSTATW** register is a 16-bit register and it is mainly used to indicate and monitor the working status of the I2C mode. The upper 8 bit, **UCBCNTx**, can be used to

15	14	13	12	11	10	9	8
Reserved							UCETXINT
7	**6**	**5**	**4**	**3**	**2**	**1**	**0**
UCCLTO		UCSTPNACK	UCSWACK	UCASTPx		UCGLITx	

Figure 8.43 Bit configurations for the UCBxCTLW1 Register.

Table 8.19 Bit value and function for UCBxCTLW1 register

Bit	Field	Type	Reset	Function
15:9	**Reserved**	RO	0	Reserved
8	**UCETXINT**	RW	0	Early **UCTXIFG0**. Only in slave mode. When this bit is set, the slave addresses defined in **UCxI2COA1** to **UCxI-2COA3** must be disabled 0: **UCTXIFGx** is set after an address match with **UCxI-2COAx** and the direction bit indicating slave transmit 1: **UCTXIFG0** is set for each **START** condition
7:6	**UCCLTO**	RW	0	Clock low timeout select **00:** Disable clock low timeout counter **01:** 135 000 SYSCLK cycles (approximately 28 ms) **10:** 150 000 SYSCLK cycles (approximately 31 ms) **11:** 165 000 SYSCLK cycles (approximately 34 ms)
5	**UCSTP-NACK**	RW	0	Allow to make the eUSCI_B master acknowledge the last byte in master receiver mode as well. This is not conform to the I2C specification and should only be used for slaves, which automatically release the SDA after a fixed packet length **0:** Send a nonacknowledge before the **STOP** condition as a master receiver **1:** All bytes are acknowledged by the eUSCI_B when configured as master receiver
4	**UCSWACK**	RW	0	Using this bit, it is possible to select, whether the eUSCI_B module triggers the sending of the ACK of the address or if it is controlled by software **0:** The address acknowledge of the slave is controlled by the eUSCI_B module **1:** The user needs to trigger the sending of the address ACK by issuing UCTXACK
3:2	**UCASTPx**	RW	0	Automatic **STOP** condition generation. In slave mode only **UCBCNTIFG** is available **00:** No automatic STOP generation. The STOP condition is generated after the user sets the **UCTXSTP** bit. The value in **UCBxTBCNT** is a don't care **01: UCBCNTIFG** is set with byte counter reaches the threshold in **UCBxTBCNT** **10:** A **STOP** condition is generated automatically after the byte counter value reached **UCBxTBCNT**. **UCBCNTIFG** is also set for this situation **11:** Reserved
1:0	**UCGLITx**	RW	0	Deglitch time **00:** 50 ns; **01:** 25 ns; **10:** 12.5 ns; **11:** 6.25 ns

count and record the real number of data bytes that have been transmitted or received. The bit-4 (**UCBBUSY**) can be used to indicate whether the I2C is busy (1) or idle (0).

The bit configuration for the **UCBxSTATW** register is shown in Figure 8.44. The bit values of this register are shown in Table 8.20.

All bits on this register are read-only bits with no writing access.

15	14	13	12	11	10	9	8
UCBCNTx							
7	**6**	**5**	**4**	**3**	**2**	**1**	**0**
Reserved	UCSCLLOW	UCGC	UCBBUSY	Reserved			

Figure 8.44 Bit configurations for the UCBxSTATW register.

The **UCBxTBCNT** register is also a 16-bit register but only the lower 8-bit are used to set up a threshold value. This threshold value is used to set the number of I2C data bytes after which the automatic **STOP** or the **UCSTPIFG** should occur. This value is evaluated only if **UCASTPx** is different from 00. This register can only be modified when the eUSCI_B module is in the reset status.

8.5.8.4 eUSCI_B module for I2C mode addresses control registers

Seven registers are involved in this group:

- The eUSCI_Bx I2C Own Address 0 (**UCBxI2COA0**)~eUSCI_Bx I2C Own Address 3 (**UCBxI2COA3**) registers
- The eUSCI_Bx I2C Received Address (**UCBxADDRX**) register
- The eUSCI_Bx I2C Address Mask (**UCBxADDMASK**) register
- The eUSCI_Bx I2C Slave Address (**UCBxI2CSA**) register

The eUSCI_B module supports two different ways of implementing multiple slave addresses at the same time:

- Hardware support for up to four different slave addresses, each with its own interrupt flag and DMA trigger
- Software support for up to 210 different slave addresses all sharing one interrupt

Table 8.20 Bit value and function for UCBxSTATW register

Bit	Field	Type	Reset	Function
15:8	**UCBCNTx**	RO	0	Hardware byte counter value Reading this register returns the number of bytes received or transmitted on the I2C bus since the last **START** or **RESTART** There is no synchronization of this register done. When reading **UCBxBCNT** during the first bit position, a faulty read-back can occur
7	**Reserved**	RO	0	Reserved
6	**UCSCLLOW**	RO	0	SCL low **0:** SCL is not held low; **1:** SCL is held low
5	**UCGC**	RO	0	General call address received. **UCGC** is automatically cleared when a **START** condition is received **0:** No general call address received; **1:** General call address received
4	**UCBBUSY**	RO	0	Bus busy **0:** Bus is idle; **1:** Bus is busy
3:0	**Reserved**	RO	0	Reserved

Four registers **UCBxI2COA0**, **UCBxI2COA1**, **UCBxI2COA2**, and **UCBxI2COA3** are used to contain four slave addresses. All of these four address registers should be compared with a received 7- or 10-bit address from the I2C bus. Each slave address must be activated by setting the **UCAOEN** bit in the corresponding **UCBxI2COAx** register.

When one of these slave registers matches the 7- or 10-bit address seen on the bus, the address is acknowledged. In the following the corresponding receive or transmit interrupt flag (**UCTXIFGx** or **UCRXIFGx**) to the received address is updated. All of these registers can only be modified when the eUSCI_B module is in the reset status.

Register **UCBxI2COA3** has the highest priority if the address received on the bus matches more than one of the slave address registers. The priority decreases with the index number of the address register, so that **UCBxI2COA0** in combination with the address mask has the lowest priority.

The bit configuration for the **UCBxI2COA0** Register is shown in Figure 8.45. The bit values of these registers are shown in Table 8.21.

The bit configuration and bit value for **UCBxI2COA1**~ **UCBxI2COA3** Registers are identical. Therefore, the bit configurations for these registers are shown in Figure 8.46.

It can be found from Figure 8.46 that only bit-10 (**UCOAEN**) and bits 9~0 (**I2COAx**) are used for these registers. Each of these address register must be enabled by setting the **UCOAEN** bit, and each 9~0 bits, **I2COAx** (**x** = 1~3) contains a valid slave address.

The received address register (**UCBxADDRX**) is a 16-bit register but only the lower 10 bits (bits 9~0) are used to contain the last received slave address on the bus. Using this register combined with the address mask register (**UCBxADDMASK**), it is possible to react on more than one slave address using one eUSCI_B module.

15	14	13	12	11	10	9	8
UCGCEN	Reserved				UCOAEN	I2COA0	
7	**6**	**5**	**4**	**3**	**2**	**1**	**0**
I2COA0							

Figure 8.45 Bit configurations for the UCBxI2COA0 Register.

Table 8.21 Bit value and function for UCBxI2COA0 register

Bit	Field	Type	Reset	Function
15	**UCGCEN**	RW	0	General call response enable. This bit is only available in **UCBxI2COA0** **0:** Do not respond to a general call; **1:** Respond to a general call
14:11	**Reserved**	RO	0	Reserved
10	**UCOAEN**	RO	0	Own address enable register **0:** The slave address defined in **I2COA0** is disabled **1:** The slave address defined in **I2COA0** is enabled
9:0	**I2COA0**	RO	0	I2C own address. The **I2COA0** bits contain the local address of the eUSCIx_B I2C controller. The address is right justified. In 7-bit addressing mode, bit-6 is the MSB and bits 9–7 are ignored. In 10-bit addressing mode, bit-9 is the MSB

15	14	13	12	11	10	9	8
Reserved					UCOAEN	I2COAx (x = 1~3)	
7	6	5	4	3	2	1	0
I2COAx (x = 1~3)							

Figure 8.46 Bit configurations for the UCBxI2COAx register.

The **UCBxADDMASK** register is a 16-bit register but only the lower 10 bits, bits 9~0, are used to facilitate the comparison between the slave address in the own address registers **UCBxI2COA0~UCBxI2COA3** and a slave address seen on the I2C bus. If the corresponding bit on one of the own address registers **UCBxI2COA0~UCBxI2COA3** is cleared to 0, this bit will be a do-not-care bit when comparing the address on the bus to the own address. By using this method, it is possible to react on more than one slave address. When all lower 10-bit on the **ADDMASK** register are set, the address mask feature is deactivated.

The **UCBxI2CSA** is also a 16-bit register but only the lower 10 bits, bits 9~0, are used to contain the slave address that will be assigned and sent to an external device by the master. It is only used in master mode. The address is right justified. In 7-bit slave addressing mode, bit-6 is the MSB and bits 9-7 are ignored. In 10-bit slave addressing mode, bit-9 is the MSB.

8.5.8.5 eUSCI_B module for I2C mode interrupt control registers

Three registers are included in this group:

- The eUSCI_Bx I2C Interrupt Enable (**UCBxIE**) register
- The eUSCI_Bx I2C Interrupt Flag (**UCBxIFG**) register
- The eUSCI_Bx I2C Interrupt Vector (**UCBxIV**) register

The bit configuration of the **UCBxIE** register is shown in Figure 8.47. Each bit (except bit-15) on this register is used to enable (1) or disable (0) one related interrupt source. The bit configuration of the **UCBxIFG** register is identical with that of **UCBxIE** register. As each data transmit or receive action is completed, a **UCTXIEx** or a **UCRXIEx** interrupt (**x** = 0~3) is generated if it is enabled and indicated in the corresponding flag bit on the **UCBxIFG** register by setting that bit to 1. Similarly, if an error, such as **CCLTO**, byte count (**UCBCNT**) hits the threshold value, an **NAK** is received, Arbitration lost (**UCAL**), a **STOP** or a **START** condition is received, a related interrupt is also generated if that is enabled and indicated in the corresponding flag bit on the **UCBxIFG** register.

The **UCBxIV** register is a 16-bit register and it holds the highest priority-enabled interrupt vector. Reading of the **UCBxIV** register automatically resets the highest-pending interrupt flag set in the **UCBxIFG** register. The bit value of the **UCBxIV** register is shown in Table 8.22.

15	14	13	12	11	10	9	8
Reserved	UCBIT9IE	UCTXIE3	UCRXIE3	UCTXIE2	UCRXIE2	UCTXIE1	UCRXIE1
7	6	5	4	3	2	1	0
UCCLTOIE	UCBCNTIE	UCNACKIE	UCALIE	UCSTPIE	UCSTTIE	UCTXIE0	UCRXIE0

Figure 8.47 Bit configurations for the UCBxIE register.

Table 8.22 Bit value and function for UCBxIV register

Bit	Field	Type	Reset	Function
15:0	**UCIVx**	RO	0	eUSCI_B interrupt vector value. It generates a value that can be used as address offset for fast interrupt service routine handling. Writing to this register clears all pending interrupt flags **0x00:** No interrupt pending **0x02:** Interrupt source: Arbitration lost; Interrupt flag: **UCALIFG**; Priority: Highest **0x04:** Interrupt source: Not acknowledgment; Interrupt flag: **UCNACKIFG** **0x06:** Interrupt source: Start condition received; Interrupt flag: **UCSTTIFG** **0x08:** Interrupt source: Stop condition received; Interrupt flag: **UCSTPIFG** **0x0A:** Interrupt source: Slave 3 data received; Interrupt flag: **UCRXIFG3** **0x0C:** Interrupt source: Slave 3 transmit buffer empty; Interrupt flag: **UCTXIFG3** **0x0E:** Interrupt source: Slave 2 data received; Interrupt flag: **UCRXIFG2** **0x10:** Interrupt source: Slave 2 transmit buffer empty; Interrupt flag: **UCTXIFG2** **0x12:** Interrupt source: Slave 1 data received; Interrupt flag: **UCRXIFG1** **0x14:** Interrupt source: Slave 1 transmit buffer empty; Interrupt flag: **UCTXIFG1** **0x16:** Interrupt source: Data received; Interrupt flag: **UCRXIFG0** **0x18:** Interrupt source: Transmit buffer empty; Interrupt flag: **UCTXIFG0** **0x1A:** Interrupt source: Byte counter zero; Interrupt flag: **UCBCNTIFG** **0x1C:** Interrupt source: Clock low timeout; Interrupt flag: **UCCLTOIFG** **0x1E:** Interrupt source: Ninth bit position; Interrupt flag: **UCBIT9IFG**; Priority: Lowest

8.5.9 eUSCI_B module for I2C mode initializations and configurations

Before using the I2C module to perform related data operations, all eUSCI_B modules must be properly initialized and configured. Depending on the different operational modes, the initialization and configuration procedures may be different. In this section, we use an example to illustrate how to initialize and configure one of the most popular operation modes, `eUSCI_B1` module for I2C mode to receive multiple bytes by a master. The system clock source is SMCLK.

This configuration process can be divided into two major parts, the configurations to the I2C-related GPIO ports and pins and the configurations to the `eUSCI_B1` module for I2C mode.

8.5.9.1 Initializations and configurations for the I2C-related GPIO pins

These initializations and configurations include the following operations (see Table 8.17):

1. Configure GPIO pins **P6.4** (**UCB1SDA**) and **P6.5** (**UCB1SCL**) as the primary function pins by setting **P6SEL1.x**:**P6SEL0.x** = 01 (**x** = 4 and 5)
2. Configure GPIO pins **P3.2** and **P3.3** as output digital function pins
3. Configure GPIO pins **P2.5** and **P2.4** as output digital function pins

Steps 2 and 3 are used to display the received data on four LEDs, **PB3~PB0** on the Trainer. Refer to Figure 4.31 in Chapter 4 to get more details about this hardware connection.

8.5.9.2 Initializations and configurations for the eUSCI_B1 module for I2C mode

These initializations and configurations include the following operations:

1. Reset the **eUSCI_B1** module by setting the bit **UCSWRST** in the **UCB1CTLW0** Register
2. Configure the **eUSCI_B1** module as: master mode + 3-pin I2C mode
3. Configure the bit clock rate as $f_{BITCLK} = f_{BRCLK}/8$ since this is smaller than the maximum bit clock rate ($f_{BITCLK} = f_{BRCLK}/4$) that can be used for the single master. Complete this configuration by setting the **UCBRx** value in the **UCB1BRW** register as 8 ($f_{BITCLK} = f_{BRCLK}/UCBRx$)
4. Specify the slave address as **0x48** and assign it to the slave address register **UCB1I2CSA**
5. Reset the bit **UCSWRST** in the **UCB1CTLW0** Register to enable the **eUSCI_B1** module

Now, let's build an example project to illustrate how to use the **eUSCI_B1** module for I2C mode to transmit and receive data between a master and a slave device.

8.5.10 Build example I2C module projects

In these projects, two devices are included, a master and a slave device. Two MSP432P401R EVBs will be used, one works as a master and the other works as a slave. The master works as a receive mode to receive a set of data from the slave, and the slave works in the transmit mode. The interrupt mechanism for both master and slave is used.

First let's have a look at the structures and hardware connection for this project.

8.5.10.1 Hardware configuration for the master and the slave device

The hardware configuration for this project is shown in Figure 8.48.

Two MSP-EXP432P401R EVBs are used in this project, one works as a master and the other works as a slave. The GPIO pins **P6.4** and **P6.5** work as **SDA** and **SCL** pins for eUSCI_B1 module. Connect these pins with two pull-up 10 K resistors. For power supply, one can use either **V3** (pin 1 in J1 connector in either EVB) or pin 4 in J32 on the EduBASE ARM Trainer. The breadboard provided by the EduBASE ARM Trainer is a good place to put these together.

Two programs should be developed for this project, the master side and the slave side program, and each one is used to control one device.

8.5.10.2 Master side program

The program in the master side works as the receive mode to ask the slave to transmit 16-byte data in the following operational sequence:

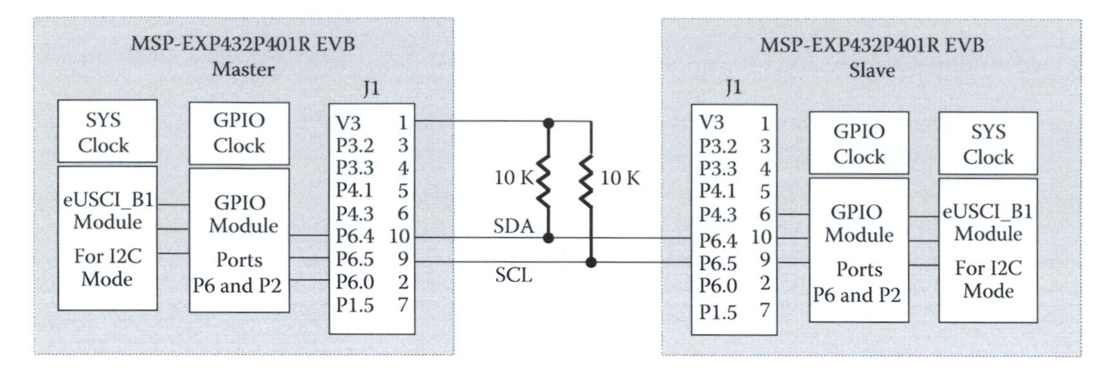

Figure 8.48 Hardware configuration for master-slave working in I2C mode.

1. The I2C-related GPIO ports and the **`eUSCI_B1`** module for I2C mode-related registers should be initialized and configured as we discussed in Section 8.5.9.2.
2. Enable the **`eUSCI_B1`** module interrupt by setting NVIC_IPR5 and NVIC_ISER0 registers.
3. The **`UCB1TBCNT`** register is set with the number of bytes (16) to be required from the slave.
4. When the 16th data byte is received, a byte counter flag is set in the **`UCBCNTIFG`** bit on the **`UCB1IFG`** register, and an interrupt would be generated if it is enabled.
5. The **`UCASTP1`** bit in the **`UCB1CTLW1`** Register should be set to **`10`** to enable a STOP condition to be generated automatically when the byte counter value equals to the threshold value set in the **`UCB1TBCNT`** register.
6. Enable the related interrupt sources, such as receive interrupt (**`UCRXIE0`**), NAK interrupt (**`UCNACKIE`**) and byte counter interrupt (**`UCBCNTIE`**), by configuring the **`UCB1IE`** register.
7. A **`while()`** loop should be used to check and send the **`STOP`** condition periodically for each data communication operation.
8. A **`START`** condition should be sent to the slave to begin the data receive operation.
9. The **`eUSCI_B1`** module ISR should be used to handle all of three interrupts enabled in step 5.
10. Finally, the received data should be collected and checked for the confirmation purpose.

Now, let's take a look at the slave side program.

8.5.10.3 Slave side program

The program in the slave side is used as the transmit mode to transmit 16-byte data to the master in the following operational sequence:

1. The I2C-related GPIO ports and the **`eUSCI_B1`** module for I2C mode-related registers should be initialized and configured as we discussed in Section 8.5.9.2.
2. Enable the **`eUSCI_B1`** module interrupt by setting NVIC_IPR5 and NVIC_ISER0 registers.
3. The **`UCB1I2COA0`** Register should be configured by setting the slave address **`0x48`** and setting the **`UCOAEN`** bit to enable this slave address.

4. Enable the related interrupt sources, such as transmit interrupt (**UCTXIE0**) and **STOP** condition interrupt (**UCSTPIE**), by configuring the **UCB1IE** register.
5. A **while()** loop should be used to wait for any interrupt to be occurred.
6. The **eUSCI_B1** module ISR should be used to handle all of two interrupts enabled in step 4.

Now, let's create and build the master side program **DRAI2CMaster**.

8.5.10.4 Create and build the master side program DRAI2CMaster

Perform the following operations to create a new project **DRAI2CMaster**:

1. Open the Windows Explorer window to create a new folder named **DRAI2CMaster** under the **C:\MSP432 Class Projects\Chapter 8** folder
2. Create a new Keil® ARM MDK µVersion5 Project **DRAI2CMaster** and save it to our new folder **DRAI2CMaster** created in step 1 above
3. On the next wizard, you need to select the device (MCU) **MSP432P401R** and click on the **OK** to close this wizard
4. Next the software components wizard is opened, and you need to set up the software development environment for your project with this wizard. Expand two icons, **CMSIS** and **Device**, and check the **CORE** and **Startup** checkboxes in the **Sel.** column, and click on the **OK** button since we need these two components to build our project

Since this project is relative simple, therefore only a C source file is good enough.

8.5.10.4.1 Create the source file DRAI2CMaster Create a new C source file **DRAI2CMaster.c** and enter the codes shown in Figure 8.49 into this file. Let's have a closer look at this piece of code to see how it works.

1. Some global variables are declared in lines 7~8.
2. The cods in lines 13~17 are used to initialize and configure related GPIO pins to work as **eUSCI_B1** I2C module and display the received data in some LEDs.
3. The GPIO pins **P2.0~P2.2** that are connected to a three-color LED are configured as the output pins in line 13. In line 14, they are set up to output 0.
4. In lines 15 and 16, the GPIO pin **P1.0** that is connected to the LED1 (red-color) in the MSP432P401R EVB is configured as an output pin and output a 0.
5. The GPIO pins **P6.4** and **P6.5** are configured as the **eUSCI_B1** for I2C mode (**UCB1SDA** and **UCB1SCL**) in line 17 by setting the **P6SEL1.x** and **P6SEL0.x** as **01** (**x** = 4 and 5).
6. The codes in lines 18~20 are used to set up and enable the interrupt related to **eUSCI_B1** module. The bits 13~15 in the **NVIC_IPR5** Register is used to set up the priority level and bit 21 in the **NVIC_ISER0** Register is used to enable the **eUSCI_B1** module's interrupt.
7. The codes in lines 22~28 are used to configure the **eUSCI_B1** module to work as the I2C mode.
8. In line 22, the **eUSCI_B1** module is first reset by setting the **UCSWRST** bit to 1.
9. Then the **eUSCI_B1** module is configured as an I2C master mode with the synchronous serial communication function in line 23 with some system macros, such

```
1  //*****************************************************************************************
2  // DRAI2CMaster.c - Main application file for project DRAI2CMaster
3  //*****************************************************************************************
4  #include <stdint.h>
5  #include <stdbool.h>
6  #include <msp.h>
7  uint8_t m = 0, RXData[16];
8  bool intFlag = true;
9  int main(void)
10 {
11    uint32_t i; uint8_t n, result[16];
12    WDTCTL = WDTPW | WDTHOLD;            // stop the watchdog timer
13    P2DIR = BIT0|BIT1|BIT2;              // set GPIO P2.0~P2.2 as output pins
14    P2OUT &= ~(BIT0|BIT1|BIT2);          // P2.0~P2.2 output 0
15    P1OUT &= ~BIT0;                      // clear P1.0 output pin
16    P1DIR |= BIT0;                       // set GPIO P1.0 as output pin
17    P6SEL0 |= BIT4|BIT5;                 // set GPIO P6.4 & P6.5 as USCIB1 IC2 pins
18    NVIC_IPR5 = 0x00006000;              // USCB1 priority level = 3 (Bits 13~15)
19    NVIC_ISER0 = 0x00200000;             // USCIB1 IRQ number = 21
20    __enable_irq();
21    // configure USCI_B1 for I2C mode
22    UCB1CTLW0 |= UCSWRST;                          // software reset enabled
23    UCB1CTLW0 |= UCMODE_3 | UCMST | UCSYNC;        // I2C mode, Master mode, sync
24    UCB1CTLW1 |= UCASTP_2;                         // automatic stop generated as UCB1TBCNT
                                                     is reached
25    UCB1BRW = 0x008;                               // baud rate = SMCLK / 8
26    UCB1TBCNT = 0x0010;                            // number of bytes to be received (16)
27    UCB1I2CSA = 0x0048;                            // set slave address as 0x48
28    UCB1CTLW0 &= ~UCSWRST;                         // enable USCIB1 module
29    UCB1IE |= UCRXIE0 | UCNACKIE | UCBCNTIE;       // interrupt sources: RX, NACK, BCNT
30    while (intFlag)                                // wait for any interrupt to be occurred
31    {
32     for (i = 2000; i > 0; i--);                   // delay program for a period of time
33     while (UCB1CTLW0 & UCTXSTP);                  // wait & ensure any previous STOP
                                                     condition has been sent
34     UCB1CTLW0 |= UCTXSTT;                         // send START condition to begin I2C
35    }
36    for (n = 0; n < 16; n++) {result[n] = RXData[n];}  // collect 16 received data from slave
37    while(1);                                      // used to check the receiving result
38 }
39 void EUSCIB1_IRQHandler(void)                     // EUSCIB1 interrupt service routine
40 {
41    if (UCB1IFG & UCNACKIFG)                       // A NAK is received
42    {
43     UCB1IFG &= ~UCNACKIFG;                        // reset the NAK
44     UCB1CTLW0 |= UCTXSTT;                         // send restart condition
45    }
46    if (UCB1IFG & UCRXIFG0)                        // a data is received from slave
47    {
48     UCB1IFG &= ~UCRXIFG0;                         // reset the data receiving flag
49     RXData[m] = UCB1RXBUF;                        // save the RX data
50     P2OUT = RXData[m]; m++;                       // display the data on P2.0~P2.2
51    }
52    if (UCB1IFG & UCBCNTIFG)                       // the UCBCNT arrived the target number
                                                     of bytes
53    {
54     UCB1IFG &= ~UCBCNTIFG;                        // reset the UCBCNT flag
55     P1OUT = BIT0;                                 // set the red LED on P1.0
56     intFlag = false;                              // release the while() loop
57    }
58 }
```

Figure 8.49 Codes for the C source file DRAI2CMaster.c.

as **UCMODE_3**, **UCMST**, and **UCSYNC**. One can use the real hexadecimal numbers to replace these macros.

10. In line 24, the **eUSCI_B1** module is configured to generate a **STOP** condition automatically as the byte number defined in the **UCB1TBCNT** register hits its threshold value. This configuration is accomplished by setting the **UCASTP1** bit field in the **UCBxCTLW1** Register.
11. The **UCB1BRW** register is configured by setting **0x8** in line 25 to generate a bit rate as **SMCLK/8**.
12. In line 26, the threshold value in the **UCB1TBCNT** register is defined as 16 (**0x10**) to allow a **STOP** condition to be generated automatically as 16-byte data have been read.
13. The slave address is defined as **0x48** and assigned to the **UCB1I2CSA** register in line 27.
14. In line 28, after the **eUSCI_B1** module has been configured, it is enabled by clearing the **UCSWRST** bit to 0.
15. In line 29, three interrupt sources, receive interrupt (**UCRXIE0**), NAK interrupt (**UCNACKIE**), and byte counter interrupt (**UCBCNTIE**), are enabled to make the module to be ready to generate these interrupts if any of them is occurred.
16. The codes in lines 30~34 are used to delay the program a period of time and check whether the previous **STOP** condition is sent. In fact, all 16 receive interrupts occurred during that time delay period and all received data can be collected during that period. The code line 33 is not exactly necessary since most systems will not send any **STOP** condition after reset. This **while()** loop is only used to check and make sure that any previous **STOP** has been sent and cleared. The **UCTXSTP** bit should be reset to 0 as soon as a **STOP** condition is sent. As that happened, a **START** condition is sent to the slave to begin or restart the data receiving operation. A point to be noted is that the condition for this **while()** loop is a Boolean variable **intFlag**, which is **True** set by the user. This **while()** loop will be terminated if this condition becomes to **False** later.
17. In line 36, if the **while()** loop is terminated, all received data from the slave are collected to a user-defined array **result[]** that can be checked later.
18. Another **while()** loop is used in line 37 to temporarily halt the program to enable us to check the received data later.
19. The codes in lines 39~57 belong to the **eUSCI_B1** ISR.
20. The first **if()** block, coding in lines 41~45, is used to check whether a NAK interrupt occurred, which means that a NAK is received from the slave. In that case, a **START** condition is sent again to the slave to ask it to restart its data transmit operation.
21. The second **if()** block, coding in lines 46~51, is used to check whether a data receiving interrupt occurred, which means that a byte has been received from the slave. If that happened, the received data byte is assigned to the global data array **RXData[]** and displayed in the three-color LED via GPIO pins **P2.0~P2.2**.
22. The third **if()** block, coding in lines 52~57, is used to check whether a byte counter interrupt occurred, which means that all 16-byte data have been successfully received. In that case, the red color LED is turned on via **P1.0** to indicate this situation. Also the global variable **intFlag** is reset to **False** to stop the **while()** loop executed in line 30.

For all three **if()** blocks, if the checked interrupt occurred, the related interrupt must be reset or released inside the **eUSCI_B1** ISR to enable the next interrupt to occur again in the future.

Now, let's continue to take care of the slave side program.

8.5.10.5 Create and build the slave side program DRAI2CSlave

Create a new folder **DRAI2CSlave** under the **C:\MSP432 Class Projects\Chapter 8** folder and create a new µVersion5 project **DRAI2CSlave**, and add it to that new folder.

Then create a new C source file **DRAI2CSlave.c** and add it into the project generated above. Enter the code shown in Figure 8.50 into this source file.

Let's have a closer look at the code in this part.

1. A global variable **TXData** is declared in line 5 and it is used to hold the data byte value to be transmitted later.
2. The GPIO pins **P6.4** and **P6.5** are configured as the **eUSCI_B1** for I2C mode (**UCB1SDA** and **UCB1SCL**) in line 10 by setting the **P6SEL1.x** and **P6SEL0.x** as **01** (**x** = 4 and 5).
3. The codes in lines 11~13 are used to set up and enable the interrupt related to **eUSCI_B1** module. The bits 13~15 in the **NVIC_IPR5** Register is used to set up the priority level and bit 21 in the **NVIC_ISER0** Register is used to enable the **eUSCI_B1** module's interrupt. Refer to Chapter 5 to get more details about these interrupt setups.
4. All interrupts are globally enabled in line 13.
5. In line 15, the **eUSCI_B1** module is first reset by setting the **UCSWRST** bit to 1.
6. The codes in lines 16~17 are used to configure the **eUSCI_B1** module as an I2C slave mode with the synchronous serial communication function with some system

```
1  //****************************************************************************************
2  // DRAI2CSlave.c - Main application for project DRAI2CSlave
3  //****************************************************************************************
4  #include <msp.h>
5  uint8_t TXData;
6  int main(void)
7  {
8     WDTCTL = WDTPW | WDTHOLD;            // stop the watchdog timer
9     // configure GPIO pins
10    P6SEL0 |= BIT4 | BIT5;               // set GPIO P6.4 & P6.5 as UCB1 pins
11    NVIC_IPR5 = 0x00006000;              // set UCB1 priority level as 3 (NVIC_TPR5 bits 13~15)
12    NVIC_ISER0 = 0x00200000;             // set UCB1 enable (IRQ number = 21)
13    enable_irq();
14    // Configure USCI_B1 for I2C mode
15    UCB1CTLW0 = UCSWRST;                 // reset eUSCI_B1 module to configure
16    UCB1CTLW0 |= UCMODE_3 | UCSYNC;      // I2C mode, sync mode
17    UCB1I2COA0 = 0x48 | UCOAEN;          // own address is 0x48 + enable
18    UCB1CTLW0 &= ~UCSWRST;               // clear reset register to enable eUSCI_B1 module
19    UCB1IE |= UCTXIE0 | UCSTPIE;         // interrupt sources: TX, STOP condition
20    while(1);                            // wait for any interrupt to be occurred...
21 }
22 void EUSCIB1_IRQHandler(void)           // UCB1 interrupt service routine
23 {
24    if (UCB1IFG & UCSTPIFG)              // STOP condition interrupt
25    {
26       UCB1IFG &= ~UCSTPIFG;             // reset the STOP flag
27       TXData = 0;                       // initiate TX data as 0
28    }
29    if (UCB1IFG & UCTXIFG0)              // TX interrupt
30    {
31       UCB1IFG &= ~UCTXIFG0;             // clear TX interrupt flag
32       UCB1TXBUF = TXData++;             // increment TX data & send to the TXBUF
33    }
34 }
```

Figure 8.50 Code for the C source file DRAI2CSlave.c.

macros, such as **UCMODE_3** and **UCSYNC**. One can use the real hexadecimal numbers to replace these macros.
7. In line 18, after the **eUSCI_B1** module has been configured, it is enabled by clearing the **UCSWRST** bit to 0.
8. In line 19, two interrupt sources, transmit interrupt (**UCTXIE0**) and STOP condition interrupt (**UCSTPIE**), are enabled to make the module to be ready to generate these interrupts if any of them is occurred.
9. An infinitive **while()** loop is executed in line 20 to wait for any interrupt to be generated.
10. The codes in lines 22~33 belong to the **eUSCI_B1** ISR.
11. The first **if()** block, coding in lines 24~28, is used to check whether a **STOP** interrupt occurred, which means that a **STOP** condition is received from the slave. In that case, the data byte to be transmitted is reset to 0 to make it to be resent.
12. The second **if()** block, coding in lines 29~33, is used to check whether a data transmit interrupt occurred, which means that a byte has been transmitted by the slave. If that happened, the transmit data byte is incremented by 1 and assigned to the **UCB1TXBUF**.

Now, let's set up the environment to build and run our project to test the I2C module function.

8.5.10.6 Set up the environment to build and run the project

Before you can build and run the project, make sure that the following two issues have been set up correctly:

- The debugger used for this project is **CMSIS-DAP Debugger**. This can be configured in the **Debug** tab under the menu item **Project|Options for Target "Target 1."**
- Also make sure that all settings for the debugger and the flash download are correct.

Now go to the **Project|Build target** menu item to build the project. Then go to the menu item **Flash|Download** to download the project image file into the flash memory.

Before you can run the projects, make sure to set a break point for the master program in line 37 (**while(1)** loop line) since we need to temporarily halt the program to check the data receiving results.

When running the projects, first run the slave program and then run the master program.

As the project runs, you can find that the three-color LED installed on the MSP432P401R EVB and the red color LED, LED1, are both on. This indicates that our projects run successfully.

Now the master program will be temporarily stopped at the break point. To check the running result, open the **Call Stack + Locals** window located at the lower right corner. Then expand the **result[]** array, you can find all 16-byte data have been successfully received, as shown in Figure 8.51.

8.5.11 I2C API functions provided by MSPWare Peripheral Driver Library

In this section, we will introduce another way to access and interface to the **eUSCI_B** modules for I2C mode to perform desired I2C serial communication tasks. In the MSPPDL, there are about 40 API functions used to interface to I2C modules. However, in this section,

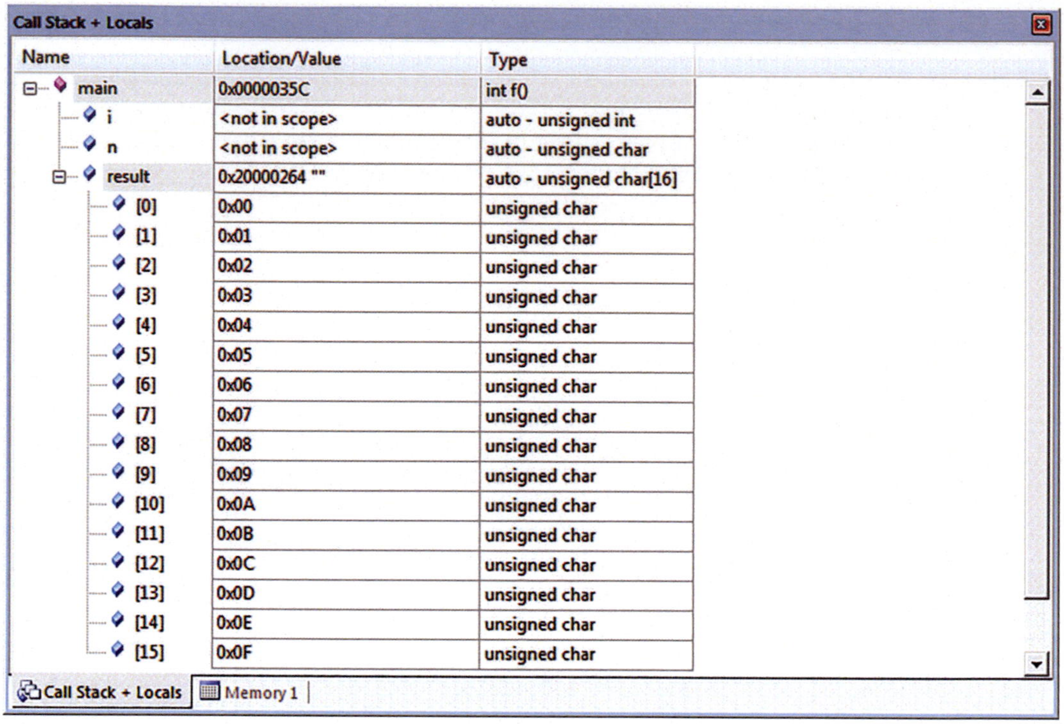

Call Stack + Locals

Name	Location/Value	Type
main	0x0000035C	int f()
i	<not in scope>	auto - unsigned int
n	<not in scope>	auto - unsigned char
result	0x20000264 ""	auto - unsigned char[16]
[0]	0x00	unsigned char
[1]	0x01	unsigned char
[2]	0x02	unsigned char
[3]	0x03	unsigned char
[4]	0x04	unsigned char
[5]	0x05	unsigned char
[6]	0x06	unsigned char
[7]	0x07	unsigned char
[8]	0x08	unsigned char
[9]	0x09	unsigned char
[10]	0x0A	unsigned char
[11]	0x0B	unsigned char
[12]	0x0C	unsigned char
[13]	0x0D	unsigned char
[14]	0x0E	unsigned char
[15]	0x0F	unsigned char

Call Stack + Locals | Memory 1

Figure 8.51 Running results for the projects (I2C master and slave).

we concentrate on some most important and popular master and slave API functions to illustrate how to use them to build sophisticated projects to efficiently access and interface to I2C modes to fulfill the desired serial data communication tasks.

Before we can get more details about these API functions, let's take a look at one important struct **eUSCI_I2C_MasterConfig** used to configure the master **eUSCI_B** modules for I2C mode.

The prototype of this struct is

```
typedef struct
{
   uint_fast8_t    selectClockSource;
   uint32_t        i2cClk;
   uint32_t        dataRate;
   uint_fast8_t    byteCounterThreshold;
   uint_fast8_t    autoSTOPGeneration;
} eUSCI_I2C_MasterConfig;
```

When initializing the master device, this struct should be used to make the configuration process easier.

These I2C API functions are provided to initialize the I2C modules, to send and receive data, obtain status, and to manage interrupts for the I2C modules. All of these API functions and struct are contained in a C file **i2c.c** with a header file **i2c.h**. Both files are located at the folder **C:\ti\msp\MSPWare_2_21_00_39\driverlib\driverlib\MSP432P4xx** in your host computer.

8.5.11.1 Master operations

When using these APIs to drive the I2C master module, the user must perform the I2C module configurations and data operations in the following sequence:

1. First initialize the I2C master module with a call to **I2C_initMaster()**. That function uses the struct **eUSCI_I2C_MasterConfig** to set and enable the master module.
2. Then the user may transmit or receive data via I2C master module. Data are transferred by first setting the slave address using **I2C_setSlaveAddress()**.
3. The function **I2C_setMode()** is used to define whether the transfer is a send (a write to the slave from the master) or a receive (a read from the slave by the master).
4. Now the I2C module can be enabled by calling the function **I2C_enableModule()**.
5. If an interrupt mechanism is used, the function **I2C_enableInterrupt()** should be called.
6. The transaction can then be initiated on the bus by calling the transmit or receive related APIs as listed below:
   a. **I2C_masterSendSingleByte()**
   b. **I2C_masterSendMultiByteStart()**
   c. **I2C_masterSendMultiByteNext()**
   d. **I2C_masterSendMultiByteFinish()**
   e. **I2C_masterSendMultiByteStop()**
   f. **I2C_masterReceiveSingleByte()**
   g. **I2C_masterReceiveStart()**
   h. **I2C_masterReceiveMultiByteNext()**
   i. **I2C_masterReceiveMultiByteFinish()**
   j. **I2C_masterReceiveMultiByteStop()**

8.5.11.2 I2C module master initialization and configuration API functions

Table 8.23 lists these master mode initialization and configuration functions.

The function **I2C_initMaster()** is an important function used to set up and configure the I2C module working in a master mode. Only after the I2C module is correctly configured, it can work properly to provide the normal functions. Before performing any data operations, the I2C bus must be checked to confirm that it is in the idle status by using **I2C_isBusBusy()** function. The **I2C_enableModule()** function must be called before the master can perform any data operation.

8.5.11.3 I2C module master sending and receiving data API functions

Table 8.24 shows some popular I2C master sending and receiving data bytes API functions.

Generally, the transaction can be processed in the following sequence:

1. For the single send and receive cases, the polling method involves looping on the return from **I2C_isBusBusy()**. Once that function indicates that the I2C master is not busy, the bus transaction has been completed and can be checked from related data destination.
2. If there are no errors, then the data have been sent or is ready to be read using the function **I2C_masterReceiveSingleByte()**.
3. For the multibyte send and receive cases, the polling method also involves calling the **I2C_isBusBusy()** function for each byte transmitted or received (using **I2C_masterSendMulti ByteNext()** or **I2C_masterReceiveMultiByteNext()**), and

Table 8.23 I2C master module initialization and configuration functions

API function	Parameter	Description
void **I2C_initMaster** **(**uint32_t mInst, const eUSCI_I2C_MasterConfig * config**)**	**mInst** is the instance of the eUSCI B (I2C) module. Valid values can include: **EUSCI_B0_MODULE** **EUSCI_B1_MODULE** **EUSCI_B2_MODULE** **EUSCI_B3_MODULE** **config** is a pointer of the struct eUSCI_I2C_Master-Config	Initialize operation of the I2C master blockUpon successful initialization of the I2C block, this function will have set the bus speed for the master; however, I2C module is still disabled till **I2C_enableModule()** is invoked **selectClockSource** is the clock source. Valid values are: **EUSCI_B_I2C_CLOCK-SOURCE_ACLK** **EUSCI_B_I2C_CLOCK-SOURCE_SMCLK** **i2cClk** is the rate of the clock supplied to the I2C module (the frequency in Hz of the clock source specified in selectClock-Source) **dataRate** set up data transfer rate. Valid values are: **EUSCI_B_I2C_SET_DATA_RATE_400KBPS** **EUSCI_B_I2C_SET_DATA_RATE_100KBPS** **byteCounterThreshold** sets threshold for automatic STOP **autoSTOPGeneration** sets up the STOP condition generation: **EUSCI_B_I2C_NO_AUTO_STOP** **EUSCI_B_I2C_SET_BYTE-COUNT_THRESHOLD_FLAG** **EUSCI_B_I2C_SEND_STOP_AUTOMATICALLY_ON_BYTE-COUNT_THRESHOLD**
void **I2C_setSlaveAddress** **(**uint32_t mInst, uint_fast16_t sAddress**)**	**mInst** is the instance of the eUSCI B (I2C) module. Valid values can include: **EUSCI_B0_MODULE** **EUSCI_B1_MODULE** **EUSCI_B2_MODULE** **EUSCI_B3_MODULE** **sAddress** is 7-bit slave address	This function will set the address that the I2C master will place on the bus when initiating a transaction. Modified register is **UCBx-I2CSA** register

(*Continued*)

Table 8.23 (Continued) I2C master module initialization and configuration functions

API function	Parameter	Description
void **I2C_setMode(**uint32_t mInst, uint_fast8_t mode**)**	**mInst** is the instance of the eUSCI B (I2C) module. Valid values are same as those in the **I2C_initMaster()** above	Set the operational mode of the I2C device **mode** is in transmit/receive mode, valid values are: **EUSCI_B_I2C_TRANSMIT_MODE** **EUSCI_B_I2C_RECEIVE_MODE** [Default value]
void **I2C_enableModule** **(**uint32_t mInst**)**	**mInst** is the instance of the eUSCI B (I2C) module. Valid values are same as those in the **I2C_initMaster()** above	This will enable operation of the I2C block. Modified bits are **UCSWRST** of **UCBxCTL1** register

for the last byte sent or received (using **`I2C_masterSendMultiByteFinish()`** or **`I2C_masterReceiveMultiByteFinish()`** functions).

4. If any error is detected during the burst transfer, the **`I2C_masterSendStart()`** function should be called to restart the failed transaction or the **`I2C_masterIsStopSent()`** is called to send a STOP condition to terminate the transaction.
5. For the interrupt-driven transaction, the user must register an interrupt handler for the I2C devices and enable the I2C master interrupt; the interrupt occurs when the master is no longer busy.

Now, let's have a closer look at those slave side API functions.

8.5.11.4 Slave operations

When working in the slave module, the APIs need to be invoked in the following order:

- **`I2C_initSlave()`**
- **`I2C_setMode()`**
- **`I2C_enableModule()`**
- **`I2C_enableInterrupt()`** (if interrupt mechanisms are used)

8.5.11.5 I2C module slave initialization and configuration API functions

Table 8.25 lists these slave mode initialization and configuration functions. Refer to Table 8.23 to get details about the functions **`I2C_setMode()`** and **`I2C_enableModule()`**.

The **`I2C_initSlave()`** function should be called first to initialize the slave module to work in the I2C mode and set the slave address. This is followed by a call to set the mode of operation, either transmit or receive. The I2C module may now be enabled by calling the **`I2C_enableModule()`**. It is recommended to enable the I2C module before enabling the interrupts. Any transmission or reception of data may be initiated at this point after interrupts are enabled.

8.5.11.6 I2C module slave sending and receiving data API functions

The transaction can then be initiated on the bus by calling the transmit or receive-related API functions as listed below.

- For a slave transmission, the API function **`I2C_slavePutData()`** should be called.
- For a slave reception, the API function **`I2C_slaveGetData()`** should be called.

Table 8.24 I2C module master sending and receiving data functions

API function	Parameter	Description
void **I2C_masterSendSingleByte** (uint32_t mInst, uint8_t txData**)**	**mInst** is the instance of the eUSCI B (I2C) module. Valid values can include: **EUSCI_B0_MODULE** **EUSCI_B1_MODULE** **EUSCI_B2_MODULE** **EUSCI_B3_MODULE** **txData** is the data byte to be transmitted	This function is used by the master module to send a single byte. This function sends **START**, transmits the byte to the slave, and then sends **STOP** Modified registers are **UCBxIE**, **UCBxCTL1**, **UCBxIFG**, **UCBxTXBUF**
void **I2C_masterSendMultiByteStart(**uint32_t mInst, uint8_t txData**)**	**mInst** is the instance of the eUSCI B (I2C) module. Valid values are same as those used above **txData** is the first data byte to be transmitted	This function is used by the master module to send the first byte. This function sends **START**, and then transmits the first data byte of a multibyte transmission to the slave Modified registers are **UCBxIE**, **UCBxCTL1**, **UCBxIFG**, **UCBxTXBUF**
void **I2C_masterSendMultiByteNext(**uint32_t mInst, uint8_t txData**)**	**mInst** is the instance of the eUSCI B (I2C) module. Valid values are same as those used above **txData** is the next data byte to be transmitted	This function is used by the master module to continue each byte of a multibyte transmission. This function transmits each data byte of a multibyte transmission to the slave Modified register is **UCBxTXBUF**
void **I2C_masterSendMultiByteFinish(**uint32_t mInst, uint8_t txData**)**	**mInst** is the instance of the eUSCI B (I2C) module. Valid values are same as those used above **txData** is the last data byte to be transmitted	This function is used by the master module to send the last byte and **STOP**. This function transmits the last data byte of a multibyte transmission to the slave, and sends the **STOP** Modified registers are **UCBxTXBUF** and **UCBxCTL0**
void **I2C_masterSendMultiByteStop(**uint32_t mInst)	**mInst** is the instance of the eUSCI B module. Valid values are same as those used above	This function is used by the master module to send **STOP** at the end of a multibyte transmission This function sends a **STOP** after current transmission is complete Modified bits are **UCTXSTP** bit of **UCBxCTL0**
uint8_t **I2C_masterReceiveSingleByte(**uint32_t mInst**)**	**mInst** is the instance of the eUSCI B module. Valid values are same as those used above	This function is used by the master module to receive a single byte. This function sends START and STOP, then waits for data reception, and receives one byte from the slave Modified registers: **UCBxIE**, **UCBxCTL0**, **UCBxIFG**, **UCBxTXBUF**
void **I2C_masterReceiveStart** **(**uint32_t mInst**)**	**mInst** is the instance of the eUSCI B module. Valid values are same as those used above	This function is used by the master module to initiate reception of a single byte. This function sends **START** Modified bits are **UCTXSTT** bit of **UCBxCTL0**

(*Continued*)

Table 8.24 (Continued) I2C module master sending and receiving data functions

API function	Parameter	Description
uint8_t **I2C_mas-terReceiveMulti ByteNext(**uint32_t mInst**)**	**mInst** is the instance of the eUSCI B module. Valid values are same as those used above.	This function is used by the master module to receive each byte of a multibyte reception. This function reads currently received byte Modified register: **UCBxRXBUF** **Returns:** Received byte at master side
uint8_t **I2C_mas-terReceiveMulti ByteFinish** **(**uint32_t mInst)	**mInst** is the instance of the eUSCI B module. Valid values are same as those used above	This function is used by the master module to initiate completion of a multibyte reception. This function receives the current byte and initiates the **STOP** from master to slave Modified bits are **UCTXSTP** bit of **UCBxCTL0**
void **I2C_master ReceiveMultiByte Stop(**uint32_t mInst**)**	**mInst** is the instance of the eUSCI B module. Valid values are same as those used above	This function is used by the master module to initiate **STOP** Modified bits are **UCTXSTP** bit of **UCBxCTL0**

Table 8.26 shows these I2C module slave sending and receiving API functions.

8.5.11.7 I2C module interrupt handling and processing API functions

The following API functions are involved in this group:

- `I2C_registerInterrupt()`
- `I2C_enableInterrupt()`
- `I2C_getInterruptStatus()`
- `I2C_clearInterruptFlag()`

Table 8.25 I2C slave module initialization and configuration functions

API function	Parameter	Description
void **I2C_ initSlave(**uint32_t mInst, uint_fast16_t sAddress, uint_fast8_t sAddrOffset, uint32_t sOwnAddrEnable**)**	**mInst** is the instance of the eUSCI B (I2C) module. Valid values can include: **EUSCI_B0_MODULE** **EUSCI_B1_MODULE** **EUSCI_B2_MODULE** **EUSCI_B3_MODULE** **sAddress:** 7-bit slave address	Initialize operation of the I2C as a slave mode. Upon successful initialization of the I2C blocks, this function will have set the slave address but I2C module is still disabled until **I2C_enable-Module()** is invoked The parameter **sAddress** is the value that will be compared with the slave address sent by an I2C master **sAddrOffset:** Own address offset referred to "x" value of UCBxI2COAx. Valid values are: **EUSCI_B_I2C_OWN_ADDRESS_OFFSET0,** **EUSCI_B_I2C_OWN_ADDRESS_OFFSET1,** **EUSCI_B_I2C_OWN_ADDRESS_OFFSET2,** **EUSCI_B_I2C_OWN_ADDRESS_OFFSET3** **sOwnAddrEnable:** Select if the specified address is enabled or disabled. Valid values are: **EUSCI_B_I2C_OWN_ADDRESS_DISABLE,** **EUSCI_B_I2C_OWN_ADDRESS_ENABLE**

Table 8.26 I2C module slave sending and receiving data functions

API function	Parameter	Description
void **I2C_slavePutData(**uint32_t mInst, uint8_t txData**)**	**mInst** is the instance of the eUSCI B (I2C) module. Valid values can include: **EUSCI_B0_MODULE** **EUSCI_B1_MODULE** **EUSCI_B2_MODULE** **EUSCI_B3_MODULE** **txData** is the data byte to be transmitted	This function will place the supplied data into I2C transmit data register to start transmission Modified register is **UCBxTXBUF** register **Returns:** None
uint8_t **I2C_slaveGetData(**uint32_t mInst**)**	**mInst** is the instance of the eUSCI B (I2C) module. Valid values are same as those used above	Read a byte of data from the I2C receive data register **Returns:** Returns the byte received from the I2C module, cast as an uint8_t. Modified bit is **UCBxRXBUF** register

Table 8.27 lists these most popular interrupt handling and processing API functions.

The points to be noted when using these API functions to set and handle any I2C-related interrupts are:

- The desired interrupt handler must be first registered by using **`I2C_registerInterrupt()`**.
- The desired interrupt sources must be enabled by using **`I2C_enableInterrupt()`**.
- To check the current interrupt source, the **`I2C_getInterruptStatus()`** should be used.
- As the desired interrupt has been responded and processed, it must be cleared by using the **`I2C_clearInterruptFlag()`** inside the interrupt handler.

Two I2C projects, **`Lab8_6(SDI2CMaster)`** and **`Lab8_7(SDI2CSlave)`**, which should be developed by students as lab projects with these I2C module API functions, can be found from the MSP432 lab projects part in the homework section of this chapter. Two MSP-EXP432P401R EVBs need to be connected with two I2C bus to run these two projects. The **`SDI2CMaster`** needs to be run at a master device to transmit six letters, **`MSP432`**, to the slave device, in which the **`SDI2CSlave`** program should be executed.

8.6 *Universal asynchronous receivers/transmitters*

The Universal Asynchronous Receivers and Transmitters (UARTs) provide asynchronous serial data communications for MCU- and UART-compatible devices. The UART is very similar to SSI in working principle and structure, but the only difference is that the former belongs to the asynchronous communications and the latter is the synchronous SCI.

The UART performs the parallel-to-serial functions in the transmitting mode and serial-to-parallel conversions in the receiving mode. It is very similar in functionality to a popular UART, but is not register compatible with any UART.

Table 8.27 I2C module interrupt handling and processing functions

API function	Parameter	Description
void **I2C_register-Interrupt(**uint32_t mInst, void(*)(void) intHandler**)**	**mInst** is the instance of the eUSCI B (I2C) module. Valid values can include: **EUSCI_B0_MODULE** **EUSCI_B1_MODULE** **EUSCI_B2_MODULE** **EUSCI_B3_MODULE** **intHandler** is a pointer to the function to be called when the interrupt occurs	Register the handler to be called when an I2C interrupt occurs. This function enables the global interrupt; specific I2C interrupts must be enabled via **I2C_enableInterrupt()**. It is the interrupt handler's responsibility to clear the interrupt source via the function **I2C_clearInterrupt-Flag()**
void **I2C_enable-Interrupt(**uint32_t mInst, uint_fast16_t mask**)**	**mInst** is the instance of the eUSCI B module. Valid values are same as those used above **mask** is the bit mask of the interrupt sources to be enabled	Enable individual I2C interrupt sources The **mask** is the logical OR of any of the following: **EUSCI_B_I2C_STOP_INTERRUPT**—STOP condition interrupt **EUSCI_B_I2C_START_INTERRUPT**—START condition interrupt **EUSCI_B_I2C_TRANSMIT_INTERRUPT0**—transmit interrupt0 **EUSCI_B_I2C_TRANSMIT_INTERRUPT1**—transmit interrupt1 **EUSCI_B_I2C_TRANSMIT_INTERRUPT2**—transmit interrupt2 **EUSCI_B_I2C_TRANSMIT_INTERRUPT3**—transmit interrupt3 **EUSCI_B_I2C_RECEIVE_INTERRUPT0**—receive interrupt0 **EUSCI_B_I2C_RECEIVE_INTERRUPT1**—receive interrupt1. **EUSCI_B_I2C_RECEIVE_INTERRUPT2**—receive interrupt2 **EUSCI_B_I2C_RECEIVE_INTERRUPT3**—receive interrupt3 **EUSCI_B_I2C_NAK_INTERRUPT**—not-acknowledge interrupt **EUSCI_B_I2C_ARBITRATIONLOST_INTER-RUPT**—arbitration lost interrupt **EUSCI_B_I2C_BIT9_POSITION_INTER-RUPT**—bit position 9 interrupt enable **EUSCI_B_I2C_CLOCK_LOW_TIMEOUT_INTERRUPT**—clock low timeout interrupt enable **EUSCI_B_I2C_BYTE_COUNTER_INTER-RUPT**— Byte counter interrupt enable

(*Continued*)

Table 8.27 (Continued) I2C module interrupt handling and processing functions

API function	Parameter	Description
uint_fast16_t **I2C_getInterruptStatus** **(**uint32_t mInst, uint16_t mask**)**	**mInst** is the instance of the eUSCI B module. Valid values are same as those used above **mask** is the masked interrupt flag status to be returned	Get the current I2C interrupt status The **mask** is the logical OR of any of those values used in the **I2C_enableInterrupt()** function above The returned value is also identical with those masks used in the **I2C_enableInterrupt()** function above
void **I2C_clearInterrupt Flag(**uint32_t mInst, uint_fast16_t mask**)**	**mInst** is the instance of the eUSCI B module. Valid values are same as those used above **mask** is a bit mask of the interrupt sources to be cleared	Clear I2C interrupt sources The **mask** parameter has the same definition as the mask parameter to **I2C_enableInterrupt()**

Some of important features of the MSP432 UART include:

- Programmable baud rate generator
- Automatic generation and stripping of start, stop, and parity bits
- Line-break generation and detection
- Separate transmit and receive buffer registers and shift registers
- LSB-first or MSB-first data transmit and receive
- Programmable serial interface with
  - 7 or 8 data bits with even, odd, or no parity bit generation and detection
  - 1 or 2 stop bits generation
  - Baud rate generation, from 0 to processor clock/16
- Built-in idle-line and address-bit communication frames for multiprocessor systems
- Status flags for error detection, error suppression, and for address detection
- Independent interrupts for receive, transmit, start bit received, and transmit complete

First let's take a look at the asynchronous serial data communication protocols.

8.6.1 MSP432 asynchronous serial communication protocols and data framing

Unlike SPI, in which both transmitter and receiver use the same clock rate as the data operation timing base to ensure both master and slave transmit and receive data correctly, in asynchronous serial data communications, the transmitter and receiver use its own clock as the timing base, and therefore the predefined data communication protocols and data framing are necessary to make this kind of data transmitting and receiving successful.

As shown in Figure 8.52, each piece of serial data to be transmitted should be configured as:

1. A logic high is appeared on the communication line when the transmission line is idle.
2. A high-to-low transaction starts an asynchronously data communication.
3. The data can be 7- or 8-bit long attached with 0 or 1 parity bit and 1 or 2 stop bits.

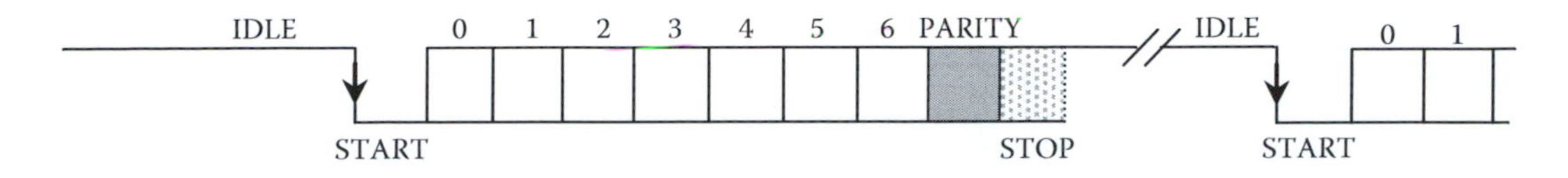

Figure 8.52 General asynchronous serial communication data framing.

Figure 8.52 shows an example of using 7 data bits with 1 parity and 1 stop bit protocol.

The MSP432P401R MCU system uses **eUSCI_A** module to provide four UART modules (**eUSCI_A0~eUSCI_A3**), and each module can work independently to perform asynchronous serial data communications. Each module can be programmed to transmit (**TX**) and receive (**RX**) data by using the different baud rates and each module can be driven by using the different system System Clock sources, such as UC0CLK, ACLK, and SMCLK.

However, in the MSP432P401R MCU system, an address-bit multiprocessor format is used as shown in Figure 8.53. In this frame, an address bit (address-bit mode) is added into this frame to enable the UART to handle the data operations with multiprocessor. The logics high and low are called **Mark** and **Space**, respectively. The **UCMSB** bit in the **UCAxCTLW0** Register controls the direction of the transfer and selects LSB- or MSB-first. LSB-first is typically required for UART.

Any UART device used in the eUSCI_A modules connect the device to an external system via two external pins, **UCAxRXD** and **UCAxTXD**, to transmit and receive data.

Generally, when two UART devices communicate asynchronously via **UCAxRXD** and **UCAxTXD** lines, the general asynchronous protocol or frame shown in Figure 8.52 is good enough and no address-bit multiprocessor format is required for the protocol. However, when three or more UART devices communicate together, the eUSCI_A module can provide supports for multi-UART devices with the following two frames:

- The idle-line and address-bit frame
- The multiprocessor communication frame

In this section, we only concentrate on the general UART device communications with the frame shown in Figure 8.52 since they are popular UART applications.

8.6.2 Architecture and organization of the eUSCI_A module for UART mode

Figure 8.54 shows a functional block diagram of the eUSCI_A module used to UART mode in the MSP432P401R MCU system. Four eUSCI_Ax modules, **eUSCI_A0~eUSCI_A3**, are

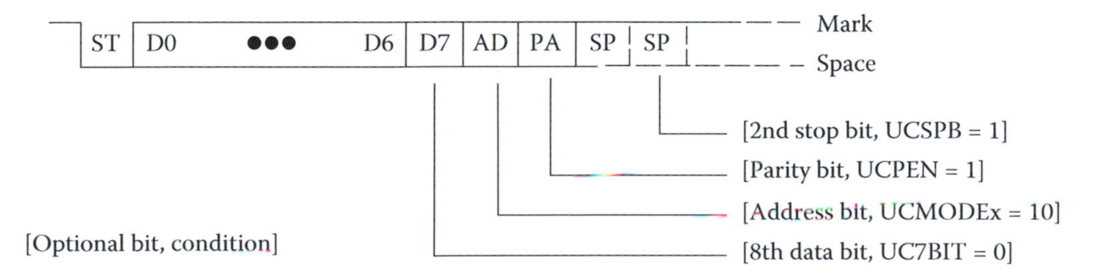

Figure 8.53 Address-bit multiprocessor serial communication data framing. (Courtesy Texas Instruments.)

configured to work as UART modes in this MCU system and each of them has the identical architecture as shown in Figure 8.54.

Each eUSCI_A module for UART mode comprises four components:

- UART data transmit control
- UART data receive control
- UART clock selection (CKSL) and bit rate generation control
- UART interrupt mechanism control

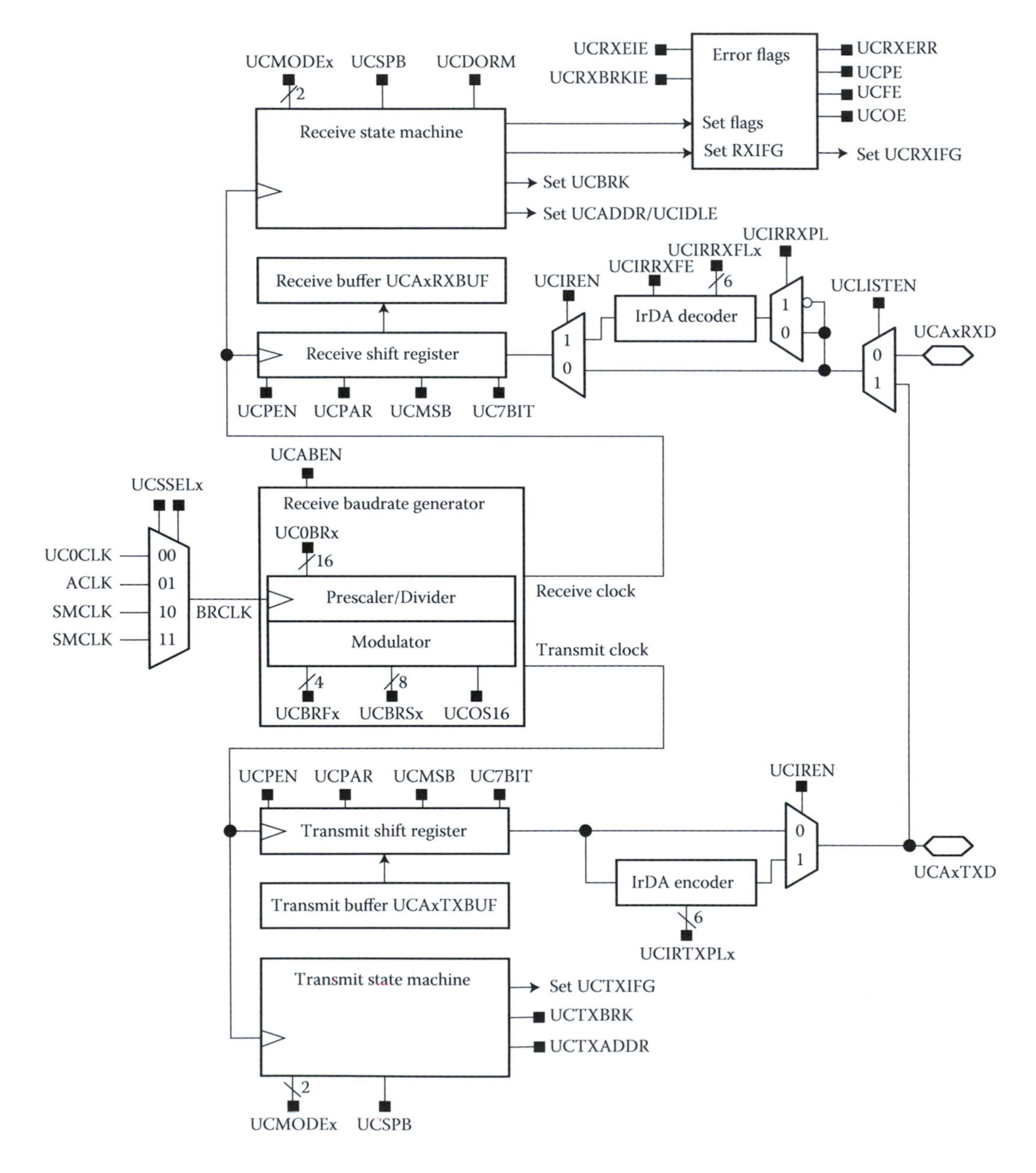

Figure 8.54 Functional block diagram for each eUSCI_Ax module as UART mode. (Courtesy Texas Instruments.)

Both UART data transmit and receive controls comprise data transmit–receive buffer and shift registers, and both actual operations are under control of the related control bits or bit fields in the eUSCI_Ax Control Word Register 0 (**UCAxCTLW0**). The UART clock source **BRCLK** can be selected from four possible system CKSEs by using the **UCSSELx** bit field in the **UCAxCTLW0** Register. The UART clock BRCLK can be further divided by the prescaler set in the eUSCI_Ax Baud Rate Control Word (**UCAxBRW**) register to get the desired baud rate used for data transmit and receive operations. Depending on the division factor **N** = BRCLK/Baud rate (>16 or < 16), which can be a noninteger value, a modulator is used to assist to determine the prescaler value. This modulator is controlled by the bits **UCBRFx**, **UCBRSx**, and **UCOS16** in the eUSCI_Ax Modulation Control Word (**UCAxMCTLW**) register.

The UART interrupt mechanism controls in the MSP432P401R MCU system include the following interrupt sources:

1. Start interrupt
2. Transmit interrupt
3. Receive interrupt
4. Transmit complete interrupt

Now, let's first have a closer look at these control registers and then discuss the operation sequence for data transmit and receive, baud rate generation, and interrupt processing.

8.6.3 Control registers used in eUSCI_A module for UART mode

All registers can be divided into the following four groups:

1. UART control and status registers
2. UART baud rate generation and control registers
3. UART data transmit and receive registers
4. UART interrupt processing registers

Let's discuss these registers one by one based on their groups.

8.6.3.1 UART control and status registers

Three registers are involved in this group:

1. The eUSCI_Ax Control Word Register 0 (**UCAxCTLW0**)
2. The eUSCI_Ax Control Word Register 1 (**UCAxCTLW1**)
3. The eUSCI_Ax Status (**UCAxSTATW**) register

The **UCAxCTLW0** Register is mainly used to configure the data transmit and receive format, eUSCI_A module operation mode, clock source selection, and some error detections. The bit configuration fort his register is shown in Figure 8.55. The bit function of this register is shown in Table 8.28.

15	14	13	12	11	10	9	8
UCPEN	UCPAR	UCMSB	UC7BIT	UCSPB	UCMODEx		UCSYNC
7	**6**	**5**	**4**	**3**	**2**	**1**	**0**
UCSSELx		UCRXEIE	UCBRKIE	UCDORM	UCTXADDR	UCTXBRK	UCSWRST

Figure 8.55 Bit configurations for the UCAxCTLW0 Register.

Table 8.28 Bit value and function for UCAxCTLW0 register

Bit	Field	Type	Reset	Function
15	**UCPEN**	RW	0	Parity enable: Parity bit is generated by **UCAxTXD** and expected by **UCAxRXD**. In address-bit multiprocessor mode, the address bit is included in the parity calculation **0:** Parity disabled; **1:** Parity enabled
14	**UCPAR**	RW	0	Parity select. **UCPAR** is not used when parity is disabled **0:** Odd parity; **1:** Even parity
13	**UCMSB**	RW	0	MSB-first select. Controls the direction of the receive and transmit shift register **0:** LSB first; **1:** MSB first
12	**UC7BIT**	RW	0	Character length. Selects 7-bit or 8-bit character length **0:** 8-bit data; **1:** 7-bit data
11	**UCSPB**	RW	0	Stop bit select. Number of stop bits **0:** One stop bit; **1:** Two stop bits
10:9	**UCMODEx**	RW	0	The eUSCI_A mode. These bits select the asynchronous mode as **UCSYNC** = 0. **00:** UART mode **01:** Idle-line multiprocessor mode **10:** Address-bit multiprocessor mode **11:** UART mode with automatic baud rate detection
8	**UCSYNC**	RW	0	Synchronous and asynchronous mode selection **0:** Asynchronous mode; **1:** Synchronous mode
7:6	**UCSSELx**	RW	0	The eUSCI_A clock source select. These bits select the BRCLK source clock **00:** UCLK; **01:** ACLK; **10:** SMCLK; **11:** SMCLK
5	**UCRXEIE**	RW	0	Receive erroneous-character interrupt enable: **0:** Error characters rejected and **UCRXIFG** bit in **UCAxIFG** register is not set **1:** Error characters received and **UCRXIFG** bit in **UCAxIFG** register is set
4	**UCBRKIE**	RW	0	Receive break character interrupt enable: **0:** Received break characters do not set **UCRXIFG** bit in **UCAxIFG** register **1:** Received break characters set **UCRXIFG** bit in **UCAxIFG** register
3	**UCDORM**	RW	0	Dormant. Puts eUSCI_A into sleep mode Only characters that are preceded by an idle-line or with address bit set UCRXIFG. In UART mode with automatic baud rate detection, only the combination of a break and synch field sets UCRXIFG **0:** Not dormant. All received characters set **UCRXIFG** bit in **UCAxIFG** register **1:** Dormant

(*Continued*)

Table 8.28 (Continued) Bit value and function for UCAxCTLW0 register

Bit	Field	Type	Reset	Function
2	**UCTXADDR**	RW	0	Transmit address. Next frame to be transmitted is marked as address, depending on the selected multiprocessor mode **0:** Next frame transmitted is data **1:** Next frame transmitted is an address
1	**UCTXBRK**	RW	0	Transmit break. Transmits a break with the next write to the transmit buffer. In UART mode with automatic baud rate detection, 055 h must be written into **UCAxTXBUF** to generate the required break/synch fields. Otherwise, 0 must be written into the transmit buffer **0:** Next frame transmitted is not a break **1:** Next frame transmitted is a break or a break/synch
0	**UCSWRST**	RW	1	Software reset enable **0:** Reset is disabled. The eUSCI_A is working in normal operation **1:** Reset is enabled. The eUSCI_A is in reset state (only in this state, initialization and configuration of the eUSCI_A can be performed)

One point to be noted is that all bits or bit fields that have a pattern background in this register can only be modified when the eUSCI_A module is in the reset status (**UCSWRST** = 1). As soon as the eUSCI_A module is enabled (**UCSWRST** = 0), these bits cannot be modified again.

Another point is that besides the eUSCI_Ax module Interrupt Enable (**UCAxIE**) register, some special errors can be enabled via this register, these errors include:

- Receive erroneous-character interrupt enable (bit **UCRXEIE**)
- Receive break character interrupt enable (bit **UCBRKIE**)

When any of these errors occurred, the **UCRXIFG** bit in the **UCAxIFG** register is set to indicate these errors. However, when a character is successfully received in the **UCAxRXBUF** register, this **UCRXIFG** bit is also set. In order to distinguish them, one needs to check the bits **UCRXEIE** and **UCBRKIE** in this register to identify whether data or an error has been received.

The **UCAxCTLW1** register is used to set up the length of the deglitch time $\mathbf{t_d}$ to prevent the eUSCI_A module from being accidentally started. Any glitch on the **UCAxRXD** pin shorter than the deglitch time $\mathbf{t_d}$ is ignored by the eUSCI_A. Only when a glitch is longer than $\mathbf{t_d}$, or a valid start bit occurs on the **UCAxRXD** pin, the eUSCI_A receive operation is started. The deglitch time $\mathbf{t_d}$ can be set to four different values using the **UCGLITx** bits, bits 1~0, in the **UCAxCTLW1** Register although it is a 16-bit register. The upper 14 bit is reserved. After a system reset, the default length of the deglitch time is set to 200 ns.

The **UCAxSTATW** register is mainly used to monitor the working or operating status of the eUSCI_A module. This is a 16-bit register, but only the lower 8 bits are used to monitor related working status of the eUSCI_A module. The bit configuration fort his register is shown in Figure 8.56 and the bit function of this register is shown in Table 8.29.

15	14	13	12	11	10	9	8
RESERVED							
7	**6**	**5**	**4**	**3**	**2**	**1**	**0**
UCLISTEN	UCFE	UCOE	UCPE	UCBRK	UCRXERR	UCADDR/UCIDLE	UCBUSY

Figure 8.56 Bit configurations for the UCAxSTATW register.

Table 8.29 Bit value and function for UCAxSTATW register

Bit	Field	Type	Reset	Function
15:8	**Reserved**	RO	0	Reserved
7	**UCLISTEN**	RW	0	Listen enable. The **UCLISTEN** bit selects loopback testing mode **0:** Disabled **1:** Enabled. **UCAxTXD** is internally fed back to the receiver pin **UCAxRXD**
6	**UCFE**	RW	0	Framing error flag. **UCFE** is cleared when **UCAxRXBUF** bit is read **0:** No error; **1:** Character received with framing error
5	**UCOE**	RW	0	Overrun error flag. This bit is set when a character is transferred into the **UCAxRXBUF** before the previous character was read. **UCOE** is cleared automatically when **UCxRXBUF** bit is read, and must not be cleared by software Otherwise, it does not function correctly **0:** No error; **1:** Overrun error occurred
4	**UCPE**	RW	0	Parity error flag. When **UCPEN** = 0, **UCPE** is read as 0. **UCPE** is cleared when **UCAxRXBUF** bit is read **0:** No error; **1:** Character received with parity error
3	**UCBRK**	RW	0	Break detect flag. **UCBRK** is cleared when **UCAxRXBUF** bit is read **0:** No break condition; **1:** Break condition occurred
2	**UCRXERR**	RW	0	Receive error flag. This bit indicates a character was received with one or more errors. When **UCRXERR** = 1, on or more error flags, **UCFE**, **UCPE**, or **UCOE** is also set. **UCRXERR** is cleared when **UCAxRXBUF** bit is read **0:** No receive errors detected; **1:** Receive error detected
1	**UCADDR/ UCIDLE**	RW	0	**UCADDR**: Address received in address-bit multiprocessor mode. **UCADDR** is cleared when **UCAxRXBUF** bit is read **UCIDLE**: Idle line detected in idle-line multiprocessor mode. **UCIDLE** is cleared when **UCAxRXBUF** bit is read **0: UCADDR**: Received character is data. **UCIDLE**: No idle line detected **1: UCADDR**: Received character is an address. **UCIDLE**: Idle line detected
0	**UCBUSY**	RW	0	eUSCI_A busy. This bit indicates if a transmit or receive operation is in progress **0:** eUSCI_A is inactive **1:** eUSCI_A is transmitting or receiving data

The bit-7, **UCLISTEN**, can be used to enable the UART to work in a loopback format to test its function. When this bit is set, the **UCAxTXD** pin and the **UCAxRXD** pin can be internally connected together to form a loop. This bit can only be modified when the eUSCI_A module is in the reset status. Once the eUSCI_A module is enabled, it cannot be changed again.

The **UCRXERR** is a special bit and it is affected by some other error conditions. This bit is set to 1 if any of bits, such as **UCFE**, **UCOE**, and **UCPE**, is set to 1. This means that if any error happens, it can be ORed with any other error condition together to set the **UCRXERR** bit.

8.6.3.2 UART baud rate generation and control registers

Three registers are involved in this group:

1. The **UCAxBRW**
2. The **UCAxMCTLW**
3. The eUSCI_Ax Auto Baud Rate Control (**UCAxABCTL**) register

The **UCAxBRW** register is a 16-bit register and all its 16 bits or bit field **UCBRx** are used to store the prescaler value. This register can only be modified when the eUSCI_A module is in the reset status (**UCSWRST** = 1). As soon as the eUSCI_A module is enabled (**UCSWRST** = 1), no modification can be made for this register.

The **UCAxMCTLW** register is mainly used to support to select and generate the baud rate for the UART. The bit configuration of this register is shown in Figure 8.57.

The eUSCI_A baud rate generator can generate standard baud rates from nonstandard source frequencies. It provides two modes of operation selected by the **UCOS16** bit.

- When the **UCOS16** = 0, the low-frequency baud rate mode is selected. This mode allows generation of baud rates from low-frequency CKSEs, such as 9600 baud from a 32768-Hz crystal. In this mode, the baud rate generator uses one prescaler and one modulator to generate bit clock timing. This combination supports fractional divisors for baud rate generation. In this mode, the maximum eUSCI_A baud rate is 1/3 the UART source clock frequency **BRCLK**. The modulation is based on the values on the bit field **Second Modulation Stage Select**, **UCBRSx**, which is arranged from **0x00** to **0xFF**, in this register. A bit value of 1 in this bit field indicates that m = 1 and the corresponding **BITCLK** period is one **BRCLK** period longer than a **BITCLK** period with m = 0.
- When the **UCOS16** = 1, the high-frequency baud rate mode or the so-called oversampling mode is selected. This mode enables sampling a UART bit stream with higher input clock frequencies. This mode uses one prescaler and one modulator to generate the **BITCLK16** clock that is 16 times faster than the **BITCLK**. An additional divider by 16 and modulator stage generates **BITCLK** from **BITCLK16**. This combination supports fractional divisions of both **BITCLK16** and **BITCLK** for baud rate generation. In this mode, the maximum eUSCI_A baud rate is 1/16 the

15	14	13	12	11	10	9	8
UCBRSx							
7	**6**	**5**	**4**	**3**	**2**	**1**	**0**
UCBRFx				RESERVED			UCOS16

Figure 8.57 Bit configurations for the UCAxMCTLW register.

UART source clock frequency **BRCLK**. The modulation for **BITCLK16** is based on the bit field **First Modulation Stage Select**, **UCBRFx** setting in this register. A bit value of 1 in this bit field indicates that the corresponding **BITCLK16** period is one **BRCLK** period longer than the periods m = 0. The modulation restarts with each new bit timing. Modulation for **BITCLK** is based on the bit field **UCBRSx** setting.

The **UCAxABCTL** register is mainly used to support the auto baud rate detection function. When the bit field **UCMODEx** = 11, the UART is selected to work with the automatic baud rate detection mode. For automatic baud rate detection, a normal data frame is preceded by a synchronization sequence that consists of a break and a synch field. A break is detected when 11 or more continuous zeros (spaces) are received. If the length of the break exceeds 21 bit times, the break timeout error flag **UCBTOE** is set. The eUSCI_A module cannot transmit data while receiving the break/sync field.

8.6.3.3 UART data transmit and receive registers

Two registers are involved in this group:

1. The eUSCI_Ax Transmit Buffer Register **(UCAxTXBUF)**
2. The eUSCI_Ax Receive Buffer Register **(UCAxRXBUF)**

The **UCAxTXBUF** register is a 16-bit register but only the lower 8 bits are used to temporarily hold the data byte to be transmitted to the receiver. Similarly, the **UCAxRXBUF** register is also a 16-bit register with lower 8 bits being used to hold the received data byte from the transmitter.

The data transmit will not start until a valid data byte is written into the **UCAxTXBUF** register. Similarly, the data receiving will not start until a valid data byte has been received in the **UCAxRXBUF** register.

8.6.3.4 UART interrupt processing registers

Three registers are involved in this group:

1. The eUSCI_Ax Interrupt Enable Register **(UCAxIE)**
2. The eUSCI_Ax Interrupt Flag Register **(UCAxIFG)**
3. The eUSCI_Ax Interrupt Vector Register **(UCAxIV)**

The **UCAxIE** register is used to enable or disable four UART-related interrupt sources. This is a 16-bit register but only the lower 4 bits are used to enable (bit value = 1) or disable (bit value = 0) four UART-related interrupt sources:

1. Transmit complete interrupt enable, **UCTXCPTIE** (bit-3)
2. Start bit interrupt enable, **UCSTTIE** (bit-2)
3. Transmit interrupt enable, **UCTXIE** (bit-1)
4. Receive interrupt enable, **UCRXIE** (bit-0)

The difference between the **UCTXIE** and the **UCTXCPTIE** is that a **UCTX** interrupt is generated as soon as a data byte is written into the **UCAxTXBUF** register and a new transmit starts, but a **UCTXCPT** interrupt will not generated until a data byte is entirely shifted out from the internal shift register and the **UCAxTXBUF** register is empty.

The **UCAxIFG** register has the same bit configuration as that of **UCAxIE** register. Only the lower 4 bits, **UCTXCPTIFG**, **UCSTTIFG**, **UCTXIFG**, and **UCRXIFG**, are used to indicate whether one of four UART-related interrupt sources has generated an interrupt request. The order of these four interrupt flags is identical with the order in the **UCAxIE** register. If one of these 4 bits is set, the corresponding interrupt request will be generated. For example, if the bit **UCRXIFG** is set, it means that a data byte has been received and written into the **UCAxRXBUF**.

The **UCAxIV** register is a 16-bit register used to hold an interrupt vector or the entry address of an interrupt source that has the highest priority level. Table 8.30 shows all possible values for four interrupt sources.

Reading of the **UCAxIV** register automatically resets the highest pending interrupt condition and related flag bit set in the **UCAxIFG** register. Disabled interrupts do not affect the **UCAxIV** value.

8.6.4 Operational procedure of the UART and some important properties

Now that we have discussed all registers used in the eUSCI_A module for UART mode, let's have a clear picture about the operations of the UART mode and its properties.

The operation procedure of a UART can be divided into the following sections:

- The UART initialization and configuration
- The UART baud rate generation
- The UART transmit enable and data transmit operation
- The UART receive enable and data receive operation
- The UART errors detection

8.6.4.1 UART initialization and configuration

Generally, the following operations should be performed to initialize and configure an eUSCI_A module for the UART mode:

1. Initialize and configure GPIO-related ports and pins used for the UART mode
2. Disable the eUSCI_A module by setting the **UCSWRST** bit in the **UCAxCTLW0** Register

Table 8.30 Bit value and function for UCAxIV register

Bit	Field	Type	Reset	Function
15:0	**UCIVx**	RO	0	The eUSCI_A interrupt vector value **00:** No interrupt pending. **02:** Interrupt source: Receive buffer full; Interrupt flag: **UCRXIFG**; Interrupt priority: Highest **04:** Interrupt source: Transmit buffer empty; Interrupt flag: **UCTXIFG** **06:** Interrupt source: Start bit received; Interrupt flag: **UCSTTIFG** **08:** Interrupt source: Transmit complete; Interrupt flag: **UCTXCPTIFG**; Interrupt priority: Lowest

3. Set up correct UART data operation protocol, which includes:
   a. The length of the data byte
   b. The parity bit
   c. The stop bits
   d. The MSB- or the LSB-first
   e. The UART mode
   f. The UART CKSE
   g. The receive error and break interrupt enables
   h. The synchronous or asynchronous mode
   i. The prescaler value in the **UCAxBRW** register
4. Enable the eUSCI_A module by clearing the **UCSWRST** bit in the **UCAxCTLW0** Register
5. Enable UART-related interrupts by configuring the **UCAxIE** register

8.6.4.2 UART baud rate generation

Since the eUSCI_A baud rate generator can generate standard baud rates from nonstandard source frequencies, therefore two possible modes may be used to calculate the correct baud rate and this makes the situation a little complicated.

For a selected **BRCLK** CKSE, first a division factor **N** should be determined based on the following equation:

$$N = f_{BRCLK}/\text{Baud rate}$$

The division factor **N** is often a noninteger value, thus at least one divider and one modulator stage is used to meet the factor as closely as possible.

After the required division factor **N** is determined, follow the steps below to complete this baud rate calculation:

1. If **N** > 16, configure the **UCAxMCTLW** register as below:
   a. **OS16** = 1
   b. **UCBRx** = INT(**N**/16)
   c. **UCBRFx** = INT([(**N**/16) – INT(**N**/16)] × 16)
2. Else If **N** ≤ 16, configure the **UCAxMCTLW** register as below:
   a. **OS16** = 0
   b. **UCBRx** = INT(**N**)
3. The **UCBRSx** value can be found by looking up the fractional part of **M** [= **N** - INT(**N**)] in a lookup table, Table 8.31.

A point to be noted is that the eUSCI module is not functional when the device is in LPM3, LPM4, or LPMx.5 modes of operation.

8.6.4.3 UART transmit enable and data transmit operation

The UART transmitter can be enabled and used to transmit data in the following way:

1. After the hardware reset, the eUSCI_A module is in the reset status (**UCSWRST** = 1). The eUSCI_A module can be enabled by clearing the **UCSWRST** bit in the **UCAxCTLW0** Register and the transmitter is ready and in an idle state. The transmit baud rate generator is ready but is neither clocked nor produces any clocks.

Table 8.31 UCBRSx settings for fractional portion of M

Fractional portion of N	UCBRSx	Fractional portion of N	UCBRSx
0.0000	0x00	0.5002	0xAA
0.0529	0x01	0.5715	0x6B
0.0715	0x02	0.6003	0xAD
0.0835	0x04	0.6254	0xB5
0.1001	0x08	0.6432	0xB6
0.1252	0x10	0.6667	0xD6
0.1430	0x20	0.7001	0xB7
0.1670	0x11	0.7147	0xBB
0.2147	0x21	0.7503	0xDD
0.2224	0x22	0.7861	0xED
0.2503	0x44	0.8004	0xEE
0.3000	0x25	0.8333	0xBF
0.3335	0x49	0.8464	0xDF
0.3575	0x4A	0.8572	0xEF
0.3753	0x52	0.8751	0xF7
0.4003	0x92	0.9004	0xFB
0.4286	0x53	0.9170	0xFD
0.4378	0x55	0.9288	0xFE

2. A transmission can be initiated by writing a data byte to the **UCAxTXBUF** register. When this occurs, the baud rate generator is enabled, and the data in the **UCAxTXBUF** register are moved to the TXS register on the next **BITCLK** after the TXS register is empty. The **UCTXIFG** bit in the **UCAxIFG** register is set when the **UCAxTXBUF** register is empty.
3. If the **UCTXIE** bit in the **UCAxIE** register is set to enable the TX interrupt, a **UCTXIE** interrupt is generated and the interrupt vector can be found from the **UCAxIV** register.
4. Transmission continues as long as a new data byte is available in the **UCAxTXBUF** register at the end of the previous data transmission. If a new data byte is not in the **UCAxTXBUF** register when the previous byte has transmitted, the transmitter returns to its idle state and the baud rate generator is turned off.

8.6.4.4 UART receive enable and data receive operation

The UART receiver can be enabled and used to receive data in the following way:

1. After the hardware reset, the eUSCI_A module is in the reset status (**UCSWRST** = 1). The eUSCI_A module can be enabled by clearing the **UCSWRST** bit and the receiver is ready and in an idle state. The receive baud rate generator is in the ready state but is neither clocked nor produces any clocks.
2. The falling edge of the start bit enables the baud rate generator and the UART state machine checks for a valid start bit. If no valid start bit is detected, the UART state machine returns to its idle state and the baud rate generator is turned off again. If a valid start bit is detected, a character should be received.
3. If the **UCRXIE** bit in the **UCAxIE** register is set to enable the RX interrupt, a **UCRXIE** interrupt is generated and the interrupt vector can be found from the **UCAxIV** register.

4. Now the received data byte can be picked up from the **UXAxRXBUF** register.
5. Reading the **UCAxRXBUF** register resets the receive error bits, the **UCADDR** or **UCIDLE** bit, and the **UCRXIFG** flag bit in the **UCAxIFG** register.
6. If a character is transferred into the **UCAxRXBUF** register before the previous character was read out, the overrun error bit (**UCOE**) is set in the **UCAxSTATW** register. The **UCOE** bit is cleared automatically when the **UCxRXBUF** register is read, and must not be cleared by software. Otherwise, it does not function correctly.

8.6.4.5 UART errors detection

The eUSCI_A module can automatically detect the following errors when receiving characters:

- Framing errors
- Parity errors
- Overrun errors
- Break conditions

The error flag bits **UCFE**, **UCPE**, **UCOE**, and **UCBRK** in the **UCAxSTATW** register are set when their respective condition is detected. When the error flags **UCFE**, **UCPE**, or **UCOE** are set, the **UCRXERR** bit in the **UCAxSTATW** register is also set. The error conditions are described in Table 8.32.

When the bit **UCRXEIE** in the **UCAxCTLW0** is 0 and a framing error or parity error is detected, no character can be received into the **UCAxRXBUF** register. When the **UCRXEIE** = 1, characters can be received into the **UCAxRXBUF** register and the **UCRXIFG** error bit is set.

When any error is detected, perform the following operation procedure to handle them:

1. When any of the **UCFE**, **UCPE**, **UCOE**, **UCBRK**, or **UCRXERR** bits is set, the bit remains set until user software resets it or the **UCAxRXBUF** register is read.
2. However, the **UCOE** bit must be reset by reading the **UCAxRXBUF**. Otherwise, it does not function properly.

Table 8.32 Received error conditions

Error condition	Error flag	Description
Framing Error	**UCFE**	A framing error occurs when a low stop bit is detected. When two stop bits are used, both stop bits are checked for framing error. When a framing error is detected, the **UCFE** bit is set
Parity Error	**UCPE**	A parity error is a mismatch between the number of 1 s in a character and the value of the parity bit. When a parity error is detected, the **UCPE** bit is set
Receive Overrun	**UCOE**	An overrun error occurs when a character is loaded into **UCAxRXBUF** before the prior character has been read. When an overrun occurs, the **UCOE** bit is set
Break Condition	**UCBRK**	When not using automatic baud rate detection, a break is detected when all data, parity, and stop bits are low. When a break condition is detected, the **UCBRK** bit is set. A break condition can also set the interrupt flag **UCRXIFG** if the break interrupt enable **UCBRKIE** bit is set

3. To detect overflows (**UCOE**) reliably, the following steps are recommended:
   a. After a character is received and the **UCAxRXIFG** is set, first read the **UCAxSTATW** register to check all error flags, including the overflow flag **UCOE**.
   b. Next, read the **UCAxRXBUF** register. This clears all error flags except **UCOE** if the **UCAxRXBUF** register is overwritten between the read access to the **UCAxSTATW** and to the **UCAxRXBUF**.
   c. Therefore, the **UCOE** flag should be checked after reading the **UCAxRXBUF** register to detect this condition. However, in this case the **UCRXERR** flag is not set.

8.6.5 *UART modem handshake serial communication mode*

A UART can work as a modem to provide the interface between a computer and a serial asynchronous device. A computer can also be considered as UART device when it is connected as data terminal equipment (DTE) and a model can be considered a data communication equipment (DCE). In general, a modem is a DCE and a computing device that connects to a modem is a DTE. Figure 8.58 shows an illustrating block diagram for this equipment.

When using UART to perform asynchronous data operations, the data communication can be divided into the following four modes based on its functionality:

- **Simplex**: Serial communication is only taking place in one direction, either from DTE to DCE or vice versa.
- **Half-duplex**: Serial communication can take place in both directions, but the communication can only take place in one direction at a moment. This means that either sender or receiver can send or receive the information in different time, but they cannot send and receive the data simultaneously.
- **Full-duplex**: Allows serial communications to take place in both directions at the same time, which means that both sender and receiver can handle and exchange the data information simultaneously.
- **Multiplex**: Allows multiple serial communications channels to occur over the same serial communication line. Multiplex operations are performed by either allocating separate frequencies or time slice to the individual serial communication channels.

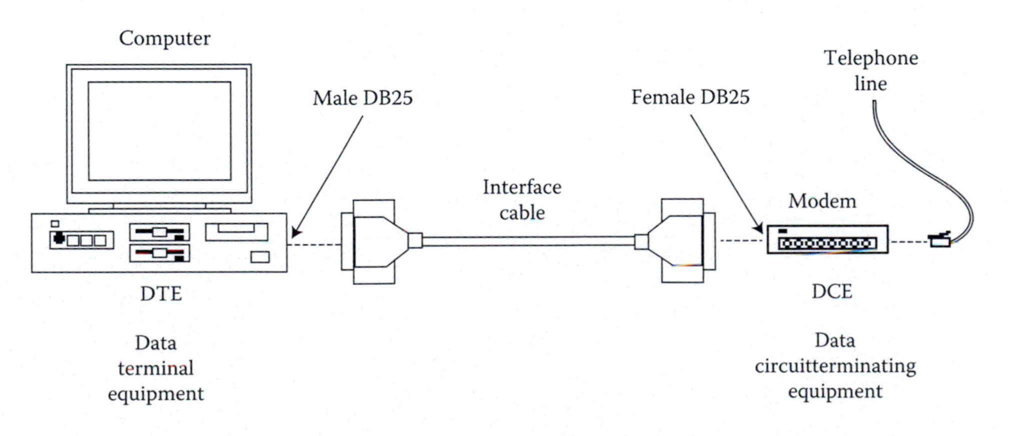

Figure 8.58 DTE and DCE connections.

A full group of modem flow control signals used for DTE and DCE is shown in Figure 8.59.

In UART module, only two signals, U1CTS and U1RTS, are used and they are defined differently based on the roles of these signals. When used as a DTE, the modem flow control signals are defined as (see Figure 8.59):

- /$\overline{\text{U1CTS}}$ is Clear To Send (CTS)
- /U1RTS is Request To Send (RTS)

When used as a DCE, the modem flow control signals are defined as:

- /$\overline{\text{U1CTS}}$ is Request To Send (RTS)
- /U1RTS is Clear To Send (CTS)

The flow control can be designed and implemented by either hardware or software. The so-called hardware flow control is to use RTS and CTS lines to get the UART status. The software flow control is to use UART interrupts to get the UART status.

8.6.6 eUSCI_A module for UART control signals and GPIO pins

In the MSP432P401R MCU system, four eUSCI_Ax modules, **eUSCI_A0**~**eUSCI_A3**, are configured to work as UART modes. All of these modules are interfaced to external UART devices via GPIO Ports **P1**~**P3** and **P9** (refer to Figure 8.4). Table 8.33 lists these GPIO pins and related UART control signals.

In MSP-EXP432P401R EVB, only limited GPIO pins, **P2.3**, **P3.2**, and **P3.3**, are exposed to the external UART devices via jumper connectors J1 and J4. These GPIO pins have been highlighted in Table 8.33.

We may use these GPIO pins (**P3.2** and **P3.3**) and related **eUSCI_A2** UART mode to build our sample project later since they are easy to access.

8.6.7 Develop an example eUSCI_A2 UART mode project

In this section, we will develop and build an example UART project to echo each character sent via the UART transmitter with the **eUSCI_A2** module for UART mode. This project can be considered as a self-testing project to check the UART transmit and receive

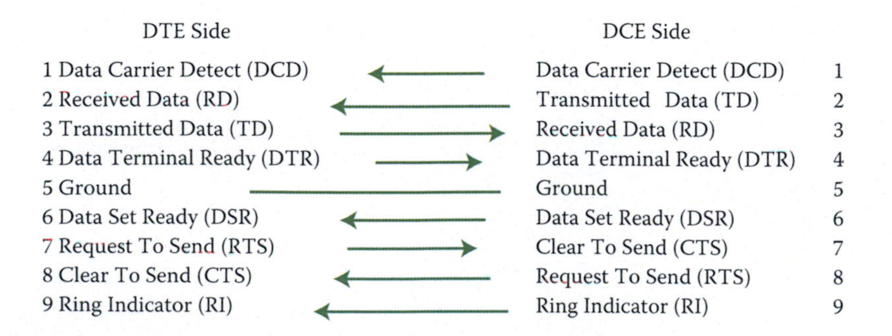

Figure 8.59 Modem flow control signals used in UART communications.

Table 8.33 UART control signals and the related GPIO pins distributions

UART pin	GPIO pin	Pin type	Buffer type	Pin function
UCA0RXD	**P1.2**	I/O	TTL	eUSCI_A0 UART mode—Data Receive
UCA0TXD	**P1.3**	I/O	TTL	eUSCI_A0 UART mode—Data Transmit
PM_UCA1RXD	**P2.2**	I/O	TTL	eUSCI_A1 UART mode—Data Receive
PM_UCA1TXD	**P2.3**	I/O	TTL	eUSCI_A1 UART mode—Data Transmit
PM_UCA2RXD	**P3.2**	I/O	TTL	eUSCI_A2 UART mode—Data Receive
PM_UCA2TXD	**P3.3**	I/O	TTL	eUSCI_A2 UART mode—Data Transmit
UCA3RXD	**P9.6**	I/O	TTL	eUSCI_A3 UART mode—Data Receive
UCA3TXD	**P9.7**	I/O	TTL	eUSCI_A3 UART mode—Data Transmit

functions. The **UCA2RXD** (**P3.2**) pin needs to be connected to the **UCA2TXD** (**P3.3**) pin via a jumper. The project design steps include the following operations:

1. Initialize and configure the eUSCI_A2 UART mode-related GPIO ports and pins
2. Initialize and configure the CKSE and baud rate for the eUSCI_A2 UART mode
3. Initialize and configure the eUSCI_A2 UART mode

Let's start from the initialization and configuration of the UART module with an assumption of using a default DCO with 3-MHz system clock SMCLK as the CKSE for this mode.

8.6.7.1 Initialize and configure the UART-related GPIO ports and pins

Perform the following operations to complete the configurations for the UART-related GPIO port 3 and related pins, **P3.2** and **P3.3**:

- Set the **P3SEL1x:P3SEL0.x** = 01 (**x** = 2 and 3) to enable pins **P3.2** and **P3.3** to work as **UCA2RXD** and **UCA2TXD** pins, respectively (refer to Table 7.11 in Chapter 7). Since GPIO port 3 **P3** is a map control port, thus both pins can be mapped to those default pins.

8.6.7.2 Initialize and configure clock source and baud rate for the UART module

In this project, we like to use a high baud rate as **115200**. Perform the following operations to complete these configurations:

- Determine the required division factor **N** = f_{BRCLK}/baud rate = **3000000/115200** = **26.042**
- Since **N** > 16, so set **OS16** = 1, **UCBR1** = INT(**N**/16) = INT(26.042/16) = **1**, **UCBRF1** = INT([(**N**/16) – INT(**N**/16)] × 16) = INT[(1.628 – 1) × 16] = **10**, **UCBRSx** value can be found from Table 8.31, which is **0x00** [**N** - INT(**N**) = 26.042 – 26 = 0.042]
- Set these parameters to the **UCA2BRW** and **UCA2MCTLW** registers as:
  - **UCA2BRW = 1**
  - **UCA2MCTLW = 0x00A1**

8.6.7.3 Initialize and configure the eUSCI_A2 for UART mode

The following asynchronous communication parameters are selected and used for this eUSCI_A2 for UART mode operations:

- Parity disabled
- LSB-first
- 8-bit data length
- One stop bit
- Asynchronous UART mode
- SMCLK is selected as the CKSE
- All other settings are default
- Enable the UART received interrupt by setting the bit **UCRXIE** in the **UCA2IE** register

Next, let's build an example UART project to illustrate how to use this peripheral to communicate to a PC to transmit and receive characters asynchronously.

8.6.7.4 Build an example UART module project

In this project, we want to perform a loop back testing by sending 16 numbers from the **UCA2TXD** pin and try to receive them from the **UCA2RXD** pin via the eUSCI_A2 for UART mode. The hardware setups and software features for this project include the following points:

1. In the MSP-EXP432P401R EVB, connect the GPIO pins **P3.2** and **P3.3** with a jumper. Both GPIO pins **P3.2** and **P3.3** can be found from the J1 connector on that EVB.
2. Configure GPIO pin **P1.0** and pin **P2.2** as output pins since these pins are connected to a red color and a blue color LED on the EVB. We need to use these LEDs to display the running status of our project. If all 16 data bytes are received successfully, the blue color LED is on, but the red color LED will be on if any receiving error is detected.

Keep all these points in mind; now, let's create a new UART project **DRAUART**.

8.6.7.4.1 Create a new UART module project DRAUART Perform the following operations to create a new project **DRAUART**:

1. Open the Windows Explorer to create a new folder **DRAUART** under the class project folder **C:\MSP432 Class Projects\Chapter 8**
2. Create a new µVersion project **DRAUART** and save it to our project folder **DRAUART** that is created in step 1 above
3. On the next wizard, you need to select the device (MCU) **MSP432P401R** and click on the **OK** to close this wizard
4. Next, the software components wizard is opened, and you need to set up the software development environment for your project with this wizard. Expand two icons, **CMSIS** and **Device**, and check the **CORE** and **Startup** checkboxes in the **Sel.** column, and click on the **OK** button since we need these two components to build our project

Since this project is relatively simple, therefore only a C source file is good enough.

8.6.7.4.2 Create a new C source file DRAUART.c Create a new C source file and name it **DRAUART.c** in the new created project. Enter the code shown in Figure 8.60 into the source file.

```
1  //****************************************************************************
2  // DRAUART.c - Main application file for project DRAUART
3  //****************************************************************************
4  #include <stdint.h>
5  #include <stdbool.h>
6  #include <msp.h>
7  bool intFlag = true;
8  uint8_t RXData[16], TXData = 0, rxMAX = 16;
9  int main(void)
10 {
11    uint8_t n, result[16];
12    WDTCTL = WDTPW | WDTHOLD;          // stop watchdog timer
13    P1DIR |= BIT0;                     // set P1.0 as output pin
14    P1OUT &= ~BIT0;                    // set P1.0 output low
15    P2DIR = BIT2;                      // set P2.2 as output pin
16    P2OUT &= ~ BIT2;                   // set P2.2 output low
17    P3SEL0 |= BIT2|BIT3;               // set P3.2 & P3.3 as UCA2RXD and
                                          UCA2TXD pins
18    // Configure UART
19    UCA2CTLW0 |= UCSWRST;              // reset & disable the USCI_A2 module
20    UCA2CTLW0 |= UCSSEL_SMCLK;         // select the clock source as SMCLK
21    // Baud Rate calculation
22    UCA2BRW = 1;                       // INT(26.042/16) = 1
23    UCA2MCTLW = 0x00A1;                // UCBRFx = INT([(N/16) - INT(N/16)] x 16) = 10
24                                       // UCBRSx value = 0x00 (See Table 8.31)
25    UCA2CTLW0 &= ~UCSWRST;             // enable the USCI_A2 module
26    UCA2IE |= UCRXIE;                  // enable USCI_A2 RX interrupt
27    NVIC_IPR4 = 0x00600000;            // USCIA2 priority level = 3 (Bits 23~21)
28    NVIC_ISER0 = 0x00040000;           // USCIA2 IRQ number = 18
29    ___enable_irq();
30    while(intFlag)
31    {
32      while(!(UCA2IFG & UCTXIFG));
33      UCA2TXBUF = TXData;              // load data into TX buffer
34      ___wfi();                        // wait for interrupt…
35    }
36    for (n = 0; n < rxMAX; n++)        // collect 16 received data bytes
37      result[n] = RXData[n];
38    while(1);
39 }
40 void EUSCIA2_IRQHandler(void)         // UART interrupt service routine
41 {
42    if (UCA2IFG & UCRXIFG)
43    {
44      UCA2IFG &=~UCRXIFG;              // clear interrupt
45      RXData[TXData] = UCA2RXBUF;      // pick up received byte & clear RX buffer
46      if(RXData[TXData] != TXData)     // check the received value
47      {
48        P1OUT |= BIT0;                 // if incorrect turn on P1.0
49        while(1);                      // trap CPU
50      }
51      TXData++;                        // increment data byte
52      if (TXData > rxMAX)              // all 16 bytes received?
53      {
54        intFlag = false;               // stop the while() loop
55        P2OUT = BIT2;                  // turn on blue LED to indicate the end
56      }
57    }
58 }
```

Figure 8.60 Code for the source file DRAUART.c.

Let's have a closer look at this piece of code to see how it works. Since most codes in this source file are straightforward and easy to understand therefore we only emphasize some important code lines.

1. The code lines 13~17 are used to initialize and configure UART-related GPIO ports and pins. The code line 17 is important since this line configures the **P3SEL1.x** and **P3SEL0.x** as **01** to enable **P3.2** and **P3.3** pins to work as the **UCA2RXD** and **UCA2TXD** pins.
2. The code lines 19~26 are used to configure the USCIA2 module to enable it to work as an UART mode. In fact, most configuration parameters used for the USCIA2 module are default except the CKSE selection parameter in the **UCA2CTLW0** Register. A system macro **UCSSEL__SMCLK** is used to select the **SMCLK** as the clock source **BRCLK**.
3. The baud rate calculations are performed in code lines 22~23. Since the required division factor **N** is greater than 16, both a prescaler and a modulator is used to set up the **UCA2BRW** and the **UCA2MCTLW** registers. The bit field **UCBRF1** is set to 10 (**0xA0**), and the bit field **UCBRS1** is set to 0 based on the setup principle and values in the Table 8.31. The bit **UCOS16** must be set to 1 if the N is greater than 16. Therefore, the combination setup value for the **UCA2MCTLW** register is **0x00A1**.
4. The code lines in 27~29 are used to set up the interrupt priority level and enable the interrupt source for the EUSCIA2 module. Refer to Sections 5.3.2.1 and 5.3.2.2 in Chapter 5 to get more details about these interrupt configurations.
5. A conditional **while()** loop is used starting at line 30. This loop is to repeatedly check the **UCTXIFG** flag to make sure that a new data byte can be transmitted into the **UCA2TXBUF** register as long as it is empty. This loop will be terminated if the loop condition intFlag becomes false, which will be set by the end of the data transmission inside the UART ISR later.
6. As soon as the **while()** loop is terminated, all 16 received data bytes are collected in the coding lines 36~37 with a **for()** loop.
7. An infinitive **while()** loop in line 38 is used to temporarily halt the program to enable us to check the receiving result. A break point should be set on this line to enable us to do that checking later.
8. The USCIA2 interrupt handler or ISR starts at line 40. The main job performed by this ISR is to check the **UCRXIFG** flag bit to make sure that a data byte has been successfully received in the **UCA2RXBUF** register.
9. Then the received data byte is reserved to the global data array **RXData[]** in line 45.
10. If the received data byte value is not equal to the byte value sent by the **UCA2TXD** pin, the red color LED is turned on by setting the **P1.0** pin in line 48 and the program is hung up in line 49 with an infinitive **while()** loop.
11. Otherwise the data to be transmitted are incremented by 1 in line 51.
12. The codes in lines 52~57 are used to check whether all 16 data bytes have been received. If that is true, the global Boolean variable **intFlag** is reset to **false** to stop the conditional **while()** loop to begin to collect received data in the main program. Also the blue color LED is turned on by setting the **P2.2** pin to indicate the end of the program.

Now, let's set up environments to build and run the project.

8.6.7.4.3 Set up the environment to build and run the project Before the project can be built, make sure that the following issues have been set up correctly:

- The debug adapter used for this project is **`CMSIS-DAP Debugger`**. This adapter can be configured in the **`Debug`** tab under the menu item **`Project|Options for Target`** "**`Target 1`**."
- Make sure that all settings for the debugger and the flash download are correct.
- Set up a break point at line 38 to enable us to check the running result later.

Now build and run the project. After the project runs, open the **`Call Stack`** + **`Locals`** window and expand the **`result[]`** array. You can find that all 16 numbers, from **`0x00`** to **`0x0F`**, have been successfully transmitted and received. Click **`Debug|Stop`** to terminate the project.

8.6.7.5 A practical example URAT lab project with two UART modes

This is a lab project to be assigned to students and readers as a part of homework. This project contains two UART modes working with two MSP-EXP432P401R MCUs that work as two DTEs. One DTE works as a sender–receiver that first transmits a sequence of data bytes to and then receives data from other DTE. The other DTE works as a receiver–sender that first receives data from and then sends data to the first DTE.

In most real UART applications, a RS-232 cable, either DB-9 or DB-25, is used to transmit–receive data between DTEs or DCEs. The voltage levels applied on the RS-232 are different with those applied in TTL and MOS circuits. Therefore, a voltage level conversion is necessary for the UART data communications.

In this lab project, we used a MAX3232 that is a 3.3~5.5 V true RS-232 transceiver to perform this voltage-level conversion. During transmission, the MAX3232 converts a digital high on the microcontroller side to –5.5 V on the RS232 cable side, and a digital low to +5.5 V. However, during the receiving, the –5.5 V on the RS-232 cable side is converted to a digital high on the microcontroller side, and the +5.5 V to a digital low. Each MAX3232 contains two drivers and two receivers. All four capacitors are 0.1 μF.

Two MSP-EXP432P401R MCUs are used to connect together with two UART modes. Both MCUs use the **`eUSCI_A2`** for UART mode, exactly the pin **`P3.2`** (**`UCA2RXD`**) works as a receive pin and the **`P3.3`** (**`UCA2TXD`**) works as a transmit pin. To perform a data transmit and receiving operation, the **`P3.2`** in the first DTE should be connected to the **`P3.3`** in the second DTE, and the **`P3.2`** in the second DTE should be connected to the **`P3.3`** in the first TDE.

In most actual applications, a RS-232 cable should be used for the data transmission and receiving. In this lab project, we just use some wires to replace that cable to make our lab easier.

The hardware configuration for this lab is shown in Figure 8.61.

Build this circuit on the breadboard provided by the EduBASE ARM Trainer. Connect the 3.3V power and the ground signals from the pin 4 and pin 5 on the J32 jumper connector on the EduBASE ARM Trainer to the breadboard. Connect two grounds on two MSP432P401R EVBs with a wire.

Two programs, **`Lab8_9 (UARTSendRecv)`** and **`Lab8_10 (UARTRecvSend)`**, will be developed by students and readers in the lab project part on the homework attached.

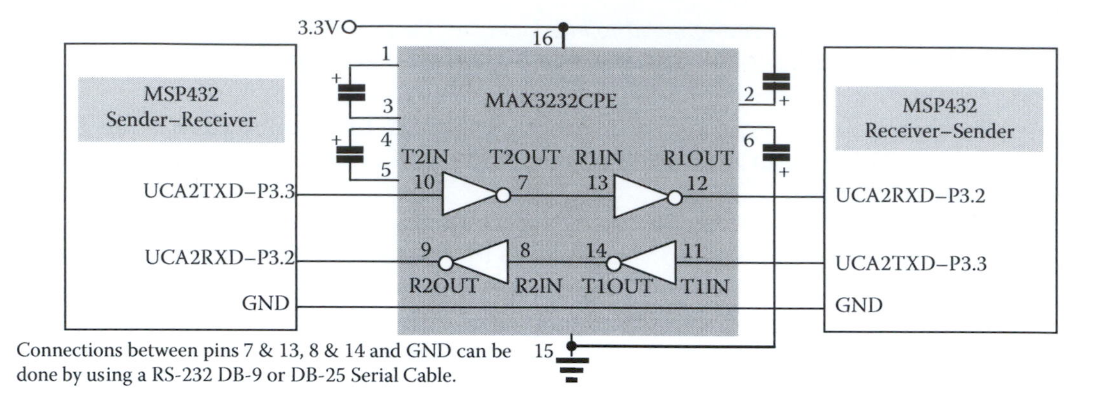

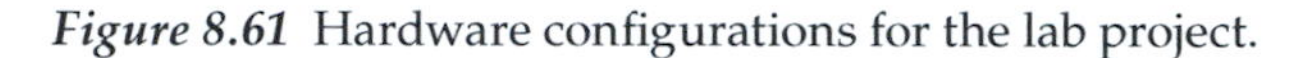
Figure 8.61 Hardware configurations for the lab project.

8.6.8 UART API functions provided by the MSPWare Peripheral Driver Library

The UART API provides a set of functions for using the eUSCI_A2 for UART mode. Functions are provided to configure and control the UART modules, to send and receive data, and to manage interrupts for the UART modules.

There are about 20 UART API functions provided by this library. In this section, we only introduce and discuss some popular and important API functions.

First let's take a look at a popular struct **_eUSCI_eUSCI_UART_Config** used in this library. When using this struct in the user's application program, a struct variable **eUSCI_UART_Config** should be generated based on this struct. This struct must be used when initializing the eUSCI_A module.

The protocol of this struct is:

```
typedef struct _eUSCI_eUSCI_UART_Config
{
   uint_fast8_t      selectClockSource;
   uint_fast16_t     clockPrescalar;
   uint_fast8_t      firstModReg;
   uint_fast8_t      secondModReg;
   uint_fast8_t      parity;
   uint_fast16_t     msborLsbFirst;
   uint_fast16_t     numberofStopBits;
   uint_fast16_t     uartMode;
   uint_fast8_t      overSampling;
} eUSCI_UART_Config;
```

The UART API function set can be divided into four groups: UART initialization and configuration, UART control and status, UART send and receive data, and UART interrupt handling.

The UART initialization and configuration functions include:

- **UART_enableModule()**
- **UART_disableModule()**
- **UART_initModule()**

The UART control and status functions include:

- **UART_queryStatusFlags()**
- **UART_selectDeglitchTime()**

The UART send and receive data API functions contain:

- **UART_transmitData()**
- **UART_receiveData()**

The UART interrupt handling functions include:

- **UART_registerInterrupt()**
- **UART_enableInterrupt()**
- **UART_getInterruptStatus()**
- **UART_clearInterruptFlag()**

Let's have a closer look at these API functions based on their groups.

8.6.8.1 UART initialization and configuration API functions

Three API functions are included in this group. Table 8.34 shows the protocol and arguments of these functions. The function **UART_initModule()** is a main function used to initialize and configure the desired UART module. The second argument is a pointer point to the UART struct **eUSCI_UART_Config**. This struct variable should be filled before this function can be called. The UART mode is still disabled after that initialization and needs to be enabled by using the API function **UART_enableModule()** to allow the UART to begin to start its normal operation. The UART mode can also be disabled by calling the **UART_disableModule()** function.

8.6.8.2 UART control and status API functions

Two popular API functions are involved in this group. Table 8.35 shows these functions.

The API function **UART_queryStatusFlags()** returns the masked status flags, and the masked flags can be any one or an OR of multiple flags shown in Table 8.35. The API function **UART_selectDeglitchTime()** can be used to set up the desired deglitch time period for the deglitch filter used to reduce the deglitch noise for the UART data communications. The default deglitch time period is 200 ns (**EUSCI_A_UART_DEGLITCH_TIME_200ns**).

When using these API functions, one point to be noted is that for eUSCI modules applied in these functions, only the eUSCI_Ax modules such as **eUSCI_A0~eUSCI_A3** can be used. Any of **eUSCI_Bx** modules such as **eUSCI_B0~eUSCI_B3** do not support the UART mode, instead they only support 3-pin or 4-pin SPI and I2C modules. The **eUSIC_Ax** modules also support 3-pin or 4-pin SPI modules.

8.6.8.3 UART send and receive data API functions

Two API functions are involved in this group. Table 8.36 shows these functions.

8.6.8.4 UART interrupt handling API functions

Four popular API functions are involved in this group. Table 8.37 shows these functions.

Generally, the operational sequence of enable and response to an interrupt is

Table 8.34 UART initialization and configuration API functions

API function	Parameter	Description
void **UART_enable-Module(**uint32_t mInst**)**	**mInst** is the instance of the eUSCI A (UART) module. Valid values include: **EUSCI_A0_MODULE**, **EUSCI_A1_MODULE**, **EUSCI_A2_MODULE**, **EUSCI_A3_MODULE**	Enable the normal operations of the UART block Modified register is **UCAxCTL1**
void **UART_disable Module(**uint32_t mInst**)**	**mInst** is the instance of the eUSCI A (UART) module. Valid values are same as above	Disable the normal operations of the UART block Modified register is **UCAxCTL1**
bool **UART_initModule (**uint32_t mInst, const eUSCI_UART_Config *config**)**	**mInst** is the instance of the eUSCI A (UART) module. Valid values are same as above **config** is the configuration structure for the UART module	Initialize the module, but the UART block still remains disabled and must be enabled with **UART_enableModule()** **selectClockSource**. Valid values are: **EUSCI_A_UART_CLOCKSOURCE_SMCLK** **EUSCI_A_UART_CLOCKSOURCE_ACLK** **clockPrescalar** is the value to be written into **UCBRx** bits **firstModReg** is first modulation stage register setting. This value is a precalculated value which can be obtained above. This value is written into **UCBRFx** bits of **UCAxMCTLW** **secondModReg** is second modulation stage register setting. This value is a precalculated value and can be obtained from above. It is written into **UCBRSx** bits of **UCAxMCTLW** **parity** is the desired parity. Valid values are: **EUSCI_A_UART_NO_PARITY** [Default Value], **EUSCI_A_UART_ODD_PARITY**, **EUSCI_A_UART_EVEN_PARITY**. **msborLsbFirst** controls direction of receive and transmit shift register. Valid values are **EUSCI_A_UART_MSB_FIRST** or **EUSCI_A_UART_LSB_FIRST** [Default Value] **numberofStopBits** indicates one/two STOP bits. Valid values are: **EUSCI_A_UART_ONE_STOP_BIT** [Default Value] or **EUSCI_A_UART_TWO_STOP_BITS**

(*Continued*)

Table 8.34 (Continued) UART initialization and configuration API functions

API function	Parameter	Description
		uartMode selects the mode of operation. Valid values are: **EUSCI_A_UART_MODE** [Default Value], **EUSCI_A_UART_IDLE_LINE_ MULTI_PROCESSOR_MODE**, **EUSCI_A_UART_ADDRESS_BIT_ MULTI_PROCESSOR_MODE**, **EUSCI_A_UART_AUTOMATIC_ BAUDRATE_DETECTION_MODE** **overSampling** indicates low frequency or oversampling baud generation. Valid values are: **EUSCI_A_UART_OVERSAMPLING_ BAUDRATE_GENERATION**, **EUSCI_A_UART_LOW_FREQUENCY_ BAUDRATE_GENERATION**

Table 8.35 UART control and status API functions

API function	Parameter	Description
uint_fast8_t **UART_ queryStatusFlags(**uint32_t mInst, uint_fast8_t mask**)**	**mInst** is the instance of the eUSCI A (UART) module. Valid values include: **EUSCI_A0_MODULE**, **EUSCI_A1_MODULE**, **EUSCI_A2_MODULE**, **EUSCI_A3_MODULE** **mask** is the masked interrupt flag status to be returned	Get the current UART running status **mask** parameter can be either any of the following selection: **EUSCI_A_UART_LISTEN_ENABLE** **EUSCI_A_UART_FRAMING_ ERROR** **EUSCI_A_UART_OVERRUN_ ERROR** **EUSCI_A_UART_PARITY_ERROR** **eUARTBREAK_DETECT** **EUSCI_A_UART_RECEIVE_ERROR** **EUSCI_A_UART_ADDRESS_ RECEIVED** **EUSCI_A_UART_IDLELINE** **EUSCI_A_UART_BUSY**
void **UART_selectDeglitch Time(**uint32_t mInst, uint32_t dTime**)**	**mInst** is the instance of the eUSCI A (UART) module. Valid values are same as above **dTime** is the selected deglitch time	Set the deglitch time Valid values for the **dTime** are: **EUSCI_A_UART_DEGLITCH_ TIME_2ns** **EUSCI_A_UART_DEGLITCH_ TIME_50ns** **EUSCI_A_UART_DEGLITCH_ TIME_100ns** **EUSCI_A_UART_DEGLITCH_ TIME_200ns**

Table 8.36 UART send and receive data API functions

API function	Parameter	Description
void **UART_transmitData (**uint32_t mInst, uint_fast8_t tData**)**	**mInst** is the instance of the eUSCI A (UART) module. Valid values include: **EUSCI_A0_ MODULE**, **EUSCI_A1_MODULE**, **EUSCI_A2_MODULE**, **EUSCI_A3_MODULE** **tData** is the data byte to be transmitted from the UART module	This function will place the supplied data into UART transmit data register to start transmission Modified register is **UCAx-TXBUF**
uint8_t **UART_receiveData (**uint32_t mInst**)**	**mInst** is the instance of the eUSCI A (UART) module. Valid values include: **EUSCI_A0_MODULE**, **EUSCI_A1_MODULE**, **EUSCI_A2_MODULE**, **EUSCI_A3_MODULE**	This function reads a byte of data from the UART receive data register Modified register is **UCAxRXBUF** **Return:** Return the byte received from by the UART module, cast as an uint8_t

- Use the **`UART_registerInterrupt()`** function to first register the interrupt's handler
- Use the **`UART_enableInterrupt()`** function to enable the interrupt source, such as the transmit or receive interrupts to be triggered
- Use the API function **`UART_getInterruptStatus()`** to get the current special interrupt source. When using this function, the special interrupt source must be clearly indicated on the second parameter that is a mask
- After the generated interrupt has been handled and responded, that interrupt source must be cleared by calling the API function **`UART_clearInterruptFlag()`** inside the ISR.

Because of the space limitation, we leave a project at the homework section to enable students to build a UART loop back project with the SD model in lab **`Lab8_8`**.

8.7 Summary

The main topics discussed in this chapter are about the MSP432 serial port programming, which include the synchronous and asynchronous serial data communications and operations. Because the MSP432P401R MCU contains quite a few peripherals that are related to serial interface and communications, we only introduced and discussed three major serial peripherals, SPI, I2C, and UART in this chapter.

We started our discussion from the synchronous peripheral interface (SPI) that is similar to SSI in this chapter. Three example projects related to SPI are introduced and discussed with line by line illustrations:

- On-Board LCD Interface Programming Project
- On-Board Seven-Segment LED Interface Programming Project
- DAC Programming Project

Table 8.37 UART interrupt handling API functions

API function	Parameter	Description
void **UART_registerInterrupt** **(**uint32_t mInst, void(*)(void) intHandler**)**	**mInst** is the instance of the eUSCI A (UART) module. Valid values include: **EUSCI_A0_MODULE**, **EUSCI_A1_MODULE**, **EUSCI_A2_MODULE**, **EUSCI_A3_MODULE** **intHandler** is a pointer to the function to be called when the interrupt occurs	This function registers the handler to be called when an UART interrupt occurs. This function enables the global interrupt in the interrupt controller; specific UART interrupts must be enabled via **UART_enableInterrupt()**. It is the interrupt handler's responsibility to clear the interrupt source via **UART_clearInterruptFlag()**
void **UART_enableInterrupt** **(**uint32_t mInst, uint_fast8_t mask**)**	**mInst** is the instance of the eUSCI A (UART) module. Valid values are same as above **mask** is the bit mask of the interrupt sources to be enabled	Enable the indicated UART interrupt sources The **mask** parameter is the logical OR of any of the following: **EUSCI_A_UART_RECEIVE_INTERRUPT**—receive interrupt **EUSCI_A_UART_TRANSMIT_INTERRUPT**—transmit interrupt **EUSCI_A_UART_RECEIVE_ERRONEOUSCHAR_INTERRUPT**—receive erroneous-character interrupt enable **EUSCI_A_UART_BREAKCHAR_INTERRUPT**—receive break character interrupt enable
uint_fast8_t **UART_getInterruptStatus** **(**uint32_t mInst, uint8_t mask**)**	**mInst** is the instance of the eUSCI A (UART) module. Valid values are same as above **mask** is the masked interrupt flag status to be returned	Get the current UART interrupt status **mask** value is the logical OR of any of the following: **EUSCI_A_UART_RECEIVE_INTERRUPT_FLAG**, **EUSCI_A_UART_TRANSMIT_INTERRUPT_FLAG**, **EUSCI_A_UART_STARTBIT_INTERRUPT_FLAG**, **EUSCI_A_UART_TRANSMIT_COMPLETE_INTERRUPT_FLAG** **Return:** The current interrupt status as an OR bit mask: **EUSCI_A_UART_RECEIVE_INTERRUPT_FLAG**—receive interrupt flag **EUSCI_A_UART_TRANSMIT_INTERRUPT_FLAG**—transmit interrupt flag
void **UART_clearInterruptFlag** **(**uint32_t mInst, uint_fast8_t mask**)**	**mInst** is the instance of the eUSCI A (UART) module. Valid values are same as above **mask** is a bit mask of the interrupt sources to be cleared	Clear UART interrupt sources The **mask** parameter has the same definition as the mask parameter to **EUSCI_A_UART_enableInterrupt()** Modified register is **UCAxIFG**

These projects are built by using the DRA model and method.

Also some important and popular SPI-related API functions provided by the MSPPDL are discussed in Section 8.4.11. An SPI-related project using the API function, **SDLCD**, is analyzed and discussed with details in that part.

Following the discussions about the SPI, we began our discussion about the I2C bus with two real projects to interface to two MSP432P401R MCUs via the I2C bus, one works as a master and the other works as a slave device in Section 8.5.10.

The I2C API functions provided by MSPPDL are discussed in Section 8.5.11. Two I2C lab projects, **Lab8_6(SDI2CMaster)** and **Lab8_7(SDI2CSlave)**, are assigned as a part of homework to the readers.

The UARTs are discussed at Section 8.6. This is an asynchronous serial data communication protocol and widely applied in most data communication implementations. A real project that is to use the loop back test for the UART mode is built with the DRA model in Section 8.6.7. Two UART lab projects, **Lab8_9 (UARTSendRecv)** and **Lab8_10 (UARTRecvSend)**, are assigned as a part of homework to the readers.

The API functions provided by MSPPDL are discussed in Section 8.6.8. An example of UART project developed by using the UART API functions is provided as a lab project **Lab8_8**, which belongs to a part of homework.

HOMEWORK

I. True/False Selections

_____1. SPI peripherals on most microcontrollers can implement SSI with external differential driver ICs and program-controlled timing.

_____2. The SPI and I2C belong to SSI or bus, but the UART is an asynchronous serial data communication system.

_____3. Generally, the SCI belongs to the asynchronous SCI but the SPI or SSI belongs to the synchronous SCI.

_____4. Both full-duplex and half-duplex communication mode use two data lines to transmit and receive data in the serial format.

_____5. A basic asynchronous serial data framing comprises start signal, number of data bits, number of stop bits, and parity bit.

_____6. For the synchronous serial data communications, both master and slave use the same CKSE, but for the asynchronous serial operations, the transmitter and receiver use the different CKSEs.

_____7. Three control signals used in SPI are: **SCLK**, **MOSI**, and **MISO**.

_____8. Before a normal SPI bus can perform its work, either a master or a slave can set up the appropriate CKPL and CKPH with respect to the data to select a frame.

_____9. Each I2C module uses two signals, **SCL** and **SDA**, to transmit and receive data between the master and the slave. The **SDA** pin must be configured as an open-drain signal.

____10. When any UART receive error occurred, the **UCRXIFG** bit in the **UCAxIFG** register is set. However, when a character is successfully received in the **UCAxRXBUF** register, this **UCRXIFG** bit is also set. In order to distinguish them, one needs to check the bits **UCRXEIE** and **UCBRKIE** to identify whether a data or an error has been received.

II. Multiple Choices

1. The following modules belong to asynchronous serial data operation modules ____________.
   a. SSI
   b. I2C
   c. SPI
   d. UART
2. The **`Full-duplex`** protocol is defined as: ________________________.
   a. Serial communications can take place in both directions at the same time
   b. Serial communications are only taking place in one direction
   c. Serial communications can take place in both directions, but the communication can only take place in one direction at a moment
   d. Multiple serial communications can occur over the same serial communication line
3. Both SPI and SSI can use either ____________ or ____________ mode to perform the data transfers between terminals and peripherals.
   a. Full-duplex, single-duplex
   b. Half-duplex, multi-duplex
   c. Half-duplex, full-duplex
   d. Multi-duplex, single-duplex
4. In the SPI mode, the transmit/receiving frame is determined by the __________and __________.
   a. **`CKSE`**, **`CKPH`**
   b. **`CKPL`**, **`CKSE`**
   c. **`CKPL`**, **`CKPH`**
   d. **`CKSL`**, **`CKPH`**
5. Every transaction on the I2C bus is ________bits long, consisting of __________bits and a single acknowledge bit.
   a. 8, 7 data
   b. 9, 8 data
   c. 10, 9 data
   d. 11, 10 data
6. The number of bytes per I2C transfer is defined as the time period between a valid ____________ and ______________ condition, is unrestricted, but each data byte has to be followed by an acknowledge bit, and data must be transferred _______ first.
   a. **`START, LSB, STOP`**
   b. **`STOP, START, MSB`**
   c. **`LSB, START, STOP`**
   d. **`START, STOP, MSB`**
7. In each I2C module, after the **`START`** condition, a slave address is first transmitted. This address is _______ bits long followed by a ______ bit.
   a. 8, START
   b. 8, STOP
   c. 7, R/S
   d. 8, R/S

8. If the **UCBxCTLW0** Register is configured as: **UCMODEx** = 11, **UCSYNC** = 1 and **UCMST** = 1, this means that the **eUSCI_Bx** module is selected to work as a ______________________.
   a. Synchronous SPI mode with master mode
   b. Asynchronous UART mode
   c. Synchronous SPI mode with slave mode
   d. Synchronous I2C mode with master mode
9. If the **UCAxCTLW0** Register is configured as: **UCMODEx** = 00, **UCSYNC** = 0 and **UCMSB** = 1, this means that the **eUSCI_Ax** module is selected to work as a ______________________.
   a. Synchronous SPI mode with MSB-first
   b. Asynchronous UART mode with MSB-first
   c. Synchronous I2C mode with MSB-first
   d. Asynchronous UART mode with LSB-first
10. If the **UCRXERR** bit in the **UCAxSTATW** register is set, which means that a __________________.
   a. **UCFE** error has been detected
   b. **UCOE** error has been detected
   c. **UCPE** error has been detected
   d. All of them

III. Exercises
1. Provide a brief description about operational procedure of the general SPI mode.
2. Provide a brief description about the eUSCI interface used for SPI mode.
3. Provide a brief description about the operational procedure of the I2C module worked in the master transmit mode.
4. Provide a brief discussion about the initializations and configurations for the eUSCI_B1 module for I2C mode.
5. Provide a brief description about some important features of the MSP432 UART.

IV. Practical Laboratory

LABORATORY 8 MSP432 GPIO serial port programming

8.0 GOALS

This laboratory exercise allows students to learn and practice MSP432P401R MCU serial ports programming by developing six labs.

1. Program **Lab8_1(DRALCDINT)** lets students build an LCD project to display **MSP432** on the first line and **2016** on the second line in the LCD panel. The first line starts from the position **05** (DDRAM address) and the second line starts at position **46**.
2. Program **Lab8_2(DRALED)** enables students to use DRA model to build a seven-segment LED project to use **eUSCI_B0** module to repeatedly display seven segments in a sequence for all four seven-segment LEDs on the EduBASE ARM® Trainer.
3. Program **Lab8_3(SDLED)** is very similar to **Lab8_2**. This lab allows students to use API functions provided by the MSPPDL to interface to the 74VHCT595 to build a seven-segment LED project to use **eUSCI_B0** module to repeatedly display seven

segments in a sequence for all four seven-segment LEDs on the EduBASE ARM® Trainer.

4. Program **Lab8_4(DRAI2CMaster)** allows students to use the **eUSCI_B1** module for I2C mode to build a master program to transmit 16-byte data to a slave device via the I2C bus. The slave program is built in **Lab8_5(DRAI2CSlave)** later.
5. Program **Lab8_5(DRAI2CSlave)** enables students to use the **eUSCI_B1** module for I2C mode to build a slave program to receive 16-byte data from a master device via the I2C bus. Both programs, **Lab8_4(DRAI2CMaster)** and **Lab8_5(DRAI2CSlave)**, should be installed on two different MSP-EXP432P401R EVBs, and two EVBs should be connected with two I2C wires to form a I2C bus (refer to Figure 8.48 for this connection).
6. Program **Lab8_6(SDI2CMaster)** enables students to use **eUSCI_B1** module for I2C mode to build a master program to transmit six letters, **MSP432**, to a slave device via the I2C bus by using the I2C API functions provided by the MSPPDL. The slave program is built in **Lab8_7(SDI2CSlave)** later.
7. Program **Lab8_7(SDI2CSlave)** enables students to use the **eUSCI_B1** module for I2C mode to build a slave program to receive six letters from a master device via the I2C bus by using the I2C API functions provided by the MSPPDL. Both programs, **Lab8_6(SDI2CMaster)** and **Lab8_7(SDI2CSlave)**, should be installed on two different MSP-EXP432P401R EVBs, and two EVBs should be connected with two I2C wires to form a I2C bus (refer to Figure 8.48 for this connection).
8. Program **Lab8_8** allows students to use **eUSCI_A2** for UART mode with the SD model to build a transmit–receive loop back testing project to test the asynchronous serial data communication functions by using UART API functions provided by the MSPPDL.
9. Programs, **Lab8_9 (UARTSendRecv)** and **Lab8_10 (UARTRecvSend)**, all ow students to build a real UART data communication system with two UART modes installed on two MSP-EXP432P401R EVBs. Both **P3.2 (UCA2RXD)** and **P3.3 (UCA2TXD)** pins on two EVBs are connected crossly via the MAX3232 RS-232 transceiver. Refer to hardware configurations shown in Figure 8.61 to complete this connection.

After completion of these programs, you should understand the basic architecture and operational procedure for most popular serial-related peripheral devices, including the SPI, I2C, and UART modules installed in the MSP432P401R MCU system. You should be able to code some sophisticated programs to access the desired peripheral devices to perform desired control operations via serial ports, either synchronously or asynchronously.

8.1 LAB8_1

8.1.1 Goal

In this project, students need to build an LCD project to display **MSP432** on the first line on the LCD panel and **2016** on the second line in the LCD panel. The first line starts from the position **05** (DDRAM address) and the second line starts at **46**. The **eUSCI_B0** module is used to enable the interrupt function to display letters on the LCD panel for this project. In fact, students can modify a class project **DRALCD** to make it as this project with the interrupt mechanism.

8.1.2 Data assignment and hardware configuration

No hardware configuration is needed in this project. Refer to Figure 8.10 to get a detailed hardware configuration about this LCD architecture and connection in the Trainer. Refer to Figure 8.13 to get a clear picture about the DDRAM address and LCD letter positions.

8.1.3 Development of the project and the header file

Both a header file and a C code file are needed in this project.

Create a new folder **Lab8_1(DRALCDINT)** under the folder **C:\MSP432 Lab Projects** with the Windows Explorer. Then create a new µVersion5 project **DRALCDINT** under the folder **Lab8_1(DRALCDINT)** created above.

Students need to create a new header file **DRALCDINT.h**. Then copy the header file **DRALCD.h** and paste it into this header file. Refer to class project **DRALCD** to complete this project.

8.1.4 Development of the source code DRALCDINT.c

Students need to create a new C source file **DRALCDINT.c** and modify a class project source file **DRALCD.c** to make it as the source file for this project. Perform the following modifications based on the **DRALCD.c** source file:

1. Include the header file **DRALCDINT.h** at the top of this C source file first
2. Create a new global variable **unsigned char sdata**
3. Inside the **main()** program, after the code line **LCD_Comd(1)**, add one code line to define the start position **05** on the first line on the LCD. Refer to Figure 8.22 to get more details for the value of the argument for the subroutine **LCD_Comd()**
4. Call the **LCD_Data()** subroutines to display **MSP432** in the first line of the LCD
5. Then use the **LCD_Comd()** subroutine to define the start position **46** on the second line of the LCD. Select the correct value for the argument of this subroutine
6. Call the **LCD_Data()** subroutines to display **2016** in the second line of the LCD
7. Inside the **LCD_Init()** subroutine, add two coding lines to set up the priority register **NVIC_IPR5** and interrupt enable register **NVIC_ISER0** with appropriate values to set up the **eUSCI_B0** module with priority level of 3 and enable this module's interrupt. These two lines can be added in any location inside this subroutine. But it had better to be under the GPIO port initialization coding lines
8. Add **__enable_irq()** to globally enable all interrupts after the above two lines
9. Modify the subroutine **SPIB_Write()** by replacing all original code with following two coding lines:

```
sdata = data;
UCB0IE |= UCTXIE;
```

10. Create a new **eUSCI_B0** interrupt handler **EUSCIB0_IRQHandler()** with the following modifications
11. Create a local variable **uint16_t res;**
12. Output low to GPIO pin **P6.7**
13. Use an **if()** selection structure to check whether a data transmit interrupt has occurred. This checking can be done by inspecting the **UCTXIFG** bit on the **UCB0IFG** register, and an **AND** operator can be sued between them

14. If a data transmit interrupt occurred, send the **sdata** to the **UCB0TXBUF** register to transmit this data to the LCD controller SPLC780 via the 74VHCT595
15. Disable the interrupt for the **eUSCI_B0** module
16. Read the **UCB0IV** register and assign the reading result to the local variable **res**. This reading is to reset the **UCTXIFG** flag bit in the **UCB0IFG** register
17. Use the code line **while (UCB0STATW & UCBUSY);** to wait for the 74VHCT595 serial data shifting to be done.
18. Set GPIO pin **P6.7** to high

8.1.5 Set up the environment to build and run the project

To build and run the project, one needs to perform the following operations to set up the environments:

- Select the **CMSIS-DAP Debugger** in the **Debug** tab under the **Project|Options for Target "Target 1"** menu item. Make sure that all settings for the debugger and the flash download are correct
- Now, let's build the project by going to **Project|Build target** menu item
- Then go to **Flash|Download** menu item to download the image file of this project into the flash memory in the MSP432P401R Evaluation Board
- Now go to **Debug|Start/Stop Debug Session** to ready to run the project
- Go to **Debug|Run** menu item to run the project

8.1.6 Demonstrate your program by checking the running result

After the project runs, check the LCD panel to make sure that the required letters are displayed in the desired positions and lines.

8.2 LAB8_2

8.2.1 Goal

This project enables students to use DRA model to build a seven-segment LED project to use **eUSCI_B0** module to repeatedly display seven segments in a sequence for all four seven-segment LEDs on the EduBASE ARM® Trainer.

8.2.2 Data assignment and hardware configuration

No hardware connections are needed for this project.

Refer to Figure 8.24 to get a detailed hardware configuration about these four seven-segment LEDs architecture and connection in the Trainer. Refer to Tables 8.10 and 8.11 to get a clear picture between the output bits of serial shift register 74VHCT595 and selected LED digit.

8.2.3 Development of the project

To make this project easy, we used DRA model with three user-defined functions, **LED_Init()**, **SPIB_Write()**, and **delay()** to support this project.

Follow the steps below to develop this project. Only a C code source file is used in this project since this project is simple. Create the project and develop the C source file with the following steps:

1. Create a new folder **Lab8_2(DRALED)** under the folder **C:\MSP432 Lab Projects\Chapter 8** in the Windows Explorer
2. Open the Keil® ARM-MDK µVersion5, create a new project named **DRALED** and save this project into the folder **Lab8_2(DRALED)** created in step 1
3. On the next wizard, you need to select the device (MCU) **MSP432P401R**. Click on the **OK** to close this wizard
4. Next, the software components wizard is opened, and you need to set up the software development environment for your project with this wizard. Expand two icons, **CMSIS** and **Device**, and check the **CORE** and **Startup** checkboxes in the **Sel.** column, and click on the **OK** button since we need these two components to build our project

8.2.4 Development of the C source file

1. In the **Project** pane, expand the **Target** folder and right click on the **Source Group 1** folder and select the **Add New Item to Group "Source Group 1."**
2. Select the **C File (.c)** and enter **DRALED** into the **Name:** box, and click on the **Add** button to add this source file into the project.
3. Include the following system header files into this source file:
   a. **#include <stdint.h>**
   b. **#include <stdbool.h>**
   c. **#include <msp.h>**
4. Declare three user-defined functions, **LED_Init(void)**, **SPIB_Write(unsigned char data)** and **delay(int time)**. All functions return **void**.
5. Inside the **main()** program, first declare three **uint8_t** local variables, **uiLoop**, **uiUpper** = **8** and **uiDigit** = **0xF7**. The first variable is used as a loop number and the second works as the upper bound for a **for()** loop, and the third one is used to provide a digit selection for the selected LED. Refer to Tables 8.10 and 8.11 to get a clear picture about the segments and the digit of the selected LED.
6. Call the **LED_Init()** function to initialize and configure the LEDs.
7. Starting with an infinitive **for()** loop or **while()** loop to repeatedly display all segments for four LEDs in a sequence.
8. Starting with a finite **for()** loop with the **uiLoop** as the loop number, starting from **0** and ending at **uiUpper** (8). In this way, we can scan and turn on all eight segments, including the decimal point segment **DP**, one by one.
9. Inside the finite **for()** loop, call the **SPIB_Write()** function to send the first segment (**a**) to all four LEDs. The argument of this function should be 1 shifted left by **uiLoop** bits. In this way, all seven segments can be turned ON from **a**, and **b**, **c**, … until **g**.
10. Call the **SPIB_Write()** function again to select the desired LED by turning on the related digit of the LEDs. The argument of this function should be the digit number of the desired LED (**uiDigit**). The first digit, DIGIT3, should be **0xF7** if we prefer to start from DIGIT3.
11. Call **delay(500)** to delay the project by 500 ms.
12. Outside the finite **for()** loop, all eight segments for one LED digit has been done. Now we need to point to the next LED digit by shifting the **uiDigit** right by 1 bit. However, in order to make sure that the MSB on the above shifted result is 1, the shift result should be **ORed** with **0xF0**.

13. Use an **if()** selection structure to check whether the **uiDigit** gets **0xFF**, which means that all four LEDs have been completed in one loop. If this is true, we need to reinitialize **uiDigit** to **0xF7** to continue for the next loop.

The following codes are used for **LED_Init()** function.

14. Set three GPIO pins, **P1.5**, **P1.6**, and **P1.7** as the second function (**eUSCI_B0**) module by setting the **P1SEL0.x** = 1 (**x** = 5~7). Since by default, **P1SEL1** = 0 after a system reset, therefore we can get **P1SEL1.x:P1SEL0.x** = 01 after this setting.
15. Set GPIO pin **P2.3** as output pin since we need to use this pin to generate a positive-going edge pulse to trigger the conversion from the serial to parallel data for 74VHCT595.
16. Perform the following operations to initialize and configure the **eUSCI_B0** module to enable it to work as SPI 3-Pin mode:
    a. Reset the **eUSCI_B0** module first to enable the initialization process
    b. Set the **eUSCI_B0** module as a master 3-pin mode
    c. Set the **eUSCI_B0** module to work on synchronous (SPI) mode
    d. Set the inactive state of the **SCLK** is high
    e. Enable the MSB is to be transmitted first
    f. Select the CKSE as **ACLK**
    g. Assign **0x1** to the **UCB0BRW** register to define the prescaler as 2 to divide the CKSE by 2 to get the SCLK
    h. Remove the reset for the **eUSCI_B0** module to enable it to begin to perform normal transmit–receive data

The following codes are used for the function **SPIB_Write**(unsigned char **data**).

17. Reset GPIO pin **P2.3** to 0 to output a low on **CS_LED**.
18. Assign the **data** to be transmitted to the **UCB0TXBUF** register to transfer it.
19. Use a **while()** loop to wait for data transmit done by checking the **UCBUSY** bit in **UCB0STATW** register. The loop condition should be **UCB0STATW & UCBUSY**.
20. Set the GPIO pin **P2.3** to high. This step is to simulate a low-to-high edge to trigger a serial-to-parallel conversion on the serial shift register 74VHCT595.

For the **delay()** function, refer to a same function used in the class project **DRALED**.

8.2.5 Set up the environment to build and run the project

Perform the following operations to make sure that the debugger you are using is the **CMSIS-DAP Debugger**. You can do this checking by:

- Going to **Project|Options for Target "Target 1"** menu item to open the Options wizard
- On the opened Options wizard, click on the **Debug** tab
- Make sure that the debugger shown in the **Use:** box is **CMSIS-DAP Debugger**. Otherwise you can click on the dropdown arrow to select this debugger from the list
- Make sure that all settings for the debugger and the flash download are correct

8.2.6 Build and demonstrate your program

As the project runs, seven segments and the decimal point for each LED are turned on one by one in a sequence. Based on on the Source Group 1 folder, answer the following questions:

- Can we modify this project to display only seven segments without the DP segment? If your answer is yes, how do we do that?
- What did you learn from this project?

8.3 LAB8_3

8.3.1 Goal

This lab is very similar to **Lab8_2** to allow students to use API functions provided by the MSPPDL to interface to the 74VHCT595 to build a seven-segment LED project to use **eUSCI_B0** module to repeatedly display seven segments in a sequence for all four seven-segment LEDs on the EduBASE ARM® Trainer. One can modify the **Lab8_2** by replacing some register accessing methods with the API functions.

8.3.2 Data assignment and hardware configuration

Refer to Section 8.4.9.2 and Figure 8.24 to get a detailed hardware configuration for the interface between the **eUSCI_B0** module and 74VHCT595.

8.3.3 Development of the project

Use the steps below to develop this project. Both a header file and a C source file are used in this project since it is a little complicated. Create a new project with the following steps:

1. Create a new folder **Lab8_3(SDLED)** under the folder **C:\MSP432 Lab Projects\Chapter 8** in the Windows Explorer
2. Create a new ARM-MDK μVersion5 project named **SDLED** and save this project in the folder **Lab8_3(SDLED)** created in step 1
3. Select the correct MCU **MSP432P401R** for this project
4. Do the necessary steps to use **CMSIS CORE** and the **Device Startup** tools for this project

8.3.4 Development of the header file SDLED.h

Create a new header file and include the following system header files and macros:

```
#include <stdint.h>
#include <stdbool.h>
#include <MSP432P4xx\gpio.h>
#include <MSP432P4xx\spi.h>
#include <MSP432P4xx\wdt_a.h>

void delay(int time);
void LED_Init(void);
void SPIB_Write(unsigned char data);

const eUSCI_SPI_MasterConfig spiMasterConfig =
{
  EUSCI_SPI_CLOCKSOURCE_ACLK,            // ACLK Clock Source
```

```
    32768,                                  // ACLK = LFXT = 32.768khz
    32768,                                  // SPICLK = 32.768khz
    EUSCI_SPI_MSB_FIRST,                    // MSB First
    EUSCI_SPI_PHASE_DATA_CHANGED_
    ONFIRST_CAPTURED_ON_NEXT,               // Phase
    EUSCI_SPI_CLOCKPOLARITY_
    INACTIVITY_HIGH,                        // High polarity
    EUSCI_SPI_3PIN                          // 3-Wire SPI Mode
};
```

The **eUSCI_SPI_MasterConfig** is the protocol of the struct used to hole all initialization parameters for the SPI module. Save this header file as **SDLED.h**.

8.3.5 Development of the C source file

One can modify the source codes for the **Lab8_2** to make it as the source codes for this lab. The main modifications can be divided into three parts:

1. Modification made in the main() program
2. Modifications made in the LED_Init() subroutine
3. Modifications made in the SPIB_Write() function

Perform the following operations to create and modify this source file:

1. Create a new C source file and name it **SDLED.c**
2. Include the header file "**SDLED.h**" into this source file
3. Use the **WDT_A_holdTimer();** to replace the original **WDTCTL = WDTPW | WDTHOLD;** instruction to stop the Watchdog timer inside the main program

Perform the following operations to modify the code inside the **LED_Init()** function:

4. Use the **GPIO_setAsPeripheralModuleFunctionOutputPin()** API function to configure GPIO pins **P1.5~P1.7** as the primary function pins
5. Use **GPIO_setAsOutputPin()** API function to set GPIO pin **P2.3** as the output pin
6. Use **SPI_disableModule()** API function to disable or reset the **eUSCI_B0** module
7. Use the API function **SPI_initMaster()** to configure the **eUSCI_B0** module. This function contains two arguments, the **eUSCI_B0** module name and the pointer of the **spiMasterConfig** struct that is defined in the **SDLED.h** header file
8. Use the API function **SPI_enableModule()** to enable the **eUSCI_B0** module and make it ready to perform data transfer jobs

Perform the following operations to modify the codes inside the **SPIB_Write()** function:

9. Use **GPIO_setOutputLowOnPin()** API function to set the GPIO pin **P2.3** to low to reset the STCP (**CS_LED**) signal to select the LED
10. Use **SPI_transmitData()** API function to transmit the data to the **UCB0SIMO** line
11. Use **while (UCB0STATW & UCBUSY);** to wait for 74VHCT595 serial data shifting to be done

12. Use **GPIO_setOutputHighOnPin()** API function to set GPIO pin **P2.3** to high to trigger the 74VHCT595 to complete its data transmit to the seven-segment LEDs

8.3.6 Set up the environment to build and run the project

This setup contains the following three operations:

1. Include the system header files by adding the include path
2. Add the MSPPDL into the project since we need to use some API functions provided by that library
3. Check and configure the correct debugger used in the project

Perform the following operations to finish these setups:

- Go to C/C++ tab in the **Project|Options for Target "Target 1"** to set up the **Include Path** as: **C:\ti\msp\MSPWare_2_21_00_39\driverlib\driverlib**
- Add the MSPPDL into this project by right clicking on the **Source Group 1** folder in the **Project** panel on the left, and select the **Add Existing Files to Group "Source Group 1"** menu item. On the opened box, select the driver library file path **C:\ti\msp\MSPWare_2_21_00_39\driverlib\driverlib\MSP432P4xx\keil** and then select the driver library file **msp432p4xx_driverlib.lib**. Click the **Add** button to add it into the project
- Select the correct debugger by clicking on the **Debug** tab and select the **CMSIS-DAP Debugger** from the **Use** box. Also make sure that all settings for the debugger and the flash download are correct by clicking on the Settings button

8.3.7 Demonstrate your program

Now you can build, download, and run your project. As the project runs, you can find that seven segment on all four LEDs are repeatedly turned on in a sequence.

Based on these results, try to answer the following questions:

- Can you modify this lab to display a number "**8**" on each LED in a sequence?
- Can you modify this lab to make each segment rotate or shift in the opposite direction?
- What did you learn from this project?

8.4 LAB8_4

8.4.1 Goal

This program **Lab8_4(DRAI2CMaster)** allows students to use the **eUSCI_B1** module for I2C mode to build a master program to transmit 16-byte data to a slave device via the I2C bus. The slave program is built in **Lab8_5(DRAI2CSlave)** later.

8.4.2 Data assignment and hardware configuration

Refer to Figure 8.48 to get a detailed hardware configuration and connection between this master device and a slave device. Both devices are MSP-EXP432P401R EVBs.

8.4.3 Development of the project

Create a new project with the following steps:

1. Create a new folder **Lab8_4(DRAI2CMaster)** under the MSP432 Lab folder **C:\MSP432 Lab Projects\Chapter 8** in the Windows Explorer
2. Create a new ARM-MDK µVersion5 project **DRAI2CMaster** and save this project in the folder **Lab8_4(DRAI2CMaster)** created in step 1
3. Do the necessary steps to use **CMSIS CORE** and the **Device Startup** tools for this project
4. Create a new C source file **DRAI2CMaster.c** and add it into this project

8.4.4 Development of the C source file

This source file is similar to one used in a class project **DRAI2CMaster.c**. Students can modify that file to make it as this source file. The difference is that this program is to use the **eUSCI_B1** module as a master working in the transmit mode, not in receive mode used in the class project.

Perform the following modifications to make this source file:

1. Replace all original global variables with a new global variable **uint8_t TXData** = **0**
2. Remove the original local variables **uint8_t n, result[16]**
3. Remove all configurations for GPIO pin **P2**
4. Add a code line to enable the transmit mode and **START** condition by **ORing** the macros **UCTR|UCTXSTT** and assigning them to the **UCB1CTLW0** Register
5. Replace one of interrupt sources, **UCRXIE0**, with the **UCTXIE0** in the assignment to the **UCB1IE** register since you need to use a transmit interrupt to transfer the data bytes to the slave device
6. Use an infinitive **while(1)** loop to replace the conditional **while(intFlag)** loop
7. Keep the first code line inside this **while()** loop, but replace the following two code lines with the code line shown below:
   a. **UCB1CTLW0 |= UCTR | UCTXSTT;**
8. Remove the original **for()** loop and an infinitive **while(1)** loop
9. Inside the ISR, replace the original RX **if()** block (**if (UCB1IFG & UCRXIFG0)**) with the following TX **if()** block:

```
if (UCB1IFG & UCTXIFG0)
{
    UCB1IFG & = ~UCTXIFG0;
    UCB1TXBUF = TXData + +;
}
```

10. Replace the code for the byte counter interrupt **if()** block with the following code:

```
UCB1IFG & = ~UCBCNTIFG;
UCB1CTLW0 |= UCTXSTP;
P1OUT = BIT0;
```

8.4.5 Set up the environment to build and run the project

Make sure that the debugger you are using is the **CMSIS-DAP**. You can do this checking by:

- Going to **Project|Options for Target "Target 1"** menu item to open the Options wizard

- On the opened Options wizard, click on the **Debug** tab
- Make sure that all settings for the debugger and the flash download are correct

8.4.6 Build your program

Perform the following operations to build your program:

- Go to **Project|Build target** to build the project
- Go to **Flash|Download** menu item to download your program into the flash memory

You cannot run this project until you finished building the next project, **Lab8_5**.

8.5 LAB8_5

8.5.1 Goal

This program **Lab8_5(DRAI2CSlave)** enables students to use the **eUSCI_B1** module for I2C mode to build a slave program to receive 16-byte data from a master device via the I2C bus. Both programs, **Lab8_4(DRAI2CMaster)** and **Lab8_5(DRAI2CSlave)**, should be installed on two different MSP-EXP432P401R EVBs, and two EVBs should be connected with two I2C wires to form a I2C bus.

8.5.2 Data assignment and hardware configuration

Refer to Figure 8.48 to get details about the hardware configuration and connection between the master and the slave device.

8.5.3 Development of the project

Create a new project with the following steps:

1. Create a new folder **Lab8_5(DRAI2CSlave)** under the MSP432 Lab folder **C:\MSP432 Lab Projects\Chapter 8** in the Windows Explorer
2. Create a new ARM MDK µVersion5 project **DRAI2CSlave** and save this project in the folder **Lab8_5(DRAI2CSlave)** created in step 1
3. Do the necessary steps to use **CMSIS CORE** and the **Device Startup** tools for this project
4. Create a new C source file **DRAI2CSlave.c** and add it into this project

8.5.4 Development of the C source file

This source file is similar to one used in a class project **DRAI2CSlave.c**. Students can modify that file to make it as this source file. The difference is that this program is to use the **eUSCI_B1** module as a slave working in the receive mode, not in transmit mode used in the class project.

Perform the following modifications to make this source file:

1. Replace the original global variable with the following variables:

```
uint8_t m = 0, RXDataCount = 16, RXData[16];
bool intFlag = true;
```

2. Add two local variables **uint8_t n, result[16]** at the beginning of the **main()** program

3. Add the following codes to configure GPIO ports and pins (**P2** and **P1**):

```
P2DIR = BIT0|BIT1|BIT2;
P2OUT &= ~(BIT0|BIT1|BIT2);
P1OUT &= ~BIT0;
P1DIR |= BIT0;
```

4. Replace all original interrupt sources with one interrupt source **UCRXIE0** in the assignment to the **UCB1IE** register since you need to use a receive interrupt to get the data bytes from the master device
5. Use a conditional **while(intFlag){}** loop to replace the original infinitive **while(1)** loop
6. Add the following code lines under that conditional **while()** loop:

```
for (n = 0; n < 16; n++) {result[n] = RXData[n];}
```

7. Add an infinitive **while(1)** loop to check the running results of the project
8. Inside the ISR, replace the original **if()** blocks with the following single RX **if()** block:

```
if (UCB1IFG & UCRXIFG0)
{
    UCB1IFG &=~UCRXIFG0;
    if (RXDataCount)
    {
        RXData[m] = UCB1RXBUF;
        P2OUT = RXData[m];
        m + +;
        RXDataCount–;
    }
    else
    {
        P1OUT = BIT0;
        intFlag = false;
    }
}
```

8.5.5 Setup the environment to build and run the project

Make sure that the debugger you are using is the **CMSIS-DAP**. You can do this checking by:

- Going to **Project|Options for Target "Target 1"** menu item to open the Options wizard
- On the opened Options wizard, click on the **Debug** tab
- Make sure that all settings for the debugger and the flash download are correct

8.5.6 Demonstrate your program

Now you can build and download this slave project.

Make sure that both the master and the slave programs are installed in two different MSP-EXP432P401R EVBs and the I2C bus connection between them are identical with one shown in Figure 8.48.

Before you can run the project, make sure to make a break point at the infinitive **while()** loop line in the slave program. In this way, we can temporarily halt the program to check the data receiving result.

Now you can run both programs on two EVBs. First run the slave program and then the master program. The interval between these two running times should be as short as possible. Ideally it should be less than 3 seconds. If the interval is greater than that, the transmitted and received data bytes may be not starting from 0.

After both red color and three-color LEDs on the slave device are on, open the **Call Stack + Locals** window and expand the receive array **result[]**. You can find that 16 data bytes, starting from **0x00** and ending at **0x0F**, which were sent by the master, have been received by the slave and displayed in this window.

Based on these results, try to answer the following questions:

- Why do we need to use a conditional **while(intFlag)** loop in the slave program to keep this project running? What will happen without this loop?
- Why do we need to use two infinitive **while(1)** loops in both the master and the slave program? What is the functional difference between these two **while()** loops?
- Why do we only use a limited number of parameters to initialize and configure the **eUSCI_B1** module as an I2C mode? According to the text, there should be more parameters to be configured; why?
- What did you learn from this project?

8.6 LAB8_6

8.6.1 Goal

The program **Lab8_6(SDI2CMaster)** enables students to use **eUSCI_B1** module for I2C mode to build a master program to transmit six letters, **MSP432**, to a slave device via the I2C bus by using the I2C API functions provided by the MSPPDL. The slave program is built in **Lab8_7(SDI2CSlave)** later.

8.6.2 Data assignment and hardware configuration

Refer to Figure 8.48 to get details about the hardware configuration and connection between the master and the slave device.

8.6.3 Development of the project

Create a new project with the following steps:

1. Create a new folder **Lab8_6(SDI2CMaster)** under the MSP432 Lab folder **C:\MSP432 Lab Projects\Chapter 8** in the Windows Explorer
2. Create a new ARM MDK µVersion5 project **SDI2CMaster** and save this project in the folder **Lab8_6(SDI2CMaster)** created in step 1
3. Do the necessary steps to use **CMSIS CORE** and the **Device Startup** tools for this project
4. Create a new C source file **SDI2CMaster.c** and add it into this project

8.6.4 Development of the C source file

1. Add the following system header files and macros into the **SDI2CMaster.c** source file:

```
#include <stdint.h>
#include <stdbool.h>
#include <MSP432P4xx\i2c.h>
```

```
#include <MSP432P4xx\gpio.h>
#include <MSP432P4xx\wdt_a.h>
#include <MSP432P4xx\interrupt.h>

#define SLAVE_ADDRESS 0x68
void EUSCIB1_ISR(void);

char txData[6] = {'M', 'S', 'P', '4', '3', '2'};
uint8_t n = 0, txDataCount = 6;
```

Three global variables are used for this project, **txData[6]**, **n**, and **txDataCount**. The I2C interrupt handler is **EUSCIB1_ISR()**.

2. Use the **eUSCI_I2C_MasterConfig** struct to create a struct variable **i2cConfig** as:

```
const eUSCI_I2C_MasterConfig i2cConfig =
{
  EUSCI_B_I2C_CLOCKSOURCE_SMCLK,       // SMCLK Clock Source
  3000000,                             // SMCLK = 3MHz
  EUSCI_B_I2C_SET_DATA_RATE_400KBPS,   // Desired I2C Clock of 400khz
  0,                                   // No byte counter threshold
  EUSCI_B_I2C_NO_AUTO_STOP             // No Autostop
};
```

3. Inside the **main** function, declare a local **uint32_t** variable **Loop**.
4. Call an API function to stop the Watchdog timer.
5. Since we need to use GPIO pin **P1.0** (connected to a red-color LED) to indicate the running status of our master program, therefore use the **GPIO_setAsOutputPin()** API function to set **P1.0** pin as an output pin, and use the **GPIO_setOutputLowOnPin()** API function to enable **P1.0** pin to output low.
6. Use **GPIO_setAsPeripheralModuleFunctionInputPin()** function to configure **P6.4** and **P6.5** pins as **UCB1SDA** and **UCB1SCL** pins.
7. Use **I2C_initMaster()** function with the struct variable **i2cConfig** to initialize and configure **EUSCI_B1_MODULE** properly.
8. Use **I2C_setSlaveAddress()** API function to set the slave device address, **SLAVE_ADDRESS**, which is **0x68** defined as a macro in the top of this source file.
9. Use **I2C_setMode()** function to set the eUSCI_B1 module (**EUSCI_B1_MODULE**) to work in the transmit mode (**EUSCI_B_I2C_TRANSMIT_MODE**).
10. Use **I2C_enableModule()** to enable the eUSCI_B1 module.
11. Call the **I2C_registerInterrupt()** API function to register the eUSCI_B1 module interrupt handler (**EUSCIB1_ISR**).
12. Call the **I2C_clearInterruptFlag()** function to clear any possible interrupt flags set before. Two flags, **EUSCI_B_I2C_TRANSMIT_INTERRUPT0** and **EUSCI_B_I2C_NAK_INTERRUPT**, can be **ORed** together as the mask to finish this clear operation.
13. Call the **I2C_enableInterrupt()** API function to enable the eUSCI_B1 module. Similarly, the macros **EUSCI_B_I2C_TRANSMIT_INTERRUPT0** and **EUSCI_B_I2C_NAK_INTERRUPT** can be **ORed** together as the mask to finish this enable function.
14. Use **Interrupt_enableInterrupt(INT_EUSCIB1)()** API function to enable all eUSCI_B1 module-related interrupt sources.
15. Use an infinitive **while()** loop to perform the following operations:
    a. Use the code line **for (Loop = 0; Loop < 2000; Loop++);** to delay the program a certain period of time.

b. Make sure that any previous transaction has been completed by using the code line: **while(I2C_masterIsStopSent(EUSCI_B1_MODULE) == EUSCI_B_I2C_SENDING_STOP);**
c. Use **I2C_masterSendMultiByteStart(EUSCI_B1_MODULE, txData[n++]);** to start a transmit.
d. Use an intrinsic function **__wfi()** to wait for any interrupt to occur.

The following codes are used for the eUSCI_B1 interrupt handler.

16. A local **uint_fast16_t** variable **status** is declared first inside this interrupt handler.
17. Call the **I2C_getInterruptStatus()** API function to get the current interrupt source. This function contains two parameters, the **EUSCI_B1_MODULE** and a **mask**. The **mask** is the transmit interrupt (**EUSCI_B_I2C_TRANSMIT_INTERRUPT0**). This function returns the current interrupt source and assigns it to the local variable **status**.
18. Use the **I2C_clearInterruptFlag()** API function to clear the accepted interrupt sources to enable them to generate interrupts in the future. This step is very important and no further interrupts could be generated if this step is missed. Two parameters are used for this function, the **EUSCI_B1_MODULE** and a **mask** represented the interrupt source to be cleared. One can directly use the **status** as the interrupt source since it did contain the interrupt source obtained from the last step (step 17).
19. Use an **if()** block with the condition **status & EUSCI_B_I2C_NAK_INTERRUPT** to check whether this interrupt is a NAK interrupt. If it is, call the **I2C_masterSendStart()** function to restart the data transmit function.
20. Use another **if**() block with the condition **status & EUSCI_B_I2C_TRANSMIT_INTERRUPT0** to check whether this interrupt is a data transmit interrupt. If it is, use another **if-else** block to transmit the next or the last data.
    a. Use an **if**() block with the condition **txDataCount** to check whether all six letters have been sent out (**txDataCount** = 0). If not, use the following code line to send next letter: **I2C_masterSendMultiByteNext(EUSCI_B1_MODULE, txData[n++]);**. Also reduce the **txDataCount** by 1.
    b. If **txDataCount** = 0, which means that all six letters have been sent out, use the API function **I2C_masterSendMultiByteStop()** to send a **STOP** to the slave. Also call the API function **GPIO_setOutputHighOnPin()** to set pin **P1.0** to high to turn on the red color LED to indicate that all data transmits have been done.

8.6.5 Set up the environment to build the project

This setup contains the following three operations:

1. Include the system header files by adding the include path
2. Add the MSPPDL into the project since we need to use some I2C-related API functions provided by that library
3. Check and configure the correct debugger used in the project

Perform the following operations to finish these setups:

- Go to **C/C++** tab in the **Project|Options for Target "Target 1"** to set up the **Include Path** as: **C:\ti\msp\MSPWare_2_21_00_39\driverlib\driverlib**

- Add the MSPPDL into this project by right clicking on the **Source Group 1** folder in the **Project** panel on the left, and select the **Add Existing Files to Group "Source Group 1"** menu item. On the opened box, select the driver library file path **C:\ti\msp\MSPWare_2_21_00_39\driverlib\driverlib\MSP432P4xx\keil** and then select the driver library file **msp432p4xx_driverlib.lib**. Click the **Add** button to add it into the project
- Select the correct debugger by clicking on the **Debug** tab and select the **CMSIS-DAP Debugger** from the **Use** box. Also make sure that all settings for the debugger and the flash download are correct by clicking on the Settings button

8.6.6 Download your program

Perform the following operations to download your program to the flash memory:

- Go to **Flash|Download** menu item to download your program into the flash memory
- Go to **Debug|Start/Stop Debug Session** to begin debugging your program. Click on the **OK** button on the 32KB memory size limitation message box to continue

You cannot run this master program until you finish building the next lab project, **SDI2CSlave** since this master program needs to communicate with that slave device to transmit data between them.

8.7 LAB8_7

8.7.1 Goal

The program **Lab8_7(SDI2CSlave)** enables students to use the **eUSCI_B1** module for I2C mode to build a slave program to receive six letters from a master device via the I2C bus by using the I2C API functions provided by the MSPPDL. Both programs, **Lab8_6(SDI2CMaster)** and **Lab8_7(SDI2CSlave)**, should be installed on two different MSP-EXP432P401R EVBs, and two EVBs should be connected with two I2C wires to form a I2C bus (refer to Figure 8.48 for this connection).

8.7.2 Data assignment and hardware configuration

Refer to Figure 8.48 to get details about the hardware configuration and connection between the master and the slave device.

8.7.3 Development of the project

Create a new project with the following steps:

1. Create a new folder **Lab8_7(SDI2CSlave)** under the MSP432 Lab folder **C:\MSP432 Lab Projects\Chapter 8** in the Windows Explorer
2. Create a new ARM MDK µVersion5 project **SDI2CSlave** and save this project in the folder **Lab8_7(SDI2CSlave)** created in step 1
3. Do the necessary steps to use **CMSIS CORE** and the **Device Startup** tools for this project
4. Create a new C source file **SDI2CSlave.c** and add it into this project

8.7.4 Development of the C source file

1. Add the following system header files and macros into the **SDI2CSlave.c** source file:

```
#include <stdint.h>
#include <stdbool.h>
#include <MSP432P4xx\i2c.h>
#include <MSP432P4xx\gpio.h>
#include <MSP432P4xx\wdt_a.h>
#include <MSP432P4xx\interrupt.h>

#define SLAVE_ADDRESS      0x68
#define NUM_OF_RX_BYTES    6

void EUSCIB1_ISR(void);
bool intFlag = true;
uint8_t n = 0, rxData[NUM_OF_RX_BYTES];
```

   Three global variables are used for this project, **rxData[6]**, **n**, and **intFlag**. The I2C interrupt handler is **EUSCIB1_ISR()**.
2. Inside the **main** function, declare an **uint8_t** local array **result[6]**.
3. Call an API function to stop the Watchdog timer.
4. Since we need to use GPIO pin **P1.0** (connected to a red-color LED) to indicate the running status of this slave program, therefore use the **GPIO_setAsOutputPin()** API function to set **P1.0** pin as an output pin, and use the **GPIO_setOutputLowOnPin()** API function to enable **P1.0** pin to output low.
5. Use **I2C_initSlave()** function to initialize the eUSCI_B1 module (**EUSCI_B1_MODULE**) as a slave device with the following parameters:
   a. Slave address: **SLAVE_ADDRESS** (defined as **0x68** at the top of this source file).
   b. Slave address offset: **EUSCI_B_I2C_OWN_ADDRESS_OFFSET0**.
   c. Slave own address enable: **EUSCI_B_I2C_OWN_ADDRESS_ENABLE**.
6. Use **I2C_setMode()** function to set the eUSCI_B1 module (**EUSCI_B1_MODULE**) to work in the receive mode (**EUSCI_B_I2C_RECEIVE_MODE**).
7. Use **I2C_enableModule()** to enable the eUSCI_B1 module.
8. Call the **I2C_registerInterrupt()** API function to register the eUSCI_B1 module interrupt handler (**EUSCIB1_ISR**).
9. Call the **I2C_clearInterruptFlag()** function to clear any possible interrupt flags set before. Only one flag, **EUSCI_B_I2C_RECEIVE_INTERRUPT0**, can be used as the mask to finish this clear operation.
10. Call the **I2C_enableInterrupt()** API function to enable the eUSCI_B1 module. Similarly, the macros **EUSCI_B_I2C_RECEIVE_INTERRUPT0** can be used as the mask to finish this enable function.
11. Use **Interrupt_enableInterrupt(INT_EUSCIB1)()** API function to enable all eUSCI_B1 module-related interrupt sources.
12. Call the **Interrupt_enableMaster()** API function to globally enable all interrupt sources.
13. Use a conditional **while()** loop with the **intFlag** as the condition to wait for all receive interrupts to be occurred. By default, the **intFlag** is set to **true** at the beginning of this file. In this way, the program will keep waiting for all receive interrupts to be occurred without proceeding to the next step. This means that the program

will not continue until all six letters transmitted from the master have been received inside the interrupt handler **EUSCIB1_ISR()**.

14. After all six letters have been received, the **while()** loop will be terminated by resetting the **intFlag** to **false** in the receive ISR. Use the following codes to collect all received letters:

```
for (n = 0; n < NUM_OF_RX_BYTES; n++)
  result[n] = rxData[n];
```

15. An infinitive **while()** loop should be added here to enable us to check the running result of the project, exactly to check the **result[]** array to confirm that all six letters have been successfully received by this slave project. To do this, a break point should be set at this **while()** loop line later.

The following codes are used for the eUSCI_B1 interrupt handler **EUSCIB1_ISR()**.

16. A local **uint_fast16_t** variable **status** is declared first inside this interrupt handler.
17. Call the **I2C_getInterruptStatus()** API function to get the current interrupt source. This function contains two parameters, the **EUSCI_B1_MODULE** and a **mask**. The **mask** is the receive interrupt (**EUSCI_B_I2C_RECEIVE_INTERRUPT0**). This function returns the current interrupt source and assigns it to the local variable **status**.
18. Use the **I2C_clearInterruptFlag()** API function to clear the accepted interrupt sources to enable them to generate interrupts in the future. This step is very important and no further interrupts could be generated if this step is missed. Two parameters are used for this function, the **EUSCI_B1_MODULE** and a **mask** represented the interrupt source to be cleared. One can directly use the **status** as the interrupt source since it did contain the interrupt source obtained from the last step (step 17).
19. Use an **if()** block with the condition **status & EUSCI_B_I2C_RECEIVE_INTERRUPT0** to check whether this interrupt is a data receive interrupt. If it is, perform the following operations to get the received letters:
    a. Get the received data: **rxData[n++] = I2C_slaveGetData(EUSCI_B1_MODULE);**
    b. Use another **if()** block with the condition **n == NUM_OF_RX_BYTES** to check whether all six letters have been received. If it is, call the function **GPIO_setOutputHighOnPin()** to set pin **P1.0** to high to turn on the red color LED to indicate that all transmitted data have been received. Also reset the **intFlag** variable to **false** to terminate the **while(intFlag)** loop to collect all received letters.

8.7.5 Set up the environment to build and run the project

Refer to Section 8.6.5 in **Lab8_6** to complete this setup.

Perform the following operations to make it ready to be run:

- Go to **Flash|Download** menu item to download your program into the flash memory
- Go to **Debug|Start/Stop Debug Session** to begin debugging your program. Click on the **OK** button on the 32 KB memory size limitation message box to continue

8.7.6 Demonstrate your program

In order to test the master and slave programs, the **SDI2CMaster.c** built in Lab8_6 and the **SDI2CSlave.c** built in this Lab8_7, should be run at the same time.

However, before you can run any program, make sure to make a break point at the infinitive **while(1)** loop line in the **SDI2CSlave.c** file since we need to temporarily halt the program to check the running result.

Now run the project by first running the **SDI2CSlave** program and then running the **SDI2CMaster** program. The interval between these two running should be as small as possible. Regularly it should be less than 3 seconds.

After both programs running, the red color LED in both MSP-EXP432P401R EVBs should be ON, which means that the project running is successful.

To check the running result, open the **Call Stack + Locals** window and expand the receive array **result[]**. You can find that all six letters sent by the master, **MSP432**, have been received by the slave and displayed in this array.

Based on these results, try to answer the following questions:

- Why can we use the **txData[n++]** and **rxData[n++]** to transmit and receive data? What is the function of the index **n++**?
- Inside the **while(1)** loop in the **SDI2CMaster** program, why do we need to use the code line: **while(I2C_masterIsStopSent(EUSCI_B1_MODULE) == EUSCI_B_I2C_SENDING_STOP);**? What is the function of this code line?
- What did you learn from this project?

8.8 LAB8_8

8.8.1 Goal

This project allows students to use eUSCI_A2 for UART mode to build a transmit–receive loop back testing project to test the asynchronous serial data communication functions by using UART API functions provided by the MSPPDL. The UART interrupt mechanism is used in this project to enable received data to be picked up immediately.

8.8.2 Data assignment and hardware configuration

Connect **P3.2** and **P3.3** pins together with a jumping wire in the J1 connector at the MSP-EXP432P401R EVB since we need to do a loop back testing for this UART mode.

Refer to Section 8.6.8 to get more details about the UART control signals and related GPIO pins distributions. The GPIO pin **P3.3** works as the **UCA2TXD** and **P3.2** works as the **UCA2RXD** signal pins.

8.8.3 Development of the project

Create a new project **SDUART** with the following steps:

1. Create a new folder **Lab8_8(SDUART)** under the folder **C:\MSP432 Lab Projects\Chapter 8** in the Windows Explorer
2. Create a new ARM MDK µVersion5 project named **SDUART** and save this project in the folder **Lab8_8(SDUART)** created in step 1
3. Do the necessary steps to use **CMSIS CORE** and the **Device Startup** tools for this project

This project is to use the eUSCI_A2 for UART mode to transmit and receive 26 characters with a loop back running mode. Two GPIO pins, **P3.3** (**UCA2TXD**) and **P3.2** (**UCA2RXD**), are used to transmit and receive characters asynchronously. The LEDs connected to **P1.0** and **P2.2** are ON when this testing is complete with error or without error. The eUSCI_A2 for UART mode receiving interrupt mechanism is used to enable the received data to be picked up immediately.

8.8.4 Development of the C source file

1. Create a new C source file named **Lab8_8.c** and add the following system header files and macros as well as the struct variable into this source file:

```
#include <stdint.h>
#include <stdbool.h>
#include <MSP432P4xx\uart.h>
#include <MSP432P4xx\gpio.h>
#include <MSP432P4xx\wdt_a.h>
#include <MSP432P4xx\interrupt.h>

bool intFlag = true;
uint8_t m = 0, rxMAX = 26, RXData[26], TXData[26];
void EUSCIA2_ISR(void);

const eUSCI_UART_Config uartConfig =
{
 EUSCI_A_UART_CLOCKSOURCE_SMCLK,               // SMCLK Clock Source
 1,                                            // BRDIV = 1
 10,                                           // UCxBRF = 10
 0,                                            // UCxBRS = 0
 EUSCI_A_UART_NO_PARITY,                       // No Parity
 EUSCI_A_UART_LSB_FIRST,                       // LSB First
 EUSCI_A_UART_ONE_STOP_BIT,                    // One stop bit
 EUSCI_A_UART_MODE,                            // UART mode
 EUSCI_A_UART_OVERSAMPLING_BAUDRATE_GENERATION // Oversampling
};
```

   Some global variables are used here, **RXData[26]**, **TXData[26]**, **num**, **rxMAX**, and **intFlag**.
2. Inside the **main** function, declare two local variables and a data array: **char s_data, uint8_t num = 0** and **result[26]**.
3. Use a **for()** loop to fill the sending data array **TXData[]** with 26 uppercase letters:

```
for(s _data = 'A'; s_data <= 'Z'; s_data = s_data + 1)
{
    TXData[num] = s_data;
    num++;
}
```

4. Use **WDT_A_holdTimer()** API function to halt the Watchdog timer.
5. Use **GPIO_setAsPeripheralModuleFunctionInputPin()** API function to set up GPIO pins **P3.2** and **P3.3** to work as primary module function (**GPIO_PRIMARY_MODULE_FUNCTION**).
6. Use **GPIO_setAsOutputPin()** API function to set GPIO pins **P1.0** and **P2.2** as output pins.
7. Use **GPIO_setOutputLowOnPin()** API function to enable pins **P1.0** and **P2.2** to output low.
8. Use **UART_initModule(EUSCI_A2_MODULE, &uartConfig)** API function to initialize the eUSCI_A2 to work as the UART mode with the preset parameters in the **uartConfig** struct defined at the top of this source file.

9. Use **UART_enableModule()** API function to enable the eUSCI_A2 module.
10. Use **UART_registerInterrupt()** API function to register the **EUSCIA2_ISR()** as the interrupt handler for this UART mode.
11. Use **UART_clearInterruptFlag()** API function to clear any previous receiving interrupt flag (**EUSCI_A_UART_RECEIVE_INTERRUPT**).
12. Call the **UART_enableInterrupt()** API function to enable the UART receiving interrupt (**EUSCI_A_UART_RECEIVE_INTERRUPT**).
13. Call the **Interrupt_enableInterrupt(INT_EUSCIA2)** API function to globally enable this UART receiving interrupt.
14. Use a conditional **while()** loop to repeatedly transmit the data bytes via the **UCA2TXBUF** register. The condition is the global Boolean variable **intFlag**, which is **true** by default.
15. Inside this conditional **while()** loop, use **UART_transmitData()** API function to transmit data item **TXData[m]** to the **UCA2TXBUF** register.
16. Call the CMSIS intrinsic function **__wfi()** to wait for the UART receiving interrupt to be occurred.
17. After the conditional **while()** loop, use a **for()** loop to collect all received 26 letters:

```
for (num = 0; num < rxMAX; num++)     // collect 26 received data bytes
     result[num] = RXData[num];
```

18. Use an infinitive **while ()** loop to temporarily halt the program to enable us to check the running result of this project later. A break point should be set at this line.

The following codes are used for the eUSCI_A2 interrupt handler (**EUSCIA2_ISR**).

19. Declare this handler as **void EUSCIA2_ISR(void)**.
20. A local 32-bit integer variable **status** is declared first inside this interrupt handler.
21. Call the **UART_getInterruptStatus()** API function to get the current interrupt source. The second argument of this function is **EUSCI_A_UART_RECEIVE_INTERRUPT_FLAG**. The returned status is assigned to the local variable **status**.
22. Use the **UART_clearInterruptFlag()** API function to clear the accepted interrupt sources to enable them to generate interrupts in the future. This step is very important and no further interrupts could be generated if this step is missed. Two parameters are used for this function, the **EUSCI_A2_MODULE** and the interrupt source to be cleared. One can directly use the **status** as the interrupt source since it did contain the interrupt source.
23. Use an **if()** statement, **if(status & EUSCI_A_UART_RECEIVE_INTERRUPT)**, to check whether this interrupt is triggered by the UART receiving interrupt. If it is, use the code line:

    RXData[m] = UART_receiveData(EUSCI_A2_MODULE); to pick up the received data.

24. Then use another **if()** statement: **if(RXData[m] != TXData[m])** to compare the received byte with the transmitted byte to see whether they are equal.
25. If they are not equal, it means that this data transmit and receive transaction is wrong, use the **GPIO_setOutputHighOnPin()** API function to set GPIO pin **P1.0** to high to turn on the red color LED to indicate this error, and use an infinitive **while()** loop to hang up the program.

26. If they are equal, it means that the data transmit and receive transaction is successful, increment the global variable **m** by 1 to point to the next data.
27. Use another **if()** statement: **if (m > rxMAX)** to check whether all 26 letters have been sent out. If it is, reset the global Boolean variable **intFlag** to **false** to stop the conditional **while()** loop set in step 14 above. Then use the **GPIO_setOutputHighOnPin()** API function to set GPIO pin **P2.2** to high to turn on the blue color LED to indicate the end of the program.

8.8.5 Set up the environment to build and run the project

Refer to Section 8.6.5 in **Lab8_6** to complete this setup.

8.8.6 Demonstrate your program

Perform the following operations to run your program and check the running results:

- Set up a break point at the infinitive **while()** loop set in step 18
- Go to **Flash|Download** menu item to download your program into the flash memory
- Go to **Debug|Start/Stop Debug Session** to begin debugging your program. Click on the **OK** button on the 32 KB memory size limitation message box to continue
- Then go to **Debug|Run** menu item to run your program

Now open the **Call Stack + Locals** window and expand the receive array **result[]**. You can find that all 26 letters sent by the UART have been received and displayed in this array.

Based on these results, try to answer the following questions:

- Why can we use the local **char** variable **s_data** as the loop counter?
- Why do we need to use a CMSIS intrinsic function **__wfi()** inside the conditional **while()** loop? What will happen without this instruction?
- What did you learn from this project?

8.9 LAB8_9

8.9.1 Goal

Two programs, **Lab8_9 (UARTSendRecv)** and **Lab8_10 (UARTRecvSend)**, will be built to enable students to design and develop a real UART data communication system with two UART modes installed on two MSP-EXP432P401R EVBs. Both **P3.2 (UCA2RXD)** and **P3.3 (UCA2TXD)** pins on two EVBs are connected crossly via the MAX3232 RS-232 transceiver. Refer to hardware configurations shown in Figure 8.61 to complete this connection.

The program **Lab8_9 (UARTSendRecv)** is built in this project and the program **Lab8_10 (UARTRecvSend)**, will be built in the next project.

8.9.2 Data assignment and hardware configuration

Refer to hardware configurations shown in Figure 8.61 to complete this connection.

8.9.3 Development of the project

Create a new project **UARTSendRecv** with the following steps:

1. Create a new folder **Lab8_9 (UARTSendRecv)** under the folder **C:\MSP432 Lab Projects\Chapter 8** in the Windows Explorer
2. Create a new ARM MDK µVersion5 project named **UARTSendRecv** and save this project in the folder **Lab8_9 (UARTSendRecv)** created in step 1
3. Do the necessary steps to use **CMSIS CORE** and the **Device Startup** tools for this project

This project is to use the eUSCI_A2 for UART mode to first transmit a data string: "**How Are You?**" to the second MSP432P401R EVB, and then receive another string: "**I am Fine!**" from the second MSP432P401R EVB. Two GPIO pins, **P3.3** (**UCA2TXD**) and **P3.2** (**UCA2RXD**), are used to transmit and receive characters asynchronously. The LEDs connected to GPIO pins **P1.0**, **P2.1**, and **P2.2** are ON when this testing is complete with error or without error. The eUSCI_A2 for UART mode receiving interrupt mechanism is used to enable the received data to be picked up immediately.

8.9.4 Development of the C source file

Create a new C source file named **UARTSendRecv.c** and add the following system header files and macros as well as some global variables into this source file:

```
#include <stdint.h>
#include <stdbool.h>
#include <msp.h>

#define d_size 11

bool intFlag = true;
uint8_t num = 0, RXData[d_size];
uint8_t TXData[13] = {'H', 'o', 'w', ' ', 'A', 'r', 'e', ' ', 'Y',
   'o', 'u', '?', '#'};
```

Some global variables are used here, **RXData[11]**, **TXData[13]**, **num**, and **intFlag**. The data string to be transmitted to the second DTE is: "**How Are You?**" The data array **RXData[11]** is used to receive the returned string: "**I am Fine!**" from the second DTE.

A flow chart for the main() program of this project is shown in Figure 8.62.

The code development for the eUSCI_A2 module interrupt handler is displayed following this block diagram.

A break point should be set on the last infinitive **while()** loop in the flow chart to enable us to check the received characters later. This is very important.

The follow codes are used for the eUSCI_A2 interrupt handler (**EUSCIA2_IRQHandler**).

1. Declare this handler as **void EUSCIA2_IRQHandler(void)**
2. Use an **if()** statement, **if (UCA2IFG & UCRXIFG)**, to check whether this interrupt is triggered by the RX interrupt
3. If it is, use **UCA2IFG &=~UCRXIFG;** to clear this interrupt flag
4. Use a nested **if()** statement, **if(UCA2STATW & UCRXERR)**, to check whether any RX error triggered this interrupt
5. If it is, set **P1.0** to high to turn on the red color LED and hang up the CPU with another infinitive **while()** loop
6. If no, toggle the pin **P2.2** to flash the blue color LED to indicate that a received data triggered this interrupt

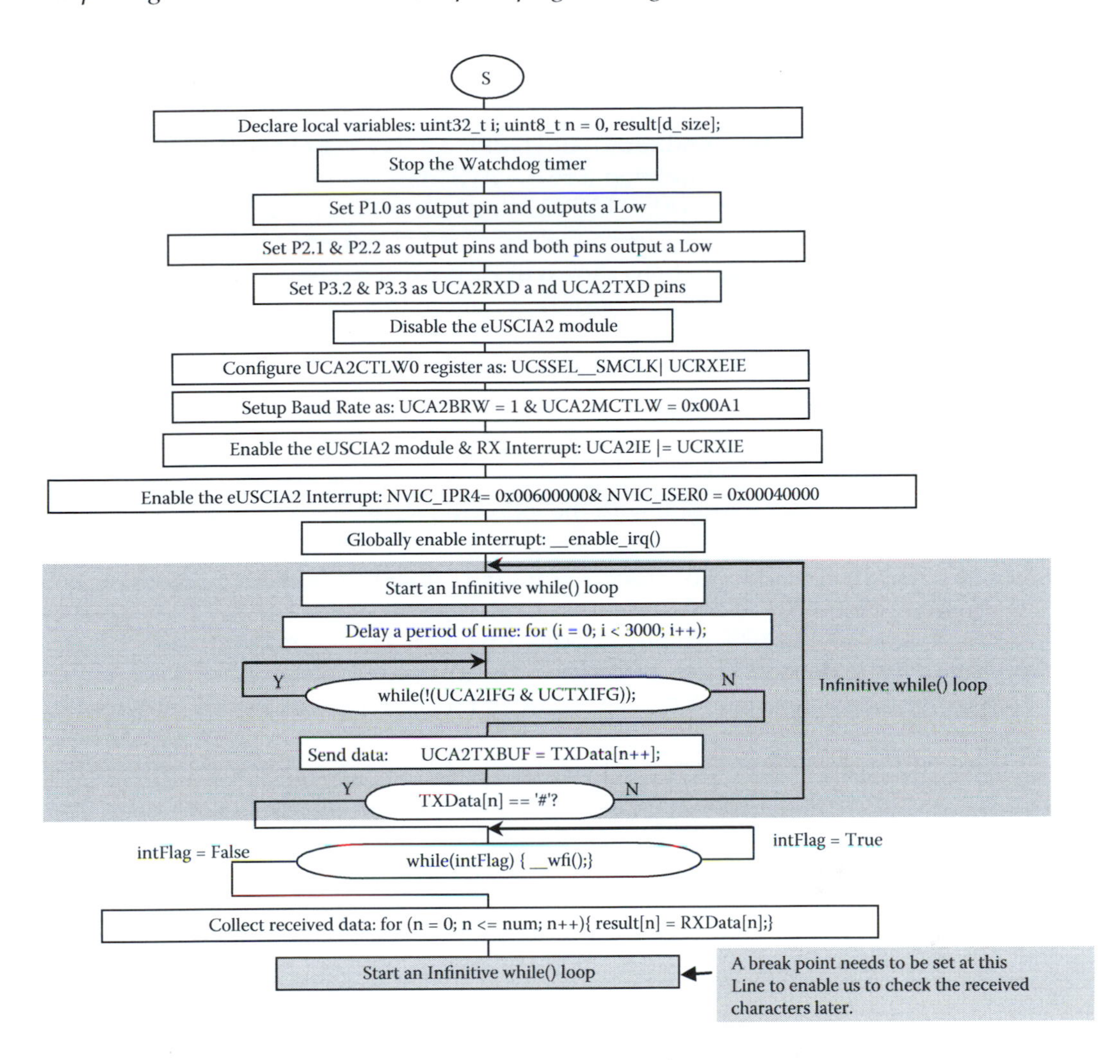

Figure 8.62 Flow chart for the main program—UARTSendRecv.

7. Use the code line: **`RXData[num] = UCA2RXBUF;`** to receive the data
8. Use another nested **`if()`** statement, **`if (RXData[num] == '!')`**, to check if the last character is received
9. If it is, reset the global Boolean variable **`intFlag`** to **`False`** to stop the **`while()`** loop in the **`main()`** program to start to collect all received characters. Set **`P2.1`** to high to turn on the green color LED to indicate the end of the program. Also disable the RX interrupt by using the code line: **`UCA2IE &=~UCRXIE;`**
10. If it is not the last character, increment the index **`num`** by 1

8.9.5 Set up the environment to build and download the project

Refer to Section 8.6.5 in **`Lab8_6`** to complete this setup. Go to **`Flash|Download`** menu item to download this program into the flash memory.

This project cannot be run until the next project, **`UARTRecvSend`**, has been built and run.

8.10 LAB8_10

8.10.1 Goal

Two programs, **Lab8_9 (UARTSendRecv)** and **Lab8_10 (UARTRecvSend)**, will be built to enable students to design and develop a real UART data communication system with two UART modes installed on two MSP-EXP432P401R EVBs. Both **P3.2 (UCA2RXD)** and **P3.3 (UCA2TXD)** pins on two EVBs are connected crossly via the MAX3232 RS-232 transceiver. Refer to hardware configurations shown in Figure 8.61 to complete this connection.

The program **Lab8_10 (UARTRecvSend)** is built in this project.

8.10.2 Data assignment and hardware configuration

Refer to hardware configurations shown in Figure 8.61 to complete this connection.

8.10.3 Development of the project

Create a new project **UARTRecvSend** with the following steps:

1. Create a new folder **Lab8_10 (UARTRecvSend)** under the folder **C:\MSP432 Lab Projects\Chapter 8** in the Windows Explorer
2. Create a new ARM MDK µVersion5 project named **UARTRecvSend** and save this project in the folder **Lab8_10 (UARTRecvSend)** created in step 1
3. Do the necessary steps to use **CMSIS CORE** and the **Device Startup** tools for this project

This project is to use the eUSCI_A2 for UART mode to first receive a data string: "**How Are You?**" from the first MSP432P401R EVB, and then transmit back another string: "**I am Fine!**" to the first MSP432P401R EVB. Two GPIO pins, **P3.3** (**UCA2TXD**) and **P3.2** (**UCA2RXD**), are used to receive and transmit characters asynchronously. The LEDs connected to GPIO pins **P1.0**, **P2.1** and **P2.2** are ON when this testing is complete with error or without error. The eUSCI_A2 for UART mode receiving interrupt mechanism is used to enable the received data to be picked up immediately.

8.10.4 Development of the C source file

Create a new C source file named **UARTRecvSend.c** and add the following system header files and macros as well as some global variables into this source file:

```
#include <stdint.h>
#include <stdbool.h>
#include <msp.h>

#define d_size 12

bool intFlag = true;
uint8_t num = 0, rxMAX = d_size, RXData[d_size];
uint8_t TXData[11] = {'I', ' ', 'a', 'm', ' ', 'F', 'i', 'n', 'e',
   '!', '#'};
```

Some global variables are used here, **RXData[12]**, **TXData[11]**, **num**, **rxMAX**, and **intFlag**. The data string to be transmitted to the first DTE is: "**I Am Fine!**." The data array **RXData[11]** is used to receive the transmitted string: "**How Are You?**" from the first DTE.

A flow chart for the **main()** program of this project is shown in Figure 8.63.

The code development for the eUSCI_A2 module interrupt handler is displayed following this block diagram.

A break point should be set on the last infinitive **while()** loop in the flow chart to enable us to check the received characters later. This is very important.

The following codes are used for the eUSCI_A2 interrupt handler (**EUSCIA2_IRQHandler**).

1. Declare this handler as **void EUSCIA2_IRQHandler(void)**
2. Use an **if()** statement, **if (UCA2IFG & UCRXIFG)**, to check whether this interrupt is triggered by the RX interrupt
3. If it is, use **UCA2IFG &= ~UCRXIFG;** to clear this interrupt flag
4. Use a nested **if()** statement, **if(UCA2STATW & UCRXERR)**, to check whether any RX error triggered this interrupt
5. If it is, set **P1.0** to high to turn on the red color LED and hang up the CPU with another infinitive **while()** loop

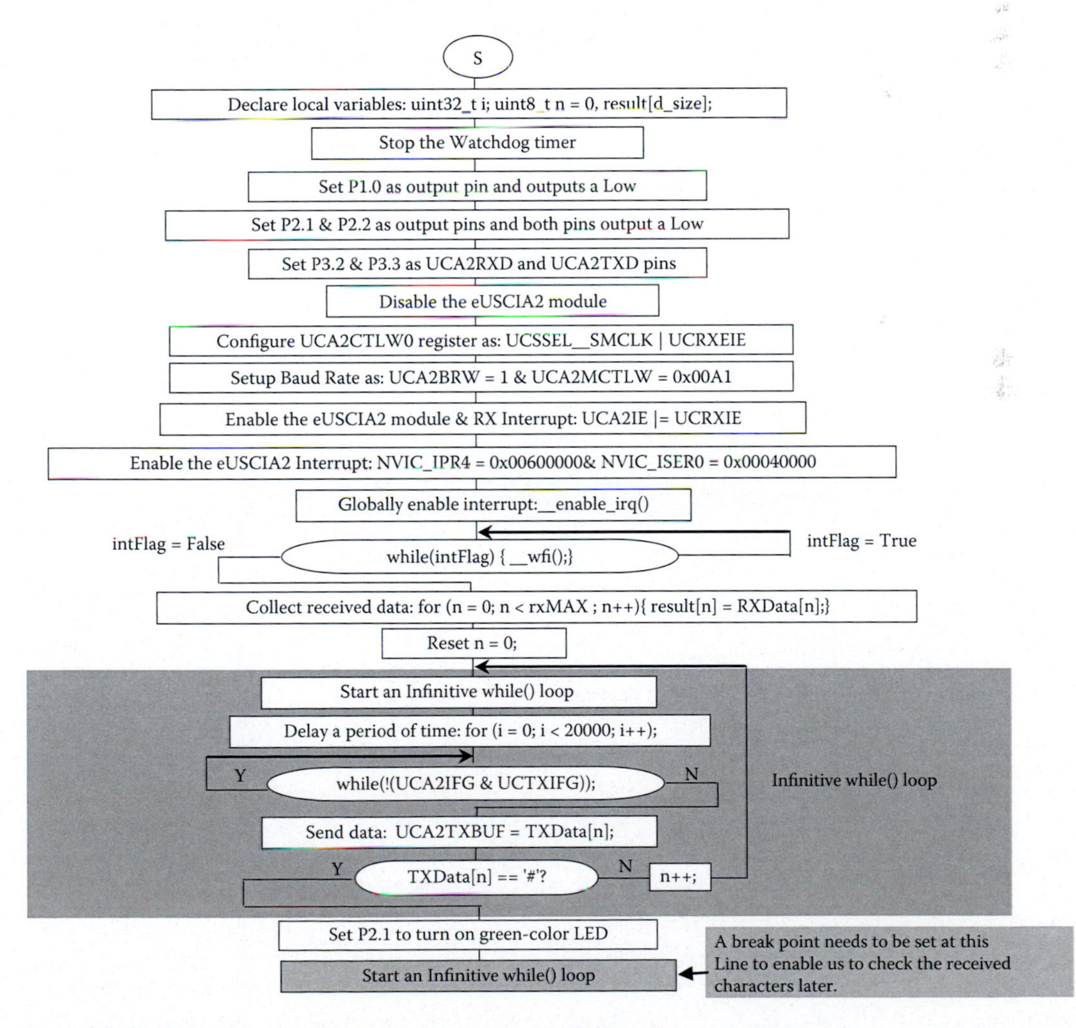

Figure 8.63 Flow chart for the main program—UARTRecvSend.

6. If no, use the code line: `RXData[num] = UCA2RXBUF;` to receive the data
7. Increment the index `num` by 1
8. Use another nested `if()` statement, `if (num > rxMAX)`, to check if the last character is received
9. If it is, reset the global Boolean variable `intFlag` to `False` to stop the `while()` loop in the `main()` program to start to collect all received characters. Set `P2.2` to high to turn on the blue color LED to indicate the end of the program. Also disable the RX interrupt by using the code line: `UCA2IE &= ~UCRXIE;`

8.10.5 Set up the environment to build and download the project

Refer to Section 8.6.5 in `Lab8_6` to complete this setup. Go to `Flash|Download` menu item to download this program into the flash memory.

This project will be run before the last project `UARTSendRecv` can be run.

(a)

Call Stack + Locals

Name	Location/Value	Type
result	0x20000270 "How Are ...	auto - unsign
[0]	0x48 'H'	unsigned cha
[1]	0x6F 'o'	unsigned cha
[2]	0x77 'w'	unsigned cha
[3]	0x20 ' '	unsigned cha
[4]	0x41 'A'	unsigned cha
[5]	0x72 'r'	unsigned cha
[6]	0x65 'e'	unsigned cha
[7]	0x20 ' '	unsigned cha
[8]	0x59 'Y'	unsigned cha
[9]	0x6F 'o'	unsigned cha
[10]	0x75 'u'	unsigned cha
[11]	0x3F '?'	unsigned cha

Call Stack + Locals | Memory 1

(b)

Call Stack + Locals

Name	Location/Value	Type
result	0x20000270 "I am Fi...	auto - u...
[0]	0x49 'I'	unsigne...
[1]	0x20 ' '	unsigne...
[2]	0x61 'a'	unsigne...
[3]	0x6D 'm'	unsigne...
[4]	0x20 ' '	unsigne...
[5]	0x46 'F'	unsigne...
[6]	0x69 'i'	unsigne...
[7]	0x6E 'n'	unsigne...
[8]	0x65 'e'	unsigne...
[9]	0x21 '!'	unsigne...
[10]	0x00	unsigne...

Call Stack + Locals | Memory 1

Figure 8.64 Running results for both projects.

8.10.6 Run and demonstrate your programs

Perform the following operations to run your program and check the running results:

- Set up a break point at the infinitive **`while()`** loop shown at the last step in Figure 8.63
- First run this **`UARTRecvSend`** project by going to **`Debug|Start/Stop Debug Session`** to begin debugging your program. Click on the **`OK`** button on the 32KB memory size limitation message box to continue. Then go to **`Debug|Run`** menu item to run this program
- Now run the program **`UARTSendRecv`** by going to **`Debug|Start/Stop Debug Session`** to begin debugging your program. Click on the **`OK`** button on the 32 KB memory size limitation message box to continue. Then go to **`Debug|Run`** menu item to run this program

First the blue color LEDs on both MSP-EXP432P401R EVBs are on and then both green color LEDs are also on in both EVBs.

Now open two **`Call Stack + Locals`** windows on both μVersion5 window, and expand two receive arrays **`result[]`**. You can find that the first string: "**`How Are You?`**" sent by the project **`UARTSendRecv`** has been received by the second project **`UARTRecvSend`** and the string: "**`I am Fine!`**," sent by the second project **`UARTRecvSend`** is received by the first project **`UARTSendRecv`** as shown in Figures 8.64a and 8.64b.

Based on these results, try to answer the following questions:

- What is the purpose of the code line **`while(intFlag) {__wfi();}`** in the **`UART SendRecv`**?
- What is the purpose of the code line **`while(intFlag) {__wfi();}`** in the **`UART RecvSend`**?
- Why do we need to delay code line: **`for (i = 0; i < 20000; i++){}`** in the project **`UARTRecvSend`**? Does this code line have any relationship to the code line: **`P2OUT ^= BIT2`** in the UART ISR at the **`UARTSebdRecv`** project?
- What did you learn from this project?

chapter nine

MSP432™ 16-bit timers and PWM modules

This chapter provides general information about an MSP432P401R microcontroller general-purpose 16-bit timer programming, including the different timer modules programming for TA0~TA3 and the PWM modules. The discussion also includes the GPIO ports related to general-purpose 16-bit timers and four modules used in the MSP432P401R MCU system. This discussion concentrates on the architectures and programming interfaces applied on GPTMs and PWM modules specially designed for the MSP432P401R MCU system. Special or alternative control functions for different GPIO pins related to different peripherals, including timers, are also discussed and introduced with example projects in this chapter.

9.1 Overview

Timer_A is a 16-bit timer/counter with up to five capture/compare registers.

Timer_A supports four timer modules, **TA0**, **TA1**, **TA2**, and **TA3**, and each of them is a 16-bit timer/counter with the Timer_A type and has up to five capture/compare registers. Each timer supports multiple capture/compares, PWM outputs, and interval timing. Each has extensive interrupt capabilities. Interrupts may be generated from the counter on overflow conditions and from each of the capture/compare registers.

A standard Timer_A provides the following features:

- Asynchronous 16-bit timer/counter with four operating modes
- Selectable and configurable clock source
- Up to five configurable capture/compare registers
- Configurable outputs with PWM capability
- Asynchronous input and output latching

When using these timer modules, the following definitions are applied:

- **CxOUT**: output from the Comparator **x** (**C0OUT~C3OUT**: the output of the comparators on the **TA0** through **TA3** modules)
- **TAx_Cy**: I/O for Timer **x** with Capture/Compare module **y**. Input is for capture but output is for compare. For example, **TA0_C0** and **TA3_C2**, the input or the output of the Timer0 with Capture/Comparator0 and the input or the output of the Timer3 with Capture/Comparator2. Note that the following equal relations exist:
    - TAx_Cy = TAx.y. For example, TA0_C0 = TA0.0, TA3_C2 = TA3.2
    - TAx_Cy = CCRy. For example, TAx_C0 = CCR0, TAx_C3 = CCR3

In this chapter, we will concentrate on the following peripherals used in the MSP432P401R MCU system:

- Timer_A timers
- PWM module

Some example projects related to those peripherals are to be developed and built in this chapter. First let's take a look at the Timer_A architecture and organization.

9.2 Timer_A architecture and functional block diagram

Figure 9.1 shows the architecture and functional block diagram of the Timer_A.

Since all four Timer_A modules have the identical architecture and organization, here only an example module, **TA4**, is shown with its capture/comparator block.

Each module has its own 16-bit timer-counter **TAxR** (**TA0R~TA3R**) and related capture/comparator block. The dash lines in Figure 9.1 indicated with **CCR0~CCR3** mean the

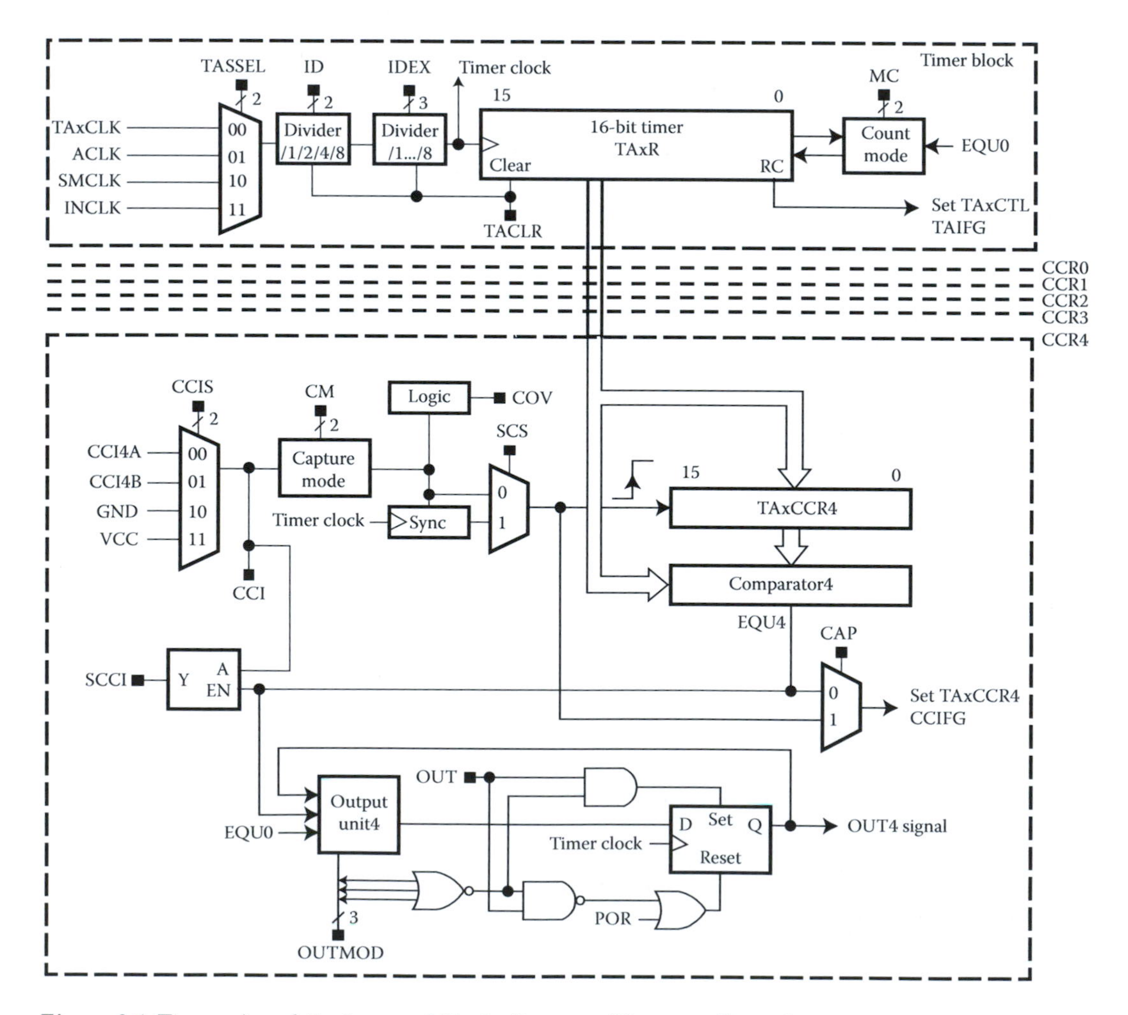

Figure 9.1 Timer_A architecture and block diagram. (Courtesy Texas Instruments.)

associated capture/comparator blocks. Based on this architecture, three control blocks can be derived for each module:

1. Global timer configuration control block
2. Each module capture/comparator control block
3. Timer_A module output control and interrupt control blocks

Each control block comprises related control registers with associated control signals. The global timer configuration control block contains the Timer_Ax Control (**TAxCTL**) register and 16-bit timer/counter (**TAxR**) register. Five module capture/comparator blocks contain the Timer_Ax Capture/Compare Control 0 (**TAxCCTL0**) to Timer_Ax Capture/Compare Control 4 (**TAxCCTL4**) registers. The related Timer_Ax Capture/Compare 0 (**TAxCCR0**) to Timer_Ax Capture/Compare 4 (**TAxCCR4**) registers are also included in this block.

Some control-related signals used for each module output and interrupt are located at some control registers discussed in the previous two blocks. For the timer interrupt enable control and flag signals, they are in the **TAxCTL** register. For all module capture/comparator interrupt enable flag signals and output controls, they can be found from related **TAxCCTL0~TAxCCTL4** registers.

9.3 Timer_A modules operations

Each Timer_A module supports four different operational modes, and each mode is determined by configuring the bit field **MC** (bits 5~4) in the **TAxCTL** register. These four operation modes are

1. Stop (**MC = 00**): the timer is stopped
2. Up count (**MC = 01**): the timer repeatedly counts from zero to the value of **TAxCCR0**
3. Continuous (**MC = 10**): the timer repeatedly counts from zero to **0xFFFF**
4. Up/down count (**MC = 11**): the timer repeatedly counts from zero up to the value of **TAxCCR0** and back down to zero

9.3.1 Timer_A module control

As shown in Figure 9.1, the global timer configuration control block provides the following control and configuration functions:

- Select the clock source as the input clock or timing base for the timer/counter by setting the bit field **TASSEL** (bits 9~8) in the **TAxCTL** register.
- The selected clock can be divided by 1, 2, 4, and 8 via an internal timer divider to allow the timer/counter to use a lower clock frequency. The divider factor such as 1, 2, 4, and 8 can be configured by setting the bit field **ID** (bits 7~6) in the **TAxCTL** register.
- The reduced input clock can further be divided by another factor, from 1 to 8, to get lower input clock to the timer/counter. By configuring the bit field **TAIDEX** (bits 2~0) in the Timer_Ax Expansion 0 (**TAxEX0**) Register, such as 000 → divided by 1, 001 → divided by 2, 010 → divided by 3, … 111 → divided by 8, the desired dividing factor can be obtained and used.
- One of the four timer/counter operation modes can be selected by configuring the bit field **MC** as above to enable the timer/counter to work as a desired mode. The timer/

counter can be cleared at any time by setting the bit **TACLR** (bit-2) in the **TAxCTL** register. This clearing also resets the timer divider logic and the count direction. After finishing this clear, the **TACLR** bit is automatically reset and is always read as zero.

Figure 9.2 shows the bit configurations of the **TAxCTL** register.

Table 9.1 lists all bit values and functions for the **TAxCTL** register.

Two bits, **TAIE** and **TAIFG**, in this register provide the functions of the Timer_A module interrupt control block. If both of these bits are set to 1, a timer/counter-related interrupt can be generated and the **TAIFG** bit will be set to indicate this interrupt.

The **TAxEX0** Register is a 16-bit register, but only the lower 3 bits, bits 2~0, are used to provide the additional divider factor **TAIDEX**. The upper 13 bits, bits 15~3, are reserved for this register. Table 9.2 shows all possible bit values and related divider factor values provided by this register.

One point to be noted when using this **TAxEX0** Register to set up an additional divider factor is after programming **TAIDEX** bits and configuration of the timer, one needs to set **TACLR** bit to ensure proper reset of the timer divider logic before it can operate properly.

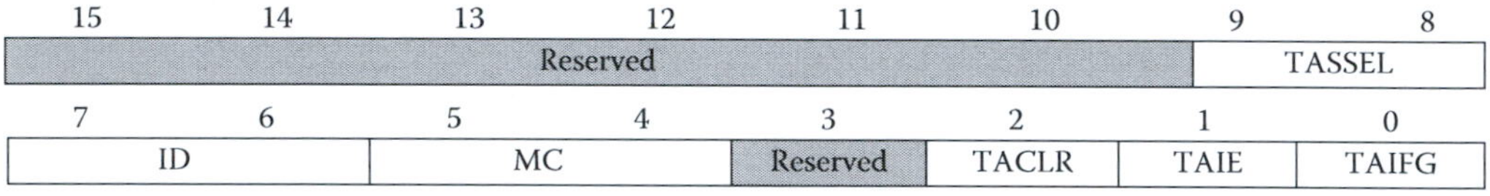

15	14	13	12	11	10	9	8
Reserved						TASSEL	
7	6	5	4	3	2	1	0
ID		MC		Reserved	TACLR	TAIE	TAIFG

Figure 9.2 Bit configurations of the TAxCTL register.

Table 9.1 Bit value and function for TAxCTL register

Bit	Field	Type	Reset	Function
15:10	Reserved	RO	0	Reserved
9:8	TASSEL	RW	0	Timer_A clock source select: **00:** TAxCLK; **01:** ACLK; **10:** SMCLK; **11:** INCLK
7:6	ID	RW	0	Input divider. These bits along with the **TAIDEX** bits select the divider for the input clock **00:** ÷1; **01:** ÷2; **10:** ÷4; **11:** ÷8
5:4	MC	RW	0	Mode control. Set **MC** = 00 when Timer_A is not in use conserves power **00:** Stop mode: timer is halted **01:** Up-count mode: timer counts up to **TAxCCR0** **10:** Continuous mode: timer counts up to **0FFFF** **11:** Up/down-count mode: timer counts up to **TAxCCR0** then down to **0000**
3	Reserved	RO	0	Reserved
2	TACLR	RW	0	Timer_A clear. Setting this bit resets **TAxR**, the timer clock divider logic, and the count direction. The **TACLR** bit is automatically reset and is always read as zero
1	TAIE	RW	0	Timer_A interrupt enable. This bit enables the **TAIFG** interrupt request **0:** Interrupt disabled; **1:** Interrupt enabled
0	TAIFG	RW	0	Timer_A interrupt flag **0:** No interrupt pending; **1:** Interrupt pending

Table 9.2 Bit value and function for TAxEX0 register

Bit	Field	Type	Reset	Function
15:3	**Reserved**	RO	0	Reserved
2:0	**TAIDEX**	RW	0	Input divider expansion. These bits along with the **ID** bits select the divider for the input clock **000:** ÷ 1; **001:** ÷ 2; **010:** ÷ 3; **011:** ÷ 4; **100:** ÷ 5; **101:** ÷ 6; **110:** ÷ 7; **111:** ÷ 8

9.3.2 *Timer_A module operation modes*

The actual timer/count function is provided by the 16-bit timer/counter register **TAxR** (**x** = 0~3). Based on the mode selection value on bits **MC** in the **TAxCTL** register, this register can work as up-count, continuous, or up–down-count mode. In fact, this timer/counter performs the counting function to count the number of periods of the input clock. Let's have a closer look at these operation modes.

9.3.2.1 *Continuous mode*

This is the simplest operation mode for this 16-bit counter. When working in this mode, the 16-bit counter just repeatedly counts from zero to **0xFFFF** and restarts from zero as shown in Figure 9.3. The capture/compare register **TAxCCR0** is not used for this mode.

When the counter counts to the final value, **0xFFFF**, exactly at the falling edge of the **0xFFFF** period as shown in Figure 9.3, an overflow interrupt occurr and it is indicated by setting the **TAIFG** flag bit in the **TAxCTL** register if the **TAIE** bit is set to enable this kind of interrupt source. This operation mode will continue until a timer/counter clear signal **TACLR** is applied.

A fixed time interval or period can be generated by using this mode. Depending on the period of the input clock and the selected divider factor, a different time interval or time period can be obtained. This kind of application can be used as a timer or fixed period signal source.

9.3.2.2 *Up-count mode*

If an application needs to use a flexible time interval or a variable time period generator, the up-count mode can be used since the upper bound of the time can be set by configuring the value in the **TAxCCR0** Register.

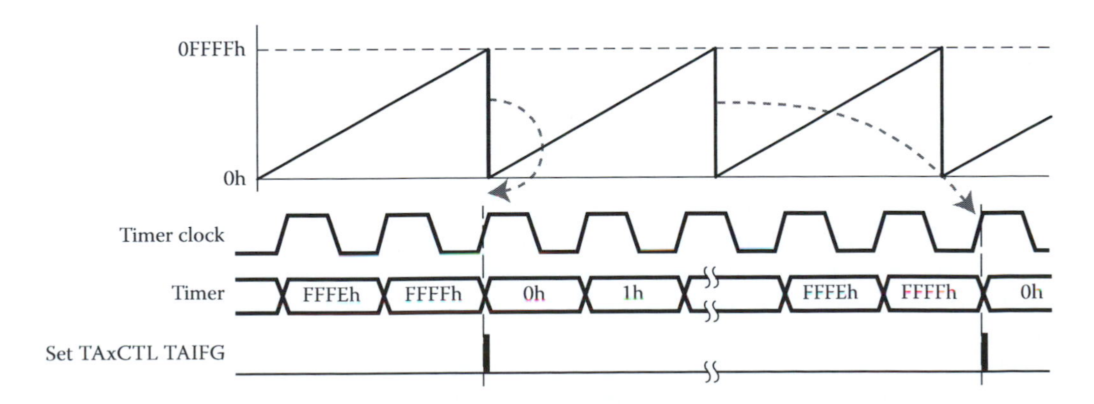

Figure 9.3 Continuous mode operation sequence and overflow interrupt. (Courtesy Texas Instruments.)

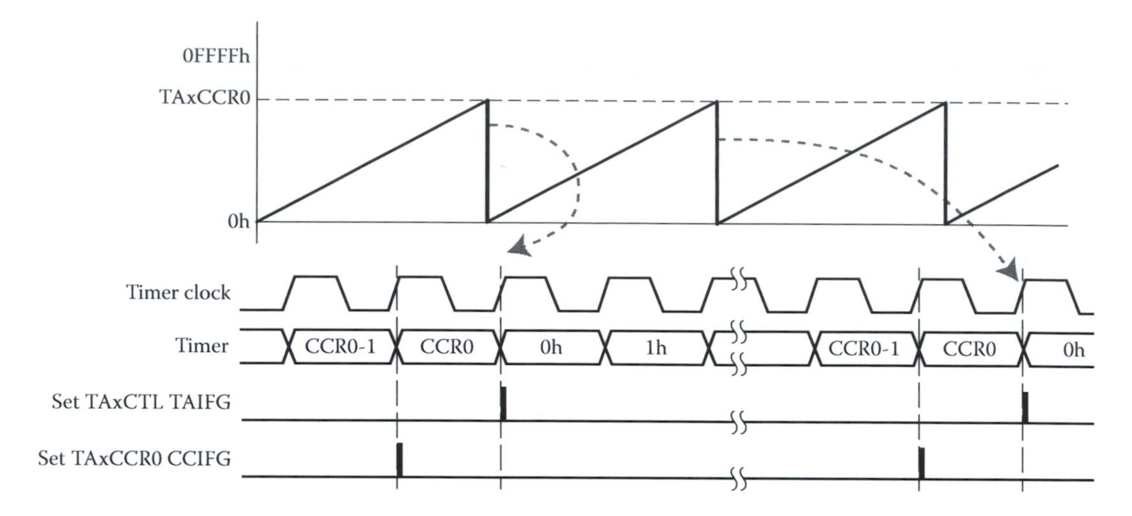

Figure 9.4 Up mode operation sequence and its interrupt. (Courtesy Texas Instruments.)

Each of five Timer_A**x** Capture/Compare**y** registers, **TAxCCRy** (**x** = 0~3 and **y** = 0~4) or TA**x.y**, can work either in the capture or in the compare mode. When working in the compare mode, an upper limit value can be feed into the **TAxCCR0** Register as the upper bound value used to compare the current counting value in the 16-bit timer/counter **TAxR** register.

Before the timer/counter can be operated in this mode, an upper limit value is loaded into the **TAxCCR0** Register. Then the timer repeatedly counts from zero up to the value of compare register **TAxCCR0**, which defines the period (see Figure 9.4). The number of timer counts in the period is **TAxCCR0 + 1**. When the timer value equals to **TAxCCR0**, the timer restarts counting from zero. If up mode is selected when the timer value is greater than **TAxCCR0**, the timer immediately restarts counting from zero.

When the counter counts to the upper limit value stored in the **TAxCCR0** Register, the **CCIFG** interrupt flag bit in the **TAxCCTL0** Register is set, and a compare equal interrupt is occurred and it is indicated by setting the **TAIFG** flag bit in the **TAxCTL** register at the falling edge of the value in the **TAxCCR0** Register, as shown in Figure 9.4. To enable this interrupt to be occurred, the **TAIE** bit in the **TAxCTL** register must be set. This operation mode will continue until a timer/counter clear signal **TACLR** is applied.

When changing the value stored in the **TAxCCR0** Register while the timer is running, if the new period is greater than or equal to the old period or greater than the current count value in the **TAxR** register, the timer counts up to the new period. If the new period is less than the current count value in the **TAxR** register, the timer rolls to zero to restart its up count. However, one additional count may occur before the counter rolls to zero.

9.3.2.3 Up/down-count mode

When working in this mode, the timer repeatedly counts up to the value stored in the **TAxCCR0** Register and back down to zero as shown in Figure 9.5. The period is twice the value in the **TAxCCR0** Register.

The count direction is latched and this allows the timer to stop and then restart in the same direction it was counting before it stopped. If this is not desired, the **TACLR** bit must be set to clear the direction.

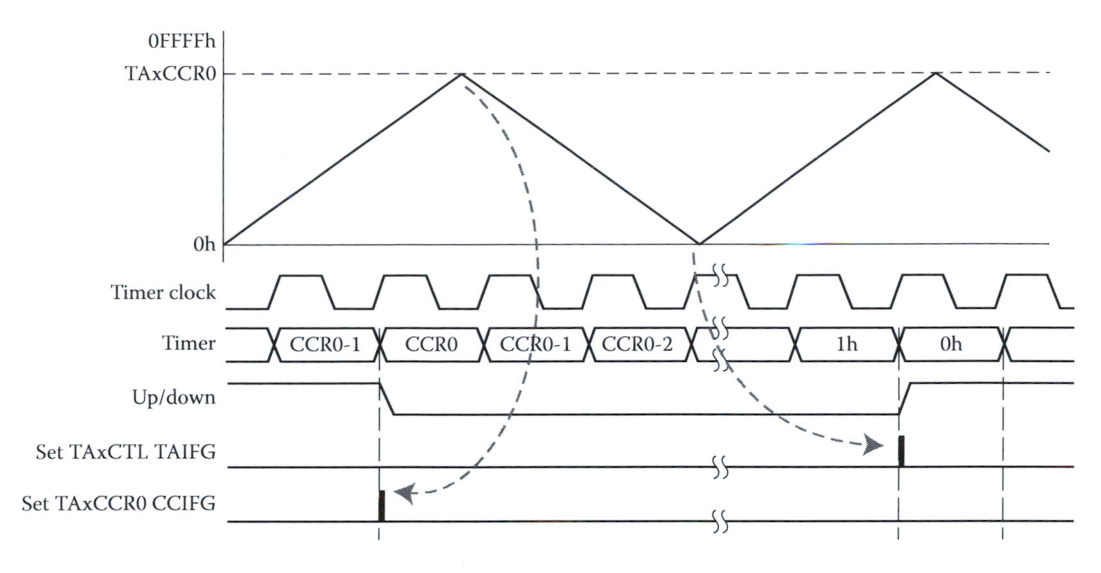

Figure 9.5 Up/down mode sequence and its interrupt. (Courtesy Texas Instruments.)

In up/down mode, the **CCIFG** interrupt flag bit in the **TAxCCTL0** Register that is used to control the **TAxCCR0** Register and the **TAIFG** interrupt flag bit are set only once during a period, separated by one-half the timer period. The **CCIFG** interrupt flag bit is set when the timer counts from **TAxCCR0-1** to **TAxCCR0**, and the **TAIFG** bit is set when the timer completes counting down from **0001** to **0000**. Figure 9.5 shows this flag set cycle.

When changing the value in the **TAxCCR0** Register but the timer is running, the following operation sequences are observed:

- When the timer is counting in the down direction, the timer continues its descent until it reaches zero. The new period takes effect after the counter counts down to zero.
- When the timer is counting in the up direction, and the new period is greater than or equal to the old period or greater than the current count value, the timer counts up to the new period before counting down.
- When the timer is counting in the up direction and the new period is less than the current count value, the timer begins counting down. However, one additional count may occur before the counter begins counting down.

Since both the up-count and up/down-count modes are under the control of the capture/compare control block, let's have a closer look at this control block.

9.3.2.4 Capture/compare control block

Each Timer_A module has five capture/compare registers used to perform either capture or compare function. The main components used for these functions are five related Timer_A**x** capture/compare **y** registers, **TAxCCR0~TAxCCR4**. All of these capture/compare registers are under control of five corresponding Timer_A**x** capture/compare control **y** registers, **TAxCCTL0~TAxCCTL4**. All of these registers are 16-bit registers. Generally, both capture and compare function need to use one or more **TAxCCRy** (**y** = 0~4) registers to perform its functions. Depending on the different functions, the **TAxCCRy** registers provide different functions:

- When a capture function is performed, the current count value in the 16-bit timer/counter register **TAxR** is loaded into the **TAxCCRy** register each time when an input capture signal is detected or captured.
- When working as a comparator, the target count value is preloaded into the **TAxCCRy** register. The current count value in the 16-bir timer/counter **TAxR** is compared with the value preloaded into the **TAxCCRy** register. If both are equal, an interrupt may be generated if that interrupt is enabled.

As we mentioned, all of these **TAxCCRy** registers are 16-bit registers and they can be used to hold the target or preloaded value or count used for capture or compare mode.

Now, let's have a closer look at these two modes.

9.3.2.4.1 Capture mode The capture mode operation is under control of the corresponding **TAxCCTLy** register. When the bit **CAP**, bit-8, in the **TAxCCTLy** register is set, the capture mode is selected. As shown in Figure 9.1, four sources can be used as the input capture signals, **CCIyA**, **CCIyB**, GND, and V_{CC}. Any of these input signals can be selected as the capture input by configuring the bit field **CCIS**, bits 13~12, in the **TAxCCTLy** register. Both **CCIyA** and **CCIyB** are two external capture signals that can be selected as any detected signals.

Figure 9.6 shows the bit configuration for the related **TAxCCTLy** registers.

The bit field **CM**, bits 15~14, is used to select the actual capture detecting mode, either the rising edge, falling edge, or both edges of the capture input signal.

A capture can occur on the selected edge of the input signal. If a capture occurs:

- The current timer value is copied into the **TAxCCRy** register
- The interrupt flag **CCIFG** is set
- A capture interrupt can be generated if the bit **CCIE** (bit-4) in the **TAxCCTLy** register is set to enable this capture interrupt

The input signal level can be read at any time via the **CCI** bit, bit-3, in the **TAxCCTLy** register. Devices may have different signals connected to **CCIyA** and **CCIyB**.

The bit **SCS**, bit-11, in the **TAxCCTLy** register can be used to work as a synchronization capture source to synchronize the capture input signal with the timer clock.

A capture overflow may occur if the second capture was performed before the value from the first capture had been read. The bit **COV**, bit-1, in the **TAxCCTLy** register is set when this occurs and this **COV** flag bit must be reset by software.

Table 9.3 lists the bit values and related functions on the **TAxCCTLy** register.

Since some control signals of the output control block are involved in this **TAxCCTLy** register, let's take care of these signals in this section.

15	14	13	12	11	10	9	8
CM		CCIS		SCS	SCCI	Reserved	CAP

7	6	5	4	3	2	1	0
OUTMOD			CCIE	CCI	OUT	COV	CCIFG

Figure 9.6 Bit configuration of the TAxCCTLy register.

Table 9.3 Bit Value and function for TAxCCTLy register

Bit	Field	Type	Reset	Function
15:14	**CM**	RW	0	Capture mode: **00:** No capture; **01:** Capture on rising edge; **10:** Capture on falling edge; **11:** Capture on both rising and falling edges
13:12	**CCIS**	RW	0	Capture/compare input select. These bits select the **TAxCCRy** input signal **00:** CCIxA; **01:** CCIxB; **10:** GND; **11:** V_{CC}
11	**SCS**	RW	0	Synchronize capture source. This bit is used to synchronize the capture input signal with the timer clock **0:** Asynchronous capture; **1:** Synchronous capture
10	**SCCI**	RW	0	Synchronized capture/compare input. The selected **CCI** input signal is latched with the **EQUx** signal and can be read via this bit
9	**Reserved**	RO	0	Reserved
8	**CAP**	RW	0	Capture/compare mode selection: **0:** Compare mode; **1:** Capture mode
7:5	**OUTMOD**	RW	0	Output mode. Modes 2, 3, 6, and 7 are not useful for **TAxCCR0** because **EQUx** = **EQU0** **000:** OUT bit value; **001:** Set; **010:** Toggle/reset; **011:** Set/reset **100:** Toggle; **101:** Reset; **110:** Toggle/set; **111:** Reset/set
4	**CCIE**	RW	0	Capture/compare interrupt enable **0:** Interrupt disabled; **1:** Interrupt enabled
3	**CCI**	RO	0	Capture/compare input. The selected input signal can be read by this bit
2	**OUT**	RW	0	Output. For output mode 0, this bit directly controls the state of the output **0:** Output low; **1:** Output high
1	**COV**	RW	0	Capture overflow. This bit **COV** must be reset with software **0:** No capture overflow occurred; **1:** Capture overflow occurred
0	**CCIFG**	RW	0	Capture/compare interrupt flag **0:** No interrupt pending; **1:** Interrupt pending

Each capture/compare block contains an output unit. The output unit is used to generate output signals, such as PWM signals. Each output unit has eight operating modes that generate signals based on the **EQU0** and **EQUn** signals.

The output modes are defined by the **OUTMOD** bits shown in Table 9.3. The **OUTn** signal is changed with the rising edge of the timer clock for all modes except mode 0. Output modes 2, 3, 6, and 7 are not useful for output unit 0 because **EQUn** = **EQU0**. A more detailed description about all eight output modes is shown in Table 9.4.

It can be found from Table 9.4 that only when the bit field **OUTMOD** = 000 (mode0), the output value is defined by the bit **OUT** in the **TAxCCTLy** register. A high output is obtained at **OUTn** pin if the **OUT** = 1, and a low output can be achieved at **OUTn** pin when the **OUT** = 0.

Table 9.4 Bit value and function for TAxCCTLy register

OUTMOD	Mode	Description
000	**Output**	The output signal **OUTn** is defined by the **OUT** bit. The **OUTn** signal updates immediately when **OUT** is updated
001	**Set**	The output is set when the timer counts to the **TAxCCRn** value. It remains set until a reset of the timer, or until another output mode is selected and affects the output
010	**Toggle/Reset**	The output is toggled when the timer counts to the **TAxCCRn** value. It is reset when the timer counts to the **TAxCCR0** value
011	**Set/Reset**	The output is set when the timer counts to the **TAxCCRn** value. It is reset when the timer counts to the **TAxCCR0** value
100	**Toggle**	The output is toggled when the timer counts to the **TAxCCRn** value. The output period is double the timer period
101	**Reset**	The output is reset when the timer counts to the **TAxCCRn** value. It remains reset until another output mode is selected and affects the output
110	**Toggle/Set**	The output is toggled when the timer counts to the **TAxCCRn** value. It is set when the timer counts to the **TAxCCR0** value
111	**Reset/Set**	The output is reset when the timer counts to the **TAxCCRn** value. It is set when the timer counts to the **TAxCCR0** value

Also when the **OUTMOD** = 001 (mode1), 100 (mode4), and 101 (mode5), the output is only changed as the timer counts to the **TAxCCRy** value, and has nothing to do with the **TAxCCR0**.

However, when the **OUTMOD** = 010 (mode2), 011 (mode3), 110 (mode6), and 111 (mode7), the output is changed with two conditions: the first change occurred when the timer counts to the **TAxCCRy** value, and the second change happened as the timer counts to the **TAxCCR0** value.

With the help of modes 2, 3, 6, and 7, various periodic waveforms with different duty cycles can be generated, and these can be used as the driving signals to generate PWM outputs.

9.3.2.4.2 Compare mode As mentioned, the function of a comparator is to compare the current count value in the 16-bit timer/counter **TAxR** register with the preloaded value in the **TAxCCRy** register. If both are equal, an interrupt may be generated if that interrupt is enabled.

The compare mode is selected by resetting the bit **CAP** to 0. The compare mode can be used to generate PWM output signals (modes 2, 3, 6, and 7) or interrupts at specific time intervals. When the current counting value in the **TAxR** register is equal to the value in a **TAxCCRy** register, the following actions happened:

- Interrupt flag **CCIFG** is set
- Internal signal **EQUn** = 1
- **EQUn** affects the output according to the output mode
- The input signal **CCI** is latched into **SCCI**

9.3.3 *Output examples of using three operational modes*

Depending on the **OUTMOD** bit field values, different output waveforms can be obtained and implemented. In this section, we want to have a closer look at these implementations.

9.3.3.1 Output implementations for Timer_A module in the up-count mode

The **OUTn** signal is changed when the timer counts up to the **TAxCCRy** value and changed again when counts up to the **TAxCCR0** Register, depending on the output mode. Figure 9.7 shows an example of using the **TAxCCR0** and **TAxCCR1** registers.

Various periodic waveforms with different duty cycles can be achieved by adjusting the values set in **TAxCCR0** and **TAxCCR1** registers.

9.3.3.2 Output implementations for Timer_A module in the continuous mode

The **OUTn** signal is changed when the timer reaches both the **TAxCCRy** and the **TAxCCR0** values, depending on the output mode. Figure 9.8 shows an example of using the **TAxCCR0** and **TAxCCR1** registers when the timer is working in the compare mode with the continuous mode.

The only difference between the up-count mode and the continuous mode is that the terminal count value is different. The terminal count value for the up-count mode is a flexible one and it can be determined by the users. However, the terminal count value for the continuous mode is a fixed one, **0xFFFF**.

9.3.3.3 Output implementations for Timer_A module in the up–down-count mode

Figure 9.9 shows an example of using the timer to work in the up–down-count mode with **TAxCCR0** and **TAxCCR2** registers.

The **OUTn** signal changes when the timer equals to **TAxCCRy** in either count direction or when the timer equals the **TAxCCR0**, depending on the output mode.

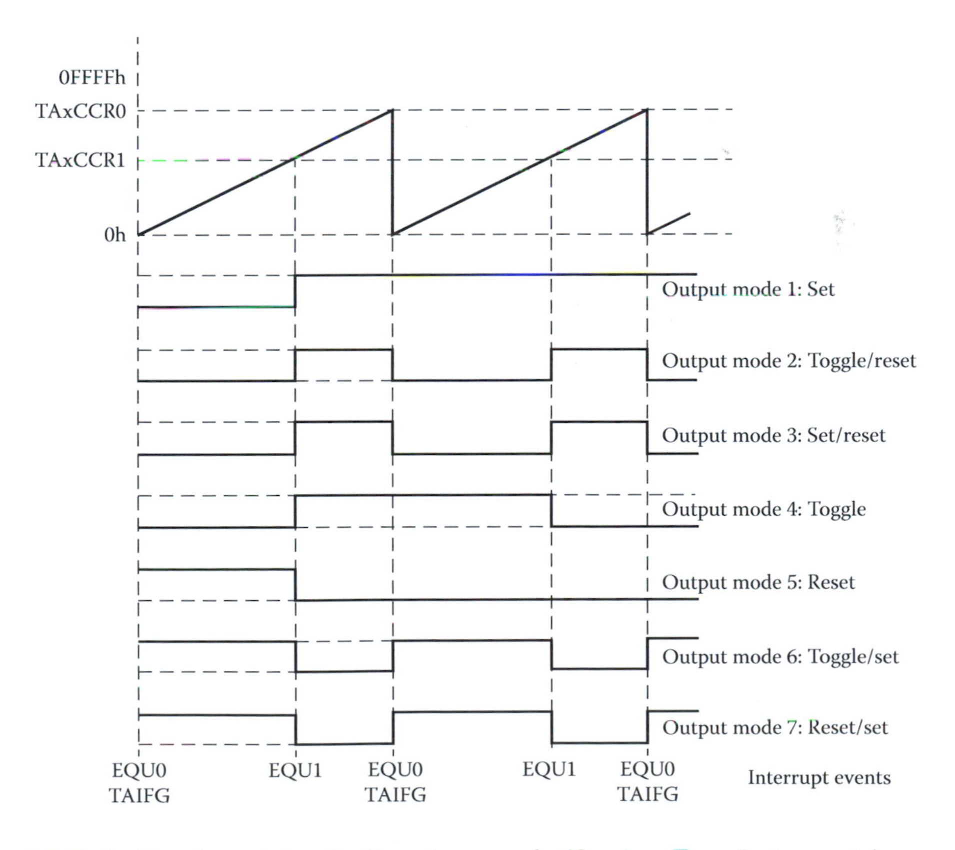

Figure 9.7 Output implementation for timer in up mode. (Courtesy Texas Instruments.)

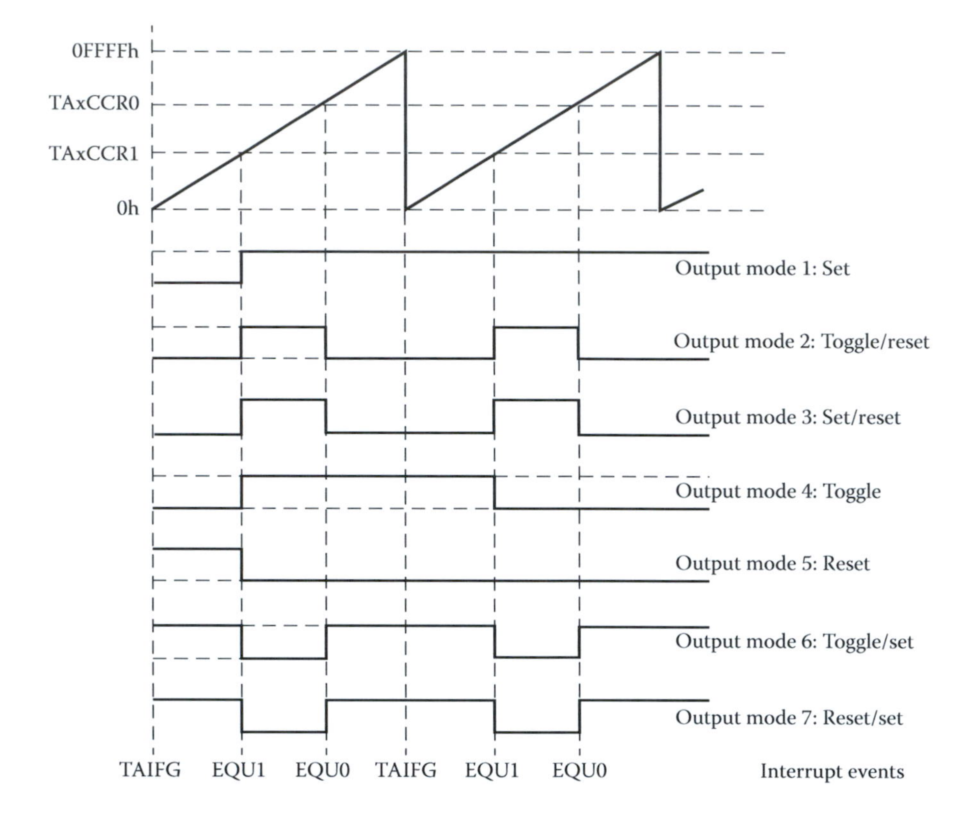

Figure 9.8 Output implementations of timer in continuous mode. (Courtesy Texas Instruments.)

A little trick in Figure 9.9 is the triggering levels for the output modes 2, 3, 6, and 7.

In mode 2, it works in the toggle/reset mode. The original output level is assumed as high. When the first **TAxCCR2** is met, the output is toggled from high to low. As the first **TAxCCR0** is hit, the output should be reset to low. However, the output is already in low, so it keeps its low. When the second **TAxCCR2** is hit, it is toggled again from low to high.

For the output mode 3, assumed that the original output level is high. In this mode, it works in the set/reset mode. When the first **TAxCCR2** is met, the output should be set to high. But it is already in high, so it keeps this high without any change. As the first **TAxCCR0** is hit, the output is reset to low. When the second **TAxCCR2** is hit, the output is set to high again.

The output modes 6 and 7 follow the similar operational principle as modes 2 and 3.

When switching between output modes, one of the **OUTMOD** bits should remain set during the transition, unless switching to mode 0. Otherwise, output glitch can occur because a NOR gate decodes output mode 0. A safe method for switching between output modes is to use output mode 7 as a transition state.

9.3.3.4 Timer_A interrupt processing

Most Timer_A-related interrupt control and flag signals are involved in the **TAxCTL** and the **TAxCCTLy** registers. Two interrupt vectors associated with the 16-bit Timer_A module are:

- **TAxCCR0** interrupt vector for **TAxCCR0 CCIFG** flag bit
- **TAxIV** interrupt vector for all other **CCIFG** flags and **TAIFG** bit

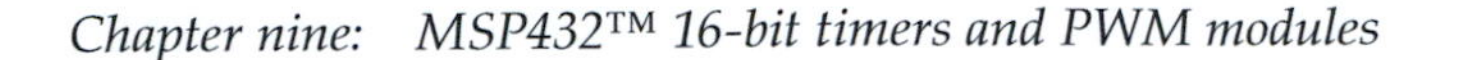

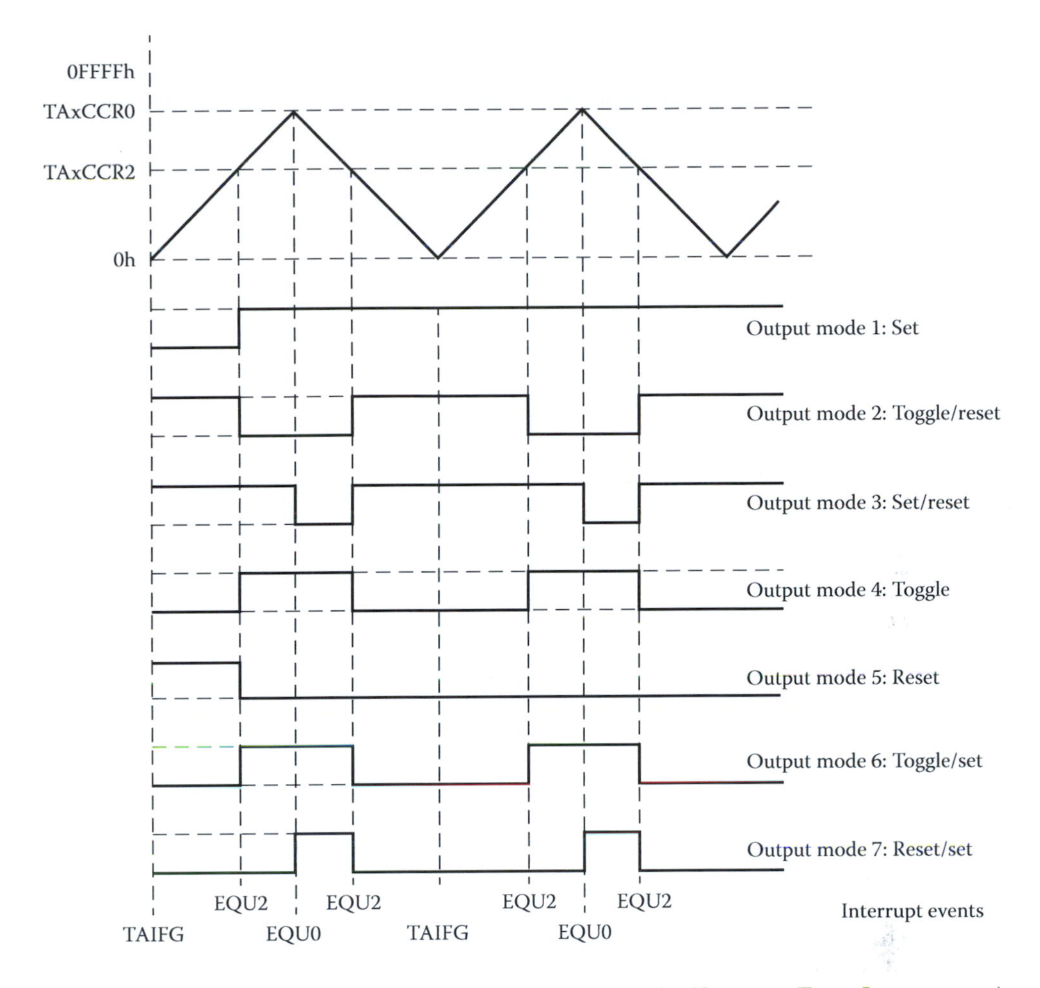

Figure 9.9 Output implementations of timer in up–down mode. (Courtesy Texas Instruments.)

When working in the capture mode, any **CCIFG** flag is set when a timer value is captured in the associated **TAxCCRy** register. In compare mode, any **CCIFG** flag is set if **TAxR** counts to the associated **TAxCCRy** value. Software may also set or clear any **CCIFG** flag. All **CCIFG** flags request an interrupt when their corresponding **CCIE** bit is set.

The **TAxCCRy CCIFG** flags and **TAIFG** flags are prioritized and combined to source a single interrupt vector. The interrupt vector register **TAxIV** is used to determine which flag requested an interrupt.

The **TAxIV** is a 16-bit register with all 16 bits to build an interrupt vector. Table 9.5 shows all possible interrupt vectors and their priority levels.

Any access, reading or writing, of the **TAxIV** register automatically resets the highest pending interrupt flag. If another interrupt flag is set, another interrupt is immediately generated after servicing the initial interrupt. For example, if the **TAxCCR1** and **TAxCCR2** **CCIFG** flags are set when the ISR accesses the **TAxIV** register, **TAxCCR1 CCIFG** is handled first and reset automatically. After the completion of **TAxCCR1 CCIFG** ISR, the **TAxCCR2** **CCIFG** flag generates another interrupt.

Table 9.5 Bit value and function for TAxIV register

Bit	Field	Type	Reset	Function
15:0	**TAIV**	RO	0	Timer_A interrupt vector value: **00:** No interrupt pending **02:** Interrupt source: Capture/compare 1; Interrupt flag: **TAxCCR1 CCIFG**; Interrupt priority: Highest **04:** Interrupt source: Capture/compare 2; Interrupt flag: **TAxCCR2 CCIFG** **06:** Interrupt source: Capture/compare 3; Interrupt flag: **TAxCCR3 CCIFG** **08:** Interrupt source: Capture/compare 4; Interrupt flag: **TAxCCR4 CCIFG** **0A:** Interrupt source: Capture/compare 5; Interrupt flag: **TAxCCR5 CCIFG** **0C:** Interrupt source: Capture/compare 6; Interrupt flag: **TAxCCR6 CCIFG** **0E:** Interrupt source: Timer overflow; Interrupt flag: **TAxCTL TAIFG**; Interrupt priority: Lowest

9.3.4 Implementations of the Timer_A modules

Based on the different operation modes, the Timer_A modules can have the following three implementation modes:

1. Input edge-count mode
2. Input edge-time mode
3. PWM mode

The function of the input edge-count mode is to calculate the total number of an external event or signal that has occurred or been detected.

When working in the input edge-count mode, the timer is selected to work in the compare mode. The input clock source to the 16-bit timer/counter **TAxR** register is an external signal (connected to the **TAxCLK**) that can be detected and counted by the timer. Each time a rising edge of the external input signal is detected, the 16-bit timer/counter **TAxR** is incremented by 1. This value is compared with the preloaded value in the **TAxCCRy** register, which is a definite value. If both of them are equal, an interrupt is generated and the value in the **TAxCCRy** register indicates the times of the input signal that has been detected.

An optional way to count the number of periods of an internal/external signal is to use the capture mode. In this mode, connect the internal/external signal to both the timer clock input **TAxCLK** and the capture input **CCIyA**. In this way, both the timer and the capturer use the same clock source to have the same count. When the capturer detects one active edge of the input signal, the timer incremented its count by 1 and sends it to the **TAxCCRy** register. Thus, both the timer and the **TAxCCRy** register have the same count. As the timer counts to the desired number of input signal's period, the program can be terminated.

The function of the input edge-time mode is to calculate the total time period or the width of a high pulse of an external event or signal.

When working in this mode, the timer is selected to work in the capture mode. The external signal works as the input to the capture input **CCIyA**. The bit field **CM** should be set to 11 to detect both edges. Each time when a rising edge of the external input signal is detected, an interrupt is generated and the current count value in the 16-bit timer/counter **TAxR** is sent to the **TAxCCR0** Register. Inside the ISR, the value in the **TAxCCR0** Register is reserved. When a falling edge of the external signal is detected, another interrupt is also generated and the count value in the **TAxR** is sent to the **TAxCCR0** again. By calculating the difference between two **TAxCCR0** values, the period or the width of the external input signal can be achieved.

The function of the PWM mode is to generate desired PWM signal to drive some DC motors via related motor amplifiers.

When working in this mode, the compare mode is selected. The value in the **TAxCCRy** and **TAxCCR0** registers can be set as different ones. By using the output mode 2, 3, 6, and 7 we discussed above, various periodic waveforms with different duty cycles can be obtained to drive the motor amplifier.

9.3.4.1 Input edge-count mode

In edge-count mode, the timer is configured as a 16-bit continuous or up-count mode. The **CAP** bit in the **TAxCCTLy** register must be reset to 0 to enable the timer/counter to work as a compare mode. The defined number of external event or signal to be detected should be loaded into the **TAxCCR0** Register as the target number of events to be occurred and detected. In this mode, the timer is capable of capturing the rising edge of the external input signal only.

Figure 9.10 shows an example of using a continuous timer to perform input edge-count detections. In this example, the timer input clock is an external signal to be detected and it should be a digital periodic signal applied on the **TAxCLK** input. The match value is set to the **TAxCCR0** Register (=**0x0006**) so that six edge events are counted. When the timer counts to **0x0006**, a matched is met, the **CCIFG** flag bit in the **TAxCCTLy** register is set and an interrupt is generated if it has been enabled by setting the **CCIE** bit in the **TAxCCTLy** register.

One issue to be noted when using this method is that the overall counting period of the 16-bit timer/counter **TAxR**, from **0x0** to **0xFFFF**, must be greater than **N** × period of the external signal to be detected (**N** = the number of periods of the external signal to be counted). In other words, the maximum number of periods of the external signal can be detected is: **65535** = $2^{16} - 1$.

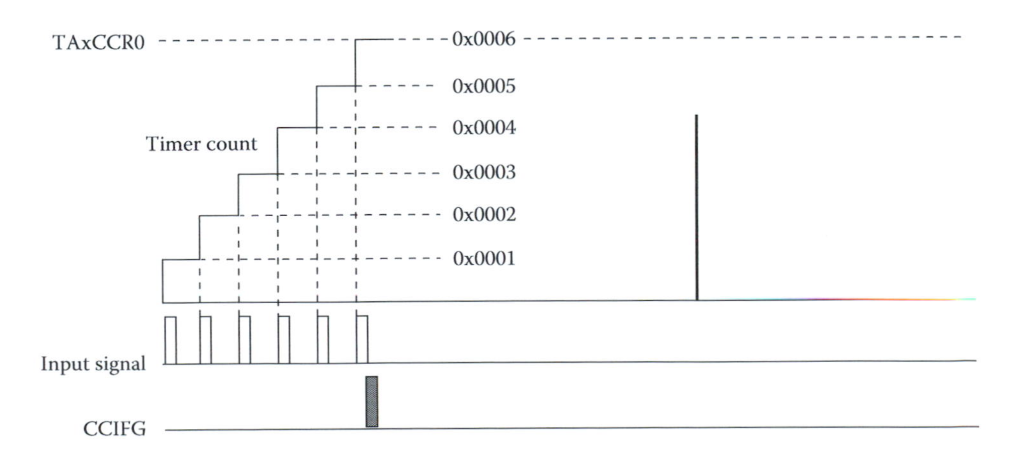

Figure 9.10 Example of using a continuous timer to detect input edge events.

9.3.4.2 Input edge-time mode

This operational mode is similar to the input edge-count mode discussed in the last section. The difference is that the timer should be configured to the capture mode by setting the **CAP** bit in the **TAxCCTLy** register. In this mode, the timer is capable of capturing the rising, falling, or both edge of the external input signal. To measure the period of the external signal, configure the capturer to capture both edges mode by setting the **CM** bits in the **TAxCCTLy** register as **11**.

Each time when a rising edge event is detected, an interrupt is generated and the current count value is loaded into the **TAxCCRy** register, and is held there until the falling edge is detected. At that point that count value should be reserved inside the interrupt handler to enable the **TAxCCRy** register to accept the falling edge count. As a falling edge of the external signal is detected, the current count in the timer is also loaded into the **TAxCCRy** register. By calculating the difference between this and the reserved count value, the period of the external signal can be obtained.

When operating in edge-time mode, the 16-bit timer/counter uses modulo 2^{16} count when working in the continuous mode. After events have been captured, the timer does not stop its counting. It continues to count until the timeout value **0xFFFF** is hit. When the timer reaches the timeout value, it is reloaded with 0 and continues its count up to another timeout value.

Figure 9.11 shows an example of using a continuous mode timer to perform the input edge timing detections. In the diagram, it is assumed that the start value of the timer is the default value of **0x0**, and the timer is configured to capture both edges events.

An issue when using this mode to measure the period of an external signal is the overlapping. As shown in Figure 9.11, the second falling edge of the input signal triggered the fourth interrupt and that moment count value is loaded into the **TAxCCR0** Register. However, that count value is incorrect since it is counted based on a new period. To avoid this overlapping mistake, that count value must be added by the whole period counts, **0xFFFF**, and then it can be used to minus the previous reserved count value to get the correct period of the detected signal. This situation can be checked in software to make sure that the current count value is always greater than the previous count value. If not, the whole period counts, **0xFFFF**, should be added into the current count to adjust the current count value. This method can be used to calculate an average value for multiple measured periods.

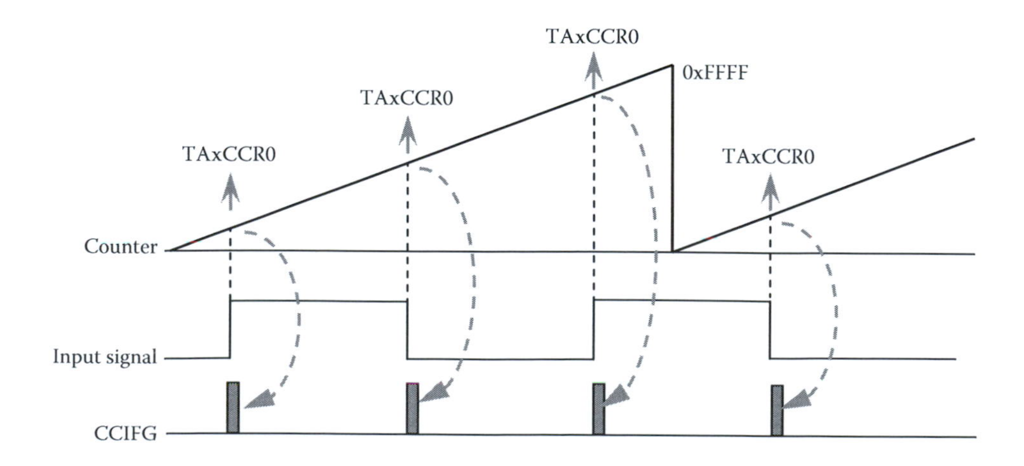

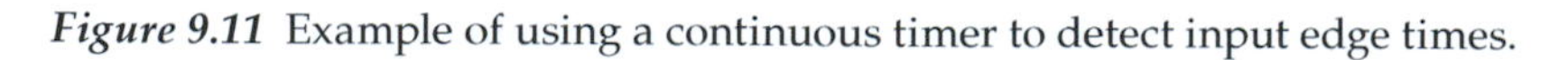

Figure 9.11 Example of using a continuous timer to detect input edge times.

9.3.4.3 PWM mode

As we discussed in Section 9.3.3, different waveforms with various duty cycles can be achieved when the timer/counter works in the compare mode by comparing the **TAxCCR0** and the **TAxCCRy** registers with desired output mode.

The period and frequency of the PWM signal is controlled by the upper limit value setup either in the **TAxCCR0** Register (up-count mode) or **0xFFFF** (continuous mode), multiplied by the input clock period or frequency. But the duty cycle is determined by the setup values in the **TAxCCR0** and the **TAxCCRy** registers. Based on selected output mode (**OUTMOD**), in most cases are output modes 2, 3, 6, and 7, different setup values in the **TAxCCR0** and the **TAxCCRy** registers can output different duty cycles. For example, in the continuous mode (refer to Figure 9.8), the setup value in the **TAxCCRy** register should be smaller than that in the **TAxCCR0** Register. When working in the output modes 2 and 3, the lower in the **TAxCCRy** and the higher in the **TAxCCR0**, the higher duty cycle can be obtained. Otherwise, the lower duty cycle can be achieved. However, if working in the output modes 6 and 7, the opposite results can be obtained.

An example operational sequence of using the Timer_A module 2 to generate a PWM signal is (Timer_A module 2 is working in the continuous mode):

1. Set the **TA2** timer to work in the compare mode with the output modes 2 or 3.
2. The start value (rising edge) is loaded into the **TA2CCR2** Register, and the terminate value (falling edge) is loaded into the **TA2CCR0** Register. The duty cycle of the PWM signal is determined based on these two setup values. The period of the PWM signal is determined by the input clock period multiplied by the 65536 (**0xFFFF**).
3. Enable the **TA2CCR2** and **TA2CCR0** interrupts by setting the **CCIE** bits in the **TA2CCTL2** and **TA2CCTL0** registers.
4. The timer is enabled and starts its continuous count until it reaches the **0xFFFF** state. Then it reloads the 0 and continues for the next cycle until disabled by software, setting the **TACLR** bit in the **TA2CTL** register.
5. As the timer counts to the value in the **TA2CCR2**, the PWM pulse rising edge is generated with outputting high. When the timer counts to the value in the **TA2CCR0** Register, the pulse of the PWM signal is terminated with outputting low.
6. The duty cycle of the PWM signal can be controlled by the setup different values in either or both **TA2CCR2** and **TA2CCR0** registers.

Figure 9.12 shows an example of using Timer_A module 2 working in the continuous mode to generate an output PWM with a 1-ms period and a 67% duty cycle. For this duty setting, the **TA2CCR2** = **0xAB85** (43909) and the **TA2CCR0** = **0xFFFF** (65535) assuming a 65.536-MHz input clock to the timer. The duty cycle could be 33% if the **TA2CCR2** = **0x547B** (21627).

9.3.5 Timer_A module GPIO-related control signals

In the MSP-432P401R MCU system, all GPIO ports, except GPIO Ports 1 and 3, provide Timer_A-related GPIO control signal pins. Figure 9.13 shows all of these GPIO pins and their control functions. All of those Timer_A-related pins have been highlighted. Table 9.6 shows these pins and their timer function assignments.

As we mentioned, all peripherals in the MSP432P401R MCU system are interfaced to the processor via GPIO ports and pins, and most GPIO pins can be configured as multiple or alternate function pins to perform multiple different functions. This is also true of the Timer_A and its four modules.

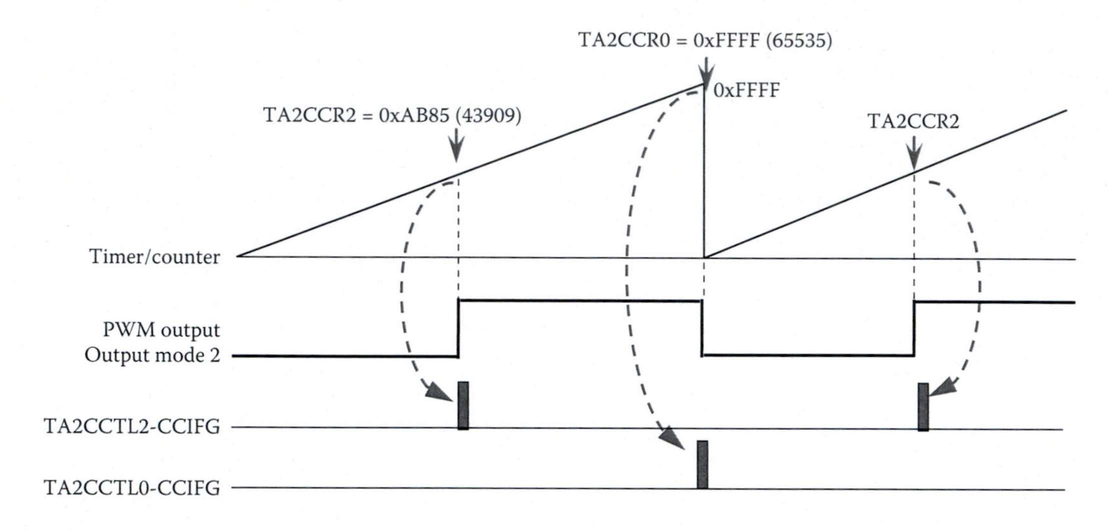

Figure 9.12 Example of using Timer_A module 2 to generate a PWM signal.

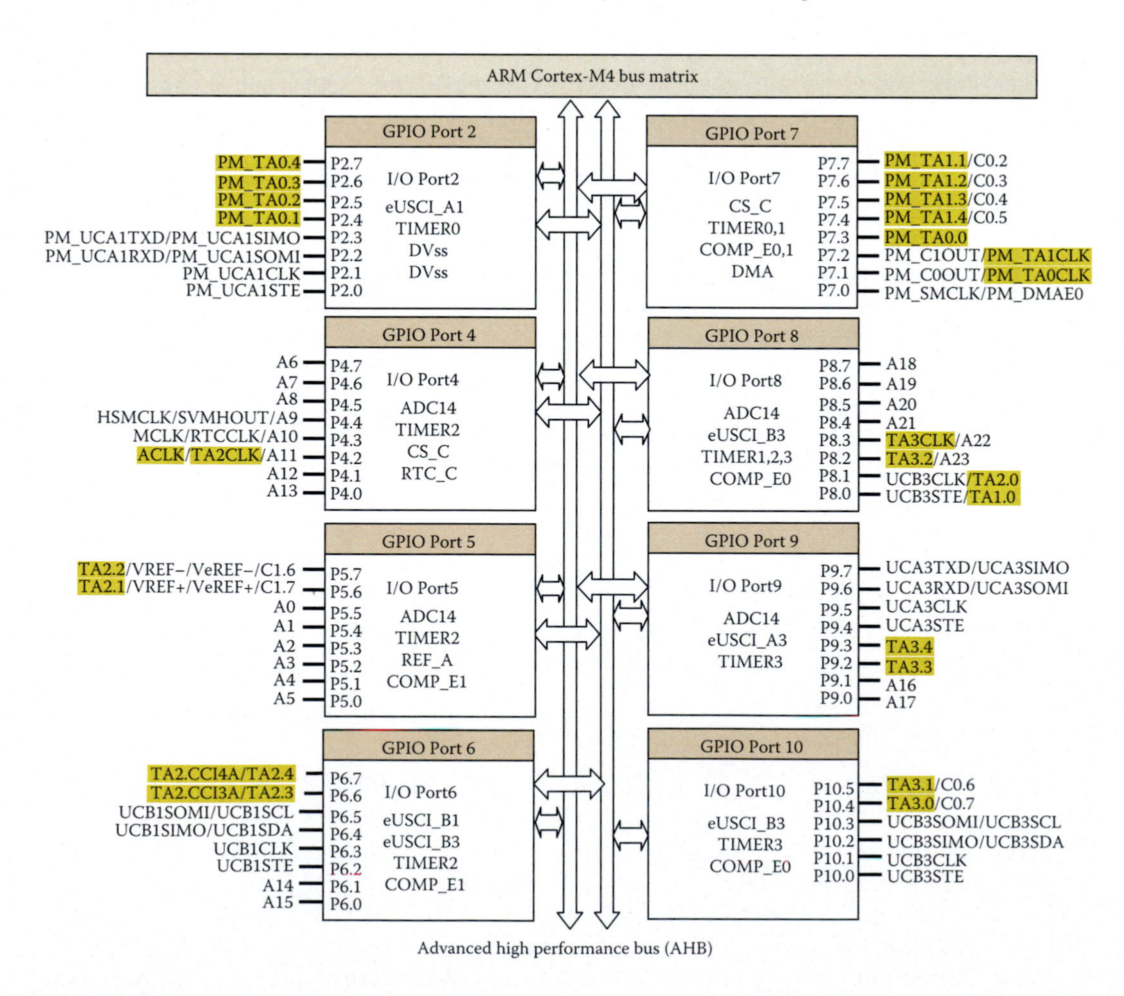

Figure 9.13 Timer_A module-related GPIO pins.

Table 9.6 Timer_A module input-output pins and their functions

Timer_A	Input Function Pin	Output Function Pin	GPIO Pins
PM_TA0.0	TA0 CCR0 capture input CCI0A	TA0 CCR0 compare output OUT0	**P7.3**
PM_TA0.1	TA0 CCR1 capture input CCI1A	TA0 CCR1 compare output OUT1	**P2.4**
PM_TA0.2	TA0 CCR2 capture input CCI2A	TA0 CCR2 compare output OUT2	**P2.5**
PM_TA0.3	TA0 CCR3 capture input CCI3A	TA0 CCR3 compare output OUT3	**P2.6**
PM_TA0.4	TA0 CCR4 capture input CCI4A	TA0 CCR4 compare output OUT4	**P2.7**
PM_TA1.0	TA1 CCR0 capture input CCI0A	TA1 CCR0 compare output OUT0	**P8.0**
PM_TA1.1	TA1 CCR1 capture input CCI1A	TA1 CCR1 compare output OUT1	**P7.7**
PM_TA1.2	TA1 CCR2 capture input CCI2A	TA1 CCR2 compare output OUT2	**P7.6**
PM_TA1.3	TA1 CCR3 capture input CCI3A	TA1 CCR3 compare output OUT3	**P7.5**
PM_TA1.4	TA1 CCR4 capture input CCI4A	TA1 CCR4 compare output OUT4	**P7.4**
TA2.0	TA2 CCR0 capture input CCI0A	TA2 CCR0 compare output OUT0	**P8.1**
TA2.1	TA2 CCR1 capture input CCI1A	TA2 CCR1 compare output OUT1	**P5.6**
TA2.2	TA2 CCR2 capture input CCI2A	TA2 CCR2 compare output OUT2	**P5.7**
TA2.3	TA2 CCR3 capture input CCI3A	TA2 CCR3 compare output OUT3	**P6.6**
TA2.4	TA2 CCR4 capture input CCI4A	TA2 CCR4 compare output OUT4	**P6.7**
TA3.0	TA3 CCR0 capture input CCI0A	TA3 CCR0 compare output OUT0	**P10.4**
TA3.1	TA3 CCR1 capture input CCI1A	TA3 CCR1 compare output OUT1	**P10.5**
TA3.2	TA3 CCR2 capture input CCI2A	TA3 CCR2 compare output OUT2	**P8.2**
TA3.3	TA3 CCR3 capture input CCI3A	TA3 CCR3 compare output OUT3	**P9.2**
TA3.4	TA3 CCR4 capture input CCI4A	TA3 CCR4 compare output OUT4	**P9.3**
PM_TA0CLK	Timer_A0 external clock input	—	**P7.1**
PM_TA1CLK	Timer_A1 external clock input	—	**P7.2**
TA2CLK	Timer_A2 external clock input	ACLK	**P4.2**
TA3CLK	Timer_A3 external clock input	—	**P8.3**

As we mentioned at the beginning of this chapter, each Timer_A-related GPIO pin is marked as **TAx.y**, which is equivalent to the **TAx_Cy** input or output signal pin. These pins have two different functions, either input or output:

- When working as an input pin, the pin is connected to an external signal and it is used to capture that signal as a detecting source.
- When working as an output pin, the pin works as the compare output pin (**OUTn**) for the selected compare register **TAxCCRy**.
- The I/O direction selection for those pins can be determined by configuring the related **PxDIR** registers.

For example, the **TA0.1** pin is the TA0 CCR1 (**TA0_C1**) capture input pin **CCI1A** when it works as an input pin. However, it will be the TA0 CCR1 compare output Out1 (**OUT1**) pin when it works as an output pin, as shown in Table 9.6.

Not all of these GPIO pins are available and connected to the J1~J4 connectors in the MSP-EXP432P401R EVB, and most of them are connected to J5 expended connector in that EVB. All pins that are connected to J1~J4 have been highlighted in Table 9.6.

Generally, if the timers are only used for the timing base or timing up indications, one does not need to use these I/O pins since these pins are mainly connected to the external signals to detect the number of events occurred or the time period events experienced. In most applications, the timer interrupts are used to indicate that the specified time period has passed and the time is up to ask users to handle desired jobs. The polling method can also be used to detect the **TAIFG** bit in the **TAxCTL** register or **CCIFG** bit in the **TAxCCTLy** register to check whether a time-out event or a time match event has occurred for the Timer_A module.

9.3.6 Build example Timer_A module projects

In this section, we use some Timer_A modules as a 16-bit timer/counter to build a few projects. The first project is to use Timer_A module 2 to periodically generate an overflow or time-out interrupt to turn on four LEDs, **PB3~PB0**, via GPIO Port 2, on the EduBASE ARM Trainer. The input clock to the timer is the 3 MHz SMCLK and the period to be counted in the **TA2** counter is 21.85 ms ($65536 \times 1/3000000$).

To perform this periodic interrupt for each 21.85 ms, one needs to:

1. Enable and clock the **TA2** to use SMCLK as clock source and to work in the continuous mode with overflow interrupt
2. Configure GPIO Port 2 and related pins, **P3.2**, **P3.3**, **P2.5**, and **P2.4** since they are connected to four red color LEDs on the EduBASE ARM Trainer
3. Use CMSIS **NVIC_IPP3** Register to set the interrupt priority level as 3 for the **TA2**
4. Use CMSIS **NVIC_ISER0** Register to enable the time-out interrupt for the **TA2**
5. Use Figure 9.14 as per set lines **__enable_irq()** intrinsic function to globally enable all interrupts
6. Use an infinitive **while()** loop to wait for any interrupt coming

In addition to the main program, one also needs to build the **TA2** interrupt handler:

1. Clear the time-out interrupt for **TA2** by resetting the bit **TAIFG** in the **TA2CTL** register to enable it to be generated in the future
2. Turn on the related LEDs via GPIO Port 2-related pins
3. Update the interrupt count number

Figure 9.14 shows the detailed codes for these steps. These codes are located in a class project, **DRATA2OVInt**, which is located at the folder: **C:\MSP432 Class Project\Chapter 9**.

For steps 17 and 18, refer to Section 5.3.2.1 (Tables 5.11 and 5.13) and Section 5.3.2.2 (Table 5.14) to get more details about these instructions.

To get a clearer picture about these interrupt numbers, one can modify code line 16 to use a lower input clock, such as **ACLK** (32 KHz), with the system macro **TASSEL_1** to get a much slower count display rate on four LEDs. In this case, the count period for each overflow is 2.0 seconds.

The next example project is to use three Timer_A compare registers, **TA0CCR0**, **TA0CCR1**, and **TA0CCR2** as well as **TA0.1** overflow to generate four independent interrupt intervals. Inside the first three ISRs, the interval is increased by different lengths to get the incremented interrupt intervals. The output mode (**OUTMOD**) is set to toggle for all these modules. When these modules work as continuous modes with the input clock as the **ACLK** = 32768 Hz, the toggle rates are:

- **TA0CCR0** = 32768/(2*2500) = 6.5536 Hz
- **TA0CCR1** = 32768/(2*5000) = 3.2768 Hz

```
1  //*****************************************************************************
2  // DRATA2OVInt.c - Main application file for project DRATA2OVInt
3  //*****************************************************************************
4  #include <stdint.h>
5  #include <stdbool.h>
6  #include <msp.h>
7  uint8_t intCount = 0;
8  void Display(uint8_t intNUM);
9  int main(void)
10 {
11  WDTCTL = WDTPW|WDTHOLD;              // stop WDT
12  P3DIR = BIT2|BIT3;                   // set P3.2 & P3.3 as output pins
                                            (LEDs PB0 & PB1)
13  P2DIR = BIT5|BIT4;                   // set P2.5 & P2.4 as output pins
                                            (LEDs PB2 & PB3)
14  P3OUT &= ~(BIT2|BIT3);               // set P3.3 & P3.3 output low
15  P2OUT &= ~(BIT5|BIT4);               // set P2.5 & P2.4 output low
16  TA2CTL =                             // SMCLK, count mode, clear TAR, enable
    TASSEL_2|MC_2|TACLR|TAIE;               interrupt
17  NVIC_IPR3 = 0x00006000;              // TA2_N priority is 3 in bits 15 ~ 13 of
                                            NVIC_IPR3
18  NVIC_ISER0 = 0x00002000;             // TA2_N IRQ = 13, set bit 13 on NVIC_ISER0
19  __enable_irq();                      // globally enable interrupts
20  while(1);                            // wait for time-out interrupt...
21 }
22 void TA2_N_IRQHandler(void)           // Timer A2 interrupt service routine
23 {
24  TA2CTL &= ~TAIFG;                    // clear TA2 overflow interrupt
25  Display(intCount);
26  intCount++;                          // call Display() to display interrupt numbers
27 }
28 void Display(uint8_t intNUM)          // Display() subroutine
29 {
30  P3OUT = (intNUM & 0x0F) << 2;
31  P2OUT = 0;                           // display intNUM on LEDs PB1-PB0
32  P2OUT |= (intNUM & BIT2) << 3;       // clean P2 output pins to 0
33  P2OUT |= (intNUM & BIT3) << 1;       // display bit2 on intNUM in PB2 (P2.5 pin)
34 }                                     // display bit3 on intNUM in PB3 (P2.4 pin)
```

Figure 9.14 Codes for the example project DRATA2OVInt.

- **TA0CCR2** = 32768/(2*10000) = 1.6384 Hz
- **TA0** overflow = 32768/(2*65536) = 0.25 Hz.

Figure 9.15 shows the detailed code for this project. This project is located in a class project, **DRATA0Toggles**, which is located in the folder: **C:\MSP432 Class Project\Chapter 9**.

Next let's take care of the most popular implementations of using the Timer_A modules.

```
1  //**************************************************************************************
2  // DRATA0Toggles.c - Main application file for project DRATA0Toggles
3  //**************************************************************************************
4  #include <stdint.h>
5  #include <stdbool.h>
6  #include <msp.h>
7  void TA0_Init(void);
8  int main(void)
9  {
10  WDTCTL = WDTPW | WDTHOLD;       // stop WDT
11  TA0_Init();                     // call TA0_Init() to initialize GPIO pins and TA0
12  NVIC_IPR2 = 0x00006060;         // set TA0.0 priority as 3 (bits 7 ~ 5 in NVIC_IPR2)
13                                  // set TA0.N priority as 3 (bits 15 ~ 13 in NVIC_IPR2)
14  NVIC_ISER0 = 0x00000300;        // TA0.0 IRQ = 8, TA0.N IRQ = 9, set bits 8 & 9 in NVIC_ISER0
15  __enable_irq();
16  while(1);                       // wait for interrupts...
17 }
18 void TA0_Init(void)
19 {
20  P1DIR |= BIT0; P1OUT &= ~BIT0; // set P1.0 as output pin & set P1.0 output low
21  P7DIR |= BIT3;                  // set P7.3 as TA0.0 pin
22  P7SEL0 |= BIT3;
23  P2DIR |= BIT2|BIT4|BIT5;        // set P2.2, P2.4 & P2.5 as output pins
24  P2SEL0 |= BIT4|BIT5;            // set P2.4 & P2.5 as TA0.1 & TA0.2 pin
25  P2OUT &= ~BIT2;                 // set P2.2 output low
26  P3DIR = BIT2|BIT3;              // set P3.2 & P3.3 as output pins
27  P3OUT &= ~(BIT2|BIT3);          // set P3.2 & P3.3 output low
28  TA0CCTL0 = OUTMOD_4|CCIE;       // TA0CCR0 toggle, interrupt enabled
29  TA0CCTL1 = OUTMOD_4|CCIE;       // TA0CCR1 toggle, interrupt enabled
30  TA0CCTL2 = OUTMOD_4|CCIE;       // TA0CCR2 toggle, interrupt enabled
31  TA0CTL = TASSEL_1|MC_2|TAIE;    // TA0 - ACLK input, continuous mode, interrupt enabled
32 }
33 void TA0_0_IRQHandler(void)     // Timer A0.0 interrupt service routine
34 {                               // TA0.0 compare interrupt occurred
35  TA0CCTL0 &= ~CCIFG;             // clear TA0.0 compare interrupt flag
36  TA0CCR0 += 2500;                // add Offset to TA0CCR0
37  P2OUT ^= BIT2;                  // toggle blue-color LED
38 }
39 void TA0_N_IRQHandler(void)     // Timer A0.N interrupt service routine
40 {
41  if(TA0CCTL1&CCIFG)              // TA0.1 compare interrupt occurred
42  {
43   TA0CCTL1 &= ~CCIFG;            // clear TA0.1 compare interrupt flag
44   TA0CCR1 += 5000; P3OUT ^=      // add Offset to TA0CCR1 & toggle P3.2 red LED
    BIT2;
45  }
46  if(TA0CCTL2&CCIFG)              // TA0.2 compare interrupt occurred
47  {
48   TA0CCTL2 &= ~CCIFG;            // clear TA0.2 compare interrupt flag
49   TA0CCR2 += 10000;              // add Offset to TA0CCR2
50   P3OUT ^= BIT3;                 // toggle P3.3 red LED
51  }
52 if(TA0CTL&TAIFG)                // TA0 overflow interrupt occurred
53 {
54  TA0CTL &= ~TAIFG; P1OUT ^=      // clear TA0 overflow interrupt flag & toggle P1.0 red-color
    BIT0;                           LED
55  }
56 }
```

Figure 9.15 Code for the example project DRATA0Toggles.

9.3.7 Popular implementations on Timer_A modules

The Timer_A system is a flexible system with many different components and functions. In this section, we only discuss some very popular and important implementations. These include:

- Using timer compare functions to detect the number of the input signal's edges
- Using timer capture functions to measure the period for periodic signals
- Using timer compare functions to generate PWM signals to control motors

First let's start from the timer capture function implementations.

9.3.7.1 Input edge-count implementations

In Section 9.3.4.1, we have provided a detailed discussion about this mode. In this section, we try to use the Timer_A module 2 or **TA2.2** compare function to measure the number of rising edges of an external input signal. The input source can be a function generator that can generate a periodic square waveform with variable frequency and period.

This implementation includes the following components and methods:

- The hardware connection is to connect the GPIO pin **P4.2** that works as a clock input to **TA2** module (**TA2CLK**) to a function generator, which can be considered as an external signal input to the timer/counter.
- Assign the **TA2CCR2** Register to 50 as the upper limit value for the number of rising edges of the external signal that is to be detected.
- As the 16-bit timer/counter detected 50 rising edges of the external signal, a compare interrupt is generated since the value in the 16-bit timer/counter is equal to the upper limit value in the **TA2CCR2** Register.
- Inside the ISR, the current count in the **TA2R** is reserved and it can be checked later.

Figure 9.16 shows the detailed code for this project **DRAEdgeCount**. The project is located in the folder: **C:\MSP432 Class Project\Chapter 9**.

The detected external signal coming from a function generator must be a standard CMOS level (0~3.3 V) digital periodic waveform. The measurable frequency range of this signal is from 1 Hz to 100 KHz.

In order to check the count result, a break point must be set in line 29, which is an infinitive **while()** loop, to enable the program to be halted at that line. Then one can open the **Call Stack + Locals** window to check the **count[]** data array for the edge-counting result, which should be 50.

9.3.7.2 Input edge-time implementations

Before we can develop this project, first let's have a closer look at some properties of the period of a signal to be measured. Two popular parameters for measuring a period are the *resolution* and *range* of the period to be measured.

Generally, the resolution is determined by the input clock frequency used by the timer. For example, if the working clock used by a timer is the **SMCLK** (3 MHz), the period for this clock is **0.33** μs. This is the minimized measurable period unit, and it can be considered as the resolution of the period that can be measured by using this clock.

The range of the period is determined by the size or length of the register to be used by the timer. For example, for a 16-bit counter, the maximized range of a period that can

```
1  //*****************************************************************************
2  // DRAEdgeCount.c - Main application fie for project DRAEdgeCount
3  //*****************************************************************************
4  #include <stdint.h>
5  #include <stdbool.h>
6  #include <msp.h>
7  uint32_t cNUM;
8  bool intFlag = true;
9  int main(void)
10 {
11   uint16_t count[1];
12   WDTCTL = WDTPW | WDTHOLD;          // stop watchdog timer
13   // Configure GPIO
14   P2DIR |= BIT2;                     // set P2.2 as output
15   P2OUT &= ~BIT2;                    // set P2.2 low
16   P4SEL1 |= BIT2;                    // set P4.2 as TA2CLK input pin, third function
17   P4DIR &= ~BIT2;
18   TA2CCR2 = 50;                      // set comparator value as 50
19   TA2CCTL2 |= CCIS_0|CCIE|SCS;       // enable compare mode
20                                      // synchronous capture, enable compare interrupt
21   TA2CTL |= TASSEL_0|MC_2|TACLR;     // use TA2CLK as clock source, clear TA2R
22                                      // start timer in continuous mode
23   NVIC_IPR3 = 0x00006000;            // TA2_N priority is 3 in bits 15 ~ 13 of
                                           NVIC_IPR3
24   NVIC_ISER0 = 0x00002000;           // TA2_N IRQ = 13, set bit 13 on NVIC_ISER0
25   __enable_irq();                    // globally enable interrupts
26   __wfi();
27   while(intFlag);                    // wait for interrupt...
28   count[0] = cNUM;                   // collect the number of detected edges
29   while(1);                          // check the result (a break point needs to be
                                           set at this line)
30 }
31 void TA2_N_IRQHandler(void)          // Timer A2 interrupt service routine
32 {
33   TA2CCTL2 &= ~CCIFG;                // clear the compare interrupt flag
34   P2OUT = BIT2;                      // set P2.2 (blue-color LED)
35   cNUM = TA2R;                       // get the number of detected edges
36   intFlag = false;                   // release the conditional while(intFlag) loop
37 }
```

Figure 9.16 Code for the example project DRAEdgeCount.

be measured is **65536 × 0.33** μs = **21.845** ms. However, for a 32-bit counter, the maximized period range can be expanded to **4294967296 × 0.33** μs = **1417.34** s.

As we discussed in Section 9.3.4.2, when working in this mode, the timer module should be configured to the capture mode by setting the **CAP** bit in the **TAxCCTLy** register. In this mode, the timer is capable of capturing the rising, falling, or both edge of the external input signal. To measure the period of the external signal, configure the capturer to capture either the rising edges or the falling edges, but for the pulse width measurement, configure the timer to measure both edges by setting the **CM** bits in the **TAxCCTLy** register as **11**.

Each time when a rising edge event is detected, an interrupt is generated and the current count value is loaded into the **TAxCCRy** register, and is held there until the next rising edge (period measuring) or the falling edge (pulse width measuring) is detected. At that point, the first count value should be reserved inside the interrupt handler to enable the **TAxCCRy** register to accept the next rising or falling edge count. As the next rising or falling edge of the external signal is detected, the current count in the timer is also loaded into the **TAxCCRy** register. Calculate the difference between this and the reserved count value, the period or the pulse width of the external signal can be obtained.

In this section, we try to use the **TA2** module (**TA2_C3**) to work as a 16-bit timer to measure the period of a periodic input signal. The input source can come from a function generator.

The working principle of using this **TA2_C3** to measure the period is:

1. The **TA2** is configured as the capture mode with the interrupt mechanism, and begins to perform the continuous counting action as it is enabled.
2. When the first rising edge of a periodic signal is detected, the current count value in the **TA2R** will be loaded into the **TA2CCR3** Register, and this value should be reserved for the further usage.
3. When the second rising edge of a periodic signal is detected, similarly the current count value can also be obtained from the **TA2CCR3** Register. The difference between these two rising edges can be considered as the period of the signal to be detected. This difference calculation can be performed and obtained in the interrupt handler.

This implementation includes the following components and methods:

- The hardware connection is to connect the GPIO pin **P6.6** (**TA2.3**) to a function generator.
- Use **TA2_C3** to capture rising edges for a periodic input signal coming from a function generator or an oscilloscope since most modern oscilloscopes have a standard built-in testing signal. Regularly this testing signal is a 1 KHz (1-ms period) square waveform.
- The **TA2** works in the 16-bit capture mode and generates a captured rising edge interrupt when a rising edge is detected.
- When a rising edge is detected, a **TA2_C3** edge-capture interrupt is generated. Inside the interrupt handler, the period can be calculated by two different count values, one is the current count value and the other is the previous count value stored in a user-defined variable. Also the red color LED connected to GPIO pin **P1.0** is toggled to indicate that an edge has been detected.
- When the period is obtained, a global Boolean flag **intFlag** can be reset to **false** to release the conditional **while()** loop in the main program to enable user to check the period measurement result.

Figure 9.17 shows the detailed code for this project named **DRATA2Period**. This project is located in the folder: **C:\MSP432 Class Projects\Chapter 9**. By using this program, the upper bound frequency can be measured for any external periodic signal is 20 KHz.

```
1  //*****************************************************************************
2  //DRATA2Period.c - Main application file for project DRATA2Period - TA2_C3 Capture Mode
3  //*****************************************************************************
4  #include <stdint.h>
5  #include <stdbool.h>
6  #include <msp.h>
7  void TA2_Init(void);
8  bool intFlag = true;
9  uint16_t period[2], count = 0;
10 int main(void)
11 {
12  uint16_t Period[1];
13  WDTCTL = WDTPW | WDTHOLD;            // stop WDT
14  TA2_Init();                          // call TA2_Init() to initialize GPIO pins and TA2
15  NVIC_IPR3 = 0x00006000;              // TA2_N priority is 3 in bits 15 ~ 13 of NVIC_IPR3
16  NVIC_ISER0 = 0x00002000;             // TA2_C3 IRQ = 13, set bit 13 in NVIC_ISER0
17  __enable_irq();
18  while(intFlag) {}                    // wait for the period to be collected
19  Period[0] = period[1];               // collect the period value
20  Period[0] = Period[0] + 0;           // useless at all! just avoid compiling warning!!!
21  while(1);                            // wait for capture interrupts...
22 }
23 void TA2_Init(void)
24 {
25  P1DIR |= BIT0;                       // set P1.0 as output pin
26  P1OUT &= ~BIT0;                      // set P1.0 output low
27  P2DIR |= BIT2;                       // set P2.2 as output pin
28  P2OUT &= ~BIT2;                      // set P2.2 output low
29  P6DIR &= ~BIT6;                      // set P6.6 as TA2_C3 capture input pin
30  P6SEL0 |= BIT6;
31  TA2CCTL3 = CCIS_0|CM_1|CAP|CCIE;     // TA2 _ C3 capture mode, CCI3A input on rising edge,
                                          interrupt enabled
32  TA2CTL = TASSEL_2|MC_2|TACLR;        // TA2 - SMCLK input, continuous mode, clear TA2R
33 }
34 void TA2_N_IRQHandler(void)          // Timer A2.N interrupt service routine
35 {
36  TA2CCTL3 &= ~CCIFG;                  // clear TA2.3 capture interrupt flag
37  if (count == 0)                      // is the first rising edge ?
38  period[0] = TA2CCR3;                 // yes, reserve the first rising edge
39  if (count == 1)                      // is the second rising edge?
40  {
41   period[1] = TA2CCR3;                // yes, save the second rising edge
42   period[1] = period[1] - period[0];  // get the difference between two rising edges
43   intFlag = false;                    // release the conditional while() loop in the
                                          main()
44   P2OUT = BIT2;                       // set blue LED to indicate the end
45  }
46  P1OUT ^= BIT0;                       // toggle red-color LED to indicate the capture
                                          interrupt
47  count++;                             // increment the count
48 }
```

Figure 9.17 Detailed code for the project DRATA2Period.

For code lines 15 and 16, refer to Tables 5.11, 5.13, and 5.14 in Chapter 5 to get more details about the coding function and values. For coding lines 29 and 30, refer to Table 7.4 in Chapter 7 to get more details about the configuring of the GPIO pin **P6.6**.

As the project runs, a function generator or an oscilloscope can be connected to the **TA2_C3** (**P6.6**) pin as the input for the edge signal to be detected. As soon as the blue LED is on, the period has been measured and collected in the user-defined variable **Period[0]**. To check this value, a break point should be set in line 21 with the **Call Stack + Locals** window.

9.3.7.3 PWM implementations

As we discussed in Section 9.3.4.3, the period and frequency of the PWM signal is controlled by the upper limit value setup either in the **TAxCCR0** Register (up-count mode) or **0xFFFF** (continuous mode), multiplied by the input clock period. But the duty cycle is determined by the setup values in the **TAxCCR0** and the **TAxCCRy** registers. Based on selected output mode (**OUTMOD**), in most cases are output modes 2, 3, 6, and 7, different setup values in the **TAxCCR0** and the **TAxCCRy** registers can output different duty cycles.

Actual period and duty cycle values depend on the PWM clock source used in the timer. For example, for an 8 MHz PWM clock source, the clock period is **125** ns. To use this clock source to generate a PWM signal with 5 ms as period and 1% duty cycle on **TA0_C2** or **TA0.2**, the following parameters and registers should be used:

The length of the high pulse = 5 ms × 1% = 0.05 ms, the length of the low pulse = 4.95 ms
TA0CCR0 = Period = 5 ms/125 ns = 5000000/125 = **40000**
TA0CCR2 = Length of low pulse = 4.95 ms/125 ns = 4950000/125 = **39600**.

In order to use the Timer_A module TA0 to generate the above PWM signal output, the following steps are needed to configure the TA0 module:

1. The input clock to the TA0 module **SMCLK** needs to be configured by tuning the DCO to output an 8 MHz clock.
2. The TA0 will work in the up-count mode with the **TA0CCR0** Register setup.
3. The Timer_A0 will work in the compare mode with the **TA0CCR2** Register setup and with the output mode as 3 (**OUTMOD** = 3).
4. The Timer_A0 will be interrupt enabled.

When the up-count counter gets the upper bound value stored in the **TA0CCR0** Register, it will reset to 0 and continue for the next counting operation until the **TACLR** bit is set.

The resolution or the minimized period of the PWM signal that can be generated by the timer is determined by the clock source. Still using an 8 MHz clock, the normal minimized period of the PWM signal is 125 ns with 0% duty cycle.

Figure 9.18 shows a functional block diagram for using **Timer_A0** as a 16-bit up-count counter to generate a PWM output signal to control a DC/Servo motor. The **TA0_C2** (**TA0.2**) pin in the **Timer_A0** module is the PWM signal output pin to transfer PWM signal to the motor driver to control the running speed of the DC/Servo motor.

In the following implementation, we use the **Timer_A0** as a 16-bit up-count counter to generate a variable duty PWM signal, which means that the duty cycle can be changed by software to control the running speed of a DC/Servo motor. No interrupt is used for this project since it is very simple.

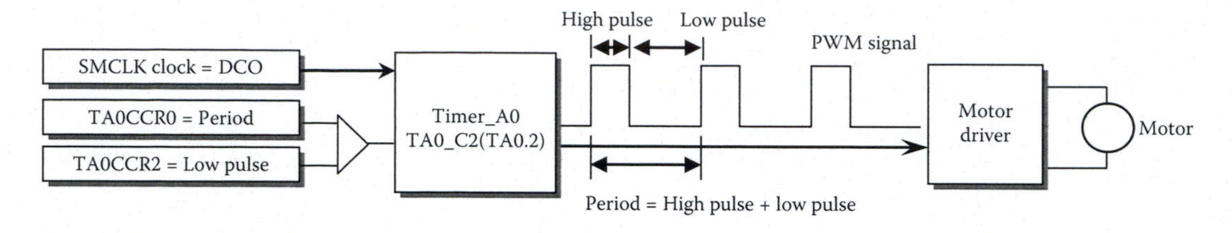

Figure 9.18 Functional block diagram of using Timer_A0 to generate PWM signals.

This implementation includes the following hardware components and methods:

1. An RC servo motor **HXT900 9GR** is used in this project as the target motor to be driven. One can buy this motor at any related site, such as **HobbyKing.com**, for less than $5.00. This is a low-voltage servo motor that can be driven by 3-V DC power, as shown in Figure 9.19a. To connect this motor to the circuit, you need to change three pins arrangement to make them as the following order:
   a. V_{CC}—red color wire to **+5 V** in pin-6 on J32 (EduBASE ARM Trainer)
   b. **GND**—brown color wire to **GND** on J2 in MSP432P401R EVB
   c. **Signal**—yellow color wire to **P2.5** on J2 in MSP432P401R EVB.
2. The **TA0.2** pin works as the PWM output pin for the Timer_A0. The order of these two pins, **GND** and **TA0.2**, on the **J2** connector in the MSP-EXP432P401R EVB is exactly same as the order of those two signal pins we just changed above. But the red color power line needs to be connected separately to the +5 V pin on the **J3** connector in that EVB, which is shown in Figure 9.19b.
3. After the hardware connection is configured and completed, you need to take care of the software control and configuration part. For this project, we want to change the duty cycle from 1% to 100% with an 8 MHz clock input. Refer to calculations we did above, for a 5 ms period, **40000** should be loaded into the **TA0CCR0** Register as the terminal value, and **39600** should be loaded into the **TA0CCR2** Register as the low pulse value for 1% duty cycle.

Before we can continue to build this example project, let's first take a closer look at the DCO frequency change and modification issue since we need to increase the SMCLK clock frequency from the default 3 to 8 MHz in this project.

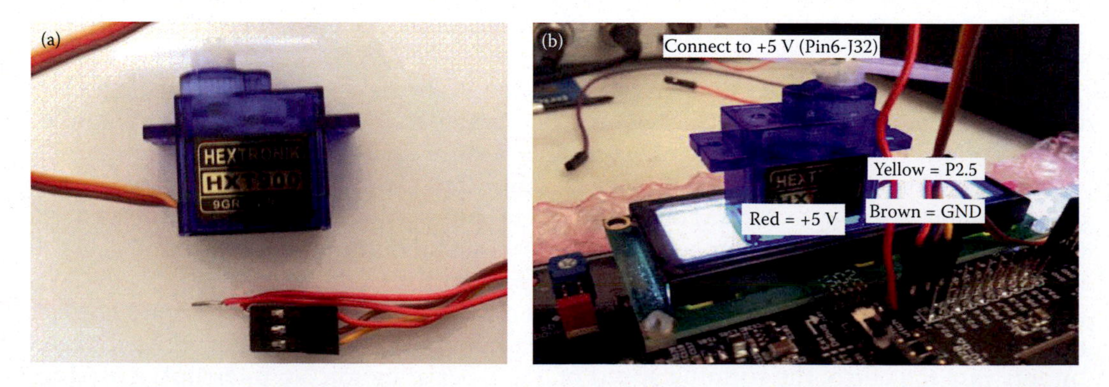

Figure 9.19 HXT900 9GR RC Servo motor and its connection.

9.3.7.3.1 Change the DCO frequency and range As we discussed in Section 2.3.5.4 in Chapter 2, the default system clock source is DCO, which can be connected to the **MCLK** or the **SMCLK** as the system clock source. The DCO clock can be selected by software via related system clock control registers, such as **CSCTL0**, **CSCTL1**, and **CSCLKEN**. By default, each DCO can be output in a certain frequency range with a center nominal frequency value (refer to Table 2.15 in Chapter 2).

To get the desired frequency, the bit fields, **DCORSEL** (bits 18~16) and **DCOTUNE** (bits 12~0), in the **CSCTL0** Register must be configured correctly. The bit field **DCORSEL** is used to determine the DCO output center frequency with a range, and the bit field **DCOTUNE** is used to define the final accurate frequency desired by the user. To obtain the accurate frequency value, the DCO tune value **N** can be calculated based on the following equation and filled into the bit field **DCOTUNE**:

$$N_{DCOTUNE} = \frac{(F_{DCO,nom} - F_{RSELx\_CTR,nom}) \times (1 + K_{DCOCONST} \times (768 - FCAL_{CSDCOxRCAL})) \times 8}{F_{DCO,nom} \times K_{DCOCONST}}$$

where

- $F_{DCO,nom}$ = target nominal frequency
- $F_{RSELx\_CTR,nom}$ = calibrated nominal center frequency for DCO frequency range x
- $K_{DCOCONST}$ = DCO constant (floating-point value)
- $N_{DCOTUNE}$ = DCO tune value in decimal
- $F_{CALCSDCOxRCAL}$ = DCO frequency calibration (FCAL) value for range x for internal resistor modes

All of the above parameters can be found from the device descriptor tag-length-value (TLV) structure for MSP432P401R MCUs or devices. For example,

- The DCO FCAL value for DCO internal resistor modes is **0x00000188** or **392**.
- The DCO constant K for **DCORSEL0~4** in internal resistor mode is **0.004944**.
- The DCO maximum allowed positive tune for DCORSEL0~4 in internal resistor mode is **0x00000600** or **1536**.

In this application, we try to set the DCO as 8-MHz clock source for **MCLK** and **SMCLK**. Based on Table 2.15 in Chapter 2, we can select the bit field **DCORSEL** as either **0x2** or **0x3** since both frequency ranges covered 8 MHz. We prefer to select **DCORSEL** = **0x2** with the nominal center frequency as 6 MHz. Now, let's calculate the **DCOTUNE** bit field value **N** and then fill it into that bit field to get our desired nominal 8-MHz frequency.

Apply all above values into the above equation, we get:

$$\begin{aligned} N_{DCOTUNE} &= \frac{(8-6) \times (1 + 0.004944 \times (768 - 392)) \times 8}{8 \times 0.004944} \\ &= 1156.53 \approx 1157 \text{ or } 0x485 \end{aligned} \tag{9.1}$$

Now, let's build our PWM project.

9.3.7.3.2 Build a PWM example project Create a new folder **DRATA0PWM5VMotor** under the class project folder **C:\MSP432 Class Projects\Chapter 9** with Windows

Explorer. Then create a new µVersion5 project **DRATA0PWM5VMotor** and add it into the above folder.

Create a new C Source file **DRATA0PWM5VMotor.c** and add it into the new project. Enter the code shown in Figure 9.20 into this C file.

All initialization and configuration processes are placed in a subroutine **PWM_Init()**. Since we use the GPIO **P2.2** pin as an indicator for the modifications of the duty cycle, the **P2.2** pin is configured as an output pin. The GPIO **P2.5** pin is connected to the **TA0.2** output to transfer the PWM signal to the motor. Refer to Table 7.9 in Chapter 7 to configure the GPIO pin **P2.5** to work as a Timer_A compare output pin.

An infinitive **while()** loop is used to tune the duty cycle for this PWM signal. The **TA0CCR2** Register is updated by subtracting a fixed number of 2 (high) periodically to reduce the low pulse value and increment the high pulse value since the value stored in the **TA0CCR2** Register is the length of the low pulse. The smaller this value, the shorter the length of the low pulse, and therefore the longer the high pulse. As the **high** value becomes higher and higher, the length of the low pulse becomes shorter and shorter. When this **high** value goes up to **39600**, the duty cycle is 100%. Then it is reduced back to 1 (duty cycle is 1%) to continue for the following loops.

9.3.7.3.3 Another professional PWM example project In this implementation, we try to use two buttons, **SW2** and **SW3**, which are installed in the EduBASE ARM Trainer as inputs to control the rotating speed of a DC motor. Refer to Figure 4.31 in Chapter 4, one can find that the button **SW2** is connected to the GPIO pin **P4.4**, and the button **SW3** is connected to the pin **P4.2**. When any of these buttons is pressed, a logic high is generated, otherwise it is a logic low.

In the EduBASE ARM® Trainer, a quad half H-bridge motor driver **TB6612FNG** is used to support and interface PWM modules to control and drive two servo or DC motors. A functional block diagram and hardware connection between the PWM modules and this motor driver TB6612FNG is shown in Figure 9.21. Table 9.7 shows the input and output relationship for this motor driver.

Each TB6612FNG driver has two control and driver blocks and each block can be used to drive one servo or DC motor. Therefore, totally two motors can be driven by using this driver.

As shown in Figure 9.21 and Table 9.7, each control block in the TB6612FNG driver has three inputs and two outputs:

- Two inputs, **IN1** and **IN2**, are used to control the rotating direction of the motor. When the input value is **IN1** = L and **IN2** = H (**01**), the motor is rotated in the counter clockwise direction when the PWM is in high pulse. When **IN1** = H and **IN2** = L (**10**), the motor is rotated in the clockwise direction as a high pulse is obtained from the PWM input.
- The PWM input is used to control the rotating speed of the motor. The longer or the wider the high pulse, the higher of the motor rotating speed.
- Two outputs, **O1** and **O2**, are directly connected to a motor to provide the driving power to the motor.
- The standby signal **STBY** should always be high to make the driver work. If this signal is low, it makes the driver to standby with high output impedances on two outputs.

In this example project, we need to use the Timer_A module 0 and the PWM driver to control a DC motor to test the control and driving ability of the PWM function. The DC

```
1  //****************************************************************************
2  // DRATA0PWM5VMotor.c - Main application file for project DRATA0PWM5VMotor - 5V DC Motor
3  //****************************************************************************
4  #include <stdint.h>
5  #include <stdbool.h>
6  #include <msp.h>
7  #define DCOTUNE 0x485
8  void PWM_Init(void);
9  void delay(int time);
10 int main(void)
11 {
12   uint32_t  high = 1;
13    WDTCTL = WDTPW | WDTHOLD;          // stop WDT
14    PWM_Init();                         // call PWM_Init() to initialize GPIO pins and
                                          TA0
15    while(1)
16   {
17     TA0CCR2 = 39600 - high;           // increment the duty cycle
18     high += 2;
19     delay(80);
20     P2OUT ^= BIT2;                    // toggle blue LED (P2.2)
21     if (high >= 39600)
22     high = 1;
23   }
24 }
25 void PWM_Init(void)
26 {
27    P2DIR |= BIT2;                     // set P2.2 as output pin
28    P2OUT &= ~BIT2;                    // set P2.2 output low
29    P2DIR |= BIT5;                     // set P2.5 as TA0.2 PWM output pin
30    P2SEL0 |= BIT5;
31    CSKEY = CSKEY_VAL;                 // unlock CS module for register access
32    CSCTL0 = 0;                        // reset tuning parameters
33    CSCTL0 = DCOEN|DCORSEL_2|DCOTUNE;// set DCO to 8 MHz (nominal, center of 4 ~ 8
                                         MHz range)
34    CSCTL1 = SELS_3|SELM_3;            // select SMCLK = MCLK = DCO
35    CSKEY = 0;                         // lock the CS module
36    TA0CCTL2 = OUTMOD_3|CCIE;          // TA0_C2 compare mode, interrupt enabled
37    TA0CTL = TASSEL_2|MC_1|TACLR;      // TA0 = SMCLK input (8MHz), up mode, clear TA0R
38    TA0CCR0 = 40000;                   // load the period to TA0CCR0
39    TA0CCR2 = 39600;                   // load the low pulse to TA0CCR2
40 }
41 void delay(int time)
42 {
43   int n;
44   for(n = 0 ; n < time; n++);
45 }
```

Figure 9.20 Complete code for the project DRATA0PWM5VMotor.

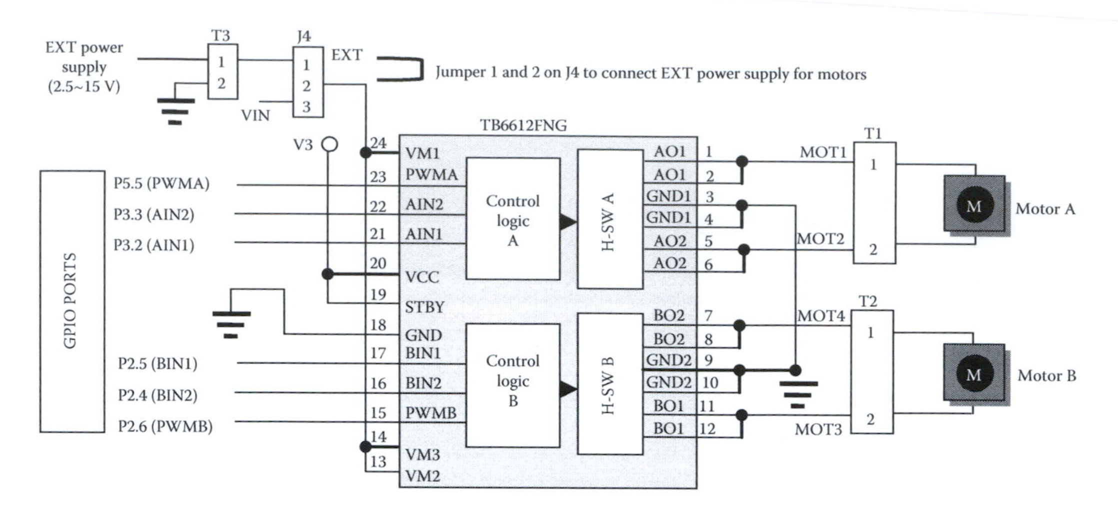

Figure 9.21 Hardware configuration of the PWM driver TB6612FNG.

motor used in this project is a DC motor fan (model number: **EE80251S2-000U-999**). The hardware connection is shown in Figure 9.22.

The hardware and software configurations for this example project include:

- We use the second control and driving block, block B, in the motor driver TB6612FNG to control and drive our target 12V DC motor and fan.
- The DC motor with a fan is a 12V DC motor with single rotating direction.
- The hardware connection and the GPIO pins to be used include:
  - **P2.6** (**TA0.3**) is used to provide the PWM driving signal to the motor (**PWMB**).
  - **P2.5** (**BIN1**) and **P2.4** (**BIN2**) are used to provide the rotating direction signal (**P2.5:P2.4** = **01** → CCW and **P2.5:P2.4** = **10** → CW) for the motor. In this project, we use the first rotating direction **P2.5:P2.4** = 01 since the motor can be rotated in only a single direction.
- Connect the DC motor to two blue interfacing blocks **M3** and **M4** in the EduBASE ARM Trainer with positive (red color) wire to **M4** and reference (black color) wire to **M3**
- Connect an external 12V DC power supply to the blue block **T3** with positive wire to the **+** terminal on the **T3**
- Change the jumper on the **J4** to connect 1 and 2 pins together to connect the external power supply to the TB6612FNG motor driver

Table 9.7 Input and Output of the TB6612FNG Driver

Input				Output		
IN1	IN2	PWM	STBY	OUT1	OUT2	MODE
H	H	H/L	H	L	L	Short brake
L	H	H	H	L	H	CCW
		L	H	L	L	Short brake
H	L	H	H	H	L	CW
		L	H	L	L	Short brake
L	L	H	H	OFF (high impedance)		Stop
H/L	H/L	H/L	L	OFF (high impedance)		Standby

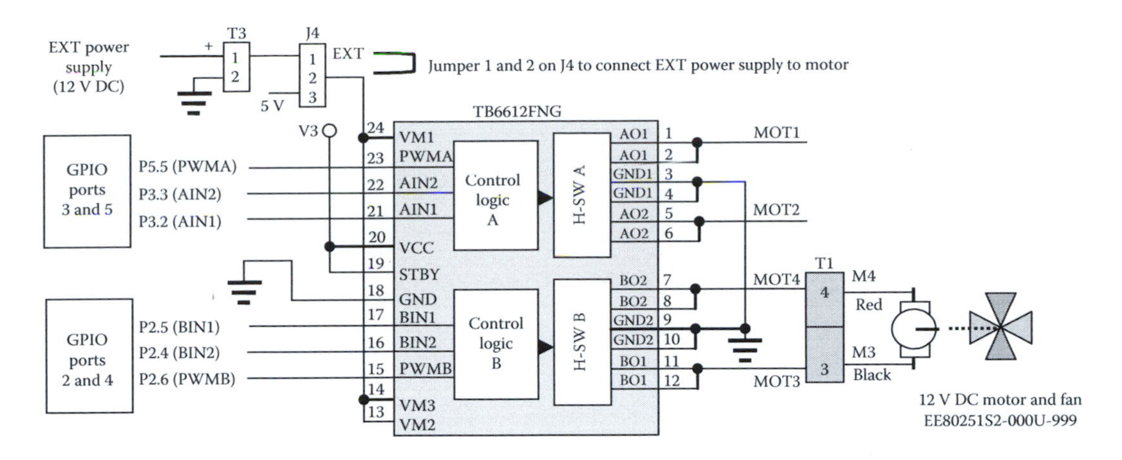

Figure 9.22 Hardware configurations for the example project.

In this example project, we use Timer_A module 0, exactly **TA0.3** to output a duty-modified PWM signal to control this DC motor via GPIO pin **P2.6**. The system input clock to the timer is 8 MHz. The output duty cycle is controlled by two user buttons, **SW2** and **SW3**, which are installed in the EduBASE ARM Trainer as inputs to control the rotating speed of a DC motor. The **SW2** and **SW3** are connected to GPIO pins **P4.4** and **P4.2**, respectively. The GPIO pins **P2.5** and **P2.4** are used to provide the rotating direction and they are fixed to **01** to allow the motor to rotate in the CCW direction since this DC motor is a single-direction rotating motor.

Now, let's start to build this project **DRATA0PWM12VMotor**.

Create a new folder **DRATA0PWM12VMotor** under the class project folder **C:\MSP432 Class Projects\Chapter 9** with the Windows Explorer. Then create a new µVersion5 project **DRATA0PWM12VMotor** and add it into the folder created above.

Create a new C file **DRATA0PWM12VMotor.c** and add it into this new project. Enter the codes shown in Figure 9.23 into this C file.

One point to be noted when using this project to generate PWM to drive a 12 V DC motor via **TA0.3** pin is how to stop the timer to disable the generation of the PWM signal to stop the motor. There are two possible ways to stop a timer:

1. To clear the timer by setting the **TACLR** bit on the **TA0CTL** register. When the timer is reset in this way, not only the timer/counter is reset but also the divider logic and count direction are reset. The timer/counter will restart from 0 next time when the timer returns to the normal counting operation as the **TACLR** is reset to 0.
2. Reset the mode control bit field **MC** (bits 5 and 4) on the **TA0CTL** register to 00. When the timer is stopped in this way, the timer/counter can continue its count from the value it stopped in the last stopping time when the **MC** bit field is modified and the resulted **MC** > 00.

Compare between these two methods, we selected the second way to stop the timer since it has nothing to do with the divider logic and the count direction for the timer. More important, it does not need to restart its count from 0 when the timer returns to the normal counting operation.

The codes in lines 23 and 24 set the maximum PWM duty cycle to 39990 when the **high** is greater than the upper bound to limit the maximum motor speed. Similarly,

```
 1 //***************************************************************************************
 2 //DRATA0PWM12VMotor.c - Main application file for project DRATA0PWM12VMotor - 12V DC Motor
 3 //***************************************************************************************
 4 #include <stdint.h>
 5 #include <stdbool.h>
 6 #include <msp.h>
 7 #define DCOTUNE 0x485
 8 void PWM_Init(void);
 9 void delay(int time);
10 int main(void)
11 {
12  uint32_t high = 1, low = 1;
13  WDTCTL = WDTPW | WDTHOLD;              // stop WDT
14  PWM_Init();                            // call PWM_Init() to initialize GPIO pins and TA0
15  while(1)
16  {
17   if (P4IN == BIT4)                     // SW2 button (P4.4) is pressed
18   {
19    TA0CTL |= MC_1;                      // enable the timer
20    TA0CCR3 = 40000 - high;              // increment the duty cycle
21    high += 4; delay(200);
22    P2OUT ^= BIT2;                       // toggle blue LED (P2.2)
23    if (high >= 39990)                   // if high > upper bound, keep high value
24     high = 39990;
25    low = 1;                             // reset the low value
26   }
27   if (P4IN == BIT2)                     // SW3 button (P4.2) is pressed
28   {
29    TA0CTL |= MC_1;                      // enable the timer
30    TA0CCR3 = 10 + low;                  // decrement the duty cycle
31    low += 4; delay(200);
32    P1OUT ^= BIT0;                       // toggle red LED (P1.0)
33    if (low >= 39990)                    // if low < lower bound, keep low value
34     low = 39990;
35    high = 1;                            // reset the high value
36   }
37   else
38    TA0CTL &= ~0x30;                     // stop the timer (MC = 00)
39  }
40 }
41 void PWM_Init(void)
42 {
43  P1DIR |= BIT0; P1OUT &= ~BIT0;         // set P1.0 as output pin & output low
44  P2DIR |= BIT2|BIT4|BIT5;               // set P2.2, P2.4 & P2.5 as output pins
45  P2OUT &= ~BIT2;                        // set P2.2 output low
46  P2OUT &= ~BIT5; P2OUT |= BIT4;         // set P2.5:P2.4 = 01(CCW rotate direction)
47  P4DIR |= ~(BIT4|BIT2);                 // set P4.2 & P4.4 as input pins
48  P2DIR |= BIT6; P2SEL0 |= BIT6;         // set P2.6 as TA0.3 PWM output pin
49  CSKEY = CSKEY_VAL; CSCTL0 = 0;         // unlock CS module for register access & reset tuning
                                           parameters
50  CSCTL0 = DCOEN|DCORSEL_2|DCOTUNE;// set DCO to 8 MHz (nominal, center of 4 ~ 8 MHz range)
51  CSCTL1 = SELS_3|SELM_3;                // select SMCLK = MCLK = DCO
52  CSKEY = 0;                             // lock the CS module
53  TA0CCTL3 = OUTMOD_3|CCIE;              // TA0_C3 compare mode, interrupt enabled
54  TA0CTL = TASSEL_2|MC_1|TACLR;          // TA0 = SMCLK input (8MHz), up mode, clear TA0R
55  TA0CCR0 = 40000;                       // load the period to TA0CCR0
56 }
57 void delay(int time) { int n; for(n = 0 ; n < time; n++); }
```

Figure 9.23 Complete code for the project DRATA0PWM12VMotor.

the codes in lines 33 and 34 set the minimum PWM duty cycle to 10 (40000–39990) when the **low** is greater than 39990. These two settings limit the motor to rotate between these two bounded speeds no matter how long you press either the **SW2** or the **SW3** buttons.

The codes in line 25 and 35 are important since they reset the **high** and the **low** to the initial values. Without these coding lines, the PWM duty cycle will keep in 99% (39990/40000) when the **SW2** button is pressed, and it will also keep in 0% (40000–40000) as the **SW3** button is pressed. The problem is that next time when the **SW3** button is pressed again, the **low** will start from 39990, not 1, to calculate its PWM cycle value, and this makes duty cycle as 0% immediately as the **SW3** button is pressed.

As the project runs, as the **SW2** button is pressed, the motor speed is going up and then keeps it in the maximum value. The motor speed will be going down as the **SW3** button is pressed until to 0. However, the motor will stop if both the **SW2** and the **SW3** button is released before the motor gets its maximum or minimum speed because of the instruction **TA0CTL &= ~0x30** under the selection statement **else**.

9.3.8 API functions used for Timer_A module

The MSPWare Peripheral Driver Library provides more than 25 API functions to support the project developments by using the Timer_A modules. Functions are provided to configure and control the timer, to modify timer/counter values, and to manage interrupt handling for the timer.

Totally four Timer_A modules, **TA0~TA3**, are available for the MSP432P401R MCU system. Also, each module can have up to five capture/compare registers, **TAxCCR0~TAxCCR4**, to perform capture and compare functions. Based on these modules and related capture/compare operations, these API functions can be divided into three major categories based on their functions:

1. Timer_A module (**TA0~TA3**) Configuration and Control
   a. Up mode
   b. Continuous mode
   c. Up–down mode
   d. Interrupt handling
2. Timer_A module five capture/compare Configuration and Control
   a. Each Module Capture/Compare Configuration and Control
   b. Each Module PWM Configuration and Control
   c. Each module capture/compare interrupt control
3. Clock System Configuration and Control

Since the **CS** is closely related to the Timer_A PWM module, some associate API functions are also included and discussed in this part.

All of these API functions are included in a C file **timer_a.c** and a header file **timer_a.h**, and both are located at the folder **C:\ti\msp\MSPWare_2_21_00_39\driverlib\driverlib\MSP432P4xx**.

In order to assist users to use these API functions to build sophisticated timer-related projects, some structs are built and provided in this library. Therefore, let's have a closer look at these structs in detail first.

9.3.8.1 Timer_A module-related structs

Six structs exist and are provided by this library:

1. struct_**Timer_A_UpModeConfig**
2. struct_**Timer_A_ContinuousModeConfig**
3. struct_**Timer_A_UpDownModeConfig**
4. struct_**Timer_A_CaptureModeConfig**
5. struct_**Timer_A_CompareModeConfig**
6. struct_**Timer_A_PWMConfig**

The details about these structs are listed below.

```
typedef struct _Timer_A_UpModeConfig
{
  uint_fast16_t clockSource;
  uint_fast16_t clockSourceDivider;
  uint_fast16_t timerPeriod;
  uint_fast16_t timerInterruptEnable_TAIE;
  uint_fast16_t captureCompareInterruptEnable_CCR0_CCIE;
  uint_fast16_t timerClear;
} Timer_A_UpModeConfig;

typedef struct _Timer_A_ContinuousModeConfig
{
  uint_fast16_t clockSource;
  uint_fast16_t clockSourceDivider;
  uint_fast16_t timerInterruptEnable_TAIE;
  uint_fast16_t timerClear;
} Timer_A_ContinuousModeConfig;
typedef struct _Timer_A_UpDownModeConfig
{
  uint_fast16_t clockSource;
  uint_fast16_t clockSourceDivider;
  uint_fast16_t timerPeriod;
  uint_fast16_t timerInterruptEnable_TAIE;
  uint_fast16_t captureCompareInterruptEnable_CCR0_CCIE;
  uint_fast16_t timerClear;
} Timer_A_UpDownModeConfig;

typedef struct _Timer_A_CaptureModeConfig
{
  uint_fast16_t captureRegister;
  uint_fast16_t captureMode;
  uint_fast16_t captureInputSelect;
  uint_fast16_t synchronizeCaptureSource;
  uint_fast8_t captureInterruptEnable;
  uint_fast16_t captureOutputMode;
} Timer_A_CaptureModeConfig;
typedef struct _Timer_A_CompareModeConfig
{
  uint_fast16_t compareRegister;
  uint_fast16_t compareInterruptEnable;
  uint_fast16_t compareOutputMode;
```

```
    uint_fast16_t compareValue;
} Timer_A_CompareModeConfig;
typedef struct _Timer_A_PWMConfig
{
    uint_fast16_t clockSource;
    uint_fast16_t clockSourceDivider;
    uint_fast16_t timerPeriod;
    uint_fast16_t compareRegister;
    uint_fast16_t compareOutputMode;
    uint_fast16_t dutyCycle;
} Timer_A_PWMConfig;
```

The data field **timerPeriod** is used in the **UpMode**, **UpDownMode**, and **PWM** mode structs. In fact, this value is used to set up the **TAxCCR0** Registers since the upper bound value of the timer is determined by the value in the **TAxCCR0** Registers when the timer is working in these modes.

*9.3.8.2 API functions for Timer_A module (**TA0~TA3**) CFGCTRL*

The following API functions are included in this group:

1. **Timer_A_configureUpMode()**—configures timer to work in the up mode
2. **Timer_A_configureContinuousMode()**—configures the timer to work in the continuous mode
3. **Timer_A_configureUpDownMode()**—configures the timer to work in the up–down mode
4. **Timer_A_startCounter()**—starts the timer to perform the normal count job
5. **Timer_A_getCounterValue()**—gets the current counter's counting value
6. **Timer_A_stopTimer()**—stops the timer and the count job
7. **Timer_A_clearTimer()**—resets the timer/counter, divider logic, and count direction
8. **Timer_A_registerInterrupt()**—registers an interrupt handler for the timer interrupt
9. **Timer_A_enableInterrupt()**—enables the timer interrupt
10. **Timer_A_getInterruptStatus()**—gets the current timer interrupt status
11. **Timer_A_disableInterrupt()**—disables the timer interrupt
12. **Timer_A_clearInterruptFlag()**—clears the timer **TAIFG** interrupt flag

Tables 9.8 and 9.9 show these Timer_A module configuration API functions and their parameters as well as available values to those parameters. Because of the similarity between the enable and disable functions, only enable function is shown in Tables 9.8 and 9.9.

When using these API functions, the following points should be noted:

- The **Timer_A_stopTimer()** and **Timer_A_clearTimer()** has different functions. The former is to stop the timer/count, but the latter only clears the timer/count. After the timer is re-enabled, it can continue its original count for the stop function. But for the clear function, it must restart its count from 0.
- The function **Timer_A_registerInterrupt()** only enables the global interrupt, but the specific Timer_A interrupts must be enabled via **Timer_A_enableInterrupt()**. Inside the interrupt handler, the **Timer_A_clearInterruptFlag()** must be used to clear the responded interrupt.

Table 9.8 Timer_A Module configuration and control API functions

API function	Parameter	Description
void **Timer_A_configureUpMode**(uint32_t timer, const Timer_A_UpModeConfig *config)	**timer** is the instance of the Timer_A module. Valid values can include: **TIMER_A0_MODULE** **TIMER_A1_MODULE** **TIMER_A2_MODULE** **TIMER_A3_MODULE** **config** is the configuration structure for Timer_A Up mode **clockSource**—valid values are: **TIMER_A_CLOCKSOURCE_EXTERNAL_TXCLK** [Default] **TIMER_A_CLOCKSOURCE_ACLK** **TIMER_A_CLOCKSOURCE_SMCLK** **TIMER_A_CLOCKSOURCE_INVERTED_EXTERNAL_TXCLK** **timerPeriod** is the specified Timer_A period. This is the value to be written into the **CCR0**. Limited to 16 bits **timerInterruptEnable_TAIE** is to enable or disable Timer_A interrupt. Valid values are: **TIMER_A_TAIE_INTERRUPT_ENABLE** and **TIMER_A_TAIE_INTERRUPT_DISABLE** [Default]	**clockSourceDivider** is the divider for clock source. Valid values are: **TIMER_A_CLOCKSOURCE_DIVIDER_1** [Default] **TIMER_A_CLOCKSOURCE_DIVIDER_2** **TIMER_A_CLOCKSOURCE_DIVIDER_4** **TIMER_A_CLOCKSOURCE_DIVIDER_8** **TIMER_A_CLOCKSOURCE_DIVIDER_3** **TIMER_A_CLOCKSOURCE_DIVIDER_5** **TIMER_A_CLOCKSOURCE_DIVIDER_6** **TIMER_A_CLOCKSOURCE_DIVIDER_7** **TIMER_A_CLOCKSOURCE_DIVIDER_10** **TIMER_A_CLOCKSOURCE_DIVIDER_12** **TIMER_A_CLOCKSOURCE_DIVIDER_14** **TIMER_A_CLOCKSOURCE_DIVIDER_16** **TIMER_A_CLOCKSOURCE_DIVIDER_20** **TIMER_A_CLOCKSOURCE_DIVIDER_24** **TIMER_A_CLOCKSOURCE_DIVIDER_28** **TIMER_A_CLOCKSOURCE_DIVIDER_32** **TIMER_A_CLOCKSOURCE_DIVIDER_40** **TIMER_A_CLOCKSOURCE_DIVIDER_48** **TIMER_A_CLOCKSOURCE_DIVIDER_56** **TIMER_A_CLOCKSOURCE_DIVIDER_64** **captureCompareInterruptEnable_CCR0_CCIE** is to enable or disable Timer_A CCR0 capture/compare interrupt. Valid values: **TIMER_A_CCIE_CCR0_INTERRUPT_ENABLE** and **TIMER_A_CCIE_CCR0_INTERRUPT_DISABLE** [Default] **timerClear** decides if Timer_A clock divider, count direction, count need to be reset. Valid values are: **TIMER_A_DO_CLEAR** **TIMER_A_SKIP_CLEAR** [Default value]
void **Timer_A_configureContinuousMode**(u`int32_t timer, const Timer_A_ContinuousModeConfig * config)	**timer** is the instance of the Timer_A module. Valid values can include: **TIMER_A0_MODULE** **TIMER_A1_MODULE** **TIMER_A2_MODULE** **TIMER_A3_MODULE** **config** is the configuration structure for Timer_A continuous mode	The struct parameters, **clockSource**, **timerClear**, **clockSourceDivider**, **timerInterruptEnable_TAIE**, are identical with those used in the above API function **Timer_A_configureUpMode**()

(Continued)

Table 9.8 (Continued) Timer_A module configuration and control API functions

API function	Parameter	Description
void **Timer_A_configureUpDownMode** (uint32_t timer, const Timer_A_UpDownMode-Config * config)	**timer** is the instance of the Timer_A module. Valid values can include: **TIMER_A0_MODULE** **TIMER_A1_MODULE** **TIMER_A2_MODULE** **TIMER_A3_MODULE** **config** is the configuration structure for Timer_A up–down mode	The struct parameters, **clockSource**, **timerClear**, **clockSourceDivider**, **timer-InterruptEnable_TAIE**, **captureCompa-reInterruptEnable_CCR0_CCIE** are identical with those used in the above API function **Timer_A_configureUpMode**()
void **Timer_A_start-Coun-ter**(uint32_t timer, uint_fast16_t timer-Mode)	**timer** is the instance of the Timer_A module. Valid values are identical with those used above	Select the timer mode. Valid values are: **TIMER_A_CONTINUOUS_MODE** [Default] **TIMER_A_UPDOWN_MODE** or **TIMER_A_UP_MODE**
uint16_t **Timer_A_get-CounterValue**(uint32_t timer)	**timer** is the instance of the Timer_A module. Valid values are identical with those used above	Return the current value of the specified timer. Note that reading the value of the counter is unreliable if the system clock is asynchronous from the timer clock
void **Timer_A_stopTimer**(uint32_t timer)	**timer** is the instance of the Timer_A module. Valid values are identical with those used above	Stop the timer Return: none
void **Timer_A_clearTimer**(uint32_t timer)	**timer** is the instance of the Timer_A module. Valid values are identical with those used above	Reset/clear the timer count, clock divider, count direction. Return: none

9.3.8.3 Timer_A module five capture/compare CFGCTRL API functions

The following API functions are included in this group:

1. `Timer_A_initCapture()`—initializes the timer to work in the capture mode
2. `Timer_A_initCompare()`—initializes the timer to work in the compare mode
3. `Timer_A_setCompareValue()`—sets the compare value to the **TAxCCRy** register
4. `Timer_A_getCaptureCompareCount()`—gets the current capture compare count
5. `Timer_A_generatePWM()`—creates a PWM with timer running in the up mode
6. `Timer_A_enableCaptureCompareInterrupt()`—enables the capture/compare interrupts
7. `Timer_A_getCaptureCompareInterruptStatus()`—gets the capture compare interrupt status
8. `Timer_A_clearCaptureCompareInterrupt()`—clears the capture-compare interrupt flag

Tables 9.10 and 9.11 show these API functions and their parameters as well as available values for those parameters.

Table 9.9 Timer_A module interrupt and control API functions

API function	Parameter	Description
void **Timer_A_registerInterrupt**(uint32_t timer, uint_fast8_t interruptSelect, void(*)(void) intHandler)	**timer** is the instance of the Timer_A module. Valid values are: **TIMER_A0_MODULE** **TIMER_A1_MODULE** **TIMER_A2_MODULE** **TIMER_A3_MODULE**	**interruptSelect** select which timer interrupt handler to register. Two interrupt handlers that can be registered: **TIMER_A_CCR0_INTERRUPT** interrupt for CCR0 **TIMER_A_CCRX_AND_OVERFLOW_INTERRUPT** interrupt for CCR1~4, and overflow interrupt **intHandler** is a pointer to the function to be called when the timer capture compare interrupt occurs
void **Timer_A_enableInterrupt**(uint32_t timer)	**timer** is the instance of the Timer_A module. Valid values are identical with those used above	Enable timer interrupt (TAIE) Return: none
uint32_t **Timer_A_getInterruptStatus**(uint32_t timer)	**timer** is the instance of the Timer_A module. Valid values are identical with those used above	Get timer interrupt status Return interrupt status. Valid values are: **TIMER_A_INTERRUPT_PENDING** **TIMER_A_INTERRUPT_NOT_PENDING**
void **Timer_A_clearInterruptFlag**(uint32_t timer)	**timer** is the instance of the Timer_A module. Valid values are identical with those used above	Clear the timer TAIFG interrupt flag Return: none

Regularly, the interrupt handling API functions for the Timer_A modules are different with those API functions used for Timer_A module capture/compare interrupt handling. Therefore, both clear functions, **`Timer_A_clearInterruptFlag()`** and **`Timer_A_clearCaptureCompareInterrupt()`**, should be called to clear the processed interrupt. The former is to clear the **`TAIFG`** in the **`TAxCTL`** register but the latter is to clear the **`CCIFG`** in the **`TAxCCTLy`** register.

9.3.8.4 API functions used for Clock System Configuration and Control

The following API functions are included in this group:

1. **`CS_initClockSignal()`**—initializes each of the clock signals
2. **`CS_setDCOCenteredFrequency()`**—sets the centered frequency of DCO operation
3. **`CS_tuneDCOFrequency()`**—tunes the DCO to a specific frequency
4. **`CS_setDCOFrequency()`**—automatically sets/tunes the DCO to the given frequency
5. **`CS_setExternalClockSourceFrequency()`**—sets the external clock sources LFXT and HFXT crystal oscillator frequency values

Table 9.10 Timer_A capture/compare configuration and control API functions

API Function	Parameter	Description
void **Timer_A_initCapture**(uint32_t timer, const Timer_A_CaptureMode-Config * config)	**timer** is the instance of the Timer_A module. Valid values can include: **TIMER_A0_MODULE** **TIMER_A1_MODULE** **TIMER_A2_MODULE** **TIMER_A3_MODULE** **config** is configuration structure for Timer_A capture mode **captureMode** is the capture mode selected. Valid values are: **TIMER_A_CAPTURE-MODE_NO_CAPTURE** [Default] **TIMER_A_CAPTURE-MODE_RISING_EDGE** **TIMER_A_CAPTURE-MODE_FALLING_EDGE** **TIMER_A_CAPTURE-MODE_RISING_AND_FALLING_EDGE** **synchronizeCapture-Source** decide if capture source should be synchro-nized with timer clock: **TIMER_A_CAPTURE_ASYNCHRONOUS** [Default] **TIMER_A_CAPTURE_SYNCHRONOUS**	**captureRegister**—select the capture register used Values are: **TIMER_A_CAPTURECOMPARE_REGISTER_0** **TIMER_A_CAPTURECOMPARE_REGISTER_1** **TIMER_A_CAPTURECOMPARE_REGISTER_2** **TIMER_A_CAPTURECOMPARE_REGISTER_3** **TIMER_A_CAPTURECOMPARE_REGISTER_4** **captureInputSelect**—decide the input capture signal: **TIMER_A_CAPTURE_INPUTSELECT_CCIxA** [Default] **TIMER_A_CAPTURE_INPUTSELECT_CCIxB** **TIMER_A_CAPTURE_INPUTSELECT_GND** **TIMER_A_CAPTURE_INPUTSELECT_Vcc** **captureInterruptEnable** is to enable or disable timer captureComapre interrupt. Valid values are: **TIMER_A_CAPTURECOMPARE_INTERRUPT_DISABLE** [Default] **TIMER_A_CAPTURECOMPARE_INTERRUPT_ENABLE** **captureOutputMode** specifies the output mode. Valid values are: **TIMER_A_OUTPUTMODE_OUTBITVALUE** [Default] **TIMER_A_OUTPUTMODE_SET** **TIMER_A_OUTPUTMODE_TOGGLE_RESET** **TIMER_A_OUTPUTMODE_SET_RESET** **TIMER_A_OUTPUTMODE_TOGGLE** **TIMER_A_OUTPUTMODE_RESET** **TIMER_A_OUTPUTMODE_TOGGLE_SET** **TIMER_A_OUTPUTMODE_RESET_SET**
void **Timer_A_initCompare**(uint32_t timer, const Timer_A_CompareMode-Config * config)	**timer** is the instance of the Timer_A module. Valid values are identical with those used above **config** is the configura-tion structure for Timer_A compare mode	**compareRegister**—select the compare register used. Values are identical with those used in **captureRegister** in above API function **compareInterruptEnable** is to enable or disable timer captureComapre interrupt. Valid values are same as those used in the **captureIn-terruptEnable** above **compareOutputMode** specifies the output mode. Valid values are identical with those used in the **captureOutputMode** above

(Continued)

Table 9.10 (Continued) Timer_A capture/compare configuration and control API functions

API Function	Parameter	Description
void **Timer_A_setCompareValue**(uint32_t timer, uint_fast16_t compareRegister, uint_fast16_t compareValue)	**timer** is the instance of the Timer_A module. Valid values are identical with those used above	**compareRegister**—select compare register being used. Valid values are identical with those used in the **captureRegister** above **compareValue** is the count to be compared with in compare mode
uint_fast16_t **Timer_A_getCaptureCompareCount** (uint32_t timer, uint_fast16_t capcomRegister)	**timer** is the instance of the Timer_A module. Valid values are identical with those used above	**capcomRegister** select the capture register being used. Valid values are identical with those used in the **captureRegister** in **Timer_A_initCapture()**
void **Timer_A_generatePWM** (uint32_t timer, const Timer_A_PWMConfig *config)	**timer** is the instance of the Timer_A module. Valid values are identical with those used above **config**—configuration structure for Timer_A PWM mode **compareOutputMode** specifies the output mode. Valid values are identical with those used in the **captureOutputMode** above	**clockSource**—valid values are identical with those used in the **Timer_A_configureUpMode** above **clockSourceDivider**—valid values are identical with those used in the **Timer_A_configureUpMode** above **timerPeriod** select the desired timer period **compareRegister**—selects the compare register used. Valid values are identical with those used in the **captureRegister** above **dutyCycle**—specifies the duty cycle for the generated waveform

Table 9.12 lists these API functions. When using these functions to set up or tune the clock sources, the following points should be noted:

1. The `CS_setDCOCenteredFrequency()` and the `CS_tuneDCOFrequency()` functions should be used together to set up a desired DCO frequency. The argument for the first function is the center frequency of the DCO, but the argument for the second function should be a calculated **DCOTUNE** value based on Equation 9.1 listed above.
2. The function `CS_setDCOFrequency()` can be used individually to set up a desired DCO target frequency. The argument for this function is the desired target frequency (Hz) users want to set up and use in the program.
3. The function `CS_setExternalClockSourceFrequency()` is needed only when an external crystal LFXT or HFXT is used.

Now, let's build an example project to illustrate how to use these API functions to perform some sophisticated Timer_A tasks.

Table 9.11 Timer_A Module capture/compare interrupt and control API functions

API Function	Parameter	Description
void **Timer_A_enableCaptureCompareInterrupt**(uint32_t timer, uint_fast16_t capcomRegister)	**timer** is the instance of the Timer_A module. Valid values are: **TIMER_A0_MODULE** **TIMER_A1_MODULE** **TIMER_A2_MODULE** **TIMER_A3_MODULE**	Enable capture compare interrupt **capcomRegister**—select the capture/compare register being used. Valid values are identical with those used in the **captureRegister** in **Timer_A_initCapture()**
uint32_t **Timer_A_getCaptureCompareInterruptStatus**(uint32_t timer, uint_fast16_t capcomRegister, uint_fast16_t mask)	**timer** is the instance of the Timer_A module. Valid values are identical with those used above **mask** is the mask for the interrupt status. Mask value is the logical OR of any of the following: **TIMER_A_CAPTURE_OVERFLOW** **TIMER_A_CAPTURECOMPARE_INTERRUPT_FLAG**	Return capture compare interrupt status **capcomRegister**—select the capture/compare register being used. Valid values are identical with those used in the **captureRegister** in **Timer_A_initCapture()** Return: the mask of the set flags. Valid values is OR of: **TIMER_A_CAPTURE_OVERFLOW** **TIMER_A_CAPTURECOMPARE_INTERRUPT_FLAG**
void **Timer_A_clearCaptureCompareInterrupt**(uint32_t timer, uint_fast16_t capcomRegister)	**timer** is the instance of the Timer_A module. Valid values are identical with those used above	Clear the capture-compare interrupt flag **capcomRegister**—select the capture/compare register being used. Valid values are: **TIMER_A_CAPTURECOMPARE_REGISTER_0** **TIMER_A_CAPTURECOMPARE_REGISTER_1** **TIMER_A_CAPTURECOMPARE_REGISTER_2** **TIMER_A_CAPTURECOMPARE_REGISTER_3** **TIMER_A_CAPTURECOMPARE_REGISTER_4**

9.3.8.5 Implementation of using Timer_A API functions to measure PWM pulses

In this section, we try to use some API functions discussed above to measure the width of a PWM pulse generated by Timer_A module 0, **TA0.3** with Timer_A module 2, **TA2.4**. The principle of pulse width measurement is to generate and capture two interrupts on **TA2_C4** for both rising and falling edge events. Each event captures a timer value and the difference between two timer values is the width of the PWM pulse.

As we discussed in the previous sections, the resolution and the range of the PWM width are determined by the frequency of the clock source and the length of the counter used. For example, when a 16-bit timer with a 16 MHz clock source, the resolution is 62.5 ns and the range is about 4.096 ms.

Table 9.12 API functions used for CS configuration and control

API function	Parameter	Description
void **CS_initClockSignal**(uint32_t selectedClockSignal, uint32_t clockSource, uint32_t clockSourceDivider)	**selectedClockSignal**—clock signal to initialize: **CS_ACLK**, **CS_MCLK**, **CS_HSMCLK** **CS_SMCLK, CS_BCLK** **clockSourceDivider**—this parameter is ignored when setting BLCK: **CS_CLOCK_DIVIDER_1** **CS_CLOCK_DIVIDER_2** **CS_CLOCK_DIVIDER_4** **CS_CLOCK_DIVIDER_8** **CS_CLOCK_DIVIDER_16** **CS_CLOCK_DIVIDER_32** **CS_CLOCK_DIVIDER_64** **CS_CLOCK_DIVIDER_128**	Initialize each of the clock signals. The user must ensure that this function is called for each clock signal. If not, the default state is assumed for the particular clock signal **clockSource**—clock source for the selectedClockSignal signal: **CS_LFXTCLK_SELECT** **CS_HFXTCLK_SELECT** **CS_VLOCLK_SELECT**, [Not available for BCLK] **CS_DCOCLK_SELECT**, [Not available for ACLK, BCLK] **CS_REFOCLK_SELECT** **CS_MODOSC_SELECT**, [Not available for ACLK, BCLK]
void **CS_setDCOCentered Frequency**(uint32_t dcoFreq)	**dcoFreq** selects between the valid frequencies: **CS_DCO_FREQUENCY_1_5**, [1 to 2 MHz] **CS_DCO_FREQUENCY_3**, [2 to 4 MHz] **CS_DCO_FREQUENCY_6**, [4 to 8 MHz] **CS_DCO_FREQUENCY_12**, [8 to 16 MHz] **CS_DCO_FREQUENCY_24**, [16 to 32 MHz] **CS_DCO_FREQUENCY_48** [32 to 64 MHz]	Sets the centered frequency of DCO operation CS_tuneDCOFrequency function
void **CS_tuneDCOFrequency**(int16_t tuneParameter)	**tuneParameter**—tuning parameter in 2's compliment representation Refer to Equation 9.1 to get this value	Tune the DCO to a specific frequency. Tuning of the DCO is based on Equation 9.1
void **CS_setDCOFrequency**(uint32_t dcoFrequency)	**dcoFrequency**—frequency in Hz that the user wants to set the DCO to	Automatically set/tune the DCO to the given frequency. Any valid value up to max frequency in the spec can be given to this function and the API will do its best to determine the correct tuning parameter
void **CS_setExternalClockSource Frequency**(uint32_t lfxt_XT_CLK_frequency, uint32_t hfxt_XT_ CLK_frequency)	**lfxt_XT_CLK_frequency** is the LFXT crystal frequencies in Hz **hfxt_XT_CLK_frequency** is the HFXT crystal frequencies in Hz	Set the external clock sources LFXT and HFXT crystal oscillator frequency values. This function must be called if an external crystal LFXT or HFXT is used

As we know, for rising edge detections, the input signal must be in high for at least two system clock periods following the rising edge. Similarly, for falling edge detections, the input signal must be in low for at least two system clock periods following the falling edge. Therefore, the maximum input signal frequency for edge detection is one-fourth of the system clock frequency. For a 16 MHz clock source, the maximized input frequency should be 4 MHz.

If we need to consider the time period spent for interrupt handling to record the current timer count for either the rising or the falling edge, the maximized input signal frequency should be further lower than 4 MHz. Given some reservations, the maximized input signal frequency should be about 2~3 MHz.

In this project, we use a 16-bit timer, **TA0** with **TA0_C3** (**TA0.3**) pin to generate and output a PWM waveform with a fixed duty cycle 70%, and detect both the rising and the falling edges of that PWM signal by using another 16-bit timer, **TA2** with **TA2_C4** (**TA2.4**) to measure its pulse width. The clock source to the timers is the 16 MHz DCO.

Actual period and duty cycle values depend on the PWM clock source used in the timer. For example, for a 16 MHz PWM clock source, the clock period is **62.5** ns. To use this clock source to generate a PWM signal with 0.5 ms as period and 70% duty cycle on **TA0_C3** or **TA0.3**, the following parameters and registers should be used:

1. The length of the high pulse = 0.5 ms × 70% = 0.35 ms, the length of the low pulse = 0.5 – 0.35 = 0.15 ms
2. **TA0CCR0** = Period = 0.5 ms/62.5 ns = 500000/62.5 = **8000**
3. **TA0CCR3** = Length of low pulse = 0.15 ms/62.5 ns = 150000/62.5 = **2400**

In order to use the Timer_A module **TA0** to generate the above PWM signal output, the following steps are needed to configure the **TA0** module:

1. The input clock signal to the **TA0** module should be selected as **SMCLK** and it needs to be equal to the clock source DCO. The DCO needs to be configured by tuning it to output a 16 MHz clock.
2. The **TA0** will work in the up-count mode. But this set up step can be skipped if the API function **Timer_A_generatePWM()** is used since by default the up-count mode is used by that PWM generation API function.
3. The **TA0_C3** (**TA0.3**) works in the compare mode and the **TA0CCR0** Register is set to **8000** and **TA0CCR3** is set to **2400**. The output mode should be set to 3 (**OUTMOD** = 3).

All of the above parameters can be added into the PWM struct **Timer_A_PWMConfig** to complete this set up and configuration process. For example, the first parameter **clockSource** can be assigned as **TIMER_A_CLOCKSOURCE_SMCLK**, the third and the sixth parameters **timerPeriod** and **dutyCycle** can be assigned by **8000** and **2400**, and the fifth parameter **compareOutputMode** can be assigned as the macro **TIMER_A_OUTPUTMODE_SET_RESET**.

The Timer_A module 2, **TA2** will work in the continuous mode to detect both rising and the falling edges of the PWM signal created above. The following setups are needed to configure the TA2 module:

1. The input clock to the **TA2** module **SMCLK** needs to be configured by tuning the DCO to output a 16 MHz clock.

2. The **TA2** will work in the continuous mode.
3. The **TA2_C4**, **TA2.4** works in the capture mode with capture interrupt enabled.

These setups and configurations can be completed by assigning related parameters to the structs **Timer_A_ContinuousModeConfig** and **Timer_A_CaptureModeConfig**, respectively.

When the **TA2R** counter gets the upper bound value **0xFFFF**, it will reset to 0 and continue for the next counting operation until the **TACLR** bit is set.

Since this project is a little complicated, therefore both a header file **SDPWMPulse.h** and a source file **SDPWMPulse.c** are used for this project **SDPWMPulse**.

Since we need to use **TA2.4** (**P6.7** pin) to capture the rising and the falling edges of a PWM signal generated at **TA0.3** (**P2.6** pin), therefore we need to connect these two GPIO pins together by using a jumper wire. Both pins are located at the J4 connector on the MSP-EXP432P401R EVB. Make sure to connect these two pins together before the project can be run and the measured result can be observed.

The code for the header file is shown in Figure 9.24.

Some project-used structs, macros, and variables are declared in this header file. All of these structs and macros can be found from the related system header files. To save compiling time and space, we moved those macros into this header file. The **TA2_C4** or **TA2.4** interrupt handler, **Timer_A2_IRQHandler()**, is declared in line 12, three global unsigned integer variables, **period**, **edgeCount**, and **count**, as well as one global Boolean variable **intFlag**, are also declared in this file.

The **8000** and **2400** are assigned to the **timerPeriod** and the **dutyCycle** in the **pwmConfig** struct to enable the **TA0.3** to generate a 70% duty cycle PWM signal. The output mode is set to mode 3 with a system macro **TIMER_A_OUTPUTMODE_SET_RESET**.

For the **TA2** module, it is configured as a continuous mode with the **continuous-ModeConfig** struct. The timer interrupt is disabled by using the macro **TIMER_A_TAIE_INTERRUPT_DISABLE**. The TA2.4 is configured as the capture mode with the **captureModeConfig** struct. The capture register is **TA2_C4** that can detect both the rising and the falling edge with the **CCI2A** as the capture input. This capture interrupt is enabled since we need to use the interrupt mechanism to reserve the detected edges to calculate the pulse width of the PWM signal.

The complete source codes for the C file **SDPWMPulse.c** are shown in Figure 9.25.

Let's have a closer look at this piece of code to see how it works.

1. The CS API function **CS_setDCOFrequency(16000000)** is executed in line 9 to automatically set and tune the DCO as 16 MHz frequency clock source. The argument of this function must be measured in Hz.
2. In code line 10, the CS API function **CS_initClockSignal()** is called to set the clock input signal to the timers as **SMCLK**, clock source as **DCOCLK**, and clock divider as **1**.
3. The GPIO API function **GPIO_setAsPeripheralModuleFunctionOutputPin()** is called in line 11 to configure the GPIO pin **P2.6** to work as an PWM output pin for **TA0.3**. Refer to Table 7.9 in Chapter 7 to get more details about this primary function configuration.
4. In code line 13, the GPIO API function **GPIO_setAsPeripheralModuleFunctionInputPin()** is used to set up the GPIO pin **P6.7** as the **TA2_CCI4A** capture input pin. Refer to Table 7.4 in Chapter 7 to get more details about this primary function configuration.

```
1  //****************************************************************************
2  //SDPWMPulse.h - Header file used for the project SDPWMPulse - TA0.3 = PWM
3  //****************************************************************************
4  #include <stdint.h>
5  #include <stdbool.h>
6  #include <MSP432P4xx\timer_a.h>
7  #include <MSP432P4xx\wdt_a.h>
8  #include <MSP432P4xx\gpio.h>
9  #include <MSP432P4xx\cs.h>
10 bool intFlag = true;
11 uint16_t  period, edgeCount, count = 0;
12 void Timer_A2_IRQHandler(void);
13 Timer_A_PWMConfig pwmConfig =
14 {
15  TIMER_A_CLOCKSOURCE_SMCLK,
16  TIMER_A_CLOCKSOURCE_DIVIDER_1,
17  8000,
18  TIMER_A_CAPTURECOMPARE_REGISTER_3,
19  TIMER_A_OUTPUTMODE_SET_RESET,
20  2400
21 };
22 Timer_A_ContinuousModeConfig continuousModeConfig =
23 {
24  TIMER_A_CLOCKSOURCE_SMCLK,                    // SMCLK Clock Source
25  TIMER_A_CLOCKSOURCE_DIVIDER_1,                // SMCLK/1 = 16MHz
26  TIMER_A_TAIE_INTERRUPT_DISABLE,               // disable Timer ISR
27  TIMER_A_DO_CLEAR                              // clear Counter
28 };
29 Timer_A_CaptureModeConfig captureModeConfig =
30 {
31  TIMER_A_CAPTURECOMPARE_REGISTER_4,            // CC2 Register 4
32  TIMER_A_CAPTUREMODE_RISING_AND_FALLING_EDGE,  // rising & falling Edges
33  TIMER_A_CAPTURE_INPUTSELECT_CCIxA,            // CCIxA Input Select
34  TIMER_A_CAPTURE_SYNCHRONOUS,                  // synchronized Capture
35  TIMER_A_CAPTURECOMPARE_INTERRUPT_ENABLE,      // enable interrupt
36  TIMER_A_OUTPUTMODE_OUTBITVALUE                // output bit value
37 };
```

Figure 9.24 Code for the header file SDPWMPulse.h.

5. The code lines 15 and 16 are used to set up the GPIO pin **P1.0** as an output pin and to output low since we need to use this pin that is connected to a red color LED to indicate the working status of our timers.
6. The API function **Timer_A_generatePWM()** is executed in line 17 to configure the **TA0.3** to generate a PWM signal with the help of the **pwmConfig** struct.
7. In line 18, the API function **Timer_A_configureContinuousMode()** is called to set up the **TA2** to work as the continuous mode.
8. The API function **Timer_A_initCapture()** is utilized in line 19 to configure the **TA2_C4** to work as a capture input pin to capture both rising and falling edges of a

```
 1 //*****************************************************************************
 2 // SDPWMPulse.c - Main file used for the project SDPWMPulse - TA0.3 = PWM
 3 //*****************************************************************************
 4 #include "SDPWMPulse.h"
 5 int main(void)
 6 {
 7   uint16_t Period[1];              // local data array used to store the PWM pulse width
 8   WDT_A_holdTimer();               // stop the watchdog timer
 9   CS_setDCOFrequency(16000000);
10   CS_initClockSignal(CS_SMCLK, CS_DCOCLK_SELECT, CS_CLOCK_DIVIDER_1);
11   GPIO_setAsPeripheralModuleFunctionOutputPin(GPIO_PORT_P2, GPIO_PIN6,
12                                    GPIO_PRIMARY_MODULE_FUNCTION);
13   GPIO_setAsPeripheralModuleFunctionInputPin(GPIO_PORT_P6, GPIO_PIN7,
14                                    GPIO_PRIMARY_MODULE_FUNCTION);
15   GPIO_setAsOutputPin(GPIO_PORT_P1, GPIO_PIN0);
16   GPIO_setOutputLowOnPin(GPIO_PORT_P1, GPIO_PIN0);
17   Timer_A_generatePWM(TIMER_A0_MODULE, &pwmConfig);
18   Timer_A_configureContinuousMode(TIMER_A2_MODULE, &continuousModeConfig);
19   Timer_A_initCapture(TIMER_A2_MODULE, &captureModeConfig);
20   Timer_A_registerInterrupt(TIMER_A2_MODULE, TIMER_A_CCRX_AND_OVERFLOW_INTERRUPT,
21                                    Timer_A2_IRQHandler);
22   Timer_A_enableCaptureCompareInterrupt(TIMER_A2_MODULE,
23                                    TIMER_A_CAPTURECOMPARE_REGISTER_4);
24   Timer_A_startCounter(TIMER_A2_MODULE, TIMER_A_CONTINUOUS_MODE);
25   while(intFlag) {}                // wait for the PWM pulse width to be calculated
26   Period[0] = period;              // collect pulse width value (16-bit)
27   Period[0] = Period[0] + 0;       // useless at all, just to avoid the compiling warning.
28   while(1);                        // make a break point at this line to check the result.
29 }
30 void Timer_A2_IRQHandler(void)    // TA2.4 interrupt handler
31 {
32   Timer_A_clearCaptureCompareInterrupt(TIMER_A2_MODULE,
33                                          TIMER_A_CAPTURECOMPARE_REGISTER_4);
34   count++;
35   if (count == 1)
36     edgeCount = Timer_A_getCaptureCompareCount(TIMER_A2_MODULE,
37                                                  TIMER_A_CAPTURECOMPARE_REGISTER_4);
38   else
39   {
40     period = Timer_A_getCaptureCompareCount(TIMER_A2_MODULE,
41                                       TIMER_A_CAPTURECOMPARE_REGISTER_4) - edgeCount;
42     Timer_A_stopTimer(TIMER_A2_MODULE);
43     GPIO_toggleOutputOnPin(GPIO_PORT_P1, GPIO_PIN0);
44     intFlag = false;
45   }
46 }
```

Figure 9.25 Complete C source code for the project SDPWMPulse.

PWM signal coming to **P6.7**. The capture interrupt mechanism is enabled to detect and calculate the detected edges and the pulse width of the PWM signal inside the interrupt handler.

9. In line 20, the function **Timer_A_registerInterrupt()** is executed to register an interrupt handler **Timer_A2_IRQHandler()** for the capture interrupts.
10. This capture interrupt is enabled by calling the **Timer_A_enableCaptureCompareInterrupt()** function in line 22.
11. The **TA2** module is started to work as the continuous mode by calling the API function **Timer_A_startCounter()** in line 24.

12. A conditional **while(intFlag)** loop starts at line 25 to wait for the PWM pulse width to be calculated. By default, the **intFlag** is set to **true** at the beginning of this program. Therefore, this **while()** loop will keep its waiting until the **intFlag** becomes **false**.
13. If the condition **intFlag** becomes **false**, the calculated PWM pulse width **period** is collected and assigned to the local array **Period[]** in line 26.
14. The code line 27 is a useless line and the only reason is to avoid a compiling warning.
15. Another infinitive **while()** loop starts at line 28. A break point should be set at this line before this project can be run to enable us to check the running and measuring result of the project, the pulse width of the PWM signal, from the **Call Stack + Locals** window after the project runs.

The following codes are for the **TA2.4** interrupt handler.

16. The Timer_A module 2 Capture 4 (**TA2_C4**) interrupt handler starts at line 30.
17. First this interrupt is cleared by calling the **Timer_A_clearCaptureCompareInterrupt()** API function in line 32 to enable this interrupt to be generated in the future. This step is critical and the same interrupt cannot be generated if the current interrupt has not been cleared.
18. The global variable **count** is incremented by 1 in line 34 (this variable is 0 as it is declared at the beginning of this program).
19. If the **count** is equal to 1, which means that this is the first rising edge of the PWM signal, the API function **Timer_A_getCaptureCompareCount()** is executed to pick up the current count on the **TA2R** register and reserve it to the global variable **edgeCount** in line 36.
20. Otherwise if the **count** is greater than 1, which means that it is equal to 2 and this is the first falling edge of the PWM signal, the difference between the current count in the **TA2R** register (**Timer_A_getCaptureCompareCount()**) and the previous count stored in the **edgeCount** is the pulse width of the measured PWM signal, which is reserved and assigned to the global variable **period** in line 40.
21. In line 42, the API function **Timer_A_stopTimer()** is called to stop the **TA2** timer.
22. Then the function **GPIO_toggleOutputOnPin()** is executed to toggle the red color LED via GPIO pin **P1.0** to indicate the end of the program in line 43.
23. Finally, in line 44, the global Boolean variable **intFlag** is reset to **false** to release the conditional **while(intFlag)** loop in the **main()** program to collect the measured result.

Now we need to set up the environment to build and run the project. Perform the following operations to finish these setups:

- Go to **C/C++** tab in the **Project|Options for Target "Target 1"** to set up the Include Path as: **C:\ti\msp\MSPWare_2_21_00_39\driverlib\driverlib**
- Add the MSPPDL into this project by right clicking on the **Source Group 1** folder in the **Project** panel on the left, and select the **Add Existing Files to Group "Source Group 1"** menu item. On the opened box, select the driver library file path **C:\ti\msp\MSPWare_2_21_00_39\driverlib\driverlib\MSP432P4xx\keil** and then select the driver library file **msp432p4xx_driverlib.lib**. Click the **Add** button to add it into the project

- Select the correct debugger by clicking on the **Debug** tab and select the **CMSIS-DAP Debugger** from the **Use** box. Also make sure that all settings for the debugger and the flash download are correct by clicking on the Settings button

Before running the project, make sure that a break point has been set in the infinitive **while()** loop at line 28. Now build, download, and run the project.

After the project runs and halts at the **while()** line, on the **Call Stack + Locals** window, expand the **Period[]** array and the pulse width of the detected PWM signal, **0x15E1**, is displayed. Convert this to decimal value, which is **5601**. This is the count difference between a rising and a falling edge of the PWM signal. Multiple this number by the period of the input clock (16 MHz), 62.5 ns or 0.0625 μs, the resulted pulse width is 0.35 ms. This is identical with our target duty cycle.

9.4 Summary

This chapter provides general information about MSP432P401R Microcontroller Timer_A modules programming. The discussion also includes the GPIO ports related to Timer_A modules used in the MSP432P401R MCU system. This discussion concentrates on the architectures and programming interfaces applied on Timer_A modules. Special or alternative control functions for different GPIO pins related to different peripherals, including Timer_A modules, are also discussed and introduced with example projects in this chapter.

Starting at Section 9.2, the Timer_A architecture and functional block diagram implemented in the MSP432P401R MCU system are introduced and discussed.

The Timer_A Module (**TA**) contains:

- Four 16-bit Timer_A modules or blocks **TA0**, **TA1**, **TA2**, and **TA3**.
- Each 16-bit module block can provide a 16-bit timer/counter **TAxR** (**x** = 0~3) and five capture/compare registers, **TAx_CCR0~TAx_CCR4**.
- Each module can be further configured to operate independently as timers or event edge counters, or event timers.
- Each 16-bit timer module can work in three modes, up-count mode, continuous mode, and up–down mode.
- Each module has timer overflow interrupt, capture on rising, falling, or both edges interrupt and compare equal interrupt to drive-related interrupt handler to process those corresponding interrupt.

Following the architecture and block diagram, the operations and the operational modes of the Timer_A blocks are discussed and analyzed in Section 9.3. These operational modes include:

1. Timer up-count mode
2. Timer continuous mode
3. Timer up–down mode
4. Input edge-count mode
5. Input edge-time mode
6. PWM mode

The Timer_A module GPIO-related control signals are introduced in Section 9.3.5. Some popular and important implementations of using those timer modules are discussed in detail in Sections 9.3.6~9.3.7.

The discussion about the change of the DCO frequency and range is given in Section 9.3.7.3.1. This discussion is very important to apply higher or different input clock to the timer modules.

The API functions provided by the MSPWare Peripheral Driver Library are discussed in Section 9.3.8 and an example project of using those Timer_A module API functions to measure PWM pulses is provided in Section 9.3.8.5.

HOMEWORK

I. True/False Selections

_____1. In MSP432P401R MCU, the Timer_A has five modules arranged from **TA0** to **TA4**.

_____2. In the MSP432P401R MCU system, each Timer_A module can have up to four capture/compare registers.

_____3. The core of each Timer_A module is a 16-bit free-running counter **TAxR** that can work in either stop, up, up–down, or continuous mode.

_____4. The **TA0.2** can play different functions. When working as an input, it is the **TA0_C2** input capture pin, but when working as an output, it is the **TA0.2** compare output pin.

_____5. Any of four Timer_A modules can generate a timer overflow interrupt when the bit **TAIE** is set in the **TAxCTL** register.

_____6. To stop a Timer_A module, one needs to set the **TACLR** bit in the **TAxCTL** register.

_____7. Each of five Timer_A module capture/compare registers, **TAxCCRy**, can be controlled and configured by the corresponding **TAxCCTLy** (**y** = 0~4) register.

_____8. The bit field **OUTMOD** in the **TAxCCTLy** register is used to configure the output mode for the capture register.

_____9. By configuring the **TAxCCR0** and the **TAxCCRy** (**y** ≠ 0) registers, different PWM signals with varied periods and duty cycles can be generated.

____10. If the bit **CCIE** in the **TA2CCTL3** Register is set, a **TA2.3** capture or a compare interrupt can be generated and the bit **CCIFG** is set when an edge is captured by the **TA2_C3** Register or the count in the **TA2R** is equal to the value in the **TA2.3** Register.

II. Multiple Choices

1. Before any Timer_A module can be configured, it must be ________ by ________________.
   a. Initialized, configuring it
   b. Set, setting the **TACLR** bit
   c. Reset, resetting the **TACLR** bit
   d. Reset, setting the **TACLR** bit
2. To stop a Timer_A module, the bit field _______ should be configured as ___________________.
   a. TACLR, 1
   b. TACLR, 0
   c. MC, 00
   d. MC, 11
3. In order to configure the **TA0** module to use the **SMCLK** as the input clock to the timer with 1.5 MHz, the **TASSEL** and **ID** bits in the **TA0CTL** register should be set to ________________.
   a. 0x0, 0x1

   b. 0x1, 0x1
   c. 0x2, 0x1
   d. 0x3, 0x1
4. In order to use **TA2.4** to capture rising edges of an input signal on **CCI2A**, the bit field **CM**, **CCIS**, and **CAP** in the __________ register must be set to ____________.
   a. **TA2CCTL4**, 00, 00, 0
   b. **TA2CCTL4**, 01, 00, 1
   c. **TA2CCTL4**, 01, 01, 1
   d. **TA2CCTL4**, 01, 02, 1
5. After a timer overflow interrupt has been responded, inside the interrupt handler, that interrupt must be _______ bit in the _____________ register.
   a. Enabled by setting the **TAIE**, **TAxCTL**
   b. Disabled by resetting the **TAIE**, **TAxCTL**
   c. Set by setting the **TAIFG**, **TAxCTL**
   d. Cleared by resetting the **TAIFG**, **TAxCTL**
6. After a timer compare interrupt has been responded, inside the interrupt handler, that interrupt must be __________ bit in the __________ register.
   a. Enabled by setting the **CCIE**, **TAxCCTLy**
   b. Disabled by resetting the **CCIE**, **TAxCCTLy**
   c. Set by setting the **CCIFG**, **TAxCCTLy**
   d. Cleared by resetting the **CCIFG**, **TAxCCTLy**
7. If the input clock to the **TA0** is the 16-MHz system clock and the **TA0** works as a 16-bit counter, the maximum period to be measured is _________ ms.
   a. 4.096
   b. 40.960
   c. 10.486
   d. 104.857
8. When an 8-MHz clock is applied to a timer module **TA3**, what is the maximum period of a PWM signal can be implemented? What is the value applied in the **TA3CCR0** Register?
   a. 8.192 ms, 65536
   b. 4.096 ms, 65535
   c. 4.096 ms, 65536
   d. 8.192 ms, 65535
9. When a 24 MHz **SMCLK** is applied to a timer module **TA1** working as the up mode and **TA1_C1** (**TA1.1**) working as output mode 7. The period of the desired PWM signal is 1 ms with a duty cycle of 25%, what values should be applied to **TA1CCR0** and **TA1CCR1**?
   a. 24000, 6000
   b. 2400, 600
   c. 12000, 3000
   d. 1200, 300
10. When the timer module **TA2** works in the continuous mode and the **TA2_C3** Register is used to measure the period of an input signal, the **TA2_C3** should be in ____________________ mode.
   a. Compare
   b. Up–down
   c. Capture
   d. Up

III. Exercises
1. Provide a brief description about the Timer_A module components.
2. Provide a brief description about the operation modes of the Timer_A modules.
3. Provide a brief description about the timer's operational procedure when it works in the capture/compare mode.
4. Provide a brief description about the timer's operational procedure when it works in the PWM mode.
5. Provide a brief explanation about dual functions on the same signal pin, such as **TAx_Cy** (**TAx.y**), used in the Timer_A module.

IV. Practical Laboratory

LABORATORY 9 MSP432 16-BIT TIMERS AND PWM MODULES PROGRAMMING

9.0 GOALS

This laboratory exercise allows students to learn and practice ARM® Cortex®-M4 General Timers, Watchdog timers, and USB programming by developing five labs.

1. Program **Lab9_1(TA3OVInt)** lets students use the Timer_A module 3 to build a project to periodically generate overflow interrupt and display the overflow times via four red color LEDs, **PB0~PB3** in EduBASE ARM Trainer.
2. Program **Lab9_2(TA1Toggle)** enables students to use three Timer_A module captures, **TA1CCR0**, **TA1CCR1**, and **TA1CCR2** as well as **TA1.1** overflow to generate four independent interrupt intervals. Inside the first three ISR, the interval is increased by different lengths to get the incremented interrupt intervals. The output mode (**OUTMOD**) is set to toggle for all of these timer modules. When these modules work as continuous modes with the input clock as the **ACLK** = 32768 Hz, different toggle frequencies can be obtained.
3. Program **Lab9_3(CaptureEdge)** enables students to use the Timer_A module 2, or TA2_C2 (**TA2.2**) to capture rising edges of an external periodic signal with the capture mode. In this project, the external signal is connected to both the **TA2_C2** capture input as the capture input and the clock input **TA2CLK** to the 16-bit timer/counter **TA2**.
4. Program **Lab9_4(CapAvePeriod)** enables students to use **TA2_C3** capture mode to capture and measure an average period of an external periodic signal. This program is similar to a class project **DRATA2Period**, so students can refer to and modify that project to build this project.
5. Program **Lab9_5(CaptureWidth)** enables students to capture both rising and falling edges to measure the pulse width of an external periodic signal. The Timer_A module 3, **TA2_C3**, is used to perform the capture function.
6. Program **Lab9_6(CaptureWidthHF)** is a modification to the program **Lab9_5(CaptureWidth)** to use a higher clock input to the timer by setting the SMCLK = MCLK = DCO = 24 MHz. With this higher input clock rate, the upper bound of frequency of the input signal that can be measured is extended to 70 KHz.
7. Program **Lab9_7(SDPWM)** enables students to use **SW2** (**P1.4**) to build a vary-duty-cycle PWM project via the API functions provided by the MSPPDL.

After completion of these programs, students should understand the basic architecture and operational procedure for most popular Timer_A modules applied in the

MSP432P401R MCU system. They should be able to code some sophisticated programs to access the desired peripheral devices to perform desired control operations via timer/counter blocks.

9.1 LAB9_1

9.1.1 Goal

In this project, students will use the DRA model with the Timer_A module 3 to build a project to periodically generate overflow interrupts and display the overflow times via four red color LEDs, **PB0~PB3** in EduBASE ARM Trainer

This lab project is very similar to one class project **DRATA2OVInt**. Refer to that project and make some necessary modifications on that project to complete this one. The difference between this lab project and that class project **DRATA2OVInt** includes:

- The Timer_A module 3, **TA3** and its overflow is used in this project, not **TA2**.
- An internal divider is used with a dividing factor of 8 to reduce the **SMCLK** frequency.
- The related **TA3** interrupt handler should be used, not TA2 interrupt handler.
- The configuration parameters for the **NVIC_IPR3** and the **NVIC_ISER0** should be modified to make them match to the **TA3** interrupt source.

Keep those in mind and now let's build this project.

9.1.2 Data assignment and hardware configuration

No hardware connection is needed in this project. Four red color LEDs, PB3~PB0, in the EduBASE ARM Trainer are connected to four GPIO pins, **PB0** → **P3.2**, **PB1** → **P3.3**, **PB2** for **P2.5** and **PB3** for **P2.4**.

9.1.3 Development of the project and the source code

1. Create a new folder **Lab9_1(TA3OVInt)** under the lab project folder: **C:/MSP432 Lab Projects/Chapter 9** in Windows Explorer
2. Create a new µVersion5 project **TA3OVInt** and save it into the folder **Lab9_1(TA3OVInt)** created above

   Perform the following modifications to the class project **DRATA2OVInt**:
3. Replace all **TA2CTL** registers with the **TA3CTL** registers
4. Replace the TA2 interrupt handler header **TA2_N_IRQHandler()** with the TA3 interrupt handler header **TA3_N_IRQHandler()**
5. Add the dividing factor macro, **ID__8**, and OR it with all other configuration parameters for the **TA3CTL** register
6. Refer to Tables 5.11, 5.13, and 5.14 in Chapter 5 to modify the setup parameters for the **NVIC_IPR3** and the **NVIC_ISER0** registers to configure TA3 interrupt mechanism

9.1.4 Set up the environment to build and run the project

Make sure that the debugger you are using is the **CMSIS-DAP Debugger** and all settings for the debugger and the flash download are correct. Now you can build the project.

9.1.5 Demonstrate your program by checking the running result

As the project runs, each time when a timer overflow occurred, an overflow interrupt is generated. This interrupt is counted and displayed on four red color LEDs **PB3~PB0**.

Based on the running result, try to answer the following questions:

- The **intCount** is a **uint8_t** type variable and it is used to count and record the number of the overflow interrupts that have occurred. The full scale of this variable is 255. However, four LEDs only displayed this number in a range of **0x0**~**0xF** as the project runs, why did this happen?
- Why do we need to build a special subroutine **Display()** to display this number?
- Can you reduce the input clock frequency to a lower value to make the display slower?
- What did you learn from this project?

9.2 LAB9_2

9.2.1 Goal

This project enables students to use three Timer_A module compare outputs, **TA1CCR0**, **TA1CCR1**, and **TA1CCR2** as well as **TA1.1** overflow to generate four independent interrupt intervals. Inside the first three ISR, the interval is increased by different lengths to get the incremented interrupt intervals. The output mode (**OUTMOD**) is set to toggle for all of these timer modules. When these modules work as continuous modes with the input clock as the **ACLK** = 32768 Hz, different toggle frequencies can be obtained.

This lab project is very similar to one class project **DRATA0Toggles**. Refer to that project and make some necessary modifications on that project to complete this one. The difference between this lab project and that class project **DRATA0Toggles** includes:

- The Timer_A module 1, **TA1** and its overflow is used in this project, not **TA0**.
- Three Timer_A compare registers, **TA1CCR0**, **TA1CCR1**, and **TA1CCR2**, are used in the project.
- The related GPIO pins should be modified to make them match to three compare register pins, **TA1.0**, **TA1.1**, and **TA1.2**.
- The related **TA1** interrupt handlers should be used, not **TA0** interrupt handlers.
- The configuration parameters for the **NVIC_IPR2** and the **NVIC_ISER0** should be modified to make them match to the **TA1** interrupt sources.
- All three control registers, such as **TA0CCTL0**~**TA0CCTL2**, should be replaced with **TA1CCTL0**~**TA1CCTL2**. The timer control register **TA0CTL** should be modified to **TA1CTL**.
- The user-defined subroutine **TA0_Init()** should be modified to become to **TA1_Init()**.

Keep those in mind and now let's build this project.

9.2.2 Data assignment and hardware configuration

No hardware connection and configuration are needed in this project.

9.2.3 Development of the project

Create a new folder **Lab9_2(TA1Toggle)** in the folder **C:\MSP432 Lab Projects\Chapter 9** with Windows Explorer. Then create a new µVersion5 project **TA1Toggle**. Save it into the folder **Lab9_2(TA1Toggle)** created above.

9.2.4 Development of the C source file

Perform the following operations to modify the class project **DRATA0Toggles** to complete this project:

1. The user-defined subroutine **TA0_Init()** should be modified to become to **TA1_Init()**
2. Configure the GPIO pin **P8.0** as **TA1.0** output pin (refer to Table 7.5 in Chapter 7)
3. Modify the code lines **P7DIR** and **P7SEL0** to make the GPIO pins **P7.7** and **P7.6** as **TA1.1** and **TA1.2** output pins (refer to Table 7.13 in Chapter 7)
4. Remove the original configurations for the GPIO pins **P2.4** and **P2.5**, but keep the GPIO pin **P2.2** as the output pin and outputs low
5. Modify the **NVIC_IPR2** assignment to make the **TA1.0** priority level as 3 (bits 23~21 in **NVIC_IPR2** Register), and **TA1.N** priority level as 3 (bits 31~29 in **NVIC_IPR2**)
6. Since the **TA1.0** IRQ = 10, **TA1.N** IRQ = 11, set bits 10 and 11 in **NVIC_ISER0** Register
7. Change three control registers, such as **TA0CCTL0~TA0CCTL2**, to **TA1CCTL0~TA1CCTL2**. The timer control register **TA0CTL** should be modified to **TA1CTL**
8. Change two interrupt handlers to **TA1_0_IRQHandler()** and **TA1_N_IRQHandler()**
9. Change all compare registers from **TA0CCR0~TA0CCR2** to **TA1CCR0~TA1CCR2**

9.2.5 Set up the environment to build and run the project

Refer to Section 9.1.4 in Lab9_1 to complete this step.

9.2.6 Build and demonstrate your program

Build and run the project. As the project runs, the related LEDs will be flashing with different intervals or frequencies.

Based on these results, try to answer the following questions:

- If you changed the input clock source from the **ACLK** to the **SMCLK**, what will happen? Can you use some method to make all LEDs flash at a similar rate as before?
- Can you reduce the input clock frequency to make the LEDs flash slower?
- What did you learn from this project?

9.3 LAB9_3

9.3.1 Goal

Program **Lab9_3(CaptureEdge)** enables students to use the Timer_A module 2, or **TA2_C2** (**TA2.2**) to capture rising edges of an external periodic signal with the capture mode. In this project, the external signal is connected to both the **CCI2A** as the capture input and the clock input **TA2CLK** to the 16-bit timer/counter **TA2**. Each time when a rising edge of the external signal is detected, a capture interrupt is occurred and the **TA2R** is incremented by 1 and this count is sent to the **TA2CCR2** Register. This means that at any time the count in the **TA2R** should be identical with the value in the **TA2CCR2** Register. After 1000 edges have been captured, the number of edges of the external signal is collected to a data array **count[0]**.

Some points to be noted when using this program to capture the edges of an external periodic signal are:

- The **TA2.2** should work in the continuous mode.
- The input external signal should be a standard CMOS periodic waveform with an amplitude range of 0~3.3 V.
- The frequency range of the external signals that can be captured is 1 Hz~10 KHz.

9.3.2 Data assignment and hardware configuration

Since the external signal works as both the capture source and the clock input to the TA2 timer/counter, therefore a connection between the **CCI2A** (**P5.7**) and the clock input **TA2CLK** (**P4.2**) is needed. Use a jumper wire to connect these two pins together on the J2 and J3 connectors on the MSP-EXP432P401R EVB.

9.3.3 Development of the project

Create a new folder **Lab9_3(CaptureEdge)** under the folder **C:\MSP432 Lab Projects\Chapter 9** in Windows Explorer. Then generate a new µVersion5 project **CaptureEdge** and save it to the folder **Lab9_3(CaptureEdge)** created above.

9.3.4 Development of the C source file

1. Declare three system header files:
   a. #include <stdint.h>
   b. #include <stdbool.h>
   c. #include <msp.h>
2. Declare the following global variables:
   a. #define MAXNUM 2000
   b. uint32_t num = 0, cNUM
   c. bool intFlag = true
3. Declare a user-defined subroutine **void TA2_Init(void)**
4. Inside the main() program, declare an **uint32_t** variable array **count[1];** and stop the Watchdog timer
5. Call the user-defined subroutine **TA2_Init()** to initialize all GPIO pins and TA2 module
6. Use **NVIC_IPR3** to set up the priority level as 3 (011) for the **TA2_N** interrupt since the priority bits should be located at bits 15~13 in the **NVIC_IPR3** Register
7. Use **NVIC_ISER0** to enable the **TA2_N** interrupt. Since the IRQ number of the **TA2_N** is 13, set bit-13 on the **NVIC_ISER0** Register
8. Use **__wfi()** intrinsic instruction to wait for any interrupt to be occurred
9. Use a conditional **while()** loop to wait for all edges to be captured and detected. The global variable **intFlag** should be working as the condition for this loop
10. If this loop is terminated, collect the number of captured edges by assigning the **cNUM** variable to the array **count[0]**
11. Use an infinitive **while()** loop to halt the program to enable us to check the detecting result. A break point should be set on this line to halt the program

The following codes are used in the user-defined subroutine **TA2_Init()**:

12. Configure GPIO pin **P1.0** as an output pin and set it to output low
13. Configure GPIO pin **P2.2** as an output pin and set it to output low

14. Configure GPIO pin **P4.2** as the **TA2CLK** input pin with secondary function (refer to Table 7.2 in Chapter 7)
15. Configure GPIO pin **P5.7** as the **TA2_CCI2A** capture input with primary function (refer to Table 7.3 in Chapter 7)
16. Configure **TA2CCTL2** Register with the following functions (all of them can be ORed):
    a. Enable capture mode with the rising edge detection function
    b. Select the **CCI2A** as the capture input source pin
    c. Enable the synchronous capture
    d. Enable the capture interrupt
17. Configure the **TA2CTL** register with the following functions (all of them can be ORed):
    a. Clear the **TA2R** register counter
    b. Select the **TA2CLK** as clock source
    c. Set the timer to work in the continuous mode

The following codes are used in the TA2_N ISR **TA2_N_IRQHandler()**:

18. Clear the capture interrupt flag **CCIFG** in the **TA2CCTL** register by ANDing this register with the inverse of the bit macro **CCIFG**
19. Toggle the GPIO pin **P1.0** to flash the red color LED to indicate the occurrence of this capture interrupt
20. Assign the current value in the **TA2CCR2** Register to the global variable **cNUM** to reserve this value
21. Clear the **TA2CCR2** Register to avoid any possible capture overflow to be occurred
22. Increment the global variable **num** by 1 to record the times of interrupts occurred
23. If this number is greater than the **MAXNUM** defined at the top of this program, set the **TACLR** bit in the **TA2CTL** register to stop the 16-bit timer/counter TA2
24. Also reset the global Boolean variable **intFlag** to **false** to release the conditional **while()** loop in the **main()** program to collect the total number of edges that have been detected
25. Also set GPIO pin **P2.2** to turn on the blue color LED to indicate the end of the program

9.3.5 Set up the environment to build the project

Make sure that the debugger you are using is the **CMSIS-DAP Debugger** and all settings for the debugger and the flash download are correct. Now you can build the project.

9.3.6 Demonstrate your program

Make sure to set a break point at the infinitive **while()** loop line. Download and run the project now. As the project runs, the red color LED will be flashing to indicate the number of the capture interrupts that have occurred. Depending on the frequency of the captured external signal, this flashing may be slow or fast. The blue LED should be ON at the end of the program.

To check the detected number of external signal, open the **Call Stack + Locals** window after the blue LED is ON, and expand the **count[0]** array, and you can find the number is presented in hexadecimal format, **0x000003E8**. Convert it to the decimal value if you want to get the actual number, which should be 1000.

Based on these results, try to answer the following questions:

- When the frequency of the external signal is greater than 80 Hz, can you still see the red color LED flashing? If not, why?
- When the frequency of the detected external signal is greater than 10 KHz, why does the detected or captured result become incorrect?
- What did you learn from this project?

9.4 LAB9_4

9.4.1 Goal

Program **Lab9_4(CapAvePeriod)** enables students to use **TA2_C3** capture mode to capture and measure an average period of an external periodic signal. This program is similar to a class project **DRATA2Period**, so students can refer to and modify that project to build this project.

9.4.2 Data assignment and hardware configuration

The hardware connection is to connect the GPIO pin **P6.6** (**TA2.3**) to a function generator since students need to use **TA2_C3** to capture rising edges for a periodic input signal coming from a function generator.

The differences between this project and the class project **DRATA2Period** include:

- This project will capture 32 rising edges for the input external periodic signal to calculate and measure an average period of the input signal based on those 31 periods.
- A timer reset-and-recount-from-0 may occur if the period of the total numbers of captured rising edges is greater than 65536 × period of the **SMCLK** = 65536 × 0.3333 μs = 21.82 ms. In that case, the captured and measured period should be compensated by adding the maximum count **0xFFFF** to make it correct.
- Totally 32 rising edges are needed to be captured, but the total periods are 31, not 32, since **period[1] = rising_edge[1] – rising_edge[0]**. Therefore, the average period should be calculated as:

Average_Period = (Σ period[n])/31, where n = 1, 2, 3, … 31

9.4.3 Development of the project

Create a new folder **Lab9_4(CapAvePeriod)** in the folder **C:\MSP432 Lab Projects\Chapter 9** with Windows Explorer. Then create a new μVersion5 project **CapAvePeriod** and save it to the folder **Lab9_4(CapAvePeriod)** created above. Add a new C source file **CapAvePeriod.c** to this new project. Following steps below to modify one of class projects, **DRATA2Period**, to complete this new project.

9.4.4 Development of the C source file

1. Change the global variable **uint16_t period[2]** to **int16_t period[32]**, **uint16_t count = 0** to **int16_t count = 0**, and add a new **int16_t** variable **edgeCount**
2. Change the local variable **uint16_t Period[1]** to **uint32_t Period[1] = {0}**

3. Replace the codes between the **while(intFlag)** and **while(1)** loops with the following codes:

```
for (count = 1; count < 32; count++)
   Period[0] += period[count];
Period[0] = Period[0]/31;
```

4. Replace all codes inside the TA2 interrupt handler **TA2_N_IRQHandler()** with the following codes:

```
TA2CCTL3 &= ~CCIFG;
if (count == 0)
   edgeCount = TA2CCR3;
if (count > 0)
{
    period[count] = TA2CCR3 - edgeCount;
    if (period[count] < 0)
       period[count] = period[count] + 0xFFFF;
    edgeCount = TA2CCR3;
}
P1OUT ^= BIT0;
count++;
if (count > 31)
{
    intFlag = false;
    P2OUT = BIT2;
}
```

9.4.5 Set up the environment to build and run the project

Make sure that the debugger you are using is the **CMSIS-DAP Debugger** and all settings for the debugger and the flash download are correct. Now you can build the project.

9.4.6 Demonstrate your program

Set up a break point at the infinitive **while(1)** loop line to enable you to check the period measuring result in the **Call Stack + Locals** window after the blue LED is ON by expanding the data array **Period[]**. The measured period is expressed in the hexadecimal format and you need to convert it to the decimal format to compare it with the input signal.

The upper bound frequency can be measured for the external periodic signal is 5 KHz when using this project. The higher frequency signal could be measured if the input clock to the timer/counter is higher, such as 16 or 48 MHz.

Based on these results, try to answer the following questions:

- In which situation do you need to compensate the measured period result? In which case do you not need to do this compensation? Why?
- Compare this project with the class project **DRATA2Period**, which one do you think is better used to measure an external periodic signal?
- How do you enable this project to measure external signals that have higher frequency? Why?
- What did you learn from this project?

9.5 LAB9_5

9.5.1 Goal

Program **Lab9_5(CaptureWidth)** enables students to capture both rising and falling edges to measure the pulse width of an external periodic signal. The Timer_A module 3, **TA2_C3**, is used to perform the capture function. This project is similar to one of class projects, **DRATA2Period**.

9.5.2 Data assignment and hardware configuration

The hardware connection is to connect the GPIO pin **P6.6** (**TA2.3**) to a function generator since students need to use **TA2_C3** to capture rising and falling edges for a periodic input signal coming from a function generator.

9.5.3 Development of the project

Create a new folder **Lab9_5(CaptureWidth)** under the folder **C:\MSP432 Lab Projects\Chapter 9** in Windows Explorer. Then a new µVersion project **CaptureWidth** and save it to the folder **Lab9_5(CaptureWidth)** created above. Add a new C source file **CaptureWidth.c** into this project.

9.5.4 Development of the C source file

The differences between this project and the class project **DRATA2Period** include:

- One more global variable **uint16_t edgeCount** is added into this new project
- One local variable **uint16_t Period[1]** is changed to **uint16_t Width[1] = {0}**
- The codes between the **while(intFlag)** and the **while(1)** loops are replaced by the following codes:

```
Width[0] = width;
Width[0] = Width[0] + 0;
```

- One configuration code line in the **TA2_Init()** subroutine, **TA2CCTL3 = CCIS_0|CM_1|CAP|CCIE**, should be modified to select the capture for both rising and the falling edges
- All codes inside the TA2 interrupt handler **TA2_N_IRQHandler()** should be replaced with the following codes:

```
TA2CCTL3 &= ~CCIFG;
if (count == 0)
   edgeCount = TA2CCR3;
if (count > 0)
{
    width = TA2CCR3 - edgeCount;
    intFlag = false;
    P2OUT = BIT2;
}
P1OUT ^= BIT0;
count++;
```

9.5.5 Set up the environment to build and run the project

Refer to Section 9.4.5 in Lab9_4 to finish this environment setup.

9.5.6 Demonstrate your program

Set up a break point at the infinitive **while(1)** loop line to enable you to check the width measuring result in the **Call Stack + Locals** window after the blue LED is ON by expanding the data array **Width[]**. Build and run the project if everything is fine.

The upper bound frequency can be measured for the external periodic signal is 5 KHz when using this project. The higher frequency signal could be measured if the input clock to the timer/counter is higher, such as 16 or 48 MHz.
Based on these results, try to answer the following questions:

- Can you modify this project to use other Timer_A modules to measure the pulse width of an external periodic signal? If so, what are the key points?
- What did you learn from this project?

9.6 LAB9_6

9.6.1 Goal

Program **Lab9_6(CaptureWidthHF)** is a modification to the program **Lab9_5 (CaptureWidth)** to use a higher clock input to the timer by setting the SMCLK = MCLK = DCO = 24 MHz. With this higher input clock rate, the upper bound of frequency of the input signal that can be measured is extended to 70 KHz.

9.6.2 Data assignment and hardware configuration

The hardware connection is to connect the GPIO pin **P6.6** (**TA2.3**) to a function generator since students need to use **TA2_C3** to capture rising and falling edges for a periodic input signal coming from a function generator.

9.6.3 Development of the project

Create a new folder **Lab9_6(CaptureWidthHF)** under the folder **C:\MSP432 Lab Projects\Chapter 9** in the Windows Explorer. Then create a new µVersion project **CaptureWidthHF** and save it to the folder **Lab9_6(CaptureWidthHF)** created above. Add a new C source file **CaptureWidthHF.c** into this project.

9.6.4 Development of the C source file

The differences between this project and the **Lab9_5(CaptureWidth)** include:

- The **TA2CCTL3** should be configured to capture both the rising and the falling edges
- Inside the **TA2_N_IRQHandler()**, change the second **if()** statement from **if (count > 0)** to **if (count == 1)**
- Add the following codes into the **TA2_Init()** subroutine, exactly after all GPIO pins have been configured, to use a higher input clock rate (SMCLK = 24 MHz):

```
CSKEY = CSKEY_VAL;            // unlock CS module for register access
CSCTL0 = 0;                   // reset tuning parameters
```

```
CSCTL0 = DCOEN|DCORSEL_4;   // set DCO to 24MHz (nominal, center of
                               16-32MHz range)
CSCTL1 = SELS_3| SELM_3;    // select SMCLK = MCLK = DCO
CSKEY = 0;
```

9.6.5 Set up the environment to build and run the project

Refer to Section 9.4.5 in Lab9_4 to finish this environment setup.

9.6.6 Demonstrate your program

Set up a break point at the infinitive **while(1)** loop line to enable you to check the width measuring result in the **Call Stack + Locals** window after the blue LED is ON by expanding the data array **Width[]**. Build and run the project if everything is fine.

The upper bound frequency can be measured for the external periodic signal which is 70 KHz when using this project.

Based on these results, try to answer the following questions:

- Can you modify this project to use other Timer_A modules to measure the period of an external periodic signal? If so, what are the key points?
- What did you learn from this project?

9.7 LAB9_7

9.7.1 Goal

Program **Lab9_7(SDPWM)** enables students to use **SW2** (**P1.4**) to build a vary-duty-cycle PWM project via the API functions provided by the MSPPDL.

9.7.2 Data assignment and hardware configuration

In this project, a 12V DC motor fan **EE80251S2-000U-999** can be used and it is driven by a vary-duty-cycle PWM signal output from **TA0.3** or the GPIO pin **P2.6**. Refer to Section 9.3.7.3.3 and Figure 9.21 to complete the hardware connection. A user switch button **SW2** that is installed in the MSP-EXP432P401R EVB is used as the input signal to change the duty cycle from 10% to 90%, and then return back to 10% as the **SW2** is pressed again. The interrupt mechanism on the GPIO Port 4, exactly on the pin **P1.4**, is implemented in this project since the **SW2** is connected to the GPIO pin **P1.4**.

9.7.3 Development of the project

Create a new folder **Lab9_7(SDPWM)** under the folder **C:\MSP432 Lab Projects\Chapter 9** in Windows Explorer. Then create a new µVersion project **SDPWM** and save it to the folder **Lab9_7(SDPWM)** created above. Add a new C source file **SDPWM.c** into this project.

9.7.4 Development of the C source file

Add the following system header files into the top of this C file:

```
#include <stdint.h>
#include <stdbool.h>
#include <MSP432P4xx\timer_a.h>
```

```
#include <MSP432P4xx\wdt_a.h>
#include <MSP432P4xx\gpio.h>
#include <MSP432P4xx\cs.h>
#include <MSP432P4xx\interrupt.h>
```

Add the PWM Configuration struct variable **pwmConfig** into this C file:

```
Timer_A_PWMConfig pwmConfig =
{
  TIMER_A_CLOCKSOURCE_SMCLK,
  TIMER_A_CLOCKSOURCE_DIVIDER_1,
  8000,
  TIMER_A_CAPTURECOMPARE_REGISTER_3,
  TIMER_A_OUTPUTMODE_SET_RESET,
  7200
};
```

The output mode is defined as 3, **TIMER_A_OUTPUTMODE_SET_RESET**, which means that the output is set when the **TA0R** matches the value in the **TA0CCR3** and reset when its count is equal to the value in the **TA0CCR0**. Refer to Figure 9.7 and Section 9.3.8.5 to get more details about these parameter values. In this project, the clock input to the timer is 16 MHz. The period of the PWM signal is 0.5 ms and the initial duty cycle is 10%.

Therefore, set **TA0CCR0** and **TA0CCR3** as:

- **TA0CCR0** = Period = 0.5 ms/62.5 ns = 500000/62.5 = **8000**
- **TA0CCR3** = 10% duty cycle = 90% × 8000 = **7200**

Refer to Figure 9.7, when working in the **SET_RESET** output mode, the width of the high pulse of a PWM signal is equal to:

The period (**8000**)—the length of the low pulse of the PWM (**7200**) = **800** = 10% of 8000.

Since the **pwmConfig** struct and the API function **Timer_A_generatePWM()** are used in this project, therefore you do not need to configure **TA0CCR3** and **TA0CCR3** registers yourself, instead you can assign these two values to the **timerPeriod** and **dutyCycle** parameters in the **pwmConfig** struct variable as it did above.

Now, let's develop the detailed codes for the **main()** program and the interrupt handler. First let's build the **main()** program as below:

1. Stop the Watchdog timer by using the function **WDT_A_holdTimer()** (**wdt_a.h**)
2. Use the following two code lines to set up the clock input to the timer as 16 MHz:

   ```
   CS_setDCOFrequency(16000000);
   CS_initClockSignal(CS_SMCLK, CS_DCOCLK_SELECT, CS_CLOCK_DIVIDER_1);
   ```

3. Use the **GPIO_setAsPeripheralModuleFunctionOutputPin()** API function to configure the GPIO pin **P2.6** as a primary module function pin to enable **TA0.3** to output a vary-duty-cycle PWM signal
4. Use the **GPIO_setAsInputPinWithPullUpResistor()** API function to set the GPIO pin **P1.4** that is connected to the user switch button **SW2** in the MSP432P401R EVB as an input pin with the pull-up resistor applying on this pin. We need to use this **SW2** as an input source to change the duty cycle of the output PWM signal from **TA0.3**

5. Use the **GPIO_clearInterruptFlag()** API function to clear any previous possible interrupt applied on the GPIO pin **P1.4**
6. Use the API function **GPIO_enableInterrupt()** to enable the GPIO pin **P1.4** as an interrupt source to generate an interrupt when the **SW2** button is pressed
7. Use the API function **GPIO_setAsOutputPin()** to configure GPIO pins **P2.2**, **P2.4**, and **P2.5** as output pins since we need to use a blue color LED that is connected to the **P2.2** as a running status indicator to show the running status of the program. The GPIO pins **P2.5** and **P2.4** are used to provide the rotating direction of the DC motor and they are fixed to **01** to allow the motor to rotate in the CCW direction since this DC motor is a single-direction rotating motor
8. The API function **GPIO_setOutputLowOnPin()** is used to reset GPIO pins **P2.2** and **P2.5** to 0 to turn off the blue color LED and set **P2.5** as low
9. Use the **GPIO_setOutputHighOnPin()** API function to set the pin **P2.4** as high to enable the combination of the pins **P2.5** and **P2.4** as **01** to rotate the DC motor in the CCW direction
10. Call the API function **Timer_A_generatePWM()** to generate a vary-duty-cycle PWM signal with the period as 0.5 ms and the initial duty cycle as 10% on **TA0.3** (**P2.6** pin)
11. Use the API function **Interrupt_enableInterrupt(INT_PORT1)** to enable the interrupt on the GPIO pin **P1.4** (**SW2** switch button)
12. Use an infinitive **while()** loop to wait for the **P1.4** interrupt to be occurred when the **SW2** button is pressed

Next let's build the codes for the **P1.4** interrupt handler.

13. Declare the **P1.4** interrupt handler as **void PORT1_IRQHandler(void)**
14. Use the **uint16_t status = GPIO_getInterruptStatus(GPIO_PORT_P1, GPIO_PIN4)** to get the current interrupt status on **P1.4** pin and assign it to a local variable **status**
15. Use the API function **GPIO_clearInterruptFlag(GPIO_PORT_P1, status)** to clear the interrupt on the GPIO pin **P1.4** to enable it to be occurred in the future
16. Call the API function **GPIO_toggleOutputOnPin()** to toggle the GPIO pin **P2.2** to flash the blue color LED to indicate that a **P1.4** interrupt has occurred and been processed
17. Use an **if()** statement with a condition **status & GPIO_PIN4** to check whether this interrupt is generated by the GPIO pin **P1.4** (**SW2** button pressing)
18. If it is, use another **if()** statement with the condition **pwmConfig.dutyCycle == 800** to check whether the duty cycle has reached the lower bound 800
19. If it is, reset the duty cycle value to **7200** by using **pwmConfig.dutyCycle = 7200**
20. If not, reduce the current duty cycle by 800 by using **pwmConfig.dutyCycle -= 800**
21. Call the API function **Timer_A_generatePWM()** to generate another PWM signal with the updated duty cycle value

9.7.5 Set up the environment to build the project

Perform the following operations to finish these setups:

- Go to **C/C++** tab in the **Project|Options for Target "Target 1"** to set up the Include Path as: **C:\ti\msp\MSPWare _ 2 _ 21 _ 00 _ 39\driverlib\driverlib**

- Add the MSPPDL into this project by right clicking on the **Source Group 1** folder in the **Project** panel on the left, and select the **Add Existing Files to Group "Source Group 1"** menu item. On the opened box, select the driver library file path **C:\ti\msp\MSPWare_2_21_00_39\driverlib\driverlib\MSP432P4xx\keil** and then select the driver library file **msp432p4xx_driverlib.lib**. Click the **Add** button to add it into the project
- Select the correct debugger by clicking on the **Debug** tab and select the **CMSIS-DAP Debugger** from the **Use** box. Also make sure that all settings for the debugger and the flash download are correct by clicking on the Settings button

9.7.6 Demonstrate your program

Now download and run the program. As the project runs, you can press the **SW2** button to see how the motor rotating speed will be changed. At the same time, an oscillator can be used to monitor the output PWM waveform on the GPIO pin **P2.6**. You can find that the duty cycle starts at 10%, and will be incremented to 20%, 30%, until 90% each time when the **SW2** button is pressed. Different motor speeds can also be heard with different duty cycles.

Based on these results, try to answer the following questions:

- Can you modify this project to use the GPIO pin **P1.1** that is connected to a user switch button **SW1** in the MSP-432P401R EVB as an interrupt source to change the duty cycle? If so, what are the key points?
- Can you reduce the input clock frequency to 8 MHz to develop a similar project? If yes, what are the key points?
- What did you learn from this project?

chapter ten

MSP432 32-bit timers and watchdog timer

This chapter provides general information about two 32/16-bit timers used in the MSP432P401R MCU system. The 32-bit Watchdog timer is also included and discussed in this chapter. Because of the similarity in architecture and operation of these two 32/16-bit timer modules, **`Timer32_0`** and **`Timer32_1`**, we concentrate on the first 32/16-bit timer module **`Timer32_0`**. This discussion concentrates on the architectures and programming interfaces applied on these timer modules specially designed for the MSP432P401R MCU system. Some example timer implementation projects and applications are also discussed and introduced in this chapter.

10.1 Overview

The MSP432P401R 32/16-bit timer module Timer32 is an ARM dual 32-bit timer module. It contains two 32-bit timers, each of which can be configured as an independent 16-bit timer. The two timers can generate independent interrupts or a combined interrupt, which can be processed according to application requirements. The key features of Timer32 module include:

1. Two independent counters and each can be configured as a 32-bit or a 16-bit counter.
2. Three different operation modes that are supported for each counter.
3. Prescaler to enable and divide the input clock by 1, 16, or 256.
4. Independent interrupts from each of the counter, as well as, a combined interrupt from both the counters.

For each of these timers, the following modes of operation are available:

- **`Free-Running Mode`**: The counter starts with the maximum value and counts down to zero, and then wraps after reaching its zero, and continues to count down from the maximum value. This is the default mode. Depending on the size of the timer, 32 bits or 16 bits, the maximum value is either **`0xFFFFFFFF`** (4294967296) or **`0xFFFF`** (65536).
- **`Periodic Timer Mode`**: The counter starts with a preloaded value and counts down to zero. An interrupt is generated when the counter gets to zero, then it reloads the original preloaded value after wrapping past zero and continues its count-down operation.
- **`One-Shot Timer Mode`**: The counter starts with the loaded value and counts down to zero. As soon as getting zero, an interrupt is generated only once. When the counter reaches zero, it halts until reprogrammed by the user. This can be achieved by either clearing the one-shot count bit in the control register, in which case the count proceeds according to the selection of free-running or periodic mode, or by writing a new value to the load value register.

The primary function of the Watchdog timer (**WDT_A**) module in the MSP432P401R MCU system is to perform a controlled system restart after a software problem occurs. This can be done by periodically checking the CPU and the system clock with a fixed-period Watchdog timer. If the selected time interval expires, a system reset can be generated to recover the CPU to return to the normal working status. If the Watchdog function is not needed in an application, the module can be stopped, or configured as an interval timer and can generate interrupts at selected time intervals.

10.2 MSP432 Timer32 modules

First, let's take a look at the architecture and function block diagram of Timer32 module.

10.2.1 Architecture and function block diagram of the Timer32 module

Because of the similarity, we only discuss the first module (module 0), **Timer32_0**. The architecture and functional block diagram of the **Timer32_0** module is shown in Figure 10.1.

The system clock can be any clock source available to the MSP432P401R MCU system. The default source is the DCO = 3 MHz clock. The **Timer32_0** module can be enabled and clocked by setting the **ENABLE** bit (bit-7) and the bit field **PRESCALE** (bits 3~2) in the Timer 1 Timer Control Register (**T32CONTROL1**), as shown in Figure 10.1. The **MODE** bit (bit-6) and the **ONESHOT** bit (bit-0) in this register can be used to determine the operation mode of this timer. The **SIZE** bit (bit-1) in the **T32CONTROL1** Register can be used to set up the counting length, either 32 or 16 bits.

Figure 10.2 shows the bit configuration of the **T32CONTROL1** Register. Table 10.1 lists the bit values and functions for this register.

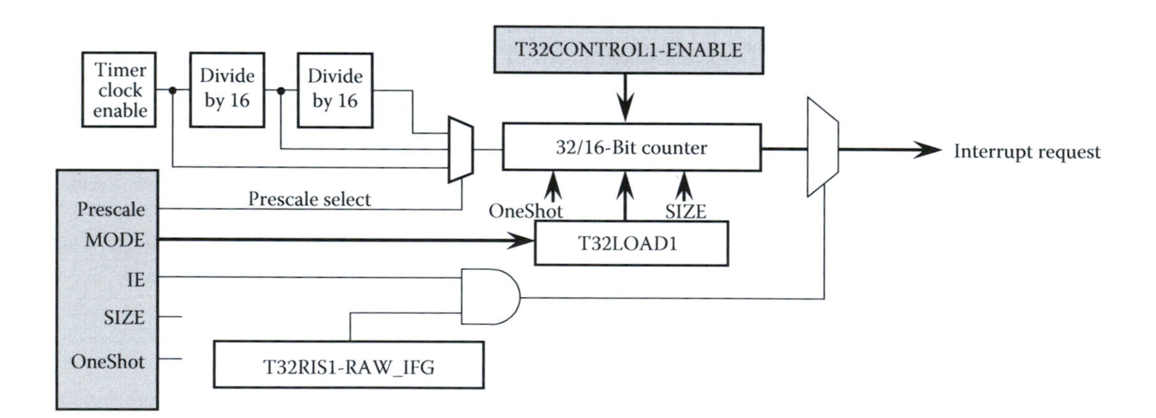

Figure 10.1 Architecture and block diagram of Timer32_0 module.

31–8
Reserved

7	6	5	4	3–2	1	0
ENABLE	MODE	IE	Reserved	PRESCALE	SIZE	ONESHOT

Figure 10.2 Bit configuration of the T32CONTROL1 Register.

Table 10.1 Bit value and function for T32CONTROL1 register

Bit	Field	Type	Reset	Function
31:8	**Reserved**	RO	0	Reserved
7	**ENABLE**	RW	0	Enable bit—**0:** Timer disabled; **1:** Timer enabled
6	**MODE**	RW	0	Mode bit—**0:** Free-running mode; **1:** Periodic mode
5	**IE**	RW	0	Interrupt enable bit—**0:** Interrupt disabled; **1:** Interrupt enabled
4	**Reserved**	RO	0	Reserved
3:2	**PRESCALE**	RW	0	Prescale bits: **00:** Clock is divided by 1; **01:** Clock is divided by 16; **10:** Clock is divided by 256; **11:** Reserved
1	**SIZE**	RW	0	Selects 16- or 32-bit counter operation: **0:** 16-bit counter; **1:** 32-bit counter
0	**ONESHOT**	RW	0	Selects one-shot or wrapping counter mode: **0:** Wrapping mode; **1:** One-shot mode

The timer input clock can be enabled by configuring the bit field **PRESCALE**. The bit **ENABLE** is then used by the counter to create a clock with a timing of one of the following:

- The system clock divided by 1 (**PRESCALE** = 00)
- The system clock divided by 16, generated by 4 bits of prescale (**PRESCALE** = 01)
- The system clock divided by 256, generated by 8 bits of prescale (**PRESCALE** = 10).

The input clock selection can be executed by using the **PRESACLE** bits to control a MUX, as shown in Figure 10.1.

10.2.2 *Operations of the Timer32 modules*

Each timer has an identical set of registers and the operation of each timer is also identical.

- When working in the free running mode, the timer starts with the maximum value and counts down to 0. Then wraps after reaching its zero, and continues to count down from the maximum value.
- When working in the periodic mode, the timer is loaded by writing to the load register **T32LOAD1**, if enabled, counts down to zero. Then wraps after reaching its zero, and continues to count down from the loaded value in the **T32LOAD1** Register.

When a counter is already running, writing to the load register causes the counter to immediately restart at the new value. Writing to the background load register **T32BGLOAD1** has no effect on the current count. The counter continues to decrement to zero, and then recommences from the new load value, if in periodic mode and one shot mode is not selected.

The Timer1 Load Register (**T32LOAD1**) is a 32-bit register containing the value from which the counter is to decrement to zero when it works in the periodic mode. When this register is written to directly, the current count is immediately reset to the new value at the next rising edge of the timer clock. The value in this register is also overwritten if the Timer 1 Background Load Register (**T32BGLOAD1**) is written to, but the current count is not immediately affected.

Both the **T32LOAD1** and the **T32BGLOAD1** are 32-bit registers and used to store loaded or background-loaded values to be used by the timer when working in the periodic mode. The difference between them is that the new loaded value into the **T32LOAD1** Register will be taken effect immediately when the next rising edge of the timer clock is coming, but the new value loaded into the **T32BGLOAD1** Register will not be taken effect until the current count gets to zero.

If values are written to both the **T32LOAD1** and the **T32BGLOAD1** registers before an enabled rising edge on the timer clock comes, the following situation occurs:

- On the next enabled **TIMCLK** edge, the value written to the **T32LOAD1** Register replaces the current count value.
- Then, each time the counter reaches zero, the current count value is reset to the value written to **T32BGLOAD1** Register.

Reading from the **T32LOAD1** Register at any time after the two writes have occurred retrieves the value written to **T32BGLOAD1**. That is, the value read from **T32LOAD1** is always the value that takes effect for periodic mode after the next time the counter reaches zero.

If one-shot mode is selected, the counter halts on reaching zero until you deselect one-shot mode, or write a new load value.

The operation mode is selected by the **MODE** bit in the **T32CONTROL1** Register. At any point, the current counter value can be read from the Current Value Register **T32VALUE1**, which is a 32-bit read-only register containing the current count value of the timer. At reset, the counter is disabled, the interrupt is cleared, and the load register is set to zero. The mode and prescale values are set to free-running mode, and clock divide is set to 1, respectively.

An interrupt is generated when the full 32-bit counter reaches zero if the **IE** bit in the **T32CONTROL1** Register is set. The bit **RAW_IFG** (bit-0) in the Timer 1 Raw Interrupt Status Register (**T32RIS1**) works as an interrupt indicator to show whether an enabled interrupt has been pended (set to 1) or not (reset to 0). Although the **T32RIS1** is a 32-bit register, but only one bit, bit-0, is used to work as an interrupt flag bit. The interrupt can be cleared by writing any value to the Timer1 Interrupt Clear Register **T32INTCLR1**. This register is a 32-bit register and holds the interrupt flag value until the interrupt is cleared. The most significant carry bit of the counter detects the counter reaching zero.

The Timer 1 Masked Interrupt Status Register (**T32MIS1**) is used to indicate the masked interrupt status from the counter. This value is the logical AND of the raw interrupt status on bit **RAW_IFG** with the timer interrupt enable bit **IE** from the **T32CONTROL1** Register. This is the same value that is passed to the interrupt output pin. Similar to **T32RIS1** Register, the **T32MIS1** Register only uses bit-0 or bit **IFG** to indicate the masked interrupt status even if it is a 32-bit register.

Both the raw interrupt status, prior to masking, and the final interrupt status, after masking, can be read from status registers, either **T32RIS1** or **T32MIS1**. The interrupts from the individual counters, after masking, are logically ORed into a combined interrupt, **TIMINTC**, which provides an additional interrupt condition from the Timer32 peripheral. Thus, the module supports three interrupts in total—**TIMINT1**, **TIMINT2**, and **TIMINTC**.

10.2.3 Mappings among the Timer32 registers and programming macros

In most real Timer32 implementation programs, all Timer32 registers discussed in the previous sections are replaced by related system macros and applied in the programs.

Table 10.2 Mapping relationships among Timer32 registers and related macros

Register address	Timer32 register	Register macro
0x4000C008	**T32CONTROL1**	**TIMER32_CONTROL1**
0x4000C000	**T32LOAD1**	**TIMER32_LOAD1**
0x4000C004	**T32VALUE1**	**TIMER32_VALUE1**
0x4000C00C	**T32INTCLR1**	**TIMER32_INTCLR1**
0x4000C010	**T32RIS1**	**TIMER32_RIS1**
0x4000C014	**T32MIS1**	**TIMER32_MIS1**
0x4000C018	**T32BGLOAD1**	**TIMER32_BGLOAD1**
0x4000C028	**T32CONTROL2**	**TIMER32_CONTROL2**
0x4000C020	**T32LOAD2**	**TIMER32_LOAD2**
0x4000C024	**T32VALUE2**	**TIMER32_VALUE2**
0x4000C02C	**T32INTCLR2**	**TIMER32_INTCLR2**
0x4000C030	**T32RIS2**	**TIMER32_RIS2**
0x4000C034	**T32MIS2**	**TIMER32_MIS2**
0x4000C038	**T32BGLOAD2**	**TIMER32_BGLOAD2**

Table 10.2 lists these mapping relationships. Users need to use those system macros located in the third column in Table 10.2 to replace the corresponding Timer32 registers to build their application programs to perform desired control functions for Timer32 modules.

There are also mappings between the value on each bit or bit fields in each Timer32 register and system macros used in the real applications. Table 10.3 shows these bit values and related macros.

A point to be noted when using this table to locate desired values or macros is

- All bit value macros shown in Table 10.3 are equivalent to setting the related bits. For any cleared bits, which mean that those bit values are 0s, no value macro is needed. For example, if the bit value macro **`TIMER32_CONTROL1_SIZE`** shows up in the program, it means that the corresponding bit, **`SIZE`** (bit-1) in the **`T32CONTROL1`** Register, is set to 1. If you want to reset this bit to 0, you do not need to show this value macro at all in your program. Another example is the bit value macro **`TIMER32_CONTROL1_MODE`**. If this macro has appeared in your program, it means that the **`MODE`** bit (bit-6) in the Register **`T32CONTROL1`** is set to enable the timer to work as a 32-bit timer.
- An exception are value macros for three prescale values, **`TIMER32_CONTROL1_PRESCALE_0`**~**`TIMER32_CONTROL1_PRESCALE_2`**. These value macros must be indicated individually.

To get more details about these mappings, refer to related TI document, exactly one of the MSP432P401R MCU system header files, **`msp432p401r.h`**. This header file is located at the folder **`C:\ti\msp\MSPWare_2_21_00_39\driverlib\inc`** at your host computer.

Now, let's build an example project to illustrate how to use these register and value macros.

Table 10.3 Mapping relationships among register bit values and related macros

Bit value macro	Register Bit(s) and value	Timer32 register
TIMER32_CONTROL1_ONESHOT	**ONESHOT** (Bit 0) = **1**	**TIMER32_CONTROL1**
TIMER32_CONTROL1_SIZE	**SIZE** (Bit 1) = **1**: 32-bit timer	**TIMER32_CONTROL1**
TIMER32_CONTROL1_PRESCALE_0	**PRESCALE** (Bits 3:2) = **0** (÷1)	**TIMER32_CONTROL1**
TIMER32_CONTROL1_PRESCALE_1	**PRESCALE** (Bits 3:2) = **4** (÷16)	**TIMER32_CONTROL1**
TIMER32_CONTROL1_PRESCALE_2	**PRESCALE** (Bits 3:2) = **8** (÷256)	**TIMER32_CONTROL1**
TIMER32_CONTROL1_IE	**IE** (Bit 5) = **1:** Enable Interrupt	**TIMER32_CONTROL1**
TIMER32_CONTROL1_MODE	**MODE** (Bit 6) = **1:** Periodic Mode	**TIMER32_CONTROL1**
TIMER32_CONTROL1_ENABLE	**ENABLE** (Bit 7) = **1:** Enable Timer	**TIMER32_CONTROL1**
TIMER32_RIS1_RAW_IFG	**IFG** (Bit 0) = **1:** Interrupt Pending	**TIMER32_RIS1**
TIMER32_MIS1_IFG	**IFG** (Bit 0) = **1:** Mask Interrupt On	**TIMER32_MIS1**
TIMER32_CONTROL2_ONESHOT	**ONESHOT** (Bit 0) = **1**	**TIMER32_CONTROL2**
TIMER32_CONTROL2_SIZE	**SIZE** (Bit 1) = **1**: 32-bit timer	**TIMER32_CONTROL2**
TIMER32_CONTROL2_PRESCALE_0	**PRESCALE** (Bits 3:2) = **0** (÷1)	**TIMER32_CONTROL2**
TIMER32_CONTROL2_PRESCALE_1	**PRESCALE** (Bits 3:2) = **4** (÷16)	**TIMER32_CONTROL2**
TIMER32_CONTROL2_PRESCALE_2	**PRESCALE** (Bits 3:2) = **8** (÷256)	**TIMER32_CONTROL2**
TIMER32_CONTROL2_IE	**IE** (Bit 5) = **1:** Enable Interrupt	**TIMER32_CONTROL2**
TIMER32_CONTROL2_MODE	**MODE** (Bit 6) = **1:** Periodic Mode	**TIMER32_CONTROL2**
TIMER32_CONTROL2_ENABLE	**ENABLE** (Bit 7) = 1: Enable Timer	**TIMER32_CONTROL2**
TIMER32_RIS2_RAW_IFG	**IFG** (Bit 0) = **1:** Interrupt Pending	**TIMER32_RIS2**
TIMER32_MIS2_IFG	**IFG** (Bit 0) = **1:** Mask Interrupt On	**TIMER32_MIS2**

10.2.4 An example Timer32 module application project

In this section, we try to build an example Timer32 project **DRATimer32** to illustrate how to control related Timer32 registers to perform some professional functions and tasks with the Timer32 module 0.

The functions of this project include:

1. The Timer32 module 0 or Timer 1 is used to work in the periodic mode. An interrupt is generated when the timer counts down to zero.
2. The **T32LOAD1** Register is loaded with an initial period values, **0x20000** (43.7 ms).
3. Inside the interrupt handler, if the number of entering the ISR is less than 16, the **T32LOAD1** is reloaded with **0x20000**. But if the number of entering the ISR is greater than or equal to 16, the **T32BGLOAD1** is loaded with a longer period, **0x80000** (174.8 ms).

```
1  //*****************************************************************************************
2  // DRATimer32.c - Main application file for project DRATimer32
3  //*****************************************************************************************
4  #include <stdint.h>
5  #include <stdbool.h>
6  #include <msp.h>
7  uint16_t count = 0;
8  int main(void)
9  {
10  WDTCTL = WDTPW | WDTHOLD;         // stop Watchdog timer
11  P1DIR |= BIT0;  P1OUT &= ~BIT0;
12  P2DIR |= BIT2;  P2OUT &= ~BIT2;
13  /* Timer32 set up in periodic mode, 32-bit, no pre-scale */
14  TIMER32_CONTROL1 = TIMER32_CONTROL1_SIZE|TIMER32_CONTROL1_MODE;
15  TIMER32_LOAD1 = 0x20000;          // load   LOAD1 with period = 0x20000
16  /* start Timer32 with interrupt enabled */
17  TIMER32_CONTROL1 |= TIMER32_CONTROL1_ENABLE|TIMER32_CONTROL1_IE;
18  NVIC_IPR6 = 0x00006000;           // set timer32 priority as 3 (bits 15 ~ 13 in NVIC_IPR6)
19  NVIC_ISER0 = 0x02000000;          // set bit-25 in NVIC_ISER0 (Timer32 IRQ number = 25)
20  __enable_irq();                   // globally enable the interrupt
21  while (1);
22 }
23 void T32_INT1_IRQHandler(void)     // Timer32 module 0 interrupt handler
24 {
25  count++;
26  TIMER32_INTCLR1 |= BIT0;          // clear Timer32 interrupt flag
27  if (count < 16)                   // if count < 16
28  {
29     P1OUT ^= BIT0;                 // toggle red-color LED
30     TIMER32_LOAD1 = 0x20000;       // load 0x20000 to T32LOAD1 as new period
31  }
32  else                              // if count >= 16
33  {
34     P2OUT ^= BIT2;                 // toggle blue-color LED
35     TIMER32_BGLOAD1 = 0x80000;     // load 0x80000 to T32BGLOAD1 as new period
36  }
37  if (count > 32) { count = 0; }    // reset count to 0
38 }
```

Figure 10.3 Complete code for the project DRATimer32.

4. Two LEDs, a red-color and a blue-color LED, work as indicators to flash itself to indicate the number of entering the ISR, either less than or greater than 16.

Open Windows Explorer and create a new folder **DRATimer32** under the class project folder **C:\MSP432 Class Projects\Chapter 10**. Then create a new µVersion5 project **DRATimer32** and save it to the folder **DRATimer32** created above. Create a new C source file **DRATimer32.c** and add it into the created project above. Enter the code shown in Figure 10.3 into this C file.

The timer is initially loaded with **0x20000** as the period to enable the Timer 1 to work as a periodic counter. When it counts down to zero, an interrupt is generated and it is directed to the interrupt handler (ISR). If the number of entering the ISR is less than 16, the **T32LOAD1** is still used with the **0x20000** as the period to continue this periodic counting function. However, if the number of entering the ISR is greater than 16, the **T32BGLOAD1** is loaded with a longer period value of **0x80000** to enable the timer to start with this as a new period.

As the project runs, one can find that both the red- and blue-color LEDs are flashing alternatively with different frequencies, the red-color LED has higher frequency but the blue-color LED has a lower frequency.

For assignments to **NVIC_IPR6** and **NVIC_ISER0** registers, refer to Tables 5.11, 5.13. and 5.14 in Chapter 5 to get more details about these setup values.

10.2.5 API functions used for Timer32 timers in the MSPWare driver library

There are twelve Timer32-related API functions provided by the MSPWare Peripheral Driver Library. These API functions can be divided into three groups based on their functions:

1. API functions used to initialize and configure Timer32 module.
2. API functions used to monitor and check the working status of the Timer32 module.
3. API functions used to configure and handle interrupts related to the Timer32 module.

10.2.5.1 API functions used to initialize and configure Timer32 module

Five API functions are included in this group:

- **Timer32_initModule()**
- **Timer32_setCount()**
- **Timer32_setCountInBackground()**
- **Timer32_startTimer()**
- **Timer32_haltTimer()**

Table 10.4 shows these API functions and their parameters as well as available values for these parameters.

A point to be noted when using these API function to configure the Timer32 module is that both the **Timer32_setCount()** and **Timer32_setCountBackground()** (if a background load register is used) must be called and used before the **Timer32_start-Timer()** API function can be called. As soon as the **Timer32_startTimer()** is executed, any configuration for the timer must have been completed prior to call that function.

Another point is that when using the **Timer32_setCount()** or **Timer32_set-CountBackground()** API functions to set up the maximum value for the up-count counter, depends on the mode of the counter to be used, either 16 bits or 32 bits, the maximum value will be different. If this setting value is greater than **UINT16_MAX** (**0xFFFF** = **65535**) for the 16-bit mode, the value will be truncated to the **UINT16_MAX**. Similar situation would happen to the 32-bit count setup.

The third point to be noted is that when using the **Timer32_haltTimer()** to stop the running of a timer module, all current count and settings are reserved temporarily, and the timer module can be resumed and recovered to work based on those settings and count if it is enabled again.

10.2.5.2 API functions used to check the working status of the Timer32 module

Only one API function is included in this group, **Timer32_getValue()**. The protocol of this function is: **uint32_t Timer32_getValue(uint32_t timer)**.

The parameter **timer** is the instance of the Timer32 module. Valid parameters must be one of the following values: **TIMER32_0_MODULE** or **TIMER32_1_MODULE**.

This function returns the current count value in the 16-bit or 32-bit counter.

Table 10.4 API functions used to initialize and configure Timer32 module

API function	Parameter	Description
void **Timer32_initModule(** uint32_t timer, uint32_t preScaler, uint32_t resolution, uint32_t mode**)**	**timer** is the instance of the Timer32 module. Valid values: **TIMER32_0_MODULE** **TIMER32_1_MODULE** **preScaler** is the prescaler to divide the clock source given to the Timer32 module. Valid values: **TIMER32_PRESCALER_1** [Default] **TIMER32_PRESCALER_16** **TIMER32_PRESCALER_256**	Initialize the Timer32 module **resolution** is the bit resolution of the Timer32 module. Valid values are: **TIMER32_1_MODULE6BIT** [Default] **TIMER32_32BIT** **mode** select either free run or periodic mode. In free run mode, the value of the timer is reset to **UINT16_MAX** (for 16-bit mode) or **UINT32_MAX** (for 32-bit mode) when the timer reaches zero. Valid values are: **TIMER32_FREE_RUN_MODE** [Default] **TIMER32_PERIODIC_MODE**
void **Timer32_setCount(** uint32_t timer, uint32_t count**)**	**timer** is the instance of the Timer32 module. Valid values are identical with those used above. **count** is the value of the timer to be set	Set the count of the timer and resets the current value to the value passed. Note that if the timer is in 16-bit mode and a **count** value is passed in that exceeds **UINT16_MAX**, the value will be truncated to **UINT16_MAX**
void **Timer32_setCountInBackground**(uint32_t timer, uint32_t count)	**timer** and **count** are identical with those used in above function **Timer32_setCount()**.	Set the count of the timer without resetting the current value. When the current value of the timer reaches zero, the value passed into this function will be set as the new count value
void **Timer32_startTimer(** uint32_t timer, bool oneShot**)**	**timer** is the instance of the Timer32 module. Valid values are identical with those used above. **oneShot** set whether the Timer32 module operates in one shot or continuous mode	Start the timer. The **Timer32_initModule()** should be called (in conjunction with **Timer32_setCount** if periodic mode is desired) prior to this function. A **true** value will cause the timer to operate in one shot mode while a **false** value will cause the timer to operate in continuous mode
void **Timer32_haltTimer(** uint32_t timer**)**	**timer** is the instance of the Timer_A module. Valid values are identical with those used above	Halt the timer. Current count and setting values are preserved

Table 10.5 API functions used to configure and handle interrupts of the Timer32 module

API function	Parameter	Description
void **Timer32_registerInterrupt**(uint32_t timerInterrupt, void(*)(void) intHandler**)**	**timerInterrupt** is the specific interrupt to register. Valid values: **TIMER32_0_INTERRUPT** **TIMER32_1_INTERRUPT** **TIMER32_COMBINED_INTERRUPT**	Registers an interrupt handler for Timer32 interrupts. **intHandler** is a pointer to the function to be called when the Timer32 interrupt occurs The **TIMER32_COMBINED_INTERRUPT** is a logic OR of the above two separate interrupts
void **Timer32_enableInterrupt**(uint32_t timer)	**timer** is the instance of the Timer32 module. Valid values: **TIMER32_0_MODULE** **TIMER32_1_MODULE**	Enable a Timer32 interrupt source
uint32_t **Timer32_getInterruptStatus**(uint32_t timer)	**timer** is the instance of the Timer32 module. Valid values are identical with those used above	Get the current Timer32 interrupt status. Return the interrupt status for the Timer32 module. A positive value indicate that an interrupt is pending while a zero value indicate that no interrupt is pending
void **Timer32_clearInterruptFlag**(uint32_t timer)	**timer** is the instance of the Timer32 module. Valid values are identical with those used above	Clear Timer32 interrupt source

10.2.5.3 *API functions used to configure and handle interrupts of the Timer32 module*

Six API functions are included in this group:

- **`Timer32_registerInterrupt()`**
- **`Timer32_unregisterInterrupt()`**
- **`Timer32_enableInterrupt()`**
- **`Timer32_disableInterrupt()`**
- **`Timer32_getInterruptStatus()`**
- **`Timer32_clearInterruptFlag()`**

Table 10.5 shows these API functions and their parameters. To save space, only the **`Timer32_registerInterrupt()`** and **`Timer32_enableInterrupt()`** API functions are shown in this table.

A project **`SDTimer32`** developed by using the API functions provided by the MSPPDL is assigned as a lab project for students in the MSP432 Lab Projects section as homework.

10.3 *MSP432 watchdog timer*

In the MSP432P401R MCU system, the Watchdog timer module **`WDT_A`** is exactly a 32-bit up-count counter. This 32-bit timer/counter can be configured to work in two modes: the Watchdog timer mode or the interval timer mode.

10.3.1 Two operational modes of the watchdog timer

The Watchdog timer **WDT_A** can work in two different modes, and they are as follows.

10.3.1.1 Watchdog timer mode

When working as a Watchdog timer, different periods or time intervals can be set up to enable the 32-bit counter to perform an up-count action. When the counter gets its maximum value, a time up event occurs or called the time interval is expired, a system reset is sent to the device to restart the whole system if any error or fault has been detected.

After a system reset, the **WDT_A** module is configured as the Watchdog mode with an initial 10.92 ms reset interval using the **SMCLK**. The user must set up, halt, or clear the Watchdog timer before this initial reset interval expires, or another system reset is generated. When the Watchdog timer is configured to operate in Watchdog mode, writing to the Watchdog timer control **WDTCTL** register with an incorrect password or expiration of the selected time interval also triggers a system reset. A system reset resets the Watchdog timer to its default condition.

10.3.1.2 Interval timer mode

When working as an interval timer, the 32-bit counter can periodically generate interrupts as the time interval expires. All of these configurations and operations are under the control of the **WDTCTL** register. This mode can be used to provide periodic interrupts. In interval timer mode, the WDT generates an interrupt at the end of each interval.

Overall, the Watchdog timer module **WDT_A** provides the following features:

- Eight software-selectable time intervals
- Watchdog mode
- Interval Timer mode
- Password-protected access to **WDTCTL** register
- Selectable clock source
- Can be stopped to conserve power

10.3.2 Architecture and functional block diagram of the WDT_A module

Figure 10.4 shows a functional block diagram of the Watchdog timer.

The 32-bit up-count counter **WDTCNT** is made of two 16-bit counters that are cascaded together with a series format. The input clock source to this counter can be selected from four possible sources; **SMCLK**, **ACLK**, **VLOCLK**, and **BCLK** via the bit field **WDTSSEL** (bits 6~5) in the **WDTCTL** register. The counting function of this 32-bit counter can be stopped or disabled by setting the bit **WDTHOLD** (bit-7) in the **WDTCTL** register via an OR gate shown in Figure 10.4.

The output of this 32-bit counter can be selected by configuring the bit **WDTIS2** (bit-2) in the **WDTCTL** register. When this bit is cleared to 0, one of the upper four time intervals can be selected. However, if this bit is set to 1, one of the lower time intervals is selected, as shown in Figure 10.4. Different combinations of values in bits **WDTIS1** and **WDTIS0** can be used to select one of four time intervals as the output from either the upper or lower 16-bit counter.

Totally, eight different time intervals can be selected by configuring the bit field **WDTIS**, which contains **WDTIS2~WDTIS0** or bits (2~0) in the **WDTCTL** register.

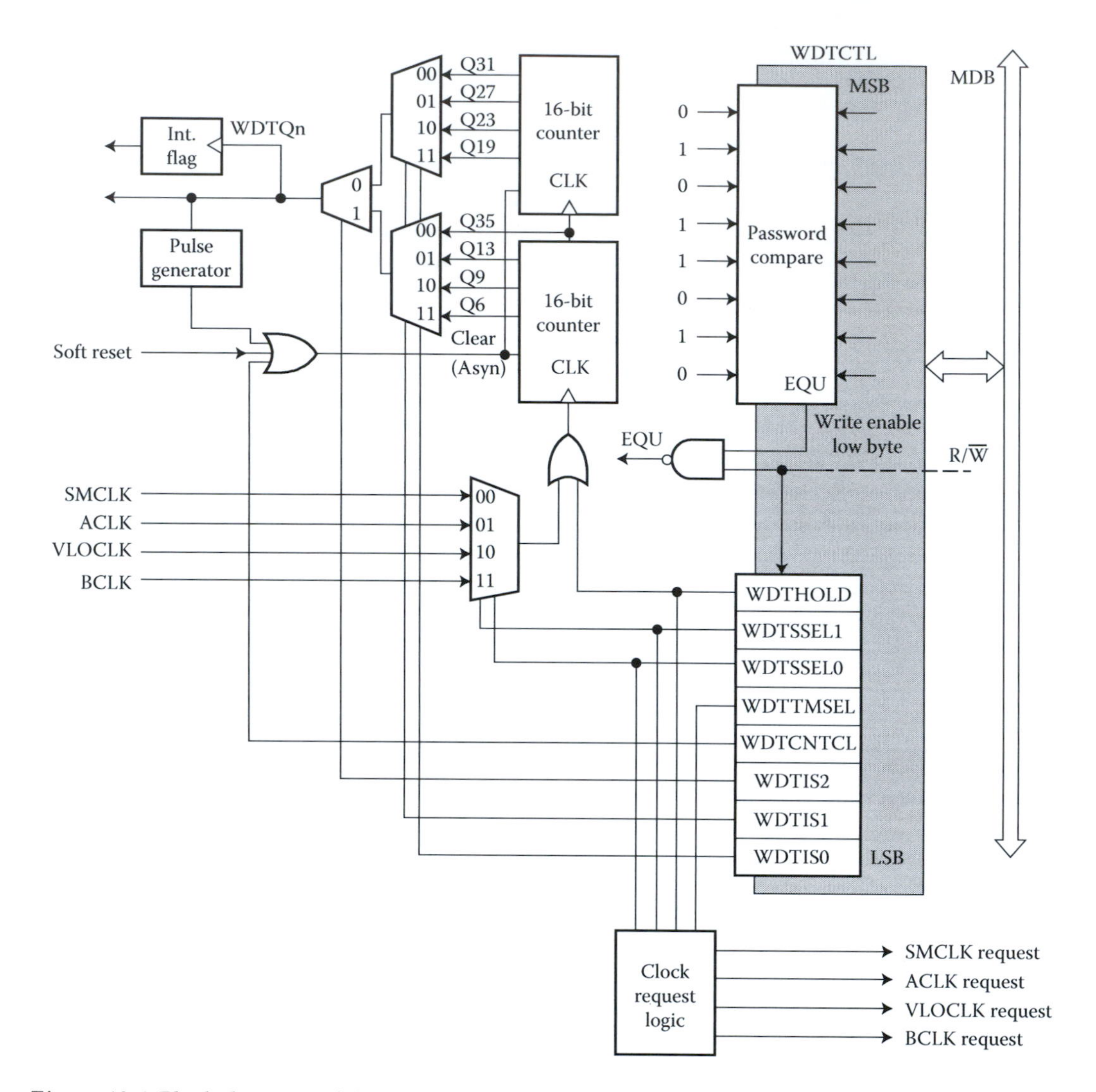

Figure 10.4 Block diagram of the watchdog timer WDT_A. (Courtesy Texas Instruments.)

When the time interval has expired, a time interval interrupt can be generated and set in the interrupt flag register. This flag can be detected by the NVIC and responded by the related WDT interrupt handler.

10.3.3 Watchdog timer-related interrupts and flags

The Watchdog timer interrupt is handled through the enable/set/clear/pending flag of the NVIC IRQ that the WDT is mapped to. When working in the Watchdog timer mode, a system reset will be generated as the time interval expires. This kind of reset can be considered as either hardware or software reset in the MSP432P401R MCU system, and it can be identified and handled by the reset controller registers. The MSP432P401R MCU supports up to 16 Hard Reset (HR) and Soft Reset (SR) sources. The reset source number of the **`WDT_A`** time-out is `1` for both HR and SR in this MCU system.

15							8
WDTPW = 0 × 69							

7	6	5	4	3	2	1	0
WDTHOLD	WDTSSEL		WDTTMSEL	WDTCNTCL	WDTIS		

Figure 10.5 Bit configuration of the WDTCTL register.

The functionality of the NVIC ensures that the related interrupt pending flag is automatically cleared when the interrupt is serviced. This is significantly different with any other interrupt that must be cleared before the ISR is completed, otherwise an error of re-entry into the same ISR will occur if the interrupt has been handled but it has not been cleared. With the help of the NVIC, the pended **`WDT_A`** interrupt flag can be cleared automatically without needing to clear it by uses' software.

When using the Watchdog timer in the Watchdog mode, a reset issued by the WDT may be mapped either as a HR or a SR of the device. In addition, flags in the reset controller can be used to determine if the WDT was indeed the cause of the reset.

Table 10.6 Bit value and function for WDTCTL register

Bit	Field	Type	Reset	Function
15:8	**WDTPW**	RW	0x69	Watchdog timer password. Always read as 069 h. Must be written as 05Ah, or the WDT will generate a system reset
7	**WDTHOLD**	RW	0	Watchdog timer hold. This bit stops the Watchdog timer. Setting WDTHOLD = 1 when the WDT is not in use conserves power **0:** Watchdog timer is not stopped **1:** Watchdog timer is stopped
6:5	**WDTSSEL**	RW	0	Watchdog timer clock source select **00:** SMCLK **01:** ACLK **10:** VLOCLK **11:** BCLK
4	**WDTTMSEL**	RW	0	Watchdog timer mode select **0:** Watchdog mode **1:** Interval timer mode
3	**WDTCNTCL**	RW	0	Watchdog timer counter clear. Setting WDTCNTCL = 1 clears the count value to 0x0000. This bit always reads 0 **0:** No action **1:** WDTCNT = 0x0000
2:0	**WDTIS**	RW	4 (**100**)	Watchdog timer interval select. These bits select the Watchdog timer interval to generate either a WDT interrupt or a WDT reset **000:** Watchdog clock source/(2^(31)) (18:12:16 at 32.768 kHz) **001:** Watchdog clock source /(2^(27)) (01:08:16 at 32.768 kHz) **010:** Watchdog clock source /(2^(23)) (00:04:16 at 32.768 kHz) **011:** Watchdog clock source /(2^(19)) (00:00:16 at 32.768 kHz) **100:** Watchdog clock source /(2^(15)) (1 s at 32.768 kHz) **101:** Watchdog clock source/(2^(13)) (250 ms at 32.768 kHz) **110:** Watchdog clock source/(2^(9)) (15.625 ms at 32.768 kHz) **111:** Watchdog clock source/(2^(6)) (1.95 ms at 32.768 kHz)

The **WDTCTL** is a 16-bit password-protected read/write register. Any read or write access must use half-word instructions, and write accesses must include the write password **0x05A** in the upper byte. A write to **WDTCTL** with any other value in the upper byte is a password violation and causes a system reset, regardless of the WDT mode of operation. Any read of **WDTCTL** register always returns **0x069** in the upper byte. Writing byte wide only to upper or lower parts of **WDTCTL** also results in a system reset, since this register must always be accessed in half-word mode.

The bit configuration of the **WDTCTL** register is shown in Figure 10.5. The bit values and functions are shown in Table 10.6.

When using the **WDTCTL** register to configure the mode, the following points need to be noted:

1. Writes to this register **MUST** be 16-bit wide, else the **WDT_A** generates a system reset.
2. Clock requests to **ACLK** and **SMCLK** clocks in Watchdog mode of operation are unconditional clock requests. Thus, disabling the corresponding clock-enable bits in CS does not stop the clock to the **WDT_A** module.
3. In the interval time mode, both the **ACLK** and **SMCLK** clock requests are always conditional from **WDT_A** module.
4. As we mentioned, eight different time intervals can be selected by configuring the bit field **WDTIS**, which contains **WDTIS2~WDTIS0** or bits (2~0) in the **WDTCTL** register. The actual time intervals are determined by the input clock frequency to the **WDT_A** module. All time intervals shown in Table 10.4 are for the default clock frequency, 32.768 KHz. For example, if **WDTIS** = **011**, the timer interval should be $2^{19} \times (1/32768) = 16$ seconds.

10.3.4 Watchdog timer operation in different power modes

The Watchdog timer module **WDT_A** can be used either in Watchdog or interval timer mode when the MCU is in one of the LPM0 modes of operation. All four clock sources, such as **SMCLK**, **ACLK**, **VLOCLK**, and **BCLK** are available.

The **WDT_A** can work in the LPM3 and LPM3.5 modes of operation. However, it may be used **only in the interval timer mode**. This is because there is no execution activity to handle the watchdog, and a reset during a LPM3 condition may result in a nondeterministic device state when it returns to the active mode. On the other hand, if the **WDT_A** is configured as an interval timer, the interval timer event can be used to wake up the MCU back to the active mode with full state retention, with the guarantee that the interval timer interrupt will be processed correctly.

Before invoking the MCU entry to LPM3 mode, the **WDT_A** must be configured to use either the **BCLK** or the **VLOCLK** as the clock source. If any of the other clock sources are used, the **WDT_A** may prevent the MCU from entering LPM3 mode and as a result the MCU will consume excess current.

The **WDT_A** cannot work in the LPM4 and the LPM4.5 modes

Before we can build an example **WDT_A** module project to illustrate how to use this timer to perform Watchdog timer and timer interval functions, first let's take a look at some system macros related to **WDT_A** registers and parameter values used in the users' program.

10.3.5 Mappings among the WDT_A bit values and programming macros

In most real **WDT_A** module implementation programs, all **WDT_A** bit values discussed in the previous sections are replaced by related system macros and applied in the programs.

Table 10.7 Mapping relationships among register bit values and related macros

Bit Value Macro	Register Bit(s) and Value	Timer32 Register
WDTPW	Bits 15~8 = **0x5A00**	**WDTCTL**
WDTHOLD	Bit 7 = **0x0080** (stop the WDT_A)	**WDTCTL**
WDTSSEL__SMCLK	Bits 6~5 = **0x0000 (SMCLK)**	**WDTCTL**
WDTSSEL__ACLK	Bits 6~5 = **0x0020 (ACLK)**	**WDTCTL**
WDTSSEL__VLOCLK	Bits 6~5 = **0x0040 (VLOCLK)**	**WDTCTL**
WDTSSEL__BCLK	Bits 6~5 = **0x0060 (BCLK)**	**WDTCTL**
WDTTMSEL	Bit 4 = **0x0010 (Interval Timer)**	**WDTCTL**
WDTCNTCL	Bit 3 = **0x0008 (WDTCNT = 0)**	**WDTCTL**
WDTIS_0	Bits 2~0 = **000 :Interval** 2^(31)	**WDTCTL**
WDTIS_1	Bits 2~0 = **001 :Interval** 2^(27)	**WDTCTL**
WDTIS_2	Bits 2~0 = **010 :Interval** 2^(23)	**WDTCTL**
WDTIS_3	Bits 2~0 = **011 :Interval** 2^(19)	**WDTCTL**
WDTIS_4	Bits 2~0 = **100 :Interval** 2^(15)	**WDTCTL**
WDTIS_5	Bits 2~0 = **101 :Interval** 2^(13)	**WDTCTL**
WDTIS_6	Bits 2~0 = **110 :Interval** 2^(9)	**WDTCTL**
WDTIS_7	Bits 2~0 = **111 :Interval** 2^(6)	**WDTCTL**

Table 10.7 lists these mapping relationships. Users must use those system macros located at the first column in Table 10.7 to replace the corresponding **WDT_A** bit values to build their application programs to perform desired control functions for **WDT_A** module.

Like system macros used for the Timer32 modules, if a system macro is not appeared, its value should be zero. For example, if users want to select the Watchdog timer to work in the timer interval mode, the macro **WDTTMSEL** can be used in the program and assigned to the **WDTCTL** register. However, if users want to select the **WDT_A** module to work as a Watchdog timer, you do not need to use any macro at all. The default mode is the Watchdog timer mode. Similar situation is true to the macro **WDTCNTCL**.

Now, let's build an example project to illustrate how to use the **WDT_A** module to work in the timer interval mode to periodically generate interrupts to toggle some LEDs.

10.3.6 *Build an example watchdog timer project*

In this example project **DRAWDT**, the **WDT_A** is configured as a timer interval producer to periodically create a sequence of interrupts to toggle some LEDs. If the numbers of interrupts are even, the red-color LED is toggled. However, if the numbers of interrupts are odd, the blue-color LED is toggled.

Create a new folder **DRAWDT** under the folder **C:\MSP432 Class Projects\Chapter 10** with Windows Explorer. Then create a new µVersion5 project **DRAWDT** and add it into the folder **DRAWDT** created above. Develop a new C source file **DRAWDT.c** and add the code shown in Figure 10.6 into this source file.

All codes in this source file are straightforward and easy to be understood. The only issue to be noted is that the **%** is a modulus operator in the C programming.

As the project runs, both the blue-color and the red-color LEDs are alternatively on and off to indicate the even and the odd numbers of entering the **WDT_A** interrupt handler.

```
1  //****************************************************************************************
2  // DRAWDT.c - Main application file for project DRAWDT
3  //****************************************************************************************
4  #include <stdint.h>
5  #include <stdbool.h>
6  #include <msp.h>
7  uint32_t count = 0;
8  int main(void)
9  {
10  /* Configure watch dog
11   * - SMCLK as clock source = 3 MHz
12   * - Interval timer mode
13   * - Clear WDT counter (initial value = 0)
14   * - Watchdog interval = 2^19 x (1/3000000) = 175 ms
15  */
16   WDTCTL = WDTPW|WDTSSEL__SMCLK|WDTTMSEL|WDTCNTCL|WDTIS_3;
17   P1DIR |= BIT0;                  // configure GPIO P1.0 and P2.2 as output pins
18   P2DIR |= BIT2;
19   NVIC_IPR0 = 0x60000000;         // set WDT_A priority level as 3 (bits 31:29 in NVIC_IPR0)
20   NVIC_ISER0 = 0x08;              // set bit 3 in NVIC_ISER0 since IRQ number of WDT_A is 3
21   __enable_irq();
22   while(1);                       // wait for WDT_A interrupts...
23 }
24 void WDT_A_IRQHandler(void)
25 {
26   count++;                        // increment count by 1
27   if (count%2 == 0)               // if count = even number
28      P2OUT ^= BIT2;               // toggle the blue-color LED
29   else                            // if count = odd number
30      P1OUT ^= BIT0;               // toggle the red-color LED
31 }
```

Figure 10.6 Complete code for the source file DRAWDT.c.

10.3.7 API functions used for WDT_A timer in the mspware driver library

There are nine WDT_A-related API functions provided by the MSPWare Peripheral Driver Library. In addition to these WDT_A directly related functions, there are two system control-related API functions that can be used to set up the reset types when either a WDT_A timer-out or a WDT_A password violation interrupt has occurred. These API functions can be divided into four groups based on their functions:

1. API functions used to initialize and configure the WDT_A module
2. API functions used to operate the WDT_A module
3. API functions used to configure and handle interrupts related to the WDT_A module
4. System control-related API functions used to set up the reset types

10.3.7.1 API functions used to initialize and configure the WDT_A module

Four API functions are included in this group:

- **WDT_A_initWatchdogTimer()**
- **WDT_A_initIntervalTimer()**
- **WDT_A_setTimeoutReset()**
- **WDT_A_setPasswordViolationReset()**

The **WDT_A_initWatchdogTimer()** function is used to initialize and configure the **WDT_A** module to work as a Watchdog timer mode, but the function

WDT_A_initIntervalTimer() can be used to configure the **WDT_A** module to work in a timer interval mode. The actual time interval applied to the **WDT_A** module is determined by the **clockDivider** multiplied by the input clock period.

The **WDT_A_setTimeoutReset()** and **WDT_A_setPasswordViolationReset()** have the similar functions as **SysCtl_setWDTTimeoutResetType()** and **SysCtl_setWDTPasswordViolationResetType()**.

Table 10.8 shows these API functions and their available parameter values.

10.3.7.2 API functions used to operate the WDT_A module

Three API functions are involved in this group:

- **WDT_A_clearTimer()**
- **WDT_A_startTimer()**
- **WDT_A_holdTimer()**

The function **WDT_A_clearTimer()** clears the Watchdog timer count to 0. The function **WDT_A_startTimer()** starts the Watchdog timer functionality to start counting.

Table 10.8 API functions used to initialize and configure the WDT_A module

API function	Parameter	Description
void **WDT_A_initWatchdogTimer**(uint_fast8_t clockSelect, uint_fast8_t clockDivider)	**clockSelect** is the clock source that the Watchdog timer will use. Valid values are: **WDT_A_CLOCKSOURCE_SMCLK** [Default] **WDT_A_CLOCKSOURCE_ACLK** **WDT_A_CLOCKSOURCE_VLOCLK** **WDT_A_CLOCKSOURCE_XCLK**	Set the clock source for the Watchdog Timer in Watchdog mode. **clockDivider** is the number of clock iterations for a Watchdog timeout. Valid values are: **WDT_A_CLOCKITERATIONS_2G** [Default] **WDT_A_CLOCKITERATIONS_128M** **WDT_A_CLOCKITERATIONS_8192K** **WDT_A_CLOCKITERATIONS_512K** **WDT_A_CLOCKITERATIONS_32K** **WDT_A_CLOCKITERATIONS_8192** **WDT_A_CLOCKITERATIONS_512** **WDT_A_CLOCKITERATIONS_64**
void **WDT_A_initIntervalTimer**(uint_fast8_t clockSelect, uint_fast8_t clockDivider)	**clockSelect** and **clockDivider** are identical with those used in above function.	Set the clock source for the Watchdog timer in timer interval mode
void **WDT_A_setTimeoutReset**(uint_fast8_t resetType)	The **resetType** is the type of reset to set. Valid values: **WDT_A_HARD_RESET** **WDT_A_SOFT_RESET**	Set the type of **RESET** that happens when a watchdog timeout occurs
void **WDT_A_setPasswordViolationReset**(uint_fast8_t resetType)	The **resetType** is the type of reset to set. Valid values: **WDT_A_HARD_RESET** **WDT_A_SOFT_RESET**	Set the type of **RESET** that happens when a Watchdog password violation occurs

The function **`WDT_A_holdTimer()`** stops the Watchdog timer from running without any interrupt.

Table 10.9 shows these API functions and their available parameter values.

10.3.7.3 API functions used to configure and handle interrupts of the WDT_A module

Two API functions are involved in this group:

- **`WDT_A_registerInterrupt()`**
- **`WDT_A_unregisterInterrupt()`**

The API function **`WDT_A_registerInterrupt()`** is used to register an interrupt handler for the Watchdog interrupt. The protocol of this function is:

- **`void WDT_A_registerInterrupt(void(*)(void) intHandler)`**

The argument **`intHandler`** is a pointer to the function to be called when the Watchdog interrupt occurs. The API function **`WDT_A_unregisterInterrupt()`** can be used to unregister the handler to be called when a Watchdog interrupt occurs. This function also masks off the interrupt in the interrupt controller so that the interrupt handler no longer can be called. The protocol of this function is

- **`void WDT_A_unregisterInterrupt(void)`**

10.3.7.4 System control-related API functions used to set up the reset types

Two API functions are involved in this group:

- **`void SysCtl_setWDTTimeoutResetType(uint_fast8_t resetType)`**
- **`void SysCtl_setWDTPasswordViolationResetType(uint_fast8_t resetType)`**

Both functions are used to set up the reset type for two events, either a Watchdog time-out or a Watchdog password violation has occurred. The first function sets the type of RESET that happens when a Watchdog timeout occurs and the second API is used to set the type of RESET that happens when a Watchdog password violation occurs.

The arguments used for these two functions are identical, either of the following values can be used for these functions:

- **`SYSCTL_HARD_RESET`**
- **`SYSCTL_SOFT_RESET`**

Table 10.9 API functions used to operate the WDT_A module

API Function	Parameter	Description
void **WDT_A_clearTimer**(void)	None	Clear the timer counter of the Watchdog timer This function can be used as a service-the-dog when operating in Watchdog mode
void **WDT_A_startTimer**(void)	None	Start the Watchdog timer
void `WDT_A_holdTimer`(void)	None	Hold the Watchdog timer

Next, let's build an example **WDT_A** module project with these API functions to illustrate how to use the SD model to develop some sophisticated projects to perform desired WDT tasks.

10.3.7.5 Build an example WDT_A project with API functions

In this example project **SDWDT**, the **WDT_A** is configured as a Watchdog timer mode to generate a Soft Reset (SR) after 4 seconds without being serviced. If this reset occurred, the red-color LED connected to the GPIO **P1.0** pin is flashing to indicate this situation. At the normal working conditions, a Timer32 module is used to generate a periodic interrupter to provide the service to the Watchdog timer for each 1 second. However, if the button switch **SW2** that is connected to the GPIO **P1.4** pin is pressed, the Timer32 module will be interrupt disabled to prevent it from providing periodic (1 second) interrupt to provide the service to the Watchdog timer. After 4 seconds, the Watchdog timer will generate an SR.

Create a new folder **SDWDT** under the folder **C:\MSP432 Class Projects\Chapter 10** with Windows Explorer. Then create a new µVersion5 project **SDWDT** and add it into the folder **SDWDT** created above. Since this project is a little big, both header and source files are needed. Develop a new header file **SDWDT.h** and enter the code shown in Figure 10.7 into this header file.

The system macro **WDT_A_TIMEOUTRESET_SRC_1** is the reset source 1 that is defined as the **WDT_A** timeout reset source in the MSP432P401R MCU system. It is equivalent to setting the bit 1 in the Hard Reset Status register (**RSTCTL_HARDRESET_STAT**) or Soft Reset Status register (**RSTCTL_SOFTRESET_STAT**), therefore the hexadecimal number **0x2** is equivalent to this macro.

Then create a new C source file **SDWDT.c** and add the codes shown in Figure 10.8 into this source file. Most codes are straightforward and easy to understand. Let's take a closer look at some key coding lines.

1. The Watchdog timer itself is first disabled to avoid any reset operation in line 8.
2. In line 10, a system API function **ResetCtl_getSoftResetSource()** is called to get the current SR source—**WDT_A** SR. Because this resource is defined as resource 1, an **AND** operation is performed between the returned SR source and the resource 1 that is defined as a macro **WDT_A_TIMEOUT**.
3. If this **WDT_A** timeout resets the system, the red-color LED is flashing periodically in line 12 with a fixed time delay executed by a **for()** loop in line 16 inside an infinitive **while()** loop in line 13.

```
1  //****************************************************************************************
2  // SDWDT.h - Header file for project SDWDT (Watchdog timer mode - Reset)
3  //****************************************************************************************
4  #include <stdint.h>
5  #include <stdbool.h>
6  #include <MSP432P4xx\timer32.h>
7  #include <MSP432P4xx\gpio.h>
8  #include <MSP432P4xx\wdt_a.h>
9  #include <MSP432P4xx\cs.h>
10 #include <MSP432P4xx\sysctl.h>
11 #include <MSP432P4xx\reset.h>
12 #include <MSP432P4xx\interrupt.h>
13 //#define WDT_A_TIMEOUT RESET_SRC_1
14 #define WDT_A_TIMEOUT  0x2
```

Figure 10.7 Complete code for the header file SDWDT.h.

```
 1 //******************************************************************************************
 2 // SDWDT.c - Main application file for porject SDWDT (Watchdog timer mode - Reset)
 3 //******************************************************************************************
 4 #include "SDWDT.h"

 5 int main(void)
 6 {
 7  uint32_t count;

 8  WDT_A_holdTimer();

 9  GPIO_setAsOutputPin(GPIO_PORT_P2, GPIO_PIN2);
10  if(ResetCtl_getSoftResetSource() & WDT_A_TIMEOUT)
11  {
12    GPIO_setAsOutputPin(GPIO_PORT_P1, GPIO_PIN0);
13    while(1)
14    {
15    GPIO_toggleOutputOnPin(GPIO_PORT_P1, GPIO_PIN0);
16    for(count = 0; count < 8000; count++) {}
17    }
18   }
19  /* set SMCLK to DCO at 2 MHz */
20  CS_setDCOFrequency(2000000);
21  CS_initClockSignal(CS_SMCLK, CS_DCOCLK_SELECT, CS_CLOCK_DIVIDER_1);

22  /* configure GPIO P1.4 as SW2 button triggering pin */
23  GPIO_setAsInputPinWithPullUpResistor(GPIO_PORT_P1, GPIO_PIN4);
24  GPIO_clearInterruptFlag(GPIO_PORT_P1, GPIO_PIN4);

25  /* configure WDT to timeout after 8192k iterations of SMCLK, at 2MHz, this will be equal 4.096 seconds */
26  SysCtl_setWDTTimeoutResetType(SYSCTL_SOFT_RESET);
27  WDT_A_initWatchdogTimer(WDT_A_CLOCKSOURCE_SMCLK, WDT_A_CLOCKITERATIONS_8192K);

28  /* configure Timer32 to 2 MHz SMCLK in periodic mode */
29  Timer32_initModule(TIMER32_0_MODULE, TIMER32_PRESCALER_1, TIMER32_32BIT,
30                                                           TIMER32_PERIODIC_MODE);
31  Timer32_setCount(TIMER32_0_MODULE, 2000000); // period = 1 second
32  Timer32_enableInterrupt(TIMER32_0_MODULE);

33  /* enable GPIO P1.4 and Timer32 interrupts */
34  GPIO_enableInterrupt(GPIO_PORT_P1, GPIO_PIN4);
35  Interrupt_enableInterrupt(INT_PORT1);
36  Interrupt_enableInterrupt(INT_T32_INT1);

37  /* start the timers */
38  Timer32_startTimer(TIMER32_0_MODULE, false);
39  WDT_A_startTimer();

40  while(1);                                         // wait for interrupts…
41 }

42 void T32_INT1_IRQHandler(void)                     /* Timer32 ISR */
43 {
44  Timer32_clearInterruptFlag(TIMER32_0_MODULE);
45  WDT_A_clearTimer();
46  GPIO_toggleOutputOnPin(GPIO_PORT_P2, GPIO_PIN2);
47 }

48 void PORT1_IRQHandler(void)                 /* GPIO ISR */
49 {
50  uint16_t status = GPIO_getInterruptStatus(GPIO_PORT_P1, GPIO_PIN4);
51  GPIO_clearInterruptFlag(GPIO_PORT_P1, status);

52  if (GPIO_PIN4 & status)
53  Timer32_disableInterrupt(TIMER32_0_MODULE);
54 }
```

Figure 10.8 Complete code for the source file SDWDT.c.

4. The clock source used for Timer32 and **WDT_A** module is configured as the **SMCLK** = DCO clock = 2 MHz in lines 20 and 21.
5. Starting with line 25, the **WDT_A** module is configured to work in the timeout mode with a time interval of 8192 K ticks, which is equivalent to **4.096** seconds = **8192000** × (**1/2000000**).
6. In code lines 28~31, the Timer32 module 0 is configured to work as a 32-bit counter in the periodic mode with the time period as 1 second.
7. Then the Timer32 module 0 is interrupt enabled in line 32 to enable it to periodically generate an interrupt for each second to provide a service to the **WDT_A** module to avoid the latter to generate any reset.
8. The GPIO **P1.4** interrupt handler that is coded between lines 48 and 54 is used to clear the **P1.4** interrupt flag and disable the Timer32 periodic interrupt to prevent the Timer32 module 0 from providing any further service to the **WDT_A** module to enable it to generate a rest after 4 seconds.

Before you can build and run the project, you need to set up the environment as below:

- Go to **C/C++** tab in the **Project|Options for Target "Target 1"** to set up the **Include Path** as: **C:\ti\msp\MSPWare_2_21_00_39\driverlib\driverlib**.
- Add the MSPPDL into this project by right clicking on the **Source Group 1** folder in the **Project** panel on the left, and select the **Add Existing Files to Group "Source Group 1"** menu item. On the opened box, select the driver library file path **C:\ti\msp\MSPWare_2_21_00_39\driverlib\driverlib\MSP432P4xx\keil** and then select the driver library file **msp432p4xx_driverlib.lib**. Click the **Add** button to add it into the project.
- Select the correct debugger by clicking on the **Debug** tab and select the **CMSIS-DAP Debugger** from the **Use** box. Also make sure that all settings for the debugger and the flash download are correct by clicking on the settings button. Then go to the **Flash|Download** menu item to download the image file of this project into the flash memory in the MSP432P401R Evaluation Board.

Now, run the project to test the Timer32 and **WDT_A** modules functions.

Each time as the Timer32 counts down to 0 with 1 second-period done, a Timer32 interrupt is generated and a clear of the **WDT_A** timer is executed to avoid any reset operation to be generated by the **WDT_A** module. This action is indicated with the flashing of the blue-color LED.

When the **SW2** button is pressed, the Timer32 interrupt is disabled inside the GPIO **P1.4** interrupt handler (ISR) to stop any further service to the **WDT_A** module. After 4 seconds, exactly 4.096 seconds, the **WDT_A** generates a system SR, and this is indicated with the flashing of the red-color LED.

We leave a similar **WDT_A** module project as the part of the homework to enable students to build a project to provide the SR function with the DRA model.

10.4 Summary

The main topics of this chapter are about 32/16-bit timer modules and Watchdog timer module used in the MSP432P401R MCU system. Because of the similarity in architecture and operation of two 32/16-bit timer modules, **Timer32_0** and **Timer32_1**, only the first 32/16-bit timer module **Timer32_0** is introduced. The discussion concentrates on the architectures and programming interfaces applied on these timer modules specially designed for the MSP432P401R MCU system. Some example timer implementation projects and applications are also discussed and introduced in this chapter.

Following the overview and introduction, the Timer32 modules are discussed first starting from Section 10.2. The architecture and function block diagram of the Timer32 module is discussed in Section 10.2.1. The detailed introduction about the Timer32 module 0 Control Register **T32CONTROL1** is also presented in this section.

The operations of the Timer32 modules are described at Section 10.2.2. Two operational modes, the free-running counter and the periodic counter, are discussed in detail. Starting at Section 10.2.3, a mapping among the Timer32 registers and system programming macros is presented with two Tables, Tables 10.2 and 10.3. With the help of these mappings, all Timer32 module registers and bit values can be replaced by the related system macros to make the Timer32 programming easier and straightforward.

An example Timer32 project is given in Section 10.2.4.

Starting at Section 10.2.5, the API functions used for Timer32 timers in the MSPWare Driver Library are introduced in detail. A real application project is assigned as the part of the homework to students.

The Watchdog timer module is presented at Section 10.3. Two operational modes, the Watchdog timer mode and the timer interval mode, are introduced in Section 10.3.1. The architecture and functional block diagram of the **WDT_A** module, including the Watchdog timer control register, **WDTCTL**, are introduced and presented in Section 10.3.2. The operations of the **WDT_A** module, including the operations in the LPM3.0, LPM3.5 modes and **WDT_A**-related interrupts, are introduced in Sections 10.3.3 and 10.3.4.

Starting at Section 10.3.5, a mapping among the **WDT_A** bit values and programming macros is provided to make the development of the **WDT_A** module application program easier and fast. An example **WDT_A** module project is given in Section 10.3.6.

The API functions used for **WDT_A** timer in the MSPWare Driver Library are discussed in Section 10.3.7 with an example project.

HOMEWORK

I. True/False Selections

____1. Each of two 32/16-bit Timer32 modules can be configured as either a half-width (16-bit) timer/counter or a full-width (32-bit) timer/counter.

____2. If configured in one-shot mode, the timer continues counting when it reaches zero when counting down.

____3. The core of each Timer32 module is a 32-bit free-running count-down counter.

____4. When counting down in one-shot or periodic modes, the prescaler acts as a true prescaler.

____5. The Timer1 Load Register (**T32LOAD1**) is the only register used to store the starting or upper bound value for a 32-bit counter.

____6. To configure a 32-bit timer as a 16-bit timer, the bit **SIZE** in the **T32CONTROL1** Register should be reset to **0**.

____7. Like a Timer32 timer module, the core of the Watchdog timer module is a 32-bit count-down counter.

____8. Because the **WDT_A** module can use different clock sources, its timeout length can be defined based on the input clock frequency and the time interval bit field **WDTIS**.

____9. After a system reset, the **WDT_A** module is configured in the Watchdog mode with an initial 10.92 ms reset interval using the **SMCLK**. The user must set up, halt, or clear the Watchdog timer before this initial reset interval expires.

____10. When using the Watchdog timer in the Watchdog mode, a reset issued by the WDT may be mapped either as an HR or an SR of the MCU device.

II. Multiple Choices

1. The input clock to a Timer32 module can be enabled by setting ____________.
   a. **ENABLE** bit in the **T32CONTROL1** Register
   b. A starting value in the **T32LOAD1** Register
   c. **PRESCALE** bits in the **T32CONTROL1** Register
   d. **MODE** bit in the **T32CONTROL1** Register
2. Two Timer32 modules can be programmed to work in the following modes: ________.
   a. Free-running, periodic, and count-up
   b. Count-down, count-up, and one-shot
   c. Periodic, count-up, and count-down
   d. Periodic, free-running, and one-shot
3. To enable a 32-bit Timer32 module to begin its count, _______ bit(s) must be set.
   a. **ENABLE** bit in the **T32CONTROL1** Register
   b. A starting value in the **T32LOAD1** Register
   c. Both a and b
   d. Either a or b
4. At reset, a Timer32 module __________, __________, and __________.
   a. Is disabled, the interrupt is cleared, the load register is set to zero
   b. The mode, and the prescale values are set to free-running, the clock divide is set to 1
   c. Either a or b
   d. Both a and b
5. After a Timer32 module 0 interrupt has been responded, inside the interrupt handler, that interrupt must be _________ by writing ________ in the _________ register.
   a. Enabled, 1, **T32CONTROL1**
   b. Disabled, 0, **T32CONTROL1**
   c. Set, 1, **T32INTCLR1**
   d. Cleared, any value, **T32INTCLR1**
6. To disable the Timer32 module 0, the bit(s) ________ in the _________ register must be _____.
   a. **IE, T32CONTROL1**, reset to 0
   b. **ENABLE, T32CONTROL1**, reset to 0
   c. **PRESCALE, T32CONTROL1**, reset to 0
   d. **MODE, T32CONTROL1**, reset to 0
7. If the input clock to the Timer32 module 0 is a 16 MHz clock and the timer works in the periodic mode with a value **80000** loaded into the **T32LOAD1** Register, the maximum time length can be counted is ________ ms.
   a. 5
   b. 6
   c. 10
   d. 20
8. To access the **WDTCTL** Register to configure the **WDT_A** module, ____________.
   a. A password **0x05A** must be written in the lower byte
   b. A password **0x05A** must be written in the upper byte

c. A password **0x069** must be written in the lower byte
d. A password **0x069** must be written in the upper byte

9. After a system reset, the **WDT_A** module is configured in the Watchdog mode with an initial 10.92 ms reset interval with the **SMCLK**. The user must _________ before this initial reset interval expires.
   a. Setup
   b. Halt
   c. Clear the WDT_A counter
   d. All of them
10. When the input clock to the **WDT_A** module that works in the Watchdog mode is 4 MHz and an SR needs to be generated after 2 seconds, the **WDTIS** should be __________.
   a. 001
   b. 010
   c. 011
   d. 100

III. Exercises

1. Provide a brief description about the Timer32 modules in the MSP432P401R.
2. Provide a brief description about the operations of the Timer32 modules.
3. Provide a brief description about the interrupt processing procedure for the Timer32 module 0.
4. Explain the difference between the API functions, **Timer32_setCount()**, and **Timer32_startTimer()**.
5. Provide a brief discussion about two operational modes of the Watchdog timer used in the MSP432P401R.
6. Design a program to enable the **WDT_A** module to generate a SR after the system starting about 5 seconds. Provide the input clock frequency and **WDTIS** value.

IV. Practical Laboratory

LABORATORY 10 MSP432 32-Bit timers and watchdog timer

10.0 GOALS

This laboratory exercise allows students to learn and practice MSP432P401R Timer32 modules and Watchdog timer programming by developing related lab projects.

1. Program **Lab10_1(SDTimer32)** lets students use the API functions provided by MSPPDL to build a one-shot 32-bit timer with the interrupt mechanism function. Each time when the button switch **SW2** that is connected to the GPIO **P1.4** pin is pressed, a GPIO Port 1 interrupt is generated and the red-color LED is toggle-on. Then the Timer32 module 0 is configured as a one-shot timer to delay 1-second period and then toggle-off the red-color LED connected to the GPIO **P1.0** pin
2. Program **Lab10_2(DRAWDT)** enables students to use the **SW2** switch button that is connected to the GPIO **P1.4** pin to trigger a **WDT_A** module to generate an SR action to the system after 4 seconds when the **WDT_A** module lost its periodic service
3. Program **Lab10_3(DRATimer32WDT)** is similar to one of class example projects, **SDWDT**. Students need to use the DRA model to replace the SD model to build that project.

After completion of these programs, students should understand the basic architecture and operational procedure of Timer32 timer modules and Watchdog timer module installed in the MSP432P401R MCU system. They should be able to code some sophisticated programs to access the desired peripheral devices to perform desired control operations via Timer32 modules and Watchdog timer module **WDT_A**.

10.1 LAB10_1

10.1.1 Goal

Program **Lab10_1(SDTimer32)** lets students use the API functions provided by MSPPDL to build a one-shot 32-bit timer with the interrupt mechanism function. Each time when the button switch **SW2** that is connected to the GPIO **P1.4** pin is pressed, a GPIO Port 1 interrupt is generated and the red-color LED is toggle-on. Then the Timer32 module 0 is configured as a one-shot timer to delay 1-second period and then toggle-off the red-color LED connected to the GPIO **P1.0** pin.

10.1.2 Data assignment and hardware configuration

No hardware configuration is needed for this project. The button switch **SW2** in the MSP-EXP432P401R EVB is already connected to the GPIO **P1.4** pin.

10.1.3 Development of the source code

Perform the following operations to complete this project:

1. Create a new folder **Lab10_1(SDTimer32)** under the folder **C:\MSP432 Lab Projects\Chapter 10** in Windows Explorer. Then create a new µVersion5 project **SDTimer32** and save it to the folder **Lab10_1(SDTimer32)** created above. Generate a new C file **SDTimer32.c**. and add it into the new created project **SDTimer32**. Build this C file with the following steps
2. Include the following system header files at the top of this source file:

```
#include <stdint.h>
#include <stdbool.h>
#include <MSP432P4xx\timer32.h>
#include <MSP432P4xx\gpio.h>
#include <MSP432P4xx\wdt_a.h>
#include <MSP432P4xx\cs.h>
#include <MSP432P4xx\interrupt.h>
```

3. Inside the **main()** program, use the **WDT_A_holdTimer()** to stop the Watchdog timer.
4. Use **CS_setDCOFrequency()** API function to set the DCO frequency as 8 MHz.
5. Use **CS_initClockSignal(CS_MCLK, CS_DCOCLK_SELECT, CS_CLOCK_DIVIDER_1)** function to configure the MCLK to use the DCO clock source.
6. Use **GPIO_setAsOutputPin()** API function to configure the GPIO **P1.0** pin as an output pin since you need to use a red-color LED that is connected to that pin to show the running status of the project later.
7. Use **GPIO_setOutputLowOnPin()** API function to set the GPIO **P1.0** pin to output low.
8. Use **GPIO_setAsInputPinWithPullUpResistor()** API function to configure the GPIO **P1.4** pin as an input pin since this pin is connected to a button switch **SW2**. You need to press the **SW2** button to trigger the **P1.4** pin to generate an interrupt as the project runs.

9. Use **GPIO_clearInterruptFlag(GPIO_PORT_P1, GPIO_PIN4)** function to clear any previous possible interrupt generated by GPIO **P1.4** pin.
10. Use **GPIO_enableInterrupt(GPIO_PORT_P1, GPIO_PIN4)** API function to enable the GPIO interrupt on the **P1.4** pin.
11. Configure Timer32 module 0 to use 8 MHz **MCLK** to work as a 32-bit periodic mode timer with the API function **Timer32_initModule()**. Use the default prescaler.
12. Enable the GPIO Port 1 interrupt with the function **Interrupt_enableInterrupt (INT_PORT1)**.
13. Enable the Timer32 module 0 interrupt with the **Interrupt_enableInterrupt (INT_T32_INT1)**.
14. Use an infinitive **while()** loop to wait for any interrupt to be occurred.

Now, using the following steps to build the interrupt handler for the GPIO **P1.4** interrupts:

15. Declare the interrupt handler as **void PORT1_IRQHandler(void)**.
16. Use **uint16_t status = GPIO_getInterruptStatus(GPIO_PORT_P1, GPIO_PIN4)** API function to get the current interrupt status.
17. Use the API function **GPIO_clearInterruptFlag(GPIO_PORT_P1, status)** to clear this interrupt flag to enable the future interrupt to be occurred.
18. Use an **if()** statement with the condition **GPIO_PIN4 & status** to check and confirm that this interrupt is generated by the GPIO **P1.4** pin (**SW2** is pressed).
19. If it is, use the API function **GPIO_disableInterrupt()** to disable that interrupt.
20. Turn on the red-color LED by using the function **GPIO_setOutputHighOnPin()** to indicate that a GPIO **P1.4** interrupt has occurred.
21. Use **Timer32_setCount(TIMER32_0_MODULE, 8000000)** function to set the target count as **8000000** for the Timer32 module 0 to enable it to count a 1-second time period.
22. Enable the Timer32 module 0 interrupt with the API function **Timer32_enableInterrupt()**.
23. Use **Timer32_startTimer(TIMER32_0_MODULE, true)** API function to start the Timer32 module 0 to work as a one-shot mode.

The following steps are used to build codes for the Timer32 module 0 interrupt handler:

24. Declare this interrupt handler as **void T32_INT1_IRQHandler(void)**.
25. Use **Timer32_clearInterruptFlag(TIMER32_0_MODULE)** API function to clear this interrupt flag to enable the future interrupt to be occurred.
26. Use the API function **GPIO_setOutputLowOnPin()** to turn off the red-color LED that is connected to the GPIO **P1.0** pin.
27. Use the **GPIO_enableInterrupt(GPIO_PORT_P1, GPIO_PIN4)** to enable the GPIO P1.4 interrupt.

10.1.4 Set up the environment to build and run the project

To build and run the project, one needs to perform the following operations to set up the environments:

- Go to **C/C++** tab in the **Project|Options for Target "Target 1"** to set up the **Include Path** as: **C:\ti\msp\MSPWare_2_21_00_39\driverlib\driverlib**.

- Add the MSPPDL into this project by right clicking on the **Source Group 1** folder in the **Project** panel on the left, and select the **Add Existing Files to Group 'Source Group 1'** menu item. On the opened box, select the driver library file path **C:\ti\msp\MSPWare_2_21_00_39\driverlib\driverlib\MSP432P4xx\keil** and then select the driver library file **msp432p4xx_driverlib.lib**. Click the **Add** button to add it into the project.
- Select the correct debugger by clicking on the **Debug** tab and select the **CMSIS-DAP Debugger** from the **Use** box. Also, make sure that all settings for the debugger and the flash download are correct by clicking on the settings button. Then go to the **Flash|Download** menu item to download the image file of this project into the flash memory in the MSP432P401R Evaluation Board.
- Now go to **Debug|Start/Stop Debug Session** to ready to run the project.
- Go to **Debug|Run** menu item to run the project.

10.1.5 Demonstrate your program by checking the running result

As the project runs, each time when the **SW2** button is pressed, a GPIO **P1.4** interrupt is generated and the red-color LED should be ON. After 1-second period is passed, the red-color LED should be OFF to indicate that 1-second has been up inside the Timer32 module 0 interrupt handler.

Based on the running result, try to answer the following questions:

- Can you modify this project by using the different input clock frequency to enable the Timer32 module 1 to count and delay a longer period?
- What did you learn from this project?

10.2 LAB10_2

10.2.1 Goal

The program **Lab10_2(DRAWDT)** enables students to use the **SW2** switch button that is connected to the GPIO **P1.4** pin to trigger a **WDT_A** module to generate an SR action to the system after 4 seconds when the **WDT_A** module has lost its periodic service.

10.2.2 Data assignment and hardware configuration

The operation sequence of this project includes:

1. Set the input clock to the **WDT_A** module as the **SMCLK** = DCO = 2 MHz.
2. The Watchdog time module **WDT_A** is configured as a Watchdog timer mode to generate a timeout SR after 4 seconds without being serviced.
3. Configure GPIO **P1.4** as an input pin to generate an interrupt as the **SW2** button is pressed.
4. Inside the GPIO **P1.4** ISR, provide a service to the **WDT_A** module by clearing the counter of the **WDT_A** module within 4 seconds to avoid it to generate an SR action. This interrupt action can be indicated by flashing the blue-color LED.
5. If after 4 seconds no service (including the clearing of the **WDT_A** counter) is provided to the **WDT_A** module, an SR is generated by the **WDT_A** module, and the red-color LED will be flashing to indicate this situation.

10.2.3 Development of the project

To make this project easy, we used DRA model

Create a new folder **Lab10_2(DRAWDT)** under the folder **C:\MSP432 Lab Projects\Chapter 10** with Windows Explorer and a new µVersion5 project **DRAWDT**, add this new project into the folder **Lab10_2(DRAWDT)** created above. Create a new C file **DRAWDT.c** and save it to the project.

10.2.4 Development of the C Source File

Perform the following operations to complete this project:

1. Include the following system header files and definition macro:

```
#include <stdint.h>
#include <stdbool.h>
#include <msp.h>
#define DCOTUNE 0x485
```

   The macro **DCOTUNE** is the tuning parameter for the center frequency of 1.5 MHz to get 2 MHz final frequency for the DCO clock source.
2. Inside the **main()** program, declare a local variable **uint32_t n** and it works as a loop count later for a **for()** loop to delay a period of time for the main program.
3. Stop the Watchdog timer.
4. Configure GPIO **P1.0** and **P2.2** pins as output pins and output low for both pins.
5. Use an **if()** statement with the condition **RSTCTL_SOFTRESET_STAT & 0x2** to check whether a **WDT_A** SR has occurred. The macro **RSTCTL_SOFTRESET_STAT** is the Soft Reset Status register reflecting the current SR source status. The bit-1 or bit **SRC1** in this register is the **WDT_A** Timeout status bit. If this bit is set (**0x2**), it indicates that a **WDT_A** Timeout reset has occurred. Refer to Section 2.3.5.1.4 in Chapter 2 to get more detail about this register and SR sources.
6. If a **WDT_A** Timeout reset happened, use an infinitive **while()** loop to toggle the red-color LED that is connected to the GPIO **P1.0** pin to indicate that a **WDT_A** reset has occurred. Inside this **while()** loop, a **for()** loop can be used with the local variable **n** as the loop counter to delay the system with **8000** loops.
7. Use the following code to set up the DCO clock frequency as 2 MHz and make **SMCLK = MCLK = DCO**:

```
CSKEY = CSKEY_VAL;                      // unlock CS module for register
                                           access
CSCTL0 = 0;                             // reset tuning parameters
CSCTL0 = DCOEN|DCORSEL_0|DCOTUNE;       // set DCO to 2 MHz
                                           (center = 1.5 MHz, 0~2 MHz range)
CSCTL1 = SELS_3|SELM_3;                 // select SMCLK = MCLK = DCO
CSKEY = 0;                              // lock the CS module
```

8. Use the following code to configure the GPIO **P1.4** pin as an input pull-up-resistor pin:

```
P1DIR &= ~BIT4;
P1REN | = BIT4;
P1OUT | = BIT4;
```

9. Use following code to enable the GPIO **P1.4** pin interrupt and clear any previous pending interrupt flag on this pin:

```
P1IE |= BIT4;
P1IFG &= ~BIT4;
```

10. Use the Watchdog Reset Control Register (**SYSCTL_WDTRESET_CTL**) to set the **WDT_A** timeout reset as a SR. The bit 0 in this register is used to control the **WDT_A** timeout reset type; a 0 in this bit indicates that a SR is applied, but a 1 means that a HR type is set.
11. Use the following code to configure the WDT to timeout after 2^{23} iterations of **SMCLK**. For a 2 MHz **SMCLK**, this will be equal 4.194 seconds ($2^{23} \times (1/2000000)$):

```
WDTCTL = WDTPW|WDTSSEL__SMCLK|WDTCNTCL|WDTIS_2;
```

12. Assign an appropriate value to the **NVIC_IPR8** Register to set the GPIO **P1.4** interrupt priority level as 3 (bits 31~29 in **NVIC_IPR8**). Refer to Table 5.13 in Chapter 5 to get more detail about this configuration.
13. Assign an appropriate value to the **NVIC_ISER1** Register to enable the interrupt on the GPIO **P1.4** pin. The IRQ number of the GPIO Port 1 is **35**. Refer to Tables 5.11 and 5.14 in Chapter 5 to get more detail about this setup.
14. Use a CMSIS intrinsic function **__enable_irq()** to globally enable all NVIC interrupt.
15. Use an infinitive **while(1)** loop to wait for any interrupts to be occurred.

The following codes are for the GPIO **P1.4** Interrupt Handler.

16. Declare this handler as: **void PORT1_IRQHandler(void)**. Refer to Table 5.11 for this step.
17. Toggle the GPIO pin **P2.2** to flash the blue-color LED to indicate that a GPIO **P1.4** interrupt has occurred.
18. Use an **if()** statement with the condition **P1IFG & BIT4** to confirm whether a GPIO **P1.4** interrupt has occurred.
19. If it is, first clear this interrupt flag.
20. Then use the code **WDTCTL |= WDTPW|WDTCNTCL** to clear the **WDT_A** counter to provide a service to the **WDT_A** module to prevent it from generating an SR to the system. One point to be noted when using this instruction is that the password macro **WDTPW** (**0x05A**) must be used with this instruction together to unlock this **WDTCTL** register to execute any desired task.

10.2.5 Set up the environment to build and run the project

Make sure that the debugger you are using is the **CMSIS-DAP Debugger** and all settings for the Debugger and the flash download are correct. Now you can build the project.

10.2.6 Build and demonstrate your program

Build and run the project. As the project runs, press the **SW2** button once within 4 seconds. The blue-color LED will be flashing to indicate that a GPIO **P1.4** interrupt has occurred. If no pressing is executed within the first and the following 4 seconds, the red-color LED will be flashing to indicate that a **WDT_A** timeout SR has been generated. To avoid any **WDT_A** timeout to be occurred, keep pressing the **SW2** button within each 4 seconds.

Based on these results, try to answer the following questions:

- Can you modify this project to enable the **WDT_A** module to generate an HR when a timeout has occurred? If so, what are the key points?
- Can you modify this project to make a longer timeout period for the **WDT_A** module?
- What did you learn from this project?

10.3 LAB10_3

10.3.1 Goal

This project, **Lab10_3(DRATimer32WDT)** is very similar to one of class example projects, **SDWDT**. Students need to use the DRA model to replace the SD model to modify that project to make it as a new project **DRATimer32WDT**.

10.3.2 Data assignment and hardware configuration

No hardware configuration is needed for this project.

One of GPIO pins, **P1.4**, which is connected to the **SW2** switch button, is used as the interrupt triggering source to start its interrupt handler to disable the Timer32 interrupt to stop the service to the **WDT_A** module to enable the **WDT_A** module to generate an SR.

10.3.3 Development of the project

Create a new folder **Lab10_3(DRATimer32WDT)** under the folder **C:\MSP432 Lab Projects\Chapter 10** with the Windows Explorer. Then create a new µVersion5 project **DRATimer32WDT** and add it into the folder **Lab10_3(DRATimer32WDT)** created above.

Create a new C source file **DRATimer32WDT.c** and save it into this new project. Copy and paste the C file **SDWDT.c** into this C file. Make the following modifications to this C source file:

10.3.4 Development and modification of the C source file

1. Change the C file name to **DRATimer32WDT.c**.
2. Replace the original header file with the following system header files and one macro:

```
#include <stdint.h>
#include <stdbool.h>
#include <msp.h>
#define DCOTUNE 0x485
```

3. Replace the code line **WDT_A_holdTimer()** with the DRA model code.
4. Replace the code line **GPIO_setAsOutputPin(GPIO_PORT_P2, GPIO_PIN2)** with the DRA model code.
5. Replace the code line **if(ResetCtl_getSoftResetSource() & WDT_A_TIMEOUT)** with the DRA code line. Refer to step 5 in the **Lab10_2** to get more details about this replacement.
6. Configure **PD2** as the **WT3CCP0** pin with **GPIOD->PCTL** register.
7. For all **WTimer3A** configurations, just prefix **W** before **TIMER3A** to make them for wide Timer3A module.

8. Replace the code line **GPIO_setAsOutputPin(GPIO_PORT_P1, GPIO_PIN0)** with the DRA code line to set the GPIO pin **P1.0** as an output pin.
9. Inside the **while(1)** loop, replace the code **GPIO_toggleOutputOnPin(GPIO_PORT_P1, GPIO_PIN0)** with the DRA model code.
10. Replace the following code lines with the DRA model code lines:

```
CS_setDCOFrequency(2000000);
CS_initClockSignal(CS_SMCLK, CS_DCOCLK_SELECT, CS_CLOCK_DIVIDER_1);
```

   Refer to step 7 in the **Lab10_2** above to get more details about these modifications.
11. Replace the code **GPIO_setAsInputPinWithPullUpResistor(GPIO_PORT_P1, GPIO_PIN4)** with the DRA model code. Refer to step 8 in the **Lab10_2** above to get more details about this modification.
12. Replace the code line **GPIO_clearInterruptFlag(GPIO_PORT_P1, GPIO_PIN4)** with the DRA model code. Refer to step 9 in the **Lab10_2** above to get more details about this modification.
13. Replace the code **SysCtl_setWDTTimeoutResetType(SYSCTL_SOFT_RESET)** with the DRA model code. Refer to step 10 in the **Lab10_2** above to get more details about this modification.
14. Replace the code **WDT_A_initWatchdogTimer(WDT_A_CLOCKSOURCE_SMCLK, WDT_A_ CLOCKITERATIONS_8192 K)** with the DRA model code. Refer to step 11 in the **Lab10_2** above to get more details about this modification.
15. Use the code **TIMER32_CONTROL1 = TIMER32_CONTROL1_SIZE|TIMER32_CONTROL1_MODE** to replace the code **Timer32_initModule(TIMER32_0_MODULE, TIMER32_PRESCALER_1, TIMER32_32BIT, TIMER32_PERIODIC_MODE)** code line.
16. Replace the code line **Timer32_setCount(TIMER32_0_MODULE, 2000000)** with the code line **TIMER32_LOAD1 = 2000000**.
17. Replace the code line **Timer32_enableInterrupt(TIMER32_0_MODULE)** with the DRA model code **TIMER32_CONTROL1 |= TIMER32_CONTROL1_IE**.
18. Delete the code line **GPIO_enableInterrupt(GPIO_PORT_P1, GPIO_PIN4)** since the GPIO pin **P1.4** interrupt has been enabled at step 12 above.
19. Replace the code line **Interrupt_enableInterrupt(INT_PORT1)** with the following DRA code lines:

```
NVIC_IPR8 = 0x60000000;   // set P1.4 priority level as 3 (bits 31~29
                              in NVIC_IPR8)
NVIC_ISER1 = 0x08;        // set bit 3 on NVIC_ISER1 since P1.4 IRQ# = 35
```

20. Replace the code line **Interrupt_enableInterrupt(INT_T32_INT1)** with the following DRA model code:

```
NVIC_IPR6 = 0x00006000;   // set timer32 priority as 3 (bits 15~13 in
                             NVIC_IPR6)
NVIC_ISER0 = 0x02000000; // set bit-25 in NVIC_ISER0 (Timer32 IRQ
                             number = 25)
__enable_irq();           // globally enable the interrupt
```

21. Replace the code line **Timer32_startTimer(TIMER32_0_MODULE, false)** with the DRA model code line **TIMER32_CONTROL1 |= TIMER32_CONTROL1_ENABLE**.

22. Delete the code line **WDT_A_startTimer()** since the **WDT_A** module has been started at step 14 above.

The following modifications are for the Timer32 interrupt handler (**T32_INT1_IRQHandler**).

23. Add a code line **P2OUT ^= BIT2;** to toggle the GPIO **P2.2** pin to flash the blue-color LED.
24. Replace the code line **Timer32_clearInterruptFlag(TIMER32_0_MODULE)** with the DRA model code **TIMER32_INTCLR1 |= BIT0**.
25. Replace the code **WDT_A_clearTimer()** with the DRA model code **WDTCTL |= WDTPW| WDTCNTCL** to clear the WDT_A counter to 0.
26. Delete the code **GPIO_toggleOutputOnPin(GPIO_PORT_P2, GPIO_PIN2)** code line.

The following modifications are for the GPIO pin **P1.4** interrupt handler (**PORT1_IRQHandler**).

27. Delete the original two code lines from this interrupt handler:

```
uint16_t status = GPIO_getInterruptStatus(GPIO_PORT_P1, GPIO_PIN4);
GPIO_clearInterruptFlag(GPIO_PORT_P1, status);
```

28. Replace the original **if()** code line **if (GPIO_PIN4 & status)** with the DRA model **if()** code line **if (P1IFG & BIT4)**.
29. Inside this modified **if()** selection structure, add a new code line **P1IFG &=~BIT4;** to clear the GPIO pin **P1.4** interrupt flag.
30. Replace the code line **Timer32_disableInterrupt(TIMER32_0_MODULE)** with the DRA model code line **TIMER32_CONTROL1 &=~TIMER32_CONTROL1_IE** to disable the Timer32 interrupt.

10.3.5 Set up the environment to build and run the project

Make sure that the debugger you are using is the **CMSIS-DAP Debugger** and all settings for the debugger and the flash download are correct. Now you can build the project.

10.3.6 Demonstrate your program

Build and run the project if everything is fine. As the project runs, the blue-color LED will be periodically flashing per second to indicate that the Timer32 interrupts are periodically occurred to clear the WDT_A counter to provide a service to the Watchdog timer to avoid any SR to be generated. Now press the **SW2** button and wait for about 4 seconds, the red-color LED is flashing to indicate that a WDT_A timeout SR is generated because the service provided by the Timer32 timer has been disabled inside the GPIO **P1.4** pin interrupt handler.

Based on these results, try to answer the following questions:

- Compare this project and the class project **SDWDT**, what can you say about the differences between them?
- Can you modify this project to make a longer timeout period for the WDT_A module?
- What did you learn from this project?

chapter eleven

MSP432™ floating-point unit (FPU)

In the MSP432P401R MCU system, an optional Floating Point Unit (FPU) is applied in the Cortex®-M4 Core to enhance the floating-point data operations for this MCU system.

As we discussed in Chapter 2, one of the most important differences between the Cortex®-M4 MCU and the Cortex®-M3 MCU is that an optional FPU is added into the Cortex®-M4 Core to enhance the floating-point data operations. The Cortex®-M4 FPU implements ARMv7E-M architecture with FPv4-SP extensions. It provides floating-point computation functionality that is compliant with the *ANSI/IEEE Standard 754-2008, IEEE Standard for Binary Floating-Point Arithmetic*. In this chapter, we will concentrate our discussions on this FPU element with more detailed analysis for this unit.

11.1 Overview

In the Cortex®-M4 MCU, an optional FPU is provided to supply different methods for manipulating the behavior of the floating-point data in an easy and convenient way. By default, the FPU is disabled and must be enabled prior to the execution of any floating-point instructions. If a floating-point instruction is executed but the FPU is disabled, a No Coprocessor (**NOCP**) usage fault is generated.

The Cortex®-M4F FPU fully supports half- and single-precision add, subtract, multiply, divide, multiply and accumulate, and square root operations. It provides conversions between fixed-point and floating-point data formats, and floating-point constant instructions. It also supports the half-precision and single-precision floating-point formats to provide different manipulations for the floating-point data with different lengths and accuracies.

The usage of FPU in the Cortex®-M4 MCU system can be divided into two categories: (1) in the main thread of the program without using any interrupt and (2) inside an interrupt handler.

When using FPU inside an interrupt handler, its use can still be further divided into the following three categories:

- It can do nothing with the floating-point data
- It can always save the floating-point data into the stack
- It can perform a lazy save/restore of the floating-point data

The default handling of the floating-point data is to perform a lazy save/restore. When an interrupt is occurs, spaces are reserved on the stack for the floating-point data but the data are not written into the stack. This method reduces the interrupt working load to a minimum because only the integer state is written to the stack. When a floating-point instruction is executed within the interrupt handler, the floating-point data are written to the stack prior to the execution of the floating-point instruction. Finally, upon return from the interrupt, the floating-point data are restored from the stack only if they were written. Lazy save/restore provides a blend between fast interrupt response and the ability to use floating-point instructions in the interrupt handler.

The FPU can generate an interrupt when one of the several exceptions occurs. These exceptions include the underflow, overflow, divide by zero, invalid operation, abnormal input, and inexact exception. The application can optionally choose to enable one or more of these interrupts and use the interrupt handler to decide which action should be taken for each case.

Let's first have a basic idea about different floating-point data.

11.2 Three types of the floating-point data

Floating-point data allow the MCU to process a much wider data range, from a huge to a very small value in a computer system. Based on the different precision requirements and formats, a floating-point data can be divided into the following three format groups:

- The half-precision floating-point format (16-bit data)
- The single-precision floating-point format (32-bit data)
- The double-precision floating-point format (64-bit data)

One point to be noted is that not all C compilers support the half-precision floating-point format numbers. Both **gcc** and **ARM**® **C** compilers support this kind of half-precision floating-point format. However, they do not support the double-precision floating-point format.

11.2.1 Half-precision floating-point data

The half-precision is exactly a binary floating-point computer number format that occupies 16 bits or 2 bytes in the computer memory.

According to the IEEE 754-2008 definition, a half-precision floating-point number can be represented as three segments as shown in Figure 11.1, which include the **Sign** bit (bit-15), **Exponent** field (5 bits, bits 14~10), and **Fraction** field (10 bits, bits 9~0).

When using this half-precision floating-point format to represent a floating-point number, the following points need to be noted:

1. The 1-bit **Sign** field is used to determine the polarity of the floating-point number. A value of **0** on this bit indicates that this floating-point number is a positive one, but a value of **1** means that this floating-point number is a negative number.
2. The 5-bit **Exponent** field is a *biased* exponent value and a biased value of **15** must be subtracted from this biased exponent, **Exponent**, to get the real or the true exponent value for the given floating-point number. The reason for using this biased exponent is to enlarge the range of the floating-point numbers.

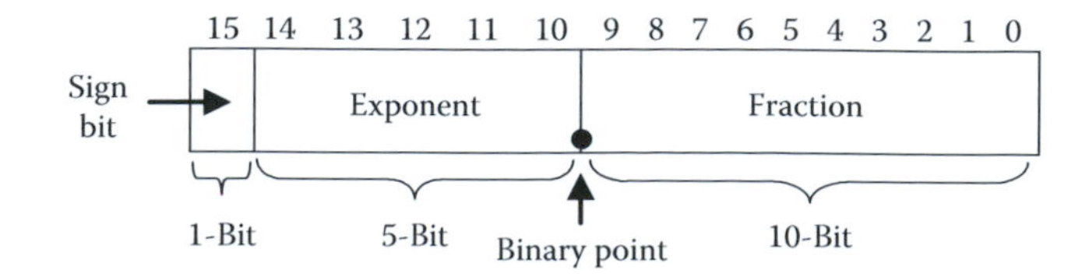

Figure 11.1 IEEE 754 half-precision floating-point format.

3. The 10-bit **Fraction** field contains 10-bit binary significant bits (significands) after the decimal point. The key is that between bit-10 and bit-9, or between the **Exponent** and the **Fraction** fields, a binary point exists and this must be kept in mind when converting this half-precision floating-point number to its real value.

Thus, the real or the true value of a floating-point number can be calculated as

$$\texttt{True Value} = (-1)^{\texttt{Sign}} \times 2^{(\texttt{Exponent}-15)} \times (1.0 + \texttt{Fraction}) \tag{11.1}$$

When using Equation 11.1 to calculate the true value for a given half-precision floating-point number, the following issues must be considered:

1. If the biased exponent, **Exponent**, is in the range **0 < Exponent < 0x1F**, the value of the true floating-point number is a normalized value and it can be calculated by using Equation 11.1.
2. If the biased exponent, **Exponent**, is 0, the following possible results exist:
   a. If the **Fraction** is 0 and the **Sign** bit is 0, then the true value is 0 (+0).
   b. If the **Fraction** is 0 but the **Sign** bit is 1, then the true value is also 0 (–0).
   c. If the **Fraction** is not 0, then the true value is a subnormal or a denormalized value and its true value should be calculated by using Equation 11.2:

$$\texttt{True Value} = (-1)^{\texttt{Sign}} \times 2^{-14} \times (0.0 + \texttt{Fraction}) \tag{11.2}$$

3. If the biased exponent, **Exponent**, is equal to **0x1F** (11111), the following possible results exist:
   a. If the **Fraction** is 0 and the **Sign** bit is also 0, then the true value is infinitive (+∞).
   b. If the **Fraction** is 0 and the **Sign** bit is 1, then the true value is also infinitive (–∞).
   c. If the **Fraction** is not 0, then it is a special code or known as **Not a Number** (**NaN**).

A NaN value can be categorized into a signaling NaN or a quiet NaN, and this is defined as:

1. If bit-9 of the **Fraction** is 0, it is a signaling NaN.
2. If bit-9 of the **Fraction** is 1, it is a quiet NaN.

Table 11.1 shows a summarization for these possible results.

Table 11.1 Possible true value for a given half-precision floating-point number

Exponent	Fraction = 0	Fraction ≠ 0	Calculation equation
00000	+0, −0	Subnormal numbers	$(-1)^{Sign} \times 2^{-14} \times (0.0 + fraction)$
00001~11110	Normalized numbers		$(-1)^{Sign} \times 2^{(Exponent-15)} \times (1.0 + fraction)$
11111	± infinitive (±∞)	NaN (signaling or quiet)	

The IEEE 754 standard specifies **binary16** as a data type for a half-precision floating-point number.

An example of using the IEEE 754 standard to present a half-precision floating-point number is **1111101101101000**. The **Sign** bit is 1, which means that this is a negative floating-point number. The biased exponent, **Exponent**, is **11110B** = 30D, and the **Fraction** is **1101101000B**. Based on Equation 11.1, the true value of this floating-point number is

$$\begin{aligned}\text{TrueValue} &= (-1)^1 \times 2^{(30-15)} \times (1.0 + 1\times 2^{-1} + 1\times 2^{-2} + 1\times 2^{-4} + 1\times 2^{-5} + 1\times 2^{-7}) \\ &= -(2^{15} \times (1.0 + 0.5 + 0.25 + 0.0625 + 0.03125 + 0.0078125)) \\ &= -(2^{15} \times 1.8515625) = -60672\end{aligned}$$

To convert **−60672** back to the IEEE 754 standard half-precision floating-point format, perform the following operations:

1. Convert **60672** to a hexadecimal or a binary number, which is **ED00** or **1110110100000000**.
2. Convert this binary number to the normalized format, which is 1.1101101×2^{15}. This is equivalent to move the binary point to the left by 15 bits.
3. Convert the normal exponent **15** to the biased exponent, **Exponent**, by adding the bias value **15**. The biased exponent, **Exponent**, is **15** + **15** = **30D** (**11110B**).
4. The **Fraction** is the significant bits after the decimal point, **1101101000**. The tailed three zeros can be added to make this **Fraction** to be 10 bits.
5. Add the **Sign** value 1 to bit-15 since this is a negative number.
6. The final IEEE 754 standard half-precision floating-point format for this number is **Sign**-Bit: **1, Exponent: 11110, Fraction: 1101101000.** Together it is: **1111101101101000.**

Next, let us take a look at the single-precision floating-point format.

11.2.2 *Single-precision floating-point data*

Similar to IEEE 754 half-precision floating-point format, the single-precision data are also a binary floating-point computer number that occupies 32 bits or 4 bytes in the computer memory.

According to the IEEE 754-2008 definition, a single-precision floating-point number can be represented as three segments as shown in Figure 11.2, which include the **Sign** bit (bit-31), **Exponent** field (8-bits, bits 30~23), and **Fraction** field (23 bits, bits 22~0).

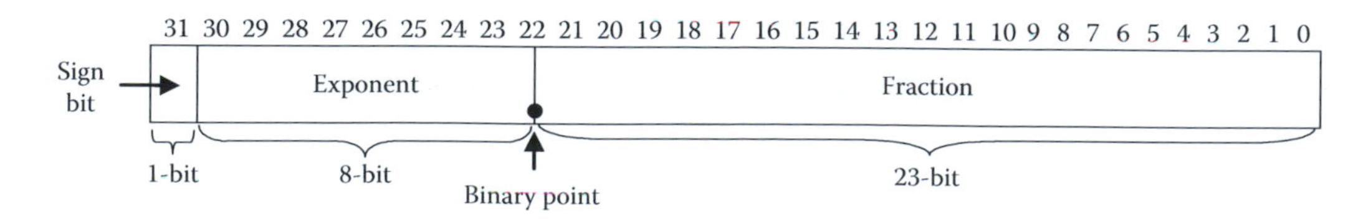

Figure 11.2 IEEE 754 single-precision floating-point format.

Similar to the half-precision format, when using this single-precision floating-point format to represent a floating-point number, the following points need to be noted:

1. The 1-bit **Sign** field is used to determine the polarity of the floating-point number. A value of **0** on this bit indicates that this floating-point number is a positive one, but a value of **1** means that this floating-point number is a negative number.
2. The 8-bit **Exponent** field is a *biased* exponent value and a biased value of **127** must be subtracted from this biased exponent, **Exponent**, to get the real or the true exponent value for the given floating-point number. The reason for using this biased exponent is to enlarge the range of floating-point numbers.
3. The 23-bit **Fraction** field contains 23-bit binary significant bits (significands) after the decimal point. The key is that between bit-23 and bit-22, or between the **Exponent** and the **Fraction** fields, a binary point exists and this must be kept in mind when converting this single-precision floating-point number to its real value.

The real or the true value of a single-precision floating-point number can be calculated as

$$\mathtt{True\ Value} = (-1)^{\mathtt{Sign}} \times 2^{(\mathtt{Exponent}-127)} \times (1.0 + \mathtt{Fraction}) \quad (11.3)$$

When using Equation 11.3 to calculate the true value for a given single-precision floating-point number, the following issues must be considered:

1. If the biased exponent, **Exponent**, is in the range **0 < Exponent < 0xFF**, the value of the true floating-point number is a normalized value and it can be calculated by using Equation 11.3.
2. If the biased exponent, **Exponent**, is 0, the following possible results exist:
   a. If the **Fraction** is 0 and the **Sign** bit is 0, then the true value is 0 (+0).
   b. If the **Fraction** is 0 but the **Sign** bit is 1, then the true value is also 0 (–0).
   c. If the **Fraction** is not 0, then the true value is a subnormal or a denormalized value and its true value should be calculated by using

$$\mathtt{True\ Value} = (-1)^{\mathtt{Sign}} \times 2^{-126} \times (0.0 + \mathtt{Fraction}) \quad (11.4)$$

3. If the biased exponent, **Exponent**, is equal to **0xFF** (11111111B), the following possible results exist:
   a. If the **Fraction** is 0 and the **Sign** bit is also 0, then the true value is infinitive (+∞).
   b. If the **Fraction** is 0 and the **Sign** bit is 1, then the true value is also infinitive (–∞).
   c. If the **Fraction** is not 0, then it is a special code or known as **NaN**.

A NaN value can be categorized into a signaling NaN or a quiet NaN, and this is defined as:

1. If bit-22 of the **Fraction** is 0, it is a signaling NaN.
2. If bit-22 of the **Fraction** is 1, it is a quiet NaN.

Table 11.2 The possible true value for a given single-precision floating-point number

Exponent	Fraction = 0	Fraction ≠ 0	Calculation equation
00000000	**+0, −0**	**Subnormal numbers**	$(-1)^{Sign} \times 2^{-126} \times (0.0 + fraction)$
01H~FEH	**Normalized numbers**		$(-1)^{Sign} \times 2^{(Exponent-127)} \times (1.0 + fraction)$
11111111	**± infinitive (±∞)**	**NaN (signaling or quiet)**	

Table 11.2 shows a summarization for these possible results.

The IEEE 754 standard specifies **`binary32`** as a data type for a single-precision floating-point number.

An example of using the IEEE 754 standard to present a single-precision floating-point number is **`11000001011011010001000000000000`**. The **`Sign`** bit is 1, which means that this is a negative floating-point number. The biased exponent, **`Exponent`**, is **`10000010B`** = 130D, and the **`Fraction`** is **`11011010001000000000000B`**. Based on Equation 11.3, the true value of this single-precision floating-point number is

$$\text{TrueValue} = (-1)^1 \times 2^{(130-127)} \times (1.0 + 1 \times 2^{-1} + 1 \times 2^{-2} + 1 \times 2^{-4} + 1 \times 2^{-5} + 1 \times 2^{-7} + 2^{-11})$$
$$= -(2^3 \times (1.0 + 0.5 + 0.25 + 0.0625 + 0.03125 + 0.0078125 + 0.00048828125))$$
$$= -(2^3 \times 1.85205078125) = -14.81640625$$

To convert −**`14.81640625`** back to the IEEE 754 standard single-precision floating-point format, perform the following operations:

1. Convert **`14.81640625`** to hexadecimal or binary number, which is **`0xE.D1`** or **`1110.11010001B`**.
2. Convert this binary number to the normalized format, which is **`1.11011010001`** × 2^3. This is equivalent to move the binary point to the left by 3 bits.
3. Convert the normal exponent **`3`** to the biased exponent, **`Exponent`**, by adding the bias value **`127`**. The biased exponent, **`Exponent`**, is **`3 + 127`** = **`130D`** (**`10000010B`**).
4. The **`Fraction`** is the significant bits after the decimal point for the normalized format value, which is **`11011010001000000000000`**. The tailed 12 zeros can be added to make this **`Fraction`** to be 23 bits.
5. Add the **`Sign`** value 1 to bit-31 since this is a negative number.
6. The final IEEE 754 standard single-precision floating-point format for this number is **`Sign`**-Bit: **`1, Exponent: 10000010, Fraction: 11011010001000000000000.`** Together it is: **`11000001011011010001000000000000.`**

Next, let's take a look at the double-precision floating-point format.

11.2.3 Double-precision floating-point data

Similar to IEEE 754 single-precision floating-point format, the double-precision data are also a binary floating-point computer number that occupies 64 bits or 8 bytes in the computer memory.

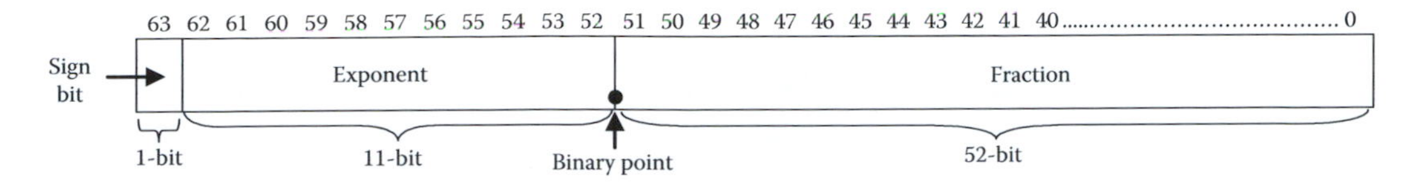

Figure 11.3 IEEE 754 double-precision floating-point format.

According to the IEEE 754 definition, a double-precision floating-point number can be represented as three segments as shown in Figure 11.3, which include the **Sign** bit (bit-63), **Exponent** field (11 bits, bits 62~52), and **Fraction** field (52 bits, bits 51~0).

Similar to the single-precision format, when using this double-precision floating-point format to represent a floating-point number, the following points need to be noted:

1. The 1-bit **Sign** field is used to determine the polarity of the floating-point number. A value of **0** on this bit indicates that this floating-point number is a positive one, but a value of **1** means that this floating-point number is a negative number.
2. The 11-bit **Exponent** field is a *biased* exponent value and a biased value of **1023** must be subtracted from this biased exponent, **Exponent**, to get the real or the true exponent value for the given floating-point number. The reason for using this biased exponent is to enlarge the range of the floating-point numbers.
3. The 52-bit **Fraction** field contains 52-bit binary significant bits (significands) after the decimal point. The key is that between bit-52 and bit-51, or between the **Exponent** and the **Fraction** fields, a binary point exists and this must be kept in mind when converting this double-precision floating-point number to its real value.

The real or the true value of a double-precision floating-point number can be calculated as

$$\mathbf{True\,Value = (-1)^{Sign} \times 2^{(Exponent-1023)} \times (1.0 + Fraction)} \quad (11.5)$$

When using Equation 11.5 to calculate the true value for a given double-precision floating-point number, the following issues must be considered:

1. If the biased exponent, **Exponent**, is in the range **0 < Exponent < 0x7FF**, the value of the true floating-point number is a normalized value and it can be calculated by using Equation 11.5.
2. If the biased exponent, **Exponent**, is 0, the following possible results exist:
   a. If the **Fraction** is 0 and the **Sign** bit is 0, then the true value is 0 (+0).
   b. If the **Fraction** is 0 but the **Sign** bit is 1, then the true value is also 0 (–0).
   c. If the **Fraction** is not 0, then the true value is a subnormal or a denormalized value and its true value should be calculated by using

$$\mathbf{True\,Value = (-1)^{Sign} \times 2^{-1022} \times (0.0 + Fraction)} \quad (11.6)$$

3. If the biased exponent, **Exponent**, is equal to **0x7FF** (011111111111B), the following possible results exist:
   a. If the **Fraction** is 0 and the **Sign** bit is also 0, then the true value is infinitive (+∞).

Table 11.3 Possible true value for a given double-precision floating-point number

Exponent	Fraction = 0	Fraction ≠ 0	Calculation equation
000H	**+0, −0**	**Subnormal numbers**	$(-1)^{Sign} \times 2^{-1022} \times (0.0 + \text{fraction})$
001H~07FEH	**Normalized numbers**		$(-1)^{Sign} \times 2^{(Exponent-1023)} \times (1.0 + \text{fraction})$
07FFH	**± infinitive (±∞)**	**NaN (signaling or quiet)**	

b. If the **Fraction** is 0 and the **Sign** bit is 1, then the true value is also infinitive (−∞).
c. If the **Fraction** is not 0, then it is a special code or known as **NaN**.

A NaN value can be categorized into a signaling NaN or a quiet NaN, and this is defined as

1. If bit-51 of the **Fraction** is 0, it is a signaling NaN.
2. If bit-51 of the **Fraction** is 1, it is a quiet NaN.

Table 11.3 shows a summarization for these possible results.

The IEEE 754 standard specifies **binary64** as a data type for a double-precision floating-point number. One point to be noted when dealing with the double-precision floating-point number is that the ARM® Cortex®-M4 MCU does not support this kind of floating-point data.

11.3 FPU in the Cortex®-M4 MCU

As we mentioned, the Cortex®-M4F FPU fully supports single-precision floating-point data operations, which include addition, subtraction, multiplication, division, multiplication and accumulation (MAA), and square root operations. It also provides conversions between fixed-point and floating-point data formats, and floating-point constant instructions.

Basically, the FPU in the Cortex®-M4F system provides the following functions:

- 32-bit instructions for single-precision (C float) data processing operations.
- Combined multiply and accumulate instructions for increased precision (Fused MAC).
- Hardware support for conversion, addition, subtraction, multiplication with optional accumulate, division, and square root operation.
- Hardware support for denormals and all IEEE rounding (R) modes.
- 32 dedicated 32-bit single-precision registers, also addressable as 16 double word registers.
- Decoupled three stage pipelines.

The Cortex®-M4F floating-point instruction set does not support all operations defined in the IEEE 754-2008 standard. Also the FPU is disabled from a system reset and you must enable it before you can use any floating-point instructions. The processor must be in privileged mode to read from and write to the Coprocessor Access Control Register (**CPACR**).

11.3.1 *Architecture of the floating-point registers*

The Cortex®-M4 MCU provides an optional FPU. Additional registers are needed to support floating data operations if this FPU is used. These registers include **FPDPR** and **FPSCR**.

The **FPDPR** are composed of 32 single-precision registers, **S0~S31**, or 16 double-precision registers, **D0~D15**, respectively. Each of the 32-bit single-precision registers S0 to S31 can be accessed using floating-point instructions. These registers can also be accessed as a pair or double-precision registers D0 to D15 (64-bit). The configuration of these registers is shown in Figure 11.4. All of these registers make up a so-called floating-point register bank.

One point to be noted is that the FPU in the Cortex®-M4 does not support double-precision floating-point calculations, but you can still use floating-point instructions to transfer double-precision data to the single-precision data via some runtime library functions.

All floating-point data calculations are under the control of the **FPSCR**. This register provides the following control functions:

- Define the floating-point operation behaviors.
- Provide status information about the floating-point operation results.

Bits functions on the **FPSCR** are shown in Figure 11.5.

The functions of bits **N**, **Z**, **C**, and **V** are identical with those in the **PSR**. The function of each other bit in the **FPSCR** is (bits 5~6, 8~21, and 27 are reserved)

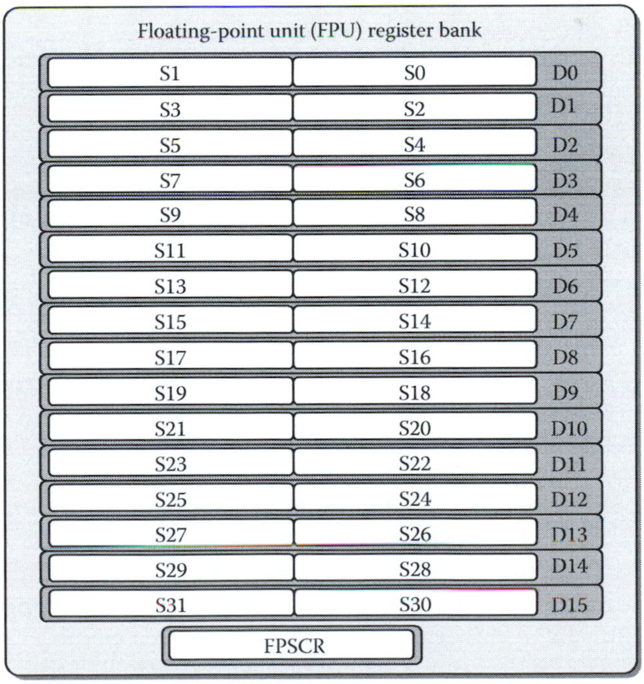

Floating-point unit (FPU) register bank

S1	S0	D0
S3	S2	D1
S5	S4	D2
S7	S6	D3
S9	S8	D4
S11	S10	D5
S13	S12	D6
S15	S14	D7
S17	S16	D8
S19	S18	D9
S21	S20	D10
S23	S22	D11
S25	S24	D12
S27	S26	D13
S29	S28	D14
S31	S30	D15

FPSCR

Figure 11.4 Configuration of the floating-point register bank.

Bits	31 30 29 28	27	26	25	24	23:22	21:8	7	6:5	4	3	2	1	0
FPSCR	N Z C V		AHP	DN	FZ	RMode		IDC		IXC	UFC	OFC	DZC	IOC

Figure 11.5 Bit function and structure on FPSCR.

- **AHP** (Bit-26): The value on this bit defines the AHP format for the floating-point operations. A 0, which is the default value on this bit, is used to define the IEEE half-precision format. A 1 is to define an AHP format.
- **DN** (Bit-25): The value on this bit is used to define the default NaN mode. A 0 means that the NaN operands propagate through to the output of a floating-point operation, and this is the default value. A 1 indicates that any operation including one or more NaNs returns the default NaN.
- **FZ** (Bit-24): The value on this bit indicates whether the flush-to-zero model is enabled or disabled. A value of 0, which is the default value, on this bit means that the FZ model is disabled, otherwise if this bit value is 1, which means that the FZ model is enabled.
- **RMode** (Bits 23 and 22): These two bits are used to set up the specified rounding mode that is used by all floating-point operational instructions. The values of these bits are:
  - 00—RN mode (default)
  - 01—RP mode
  - 10—RM mode
  - 11—RZ mode
- **IDC** (Bit-7): This bit is used to monitor whether a floating-point exception has occurred or not. A 1 indicated that a floating-point exception has happened, and the result is not within the normalized value range. A 0 means that no floating-point exception occurred. This bit can be cleared by writing 0 to it.
- **IXC** (Bit-4): This bit is used to detect whether an inexact cumulative exception occurred or not. A 1 in this bit indicated that a floating exception has occurred, otherwise a 0 means that no floating-point exception occurred. This bit can be cleared by writing 0 to it.
- **UFC** (Bit-3): This bit is the underflow cumulative exception status bit. A 1 in this bit indicated that an underflow cumulative exception has occurred. Otherwise if this bit is 0, which means that no underflow cumulative exception has occurred. This bit can be cleared by writing 0 to it.
- **OFC** (Bit-2): This bit is the overflow cumulative exception status bit. A 1 in this bit indicated that an overflow cumulative exception has occurred. Otherwise if this bit is 0, which means that no overflow cumulative exception has occurred. This bit can be cleared by writing 0 to it.
- **DZC** (Bit-1): This bit is the divided by zero cumulative exception status bit. A 1 in this bit indicated that a divided by zero cumulative exception has occurred. Otherwise if this bit is 0, which means that no divided by zero cumulative exception has occurred. This bit can be cleared by writing 0 to it.
- **IOC** (Bit-0): This bit is the invalid operation cumulative exception status bit. A 1 in this bit indicated that an invalid operation cumulative exception has occurred. Otherwise if this bit is 0, which means that no invalid operation cumulative exception has occurred. This bit can be cleared by writing 0 to it.

In addition to this register bank, some additional registers are provided to support the floating-point data operations, these registers include:

- Coprocessor Access Control Register (**CPACR**) in the System Control Block (SCB).
- Floating-Point Context Control Register (**FPCCR**).
- Floating-Point Context Address Register (**FPCAR**).
- Floating-Point Default Status Control Register (**FPDSCR**).
- Media and FP Feature Register 0 (**MVFR0**).
- Media and FP Feature Register 1 (**MVFR1**).

In the ARM® Cortex®-M4 system, a set of floating-point instructions are used to perform floating-point operations and floating-point data transfers, instead of using the coprocessor access instructions.

The **CPACR** allows the users to enable or disable the FPU by using **SCB → CPACR** instruction in the CMSIS Core. Since the Cortex®-M4 MCU system does not use any coprocessor-related instruction to access and control the floating-point operations, we will not provide very detailed discussions about these registers in this section. Refer to related documents to get more details for these registers.

Now, let's have a closer look at the operation modes of the FPU.

11.3.2 FPU operational modes

In the ARM® Cortex®-M4 MCU system, the FPU is considered as a coprocessor and all operations related to this FPU belong to coprocessor data operations (CDP). This coprocessor provides three operation modes to handle and accommodate a variety of operations.

- **Full-compliance mode:** In this operation mode, the FPU processes all operations according to the IEEE 754 standard in hardware.
- **Flush-to-zero mode:** Setting the FZ bit of the **FPSCR** to enable flush-to-zero mode. In this mode, the FPU treats all subnormal input operands of arithmetic CDP operations as zeros in the operation. Exceptions that result from a zero operand are signaled appropriately. Some floating-point operations, such as absolute (**VABS**), negative (**VNEG**), and move immediate (**VMOV**), are not considered arithmetic CDP operations and are not affected by flush-to-zero mode. A result that is tiny, as described in the IEEE 754 standard, where the destination precision is smaller in magnitude than the minimum normal value before rounding, is replaced with a zero. The **IDC** bit in **FPSCR** indicates when an input flush occurs. The **UFC** bit in **FPSCR** indicates when a result flush occurs.
- **Default NaN mode:** Setting the **DN** bit in the **FPSCR** enables default NaN mode. In this mode, the result of any arithmetic data processing operation that involves an input NaN, or that generates a NaN result, returns the default NaN. Propagation of the fraction bits is maintained only by VABS, VNEG, and VMOV operations. All other CDP operations ignore any information in the fraction bits of an input NaN.

Next, let's take care of the implementations of the FPU in the Cortex®-M4 system.

11.4 Implementing the FPU

In the ARM® Cortex®-M4F MCU system, all FPU-related operations are performed by using a set of floating-point instructions. The Cortex®-M4F floating-point instruction set does not support all operations defined in the IEEE 754-2008 standard. Most unsupported operations include, but are not limited to the following operations:

- Floating-point remainder, such as z = fmod(x, y).
- Round floating-point number to integer-valued floating-point number
- Binary-to-decimal conversions
- Decimal-to-binary conversions
- Direct comparison of single-precision and double-precision values

The Cortex®-M4 FPU supports fused MAC operations as described in the IEEE standard. For complete implementation of the IEEE 754-2008 standard, floating-point functionality must be cooperated with library runtime functions.

11.4.1 Floating-point support in CMSIS-Core

The CMSIS-Core provides a set of macros, data structure, and instructions to support the FPU-related operations in the Cortex®-M4 MCU system. Two often used macros are

- __FPU_PRESENT
- __FPU_USED

If the FPU is present in the Cortex®-M4 MCU system, the macro **__FPU_PRESENT** is set to 1 by the device-specific header file. If the **__FPU_PRESENT** is 0, the **__FPU_USED** must also be cleared to 0 to indicate that no FPU is presented and used in this Cortex®-M4 system.

All FPU-related data structures are available if **__FPU_PRESENT** is set to 1. The system initialization function **SystemInit()** is executed to enable the FPU if the **__FPU_USED** is set to 1 by writing to the **CPACR** when a reset is generated and the reset handler is executed.

Some popular FPU-related instructions supported by CMSIS-Core include the **SCB→CPACR** that is used to access the **CPACR**, **FPU→FPSCR** that is used to access the **FPSCR**, **FPU→FPCCR** that is used to access the **FPCCR**, **FPU→FPCAR** that is used to access the **FPCAR**, and **FPU→FPDSCR** for accessing Media and FP Feature Registers (**MVFR0** and **MVFR1**).

By default, the FPU is presented and used in most Cortex®-M4-related IDEs. For example, in Keil® MDK-ARM µVersion5 IDE, a single-precision FPU is automatically selected and used as this IDE is opened, as shown in Figure 11.6.

The **use single-precision** is selected in the **floating-point hardware** combo box in the target wizard when the Keil® MDK-ARM µVersion5 IDE is opened.

For the **gcc** users, you need to type some special command with some options to activate the FPU in that IDE.

11.4.2 Floating-point programming in the MSP432P401R MCU system

In the MSP432P401R MCU system, several control registers are used to support the FPU operations and these registers are:

- Coprocessor Access Control Register (**CPACR**)
- Floating Point Context Control Register (**FPCCR**)

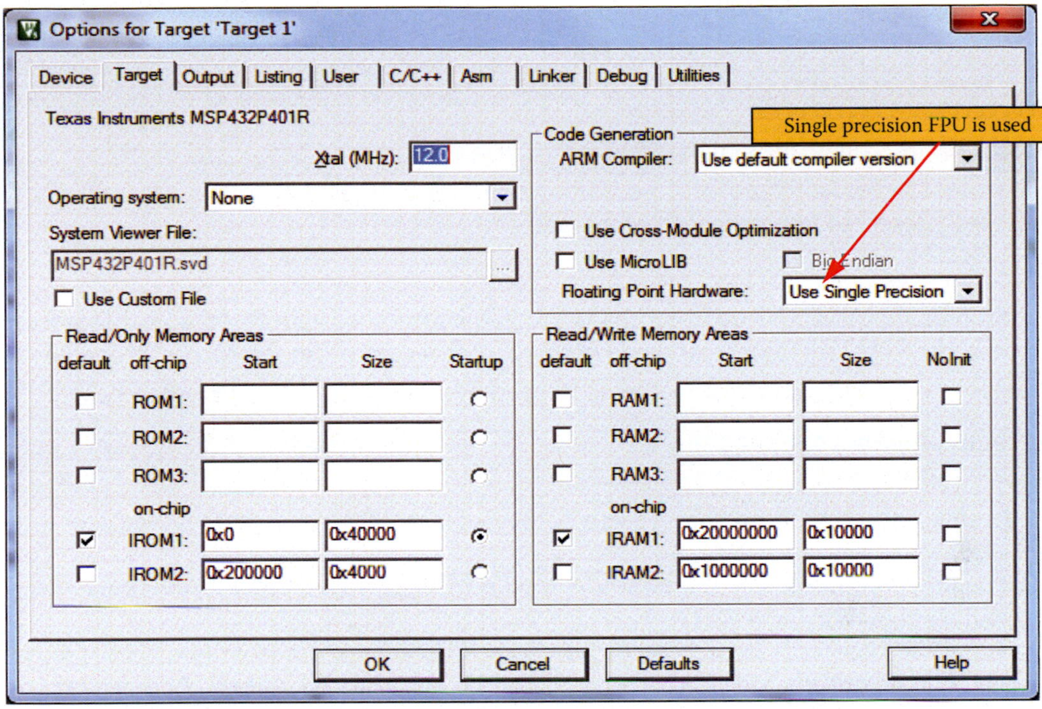

Figure 11.6 Single-precision FPU is automatically used in MDK-ARM µVersion5 IDE.

- Floating Point Context Address Register (**FPCAR**)
- Floating Point Default Status Control Register (**FPDSCR**)

After a system reset operation, the FPU is disabled. You must enable it before you can use any floating-point instructions. The processor must be in privileged mode to read from and write to the **CPACR**.

Ideally, in total, about 16 coprocessors can be defined in an ARM® MCU system, but the FPU is defined as coprocessor 10 (**CP10**) and coprocessor 11 (**CP11**) in the Cortex®-M4 MCU system by using this **CPACR**. Since no other coprocessor is available, therefore both **CP10** and **CP11** are used for the FPU. When programming this register, the coding for **CP11** and **CP10** must be identical to make sure to select and use the FPU. As we mentioned, one can use **SCB** → **CPACR** instruction to access this register. Let's have a closer look at these registers in the next section.

11.4.2.1 FPU in the DRA model

The so-called direct register model (**DRA**) is to access and assign the values to these FPU-related control registers directly.

First let's take care of the **CPACR**.

Figure 11.7 shows the bit configuration for this register.

Two bit fields, **CP11** (bits 23:22) and **CP10** (bits 21:20), have the same bit values and functions:

1. **0x0**: Access denied (no access to this FPU is allowed).
2. **0x1**: Privileged access only.

Coprocessor Access Control Register (CPACR)

31–24	23–22	21–20	19–0
Reserved bits 31~24	CP11	CP10	Reserved bits 19~0

Figure 11.7 Bit configuration for the CPAC register.

3. **0x2**: Reserved (equivalent to no access).
4. **0x3**: Full access.

By default, **CP11** and **CP10** bit fields are zeros after a system reset, and this setting disables the FPU functions to allow lower power dissipations on the MCU. To enable the FPU operations, one can set **CP11** and **CP10** fields to enable the FPU to be fully accessed as:

```
SCB→CPACR = 0x00F00000
```

In this setting, both **CP11** and **CP10** fields are set to 11 to enable the FPU to be fully accessed.

The bit configuration of the **FPCCR** is shown in Figure 11.8. The bit functions of this register are shown in Table 11.4.

The **FPCC** register is used to set or return the FPU-related control data. These control data include the top six settings, which contain the **Automatic State Preservation Enable** to automatically preserve and restore the hardware state for floating-point context on exception and exit, the **Lazy State Preservation Enable** to allow lazy state to be preserved automatically for floating-point context, the **Monitor Ready** to enable the debug monitor to set the **MON_PEND** bit when the floating-point stack is allocated, the BusFault Ready to set the BusFault bit to pending the BusFault handler as the floating-point stack is allocated, the **Memory Management Fault Ready** to set the MemManage handler to the pending state as the floating-point stack is allocated, the Hard Fault Ready to set the Hard Fault handler to the pending state when the floating-point stack is allocated.

The lower three data are used to show the running status for the FPU operations. These include the **Thread Mode** that is used to indicate whether the current running mode is the thread mode or not, the **User Privilege Level** that is used to indicate whether the user has the privilege level to run the user's program, the **Lazy State Preservation Active** that is used to indicate whether the lazy preservation is active or not.

The **FPCAR** holds the address of the unpopulated floating-point register space allocated on an exception stack frame. This is a 32-bit register but only the top 29 bits (31:3) are used to hold the address of the unpopulated floating-point register space used in the stack space.

The **FPDSCR** is used to hold the default values for the **FPSCR**. The bit configuration of the **FPDSCR** is shown in Figure 11.9.

For the upper three bits, **AHP** (bit-26), **DN** (bit-25), and **FZ** (bit-24), each bit holds the default value for the corresponding bit, **AHP**, **DN**, and **FZ**, in the **FPSCR**. The bit field **RMODE** (bits

Floating-Point Context Control Register (FPCCR)

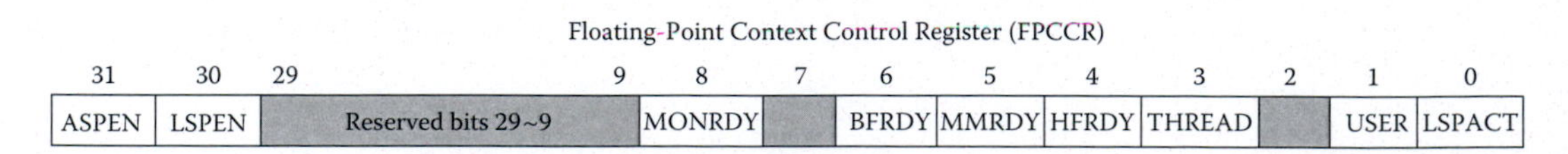

Figure 11.8 Bit configuration for the FPCC register.

Table 11.4 Bit value and its function for FPCC register

Bit	Name	Reset	Function
31	**ASPEN**	1	Automatic state preservation enable: When set, enables the use of the FRACTV bit in the **CONTROL** register on execution of a floating-point instruction. This results in automatic hardware state preservation and restoration, for floating-point context, on exception entry and exit
30	**LSPEN**	1	Lazy state preservation enable: When set, enables automatic lazy state preservation for floating-point context
29:9	**Reserved**	0x00	Reserved
8	**MONRDY**	0	Monitor ready: When set, the debug monitor is enabled and priority permits setting **MON_PEND** when the floating-point stack frame was allocated
7	**Reserved**	0	Reserved
6	**BFRDY**	0	BusFault ready: When set, the **BusFault** is enabled and priority permitted setting the **BusFault** handler to the pending state when the floating-point stack frame was allocated
5	**MMRDY**	0	Memory management fault ready: When set, the **MemManage** is enabled and priority permitted setting the **MemManage** handler to the pending state when the floating-point stack frame was allocated
4	**HFRDY**	0	Hard Fault ready: When set, priority permitted setting the **HardFault** handler to the pending state when the floating-point stack frame was allocated
3	**THREAD**	0	Thread mode: When set, mode was thread mode when the floating-point stack frame was allocated
2	**Reserved**	0	Reserved
1	**USER**	0	User privilege level: When set, privilege level was user when the floating-point stack frame was allocated.
0	**LSPACT**	0	Lazy state preservation active: When set, lazy state preservation is active. Floating-point stack frame has been allocated but saving state to it has been deferred

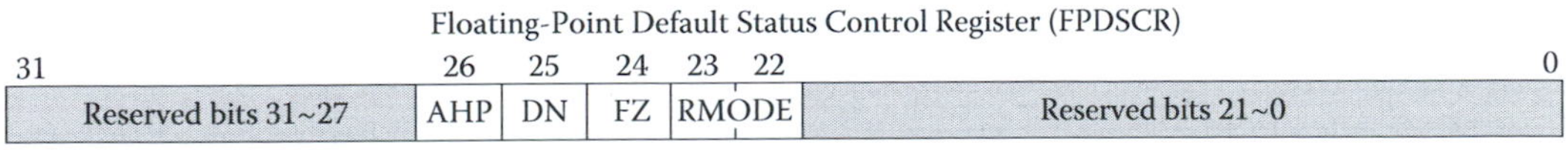

Floating-Point Default Status Control Register (FPDSCR)

31	26	25	24	23 22	0
Reserved bits 31~27	AHP	DN	FZ	RMODE	Reserved bits 21~0

Figure 11.9 Bit configuration for the FPDSC register.

23:22) in this register also holds the default values for the corresponding bit field **RMODE** in the **FPSCR**. The available bit values and the function values for this bit field are:

- **0x0:** Round to Nearest (**RN**) mode.
- **0x1:** Round towards Plus Infinity (**RP**) mode.
- **0x2:** Round towards Minus Infinity (**RM**) mode.
- **0x3:** Round towards Zero (**RZ**) mode.

Next, let's have a closer look at the FPU-related API functions provided by the MSPW Peripheral Driver Library.

11.4.2.2 FPU in the SD model

In the MSP432P401R MCU system, the MSPW Peripheral Driver Library provides nine API functions for manipulating the behavior of the FPU in the Cortex®-M4 microprocessor.

The behavior of the FPU can also be adjusted and specified to the different formats, such as the half-precision and single-precision floating-point values, the handle of NaN values, the flush-to-zero mode that sacrifices the full IEEE 754 compliance for execution speed, and the rounding mode for results.

This driver is contained in a C source file **fpu.c** and a header file **fpu.h**, and both files are located at the default folder **C:\ti\msp\MSPWrae_2_21_00_39\driverlib\driverlib\MSP432P4xx**.

Table 11.5 shows these API functions with the related arguments. All FPU interrupt-related API functions are not provided in this library and therefore they are not supported for these API functions.

These FPU API functions can be divided into the following three groups based on their functions:

1. Enable and disable the FPU: **FPU_enableModule()** and **FPU_disableModule()**.
2. Control how the floating-point state is stored on the stack when interrupts occur: **FPU_enableStacking()**, **FPU_enableLazyStacking()**, and **FPU_disableStacking()**.
3. Adjust the operation of the FPU: **FPU_setHalfPrecisionMode(), FPU_setNaNMode(), FPU_setFlushToZeroMode(),** and **FPU_setRoundingMode()**.

11.4.3 A FPU example project using the DRA model

In this section, we build an example FPU project to access the FPU-related registers directly to generate a sinusoidal sequence with floating-point data operations.

Create a new folder **DRAFPU** under the folder **C:\MSP432 Class Projects\Chapter 11** with Windows Explorer. Then create a new µVersion5 project **DRAFPU** and save it into the new folder **DRAFPU** created above. Create a new C source file **DRAFPU.c** and enter the codes shown in Figure 11.10 into the source file. Let's have a closer look at this piece of code to see how it works.

1. The codes in lines 4~7 are used to include all necessary system header files to be used in this project. The **<math.h>** header file is specially used for the floating-point function **sinf()**.
2. The codes in lines 8 and 9 are used to declare two user-defined constants, **M_PI** and the length of the floating-point array **gSData[]**.
3. The **main()** program starts at line 10.
4. Two local variables, **count**, **fRadians,** and a data array **gSData[]**, are declared in lines 12 and 13. These variables are used as the loop counter and intermediate value holder as well as the data collector of the calculated results for the floating-point data.
5. The **ASPEN** and the **LSPEN** bits on the **FPCCR** are set to 1 to enable the hardware state and the lazy state to be automatically preserved and restored for the floating-point context in line 14. A pointer structure **FPU → FPCCR** is used in this instruction to access the **FPCCR** since the **FPU** is defined as a pointer point to a structure **FPU_Type**,

Table 11.5 FPU operation API functions

API Function	Parameter	Description
void **FPU_enableModule(**void**)**	**None**	Enables the floating-point unit, allowing the floating-point instructions to be executed. This function must be called prior to performing any hardware floating-point operations
void **FPU_disableModule(**void**)**	**None**	Disables the floating-point unit, preventing floating-point instructions from executing
void **FPU_enableStacking(**void**)**	**None**	Enables the stacking of floating-point registers **s0–s15** when an interrupt is generated and handled When enabled, space is reserved on the stack for the floating-point context and the floating-point state is saved into this stack space. Upon return from the interrupt, the floating-point context is restored
void **FPU_enableLazyStacking** **(**void**)**	**None**	Enables the lazy stacking of floating-point registers **s0–s15** when an interrupt is handled. When lazy stacking is enabled, space is reserved on the stack for the floating-point context, but the floating-point state is not saved. If a floating-point instruction is executed from within the interrupt context, the floating-point context is first saved into the space reserved on the stack. On completion of the interrupt handler, the floating-point context is only restored if it was saved (as the result of executing a floating-point instruction)
void **FPU_disableStacking(**void**)**	**None**	Disables the stacking of floating-point registers **s0–s15** when an interrupt is generated and handled When floating-point context stacking is disabled, floating-point operations performed in an interrupt handler destroy the floating-point context of the main thread of execution
void **FPU_setHalfPrecisionMode** **(**uint32_t ui32Mode**)**	**ui32Mode** is the format for half-precision floating-point value, which is either **FPU_HALF_IEEE** or **FPU_HALF_ALTERNATE**	Selects between the IEEE half-precision floating-point representation and the Cortex-M processor alternative representation. The alternative representation has a larger range but does not have a way to encode infinity (positive or negative) or NaN (quiet or signaling) The default setting is the IEEE format

(*Continued*)

Table 11.5 (Continued) FPU operation API functions

API Function	Parameter	Description
void **FPU_setNaNMode** **(**uint32_t ui32Mode**)**	**ui32Mode** is the mode for **NaN** results; which is either **FPU_NAN_PROPAGATE** or **FPU_NAN_DEFAULT**	Selects the handling of **NaN** results during floating-point computations. **NaNs** can either propagate (the default), or they can return the default **NaN**
void **FPU_setFlushToZeroMode** **(**uint32_t ui32Mode**)**	**ui32Mode** is the flush-to-zero mode; which is either **FPU_FLUSH_TO_ZERO_DIS** or **FPU_FLUSH_TO_ZERO_EN**	Enables or disables the **flush-to-zero** mode of the floating-point unit. When disabled (the default), the floating-point unit is fully IEEE compliant. When enabled, values close to zero are treated as zero, greatly improving the execution speed at the expense of some accuracy (as well as IEEE compliance)
Void **FPU_setRoundingMode** **(**uint32_t ui32Mode**)**	**ui32Mode** is the rounding mode	Selects the rounding mode for floating-point results. After a floating-point operation, the result is rounded toward the specified value. The default mode is **FPU_ROUND_NEAREST** The following rounding modes are available (as specified by **ui32Mode**): **FPU_ROUND_NEAREST**—round toward the nearest value **FPU_ROUND_POS_INF**—round toward positive infinity **FPU_ROUND_NEG_INF**—round toward negative infinity **FPU_ROUND_ZERO**—round toward zero

which is defined in the **core_cm4.h** header file. If you like, you can use the macro **FPU_FPCCR** that is defined in the **msp.h** header file to replace that pointer structure to access the **FPCCR**.

6. In line 15, the bit fields **CP11** and **CP10** in the **CPACR** are set to **0xFF** to enable the coprocessor (FPU) to begin its normal operations. Similarly, a pointer structure **SCB → CPACR** is used to access the **CPACR**. You can use the system macro **SCB_CPACR** that is defined in the **msp.h** header file to replace that pointer structure if you like.
7. The unit radian degree **fRadians** is calculated as **2π/sLength**, which is **0.2π**, in line 16.
8. A conditional **while()** loop starts at line 17. This loop is to repeatedly update the unit radian degree **fRadians** by incrementing it by 0.2π and calculating the sinusoidal result for that updated **fRadians**, and assign them to the data array **gSData[]** in lines 17~19.
9. The code in line 20 has no usage in programming function. Instead, we just try to avoid a compiling warning for the data array **gSData[]** since no operation has been performed by this data array except the data assignments.

```
1  //*****************************************************************************************
2  // DRAFPU.c - Main Application File for the DRAFPU Project
3  //*****************************************************************************************
4  #include <stdint.h>
5  #include <stdbool.h>
6  #include <math.h>
7  #include <msp.h>
8  #define M_PI    3.14159265358979323846
9  #define sLength 100
10 int main(void)
11 {
12   int32_t count = 0;
13   float fRadians,  gSData[sLength];
14   FPU->FPCCR |= 0xC0000000;            // set ASPEN & LSPEN bits to enable auto HW & Lazy set/restore
15   SCB->CPACR = 0x00F00000;             // enable FPU
16   fRadians = (2 * M_PI)/sLength;
17   while(count < sLength)
18   {
19     gSData[count] = sinf(fRadians * count);
20     gSData[count] = gSData[count] + 0; // useless at all - just avoid compiler warning
21     count++;
22   }
23   while(1){}
24 }
```

Figure 11.10 Detailed code for the project DARFPU.

10. Then the loop counter **count** is updated by 1 in line 21. This loop can be replaced by a **for()** loop. The difference between the **sin()** and the **sinf()** functions in the C codes is that the former accepts only a double-precision value and returns a double-precision result. However, the latter only accepts the single floating-point value and returns a single floating-point result.
11. In line 23, an infinitive **while()** loop is used to enable users to check the calculation results stored in the data array **gSData[]** from the **Call Stack + Locals** window.

Before the project can be run, appropriate environments need to be set to enable the project to be built and downloaded into the flash memory in the MSP-EXP432P401R EVB. These environment setups include the selection of the correct debugger for this project, and making the correct setups for the debugger and the flash downloading operations. Perform the following operations to finish these setups:

- Going to **Project|Options for Target "Target 1"** menu item to open the **Options** wizard.
- On the opened Options wizard, click on the **Debug** tab.
- Make sure that the debugger shown in the **Use:** box is **CMSIS-DAP Debugger**. Otherwise you can click on the dropdown arrow to select this debugger from the list.

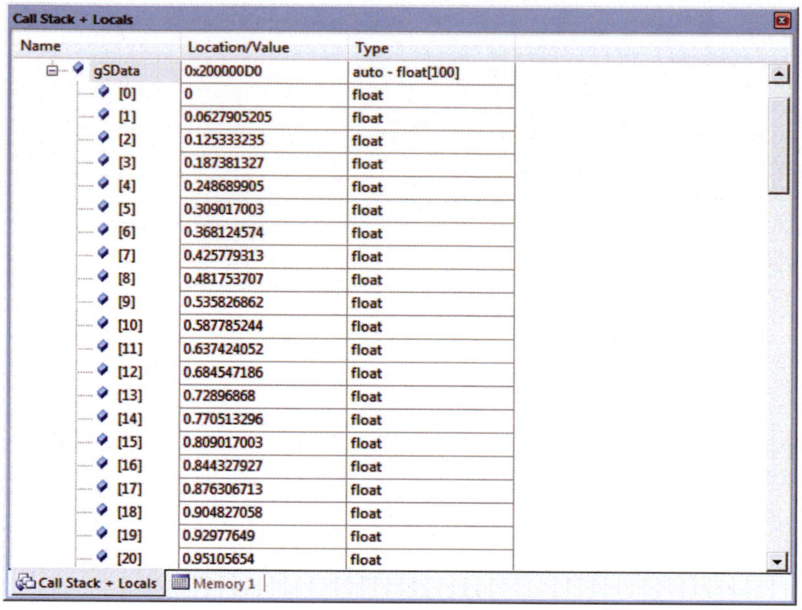

Call Stack + Locals

Name	Location/Value	Type
gSData	0x200000D0	auto - float[100]
[0]	0	float
[1]	0.0627905205	float
[2]	0.125333235	float
[3]	0.187381327	float
[4]	0.248689905	float
[5]	0.309017003	float
[6]	0.368124574	float
[7]	0.425779313	float
[8]	0.481753707	float
[9]	0.535826862	float
[10]	0.587785244	float
[11]	0.637424052	float
[12]	0.684547186	float
[13]	0.72896868	float
[14]	0.770513296	float
[15]	0.809017003	float
[16]	0.844327927	float
[17]	0.876306713	float
[18]	0.904827058	float
[19]	0.92977649	float
[20]	0.95105654	float

Call Stack + Locals | Memory 1

Figure 11.11 Running result of the project DRAFPU.

- Click on the **Settings** button on the right of this Debugger box and make sure both checkboxes are checked:
  - **Verify Code Download**
  - **Download to Flash**
- Click on the **Flash Download** tab to make sure that the radio button **Erase Full Chip** is checked.

Now run the project by going to **Debug|Start/Stop Debug Session** and the **Debug|Run** menu items. In order to check the running result, one needs to go to **Debug|Stop** menu item to stop the project. On the opened **Call Stack + Locals** window, expand the data array **gSData[]** and you can find all 100 floating-point results, as shown in Figure 11.11.

A point to be noted when you expand the **gSData[]** array to check the result is that you have to wait for a couple of seconds to get the results since this is a floating-point data operation and needs some time to complete. Keep your patience for this issue.

If plotting the data array **gSData[]**, the graphic result is a sinusoidal signal with one cycle time of 2π, as shown in Figure 11.12.

11.5 Summary

One of the most important differences between the Cortex®-M4 MCU and Cortex®-M3 MCU is that an optional FPU is added into the Cortex®-M4 Core to enhance the floating-point data operations. The Cortex®-M4 FPU implements ARMv7E-M architecture with FPv4-SP extensions. It provides floating-point computation functionality that is compliant with the *ANSI/IEEE Standard 754-2008, IEEE Standard for Binary Floating-Point Arithmetic*. In this chapter, we concentrated our discussions on this FPU element with a completed and detailed analysis for this unit.

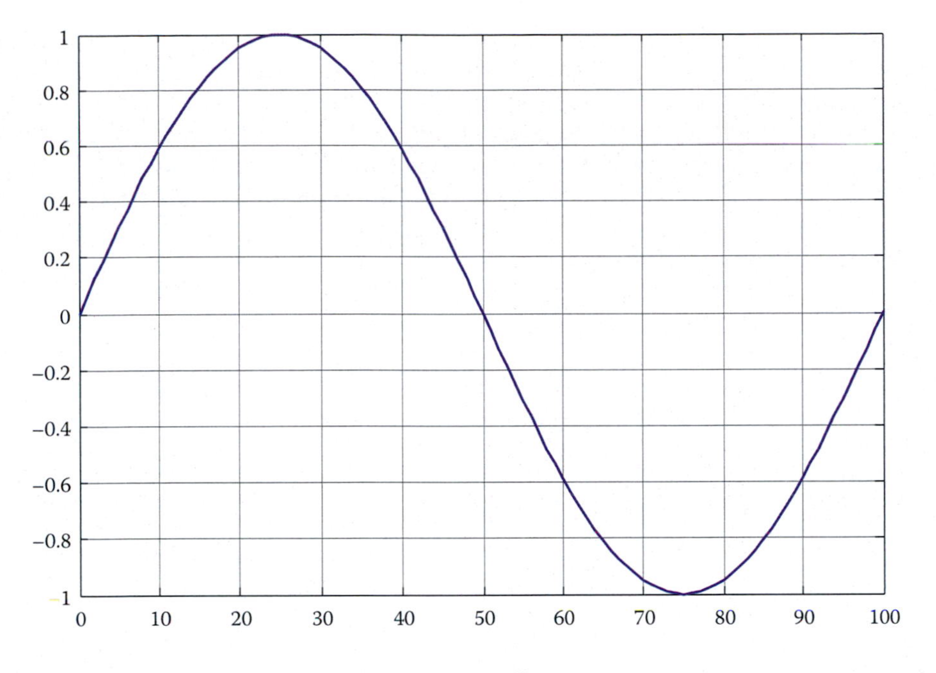

Figure 11.12 Plotting result for the data array gSData[].

An introduction and overview about the FPU are given in Section 11.1. Three types of the floating-point data formats, half-precision, single-precision, and double-precision, are discussed in Section 11.2 with some real examples.

A detailed discussion about the FPU applied in the Cortex®-M4 MCU is provided in Section 11.3. This discussion includes the architecture of most popular FPU-related registers and FPU operational modes. The implementations of the FPU are given in Section 11.4. This section contains the floating-point support in CMSIS-Core, floating-point programming in the MSP432P401R MCU system, FPU API functions provided by the MSPW are Peripheral Driver Library, and a real FPU example project using the DRA model.

HOMEWORK

I. True/False selections

____ 1. One of the most important differences between the Cortex®-M4 MCU and Cortex®-M3 MCU is that an optional FPU is added into the Cortex®-M4 Core to enhance the floating-point data operations.

____ 2. If a floating-point instruction is executed but the FPU is disabled, no fault would be generated.

____ 3. The Cortex®-M4F FPU fully supports half-precision, single-precision, and double-precision floating-point data operations.

____ 4. The FPU can generate an interrupt when one of the several exceptions occurs. These exceptions include the underflow, overflow, divide by zero, invalid operation, abnormal input, and inexact exception.

____ 5. In the half-precision floating-point format, it includes one **`Sign`** bit (bit-15), 5 **`Exponent`** bits (bits 14~10), and 10 **`Fraction`** bits (bits 9~0).

____ 6. If the biased exponent, **`Exponent`**, is 0, and if the **`Fraction`** is not 0, then the true value for a half-precision number is: **`True Value`** = $(-1)^{Sign} \times 2^{(Exponent-15)} \times (1.0 + Fraction)$.

____ 7. If the biased exponent, **`Exponent`**, is equal to **`0xFF`** (11111111B), and if the **`Fraction`** is not 0, then it is a special code or known as **`NaN`**.

____ 8. The FPU is disabled from a system reset and you must enable it before you can use any floating-point instructions. The processor must be in privileged mode to read from and write to the **`CPACR`**.

____ 9. In the Cortex®-M4 MCU system, the FPU is defined as **`CP10`** and **`CP11`**. The coding for **`CP11`** and **`CP10`** bit fields in the **`CPACR`** can be different to make sure to select and use the FPU.

____ 10. For complete implementations of the IEEE 754-2008 standard, the floating-point functionality must be cooperated with some library runtime functions.

II. Multiple choices

1. Which one of the following statements is true? ____________.
   a. An optional FPU in the Cortex®-M4 is to enhance the floating-point data operations.
   b. The Cortex®-M4 FPU implements ARMv7E-M architecture with FPv4-SP extensions.
   c. It is compliant with the *ANSI/IEEE Standard 754-2008.*
   d. All of them
2. The Cortex®-M4F FPU fully supports ____________.
   a. Half-precision and single-precision floating-point formats.
   b. Single-precision and double-precision floating-point formats.
   c. Only a
   d. Both a and b
3. The usage of the FPU in the Cortex®-M4 MCU system can be divided into two categories ____________.
   a. Full-compliance mode, flush-to-zero mode.
   b. Default NaN mode, Normal mode.
   c. In the main thread of the program, inside an interrupt handler.
   d. Inside an interrupt handler, in the normal mode.
4. The FPU can generate an interrupt when one of the several exceptions occurs. These exceptions include the ____________.
   a. A **`NaN`** result is obtained.
   b. The exponent is 0.
   c. The fraction is 0.
   d. An overflow or an underflow occurred.
5. A single-precision floating-point number is composed of ____________.
   a. 1 Sign bit, 8 Exponent bits, and 10 Fraction bits
   b. 1 Sign bit, 5 Exponent bits, and 10 Fraction bits
   c. 1 Sign bit, 8 Exponent bits, and 23 Fraction bits
   d. 1 Sign bit, 8 Exponent bits, and 52 Fraction bits
6. The true value for a half-precision floating-point number **`0011100010010100`** is ____________.
   a. **`0.576256256`**
   b. **`0.572265625`**

   c. **0.672256256**
   d. **0.472256256**
7. The half-precision floating-point presentation for the number 15.5 is ___________.
   a. 1100101111000000
   b. 0100101111111111
   c. 1100101111111111
   d. 0100101111000000
8. The FPU can perform ____________________ operation modes and they are ______________.
   a. 2, Full-compliance mode and flush-to-zero mode.
   b. 3, Full-compliance mode, default NaN mode, and flush-to-zero mode.
   c. 4, Full-compliance mode, NaN mode, normal mode, and flush-to-zero mode.
   d. 5, Full mode, Default mode, NaN mode, normal mode, and flush-to-zero mode.
9. Two often used macros defined in the Cortex®-M4 FPU are ______________.
   a. __FPU_USED, __FPU_PRESENT
   b. __FPU_NOT_USED, __FPU_SHOW
   c. __FPU_WORK, __FPU_ON
   d. __FPU_DONE, __FPU_OK
10. Set **CP11** and **CP10** bit fields to enable the FPU to be fully accessed as ________.
   a. 0x0
   b. 0x1
   c. 0x2
   d. 0x3

III. Exercises
1. Provide a brief description about the FPU used in the Cortex®-M4 Core.
2. Provide a brief description about the half-precision format.
3. Provide a brief description about the single-precision format.
4. Provide a brief discussion about the FPU operation modes.
5. Provide a brief discussion about the popular control registers used for the FPU.

IV. Practical laboratory

LABORATORY 11 MSP432 FPU

11.0 GOALS

This laboratory exercise allows students to learn and practice ARM® Cortex®-M4 FPU and related operations. Some FPU-related API functions are used in this lab to enable students to make them familiar with these functions.

1. Program **Lab11 _1(SDFPU)** lets students use some FPU-related API functions to access some FPU registers to perform desired floating-point data operations.

After completion of this program, students should understand the basic architecture and operational procedure for the FPU installed in the MSP432P401R MCU system. They should be able to code some sophisticated programs to perform desired FPU-related operations with specified algorithms.

11.1 LAB11_1

11.1.1 Goal

In this project, students will use the SD model to build a floating-point data program to perform desired floating-point data operations. Some FPU-related API functions are to be used in this lab project to enable students to make them familiar with these functions and apply them in the real applications.

The advantage of using these API functions to access and operate the FPU-related registers is that some complicated registers' structures and bit field values can be avoided to improve the coding efficiency and coding process seed.

11.1.2 Data assignment and hardware configuration

No data assignment and hardware configuration are needed for this lab project.

11.1.3 Development of the project and the source code

Perform the following operations to complete this project:

1. Create a new folder **Lab11_1(SDFPU)** in the folder **C:\MSP432 Lab Projects\Chapter 11** with Windows Explorer.
2. Create a new µVersion5 project **SDFPU** and save it into the folder **Lab11_1(SDFPU)** created above.
3. Create a new C source file **SDFPU.c** and add it into the project.
4. Place the following system header files and macros into this source file:

```
#include <stdint.h>
#include <stdbool.h>
#include <math.h>
#include <MSP432P4xx\fpu.h>
#include <MSP432P4xx\wdt_a.h>
```

5. Use **#define** to define **M_PI** as **3.14159265358979323846**.
6. Use **#define** to define a variable **sLength** as 100.
7. Inside the **main()** program, declare one **unit32_t** variable **count** and one **float** variable **fRadian**. Then declare another float data array **gSData**[100];
8. Use an API function to stop the Watchdog timer.
9. Use the API function **FPU_enableLazyStacking()** to set the **ASPEN** and **LSPEN** bits on the **FPCCR** to enable the hardware state and the lazy state to be automatically preserved and restored for the floating-point context.
10. Call the API function **FPU_enableModule()** to set the bit fields **CP11** and **CP10** in the **CPACR** to **0xFF** to enable the coprocessor (FPU) to begin its normal operations.
11. Use the code line **fRadians** = **(2*M_PI)/sLength** to get the unit radian degree.
12. Use a **for()** loop with the variable **count** as the loop counter to repeatedly calculate the sinusoidal value for each radian degree unit, and assign them to the data array **gSData[]**.
13. Use an infinitive **while()** loop to check the running result of this project.

11.1.4 Set up the environment to build and run the project

To build and run the project, one needs to perform the following operations to set up the environments:

- Go to **C/C++** tab in the **Project|Options for Target "Target 1"** to set up the **Include Path** as: **C:\ti\msp\MSPWare_2_21_00_39\driverlib\driverlib**.
- Add the MSPWare Peripheral Driver Library into this project by right clicking on the **Source Group 1** folder in the **Project** panel on the left, and select the **Add Existing Files to Group "Source Group 1"** menu item. On the opened box, select the driver library file path **C:\ti\msp\MSPWare_2_21_00_39\driverlib\driverlib\MSP432P4xx\keil** and then select the driver library file **msp432p4xx_driverlib.lib**. Click the **Add** button to add it into the project.
- Select the correct debugger by clicking on the **Debug** tab and select the **CMSIS-DAP Debugger** from the **Use** box. Also make sure that all settings for the debugger and the flash download are correct by clicking on the settings button. Then go to the **Flash|Download** menu item to download the image file of this project into the flash memory in the MSP432P401R EVB.
- Now go to **Debug|Start/Stop Debug Session** to be ready to run the project.
- Go to **Debug|Run** menu item to run the project.

After the project runs, go to **Debug|Stop** to stop project. Then you can check the running result of this project by expanding the **gSData[]** array from the **Call Stack + Locals** window.

chapter twelve

MSP432™ memory protection unit (MPU)

In the MSP432P401R MCU system, an optional MPU is applied in the Cortex®-M4 Core to enhance the memory security for this MCU system.

Both Cortex®-M3 and Cortex®-M4 MCU support an optional feature called the MPU. In fact, the MPU can be considered as a programmable device used to:

- Divide the memory space into several (eight) regions that can be used for the different applications with different access levels.
- Define the memory access permissions to enable different memory regions to be accessed in either privileged level or full access level.
- Define the memory attributes as bufferable or catcheable regions.

Each memory region has its own programmable starting addresses, sizes, and settings. The MPU also supports the background region feature.

One point to be noted is that there is no need to set up the memory regions for the **PPB** address ranges and the vector table including the **SCS**. Accessing to PPB, which includes the MPU, NVIC, SysTick, and ITM, is always allowed in the privileged state and the vector fetches are always permitted by the MPU.

The MPU can make an embedded system more robust and secure by:

- Preventing applications from corrupting stack or data memory used by other tasks
- Preventing unprivileged tasks from accessing some peripherals that can be critical to the system applications
- Defining the SRAM or RAM regions as nonexecutable to prevent code injection attacks

In summary, the MPU is a programmable security device used to protect the system memory and the user's memory spaces from corrupting and attacking by undesired tasks.

12.1 Overview

The MPU is an optional component in the ARM® Cortex®-M4 MCU. The main purpose of using this MCU is to protect memory regions by defining different access permissions in privileged and unprivileged access levels for some embedded OS.

The MPU is a programmable unit and can be programmed up to eight regions. In some simple applications, the MPU can be programmed to protect certain memory regions only, for example, to make some memory regions read only.

In the MSPWare Series LaunchPad™ EVB, MSP-EXP432P401R, the MPU has been defined with the following protection functions:

- The memory attributes affect the behavior of memory accesses to the region. The Cortex®-M4 MPU defines eight separate memory regions, 0~7, and a background region accessible only from privileged mode. The background region has the same memory access attributes as the default memory map.

- When memory regions overlap, a memory access is affected by the attributes of the region with the highest number. For example, if a transfer address is within the address region defined for region 1 and region 4, the region 4 settings will be used.
- Regions of 256 bytes or more are divided into eight equal-sized subregions.
- MPU definitions for all regions include:
  - Location
  - Size
  - Access permissions
  - Memory attributes
- Accessing a prohibited region causes a memory management fault.
- The Cortex®-M4 MPU memory map is unified, meaning that instruction accesses and data accesses have the same region settings.
- If a program accesses a memory location that is prohibited by the MPU, the processor generates a memory management fault, causing a fault exception and possibly causing termination of the process in an OS environment.

To make the MPU work effectively, one also needs to define the fault handler for either the HardFault or MemManage (Memory Management) fault. The MemManage exception by default is disabled, and it can be enabled by setting the **MEMFAULTENA** bit in the System Handler Control and State Register (**SHCSR**) with the CMSIS-Core supported instructions.

The MemManage handler, **void MemManage_ Handler(void)**, should be defined in the user's program if the MemManage exception is enabled. Also the HardFault EXH, **void HardFault_Handler(void)**, should always be defined even if the MemManage exception is enabled.

The MPU needs to be programmed and enabled before it can be used in any application. If the MPU is not enabled or programmed, all MPU regions can overlap.

12.2 Implementation of the MPU

Depending on the different applications, the MPU can be used in several different ways.

For a system without any embedded OS, the MPU can be configured to have a static feature, and this includes the following functions:

1. Configure a RAM or SRAM region as a read-only region to prevent important program data from attacking or corrupting.
2. Make a part or a portion of a RAM or SRAM region at the bottom of the stack inaccessible to detect stack overflow.
3. Set a SRAM or RAM region as nonexecutable to prevent code injection attacks.
4. Define the memory attributes that can be used by system level catch or the memory controllers.

For a system with an embedded OS, the MPU can be configured to make each application to have its own MPU settings, which includes:

1. Set memory access permissions to the different levels for the stack operations of all applications to enable each application to access its own stack space to prevent stack corruptions of other stacks if the stack leaking situation occurred.
2. Set memory access permissions to limit any application to only access to a limited number of peripherals to avoid any conflict among peripherals.

3. Set memory access permissions to limit each application to access its own data to prevent any possible data corruption.

The static configuration is not limited to be used by a system without an embedded OS. A system with an embedded OS can also be configured with a static feature.

In order to use MPU to protect a memory system, we need to have the basic knowledge about the memory requirements and properties.

12.2.1 Memory regions, types, and attributes

In MSP432P401R MCU system, the memory map and the programming split the memory map into different regions, as shown in Figure 6.34 and Table 6.21 in Chapter 6. Each region has a defined memory type, and some regions have additional memory attributes. The memory type and attributes determine the behavior of accesses to the region.

In order to correctly and effectively access the desired memory region to perform the selected operations, we need to have a clear and complete picture about the memory requirements and some important memory properties.

We have provided detailed discussions about these memory requirements and properties in Section 6.8 in Chapter 6. Refer to that section to get more details for this issue.

In fact, the MPU is configured and controlled by a set of registers. Let's first have a closer look at these registers and their functions. The MPU registers can only be accessed from the privileged mode.

12.2.2 MPU Configuration and Control Registers

The MPU contains 11 registers and all of these registers are located at the **System Control Space (SCS)** with a starting memory address of **0xE000E000** (Figure 6.34 in Chapter 6). In fact, only the top 5 registers in those 11 registers are important, and the other 6 registers are only the alias registers used to duplicate some of top 5 registers.

The top five registers include:

- MPU Type Register (**MPUTYPE**)—Read-only register
- MPU Control Register (**MPUCTRL**)
- MPU Region Number Register (**MPUNUMBER**)
- MPU Region Base Address Register (**MPUBASE**)
- MPU Region Attribute and Size Register (**MPUATTR**)

The following six alias registers include:

- MPU Region Base Address Alias 1 Register (**MPUBASE1**)
- MPU Region Attribute and Size Alias 1 Register (**MPUATTR1**)
- MPU Region Base Address Alias 2 Register (**MPUBASE2**)
- MPU Region Attribute and Size Alias 2 Register (**MPUATTR2**)
- MPU Region Base Address Alias 3 Register (**MPUBASE3**)
- MPU Region Attribute and Size Alias 3 Register (**MPUATTR3**)

To use any alias register, such as **MPUBASE1**, it is exactly to use the **MPUBASE** Register itself. The reason to use these alias registers is to allow multiple MPU regions to be accessed and programmed in one instruction, such as store-multiple (**STM**), to save time and speed up the execution of the programs.

12.2.2.1 MPU type register (MPUTYPE)

This is a 32-bit register but only some bits or bit fields are used to indicate whether the MPU is present and how many regions it supports. The bit fields and bit configurations of this register are shown in Figure 12.1.

The bit field **IREGION** contains eight bits (bits 23~16) and this field is used to indicate how many memory regions can be used for the instructions. The default value for this field is **0x00** after a system reset.

The bit field **DREGION** also contains eight bits (bits 15~8) and this field is used to indicate how many memory regions can be used for the data. The default value for this field is **0x08** after a system reset.

The bit **SEPARATE** (bit 0) is used to indicate whether the MPU is unified (**0**) or separate (**1**). The default value for this bit is 0 (unified) after a system reset.

One point to be noted for using this register is that this register is a read-only register and it can be accessed from privileged mode.

12.2.2.2 MPU control register (MPUCTRL)

This is a 32-bit register but only the lowest three bits (bits 2~0) are used to enable the MPU, enable the default memory map background region, and enable use of the MPU when in the hard fault, NMI, and **FAULTMASK** Register escalated handlers.

The bit configuration of this register is shown in Figure 12.2. The bit function of this register is shown in Table 12.1. This register can only be accessed from the privileged mode, and these three bits are readable and writable.

When using this register to enable or disable the MPU, the following points should be noted:

- When the **ENABLE** and the **PRIVDEFEN** bits are both set:
  - For privileged accesses, the default memory map is as described in Figure 6.34 in Chapter 6. Any access by privileged software that does not address an enabled memory region behaves as defined by the default memory map.
  - Any access by unprivileged software that does not address an enabled memory region causes a memory management fault.
- Execute never (**XN**) and strongly-ordered rules always apply to the SCS regardless of the value of the **ENABLE** bit.
- When the **ENABLE** bit is set, at least one region of the memory map must be enabled for the system to function unless the **PRIVDEFEN** bit is set. If the **PRIVDEFEN** bit is set and no regions are enabled, then only privileged software can operate.

MPU Type Register (MPUTYPE)

31 30 29 28 27 26 25 24	23 … 16	15 … 8	7 … 1	0
Reserved bits 31 ~ 24	IREGION	DREGION	Reserved bits 7 ~ 1	SEPARATE

Figure 12.1 Bit fields of the MPUTYPE register.

MPU Control Register (MPUCTRL)

31 … 3	2	1	0
Reserved bits 31 ~ 3	PRIVDEFEN	HFNMIENA	ENABLE

Figure 12.2 Bit configuration of the MPUCTRL register.

Table 12.1 Bit value and its function for MPUCTRL register

Bit	Name	Reset	Function
31:3	**Reserved**	0x00	Reserved
2	**PRIVDEFEN**	0	MPU default region: This bit enables privileged software to access the default memory map **0:** If the MPU is enabled, this bit disables use of the default memory map. Any memory access to a location not covered by any enabled region causes a fault **1:** If the MPU is enabled, this bit enables use of the default memory map as a background region for privileged software accesses When this bit is set, the background region acts as if it is region number −1. Any region that is defined and enabled has priority over this default map. If the MPU is disabled, the processor ignores this bit
1	**HFNMIENA**	0	MPU Enabled During Faults: This bit controls the operation of the MPU during hard fault, NMI, and **FAULTMASK** handlers **0:** The MPU is disabled during hard fault, NMI, and **FAULTMASK** handlers, regardless of the value of the **ENABLE** bit **1:** The MPU is enabled during hard fault, NMI, and **FAULTMASK** handlers When the MPU is disabled and this bit is set, the resulting behavior is unpredictable
0	**ENABLE**	0	MPU Enable: **0:** The MPU is disabled; **1:** The MPU is enabled When the MPU is disabled and the **HFNMIENA** bit is set, the behavior is unpredictable

- When the **ENABLE** bit is clear, the system uses the default memory map, which has the same memory attributes as if the MPU is not implemented. The default memory map applies to accesses from both privileged and unprivileged software.
- When the MPU is enabled, accesses to the System Control Space and vector table are always permitted. Other areas are accessible based on regions and whether **PRIVDEFEN** is set.

12.2.2.3 MPU region number register (MPUNUMBER)

This is a 32-bit register but only the lowest three bits (bits 2~0) are used to indicate which memory region (0~7) is used currently. In fact, this register is used to select which memory region is referenced by the **MPUBASE** and MPU Region Attribute and Size (**MPUATTR**) registers. Normally, the required region number should be written to this register before accessing the **MPUBASE** or the **MPUATTR** register. However, the region number can be changed by writing to the **MPUBASE** Register with the **VALID** bit set. This write updates the value of the **REGION** field in that register.

In the MSP432P401R MCU system, the MPU supports up to 8 (0~7) memory regions.

12.2.2.4 MPU region base address register (MPUBASE)

This is a 32-bit register used to hold the valid base address of the MPU region selected by the MPU Region Number (**MPUNUMBER**) Register and can update the value stored in the

MPUNUMBER Register. To change the current region number and update the **MPUNUMBER** Register, write the **MPUBASE** Register with the **VALID** bit set.

Although the default base address, **ADDR** bit field, takes 27 bits (bits 31~5), in real applications, this **ADDR** field takes only bits **31:N** on the **MPUBASE** Register. Bits **(N-1):5** are reserved. The actual value **N** is determined by the memory region size specified by the **SIZE** field in the MPU Region Attribute and Size (**MPUATTR**) Register as:

$$N = \text{Log}_2(\text{SIZE})$$

If the region size is configured to 4 GB in the **MPUATTR** Register, there is no valid **ADDR** field. In this case, the region occupies the complete memory map, and the base address is **0x0000.0000**. The base address should be aligned to the size of the region. For example, a 64 KB region must be aligned on a multiple of 64 KB, for example, at **0x0001.0000** or **0x0002.0000**.

The bit configuration of this register is shown in Figure 12.3.

The bit **VALID** is used to indicate whether the region number stored in the **MPUNUMBER** Register is updated and equal to the number stored in the **REGION** bit field in this register or not. If this **VALID** is 1, the region number in the **MPUNUMBER** Register is updated and equals to the value in the **REGION** bit field in this register. Otherwise if this **VALID** is 0, which means that the region number in the **MPUNUMBER** Register is not updated and the processor updates the base address based on the nonupdated region number stored in the **MPUNUMBER** Register and ignores the number in the **REGION** bit field in this register.

The value in the bit field **REGION** contains the current region number. When writing, this value is written into the **MPUNUMBER** Register, when reading, it returns the current region number stored in the **MPUNUMBER** Register.

Three MPU Region Base Address Alias Registers (**MPUBASE1~MPUBASE3**) have the same bit fields and functions.

12.2.2.5 MPU region attribute and size register (MPUATTR)

This is a 32-bit register used to define the region size and memory attributes of the MPU region specified by the MPU Region Number (**MPUNUMBER**) Register and enables that region and any subregions.

The bit configuration of this register is shown in Figure 12.4.

The **MPUATTR** Register is accessible using word or half-word accesses with the most significant half-word holding the region attributes and the least significant half-word holds the region size and the region and subregion enable bits.

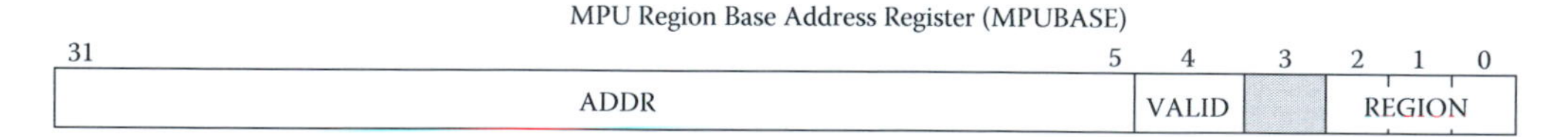

Figure 12.3 Bit configuration of the MPUBASE register.

Figure 12.4 Bit configuration of the MPUATTR register.

Table 12.2 Bit value and its function for MPUATTR register

Bit	Name	Reset	Function
31:29	**Reserved**	0x00	Reserved
28	**XN**	0	Instruction access disable: **0:** Instruction fetches are enabled; **1:** Instruction fetches are disabled
27	**Reserved**	0	Reserved
26:24	**AP**	0x00	Access Privilege: **000:** No access; **001–010:** Access from privileged software only; **011:** Full access **100:** Reserved; **101:** Reads by privileged software only; **110–111:** Read-only by privileged or unprivileged software
23:22	**Reserved**	0x00	Reserved
21:19	**TEX**	0	Type extension mask. For MPU access permissions, see Table 12.4
18	**S**	0	Shareable. For MPU access permissions, see Table 12.4
17	**C**	0	Cacheable. For MPU access permissions, see Table 12.4
16	**B**	0	Bufferable. For MPU access permissions, see Table 12.4
15:8	**SRD**	0x00	Subregion disable bits
			0: The corresponding subregion is enabled **1:** The corresponding subregion is disabled
7:6	**Reserved**	0x0	Reserved
5:1	**SIZE**	0x00	Region size mask: The **SIZE** field defines the size of the MPU memory region specified by the **MPUNUMBER** Register
0	**ENABLE**	0	Region enable: **0:** The region is disabled; **1:** The region is enabled

The MPU access permission attribute bits, **XN**, **AP**, **TEX**, **S**, **C**, and **B**, are used to control access to the corresponding memory region. If an access is made to an area of memory without the required permissions, then the MPU generates a permission fault. The bit functions of this register are shown in Table 12.2. The MPU access permission attributes bits, **XN**, **AP**, **TEX**, **S**, **C**, and **B**, as well as their functions are shown in Table 12.4.

The **SIZE** field defines the size of the MPU memory region specified by the **MPUNUMBER** Register as follows:

```
(Region Size in Bytes) = 2^(SIZE+1)
```

The smallest permitted region size is 32 bytes, corresponding to a **SIZE** value of 4. Table 12.3 shows some popular example **SIZE** values with the corresponding region size and value of **N** in the MPU Region Base Address (**MPUBASE**) Register.

All encodings shown in Table 12.4 are completeness. However, the current implementation of the Cortex®-M4 does not support the concept of cacheability or shareability.

Table 12.5 shows the cache policy for memory attribute encodings with a **TEX** value in the range of **0x4~0x7**.

Three MPU Region Attribute and Size Alias Registers (**MPUATTR1~MPUATTR3**) have the same bit fields and functions.

Now that all MPU-related control and status registers have been discussed, let's take a look at the MPU setups and configurations.

Table 12.3 Some popular SIZE values and related region sizes

SIZE	Region size	N	Description
0x4	**32B**	**5**	The Min permitted size
0x9	**1 KB**	**10**	
0x13	**1 MB**	**20**	
0x1D	**1 GB**	**30**	
0x1F	**4 GB**	No valid **ADDR** field in **MPUBASE**; the Max possible size region occupies the complete memory map	The Max possible size

Table 12.4 TEX, S, C, and B bit field encoding

TEX	S	C	B	Memory type	Shareable	Others
000	**X**	**0**	**0**	Strongly ordered	Shareable	—
000	**X**	**0**	**1**	Device	Shareable	—
000	**0**	**1**	**0**	Normal	Not shareable	Outer and inner write-through No write allocate
000	**1**	**1**	**0**	Normal	Shareable	
000	**0**	**1**	**1**	Normal	Not shareable	
000	**1**	**1**	**1**	Normal	Shareable	
001	**0**	**0**	**0**	Normal	Not shareable	Outer and inner noncacheable
001	**1**	**0**	**0**	Normal	Shareable	
001	**X**	**0**	**1**	Reserved encoding	—	—
001	**X**	**1**	**0**	Reserved encoding	—	—
001	**0**	**1**	**1**	Normal	Not shareable	Outer and inner write-back Write and read allocate
001	**1**	**1**	**1**	Normal	Shareable	
010	**X**	**0**	**0**	Device	Not shareable	Not shareable device
010	**X**	**0**	**1**	Reserved encoding	—	—
010	**X**	**1**	**X**	Reserved encoding	—	—
1BB	**0**	**A**	**A**	Normal	Not shareable	Cached memory (BB = outer policy, AA = inner policy) See Table 12.5 for the encoding of the AA and BB bits
1BB	**1**	**A**	**A**	Normal	Shareable	

Table 12.5 Cache policy for memory attribute encoding

Encoding (AA or BB)	Cache policy
00	Noncacheable
01	Write back, write, and read allocate
10	Write through, no write allocate
11	Write back, no write allocate

12.3 Initialization and configuration of the MPU

In most real applications, the MPU is not required. By default, the MPU is disabled and the MCU system works as if the MPU does not exists. In order to use MPU, one needs to configure out what memory regions the user's program needs to access.

In a simple application, the goal is just to prevent unprivileged tasks from accessing certain memory regions. In that case, the background region feature is a very useful tool since it reduced the memory setup requirements. One only needs to set up the region setting for unprivileged tasks, and the privileged tasks and interrupt handlers have full access abilities to other memory spaces using the background region.

Basically, the MPU setup procedure can be completed by the following steps:

1. Check the MPU Type register **(MPUTYPE)** to see whether the MPU is presented and how many regions are supported.
2. If the MPU exists, disable the MPU before it can be configured and initialized.
3. Select all regions one by one and program region base address and configuration. This step can be performed in a loop with **n** times, and the **n** is the number of regions to be selected and configured.
4. Select the configured region and disable unused regions. This step can be performed as a loop with 8 – **n** times, and the **n** is the number of used regions.
5. Enable the MPU.

One also needs to define the fault handler for either the HardFault or MemManage (Memory Management) fault. The MemManage exception by default is disabled, and one can enable this by setting the **MEMFAULTENA** bit in the System Handler Control (**SCB**) and State Register (**SHCSR**) with the CMSIS-Core supported instruction **SCB → SHCSR 0= 0x00010000** (set bit 16).

The MemManage handler, **void MemManage_Handler(void)**, should be defined in the user's program if the MemManage exception is enabled. Also the HardFault EXH, **void HardFault_Handler(void)**, should always be defined even if the MemManage exception is enabled.

The MPU interrupt handler should also be registered and enabled before any exception can be responded and handled.

Prior to enabling the MPU, at least one region must be set or else by enabling the default region for privileged mode. Once the MPU is enabled, a memory management fault is generated for memory access violations.

Now let's build a real example MPU project to illustrate how to set up and use the MPU to protect the memory regions from being illegally accessed based on our above discussions.

12.4 Building a practical example MPU project

In this project, we use the DRA model to build a MPU project to divide the entire memory map into three regions with the base addresses and sizes as below:

- **Region 1**: A 28-KB region of flash from **0x00000000** to **0x00007000**. The region is executable and read-only for both privileged and user modes. To set up the region, a 32-KB region (**#0**) is defined starting at address 0, and then a 4-KB hole is removed at the end by disabling the last subregion. The region is initially enabled.

- **Region 2**: A 32-KB region (**#1**) of RAM from **0x20000000** to **0x20008000**. The region is not executable, and is R/W access for privileged and user modes.
- **Region 3**: An additional 8-KB region (**#2**) in RAM from **0x20008000** to **0x2000A000** that is R/W accessible only from privileged mode. This region is initially disabled and it can be enabled later.

In order to test the MPU function, we develop a function **WriteFlash()** to try to access the flash memory at **0x1000**, which has been defined as a read-only region (**Region 1**), to trigger the MPU to start its protection function. This protection function is performed by calling and executing two handlers, **HardFault_Handler()** and **MemManage_Handler()**, respectively.

The unused five regions are also disabled in this project by using a loop function.

12.4.1 Create a new DRA model MPU project DRAMPU

Since this project is a little complicated, a header file and a C source file are needed.

Create a new folder **DRAMPU** under the folder **C:\MSP432 Class Projects\Chapter 12** with Windows Explorer. Create a new µVersion5 project **DRAMPU** and add it into the folder **DRAMPU** created above.

Add a new header file **DRAMPU.h** into this project and enter the code shown in Figure 12.5 into this file. Let's have a closer look at this header file to see how it works.

1. The codes in lines 4~7 are used to include all system header files to be used in this project. One of the MSPPDL header file **mpu.h** is also included here since we need to use some memory-related macros defined in that header file.
2. Two user-defined function, **SetupMPU()** and **WriteFlash()**, are declared in lines 8 and 9. These two functions are used to configure the MPU before it can be operated and try to access a read-only region to program some flash memory units to test the MPU later.
3. In line 10, a data array **ui32RegionAddr[]** is declared and initialized with three starting base addresses, **0**, **0x20000000** and **0x20008000**, for three regions to be divided by this project.

```
1   //*****************************************************************************
2   // DRAMPU.h - The Header File for project DRAMPU -
3   //*****************************************************************************
4   #include <stdint.h>
5   #include <stdbool.h>
6   #include <msp.h>
7   #include <MSP432P4xx\mpu.h>

8   int SetupMPU(void);
9   int WriteFlash(void);

10  uint32_t ui32RegionAddr[3] = {0, 0x20000000, 0x20008000};
11  uint32_t ui32RegionAttr[3] = {(MPU_RGN_SIZE_32K|MPU_RGN_PERM_EXEC|
12                 MPU_RGN_PERM_PRV_RO_USR_RO|MPU_SUB_RGN_DISABLE_7|MPU_RGN_ENABLE),
13                 (MPU_RGN_SIZE_32K|MPU_RGN_PERM_NOEXEC|
14                 MPU_RGN_PERM_PRV_RW_USR_RW|MPU_RGN_ENABLE),
15                 (MPU_RGN_SIZE_8K|MPU_RGN_PERM_NOEXEC|
16                 MPU_RGN_PERM_PRV_RW_USR_NO|MPU_RGN_DISABLE)};
```

Figure 12.5 Header file for the project DRAMPU.

4. Similarly in line 11, another data array **ui32RegionAttr[]** is declared and initialized with three groups of configuration data for the corresponding attributes of three regions created above. The reason we use these data arrays is to save the space for the set up of three memory regions. These attributes can be **ORed** together to get the target attributes.
5. The attributes for the first region are defined by the following macros (lines 11~12):
   a. **MPU_RGN_SIZE_32K:** The Initial region size is 32 KB
   b. **MPU_RGN_PERM_EXEC:** The region is executable
   c. **MPU_RGN_PERM_PRV_RO_USR_RO:** Read-only for privileged and user modes
   d. **MPU_SUB_RGN_DISABLE_7:** The last 4-KB hole are removed to get 28 KB
   e. **MPU_RGN_ENABLE:** This region is initially enabled
6. The attributes for the second region are defined by the following macros (lines 13~14):
   a. **MPU_RGN_SIZE_32K:** The region size is 32 KB
   b. **MPU_RGN_PERM_NOEXEC:** The region is nonexecutable
   c. **MPU_RGN_PERM_PRV_RW_USR_RW:** R/W for privileged and user modes
   d. **MPU_RGN_ENABLE:** This region is initially enabled
7. The attributes for the third region are defined by the following macros (lines 15~16):
   a. **MPU_RGN_SIZE_8K:** The region size is 8 KB
   b. **MPU_RGN_PERM_NOEXEC:** The region is nonexecutable
   c. **MPU_RGN_PERM_PRV_RW_USR_NO:** R/W for privileged mode only
   d. **MPU_RGN_DISABLE:** This region is initially disabled

Now create a new C source file **DRAMPU.c** and add it into this project. Enter the code shown in Figure 12.6 into this source file.

Let's have a closer look at this source file to see how it works.

1. The main program starts at line 5.
2. The codes in lines 7~8 are used to set the GPIO pins **P2.0** and **P2.2** as output pins since we need to use these two pins, exactly two-color LEDs, red and blue color, to indicate the running status of this project later. Both GPIO pins are configured to output low.
3. In line 9, the user-defined function **SetupMPU()** is called to perform the set up and configuration process for the MPU. A **0** is returned if this function execution is successful. Otherwise a nonzero value is returned if any error is encountered. A 1 is returned to the OS if any error occurred while executing this function.
4. If the function **SetupMPU()** is executed successfully, the **SCB->SHCSR** instruction is executed to set bit **MEMFAULTENA** (bit 16) on the **SHCSR** to enable the MemManage exception in line 10.
5. Another user-defined function **WriteFlash()** is called in line 11 to try to access the flash memory region at addresses starting at **0x1000**, which has been defined as a read-only region by the MPU, to write some bytes. The purpose of calling this function is to test the MPU protection function. A HardFault and a MemManage exception should be triggered if this writing happened and the handlers for both exceptions should be excited and performed to handle this violation.
6. The first user-defined function **SetupMPU()** starts at line 13.
7. A local **uint32_t** variable **index** is declared first and this variable works as a loop counter.
8. In line 16, the MPU Type register **MPUTYPE** is checked first to make sure that a MPU is presented in the current system. Otherwise a 1 would be returned to the main program to indicate a MPU nonpresent error if all-bit on the **MPUTYPE** Register is zero.

```
1  ***********************************************************************************
2  // DRAMPU.c - Main Application File for project DRAMPU -
3  //*********************************************************************************
4  #include "DRAMPU.h"
5  int main(void)
6  {
7    P2DIR = BIT0|BIT2;                         // set P2.0 & P2.2 as output pins
8    P2OUT &= ~(BIT0|BIT2);                     // enable P2.0 & P2.2 to output low
9    if (SetupMPU()) {return 1;}                // setup the MPU. if error, stop program
10   SCB->SHCSR = 0x00010000;                   // enable the MemManage exception
11   if (WriteFlash()) {return 1;}              // try to access flash memory at 0x1000 (read-
                                                   only region)
12 }
13 int SetupMPU(void)
14 {
15   uint32_t index;
16   if (MPU->TYPE == 0) {return 1;}            // if MPU is not presented, returns an error
17   MPU->CTRL = 0;                             // disable MPU to setup it
18   for (index = 0; index < 3; index++)        // configure 3 regions (0, 1, 2)
19   {
20    MPU->RNR = index;
21    MPU->RBAR = ui32RegionAddr[index];
22    MPU->RASR = ui32RegionAttr[index];
23   }
24   for (index = 3; index < 8; index++)        // disable unused 5 regions
25   {
26    MPU->RNR = index;
27    MPU->RBAR = 0;
28    MPU->RASR = 0;
29   }
30   MPU->CTRL = 0x1;                           // enable the MPU
31   return 0;
32 }
33 int WriteFlash(void)                         // try to access a read-only region
34 {
35   uint32_t ret;
36   unsigned long ulCount = 3, ulAddress = 0x1000;
37   unsigned long Data[3] = {0x78563412, 0x8B674523, 0xA3456789};
38   ret = FLCTL_PRG_CTLSTAT;                       // get the current content of FLCTL_PRG_
                                                       CTLSTAT register
39   while(!(ret & 0xFFFCFFFF)){}                   // bits 17:16 are zero (idle)? If not, wait
                                                       here......
40   FLCTL_PRG_CTLSTAT = 0x0D;                      // Immediate mode with auto-verify (enable
                                                       the write)
41   FLCTL_PRGBRST_CTLSTAT &= ~0x000000C0;          // clear burst program auto verify functions
42   while(ulCount)                                 // loop to perform 3 words programming
43   {
44    FLCTL_BANK0_MAIN_WEPROT = 0;                  // clear all write/erase protection functions
45    FLCTL_CLRIFG = 0x0000020E;                    // clear IFG errors (PRG Done |PreV | PostV
                                                       |PRG Error)
46    HWREG32(ulAddress) = Data[ulCount];           // Write a 32-bit word to destination in
                                                       flash memory
47    while(FLCTL_IFG & 0x8){}                      // Wait for the PRG to be done......
48    if (FLCTL_IFG & 0x00000206)                   // check all possible errors (PreV|PostV|PRG
                                                       Error)
49    return 1;                                     // if any error occurred, return non-zero
50    ulAddress += 4;                               // update to the next word
51    ulCount--;
52   }
53   return 0;
54 }
55 void HardFault_Handler(void) { P2OUT = BIT0; }      // turn on red-color LED
56 void MemManage_Handler(void) { P2OUT = BIT2;}       // turn on blue-color LED & trigger
                                                       the HardFault_Handler()
```

Figure 12.6 Complete code for the project DRAMPU.

9. If a MPU is presented, the **MPUCTRL**, is reset to disable the MPU to enable the set up and configuration of the MPU to start in line 17.
10. A **for()** loop is used starting at line 18 to repeatedly set up three memory regions by assigning the region numbers (**0~2**) to the MPUNUMBER (**MPU→RNR**) via the **index**, by assigning three memory starting or base addresses, **0**, **0x20000000** and **0x20008000**, to the MPUBASE (**MPU→RBAR**) via the data array **ui32Region-Addr[]**, by assigning three memory attributes to the MPUATTR (**MPU→RASR**) via the data array **ui32RegionAttr[]**.
11. Similarly, another **for()** loop is used for the codes in lines 24~29 to disable the rest five unused memory regions by assigning the corresponding region numbers to the MPUNUMBER registers, and **0** to those MPUBASE registers and MPU Region Attribute and Size registers.
12. When the MPU is configured, it is enabled in line 30 by assigning **0x1** to the MPUCTRL (**MPU→CTRL**).
13. A **0** is returned to the main program if no error is encountered in line 31.
14. The user-defined function **WriteFlash()** starts at line 33.
15. Some local variables are declared and initialized first. These variables include the status holder variable **ret**, a data count **ulCount**, flash memory destination starting address **ulAddress**, and data array **Data[]**. All of these variables are initialized with given values.
16. In line 38, the **FLCTL_PRG_CTLSTAT** Register is read to get the current running status of the flash controller.
17. A **while()** loop is used in line 39 to check whether the bits 17–16 in that register are 00, which means that the flash controller is in the idle status. If both bits are 0, which means that the flash controller is in the idle status and the next instruction can be executed. Otherwise the **while()** loop continues this checking and waits until both bits are 0.
18. If the flash controller is in the idle status, the **FLCTL_PRG_CTLSTAT** Register is configured to start an Immediate Word programming operation in line 40 with the Post and Pre Verify function to be activated by assigning **0x0D** (1101)—refer to Table 6.11 in Chapter 6.
19. In line 41, the Burst Program Auto-Verify bits (**AUTO_PST** and **AUTO_PRE**) in the **FLCTL_PRGBRST_CTLSTAT** Register are cleared to avoid any possible disturbance for those bits' functions (refer to Table 6.12 in Chapter 6).
20. The codes in lines 42 through 52 are used to continuously write three 32-bit words into the flash memory starting at **0x1000**.
21. Before any WE operation can be performed, the related Flash **WE** Protection Register must be cleared to remove those protections since all 32-bit on those protection registers are set to 1 (protected) after a system reset. Since we try to write data into the Bank0 in the main flash memory, the related **WE** protection register for Bank0, **FLCTL_BANK0_MAIN_WEPROT**, is cleared in line 44.
22. In order to check this writing result, we first need to clear all previous setup bits on the Flash Interrupt Flag Register (**FLCTL_IFG**) in line 45 by setting the related bits on the **FLCTL_CLRIFG** Register. Bit 3 (**PRG**) on the **FLCTL_IFG** Register will be set to 1 when a program operation is complete. Also bits 9 (**PRG_ERR**), 2 (**AVPST**), and 1 (**AVPRE**) on the **FLCTL_IFG** Register would be set to 1 if any error related to programming operation (**PRG_ERR** = 1), or any error related to either post auto verify (**AVPST** = 1) or pre auto verify (**AVPRE** = 1) occurred (refer to Table 6.14 in Chapter 6)

23. In line 46, the source data stored in the **Data[]** array is written into the flash memory starting at the address of **ulAddress**(**0x1000**). The address constant **ulAddress** is converted to the pointer by using the system macro **HWREG32()** to facilitate this data writing.
24. Then a conditional **while()** loop is used in line 47 to check whether this data has been written into the flash by inspecting bit 3 in the **FLCTL_IFG** Register. If not, this **while()** loop will wait there until this programming operation is complete. In fact, this writing operation can never be executed since this flash memory region is a protected region with the read-only accessing protection, and a MemManage and HardFault exception would occurr when this writing operation begins.
25. If this writing is complete, we need to check if any error has been encountered by inspecting bits 9 (**PRG_ERR**), 2 (**AVPST**), and 1 (**AVPRE**) on the **FLCTL_IFG** Register in line 48.
26. If any error occurs, a nonzero value is returned to the main program in line 49.
27. Otherwise the destination address **ulAddress** and the data count **ulCount** are updated to point to the next address and the next data to make it ready for the next writing in lines 50 and 51.
28. A 0 is returned if no error has occurred for this writing operation in line 53.
29. The codes in lines 55~56 are two EXHs. These two handlers must be defined in this project to enable them to be executed. Two LEDs connected to **P2.0** (red color) and **P2.2** (blue color) are turned on when these handlers are triggered and responded. The execution order of these handlers is: the **MemManage_Handler()** is executed first, and the **HardFault_Handler()** is executed second.

Now let's set up the environment to build and run this project to test the MPU function.

12.4.2 *Set up the environment to build and run the project*

This setup contains the following two steps:

1. Include the system header files by adding the include path.
2. Check and configure the correct debugger used in the project.

Perform the following operations to include the header file path in the project:

- Go to **Project**|**Options for Target "Target 1"** menu item. Then click on the **C/C++** tab.
- Go to **Include Paths** box and browse to the folder where our header files are located, it is **C:\ti\msp\MSPWare_2_21_00_39\driverlib\driverlib**. Select this folder and click on the **OK** button.

Perform the following operations to select the correct debugger driver:

- Go to **Project**|**Options for Target "Target 1"** menu item to open the **Options** wizard.
- On the opened Options wizard, click on the **Debug** tab.
- Make sure that the debugger shown in the **Use:** box is **CMSIS-DAP Debugger**. Otherwise you can click on the dropdown arrow to select this debugger from the list.
- Make sure that all settings for the Debugger and the Flash Download are correct.

Now let's build and run the project if everything is fine. After the project runs, the red color LED is on to indicate that a memory access violation has occurred and the related EXHs have been triggered and responded to handle this error. This indicates that our MPU works fine.

A small question is why is only the red color LED on but the blue color LED is never on? You can figure out this question and find out the solution yourself easily.

Next, let's take care of the API functions provided by the MSPPDL. By using these API functions, one can directly call them to avoid direct accessing to those MPU registers to make the memory protection jobs simple and easy.

12.5 API functions provided by the MSPWare Peripheral Driver Library

The MSPWare Peripheral Driver Library provides 11 MPU-related API functions to configure, enable, set, and get required memory regions for the MPU. The MPU is tightly coupled to the Cortex®-M4 processor core and provides a means to establish access permissions on regions of memory.

The entire memory map can be divided into eight regions and each region can be configured for read-only access, R/W access, or no access for both privileged and user modes. Access permissions can be used to create an environment where only kernel or system code can access certain hardware registers or sections of code.

The MPU can create eight subregions within each region. Any subregion or combination of subregions can be disabled, allowing creation of holes or complex overlaying regions with different permissions. The subregions can also be used to create an unaligned beginning or ending of a region by disabling one or more of the leading or trailing subregions.

Once the regions are defined and the MPU is enabled, any access violation of a region causes a memory management fault, and the fault handlers, including the **MemManage_Handler()** and **HardFault_Handler()**, are activated.

These 11 API functions can be divided into three groups based on their functions and purposes as below:

- The MPU Setup and Status Group
  - **MPU_setRegion()**
  - **MPU_getRegion()**
  - **MPU_getRegionCount()**
  - **MPU_enableRegion()**
  - **MPU_disableRegion()**
- The MPU Enable and Disable Group
  - **MPU_enableModule()**
  - **MPU_disableModule()**
- The MPU Interrupt Handler Control Group
  - **MPU_registerInterrupt()**
  - **MPU_enableInterrupt()**
  - **MPU_disableInterrupt()**

Let's have a closer look at these API functions one by one based on their groups.

12.5.1 MPU setup and status API functions

As we mentioned, the MPU must be set up and configured before it can be enabled. This setup and configuration process includes the memory regions defined (region number, region base address, and region attributes). The regions can be configured by calling the API function **MPU_setRegion()** once for each region to be configured.

When each region is set up and configured by the function **MPU_setRegion()**, it can be initially enabled or disabled. If a region is not initially enabled, it can be enabled later by calling the API function **MPU_enableRegion()**. An enabled region can be disabled by calling another API function **MPU_disableRegion()**. When a region is disabled, its configuration is preserved as long as it is not overwritten. In this case, it can be enabled again with **MPU_enableRegion()** without the need to reconfigure the region.

After a region has been configured, the attributes of a region can be retrieved and saved using the **MPU_getRegion()** function. This function is used to save the attributes in a format that can be used later to reload the region using the **MPU_setRegion()** function. Note that the enable state of the region is saved with the attributes and takes effect when the region is reloaded.

The function **MPU_getRegionCount()** can be used to get the total number of regions that are supported by the MPU, including regions that are already programmed and configured.

When one or more memory regions have been defined, the MPU can be enabled by calling the API function **MPU_enableModule()**. This function turns on the MPU and defines the behavior in privileged mode with the HardFault and NMI fault handlers. The MPU can be configured so that when in privileged mode and no regions are enabled, a default memory map is applied. If this feature is not enabled, then a memory management fault is generated if the MPU is enabled and no regions are configured and enabled. The MPU can also be set to use a default memory map when in the HardFault or NMI handlers, instead of using the configured regions. All of these features are selected when calling the function **MPU_enableModule()**. When the MPU is enabled, it can be disabled by calling the API function **MPU_disableModule()**.

If an application is using the run-time interrupt registration API **Interrupt_registerInterrupt()**, then the function **MPU_registerInterrupt()** can be used to install the Fault Handler which is called whenever a memory protection violation occurs. This function also enables the Fault Handler. If compile-time interrupt registration is used, then the **Interrupt_enableInterrupt()** function with the parameter **FAULT_MPU** must be used to enable the memory management fault handler.

12.5.1.1 API function MPU_setRegion()

This function is used to set up the protection rules for a region. The region has a base address and a set of attributes including the size. The base address parameter, **addr**, must be aligned according to the size, and the size must be a power of 2.

The protocol of this function is:

```
void MPU_setRegion(uint32_t region, uint32_t addr, uint32_t flags)
```

- **region** is the region number to set up and it is ranged from 0 to 7.
- **addr** is the base address of the region. It must be aligned according to the size of the region specified in **flags**.
- **flags** is a set of flags to define the attributes of the region.

The **flags** parameter is the logical **OR** of all of the attributes of the region. It is a combination of region size, execute permission, R/W permissions, disabled subregions, and a flag to determine if the region is enabled.

The size flag determines the size of a region and must be one of the following:

```
MPU_RGN_SIZE_32B;       MPU_RGN_SIZE_64B;       MPU_RGN_SIZE_128B;
MPU_RGN_SIZE_256B;      MPU_RGN_SIZE_512B;      MPU_RGN_SIZE_1K;
MPU_RGN_SIZE_2K;        MPU_RGN_SIZE_4K;        MPU_RGN_SIZE_8K;
MPU_RGN_SIZE_16K;       MPU_RGN_SIZE_32K;       MPU_RGN_SIZE_64K;
MPU_RGN_SIZE_128K;      MPU_RGN_SIZE_256K;      MPU_RGN_SIZE_512K;
MPU_RGN_SIZE_1M;        MPU_RGN_SIZE_2M;        MPU_RGN_SIZE_4M;
MPU_RGN_SIZE_8M;        MPU_RGN_SIZE_16M;       MPU_RGN_SIZE_32M;
MPU_RGN_SIZE_64M;       MPU_RGN_SIZE_128M;      MPU_RGN_SIZE_256M;
MPU_RGN_SIZE_512M;      MPU_RGN_SIZE_1G;        MPU_RGN_SIZE_2G;
MPU_RGN_SIZE_4G
```

The execute permission flag must be one of the following:

```
MPU_RGN_PERM_EXEC:      Enables the region for execution of code.
MPU_RGN_PERM_NOEXEC:    Disables the region for execution of code.
```

The R/W access permissions are applied separately for the privileged and user modes. The R/W access flags must be one of the following:

```
MPU_RGN_PERM_PRV_NO_USR_NO:    No access in privileged or user mode.
MPU_RGN_PERM_PRV_RW_USR_NO:    Privileged read/write, user no access.
MPU_RGN_PERM_PRV_RW_USR_RO:    Privileged read/write, user read-only.
MPU_RGN_PERM_PRV_RW_USR_RW:    Privileged read/write, user read/write.
MPU_RGN_PERM_PRV_RO_USR_NO:    Privileged read-only, user no access.
MPU_RGN_PERM_PRV_RO_USR_RO:    Privileged read-only, user read-only.
```

Each region is automatically divided into eight equally-sized subregions by the MPU. Subregions can only be used in regions of size 256 bytes or larger. Any of these eight subregions can be disabled, allowing for creation of holes in a region which can be left open, or overlaid by another region with different attributes. Any of the eight subregions can be disabled with a logical **OR** of any of the following flags:

```
MPU_SUB_RGN_DISABLE_0
MPU_SUB_RGN_DISABLE_1
MPU_SUB_RGN_DISABLE_2
MPU_SUB_RGN_DISABLE_3
MPU_SUB_RGN_DISABLE_4
MPU_SUB_RGN_DISABLE_5
MPU_SUB_RGN_DISABLE_6
MPU_SUB_RGN_DISABLE_7
```

Finally, each region can be initially enabled or disabled with one of the following flags:

```
MPU_RGN_ENABLE
MPU_RGN_DISABLE
```

Table 12.6 shows all other MPU setup and status API functions.

Table 12.6 MPU setup and status API functions

API function	Parameter	Description
void **MPU_getRegion (**uint32_t region, uint32_t *addr, uint32_t *pflags**)**	**region** is the region number to get **addr** points to storage for the base address of the region **pflags** points to the attribute flags for the region	This function retrieves the configuration of a specific region. The meanings and format of the parameters is the same as that of the **MPU_setRegion()** function This function can be used to save the configuration of a region for later use with the **MPU_setRegion()** function. The region's enable state is preserved in the attributes that are saved
uint32_t **MPU_getRegionCount(**void**)**	**None**	This function is used to get the total number of regions that are supported by the MPU, including regions that are already programmed and configured The function returns the number of memory protection regions that are available for programming using **MPU_setRegion()**
void **MPU_enableRegion (**uint32_t region**)**	**region** is the region number to enable	This function is used to enable a memory protection region. The region should already be configured with the **MPU_setRegion()** function. Once enabled, the memory protection rules of the region are applied and access violations cause a memory management fault
void **MPU_disableRegion (**uint32_t region**)**	**region** is the region number to disable	This function is used to disable a previously enabled memory protection region. The region remains configured if it is not overwritten with another call to **MPU_setRegion()**, and can be enabled again by calling **MPU_enableRegion()**

12.5.2 *MPU module enable and disable API functions*

After one or more regions are configured, the MPU can be enabled by calling the API function **MPU_enableModule()**. This function turns on the MPU and also defines the behavior in privileged mode, in the HardFault and NMI fault handlers. The MPU can be configured to use the default memory map or the background region when in privileged mode and no regions are enabled. If this feature is not enabled, then a memory management fault is generated if the MPU is enabled and no regions are configured and enabled.

The protocol of the **MPU_enableModule()** function is

void MPU_enableModule(uint32_t **mpuConfig)**

- **mpuConfig** is the logical **OR** of the possible configurations

This function is used to enable the Cortex®-M4 MPU. It also configures the default behavior when in privileged mode and while handling a HardFault or NMI. Before any MPU can be enabled, at least one region must be set by calling the API function

MPU_setRegion() or by enabling the default region for privileged mode by passing the **MPU_CONFIG_PRIV_DEFAULT** flag to the API function **MPU_enableModule()**.

Once the MPU is enabled, a memory management fault is generated for any memory access violations. The **mpuConfig** parameter should be the logical **OR** of any of the following:

- **MPU_CONFIG_PRIV_DEFAULT**: Enables the default memory map when in privileged mode and when no other regions are defined. If this option is not enabled, there must be at least one valid region that has been already defined when the MPU is enabled.
- **MPU_CONFIG_HARDFLT_NMI**: Enables the MPU while in a HardFault or NMI EXH. If this option is not enabled, the MPU would be disabled when one of these EXHs is triggered and the default memory map is applied.
- **MPU_CONFIG_NONE**: Chooses none of the above options. In this case, no default memory map is provided in privileged mode, and the MPU is not enabled in the fault handlers.

The protocol of the **MPU_disableModule()** function is

```
void MPU_disableModule(void)
```

This function disables the Cortex®-M4 MPU. When the MPU is disabled, the default memory map is used and memory management faults are not generated.

12.5.3 MPU interrupt handler control API functions

Table 12.7 shows these MPU interrupt handler control functions.

At this point, we have completed our discussions about the MPU API functions provided by the MSPPDL. We need to build a real MPU project by using these API functions to test the MPU functions. We prefer to leave this as a lab project for the readers to allow students to do something themselves.

12.6 Summary

The main topic of this chapter is the MPU applied in the MSP432P401R MCU system. In fact, an optional MPU is included in the ARM® Cortex®-M4 MCU system. The purpose of

Table 12.7 The MPU interrupt handler control API functions

API function	Parameter	Description
void **MPU_registerInterrupt(**void (*) (void) intHandler**)**	**intHandler** is a pointer to the function to be called when the memory management fault occurs	This function sets and enables the handler to be called when the MPU generates a memory management fault due to a protection region access violation
void **MPU_enableInterrupt(**void**)**	**None**	This function enables the interrupt for the memory management fault
void **MPU_disableInterrupt(**void**)**	**None**	This function disables the interrupt for the memory management fault

the MCU is to protect the memory regions in the Cortex®-M4 system from illegal accessing by some other programs and devices. This protection is performed by dividing the entire memory map into eight regions, and each region has its own region base address, region size, and related attributes. In fact, the MPU can be considered as a programmable device used to:

- Divide the memory space into several (eight) regions that can be used for the different applications with different access levels.
- Define the memory access permissions to enable different memory regions to be accessed in either privileged level or full access level.
- Define the memory attributes as bufferable or catcheable regions.

An introduction and overview for the MPU applied in the MSP432P401R MCU is provided in Section 12.1. Starting in Section 12.2, the implementation of the MPU is discussed in detailed with all aspects of applications of a MPU in real systems and applications. These include the memory regions, types, and attributes and are discussed in Section 12.2.1. The MPU-related CFGCTRL registers are given in Section 12.2.2.

Some major and useful MPU registers, such as **MPUNUMBER**, **MPUTYPE**, **MPUCTRL**, **MPUBASE**, and **MPUATTR**, are discussed in this section.

The initialization and configuration of the MPU is provided in Section 12.3. The detailed operational procedure of the set up and configuration of a MPU is discussed. Starting Section 12.4, a practical example MPU project is provided step by step to illustrate how to use these MPU registers to perform memory regions protection functions.

The API functions provided by the MSPPDL are introduced in Section 12.5.

HOMEWORK

I. True/False Selections

____1. Both Cortex®-M3 and Cortex®-M4 MCU support an optional feature called the MPU.

____2. Each memory region shares a common set of base address, size, and setting.

____3. The MPU is a programmable security device used to protect the system memory and the user's memory spaces from corrupting and attacking by undesired tasks.

____4. Accessing to PPB, which includes the MPU, NVIC, SysTick, and ITM, is always prohibited in the privileged state and the vector fetches are blocked by the MPU.

____5. The Cortex®-M4 MPU defines eight separate memory regions, 0–7, and a background region accessible only from privileged mode.

____6. If a program accesses a memory location that is prohibited by the MPU, the processor generates a memory management fault, causing a fault exception.

____7. The MPU does not need to be programmed and enabled before it can be used in any application. If the MPU is not enabled or programmed, it still can work.

____8. In order to check whether a MPU is presented in a Cortex®-M4 system, one can use the **MPUCTRL** to do this checking.

____9. In order to enable or disable a MPU, one needs to use the **MPUCTRL** to do this job.

____10. The **SIZE** field in the **MPUATTR** Register defines the size of the MPU memory region specified by the **MPUNUMBER** Register as: **(Region Size in Bytes) =** $2^{(SIZE+1)}$.

II. Multiple Choices

1. A MPU can be considered as a programmable device used to ______________.
   a. Divide the memory space into several regions that can be used for the different applications with different access levels
   b. Define the memory access permissions to enable different memory regions to be accessed in either privileged level or full access level
   c. Define the memory attributes as bufferable or catcheable regions
   d. All of them
2. You may set up the memory regions for the device __________________.
   a. NVIC
   b. ITM
   c. GPIO
   d. SysTick
3. Each memory region has its own programmable ___________.
   a. Starting or base address
   b. Size
   c. Attributes
   d. All of them
4. A MPU region includes all definitions except _________.
   a. Base address
   b. Size
   c. Function
   d. Attributes
5. The HardFault EXH should always be defined even if the ______ is enabled.
   a. HardFault exception
   b. MemManage exception
   c. NMI exception
   d. MPU
6. To check whether a MPU is presented in a Cortex®-M4 system, one needs to use ____________________________.
   a. `MPUATTR`
   b. `MPUTYPE`
   c. `MPUNUMBER`
   d. `MPUCTRL`
7. Basically, the MPU setup procedure can be completed by _________________.
   a. Check MPU, enable MPU, setup used regions, disable unused regions
   b. Check MPU, disable MPU, setup used region, disable unused regions, enable MPU
   c. Disable MPU, check MPU, setup used regions, disable unused regions, enable MPU
   d. Enable MPU, setup used regions, disable unused regions
8. Prior to enabling the MPU, at least _____ region(s) must be set or else by enabling the default region for privileged mode. Once the MPU is enabled, a ____________ fault is generated for memory access violations.
   a. 1, Memory Management
   b. 2, HardFault exception
   c. 3, NMI exception
   d. 4, Memory Overlapping exception

9. Each region is automatically divided into _____ equally-sized subregions by the MPU. Subregions can only be used in regions of size _______ bytes or larger.
   a. 2, 64
   b. 4, 128
   c. 8, 256
   d. 16, 512
10. In the **`MPU_setRegion()`** API function, the base address parameter, **`addr`**, must be aligned according to the ________, and the size must be a power of ______.
   a. Starting address, 4
   b. Region number, 8
   c. IREGION, 16
   d. Size, 2

III. Exercises
1. Provide a brief description about the MPU used in the Cortex®-M4 MCU.
2. Provide a brief description about the HardFault and MemManage exceptions and how to define both handlers to handle these exceptions.
3. Provide a brief description about the configuration for a MPU without embedded OS.
4. Provide a brief discussion about the MPU setup procedure.

IV. Practical Laboratory

LABORATORY 12 MSP432 MPU

12.0 GOALS

This laboratory exercise allows students to learn and practice ARM® Cortex®-M4 MPU and related operations. Some MPU-related API functions are used in this lab to make students familiar with these functions.

1. Program **`Lab12_1(SDMPU)`** lets students use some MPU-related API functions to configure and set up the entire memory map with eight memory regions. Furthermore, to enable students to use MPU to set up protection mechanism to limit other programs or devices to access the protected memory regions.

After completion of this program, students should understand the basic architecture and operational procedure for the MPU installed in the MSP432P401R MCU system. They should be able to code some sophisticated programs to perform the desired MPU protection functions with specified memory regions.

12.1 LAB12_1

12.1.1 Goal

In this project, students will use the SD model with MPU-related API functions to build a MPU program to perform desired memory regions protection operations. Some popular MPU-related API functions are to be used in this lab project to enable students to make them familiar with these functions and apply them in the real applications.

The advantage of using these API functions to access and operate the MPU-related registers is that some complicated register structures and bit field values can be avoided to improve the coding efficiency and coding process seed.

12.1.2 Data assignment and hardware configuration

No data assignment and hardware configuration are needed for this lab project.

In this project, the entire memory map is divided into eight different regions. The first three regions with region numbers 0~2, have the base addresses as **0x00000000**, **0x20000000**, and **0x20008000**, and the following sizes and attributes:

1. A 32-KB region of flash from **0x00000000** to **0x00008000** is set up. The region is executable and read-only for both privileged and user modes. To set up this region, a 32-KB region (**#0**) is defined starting at address 0 and the region is initially enabled.
2. Another 32-KB region (**#1**) of RAM from **0x20000000** to **0x20008000** is configured. The region is not executable and is R/W access for both privileged and user modes.
3. An additional 8-KB region (**#2**) in RAM from **0x20008000** to **0x2000A000** is defined and this region is R/W accessible only from privileged mode. This region is initially disabled, and can be enabled later.

To set up and configure the MPU with the above requirements and attributes, you need to use two data structures, **ui32RegionAddr[]** and **ui32RegionAttr[]**, to store these base addresses and attributes.

This project needs a header file and a source file.

12.1.3 Development of the project and the header file

Perform the following operations to create this project and the header file:

1. Create a new folder **Lab12_1(SDMPU)** under the folder **C:\MSP432 Lab Projects\Chapter 12** with Windows Explorer.
2. Create a new µVersion5 project **SDMPU** and save it into the folder **Lab12_1(SDMPU)** created above.
3. Then create a new header file **SDMPU.h**.
4. Place the following system header files and macros into this header file:

```
#include <stdint.h>
#include <stdbool.h>
#include <msp.h>
#include <MSP432P4xx\flash.h>
#include <MSP432P4xx\mpu.h>
#include <MSP432P4xx\gpio.h>
#include <MSP432P4xx\wdt_a.h>
#include <MSP432P4xx\interrupt.h>

int SetupMPU(void);
int WriteFlash(void);
uint32_t ui32RegionAddr[3] = {0, 0x20000000, 0x20008000};
uint32_t ui32RegionAttr[3] = {(MPU_RGN_SIZE_32K|MPU_RGN_PERM_EXEC|
         MPU_RGN_PERM_PRV_RO_USR_RO|MPU_RGN_ENABLE),
         (MPU_RGN_SIZE_32K|MPU_RGN_PERM_NOEXEC|
         MPU_RGN_PERM_PRV_RW_USR_RW|MPU_RGN_ENABLE),
         (MPU_RGN_SIZE_8K|MPU_RGN_PERM_NOEXEC|
         MPU_RGN_PERM_PRV_RW_USR_NO|MPU_RGN_DISABLE)};
```

Now, let's build the source file **SDMPU.c**.

12.1.4 Development of the source file

Perform the following operations to complete this project:

1. Create a new C source file **SDMPU.c** and add it into the project.
2. Place the following system header files and macros into this source file:

```
#include "SDMPU.h"
```

3. Start the **main()** program and the main program needs to return an integer variable.
4. Use **WDT_A_holdTimer()** to stop the Watchdog timer.
5. Use **GPIO_setAsOutputPin()** to set **P2.0** and **P2.2** pins as output pins.
6. Use **GPIO_setOutputLowOnPin()** to enable **P2.0** and **P2.2** pins to output low.
7. Call the **SetupMPU()** function to set up and configure the MPU. A 1 is returned to the OS if this function calling encountered any error.
8. Use **MPU_enableRegion()** to enable the second memory region (**2**).
9. Use **Interrupt_enableInterrupt(FAULT_MPU)** to enable the MPU fault exception.
10. Call the function **WriteFlash()** to try to access a read-only region starting at **0x1000**. A 1 is returned to the OS if this function calling encountered any error. This access would generate a memory fault exception to trigger the **HardFault_Handler()** and the **MemManage_Handler()** to handle this exception.

The following codes are for the user-defined function **SetupMPU()**:

11. Declare a **uint32_t** local variable **index**.
12. Call the API function **MPU_disableModule()** to disable the MPU to enable it to be configured.
13. Use a **for()** loop to call the API function **MPU_setRegion()** three times to configure three memory regions. Three arguments of this function are: **index**, **ui32RegionAddr[index]** and **ui32RegionAttr[index]**, which have been defined in the header file.
14. Still use another **for()** loop to call the API function **MPU_setRegion()** five times to disable the rest five memory regions with index starting from three but less than eight. Three parameters of this function are: **index**, **0**, and **0**.
15. Use the API function **MPU_enableModule(MPU_CONFIG_NONE)** to enable the MPU.
16. Return a 0 to the main program to indicate that this function is executed successfully.

The following codes are for the user-defined function **WriteFlash()**:

17. Declare one local variable **uint32_t ret**, which is used as the status holder later.
18. Use the following two code lines to define the number of words to be written into the flash memory (**ulCount**) starting at **0x1000** (**ulAddress**) and three source words data stored in a data array **Data[3]**:

```
unsigned long ulCount = 3, ulAddress = 0x1000;
unsigned long Data[3] = {0x78563412, 0x8B674523, 0xA3456789};
```

19. Use the **FlashCtl_enableWordProgramming(FLASH_IMMEDIATE_WRITE_MODE)** function to enable the flash memory controller to perform words programming operations in the immediate-write-mode.

20. Use the code line **ret = FlashCtl_programMemory(Data, &ulAddress, ulCount)** to start this word programming by giving the source data address **Data**, destination address **&ulAddress** and the number of words (**ulCount**) to be written into the flash memory.
21. This function returns a Boolean value. If a **false** is returned (**if (!ret)**), it means that this programming operation is failed and a 1 is returned to the main program to indicate this error (**return 1**).
22. Otherwise, if a **true** is returned, it means that the writing is successful, a 0 is returned to the main program (**return 0**).

The following codes are for two exception handlers, **HardFault_Handler()** and the **MemManage_Handler()**:

23. Inside the **HardFault_Handler()**, use **GPIO_setOutputHighOnPin()** to set the GPIO pin **P2.0** to high to turn on the red color LED.
24. Inside the **MemManage_Handler()**, use **GPIO_setOutputHighOnPin()** to set the GPIO pin **P2.2** to High to turn on the blue color LED.

12.1.5 Set up the environment to build and run the project

To build and run the project, one needs to perform the following operations to set up the environments:

- Go to **C/C++** tab in the **Project|Options for Target "Target 1"** to set up the **Include Path** as: **C:\ti\msp\MSPWare_2_21_00_39\driverlib\driverlib.**
- Add the MSPPDL into this project by right clicking on the **Source Group 1** folder in the **Project** panel on the left, and select the **Add Existing Files to Group 'Source Group 1'** menu item. On the opened box, select the driver library file path **C:\ti\msp\MSPWare_2_21_00_39\driverlib\driverlib\MSP432P4xx\keil** and then select the driver library file **msp432p4xx_driverlib.lib**. Click on the **Add** button to add it into the project.
- Select the correct debugger by clicking on the **Debug** tab and select the **CMSIS-DAP Debugger** from the **Use** box. Also make sure that all settings for the Debugger and the Flash Download are correct by clicking on the Settings button. Then go to the **Flash|Download** menu item to download the image file of this project into the flash memory in the MSP432P401R EVB.
- Now go to **Debug|Start/Stop Debug Session** to ready to run the project.
- Go to **Debug|Run** menu item to run the project.

Now you can build and run the project. Then you can check the MPU protection function by monitoring the red color LED.

Based on the lab running result, try to answer the following questions:

- Why do you need to use the function **MPU_enableRegion()** to enable the second memory region (2)? Why do all other memory regions, such as 0 and 1, not need to do this enabling function?
- What is the difference between the function **MPU_enableRegion()** and the API function **MPU_enableModule()**?
- What did you learn from this lab project?

Appendix A: Download and install Keil MDK-ARM 5.15 IDE and MSP432 DFP

1. Go to the URL: **https://www.keil.com/download/product** to begin this process. Click on the **MDK-ARM v5** button to begin this download process (Figure A.1).
2. Enter your information to finish this registration process. Then click on the **Submit** button to go to the next screen (Figure A.2).
3. Click on **MDK_515.EXE** link to start the download process (Figure A.2).
4. Click on the **Save** button to save this file to your C: drive. Then click on the **Run** and **Yes** buttons to download this software.
5. A confirmation page is shown in Figure A.3. Click on the **Next** button to go to the next screen.
6. Check on **I agree to all the terms of the preceding License Agreement** checkbox, and click on the **Next** button.
7. Click on the **Next** button on the next screen if you want to keep the default location to install this software (**C:\Keil_v5**).
8. On the next screen, enter your registration information again, and then click on the **Next** button. The installation starts as shown in Figure A.4. Meantime, a command window will be displayed to show that the debug driver, the **ULINK Driver**, will be installed (Figure A.5).
9. Click on the **Install** button on the next screen to confirm this installation (Figure A.6).
10. A final screen shown in Figure A.7 will be displayed when this installation is completed. Click on the **Finish** button to close this process.
11. The MDK-ARM will be initialized and the most related Device Family Packs (**DFP**) are installed and displayed as the IDE starts (Figure A.8). Click on the **OK** button to close that message box.
12. In the opened **Pack Installer** window shown in Figure A.9, all available DFPs are displayed in the left pane under the **Devices** tab. These DFPs are developed by the Keil to assist various MCU vendors to build related developments based on different types of MCUs via this IDE. One can select any desired pack to install that pack by clicking on the related **Install** button. The **Pack Installer** is a utility for installing, updating, and removing Software Packs, and can be launched from within µVision or standalone, outside of µVision.
13. Since we will use *TI MSP432P401 Launchpad—MSP-EXP432P401R* Evaluation Kit as our development tool, therefore we need to install the ***MSP432 Device Family Pack*** and related examples in this Pack Installer. Expand the **Texas Instruments** item and click on the **MSP432 Family** under the **Devices** tab in the left pane, as shown

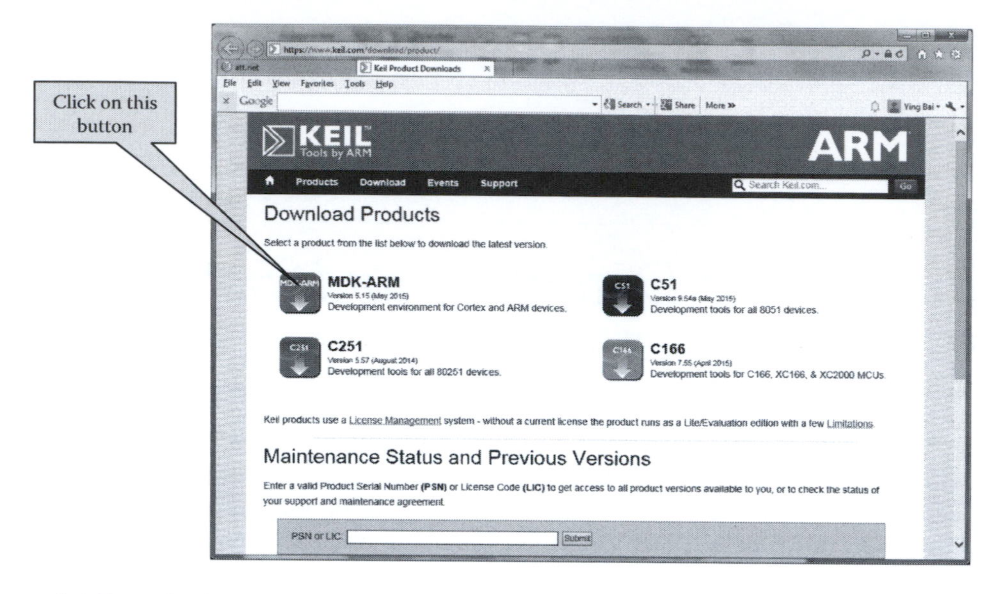

Figure A.1 Download starting page.

in Figure A.9. Then go to right pane and click on the **`Install`** button located on the right side of **`TexasInstruments:MSP432`** under the **`Packs`** tab, as shown in Figure A.9.

14. Check on **`I agree to all the terms of the preceding License Agreement`** checkbox, and click on the **`Next`** button for the pop-up message box.
15. When the MSP432 DFP is installed, the **`Install`** button under the **`Action`** tab becomes **`Up to date`**, as shown in Figure A.10. This indicates that the installed

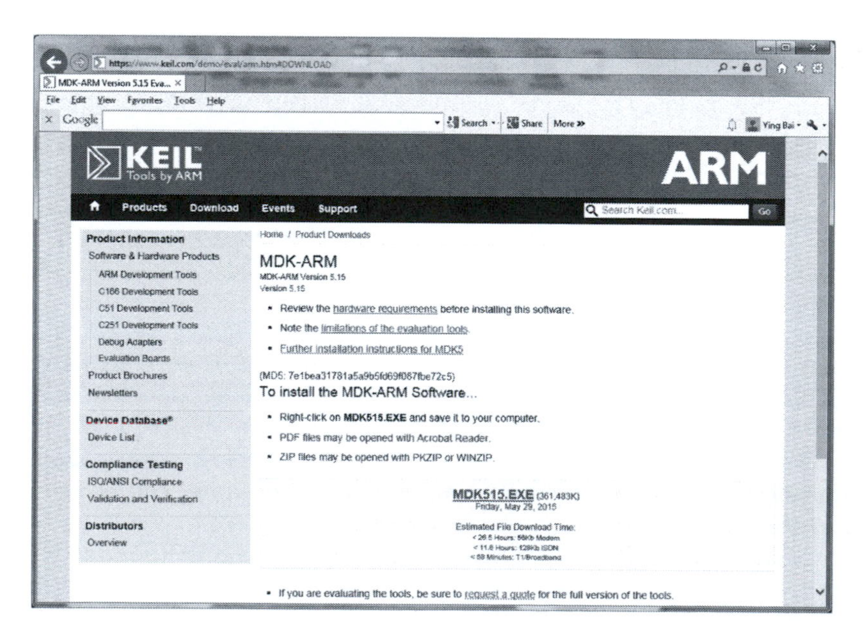

Figure A.2 Information page.

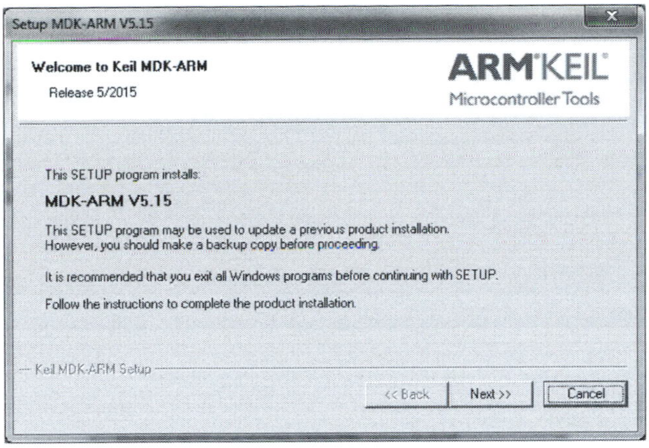

Figure A.3 Confirmation page.

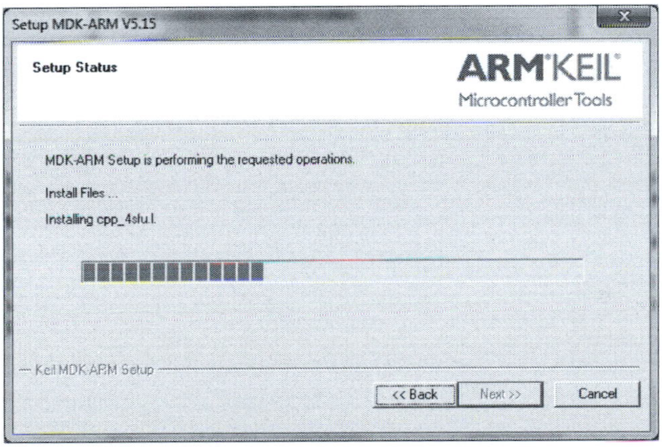

Figure A.4 Installation process.

Figure A.5 Installation process of the ULINK driver.

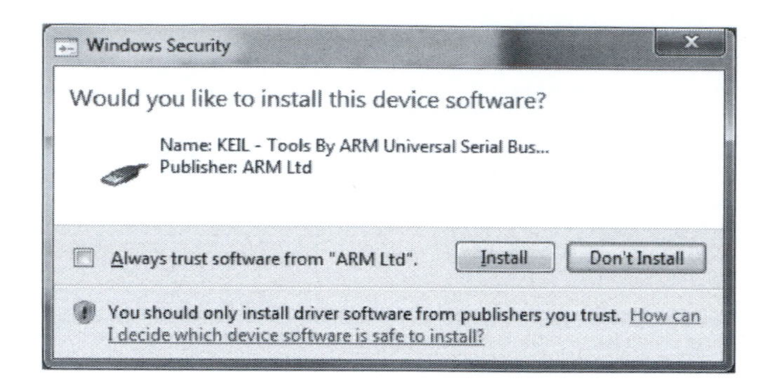

Figure A.6 Confirmation page for installing the ULINK Driver.

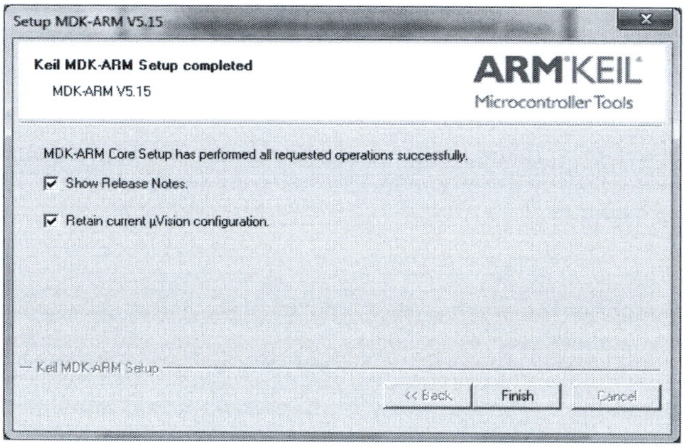

Figure A.7 Finishing page.

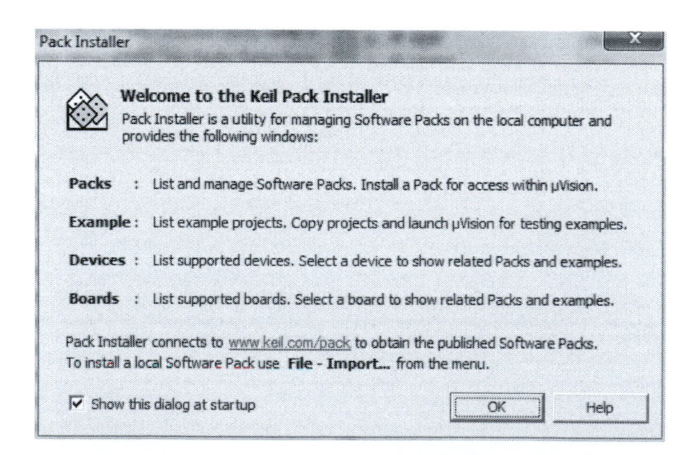

Figure A.8 Starting page of MDK-ARM.

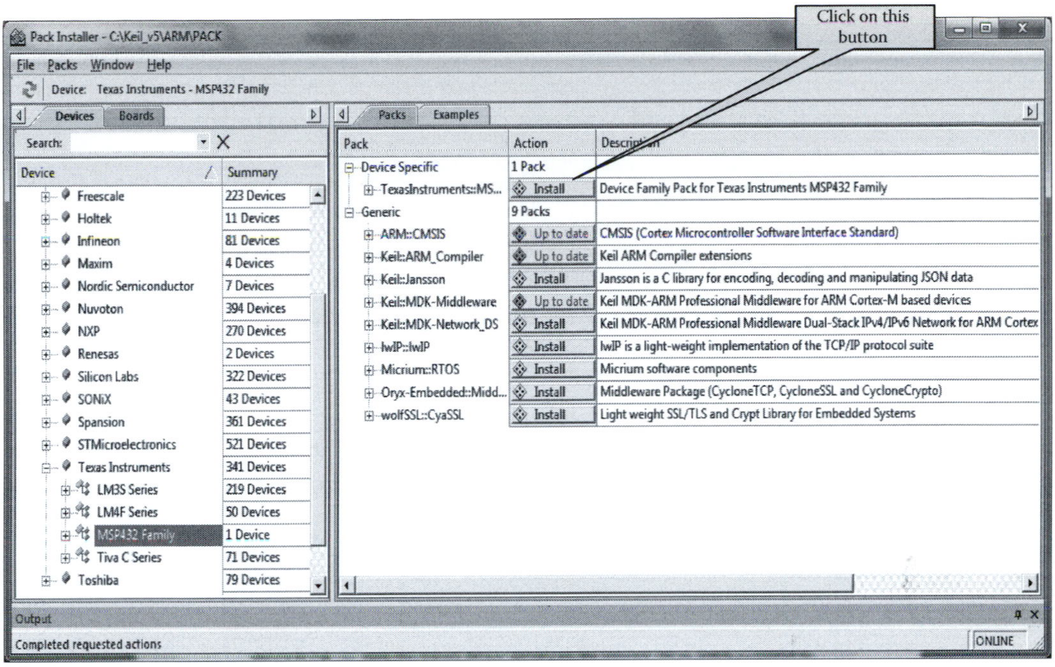

Click on this button

Pack Installer - C:\Keil_v5\ARM\PACK

File Packs Window Help

Device: Texas Instruments - MSP432 Family

Devices Boards

Search:

Device	Summary
Freescale	223 Devices
Holtek	11 Devices
Infineon	81 Devices
Maxim	4 Devices
Nordic Semiconductor	7 Devices
Nuvoton	394 Devices
NXP	270 Devices
Renesas	2 Devices
Silicon Labs	322 Devices
SONiX	43 Devices
Spansion	361 Devices
STMicroelectronics	521 Devices
Texas Instruments	341 Devices
LM3S Series	219 Devices
LM4F Series	50 Devices
MSP432 Family	1 Device
Tiva C Series	71 Devices
Toshiba	79 Devices

Packs Examples

Pack	Action	Description
Device Specific	1 Pack	
TexasInstruments::MS...	Install	Device Family Pack for Texas Instruments MSP432 Family
Generic	9 Packs	
ARM::CMSIS	Up to date	CMSIS (Cortex Microcontroller Software Interface Standard)
Keil::ARM_Compiler	Up to date	Keil ARM Compiler extensions
Keil::Jansson	Install	Jansson is a C library for encoding, decoding and manipulating JSON data
Keil::MDK-Middleware	Up to date	Keil MDK-ARM Professional Middleware for ARM Cortex-M based devices
Keil::MDK-Network_DS	Install	Keil MDK-ARM Professional Middleware Dual-Stack IPv4/IPv6 Network for ARM Cortex
lwIP::lwIP	Install	lwIP is a light-weight implementation of the TCP/IP protocol suite
Micrium::RTOS	Install	Micrium software components
Oryx-Embedded::Midd...	Install	Middleware Package (CycloneTCP, CycloneSSL and CycloneCrypto)
wolfSSL::CyaSSL	Install	Light weight SSL/TLS and Crypt Library for Embedded Systems

Output

Completed requested actions

ONLINE

Figure A.9 Opened Pack Installer window.

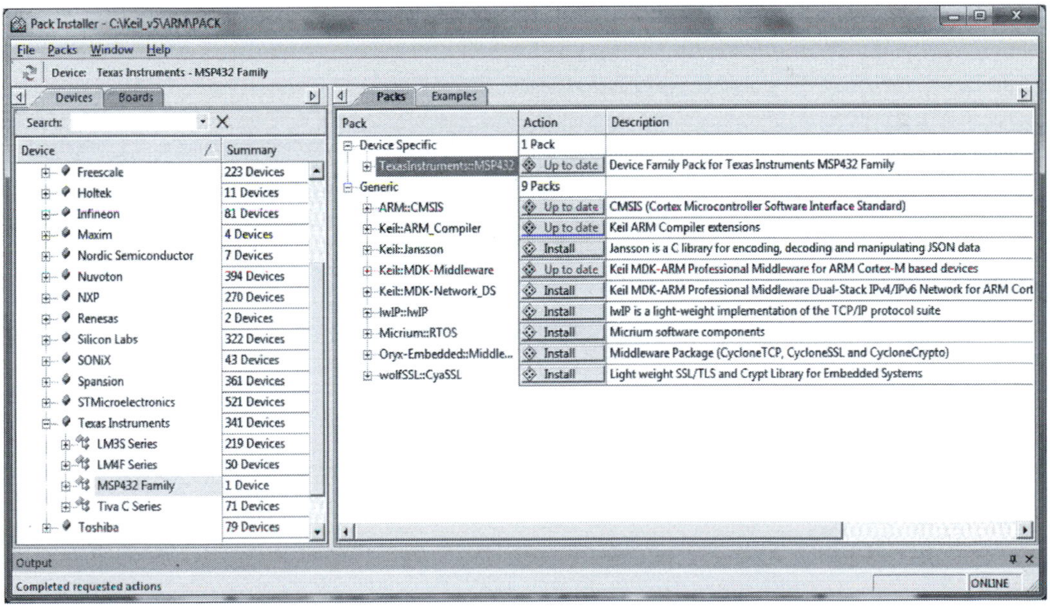

Pack Installer - C:\Keil_v5\ARM\PACK

File Packs Window Help

Device: Texas Instruments - MSP432 Family

Devices Boards

Search:

Device	Summary
Freescale	223 Devices
Holtek	11 Devices
Infineon	81 Devices
Maxim	4 Devices
Nordic Semiconductor	7 Devices
Nuvoton	394 Devices
NXP	270 Devices
Renesas	2 Devices
Silicon Labs	322 Devices
SONiX	43 Devices
Spansion	361 Devices
STMicroelectronics	521 Devices
Texas Instruments	341 Devices
LM3S Series	219 Devices
LM4F Series	50 Devices
MSP432 Family	1 Device
Tiva C Series	71 Devices
Toshiba	79 Devices

Packs Examples

Pack	Action	Description
Device Specific	1 Pack	
TexasInstruments::MSP432	Up to date	Device Family Pack for Texas Instruments MSP432 Family
Generic	9 Packs	
ARM::CMSIS	Up to date	CMSIS (Cortex Microcontroller Software Interface Standard)
Keil::ARM_Compiler	Up to date	Keil ARM Compiler extensions
Keil::Jansson	Install	Jansson is a C library for encoding, decoding and manipulating JSON data
Keil::MDK-Middleware	Up to date	Keil MDK-ARM Professional Middleware for ARM Cortex-M based devices
Keil::MDK-Network_DS	Install	Keil MDK-ARM Professional Middleware Dual-Stack IPv4/IPv6 Network for ARM Cort
lwIP::lwIP	Install	lwIP is a light-weight implementation of the TCP/IP protocol suite
Micrium::RTOS	Install	Micrium software components
Oryx-Embedded::Middle...	Install	Middleware Package (CycloneTCP, CycloneSSL and CycloneCrypto)
wolfSSL::CyaSSL	Install	Light weight SSL/TLS and Crypt Library for Embedded Systems

Output

Completed requested actions

ONLINE

Figure A.10 Installed pack window.

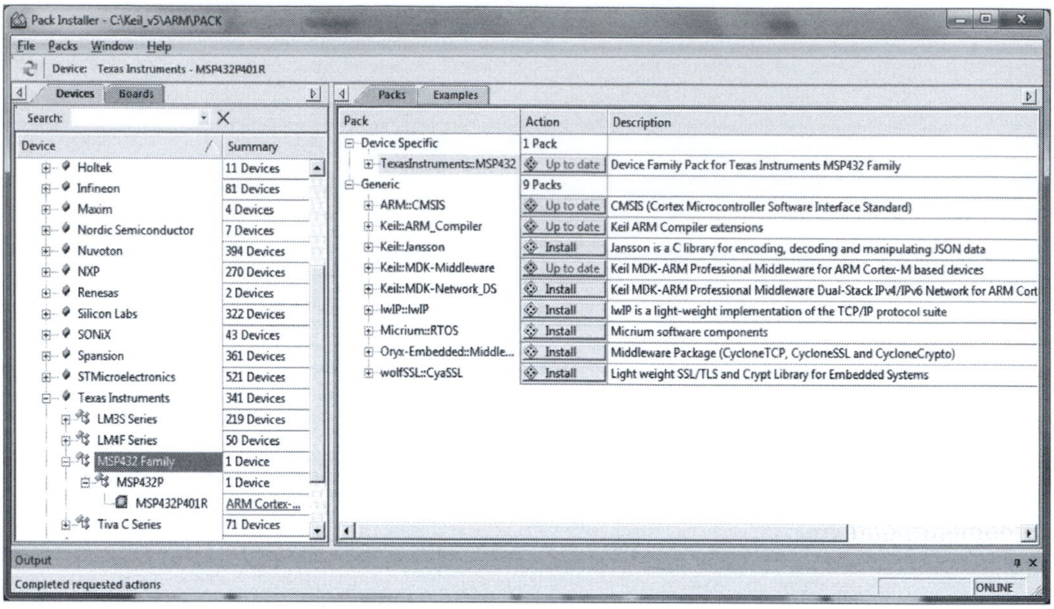

Figure A.11 Filtered MCU or device for the selected pack.

DFP for the selected MCU, **MSP432P401R**, which is an ARM Cortex-M4F-based 32-bit microcontroller, is up-to-date.

16. If one expands the **MSP432 Family** and **MSP432P** items under the **Texas Instruments** item under the **Devices** tab in the left pane, you can find the MCU we will use is **MSP432P401R**, as shown in Figure A.11
17. Three default CMSIS-related support tools, **ARM::CMSIS**, **Keil::ARM_Compiler**, and **Keil::MDK-Middleware**, are also indicated as **Up to date** as shown in Figures A.10 and A.11. We need these tools to build our application projects later.

Appendix B: Download and install MSPWare software package

MSPWare is a collection of MSP430 and MSP432 software and tools. Exactly, it is a collection of system software, including user's guides, MCU registers-related macro definitions, hardware interfacing protocols, interfacing API functions and code examples as well as other design resources for ALL MSP MCUs delivered in a convenient package—essentially everything developers need to become MSP432 experts!

In addition to providing a complete collection of existing MSP430 and MSP432 design resources, MSPWare also provides a wide selection of highly abstracted software libraries ranging from device- and peripheral-specific such as MSP Driver Library or USB, to application-specific such as Graphics Library or Capacitive Touch Library. MSP Driver Library, in particular, is an essential and independent or standalone library to help software developers leverage high-level APIs to control low-level and intricate hardware peripherals. As of today, MSP Driver Library supports MSP430F5x/6x and MSP432P4x series devices.

The developers can use either the MSPWare package or just use the MSP Driver Library to build their applications. Generally the former contains all software tools, including the latter, but the latter provides more specific support for a special MCU, such as MSP432P401R, with a smaller volume.

You need to download this package if you have already installed a supported third-party integrated development environment (IDE), such as **Keil MDK**, on your system.

This MSPWare package contains all software sources as well as installation files for MSP430 and MSP432 MCUs and the selected IDE. This package also contains several documents to help you get started with using the **MSP-EXP432P401R** LaunchPad.

1. Go to http://www.ti.com/tool/mspware#descriptionArea to begin this process. Click on **Get Software** button on the right hand side of **MSPWare** to begin the download process (Figure B.1).
2. On the next page, go to the bottom and click on the link **MSPWare_2_21_00_39_setup.exe** to download a **Standalone MSPWare 2.21.00.39 for Windows** OS.
3. If you are a new user, you need to create a new account in the next page by entering the necessary information and click on the **Create account** to go to the next page. Otherwise you can enter your username and password to login to this system.
4. In the next page, you need to enter related information to complete this US Government Approval process. Two points you need to pay attention to are: (1) the purpose of using this software and (2) confirmation to your declaration on this page. Select one of purposes by checking the related checkbox (using **Other** and putting a description into the textbox if you cannot find a matched field). Then check the **Yes**

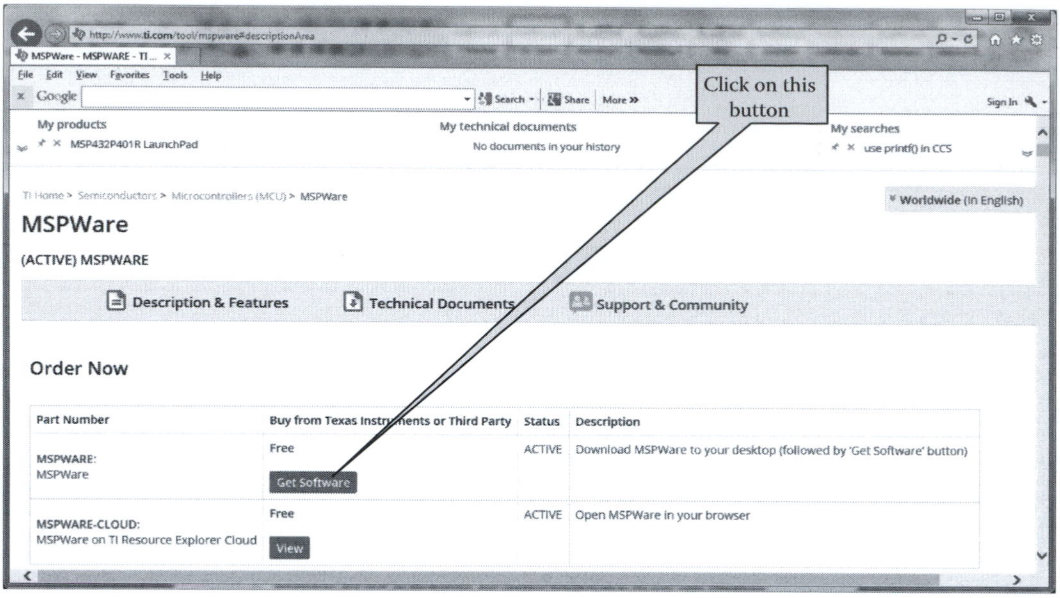

Figure B.1 Download and install Tiva C Series Evaluation Software window.

radio button for the I CERTIFY ALL THE ABOVE IS TRUE. Then click on the **Submit** button.

5. Click on the **Download** button on the next page to begin the download process. You may click the **Save** button to save this file to your PC. Check your system to make sure that your computer did not block some popup file if this download process cannot start automatically.
6. An executable file, **MSPWare_2_21_00_39_setup.exe**, should have been stored into your computer under the **Downloads** folder as this downloading process is done. Double click on this file to run it. A **Setup** wizard is displayed as shown in Figure B.2. Click on the **Next** button to start this installation.

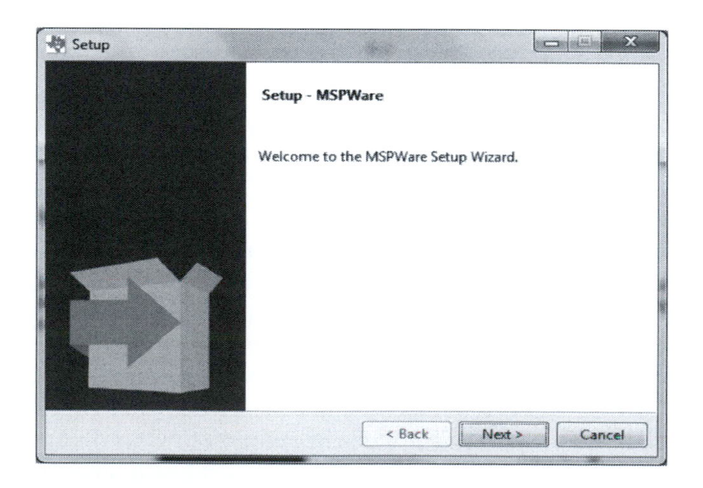

Figure B.2 Setup page for MSPWare package.

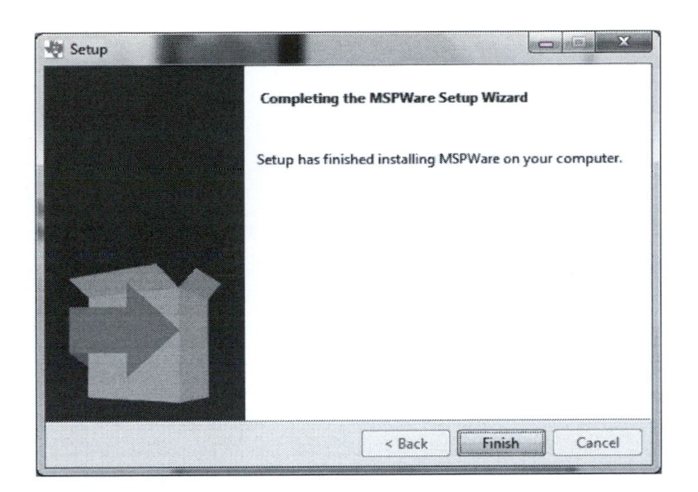

Figure B.3 Installation finish page.

7. On the next page, check on **`I accept the agreement`** radio button and the **`Next`** button to continue.
8. Keep the target location, such as **`C:\ti\msp`**, to install this package on your computer and click on the **`Next`** button if you want to use this default location. Otherwise you can change it to any folder you want in your computer.
9. Click on the **Next** button on the next page to officially begin this installation.
10. Then the installation process starts. When this is done, a **`Finished`** wizard is displayed and shown in Figure B.3.
11. Click on the **`Finish`** button to complete this installation.
12. When this package is installed, you can find it in the folder **`C:\ti\msp`** in your host computer. Exactly the **`MSP432 Driver Library`** for Keil ARM-MDK µ5 IDE is installed at the folder **`C:\ti\msp\MSPWare_2_ 21_00_39\driverlib\driverlib\MSP432P4xx\keil.`**
13. Unlike the other MCUs, such as **`TM4C123GH6PM`**, both MSP432 header files, Register Driver Definition header file and CMSIS Cortex-M4 Peripheral Layer header file, are combined together to make one header file, **`msp432p401r.h`**, which is located in your host computer at the folder **`C:\ti\msp\MSPWare_2_21_00_39\driverlib\inc.`** This combined header file provides all definitions, including the register addresses definitions and register structure definitions, to enable users to use different ways to access these registers.

Appendix C: MSP432P401R-based EVB hardware setup

To connect the MSP432P401R LaunchPad evaluation board MSP-EXP432P401R EVB to the host computer, follow the operations listed below:

1. Set up the MSP-EXP432P401R evaluation board by changing the `Power Select` switch to the left for the `XDS-ET` (`Debug Mode`).
2. Connect the USB-to-PC cable from the USB port in the host PC to the `Debug USB Port` on the MSP-EXP432P401R board, as shown in Figure C.1.

After this connection, the windows may display a Find New Hardware dialog to install the driver.

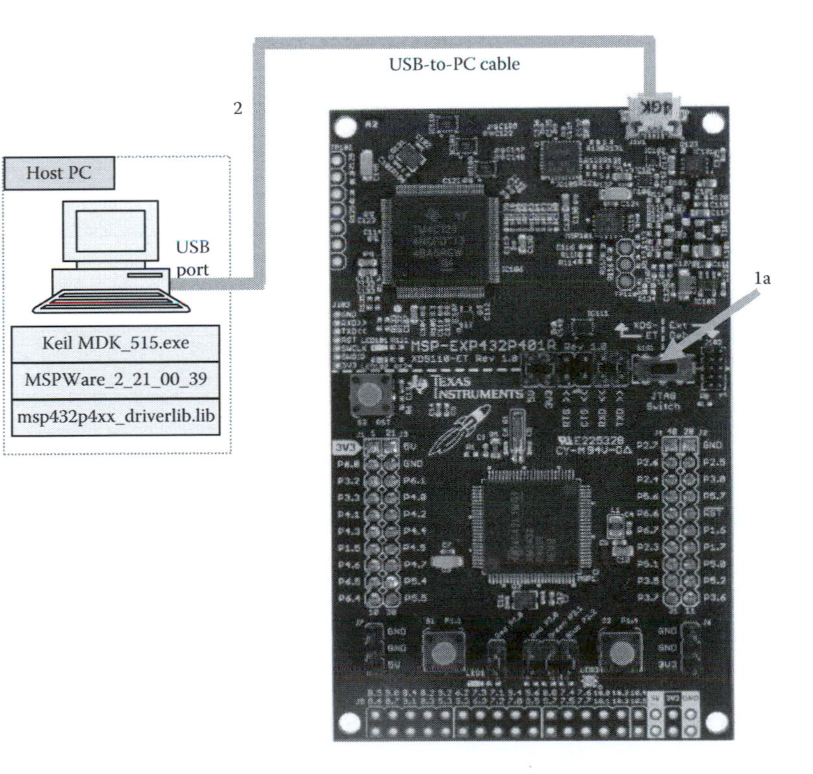

Figure C.1 MSP-EXP432P401R EVB setup.

Appendix D: The CMSIS core-specific intrinsic functions

Table D.1 lists all popular intrinsic functions provided by the CMSIS Core.

Table D.1 All popular intrinsic functions provided by CMSIS core

Instructions	CMSIS core intrinsic function	Functions
NOP	void __NOP (void);	No operation
SEV	void __SEV(void);	Send event
WFI	void __WFI(void);	Wait for interrupt (enter sleep mode)
WFE	void __WFE(void);	Wait for event
BKPT	void __BKPT(uint8_t value);	Set a software breakpoint
LDREXB	uint8_t __LDREXB (volatile uint8_t addr*);	Exclusive load byte
LDREX	uint32_t __LDREXW (volatile uint32_t addr*);	Exclusive load word
STREXB	uint32_t __STREXB (uint8_t value, volatile uint8_t addr*);	Exclusive store byte
CLZ	uint8_t __CLZ (unsigned int val);	Count leading zeros
RBIT	uint32_t __RBIT (uint32_t val);	Reverse bits order in word
ROR	unit32_t __ROR (uint32_t value, uint32_t shift);	Rotate shift right by n bits
SADD8	uint32_t __SADD8 (uint32_t val1, uint32_t val2);	Perform four 8-bit signed addition
SSUB8	uint32_t __SSUB8 (uint32_t val1, uint32_t val2);	Perform four 8-bit signed subtraction
MRS	uint32_t __get_CONTROL (void);	Read the CONTROL register
MSR	uint32_t __set_CONTROL (uint32_t control);	Set the CONTROL register
MRS	uint32_t __get_APSR (void);	Read the APSR register
MSR	uint32_t __set_PRIMASK (uint32_t priMask);	Set the PRIMASK register
CPSIE I	void __enable_irq (void);	Globally enable the IRQ interrupts
CPSID I	void __disable_irq (void);	Globally disable the IRQ interrupts

Index

D

G

J

K

R

S

T

U

V

W

X–Z